Cutnell & Johnson Physics

Twelfth Edition

DAVID YOUNG

SHANE STADLER

Louisiana State University

WILEY

VICE PRESIDENT AND GENERAL MANAGER	Aurora Martinez
EXECUTIVE EDITOR	John LaVacca
SENIOR EDITOR	Jennifer Yee
ASSOCIATE EDITOR	Georgia Larsen
ASSISTANT EDITOR	Samantha Hart
SENIOR MANAGING EDITOR	Mary Donovan
MARKETING MANAGER	Sean Willey
SENIOR MANAGER, COURSE DEVELOPMENT AND PRODUCTION	Svetlana Barskaya
SENIOR COURSE PRODUCTION OPERATIONS SPECIALIST	Patricia Gutierrez
COURSE CONTENT DEVELOPER	Corrina Santos
DESIGNER	Wendi Lai
COVER IMAGES CREDIT	Digital Composite Image Of Iris © Erwin Matuschat/EyeEm/Getty Images, DNA structure © KTSDESIGN/ SCIENCE PHOTO LIBRARY/Getty Images

This book was set in 9.5/12 STIX Two Text by Lumina Datamatics®, Inc.

Library of Congress Cataloging-in-Publication Data

Names: Young, David (David P.) (Physics professor), author. | Stadler, Shane, author.
Title: Physics / David Young, Shane Stadler, Louisiana State University.
Description: Twelfth edition. | Hoboken, NJ : Wiley, [2022] | Revision of: Physics / John D. Cutnell & Kenneth W. Johnson, David Young, Shane Stadler, Louisiana State University. c2015. 10th ed.
Identifiers: LCCN 2021039453 (print) | LCCN 2021039454 (ebook) | ISBN 9781119773610 (hardback) | ISBN 9781119788591 (adobe pdf) | ISBN 9781119773535 (epub)
Subjects: LCSH: Physics—Textbooks.
Classification: LCC QC23.2 .C87 2022 (print) | LCC QC23.2 (ebook) | DDC 530—dc23
LC record available at https://lccn.loc.gov/2021039453
LC ebook record available at https://lccn.loc.gov/2021039454

The inside back cover will contain printing identification and country of origin if omitted from this page. In addition, if the ISBN on the cover differs from the ISBN on this page, the one on the cover is correct.

Printed in the United States of America

SKY10046639_042823

About the Authors

DAVID YOUNG received his Ph.D. in experimental condensed matter physics from Florida State University in 1998. He then held a post-doc position in the Department of Chemistry and the Princeton Materials Institute at Princeton University before joining the faculty in the Department of Physics and Astronomy at Louisiana State University in 2000. His research focuses on the synthesis and characterization of high-quality single crystals of novel electronic and magnetic materials. The goal of his research group is to understand the physics of electrons in materials under extreme conditions, i.e., at temperatures close to absolute zero, in high magnetic fields, and under high pressure. He is the coauthor of over 250 research publications that have appeared in peer-reviewed journals, such as *Physical Review B, Physical Review Letters*, and *Nature*. Professor Young has taught introductory physics with the Cutnell & Johnson text since he was a senior undergraduate almost 30 years ago. He routinely lectures to large sections, often in excess of 300 students. To engage such a large number of students, he uses *WileyPLUS*, electronic response systems, tutorial-style recitation sessions, and in-class demonstrations. Professor Young has received multiple awards for outstanding teaching of undergraduates. David enjoys spending his free time with his family, playing basketball, and working on his house.

I would like to thank my family for their continuous love and support. —David Young

Courtesy Shane Stadler

SHANE STADLER Shane Stadler earned a Ph.D. in experimental condensed matter physics from Tulane University in 1998. Afterwards, he accepted a National Research Council Postdoctoral Fellowship with the Naval Research Laboratory in Washington, DC, where he conducted research on artificially structured magnetic materials. Three years later, he joined the faculty in the Department of Physics at Southern Illinois University (the home institution of John Cutnell and Ken Johnson, the original authors of this textbook), before joining the Department of Physics and Astronomy at Louisiana State University in 2008. His research group studies novel magnetic materials for applications in the areas of spintronics and magnetic cooling.

Over the past twenty years, Professor Stadler has taught the full spectrum of physics courses, from physics for students outside the sciences, to graduate-level physics courses, such as classical electrodynamics. He teaches classes that range from fewer than ten students to those with enrollments of over 300. His educational interests are focused on developing teaching tools and methods that apply to both small and large classes, and which are applicable to emerging teaching strategies, such as "flipping the classroom."

In his spare time, Shane writes science fiction/thriller novels.

I would like to thank my parents, George and Elissa, for their constant support and encouragement. —Shane Stadler

Dear Students and Instructors:

Welcome to college physics! To the students: We know there is a negative stigma associated with physics, and you yourself may harbor some trepidation as you begin this course. But fear not! We are here to help. Whether you are worried about your math proficiency, understanding the concepts, or developing your problem-solving skills, the resources available to you are designed to address all of these areas and more. When we were students and had to take introductory physics, we had a printed textbook, a pencil, and some paper. That was it! Can you learn physics this way? You bet! We did! But research has shown that learning styles vary greatly among students. Maybe some of you have a more visual preference, or auditory preference, or some other preferred learning modality. In any case, the resources available to you in this course will satisfy all of these preferences and improve your chance of success. Take a moment to explore below what the textbook and online course have to offer. We suspect that, as you continue to improve throughout the course, some of that initial trepidation will be replaced with confidence and excitement.

To start, 12e will continue to offer a new learning medium unique to this book in the form of a comprehensive set of lecture videos—one for every section (259 in all). These animated lectures (created and narrated by the authors) are 2–10 minutes in length, and explain the basic concepts and learning objectives of each section. They are assignable within WileyPLUS and can be paired with follow-up questions that are gradable. In addition to supplementing traditional lecturing, the videos can be used in a variety of ways, including, flipping the classroom, lectures for online courses, and reviewing for exams.

Next, we created a new end-of-chapter section of problems called *Physics in Biology, Medicine, and Sports*. The text has always offered real-world examples of physics applications in the life sciences, not only to generate interest for the many students in one of the premedical fields, but also to underscore the fact that the human body, like any biological system, is governed by the laws of physics. Overall, in this edition, we have more than doubled the number of bio-inspired examples and problems to more than 300 total for the text, which reflects our commitment to showing students how relevant physics is in their lives. Although they are of general instructional value, many of the end-of-chapter problems are also similar to what premed students will encounter in the Chemical and Physical Foundations of Biological Systems Passages section of the MCAT.

Building on the success of the *Team Problems* introduced in the previous edition, we have doubled their number in each chapter. These are context-rich end-of-chapter problems of variable difficulty designed for group cooperation, but they may also be tackled by the individual student. We have also added a "worksheet" activity to each new Team Problem, which helps to guide the team of students along the way through the solution. These problems offer a great way to promote peer-to-peer instruction in the classroom, as well as serve as the focal point of a group activity during a breakout session as part of an online course.

One of the great strengths of this text is the synergistic relationship it develops between problem solving and conceptual understanding. For instance, available in WileyPLUS are animated *Chalkboard Videos*, which consist of short (2–3 min) videos narrated by the authors that demonstrate step-by-step practical solutions to typical homework problems. We have more than doubled the number of *Chalkboard Videos* in the new edition, so that every section of the book has at least one, which provides comprehensive coverage of the entire course material in this format.

One of the most important techniques developed in the text for solving problems involving multiple forces is the *free-body diagram (FBD)*. Many problems in the force-intensive chapters, such as Chapters 4 and 18, take advantage of the new FBD drawing tools now available online in WileyPLUS, where students can construct the FBDs for a select number of problems and be graded on them. Also available are numerous *Guided Online (GO) Tutorials* that implement a step-by-step pedagogical approach, which provides students a low-stakes environment for refining their problem-solving skills. Finally, new to this edition will be *Adaptive Assignments*. Seamlessly integrated into WileyPLUS, *Adaptive Assignments,* which are powered by Knewton's world-class recommendation engine, offer the most targeted, efficient way for instructors to tailor practice and preparation. This new assignment type pinpoints learner knowledge gaps and offers personalized, just-in-time support, targeted instruction, and detailed answer explanations.

Over the last 20 years nothing has had a more significant impact on the way students learn than the World Wide Web. Students essentially have 24/7 access to countless sources of digital multimedia. They complete homework assignments online with their PC's, tablets, and smart phones. Online homework systems are no longer "in the future," but are now the norm. Physics is no exception. Unfortunately, having all of this information readily available comes at a price. Students have fundamentally changed the way in which they approach their homework assignments. Instead of struggling through the entire solution to a problem from scratch, where much of the learning process takes place, they default to online resources, where they can pay for access to written solutions to the end-of-chapter problems. Alternatively, many students find solutions by simply searching the questions on Google, for example. As a result, a student's online-homework grade has become a rather poor measure of their knowledge of the course material. What's even worse is the false sense of security the students feel as a result of their inflated homework grades. They feel confident and prepared for exams, because they have 95%–100% on their homework. Unfortunately, a poor performance on the first exam is often the initial indicator of their level of misunderstanding. The content outlined below, and the functionality of WileyPLUS and the *Adaptive Assignments,* will provide students with all the resources they need to be successful in the course:

- The *Lecture Videos* created by the authors for each section include questions with intelligent feedback when a student enters the wrong answer.
- The multistep *GO Tutorial* problems created in WileyPLUS are designed to provide targeted, intelligent feedback.
- The *Free-body Diagram* vector drawing tools provide students an easy way to enter answers requiring vector drawing, and also provide enhanced feedback.
- *Chalkboard Video Solutions* take the students step-by-step through the solution and the thought process of the authors. Problem-solving strategies are discussed, and common misconceptions and potential pitfalls are addressed. The students can then apply these techniques to solve similar, but different, problems.

All of these features are designed to encourage students to remain within the WileyPLUS environment, as opposed to pursuing the "pay-for solutions" websites that short circuit the learning process.

To the students—We strongly recommend that you take this honest approach to the course. Take full advantage of the many features and learning resources that accompany the text and the online content. Be engaged with the material and push yourself to work through the exercises. Physics may not be the easiest subject to understand initially, but with the Wiley resources at your disposal, and your hard work, you CAN be successful.

We are immensely grateful to all of you who have provided feedback as we've worked on this new edition, and to our students who have taught us how to teach. Thank you for your guidance, and keep the feedback coming. Best wishes for success in this course and wherever your major may take you!

Brief Contents

Contents

27 Interference and the Wave Nature of Light 860

28 Special Relativity 893

29 Particles and Waves 920

30 The Nature of the Atom 945

31 Nuclear Physics and Radioactivity 979

32 Ionizing Radiation, Nuclear Energy, and Elementary Particles 1007

Note: Chapter sections marked with an asterisk (*) can be omitted with little impact to the overall development of the material.

Appendixes A-1

Our Vision and WileyPLUS with Adaptive Assignments

Our Vision

Our goal is to provide students with the skills they need to succeed in this course, and instructors with the tools they need to develop those skills.

Skills Development

One of the great strengths of this text is the synergistic relationship between conceptual understanding, problem solving, and establishing relevance. We identify here some of the core features of the text that support these synergies.

Conceptual Understanding Students often regard physics as a collection of equations that can be used blindly to solve problems. However, a good problem-solving technique does not begin with equations. It starts with a firm grasp of physics concepts and how they fit together to provide a coherent description of natural phenomena. Helping students develop a conceptual understanding of physics principles is a primary goal of this text. The features in the text that work toward this goal are:

- *Lecture Videos (one for each section of the text)*
- *Conceptual Examples*
- *Concepts & Calculations* problems (now with video solutions)
- *Focus on Concepts* homework material
- *Check Your Understanding* questions
- *Concept Simulations* (an online feature)

Problem Solving The ability to reason in an organized and mathematically correct manner is essential to solving problems, and helping students to improve their reasoning skills is also one of our primary goals. To this end, we have included the following features:

- *Math Skills boxes* for just-in-time delivery of math support
- *Explicit reasoning steps* in all examples
- *Reasoning Strategies* for solving certain classes of problems
- *Analyzing Multiple-Concept Problems*
- *Video Support and Tutorials* (in *WileyPLUS*)
  - Chalkboard Video Solutions
  - Physics Demonstration Videos
  - Video Help
  - Concept Simulations
- *Problem Solving Insights*

Relevance Since it is always easier to learn something new if it can be related to day-to-day living, we want to show students that physics principles come into play over and over again in their lives. To emphasize this goal, we have included a wide range of applications of physics principles. Many of these applications are biomedical in nature (for example, wireless capsule endoscopy). Others deal with modern technology (for example, 3-D movies). Still others focus on things that we take for granted in our lives (for example, household plumbing). To call attention to the applications we have used the label The Physics of.

WileyPLUS with Adaptive Assignments

WileyPLUS is an innovative, research-based online environment for effective teaching and learning. The hallmark of *WileyPLUS* with Adaptive Assignments for this text is that the media- and text-based resources are all created by the authors of the project, providing a seamless presentation of content.

WileyPLUS builds students' confidence because it takes the guesswork out of studying by providing students with a clear roadmap: **what to do, how to do it, if they did it right**.

With *WileyPLUS*, our efficacy research shows that students improve their outcomes by as much as one letter grade. *WileyPLUS* helps students take more initiative, so you'll have greater impact on their achievement in the classroom and beyond.

With WileyPLUS, instructors receive:

- **Breadth and Depth of Assessment:** *WileyPLUS* contains a wealth of online questions and problems for creating online homework and assessment including:
  - ALL end-of-chapter questions, plus favorites from past editions not found in the printed text, coded algorithmically, each with at least one form of instructor-controlled question assistance (GO tutorials, hints, link to text, video help)
  - Simulation, animation, and video-based questions
  - Free body and vector drawing questions
  - Test bank questions

- **Gradebook:** *WileyPLUS* provides instant access to reports on trends in class performance, student use of course materials, and progress toward learning objectives, thereby helping instructors' decisions and driving classroom discussion.

With WileyPLUS, students receive:

- The complete digital textbook, including the Lecture Videos, saving students up to 60% off the cost of a printed text
- Question assistance, including links to relevant sections in the online digital textbook
- Immediate feedback and proof of progress, 24/7
- Integrated, multimedia resources—including animations, simulations, video demonstrations, and much more—that provide multiple study paths and encourage more active learning
- GO Tutorials
- Chalkboard Videos
- Free Body Diagram/Vector Drawing Questions

New to WileyPLUS for the Twelfth Edition

Team Problems In each chapter, the end-of-chapter problems contain four *Team Problems* that are designed for group problem-solving exercises. These are context-rich problems of medium difficulty designed for group cooperation, but may also be tackled by the individual student. Accompanying some of these team problems are "worksheets" that serve to guide the students through the solution in a tutorial fashion. Many of these problems read like parts of an adventure story, where the student (or their team) is the main character. The motivation for each problem is clear and personal—the pronoun "you" is used throughout, and the problem statements often start with "You and your team need to ...". Pictures and diagrams are not given with these problems except in rare cases. Students must visualize the problems and discuss strategies with their team members to solve them. The problems require two or more steps/multiple concepts (hence the "medium" difficulty level) and may require basic principles learned earlier. Sometimes, there is no specific target variable given, but rather questions like *Will it work?* or *Is it safe?* Suggested solutions are given in the Instructor Solutions Manual.

"The Physics of . . ." Examples The text now contains over 300 real-world application examples that reflect our commitment to showing students how relevant physics is in their lives. Each application is identified in the text with the label The Physics of. A subset of these examples focuses on biomedical applications, and we have more than doubled their number in the new edition. In fact, we have created a new end-of-chapter problem section called the *Physics in Biology, Medicine, and Sports*. Students majoring in biomedical and life sciences will find new examples in every chapter covering topics such as cooling the human brain, abdominal aortic aneurysms, the mechanical properties of bone, and many more! The application of physics principles to biomedical problems in these examples is similar to what premed students will encounter in the *Chemical and Physical Foundations of Biological Systems Passages* section of the MCAT. All biomedical examples and end-of-chapter problems will be marked with the BIO icon.

EXAMPLE 8 | BIO The Physics of Hearing Loss—Standing Waves in the Ear

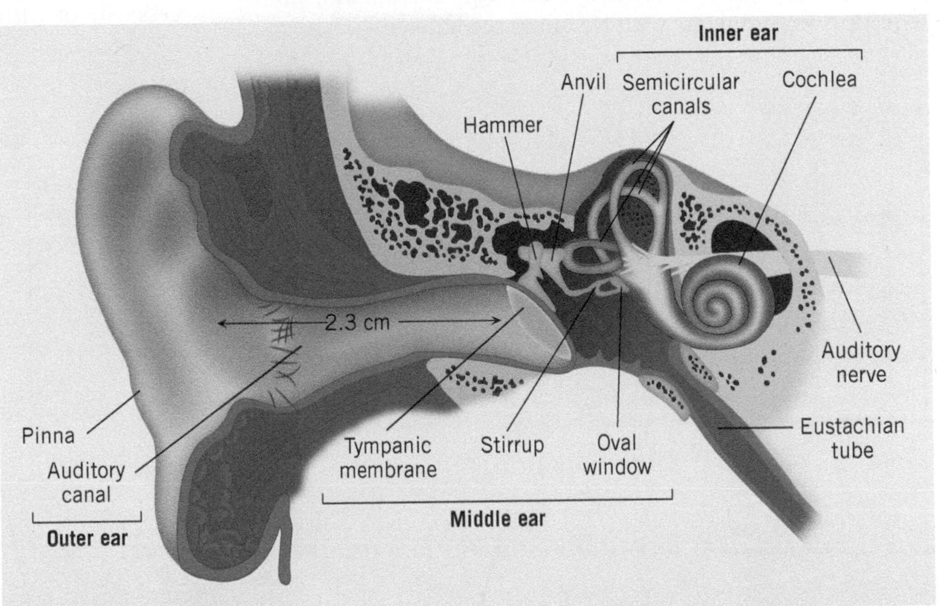

Interactive Graphics The online reading experience within WileyPLUS has been enhanced with the addition of "Interactive Graphics." Several static figures in each chapter have been transformed to include interactive elements. These graphics drive students to be more engaged with the extensive art program and allow them to more easily absorb complex and/or long multi-part figures.

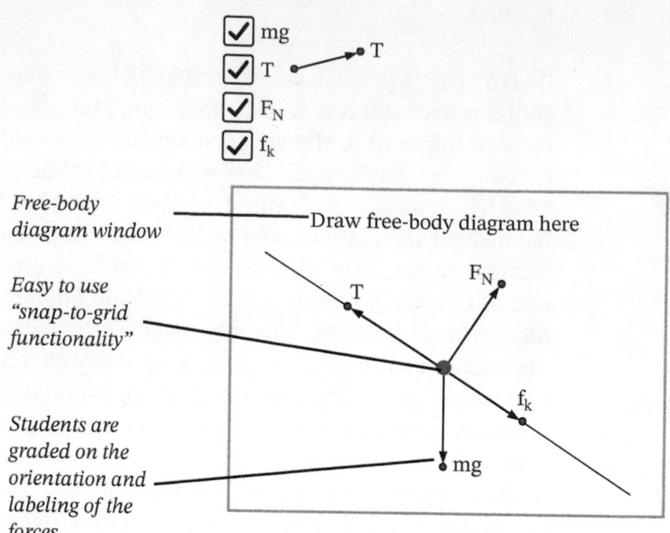

Also Available in WileyPLUS

Free-Body Diagram (FBD) Tools For many problems involving multiple forces, an interactive free-body diagram tool in *WileyPLUS* is used to construct the diagram. It is essential for students to practice drawing FBDs, as that is the critical first step in solving many equilibrium and non-equilibrium problems with Newton's second law.

GO Tutorial Problems Some of the homework problems found in the collection at the end of each chapter are marked with a special GO icon. All of these problems (550 of them) are available for assignment online via *WileyPLUS*. Each of these problems in *WileyPLUS* includes a guided tutorial option that instructors can make available for student access with or without penalty.

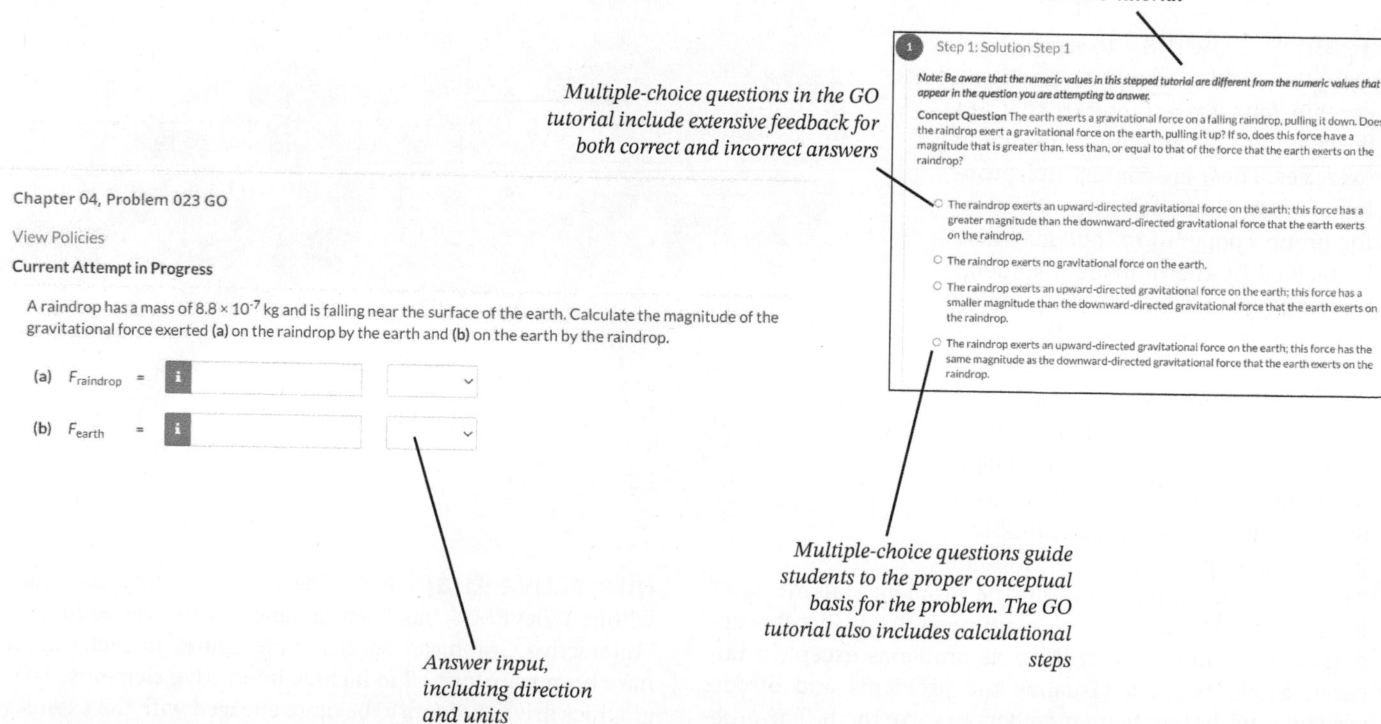

Adaptive Assignments

Seamlessly integrated into WileyPLUS and based on learning science, Adaptive Assignments are the most targeted, efficient way for instructors to tailor practice and preparation. This new assignment type pinpoints learner knowledge gaps and offers personalized, just-in-time support, targeted instruction and detailed answer explanations.

Acknowledgments

The effective revision of a textbook in this age goes well beyond the efforts of the authors and requires the expertise of a great number of people. Their abilities must range from brainstorming revision concepts to editing written and digital content to creating marketing strategies. They have to learn how the needs of our readers have changed and to anticipate how they might transform in the future. As the authors, we are extremely grateful to the people who have provided direction, talent, and effort as a part of the team that made this revision possible. Well done and thank you!

We would especially like to acknowledge John LaVacca III, our Senior Editor. His knowledge of the textbook landscape has offered crucial guidance in developing this revision in order to meet the needs of students and instructors alike. He has provided us ample encouragement and flexibility to pursue the creative aspects of this work within the constraints of a well-planned strategy.

It was a great pleasure working with Jennifer Yee, Senior Editor, on this edition, and we would like to thank her for keeping this edition on track and coordinating our efforts with the other facets of this production. Keeping us informed of the timelines for all aspects of the project made the process smooth and devoid of stress (at least on the part of authors!). We wish her well at her new position in Wiley's Professional Learning Group! Congratulations Jen!

We owe our gratitude to Corrina Santos, Course Content Developer, and Samantha Hart, Editorial Assistant for their careful quality control and editing of the new content, and Anju Joshi and Wendi Lai for layout and design, as well as keeping the production process running smoothly. With the lack of direct interaction due to travel restrictions during the past year, we were impressed with the cohesive team effort, despite the lack of in-person interactions, and we are grateful for and recognize the individual contributions from Georgia Larson, Assistant Editor, Svetlana Barskaya, Senior Manager, Course Development and Production, Mary Donovan,Senior Managing Editor, and Patricia Gutierrez, Senior Course Production Specialist. We would also like to welcome our new Marketing Manager, Sean Willey, who has hit the ground running on getting the word out about the new features of this edition.

We think all would agree that the sales representatives of John Wiley & Sons, Inc. form the backbone of this entire effort. Without you, and your expertise and understanding of the book's features, its online content, its integration into WileyPLUS, and the needs of students and instructors in higher education, it would never have had the success that is has over the past 30 years! And you should know that it was one of your own, Dayna Leaman, our good friend, who got us into this business in the first place! Thank you, Dayna!

Finally, we would like to thank our physics colleagues and their students who have kindly shared their ideas about effective pedagogy, and improved the text by identifying errors. We are grateful for all of their suggestions as they have helped us to produce a more clear and accurate work. To the reviewers of this and previous editions, we especially owe a large debt of gratitude. Specifically, we thank:

Lai Cao, *Baton Rouge Magnet High School*
Candee Chambers-Colbeck, *Maryville University*
Diana Driscoll, *Case Western Reserve University*
Costas Efthimiou, *University of Central Florida*
Robert Egler, *North Carolina State University*
Sambandamurthy Ganapathy, *The State University of New York at Buffalo*
Joseph Ganem, *Loyola University, Maryland*
Jasper Halekas, *University of Iowa*
Lilit Haroyan, *East Los Angeles College*
Richard Holland, *Rend Lake College and SIUC*
Klaus Honscheid, *Ohio State University*
Shyang Huang, *Missouri State University*
Craig Kleitzing, *University of Iowa*
Wayne Mathe, *University of South Florida, Sarasota-Manatee Campus*
Wayne Mathé, *University of South Florida Tampa*
Samuel Mensah, *University of Memphis*
Mark Morgan-Tracy, *University of New Mexico*
Kriton Papavasiliou, *Virginia Polytechnic Institute and State University*
Kriton Papavasiliou, *Virginia Tech*
Payton Parker, *Midlothian Heritage High School*
Christian Prewitt, *Midlothian Heritage High School*
Joshua Ravenscraft, *Vernon Hills High School and College of Lake County*
Bill Schmidt, *Meredith College*
Brian Schuft, *North Carolina A&T State University*
Andreas Shalchi, *University of Manitoba*
Deepshikha Shukla, *Rockford University*
Jennifer Snyder, *San Diego Mesa College*
Paul Sokol, *Indiana University Bloomington*
Richard Taylor, *University of Oregon*
Beth Thacker, *Texas Tech University*
Anne Topper, *Queen's University*
David Ulrich, *Portland Community College*
Luc T. Wille, *Florida Atlantic University*

About the cover: One of the great successes of physics as a science is its ability to accurately predict the behavior of the natural world across all length scales, from the atomically small to the astronomically large. The cover image represents this achievement with an artist's rendition of a human eye drifting in the cosmos. At the center of the eye's pupil is the famous double helix of DNA. The motion of the planets and stars is largely determined by the gravitational force. The formation of an image by our eyes is governed by the laws of optics, and the amino acid base pairs in DNA are held together by electrostatic forces. In this course, you will study all of these topics and learn how to apply them to solve a variety of problems.

In spite of our best efforts to produce an error-free book, errors no doubt remain. They are solely our responsibility, and we would appreciate hearing of any that you find. We hope that this text makes learning and teaching physics easier and more enjoyable, and we look forward to hearing about your experiences with it. Please feel free to write us care of Physics Editor, Global Education, John Wiley & Sons, Inc., 111 River Street, Hoboken, NJ 07030, or contact the authors at **dyoun14@gmail.com** or **sstadler23@gmail.com**.

Rachel Hopper/Dreamstime.com

We have all experienced static electricity in our hair and on our clothes, and have been zapped on occasion when touching a doorknob after walking on carpet. These phenomena occur when electric charges, one of the fundamental building blocks of atoms, separate, and one type (either positive or negative) becomes more abundant than the other. As we will see in this chapter, like charges repel, which is why the toddler's hair is standing on end.

Electric Forces and Electric Fields

18.1 The Origin of Electricity

The electrical nature of matter is inherent in atomic structure. An atom consists of a small, relatively massive nucleus that contains particles called protons and neutrons. A proton has a mass of 1.673×10^{-27} kg, and a neutron has a slightly greater mass of 1.675×10^{-27} kg. Surrounding the nucleus is a diffuse cloud of orbiting particles called electrons, as **Figure 18.1** suggests. An electron has a mass of 9.11×10^{-31} kg. Like mass, **electric charge** is an intrinsic property of protons and electrons, and only two types of charge have been discovered, positive and negative. A proton has a positive charge, and an electron has a negative charge. A neutron has no net electric charge.

Experiment reveals that the magnitude of the charge on the proton *exactly equals* the magnitude of the charge on the electron; the proton carries a charge $+e$, and the electron carries a charge $-e$. The SI unit for measuring the magnitude of an electric charge is the **coulomb*** (C), and e has been determined experimentally to have the value

$$e = 1.60 \times 10^{-19}\ \text{C}$$

*The definition of the coulomb depends on electric currents and magnetic fields, concepts that will be discussed later. Therefore, we postpone its definition until Section 21.7.

LEARNING OBJECTIVES

After reading this module, you should be able to...

18.1 Define electric charge.

18.2 Describe the electric force between charged particles.

18.3 Distinguish between conductors and insulators.

18.4 Explain charging by contact and charging by induction.

18.5 Use Coulomb's law to calculate the force on a point charge due to other point charges.

18.6 Calculate the net electric field due to a configuration of point charges.

18.7 Draw electric field lines.

18.8 Describe the electric field inside a conductor.

18.9 Use Gauss' law to obtain the value of the electric field due to charge distributions.

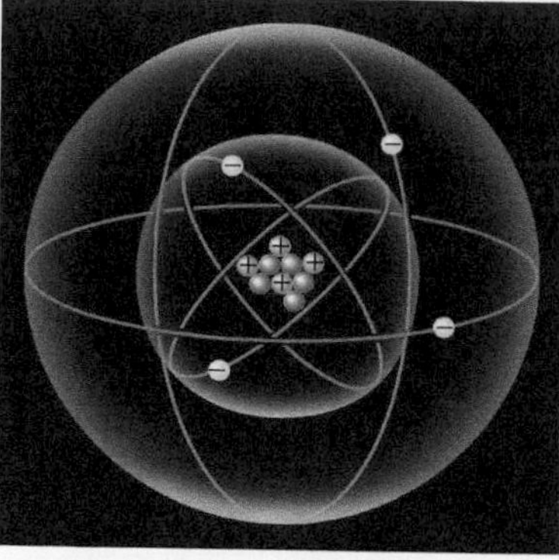

FIGURE 18.1 An atom contains a small, positively charged nucleus, about which the negatively charged electrons move. The closed-loop paths shown here are symbolic only. In reality, the electrons do not follow discrete paths, as Section 30.5 discusses.

The symbol e represents only the magnitude of the charge on a proton or an electron and does not include the algebraic sign that indicates whether the charge is positive or negative.

In nature, atoms are normally found with equal numbers of protons and electrons. Usually, then, an atom carries no net charge because the algebraic sum of the positive charge of the nucleus and the negative charge of the electrons is zero. When an atom, or any object, carries no net charge, the object is said to be *electrically neutral*. The neutrons in the nucleus are electrically neutral particles.

Charges of larger magnitude than the charge on an electron or on a proton are built up on an object by adding or removing electrons. Thus, any charge of magnitude q is an integer multiple of e; that is, $q = Ne$, where N is an integer. Because any electric charge q occurs in integer multiples of elementary, indivisible charges of magnitude e, electric charge is said to be *quantized*. In Section 19.7, we will discuss the experiment that determined the value of e. Example 1 emphasizes the quantized nature of electric charge.

EXAMPLE 1 | A Lot of Electrons

How many electrons are there in one coulomb of negative charge?

Reasoning The negative charge is due to the presence of excess electrons, since they carry negative charge. Because each electron has a charge whose magnitude is $e = 1.60 \times 10^{-19}$ C, the number of electrons is equal to the charge magnitude of one coulomb (1.00 C) divided by e.

Solution The number N of electrons is

$$N = \frac{1.00\text{ C}}{e} = \frac{1.00\text{ C}}{1.60 \times 10^{-19}\text{ C}} = \boxed{6.25 \times 10^{18}}$$

18.2 Charged Objects and the Electric Force

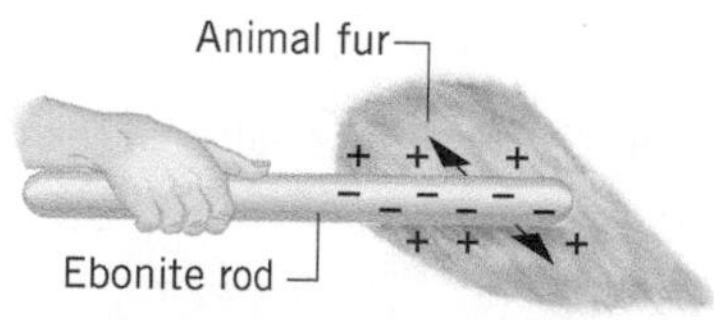

FIGURE 18.2 When an ebonite rod is rubbed against animal fur, electrons from atoms of the fur are transferred to the rod. This transfer gives the rod a negative charge (−) and leaves a positive charge (+) on the fur.

Electricity has many useful applications that have come about because it is possible to transfer electric charge from one object to another. Usually electrons are transferred, and the body that gains electrons acquires an excess of negative charge. The body that loses electrons has an excess of positive charge. Such separation of charge occurs often when two unlike materials are rubbed together. For example, when an ebonite (hard, black rubber) rod is rubbed against animal fur, some of the electrons from atoms of the fur are transferred to the rod. The ebonite becomes negatively charged, and the fur becomes positively charged, as **Figure 18.2** indicates. Similarly, if a glass rod is rubbed with a silk cloth, some of the electrons are removed from the atoms of the glass and deposited on the silk, leaving the silk negatively charged and the glass positively charged. There are many familiar examples of charge separation, as when you walk across a nylon rug or run a comb through dry hair. In each case, objects become "electrified" as surfaces rub against one another.

When an ebonite rod is rubbed with animal fur, the rubbing process serves only to separate electrons and protons already present in the materials. No electrons or protons are created or destroyed. Whenever an electron is transferred to the rod, a proton is left behind on the fur. Since the charges on the electron and proton have identical magnitudes but opposite signs, the algebraic sum of the two charges is zero, and the transfer does not change the net charge of the fur/rod system. If each material contains an equal number of protons and electrons to begin with, the net charge of the system is zero initially and remains zero at all times during the rubbing process.

Electric charges play a role in many situations other than rubbing two surfaces together. They are involved, for instance, in chemical reactions, electric circuits, and radioactive decay. A great number of experiments have verified that in any situation, the **law of conservation of electric charge** is obeyed.

LAW OF CONSERVATION OF ELECTRIC CHARGE

During any process, the net electric charge of an isolated system remains constant (is conserved).

It is easy to demonstrate that two electrically charged objects exert a force on one another. Consider **Interactive Figure 18.3*a***, which shows two small balls that have been *oppositely charged* and are light and free to move. The balls attract each other. On the other hand, balls with the *same* type of charge, either both positive or both negative, repel each other, as parts *b* and *c* of the drawing indicate. The behavior depicted in **Interactive Figure 18.3** illustrates the following fundamental characteristic of electric charges:

Like charges repel and unlike charges attract each other.

Like other forces that we have encountered, the **electric force** (also sometimes called the **electrostatic force**) can alter the motion of an object. It can do so by contributing to the net external force $\sum \vec{\mathbf{F}}$ that acts on the object. Newton's second law, $\sum \vec{\mathbf{F}} = m\vec{\mathbf{a}}$, specifies the acceleration $\vec{\mathbf{a}}$ that arises because of the net external force. Any external electric force that acts on an object must be included when determining the net external force to be used in the second law.

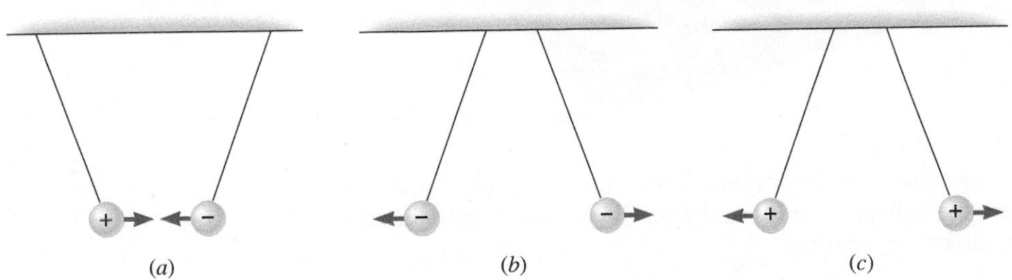

INTERACTIVE FIGURE 18.3 (*a*) A positive charge (−) and a negative charge (+) attract each other. (*b*) Two negative charges repel each other. (*c*) Two positive charges repel each other.

THE PHYSICS OF . . . electronic ink. The development of this technology began in the late 1990's and, since 2015, has been used in many applications. In fact, electronic ink, which utilizes electric forces, has revolutionized the way books and other printed matter are made, including popular devices used to read electronic books. It allows letters and graphics on a page to be changed instantly, much like the symbols displayed on a computer monitor. **Figure 18.4*a*** illustrates the essential features of electronic ink. It consists of millions of clear microcapsules, each having the diameter of a human hair and filled with a dark, inky liquid. Inside each microcapsule are several dozen extremely tiny white beads that carry a slightly negative charge. The microcapsules are sandwiched between two sheets, an opaque base layer and a transparent top layer, at which the reader looks. When a positive charge is applied to a small region of the base layer, as shown in part *b* of the drawing, the negatively charged white beads are drawn to it, leaving dark ink at the top layer. Thus, a viewer sees only the dark liquid. When a negative charge is applied to a region of the base layer, the negatively charged white beads are repelled from it and are forced to the top of the microcapsules; now a viewer sees a white area due to the beads. Thus, electronic ink is based on the principle that like charges repel and unlike charges attract each other; a positive charge causes one color to appear, and a negative charge causes another color to appear. Each small region, whether dark or light, is known as a *pixel* (short for "picture element"). Computer chips provide the

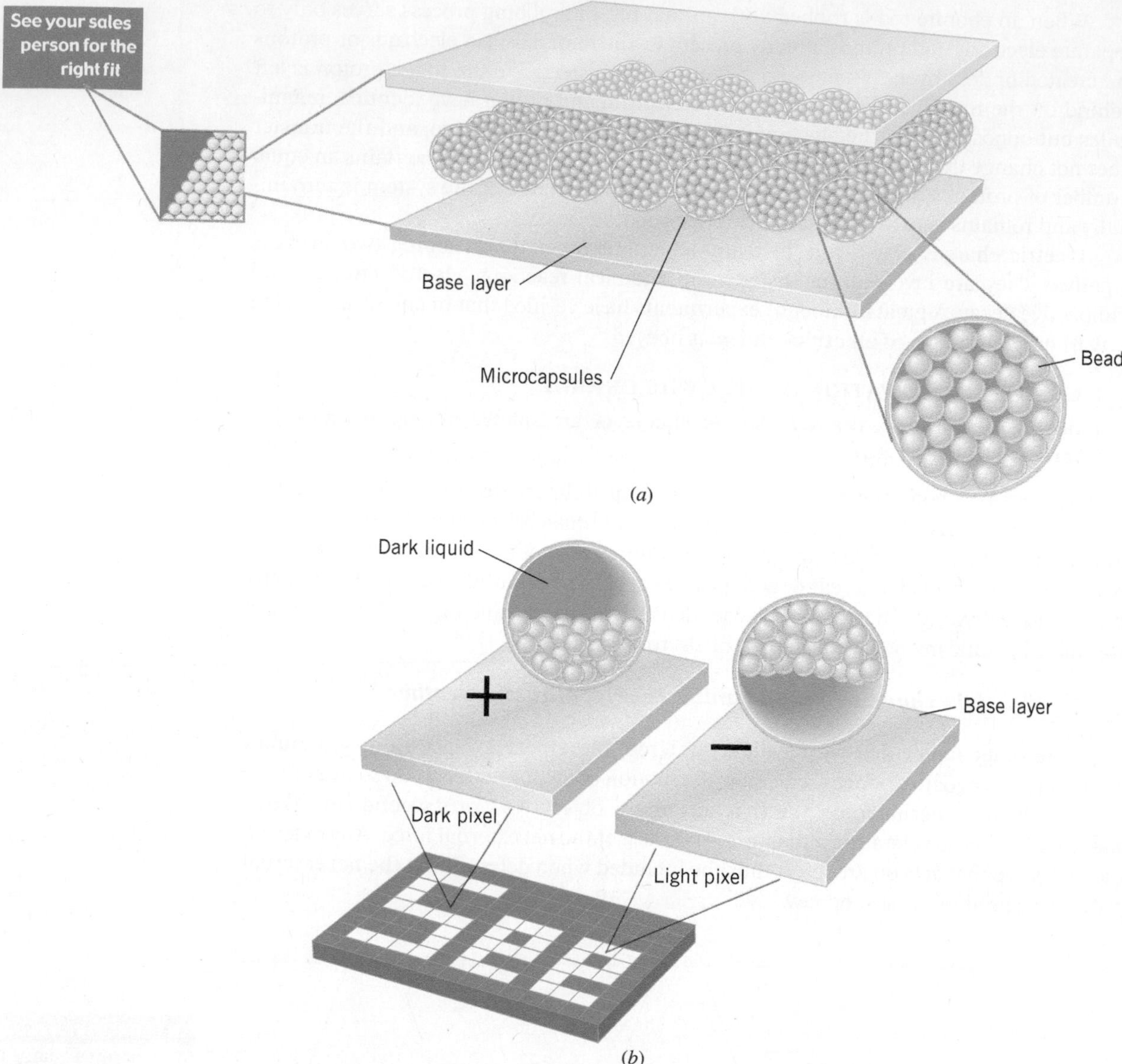

FIGURE 18.4 (*a*) Electronic ink consists of microcapsules filled with a dark, inky liquid and dozens of white beads. (*b*) Dark and light pixels are formed when positive and negative charges are placed in the base layer by electronic circuitry.

instructions to produce the negative and positive charges on the base layer of each pixel. Letters and graphics are produced by the patterns generated with the two colors.

Check Your Understanding

(*The answers are given at the end of the book.*)

1. An electrically neutral object acquires a net electric charge. Which one of the following statements concerning the mass of the object is true? **(a)** The mass does not change. **(b)** The mass increases if the charge is positive and decreases if it is negative. **(c)** The mass increases if the charge is negative and decreases if it is positive.

2. Object A and object B are each electrically neutral. Two million electrons are removed from object A and placed on object B. Expressed in coulombs, what is the resulting charge (algebraic sign and magnitude) on object A and on object B?

3. Object A has a charge of -1.6×10^{-13} C, and object B is electrically neutral. Two million electrons are removed from object A and placed on object B. Expressed in coulombs, what is the resulting charge (algebraic sign and magnitude) on object A and on object B?

18.3 Conductors and Insulators

Electric charge can not only exist *on an object,* but it can also move *through an object.* However, materials differ vastly in their abilities to allow electric charge to move or be conducted through them. To help illustrate such differences in conductivity, **Figure 18.5*a*** recalls the conduction of heat through a bar of material whose ends are maintained at different temperatures. As Section 13.2 discusses, metals conduct heat readily and, therefore, are known as thermal conductors. On the other hand, substances that conduct heat poorly are referred to as thermal insulators.

A situation analogous to the conduction of heat arises when a metal bar is placed between two charged objects, as in **Figure 18.5*b***. Electrons are conducted through the bar from the negatively charged object toward the positively charged object. Substances that readily conduct electric charge are called **electrical conductors**. Although there are exceptions, good thermal conductors are generally good electrical conductors. Metals such as copper, aluminum, silver, and gold are excellent electrical conductors and, therefore, are used in electrical wiring. Materials that conduct electric charge *poorly* are known as **electrical insulators**. In many cases, thermal insulators are also electrical insulators. Common electrical insulators are rubber, many plastics, and wood. Insulators, such as the rubber or plastic that coats electrical wiring, prevent electric charge from going where it is not wanted.

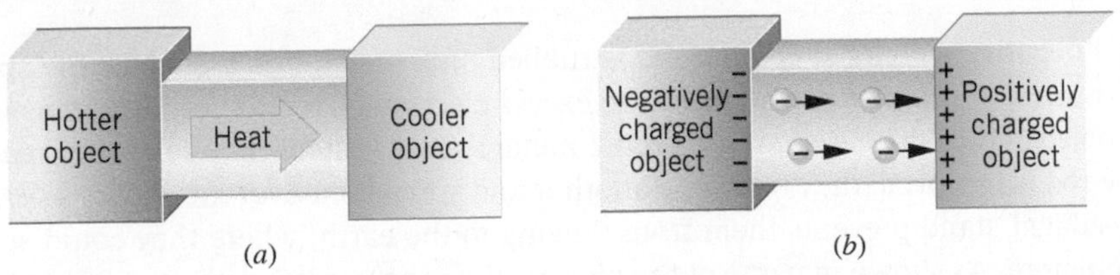

FIGURE 18.5 (*a*) Heat is conducted from the hotter end of the metal bar to the cooler end. (*b*) Electrons are conducted from the negatively charged end of the metal bar to the positively charged end.

The difference between electrical conductors and insulators is related to atomic structure. As electrons orbit the nucleus, those in the outer orbits experience a weaker force of attraction to the nucleus than do those in the inner orbits. Consequently, the outermost electrons (also called the valence electrons) can be dislodged more easily than the inner ones. In a good conductor, some valence electrons become detached from a parent atom and wander more or less freely throughout the material, belonging to no one atom in particular. The exact number of electrons detached from each atom depends on the nature of the material, but is usually between one and three. When one end of a conducting bar is placed in contact with a negatively charged object and the other end in contact with a positively charged object, as in **Figure 18.5*b***, the "free" electrons are able to move readily away from the negative end and toward the positive end. The ready movement of electrons is the hallmark of a good conductor. In an insulator the situation is different, for there are very few electrons free to move throughout the material. Virtually every electron remains bound to its parent atom. Without the "free" electrons, there is very little flow of charge when the material is placed between two oppositely charged bodies, so the material is an electrical insulator.

THE PHYSICS OF . . . photoconductivity. The conducting properties of some materials can change depending on their environments or external stimuli. For instance, there are materials that change from insulators to conductors depending on their temperature. An example of such a material that exhibits this behavior is vanadium dioxide (VO_2), which is a conductor above 341 K and an insulator below that temperature. Some materials, however, such as selenium, transform from poor conductors to good conductors when exposed to light. Such a material is said to be a photoconductor. At the atomic level, the light provides energy to the least tightly bound electrons so that they become more mobile, and can therefore more readily conduct through the material. The electrical conductivity of selenium can increase by a factor of a thousand when exposed to light!

Some of the earliest and most influential technological applications of the photoconducting properties of selenium are seen in the photocopying process of xerography and in the early development of television. The word "xerography" derives from the Greek *xeros* and *graphos*, meaning "dry writing." At the heart of a xerographic copier (an old technology that has since been replaced by modern printers) is a xerographic drum, an aluminum cylinder coated with a layer of selenium. An image of the document to be copied is focused on the selenium-coated cylinder that initially carries a uniform positive charge density. The portions exposed to the light become conducting and are then electrically neutralized by conducting electrons. This process creates an "image" in positive charge density on the surface of the cylinder that then attracts a negatively charged toner powder, which can then be transferred to positively charged paper. The paper with its toner image is then sent through heated rollers to form the final copy.

The origin of early television can be traced back to the development of selenium photoconducting cells that could be used to convert light into electrical signals. These electrical signals could then be transmitted using radio waves and then detected, processed, and displayed by a receiver.

18.4 Charging by Contact and by Induction

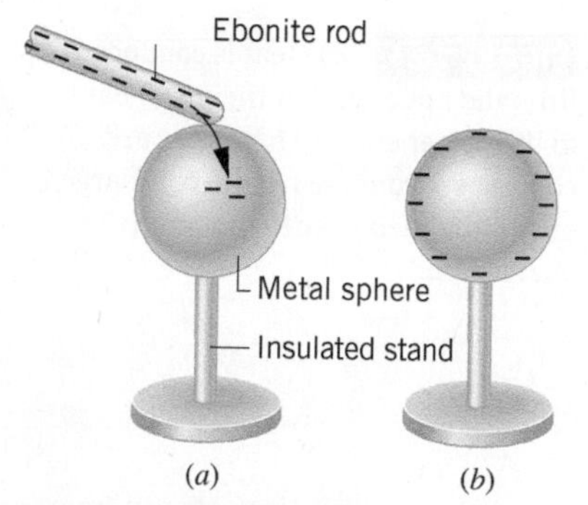

ANIMATED FIGURE 18.6 (*a*) Electrons are transferred by rubbing the negatively charged rod on the metal sphere. (*b*) When the rod is removed, the electrons distribute themselves over the surface of the sphere.

When a negatively charged ebonite rod is rubbed on a metal object, such as the sphere in **Animated Figure 18.6*a***, some of the excess electrons from the rod are transferred to the object. Once the electrons are on the metal sphere (where they can move readily) and the rod is removed, they repel one another and spread out over the sphere's surface. The insulated stand prevents them from flowing to the earth, where they could spread out even more. As shown in part *b* of the picture, the sphere is left with a negative charge distributed over its surface. In a similar manner, the sphere would be left with a positive charge after being rubbed with a positively charged rod. In this case, electrons from the sphere would be transferred to the rod. The process of giving one object a net electric charge by placing it in contact with another object that is already charged is known as **charging by contact.**

It is also possible to charge a conductor in a way that does not involve contact. In **Animated Figure 18.7**, a negatively charged rod is brought close to, *but does not touch,* a metal sphere. In the sphere, the free electrons closest to the rod move to the other side, as part *a* of the drawing indicates. As a result, the part of the sphere nearest the rod becomes positively charged and the part farthest away becomes negatively charged. These positively and negatively charged regions have been "induced" or "persuaded" to form because of the repulsive force between the negative rod and the free electrons in the sphere. If the rod were removed, the free electrons would return to their original places, and the charged regions would disappear.

Under most conditions the earth is a good electrical conductor. So when a metal wire is attached between the sphere and the ground, as in **Animated Figure 18.7*b***, some of the free electrons leave the sphere and distribute themselves over the much larger earth. If the grounding wire is then removed, followed by the ebonite rod, the sphere is left with

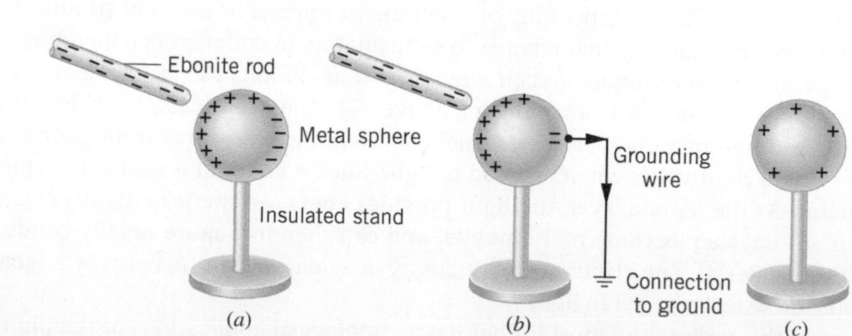

ANIMATED FIGURE 18.7 (*a*) When a charged rod is brought near the metal sphere without touching it, some of the positive and negative charges in the sphere are separated. (*b*) Some of the electrons leave the sphere through the grounding wire, with the result (*c*) that the sphere acquires a positive net charge.

a positive net charge, as part *c* of the picture shows. The process of giving one object a net electric charge *without touching* the object to a second charged object is called **charging by induction**. The process could also be used to give the sphere a negative net charge, if a positively charged rod were used. Then, electrons would be drawn up from the ground through the grounding wire and onto the sphere.

If the sphere in **Animated Figure 18.7** were made from an insulating material like plastic, instead of metal, the method of producing a net charge by induction would not work, because very little charge would flow through the insulating material and down the grounding wire. However, the electric force of the charged rod would have some effect on the insulating material. The electric force would cause the positive and negative charges in the molecules of the material to separate slightly, with the negative charges being "pushed" away from the negative rod, as **Figure 18.8** illustrates. Although no net charge is created, the surface of the plastic does acquire a slight induced positive charge and is attracted to the negative rod. It is attracted in spite of the repulsive force between the negative rod and the negative charges in the plastic. This is because the negative charges in the plastic are further away from the rod than the positive charges are. For a similar reason, one piece of cloth can stick to another in the phenomenon known as "static cling," which occurs when an article of clothing has acquired an electric charge while being tumbled about in a clothes dryer.

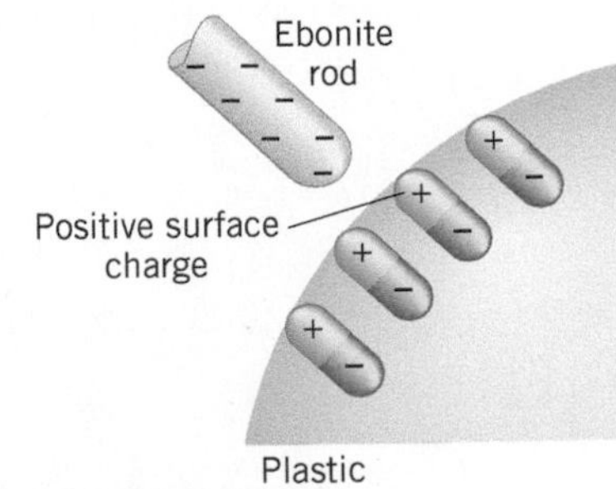

FIGURE 18.8 The negatively charged rod induces a slight positive surface charge on the plastic.

Check Your Understanding

(The answers are given at the end of the book.)

4. Two metal spheres are identical. They are electrically neutral and are touching. An electrically charged ebonite rod is then brought near the spheres without touching them, as **CYU Figure 18.1** shows. After a while, with the rod held in place, the spheres are separated, and the rod is then removed. The following statements refer to the masses m_A and m_B of the spheres after they are separated and the rod is removed. Which one or more of the statements is true? **(a)** $m_A = m_B$ **(b)** $m_A > m_B$ if the rod is positive **(c)** $m_A < m_B$ if the rod is positive **(d)** $m_A > m_B$ if the rod is negative **(e)** $m_A < m_B$ if the rod is negative

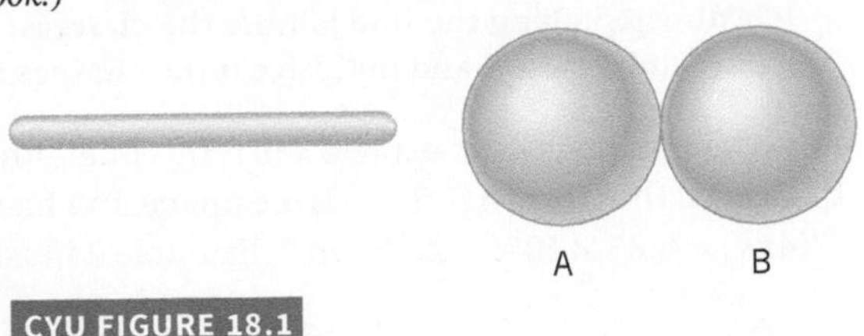

CYU FIGURE 18.1

5. Blow up a balloon, tie it shut, and rub it against your shirt a number of times, so that the balloon acquires a net electric charge. Now touch the balloon to the ceiling. When released, will the balloon remain stuck to the ceiling?

6. A rod made from insulating material carries a net charge (which may be positive or negative), whereas a copper sphere is electrically neutral. The rod is held close to the sphere but does not touch it. Which one of the following statements concerning the forces that the rod and sphere exert on each other is true? **(a)** The forces are always attractive. **(b)** The forces are always repulsive. **(c)** The forces are attractive when the rod is negative and repulsive when it is positive. **(d)** The forces are repulsive when the rod is negative and attractive when it is positive. **(e)** There are no forces.

18.5 Coulomb's Law

The Force That Point Charges Exert on Each Other

The electrostatic force that stationary charged objects exert on each other depends on the amount of charge on the objects and the distance between them. Experiments reveal that the greater the charge and the closer together they are, the greater is the force. To set the stage for explaining these features in more detail, **Figure 18.9** shows two charged bodies. These objects are so small, compared to the distance r between them, that they

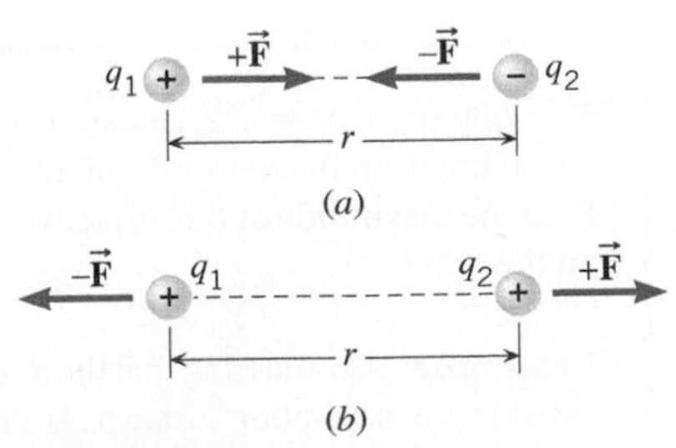

FIGURE 18.9 Each point charge exerts a force on the other. Regardless of whether the forces are (*a*) attractive or (*b*) repulsive, they are directed along the line between the charges and have equal magnitudes.

can be regarded as mathematical points. The "point charges" have magnitudes* $|q_1|$ and $|q_2|$. If the charges have *unlike* signs, as in part *a* of the picture, each object is *attracted* to the other by a force that is directed along the line between them; $+\vec{\mathbf{F}}$ is the electric force exerted on object 1 by object 2 and $-\vec{\mathbf{F}}$ is the electric force exerted on object 2 by object 1. If, as in part *b*, the charges have the *same* sign (both positive or both negative), each object is *repelled* from the other. The repulsive forces, like the attractive forces, act along the line between the charges. Whether attractive or repulsive, the two forces are equal in magnitude but opposite in direction. These forces always exist as a pair, each one acting on a different object, in accord with Newton's action–reaction law.

The French physicist Charles Augustin de Coulomb (1736–1806) carried out a number of experiments to determine how the electric force that one point charge applies to another depends on the amount of each charge and the separation between them. His result, now known as **Coulomb's law**, is stated as follows:

COULOMB'S LAW

The magnitude F of the electrostatic force exerted by one point charge q_1 on another point charge q_2 is directly proportional to the magnitudes $|q_1|$ and $|q_2|$ of the charges and inversely proportional to the square of the distance r between them:

$$F = k\frac{|q_1|\,|q_2|}{r^2} \tag{18.1}$$

where k is a proportionality constant: $k = 8.99 \times 10^9\ \mathrm{N \cdot m^2/C^2}$ in SI units. Equation 18.1 gives only the magnitude of the electrostatic force that each point charge exerts on the other; it does *not* give the direction. The electrostatic force is directed along the line joining the charges, and it is attractive if the charges have unlike signs and repulsive if the charges have like signs.

It is common practice to express k in terms of another constant ε_0, by writing $k = 1/(4\pi\varepsilon_0)$; ε_0 is called the **permittivity of free space** and has a value that is given according to $\varepsilon_0 = 1/(4\pi k) = 8.85 \times 10^{-12}\ \mathrm{C^2/(N \cdot m^2)}$. Example 2 illustrates the use of Coulomb's law.

Math Skills When using Equation 18.1, substitute only the charge magnitudes (without algebraic signs) for $|q_1|$ and $|q_2|$. Do not substitute negative numbers for these symbols. This is because the equation gives only the magnitude of the electrostatic force, and the magnitude of a force cannot be negative. For example, suppose that $q_1 = -5.0 \times 10^{-6}$ C and $q_2 = +7.0 \times 10^{-6}$ C. Then, for use in Equation 18.1, we would have the following values:

$$|q_1| = |-5.0 \times 10^{-6}\ \mathrm{C}| = 5.0 \times 10^{-6}\ \mathrm{C} \quad (\textit{not} - 5.0 \times 10^{-6}\ \mathrm{C})$$

$$|q_2| = |+7.0 \times 10^{-6}\ \mathrm{C}| = 7.0 \times 10^{-6}\ \mathrm{C}$$

EXAMPLE 2 | A Large Attractive Force

Two objects, whose charges are +1.0 and −1.0 C, are separated by 1.0 km. Compared to 1.0 km, the sizes of the objects are small. Find the magnitude of the attractive force that either charge exerts on the other.

Reasoning Considering that the sizes of the objects are small compared to the separation distance, we can treat the charges as point charges. Coulomb's law may then be used to find the magnitude of the attractive force, provided that only the *magnitudes of the charges* are used for the symbols $|q_1|$ and $|q_2|$ that appear in the law.

Solution The magnitude of the force is

$$F = k\frac{|q_1|\,|q_2|}{r^2} = \frac{(8.99 \times 10^9\ \mathrm{N \cdot m^2/C^2})(1.0\ \mathrm{C})(1.0\ \mathrm{C})}{(1.0 \times 10^3\ \mathrm{m})^2} \tag{18.1}$$

$$= \boxed{9.0 \times 10^3\ \mathrm{N}}$$

*The magnitude of a variable is sometimes called the absolute value and is symbolized by a vertical bar to the left and to the right of the variable. Thus, $|q|$ denotes the magnitude or absolute value of the variable q, which is the value of q without its algebraic plus or minus sign. For example, if $q = -2.0$ C, then $|q| = 2.0$ C.

The force calculated in Example 2 corresponds to about 2000 pounds and is so large because charges of ± 1.0 C are enormous. Such large charges are rare and are encountered only in the most severe conditions, as in a lightning bolt, where as much as 25 C can be transferred between the cloud and the ground. The typical charges produced in the laboratory are much smaller and are measured conveniently in microcoulombs (1 microcoulomb $= 1\ \mu C = 10^{-6}$ C).

Coulomb's law has a form that is remarkably similar to Newton's law of gravitation ($F = Gm_1m_2/r^2$). The force in both laws depends on the inverse square ($1/r^2$) of the distance between the two objects and is directed along the line between them. In addition, the force is proportional to the product of an intrinsic property of each of the objects, the magnitudes of the charges $|q_1|$ and $|q_2|$ in Coulomb's law and the masses m_1 and m_2 in the gravitation law. However, there is a major difference between the two laws. The electrostatic force can be either repulsive or attractive, depending on whether or not the charges have the same sign; in contrast, the gravitational force is *always* an attractive force.

Section 5.5 discusses how the gravitational attraction between the earth and a satellite provides the centripetal force that keeps a satellite in orbit. Example 3 illustrates that the electrostatic force of attraction plays a similar role in a famous model of the atom created by the Danish physicist Niels Bohr (1885–1962).

Analyzing Multiple-Concept Problems

EXAMPLE 3 | A Model of the Hydrogen Atom

In the Bohr model of the hydrogen atom, the electron (charge $= -e$) is in a circular orbit about the nuclear proton (charge $= +e$) at a radius of 5.29×10^{-11} m, as **Figure 18.10** shows. The mass of the electron is 9.11×10^{-31} kg. Determine the speed of the electron.

Reasoning Recall from Section 5.3 that a net force is required to keep an object such as an electron moving on a circular path. This net force is called the centripetal force and always points toward the center of the circle. The centripetal force has a magnitude given by $F_c = mv^2/r$, where m and v are, respectively, the mass and speed of the electron and r is the radius of the orbit. This equation can be solved for the speed of the electron. Since the mass and orbital radius are known, we can calculate the electron's speed provided that a value for the centripetal force can be found. For the electron in the hydrogen atom, the centripetal force is provided almost exclusively by the electrostatic force that the proton exerts on the electron. This attractive force points toward the center of the circle, and its magnitude is given by Coulomb's law. The electron is also pulled toward the proton by the gravitational force. However, the gravitational force is negligible in comparison to the electrostatic force.

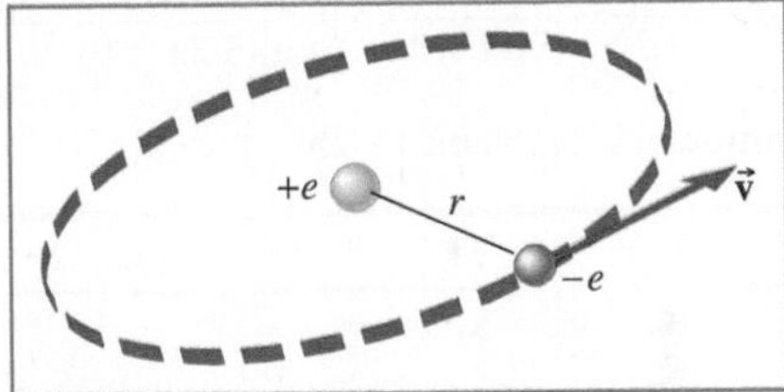

FIGURE 18.10 In the Bohr model of the hydrogen atom, the electron ($-e$) orbits the proton ($+e$) at a distance that is $r = 5.29 \times 10^{-11}$ m. The velocity of the electron is $\vec{\mathbf{v}}$.

Knowns and Unknowns The data for this problem are:

Description	Symbol	Value
Electron charge	$-e$	-1.60×10^{-19} C
Electron mass	m	9.11×10^{-31} kg
Proton charge	$+e$	$+1.60 \times 10^{-19}$ C
Radius of orbit	r	5.29×10^{-11} m
Unknown Variable		
Orbital speed of electron	v	?

Modeling the Problem

STEP 1 Centripetal Force An electron of mass m that moves with a constant speed v on a circular path of radius r experiences a net force, called the centripetal force. The magnitude F_c of this force is given by $F_c = mv^2/r$ (Equation 5.3). By solving this equation for the speed, we obtain Equation 1 at the right. The mass and radius in this expression are known. However, the magnitude of the centripetal force is not known, so we will evaluate it in Step 2.

$$v = \sqrt{\frac{rF_c}{m}} \quad (1)$$

?

STEP 2 Coulomb's Law As the electron orbits the proton in the hydrogen atom, it is attracted to the proton by the electrostatic force. The magnitude F of the electrostatic force is given by Coulomb's law as $F = k|q_1|\,|q_2|/r^2$ (Equation 18.1), where $|q_1|$ and $|q_2|$ are the magnitudes of the charges, r is the orbital radius, and $k = 8.99 \times 10^9\ \mathrm{N \cdot m^2/C^2}$. Since the centripetal force is provided almost entirely by the electrostatic force, it follows that $F_c = F$. Furthermore, $|q_1| = |-e|$ and $|q_2| = |+e|$. With these substitutions, Equation 18.1 becomes

$$F_c = k\frac{|-e|\,|+e|}{r^2}$$

All the variables on the right side of this expression are known, so we substitute it into Equation 1, as indicated in the right column.

$$v = \sqrt{\frac{rF_c}{m}} \quad (1)$$

$$F_c = k\frac{|-e|\,|+e|}{r^2}$$

Solution Algebraically combining the results of the modeling steps, we have

STEP 1 STEP 2

$$v = \sqrt{\frac{rF_c}{m}} = \sqrt{\frac{r\left(k\frac{|-e|\,|+e|}{r^2}\right)}{m}} = \sqrt{\frac{k|-e|\,|+e|}{mr}}$$

The speed of the orbiting electron is

$$v = \sqrt{\frac{k|-e|\,|+e|}{mr}}$$

$$= \sqrt{\frac{(8.99 \times 10^9\ \mathrm{N \cdot m^2/C^2})|-1.60 \times 10^{-19}\ \mathrm{C}|\,|+1.60 \times 10^{-19}\ \mathrm{C}|}{(9.11 \times 10^{-31}\ \mathrm{kg})(5.29 \times 10^{-11}\ \mathrm{m})}} = \boxed{2.19 \times 10^6\ \mathrm{m/s}}$$

Related Homework: Problems 19, 23

THE PHYSICS OF . . . adhesion. Since the electrostatic force depends on the inverse square of the distance between the charges, it becomes larger for smaller distances, such as those involved when a strip of adhesive tape is stuck to a smooth surface. Electrons shift over the small distances between the tape and the surface. As a result, the materials become oppositely charged. Since the distance between the charges is relatively small, the electrostatic force of attraction is large enough to contribute to the adhesive bond. **Figure 18.11** shows an image of the sticky surface of a piece of tape after it has been pulled off a metal surface. The image was obtained using an atomic-force microscope and reveals the tiny pits left behind when microscopic portions of the adhesive remain stuck to the metal because of the strong adhesive bonding forces.

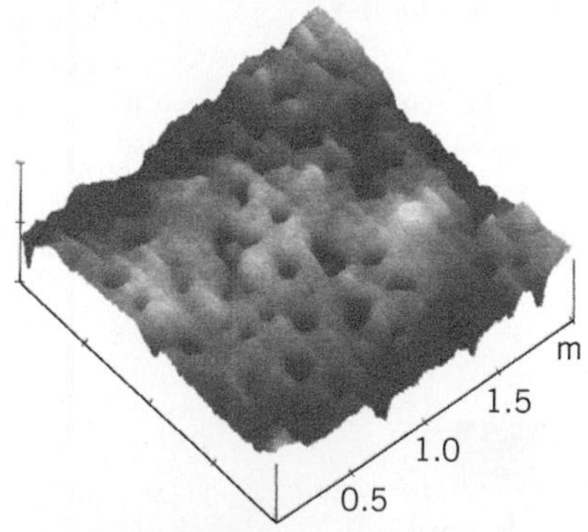

FIGURE 18.11 After a strip of tape has been pulled off a metal surface, there are tiny pits in the sticky surface of the tape, as this image shows. It was obtained using an atomic-force microscope. (Courtesy Louis Scudiero and J.Thomas Dickinson, Washington State University)

The Force on a Point Charge Due to Two or More Other Point Charges

Up to now, we have been discussing the electrostatic force on a point charge (magnitude $|q_1|$) due to another point charge (magnitude $|q_2|$). Suppose that a third point charge (magnitude $|q_3|$) is also present. What would be the net force on q_1 due to both q_2 and q_3? It is convenient to deal with such a problem in parts. First, find the magnitude and direction of the force exerted on q_1 by q_2 (ignoring q_3). Then, determine the force exerted on q_1 by q_3 (ignoring q_2). The *net force* on q_1 is the *vector sum* of these forces. Examples 4 and 5 illustrate this approach when the charges lie along a straight line and on a plane, respectively.

EXAMPLE 4 | Three Charges on a Line

Figure 18.12*a* shows three point charges that lie along the *x* axis in a vacuum. Determine the magnitude and direction of the net electrostatic force on q_1.

Reasoning Part *b* of the drawing shows a free-body diagram of the forces that act on q_1. Since q_1 and q_2 have opposite signs, they attract one another. Thus, the force exerted on q_1 by q_2 is $\vec{\mathbf{F}}_{12}$, and it points to the left. Similarly, the force exerted on q_1 by q_3 is $\vec{\mathbf{F}}_{13}$ and is also an attractive force. It points to the right in **Figure 18.12*b***. The magnitudes of these forces can be obtained from Coulomb's law. The net force is the vector sum of $\vec{\mathbf{F}}_{12}$ and $\vec{\mathbf{F}}_{13}$.

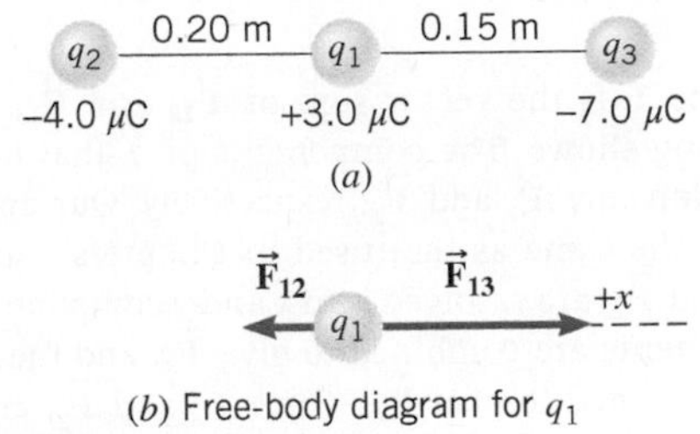

q_2 0.20 m q_1 0.15 m q_3

−4.0 μC +3.0 μC −7.0 μC

(*a*)

$\vec{\mathbf{F}}_{12}$ q_1 $\vec{\mathbf{F}}_{13}$ $+x$

(*b*) Free-body diagram for q_1

FIGURE 18.12 (*a*) Three charges lying along the *x* axis. (*b*) The force exerted on q_1 by q_2 is $\vec{\mathbf{F}}_{12}$, while the force exerted on q_1 by q_3 is $\vec{\mathbf{F}}_{13}$.

Solution In calculating the magnitudes of the individual forces with the aid of Equation 18.1, we use only the magnitudes of the charges (without algebraic signs):

$$F_{12} = k\frac{|q_1|\,|q_2|}{r_{12}^2} = \frac{(8.99\times10^9\ \text{N}\cdot\text{m}^2/\text{C}^2)(3.0\times10^{-6}\ \text{C})(4.0\times10^{-6}\ \text{C})}{(0.20\ \text{m})^2}$$

$$= 2.7\ \text{N}$$

$$F_{13} = k\frac{|q_1|\,|q_3|}{r_{13}^2} = \frac{(8.99\times10^9\ \text{N}\cdot\text{m}^2/\text{C}^2)(3.0\times10^{-6}\ \text{C})(7.0\times10^{-6}\ \text{C})}{(0.15\ \text{m})^2}$$

$$= 8.4\ \text{N}$$

Since $\vec{\mathbf{F}}_{12}$ points in the negative *x* direction, and $\vec{\mathbf{F}}_{13}$ points in the positive *x* direction, the net force $\vec{\mathbf{F}}$ is

$$\vec{\mathbf{F}} = \vec{\mathbf{F}}_{12} + \vec{\mathbf{F}}_{13} = (-2.7\ \text{N}) + (8.4\ \text{N}) = \boxed{+5.7\ \text{N}}$$

The plus sign in the answer indicates that the net force points to the right in the drawing.

EXAMPLE 5 | Three Charges in a Plane

Figure 18.13*a* shows three point charges that lie in the *x, y* plane in a vacuum. Find the magnitude and direction of the net electrostatic force on q_1.

Reasoning The force exerted on q_1 by q_2 is $\vec{\mathbf{F}}_{12}$ and is an attractive force because the two charges have opposite signs. It points along the line between the charges. The force exerted on q_1 by q_3 is $\vec{\mathbf{F}}_{13}$ and is also an attractive force. It points along the line between q_1 and q_3. Coulomb's law specifies the magnitudes of these forces. Since the forces point in different directions (see **Figure 18.13*b***), we will use vector components to find the net force.

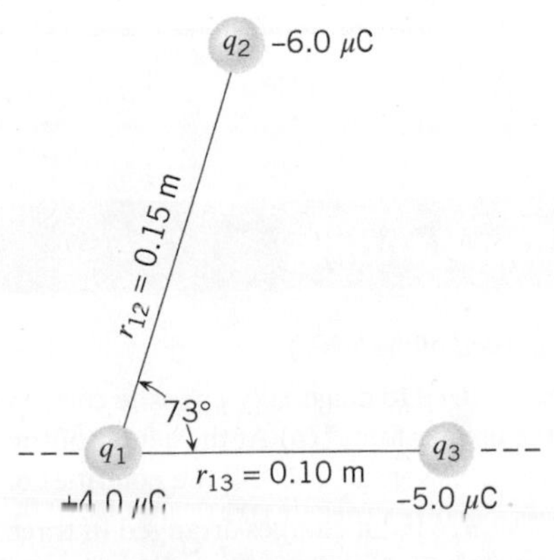

$\vec{\mathbf{F}}_{12}$ F_{12} sin 73°   73° F_{12} cos 73° $+y$ $\vec{\mathbf{F}}$ θ q_1 $\vec{\mathbf{F}}_{13}$ $+x$

(*b*) Free-body diagram for q_1

FIGURE 18.13 (*a*) Three charges lying in a plane. (*b*) The net force acting on q_1 is $\vec{\mathbf{F}} = \vec{\mathbf{F}}_{12} + \vec{\mathbf{F}}_{13}$. The angle that $\vec{\mathbf{F}}$ makes with the $+x$ axis is θ.

Solution The magnitudes of the forces are

$$F_{12} = k\frac{|q_1|\,|q_2|}{r_{12}^2} = \frac{(8.99 \times 10^9\ \text{N}\cdot\text{m}^2/\text{C}^2)(4.0 \times 10^{-6}\ \text{C})(6.0 \times 10^{-6}\ \text{C})}{(0.15\ \text{m})^2} = 9.6\ \text{N}$$

$$F_{13} = k\frac{|q_1|\,|q_3|}{r_{13}^2} = \frac{(8.99 \times 10^9\ \text{N}\cdot\text{m}^2/\text{C}^2)(4.0 \times 10^{-6}\ \text{C})(5.0 \times 10^{-6}\ \text{C})}{(0.10\ \text{m})^2} = 18\ \text{N}$$

The net force $\vec{\mathbf{F}}$ is the vector sum of $\vec{\mathbf{F}}_{12}$ and $\vec{\mathbf{F}}_{13}$, as part *b* of the drawing shows. The components of $\vec{\mathbf{F}}$ that lie in the *x* and *y* directions are $\vec{\mathbf{F}}_x$ and $\vec{\mathbf{F}}_y$, respectively. Our approach to finding $\vec{\mathbf{F}}$ is the same as that used in Chapters 1 and 4. The forces $\vec{\mathbf{F}}_{12}$ and $\vec{\mathbf{F}}_{13}$ are resolved into *x* and *y* components. Then, the *x* components are combined to give $\vec{\mathbf{F}}_x$, and the *y* components are combined to give $\vec{\mathbf{F}}_y$. Once $\vec{\mathbf{F}}_x$ and $\vec{\mathbf{F}}_y$, are known, the magnitude and direction of $\vec{\mathbf{F}}$ can be determined using trigonometry.

Force	*x* component	*y* component
$\vec{\mathbf{F}}_{12}$	$+(9.6\ \text{N}) \cos 73° = +2.8\ \text{N}$	$+(9.6\ \text{N}) \sin 73° = +9.2\ \text{N}$
$\vec{\mathbf{F}}_{13}$	$+18\ \text{N}$	$0\ \text{N}$
$\vec{\mathbf{F}}$	$\vec{\mathbf{F}}_x = +21\ \text{N}$	$\vec{\mathbf{F}}_y = +9.2\ \text{N}$

The magnitude *F* and the angle θ of the net force are

$$F = \sqrt{F_x^2 + F_y^2} = \sqrt{(21\ \text{N})^2 + (9.2\ \text{N})^2} = \boxed{23\ \text{N}}$$

$$\theta = \tan^{-1}\left(\frac{F_y}{F_x}\right) = \tan^{-1}\left(\frac{9.2\ \text{N}}{21\ \text{N}}\right) = \boxed{24°}$$

Math Skills As discussed in Section 1.7, the vector components of the force $\vec{\mathbf{F}}$ (magnitude = *F*) are two perpendicular vectors $\vec{\mathbf{F}}_x$ and $\vec{\mathbf{F}}_y$ that are parallel to the *x* and *y* axes, respectively, and add vectorially so that $\vec{\mathbf{F}} = \vec{\mathbf{F}}_x + \vec{\mathbf{F}}_y$. The scalar component F_x has a magnitude equal to that of the vector component $\vec{\mathbf{F}}_x$ and is given a positive sign if $\vec{\mathbf{F}}_x$ points along the $+x$ axis and a negative sign if $\vec{\mathbf{F}}_x$ points along the $-x$ axis. The scalar component F_y is defined similarly. **Figure 18.14** shows the vector $\vec{\mathbf{F}}$ and its components. Since the components are perpendicular, the shaded triangle is a right triangle, and we can use the Pythagorean theorem (Equation 1.7). This theorem states that the square of the length of the hypotenuse of a right triangle is equal to the sum of the squares of the lengths of the other two sides. The hypotenuse of the shaded triangle is *F*, and the other two sides are F_x and F_y. Thus, we have that

$$F^2 = F_x^2 + F_y^2 \quad \text{or} \quad F = \sqrt{F_x^2 + F_y^2}$$

To determine the angle θ we apply the tangent function ($\tan \theta$). According to Equation 1.3, $\tan \theta$ is the length of the side of the triangle opposite the angle θ divided by the length of the side adjacent to the angle θ. Thus, referring to **Figure 18.14** to identify these lengths, we see that $\tan \theta = \frac{F_y}{F_x}$. This equation means that θ is the angle whose tangent is $\frac{F_y}{F_x}$, a result that can be expressed by using the inverse tangent function ($\tan^{-1}$):

$$\theta = \tan^{-1}\left(\frac{F_y}{F_x}\right)$$

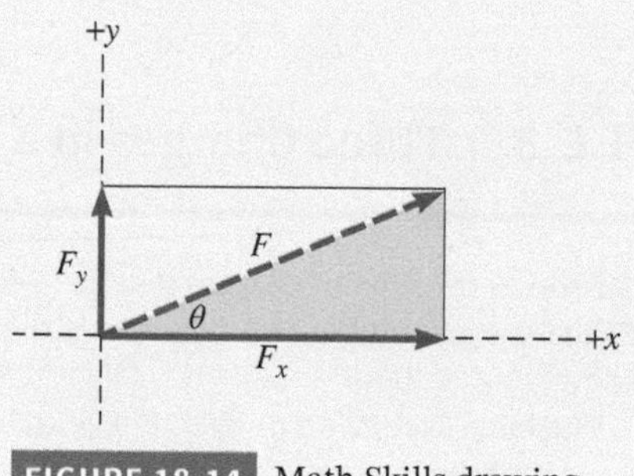

FIGURE 18.14 Math Skills drawing.

Check Your Understanding

(The answers are given at the end of the book.)

7. Identical point charges are fixed to diagonally opposite corners of a square. Where does a third point charge experience the greater force? **(a)** At the center of the square **(b)** At one of the empty corners **(c)** The question is unanswerable because the polarities of the charges are not given.

8. **CYU Figure 18.2** shows three point charges arranged in three different ways. The charges are $+q$, $-q$, and $-q$; each has the same magnitude, with one positive and the other two negative. In each of the arrangements the distance *d* is the same. Rank the arrangements in descending order (largest first) according to the magnitude of the net electrostatic force that acts on the positive charge.

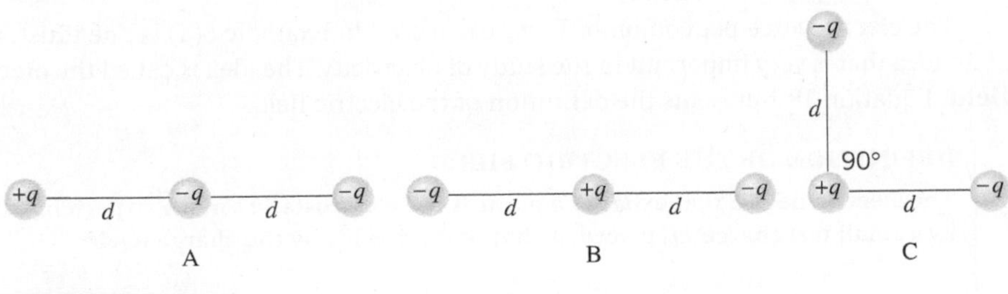

CYU FIGURE 18.2

9. A proton and an electron are held in place on the x axis. The proton is at $x = -d$, while the electron is at $x = +d$. They are released simultaneously, and the only force that affects their motions significantly is the electrostatic force of attraction that each applies to the other. Which particle reaches the origin first?

10. A particle is attached to one end of a horizontal spring, and the other end of the spring is attached to a wall. When the particle is pushed so that the spring is compressed more and more, the particle experiences a greater and greater force from the spring. Similarly, a charged particle experiences a greater and greater force when pushed closer and closer to another particle that is fixed in position and has a charge of the same polarity. Considering this similarity, will the charged particle exhibit simple harmonic motion on being released, as will the particle on the spring?

18.6 The Electric Field

Definition

As we know, a charge can experience an electrostatic force due to the presence of other charges. For instance, the positive charge q_0 in **Figure 18.15** experiences a force $\vec{\mathbf{F}}$, which is the vector sum of the forces exerted by the charges on the rod and the two spheres. It is useful to think of q_0 as a **test charge** for determining the extent to which the surrounding charges generate a force. However, in using a test charge, we must be careful to select one with a very small magnitude, so that it does not alter the locations of the other charges. The next example illustrates how the concept of a test charge is applied.

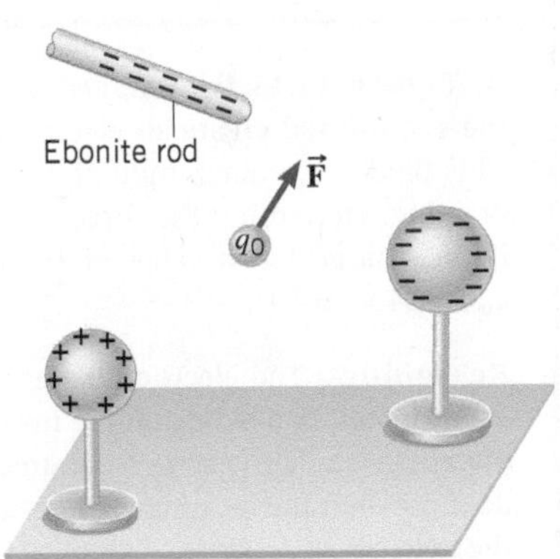

FIGURE 18.15 A positive charge q_0 experiences an electrostatic force $\vec{\mathbf{F}}$ due to the surrounding charges on the ebonite rod and the two spheres.

EXAMPLE 6 | A Test Charge

The positive test charge shown in **Figure 18.15** is $q_0 = +3.0 \times 10^{-8}$ C and experiences a force $F = 6.0 \times 10^{-8}$ N in the direction shown in the drawing. **(a)** Find the *force per coulomb* that the test charge experiences. **(b)** Using the result of part (a), predict the force that a charge of $+12 \times 10^{-8}$ C would experience if it replaced q_0.

Reasoning The charges in the environment apply a force $\vec{\mathbf{F}}$ to the test charge q_0. The force per coulomb experienced by the test charge is $\vec{\mathbf{F}}/q_0$. If q_0 is replaced by a new charge q, then the force on this new charge is the force per coulomb times q.

Solution **(a)** The force per coulomb of charge is

$$\frac{\vec{\mathbf{F}}}{q_0} = \frac{6.0 \times 10^{-8}\ \text{N}}{3.0 \times 10^{-8}\ \text{C}} = \boxed{2.0\ \text{N/C}}$$

The direction of the force per coulomb is the same as the direction of $\vec{\mathbf{F}}$ in **Figure 18.15**.

(b) The result from part (a) indicates that the surrounding charges can exert 2.0 newtons of force per coulomb of charge. Thus, a charge of $+12 \times 10^{-8}$ C would experience a force whose magnitude is

$$F = (2.0\ \text{N/C})(12 \times 10^{-8}\ \text{C}) = \boxed{24 \times 10^{-8}\ \text{N}}$$

The direction of this force would be the same as the direction of the force experienced by the test charge, since both have the same positive sign.

The electric force per coulomb, $\vec{F}/q_0$, calculated in Example 6(a) is one illustration of an idea that is very important in the study of electricity. The idea is called the **electric field**. Equation 18.2 presents the definition of the electric field.

DEFINITION OF THE ELECTRIC FIELD

The electric field $\vec{E}$ that exists at a point is the electrostatic force $\vec{F}$ experienced by a small test charge* q_0 placed at that point divided by the charge itself:

$$\vec{E} = \frac{\vec{F}}{q_0} \tag{18.2}$$

The electric field is a vector, and its direction is the same as the direction of the force $\vec{F}$ on a positive test charge.

SI Unit of Electric Field: newton per coulomb (N/C)

Equation 18.2 indicates that the unit for the electric field is that of force divided by charge, which is a newton/coulomb (N/C) in SI units.

It is the surrounding charges that create an electric field at a given point. Any positive or negative charge placed at the point interacts with the field and, as a result, experiences a force, as the next example indicates.

EXAMPLE 7 | An Electric Field Leads to a Force

In **Figure 18.16** the charges on the two metal spheres and the ebonite rod create an electric field $\vec{E}$ at the spot indicated. This field has a magnitude of 2.0 N/C and is directed as in the drawing. Determine the force on a charge placed at that spot, if the charge has a value of **(a)** $q_0 = +18 \times 10^{-8}$ C and **(b)** $q_0 = -24 \times 10^{-8}$ C.

Reasoning The electric field at a given spot can exert a variety of forces, depending on the magnitude and sign of the charge placed there. The charge is assumed to be small enough that it does not alter the locations of the surrounding charges that create the field.

Solution (a) The magnitude of the force is the product of the magnitudes of q_0 and $\vec{E}$:

$$F = |q_0|\,E = (18 \times 10^{-8}\ \text{C})(2.0\ \text{N/C}) = \boxed{36 \times 10^{-8}\ \text{N}} \tag{18.2}$$

Since q_0 is positive, the force points in the same direction as the electric field, as part *a* of the drawing indicates.

(b) In this case, the magnitude of the force is

$$F = |q_0|\,E = (24 \times 10^{-8}\ \text{C})(2.0\ \text{N/C}) = \boxed{48 \times 10^{-8}\ \text{N}} \tag{18.2}$$

The force on the negative charge points in the direction *opposite* to the force on the positive charge—that is, opposite to the electric field (see part *b* of the drawing).

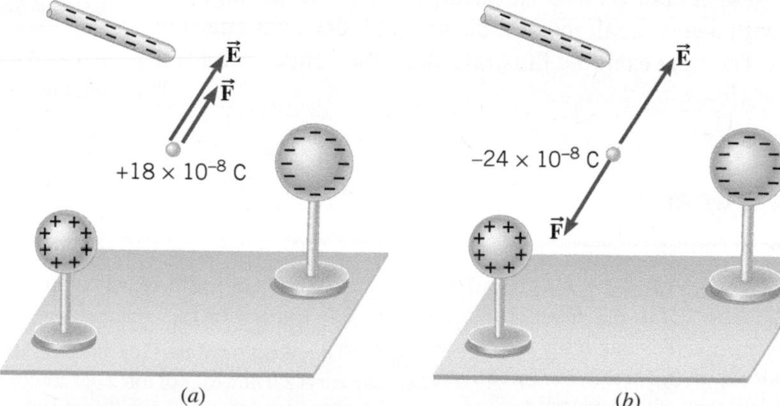

FIGURE 18.16 The electric field $\vec{E}$ that exists at a given spot can exert a variety of forces. The force exerted depends on the magnitude and sign of the charge placed at that spot. (*a*) The force on a positive charge points in the same direction as $\vec{E}$, while (*b*) the force on a negative charge points opposite to $\vec{E}$.

*As long as the test charge is small enough that it does not disturb the surrounding charges, it may be either positive or negative. Compared to a positive test charge, a negative test charge of the same magnitude experiences a force of the same magnitude that points in the opposite direction. However, the same electric field is given by Equation 18.2, in which $\vec{F}$ is replaced by $-\vec{F}$ and q_0 is replaced by $-q_0$.

At a particular point in space, each of the surrounding charges contributes to the net electric field that exists there. To determine the net field, it is necessary to obtain the various contributions separately and then find the vector sum of them all. Such an approach is an illustration of the principle of linear superposition, as applied to electric fields. (This principle is introduced in Section 17.1, in connection with waves.) Example 8 emphasizes the vector nature of the electric field.

EXAMPLE 8 | Electric Fields Add as Vectors Do

Animated Figure 18.17 shows two charged objects, A and B. Each contributes as follows to the net electric field at point *P*: $\vec{\mathbf{E}}_A$ = 3.00 N/C directed to the right, and $\vec{\mathbf{E}}_B$ = 2.00 N/C directed downward. Thus, $\vec{\mathbf{E}}_A$ and $\vec{\mathbf{E}}_B$ are perpendicular. What is the net electric field at *P*?

Reasoning The net electric field $\vec{\mathbf{E}}$ is the vector sum of $\vec{\mathbf{E}}_A$ and $\vec{\mathbf{E}}_B$: $\vec{\mathbf{E}} = \vec{\mathbf{E}}_A + \vec{\mathbf{E}}_B$. As illustrated in **Animated Figure 18.17**, $\vec{\mathbf{E}}_A$ and $\vec{\mathbf{E}}_B$ are perpendicular, so $\vec{\mathbf{E}}$ is the diagonal of the rectangle shown in the drawing. Thus, we can use the Pythagorean theorem to find the magnitude of $\vec{\mathbf{E}}$ and trigonometry to find the directional angle θ.

Solution The magnitude of the net electric field is

$$E = \sqrt{E_A^2 + E_B^2} = \sqrt{(3.00\ \text{N/C})^2 + (2.00\ \text{N/C})^2} = \boxed{3.61\ \text{N/C}}$$

The direction of $\vec{\mathbf{E}}$ is given by the angle θ in the drawing:

$$\theta = \tan^{-1}\left(\frac{E_B}{E_A}\right) = \tan^{-1}\left(\frac{2.00\ \text{N/C}}{3.00\ \text{N/C}}\right) = \boxed{33.7°}$$

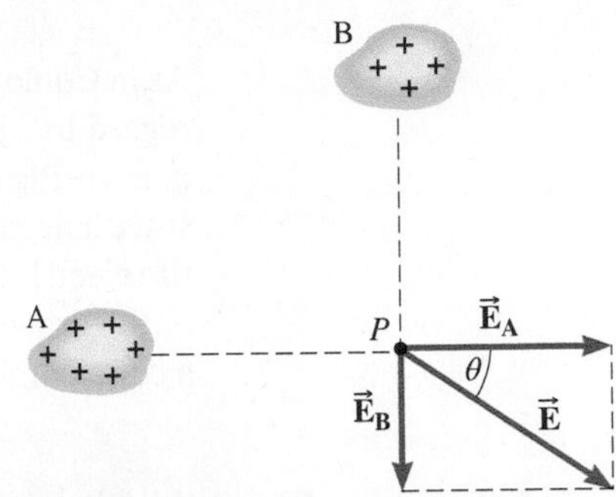

ANIMATED FIGURE 18.17 The electric field contributions $\vec{\mathbf{E}}_A$ and $\vec{\mathbf{E}}_B$, which come from the two charge distributions, are added vectorially to obtain the net field $\vec{\mathbf{E}}$ at point *P*.

Point Charges

A more complete understanding of the electric field concept can be gained by considering the field created by a point charge, as in the following example.

EXAMPLE 9 | The Electric Field of a Point Charge

There is an isolated point charge of $q = +15\ \mu\text{C}$ in a vacuum at the left in **Figure 18.18*a***. Using a test charge of $q_0 = +0.80\ \mu\text{C}$, determine the electric field at point *P*, which is 0.20 m away.

Reasoning Following the definition of the electric field, we place the test charge q_0 at point *P*, determine the force acting on the test charge, and then divide the force by the test charge.

Solution Coulomb's law (Equation 18.1) gives the magnitude of the force:

$$F = k\frac{|q_0|\,|q|}{r^2} = \frac{(8.99 \times 10^9\ \text{N}\cdot\text{m}^2/\text{C}^2)(0.80 \times 10^{-6}\ \text{C})(15 \times 10^{-6}\ \text{C})}{(0.20\ \text{m})^2}$$

$$= \boxed{2.7\ \text{N}}$$

Equation 18.2 gives the magnitude of the electric field:

$$E = \frac{F}{|q_0|} = \frac{2.7\ \text{N}}{0.80 \times 10^{-6}\ \text{C}} = \boxed{3.4 \times 10^6\ \text{N/C}}$$

The electric field $\vec{\mathbf{E}}$ points in the *same direction* as the force $\vec{\mathbf{F}}$ on the positive test charge. Since the test charge experiences a force of repulsion directed to the right, the electric field vector also points to the right, as **Figure 18.18*b*** shows.

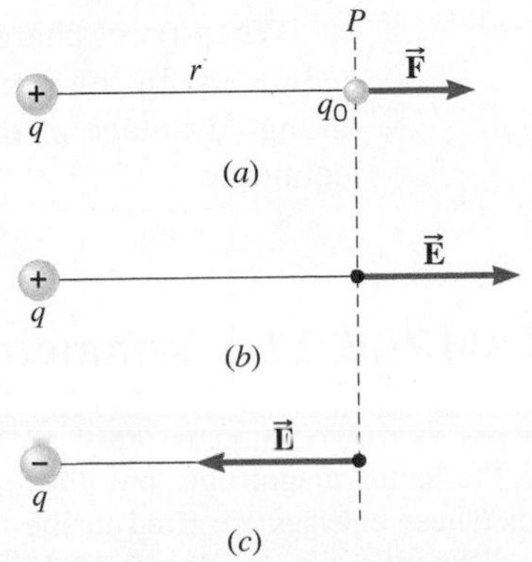

FIGURE 18.18 (*a*) At location *P*, a positive test charge q_0 experiences a repulsive force $\vec{\mathbf{F}}$ due to the positive point charge *q*. (*b*) At *P*, the electric field $\vec{\mathbf{E}}$ is directed to the right. (*c*) If the charge *q* were negative rather than positive, the electric field would have the same magnitude as in (*b*) but would point to the left.

The electric field produced by a point charge q can be obtained in general terms from Coulomb's law. First, note that the magnitude of the force exerted by the charge q on a test charge q_0 is $F = k|q||q_0|/r^2$. Then, divide this value by $|q_0|$ to obtain the magnitude of the field. Since $|q_0|$ is eliminated algebraically from the result, *the electric field does not depend on the test charge:*

Point charge q
$$E = \frac{k|q|}{r^2} \tag{18.3}$$

Problem-Solving Insight **Equation 18.3 gives only the magnitude of the electric field produced by a point charge. Therefore, do not use negative numbers for the symbol $|q|$ in this equation.**

As in Coulomb's law, the symbol $|q|$ denotes the magnitude of q in Equation 18.3, without regard to whether q is positive or negative. If q is positive, then $\vec{\mathbf{E}}$ is directed away from q, as in **Figure 18.18*b***. On the other hand, if q is negative, then $\vec{\mathbf{E}}$ is directed toward q, since a negative charge attracts a positive test charge. For instance, **Figure 18.18*c*** shows the electric field that would exist at P if there were a charge of $-q$ instead of $+q$ at the left of the drawing. Example 10 reemphasizes the fact that all the surrounding charges make a contribution to the electric field that exists at a given place.

EXAMPLE 10 | The Electric Fields from Separate Charges May Cancel

Two positive point charges, $q_1 = +16\ \mu C$ and $q_2 = +4.0\ \mu C$, are separated in a vacuum by a distance of 3.0 m, as **Figure 18.19** illustrates. Find the spot on the line between the charges where the net electric field is zero.

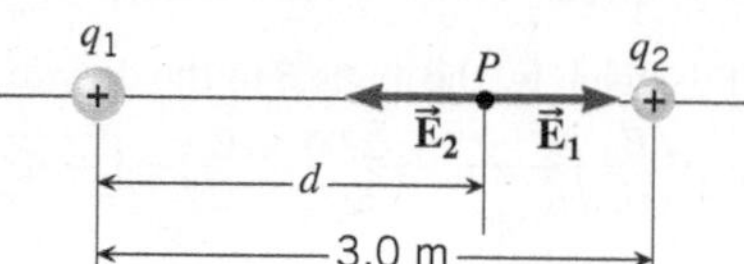

FIGURE 18.19 The two point charges q_1 and q_2 create electric fields $\vec{\mathbf{E}}_1$ and $\vec{\mathbf{E}}_2$ that cancel at a location P on the line between the charges.

Reasoning Between the charges the two field contributions have opposite directions, and the net electric field is zero at the place where the magnitude of $\vec{\mathbf{E}}_1$ equals the magnitude of $\vec{\mathbf{E}}_2$. However, since q_2 is smaller than q_1, this location must be *closer* to q_2, in order that the field of the smaller charge can balance the field of the larger charge. In the drawing, the cancellation spot is labeled P, and its distance from q_1 is d.

Solution At P, $E_1 = E_2$, and using the expression $E = k|q|/r^2$ (Equation 18.3), we have

$$\frac{k(16 \times 10^{-6}\ \text{C})}{d^2} = \frac{k(4.0 \times 10^{-6}\ \text{C})}{(3.0\ \text{m} - d)^2}$$

Rearranging this expression shows that $4.0(3.0\ \text{m} - d)^2 = d^2$. Taking the square root of each side of this equation reveals that

$$2.0(3.0\ \text{m} - d) = \pm d$$

The plus and minus signs on the right occur because either the positive or negative root can be taken. Therefore, there are two possible values for d: +2.0 m and +6.0 m. The value +6.0 m corresponds to a location off to the right of both charges, where the magnitudes of $\vec{\mathbf{E}}_1$ and $\vec{\mathbf{E}}_2$ are equal, but where the directions are the same. Thus, $\vec{\mathbf{E}}_1$ and $\vec{\mathbf{E}}_2$ do not cancel at this spot. The other value for d corresponds to the location shown in the drawing and is the zero-field location: $\boxed{d = +2.0\ \text{m}}$.

When point charges are arranged in a symmetrical fashion, it is often possible to deduce useful information about the magnitude and direction of the electric field by taking advantage of the symmetry. Conceptual Example 11 illustrates the use of this technique.

CONCEPTUAL EXAMPLE 11 | Symmetry and the Electric Field

Four point charges all have the same magnitude, but they do not all have the same sign. These charges are fixed to the corners of a rectangle in two different ways, as **Figure 18.20** shows. Consider the net electric field at the center C of the rectangle in each case. In which case, if either, is the net electric field greater? **(a)** It is greater in **Figure 18.20*a***. **(b)** It is greater in **Figure 18.20*b***. **(c)** The field has the same magnitude in both cases.

Reasoning The net electric field at C is the vector sum of the individual fields created there by the charges at each corner. Each of the individual fields has the same magnitude, since the charges all have the same magnitude and are equidistant from C. The directions of the individual fields are different, however. The field created by a positive charge points away from the charge, and the field created by a negative charge points toward the charge.

Answers (a) and (c) are incorrect. To see why these answers are incorrect, note that the charges on corners 2 and 4 are identical in both parts of **Figure 18.20**. Moreover, in **Figure 18.20*a*** the charges at corners 1 and 3 are both $+q$, so they contribute individual fields of the same magnitude at C that have *opposite* directions and, therefore, cancel. However, in **Figure 18.20*b*** the charges at corners 1 and 3 are $-q$ and $+q$, respectively. They contribute individual fields of the same magnitude at C that have the *same* directions and do not cancel, but combine to produce the field $\vec{\mathbf{E}}_{13}$ shown in Figure 18.20*b*. The fact that this contribution to the net field at C is present in **Figure 18.20*b*** but not in **Figure 18.20*a*** means that the net fields in the two cases are different and that the net field in **Figure 18.20*a*** is less than (not greater than) the net field in **Figure 18.20*b***.

Answer (b) is correct. To assess the net field at C, we need to consider the contribution from the charges at corners 2 and 4, which are $-q$ and $+q$, respectively, in both cases. This is just like the arrangement on corners 1 and 3, which was discussed previously. It leads to a contribution to the net field at C that is shown as $\vec{\mathbf{E}}_{24}$ in both parts of the figure. In **Figure 18.20*a*** the net field at C is just $\vec{\mathbf{E}}_{24}$, but in **Figure 18.20*b*** it is the vector sum of $\vec{\mathbf{E}}_{13}$ and $\vec{\mathbf{E}}_{24}$, which is clearly greater than either of these two values alone.

Related Homework: Problem 37

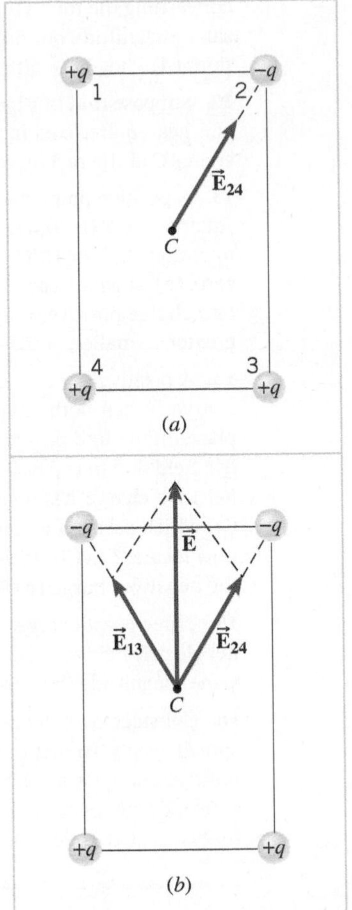

FIGURE 18.20 Charges of identical magnitude are placed at the corners of a rectangle. However, the charges do not all have the same sign and give rise to different electric fields at the center C of the rectangle, depending on the signs that they have.

The Parallel Plate Capacitor

Equation 18.3, which gives the electric field of a point charge, is a very useful result. With the aid of integral calculus, this equation can be applied in a variety of situations where point charges are distributed over one or more surfaces. One such example that has considerable practical importance is the **parallel plate capacitor**. As **Figure 18.21** shows, this device consists of two parallel metal plates, each with area A. A charge $+q$ is spread uniformly over one plate, while a charge $-q$ is spread uniformly over the other plate. In the region between the plates and away from the edges, the electric field points from the positive plate toward the negative plate and is perpendicular to both. It can be shown (see Example 16 in Section 18.9) that this electric field has a magnitude of

Parallel plate capacitor
$$E = \frac{q}{\varepsilon_0 A} = \frac{\sigma}{\varepsilon_0} \tag{18.4}$$

where ε_0 is the permittivity of free space. In this expression the Greek symbol sigma (σ) denotes the charge per unit area ($\sigma = q/A$) and is sometimes called the **charge density**. Except in the region near the edges, the field has the same value at all places between the plates. The field does *not* depend on the distance from the charges, in distinct contrast to the field created by an isolated point charge.

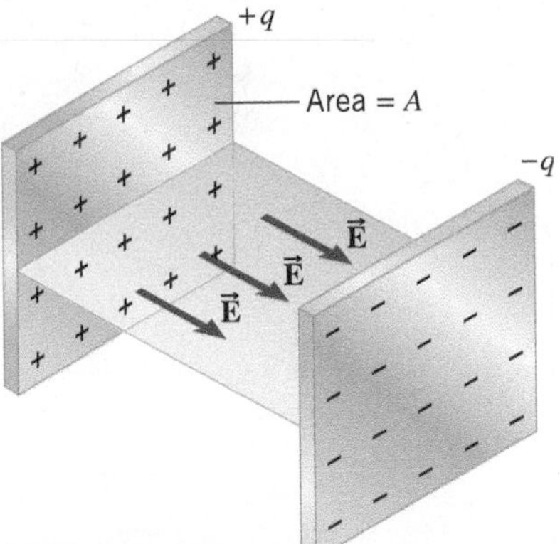

FIGURE 18.21 A parallel plate capacitor.

Check Your Understanding

(The answers are given at the end of the book.)

11. There is an electric field at point P. A very small positive charge is placed at this point and experiences a force. Then the positive charge is replaced by a very small negative charge that has a magnitude different from that of the positive charge. Which one of the following statements is true concerning the forces that these charges experience at *P*? **(a)** They are identical. **(b)** They have the same magnitude but different directions. **(c)** They have different magnitudes but the same direction. **(d)** They have different magnitudes and different directions.

12. Suppose that in **Figure 18.20*a*** point charges $-q$ are fixed in place at corners 1 and 3 and point charges $+q$ are fixed in place at corners 2 and 4. What then would be the net electric field at the center *C* of the rectangle?

13. A positive point charge $+q$ is fixed in position at the center of a square, as **CYU Figure 18.3** shows. A second point charge is fixed to corner B, C, or D. The net electric field that results at corner A is zero. **(a)** At which corner is the second charge located? **(b)** Is the second charge positive or negative? **(c)** Does the second charge have a greater, a smaller, or the same magnitude as the charge at the center?

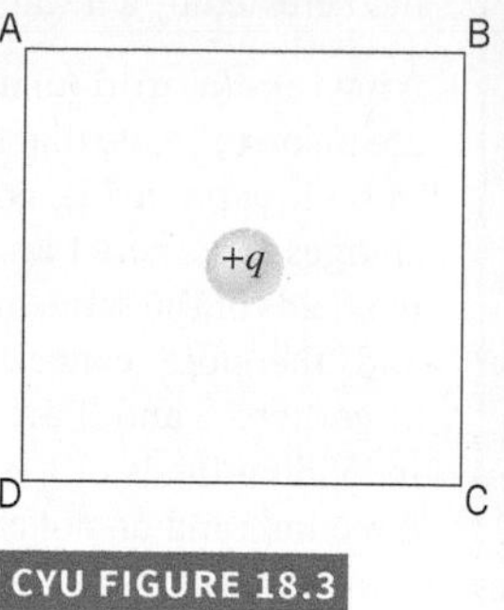

CYU FIGURE 18.3

14. A positive point charge is located to the left of a negative point charge. When both charges have the same magnitude, there is no place on the line passing through both charges where the net electric field due to the two charges is zero. Suppose, however, that the negative charge has a greater magnitude than the positive charge. On which part of the line, if any, is a place of zero net electric field now located? **(a)** To the left of the positive charge **(b)** Between the two charges **(c)** To the right of the negative charge **(d)** There is no zero place.

15. Three point charges are fixed to the corners of a square, one to a corner, in such a way that the net electric field at the empty corner is zero. Do these charges all have **(a)** the same sign and **(b)** the same magnitude (but, possibly, different signs)?

16. Consider two identical, thin, and nonconducting rods, A and B. On rod A, positive charge is spread evenly, so that there is the same amount of charge per unit length at every point. On rod B, positive charge is spread evenly over only the left half, and the same amount of negative charge is spread evenly over the right half. For each rod deduce the *direction* of the electric field at a point that is located directly above the midpoint of the rod.

18.7 Electric Field Lines

As we have seen, electric charges create an electric field in the space around them. It is useful to have a kind of "map" that gives the direction and strength of the field at various places. The great English physicist Michael Faraday (1791–1867) proposed an idea that provides such a "map"—the idea of **electric field lines**. Since the electric field is the electric force per unit charge, field lines are also called **lines of force**.

To introduce the electric field line concept, **Figure 18.22*a*** shows a positive point charge $+q$. At the locations numbered 1–8, a positive test charge would experience a repulsive force, as the arrows in the drawing indicate. Therefore, the electric field created by the charge $+q$ is directed radially outward. The electric field lines are lines drawn

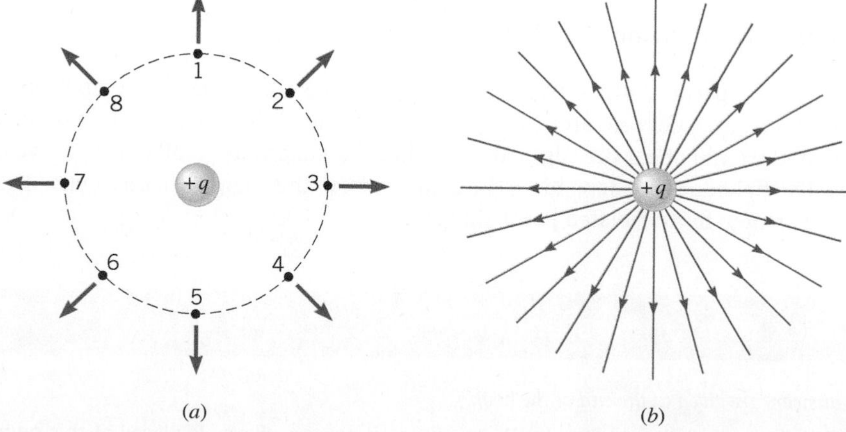

FIGURE 18.22 (*a*) At any of the eight marked spots around a positive point charge $+q$, a positive test charge would experience a repulsive force directed radially outward. (*b*) The electric field lines are directed radially outward from a positive point charge $+q$.

to show this direction, as Figure 18.22*b* illustrates. They begin on the charge $+q$ and point radially *outward*. Figure 18.23 shows the field lines in the vicinity of a negative charge $-q$. In this case they are directed radially *inward* because the force on a positive test charge is one of attraction, indicating that the electric field points inward. In general, ***electric field lines are always directed away from positive charges and toward negative charges.***

The electric field lines in Figures 18.22 and 18.23 are drawn in only two dimensions, as a matter of convenience. Field lines radiate from the charges in three dimensions, and an infinite number of lines could be drawn. However, for clarity only a small number are ever included in pictures. The number is chosen to be proportional to the magnitude of the charge; thus, five times as many lines would emerge from a $+5q$ charge as from a $+q$ charge.

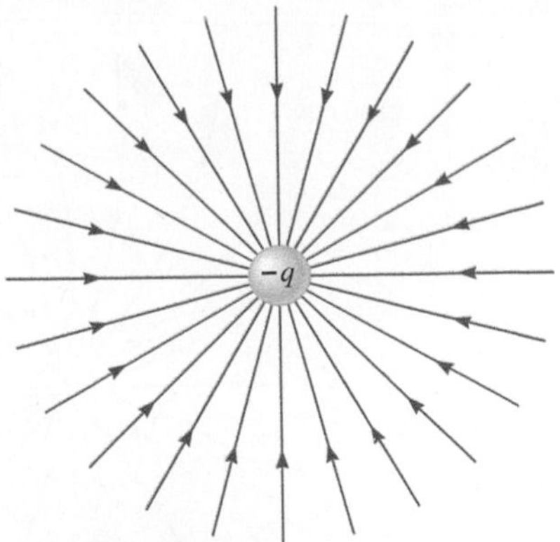

FIGURE 18.23 The electric field lines are directed radially inward toward a negative point charge $-q$.

The pattern of electric field lines also provides information about the magnitude or strength of the field. Notice that in Figures 18.22 and 18.23, the lines are closer together near the charges, where the electric field is stronger. At distances far from the charges, where the electric field is weaker, the lines are more spread out. It is true in general that the electric field is stronger in regions where the field lines are closer together. In fact, no matter how many charges are present, the number of lines per unit area passing perpendicularly through a surface is proportional to the magnitude of the electric field.

In regions where the electric field lines are equally spaced, there is the same number of lines per unit area everywhere, and the electric field has the same strength at all points. For example, Figure 18.24 shows that the field lines between the plates of a parallel plate capacitor are parallel and equally spaced, except near the edges where they bulge outward. The equally spaced, parallel lines indicate that the electric field has the same magnitude and direction at all points in the central region of the capacitor.

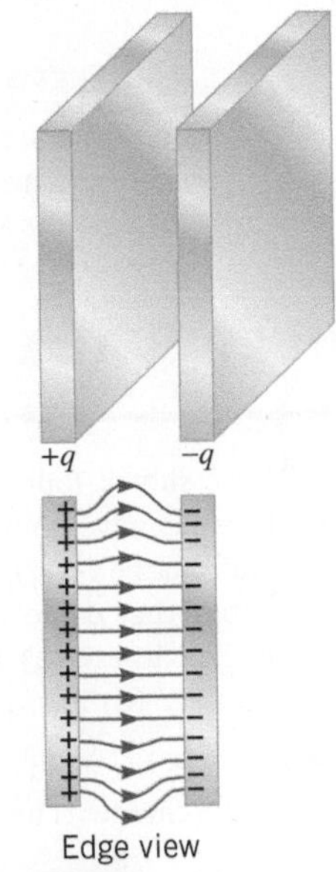

FIGURE 18.24 In the central region of a parallel plate capacitor, the electric field lines are parallel and evenly spaced, indicating that the electric field there has the same magnitude and direction at all points.

Often, electric field lines are curved, as in the case of an **electric dipole**. An electric dipole consists of two separated point charges that have the same magnitude but opposite signs. The electric field of a dipole is proportional to the product of the magnitude of one of the charges and the distance between the charges. This product is called the **dipole moment**. Many molecules, such as H_2O and HCl, have dipole moments. Figure 18.25 depicts the field lines in the vicinity of a dipole. For a curved field line, the electric field vector at a point is *tangent* to the line at that point (see points 1, 2, and 3 in the drawing). The pattern of the lines for the dipole indicates that the electric field is greatest in the region between and immediately surrounding the two charges, since the lines are closest together there.

Notice in Figure 18.25 that any given field line starts on the positive charge and ends on the negative charge. This is a fundamental characteristic of electric field lines:

Problem-Solving Insight **Electric field lines always begin on a positive charge and end on a negative charge and do not start or stop in midspace. Furthermore, the number of lines leaving a positive charge or entering a negative charge is proportional to the magnitude of the charge.**

This means, for example, that if 100 lines are drawn leaving a $+4\ \mu C$ charge, then 75 lines would have to end on a $-3\ \mu C$ charge and 25 lines on a $-1\ \mu C$ charge. Thus, 100 lines leave the charge of $+4\ \mu C$ and end on a total charge of $-4\ \mu C$, so the lines begin and end on equal amounts of total charge.

The electric field lines are also curved in the vicinity of two identical charges. Figure 18.26 shows the pattern associated with two positive point charges and reveals that there is an absence of lines in the region between the charges. The absence of lines indicates that the electric field is relatively weak between the charges.

Some of the important properties of electric field lines are considered in Conceptual Example 12, and the role of the electric field in the electrophoresis of DNA is explored in Example 13.

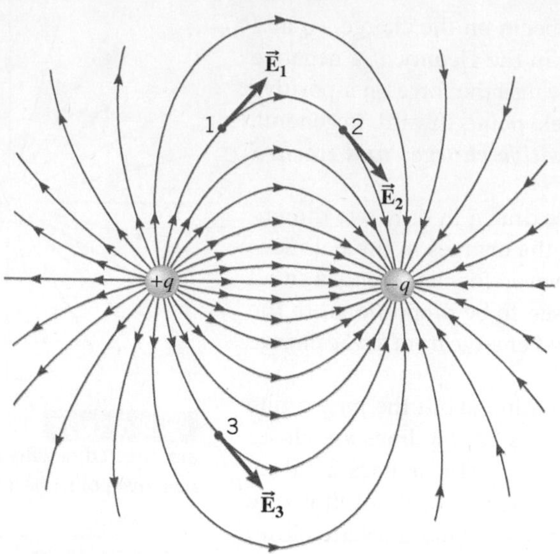

FIGURE 18.25 The electric field lines of an electric dipole are curved and extend from the positive to the negative charge. At any point, such as 1, 2, or 3, the field created by the dipole is tangent to the line through the point.

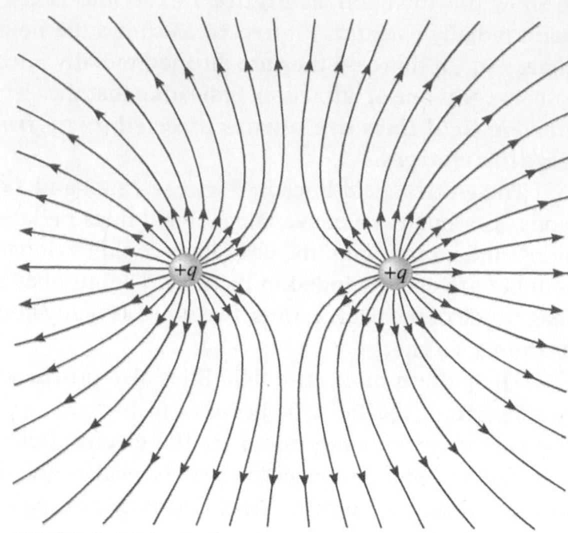

FIGURE 18.26 The electric field lines for two identical positive point charges. If the charges were both negative, the directions of the lines would be reversed.

CONCEPTUAL EXAMPLE 12 | Drawing Electric Field Lines

Figure 18.27 shows four choices for the electric field lines between three negative point charges ($-q$, $-q$, and $-2q$) and one positive point charge ($+4q$). Which of these choices is the only one of the four that could possibly show a correct representation of the field lines? **(a) Figure 18.27*a*** **(b) Figure 18.27*b*** **(c) Figure 18.27*c*** **(d) Figure 18.27*d***

Reasoning Electric field lines begin on positive charges and end on negative charges. The tangent to a field line at a point gives the direction of the electric field at that point. Equally spaced parallel field lines indicate that the field has a constant value (magnitude and direction) in the corresponding region of space.

Answer (a) is incorrect. Field lines can never cross, as they do at point *P* in **Figure 18.27*a***. If two field lines were to intersect, there would be two electric fields at the point of intersection, one associated with each line. However, there can only be one value of the electric field at a given point.

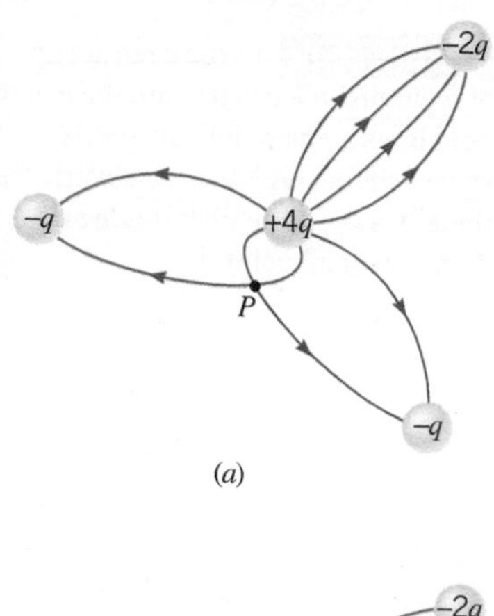

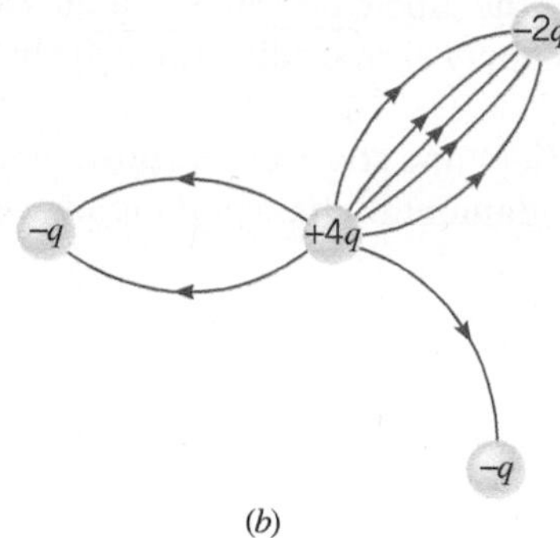

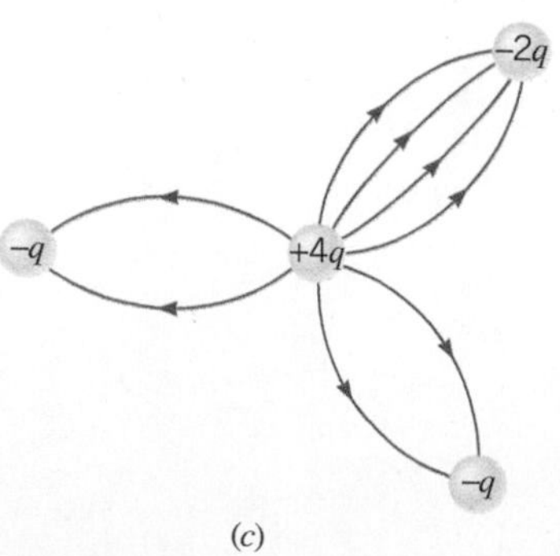

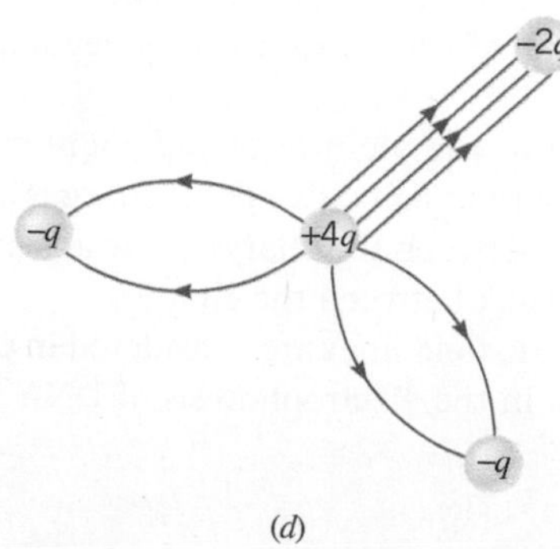

FIGURE 18.27 Only one of these drawings shows a representation of the electric field lines between the charges that could be correct. (See Conceptual Example 12.)

Answer (b) is incorrect. The number of field lines that leave a positive charge or end on a negative charge is proportional to the magnitude of the charge. Since 8 lines leave the $+4q$ charge, one-half of them (or 4) must end on the $-2q$ charge, and one-fourth of them (or 2) must end on each of the $-q$ charges. **Figure 18.27*b*** incorrectly shows 5 lines ending on the $-2q$ charge and 1 line ending on one of the $-q$ charges.

Answer (d) is incorrect. **Figure 18.27*d*** is incorrect because the field lines between the $+4q$ charge and the $-2q$ charge are parallel and evenly spaced, which would indicate that the field everywhere in this region has a constant magnitude and direction. However, the field between the $+4q$ charge and the $-2q$ charge certainly is stronger in places close to either of the charges. The field lines should, therefore, have a curved nature, similar (but not identical) to the field lines that surround a dipole.

Answer (c) is correct. **Figure 18.27*c*** contains none of the errors discussed previously and, therefore, is the only drawing that could be correct.

Related Homework: Problem 30

EXAMPLE 13 | BIO Electrophoresis of DNA

Gel electrophoresis is a method used to electrically separate charged macromolecules such as those in viruses, bacterial cells, proteins, and DNA. When subjected to an electric field, these objects experience a force and migrate. Due to the viscous properties of the medium in which they travel (a gel), smaller objects move with a greater speed than larger ones, assuming they have equal charge densities. This method is used to carry out DNA tests. In this case, restriction enzymes are first used to cut the DNA into pieces related to the specific DNA bases, that is, adenine (A), cytosine (C), guanine (G), and thymine (T). What remains are DNA fragments that can now be sorted according to size (length). The size distribution of these fragments is what is measured in a gel electrophoresis DNA test, and displayed as a series of lines (**Figure 18.28**), with shorter segments on one side and longer segments on the other.

Suppose a DNA fragment has a net negative charge with a magnitude equal to that of one electron ($e = 1.60 \times 10^{-19}$ C) per nanometer of its length. **(a)** If a set of electrodes in a gel sets up a uniform electric field of magnitude $E = 125$ N/C, determine the magnitude of the net electric force acting on a DNA fragment of length 0.220 μm. **(b)** Toward which electrode does the DNA fragment migrate (i.e., toward the negative or positive electrode) if the electric field points from the positive (+) electrode toward the negative (−) electrode?

Reasoning **(a)** From the linear charge density and the DNA fragment length, we can calculate the magnitude of the net charge (q_0) of a fragment. Since we know the electric field E, the force is given by Equation 18.2 as $F = q_0E$. **(b)** The net charge of the fragment is negative. Therefore, from the vector nature of the force on a charge due to an electric field expressed by Equation 18.2, the electric force on the fragment is antiparallel to the electric field.

Solution **(a)** The magnitude of the net charge on the DNA fragment is given by

$$|q_0| = (\text{length}) \times (\text{charge per unit length})$$

$$= (0.220\ \mu\text{m})\left(\frac{e}{1\ \text{nm}}\right) = (0.220 \times 10^{-6}\ \text{m})\left(\frac{1.60 \times 10^{-19}\ \text{C}}{1.00 \times 10^{-9}\ \text{m}}\right)$$

$$= 3.52 \times 10^{-17}\ \text{C}$$

Applying Equation 18.2 to find the magnitude of the total force, we have

$$F = |q_0|E = (3.52 \times 10^{-17}\ \text{C})(125\ \text{N/C}) = \boxed{4.40 \times 10^{-15}\ \text{N}}$$

(b) Since the electric field points from positive charges toward negative charges, the electric field points from the positive electrode toward the negative electrode. Since the force on a negatively charged particle points in a direction opposite to that of the field, the DNA fragment migrates toward the positive electrode.

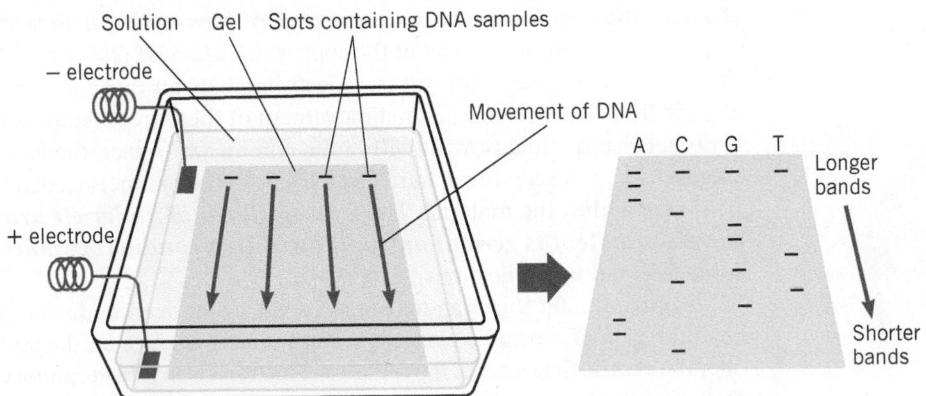

FIGURE 18.28 DNA treated with restriction enzymes is placed into slots called wells, and then subjected to an electric field set up by electrodes in an electrolytic gel. The fragments then migrate and separate according to length. (Modified from *Encyclopedia Britannica* 2007)

Check Your Understanding

(The answers are given at the end of the book.)

17. Drawings A and B in **CYU Figure 18.4** show two examples of electric field lines. Which (one or more) of the following statements are true, and which (one or more) are false? **(a)** In both A and B the electric field is the same everywhere. **(b)** As you move from left to right in each case, the electric field becomes stronger. **(c)** The electric field in A is the same everywhere, but it becomes stronger in B as you move from left to right. **(d)** The electric fields in both A and B could be created by negative charges located somewhere on the left and positive charges somewhere on the right. **(e)** Both A and B arise from a single positive point charge located somewhere on the left.

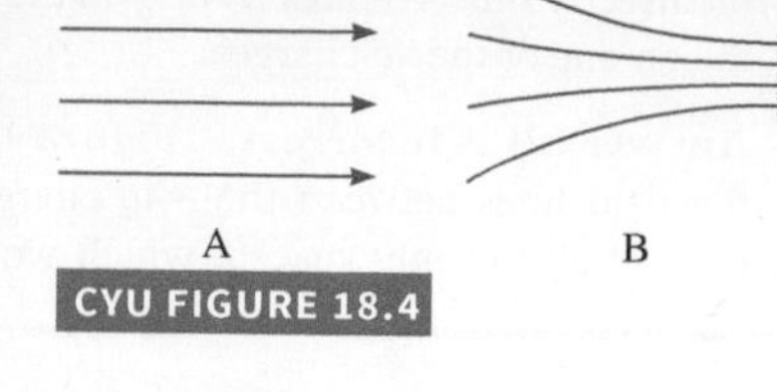

CYU FIGURE 18.4

18. A positively charged particle is moving horizontally when it enters the region between the plates of a parallel plate capacitor, as **CYU Figure 18.5** illustrates. When the particle is within the capacitor, which of the following vectors, if any, are *parallel* to the electric field lines inside the capacitor? **(a)** The particle's displacement **(b)** Its velocity **(c)** Its linear momentum **(d)** Its acceleration

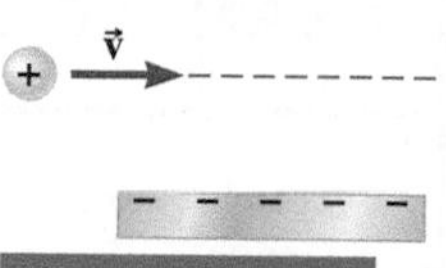

CYU FIGURE 18.5

18.8 The Electric Field Inside a Conductor: Shielding

(*a*)

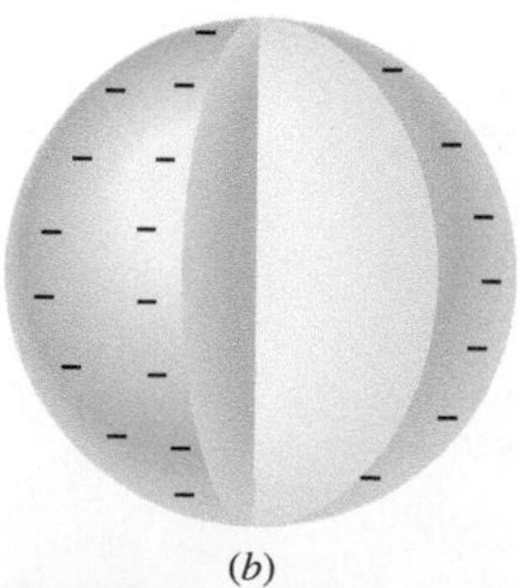

(*b*)

FIGURE 18.29 (*a*) Excess charge within a conductor (copper) moves quickly (*b*) to the surface.

In conducting materials such as copper, electric charges move readily in response to the forces that electric fields exert. This property of conducting materials has a major effect on the electric field that can exist within and around them. Suppose that a piece of copper carries a number of excess electrons somewhere within it, as in **Figure 18.29*a***. Each electron would experience a force of repulsion because of the electric field of its neighbors. And, since copper is a conductor, the excess electrons move readily in response to that force. In fact, as a consequence of the $1/r^2$ dependence on distance in Coulomb's law, they rush to the surface of the copper. Once static equilibrium is established with all of the excess charge on the surface, no further movement of charge occurs, as part *b* of the drawing indicates. Similarly, excess positive charge also moves to the surface of a conductor. In general, ***at equilibrium under electrostatic conditions, any excess charge resides on the surface of a conductor.***

Now consider the interior of the copper in **Figure 18.29*b***. The interior is electrically neutral, although there are still free electrons that can move under the influence of an electric field. The absence of a net movement of these free electrons indicates that there is no net electric field present within the conductor. In fact, the excess charges arrange themselves on the conductor surface precisely in the manner needed to make the electric field zero within the material. Thus, ***at equilibrium under electrostatic conditions, the electric field is zero at any point within a conducting material.*** This fact has some fascinating implications.

Figure 18.30*a* shows an uncharged, solid, cylindrical conductor at equilibrium in the central region of a parallel plate capacitor. Induced charges on the surface of the cylinder alter the electric field lines of the capacitor. Since an electric field cannot exist within the conductor under these conditions, the electric field lines do not penetrate the cylinder. Instead, they end or begin on the induced charges. Consequently, a test charge placed *inside* the conductor would feel no force due to the presence of the charges on the capacitor. In other words, ***the conductor shields any charge within it from electric fields created outside the conductor.*** The shielding results from the induced charges on the conductor surface.

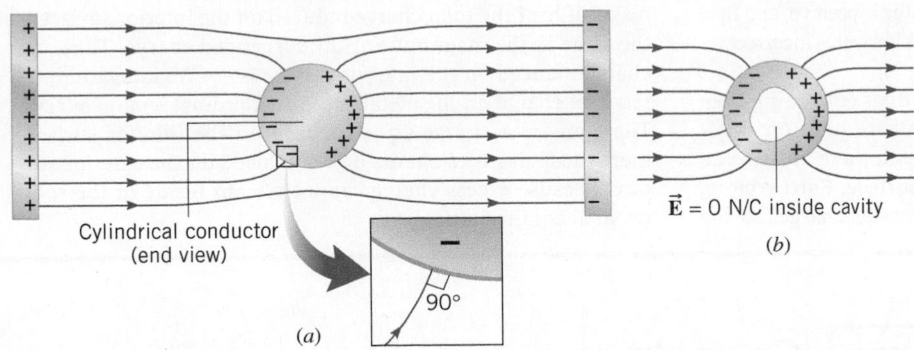

FIGURE 18.30 (*a*) A cylindrical conductor (shown as an end view) is placed between the oppositely charged plates of a capacitor. The electric field lines do not penetrate the conductor. The blowup shows that, just outside the conductor, the electric field lines are perpendicular to its surface. (*b*) The electric field is zero in a cavity within the conductor.

THE PHYSICS OF . . . shielding electronic circuits. Since the electric field is zero inside the conductor, nothing is disturbed if a cavity is cut from the interior of the material, as in **Figure 18.30*b***. Thus, the interior of the cavity is also shielded from external electric fields, a fact that has important applications, particularly for shielding electronic circuits. "Stray" electric fields are produced by various electrical appliances (e.g., hair dryers, blenders, and vacuum cleaners), and these fields can interfere with the operation of sensitive electronic circuits, such as those in stereo amplifiers, televisions, and computers. To eliminate such interference, circuits are often enclosed within metal boxes that provide shielding from external fields.

The blowup in **Figure 18.30*a*** shows another aspect of how conductors alter the electric field lines created by external charges. The lines are altered because ***the electric field just outside the surface of a conductor is perpendicular to the surface at equilibrium under electrostatic conditions.*** If the field were not perpendicular, there would be a component of the field parallel to the surface. Since the free electrons on the surface of the conductor can move, they would do so under the force exerted by that parallel component. In reality, however, no electron flow occurs at equilibrium. Therefore, there can be no parallel component, and the electric field is perpendicular to the surface.

The preceding discussion deals with features of the electric field within and around a conductor at equilibrium under electrostatic conditions. These features are related to the fact that conductors contain mobile free electrons and *do not apply to insulators,* which contain very few free electrons. Conceptual Example 14 further explores the behavior of a conducting material in the presence of an electric field.

CONCEPTUAL EXAMPLE 14 | A Conductor in an Electric Field

A charge $+q$ is suspended at the center of a hollow, electrically neutral, spherical, metallic conductor, as **Figure 18.31** illustrates. The table shows a number of possibilities for the charges that this suspended charge induces on the interior and exterior surfaces of the conductor. Which one of the possibilities is correct?

	Interior Surface	Exterior Surface
(a)	$-q$	0
(b)	$-\frac{1}{2}q$	$-\frac{1}{2}q$
(c)	$+q$	$-q$
(d)	$-q$	$+q$

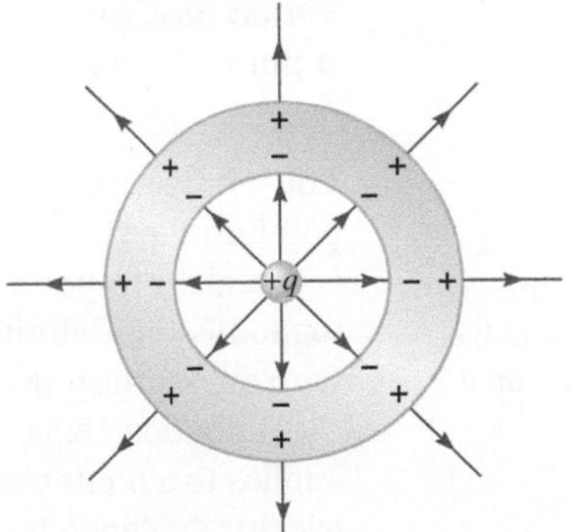

FIGURE 18.31 A positive charge $+q$ is suspended at the center of a hollow spherical conductor that is electrically neutral. Induced charges appear on the inner and outer surfaces of the conductor. The electric field within the conductor itself is zero.

Reasoning Three facts will guide our analysis. First: Since the suspended charge does not touch the conductor, the net charge on the conductor must remain zero, because it is electrically neutral to begin with. Second: Electric field lines begin on positive charges and end on negative charges. Third: At equilibrium under electrostatic conditions, there is no electric field and, hence, no field lines inside the solid material of the metallic conductor.

Answers (a) and (b) are incorrect. The net charge on the conductor in each of these answers is $-q$, which cannot be, since the conductor's net charge must remain zero.

Answer (c) is incorrect. Electric field lines emanate from the suspended positive charge $+q$ and must end on the interior surface of the metallic conductor, since they do not penetrate the metal. However, the charge on the interior surface in this answer is positive, and field lines must end on negative charges. Thus, this answer is incorrect.

Answer (d) is correct. Since the field lines emanating from the suspended positive charge $+q$ terminate only on negative charges and do not penetrate the metal, there must be an induced *negative* charge on the interior surface. Furthermore, the lines begin and end on equal amounts of charge, so the magnitude of the total charge induced on the interior surface is the same as the magnitude of the suspended charge. Thus, the charge induced on the interior surface is $-q$. We know that the total net charge on the metallic conductor must remain at zero. Therefore, if a charge $-q$ is induced on the interior surface, there must also be a charge of $+q$ induced on the exterior surface, because excess charge cannot remain inside of the solid metal at equilibrium.

18.9 Gauss' Law

Section 18.6 discusses how a point charge creates an electric field in the space around the charge. There are also many situations in which an electric field is produced by charges that are spread out over a region, rather than by a single point charge. Such an extended collection of charges is called a *charge distribution*. For example, the electric field within the parallel plate capacitor in **Figure 18.21** is produced by positive charges spread uniformly over one plate and an equal number of negative charges spread over the other plate. As we will see, Gauss' law describes the relationship between a charge distribution and the electric field it produces. This law was formulated by the German mathematician and physicist Carl Friedrich Gauss (1777–1855).

In presenting Gauss' law, it will be necessary to introduce a new idea called **electric flux**. The idea of flux involves both the electric field and the surface through which it passes. By bringing together the electric field and the surface through which it passes, we will be able to define electric flux and then present Gauss' law.

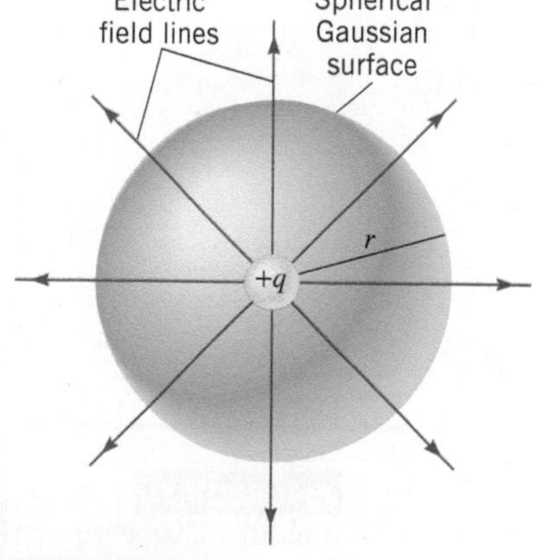

FIGURE 18.32 A positive point charge is located at the center of an imaginary spherical surface of radius r. Such a surface is one example of a Gaussian surface. Here the electric field is perpendicular to the surface and has the same magnitude everywhere on it.

We begin by developing a version of Gauss' law that applies only to a point charge, which we assume to be positive. The electric field lines for a positive point charge radiate outward in all directions from the charge, as **Figure 18.22b** indicates. The magnitude E of the electric field at a distance r from the charge is $E = kq/r^2$, according to Equation 18.3, in which we have replaced the symbol $|q|$ with the symbol q since we are assuming that the charge is positive. As mentioned in Section 18.5, the constant k can be expressed as $k = 1/(4\pi\varepsilon_0)$, where ε_0 is the permittivity of free space. With this substitution, the magnitude of the electric field becomes $E = q/(4\pi\varepsilon_0 r^2)$. We now place this point charge at the center of an imaginary spherical surface of radius r, as **Figure 18.32** shows. Such a hypothetical closed surface is called a **Gaussian surface**, although in general it need not be spherical. The surface area A of a sphere is $A = 4\pi r^2$, and the magnitude of the electric field can be written in terms of this area as $E = q/(A\varepsilon_0)$, or

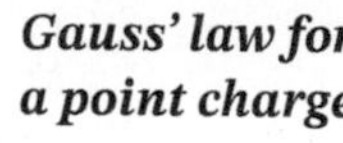
Gauss' law for a point charge

$$\underbrace{EA}_{\text{Electric flux } \Phi_E} = \frac{q}{\varepsilon_0} \tag{18.5}$$

The left side of Equation 18.5 is the product of the magnitude E of the electric field at any point on the Gaussian surface and the area A of the surface. In Gauss' law this product is especially important and is called the **electric flux**, Φ_E: $\Phi_E = EA$. (It will be necessary to modify this definition of flux when we consider the general case of a Gaussian surface with an arbitrary shape.)

Equation 18.5 is the result we have been seeking, for it is the form of Gauss' law that applies to a point charge. This result indicates that, aside from the constant ε_0, the electric flux Φ_E depends only on the charge q within the Gaussian surface and is independent of the radius r of the surface. We will now see how to generalize Equation 18.5 to account for distributions of charges and Gaussian surfaces with arbitrary shapes.

Interactive Figure 18.33 shows a charge distribution whose net charge is labeled Q. The charge distribution is surrounded by a Gaussian surface—that is, an imaginary closed surface. The surface can have *any arbitrary shape* (it need not be spherical), but *it must be closed* (an open surface would be like that of half an eggshell). The direction of the electric field is not necessarily perpendicular to the Gaussian surface. Furthermore, the magnitude of the electric field need not be constant on the surface but can vary from point to point.

To determine the electric flux through such a surface, we divide the surface into many tiny sections with areas ΔA_1, ΔA_2, and so on. Each section is so small that it is essentially flat and the electric field $\vec{\mathbf{E}}$ is a constant (both in magnitude and direction) over it. For reference, a dashed line called the "normal" is drawn perpendicular to each section on the outside of the surface. To determine the electric flux for each of the sections, we use only the component of $\vec{\mathbf{E}}$ that is perpendicular to the surface—that is, the component of the electric field that passes through the surface. From **Interactive Figure 18.33** it can be seen that this component has a magnitude of $E \cos \phi$, where ϕ is the angle between the electric field and the normal. The electric flux through any one section is then $(E \cos \phi) \Delta A$. The electric flux Φ_E that passes through the entire Gaussian surface is the sum of all of these individual fluxes: $\Phi_E = (E_1 \cos \phi_1)\Delta A_1 + (E_2 \cos \phi_2)\Delta A_2 + \cdots$, or

$$\Phi_E = \Sigma(E \cos \phi)\Delta A \quad \textbf{(18.6)}$$

where, as usual, the symbol Σ means "the sum of." Gauss' law relates the electric flux Φ_E to the net charge Q enclosed by the arbitrarily shaped Gaussian surface.

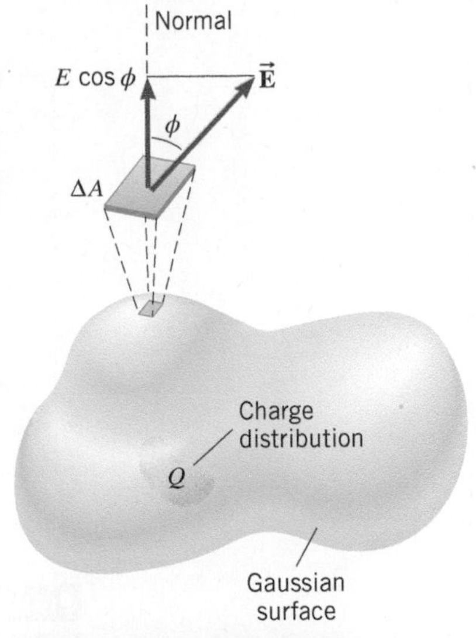

INTERACTIVE FIGURE 18.33 The charge distribution Q is surrounded by an arbitrarily shaped Gaussian surface. The electric flux Φ through any tiny segment of the surface is the product of $E \cos \phi$ and the area ΔA of the segment: $\Phi = (E \cos \phi)\Delta A$. The angle ϕ is the angle between the electric field and the normal to the surface.

GAUSS' LAW

The electric flux Φ_E through a Gaussian surface is equal to the net charge Q enclosed by the surface divided by ε_0, the permittivity of free space:

$$\underbrace{\Sigma(E \cos \phi)\Delta A}_{\text{Electric flux, } \Phi_E} = \frac{Q}{\varepsilon_0} \quad (18.7)$$

SI Unit of Electric Flux: $\text{N} \cdot \text{m}^2/\text{C}$

Although we arrived at Gauss' law by assuming the net charge Q was positive, Equation 18.7 also applies when Q is negative. In this case, the electric flux Φ_E is also negative. Gauss' law is often used to find the magnitude of the electric field produced by a distribution of charges. The law is most useful when the distribution is uniform and symmetrical. In the next two examples we will see how to apply Gauss' law in such situations.

EXAMPLE 15 | The Electric Field of a Charged Thin Spherical Shell

Figures 18.34*a* and ***b*** show a thin spherical shell of radius R (for clarity, only half of the shell is shown). A positive charge q is spread uniformly over the shell. Find the magnitude of the electric field at any point **(a)** outside the shell and **(b)** inside the shell.

Reasoning Because the charge is distributed uniformly over the spherical shell, the electric field is symmetrical. This means that the electric field is directed radially outward in all directions, and its magnitude is the same at all points that are equidistant from the shell. All such points lie on a sphere, so the symmetry is called *spherical symmetry.* With this symmetry in mind, we will use a spherical Gaussian surface to evaluate the electric flux Φ_E. We will then use Gauss' law to determine the magnitude of the electric field.

Solution **(a)** To find the magnitude of the electric field outside the charged shell, we evaluate the electric flux $\Phi_E = \Sigma(E \cos \phi)\Delta A$ by using a spherical Gaussian surface of radius r ($r > R$) that is concentric with the shell. See the blue surface labeled S in **Figure 18.34*a***. Since the electric field $\vec{\mathbf{E}}$ is everywhere perpendicular to the Gaussian surface, $\phi = 0°$ and $\cos \phi = 1$. In addition, E has the same value at all points on the surface, since they are equidistant from the charged shell. Being constant over the surface, E can be factored outside the summation, with the result that

$$\Phi_E = \Sigma(E \cos 0°)\,\Delta A = E(\underbrace{\Sigma \Delta A}_{\substack{\text{Area of}\\ \text{Gaussian}\\ \text{surface}}}) = E\,(\underbrace{4\pi r^2}_{\substack{\text{Surface area}\\ \text{of sphere}}})$$

The term $\Sigma \Delta A$ is just the sum of the tiny areas that make up the Gaussian surface. Since the area of a spherical surface is $4\pi r^2$, we have $\Sigma \Delta A = 4\pi r^2$. Setting the electric flux equal to Q/ε_0, as specified by Gauss' law, yields $E(4\pi r^2) = Q/\varepsilon_0$. Since the only charge within the Gaussian surface is the charge q on the shell, it follows that the net charge within the Gaussian surface is $Q = q$. Thus, we can solve for E and find that

$$\boxed{E = \frac{q}{4\pi\varepsilon_0 r^2} \quad (\text{for } r > R)}$$

This is a surprising result, for it is the same as that for a point charge (see Equation 18.3 with $|q| = q$). Thus, the electric field outside a uniformly charged spherical shell is the same as if all the charge q were concentrated as a point charge at the center of the shell.

(b) To find the magnitude of the electric field inside the charged shell, we select a spherical Gaussian surface that lies inside the

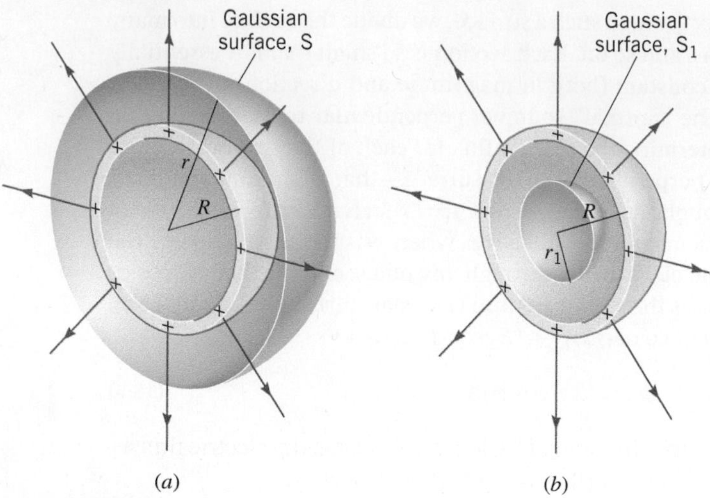

FIGURE 18.34 A uniform distribution of positive charge resides on a thin spherical shell of radius R. The spherical Gaussian surfaces S and S_1 are used in Example 15 to evaluate the electric flux (*a*) outside and (*b*) inside the shell, respectively. For clarity, only half the shell and the Gaussian surfaces are shown.

shell and is concentric with it. See the blue surface labeled S_1 in **Figure 18.34*b***. Inside the charged shell, the electric field (if it exists) must also have spherical symmetry. Therefore, using reasoning like that in part (a), the electric flux through the Gaussian surface is $\Phi_E = \Sigma(E \cos \phi)\Delta A = E(4\pi r_1^2)$. In accord with Gauss' law, the electric flux must be equal to Q/ε_0, where Q is the net charge *inside* the Gaussian surface. But now $Q = 0$ C, since all the charge lies on the shell that is *outside* the surface S_1. Consequently, we have $E(4\pi r_1^2) = Q/\varepsilon_0 = 0$, or

$$\boxed{E = 0 \text{ N/C} \quad (\text{for } r < R)}$$

Gauss' law allows us to deduce that there is no electric field inside a uniform spherical shell of charge. An electric field exists only on the outside.

EXAMPLE 16 | The Electric Field Inside a Parallel Plate Capacitor

According to Equation 18.4, the electric field inside a parallel plate capacitor, and away from the edges, is constant and has a magnitude of $E = \sigma/\varepsilon_0$, where σ is the charge density (the charge per unit area) on a plate. Use Gauss' law to obtain this result.

Reasoning **Figure 18.35*a*** shows the electric field inside a parallel plate capacitor. Because the positive and negative charges are distributed uniformly over the surfaces of the plates, symmetry requires that the electric field be perpendicular to the plates. We will take advantage of this symmetry by choosing our Gaussian surface to be a small cylinder whose axis is perpendicular to the plates (see part *b* of the figure). With this choice, we will be able to evaluate the electric flux and then, with the aid of Gauss' law, determine E.

Solution **Figure 18.35*b*** shows that we have placed our Gaussian cylinder so that its left end is inside the positive metal plate, and the right end is in the space between the plates. To determine the electric flux through this Gaussian surface, we evaluate the flux through each of the three parts—labeled 1, 2, and 3 in the drawing—that make up the total surface of the cylinder and then add up the fluxes.

Surface 1—the flat left end of the cylinder—is embedded inside the positive metal plate. As discussed in Section 18.8, the electric field is zero everywhere inside a conductor that is in equilibrium under electrostatic conditions. Since $E = 0$ N/C, the electric flux through this surface is also zero:

$$\Phi_1 = \Sigma(E \cos \phi)\,\Delta A = \Sigma[(0 \text{ N/C}) \cos \phi]\,\Delta A = 0$$

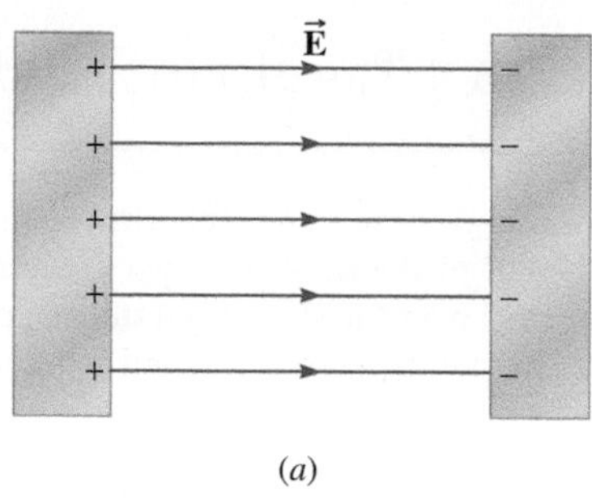

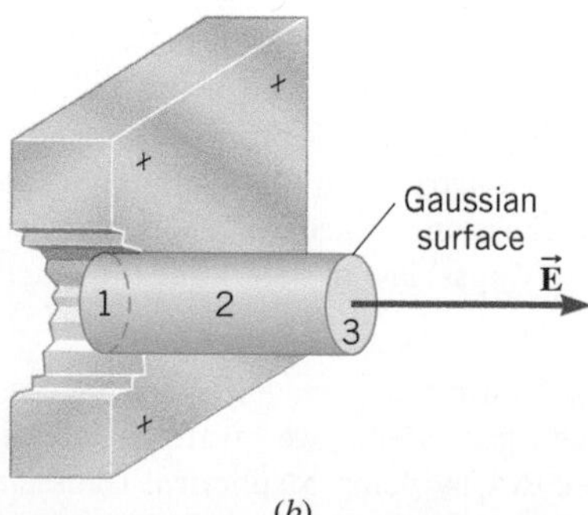

FIGURE 18.35 (*a*) A side view of a parallel plate capacitor, showing some of the electric field lines. (*b*) The Gaussian surface is a cylinder oriented so its axis is perpendicular to the positive plate and its left end is inside the plate.

Surface 2—the curved wall of the cylinder—is everywhere parallel to the electric field between the plates, so that $\cos\phi = \cos 90° = 0$. Therefore, the electric flux through this surface is also zero:

$$\Phi_2 = \Sigma(E\cos\phi)\,\Delta A = \Sigma(E\cos 90°)\,\Delta A = 0$$

Surface 3—the flat right end of the cylinder—is perpendicular to the electric field between the plates, so $\cos\phi = \cos 0° = 1$. The electric field is constant over this surface, so E can be taken outside the summation in Equation 18.6. Noting that $\Sigma\Delta A = A$ is the area of surface 3, we find that the electric flux through this surface is

$$\Phi_3 = \Sigma(E\cos\phi)\,\Delta A = \Sigma(E\cos 0°)\,\Delta A = E(\Sigma\Delta A) = EA$$

The electric flux through the entire Gaussian cylinder is the sum of the three fluxes determined above:

$$\Phi_E = \Phi_1 + \Phi_2 + \Phi_3 = 0 + 0 + EA = EA$$

According to Gauss' law, we set the electric flux equal to Q/ε_0, where Q is the net charge *inside* the Gaussian cylinder: $EA = Q/\varepsilon_0$. But Q/A is the charge per unit area, σ, on the plate. Therefore, we arrive at the value of the electric field inside a parallel plate capacitor: $\boxed{E = \sigma/\varepsilon_0.}$ The distance of the right end of the Gaussian cylinder from the positive plate does not appear in this result, indicating that the electric field is the same everywhere between the plates.

Check Your Understanding

(*The answers are given at the end of the book.*)

19. A Gaussian surface contains a single charge within it, and as a result an electric flux passes through the surface. Suppose that the charge is then moved to another spot within the Gaussian surface. Does the flux through the surface change?

20. **CYU Figure 18.6** shows an arrangement of three charges. In parts *a* and *b* different Gaussian surfaces (both in blue) are shown. Through which surface, if either, does the greater electric flux pass?

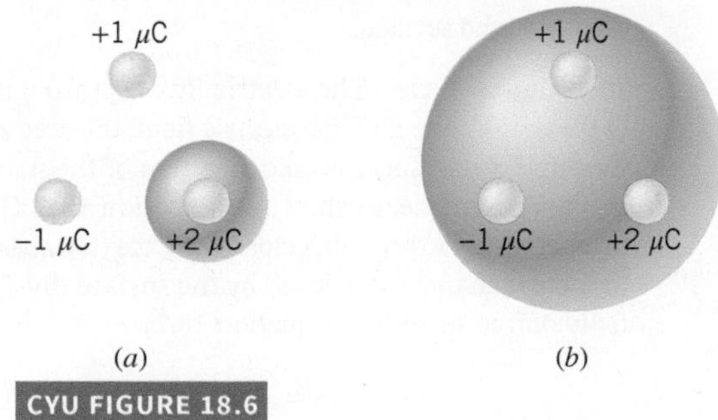

CYU FIGURE 18.6

21. **CYU Figure 18.7** shows three charges, labeled q_1, q_2, and q_3. A Gaussian surface (in blue) is drawn around q_1 and q_2. **(a)** Which charges determine the electric flux through the Gaussian surface? **(b)** Which charges produce the electric field that exists at the point *P*?

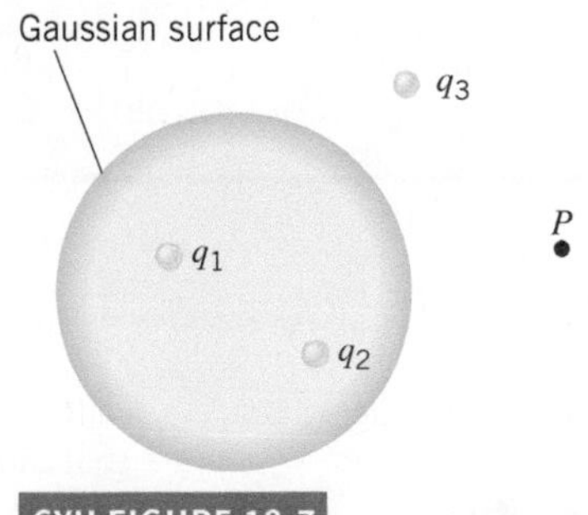

CYU FIGURE 18.7

Concept Summary

18.1 The Origin of Electricity There are two kinds of electric charge: positive and negative. The SI unit of electric charge is the coulomb (C). The magnitude of the charge on an electron or a proton is $e = 1.60 \times 10^{-19}$ C. Since the symbol e denotes a magnitude, it has no algebraic sign. Thus, the electron carries a charge of $-e$, and the proton carries a charge of $+e$. The charge on any object, whether positive or negative, is quantized, in the sense that the charge consists of an integer number of protons or electrons.

18.2 Charged Objects and the Electric Force The law of conservation of electric charge states that the net electric charge of an isolated system remains constant during any process.

Like charges repel and unlike charges attract each other.

18.3 Conductors and Insulators An electrical conductor is a material, such as copper, that conducts electric charge readily. An electrical insulator is a material, such as rubber, that conducts electric charge poorly.

18.4 Charging by Contact and by Induction Charging by contact is the process of giving one object a net electric charge by placing it in contact with an object that is already charged. Charging by induction is the process of giving an object a net electric charge without touching it to a charged object.

18.5 Coulomb's Law A point charge is a charge that occupies so little space that it can be regarded as a mathematical point. Coulomb's law gives the magnitude F of the electric force that two point charges q_1 and q_2 exert on each other, according to Equation 18.1, where $|q_1|$ and $|q_2|$ are the magnitudes of the charges and have no algebraic sign. The term k is a constant and has the value $k = 8.99 \times 10^9\ \text{N} \cdot \text{m}^2/\text{C}^2$. The force specified by Equation 18.1 acts along the line between the two charges.

$$F = k\frac{|q_1|\,|q_2|}{r^2} \quad (18.1)$$

The permittivity of free space ε_0 is defined by the relation

$$k = \frac{1}{4\pi\varepsilon_0}$$

18.6 The Electric Field The electric field $\vec{\mathbf{E}}$ at a given spot is a vector and is the electrostatic force $\vec{\mathbf{F}}$ experienced by a very small test charge q_0 placed at that spot divided by the charge itself, as given by Equation 18.2. The direction of the electric field is the same as the direction of the force on a positive test charge. The SI unit for the electric field is the newton per coulomb (N/C). The source of the electric field at any spot is the collection of charged objects surrounding that spot.

$$\vec{\mathbf{E}} = \frac{\vec{\mathbf{F}}}{q_0} \quad (18.2)$$

The magnitude of the electric field created by a point charge q is given by Equation 18.3, where $|q|$ is the magnitude of the charge and has no algebraic sign and r is the distance from the charge. The electric field $\vec{\mathbf{E}}$ points away from a positive charge and toward a negative charge.

$$E = \frac{k|q|}{r^2} \quad (18.3)$$

For a parallel plate capacitor that has a charge per unit area of σ on each plate, the magnitude of the electric field between the plates is given by Equation 18.4.

$$E = \frac{\sigma}{\varepsilon_0} \quad (18.4)$$

18.7 Electric Field Lines Electric field lines are lines that can be thought of as a "map," insofar as the lines provide information about the direction and strength of the electric field. The lines are directed away from positive charges and toward negative charges. The direction of the lines gives the direction of the electric field, since the electric field vector at a point is tangent to the line at that point. The electric field is strongest in regions where the number of lines per unit area passing perpendicularly through a surface is the greatest—that is, where the lines are packed together most tightly.

18.8 The Electric Field Inside a Conductor: Shielding Excess negative or positive charge resides on the surface of a conductor at equilibrium under electrostatic conditions. In such a situation, the electric field at any point within the conducting material is zero, and the electric field just outside the surface of the conductor is perpendicular to the surface.

18.9 Gauss' Law The electric flux Φ_E through a surface is related to the magnitude E of the electric field, the area A of the surface, and the angle ϕ that specifies the direction of the field relative to the normal to the surface, as shown in Equation 18.6. Gauss' law states that the electric flux through a closed surface (a Gaussian surface) is equal to the net charge Q enclosed by the surface divided by ε_0, the permittivity of free space (see Equation 18.7).

$$\Phi_E = \Sigma(E \cos \phi)\Delta A \quad (18.6)$$

$$\Phi_E = \Sigma(E \cos \phi)\Delta A = \frac{Q}{\varepsilon_0} \quad (18.7)$$

Focus on Concepts

Online

Additional questions are available for assignment in WileyPLUS.

Section 18.1 The Origin of Electricity

1. An object carries a charge of $-8.0\ \mu\text{C}$, while another carries a charge of $-2.0\ \mu\text{C}$. How many electrons must be transferred from the first object to the second so that both objects have the same charge?

Section 18.2 Charged Objects and the Electric Force

2. Each of three objects carries a charge. As the drawing shows, objects A and B attract each other, and objects C and A also attract each other. Which one of the following statements concerning objects B and C is true? (a) They attract each other. (b) They repel each other. (c) They neither attract nor repel each other. (d) This question cannot be answered without additional information.

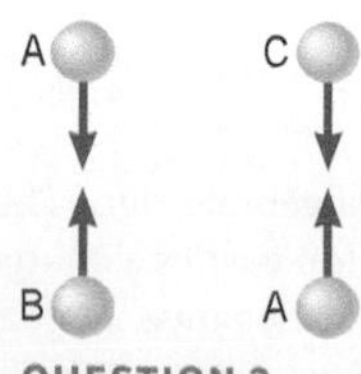

QUESTION 2

Section 18.4 Charging by Contact and by Induction

3. Each of two identical objects carries a net charge. The objects are made from conducting material. One object is attracted to a positively charged ebonite rod, and the other is repelled by the rod. After the objects are touched together, it is found that they are each repelled by the rod. What can be concluded about the initial charges on the objects? (a) Initially both objects are positive, with both charges having the same magnitude. (b) Initially both objects are negative, with both charges having the same magnitude. (c) Initially one object is positive and one is negative, with the negative charge having a greater magnitude than the positive charge. (d) Initially one object is positive and one is negative, with the positive charge having a greater magnitude than the negative charge.

4. Only one of three balls A, B, and C carries a net charge q. The balls are made from conducting material and are identical. One of the uncharged balls can become charged by touching it to the charged ball and then separating the two. This process of touching one ball to another and then separating the two balls can be repeated over and over again, with the result that the three balls can take on a variety of charges. Which one of the following distributions of charges could not possibly be achieved in this fashion, even if the process were repeated a large number of times?

(a) $q_A = \frac{1}{3}q, q_B = \frac{1}{3}q, q_C = \frac{1}{3}q$ (c) $q_A = \frac{1}{2}q, q_B = \frac{3}{8}q, q_C = \frac{1}{4}q$
(b) $q_A = \frac{1}{2}q, q_B = \frac{1}{4}q, q_C = \frac{1}{4}q$ (d) $q_A = \frac{3}{8}q, q_B = \frac{3}{8}q, q_C = \frac{1}{4}q$

Section 18.5 Coulomb's Law

5. Three point charges have equal magnitudes and are located on the same line. The separation d between A and B is the same as the separation between B and C. One of the charges is positive and two are negative, as the drawing shows. Consider the net electrostatic force that each charge experiences due to the other two charges. Rank the net forces in descending order (greatest first) according to magnitude. (a) A, B, C (b) B, C, A (c) A, C, B (d) C, A, B (e) B, A, C

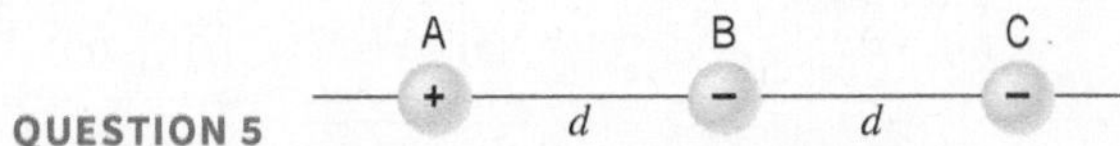

QUESTION 5

6. Three point charges have equal magnitudes and are fixed to the corners of an equilateral triangle. Two of the charges are positive and one is negative, as the drawing shows. At which one of the corners is the net force acting on the charge directed parallel to the x axis? (a) A (b) B (c) C

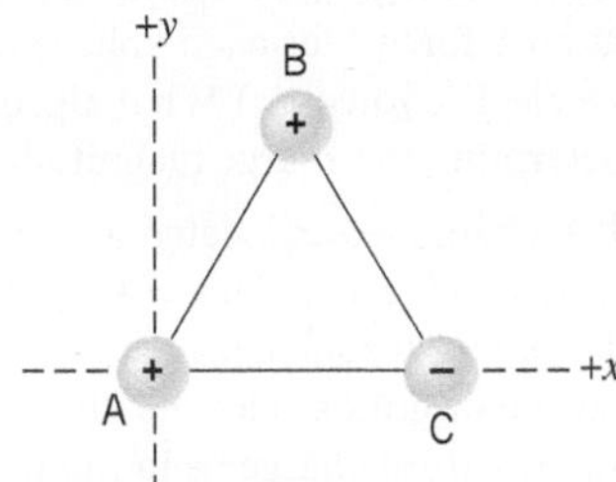

QUESTION 6

Section 18.6 The Electric Field

7. A positive point charge q_1 creates an electric field of magnitude E_1 at a spot located at a distance r_1 from the charge. The charge is replaced by another positive point charge q_2, which creates a field of magnitude $E_2 = E_1$ at a distance of $r_2 = 2r_1$. How is q_2 related to q_1? (a) $q_2 = \frac{1}{2}q_1$ (b) $q_2 = \frac{1}{2}q_1$ (c) $q_2 = 4q_1$ (d) $q_2 = \frac{1}{4}q_1$ (e) $q_2 = \sqrt{2}\,q_1$

8. The drawing shows a positive and a negative point charge. The negative charge has the greater magnitude. Where on the line that passes through the charges is the one spot where the total electric field is zero? (a) To the right of the negative charge (b) To the left of the positive charge (c) Between the charges, to the left of the midpoint (d) Between the charges, to the right of the midpoint

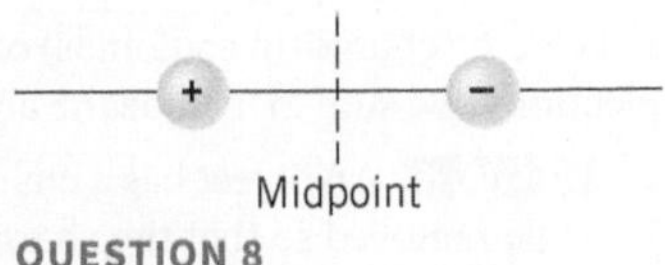

QUESTION 8

Section 18.7 Electric Field Lines

9. The drawing shows some electric field lines. For the points indicated, rank the magnitudes of the electric field in descending order (largest first). (a) B, C, A (b) B, A, C (c) A, B, C (d) A, C, B (e) C, A, B

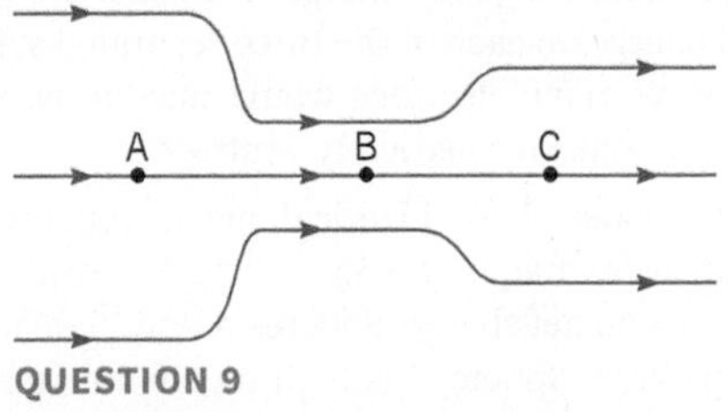

QUESTION 9

Section 18.8 The Electric Field Inside a Conductor: Shielding

10. The drawings show (in cross section) two solid spheres and two spherical shells. Each object is made from copper and has a net charge, as the plus and minus signs indicate. Which drawing correctly shows where the charges reside when they are in equilibrium? (a) A (b) B (c) C (d) D

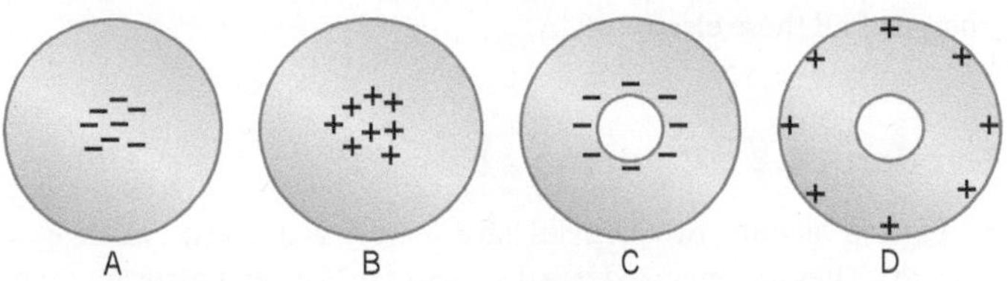

QUESTION 10

Section 18.9 Gauss' Law

11. A cubical Gaussian surface surrounds two charges, $q_1 = +6.0 \times 10^{-12}$ C and $q_2 = -2.0 \times 10^{-12}$ C, as the drawing shows. What is the electric flux passing through the surface?

Problems

Online

Additional questions are available for assignment in WileyPLUS.

SSM Student Solutions Manual
MMH Problem-solving help
GO Guided Online Tutorial
V-HINT Video Hints
CHALK Chalkboard Videos
BIO Biomedical application

E Easy
M Medium
H Hard
WS Worksheet
T Team Problem

Section 18.1 The Origin of Electricity, Section 18.2 Charged Objects and the Electric Force, Section 18.3 Conductors and Insulators, Section 18.4 Charging by Contact and by Induction

1. E SSM Iron atoms have been detected in the sun's outer atmosphere, some with many of their electrons stripped away. What is the

net electric charge (in coulombs) of an iron atom with 26 protons and 7 electrons? Be sure to include the algebraic sign (+ or −) in your answer.

2. **E** **CHALK** An object has a charge of $-2.0\ \mu C$. How many electrons must be removed so that the charge becomes $+3.0\ \mu C$?

3. **E** **CHALK** Four identical metallic objects carry the following charges: +1.6, +6.2, −4.8, and −9.4 μC. The objects are brought simultaneously into contact, so that each touches the others. Then they are separated. **(a)** What is the final charge on each object? **(b)** How many electrons (or protons) make up the final charge on each object?

4. **E** **GO** Four identical metal spheres have charges of $q_A = -8.0\ \mu C$, $q_B = -2.0\ \mu C$, $q_C = +5.0\ \mu C$, and $q_D = +12.0\ \mu C$. **(a)** Two of the spheres are brought together so they touch, and then they are separated. Which spheres are they, if the final charge on each one is $+5.0\ \mu C$? **(b)** In a similar manner, which three spheres are brought together and then separated, if the final charge on each of the three is $+3.0\ \mu C$? **(c)** The final charge on each of the three separated spheres in part (b) is $+3.0\ \mu C$. How many electrons would have to be added to one of these spheres to make it electrically neutral?

5. **E** **SSM** Consider three identical metal spheres, A, B, and C. Sphere A carries a charge of $+5q$. Sphere B carries a charge of $-q$. Sphere C carries no net charge. Spheres A and B are touched together and then separated. Sphere C is then touched to sphere A and separated from it. Last, sphere C is touched to sphere B and separated from it. **(a)** How much charge ends up on sphere C? What is the total charge on the three spheres **(b)** before they are allowed to touch each other and **(c)** after they have touched?

6. **E** **GO** **CHALK** A plate carries a charge of $-3.0\ \mu C$, while a rod carries a charge of $+2.0\ \mu C$. How many electrons must be transferred from the plate to the rod, so that both objects have the same charge?

7. **M** **V-HINT** **CHALK** Water has a mass per mole of 18.0 g/mol, and each water molecule (H_2O) has 10 electrons. **(a)** How many electrons are there in one liter ($1.00 \times 10^{-3}\ m^3$) of water? **(b)** What is the net charge of all these electrons?

Section 18.5 Coulomb's Law

8. **E** In a vacuum, two particles have charges of q_1 and q_2, where $q_1 = +3.5\ \mu C$. They are separated by a distance of 0.26 m, and particle 1 experiences an attractive force of 3.4 N. What is q_2 (magnitude and sign)?

9. **E** **SSM** Two spherical objects are separated by a distance that is 1.80×10^{-3} m. The objects are initially electrically neutral and are very small compared to the distance between them. Each object acquires the same negative charge due to the addition of electrons. As a result, each object experiences an electrostatic force that has a magnitude of 4.55×10^{-21} N. How many electrons did it take to produce the charge on one of the objects?

10. **E** Two tiny conducting spheres are identical and carry charges of $-20.0\ \mu C$ and $+50.0\ \mu C$. They are separated by a distance of 2.50 cm. **(a)** What is the magnitude of the force that each sphere experiences, and is the force attractive or repulsive? **(b)** The spheres are brought into contact and then separated to a distance of 2.50 cm. Determine the magnitude of the force that each sphere now experiences, and state whether the force is attractive or repulsive.

11. **E** **SSM** Two very small spheres are initially neutral and separated by a distance of 0.50 m. Suppose that 3.0×10^{13} electrons are removed from one sphere and placed on the other. **(a)** What is the magnitude of the electrostatic force that acts on each sphere? **(b)** Is the force attractive or repulsive? Why?

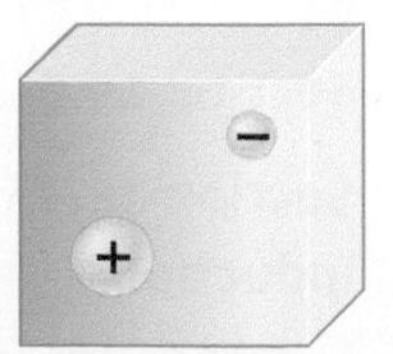

QUESTION 11

12. **E** Two charges attract each other with a force of 1.5 N. What will be the force if the distance between them is reduced to one-ninth of its original value?

13. **E** Two point charges are fixed on the y axis: a negative point charge $q_1 = -25\ \mu C$ at $y_1 = +0.22$ m and a positive point charge q_2 at $y_2 = +0.34$ m. A third point charge $q = +8.4\ \mu C$ is fixed at the origin. The net electrostatic force exerted on the charge q by the other two charges has a magnitude of 27 N and points in the $+y$ direction. Determine the magnitude of q_2.

14. **E** **GO** The drawings show three charges that have the same magnitude but may have different signs. In all cases the distance d between the charges is the same. The magnitude of the charges is $|q| = 8.6\ \mu C$, and the distance between them is $d = 3.8$ mm. Determine the magnitude of the net force on charge 2 for each of the three drawings.

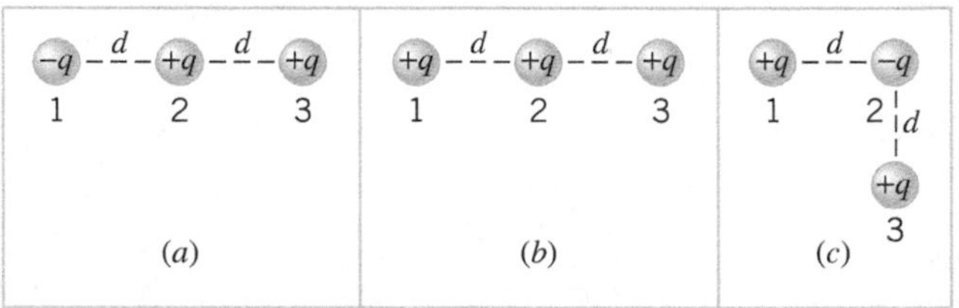

PROBLEM 14

15. **E** **SSM** **MMH** Two tiny spheres have the same mass and carry charges of the same magnitude. The mass of each sphere is 2.0×10^{-6} kg. The gravitational force that each sphere exerts on the other is balanced by the electric force. **(a)** What algebraic signs can the charges have? **(b)** Determine the charge magnitude.

16. **E** **GO** A charge $+q$ is located at the origin, while an identical charge is located on the x axis at $x = +0.50$ m. A third charge of $+2q$ is located on the x axis at such a place that the net electrostatic force on the charge at the origin doubles, its direction remaining unchanged. Where should the third charge be located?

17. **E** **SSM** Two particles, with identical positive charges and a separation of 2.60×10^{-2} m, are released from rest. Immediately after the release, particle 1 has an acceleration $\vec{\mathbf{a}}_1$ whose magnitude is $4.60 \times 10^3\ m/s^2$, while particle 2 has an acceleration $\vec{\mathbf{a}}_2$ whose magnitude is $8.50 \times 10^3\ m/s^2$. Particle 1 has a mass of 6.00×10^{-6} kg. Find **(a)** the charge on each particle and **(b)** the mass of particle 2.

18. **E** A charge of $-3.00\ \mu C$ is fixed at the center of a compass. Two additional charges are fixed on the circle of the compass, which has a radius of 0.100 m. The charges on the circle are $-4.00\ \mu C$ at the position due north and $+5.00\ \mu C$ at the position due east. What are the magnitude and direction of the net electrostatic force acting on the charge at the center? Specify the direction relative to due east.

19. **M** Multiple-Concept Example 3 provides some pertinent background for this problem. Suppose a single electron orbits about a nucleus containing two protons ($+2e$), as would be the case for a helium atom from which one of the two naturally occurring electrons is removed. The radius of the orbit is 2.65×10^{-11} m. Determine the magnitude of the electron's centripetal acceleration.

20. **M** **V-HINT** The drawing shows an equilateral triangle, each side of which has a length of 2.00 cm. Point charges are fixed to each corner, as shown. The $4.00\ \mu C$ charge experiences a net force due to the charges q_A and q_B. This net force points vertically downward and has a magnitude of 405 N. Determine the magnitudes and algebraic signs of the charges q_A and q_B.

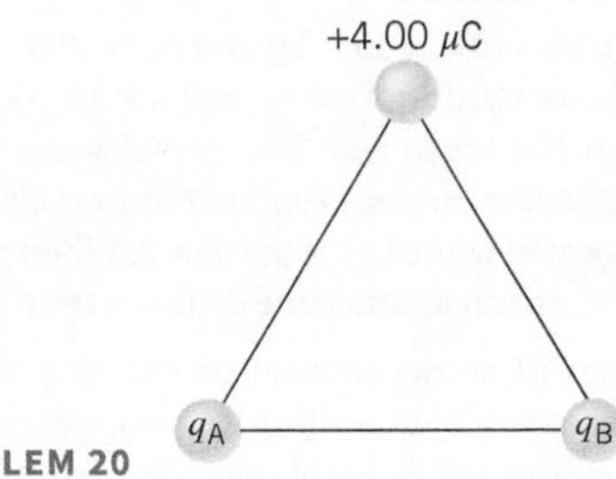

PROBLEM 20

21. M V-HINT MMH The drawing shows three point charges fixed in place. The charge at the coordinate origin has a value of $q_1 =$ +8.00 μC; the other two charges have identical magnitudes, but opposite signs: $q_2 = -5.00$ μC and $q_3 = +5.00$ μC. **(a)** Determine the net force (magnitude and direction) exerted on q_1 by the other two charges. **(b)** If q_1 had a mass of 1.50 g and it were free to move, what would be its acceleration?

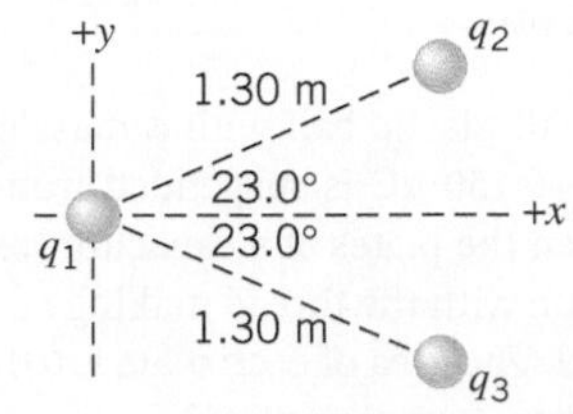

PROBLEM 21

22. M GO An electrically neutral model airplane is flying in a horizontal circle on a 3.0-m guideline, which is nearly parallel to the ground. The line breaks when the kinetic energy of the plane is 50.0 J. Reconsider the same situation, except that now there is a point charge of $+q$ on the plane and a point charge of $-q$ at the other end of the guideline. In this case, the line breaks when the kinetic energy of the plane is 51.8 J. Find the magnitude of the charges.

23. M V-HINT Multiple-Concept Example 3 illustrates several of the concepts that come into play in this problem. A single electron orbits a lithium nucleus that contains three protons ($+3e$). The radius of the orbit is 1.76×10^{-11} m. Determine the kinetic energy of the electron.

24. M GO An unstrained horizontal spring has a length of 0.32 m and a spring constant of 220 N/m. Two small charged objects are attached to this spring, one at each end. The charges on the objects have equal magnitudes. Because of these charges, the spring stretches by 0.020 m relative to its unstrained length. Determine **(a)** the possible algebraic signs and **(b)** the magnitude of the charges.

25. M SSM In the rectangle in the drawing, a charge is to be placed at the empty corner to make the net force on the charge at corner A point along the vertical direction. What charge (magnitude and algebraic sign) must be placed at the empty corner?

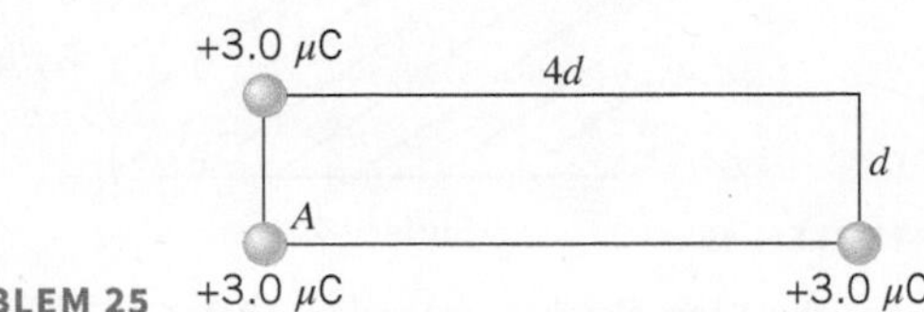

PROBLEM 25

26. H CHALK SSM A small spherical insulator of mass 8.00×10^{-2} kg and charge +0.600 μC is hung by a thread of negligible mass. A charge of −0.900 μC is held 0.150 m away from the sphere and directly to the right of it, so the thread makes an angle θ with the vertical (see the drawing). Find **(a)** the angle θ and **(b)** the tension in the thread.

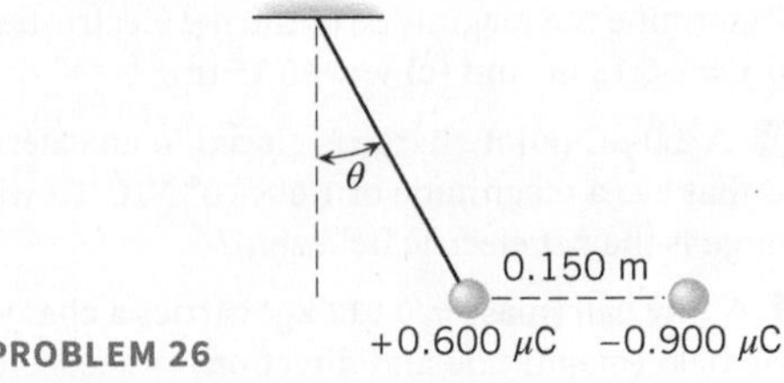

PROBLEM 26

Section 18.6 The Electric Field, Section 18.7 Electric Field Lines, Section 18.8 The Electric Field Inside a Conductor: Shielding

27. E SSM At a distance r_1 from a point charge, the magnitude of the electric field created by the charge is 248 N/C. At a distance r_2 from the charge, the field has a magnitude of 132 N/C. Find the ratio r_2/r_1.

28. E GO Suppose you want to determine the electric field in a certain region of space. You have a small object of known charge and an instrument that measures the magnitude and direction of the force exerted on the object by the electric field. **(a)** The object has a charge of +20.0 μC and the instrument indicates that the electric force exerted on it is 40.0 μN, due east. What are the magnitude and direction of the electric field? **(b)** What are the magnitude and direction of the electric field if the object has a charge of −10.0 μC and the instrument indicates that the force is 20.0 μN, due west?

29. E An electric field of 260 000 N/C points due west at a certain spot. What are the magnitude and direction of the force that acts on a charge of −7.0 μC at this spot?

30. E MMH Review the important features of electric field lines discussed in Conceptual Example 12. Three point charges ($+q$, $+2q$, and $-3q$) are at the corners of an equilateral triangle. Sketch in six electric field lines between the three charges.

31. E Four point charges have the same magnitude of 2.4×10^{-12} C and are fixed to the corners of a square that is 4.0 cm on a side. Three of the charges are positive and one is negative. Determine the magnitude of the net electric field that exists at the center of the square.

32. E GO The drawing shows two situations in which charges are placed on the x and y axes. They are all located at the same distance of 6.1 cm from the origin O. For each of the situations in the drawing, determine the magnitude of the net electric field at the origin.

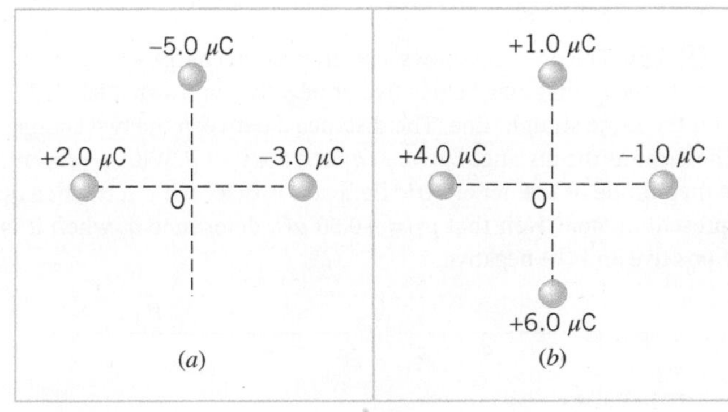

PROBLEM 32

33. E MMH A uniform electric field exists everywhere in the x, y plane. This electric field has a magnitude of 4500 N/C and is directed in the positive x direction. A point charge -8.0×10^{-9} C is placed at

the origin. Determine the magnitude of the net electric field at (a) $x = -0.15$ m, (b) $x = +0.15$ m, and (c) $y = +0.15$ m.

34. **E** **MMH** A 3.0 μC point charge is placed in an external uniform electric field that has a magnitude of 1.6×10^4 N/C. At what distance from the charge is the net electric field zero?

35. **E** **SSM** A tiny ball (mass = 0.012 kg) carries a charge of −18 μC. What electric field (magnitude and direction) is needed to cause the ball to float above the ground?

36. **E** **GO** A proton and an electron are moving due east in a constant electric field that also points due east. The electric field has a magnitude of 8.0×10^4 N/C. Determine the magnitude of the acceleration of the proton and the electron.

37. **E** Review Conceptual Example 11 before attempting to work this problem. The magnitude of each of the charges in **Figure 18.20** is 8.60×10^{-12} C. The lengths of the sides of the rectangles are 3.00 cm and 5.00 cm. Find the magnitude of the electric field at the center of the rectangle in **Figures 18.20*a*** and ***b***.

38. **E** Two charges are placed between the plates of a parallel plate capacitor. One charge is $+q_1$ and the other is $q_2 = +5.00$ μC. The charge per unit area on each of the plates has a magnitude of $\sigma = 1.30 \times 10^{-4}$ C/m^2. The magnitude of the force on q_1 due to q_2 equals the magnitude of the force on q_1 due to the electric field of the parallel plate capacitor. What is the distance r between the two charges?

39. **M** **MMH** A small object has a mass of 3.0×10^{-3} kg and a charge of −34 μC. It is placed at a certain spot where there is an electric field. When released, the object experiences an acceleration of 2.5×10^3 m/s^2 in the direction of the $+x$ axis. Determine the magnitude and direction of the electric field.

40. **M** **SSM** Two point charges are located along the x axis: $q_1 = +6.0$ μC at $x_1 = +4.0$ cm, and $q_2 = +6.0$ μC at $x_2 = -4.0$ cm. Two other charges are located on the y axis: $q_3 = +3.0$ μC at $y_3 = +5.0$ cm, and $q_4 = -8.0$ μC at $y_4 = +7.0$ cm. Find the net electric field (magnitude and direction) at the origin.

41. **M** **GO** The total electric field $\vec{\mathbf{E}}$ consists of the vector sum of two parts. One part has a magnitude of $E_1 = 1200$ N/C and points at an angle $\theta_1 = 35°$ above the $+x$ axis. The other part has a magnitude of $E_2 = 1700$ N/C and points at an angle $\theta_2 = 55°$ above the $+x$ axis. Find the magnitude and direction of the total field. Specify the directional angle relative to the $+x$ axis.

42. **M** **V-HINT** **CHALK** A particle of charge +12 μC and mass 3.8×10^{-5} kg is released from rest in a region where there is a constant electric field of +480 N/C. What is the displacement of the particle after a time of 1.6×10^{-2} s?

43. **M** **GO** The drawing shows a positive point charge $+q_1$, a second point charge q_2 that may be positive or negative, and a spot labeled P, all on the same straight line. The distance d between the two charges is the same as the distance between q_1 and the spot P. With q_2 present, the magnitude of the net electric field at P is twice what it is when q_1 is present alone. Given that $q_1 = +0.50$ μC, determine q_2 when it is (a) positive and (b) negative.

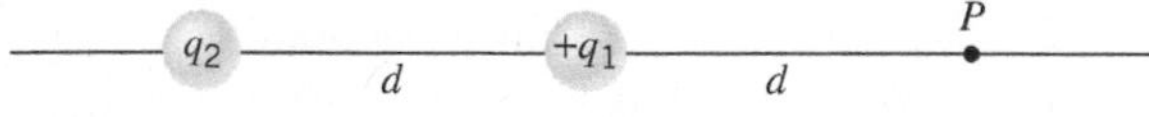

PROBLEM 43

44. **M** **CHALK** An electron is released from rest at the negative plate of a parallel plate capacitor. The charge per unit area on each plate is $\sigma = 1.8 \times 10^{-7}$ C/m^2, and the plate separation is 1.5×10^{-2} m. How fast is the electron moving just before it reaches the positive plate?

45. **H** Two particles are in a uniform electric field that points in the $+x$ direction and has a magnitude of 2500 N/C. The mass and charge of particle 1 are $m_1 = 1.4 \times 10^{-5}$ kg and $q_1 = -7.0$ μC, while the corresponding values for particle 2 are $m_2 = 2.6 \times 10^{-5}$ kg and $q_2 = +18$ μC. Initially the particles are at rest. The particles are both located on the same electric field line but are separated from each other by a distance d. Particle 1 is located to the left of particle 2. When released, they accelerate but always remain at this same distance from each other. Find d.

46. **H** The drawing shows an electron entering the lower left side of a parallel plate capacitor and exiting at the upper right side. The initial speed of the electron is 7.00×10^6 m/s. The capacitor is 2.00 cm long, and its plates are separated by 0.150 cm. Assume that the electric field between the plates is uniform everywhere and find its magnitude.

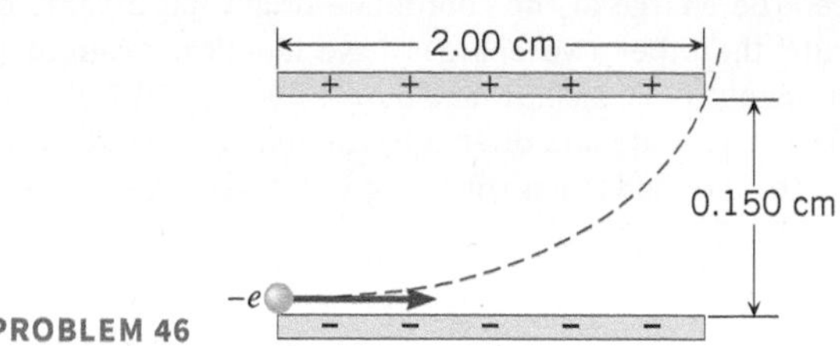

PROBLEM 46

47. **H** A small plastic ball with a mass of 6.50×10^{-3} kg and with a charge of +0.150 μC is suspended from an insulating thread and hangs between the plates of a capacitor (see the drawing). The ball is in equilibrium, with the thread making an angle of 30.0° with respect to the vertical. The area of each plate is 0.0150 m^2. What is the magnitude of the charge on each plate?

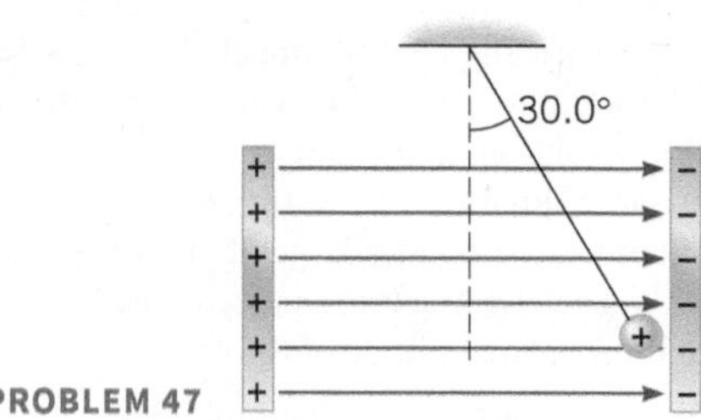

PROBLEM 47

Section 18.9 Gauss' Law

48. **E** A spherical surface completely surrounds a collection of charges. Find the electric flux through the surface if the collection consists of (a) a single $+3.5 \times 10^{-6}$ C charge, (b) a single -2.3×10^{-6} C charge, and (c) both of the charges in (a) and (b).

49. **E** **SSM** The drawing shows an edge-on view of two planar surfaces that intersect and are mutually perpendicular. Surface 1 has an area of 1.7 m^2, while surface 2 has an area of 3.2 m^2. The electric field $\vec{\mathbf{E}}$ in the drawing is uniform and has a magnitude of 250 N/C. Find the magnitude of the electric flux through (a) surface 1 and (b) surface 2.

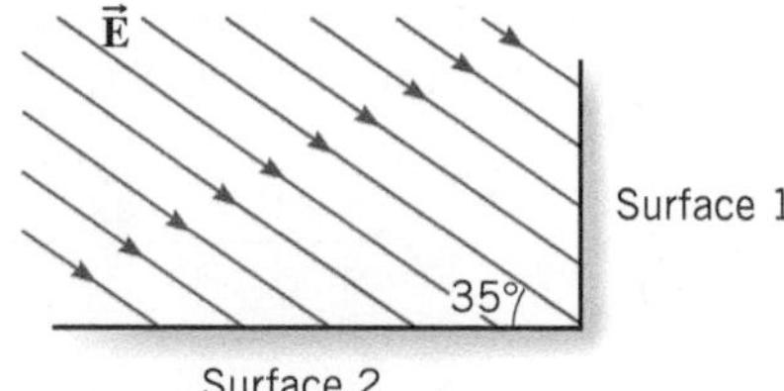

PROBLEM 49

50. **E** A surface completely surrounds a $+2.0 \times 10^{-6}$ C charge. Find the electric flux through this surface when the surface is (a) a sphere with a radius of 0.50 m, (b) a sphere with a radius of 0.25 m, and (c) a cube with edges that are 0.25 m long.

51. **E** A circular surface with a radius of 0.057 m is exposed to a uniform external electric field of magnitude 1.44×10^4 N/C. The magnitude of the electric flux through the surface is 78 N · m^2/C. What is the angle (less than 90°) between the direction of the electric field and the normal to the surface?

52. E V-HINT A charge Q is located inside a rectangular box. The electric flux through each of the six surfaces of the box is: $\Phi_1 = +1500\ \text{N} \cdot \text{m}^2/\text{C}$, $\Phi_2 = +2200\ \text{N} \cdot \text{m}^2/\text{C}$, $\Phi_3 = +4600\ \text{N} \cdot \text{m}^2/\text{C}$, $\Phi_4 = -1800\ \text{N} \cdot \text{m}^2/\text{C}$, $\Phi_5 = -3500\ \text{N} \cdot \text{m}^2/\text{C}$, and $\Phi_6 = -5400\ \text{N} \cdot \text{m}^2/\text{C}$. What is Q?

53. M GO MMH CHALK Two spherical shells have a common center. A -1.6×10^{-6} C charge is spread uniformly over the inner shell, which has a radius of 0.050 m. A $+5.1 \times 10^{-6}$ C charge is spread uniformly over the outer shell, which has a radius of 0.15 m. Find the magnitude and direction of the electric field at a distance (measured from the common center) of **(a)** 0.20 m, **(b)** 0.10 m, and **(c)** 0.025 m.

Additional Problems

Online

54. E GO The masses of the earth and moon are 5.98×10^{24} and 7.35×10^{22} kg, respectively. Identical amounts of charge are placed on each body, such that the net force (gravitational plus electrical) on each is zero. What is the magnitude of the charge placed on each body?

55. E A small drop of water is suspended motionless in air by a uniform electric field that is directed upward and has a magnitude of 8480 N/C. The mass of the water drop is 3.50×10^{-9} kg. **(a)** Is the excess charge on the water drop positive or negative? Why? **(b)** How many excess electrons or protons reside on the drop?

56. E SSM Two charges are placed on the x axis. One of the charges ($q_1 = +8.5\ \mu\text{C}$) is at $x_1 = +3.0$ cm and the other ($q_2 = -21\ \mu\text{C}$) is at $x_1 = +9.0$ cm. Find the net electric field (magnitude and direction) at **(a)** $x = 0$ cm and **(b)** $x = +6.0$ cm.

57. E When point charges $q_1 = +8.4\ \mu\text{C}$ and $q_2 = +5.6\ \mu\text{C}$ are brought near each other, each experiences a repulsive force of magnitude 0.66 N. Determine the distance between the charges.

58. M CHALK MMH The drawing shows two positive charges q_1 and q_2 fixed to a circle. At the center of the circle they produce a net electric field that is directed upward along the vertical axis. Determine the ratio $|q_2|/|q_1|$ of the charge magnitudes.

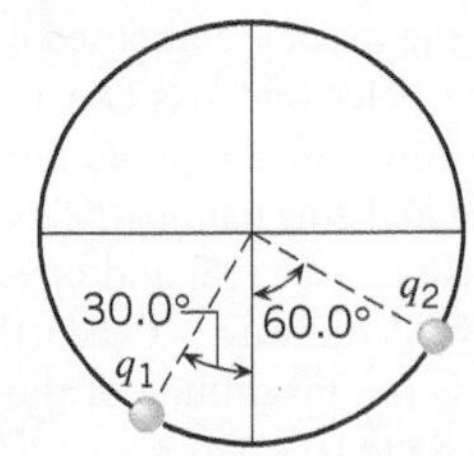

PROBLEM 58

59. M GO Three point charges have equal magnitudes, two being positive and one negative. These charges are fixed to the corners of an equilateral triangle, as the drawing shows. The magnitude of each of the charges is 5.0 μC, and the lengths of the sides of the triangle are 3.0 cm. Calculate the magnitude of the net force that each charge experiences.

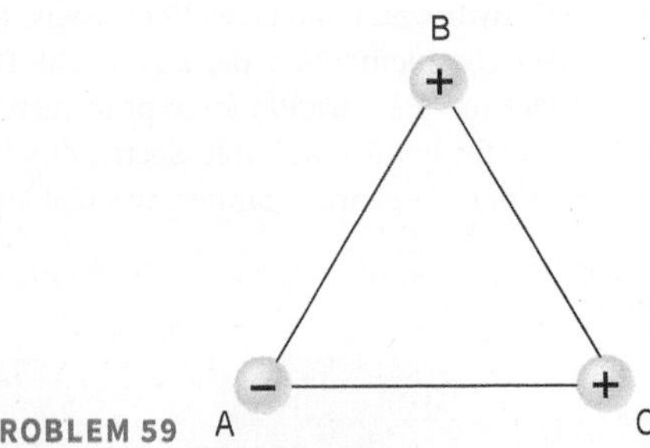

PROBLEM 59

60. M GO Four point charges have equal magnitudes. Three are positive, and one is negative, as the drawing shows. They are fixed in place on the same straight line, and adjacent charges are equally separated by a distance d. Consider the net electrostatic force acting on each charge. Calculate the ratio of the largest to the smallest net force.

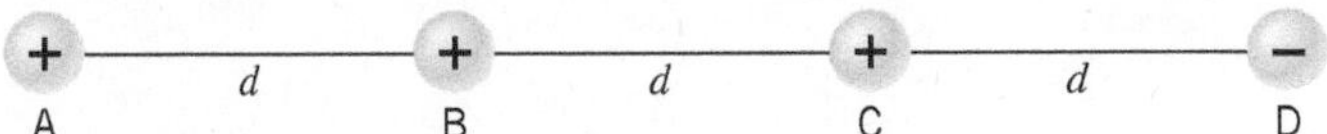

PROBLEM 60

61. H Two identical small insulating balls are suspended by separate 0.25-m threads that are attached to a common point on the ceiling. Each ball has a mass of 8.0×10^{-4} kg. Initially the balls are uncharged and hang straight down. They are then given identical positive charges and, as a result, spread apart with an angle of 36° between the threads. Determine **(a)** the charge on each ball and **(b)** the tension in the threads.

62. **(animated)** M GO In a vacuum, a proton (charge $= +e$, mass $= 1.67 \times 10^{-27}$ kg) is moving parallel to a uniform electric field that is directed along the $+x$ axis (see the figure below). The proton starts with a velocity of $+2.5 \times 10^4$ m/s and accelerates in the same direction as the electric field, which has a value of $+2.3 \times 10^3$ N/C. Find the velocity of the proton when its displacement is $+2.0$ mm from the starting point.

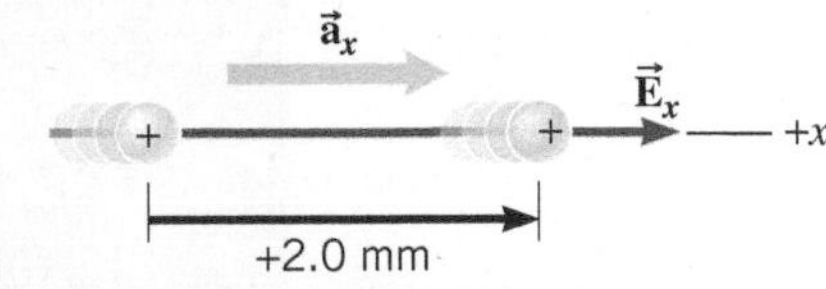

ANIMATED PROBLEM 62

Physics in Biology, Medicine, and Sports

63. E BIO GO **18.6** The membrane surrounding a living cell consists of an inner and an outer wall that are separated by a small space. Assume that the membrane acts like a parallel plate capacitor in which the effective charge density on the inner and outer walls has a magnitude of $7.1 \times 10^{-6}\ \text{C/m}^2$. **(a)** What is the magnitude of the electric field within the cell membrane? **(b)** Find the magnitude of the electric force that would be exerted on a potassium ion (K^+; charge $= +e$) placed inside the membrane.

64. M BIO **18.5** The average human body contains 7 billion billion billion atoms, or 7.0×10^{27} atoms. Of these, 65% are hydrogen. Each hydrogen atom is electrically neutral, as the positive charge on its proton is balanced by the negative charge on its electron. Imagine if just 1.0% of your hydrogen atoms were missing their corresponding electron. This would leave you with a net positive charge. Assume your best friend, who is standing 1.0 cm away from you, also has the same net positive charge. What would be the repulsive force between you? Treat you and your friend as point charges. For comparison, the force required to lift the weight of the entire earth is about 6.0×10^{25} N.

65. E BIO **18.5** Electrostatic forces play an important role in biological systems, as many macromolecules, like DNA, are heavily

charged (see Example 13). In fact, the electrostatic attractive force holds the two strands of DNA together, giving the molecule its strength and the famous structure of its double helix (see the image). The four organic bases of DNA (cytosine [C], guanine [G], adenine [A], and thymine [T]) hold the two strands together by forming pairs of bases that bond through these electrostatic forces. The order of the bases varies along the strand, but the pairing between bases is always the same; C always pairs with G, and A always pairs with T. In order for the pairs of bases to bond, they must be sufficiently close together. A guanine and cytosine, for example (see the image), pair by forming three hydrogen bonds (noncovalent bonds) between oxygen and nitrogen atoms. While a single hydrogen bond is relatively weak, the large number of them between complementary pairs gives the DNA molecule its strength. Calculate the electrostatic force of attraction within one hydrogen bond. Model the bond as a simple electric dipole, where equal and opposite charges of ($+/-e$) are separated by a distance of 0.30 nm.

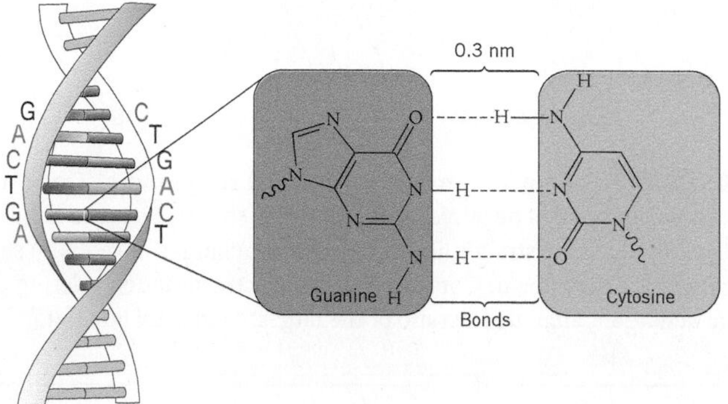

PROBLEM 65

66. E BIO 18.4 Many of a computer's internal components, such as the random access memory (RAM) and central processing unit (CPU), are very sensitive to electrostatic discharge (ESD). You have probably experienced ESDs yourself, if you have ever shuffled your feet across a carpeted floor and then touched a metal doorknob. You experienced a small shock, as the excess electrons you stripped from the carpet are transferred to the door knob. If the relative humidity in the room is low, the buildup of charge on the surface of your body, which acts like an insulator, can be significant—more than enough to damage a computer's components. To prevent ESDs, people assembling computers will wear antistatic equipment, such as that shown in the photo. The wrist strap is said to be grounded, which gives the static charge a safe conducting path away from the body, thereby protecting the computer components. In walking across a carpeted hallway, a computer technician acquires a charge of -4.79×10^{-9} C. Assume he touches one of the contact pins of a RAM module without wearing his antistatic equipment and completely delivers his excess charge through the component. How many electrons enter the RAM module during the discharge? Round your answer to the nearest whole number.

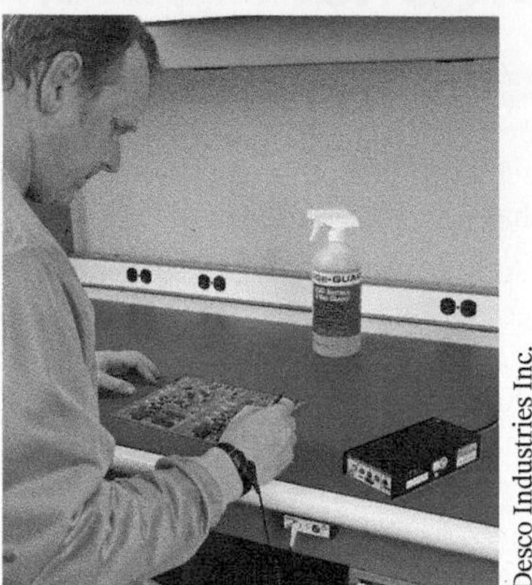
Desco Industries Inc.

PROBLEM 66

67. M BIO 18.6 During the COVID-19 pandemic in 2020, social distancing and the wearing of protective face masks became a way of life. Normal cloth masks provide protection against the spread of the disease by reducing the transmission of water droplets that contain the virus. However, researchers have developed "smartmasks" that use electric fields to destabilize the coronavirus particles, and while still in the early stages of development, they have shown promise in damaging structures the virus needs to infect host cells. The electroceutical fabric of which the mask is composed consists of a dot matrix of silver and zinc nanoparticles that acts like an array of electric dipoles when exposed to moisture (see the photo and figures below). Consider just one pair of silver and zinc nanoparticles extracted from the mask. Treat each nanoparticle as an equal and opposite point charge. If the value of the electric field midway between them (point P in the figure) is 120 N/C, what is the magnitude of the charge on each nanoparticle? Take the distance r to be 2.0 mm.

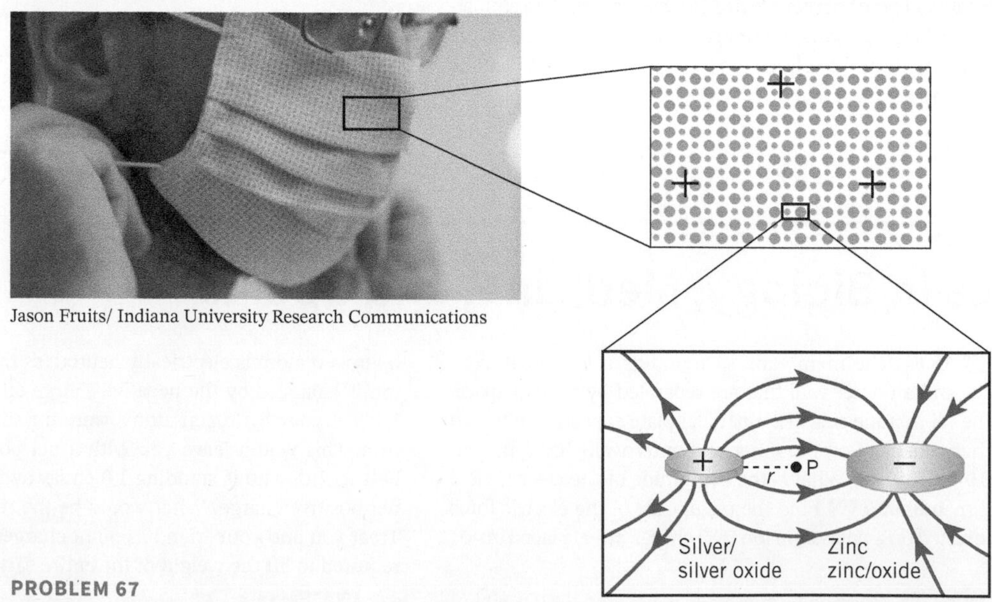

Jason Fruits/ Indiana University Research Communications

PROBLEM 67

68. **M** **BIO** **18.6** Natural diffusion of positive sodium ions and negative chlorine ions through the epithelial cells creates an endogenous electric field whenever you cut or wound your skin. Research has shown that electricity has a major impact on the healing of wounds. Certain proteins and genes within our cells respond to the natural electric fields that are produced at the site of a wound, which aids in the healing process (see the figures). In fact, research has shown that patients with slow-healing wounds, like those with diabetes, have smaller endogenous electric fields at the wound site. By exposing the wound to an external electric field and varying the intensity, the rate of cell migration to the wound can be increased, which promotes healing. One of the most important cells involved in wound healing are basal keratinocytes, which make up 90% of the epidermis. Consider a single keratinocyte with a net charge of $+5e$ and a mass of 1.0×10^2 picograms. Assume it is accelerated from rest through a wound gap by a uniform endogenous electric field of 55 N/C. How long will it take the cell to move through a distance of 1.0 mm, which is half the width of the wound? The cell is moving through a liquid medium of blood plasma and serous fluid, but we will ignore these effects.

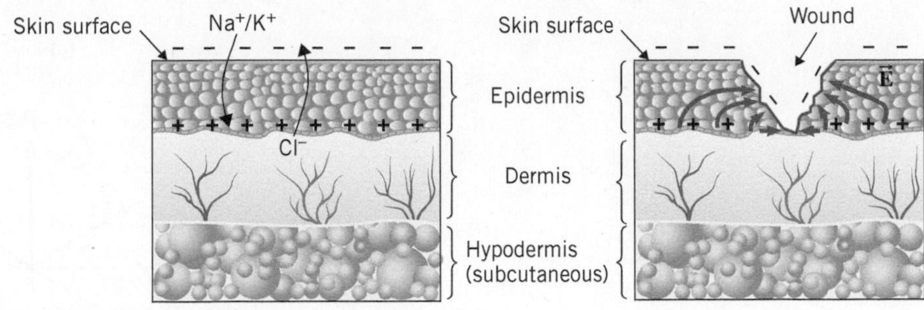

PROBLEM 68

69. **E** **BIO** **18.6** Electroreception is an animal's ability to detect electrical stimuli (i.e., electric fields in its environment). Certain species of fish, sharks, and dolphins use this ability to locate prey and other objects around them. Studies suggest that sharks are the most electrically sensitive animals known, being able to detect electric fields as low as 5.0×10^{-7} N/C. If a small fish carrying a charge of −0.1 nC is creating this electric field (like a point charge), what is the maximum distance the shark could be away from the fish and still detect it? Note, we are neglecting the effect of the water, which can significantly reduce the electric field. See Section 19.5.

Concepts and Calculations Problems

Online

In this chapter we have studied electric forces and electric fields. The format of the following problems stresses the role of conceptual understanding in problem solving. The purpose of the conceptual question section is to review the important concepts and to build intuition to help in anticipating some of the characteristics of the numerical answers.

70. **M** **CHALK** The charges on three identical metal spheres are −12 μC, +4.0 μC, and +2.0 μC. The spheres are brought together so they simultaneously touch each other. They are then separated and placed on the x and y axes, as shown in the figure. Treat the spheres as if they were particles. *Concepts:* (i) Is the net charge on the system comprising the three spheres the same before and after touching? Why? (ii) After the spheres touch and are separated, how is the charge distributed, and what is the value of the charge on each sphere? (iii) Do q_2 and q_3 exert forces of equal magnitude on q_1? (iv) Is the magnitude of the net force exerted on q_1 equal to $2F$, where F is the magnitude of the force that either q_2 or q_3 exerts on q_1? *Calculations:* What is the net force (magnitude and direction) exerted on the sphere at the origin?

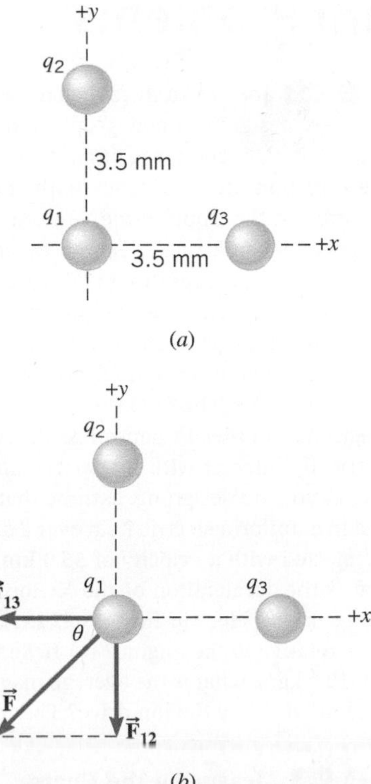

PROBLEM 70 (a) Three equal charges lie on the x and y axes. (b) The net force exerted by the two charges is $\vec{\mathbf{F}}$.

71. M CHALK Two point charges are lying on the y axis as in the figure: $q_1 = -4.00\ \mu C$ and $q_2 = +4.00\ \mu C$. They are equidistant from the point P, which lies on the x axis. *Concepts:* (i) There is no charge at P in part a of the figure. Is there an electric field at P? (ii) What is the direction of the electric field at point P due to charge q_2? (iii) Is the magnitude of the electric field at P equal to $E_1 + E_2$, where E_1 and E_2 are the magnitudes of the electric fields produced by q_1 and q_2? Explain. *Calculations:* (a) What is the net electric field at P? (b) A small object of charge $q_0 = +8.00\ \mu C$ and mass $m = 1.2$ g is placed at P. When it is released, what is its acceleration?

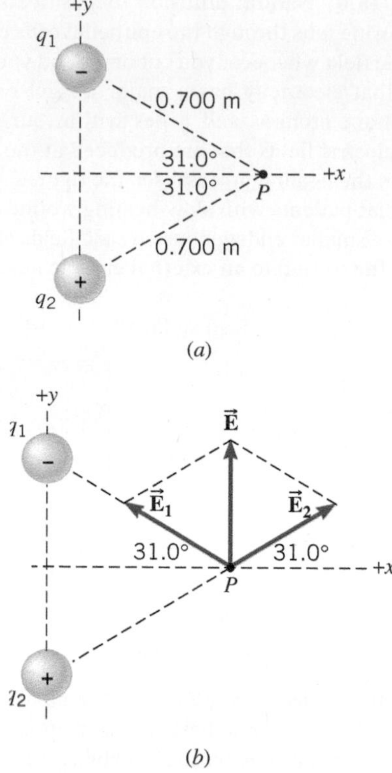

(a) Charges q_1 and q_2 lie on the y axis, and point P lies on the x axis. (b) The net electric field at point P is $\vec{E}$.

PROBLEM 71

Team Problems

Online

72. M T WS **Ion Thrusters I: The Electric Field.** You and your team are designing a new propulsion engine that ejects xenon (Xe) ions in one direction, causing a spacecraft to accelerate in the opposite direction, in accordance with the conservation of linear momentum and the impulse-momentum theorem. This is exactly what happens when a rocket engine expels high velocity particles when burning fuel, except that in the case of the ion drive the ions are accelerated with an electric field. In this process, Xe gas atoms are ionized so that they each acquire a net charge of $+e$. They are then subjected to an electric field, which accelerates and ejects the ions from the engine. The final step is to infuse the ejected ion beam with electrons in order to neutralize the ions so that they no longer electrically interact with the spacecraft, or with each other. In the thruster you are designing, assume that the Xe atoms accelerate from rest in a uniform electric field over a distance of 0.110 m before they are ejected with a velocity of 55.0 km/s relative to the engine. (a) What is the acceleration of the Xe ions? (b) What must be the magnitude of the electric field such that the ejection velocity is 55.0 km/s relative to the engine? (c) If Xe ions are ejected at a rate of 9.05×10^{-6} kg/s, what is the average magnitude of the thrust (i.e., the force) provided by the ion drive? The mass of a Xe ion is 2.18×10^{-25} kg.

73. E T WS **Measuring the Charge.** You and your team are designing an experiment where two spherical insulating beads of diameter 0.45 cm and mass 0.15 g are strung together on a fine thread that passes through microscopic holes drilled through their centers. When hung vertically, the lower bead rests on a knot at the bottom of the thread and the upper bead rests upon the lower bead. When placed inside a vacuum chamber and exposed to X-rays (highly energetic electromagnetic radiation), the beads acquire equal, uniform distributions of positive charge on their surfaces due to the loss of electrons ejected when struck by the X-rays. During this experiment, the beads are hung vertically, and a gap forms between them as if the top bead were levitating above the lower one, presumably due to the repulsive electrostatic force between them. The gap, that is, the vertical separation between the top of the lower bead and the bottom of the upper bead, is 0.27 cm. (a) How much excess charge is on each bead? You can treat the net excess charge on each bead as a point charge located at its center, and may neglect friction between the beads and the thread. (b) Determine the mass of a replacement upper bead of the same size and charge as those of the original upper bead that would reduce the gap between the beads to 0.15 cm.

74. M T **An Electron Gun.** You and your team are tasked with designing an "electron gun" that operates in a vacuum chamber, the purpose of which is to direct a beam of electrons toward a tiny metallic plate in order to heat it. The electrons in the beam have a speed of 3.50×10^7 m/s and travel in the positive x direction through the center of a set of deflecting plates (a parallel-plate capacitor) that sets up a uniform electric field in the region between the plates. The target is located 22.0 cm along the x-axis from the trailing edge

of plates (i.e., the edge closest to the target), and 11.5 cm above the horizontal (i.e., in the $+y$ direction). The length of the plates (in the x direction) is 2.50 cm. **(a)** In which direction should the electric field between the plates point in order to deflect the electrons toward the target? **(b)** To what magnitude should you set the electric field so that the electron beam hits the target? **(c)** After successfully striking the target using your results from (a) and (b), you realized that the target is not heating up to the required temperature. Since the degree of heating depends on the speed of the electrons, you increase the electron speed to 5.20×10^7 m/s. With the electric field setting from **(b)**, will the electrons still be on target? If not, to what value should you set the electric field?

75. M T **An Electrostatic Positioner.** You and your team are designing a device that can be used to position a small, plastic object in the region between the plates of a parallel-plate capacitor. A small plastic sphere of mass $m = 1.20 \times 10^{-2}$ kg carries a charge $q = +0.200$ μC and hangs vertically (along the y direction) from a massless, insulating thread (length $l = 10.0$ cm) between two vertical capacitor plates. When there is no electric field, the object resides at the midpoint between the plates (at $x = 0$). However, when there is a field between plates (in the $\pm x$ direction) the object moves to a new equilibrium position. **(a)** To what value should you set the field if you want the object to be located at $x = 2.10$ cm? **(b)** To what value should you set the field if you want the object to be located at $x = -3.30$ cm?

CHAPTER 19

Lightning permeates the sky around the ash plume above the Puyehue-Cordon Caulle volcano in south-central Chile in 2011. Unlike normal lightning associated with rain clouds, where static charges are produced by colliding ice particles, volcanic lightning, or "dirty thunderstorms," can result from frictional charging between colliding ash and dust particles. The natural convective thermal currents in the hot ash cloud aid in the separation of charges. This creates extremely high differences in voltage, or potential, between different parts of the dust cloud or between the cloud and the ground. If the voltage difference is sufficiently large, the insulating properties of the air break down, and it conducts electricity in spectacular fashion. The electric potential, and its relationship to charge, will be one of the topics we study in this chapter.

Carlos Gutierrez/Reuters

LEARNING OBJECTIVES

After reading this module, you should be able to...

19.1 Define electrical potential energy.

19.2 Solve problems involving electric potential and electric potential energy.

19.3 Calculate electric potential created by point charges.

19.4 Relate equipotential surfaces to the electric field.

19.5 Solve problems involving capacitors.

19.6 Describe biomedical applications of electric potential.

19.7 Describe the Millikan oil-drop experiment.

Electric Potential Energy and the Electric Potential

19.1 Potential Energy

In Chapter 18 we discussed the electrostatic force that two point charges exert on each other, the magnitude of which is $F = k|q_1||q_2|/r^2$. The form of this equation is similar to the form for the gravitational force that two particles exert on each other, which is $F = Gm_1m_2/r^2$, according to Newton's law of universal gravitation (see Section 4.7). Both of these forces are conservative and, as Section 6.4 explains, a potential energy can be associated with a conservative force. Thus, an electric potential energy exists that is analogous to the gravitational potential energy. To set the stage for a discussion of the electric potential energy, let's review some of the important aspects of the gravitational counterpart.

Figure 19.1, which is essentially Figure 6.10, shows a basketball of mass m falling from point A to point B. The gravitational force, $m\vec{g}$, is the only force acting on the ball, where g is the magnitude of the acceleration due to gravity. As Section 6.3 discusses, the work W_{AB} done by the gravitational force when the ball falls from a height of h_A to a height of h_B is

$$W_{AB} = \underbrace{mgh_A}_{\substack{\text{Initial gravitational}\\ \text{potential energy,}\\ \text{GPE}_A}} - \underbrace{mgh_B}_{\substack{\text{Final gravitational}\\ \text{potential energy,}\\ \text{GPE}_B}} = \text{GPE}_A - \text{GPE}_B \tag{6.4}$$

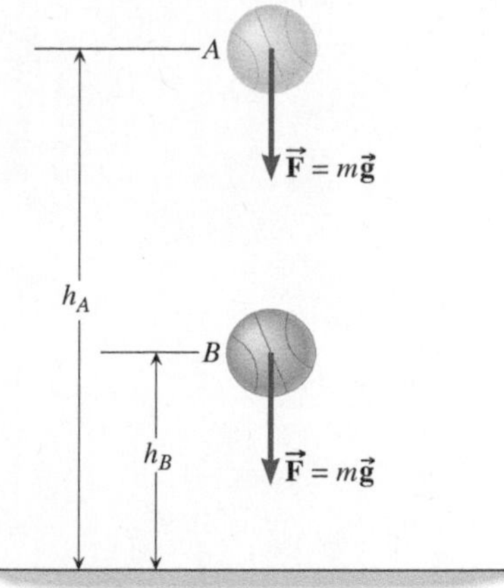

FIGURE 19.1 Gravity exerts a force, $\vec{F} = m\vec{g}$, on the basketball of mass m. Work is done by the gravitational force as the ball falls from A to B.

Recall that the quantity mgh is the gravitational potential energy* of the ball, GPE = mgh (Equation 6.5), and represents the energy that the ball has by virtue of its position relative to the surface of the earth. Thus, the work done by the gravitational force equals the initial gravitational potential energy minus the final gravitational potential energy.

Interactive Figure 19.2 clarifies the analogy between electric and gravitational potential energies. In this drawing a positive test charge $+q_0$ is situated at point A between two oppositely charged plates. Because of the charges on the plates, an electric field $\vec{E}$ exists in the region between the plates. Consequently, the test charge experiences an electric force, $\vec{F} = q_0\vec{E}$ (Equation 18.2), that is directed downward, toward the lower plate. (The gravitational force is being neglected here.) As the charge moves from A to B, work is done by this force, in a fashion analogous to the work done by the gravitational force in **Figure 19.1**. The work W_{AB} done by the electric force equals the difference between the electric potential energy EPE at A and the electric potential energy at B:

$$W_{AB} = \text{EPE}_A - \text{EPE}_B \tag{19.1}$$

This expression is similar to Equation 6.4. The path along which the test charge moves from A to B is of no consequence because the electric force, like the gravitational force, is conservative. For such forces, the work W_{AB} is the same for all paths (see Section 6.4).

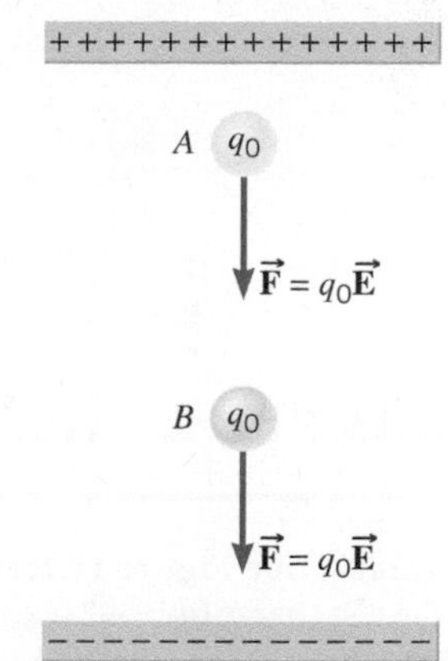

INTERACTIVE FIGURE 19.2 Because of the electric field $\vec{E}$, an electric force, $\vec{F} = q_0\vec{E}$, is exerted on a positive test charge $+q_0$. Work is done by the force as the charge moves from A to B.

19.2 The Electric Potential Difference

Since the electric force is $\vec{F} = q_0\vec{E}$, the work that it does as the charge moves from A to B in **Interactive Figure 19.2** depends on the charge q_0. It is useful, therefore, to express this work on a per-unit-charge basis, by dividing both sides of Equation 19.1 by the charge:

$$\frac{W_{AB}}{q_0} = \frac{\text{EPE}_A}{q_0} - \frac{\text{EPE}_B}{q_0} \tag{19.2}$$

Notice that the right-hand side of this equation is the difference between two terms, each of which is an electric potential energy divided by the test charge, EPE/q_0. The quantity EPE/q_0 is the electric potential energy per unit charge and is an important concept in electricity. It is called the **electric potential** or, simply, the **potential** and is referred to with the symbol V, as in Equation 19.3.

DEFINITION OF ELECTRIC POTENTIAL

The electric potential V at a given point is the electric potential energy EPE of a small test charge q_0 situated at that point divided by the charge itself:

$$V = \frac{\text{EPE}}{q_0} \tag{19.3}$$

SI Unit of Electric Potential: joule/coulomb = volt (V)

*The gravitational potential energy is now being denoted by GPE to distinguish it from the electric potential energy EPE.

The SI unit of electric potential is a joule per coulomb, a quantity known as a *volt*. The name honors Alessandro Volta (1745–1827), who invented the voltaic pile, the forerunner of the battery. In spite of the similarity in names, the electric potential energy EPE and the electric potential V are *not* the same. The electric potential energy, as its name implies, is an *energy* and, therefore, is measured in joules. In contrast, the electric potential is an *energy per unit charge* and is measured in joules per coulomb, or volts.

We can now relate the work W_{AB} done by the electric force when a charge q_0 moves from A to B to the potential difference $V_B - V_A$ between the points. Combining Equations 19.2 and 19.3, we have:

$$V_B - V_A = \frac{\text{EPE}_B}{q_0} - \frac{\text{EPE}_A}{q_0} = \frac{-W_{AB}}{q_0} \quad \textbf{(19.4)}$$

Often, the "delta" notation is used to express the difference (final value minus initial value) in potentials and in potential energies: $\Delta V = V_B - V_A$ and $\Delta(\text{EPE}) = \text{EPE}_B - \text{EPE}_A$. In terms of this notation, Equation 19.4 takes the following more compact form:

$$\Delta V = \frac{\Delta(\text{EPE})}{q_0} = \frac{-W_{AB}}{q_0} \quad \textbf{(19.4)}$$

Neither the potential V nor the potential energy EPE can be determined in an absolute sense, because only the *differences* ΔV and $\Delta(\text{EPE})$ are measurable in terms of the work W_{AB}. The gravitational potential energy has this same characteristic, since only the value at one height relative to the value at some reference height has any significance. Example 1 emphasizes the relative nature of the electric potential.

EXAMPLE 1 | Work, Electric Potential Energy, and Electric Potential

In **Interactive Figure 19.2**, the work done by the electric force as the test charge ($q_0 = +2.0 \times 10^{-6}$ C) moves from A to B is $W_{AB} = +5.0 \times 10^{-5}$ J. **(a)** Find the value of the difference, $\Delta(\text{EPE}) = \text{EPE}_B - \text{EPE}_A$, in the electric potential energies of the charge between these points. **(b)** Determine the potential difference, $\Delta V = V_B - V_A$, between the points.

Reasoning The work done by the electric force when the charge travels from A to B is $W_{AB} = \text{EPE}_A - \text{EPE}_B$, according to Equation 19.1. Therefore, the difference in the electric potential energies (final value minus initial value) is $\Delta(\text{EPE}) = \text{EPE}_B - \text{EPE}_A = -W_{AB}$. The potential difference, $\Delta V = V_B - V_A$, is the difference in the electric potential energies divided by the charge q_0, according to Equation 19.4.

Solution **(a)** The difference in the electric potential energies of the charge between the points A and B is

$$\underbrace{\text{EPE}_B - \text{EPE}_A}_{=\Delta(\text{EPE})} = -W_{AB} = \boxed{-5.0 \times 10^{-5}\ \text{J}} \quad \textbf{(19.1)}$$

The fact that $\text{EPE}_B - \text{EPE}_A$ is negative means that the charge has a higher electric potential energy at A than at B.

(b) The potential difference ΔV between A and B is

$$\underbrace{V_B - V_A}_{=\Delta V} = \frac{\text{EPE}_B - \text{EPE}_A}{q_0} = \frac{-5.0 \times 10^{-5}\ \text{J}}{2.0 \times 10^{-6}\ \text{C}} = \boxed{-25\ \text{V}} \quad \textbf{(19.4)}$$

The fact that $V_B - V_A$ is negative means that the electric potential is higher at A than at B.

The potential difference between two points is measured in volts and, therefore, is often referred to as a "voltage." Everyone has heard of "voltage" because, as we will see in Chapter 20, it is frequently used in connection with everyday devices. For example, your TV requires a "voltage" of 120 V (which is applied between the two prongs of the plug on the power cord when it is inserted into an electrical wall outlet), and your cell phone and laptop computer use batteries that provide, for example, "voltages" of 1.5 V or 9 V (which exist between the two battery terminals).

In **Figure 19.1** the speed of the basketball increases as it falls from A to B. Since point A has a greater gravitational potential energy than point B, we see that an object of mass m accelerates when it moves from a region of higher potential energy toward a region of lower potential energy. Likewise, the positive charge in **Interactive Figure 19.2** accelerates as it moves from A to B because of the electric repulsion from the upper plate and the attraction to the lower plate. Since point *A* has a higher electric potential than point *B*, we conclude the following:

Problem-Solving Insight **A positive charge accelerates from a region of higher electric potential toward a region of lower electric potential.**

On the other hand, a negative charge placed between the plates in **Interactive Figure 19.2** behaves in the opposite fashion, since the electric force acting on the negative charge is directed opposite to the electric force acting on the positive charge.

Problem-Solving Insight **A negative charge accelerates from a region of lower potential toward a region of higher potential.**

The next example illustrates the way positive and negative charges behave.

CONCEPTUAL EXAMPLE 2 | How Positive and Negative Charges Accelerate

Three points, *A*, *B*, and *C*, are located along a horizontal line, as **Figure 19.3** illustrates. A positive test charge is released from rest at *A* and accelerates toward *B*. Upon reaching *B*, the test charge continues to accelerate toward *C*. Assuming that only motion along the line is possible, what will a negative test charge do when it is released from rest at *B*? A negative test charge will **(a)** accelerate toward *C*, **(b)** remain stationary, **(c)** accelerate toward *A*.

Reasoning A positive charge accelerates from a region of higher potential toward a region of lower potential. A negative charge behaves in an opposite manner, because it accelerates from a region of lower potential toward a region of higher potential.

Answers (a) and (b) are incorrect. The positive charge accelerates from *A* to *B* and then from *B* to *C*. A negative charge placed at *B* also accelerates, but in a direction opposite to that of the positive charge. Therefore, a negative charge placed at *B* will not remain stationary, nor will it accelerate toward *C*.

Answer (c) is correct. Since the positive charge accelerates from *A* to *B*, the potential at *A* must exceed the potential at *B*. And since the positive test charge accelerates from *B* to *C*, the potential at *B* must exceed the potential at *C*. The potential at point *B*, then, must lie between the potential at points *A* and *C*, as **Figure 19.3** illustrates. When the negative test charge is released from rest at *B*, it will accelerate toward the region of higher potential, so it will begin moving toward *A*.

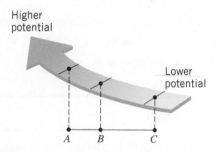

FIGURE 19.3 The electric potentials at points *A*, *B*, and *C* are different. Under the influence of these potentials, positive and negative charges accelerate in opposite directions.

As a familiar application of electric potential energy and electric potential, **Figure 19.4** shows a 12-V automobile battery with a headlight connected between its terminals. The positive terminal, point *A*, has a potential that is 12 V higher than the potential at the negative terminal, point *B*; in other words, $V_A - V_B = 12$ V. Positive charges would be repelled from the positive terminal and would travel through the wires and headlight toward the negative terminal.* As the charges pass through the headlight, virtually all their potential energy is converted into heat, which causes the filament to glow "white hot" and emit light. When the charges reach the negative terminal, they no longer have any potential energy. The battery then gives the charges an additional "shot" of potential energy by moving them to the higher-potential positive terminal, and the cycle is repeated. In raising the potential energy of the charges, the battery does work on them and draws from its reserve of chemical energy to do so. Example 3 illustrates the concepts of electric potential energy and electric potential as applied to a battery.

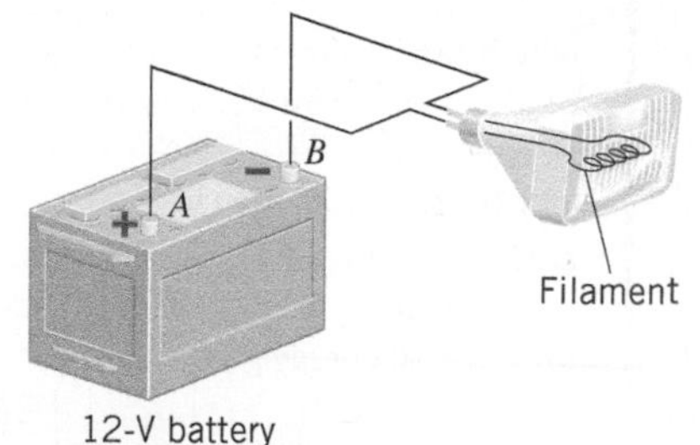

FIGURE 19.4 A headlight connected to a 12-V battery.

*Historically, it was believed that positive charges flow in the wires of an electric circuit. Today, it is known that negative charges flow in wires from the negative toward the positive terminal. Here, however, we follow the customary practice of describing the flow of negative charges by specifying the opposite but equivalent flow of positive charges. This hypothetical flow of positive charges is called the "conventional electric current," as we will see in Section 20.1.

Analyzing Multiple-Concept Problems

EXAMPLE 3 | Operating a Headlight

The wattage of the headlight in **Figure 19.4** is 60.0 W. Determine the number of particles, each carrying a charge of 1.60×10^{-19} C (the magnitude of the charge on an electron), that pass between the terminals of the 12-V car battery when the headlight burns for one hour.

Reasoning The number of particles is the total charge that passes between the battery terminals in one hour divided by the magnitude of the charge on each particle. The total charge is the amount of charge needed to convey the energy used by the headlight in one hour. This energy is related to the wattage of the headlight, which specifies the power or rate at which energy is used, and the time the light is on.

Knowns and Unknowns The following table summarizes the data provided:

Description	Symbol	Value	Comment
Wattage of headlight	P	60.0 W	
Charge magnitude per particle	e	1.60×10^{-19} C	
Electric potential difference between battery terminals	$V_A - V_B$	12 V	See **Figure 19.4**.
Time headlight is on	t	3600 s	One hour.
Unknown Variable			
Number of particles	n	?	

Modeling the Problem

STEP 1 The Number of Particles The number n of particles is the total charge q_0 that passes between the battery terminals in one hour divided by the magnitude e of the charge on each particle, as expressed by Equation 1 at the right. The value of e is given. To evaluate q_0, we proceed to Step 2.

$$n = \frac{q_0}{e} \quad (1)$$

?

STEP 2 The Total Charge Provided by the Battery The battery must supply the total energy used by the headlight in one hour. The battery does this by supplying the charge q_0 to convey this energy. The energy is the difference between the electric potential energy EPE_A at terminal A and the electric potential energy EPE_B at terminal B of the battery (see **Figure 19.4**). According to Equation 19.4, this total energy is $\text{EPE}_A - \text{EPE}_B = q_0(V_A - V_B)$, where $V_A - V_B$ is the electric potential difference between the battery terminals. Solving this expression for q_0 gives

$$q_0 = \frac{\text{EPE}_A - \text{EPE}_B}{V_A - V_B}$$

which can be substituted into Equation 1, as shown at the right. As the data table shows, a value is given for $V_A - V_B$. In Step 3, we determine a value for $\text{EPE}_A - \text{EPE}_B$.

$$n = \frac{q_0}{e} \quad (1)$$

$$q_0 = \frac{\text{EPE}_A - \text{EPE}_B}{V_A - V_B} \quad (2)$$

?

STEP 3 The Energy Used by the Headlight The rate at which the headlight uses energy is the power P or wattage of the headlight. According to Equation 6.10b, the power is the total energy $\text{EPE}_A - \text{EPE}_B$ divided by the time t, so that $P = (\text{EPE}_A - \text{EPE}_B)/t$. Solving for the total energy gives

$$\text{EPE}_A - \text{EPE}_B = Pt$$

Since P and t are given, we substitute this result into Equation 2, as indicated at the right.

$$n = \frac{q_0}{e} \quad (1)$$

$$q_0 = \frac{\text{EPE}_A - \text{EPE}_B}{V_A - V_B} \quad (2)$$

$$\text{EPE}_A - \text{EPE}_B = Pt$$

Solution Combining the results of each step algebraically, we find that

$$n \overset{\text{STEP 1}}{=} \frac{q_0}{e} \overset{\text{STEP 2}}{=} \frac{(\text{EPE}_A - \text{EPE}_B)/(V_A - V_B)}{e} \overset{\text{STEP 2}}{=} \frac{Pt/(V_A - V_B)}{e}$$

The number of particles that pass between the battery terminals in one hour is

$$n = \frac{Pt}{e(V_A - V_B)} = \frac{(60.0\text{ W})(3600\text{ s})}{(1.60 \times 10^{-19}\text{ C})(12\text{ V})} = \boxed{1.1 \times 10^{23}}$$

Related Homework: Problem 46

As used in connection with batteries, the volt is a familiar unit for measuring electric potential difference. The word "volt" also appears in another context, as part of a unit that is used to measure energy, particularly the energy of an atomic particle, such as an electron or a proton. This energy unit is called the *electron volt* (eV).

Problem-Solving Insight One electron volt is the magnitude of the amount by which the potential energy of an electron changes when the electron moves through a potential difference of one volt.

Since the magnitude of the change in potential energy is $|q_0\Delta V| = |(-1.60 \times 10^{-19}\text{ C}) \times (1.00\text{ V})| = 1.60 \times 10^{-19}$ J, it follows that

$$1\text{ eV} = 1.60 \times 10^{-19}\text{ J}$$

One million (10^{+6}) electron volts of energy is referred to as one MeV, and one billion (10^{+9}) electron volts of energy is one GeV, where the "G" stands for the prefix "giga."

In Equation 19.3, we have seen that the electric potential is the electric potential energy per unit charge. In previous chapters, we have seen that the total energy of an object, which is the sum of its kinetic and potential energies, is an important concept. Its significance lies in the fact that the total energy remains the same (is conserved) during the object's motion, provided that nonconservative forces, such as friction, are either absent or do no net work. While the sum of the energies at each instant remains constant, energy may be converted from one form to another; for example, gravitational potential energy is converted into kinetic energy as a ball falls. We now include the electric potential energy EPE as part of the total energy that an object can have:

$$\underbrace{E}_{\substack{\text{Total}\\\text{energy}}} = \underbrace{\tfrac{1}{2}mv^2}_{\substack{\text{Translational}\\\text{kinetic energy}}} + \underbrace{\tfrac{1}{2}I\omega^2}_{\substack{\text{Rotational}\\\text{kinetic energy}}} + \underbrace{mgh}_{\substack{\text{Gravitational}\\\text{potential}\\\text{energy}}} + \underbrace{\tfrac{1}{2}kx^2}_{\substack{\text{Elastic}\\\text{potential}\\\text{energy}}} + \underbrace{\text{EPE}}_{\substack{\text{Electric}\\\text{potential}\\\text{energy}}}$$

If the total energy is conserved as the object moves, then its final energy E_f is equal to its initial energy E_0, or $E_f = E_0$. Example 4 illustrates how the conservation of energy is applied to a charge moving in an electric field.

Analyzing Multiple-Concept Problems

EXAMPLE 4 | The Conservation of Energy

A particle has a mass of 1.8×10^{-5} kg and a charge of $+3.0 \times 10^{-5}$ C. It is released from rest at point A and accelerates until it reaches point B, as **Figure 19.5a** shows. The particle moves on a horizontal straight line and does not rotate. The only forces acting on the particle are the gravitational force and an electrostatic force (neither is shown in the drawing). The electric potential at A is 25 V greater than that at B; in other words, $V_A - V_B = 25$ V. What is the translational speed of the particle at point B?

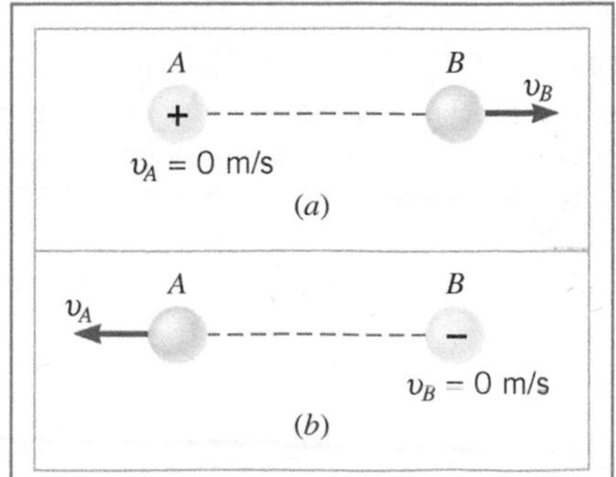

FIGURE 19.5 (*a*) A positive charge starts from rest at point A and accelerates toward point B. (*b*) A negative charge starts from rest at B and accelerates toward A.

Reasoning The translational speed of the particle is related to the particle's translational kinetic energy, which forms one part of the total energy that the particle has. The total energy is conserved, because only the gravitational force and an electrostatic force, both of which are conservative forces, act on the particle (see Section 6.5). Thus, we will determine the speed at point B by utilizing the principle of conservation of energy.

Knowns and Unknowns We have the following data:

Description	Symbol	Value	Comment
Explicit Data			
Mass of particle	m	1.8×10^{-5} kg	
Electric charge of particle	q_0	$+3.0 \times 10^{-5}$ C	
Electric potential difference between points A and B	$V_A - V_B$	25 V	See **Figure 19.5a**.
Implicit Data			
Speed at point A	v_A	0 m/s	Particle released from rest.
Vertical height above ground	h	Remains constant	Particle travels horizontally.
Angular speed	ω	0 rad/s	Particle does not rotate during motion.
Elastic force	F_{elastic}	0 N	No elastic force acts on particle.
Unknown Variable			
Speed at point B	v_B	?	

Modeling the Problem

STEP 1 Conservation of Total Energy The particle's total energy E is

$$E = \underbrace{\tfrac{1}{2}mv^2}_{\substack{\text{Translational}\\ \text{kinetic energy}}} + \underbrace{\tfrac{1}{2}I\omega^2}_{\substack{\text{Rotational}\\ \text{kinetic energy}}} + \underbrace{mgh}_{\substack{\text{Gravitational}\\ \text{potential}\\ \text{energy}}} + \underbrace{\tfrac{1}{2}kx^2}_{\substack{\text{Elastic}\\ \text{potential}\\ \text{energy}}} + \underbrace{\text{EPE}}_{\substack{\text{Electric}\\ \text{potential}\\ \text{energy}}}$$

Since the particle does not rotate, the angular speed ω is always zero (see the data table) and since there is no elastic force (see the data table), we may omit the terms $\frac{1}{2}I\omega^2$ and $\frac{1}{2}kx^2$ from this expression. With this in mind, we express the fact that $E_B = E_A$ (energy is conserved) as follows:

$$\tfrac{1}{2}mv_B^2 + mgh_B + \text{EPE}_B = \tfrac{1}{2}mv_A^2 + mgh_A + \text{EPE}_A$$

This equation can be simplified further, since the particle travels horizontally, so that $h_B = h_A$ (see the data table), with the result that

$$\tfrac{1}{2}mv_B^2 + \text{EPE}_B = \tfrac{1}{2}mv_A^2 + \text{EPE}_A$$

Solving for v_B gives Equation 1 at the right. Values for v_A and m are available, and we turn to Step 2 in order to evaluate $\text{EPE}_A - \text{EPE}_B$.

$$v_B = \sqrt{v_A^2 + \frac{2(\text{EPE}_A - \text{EPE}_B)}{m}} \quad (1)$$

?

STEP 2 The Electric Potential Difference According to Equation 19.4, the difference in electric potential energies $\text{EPE}_A - \text{EPE}_B$ is related to the electric potential difference $V_A - V_B$:

$$\boxed{\text{EPE}_A - \text{EPE}_B = q_0(V_A - V_B)}$$

The terms q_0 and $V_A - V_B$ are known, so we substitute this expression into Equation 1, as shown at the right.

$$v_B = \sqrt{v_A^2 + \frac{2(\text{EPE}_A - \text{EPE}_B)}{m}} \quad (1)$$

$$\boxed{\text{EPE}_A - \text{EPE}_B = q_0(V_A - V_B)}$$

Problem-Solving Insight A positive charge accelerates from a region of higher potential toward a region of lower potential. In contrast, a negative charge accelerates from a region of lower potential toward a region of higher potential.

Solution Combining the results of each step algebraically, we find that

STEP 1 STEP 2

$$v_B = \sqrt{v_A^2 + \frac{2(\text{EPE}_A - \text{EPE}_B)}{m}} = \sqrt{v_A^2 + \frac{2q_0(V_A - V_B)}{m}}$$

The speed of the particle at point B is

$$v_B = \sqrt{v_A^2 + \frac{2q_0(V_A - V_B)}{m}} = \sqrt{(0\text{ m/s})^2 + \frac{2(+3.0 \times 10^{-5}\text{ C})(25\text{ V})}{1.8 \times 10^{-5}\text{ kg}}} = \boxed{9.1\text{ m/s}}$$

Note that if the particle had a negative charge of -3.0×10^{-5} C and were released from rest at point B, it would arrive at point A with the same speed of 9.1 m/s (see **Figure 19.5b**). This result can be obtained by returning to Modeling Step 1 and solving for v_A instead of v_B.

Related Homework: Problems 4, 5

Check Your Understanding

(*The answers are given at the end of the book.*)

1. An ion, starting from rest, accelerates from point A to point B due to a potential difference between the two points. Does the electric potential energy of the ion at point B depend on **(a)** the magnitude of its charge and **(b)** its mass? Does the speed of the ion at B depend on **(c)** the magnitude of its charge and **(d)** its mass?

2. **CYU Figure 19.1** shows three possibilities for the potentials at two points, A and B. In each case, the same positive charge is moved from A to B. In which case, if any, is the most work done on the positive charge by the electric force?

A	B	A	B	A	B
150 V	100 V	25 V	−25 V	−10 V	−60 V
Case 1		Case 2		Case 3	

CYU FIGURE 19.1

3. A proton and an electron are released from rest at the midpoint between the plates of a charged parallel plate capacitor (see Section 18.6). Except for these particles, nothing else is between the plates. Ignore the attraction between the proton and the electron, and decide which particle strikes a capacitor plate first.

19.3 The Electric Potential Difference Created by Point Charges

A positive point charge $+q$ creates an electric potential in a way that **Figure 19.6** helps explain. This picture shows two locations A and B, at distances r_A and r_B from the charge. At any position between A and B an electrostatic force of repulsion $\vec{\mathbf{F}}$ acts on a positive test charge $+q_0$. The magnitude of the force is given by Coulomb's law as $F = kq_0q/r^2$, where we assume for convenience that q_0 and q are positive, so that $|q_0| = q_0$ and $|q| = q$. When the test charge moves from A to B, work is done by this force. Since r varies between r_A and r_B, the force F also varies, and the work is not the product of the force and the distance between the points. (Recall from Section 6.1 that work is force times distance only if the force is constant.) However, the work W_{AB} can be found with the methods of integral calculus. The result is

$$W_{AB} = \frac{kqq_0}{r_A} - \frac{kqq_0}{r_B}$$

This result is valid whether q is positive or negative, and whether q_0 is positive or negative. The potential difference, $V_B - V_A$, between A and B can now be determined by substituting this expression for W_{AB} into Equation 19.4:

$$V_B - V_A = \frac{-W_{AB}}{q_0} = \frac{kq}{r_B} - \frac{kq}{r_A} \tag{19.5}$$

As point B is located farther and farther from the charge q, r_B becomes larger and larger. In the limit that r_B is infinitely large, the term kq/r_B becomes zero, and it is customary to set V_B equal to zero also. In this limit, Equation 19.5 becomes $V_A = kq/r_A$, and it is standard convention to omit the subscripts and write the potential in the following form:

Potential of a point charge
$$V = \frac{kq}{r} \tag{19.6}$$

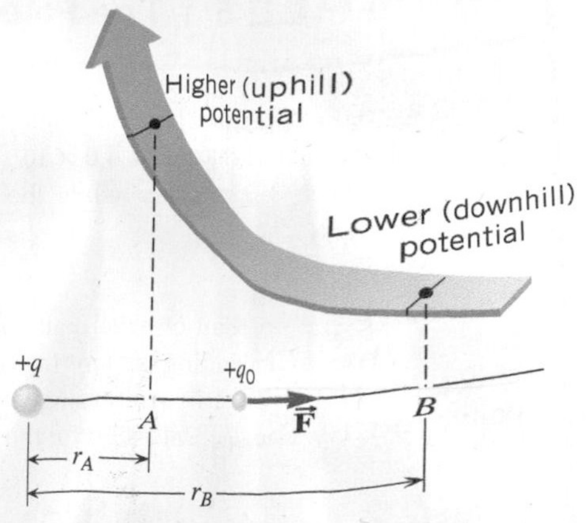

FIGURE 19.6 The positive test charge $+q_0$ experiences a repulsive force $\vec{\mathbf{F}}$ due to the positive point charge $+q$. As a result, work is done by this force when the test charge moves from A to B. Consequently, the electric potential is higher (uphill) at A and lower (downhill) at B.

Math Skills *In Equation 19.6 the symbol q denotes the value of the point charge, including both the magnitude and the algebraic sign of the charge.* Note especially that the symbol q does not have the same meaning as the symbol $|q|$, which denotes only the magnitude of the charge. When you use Equation 19.6 to solve problems dealing with the potential of a point charge, both the magnitude and the algebraic sign of the charge must be taken into account, and the presence of q (rather than $|q|$) in the equation ensures that this will be so. Using only the magnitude of a point charge in Equation 19.6 would ignore the important fact that positive and negative point charges create different potentials, even when the charge magnitudes are the same.

The symbol V in this equation does not refer to the potential in any absolute sense. Rather, $V = kq/r$ stands for the amount by which the potential at a distance r from a point charge differs from the potential at an infinite distance away. In other words, V refers to a potential difference with the arbitrary assumption that the potential at infinity is zero.

With the aid of Equation 19.6, we can describe the effect that a point charge q has on the surrounding space. When q is positive, the value of $V = kq/r$ is also positive, indicating that the positive charge has everywhere raised the potential above the zero reference value. Conversely, when q is negative, the potential V is also negative, indicating that the negative charge has everywhere decreased the potential below the zero reference value. The next example deals with these effects quantitatively.

EXAMPLE 5 | The Potential of a Point Charge

Using a zero reference potential at infinity, determine the amount by which a point charge of 4.0×10^{-8} C alters the electric potential at a spot 1.2 m away when the charge is **(a)** positive and **(b)** negative.

Reasoning A point charge q alters the potential at every location in the surrounding space. In the expression $V = kq/r$, the effect of the charge in increasing or decreasing the potential is conveyed by the algebraic sign for the value of q.

Solution **(a)** **Figure 19.7*a*** shows the potential when the charge is positive:

$$V = \frac{kq}{r} = \frac{(8.99 \times 10^9 \text{ N} \cdot \text{m}^2/\text{C}^2)(+4.0 \times 10^{-8} \text{ C})}{1.2 \text{ m}} = \boxed{+300 \text{ V}} \quad \textbf{(19.6)}$$

(b) Part *b* of the drawing illustrates the results when the charge is negative. A calculation similar to the one in part (a) shows that the potential is now negative: $\boxed{-300 \text{ V}}$.

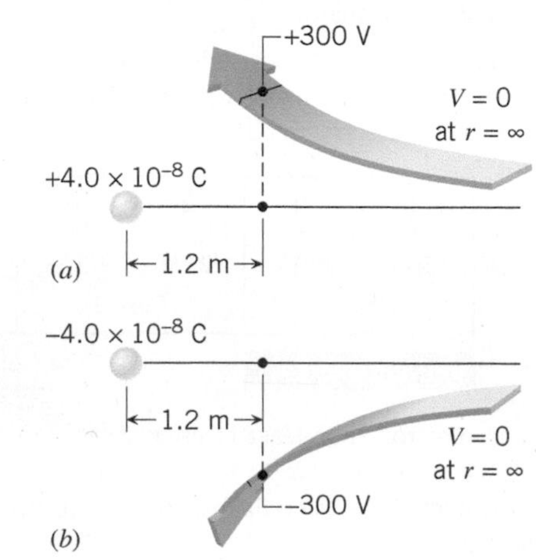

FIGURE 19.7 A point charge of 4.0×10^{-8} C alters the potential at a spot 1.2 m away. The potential is (*a*) increased by 300 V when the charge is positive and (*b*) decreased by 300 V when the charge is negative, relative to a zero reference potential at infinity.

A single point charge raises or lowers the potential at a given location, depending on whether the charge is positive or negative.

Problem-Solving Insight **When two or more charges are present, the potential due to all the charges is obtained by adding together the individual potentials.**

The next two examples will demonstrate this point.

EXAMPLE 6 | The Total Electric Potential

At locations A and B in **Figure 19.8**, find the total electric potential due to the two point charges.

Reasoning At each location, each charge contributes to the total electric potential. We obtain the individual contributions by using $V = kq/r$ and find the total potential by adding the individual contributions algebraically. The two charges have the same magnitude, but different signs. Thus, at A the total potential is positive because this spot is closer to the positive charge, whose effect dominates over that of the more distant negative charge. At B, midway between the charges, the total potential is zero, since the potential of one charge exactly offsets that of the other.

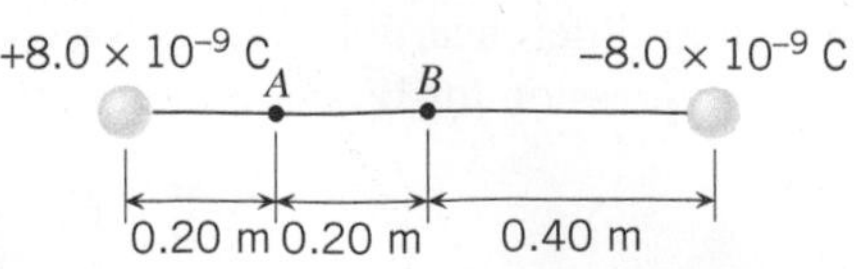

FIGURE 19.8 Both the positive and negative charges affect the electric potential at locations A and B.

Solution

Location	Contribution from + Charge		Contribution from − Charge	Total Potential
A	$\frac{(8.99 \times 10^9\ \text{N} \cdot \text{m}^2/\text{C}^2)(+8.0 \times 10^{-9}\ \text{C})}{0.20\ \text{m}}$	+	$\frac{(8.99 \times 10^9\ \text{N} \cdot \text{m}^2/\text{C}^2)(-8.0 \times 10^{-9}\ \text{C})}{0.60\ \text{m}}$ =	+240 V
B	$\frac{(8.99 \times 10^9\ \text{N} \cdot \text{m}^2/\text{C}^2)(+8.0 \times 10^{-9}\ \text{C})}{0.40\ \text{m}}$	+	$\frac{(8.99 \times 10^9\ \text{N} \cdot \text{m}^2/\text{C}^2)(-8.0 \times 10^{-9}\ \text{C})}{0.40\ \text{m}}$ =	0 V

CONCEPTUAL EXAMPLE 7 | Where Is the Potential Zero?

Two point charges are fixed in place, as in **Figure 19.9**. The positive charge is $+2q$ and has twice the magnitude of the negative charge, which is $-q$. On the line that passes through the charges, three spots are identified, P_1, P_2, and P_3. At which of these spots could the potential be equal to zero? **(a)** P_2 and P_3 **(b)** P_1 and P_3 **(c)** P_1 and P_2

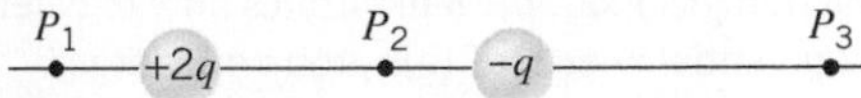

FIGURE 19.9 Two point charges, one positive and one negative. The positive charge, $+2q$, has twice the magnitude of the negative charge, $-q$.

Reasoning The total potential is the algebraic sum of the individual potentials created by each charge. It will be zero if the potential due to the positive charge is exactly offset by the potential due to the negative charge. The potential of a point charge is directly proportional to the charge and inversely proportional to the distance from the charge.

Answers (b) and (c) are incorrect. The total potential at P_1 cannot be zero. The positive charge has the larger magnitude and is closer to P_1 than is the negative charge. As a result, the potential of the positive charge at P_1 dominates over the potential of the negative charge, so the total potential cannot be zero.

Answer (a) is correct. Between the charges there is a location at which the individual potentials cancel each other. We saw a similar situation in Example 6, where the canceling occurred at the midpoint between the two charges that had equal magnitudes. Now the charges have unequal magnitudes, so the cancellation point does not occur at the midpoint. Instead, it occurs at the location P_2, which is closer to the charge with the smaller magnitude—namely, the negative charge. At P_2, since the potential of a point charge is inversely proportional to the distance from the charge, the effect of the smaller charge will be able to offset the effect of the more distant larger charge.

To the right of the negative charge there is also a location at which the individual potentials exactly cancel each other. All places on this section of the line are closer to the negative charge than to the positive charge. Therefore, there is a location P_3 in this region at which the potential of the smaller negative charge exactly cancels the potential of the more distant and larger positive charge.

In Example 6 we determined the total potential at a spot due to two point charges. In Example 8 we now extend this technique to find the total potential *energy* of three charges.

EXAMPLE 8 | The Potential Energy of a Group of Charges

Three point charges initially are infinitely far apart. Then, as **Figure 19.10** shows, they are brought together and placed at the corners of an equilateral triangle. Each side of the triangle has a length of 0.50 m. Determine the electric potential energy of the triangular group. In other words, determine the amount by which the electric potential energy of the group differs from that of the three charges in their initial, infinitely separated, locations.

Reasoning We will proceed in steps by adding charges to the triangle one at a time, and then determining the electric potential energy at each step. According to Equation 19.3, $\text{EPE} = q_0V$, the electric potential energy is the product of the charge and the electric potential at the spot where the charge is placed. The total electric potential energy of the triangular group is the sum of the energies of each step in assembling the group.

Solution The order in which the charges are put on the triangle does not matter; we begin with the charge $q_1 = +5.0\ \mu\text{C}$. When this charge is placed at a corner of the triangle, it has no electric potential energy, according to $\text{EPE}_1 = q_1(V_2 + V_3) = 0\ \text{J}$ (Equation 19.3). This is because the total potential $V_2 + V_3$ produced by the other two charges is zero at this corner, since they are infinitely far away. Once the charge q_1 is in place, the potential it creates at either empty corner ($r = 0.50$ m) is

$$V_1 = \frac{kq_1}{r} = \frac{(8.99 \times 10^9\ \text{N} \cdot \text{m}^2/\text{C}^2)(+5.0 \times 10^{-6}\ \text{C})}{0.50\ \text{m}}$$
$$= +9.0 \times 10^4\ \text{V} \qquad \textbf{(19.6)}$$

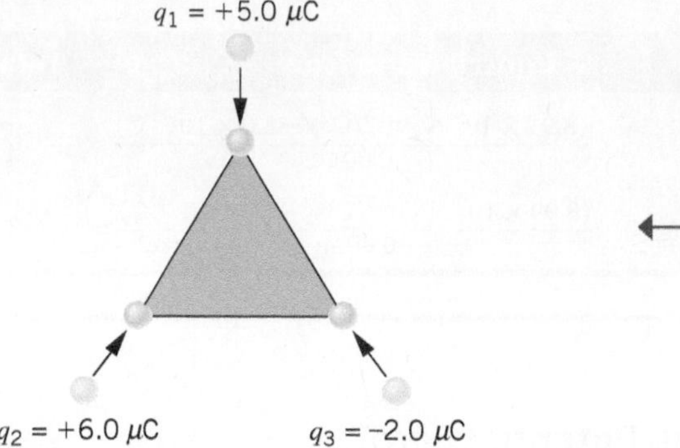

FIGURE 19.10 Three point charges are placed on the corners of an equilateral triangle. Example 8 illustrates how to determine the total electric potential energy of this group of charges.

Therefore, when the charge $q_2 = +6.0\ \mu C$ is placed at the second corner of the triangle, its electric potential energy is

$$\text{EPE}_2 = q_2V_1 = \frac{q_2kq_1}{r} = (+6.0 \times 10^{-6}\ \text{C})(+9.0 \times 10^4\ \text{V})$$

$$= +0.54\ \text{J} \qquad \textbf{(19.3)}$$

The electric potential at the remaining empty corner is the sum of the potentials due to the two charges that are already in place:

$$V_1 + V_2 = \frac{kq_1}{r} + \frac{kq_2}{r}$$

$$= \frac{(8.99 \times 10^9\ \text{N}\cdot\text{m}^2/\text{C}^2)(+5.0 \times 10^{-6}\ \text{C})}{0.50\ \text{m}}$$

$$+ \frac{(8.99 \times 10^9\ \text{N}\cdot\text{m}^2/\text{C}^2)(+6.0 \times 10^{-6}\ \text{C})}{0.50\ \text{m}}$$

$$= +2.0 \times 10^5\ \text{V}$$

When the third charge $q_3 = -2.0\ \mu C$ is placed at the remaining empty corner, its electric potential energy is

$$\text{EPE}_3 = q_3(V_1 + V_2) = \frac{q_3kq_1}{r} + \frac{q_3kq_2}{r}$$

$$= (-2.0 \times 10^{-6}\ \text{C})(+2.0 \times 10^5\ \text{V}) = -0.40\ \text{J} \qquad \textbf{(19.3)}$$

Problem-Solving Insight **Be careful to distinguish between the concepts of potential V and electric potential energy EPE. Potential is electric potential energy per unit charge: $V = \text{EPE}/q$.**

Math Skills To illustrate that the order in which the charges are put on the triangle does not matter, let us obtain the total potential energy of the group when the order begins with q_2, continues with q_3, and concludes with q_1. When q_2 is placed at a corner of the triangle, it has no electric potential energy EPE_2, since it experiences no electric potential due to the other charges, which are infinitely far away. Thus, we have

$$\text{EPE}_2 = 0\ \text{J}$$

Once q_2 is in place, it creates a potential $V_2 = \frac{kq_2}{r}$ (Equation 19.6) at either empty corner located a distance r away. When the charge q_3 is placed at an empty corner, it experiences this potential and has an electric potential energy EPE_3, as specified by Equation 19.3:

$$\text{EPE}_3 = q_3V_2 = \frac{q_3kq_2}{r}$$

Each of the charges in place creates a potential at the remaining empty corner. Using Equation 19.6 for each charge, we see that the total potential at the remaining empty corner is $V_2 + V_3 = \frac{kq_2}{r} + \frac{kq_3}{r}$. When charge q_1 is placed on this corner, therefore, it experiences this potential and has an electric potential energy EPE_1, as given by Equation 19.3:

$$\text{EPE}_1 = q_1(V_2 + V_3) = \frac{q_1kq_2}{r} + \frac{q_1kq_3}{r}$$

Thus, we obtain the following total electric potential energy of the group:

$$\text{Total potential energy} = 0\ \text{J} + \frac{q_3kq_2}{r} + \frac{q_1kq_2}{r} + \frac{q_1kq_3}{r}$$

This result is exactly the same as that determined in Example 8, where a different order of adding the charges is used: first q_1, second q_2, and third q_3.

The total potential energy of the triangular group differs from that of the widely separated charges by an amount that is the sum of the potential energies calculated previously:

$$\text{Total potential energy} = \underbrace{0\ \text{J}}_{\text{EPE}_1} + \underbrace{\frac{q_2kq_1}{r}}_{\text{EPE}_2} + \underbrace{\frac{q_3kq_1}{r} + \frac{q_3kq_2}{r}}_{\text{EPE}_3}$$

$$= 0\ \text{J} + 0.54\ \text{J} - 0.40\ \text{J} = \boxed{+0.14\ \text{J}}$$

This energy originates in the work done to bring the charges together.

Check Your Understanding

(The answers are given at the end of the book.)

4. **CYU Figure 19.2** shows four arrangements (A–D) of two point charges. In each arrangement consider the total electric potential that the charges produce at location P. Rank the arrangements (largest to smallest) according to the total potential. **(a)** B, C, A and D (a tie) **(b)** D, C, A, B **(c)** A and C (a tie), B, D **(d)** C, D, A, B

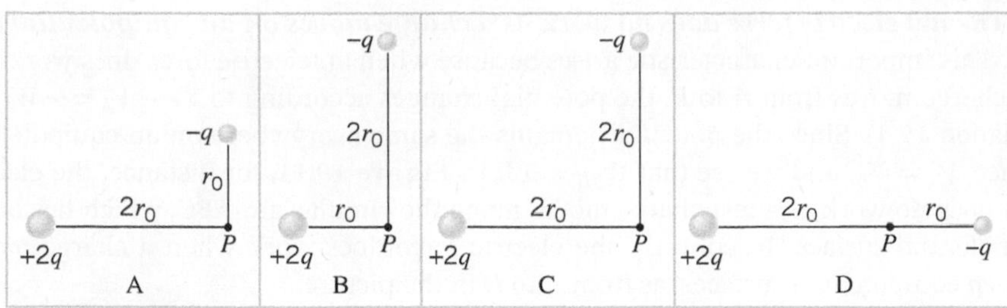

CYU FIGURE 19.2

5. A positive point charge and a negative point charge have equal magnitudes. One charge is fixed to one corner of a square, and the other is fixed to another corner. On which corners should the charges be placed, so that the same potential exists at the empty corners? The charges should be placed at **(a)** adjacent corners, **(b)** diagonally opposite corners.

6. Three point charges have identical magnitudes, but two of the charges are positive and one is negative. These charges are fixed to the corners of a square, one to a corner. No matter how the charges are arranged, the potential at the empty corner is always **(a)** zero, **(b)** negative, **(c)** positive.

7. Consider a spot that is located midway between two identical point charges. Which one of the following statements concerning the electric field and the electric potential at this spot is true? **(a)** The electric field is zero, but the electric potential is not zero. **(b)** The electric field is not zero, but the electric potential is zero. **(c)** Both the electric field and the electric potential are zero. **(d)** Neither the electric field nor the electric potential is zero.

8. Four point charges have the same magnitude (but they may have different signs) and are placed at the corners of a square, as **CYU Figure 19.3** shows. What must be the sign (+ or −) of each charge so that *both* the electric field *and* the electric potential are zero at the center of the square? Assume that the potential has a zero value at infinity.

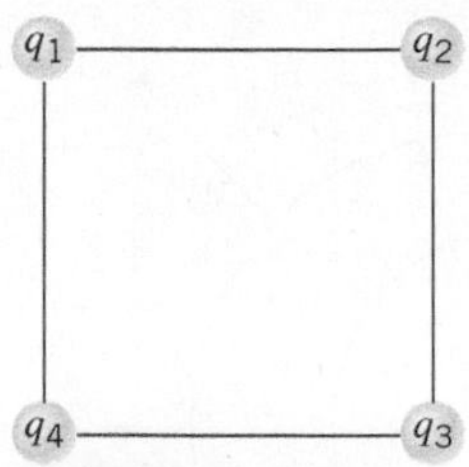

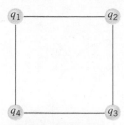

	q_1	q_2	q_3	q_4
(a)	−	−	−	−
(b)	+	+	−	−
(c)	+	+	+	+
(d)	+	−	+	−

CYU FIGURE 19.3

9. An electric potential energy exists when two protons are separated by a certain distance. Does the electric potential energy increase, decrease, or remain the same **(a)** when both protons are replaced by electrons, and **(b)** when only one of the protons is replaced by an electron?

10. A proton is fixed in place. An electron is released from rest and allowed to collide with the proton. Then the roles of the proton and electron are reversed, and the same experiment is repeated. Which, if either, is traveling faster when the collision occurs, the proton or the electron?

19.4 Equipotential Surfaces and Their Relation to the Electric Field

An **equipotential surface** is a surface on which the electric potential is the same everywhere. The easiest equipotential surfaces to visualize are those that surround an isolated point charge. According to Equation 19.6, the potential at a distance r from a point charge q is $V = kq/r$. Thus, wherever r is the same, the potential is the same, and the equipotential surfaces are spherical surfaces centered on the charge. There are an infinite number of such surfaces, one for every value of r, and **Figure 19.11** illustrates two of them. The larger the distance r, the smaller is the potential of the equipotential surface.

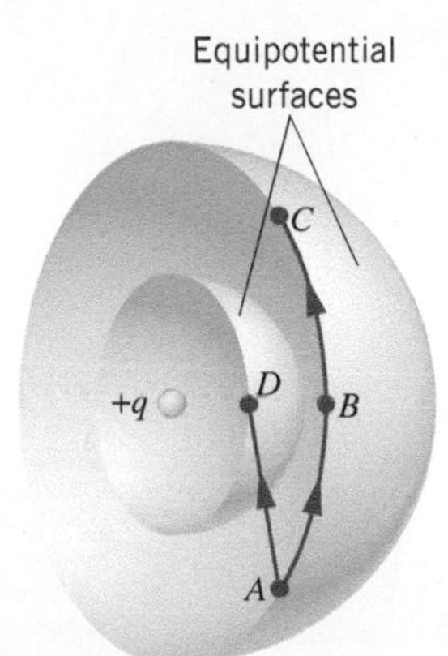

FIGURE 19.11 The equipotential surfaces that surround the point charge $+q$ are spherical. The electric force does no work as a charge moves on a path that lies on an equipotential surface, such as the path ABC. However, work is done by the electric force when a charge moves between two equipotential surfaces, as along the path AD.

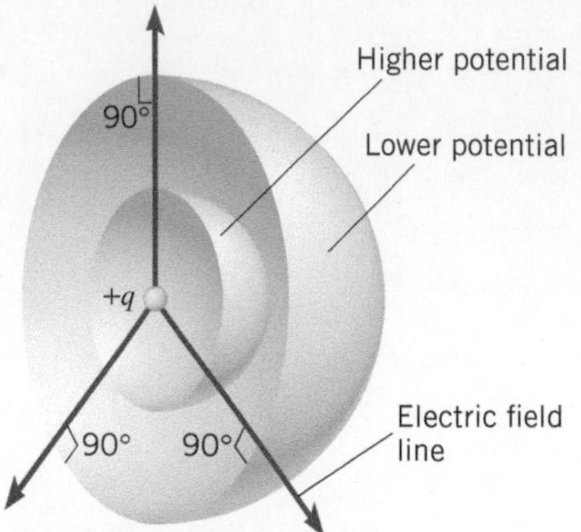

FIGURE 19.12 The radially directed electric field of a point charge is perpendicular to the spherical equipotential surfaces that surround the charge. The electric field points in the direction of *decreasing* potential.

The net electric force does no work as a charge moves on an equipotential surface. This important characteristic arises because when an electric force does work W_{AB} as a charge moves from A to B, the potential changes according to $V_B - V_A = -W_{AB}/q_0$ (Equation 19.4). Since the potential remains the same everywhere on an equipotential surface, $V_A = V_B$, and we see that $W_{AB} = 0$ J. In **Figure 19.11**, for instance, the electric force does no work as a test charge moves along the circular arc ABC, which lies on an equipotential surface. In contrast, the electric force does work when a charge moves *between* equipotential surfaces, as from A to D in the picture.

The spherical equipotential surfaces that surround an isolated point charge illustrate another characteristic of all equipotential surfaces. **Figure 19.12** shows two of the surfaces around a positive point charge, along with some electric field lines. The electric field lines give the direction of the electric field, and for a positive point charge the electric field is directed radially outward. Therefore, at each location on an equipotential sphere the electric field is perpendicular to the surface and points outward in the direction of decreasing potential, as the drawing emphasizes. This perpendicular relation is valid whether or not the equipotential surfaces result from a positive charge or have a spherical shape.

Problem-Solving Insight The electric field created by any charge or group of charges is everywhere perpendicular to the associated equipotential surfaces and points in the direction of decreasing potential.

For example, **Animated Figure 19.13** shows the electric field lines (in red) around an electric dipole, along with some equipotential surfaces (in blue), shown in cross section. Since the field lines are not simply radial, the equipotential surfaces are no longer spherical but, instead, have the shape necessary to be everywhere perpendicular to the field lines.

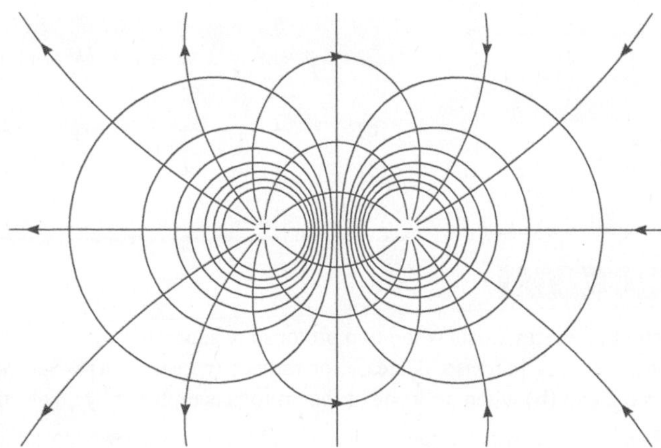

ANIMATED FIGURE 19.13 A cross-sectional view of the equipotential surfaces (in blue) of an electric dipole. The surfaces are drawn to show that at every point they are perpendicular to the electric field lines (in red) of the dipole.

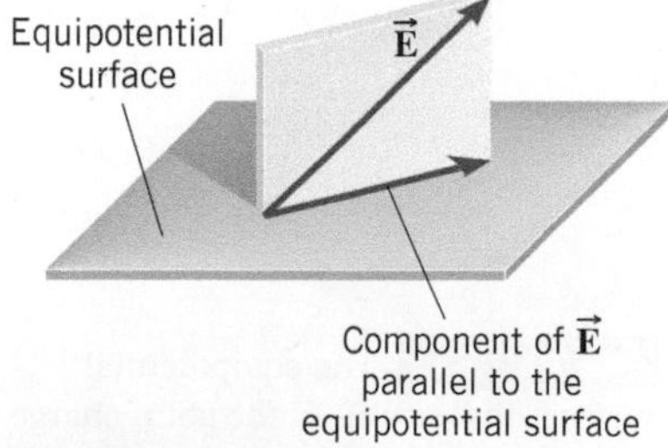

FIGURE 19.14 In this hypothetical situation, the electric field $\vec{E}$ is not perpendicular to the equipotential surface. As a result, there is a component of $\vec{E}$ parallel to the surface.

To see why an equipotential surface must be perpendicular to the electric field, consider **Figure 19.14**, which shows a hypothetical situation in which the perpendicular relation does *not* hold. If $\vec{E}$ were not perpendicular to the equipotential surface, there would be a component of $\vec{E}$ parallel to the surface. This field component would exert an electric force on a test charge placed on the surface. As the charge moved along the surface, work would be done by this component of the electric force. The work, according to Equation 19.4, would cause the potential to change, and, thus, the surface could not be an equipotential surface as assumed. The only way out of the dilemma is for the electric field to be perpendicular to the surface, so there is no component of the field parallel to the surface.

We have already encountered one equipotential surface. In Section 18.8, we found that the direction of the electric field just outside an electrical conductor is perpendicular to the conductor's surface, when the conductor is at equilibrium under electrostatic conditions. Thus, the surface of any conductor is an equipotential surface under such conditions. In fact, since the electric field is zero everywhere inside a conductor whose charges are in equilibrium, the entire conductor can be regarded as an equipotential volume.

There is a quantitative relation between the electric field and the equipotential surfaces. One example that illustrates this relation is the parallel plate capacitor in **Figure 19.15**. As Section 18.6 discusses, the electric field $\vec{\mathbf{E}}$ between the metal plates is perpendicular to them and is the same everywhere, ignoring fringe fields at the edges. To be perpendicular to the electric field, the equipotential surfaces must be planes that are parallel to the capacitor plates, which themselves are equipotential surfaces. The potential difference between the plates is given by Equation 19.4 as $\Delta V = V_B - V_A = -W_{AB}/q_0$, where A is a point on the positive plate and B is a point on the negative plate. The work done by the electric force as a positive test charge q_0 moves from A to B is $W_{AB} = F\,\Delta s$, where F refers to the electric force and Δs to the displacement along a line perpendicular to the plates. The force equals the product of the charge and the electric field E ($F = q_0 E$), so the work becomes $W_{AB} = F\,\Delta s = q_0 E\,\Delta s$. Therefore, the potential difference between the capacitor plates can be written in terms of the electric field as $\Delta V = -W_{AB}/q_0 = -q_0 E\,\Delta s/q_0$, or

$$E = -\frac{\Delta V}{\Delta s} \tag{19.7a}$$

The quantity $\Delta V/\Delta s$ is referred to as the **potential gradient** and has units of volts per meter. In general, the relation $E = -\Delta V/\Delta s$ gives only the component of the electric field along the displacement Δs; it does not give the perpendicular component. The next example deals further with the equipotential surfaces between the plates of a capacitor.

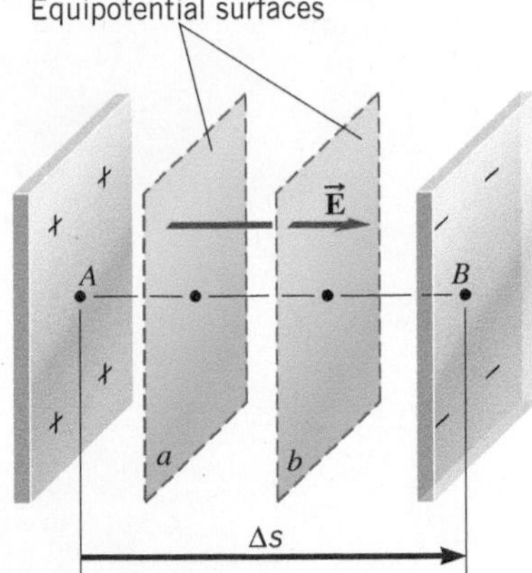

FIGURE 19.15 The metal plates of a parallel plate capacitor are equipotential surfaces. Two additional equipotential surfaces are shown between the plates. These two equipotential surfaces are parallel to the plates and are perpendicular to the electric field $\vec{\mathbf{E}}$ between the plates.

EXAMPLE 9 | The Electric Field and Potential Are Related

The plates of the capacitor in **Figure 19.15** are separated by a distance $\Delta s = (\Delta s)_{\text{plates}} = 0.032$ m, and the potential difference between them is $(\Delta V)_{\text{plates}} = V_B - V_A = -64$ V. Between the two blue equipotential surfaces there is a potential difference of $(\Delta V)_{\text{blue}} = V_b - V_a = -3.0$ V. Find the spacing between the two blue surfaces.

Reasoning To find the spacing between the blue surfaces, we will solve Equation 19.7a for $(\Delta s)_{\text{blue}}$ with $(\Delta V)_{\text{blue}} = -3.0$ V and E equal to the electric field between the plates of the capacitor. The electric field E is the same everywhere between the plates (ignoring the fringe fields). A value for E can be obtained by using Equation 19.7a with the data given for the distance between the plates and the potential difference between them [$(\Delta s)_{\text{plates}} = 0.032$ m and $(\Delta V)_{\text{plates}} = -64$ V].

Solution Solving Equation 19.7a for the spacing between the blue equipotential surfaces gives

$$(\Delta s)_{\text{blue}} = -\frac{(\Delta V)_{\text{blue}}}{E}$$

Using Equation 19.7a to determine the electric field between the plates, we find that

$$E = -\frac{(\Delta V)_{\text{plates}}}{(\Delta s)_{\text{plates}}} = -\frac{-64\text{ V}}{0.032\text{ m}} = 2.0 \times 10^3\text{ V/m}$$

Substituting this result into the equation for $(\Delta s)_{\text{blue}}$ gives

$$(\Delta s)_{\text{blue}} = -\frac{(\Delta V)_{\text{blue}}}{E} = -\frac{-3.0\text{ V}}{2.0 \times 10^3\text{ V/m}} = \boxed{1.5 \times 10^{-3}\text{ m}}$$

Equation 19.7a gives the relationship between the electric field and the electric potential. It gives the component of the electric field along the displacement Δs in a region of space where the electric potential changes from place to place and applies to a wide variety of situations. When applied strictly to a parallel plate capacitor, however, this expression is often used in a slightly different form. In **Figure 19.15**, the metal plates of the capacitor are marked A (higher potential) and B (lower potential). Traditionally, in discussions of such a capacitor, the potential difference between the plates is referred to by using the symbol V to denote the amount by which the higher potential exceeds the lower potential ($V = V_A - V_B$). In this tradition, the symbol V is often referred to as simply the "voltage." For example, if the potential difference between the plates of a capacitor is 5 volts, it is common to say that the "voltage" of the capacitor is 5 volts. In addition, the displacement from plate A to plate B is expressed in terms of the separation d between the plates ($d = s_B - s_A$). With this nomenclature, Equation 19.7a becomes

$$E = -\frac{\Delta V}{\Delta s} = -\frac{V_B - V_A}{s_B - s_A} = \frac{V_A - V_B}{s_B - s_A} = \frac{V}{d} \quad \text{(parallel plate capacitor)} \tag{19.7b}$$

Check Your Understanding

(*The answers are given at the end of the book.*)

11. **CYU Figure 19.4** shows a cross-sectional view of two spherical equipotential surfaces and two electric field lines that are perpendicular to these surfaces. When an electron moves from point A to point B (against the electric field), the electric force does $+3.2 \times 10^{-19}$ J of work. What are the electric potential differences **(a)** $V_B - V_A$, **(b)** $V_C - V_B$, and **(c)** $V_C - V_A$?

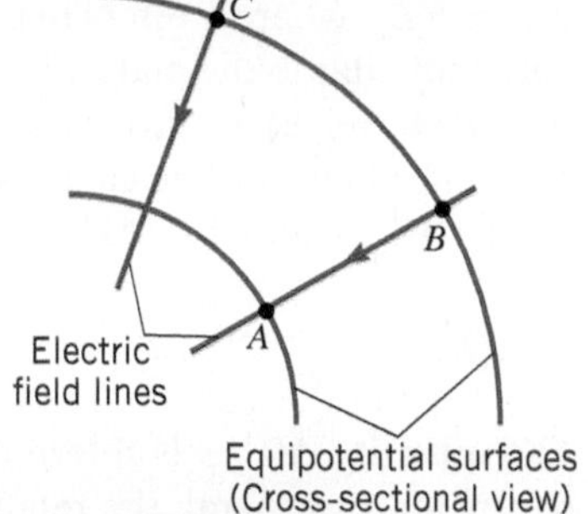

CYU FIGURE 19.4

12. The electric potential is constant throughout a given region of space. Is the electric field zero or nonzero in this region?

13. In a region of space where the electric field is constant everywhere, as it is inside a parallel plate capacitor, is the potential constant everywhere? **(a)** Yes. **(b)** No, the potential is greatest at the positive plate. **(c)** No, the potential is greatest at the negative plate.

14. A positive test charge is placed in an electric field. In what direction should the charge be moved relative to the field, so that the charge experiences a constant electric potential? The charge should be moved **(a)** perpendicular to the electric field, **(b)** in the same direction as the electric field, **(c)** opposite to the direction of the electric field.

15. The location marked P in **CYU Figure 19.5** lies midway between the point charges $+q$ and $-q$. The blue lines labeled A, B, and C are edge-on views of three planes. Which of the planes is an equipotential surface? **(a)** A and C **(b)** A, B, and C **(c)** Only B **(d)** None of the planes is an equipotential surface.

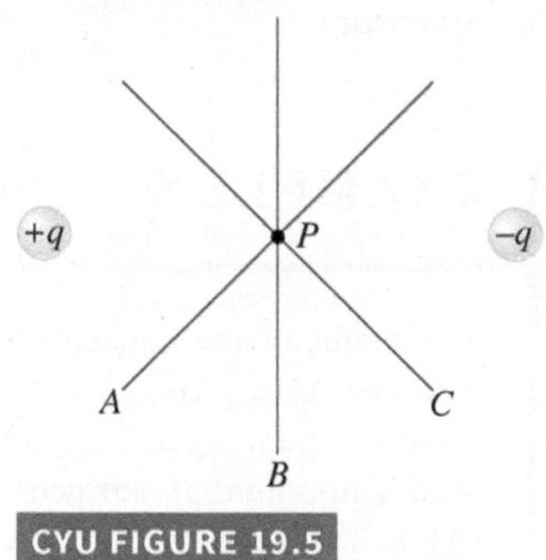

CYU FIGURE 19.5

16. Imagine that you are moving a positive test charge along the line between two identical point charges. With regard to the electric potential, is the midpoint on the line analogous to the top of a mountain or the bottom of a valley when the two point charges are **(a)** positive and **(b)** negative?

19.5 Capacitors and Dielectrics

The Capacitance of a Capacitor

In Section 18.6 we saw that a parallel plate capacitor consists of two parallel metal plates placed near one another but not touching. This type of capacitor is only one among many. In general, a **capacitor** consists of two conductors of any shape placed near one another without touching. For a reason that will become clear later on, it is common practice to fill the region between the conductors or plates with an electrically insulating material called a **dielectric**, as **Figure 19.16** illustrates.

A capacitor stores electric charge. Each capacitor plate carries a charge of the *same magnitude*, one positive and the other negative. Because of the charges, the electric potential of the positive plate exceeds that of the negative plate by an amount V, as **Figure 19.16** indicates. Experiment shows that when the magnitude q of the charge on each plate is doubled, the magnitude V of the electric potential difference is also doubled, so q is proportional to V: $q \propto V$. Equation 19.8 expresses this proportionality with the aid of a proportionality constant C, which is the **capacitance** of the capacitor.

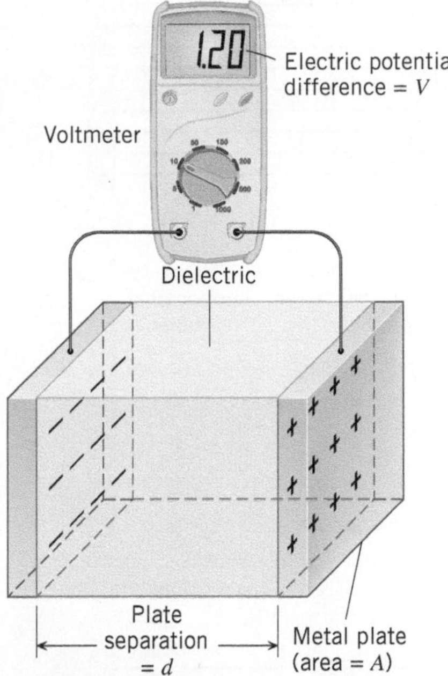

FIGURE 19.16 A parallel plate capacitor consists of two metal plates, one carrying a charge $+q$ and the other a charge $-q$. The potential of the positive plate exceeds that of the negative plate by an amount V. The region between the plates is filled with a dielectric.

THE RELATION BETWEEN CHARGE AND POTENTIAL DIFFERENCE FOR A CAPACITOR

The magnitude q of the charge on each plate of a capacitor is directly proportional to the magnitude V of the potential difference between the plates:

$$q = CV \tag{19.8}$$

where C is the capacitance.

SI Unit of Capacitance: coulomb/volt = farad (F)

Equation 19.8 shows that the SI unit of capacitance is the coulomb per volt (C/V). This unit is called the *farad* (F), named after the English scientist Michael Faraday (1791–1867). One farad is an enormous capacitance. Usually smaller amounts, such as a microfarad (1 μF = 10^{-6} F) or a picofarad (1 pF = 10^{-12} F), are used in electric circuits. The capacitance reflects the ability of the capacitor to store charge, in the sense that a larger capacitance C allows more charge q to be put onto the plates for a given value of the potential difference V.

THE PHYSICS OF . . . random-access memory (RAM) chips. The ability of a capacitor to store charge lies at the heart of the random-access memory (RAM) chips used in computers, where information is stored in the form of the "ones" and "zeros" that comprise binary numbers. **Figure 19.17** illustrates the role of a capacitor in a RAM chip. The capacitor is connected to a transistor switch, to which two lines are connected, an address line and a data line. A single RAM chip often contains millions of such transistor–capacitor units. The address line is used by the computer to locate a particular transistor–capacitor combination, and the data line carries the data to be stored. A pulse on the address line turns on the transistor switch. With the switch turned on, a pulse coming in on the data line can cause the capacitor to charge. A charged capacitor means that a "one" has been stored, whereas an uncharged capacitor means that a "zero" has been stored.

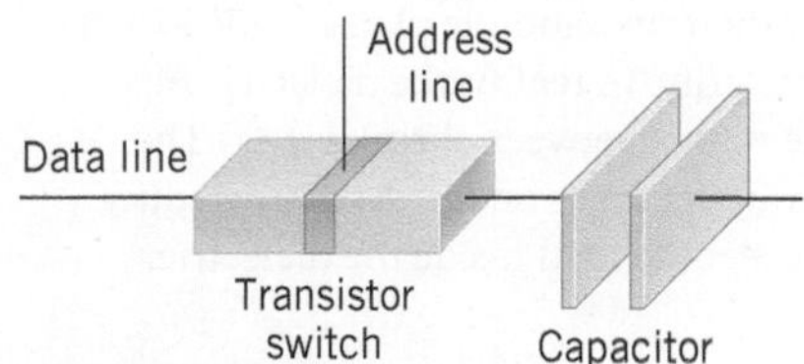

FIGURE 19.17 A transistor–capacitor combination is part of a RAM chip used in computer memories.

The Dielectric Constant

If a dielectric is inserted between the plates of a capacitor, the capacitance can increase markedly because of the way in which the dielectric alters the electric field between the plates. **Interactive Figure 19.18** shows how this effect comes about. In part *a*, the region between the charged plates is empty. The field lines point from the positive toward the negative plate. In part *b*, a dielectric has been inserted between the plates. Since the capacitor is not connected to anything, the charge on the plates remains constant as the dielectric is inserted. In many materials (e.g., water) the molecules possess permanent dipole moments, even though the molecules are electrically neutral. The dipole moment exists because one end of a molecule has a slight excess of negative charge while the other end has a slight excess of positive charge. When such molecules are placed between the charged plates of the capacitor, the negative ends are attracted to the positive plate and the positive ends are attracted to the negative plate. As a result, the dipolar molecules tend to orient themselves end to end, as in part *b*. Whether or not a molecule has a permanent dipole moment, the

$-q$ $+q$

(a)

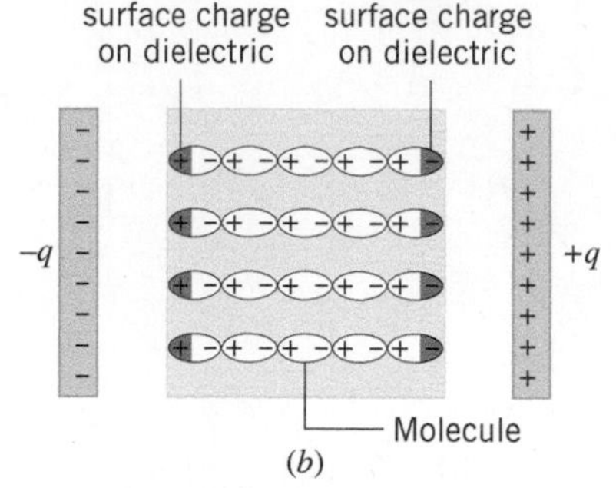

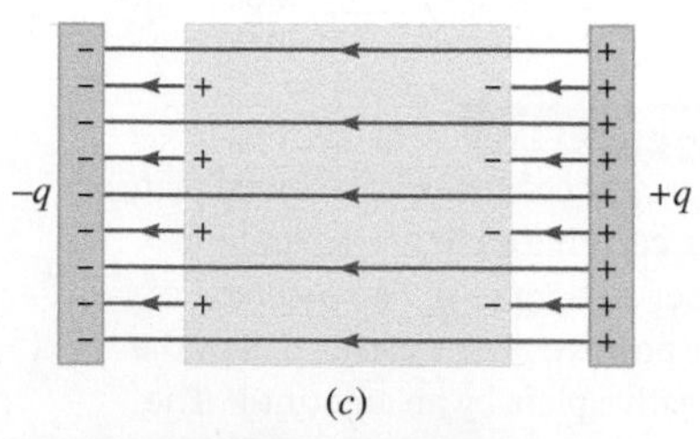

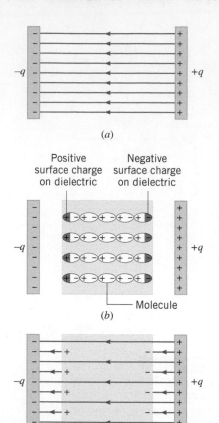

INTERACTIVE FIGURE 19.18 (*a*) The electric field lines inside an empty capacitor. (*b*) The electric field produced by the charges on the plates aligns the molecular dipoles within the dielectric end to end. The space between the dielectric and the plates is added for clarity. In reality, the dielectric fills the region between the plates. (*c*) The surface charges on the dielectric reduce the electric field inside the dielectric.

electric field can cause the electrons to shift position within a molecule, making one end slightly negative and the opposite end slightly positive. Because of the end-to-end orientation, the left surface of the dielectric becomes positively charged, and the right surface becomes negatively charged. The surface charges are shown in red in the picture.

Because of the surface charges on the dielectric, not all the electric field lines generated by the charges on the plates pass through the dielectric. As **Interactive Figure 19.18c** shows, some of the field lines end on the negative surface charges and begin again on the positive surface charges. Thus, the electric field inside the dielectric is less strong than the electric field inside the empty capacitor, assuming the charge on the plates remains constant. This reduction in the electric field is described by the **dielectric constant** κ, which is the ratio of the field magnitude E_0 without the dielectric to the field magnitude E inside the dielectric:

$$\kappa = \frac{E_0}{E} \quad \textbf{(19.9)}$$

Being a ratio of two field strengths, the dielectric constant is a number without units. Moreover, since the field $\vec{\mathbf{E}}_0$ without the dielectric is greater than the field $\vec{\mathbf{E}}$ inside the dielectric, the dielectric constant is greater than unity. The value of κ depends on the nature of the dielectric material, as **Table 19.1** indicates.

TABLE 19.1 Dielectric Constants of Some Common Substances[a]

Substance	Dielectric Constant, κ
Vacuum	1
Air	1.000 54
Teflon	2.1
Benzene	2.28
Paper (royal gray)	3.3
Ruby mica	5.4
Neoprene rubber	6.7
Methyl alcohol	33.6
Water	80.4

[a]Near room temperature.

The Capacitance of a Parallel Plate Capacitor

The capacitance of a capacitor is affected by the geometry of the plates and the dielectric constant of the material between them. For example, **Figure 19.16** shows a parallel plate capacitor in which the area of each plate is A and the separation between the plates is d. The magnitude of the electric field inside the dielectric is given by Equation 19.7b as $E = V/d$, where V is the magnitude of the potential difference between the plates. If the charge on each plate is kept fixed, the electric field inside the dielectric is related to the electric field in the absence of the dielectric via Equation 19.9. Therefore,

$$E = \frac{E_0}{\kappa} = \frac{V}{d}$$

Since the electric field within an empty capacitor is $E_0 = q/(\varepsilon_0 A)$ (see Equation 18.4), it follows that $q/(\kappa\varepsilon_0 A) = V/d$, which can be solved for q to give

$$q = \left(\frac{\kappa\varepsilon_0 A}{d}\right)V$$

A comparison of this expression with $q = CV$ (Equation 19.8) reveals that the capacitance C is

Parallel plate capacitor filled with a dielectric

$$C = \frac{\kappa\varepsilon_0 A}{d} \quad \textbf{(19.10)}$$

Only the geometry of the plates (A and d) and the dielectric constant κ affect the capacitance. With C_0 representing the capacitance of the empty capacitor ($\kappa = 1$), Equation 19.10 shows that $C = \kappa C_0$. In other words, the capacitance with the dielectric present is increased by a factor of κ over the capacitance without the dielectric. It can be shown that the relation $C = \kappa C_0$ applies to any capacitor, not just to a parallel plate capacitor. One reason, then, that capacitors are filled with dielectric materials is to increase the capacitance.

CONCEPTUAL EXAMPLE 10 | The Effect of a Dielectric When a Capacitor Has a Constant Charge

An empty capacitor is connected to a battery and charged up. The capacitor is then disconnected from the battery, and a slab of dielectric material is inserted between the plates. Does the potential difference across the plates **(a)** increase, **(b)** remain the same, or **(c)** decrease?

Reasoning Our reasoning is guided by the following fact: Once the capacitor is disconnected from the battery, the charge on its plates remains constant, for there is no longer any way for charge to be added or removed. According to Equation 19.8, the magnitude q of the charge stored by the capacitor is $q = CV$, where C is its capacitance and V is the magnitude of the potential difference between the plates.

Answers (a) and (b) are incorrect. Placing a dielectric between the plates of a capacitor reduces the electric field in that region (see **Interactive Figure 19.18c**). The magnitude V of the potential difference is related to the magnitude E of the electric field by $V = Ed$, where d is the distance between the capacitor plates. Since E decreases when the dielectric is inserted and d is unchanged, the potential difference does not increase or remain the same.

Answer (c) is correct. Inserting the dielectric causes the capacitance C to increase. Since $q = CV$ and q is fixed, the potential difference V across the plates must decrease in order for q to remain unchanged. The amount by which the potential difference decreases from the value initially established by the battery depends on the dielectric constant of the slab.

Related Homework: Problem 44

Capacitors are used often in electronic devices, and Example 11 deals with one familiar application.

EXAMPLE 11 | The Physics of a Computer Keyboard

One kind of computer keyboard is based on the idea of capacitance. Each key is mounted on one end of a plunger, and the other end is attached to a movable metal plate (see **Figure 19.19**). The movable plate is separated from a fixed plate, the two plates forming a capacitor. When the key is pressed, the movable plate is pushed closer to the fixed plate, and the capacitance increases. Electronic circuitry enables the computer to detect the *change* in capacitance, thereby recognizing which key has been pressed. The separation of the plates is normally 5.00×10^{-3} m but decreases to 0.150×10^{-3} m when a key is pressed. The plate area is 9.50×10^{-5} m^2, and the capacitor is filled with a material whose dielectric constant is 3.50. Determine the change in capacitance that is detected by the computer.

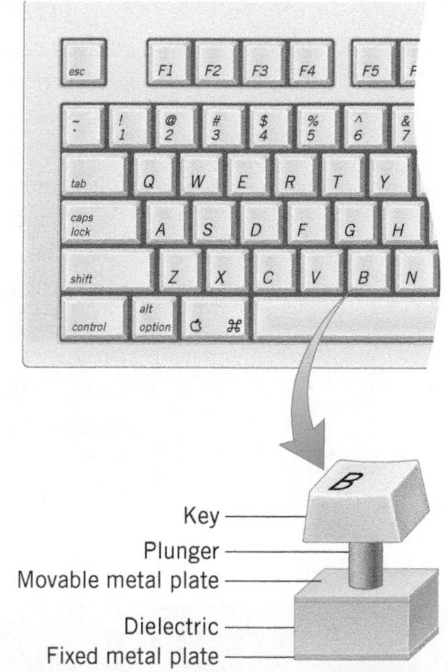

FIGURE 19.19 In one kind of computer keyboard, each key, when pressed, changes the separation between the plates of a capacitor.

Reasoning We can use Equation 19.10 directly to find the capacitance of the key, since the dielectric constant κ, the plate area A, and the plate separation d are known. We will use this relation twice, once to find the capacitance when the key is pressed and once when it is not pressed. The change in capacitance will be the difference between these two values.

Solution When the key is pressed, the capacitance is

$$C = \frac{\kappa\varepsilon_0 A}{d} = \frac{(3.50)[8.85 \times 10^{-12}\ \text{C}^2/(\text{N}\cdot\text{m}^2)](9.50 \times 10^{-5}\ \text{m}^2)}{0.150 \times 10^{-3}\ \text{m}}$$

$$= 19.6 \times 10^{-12}\ \text{F} \quad (19.6\ \text{pF}) \qquad \textbf{(19.10)}$$

A calculation similar to the one above reveals that when the key is *not* pressed, the capacitance has a value of 0.589×10^{-12} F (0.589 pF). The *change* in capacitance is an increase of $\boxed{19.0 \times 10^{-12}\ \text{F}\ (19.0\ \text{pF})}$. The *change* in the capacitance is greater with the dielectric present, which makes it easier for the circuitry within the computer to detect it.

Energy Storage in a Capacitor

When a capacitor stores charge, it also stores energy. In charging up a capacitor, for example, a battery does work in transferring an increment of charge from one plate of the capacitor to the other plate. The work done is equal to the product of the charge increment and the potential difference between the plates. However, as each increment of charge is moved, the potential difference increases slightly, and a larger amount of work is needed to move the next increment. The total work W done in completely charging the capacitor is the product of the total charge q transferred and the average potential difference $\overline{V}$; $W = q\overline{V}$. Since the average potential difference is one-half the final potential V, or $\overline{V} = \frac{1}{2}V$, the total work done by the battery is $W = \frac{1}{2}qV$. This work does not disappear but is stored as electric potential energy in the capacitor, so that Energy $= \frac{1}{2}qV$. Equation 19.8 indicates that $q = CV$ or, equivalently, that $V = q/C$. We can see, then, that our expression for the energy can be cast into two additional equivalent forms by substituting for q or for V. Equations 19.11a–c summarize these results:

$$\text{Energy} = \tfrac{1}{2}qV \tag{19.11a}$$

$$\text{Energy} = \tfrac{1}{2}(CV)V = \tfrac{1}{2}CV^2 \tag{19.11b}$$

$$\text{Energy} = \tfrac{1}{2}q\left(\frac{q}{C}\right) = \frac{q^2}{2C} \tag{19.11c}$$

It is also possible to regard the energy as being stored in the electric field between the plates. The relation between energy and field strength can be obtained for a parallel plate capacitor by substituting $V = Ed$ (Equation 19.7b) and $C = \kappa\varepsilon_0 A/d$ (Equation 19.10) into Equation 19.11b:

$$\text{Energy} = \tfrac{1}{2}CV^2 = \tfrac{1}{2}\left(\frac{\kappa\varepsilon_0 A}{d}\right)(Ed)^2$$

Since the area A times the separation d is the volume between the plates, the energy per unit volume or **energy density** is

$$\text{Energy density} = \frac{\text{Energy}}{\text{Volume}} = \tfrac{1}{2}\kappa\varepsilon_0 E^2 \tag{19.12}$$

It can be shown that this expression is valid for any electric field strength, not just that between the plates of a capacitor.

THE PHYSICS OF . . . an electronic flash attachment for a camera. The energy-storing capability of a capacitor is often put to good use in electronic circuits. For example, in an electronic flash attachment for a camera, energy from the battery pack is stored in a capacitor. The capacitor is then discharged between the electrodes of the flash tube, which converts the energy into light. Flash duration times range from 1/200 to 1/1 000 000 second or less, with the shortest flashes being used in high-speed photography (see **Figure 19.20**). Some flash attachments automatically control the flash duration by monitoring the light reflected from the photographic subject and quickly stopping or quenching the capacitor discharge when the reflected light reaches a predetermined level.

BIO THE PHYSICS OF . . . a defibrillator. During a heart attack, the heart produces a rapid, unregulated pattern of beats, a condition known as cardiac fibrillation. Cardiac fibrillation can often be stopped by sending a very fast discharge of electrical energy through the heart. For this purpose, emergency medical personnel use defibrillators, such as the one being used in **Figure 19.21**. A paddle is connected to each plate

Kyodo News/Getty Images

FIGURE 19.20 This time-lapse photo of a figure skater was obtained using a camera with an electronic flash attachment. The energy for each flash of light comes from the electrical energy stored in a capacitor.

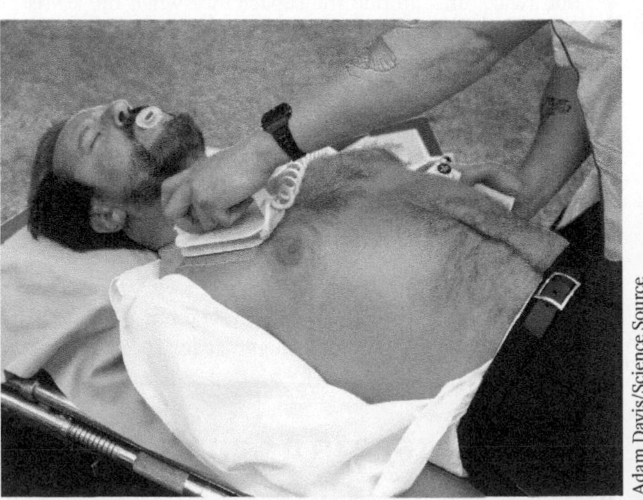

Adam Davis/Science Source

FIGURE 19.21 A paramedic is using a portable defibrillator in an attempt to revive a heart attack victim. A defibrillator uses the electrical energy stored in a capacitor to deliver a controlled electric current that can restore normal heart rhythm.

of a large capacitor, and the paddles are placed on the chest near the heart. The capacitor is charged to a potential difference of about a thousand volts. The capacitor is then discharged in a few thousandths of a second; the discharge current passes through a paddle, the heart, and the other paddle. Within a few seconds, the heart often returns to its normal beating pattern.

Check Your Understanding

(*The answers are given at the end of the book.*)

17. An empty parallel plate capacitor is connected to a battery that maintains a constant potential difference between the plates. With the battery connected, a dielectric is then inserted between the plates. Do the following quantities decrease, remain the same, or increase when the dielectric is inserted? **(a)** The electric field between the plates **(b)** The capacitance **(c)** The charge on the plates **(d)** The energy stored by the capacitor

18. A parallel plate capacitor is charged up by a battery. The battery is then disconnected, but the charge remains on the plates. The plates are then pulled apart. Do the following quantities decrease, remain the same, or increase as the distance between the plates increases? **(a)** The capacitance of the capacitor **(b)** The potential difference between the plates **(c)** The electric field between the plates **(d)** The electric potential energy stored by the capacitor

19.6 *Biomedical Applications of Electric Potential Differences

Conduction of Electrical Signals in Neurons

The human nervous system is remarkable for its ability to transmit information in the form of electrical signals. These signals are carried by the nerves, and the concept of electric potential difference plays an important role in the process. For example, sensory information from our eyes and ears is carried to the brain by the optic nerves and auditory nerves, respectively. Other nerves transmit signals from the brain or spinal column to muscles, causing them to contract. Still other nerves carry signals within the brain.

A nerve consists of a bundle of *axons,* and each axon is one part of a nerve cell, or *neuron.* As **Interactive Figure 19.22** illustrates, a typical neuron consists of a cell body with numerous extensions, called *dendrites,* and a single axon. The dendrites convert stimuli, such as pressure or heat, into electrical signals that travel through the neuron. The axon sends the signal to the nerve endings, which transmit the signal across a gap (called a *synapse*) to the next neuron or to a muscle.

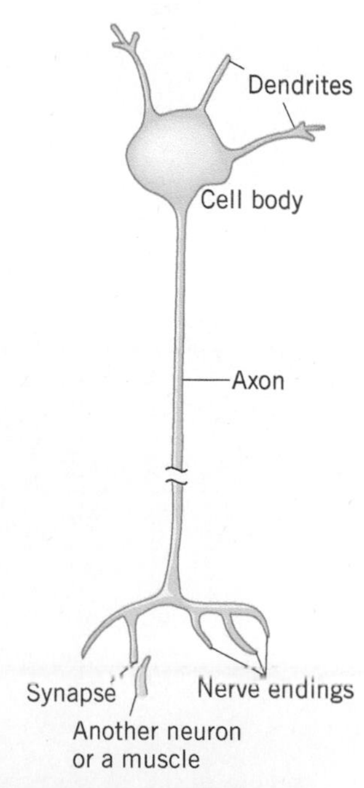

INTERACTIVE FIGURE 19.22 The anatomy of a typical neuron.

The fluid inside a cell, the intracellular fluid, is quite different from that outside the cell, the extracellular fluid. Both fluids contain concentrations of positive and negative ions. However, the extracellular fluid is rich in sodium (Na^+) and chlorine (Cl^-) ions, whereas the intracellular fluid is rich in potassium (K^+) ions and negatively charged proteins. These concentration differences between the fluids are extremely important to the life of the cell. If the cell membrane were freely permeable, the ions would diffuse across it until the concentrations on both sides were equal. (See Section 14.4 for a review of diffusion.) This does not happen, because a living cell has a selectively permeable membrane. Ions can enter or leave the cell only through membrane channels, and the permeability of the channels varies markedly from one ion to another. For example, it is much easier for K^+ ions to diffuse out of the cell than it is for Na^+ to enter the cell. As a result of selective membrane permeability, there is a small buildup of negative charges just on the

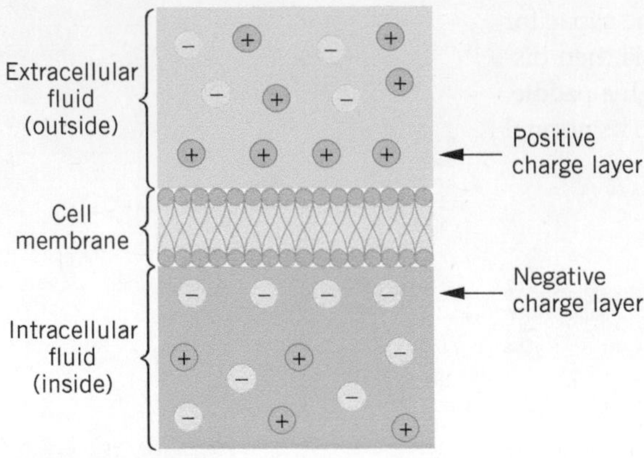

FIGURE 19.23 Positive and negative charge layers form on the outside and inside surfaces of a membrane during its resting state.

inner side of the membrane and an equal amount of positive charges on the outer side (see Figure 19.23). The buildup of charge occurs very close to the membrane, so the membrane acts like a capacitor (see Problems 56 and 57). Elsewhere in the intracellular and extracellular fluids, there are equal numbers of positive and negative ions, so the fluids are overall electrically neutral. Such a separation of positive and negative charges gives rise to an electric potential difference across the membrane, called the *resting membrane potential*. In neurons, the resting membrane potential ranges from −40 to −90 mV, with a typical value of −70 mV. The minus sign indicates that the inner side of the membrane is negative relative to the outer side.

BIO THE PHYSICS OF . . . an action potential. A "resting" neuron is one that is not conducting an electrical signal. The *change* in the resting membrane potential is the key factor in the initiation and conduction of a signal. When a sufficiently strong stimulus is applied to a given point on the neuron, "gates" in the membrane open and sodium ions flood into the cell, as Figure 19.24 illustrates. The sodium ions are driven into the cell by attraction to the negative ions on the inner side of the membrane as well as by the relatively high concentration of sodium ions outside the cell. The large influx of Na^+ ions first neutralizes the negative ions on the interior of the membrane and then causes it to become positively charged. As a result, the membrane potential in this localized region goes from −70 mV, the resting potential, to about +30 mV in a very short time (see Figure 19.25). The sodium gates then close, and the cell membrane quickly returns to its normal resting potential. This change in potential, from −70 mV to +30 mV and back to −70 mV, is known as the *action potential*. The action potential lasts for a few milliseconds, and it is the electrical signal that propagates down the axon, typically at a speed of about 50 m/s, to the next neuron or to a muscle cell.

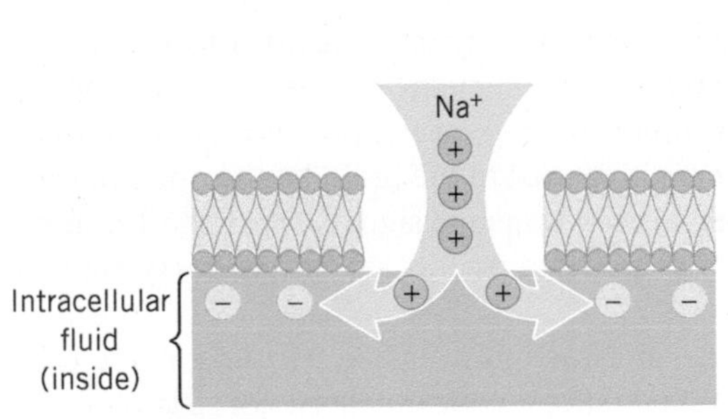

FIGURE 19.24 When a stimulus is applied to the cell, positive sodium ions (Na^+) rush into the cell, causing the interior surface of the membrane to become momentarily positive.

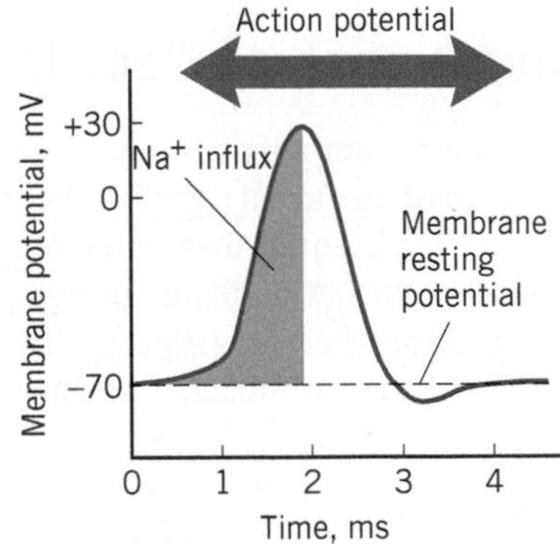

FIGURE 19.25 The action potential is caused by the rush of positive sodium ions into the cell and a subsequent return of the cell to its resting potential.

Medical Diagnostic Techniques

Several important medical diagnostic techniques depend on the fact that the surface of the human body is *not* an equipotential surface. Between various points on the body there are small potential differences (approximately 30–500 μV), which provide the basis for electrocardiography, electroencephalography, and electroretinography. The potential differences can be traced to the electrical characteristics of muscle cells and nerve cells. In carrying out their biological functions, these cells utilize positively charged sodium and potassium ions and negatively charged chlorine ions that exist within the cells and in the extracellular fluid. As a result of such charged particles, electric fields are generated that extend to the surface of the body and lead to the small potential differences.

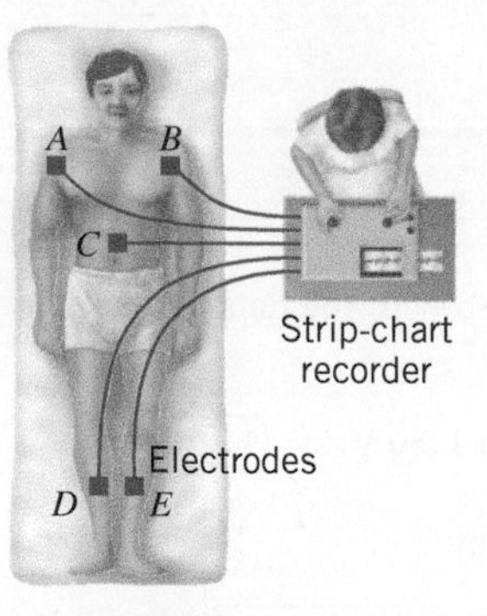

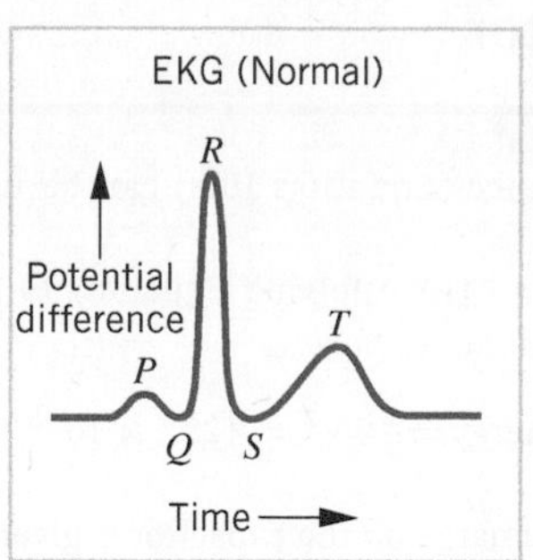

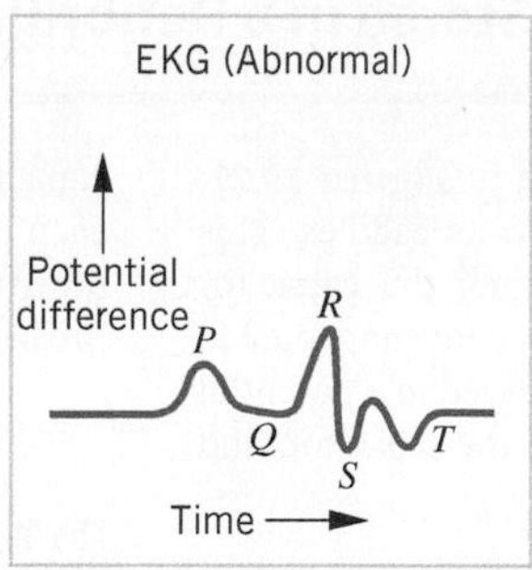

FIGURE 19.26 The potential differences generated by heart muscle activity provide the basis for electrocardiography. The normal and abnormal EKG patterns correspond to one heartbeat.

BIO THE PHYSICS OF . . . electrocardiography. Figure 19.26 shows electrodes placed on the body to measure potential differences in electrocardiography. The potential difference between two locations changes as the heart beats and forms a repetitive pattern. The recorded pattern of potential difference versus time is called an electrocardiogram (ECG or EKG), and its shape depends on which pair of points in the picture (*A* and *B*, *B* and *C*, etc.) is used to locate the electrodes. The figure also shows some EKGs and indicates the regions (*P*, *Q*, *R*, *S*, and *T*) associated with specific parts of the heart's beating cycle. The differences between the EKGs of normal and abnormal hearts provide physicians with a valuable diagnostic tool.

BIO THE PHYSICS OF . . . electroencephalography. In electroencephalography the electrodes are placed at specific locations on the head, as Figure 19.27 indicates, and they record the potential differences that characterize brain behavior. The graph of potential difference versus time is known as an electroencephalogram (EEG). The various parts of the patterns in an EEG are often referred to as "waves" or "rhythms." The drawing shows an example of the main resting rhythm of the brain, the so-called alpha rhythm, and also illustrates the distinct differences that are found between the EEGs generated by healthy (normal) and diseased (abnormal) tissue.

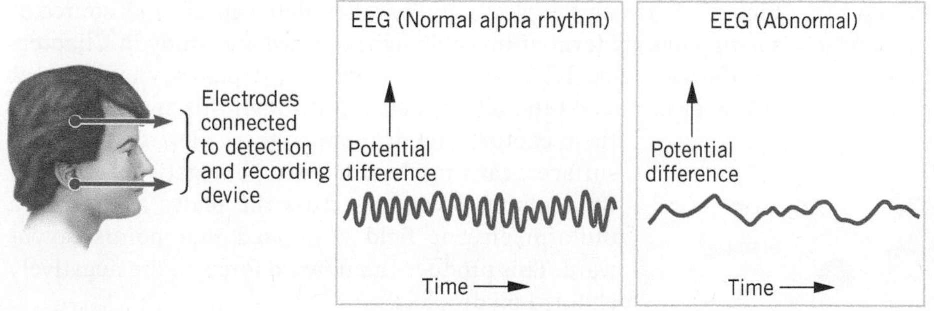

FIGURE 19.27 In electroencephalography the potential differences created by the electrical activity of the brain are used for diagnosing abnormal behavior.

BIO THE PHYSICS OF . . . electroretinography. The electrical characteristics of the retina of the eye lead to the potential differences measured in electroretinography. Figure 19.28 shows a typical electrode placement used to record the pattern of potential difference versus time that occurs when the eye is stimulated by a flash of light. One electrode is mounted on a contact lens, while the other is often placed on the forehead. The recorded pattern is called an electroretinogram (ERG), and parts of the pattern are referred to as the "*A* wave" and the "*B* wave." As the graphs show, the ERGs of normal and diseased (abnormal) eyes can differ markedly.

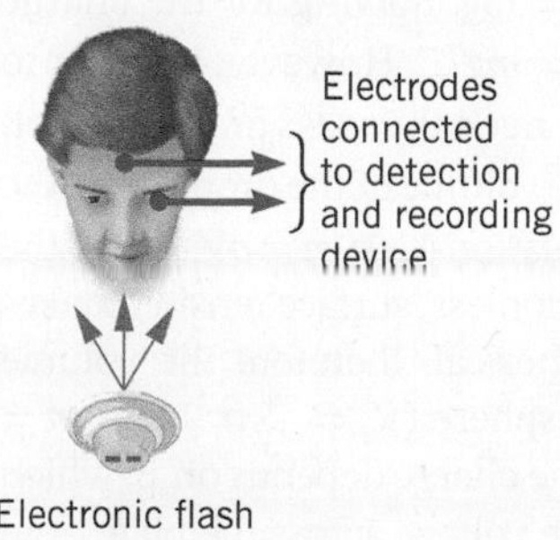

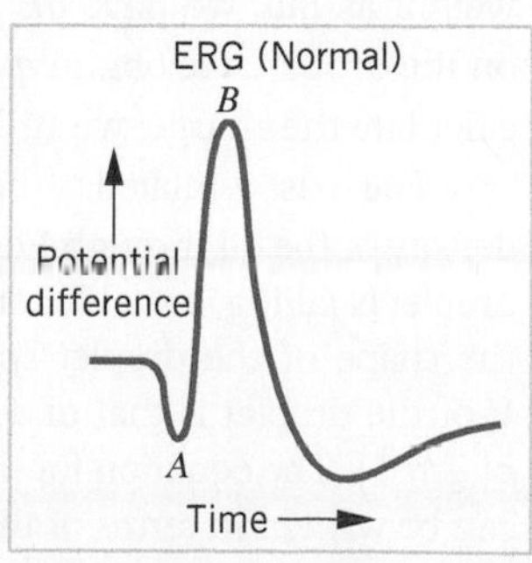

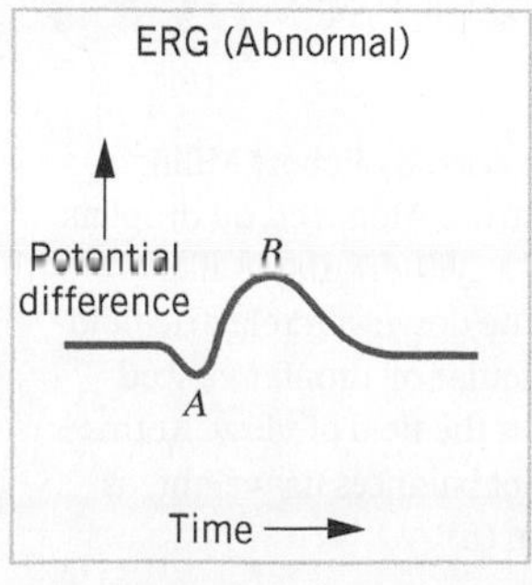

FIGURE 19.28 The electrical activity of the retina of the eye generates the potential differences used in electroretinography.

EXAMPLE 12 | BIO Energy Stored in a Defibrillator

A cardiac defibrillator, such as the one shown in **Figure 19.21**, stores energy in a large capacitor connected to its paddles. This energy is delivered to the heart by discharging the capacitor through the chest wall of the patient. If the capacitor connected to the paddles has a value of 22.5 μF, and it is charged to a potential of 1250 V, what is **(a)** the energy delivered by the capacitor, and **(b)** the charge on its plates?

Reasoning **(a)** To find the energy stored in the capacitor, we can simply apply Equation 19.11b. **(b)** The definition of capacitance (Equation 19.8) can be used to find the charge on each plate.

Solution **(a)** Applying Equation 19.11b with the data given in the problem, we have

$$\text{Energy} = \tfrac{1}{2}CV^2 = \tfrac{1}{2}(22.5 \times 10^{-6}\ \text{F})(1250\ \text{V})^2 = \boxed{17.6\ \text{J}}$$

(b) The charge on the capacitor is given by

$$Q = CV = (22.5 \times 10^{-6}\ \text{F})(1250\ \text{V}) = \boxed{0.0280\ \text{C}}$$

19.7 *The Millikan Oil-drop Experiment

In Section 18.1 we considered objects with a charge greater in magnitude than the charge on a single electron. Charge builds up on an object by adding or removing electrons, and therefore the total charge is quantized, that is, it consists of an integer multiple of the fundamental electron charge, or -1.602×10^{-19} C. However, we never discussed how this value of the fundamental charge had been determined. An American physicist, Robert A. Millikan (1868–1953), is credited with the first accurate experimental determination of the charge on a single electron.

In 1909, Millikan and his graduate student, Harvey Fletcher, analyzed a series of oil-drop experiments to measure the electron's charge. **Figure 19.29*a*** shows a schematic of the equipment they used for their measurements. The apparatus consists of a cylindrical container in which an atomizer produces a fine mist of oil droplets, some of which fall between the plates of a parallel plate capacitor. A source of X-rays, which is a high energy form of invisible light that we will study in Chapters 24 and 30, ionizes the air molecules in the space between the plates, which causes the molecules to lose electrons. As the oil droplets fall through this space, some of them capture one or more of these electrons on their surfaces, causing them to become negatively charged. By applying a voltage across the plates as shown, a uniform electric field is created that points downward. This produces an upward force on the negatively charged oil droplets.

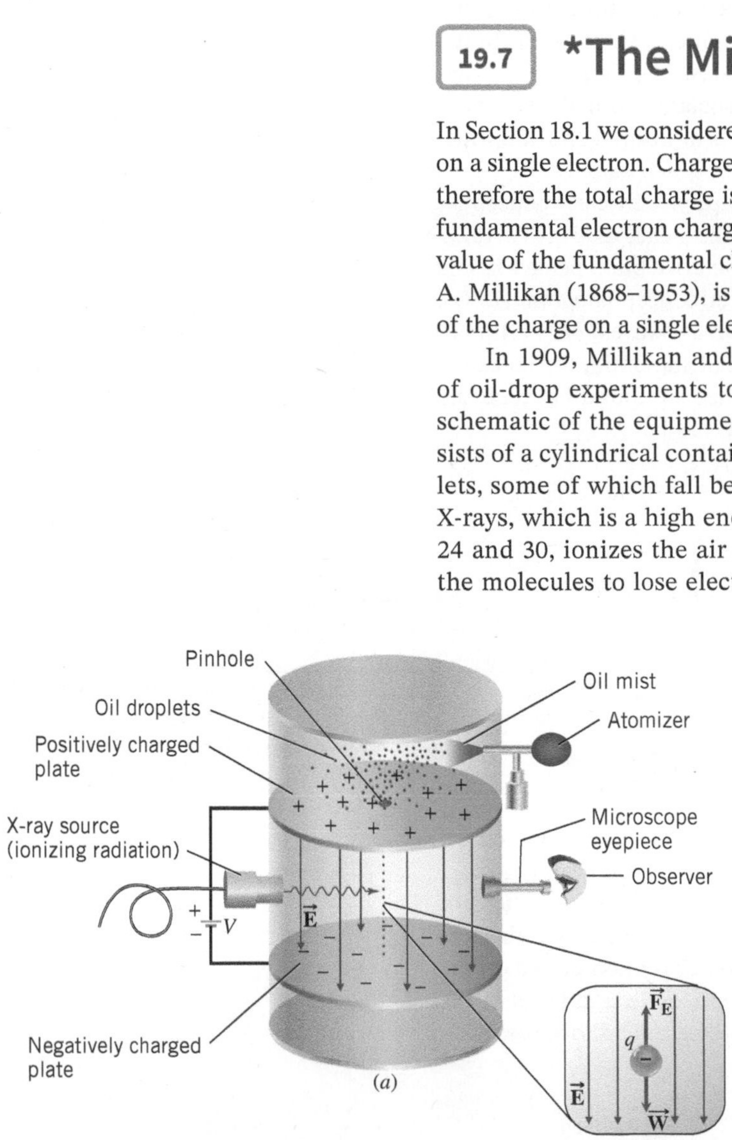

Figure 19.29*b* shows the free-body diagram for one of the oil droplets. The force due to the electric field is upward, and the droplet's weight acts downward. The voltage (V) between the plates, and therefore the electric field, can be varied, such that the upward electrostatic force can be made to balance the weight. At this point, the droplet will appear motionless in the eyepiece. Setting the electrostatic force equal to the weight, we have $F_E = W$. Since $F_E = qE$ (Equation 18.2), and writing the weight as mg, we have $qE = mg$. Solving for the charge on the droplet, we obtain $q = mg/E$. However, in order to calculate the charge, we still need the mass of the droplet.

The mass is related to the volume of the droplet and the density of the oil through Equation 11.1: $m = \rho V$. While the droplet is falling, or held motionless, surface tension makes the shape of the droplet spherical, therefore the volume V of the droplet is that of a sphere ($V_s = \frac{4}{3}\pi r^3$) and $m = \rho\left(\frac{4}{3}\pi r^3\right)$. The equation for the charge depends on E, which can be written in terms of the voltage across the plates and

FIGURE 19.29 (*a*) The experimental apparatus used by Robert Millikan to determine the fundamental charge on the electron. Atomized oil droplets fall through the ionized air between the plates of a cylindrical, parallel plate capacitor, collecting electrons as they fall. The downward electric field between the plates can be adjusted so that a particular oil droplet viewed through the microscope can be made stationary in the field of view. At this point, the upward electrostatic force on the droplet balances its weight, as shown by the free-body diagram for the droplet in (*b*).

the distance (d) between them: $E = \frac{V}{d}$. Substituting these results for m and E into the equation for the charge gives the following*:

$$q = \frac{mg}{E} = \frac{\rho\left(\frac{4}{3}\pi r^3\right)gd}{V} = \frac{4\pi\rho g d r^3}{3V} \tag{19.13}$$

All of the quantities in Equation 19.13 are known or easily measured, except for the radius of the drop. To determine the radius, a second measurement is made. Here, the voltage across the plates is set to zero, so that the electric field, and thus F_E, vanishes. The droplet then falls due to its weight, but quickly reaches terminal velocity due to the upward drag force from the air in the chamber (review Problem 91 in Chapter 4.). The terminal velocity of the droplet can be measured, and this depends on the cross-sectional area of the drop, and thus its radius. Once the radius is determined, the charge on the droplet can be calculated using Equation 19.13.

After measuring many thousands of drops, Millikan and Fletcher determined, to within their experimental uncertainty, that the charge on any drop consisted of a small integer multiple of the same factor: $e = -1.5924 \times 10^{-19}$ C. For example, each drop had a charge of $-e$ (1 electron), $-3e$ (3 electrons), or $-8e$ (8 electrons), etc., but a droplet never had a non-integer multiple of a charge, such as $-1.5e$. Millikan's experiment showed that electric charge is indeed quantized, and his careful measurement produced a value within 1% of the present day accepted value of -1.602×10^{-19} C. Millikan was awarded the Nobel Prize in Physics in 1923 for determining the fundamental charge on the electron, as well as for his work on the photoelectric effect (see Section 29.3).

Concept Summary

19.1 Potential Energy When a positive test charge $+q_0$ moves from point A to point B in the presence of an electric field, work W_{AB} is done by the electric force. The work equals the electric potential energy (EPE) at A minus that at B, as given by Equation 19.1. The electric force is a conservative force, so the path along which the test charge moves from A to B is of no consequence, for the work W_{AB} is the same for all paths.

$$W_{AB} = \text{EPE}_A - \text{EPE}_B \tag{19.1}$$

19.2 The Electric Potential Difference The electric potential V at a given point is the electric potential energy of a small test charge q_0 situated at that point divided by the charge itself, as shown in Equation 19.3. The SI unit of electric potential is the joule per coulomb (J/C), or volt (V). The electric potential difference between two points A and B is given by Equation 19.4. The potential difference between two points (or between two equipotential surfaces) is often called the "voltage."

$$V = \frac{\text{EPE}}{q_0} \tag{19.3}$$

$$V_B - V_A = \frac{\text{EPE}_B}{q_0} - \frac{\text{EPE}_A}{q_0} = \frac{-W_{AB}}{q_0} \tag{19.4}$$

A positive charge accelerates from a region of higher potential toward a region of lower potential. Conversely, a negative charge accelerates from a region of lower potential toward a region of higher potential.

An electron volt (eV) is a unit of energy. The relationship between electron volts and joules is 1 eV = 1.60×10^{-19} J.

The total energy E of a system is the sum of its translational $\left(\frac{1}{2}mv^2\right)$ and rotational $\left(\frac{1}{2}I\omega^2\right)$ kinetic energies, gravitational potential energy (mgh), elastic potential energy $\left(\frac{1}{2}kx^2\right)$, and electric potential energy (EPE), as indicated by Equation 1. If external nonconservative forces like friction do no net work, the total energy of the system is conserved. That is, the final total energy E_f is equal to the initial total energy E_0; $E_f = E_0$.

$$E = \tfrac{1}{2}mv^2 + \tfrac{1}{2}I\omega^2 + mgh + \tfrac{1}{2}kx^2 + \text{EPE} \tag{1}$$

19.3 The Electric Potential Difference Created by Point Charges The electric potential V at a distance r from a point charge q is given by Equation 19.6, where $k = 8.99 \times 10^9$ N · m²/C². This expression for V assumes that the electric potential is zero at an infinite distance away from the charge. The total electric potential at a given location due to two or more charges is the algebraic sum of the potentials due to each charge.

$$V = \frac{kq}{r} \tag{19.6}$$

The total potential energy of a group of charges is the amount by which the electric potential energy of the group differs from its initial value when the charges are infinitely far apart and far away. It is also equal to the work required to assemble the group, one charge at a time, starting with the charges infinitely far apart and far away.

19.4 Equipotential Surfaces and Their Relation to the Electric Field An equipotential surface is a surface on which the electric potential is the same everywhere. The electric force does no work as a charge moves on an equipotential surface, because the force is always perpendicular to the displacement of the charge.

The electric field created by any group of charges is everywhere perpendicular to the associated equipotential surfaces and points in

*The weight of the droplet that has to be balanced by the electrostatic force is less than the actual weight due to an upward buoyant force on the droplet. The buoyant force, according to Archimedes' principle, will be equal to the weight of the air displaced by the droplet. Thus, the density of the oil (ρ) in Equation 19.13 should be replaced by ($\rho - \rho_{air}$), where ρ_{air} is the density of the air. This is a small correction, as the density of the oil is typically several hundred times larger than the density of air.

the direction of decreasing potential. The electric field is related to two equipotential surfaces by Equation 19.7a, where ΔV is the potential difference between the surfaces and Δs is the displacement. The term $\Delta V/\Delta s$ is called the potential gradient.

$$E = -\frac{\Delta V}{\Delta s} \quad (19.7a)$$

19.5 Capacitors and Dielectrics A capacitor is a device that stores charge and energy. It consists of two conductors or plates that are near one another, but not touching. The magnitude q of the charge on each plate is given by Equation 19.8, where V is the magnitude of the potential difference between the plates and C is the capacitance. The SI unit for capacitance is the coulomb per volt (C/V), or farad (F).

$$q = CV \quad (19.8)$$

The insulating material included between the plates of a capacitor is called a dielectric. The dielectric constant κ of the material is defined as shown in Equation 19.9, where E_0 and E are, respectively, the magnitudes of the electric fields between the plates without and with a dielectric, assuming the charge on the plates is kept fixed.

$$\kappa = \frac{E_0}{E} \quad (19.9)$$

The capacitance of a parallel plate capacitor filled with a dielectric is given by Equation 19.10, where $\varepsilon_0 = 8.85 \times 10^{-12}\ C^2/(N \cdot m^2)$ is the permittivity of free space, A is the area of each plate, and d is the distance between the plates.

$$C = \frac{\kappa \varepsilon_0 A}{d} \quad (19.10)$$

The electric potential energy stored in a capacitor is given by Equations 19.11a–c. The energy density is the energy stored per unit volume and is related to the magnitude E of the electric field, as indicated in Equation 19.12.

$$\text{Energy} = \tfrac{1}{2}qV = \tfrac{1}{2}CV^2 = q^2/(2C) \quad (19.11a\text{–}c)$$

$$\text{Energy density} = \tfrac{1}{2}\kappa\varepsilon_0 E^2 \quad (19.12)$$

Focus on Concepts

Online

Additional questions are available for assignment in WileyPLUS.

Section 19.2 The Electric Potential Difference

1. Two different charges, q_1 and q_2, are placed at two different locations, one charge at each location. The locations have the same electric potential V. Do the charges have the same electric potential energy? **(a)** Yes. If the electric potentials at the two locations are the same, the electric potential energies are also the same, regardless of the type (+ or −) and magnitude of the charges placed at these locations. **(b)** Yes, because electric potential and electric potential energy are just different names for the same concept. **(c)** No, because the electric potential V at a given location depends on the charge placed at that location, whereas the electric potential energy EPE does not. **(d)** No, because the electric potential energy EPE at a given location depends on the charge placed at that location as well as the electric potential V.

2. A proton is released from rest at point A in a constant electric field and accelerates to point B (see part a of the drawing). An electron is released from rest at point B and accelerates to point A (see part b of the drawing). How does the change in the proton's electric potential energy compare with the change in the electron's electric potential energy? **(a)** The change in the proton's electric potential energy is the same as the change in the electron's electric potential energy. **(b)** The proton experiences a greater change in electric potential energy, since it has a greater charge magnitude. **(c)** The proton experiences a smaller change in electric potential energy, since it has a smaller charge magnitude. **(d)** The proton experiences a smaller change in electric potential energy, since it has a smaller speed at B than the electron has at A. This is due to the larger mass of the proton. **(e)** One cannot compare the change in electric potential energies because the proton and electron move in opposite directions.

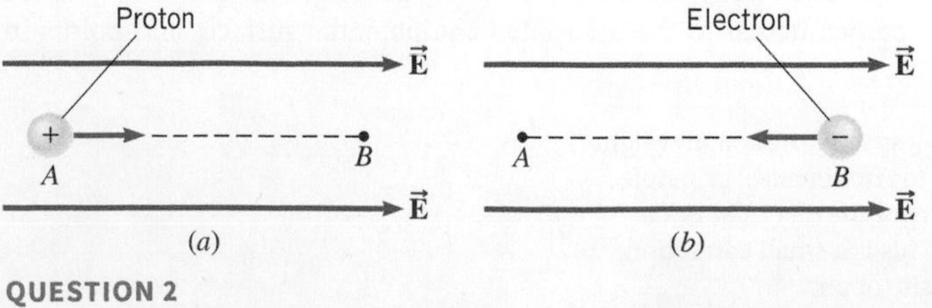

QUESTION 2

Section 19.3 The Electric Potential Difference Created by Point Charges

3. The drawing shows three arrangements of charged particles, all the same distance from the origin. Rank the arrangements, largest to smallest, according to the total electric potential V at the origin.

(a) A, B, C
(b) B, A, C
(c) B, C, A
(d) A, B and C (a tie)
(e) A and C (a tie), B

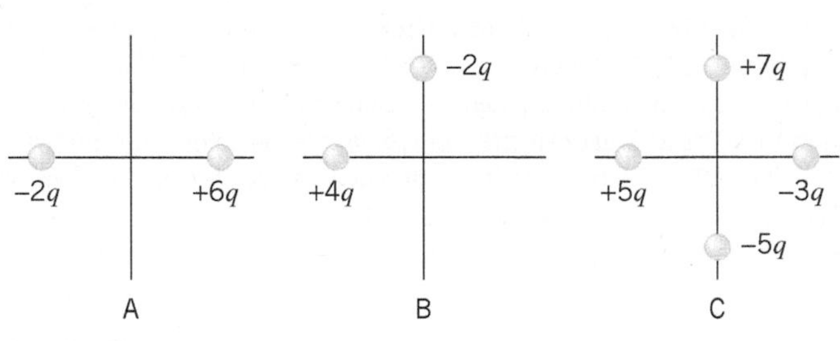

QUESTION 3

4. Four pairs of charged particles with identical separations are shown in the drawing. Rank the pairs according to their electric potential energy EPE, greatest (most positive) first.

(a) A and C (a tie), B and D (a tie)
(b) A, B, C, D
(c) C, B, D, A
(d) B, A, C and D (a tie)
(e) A and B (a tie), C, D

+4q +3q A −2q +6q B −12q −q C −4q +3q D

QUESTION 4

Section 19.4 Equipotential Surfaces and Their Relation to the Electric Field

5. The drawing shows edge-on views of three parallel plate capacitors with the same separation between the plates. The potential of each plate is indicated above it. Rank the capacitors as to the *magnitude* of the electric field inside them, largest to smallest.

(a) A, B, C (b) A, C, B
(c) C, B, A (d) C, A, B
(e) B, C, A

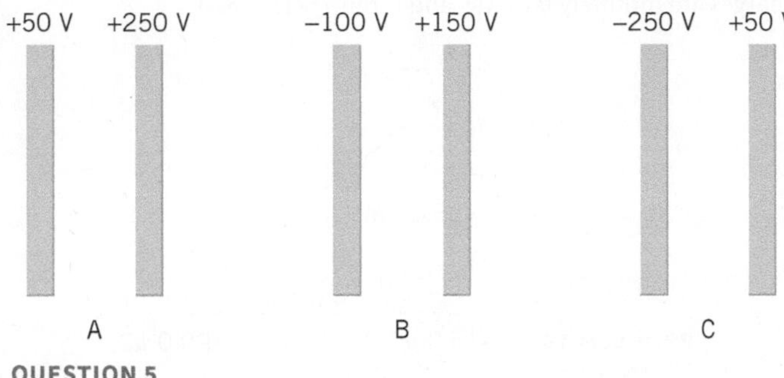

QUESTION 5

6. The drawing shows a plot of the electric potential V versus the displacement s. The plot consists of four segments. Rank the *magnitude* of the electric fields for the four segments, largest to smallest.

(a) D, C, B, A
(b) A and C (a tie), B and D (a tie)
(c) A, B, D, C
(d) B, D, C, A
(e) D, B, A and C (a tie)

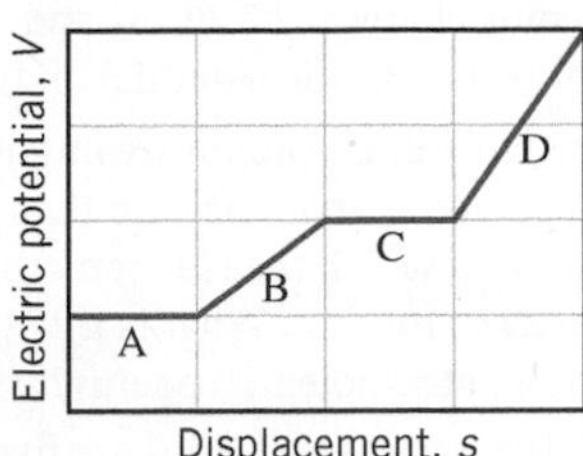

QUESTION 6

Section 19.5 Capacitors and Dielectrics

7. Which two or more of the following actions would increase the energy stored in a parallel plate capacitor when a constant potential difference is applied across the plates?

1. Increasing the area of the plates
2. Decreasing the area of the plates
3. Increasing the separation between the plates
4. Decreasing the separation between the plates
5. Inserting a dielectric between the plates

(a) 2, 4 (b) 2, 3, 5
(c) 1, 4, 5 (d) 1, 3

Problems

Online

Additional questions are available for assignment in WileyPLUS.

SSM Student Solutions Manual
MMH Problem-solving help
GO Guided Online Tutorial
V-HINT Video Hints
CHALK Chalkboard Videos
BIO Biomedical application
E Easy
M Medium
H Hard
WS Worksheet
T Team Problem

Section 19.1 Potential Energy

Section 19.2 The Electric Potential Difference

1. E During a particular thunderstorm, the electric potential difference between a cloud and the ground is $V_{\text{cloud}} - V_{\text{ground}} = 1.3 \times 10^8$ V, with the cloud being at the higher potential. What is the change in an electron's electric potential energy when the electron moves from the ground to the cloud?

2. E A particle with a charge of $-1.5\ \mu$C and a mass of 2.5×10^{-6} kg is released from rest at point A and accelerates toward point B, arriving there with a speed of 42 m/s. The only force acting on the particle is the electric force. (a) Which point is at the higher potential? Give your reasoning. (b) What is the potential difference $V_B - V_A$ between A and B?

3. E GO CHALK A particle has a charge of $+1.5\ \mu$C and moves from point A to point B, a distance of 0.20 m. The particle experiences a constant electric force, and its motion is along the line of action of the force. The difference between the particle's electric potential energy at A and at B is $\text{EPE}_A - \text{EPE}_B = +9.0 \times 10^{-4}$ J. (a) Find the magnitude and direction of the electric force that acts on the particle. (b) Find the magnitude and direction of the electric field that the particle experiences.

4. E GO Review Multiple-Concept Example 4 to see the concepts that are pertinent here. In a television picture tube, electrons strike the screen after being accelerated from rest through a potential difference of 25 000 V. The speeds of the electrons are quite large, and for accurate calculations of the speeds, the effects of special relativity must be taken into account. Ignoring such effects, find the electron speed just before the electron strikes the screen.

5. E SSM Multiple-Concept Example 4 deals with the concepts that are important in this problem. As illustrated in **Figure 19.5*b***, a negatively charged particle is released from rest at point B and accelerates until it reaches point A. The mass and charge of the particle are 4.0×10^{-6} kg and -2.0×10^{-5} C, respectively. Only the gravitational force and the electrostatic force act on the particle, which moves on a horizontal straight line without rotating. The electric potential at A is 36 V greater than that at B; in other words, $V_A - V_B = 36$ V. What is the translational speed of the particle at point A?

6. E GO An electron and a proton, starting from rest, are accelerated through an electric potential difference of the same magnitude. In the process, the electron acquires a speed v_e, while the proton acquires a speed v_p. Find the ratio v_e/v_p.

7. M V-HINT CHALK A moving particle encounters an external electric field that decreases its kinetic energy from 9520 eV to 7060 eV as the particle moves from position A to position B. The electric potential at A is −55.0 V, and the electric potential at B is +27.0 V. Determine the charge of the particle. Include the algebraic sign (+ or −) with your answer.

8. M GO During a lightning flash, there exists a potential difference of $V_{\text{cloud}} - V_{\text{ground}} = 1.2 \times 10^9$ V between a cloud and the ground. As a result, a charge of −25 C is transferred from the ground to the cloud. (a) How much work $W_{\text{ground-cloud}}$ is done on the charge by the electric force? (b) If the work done by the electric force were used to accelerate a 1100-kg automobile from rest, what would be its final

speed? (c) If the work done by the electric force were converted into heat, how many kilograms of water at 0 °C could be heated to 100 °C?

Section 19.3 The Electric Potential Difference Created by Point Charges

9. **E** Two point charges, $+3.40\ \mu C$ and $-6.10\ \mu C$, are separated by 1.20 m. What is the electric potential midway between them?

10. **E** An electron and a proton are initially very far apart (effectively an infinite distance apart). They are then brought together to form a hydrogen atom, in which the electron orbits the proton at an average distance of 5.29×10^{-11} m. What is $EPE_{final} - EPE_{initial}$, which is the change in the electric potential energy?

11. **E SSM** Two charges A and B are fixed in place, at different distances from a certain spot. At this spot the potentials due to the two charges are equal. Charge A is 0.18 m from the spot, while charge B is 0.43 m from it. Find the ratio q_B/q_A of the charges.

12. **E GO** The drawing shows a square, each side of which has a length of $L = 0.25$ m. On two corners of the square are fixed different positive charges, q_1 and q_2. Find the electric potential energy of a third charge $q_3 = -6.0 \times 10^{-9}$ C placed at corner A and then at corner B.

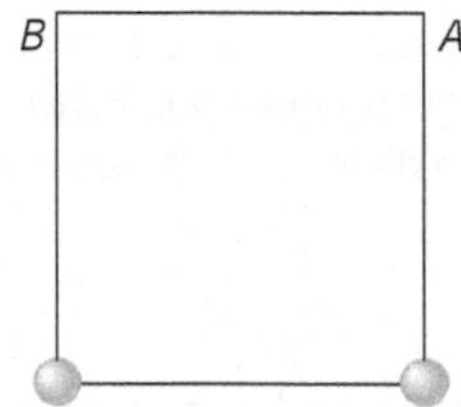

PROBLEM 12 $q_1 = +1.5 \times 10^{-9}$ C $q_2 = +4.0 \times 10^{-9}$ C

13. **E SSM** The drawing shows four point charges. The value of q is 2.0 μC, and the distance d is 0.96 m. Find the total potential at the location P. Assume that the potential of a point charge is zero at infinity.

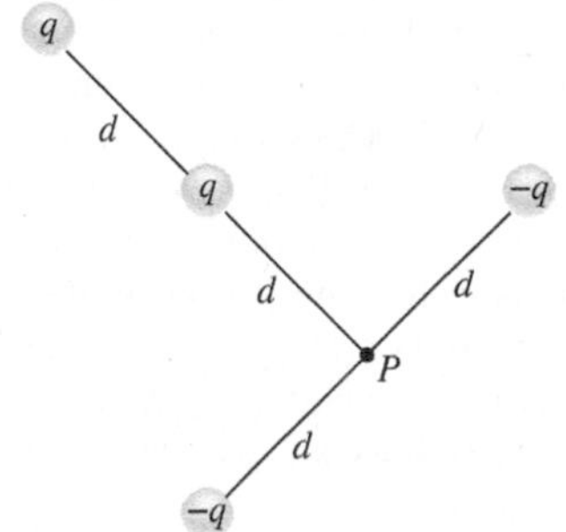

PROBLEM 13

14. **E** A charge of $+125\ \mu C$ is fixed at the center of a square that is 0.64 m on a side. How much work is done by the electric force as a charge of $+7.0\ \mu C$ is moved from one corner of the square to any other empty corner? Explain.

15. **E CHALK** The drawing shows six point charges arranged in a rectangle. The value of q is 9.0 μC, and the distance d is 0.13 m. Find the total electric potential at location P, which is at the center of the rectangle.

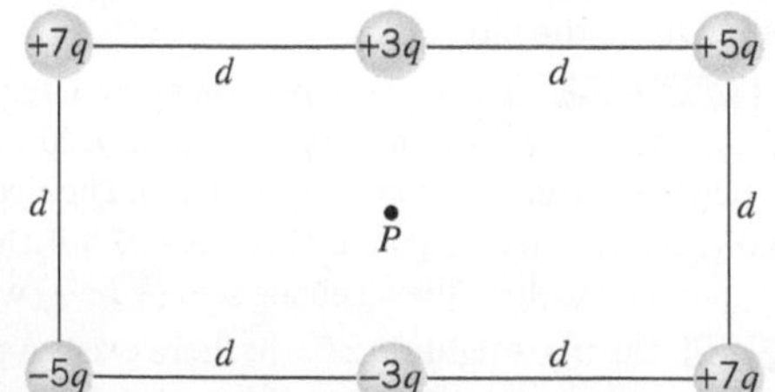

PROBLEM 15

16. **E** Location A is 3.00 m to the right of a point charge q. Location B lies on the same line and is 4.00 m to the right of the charge. The potential difference between the two locations is $V_B - V_A = 45.0$ V. What are the magnitude and sign of the charge?

17. **E** Identical $+1.8\ \mu C$ charges are fixed to adjacent corners of a square. What charge (magnitude and algebraic sign) should be fixed to one of the empty corners, so that the total electric potential at the remaining empty corner is 0 V?

18. **E GO** Charges of $-q$ and $+2q$ are fixed in place, with a distance of 2.00 m between them. A dashed line is drawn through the negative charge, perpendicular to the line between the charges. On the dashed line, at a distance L from the negative charge, there is at least one spot where the total potential is zero. Find L.

19. **M CHALK SSM** Determine the electric potential energy for the array of three charges in the drawing, relative to its value when the charges are infinitely far away and infinitely far apart.

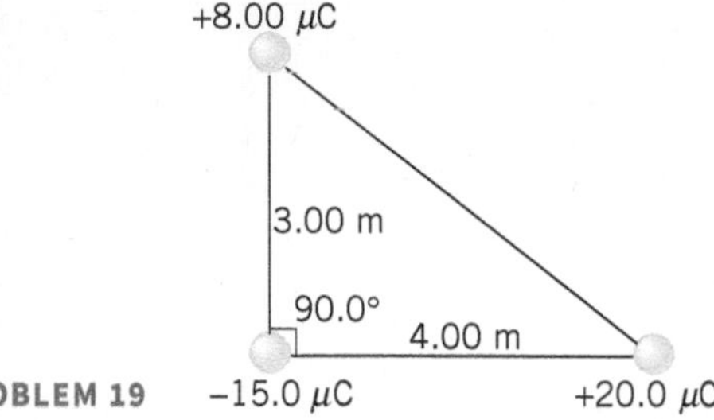

PROBLEM 19

20. **M V-HINT** Two identical point charges ($q = +7.20 \times 10^{-6}$ C) are fixed at diagonally opposite corners of a square with sides of length 0.480 m. A test charge ($q_0 = -2.40 \times 10^{-8}$ C), with a mass of 6.60×10^{-8} kg, is released from rest at one of the empty corners of the square. Determine the speed of the test charge when it reaches the center of the square.

21. **M MMH** Two protons are moving directly toward one another. When they are very far apart, their initial speeds are 3.00×10^6 m/s. What is the distance of closest approach?

22. **M V-HINT** Four identical charges ($+2.0\ \mu C$ each) are brought from infinity and fixed to a straight line. The charges are located 0.40 m apart. Determine the electric potential energy of this group.

23. **M SSM** A charge of $-3.00\ \mu C$ is fixed in place. From a horizontal distance of 0.0450 m, a particle of mass 7.20×10^{-3} kg and charge $-8.00\ \mu C$ is fired with an initial speed of 65.0 m/s directly toward the fixed charge. How far does the particle travel before its speed is zero?

24. **M GO** Identical point charges of $+1.7\ \mu C$ are fixed to diagonally opposite corners of a square. A third charge is then fixed at the center of the square, such that it causes the potentials at the empty corners to change signs without changing magnitudes. Find the sign and magnitude of the third charge.

25. **H** One particle has a mass of 3.00×10^{-3} kg and a charge of $+8.00\ \mu C$. A second particle has a mass of 6.00×10^{-3} kg and the same charge. The two particles are initially held in place and then released. The particles fly apart, and when the separation between them is 0.100 m, the speed of the 3.00×10^{-3} kg particle is 125 m/s. Find the initial separation between the particles.

Section 19.4 Equipotential Surfaces and Their Relation to the Electric Field

26. **E** Two equipotential surfaces surround a $+1.50 \times 10^{-8}$ C point charge. How far is the 190-V surface from the 75.0-V surface?

27. **E** An equipotential surface that surrounds a point charge q has a potential of 490 V and an area of 1.1 m². Determine q.

28. **E GO** A positive point charge ($q = +7.2 \times 10^{-8}$ C) is surrounded by an equipotential surface A, which has a radius of $r_A = 1.8$ m. A positive test charge ($q_0 = +4.5 \times 10^{-11}$ C) moves from surface A to another equipotential surface B, which has a radius r_B. The work done as the test charge moves from surface A to surface B is $W_{AB} = -8.1 \times 10^{-9}$ J. Find r_B.

29. E SSM A spark plug in an automobile engine consists of two metal conductors that are separated by a distance of 0.75 mm. When an electric spark jumps between them, the magnitude of the electric field is 4.7×10^7 V/m. What is the magnitude of the potential difference ΔV between the conductors?

30. E The drawing that accompanies Problem 47 shows a graph of a set of equipotential surfaces in cross section. The grid lines are 2.0 cm apart. Determine the magnitude and direction of the electric field at position *D*. Specify whether the electric field points toward the top or the bottom of the drawing.

31. M SSM MMH An electric field has a constant value of 4.0×10^3 V/m and is directed downward. The field is the same everywhere. The potential at a point *P* within this region is 155 V. Find the potential at the following points: (a) 6.0×10^{-3} m directly above *P*, (b) 3.0×10^{-3} m directly below *P*, (c) 8.0×10^{-3} m directly to the right of *P*.

32. M CHALK GO An electron is released from rest at the negative plate of a parallel plate capacitor and accelerates to the positive plate (see the drawing). The plates are separated by a distance of 1.2 cm, and the electric field within the capacitor has a magnitude of 2.1×10^6 V/m. What is the kinetic energy of the electron just as it reaches the positive plate?

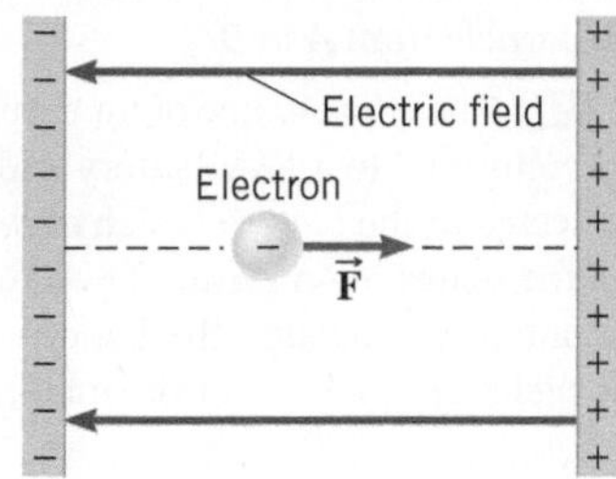

PROBLEM 32

33. M V-HINT The drawing shows the electric potential as a function of distance along the *x* axis. Determine the magnitude of the electric field in the region (a) *A* to *B*, (b) *B* to *C*, and (c) *C* to *D*.

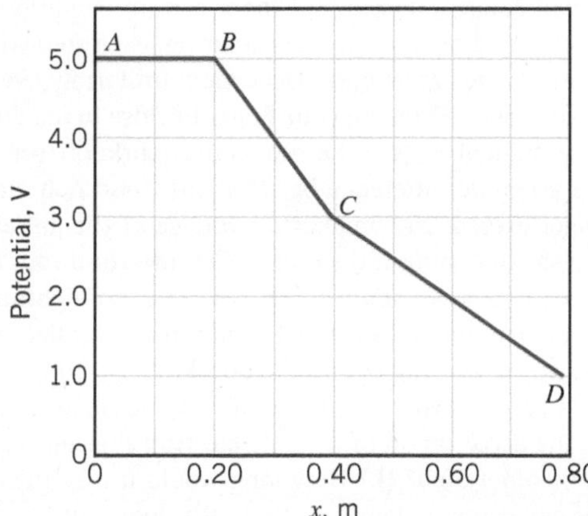

PROBLEM 33

34. M GO At a distance of 1.60 m from a point charge of $+2.00\ \mu$C, there is an equipotential surface. At greater distances there are additional equipotential surfaces. The potential difference between any two successive surfaces is 1.00×10^3 V. Starting at a distance of 1.60 m and moving radially outward, how many of the additional equipotential surfaces are crossed by the time the electric field has shrunk to one-half of its initial value? Do not include the starting surface.

35. M V-HINT The drawing shows a uniform electric field that points in the negative *y* direction; the magnitude of the field is 3600 N/C. Determine the electric potential difference (a) $V_B - V_A$ between points *A* and *B*, (b) $V_C - V_B$ between points *B* and *C*, and (c) $V_A - V_C$ between points *C* and *A*.

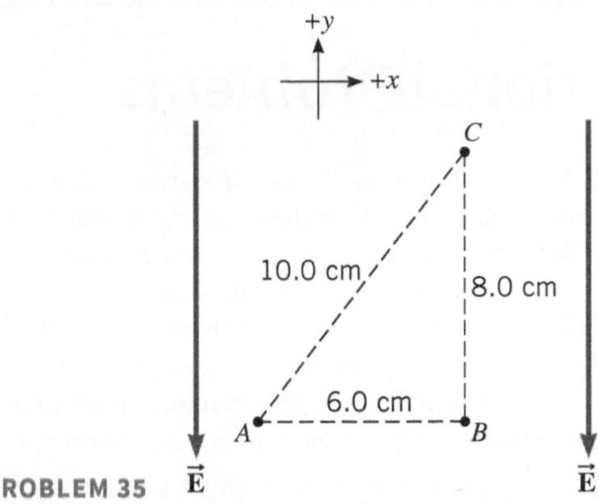

PROBLEM 35

Section 19.5 Capacitors and Dielectrics

36. E What is the capacitance of a capacitor that stores 4.3 μC of charge on its plates when a voltage of 1.5 V is applied between them?

37. E Two identical capacitors store different amounts of energy: capacitor A stores 3.1×10^{-3} J, and capacitor B stores 3.4×10^{-4} J. The voltage across the plates of capacitor B is 12 V. Find the voltage across the plates of capacitor A.

38. E MMH The electronic flash attachment for a camera contains a capacitor for storing the energy used to produce the flash. In one such unit, the potential difference between the plates of an 850-μF capacitor is 280 V. (a) Determine the energy that is used to produce the flash in this unit. (b) Assuming that the flash lasts for 3.9×10^{-3} s, find the effective power or "wattage" of the flash.

39. E GO The same voltage is applied between the plates of two different capacitors. When used with capacitor A, this voltage causes the capacitor to store 11 μC of charge and 5.0×10^{-5} J of energy. When used with capacitor B, which has a capacitance of 6.7 μF, this voltage causes the capacitor to store a charge that has a magnitude of q_B. Determine q_B.

40. E SSM A parallel plate capacitor has a capacitance of 7.0 μF when filled with a dielectric. The area of each plate is 1.5 m^2 and the separation between the plates is 1.0×10^{-5} m. What is the dielectric constant of the dielectric?

41. E GO Two capacitors are identical, except that one is empty and the other is filled with a dielectric ($\kappa = 4.50$). The empty capacitor is connected to a 12.0-V battery. What must be the potential difference across the plates of the capacitor filled with a dielectric so that it stores the same amount of electrical energy as the empty capacitor?

42. M GO CHALK Capacitor A and capacitor B both have the same voltage across their plates. However, the energy of capacitor A can melt *m* kilograms of ice at 0 °C, while the energy of capacitor B can boil away the same amount of water at 100 °C. The capacitance of capacitor A is 9.3 μF. What is the capacitance of capacitor B?

43. M SSM MMH What is the potential difference between the plates of a 3.3-F capacitor that stores sufficient energy to operate a 75-W light bulb for one minute?

44. M Review Conceptual Example 10 before attempting this problem. An empty capacitor is connected to a 12.0-V battery and charged up. The capacitor is then disconnected from the battery, and a slab of dielectric material ($\kappa = 2.8$) is inserted between the plates. Find the amount by which the potential difference across the plates changes. Specify whether the change is an increase or a decrease.

45. M V-HINT An empty parallel plate capacitor is connected between the terminals of a 9.0-V battery and charged up. The capacitor is then disconnected from the battery, and the spacing between the capacitor plates is doubled. As a result of this change, what is the new voltage between the plates of the capacitor?

Additional Problems

Online

46. E GO Refer to Multiple-Concept Example 3 to review the concepts that are needed here. A cordless electric shaver uses energy at a rate of 4.0 W from a rechargeable 1.5-V battery. Each of the charged particles that the battery delivers to the shaver carries a charge that has a magnitude of 1.6×10^{-19} C. A fully charged battery allows the shaver to be used for its maximum operation time, during which 3.0×10^{22} of the charged particles pass between the terminals of the battery as the shaver operates. What is the shaver's maximum operation time?

47. E V-HINT The drawing shows a graph of a set of equipotential surfaces seen in cross section. Each is labeled according to its electric potential. A $+2.8 \times 10^{-7}$ C point charge is placed at position *A*. Find the work that is done on the point charge by the electric force when it is moved (a) from *A* to *B*, and (b) from *A* to *C*.

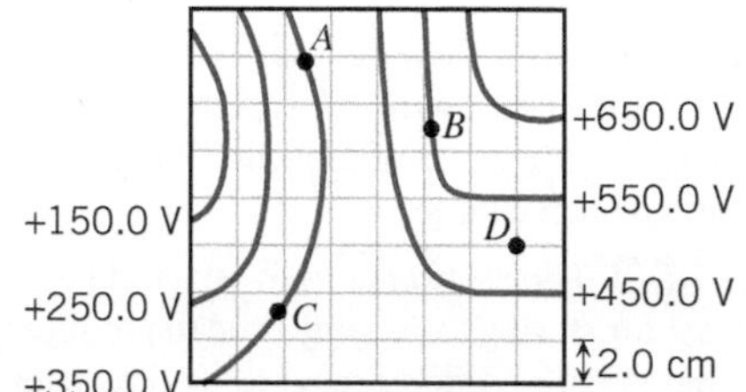

PROBLEM 47

48. E SSM The work done by an electric force in moving a charge from point *A* to point *B* is 2.70×10^{-3} J. The electric potential difference between the two points is $V_A - V_B = 50.0$ V. What is the charge?

49. E GO Two capacitors have the same plate separation, but one has square plates and the other has circular plates. The square plates are a length *L* on each side, and the diameter of the circular plate is *L*. The capacitors have the same capacitance because they contain different dielectric materials. The dielectric constant of the material between the square plates has a value of $\kappa_{\text{square}} = 3.00$. What is the dielectric constant κ_{circle} of the material between the circular plates?

50. E Three point charges, -5.8×10^{-9} C, -9.0×10^{-9} C, and $+7.3 \times 10^{-9}$ C, are fixed at different positions on a circle. The total electric potential at the center of the circle is −2100 V. What is the radius of the circle?

51. M GO Equipotential surface *A* has a potential of 5650 V, while equipotential surface *B* has a potential of 7850 V. A particle has a mass of 5.00×10^{-2} kg and a charge of $+4.00 \times 10^{-5}$ C. The particle has a speed of 2.00 m/s on surface *A*. A nonconservative outside force is applied to the particle, and it moves to surface *B*, arriving there with a speed of 3.00 m/s. How much work is done by the outside force in moving the particle from *A* to *B*?

52. M GO SSM The capacitance of an empty capacitor is 1.2 μF. The capacitor is connected to a 12-V battery and charged up. With the capacitor connected to the battery, a slab of dielectric material is inserted between the plates. As a result, 2.6×10^{-5} C of *additional* charge flows from one plate, through the battery, and onto the other plate. What is the dielectric constant of the material?

Physics in Biology, Medicine, and Sports

53. E BIO SSM **19.2** Suppose that the electric potential outside a living cell is higher than that inside the cell by 0.070 V. How much work is done by the electric force when a sodium ion (charge = $+e$) moves from the outside to the inside?

54. E BIO SSM **19.4** The inner and outer surfaces of a cell membrane carry a negative and a positive charge, respectively. Because of these charges, a potential difference of about 0.070 V exists across the membrane. The thickness of the cell membrane is 8.0×10^{-9} m. What is the magnitude of the electric field in the membrane?

55. E BIO SSM **19.5** The electric potential energy stored in the capacitor of a defibrillator is 73 J, and the capacitance is 120 μF. What is the potential difference that exists across the capacitor plates?

56. E BIO **19.5** The membrane that surrounds a certain type of living cell has a surface area of 5.0×10^{-9} m^2 and a thickness of 1.0×10^{-8} m. Assume that the membrane behaves like a parallel plate capacitor and has a dielectric constant of 5.0. (a) The potential on the outer surface of the membrane is +60.0 mV greater than that on the inside surface. How much charge resides on the outer surface? (b) If the charge in part (a) is due to positive ions (charge $+e$), how many such ions are present on the outer surface?

57. E BIO SSM **19.5** An axon is the relatively long tail-like part of a neuron, or nerve cell. The outer surface of the axon membrane (dielectric constant = 5, thickness = 1×10^{-8} m) is charged positively, and the inner portion is charged negatively. Thus, the membrane is a kind of capacitor. Assuming that the membrane acts like a parallel plate capacitor with a plate area of 5×10^{-6} m^2, what is its capacitance?

58. E BIO **19.4** Probably all of us have experienced small electric shocks after strolling across a carpeted floor and touching something metallic, like a doorknob. Under certain conditions, and depending on the kind of shoes you are wearing, each step you take strips electrons off the carpet and places them uniformly over the surface area of your body. Bringing your finger in close to the doorknob induces a separation of surface charges on the doorknob (see the figure). When the potential difference between the doorknob and the tip of your finger exceeds the breakdown voltage of the air, a spark will occur, transferring some of the excess electrons from your finger to the doorknob. This is essentially a very small form of lightning. Consider that, at a separation distance of 1.0 mm, the potential difference between the tip of your finger and the doorknob is 7500 V. (a) If we treat the field between your finger and the doorknob as uniform, what would be the acceleration of a single electron that moves from your finger to the doorknob? (b) How long would it take the electron to bridge the gap between your finger and the doorknob? Neglect the potential collisions between the electron and air molecules.

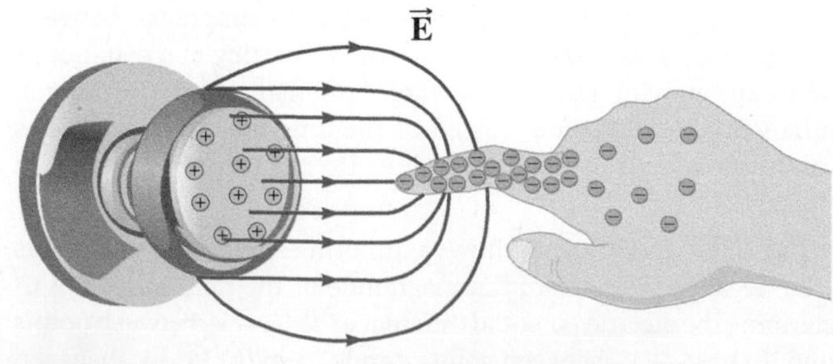

PROBLEM 58

59. E BIO 19.5 Quite often the best course of treatment for breast cancer is a mastectomy, where the breast and surrounding tissue are surgically removed. This often involves the removal of lymph nodes in the breast area, which can lead to a condition called lymphedema. This is a chronic condition that consists of a buildup of lymphatic fluid that can result in pain, swelling, decreased mobility, and recurring infections. One technique that shows promise in early detection of lymphedema is comparative tissue dielectric constant measurements, or TDC. Here, a probe is placed against the skin and transmits a low power electromagnetic wave (see Chapter 24). The wave reflects back from the skin to a sensor in the probe head that determines information about the average dielectric constant of the skin and tissue beneath the probe. This value strongly depends on how much water is located beneath the skin, which is a strong indicator of lymphedema. As part of the probe's calibration, a parallel plate capacitor is used. The capacitor is connected to a 1.0×10^2 V power source, charged, and then disconnected from the source. Different tissue samples with varying water content are then placed between the plates, and then the voltage across the plates is measured. **(a)** When the samples are inserted between the plates, do you expect the voltage across the plates to increase or decrease? **(b)** For one particular tissue sample, the voltage across the plates was measured to be 2.8 V. What is the dielectric constant of this tissue sample?

60. E BIO 19.5 In previous examples, we have discussed how a cardiac defibrillator uses a capacitor to deliver electrical energy to the heart. However, a battery, or other voltage source, could accomplish the same thing. So why does the defibrillator use a capacitor to deliver the charge, instead of a battery? Restarting a heart often requires a large charge delivered over a short burst, and because many defibrillators need to be portable, they get their power from batteries. However, batteries can only produce a continuous low flow of current. A capacitor solves this problem, as it is able to deliver a large charge very quickly. It is true that the battery will have to charge the capacitor, and this can take some time, but this is necessary to achieve the required charge. Also, by varying the charging time, the energy stored in the capacitor, and subsequently delivered to the heart, can be varied. Defibrillator capacitors come in a variety of values and sizes (see the photo). One of the larger capacitors available is 32 μF. If it is charged until the voltage across its plates is 6.0×10^2 V, how much energy is stored in the capacitor?

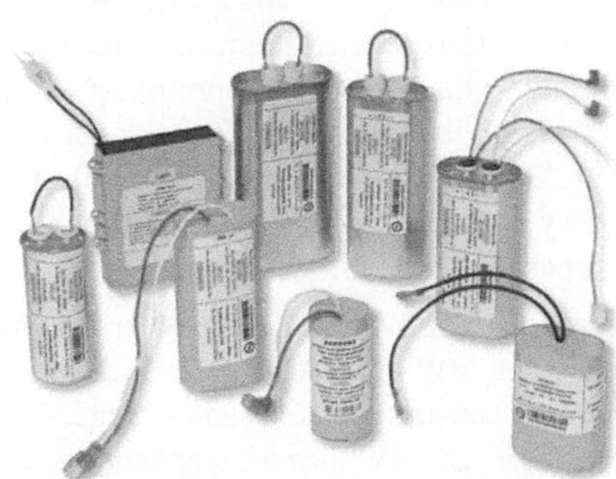

PROBLEM 60 2010-2019 CD Aero, LLC (Formerly Aerovox Corp.)

61. M BIO 19.2 When lightning strikes a tree, the effects can be explosive (see the photo). A typical flash can transport 30.0 C of charge through a potential difference of 125 MV. **(a)** How much energy would be delivered to a tree by this strike? **(b)** What mass of water in the tree would be converted to steam by this energy? Assume the initial temperature of the water is 20.0 °C.

PROBLEM 61 Photo courtesy of Professor James Klett, Colorado State University

62. BIO 19.5 As part of our natural digestive process, our bodies rely on peristalsis—the involuntary muscular contractions that keep food moving through our digestive tract. If these muscular contractions are weak, food may move too slowly, or not at all, which can lead to a condition called *gastroparesis*. Symptoms include nausea, vomiting, abdominal pain, and bloating. Dietary changes and drug therapies are often prescribed as initial treatments. However, if these prove ineffective, implantable gastric stimulation may be used. Here, a small battery-powered electronic device, often referred to as a gastric pacer, is inserted beneath the skin in the lower abdomen (see the figure). Conducting leads from the device provide controlled electrical pulses to the muscles in the lower stomach wall, which cause contractions that aid in peristalsis. The capacitor in one device delivers a short pulse containing 52 μC of charge. **(a)** If the charging voltage across the plates was 5.0 V, what is the value of the device's capacitor? **(b)** How much energy was delivered during the pulse?

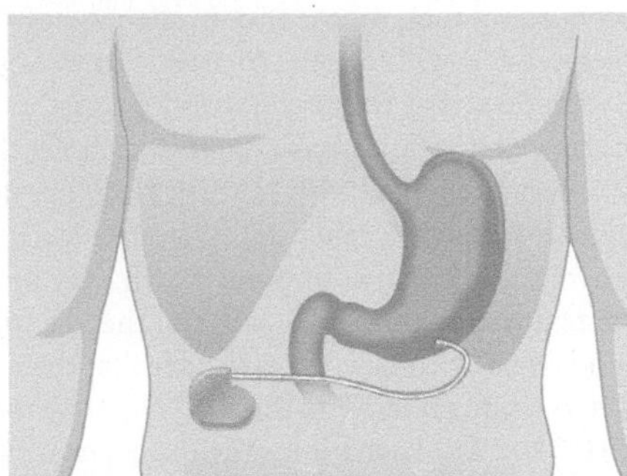

PROBLEM 62 © 2020 Henry Ford Health System

Concepts and Calculations Problems

Online

The conservation of energy (Chapter 6) and the conservation of linear momentum (Chapter 7) are two of the most broadly applicable principles in all of science. In this chapter, we have seen that electrically charged particles obey the conservation-of-energy principle, provided that the electric potential is taken into account. The behavior of electrically charged particles, however, must also be consistent with the conservation-of-momentum principle, as Problem 63 emphasizes. Chapter 18 introduces the electric field, and the present chapter introduces the electric potential. These two concepts are central to the study of electricity, and it is important to distinguish between them, as Problem 64 illustrates.

63. M CHALK Particle 1 has a mass of $m_1 = 3.6 \times 10^{-6}$ kg, while particle 2 has a mass of $m_2 = 6.2 \times 10^{-6}$ kg. Each has the same elec-

tric charge. These particles are initially held at rest, and the two-particle system has an initial electric potential energy of 0.150 J. Suddenly, the particles are released and fly apart because of the repulsive electric force that acts on each one (see the figure). The effects of the gravitational force are negligible, and no other forces act on the particles. *Concepts:* (i) What types of energy does the two-particle system have initially? (ii) What types of energy does the two-particle system have at the instant illustrated in part *b* of the drawing? (iii) Does the principle of conservation of energy apply to this problem? Explain. (iv) Does the conservation of linear momentum apply to the two particles as they fly apart? Explain. *Calculations:* At one instant following the release, the speed of particle 1 is measured to be $v_1 = 170$ m/s. What is the electric potential energy at this instant?

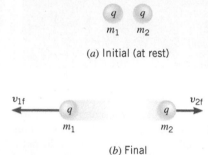

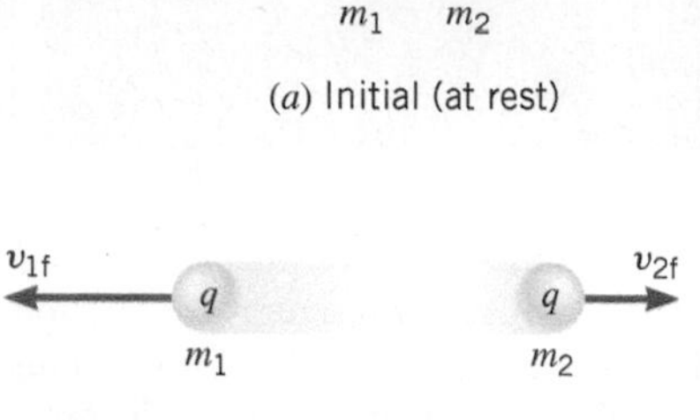

(*a*) Two particles have different masses, but the same electrical charge *q*. They are initially at rest. (*b*) At the instant following the release of the particles, they are flying apart due to the mutual force of electric repulsion.

PROBLEM 63

64. M CHALK SSM Two identical point charges ($+2.4 \times 10^{-9}$ C) are fixed in place, separated by 0.50 m (see the figure). *Concepts:* (i) The electric field is a vector and has a direction. At the midpoint, what are the directions of the individual electric-field contributions from q_A and q_B? (ii) Is the magnitude of the net electric field at the midpoint greater than, less than, or equal to zero? (iii) Is the total electric potential at the midpoint positive, negative, or zero? (iv) Does the electric potential have a direction associated with it? Explain. *Calculations:* Find the electric field and the electric potential at the midpoint of the line between the charges q_A and q_B.

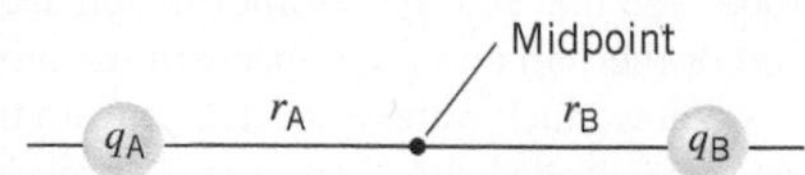

Problem 71 determines the electric field and the electric potential at the midpoint ($r_A = r_B$) between the identical charges ($q_A = q_B$).

PROBLEM 64

Team Problems

Online

65. E T WS **Ion Thrusters II: Electrostatic Potential Energy.** You and your team are designing a new propulsion engine that ejects xenon (Xe) ions in one direction, causing a spacecraft to accelerate in the opposite direction, in accordance with the conservation of linear momentum and the impulse-momentum theorem. This is exactly what happens when a rocket engine expels high velocity particles when burning fuel, except that in the case of the ion drive the ions are accelerated with an electric field. In this process, Xe gas atoms are ionized so that they each acquire a net charge of $+e$. They are then subjected to an electric field, which accelerates and ejects the ions from the engine. In the thruster you are designing, the Xe ions accelerate through a potential difference between an electrode and negatively charged grid through which they are ejected. **(a)** What is the potential difference (V) between the electrode (held at 0 V) and the grid (held at voltage $-V$) in order that the ejection velocity of the ions is 55.0 km/s relative to the grid? **(b)** If Xe ions are ejected at a rate of 9.05×10^{-6} kg/s, what must be the average "wattage" of the power source that accelerates them? The mass of a Xe ion is 2.18×10^{-25} kg.

66. M T WS **A Capacitive Scale.** You and your team are tasked with finding a way to electronically measure weight. You get the idea to measure the capacitance between two circular conducting plates of mass $m = 0.112$ kg and radius $r = 11.5$ cm separated by a nonconducting vertical spring of spring constant $k = 1.35 \times 10^4$ N/m and unstrained length $d_0 = 3.55$ cm. The idea is that the object whose weight is to be determined is placed on the upper plate, thereby compressing the spring and bringing the two plates closer together. This, in turn, changes the capacitance. An electronic instrument that measures capacitance is interfaced with a computer that records the capacitance of the plates and converts it to weight. **(a)** Your task is to derive an equation for the weight in terms of the capacitance that the computer will use to convert the measured capacitance to weight (in N). You can neglect any electrostatic forces between the plates of the capacitor. **(b)** What is the weight of an object placed on the upper plate that results in a measured capacitance of 16.3 pF?

67. M T **A Proton Gun.** You and your team are designing a device that creates a beam of high-velocity protons. The device consists of a parallel-plate capacitor in which protons start at rest at the positive plate. The protons accelerate toward the negative plate, which has a hole through which some of the protons pass. The plates are separated by 6.0 cm, and the potential difference between the plates is 5.0 kV. What are **(a)** the kinetic energies and **(b)** the speeds of the protons when they pass through the hole? **(c)** What is the acceleration of the protons between the plates?

68. M T **Storing Wind Energy.** A certain windmill produces an average of 0.500 MW of electrical power during the late evening hours when energy consumption is at a low (a four-hour span). An industrial capacitor (the "BC125") has a capacitance of $C = 63.0$ F and operates with a maximum voltage of 125 V between its plates. **(a)** You and your team are asked to determine the minimum number of these capacitors that will be needed to store the energy produced during the four off-peak hours in the late evening so that it can be used during the peak hours of the early afternoon. **(b)** The same company that makes the BC125 storage capacitor has discovered a breakthrough dielectric material that they use in their new model of storage capacitor, the SBC240. The new dielectric does two things: (1) Its dielectric constant is 2.5 times that of the dielectric in the BC125, and (2) it can operate at 240 V. What is the minimum number of SBC240 capacitors that will be needed to store the off-peak windmill energy?

CHAPTER 20

The earth at night from space. The earth's major metropolitan areas and modern industrialized regions twinkle brilliantly with electric lights. Without electric circuits, lighting of the night sky would not be possible. Virtually every aspect of life in modern society utilizes or depends upon electric circuits in some way, which is the topic of this chapter.

NASA/NOAA

Electric Circuits

LEARNING OBJECTIVES

After reading this module, you should be able to...

- **20.1** Define electromotive force and current.
- **20.2** Solve problems using Ohm's law for a simple series circuit.
- **20.3** Relate resistance and resistivity.
- **20.4** Solve problems involving electric power.
- **20.5** Solve ac circuit problems for current and power.
- **20.6** Analyze resistor circuits with series connections.
- **20.7** Analyze resistor circuits with parallel connections.
- **20.8** Analyze resistor circuits with both series and parallel connections.
- **20.9** Solve circuit problems that include internal resistances of batteries.
- **20.10** Solve complex circuit problems by applying Kirchhoff's rules.
- **20.11** Describe how ammeters and voltmeters operate.
- **20.12** Analyze capacitor circuits.
- **20.13** Analyze *RC* circuits.
- **20.14** Explain why electrical grounding is important.

20.1 Electromotive Force and Current

Look around you. Chances are that there is an electrical device nearby—a radio, a hair dryer, a computer—something that uses electrical energy to operate. The energy needed to run an MP3 player, for instance, comes from batteries, as **Figure 20.1** illustrates. The transfer of energy takes place via an electric circuit, in which the energy source (the battery pack) and the energy-consuming device (the MP3 player) are connected by conducting wires, through which electric charges move.

Within a battery, a chemical reaction occurs that transfers electrons from one terminal (leaving it positively charged) to another terminal (leaving it negatively charged). **Figure 20.2** shows the two terminals of a car battery and a flashlight battery. The drawing also illustrates the symbol (+|−) used to represent a battery in circuit drawings. Because of the positive and negative charges on the battery terminals, an electric potential difference exists between them. The maximum potential difference is called the **electromotive force* (emf)** of the battery, for which the symbol $\mathscr{E}$ is used. In a typical car battery, the chemical reaction maintains the potential of the positive terminal at a maximum of 12 volts (12 joules/coulomb) higher than the potential of the negative terminal,

*The word "force" appears in this context for historical reasons, even though it is incorrect. As we have seen in Section 19.2, electric potential is energy per unit charge, which is not force.

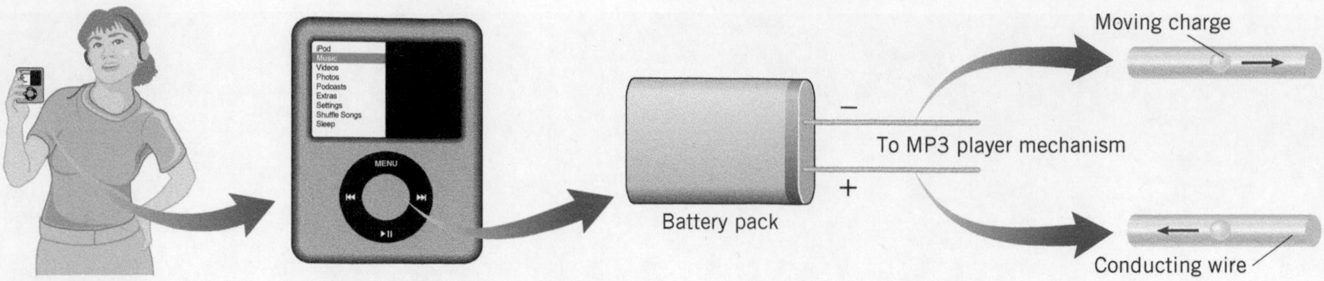

FIGURE 20.1 In an electric circuit, energy is transferred from a source (the battery pack) to a device (the MP3 player) by charges that move through a conducting wire.

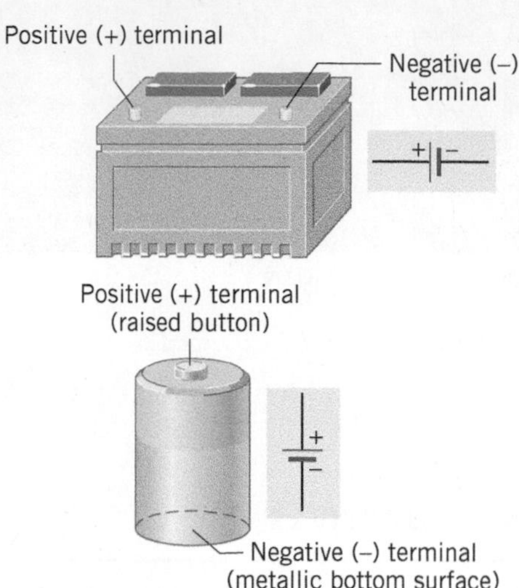

FIGURE 20.2 Typical batteries and the symbol (+|−) used to represent them in electric circuits.

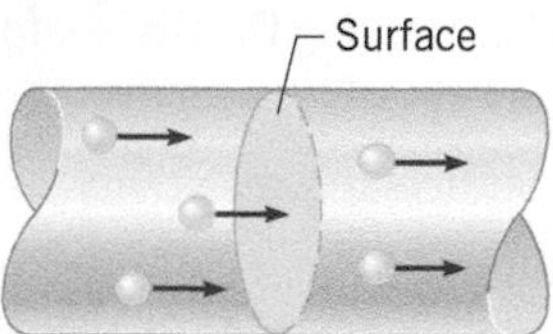

FIGURE 20.3 The electric current is the amount of charge per unit time that passes through an imaginary surface that is perpendicular to the motion of the charges.

so the emf is $\mathscr{E} = 12$ V. Thus, one coulomb of charge emerging from the battery and entering a circuit has at most 12 joules of energy. In a typical flashlight battery the emf is 1.5 V. In reality, the potential difference between the terminals of a battery is somewhat less than the maximum value indicated by the emf, for reasons that Section 20.9 discusses.

In a circuit such as the one shown in **Figure 20.1**, the battery creates an electric field within and parallel to the wire, directed from the positive toward the negative terminal. The electric field exerts a force on the free electrons in the wire, and they respond by moving. **Figure 20.3** shows charges moving inside a wire and crossing an imaginary surface that is perpendicular to their motion. This flow of charge is known as an **electric current**. The electric current I is defined as the amount of charge per unit time that crosses the imaginary surface, as in **Figure 20.3**, in much the same sense that a river current is the amount of water per unit time that is flowing past a certain point. If the rate is constant, the current is

$$I = \frac{\Delta q}{\Delta t} \tag{20.1}$$

If the rate of flow is not constant, then Equation 20.1 gives the average current. Since the units for charge and time are the coulomb (C) and the second (s), the SI unit for current is a coulomb per second (C/s). One coulomb per second is referred to as an **ampere** (A), after the French mathematician André-Marie Ampère (1775–1836).

If the charges move around a circuit in the same direction at all times, the current is said to be **direct current (dc)**, which is the kind produced by batteries. In contrast, the current is said to be **alternating current (ac)** when the charges move first one way and then the opposite way, changing direction from moment to moment. Many energy sources produce alternating current—for example, generators at power companies and microphones. Example 1 deals with direct current.

EXAMPLE 1 | A Pocket Calculator

The battery pack of a pocket calculator has a voltage* of 3.0 V and delivers a current of 0.17 mA. In one hour of operation, **(a)** how much charge flows in the circuit and **(b)** how much energy does the battery deliver to the calculator circuit?

Reasoning Since current is defined as charge per unit time, the charge that flows in one hour is the product of the current and the time (3600 s). The charge that leaves the 3.0-V battery pack has 3.0 joules of energy per coulomb of charge. Thus, the total energy delivered to the calculator circuit is the charge (in coulombs) times the energy per unit charge (in volts or joules/coulomb).

Solution **(a)** The charge that flows in one hour can be determined from Equation 20.1:

$$\Delta q = I(\Delta t) = (0.17 \times 10^{-3}\,\text{A})(3600\text{ s}) = \boxed{0.61\text{ C}}$$

(b) The energy delivered to the calculator circuit is

$$\text{Energy} = \text{Charge} \times \underbrace{\frac{\text{Energy}}{\text{Charge}}}_{\substack{\text{Battery}\\ \text{voltage}}} = (0.61\text{ C})(3.0\text{ V}) = \boxed{1.8\text{ J}}$$

*The potential difference between two points, such as the terminals of a battery, is commonly called the voltage between the points.

Today, it is known that electrons flow in metal wires. **Figure 20.4** shows the negative electrons emerging from the negative terminal of the battery and moving counterclockwise around the circuit toward the positive terminal. It is customary, however, *not* to use the flow of electrons when discussing circuits. Instead, a so-called **conventional current** is used, for reasons that date to the time when it was believed that positive charges moved through metal wires. Conventional current is the hypothetical flow of positive charges that would have the same effect in the circuit as the movement of negative charges that actually does occur. In **Figure 20.4**, negative electrons leave the negative terminal of the battery, pass through the device, and arrive at the positive terminal. The same effect would have been achieved if an equivalent amount of positive charge had left the positive terminal, passed through the device, and arrived at the negative terminal. Therefore, the drawing shows the conventional current originating from the positive terminal and moving clockwise around the circuit. A conventional current of hypothetical positive charges is consistent with our earlier use of a positive test charge for defining electric fields and potentials. The direction of conventional current is always from a point of higher potential toward a point of lower potential—that is, from the positive toward the negative terminal. In this text, the symbol I stands for conventional current.

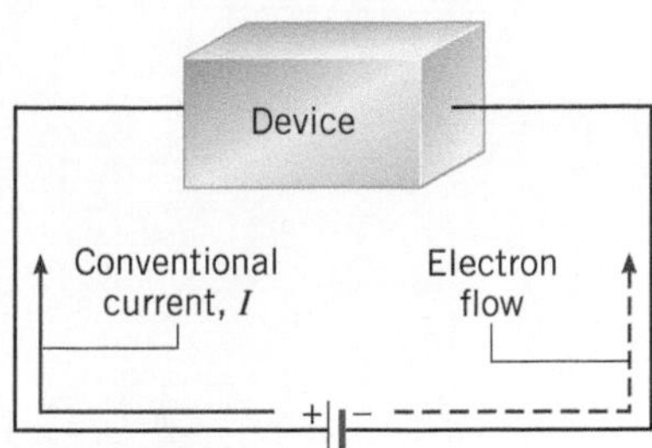

FIGURE 20.4 In a circuit, electrons actually flow through the metal wires. However, it is customary to use a conventional current I to describe the flow of charges.

20.2 Ohm's Law

The current that a battery can push through a wire is analogous to the water flow that a pump can push through a pipe. Greater pump pressures lead to larger water flow rates, and, similarly, greater battery voltages lead to larger electric currents. In the simplest case, the current I is directly proportional to the voltage V; that is, $I \propto V$. Thus, a voltage of 12 V leads to twice as much current as a voltage of 6 V, when each is connected to the same circuit.

In a water pipe, the flow rate is not only determined by the pump pressure but is also affected by the length and diameter of the pipe. Longer and narrower pipes offer higher resistance to the moving water and lead to smaller flow rates for a given pump pressure. A similar situation exists in electric circuits, and to deal with it we introduce the concept of electrical resistance. Electrical resistance is defined in terms of two ideas that have already been discussed—the electric potential difference, or voltage (see Section 19.2), and the electric current (see Section 20.1).

The **resistance** R is defined as the ratio of the voltage V applied across a piece of material to the current I through the material, or $R = V/I$. When only a small current results from a large voltage, there is a high resistance to the moving charge. For many materials (e.g., metals), the ratio V/I is the same for a given piece of material over a wide range of voltages and currents. In such a case, the resistance is a constant. Then, the relation $R = V/I$ is referred to as **Ohm's law**, after the German physicist Georg Simon Ohm (1789–1854), who discovered it.

OHM'S LAW

The ratio V/I is a constant, where V is the voltage applied across a piece of material (such as a wire) and I is the current through the material:

$$\frac{V}{I} = R = \text{constant or } V = IR \tag{20.2}$$

R is the resistance of the piece of material.

SI Unit of Resistance: volt/ampere (V/A) = ohm (Ω)

The SI unit for resistance is a volt per ampere, which is called an *ohm* and is represented by the Greek capital letter omega (Ω). Ohm's law is not a fundamental law of nature like Newton's laws of motion. It is only a statement of the way certain materials behave in electric circuits.

To the extent that a wire or an electrical device offers resistance to the flow of charges, it is called a **resistor**. The resistance can have a wide range of values. The copper wires in a television set, for instance, have a very small resistance. On the other hand, commercial resistors can have resistances up to many kilohms (1 kΩ = 10^3 Ω) or megohms (1 MΩ = 10^6 Ω). Such resistors play an important role in electric circuits, where they are typically used to limit the amount of current and establish the desired voltage levels.

In drawing electric circuits we follow the usual conventions: (1) a zigzag line (—ww—) represents a resistor and (2) a straight line (———) represents an ideal conducting wire, or one with a negligible resistance. Example 2 illustrates an application of Ohm's law to the circuit in a flashlight.

EXAMPLE 2 | A Flashlight

The filament in a light bulb is a resistor in the form of a thin piece of wire. The wire becomes hot enough to emit light because of the current in it. **Interactive Figure 20.5** shows a flashlight that uses two 1.5-V batteries (effectively a single 3.0-V battery) to provide a current of 0.40 A in the filament. Determine the resistance of the glowing filament.

Reasoning The filament resistance is assumed to be the only resistance in the circuit. The potential difference applied across the filament is that of the 3.0-V battery. The resistance, given by Equation 20.2, is equal to this potential difference divided by the current.

Solution The resistance of the filament is

$$R = \frac{V}{I} = \frac{3.0\ \text{V}}{0.40\ \text{A}} = \boxed{7.5\ \Omega} \qquad \textbf{(20.2)}$$

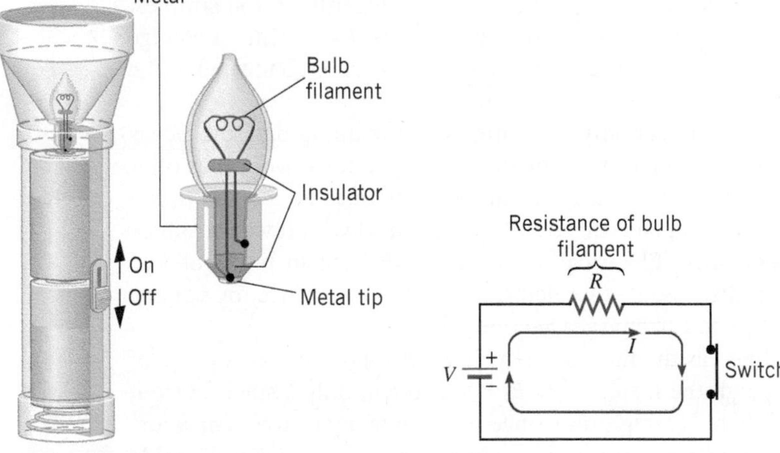

INTERACTIVE FIGURE 20.5 The circuit in this flashlight consists of a resistor (the filament of the light bulb) connected to a 3.0-V battery (two 1.5-V batteries).

Check Your Understanding

(The answers are given at the end of the book.)

1. In circuit A the battery that supplies energy has twice as much voltage as the battery in circuit B. However, the current in circuit A is only one-half the current in circuit B. Circuit A presents _____ the resistance to the current that circuit B does. **(a)** twice **(b)** one-half **(c)** the same **(d)** four times **(e)** one-fourth

2. Two circuits present the same resistance to the current. In one circuit the battery causing the flow of current has a voltage of 9.0 V, and the current is 3.0 A. In the other circuit the battery has a voltage of 1.5 V. What is the current in this other circuit?

20.3 Resistance and Resistivity

In a water pipe, the length and cross-sectional area of the pipe determine the resistance that the pipe offers to the flow of water. Longer pipes with smaller cross-sectional areas offer greater resistance. Analogous effects are found in the electrical case. For a wide range of materials, the resistance of a piece of material of length L and cross-sectional area A is

$$R = \rho \frac{L}{A} \tag{20.3}$$

where ρ is a proportionality constant known as the **resistivity** of the material. It can be seen from Equation 20.3 that the unit for resistivity is the ohm · meter ($\Omega \cdot$ m), and **Table 20.1** lists values for various materials. All the conductors in the table are metals and have small resistivities. Insulators such as rubber have large resistivities. Materials like germanium and silicon have intermediate resistivity values and are, accordingly, called *semiconductors*.

TABLE 20.1 Resistivities[a] of Various Materials

Material	Resistivity ρ ($\Omega \cdot$ m)	Material	Resistivity ρ ($\Omega \cdot$ m)
Conductors		**Semiconductors**	
Aluminum	2.82×10^{-8}	Carbon	3.5×10^{-5}
Copper	1.72×10^{-8}	Germanium	0.5[b]
Gold	2.44×10^{-8}	Silicon	20–2300[b]
Iron	9.7×10^{-8}	**Insulators**	
Mercury	95.8×10^{-8}	Mica	10^{11}–10^{15}
Nichrome (alloy)	100×10^{-8}	Rubber (hard)	10^{13}–10^{16}
Silver	1.59×10^{-8}	Teflon	10^{16}
Tungsten	5.6×10^{-8}	Wood (maple)	3×10^{10}

[a]The values pertain to temperatures near 20 °C.
[b]Depending on purity.

Resistivity is an inherent property of a material, inherent in the same sense that density is an inherent property. Resistance, on the other hand, depends on both the resistivity and the geometry of the material. Thus, two wires can be made from copper, which has a resistivity of $1.72 \times 10^{-8}\ \Omega \cdot$ m, but Equation 20.3 indicates that a short wire with a large cross-sectional area has a smaller resistance than does a long, thin wire. Wires that carry large currents, such as main power cables, are thick rather than thin so that the resistance of the wires is kept as small as possible. Similarly, electric tools that are to be used far away from wall sockets require thicker extension cords, as Example 3 illustrates.

EXAMPLE 3 | The Physics of Electrical Extension Cords

The instructions for an electric lawn mower suggest that a 20-gauge extension cord can be used for distances up to 35 m, but a thicker 16-gauge cord should be used for longer distances, to keep the resistance of the wire as small as possible. The cross-sectional area of 20-gauge wire is 5.2×10^{-7} m^2, while that of 16-gauge wire is 13×10^{-7} m^2. Determine the resistance of **(a)** 35 m of 20-gauge copper wire and **(b)** 75 m of 16-gauge copper wire.

Reasoning According to Equation 20.3, the resistance of a copper wire depends on the resistivity of copper and the length and cross-sectional area of the wire. The resistivity can be obtained from **Table 20.1**, while the length and cross-sectional area are given in the problem statement.

Solution According to **Table 20.1** the resistivity of copper is $1.72 \times 10^{-8}\ \Omega \cdot$ m. The resistance of the wires can be found using Equation 20.3:

20-gauge wire $$R = \frac{\rho L}{A} = \frac{(1.72 \times 10^{-8}\ \Omega \cdot \text{m})(35\ \text{m})}{5.2 \times 10^{-7}\ \text{m}^2} = \boxed{1.2\ \Omega}$$

16-gauge wire $$R = \frac{\rho L}{A} = \frac{(1.72 \times 10^{-8}\ \Omega \cdot \text{m})(75\ \text{m})}{13 \times 10^{-7}\ \text{m}^2} = \boxed{0.99\ \Omega}$$

Even though it is more than twice as long, the thicker 16-gauge wire has less resistance than the thinner 20-gauge wire. It is necessary to keep the resistance as low as possible to minimize heating of the wire, thereby reducing the possibility of a fire, as Conceptual Example 7 in Section 20.5 emphasizes.

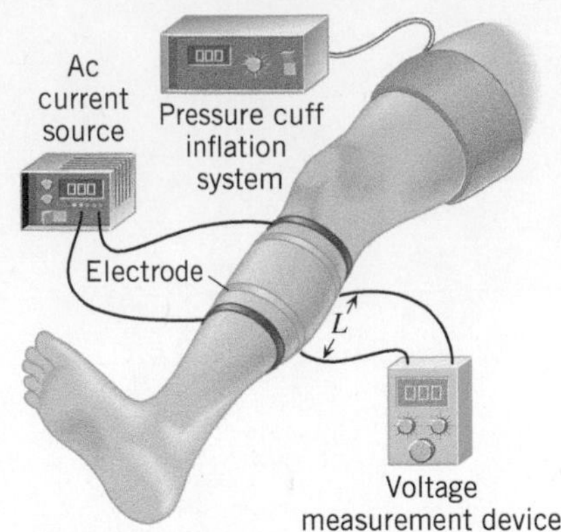

FIGURE 20.6 Using the technique of impedance plethysmography, the electrical resistance of the calf can be measured to diagnose deep venous thrombosis (blood clotting in the veins).

BIO THE PHYSICS OF . . . impedance plethysmography. Equation 20.3 provides the basis for an important medical diagnostic technique known as impedance (or resistance) plethysmography. **Figure 20.6** shows how the technique is applied to diagnose blood clotting in the veins (deep venous thrombosis) near the knee. A pressure cuff, like that used in blood pressure measurements, is placed around the midthigh, while electrodes are attached around the calf. The two outer electrodes are connected to a source that supplies a small amount of ac current. The two inner electrodes are separated by a distance L, and the voltage between them is measured. The voltage divided by the current gives the resistance. The key to this technique is the fact that resistance can be related to the volume V_{calf} of the calf between the inner electrodes. The volume is the product of the length L and the calf's cross-sectional area A, or $V_{\text{calf}} = LA$. Solving for A and substituting in Equation 20.3 shows that

$$R = \rho\frac{L}{A} = \rho\frac{L}{V_{\text{calf}}/L} = \rho\frac{L^2}{V_{\text{calf}}}$$

Thus, resistance is inversely proportional to volume, a fact that is exploited in diagnosing deep venous thrombosis. Blood flows from the heart into the calf through arteries in the leg and returns through the system of veins. The pressure cuff in **Figure 20.6** is inflated to the point where it cuts off the venous flow but does not alter the arterial flow. As a result, more blood enters than leaves the calf. Therefore, the volume of the calf increases, and the electrical resistance decreases. When the cuff pressure is removed suddenly, the volume returns to a normal value, and so does the electrical resistance. With healthy (unclotted) veins, there is a rapid return to normal values. A slow return, however, reveals the presence of clotting.

The resistivity of a material depends on temperature. In metals, the resistivity increases with increasing temperature, whereas in semiconductors the reverse is true. For many materials and limited temperature ranges, it is possible to express the temperature dependence of the resistivity as follows:

$$\rho = \rho_0[1 + \alpha(T - T_0)] \tag{20.4}$$

In this expression ρ and ρ_0 are the resistivities at temperatures T and T_0, respectively. The term α has the unit of reciprocal temperature and is the **temperature coefficient of resistivity**. When the resistivity increases with increasing temperature, α is positive, as it is for metals. When the resistivity decreases with increasing temperature, α is negative, as it is for the semiconductors carbon, germanium, and silicon. Since resistance is given by $R = \rho L/A$, both sides of Equation 20.4 can be multiplied by L/A to show that resistance depends on temperature according to

$$R = R_0[1 + \alpha(T - T_0)] \tag{20.5}$$

The next example illustrates the role of the resistivity and its temperature coefficient in determining the electrical resistance of a piece of material.

EXAMPLE 4 | The Physics of a Heating Element on an Electric Stove

Figure 20.7a shows a cherry-red heating element on an electric stove. The element contains a wire (length = 1.1 m, cross-sectional area = 3.1×10^{-6} m²) through which electric charge flows. As **Figure 20.7b** shows, this wire is embedded within an electrically insulating material that is contained within a metal casing. The wire becomes hot in response to the flowing charge and heats the casing. The material of the wire has a resistivity of $\rho_0 = 6.8 \times 10^{-5}\ \Omega \cdot \text{m}$ at $T_0 = 320$ °C and a temperature coefficient of resistivity of $\alpha = 2.0 \times 10^{-3}\ (\text{C}°)^{-1}$. Determine the resistance of the heater wire at an operating temperature of 420 °C.

Reasoning The expression $R = \rho L/A$ (Equation 20.3) can be used to find the resistance of the wire at 420 °C, once the resistivity ρ is determined at this temperature. Since the resistivity at 320 °C is given, Equation 20.4 can be employed to find the resistivity at 420 °C.

Solution At the operating temperature of 420 °C, the material of the wire has a resistivity of

$$\rho = \rho_0[1 + \alpha(T - T_0)]$$
$$\rho = (6.8 \times 10^{-5}\ \Omega \cdot \text{m})\{1 + [2.0 \times 10^{-3}(\text{C}°)^{-1}](420\ °\text{C} - 320\ °\text{C})\}$$
$$= 8.2 \times 10^{-5}\ \Omega \cdot \text{m}$$

This value of the resistivity can be used along with the given length and cross-sectional area to find the resistance of the heater wire:

$$R = \frac{\rho L}{A} = \frac{(8.2 \times 10^{-5}\,\Omega \cdot \text{m})(1.1\ \text{m})}{3.1 \times 10^{-6}\ \text{m}^2} = \boxed{29\ \Omega} \quad \textbf{(20.3)}$$

(a)

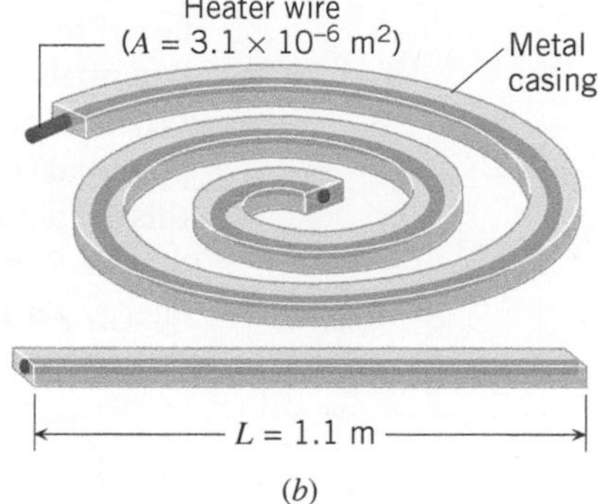

(b)

FIGURE 20.7 A heating element on an electric stove.

There is an important class of materials whose resistivity suddenly goes to zero below a certain temperature T_c, which is called the *critical temperature* and is commonly a few degrees above absolute zero. Below this temperature, such materials are called *superconductors.* The name derives from the fact that with zero resistivity, these materials offer no resistance to electric current and are, therefore, perfect conductors. Once a current is established in a superconducting ring, the current continues indefinitely without the need for an emf. Currents have persisted in superconductors for many years without measurable decay. In contrast, the current in a nonsuperconducting material drops to zero almost immediately after the emf is removed.

Many metals become superconductors only at very low temperatures, such as aluminum ($T_c = 1.18$ K), tin ($T_c = 3.72$ K), lead ($T_c = 7.20$ K), and niobium ($T_c = 9.25$ K). Materials involving copper oxide complexes have been made that undergo the transition to the superconducting state at 175 K. Superconductors have many technological applications, including magnetic resonance imaging (Section 21.7), magnetic levitation of trains (Section 21.9), cheaper transmission of electric power, powerful (yet small) electric motors, and faster computer chips.

Check Your Understanding

(*The answers are given at the end of the book.*)

3. Two materials have different resistivities. Two wires of the same length are made, one from each of the materials. Is it possible for each wire to have the same resistance? **(a)** Yes, if the material with the greater resistivity is used for a thinner wire. **(b)** Yes, if the material with the greater resistivity is used for a thicker wire. **(c)** It is not possible.

4. How does the resistance of a copper wire change when both the length and diameter of the wire are doubled? **(a)** It decreases by a factor of two. **(b)** It increases by a factor of two. **(c)** It increases by a factor of four. **(d)** It decreases by a factor of four. **(e)** It does not change.

5. A resistor is connected between the terminals of a battery. This resistor is a wire, and the following five choices give possibilities for its length and radius in multiples of L_0 and r_0, respectively. For which one or more of the possibilities is the current in the resistor a minimum? **(a)** L_0 and r_0 **(b)** $\frac{1}{2}L_0$ and $\frac{1}{2}r_0$ **(c)** $2L_0$ and $2r_0$ **(d)** $2L_0$ and r_0 **(e)** $8L_0$ and $2r_0$

20.4 Electric Power

One of the most important functions of current in an electric circuit is to transfer energy from a source (a battery or a generator) to an electrical device (MP3 player, cell phone, etc.), as **Figure 20.8** illustrates. Note that the positive (+) terminal of the battery is connected by a wire to the terminal labeled A on the device; likewise, the negative terminal (−) of the battery is connected to the B terminal. Thus, the battery maintains a constant potential difference between the terminals A and B, with A being at the higher potential. When an amount of positive charge Δq moves from the higher potential (A) to the lower potential (B), its electric potential energy decreases. In accordance with Equation 19.4, this decrease is $(\Delta q)V$, where V is the amount by which the electric potential at A exceeds that at B or, in other words, the voltage between the two points. Since the change in energy per unit time is the power P (Equation 6.10b), the electric power associated with this change in energy is

$$P = \frac{\text{Change in energy}}{\text{Time interval}} = \frac{(\Delta q)V}{\Delta t} = \underbrace{\frac{\Delta q}{\Delta t}}_{\text{Current } I} V$$

The term $\Delta q/\Delta t$ is the charge per unit time, or the current I in the device, according to Equation 20.1. It follows, then, that the electric power is the product of the current and the voltage.

ELECTRIC POWER

When electric charge flows from point A to point B in a circuit, leading to a current I, and the voltage between the points is V, the electric power associated with this current and voltage is

$$P = IV \tag{20.6}$$

SI Unit of Power: watt (W)

Power is measured in watts, and Equation 20.6 indicates that the product of an ampere and a volt is equal to a watt.

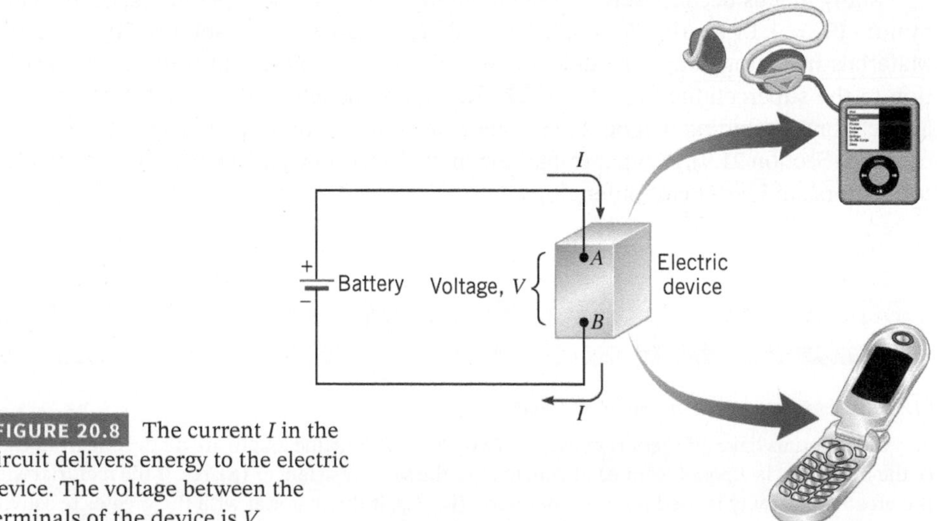

FIGURE 20.8 The current I in the circuit delivers energy to the electric device. The voltage between the terminals of the device is V.

When the charge moves through the device in **Figure 20.8**, the charge *loses* electric potential energy. The principle of conservation of energy tells us that the decrease in potential energy must be accompanied by a transfer of energy to some other form (or forms). In a cell phone, for example, the energy transferred appears as light energy (coming from the display screen), sound energy (emanating from the speaker), thermal energy (due to heating of the internal circuitry), and so on.

The charge in a circuit can also gain electrical energy. For example, when it moves through the battery in **Figure 20.8**, the charge goes from a lower to a higher potential, just the opposite of what happens in the electrical device. In this case, the charge *gains* electric potential energy. Consistent with the conservation of energy, this increase in potential energy must come from somewhere; in this case it comes from the chemical energy stored in the battery. Thus, the charge regains the energy it lost to the device, at the expense of the chemical energy of the battery.

Many electrical devices are essentially resistors that become hot when provided with sufficient electric power: toasters, irons, space heaters, heating elements on electric stoves, and incandescent light bulbs, to name a few. In such cases, it is convenient to have additional expressions that are equivalent to the power $P = IV$ but which include the resistance R explicitly. We can obtain two such equations by substituting $V = IR$, or equivalently $I = V/R$, into the relation $P = IV$:

$$P = IV \qquad \textbf{(20.6a)}$$

$$P = I(IR) = I^2R \qquad \textbf{(20.6b)}$$

$$P = \left(\frac{V}{R}\right)V = \frac{V^2}{R} \qquad \textbf{(20.6c)}$$

Example 5 deals with the electric power delivered to the bulb of a flashlight.

EXAMPLE 5 | The Power and Energy Used in a Flashlight

In the flashlight in **Interactive Figure 20.5**, the current is 0.40 A, and the voltage is 3.0 V. Find **(a)** the power delivered to the bulb and **(b)** the electrical energy dissipated in the bulb in 5.5 minutes of operation.

Reasoning The electric power delivered to the bulb is the product of the current and voltage. Since power is energy per unit time, the energy delivered to the bulb is the product of the power and time.

Solution **(a)** The power is

$$P = IV = (0.40\ \text{A})(3.0\ \text{V}) = \boxed{1.2\ \text{W}} \qquad \textbf{(20.6a)}$$

The "wattage" rating of this bulb would, therefore, be 1.2 W.

(b) The energy consumed in 5.5 minutes (330 s) follows from the definition of power as energy per unit time:

$$\text{Energy} = P\,\Delta t = (1.2\ \text{W})(330\ \text{s}) = \boxed{4.0 \times 10^2\ \text{J}}$$

Monthly electric bills specify the cost for the energy consumed during the month. Energy is the product of power and time, and electric companies compute your energy consumption by expressing power in kilowatts and time in hours. Therefore, a commonly used unit for energy is the *kilowatt · hour* (kWh). For instance, if you used an average power of 1440 watts (1.44 kW) for 30 days (720 h), your energy consumption would be (1.44 kW) (720 h) = 1040 kWh. At a cost of \$0.12 per kWh, your monthly bill would be \$125. As shown in Example 11 in Chapter 12, 1 kWh = 3.60×10^6 J of energy.

Check Your Understanding

(*The answers are given at the end of the book.*)

6. A toaster is designed to operate with a voltage of 120 V, and a clothes dryer is designed to operate with a voltage of 240 V. Based solely on this information, which appliance uses more power? **(a)** The toaster **(b)** The dryer **(c)** Insufficient information is given for an answer.

7. When an incandescent light bulb is turned on, a constant voltage is applied across the tungsten filament, which then becomes white hot. The temperature coefficient of resistivity for tungsten is a positive number. What happens to the power delivered to the bulb as the filament heats up? **(a)** It decreases. **(b)** It increases. **(c)** It remains constant.

8. CYU Figure 20.1 shows a circuit that includes a bimetallic strip (made from brass and steel; see Section 12.4) with a resistance heater wire wrapped around it. When the switch is initially closed, a current appears in the circuit because charges flow through the heater wire (which becomes hot), the strip itself, the contact point, and the light bulb. The bulb glows in response. As long as the switch remains closed, does the bulb **(a)** continue to glow, **(b)** eventually turn off permanently, or **(c)** flash on and off?

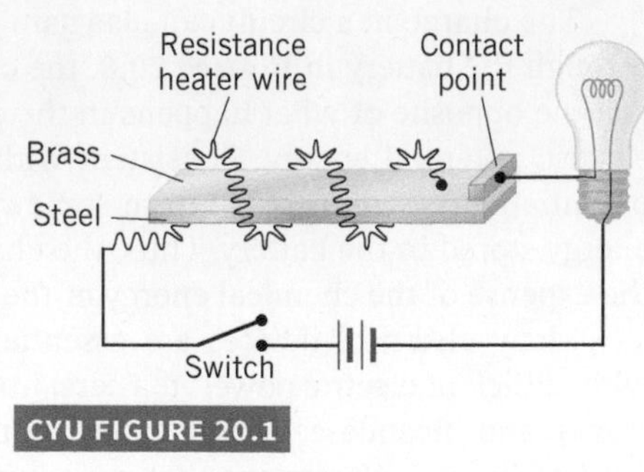

CYU FIGURE 20.1

20.5 Alternating Current

Many electric circuits use batteries and involve direct current (dc). However, there are considerably more circuits that operate with alternating current (ac), in which the charge flow reverses direction periodically. The common generators that create ac electricity depend on magnetic forces for their operation and are discussed in Chapter 22. In an ac circuit, these generators serve the same purpose as a battery serves in a dc circuit; that is, they give energy to the moving electric charges. This section deals with ac circuits that contain only resistance.

Since the electrical outlets in a house provide alternating current, we all use ac circuits routinely. For example, the heating element of a toaster is essentially a thin wire of resistance R and becomes red hot when electrical energy is dissipated in it. **Figure 20.9** shows the ac circuit that is formed when a toaster is plugged into a wall socket. The circuit schematic in the picture introduces the symbol ⊝ that is used to represent the generator. In this case, the generator is located at the electric power company.

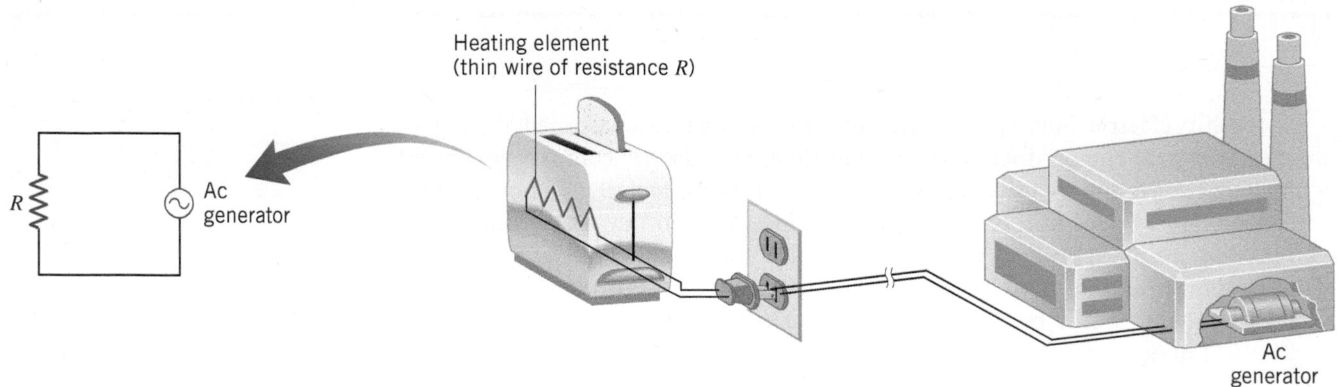

FIGURE 20.9 This circuit consists of a toaster (resistance = R) and an ac generator at the electric power company.

Figure 20.10 shows a graph that records the voltage V produced between the terminals of the ac generator in **Figure 20.9** at each moment of time t. This is the most common type of ac voltage. It fluctuates sinusoidally between positive and negative values:

$$V = V_0 \sin 2\pi ft \qquad \textbf{(20.7)}$$

where V_0 is the maximum or peak value of the voltage, and f is the frequency (in cycles/s or Hz) at which the voltage oscillates. The angle $2\pi ft$ in Equation 20.7 is expressed in radians, so a calculator must be set to its radian mode before the sine of this angle is evaluated. In the United States, the voltage at most home wall outlets has a *peak value* of approximately $V_0 = 170$ volts and oscillates with a frequency of $f = 60$ Hz. Thus, the period of each cycle is $\frac{1}{60}$ s, and the polarity of the generator terminals reverses twice during each cycle, as **Figure 20.10** indicates.

The current in an ac circuit also oscillates. In circuits that contain only resistance, the current reverses direction each time the polarity of the generator terminals reverses. Thus, the current in a circuit like that in **Figure 20.9** would have a frequency of 60 Hz and would change direction twice during each cycle. Substituting $V = V_0 \sin 2\pi ft$ into $V = IR$ shows that the current can be represented as

$$I = \frac{V_0}{R}\sin 2\pi ft = I_0 \sin 2\pi ft \qquad \textbf{(20.8)}$$

The peak current is given by $I_0 = V_0/R$, so it can be determined if the peak voltage and the resistance are known.

The power delivered to an ac circuit by the generator is given by $P = IV$, just as it is in a dc circuit. However, since both I and V depend on time, the power fluctuates as time passes. Substituting Equations 20.7 and 20.8 for V and I into $P = IV$ gives

$$P = I_0 V_0 \sin^2 2\pi ft \qquad \textbf{(20.9)}$$

This expression is plotted in **Figure 20.11**.

Since the power fluctuates in an ac circuit, it is customary to consider the average power $\overline{P}$, which is one-half the peak power, as **Figure 20.11** indicates:

$$\overline{P} = \tfrac{1}{2} I_0 V_0 \qquad \textbf{(20.10)}$$

On the basis of this expression, a kind of average current and average voltage can be introduced that are very useful when discussing ac circuits. A rearrangement of Equation 20.10 reveals that

$$\overline{P} = \left(\frac{I_0}{\sqrt{2}}\right)\left(\frac{V_0}{\sqrt{2}}\right) = I_{\text{rms}} V_{\text{rms}} \qquad \textbf{(20.11)}$$

I_{rms} and V_{rms} are called the **root-mean-square (rms)** current and voltage, respectively, and may be calculated from the peak values by dividing them by $\sqrt{2}$:*

$$I_{\text{rms}} = \frac{I_0}{\sqrt{2}} \qquad \textbf{(20.12)}$$

$$V_{\text{rms}} = \frac{V_0}{\sqrt{2}} \qquad \textbf{(20.13)}$$

Problem-Solving Insight The rms values of the ac current and the ac voltage, I_{rms} and V_{rms}, respectively, are not the same as the peak values I_0 and V_0. The rms values are always smaller than the peak values by a factor of $\sqrt{2}$.

Since the normal maximum ac voltage at a home wall socket in the United States is $V_0 = 170$ volts, the corresponding rms voltage is $V_{\text{rms}} = (170 \text{ volts})/\sqrt{2} = 120$ volts. Instructions for electrical devices usually specify this rms value. Similarly, when we specify an ac voltage or current in this text, it is an rms value, unless indicated otherwise. When we specify ac power, it is an average power, unless stated otherwise.

Except for dealing with average quantities, the relation $\overline{P} = I_{\text{rms}} V_{\text{rms}}$ has the same form as Equation 20.6a ($P = IV$). Moreover, Ohm's law can be written conveniently in terms of rms quantities:

$$V_{\text{rms}} = I_{\text{rms}} R \qquad \textbf{(20.14)}$$

Substituting Equation 20.14 into $\overline{P} = I_{\text{rms}} V_{\text{rms}}$ shows that the average power can be expressed in the following ways:

$$\overline{P} = I_{\text{rms}} V_{\text{rms}} \qquad \textbf{(20.15a)}$$

$$\overline{P} = I_{\text{rms}}^2 R \qquad \textbf{(20.15b)}$$

$$\overline{P} = \frac{V_{\text{rms}}^2}{R} \qquad \textbf{(20.15c)}$$

*This applies only for sinusoidal current and voltage.

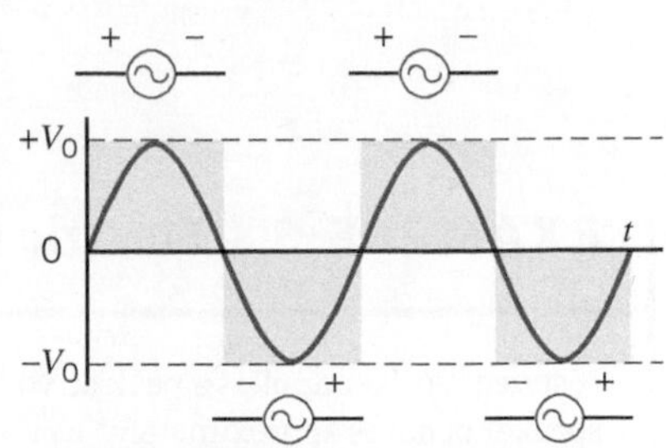

FIGURE 20.10 In the most common case, the ac voltage is a sinusoidal function of time. The relative polarity of the generator terminals during the positive and negative parts of the sine wave is indicated.

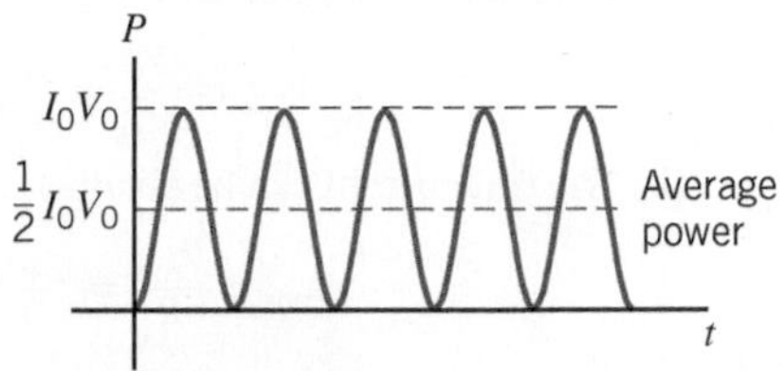

FIGURE 20.11 In an ac circuit, the power P delivered to a resistor oscillates between zero and a peak value of I_0V_0, where I_0 and V_0 are the peak current and voltage, respectively.

These expressions are completely analogous to $P = IV = I^2R = V^2/R$ (Equations 20.6a–c) for dc circuits. Example 6 deals with the average power in one familiar ac circuit.

EXAMPLE 6 | Electric Power Sent to a Loudspeaker

A stereo receiver applies a peak ac voltage of 34 V to a speaker. The speaker behaves approximately* as if it had a resistance of 8.0 Ω, as the circuit in **Figure 20.12** indicates. Determine **(a)** the rms voltage, **(b)** the rms current, and **(c)** the average power for this circuit.

Reasoning The rms voltage is, by definition, equal to the peak voltage divided by $\sqrt{2}$. Furthermore, we are assuming that the circuit contains only a resistor. Therefore, we can use Ohm's law (Equation 20.14) to calculate the rms current as the rms voltage divided by the resistance, and we can then determine the average power as the rms current times the rms voltage (Equation 20.15a).

Solution **(a)** The peak value of the voltage is $V_0 = 34$ V, so the corresponding rms value is

$$V_{\text{rms}} = \frac{V_0}{\sqrt{2}} = \frac{34\text{ V}}{\sqrt{2}} = \boxed{24\text{ V}} \qquad \textbf{(20.13)}$$

(b) The rms current can be obtained from Ohm's law:

$$I_{\text{rms}} = \frac{V_{\text{rms}}}{R} = \frac{24\text{ V}}{8.0\ \Omega} = \boxed{3.0\text{ A}} \qquad \textbf{(20.14)}$$

(c) The average power is

$$\overline{P} = I_{\text{rms}}V_{\text{rms}} = (3.0\text{ A})(24\text{ V}) = \boxed{72\text{ W}} \qquad \textbf{(20.15a)}$$

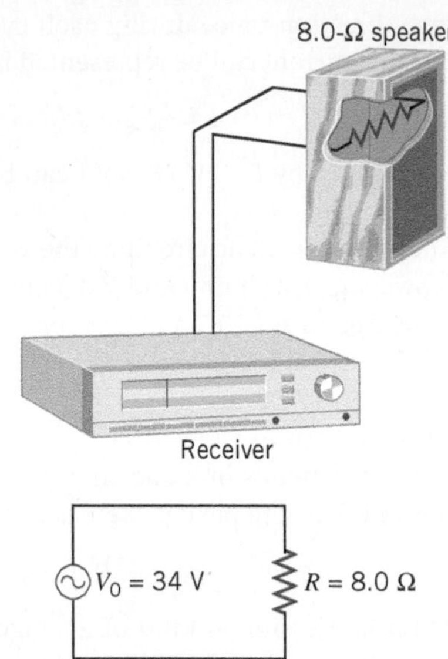

FIGURE 20.12 A receiver applies an ac voltage (peak value = 34 V) to an 8.0-Ω speaker.

The electric power dissipated in a resistor causes the resistor to heat up. Excessive power can lead to a potential fire hazard, as Conceptual Example 7 discusses.

CONCEPTUAL EXAMPLE 7 | Extension Cords and a Potential Fire Hazard

During the winter, many people use portable electric space heaters to keep warm. When the heater is located far from a 120-V wall receptacle, an extension cord is necessary (see **Figure 20.13**). To prevent fires, however, manufacturers sometimes caution about using extension cords. To minimize the risk of a fire, should a long extension cord used with a space heater be made from **(a)** larger-gauge or **(b)** smaller-gauge wire?

Reasoning An electric space heater contains a heater element that is a piece of wire of resistance R, which is heated to a high temperature because of the power $I_{\text{rms}}^2 R$ dissipated in it. A typical heater uses a relatively large current I_{rms} of about 12 A. On its way to the heater, this current passes through the wires of the extension cord. Since these additional wires offer resistance to the current, the extension cord can also heat up, just as the heater element does. This unwanted heating depends on the resistance of the wire in the extension cord and could lead to a fire. As Example 3 discusses, the larger-gauge wire is the one that has the smaller cross-sectional area. The cross-sectional area is important because it is one of the factors that determine the resistance of the wire in the extension cord.

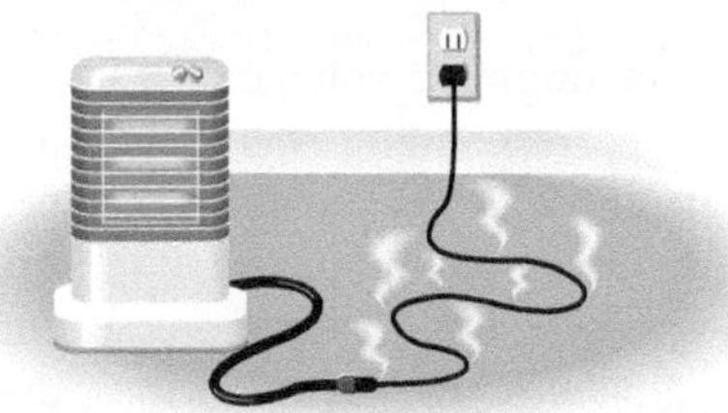

FIGURE 20.13 When an extension cord is used with a space heater, the cord must have a resistance that is sufficiently small to prevent overheating of the cord.

Answer (a) is incorrect. To keep the heating of the extension cord to a safe level, the resistance of the wire must be kept small. Recall from Section 20.3 that the resistance of a wire depends inversely on its cross-sectional area. A larger-gauge wire has a

*Other factors besides resistance can affect the current and voltage in ac circuits; they are discussed in Chapter 23.

smaller cross-sectional area and, therefore, has a greater resistance $R_{\text{extension cord}}$. This greater resistance would lead to more (not less) heating of the extension cord, because of the power $I_{\text{rms}}^2 R_{\text{extension cord}}$ dissipated in it.

Answer (b) is correct. Since the wire's resistance depends inversely on its cross-sectional area, a smaller-gauge wire, with its larger cross-sectional area, has a smaller resistance $R_{\text{extension cord}}$. This smaller resistance would lead to less heating of the extension cord due to the power $I_{\text{rms}}^2 R_{\text{extension cord}}$ dissipated in it, thus minimizing the risk of a fire.

Related Homework: Problem 30

Check Your Understanding

(The answers are given at the end of the book.)

9. CYU Figure 20.2 shows a circuit in which a light bulb is connected to the household ac voltage via two switches, S_1 and S_2. This is the kind of wiring, for example, that allows you to turn on a carport light from either inside the house or out in the carport. Which one or more of the combinations of the switch positions will turn on the light? **(a)** S_1 set to A and S_2 set to B **(b)** S_1 set to B and S_2 set to B **(c)** S_1 set to B and S_2 set to A **(d)** S_1 set to A and S_2 set to A

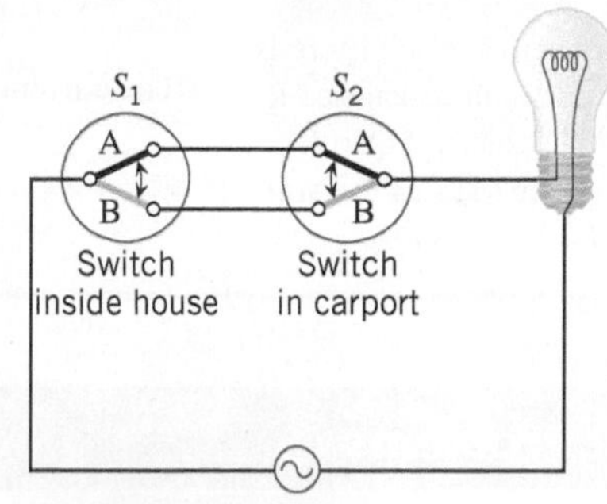

CYU FIGURE 20.2

10. Two light bulbs are designed for use with an ac voltage of 120 V and are rated at 75 W and 150 W. Which bulb, if either, has the greater filament resistance?

11. An ac circuit contains only a generator and a resistor. Which one of the following changes leads to the greatest average power being delivered to the circuit? **(a)** Double the peak voltage of the generator and double the resistance. **(b)** Double the resistance. **(c)** Double the peak voltage of the generator. **(d)** Double the peak voltage of the generator and reduce the resistance by a factor of two.

20.6 Series Wiring

Thus far, we have dealt with circuits that include only a single device, such as a light bulb or a loudspeaker. There are, however, many circuits in which more than one device is connected to a voltage source. This section introduces one method by which such connections may be made—namely, series wiring. ***Series wiring means that the devices are connected in such a way that there is the same electric current through each device.*** Figure 20.14 shows a circuit in which two different devices, represented by resistors R_1 and R_2, are connected in series with a battery. Note that if the current in one resistor is interrupted, the current in the other is too. This could occur, for example, if two light bulbs were connected in series and the filament of one bulb broke. Because of the series wiring, the voltage V supplied by the battery is divided between the two resistors. The drawing indicates that the portion of the voltage across R_1 is V_1, while the portion across R_2 is V_2, so $V = V_1 + V_2$. For the individual resistances, the definition of resistance indicates that $R_1 = V_1/I$ and $R_2 = V_2/I$, so that $V_1 = IR_1$ and $V_2 = IR_2$. Therefore, we have

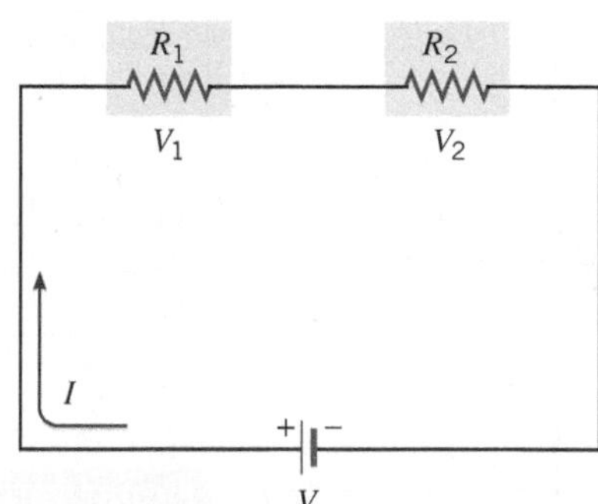

FIGURE 20.14 When two resistors are connected in series, the same current I is in both of them.

$$V = V_1 + V_2 = IR_1 + IR_2 = I(R_1 + R_2) = IR_S$$

where R_S is called the **equivalent resistance** of the series circuit. Thus, two resistors in series are equivalent to a single resistor whose resistance is $R_S = R_1 + R_2$, in the sense that there is the same current through R_S as there is through the series combination of R_1 and R_2. This line of reasoning can be extended to any number of resistors in series if we note the following:

Problem-Solving Insight **The voltage across all the resistors in series is the sum of the individual voltages across each resistor.**

The result for the equivalent resistance is

Series resistors $$R_S = R_1 + R_2 + R_3 + \cdots \quad (20.16)$$

Examples 8 and 9 illustrate the concept of equivalent resistance in series circuits.

EXAMPLE 8 | A Series Circuit

Suppose that the resistances in **Figure 20.14** are $R_1 = 47\ \Omega$ and $R_2 = 86\ \Omega$, and the battery voltage is 24 V. Determine the equivalent resistance of the two resistors and the current in the circuit.

Reasoning The two resistors are wired in series, since there is the same current through each one. The equivalent resistance R_S of a series circuit is the sum of the individual resistances, so $R_S = R_1 + R_2$. The current I can be obtained from Ohm's law as the voltage V divided by the equivalent resistance: $I = V/R_S$.

Solution The equivalent resistance is

$$R_s = R_1 + R_2 = 47\ \Omega + 86\ \Omega = \boxed{133\ \Omega} \quad (20.16)$$

The current in the circuit is

$$I = \frac{V}{R_s} = \frac{24\ \text{V}}{133\ \Omega} = \boxed{0.18\ \text{A}} \quad (20.2)$$

Analyzing Multiple-Concept Problems

EXAMPLE 9 | Power Delivered to a Series Circuit

A 6.00-Ω resistor and a 3.00-Ω resistor are connected in series with a 12.0-V battery, as **Figure 20.15** indicates. Assuming that the battery contributes no resistance to the circuit, find the power delivered to each of the resistors.

Reasoning The power P delivered to each resistor is the product of the current squared (I^2) and the corresponding resistance R, or $P = I^2R$. The resistances are known, and Ohm's law can be used to find the current. Ohm's law states that the current in the circuit (which is also the current through each of the resistors) equals the voltage V of the battery divided by the equivalent resistance R_S of the two resistors: $I = V/R_S$. Since the resistors are connected in series, we can obtain the equivalent resistance by adding the two resistances (see **Figure 20.15**).

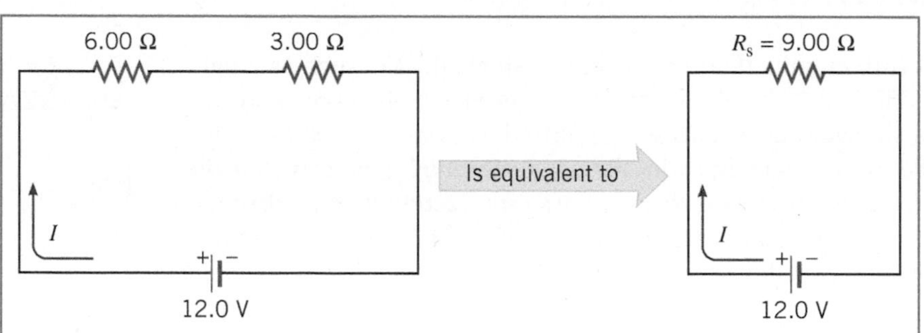

FIGURE 20.15 A 6.00-Ω and a 3.00-Ω resistor connected in series are equivalent to a single 9.00-Ω resistor.

Knowns and Unknowns The data for this problem are:

Description	Symbol	Value
Resistance of 6.00-Ω resistor	R_1	6.00 Ω
Resistance of 3.00-Ω resistor	R_2	3.00 Ω
Battery voltage	V	12.0 V
Unknown Variables		
Power delivered to 6.00-Ω resistor	P_1	?
Power delivered to 3.00-Ω resistor	P_2	?

Modeling the Problem

STEP 1 Power The power P_1 delivered to the 6.00-Ω resistor is given by $P_1 = I^2R_1$ (Equation 20.6b), where I is the current through the resistor and R_1 is the resistance. In this expression R_1 is a known quantity, and the current I will be determined in Step 2.	$P_1 = I^2R_1$ ↑ ?	**(20.6b)**
STEP 2 Ohm's Law The current I in the circuit depends on the voltage V of the battery and the equivalent resistance R_S of the two resistors in series (see **Figure 20.15**). This dependence is given by Ohm's law (Equation 20.2) as $\boxed{I = \frac{V}{R_S}}$ This result for the current can be substituted into Equation 20.6b, as indicated at the right. Note from the data table that the voltage is given. In Step 3 we will evaluate the equivalent resistance from the individual resistances R_1 and R_2.	$P_1 = I^2R_1$ ↑ $I = \frac{V}{R_S}$ ↑ ?	**(20.6b)** **(1)**
STEP 3 Equivalent Resistance Since the two resistors are wired in series, the equivalent resistance R_S is the sum of the two resistances (Equation 20.16): $\boxed{R_S = R_1 + R_2}$ The resistances R_1 and R_2 are known. We substitute this expression for R_S into Equation 1, as shown in the right column.	$P_1 = I^2R_1$ ↑ $I = \frac{V}{R_S}$ ↑ $R_S = R_1 + R_2$	**(20.6b)** **(1)** **(20.16)**

Solution Algebraically combining the results of the three steps, we have

$$P_1 \overset{\text{STEP 1}}{=} I^2R_1 \overset{\text{STEP 2}}{=} \left(\frac{V}{R_S}\right)^2 R_1 \overset{\text{STEP 2}}{=} \left(\frac{V}{R_1 + R_2}\right)^2 R_1$$

The power delivered to the 6.00-Ω resistor is

$$P_1 = \left(\frac{V}{R_1 + R_2}\right)^2 R_1 = \left(\frac{12.0\ \text{V}}{6.00\ \Omega + 3.00\ \Omega}\right)^2 (6.00\ \Omega) = \boxed{10.7\ \text{W}}$$

In a similar fashion, it can be shown that the power delivered to the 3.00-Ω resistor is

$$P_2 = \left(\frac{V}{R_1 + R_2}\right)^2 R_2 = \left(\frac{12.0\ \text{V}}{6.00\ \Omega + 3.00\ \Omega}\right)^2 (3.00\ \Omega) = \boxed{5.3\ \text{W}}$$

Related Homework: Problems 37, 39

In Example 9 the total power sent to the two resistors is $P = 10.7\ \text{W} + 5.3\ \text{W} = 16.0\ \text{W}$. Alternatively, the total power could have been obtained by using the voltage across the two resistors (the battery voltage) and the equivalent resistance R_S:

$$P = \frac{V^2}{R_S} = \frac{(12.0\ \text{V})^2}{6.00\ \Omega + 3.00\ \Omega} = 16.0\ \text{W} \qquad \textbf{(20.6c)}$$

Problem-Solving Insight The total power delivered to any number of resistors in series is equal to the power delivered to the equivalent resistance.

THE PHYSICS OF . . . personal digital assistants. Pressure-sensitive pads form the heart of computer input devices that function as personal digital assistants, or PDAs, and they offer an interesting application of series resistors. These devices are simple to use. You write directly on the pad with a plastic stylus that itself contains no electronics (see **Figure 20.16**). The writing appears as the stylus is moved, and recognition software interprets it as input for the built-in computer. The pad utilizes two transparent conductive layers that are separated by a small distance, except where pressure from the stylus brings them into contact (see point P in the drawing). Current I enters the positive side of the top layer, flows into the bottom layer through point P, and exits that layer through its negative side. Each layer provides resistance to the current, the amounts depending on where the point P is located. As the right side of the drawing indicates, the resistances from the layers are in series, since the same current exists in both of

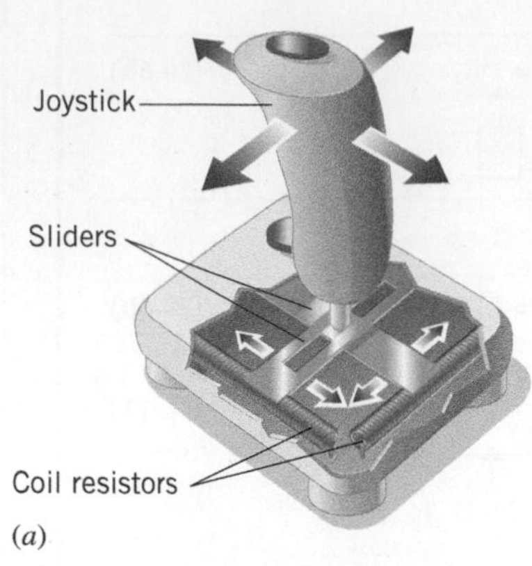

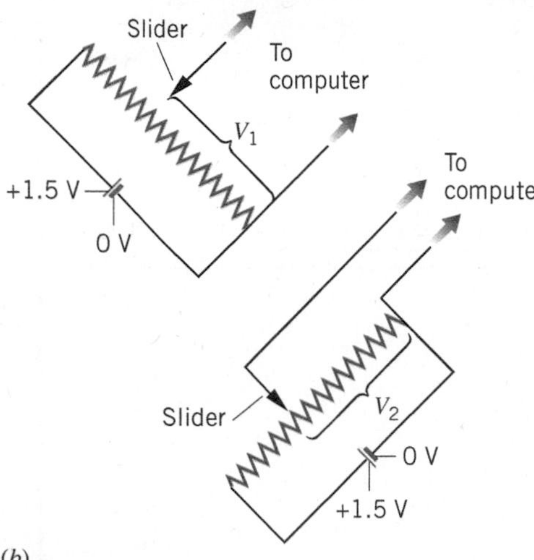

FIGURE 20.17 (*a*) A joystick uses two perpendicular movable sliders, and each makes contact with a coil resistor. (*b*) The sliders allow detection of the voltages V_1 and V_2, which a computer translates into positional data.

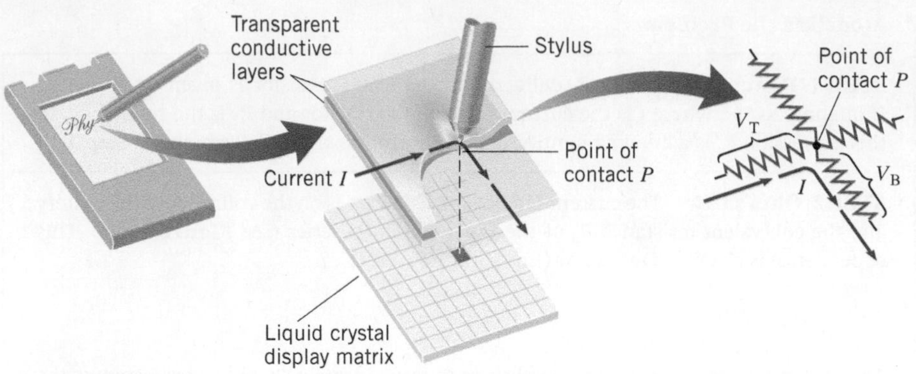

FIGURE 20.16 The pressure pad on which the user writes in a personal digital assistant is based on the use of resistances that are connected in series.

them. The voltage across the top-layer resistance is V_T, and the voltage across the bottom-layer resistance is V_B. In each case, the voltage is the current times the resistance. These two voltages are used to locate the point *P* and to activate (darken) one of the elements or pixels in a liquid crystal display matrix that lies beneath the transparent layers (see Section 24.6 for a discussion of liquid crystal displays). As the stylus is moved, the writing becomes visible as one element after another in the display matrix is activated.

THE PHYSICS OF . . . a joystick. The joystick, found in computer games, also takes advantage of resistors connected in series. A joystick contains two straight coils of resistance wire that are oriented at 90° to each other (see Figure 20.17*a*). When the joystick is moved, it repositions the metallic slider on each of the coils. As part *b* of the drawing illustrates, each coil is connected across a 1.5-V battery.* Because one end of a coil is at 1.5 V and the other at 0 V, the voltage at the location of the slider is somewhere between these values; the voltage of the left slider in the drawing is labeled V_1 and the voltage of the right slider is V_2. The slider voltages are sent via wires to a computer, which translates them into positional data. In effect, the slider divides each resistance coil into two smaller resistance coils wired in series, and allows the voltage at the point where they are joined together to be detected.

Check Your Understanding

(*The answer is given at the end of the book.*)

12. The power rating of a 1000-W heater specifies the power that the heater uses when it is connected to an ac voltage of 120 V. What is the total power used by two of these heaters when they are connected in series with a single ac voltage of 120 V? **(a)** 4000 W **(b)** 3000 W **(c)** 2000 W **(d)** 1000 W **(e)** 500 W

20.7 Parallel Wiring

Parallel wiring is another method of connecting electrical devices. ***Parallel wiring means that the devices are connected in such a way that the same voltage is applied across each device.*** Figure 20.18 shows two resistors connected in parallel between the terminals of a battery. Part *a* of the picture is drawn so as to emphasize that the entire voltage of the battery is applied across each resistor. Actually, parallel connections are

*For clarity, two batteries are shown in Figure 20.17*b*, one associated with each resistance coil. In reality, both coils are connected across a single battery.

rarely drawn in this manner; instead they are drawn as in part *b*, where the dots indicate the points where the wires for the two branches are joined together. **Figures 20.18*a*** and ***b*** are equivalent representations of the same circuit.

Parallel wiring is very common. For example, when an electrical appliance is plugged into a wall socket, the appliance is connected in parallel with other appliances, as in **Figure 20.19**, where the entire voltage of 120 V is applied across each one of the devices: the television, the stereo, and the light bulb (when the switch is turned on). The presence of the unused socket or other devices that are turned off does not affect the operation of those devices that are turned on. Moreover, if the current in one device is interrupted (perhaps by an opened switch or a broken wire), the current in the other devices is not interrupted. In contrast, if household appliances were connected in series, there would be no current through any appliance if the current in the circuit were halted at any point.

When two resistors R_1 and R_2 are connected as in **Figure 20.18**, each receives current from the battery as if the other were not present. Therefore, R_1 and R_2 together draw more current from the battery than does either resistor alone. According to the definition of resistance, $R = V/I$, a larger current implies a smaller resistance. Thus, the two parallel resistors behave as a single equivalent resistance that is *smaller* than either R_1 or R_2. **Figure 20.20** returns to the water-flow analogy to provide additional insight into this important feature of parallel wiring. In part *a*, two sections of pipe that have the same length are connected in parallel with a pump. In part *b* these two sections have been replaced with a single pipe of the same length, whose cross-sectional area equals the combined cross-sectional areas of section 1 and section 2. The pump (analogous to a voltage source) can push more water per second (analogous to current) through the wider pipe in part *b* (analogous to a wider wire) than it can through *either* of the narrower pipes (analogous to narrower wires) in part *a*. In effect, the wider pipe offers less resistance to the flow of water than either of the narrower pipes offers individually.

As in a series circuit, it is possible to replace a parallel combination of resistors with an equivalent resistor that results in the same total current and power for a given voltage as the original combination. To determine the equivalent resistance for the two resistors in **Figure 20.18*b***, note that the total current I from the battery is the sum of I_1 and I_2, where I_1 is the current in resistor R_1 and I_2 is the current in resistor R_2: $I = I_1 + I_2$. Since the same voltage V is applied across each resistor, the definition of resistance indicates that $I_1 = V/R_1$ and $I_2 = V/R_2$. Therefore,

$$I = I_1 + I_2 = \frac{V}{R_1} + \frac{V}{R_2} = V\left(\frac{1}{R_1} + \frac{1}{R_2}\right) = V\left(\frac{1}{R_P}\right)$$

where R_P is the equivalent resistance. Hence, when two resistors are connected in parallel, they are equivalent to a single resistor whose resistance R_P can be obtained from $1/R_P = 1/R_1 + 1/R_2$.

Problem-Solving Insight For any number of resistors wired in parallel, the total current from the voltage source is the sum of the currents in the individual resistors.

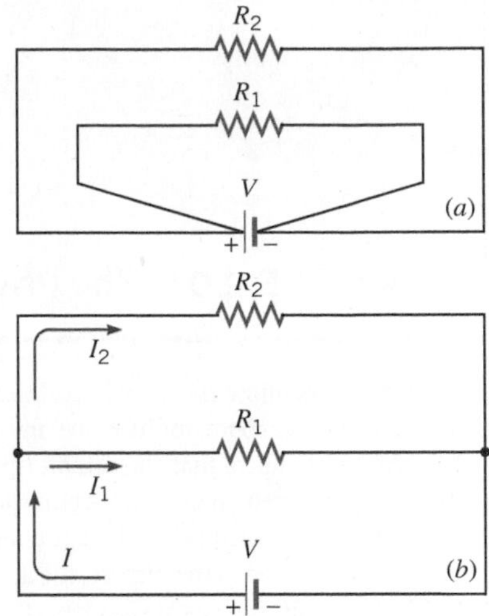

FIGURE 20.18 (*a*) When two resistors are connected in parallel, the same voltage V is applied across each resistor. (*b*) This drawing is equivalent to part *a*. I_1 and I_2 are the currents in R_1 and R_2.

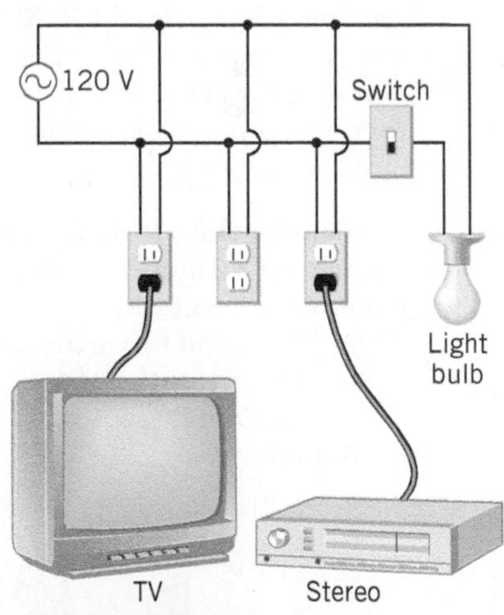

FIGURE 20.19 This drawing shows some of the parallel connections found in a typical home. Each wall socket provides 120 V to the appliance connected to it. In addition, 120 V is applied to the light bulb when the switch is turned on.

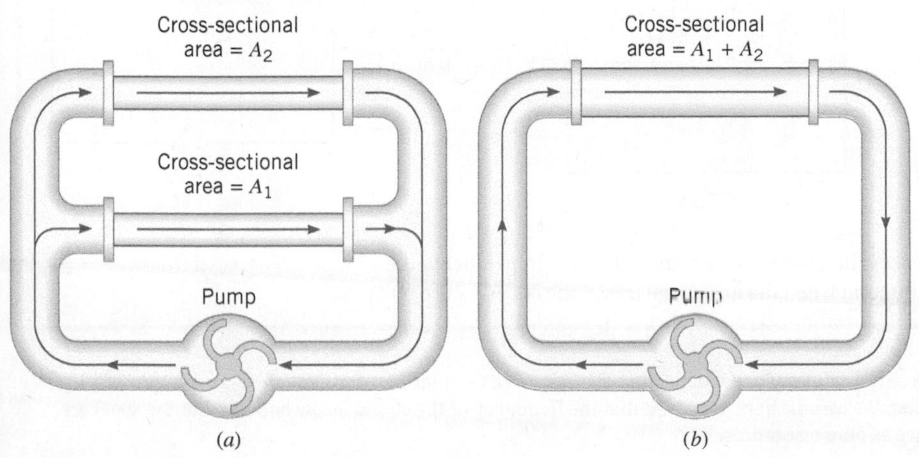

FIGURE 20.20 (*a*) Two equally long pipe sections, with cross-sectional areas A_1 and A_2, are connected in parallel to a water pump. (*b*) The two parallel pipe sections in part *a* are equivalent to a single pipe of the same length whose cross-sectional area is $A_1 + A_2$.

Thus, a similar line of reasoning reveals that the equivalent resistance is

Parallel resistors $$\frac{1}{R_P} = \frac{1}{R_1} + \frac{1}{R_2} + \frac{1}{R_3} + \cdots \quad \textbf{(20.17)}$$

The next example deals with a parallel combination of resistors that occurs in a stereo system.

EXAMPLE 10 | The Physics of Main and Remote Stereo Speakers

Most receivers allow the user to connect "remote" speakers (to play music in another room, for instance) in addition to the main speakers. **Figure 20.21** shows that the remote speaker and the main speaker for the right stereo channel are connected to the receiver in parallel (for clarity, the speakers for the left channel are not shown). At the instant represented in the picture, the ac voltage across the speakers is 6.00 V. The main-speaker resistance is 8.00 Ω, and the remote-speaker resistance is 4.00 Ω.* Determine **(a)** the equivalent resistance of the two speakers, **(b)** the total current supplied by the receiver, **(c)** the current in each speaker, and **(d)** the power dissipated in each speaker.

Reasoning The total current supplied to the two speakers by the receiver can be calculated as $I_{rms} = V_{rms}/R_P$, where R_P is the equivalent resistance of the two speakers in parallel and can be obtained from $1/R_P = 1/R_1 + 1/R_2$. The current in each speaker is different, however, since the speakers have different resistances. The average power delivered to a given speaker is the product of its current and voltage. In the parallel connection the same voltage is applied to each speaker.

Problem-Solving Insight The equivalent resistance R_P of a number of resistors in parallel has a reciprocal given by $R_P^{-1} = R_1^{-1} + R_2^{-1} + R_3^{-1} + \ldots$, where R_1, R_2, and R_3 are the individual resistances. After adding together the reciprocals R_1^{-1}, R_2^{-1}, and R_3^{-1}, do not forget to take the reciprocal of the result to find R_P.

Solution **(a)** According to Equation 20.17, the equivalent resistance of the two speakers is given by

$$\frac{1}{R_P} = \frac{1}{8.00\ \Omega} + \frac{1}{4.00\ \Omega} = \frac{3}{8.00\ \Omega} \quad \text{or} \quad R_P = \frac{8.00\ \Omega}{3} = \boxed{2.67\ \Omega}$$

This result is illustrated in part *b* of the drawing.

(b) Using the equivalent resistance in Ohm's law shows that the total current is

$$I_{rms} = \frac{V_{rms}}{R_P} = \frac{6.00\ \text{V}}{2.67\ \Omega} = \boxed{2.25\ \text{A}} \quad \textbf{(20.14)}$$

(c) Applying Ohm's law to each speaker gives the individual speaker currents:

8.00-Ω speaker $$I_{rms} = \frac{V_{rms}}{R} = \frac{6.00\ \text{V}}{8.00\ \Omega} = \boxed{0.750\ \text{A}}$$

4.00-Ω speaker $$I_{rms} = \frac{V_{rms}}{R} = \frac{6.00\ \text{V}}{4.00\ \Omega} = \boxed{1.50\ \text{A}}$$

The sum of these currents is equal to the total current obtained in part (b).

(d) The average power dissipated in each speaker can be calculated using $\overline{P} = I_{rms}V_{rms}$ with the individual currents obtained in part (c):

8.00-Ω speaker $$\overline{P} = (0.750\ \text{A})(6.00\ \text{V}) = \boxed{4.50\ \text{W}} \quad \textbf{(20.15a)}$$

4.00-Ω speaker $$\overline{P} = (1.50\ \text{A})(6.00\ \text{V}) = \boxed{9.00\ \text{W}} \quad \textbf{(20.15a)}$$

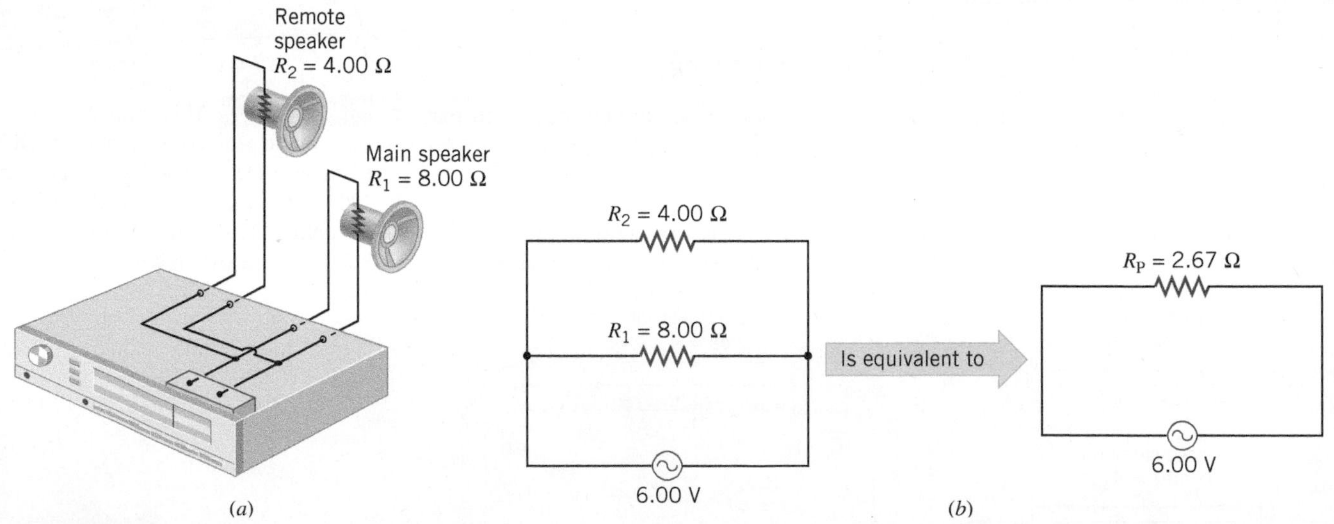

FIGURE 20.21 (*a*) The main and remote speakers in a stereo system are connected in parallel to the receiver. (*b*) The circuit schematic shows the situation when the ac voltage across the speakers is 6.00 V.

*In reality, frequency-dependent characteristics (see Chapter 23) play a role in the operation of a loudspeaker. We assume here, however, that the frequency of the sound is low enough that the speakers behave as pure resistances.

In Example 10, the total power delivered by the receiver is the sum of the individual values that were found in part (d), $\overline{P} = 4.50\text{ W} + 9.00\text{ W} = 13.5\text{ W}$. Alternatively, the total power can be obtained from the equivalent resistance $R_P = 2.67\ \Omega$ and the total current in part (b):

$$\overline{P} = I_{rms}^2 R_P = (2.25\text{ A})^2(2.67\ \Omega) = 13.5\text{ W} \quad \textbf{(20.15b)}$$

Problem-Solving Insight The total power delivered to any number of resistors in parallel is equal to the power delivered to the equivalent resistor.

In a parallel combination of resistances, it is the *smallest* resistance that has the largest impact in determining the equivalent resistance. In fact, if one resistance approaches zero, then according to Equation 20.17, the equivalent resistance also approaches zero. In such a case, the near-zero resistance is said to *short out* the other resistances by providing a near-zero resistance path for the current to follow as a shortcut around the other resistances.

An interesting application of parallel wiring occurs in a three-way light bulb, as Conceptual Example 11 discusses.

CONCEPTUAL EXAMPLE 11 | The Physics of a Three-Way Light Bulb

Three-way light bulbs are popular because they can provide three levels of illumination (e.g., 50 W, 100 W, and 150 W) using a 120-V socket. The socket, however, must be equipped with a special three-way switch that enables one to select the illumination level. This switch does not select different voltages, because a three-way bulb uses a single voltage of 120 V. Within the bulb are two separate filaments. When the bulb is producing its highest illumination level and one of the filaments burns out (i.e., vaporizes), the bulb shines at one of the other illumination levels (either the lowest or the intermediate one). When the bulb is set to its highest level of illumination, how are the two filaments connected, **(a)** in parallel or **(b)** in series?

Reasoning In a series connection, the filaments would be connected in such a way that there is the same current through each one. The current would enter one filament and then leave that filament and enter into the other one. In a parallel connection, the same voltage would be applied across each filament, but the current through each would, in general, be different, the two currents existing independently of one another.

Answer (b) is incorrect. If the filaments were wired in series and one of them burned out, no current would pass through the bulb and none of the illumination levels would be available, contrary to what is observed. Therefore, the filaments are not wired in series.

Answer (a) is correct. Since the filaments are not in series, they must be in parallel, as **Interactive Figure 20.22** helps to explain. The power dissipated in a resistance R is $\overline{P} = V_{rms}^2/R$, according to Equation 20.15c. With a single value of 120 V for the voltage V_{rms}, three different power ratings for the bulb can be obtained only if three different values for the resistance R are available. In a 50-W/100-W/150-W bulb, for example, one resistance R_{50} is provided by the 50-W filament, and the second resistance R_{100} comes from the 100-W filament. The third resistance R_{150} is the parallel combination of the other two and can be obtained from $1/R_{150} = 1/R_{50} + 1/R_{100}$. **Interactive Figure 20.22** illustrates a simplified version of how the three-way switch operates in such a bulb. The first position of the switch closes contact A and leaves contact B open, energizing only the 50-W filament. The second position closes contact B and leaves contact A open, energizing only the 100-W filament. The third position closes both contacts A and B, so that both filaments light up to give the highest level of illumination.

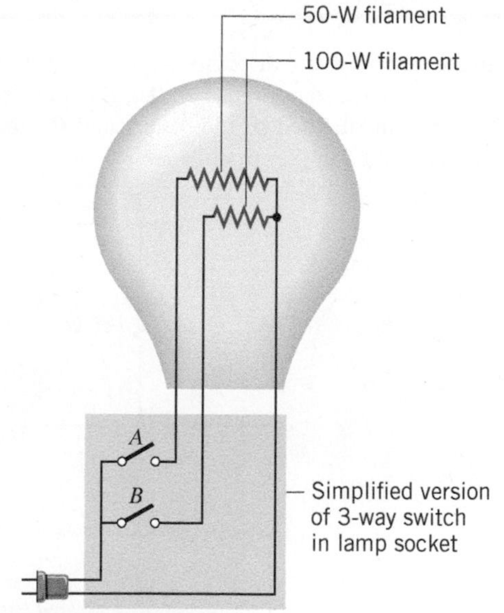

INTERACTIVE FIGURE 20.22 A three-way light bulb uses two connected filaments. The filaments can be turned on one at a time or both together in parallel.

Related Homework: Problem 44

Check Your Understanding

(The answers are given at the end of the book.)

13. A car has two headlights, and their power is derived from the car's battery. The filament in one burns out, but the other headlight stays on. Are the headlights connected in series or in parallel?

14. Two identical light bulbs are connected to identical batteries in two different ways. In method A the bulbs are connected in parallel, and the parallel combination is connected between the one battery's terminals. In method B the bulbs are connected in series, and the series combination is connected between the other battery's terminals. What is the ratio of the power supplied by the battery in method A to the power supplied in method B? **(a)** $\frac{1}{4}$ **(b)** 4 **(c)** $\frac{1}{2}$ **(d)** 2 **(e)** 1

20.8 Circuits Wired Partially in Series and Partially in Parallel

Often an electric circuit is wired partially in series and partially in parallel. The key to determining the current, voltage, and power in such a case is to deal with the circuit in parts, with the resistances in each part being either in series or in parallel with each other. Example 12 shows how such an analysis is carried out.

EXAMPLE 12 | A Four-Resistor Circuit

Figure 20.23*a* shows a circuit composed of a 24-V battery and four resistors, whose resistances are 110, 180, 220, and 250 Ω. Find **(a)** the total current supplied by the battery and **(b)** the voltage between points *A* and *B* in the circuit.

Reasoning The total current that is supplied by the battery can be obtained from Ohm's law, $I = V/R$, where R is the equivalent resistance of the four resistors. The equivalent resistance can be calculated by dealing with the circuit in parts. The voltage V_{AB} between the two

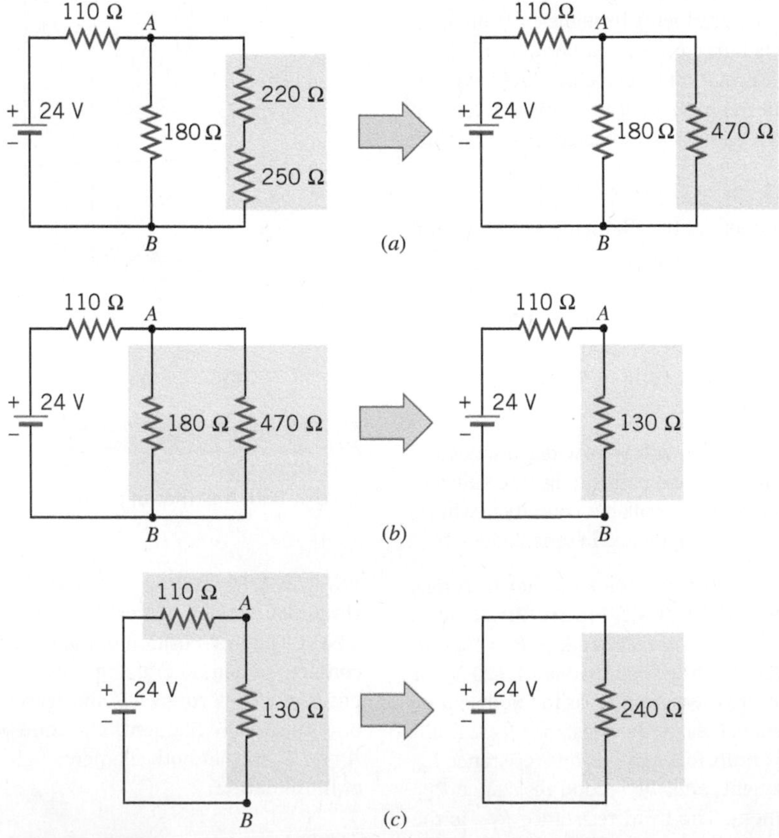

FIGURE 20.23 The circuits shown in this picture are equivalent.

points A and B is also given by Ohm's law, $V_{AB} = IR_{AB}$, where I is the current and R_{AB} is the equivalent resistance between the two points.

Solution **(a)** The 220-Ω resistor and the 250-Ω resistor are in series, so they are equivalent to a single resistor whose resistance is 220 Ω + 250 Ω = 470 Ω (see **Figure 20.23*a***). The 470-Ω resistor is in parallel with the 180-Ω resistor. Their equivalent resistance can be obtained from Equation 20.17:

$$\frac{1}{R_{AB}} = \frac{1}{470\ \Omega} + \frac{1}{180\ \Omega} = 0.0077\ \Omega^{-1}$$

$$R_{AB} = \frac{1}{0.0077\ \Omega^{-1}} = 130\ \Omega$$

The circuit is now equivalent to a circuit containing a 110-Ω resistor in series with a 130-Ω resistor (see **Figure 20.23*b***). This combination behaves like a single resistor whose resistance is $R = 110\ \Omega + 130\ \Omega = 240\ \Omega$ (see **Figure 20.23*c***). The total current from the battery is, then,

$$I = \frac{V}{R} = \frac{24\ \text{V}}{240\ \Omega} = \boxed{0.10\ \text{A}}$$

(b) The current $I = 0.10$ A passes through the resistance between points A and B. Therefore, Ohm's law indicates that the voltage across the 130-Ω resistor between points A and B is

$$V_{AB} = IR_{AB} = (0.10\ \text{A})(130\ \Omega) = \boxed{13\ \text{V}}$$

Check Your Understanding

(The answers are given at the end of the book.)

15. In one of the circuits in **CYU Figure 20.3** none of the resistors is in series or in parallel with any of the other resistors. Which circuit is it?

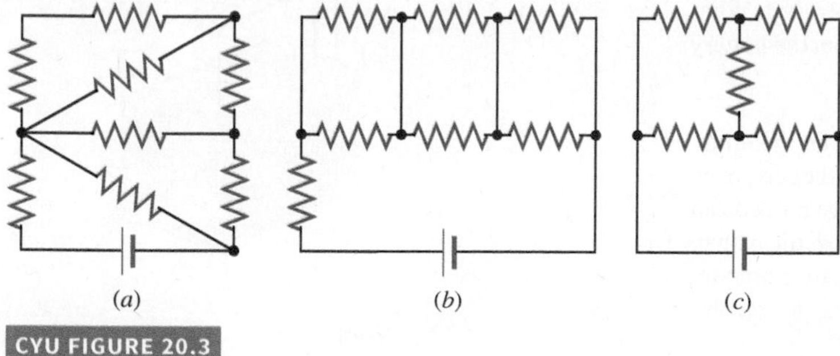

CYU FIGURE 20.3

16. You have three resistors, each of which has a resistance R. By connecting all three together in various ways, which one or more of the following resistance values can you obtain? **(a)** $3R$ **(b)** $\frac{3}{2}R$ **(c)** R **(d)** $\frac{2}{3}R$ **(e)** $\frac{1}{3}R$

17. You have four resistors, each of which has a resistance R. It is possible to connect these four together so that the equivalent resistance of the combination is also R. How many ways can you do it? There is more than one way.

20.9 Internal Resistance

So far, the circuits we have considered include batteries or generators that contribute only their emfs to a circuit. In reality, however, such devices also add some resistance. This resistance is called the **internal resistance** of the battery or generator because it is located inside the device. In a battery, the internal resistance is due to the chemicals within the battery. In a generator, the internal resistance is the resistance of wires and other components within the generator.

Figure 20.24 presents a schematic representation of the internal resistance r of a battery. The drawing emphasizes that when an external resistance R is connected to the battery, the resistance is connected *in series* with the internal resistance. The internal resistance of a functioning battery is typically small (several thousandths of an ohm for a new car battery). Nevertheless, the effect of the internal resistance may not be negligible. Example 13 illustrates that when current is drawn from a battery, the internal resistance causes the voltage between the terminals to drop below the maximum value specified by the battery's emf. The actual voltage between the terminals of a battery is known as the **terminal voltage**.

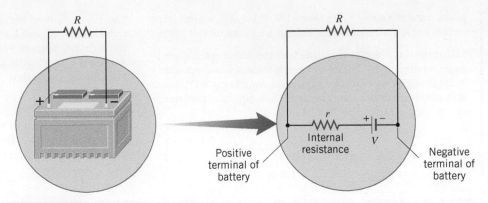

FIGURE 20.24 When an external resistance R is connected between the terminals of a battery, the resistance is connected in series with the internal resistance r of the battery.

EXAMPLE 13 | The Physics of Automobile Batteries

Figure 20.25 shows a car battery whose emf is 12.0 V and whose internal resistance is 0.010 Ω. This resistance is relatively large because the battery is old and the terminals are corroded. What is the terminal voltage when the current I drawn from the battery is **(a)** 10.0 A and **(b)** 100.0 A?

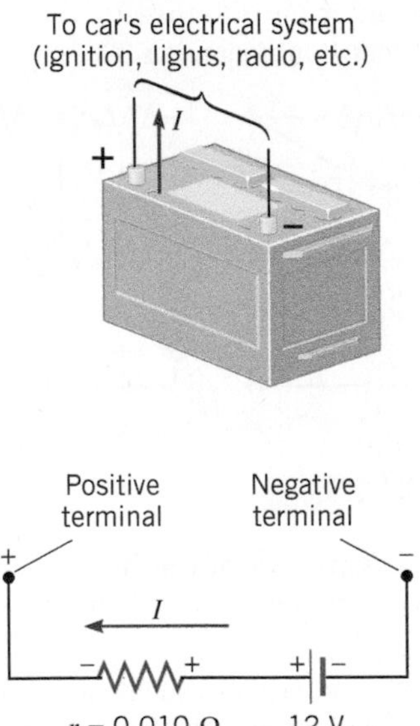

FIGURE 20.25 A car battery whose emf is 12 V and whose internal resistance is r.

Reasoning The voltage between the terminals is not the entire 12.0-V emf, because part of the emf is needed to make the current go through the internal resistance. The amount of voltage needed can be determined from Ohm's law as the current I through the battery times the internal resistance r. For larger currents, a larger amount of voltage is needed, leaving less of the emf between the terminals.

Solution **(a)** The voltage needed to make a current of $I = 10.0$ A go through an internal resistance of $r = 0.010\ \Omega$ is

$$V = Ir = (10.0\ \text{A})(0.010\ \Omega) = 0.10\ \text{V}$$

To find the terminal voltage, remember that the direction of conventional current is always from a higher toward a lower potential. To emphasize this fact in the drawing, plus and minus signs have been included at the right and left ends, respectively, of the resistance r. The terminal voltage can be calculated by starting at the negative terminal and keeping track of how the voltage increases and decreases as we move toward the positive terminal. The voltage rises by 12.0 V due to the battery's emf. However, the voltage drops by 0.10 V because of the potential difference across the internal resistance. Therefore, the terminal voltage is $12.0\ \text{V} - 0.10\ \text{V} = \boxed{11.9\ \text{V}}$.

(b) When the current through the battery is 100.0 A, the voltage needed to make the current go through the internal resistance is

$$V = Ir = (100.0\ \text{A})(0.010\ \Omega) = 1.0\ \text{V}$$

The terminal voltage now decreases to $12.0\ \text{V} - 1.0\ \text{V} = \boxed{11.0\ \text{V}}$.

Example 13 indicates that the terminal voltage of a battery is smaller when the current drawn from the battery is larger, an effect that any car owner can demonstrate. Turn the headlights on before starting your car, so that the current through the battery is about 10 A, as in part (a) of Example 13. Then start the car. The starter motor draws a large amount of additional current from the battery, momentarily increasing the total current by an appreciable amount. Consequently, the terminal voltage of the battery decreases, causing the headlights to become dimmer.

20.10 Kirchhoff's Rules

Electric circuits that contain a number of resistors can often be analyzed by combining individual groups of resistors in series and parallel, as Section 20.8 discusses. However, there are many circuits in which no two resistors are in series or in parallel. To deal with such circuits it is necessary to employ methods other than the series–parallel method. One alternative is to take advantage of Kirchhoff's rules, named after their developer Gustav Kirchhoff (1824–1887). There are two rules, the **junction rule** and the **loop rule**, and both arise from principles and ideas that we have encountered earlier. The junction rule is an application of the law of conservation of electric charge (see Section 18.2) to the electric current in a circuit. The loop rule is an application of the principle of conservation of energy (see Section 6.8) to the electric potential (see Section 19.2) that exists at various places in a circuit.

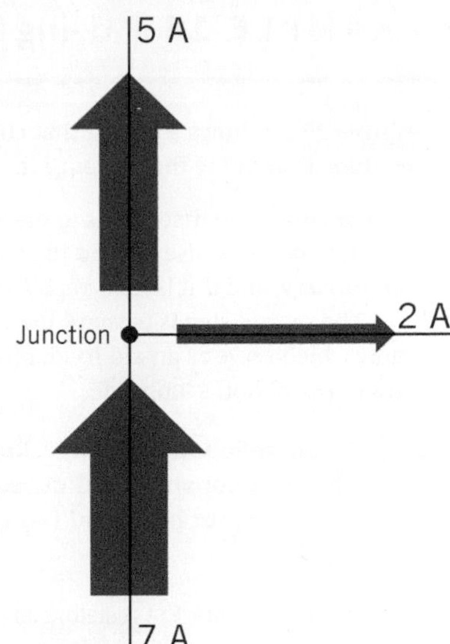

FIGURE 20.26 A junction is a point in a circuit where a number of wires are connected together. If 7 A of current is directed into the junction, then a total of 7 A (5 A + 2 A) of current must be directed out of the junction.

Figure 20.26 illustrates in greater detail the basic idea behind Kirchhoff's junction rule. The picture shows a junction where several wires are connected together. As Section 18.2 discusses, electric charge is conserved. Therefore, since there is no accumulation of charges at the junction itself, the total charge per second flowing into the junction must equal the total charge per second flowing out of it. In other words, ***the junction rule states that the total current directed into a junction must equal the total current directed out of the junction,*** or 7 A = 5 A + 2 A for the specific case shown in the picture.

To help explain Kirchhoff's loop rule, **Figure 20.27** shows a circuit in which a 12-V battery is connected to a series combination of a 5-Ω and a 1-Ω resistor. The plus and minus signs associated with each resistor remind us that, outside a battery, conventional current is directed from a higher toward a lower potential. From left to right, there is a potential drop of 10 V across the first resistor and another drop of 2 V across the second resistor. Keeping in mind that potential is the electric potential energy per unit charge, let us follow a positive test charge clockwise* around the circuit. Starting at the negative terminal of the battery, we see that the test charge gains potential energy because of the 12-V rise in potential due to the battery. The test charge then loses potential energy because of the 10-V and 2-V drops in potential across the resistors, ultimately arriving back at the negative terminal. In traversing the closed-circuit loop, the test charge is like a skier gaining gravitational potential energy in going up a hill on a chair lift and then losing it to friction in coming down and stopping. When the skier returns to the starting point, the gain equals the loss, so there is no net change in potential energy. Similarly, when the test charge arrives back at its starting point, there is no net change in electric potential energy, the gains matching the losses. ***The loop rule expresses this example of energy conservation in terms of the electric potential and states that for a closed-circuit loop, the total of all the potential drops is the same as the total of all the potential rises,*** or 10 V + 2 V = 12 V for the specific case in **Figure 20.27**.

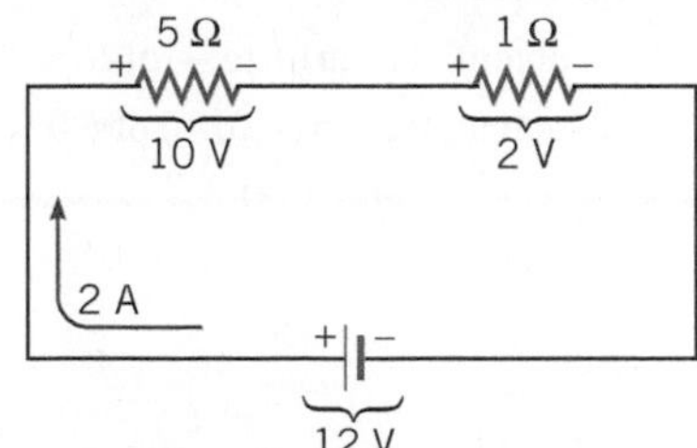

FIGURE 20.27 Following a positive test charge clockwise around the circuit, we see that the total voltage drop of 10 V + 2 V across the two resistors equals the voltage rise of 12 V due to the battery. The plus and minus signs on the resistors emphasize that, outside a battery, conventional current is directed from a higher potential (+) toward a lower potential (−).

Kirchhoff's rules can be applied to any circuit, even when the resistors are not in series or in parallel. The two rules are summarized below, and Examples 14 and 15 illustrate how to use them.

> KIRCHHOFF'S RULES
>
> **Junction rule. The sum of the magnitudes of the currents directed into a junction equals the sum of the magnitudes of the currents directed out of the junction.**
>
> **Loop rule. Around any closed-circuit loop, the sum of the potential drops equals the sum of the potential rises.**

*The choice of the clockwise direction is arbitrary.

EXAMPLE 14 | Using Kirchhoff's Loop Rule

Figure 20.28 shows a circuit that contains two batteries and two resistors. Determine the current I in the circuit.

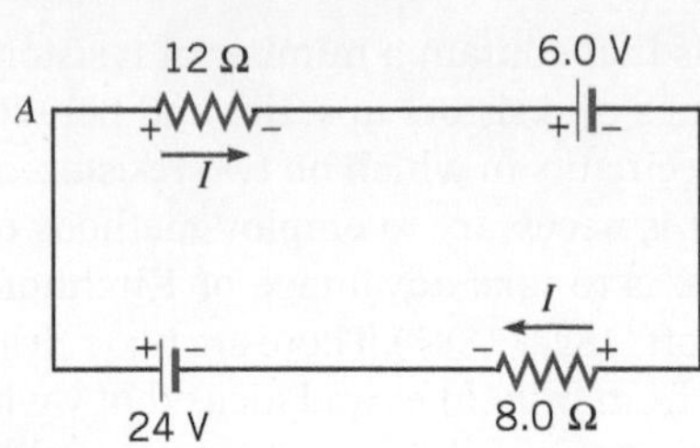

FIGURE 20.28 A single-loop circuit that contains two batteries and two resistors.

Reasoning The first step is to draw the current, which we have chosen to be clockwise around the circuit. The choice of direction is ***arbitrary***, and if it is incorrect, I will turn out to be negative.

The second step is to mark the resistors with plus and minus signs, which serve as an aid in identifying the potential drops and rises for Kirchhoff's loop rule.

> **Problem-Solving Insight** **Remember that, outside a battery, conventional current is always directed from a higher potential (+) toward a lower potential (–).**

Thus, we ***must*** mark the resistors as indicated in **Figure 20.28**, to be consistent with the clockwise direction chosen for the current.

We may now apply Kirchhoff's loop rule to the circuit, starting at corner A, proceeding clockwise around the loop, and identifying the potential drops and rises as we go. The potential across each resistor is given by Ohm's law as $V = IR$. The clockwise direction is arbitrary, and the same result is obtained with a counterclockwise path.

Solution Starting at corner A and moving clockwise around the loop, there is

1. A potential drop (+ to –) of $IR = I(12\ \Omega)$ across the 12-Ω resistor
2. A potential drop (+ to –) of 6.0 V across the 6.0-V battery
3. A potential drop (+ to –) of $IR = I(8.0\ \Omega)$ across the 8.0-Ω resistor
4. A potential rise (– to +) of 24 V across the 24-V battery

Setting the sum of the potential drops equal to the sum of the potential rises, as required by Kirchhoff's loop rule, gives

$$\underbrace{I(12\ \Omega) + 6.0\ \text{V} + I(8.0\ \Omega)}_{\text{Potential drops}} = \underbrace{24\ \text{V}}_{\text{Potential rises}}$$

Solving this equation for the current yields $\boxed{I = 0.90\ \text{A}}$. The current is a positive number, indicating that our initial choice for the direction (clockwise) of the current was correct.

EXAMPLE 15 | The Physics of an Automobile Electrical System

In a car, the headlights are connected to the battery and would discharge the battery if it were not for the alternator, which is run by the engine. **Interactive Figure 20.29** indicates how the car battery, headlights, and alternator are connected. The circuit includes an internal resistance of 0.0100 Ω for the car battery and its leads and a resistance of 1.20 Ω for the headlights. For the sake of simplicity, the alternator is approximated as an additional 14.00-V battery with an internal resistance of 0.100 Ω. Determine the currents through the car battery (I_B), the headlights (I_H), and the alternator (I_A).

Reasoning The drawing shows the directions chosen arbitrarily for the currents I_B, I_H, and I_A. If any direction is incorrect, the analysis will reveal a negative value for the corresponding current.

Next, we mark the resistors with the plus and minus signs that serve as an aid in identifying the potential drops and rises for

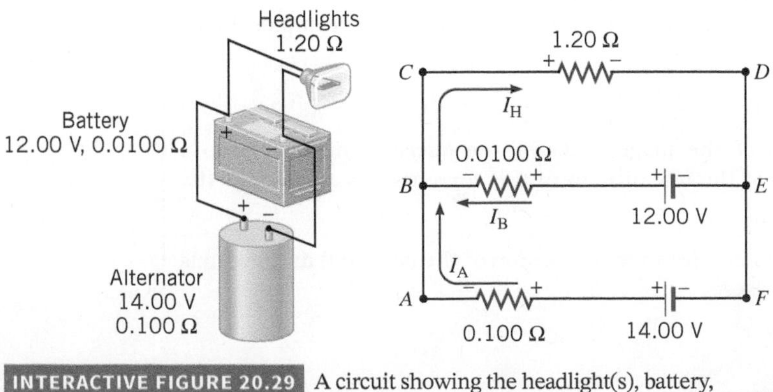

INTERACTIVE FIGURE 20.29 A circuit showing the headlight(s), battery, and alternator of a car.

the loop rule, recalling that, outside a battery, conventional current is always directed from a higher potential (+) toward a lower potential (−). Thus, given the directions selected for I_B, I_H, and I_A, the plus and minus signs ***must*** be those indicated in **Interactive Figure 20.29**.

Kirchhoff's junction and loop rules can now be used.

> **Problem-Solving Insight In problems involving Kirchhoff's rules, it is always helpful to mark the resistors with plus and minus signs to keep track of the potential rises and drops in the circuit.**

Solution The junction rule can be applied to junction *B* or junction *E* to give the same result:

Junction rule applied at B

$$\underbrace{I_A + I_B}_{\text{Into junction}} = \underbrace{I_H}_{\text{Out of junction}}$$

In applying the loop rule to the lower loop *BEFA*, we start at point *B*, move clockwise around the loop, and identify the potential drops and rises. There is a potential rise (− to +) of I_B (0.0100 Ω) across the 0.0100-Ω resistor and then a drop (+ to −) of 12.00 V due to the car battery. Continuing around the loop, we find a 14.00-V rise (− to +) across the alternator, followed by a potential drop (+ to −) of I_A (0.100 Ω) across the 0.100-Ω resistor. Setting the sum of the potential drops equal to the sum of the potential rises gives the following result:

Loop rule: BEFA

$$\underbrace{I_A(0.100\ \Omega) + 12.00\ \text{V}}_{\text{Potential drops}} = \underbrace{I_B(0.0100\ \Omega) + 14.00\ \text{V}}_{\text{Potential rises}}$$

Since there are three unknown variables in this problem, I_B, I_H, and I_A, a third equation is needed for a solution. To obtain the third equation, we apply the loop rule to the upper loop *CDEB*, choosing a clockwise path for convenience. The result is

Loop rule: CDEB

$$\underbrace{I_B(0.0100\ \Omega) + I_H(1.20\ \Omega)}_{\text{Potential drops}} = \underbrace{12.00\ \text{V}}_{\text{Potential rises}}$$

These three equations can be solved simultaneously to show that

$$\boxed{I_A = 19.1\ \text{A},\quad I_B = -9.0\ \text{A},\quad I_H = 10.1\ \text{A}}$$

The negative answer for I_B indicates that the current through the battery is not directed from right to left, as drawn in **Interactive Figure 20.29**. Instead, the 9.0-A current is directed from left to right, opposite to the way current would be directed if the alternator were not connected. It is the left-to-right current created by the alternator that keeps the battery charged.

Note that we can check our results by applying Kirchhoff's loop rule to the outer loop *ABCDEF*. If our results are correct, then the sum of the potential drops around this loop will be equal to the sum of the potential rises.

Math Skills The three equations that must be solved simultaneously for I_A, I_B, and I_H are

$$I_A + I_B = I_H \qquad (1)$$

$$I_A(0.100) + 12.00 = I_B(0.0100) + 14.00 \qquad (2)$$

$$I_B(0.0100) + I_H(1.20) = 12.00 \qquad (3)$$

For clarity, units have been omitted. Substituting Equation 1 for I_H into Equation 3 gives

$$I_B(0.0100) + \underbrace{(I_A + I_B)}_{I_H}(1.20) = 12.00 \text{ or}$$

$$I_B(1.21) + I_A(1.20) = 12.00 \qquad (4)$$

Solving Equation 4 for I_B shows that

$$I_B = \frac{12.00 - I_A(1.20)}{1.21} = 9.92 - I_A(0.992) \qquad (5)$$

Substituting Equation 5 for I_B into Equation 2, we find that

$$I_A(0.100) + 12.00 = \underbrace{[9.92 - I_A(0.992)](0.0100)}_{I_B} + 14.00$$

This result contains only the unknown variable I_A and can be solved to show that $\boxed{I_A = 19.1\ \text{A}}$. Using this value for I_A in Equation 5 reveals that

$$I_B = 9.92 - \underbrace{(19.1)}_{I_A}(0.992) = \boxed{-9.0\ \text{A}}$$

Finally, substituting the values for both I_A and I_B into Equation 1 yields

$$I_H = \underbrace{19.1}_{I_A} + \underbrace{(-9.0)}_{I_B} = \boxed{10.1\ \text{A}}$$

REASONING STRATEGY **Applying Kirchhoff's Rules**

1. **Draw the current in each branch of the circuit. Choose any direction. If your choice is incorrect, the value obtained for the current will turn out to be a negative number.**
2. **Mark each resistor with a plus sign at one end and a minus sign at the other end, in a way that is consistent with your choice for the current direction in Step 1. Outside a battery, conventional current is always directed from a higher potential (the end marked +) toward a lower potential (the end marked −).**
3. **Apply the junction rule and the loop rule to the circuit, obtaining in the process as many independent equations as there are unknown variables.**
4. **Solve the equations obtained in Step 3 simultaneously for the unknown variables. (See Appendix C.3.)**

Check Your Understanding

(The answer is given at the end of the book.)

18. CYU Figure 20.4 shows a circuit containing three resistors and three batteries. In preparation for applying Kirchhoff's rules, the currents in each resistor have been drawn. For these currents, write down the equations that result from applying Kirchhoff's junction rule and loop rule. Apply the loop rule to loops *ABCD* and *BEFC*.

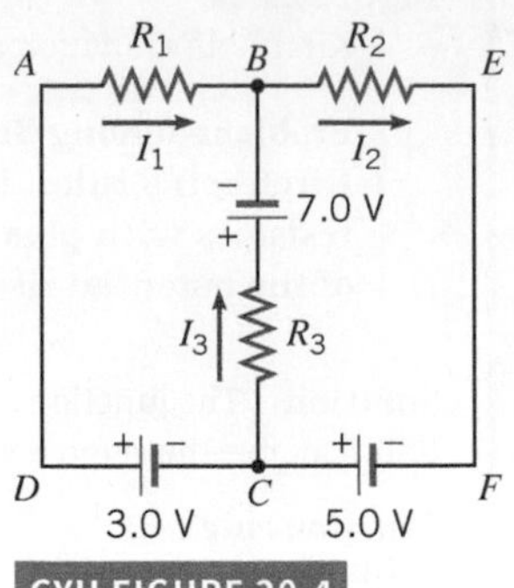

CYU FIGURE 20.4

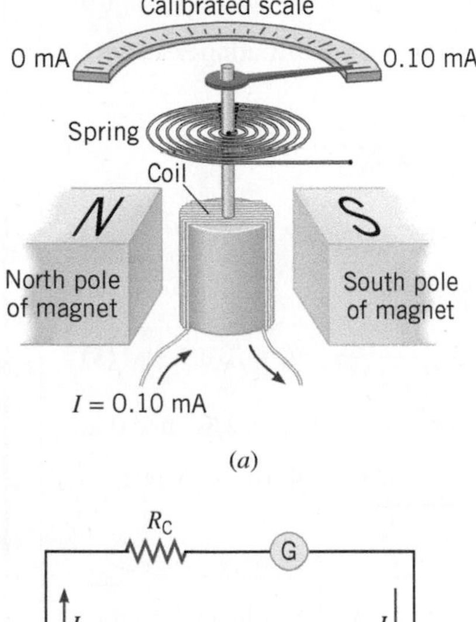

(*a*)

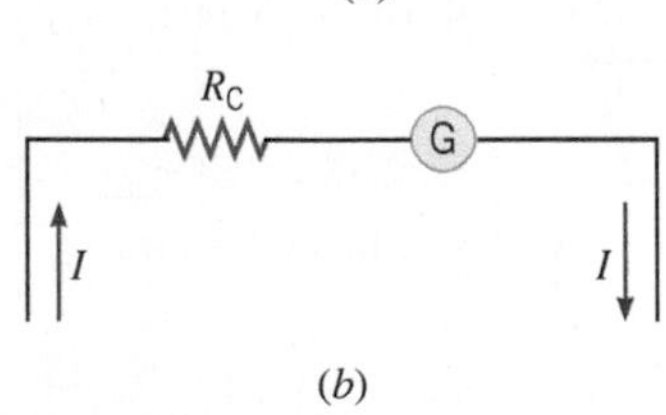

(*b*)

FIGURE 20.30 (*a*) A dc galvanometer. The coil of wire and pointer rotate when there is a current in the wire. (*b*) A galvanometer with a coil resistance of R_C is represented in a circuit diagram as shown here.

20.11 The Measurement of Current and Voltage

Current and voltage can be measured with devices known, respectively, as ammeters and voltmeters. There are two types of such devices: those that use digital electronics and those that do not. The essential feature of nondigital devices is the dc *galvanometer.* As **Figure 20.30*a*** illustrates, a galvanometer consists of a magnet, a coil of wire, a spring, a pointer, and a calibrated scale. The coil is mounted so that it can rotate, which causes the pointer to move in relation to the scale. The coil rotates in response to the torque applied by the magnet when there is a current in the coil (see Section 21.6). The coil stops rotating when this torque is balanced by the torque of the spring.

Two characteristics of a galvanometer are important when it is used as part of a measurement device. First, the amount of dc current that causes full-scale deflection of the pointer indicates the sensitivity of the galvanometer. For instance, **Figure 20.30*a*** shows an instrument that deflects full scale when the current in the coil is 0.10 mA. The second important characteristic is the resistance R_C of the wire in the coil. **Figure 20.30*b*** shows how a galvanometer with a coil resistance of R_C is represented in a circuit diagram.

THE PHYSICS OF . . . an ammeter. Since an **ammeter** is an instrument that measures current, it must be inserted in the circuit so the current passes directly through it, as **Figure 20.31** shows. (This is true for both digital and nondigital ammeters; **Figure 20.31** shows a digital instrument.)

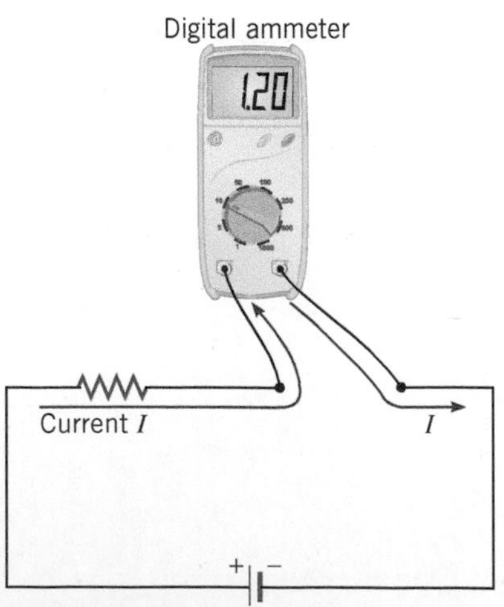

FIGURE 20.31 An ammeter must be inserted into a circuit so that the current passes directly through it.

A nondigital ammeter includes a galvanometer and one or more *shunt resistors,* which are connected in parallel with the galvanometer and provide a bypass for current in excess of the galvanometer's full-scale limit. The bypass allows the ammeter to be used to measure a current exceeding the full-scale limit. In **Figure 20.32**, for instance, a current of 60.0 mA enters terminal *A* of an ammeter (nondigital), which uses a galvanometer with a full-scale current of 0.100 mA. The shunt resistor *R* can be selected so that 0.100 mA of current enters the galvanometer, while 59.9 mA bypasses it. In such a case, the measurement scale on the ammeter would be labeled 0 to 60.0 mA. To determine the shunt resistance, it is necessary to know the coil resistance R_C (see Problem 76).

When an ammeter is inserted into a circuit, the equivalent resistance of the ammeter adds to the circuit resistance. Any increase in circuit resistance causes a reduction in current, and this is a problem, because an ammeter should only measure the current, not change it. Therefore, an *ideal* ammeter would have zero resistance. In practice, a good ammeter is designed with an equivalent resistance small enough so there is only a negligible reduction of the current in the circuit when the ammeter is inserted.

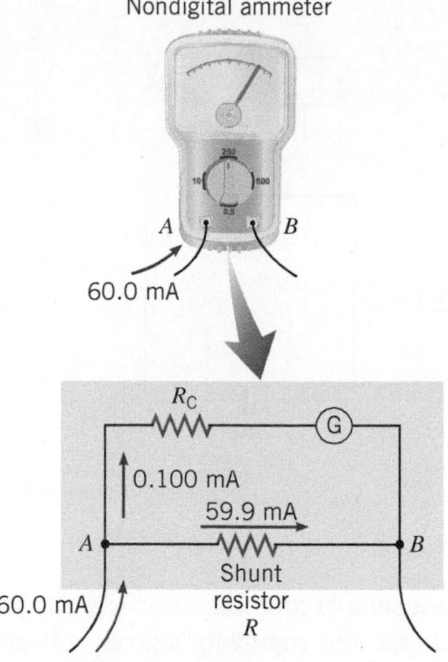

FIGURE 20.32 If a galvanometer with a full-scale limit of 0.100 mA is to be used to measure a current of 60.0 mA, a shunt resistance R must be used, so the excess current of 59.9 mA can detour around the galvanometer coil.

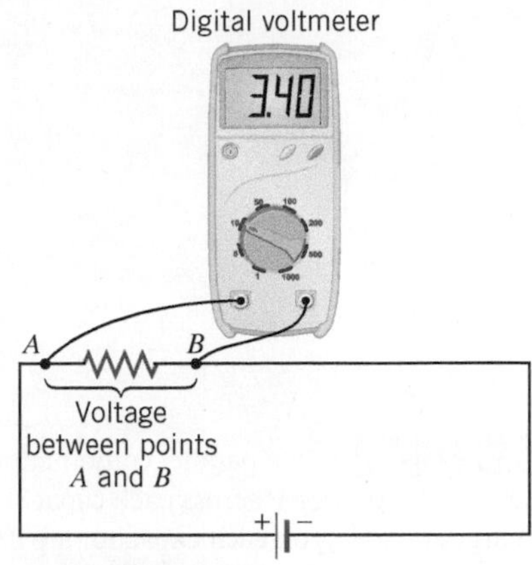

FIGURE 20.33 To measure the voltage between two points A and B in a circuit, a voltmeter is connected between the points.

THE PHYSICS OF . . . a voltmeter. A **voltmeter** is an instrument that measures the voltage between two points, A and B, in a circuit. **Figure 20.33** shows that the voltmeter must be connected between the points and is *not* inserted into the circuit as an ammeter is. (This is true for both digital and nondigital voltmeters; **Figure 20.33** shows a digital instrument.)

A nondigital voltmeter includes a galvanometer whose scale is calibrated in volts. Suppose, for instance, that the galvanometer in **Figure 20.34** has a full-scale current of 0.1 mA and a coil resistance of 50 Ω. Under full-scale conditions, the voltage across the coil would, therefore, be $V = IR_C = (0.1 \times 10^{-3}\ \text{A})(50\ \Omega) = 0.005\ \text{V}$. Thus, this galvanometer could be used to register voltages in the range 0–0.005 V. A nondigital voltmeter, then, is a galvanometer used in this fashion, along with some provision for adjusting the range of voltages to be measured. To adjust the range, an additional resistance R is connected in series with the coil resistance R_C (see Problem 77).

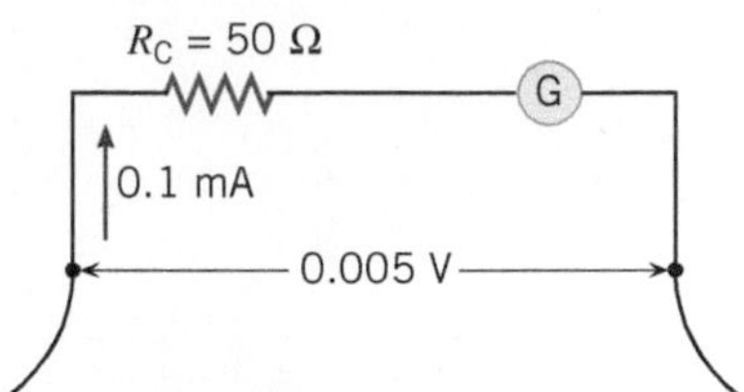

FIGURE 20.34 The galvanometer shown has a full-scale deflection of 0.1 mA and a coil resistance of 50 Ω.

Ideally, the voltage registered by a voltmeter should be the same as the voltage that exists when the voltmeter is not connected. However, a voltmeter takes some current from a circuit and, thus, alters the circuit voltage to some extent. An *ideal* voltmeter would have infinite resistance and would draw away only an infinitesimal amount of current. In reality, a good voltmeter is designed with a resistance that is large enough so the unit does not appreciably alter the voltage in the circuit to which it is connected.

Check Your Understanding

(*The answers are given at the end of the book.*)

19. An ideal ammeter has ________ resistance, whereas an ideal voltmeter has ________ resistance. **(a)** zero, zero **(b)** infinite, infinite **(c)** zero, infinite **(d)** infinite, zero

20. What would happen to the current in a circuit if a voltmeter, inadvertently mistaken for an ammeter, were inserted into the circuit? The current would **(a)** increase markedly **(b)** decrease markedly **(c)** remain the same.

20.12 Capacitors in Series and in Parallel

Figure 20.35 shows two different capacitors connected in parallel to a battery. Since the capacitors are in parallel, they have the same voltage V across their plates. However, the capacitors *contain different amounts of charge.* The charge stored by a capacitor is $q = CV$ (Equation 19.8), so $q_1 = C_1V$ and $q_2 = C_2V$.

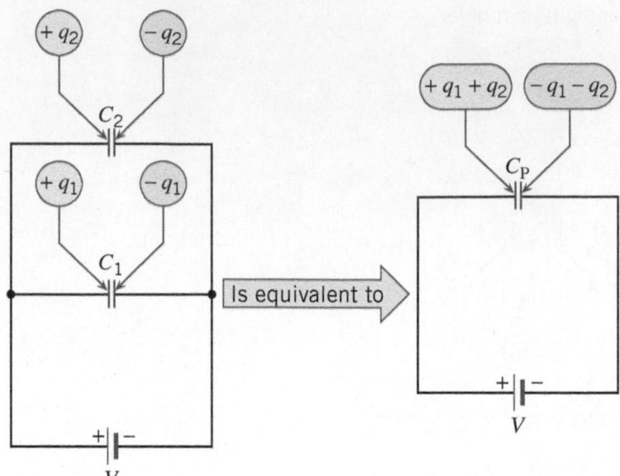

FIGURE 20.35 In a parallel combination of capacitances C_1 and C_2, the voltage V across each capacitor is the same, but the charges q_1 and q_2 on each capacitor are different.

As with resistors, it is always possible to replace a parallel combination of capacitors with an *equivalent capacitor* that stores the same charge and energy for a given voltage as the combination does. To determine the equivalent capacitance C_P, note that the total charge q stored by the two capacitors is

$$q = q_1 + q_2 = C_1 V + C_2 V = (C_1 + C_2)V = C_P V$$

This result indicates that two capacitors in parallel can be replaced by an equivalent capacitor whose capacitance is $C_P = C_1 + C_2$. For any number of capacitors in parallel, the equivalent capacitance is

Parallel capacitors $$C_P = C_1 + C_2 + C_3 + \cdots \tag{20.18}$$

Capacitances in parallel simply add together to give an equivalent capacitance. This behavior contrasts with that of resistors in parallel, which combine as reciprocals, according to Equation 20.17. The reason for this difference is that the charge q on a capacitor is directly proportional to the capacitance C $(q = CV)$, whereas the current I in a resistor is inversely proportional to the resistance R $(I = V/R)$.

The equivalent capacitor not only stores the same amount of charge as the parallel combination of capacitors, but also stores the same amount of energy. For instance, the energy stored in a single capacitor is $\frac{1}{2}CV^2$ (Equation 19.11b), so the total energy stored by two capacitors in parallel is

$$\text{Total energy} = \tfrac{1}{2}C_1 V^2 + \tfrac{1}{2}C_2 V^2 = \tfrac{1}{2}(C_1 + C_2)V^2 = \tfrac{1}{2}C_P V^2$$

which is equal to the energy stored in the equivalent capacitor C_P.

When capacitors are connected in series, the equivalent capacitance is different from when they are in parallel. As an example, **Figure 20.36** shows two capacitors in series and reveals the following important fact.

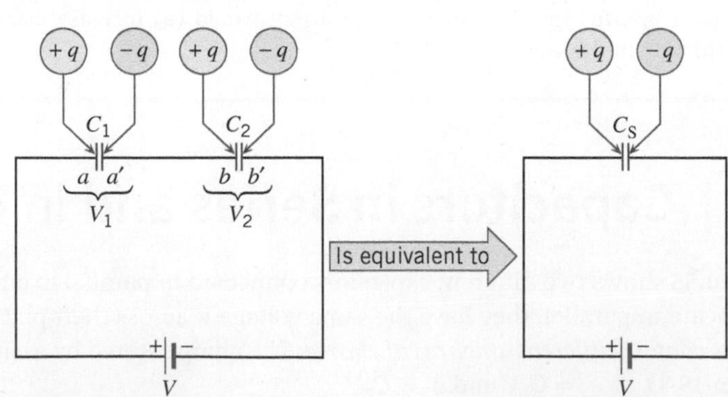

FIGURE 20.36 In a series combination of capacitances C_1 and C_2, the same amount of charge q is on the plates of each capacitor, but the voltages V_1 and V_2 across each capacitor are different.

Problem-Solving Insight All capacitors in series, regardless of their capacitances, contain charges of the same magnitude, $+q$ and $-q$, on their plates.

The battery places a charge of $+q$ on plate a of capacitor C_1, and this charge induces a charge of $+q$ to depart from the opposite plate a', leaving behind a charge $-q$. The $+q$ charge that leaves plate a' is deposited on plate b of capacitor C_2 (since these two plates are connected by a wire), where it induces a $+q$ charge to move away from the opposite plate b', leaving behind a charge of $-q$. Thus, all capacitors in series contain charges of the same magnitude on their plates. Note the difference between charging capacitors in parallel and in series. When charging parallel capacitors, the battery moves a charge q that is the sum of the charges moved for each of the capacitors: $q = q_1 + q_2 + q_3 + \cdots$. In contrast, when charging a series combination of n capacitors, the battery only moves a charge q, not nq, because the charge q passes by induction from one capacitor directly to the next one in line.

The equivalent capacitance C_S for the series connection in **Figure 20.36** can be determined by observing that the battery voltage V is shared by the two capacitors. The drawing indicates that the voltages across C_1 and C_2 are V_1 and V_2, so that $V = V_1 + V_2$. Since the voltages across the capacitors are $V_1 = q/C_1$ and $V_2 = q/C_2$, we find that

$$V = V_1 + V_2 = \frac{q}{C_1} + \frac{q}{C_2} = q\left(\frac{1}{C_1} + \frac{1}{C_2}\right) = q\left(\frac{1}{C_S}\right)$$

Thus, two capacitors in series can be replaced by an equivalent capacitor, which has a capacitance C_S that can be obtained from $1/C_S = 1/C_1 + 1/C_2$. For any number of capacitors connected in series the equivalent capacitance is given by

Series capacitors

$$\frac{1}{C_S} = \frac{1}{C_1} + \frac{1}{C_2} + \frac{1}{C_3} + \cdots \qquad \textbf{(20.19)}$$

Equation 20.19 indicates that capacitances in series combine as reciprocals and do not simply add together as resistors in series do. It is left as an exercise (Problem 87) to show that the equivalent series capacitance stores the same electrostatic energy as the sum of the energies of the individual capacitors.

It is possible to simplify circuits containing a number of capacitors in the same general fashion as that outlined for resistors in Example 12 and **Figure 20.23**. The capacitors in a parallel grouping can be combined according to Equation 20.18, and those in a series grouping can be combined according to Equation 20.19, as the following example shows.

EXAMPLE 16 | BIO The Physics of Projective Capacitive Touchscreens

Chances are you are the owner of a smartphone, such as an iPhone, and you use your fingers to communicate with the device by touching its front panel. Touchscreens on modern smartphones use capacitive sensing to locate the position of your fingers on the screen. If that position is located on the screen above the icon of your favorite app, then the phone will launch that app. **Figure 20.37a** shows the essential components of the touchscreen, which utilizes a grid of parallel plate capacitors composed of sensing lines and driving lines that are etched into a glass substrate. This is referred to as **projective capacitive touch (PCT) technology**. A voltage is applied across the sensing and driving lines, which places a small charge on their surfaces, effectively creating a grid of capacitors, with positive charges on one grid and negative charges on the other. When a finger, or other conductor, is brought near the surface of the screen, it is conductive enough to disrupt the electric field lines around the grid. Negative charges on your skin are attracted to the positive charges on the driving lines and move toward the tip of your finger. The positive charges on the driving lines are maintained by a constant current. The negative charges in your finger repel some of the negative charges from the sensing electrode, thereby decreasing the effective capacitance between the electrodes (**Figure 20.37b**).

The electrical conductivity of your skin is essential for the operation of the touchscreen. This is why your smartphone screen doesn't work when you wear gloves, as cloth is a good electrical insulator. However, special touchscreen gloves will work, as they use conductive thread in their fingertips. The drop in capacitance is measured by the phone's sensing electronics, which pinpoint the location of your finger on the screen. By continuously monitoring the mutual capacitance of all the grid crossing points, swiping motions and two-fingered movements, such as zooming in or out, can also be detected.

We can model the finger touch on the screen as a series of capacitors, as shown in **Figure 20.37b**, where C_{E0} is the capacitance between the grid electrodes before the screen is touched, C_{EF} is the capacitance between the grid electrodes after the screen is touched, and C_F is the capacitance between the sensing electrode and the finger. Suppose that, after the screen is touched, the sensing circuit measures the two capacitors in **Figure 20. 38b** in series (C_S) to be 0.19 pF. If the value of $C_{EF} = 0.25$ pF, what is the value of the capacitance between the finger and the sensing electrode (C_F)?

Reasoning The equivalent capacitance for capacitors connected in series is less than the individual capacitance values, as the measurement indicates. We can use Equation 20.19 to calculate the equivalent capacitance in terms of C_{EF} and C_F, which we can then solve for C_F.

Solution Applying Equation 20.19 to the two capacitors in series in **Figure 20. 37b**, we have the following:

$$\frac{1}{C_S} = \frac{1}{C_{EF}} + \frac{1}{C_F} \Rightarrow \frac{1}{C_F} = \frac{1}{C_S} - \frac{1}{C_{EF}} \Rightarrow C_F = \left(\frac{1}{C_S} - \frac{1}{C_{EF}}\right)^{-1}$$

$$\Rightarrow C_F = \left(\frac{1}{0.19\ \text{pF}} - \frac{1}{0.25\ \text{pF}}\right)^{-1} = \boxed{0.79\ \text{pF}}$$

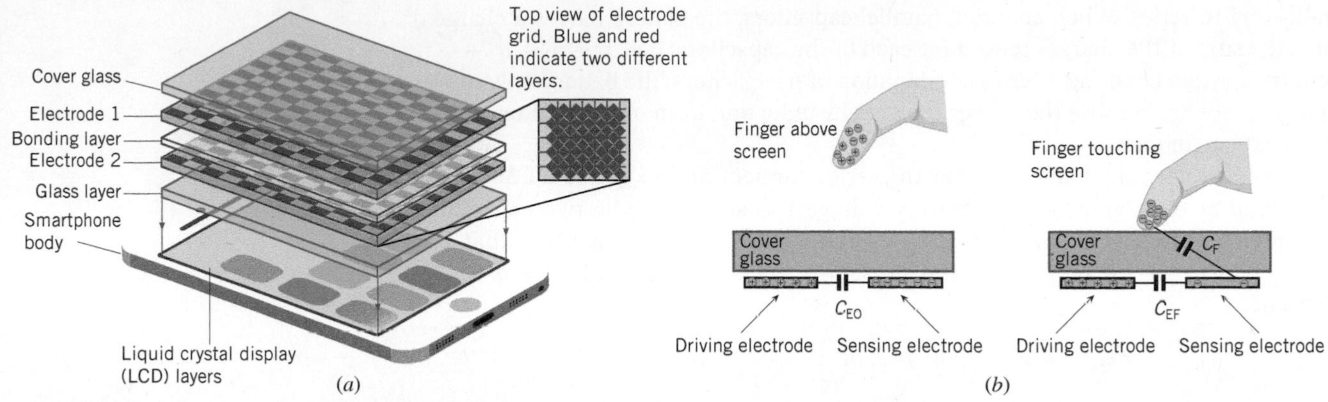

FIGURE 20.37 (*a*) The main components of a smartphone touchscreen. The driving and sensing electrodes are composed of transparent conducting layers of indium tin oxide (ITO). (*b*) Bringing a conducting object, like a bare finger, in contact with the screen changes the capacitance near the finger. The phone's circuitry detects this change, which pinpoints the finger's location.

20.13 *RC* Circuits

Many electric circuits contain both resistors and capacitors. **Figure 20.38** illustrates an example of a resistor–capacitor circuit, or *RC* circuit. Part *a* of the drawing shows the circuit at a time *t* after the switch has been closed and the battery has begun to charge up the capacitor plates. The charge on the plates builds up gradually to its equilibrium value of $q_0 = CV_0$, where V_0 is the voltage of the battery. Assuming that the capacitor is uncharged at time $t = 0$ s when the switch is closed, it can be shown that the magnitude q of the charge on the plates at time t is

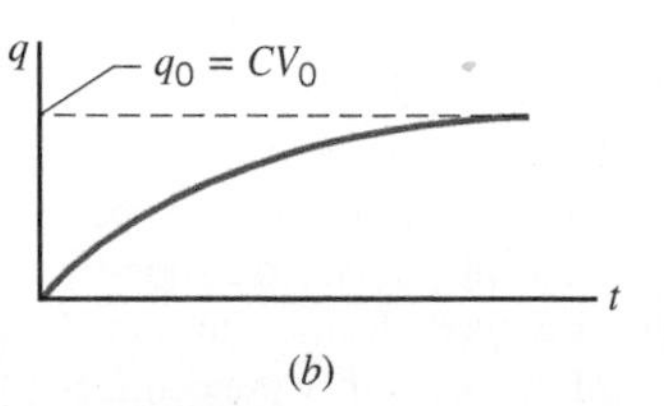

Capacitor charging $$q = q_0\left[1 - e^{-t/(RC)}\right] \tag{20.20}$$

FIGURE 20.38 Charging a capacitor.

Math Skills If the need arises, you can solve Equation 20.20 for the time *t* with the aid of natural logarithms. The first step is to rearrange the equation so as to isolate the exponential *e* on one side of the equals sign:

$$q = q_0\left[1 - e^{-t/(RC)}\right] \quad \text{or} \quad \frac{q}{q_0} = 1 - e^{-t/(RC)} \quad \text{or} \quad 1 - \frac{q}{q_0} = e^{-t/(RC)}$$

According to Equation D-9 in Appendix D, the natural logarithm of e^Z is $\ln e^Z = Z$. Therefore, we can take the natural logarithm of both sides of the rearranged equation and obtain

$$\ln\left(1 - \frac{q}{q_0}\right) = \ln e^{-t/(RC)} = -\frac{t}{RC} \quad \text{or} \quad t = -RC\left[\ln\left(1 - \frac{q}{q_0}\right)\right]$$

To illustrate this result, suppose that $RC = 5.00$ s and that the capacitor has acquired a charge q that is one-fourth of its final equilibrium value q_0, so that $q = 0.250\ q_0$. The time required for the capacitor to acquire the charge is

$$t = -(5.00\ \text{s})\left[\ln\left(1 - \frac{0.250\ q_0}{q_0}\right)\right]$$

$$= -(5.00\ \text{s})\ \ln 0.750 = -(5.00\ \text{s})(-0.288) = 1.44\ \text{s}$$

where the exponential e has the value of 2.718. . . . Part b of the drawing shows a graph of this expression, which indicates that the charge is $q = 0$ C when $t = 0$ s and increases gradually toward the equilibrium value of $q_0 = CV_0$. The voltage V across the capacitor at any time can be obtained from Equation 20.20 by dividing the charges q and q_0 by the capacitance C, since $V = q/C$ and $V_0 = q_0/C$.

The term RC in the exponent in Equation 20.20 is called the **time constant** τ of the circuit:

$$\tau = RC \quad (20.21)$$

The time constant is measured in seconds; verification of the fact that an ohm times a farad is equivalent to a second is left as an exercise (see Check Your Understanding Question 21). The time constant is the amount of time required for the capacitor to accumulate 63.2% of its equilibrium charge, as can be seen by substituting $t = \tau = RC$ in Equation 20.20; $q_0(1 - e^{-1}) = q_0(0.632)$. The charge approaches its equilibrium value rapidly when the time constant is small and slowly when the time constant is large.

Figure 20.39a shows a circuit at a time t after a switch is closed to allow a charged capacitor to begin discharging. There is no battery in this circuit, so the charge $+q$ on the left plate of the capacitor can flow counterclockwise through the resistor and neutralize the charge $-q$ on the right plate. Assuming that the capacitor has a charge q_0 at time $t = 0$ s when the switch is closed, it can be shown that

Capacitor discharging

$$q = q_0 e^{-t/(RC)} \quad (20.22)$$

where q is the amount of charge remaining on either plate at time t. The graph of this expression in part b of the drawing shows that the charge begins at q_0 when $t = 0$ s and decreases gradually toward zero. Smaller values of the time constant RC lead to a more rapid discharge. Equation 20.22 indicates that when $t = \tau = RC$, the magnitude of the charge remaining on each plate is $q_0 e^{-1} = q_0(0.368)$. Therefore, the time constant is also the amount of time required for a charged capacitor to *lose* 63.2% of its charge.

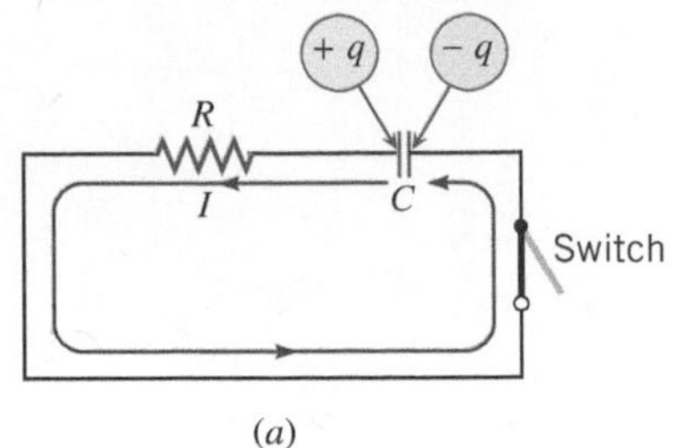

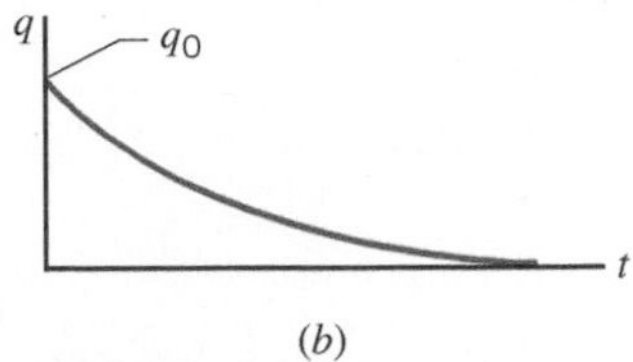

FIGURE 20.39 Discharging a capacitor.

EXAMPLE 17 | BIO The Physics of Pacemakers

THE PHYSICS OF . . . heart pacemakers. The charging/discharging of a capacitor has many applications. Heart pacemakers, for instance, incorporate RC circuits to control the timing of voltage pulses that are delivered to a malfunctioning heart to regulate its beating cycle. The pulses are delivered by electrodes attached externally to the chest or located internally near the heart when the pacemaker is implanted surgically (see Figure 20.40). A voltage pulse is delivered when the capacitor discharges to a preset level, following which the capacitor is recharged rapidly and the cycle repeats. The value of the time constant RC controls the pulsing rate, which is about once per second.

One particular pacemaker uses an RC circuit that discharges when the charge on the capacitor reaches 75% of its equilibrium charge value. If this occurs every 1.2 seconds, and the capacitance of the RC circuit is 110 μF, what is the resistance of the resistor?

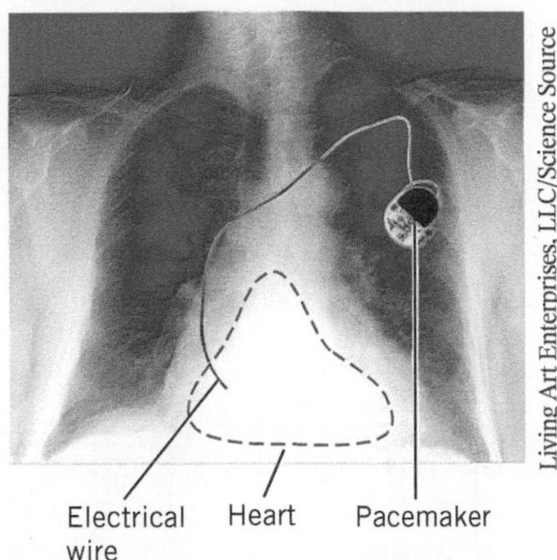

FIGURE 20.40 This color-enhanced frontal X-ray photograph of a human chest shows an enlarged heart and a heart pacemaker that has been implanted surgically.

Reasoning To calculate the resistance of the RC circuit we can use Equation 20.20, which is the charging equation for the capacitor in the circuit.

Solution Beginning with Equation 20.20, we have: $q = q_0[1 - e^{-t/(RC)}]$. We rearrange this equation and solve for the resistance, R:

$$1 - \frac{q}{q_0} = e^{-t/(RC)} \Rightarrow \ln\left(1 - \frac{q}{q_0}\right) = \frac{-t}{RC} \Rightarrow R = \frac{-t}{C\left[\ln\left(1 - \frac{q}{q_0}\right)\right]}$$

Using the values given in the problem, we can compute the resistance as follows:

$$R = \frac{-1.2\ \text{s}}{(110 \times 10^{-6}\ \text{F})\,[\ln(1 - 0.75)]} = \boxed{7.9 \times 10^3\ \Omega}$$

THE PHYSICS OF . . . windshield wipers. The charging/discharging of a capacitor is also used in automobiles that have windshield wipers equipped for intermittent operation during a light drizzle. In this mode of operation, the wipers remain off for a while and then turn on briefly. The timing of the on–off cycle is determined by the time constant of a resistor–capacitor combination.

Check Your Understanding

(*The answers are given at the end of the book.*)

21. The time constant τ of a series RC circuit is measured in seconds (s) and is given by $\tau = RC$, where the resistance R is measured in ohms (Ω) and the capacitance C is measured in farads (F). Verify that an ohm times a farad is equivalent to a second.

22. **CYU Figure 20.5** shows two different resistor–capacitor arrangements. The time constant for arrangement A is 0.20 s. What is the time constant for arrangement B? **(a)** 0.050 s **(b)** 0.10 s **(c)** 0.20 s **(d)** 0.40 s **(e)** 0.80 s

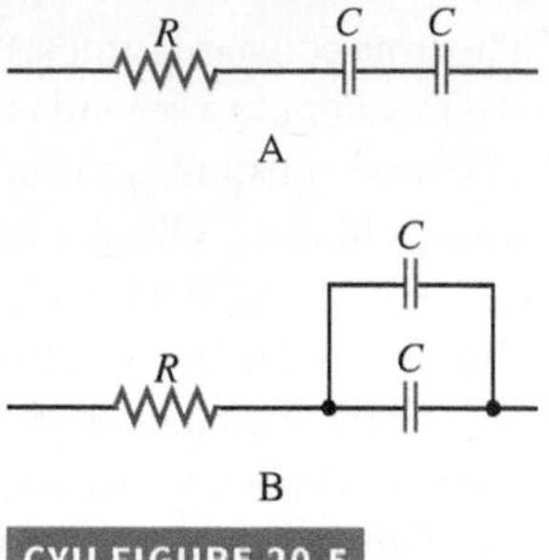

CYU FIGURE 20.5

20.14 Safety and the Physiological Effects of Current

Electric circuits, although very useful, can also be hazardous. To reduce the danger inherent in using circuits, proper **electrical grounding** is necessary. The next two figures help to illustrate what electrical grounding means and how it is achieved.

Figure 20.41*a* shows a clothes dryer connected to a wall socket via a two-prong plug. The dryer is operating normally; that is, the wires inside are insulated from the metal casing of the dryer, so no charge flows through the casing itself. Notice that one terminal of the ac generator is customarily connected to ground (⏚) by the electric power company. Part *b* of the drawing shows the hazardous result that occurs if a wire comes loose and contacts the metal casing. A person touching it receives a shock, since electric charge flows through the casing, the person's body, and the ground on the way back to the generator.

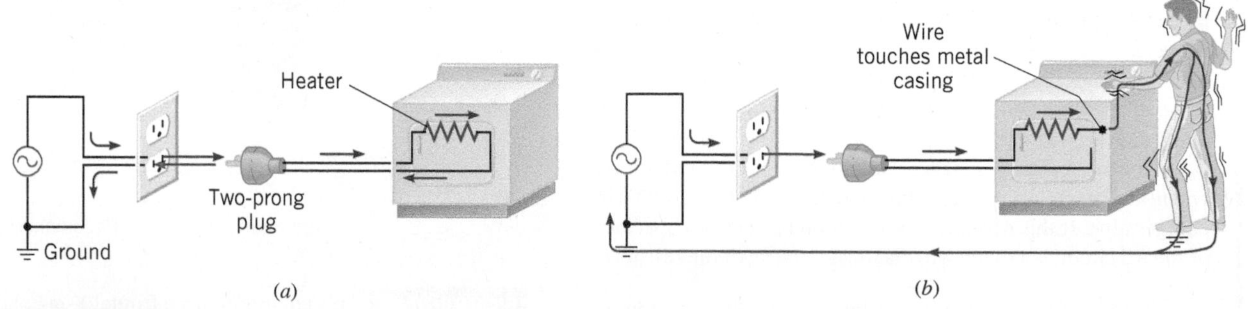

FIGURE 20.41 (*a*) A normally operating clothes dryer that is connected to a wall socket via a two-prong plug. (*b*) An internal wire accidentally touches the metal casing, and a person who touches the casing receives an electrical shock.

THE PHYSICS OF . . . safe electrical grounding. **Figure 20.42** shows the same appliance connected to a wall socket via a three-prong plug that provides safe electrical grounding. The third prong connects the metal casing directly to a copper rod driven into the ground or to a copper water pipe that is in the ground. This arrangement protects against electrical shock in the event that a broken wire touches the metal casing. In this event, charge would flow through the casing, through the third prong, and into

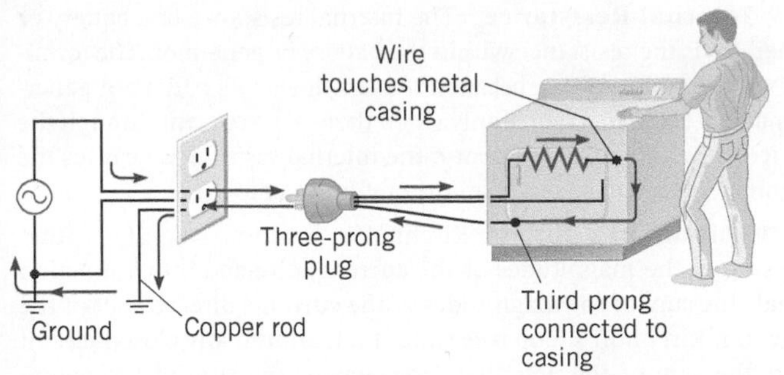

FIGURE 20.42 A safely connected dryer. If the dryer malfunctions, a person touching it receives no shock, since electric charge flows through the third prong and into the ground via a copper rod, rather than through the person's body.

the ground, returning eventually to the generator. No charge would flow through the person's body, because the copper rod provides much less electrical resistance than does the body.

BIO THE PHYSICS OF . . . the physiological effects of current. Serious and sometimes fatal injuries can result from electrical shock. The severity of the injury depends on the magnitude of the current and the parts of the body through which the moving charges pass. The amount of current that causes a mild tingling sensation is about 0.001 A. Currents on the order of 0.01–0.02 A can lead to muscle spasms, in which a person "can't let go" of the object causing the shock. Currents of approximately 0.2 A are potentially fatal because they can make the heart fibrillate, or beat in an uncontrolled manner. Substantially larger currents stop the heart completely. However, since the heart often begins beating normally again after the current ceases, the larger currents can be less dangerous than the smaller currents that cause fibrillation.

Concept Summary

20.1 Electromotive Force and Current There must be at least one source or generator of electrical energy in an electric circuit. The electromotive force (emf) of a generator, such as a battery, is the maximum potential difference (in volts) that exists between the terminals of the generator.

The rate of flow of charge is called the electric current. If the rate is constant, the current I is given by Equation 20.1, where Δq is the magnitude of the charge crossing a surface in a time Δt, the surface being perpendicular to the motion of the charge. The SI unit for current is the coulomb per second (C/s), which is referred to as an ampere (A). When the charges flow only in one direction around a circuit, the current is called direct current (dc). When the direction of charge flow changes from moment to moment, the current is known as alternating current (ac). Conventional current is the hypothetical flow of positive charges that would have the same effect in a circuit as the movement of negative charges that actually does occur.

$$I = \frac{\Delta q}{\Delta t} \tag{20.1}$$

20.2 Ohm's Law The definition of electrical resistance R is $R = V/I$, where V (in volts) is the voltage applied across a piece of material and I (in amperes) is the current through the material. Resistance is measured in volts per ampere, a unit called an ohm (Ω). If the ratio of the voltage to the current is constant for all values of voltage and current, the resistance is constant. In this event, the definition of resistance becomes Ohm's law, Equation 20.2.

$$\frac{V}{I} = R = \text{constant} \quad \text{or} \quad V = IR \tag{20.2}$$

20.3 Resistance and Resistivity The resistance of a piece of material of length L and cross-sectional area A is given by Equation 20.3, where ρ is the resistivity of the material. The resistivity of a material depends on the temperature. For many materials and limited temperature ranges, the temperature dependence is given by Equation 20.4, where ρ and ρ_0 are the resistivities at the temperatures T and T_0, respectively, and α is the temperature coefficient of resistivity. The temperature dependence of the resistance R is given by Equation 20.5, where R and R_0 are the resistances at the temperatures T and T_0, respectively.

$$R = \rho\frac{L}{A} \tag{20.3}$$

$$\rho = \rho_0[1 + \alpha(T - T_0)] \tag{20.4}$$

$$R = R_0[1 + \alpha(T - T_0)] \tag{20.5}$$

20.4 Electric Power When electric charge flows from point A to point B in a circuit, leading to a current I, and the voltage between the points is V, the electric power associated with this current and voltage is given by Equation 20.6a. For a resistor, Ohm's law applies, and it follows that the power delivered to the resistor is also given by either Equation 20.6b or 20.6c.

$$P = IV \tag{20.6a}$$

$$P = I^2R \tag{20.6b}$$

$$P = \frac{V^2}{R} \tag{20.6c}$$

20.5 Alternating Current The alternating voltage between the terminals of an ac generator can be represented by Equation 20.7, where V_0 is the peak value of the voltage, t is the time, and f is the frequency (in hertz) at which the voltage oscillates. Correspondingly, in a circuit containing only resistance, the ac current is given by Equation 20.8, where I_0 is the peak value of the current and is related to the peak voltage via $I_0 = V_0/R$.

$$V = V_0 \sin 2\pi ft \quad (20.7)$$

$$I = I_0 \sin 2\pi ft \quad (20.8)$$

For sinusoidal current and voltage, the root mean square (rms) current and voltage are related to the peak values according to Equations 20.12 and 20.13.

$$I_{rms} = \frac{I_0}{\sqrt{2}} \quad (20.12)$$

$$V_{rms} = \frac{V_0}{\sqrt{2}} \quad (20.13)$$

The power in an ac circuit is the product of the current and the voltage and oscillates in time. The average power is given by Equation 20.15a. For a resistor, Ohm's law applies, so that $V_{rms} = I_{rms}R$, and the average power delivered to the resistor is also given by Equations 20.15b and 20.15c.

$$\overline{P} = I_{rms} V_{rms} \quad (20.15a)$$

$$\overline{P} = I_{rms}^2 R \quad (20.15b)$$

$$\overline{P} = \frac{V_{rms}^2}{R} \quad (20.15c)$$

20.6 Series Wiring When devices are connected in series, there is the same current through each device. The equivalent resistance R_S of a series combination of resistances (R_1, R_2, R_3, etc.) is given by Equation 20.16. The power delivered to the equivalent resistance is equal to the total power delivered to any number of resistors in series.

$$R_S = R_1 + R_2 + R_3 + \cdots \quad (20.16)$$

20.7 Parallel Wiring When devices are connected in parallel, the same voltage is applied across each device. In general, devices wired in parallel carry different currents. The reciprocal of the equivalent resistance R_P of a parallel combination of resistances (R_1, R_2, R_3, etc.) is given by Equation 20.17. The power delivered to the equivalent resistance is equal to the total power delivered to any number of resistors in parallel.

$$\frac{1}{R_P} = \frac{1}{R_1} + \frac{1}{R_2} + \frac{1}{R_3} + \cdots \quad (20.17)$$

20.8 Circuits Wired Partially in Series and Partially in Parallel Sometimes, one section of a circuit is wired in series, while another is wired in parallel. In such cases the circuit can be analyzed in parts, according to the respective series and parallel equivalent resistances of the various sections.

20.9 Internal Resistance The internal resistance of a battery or generator is the resistance within the battery or generator. The terminal voltage is the voltage between the terminals of a battery or generator and is equal to the emf only when there is no current through the device. When there is a current I, the internal resistance r causes the terminal voltage to be less than the emf by an amount Ir.

20.10 Kirchhoff's Rules Kirchhoff's junction rule states that the sum of the magnitudes of the currents directed into a junction equals the sum of the magnitudes of the currents directed out of the junction. Kirchhoff's loop rule states that, around any closed-circuit loop, the sum of the potential drops equals the sum of the potential rises. The Reasoning Strategy given at the end of Section 20.10 explains how these two rules are applied to analyze any circuit.

20.11 The Measurement of Current and Voltage A galvanometer is a device that responds to electric current and is used in nondigital ammeters and voltmeters. An ammeter is an instrument that measures current and must be inserted into a circuit in such a way that the current passes directly through the ammeter. A voltmeter is an instrument for measuring the voltage between two points in a circuit. A voltmeter must be connected between the two points and is not inserted into a circuit as an ammeter is.

20.12 Capacitors in Series and in Parallel The equivalent capacitance C_P for a parallel combination of capacitances (C_1, C_2, C_3, etc.) is given by Equation 20.18. In general, each capacitor in a parallel combination carries a different amount of charge. The equivalent capacitor carries the same total charge and stores the same total energy as the parallel combination.

$$C_P = C_1 + C_2 + C_3 + \cdots \quad (20.18)$$

The reciprocal of the equivalent capacitance C_S for a series combination (C_1, C_2, C_3, etc.) of capacitances is given by Equation 20.19. The equivalent capacitor carries the same amount of charge as *any one* of the capacitors in the combination and stores the same total energy as the series combination.

$$\frac{1}{C_S} = \frac{1}{C_1} + \frac{1}{C_2} + \frac{1}{C_3} + \cdots \quad (20.19)$$

20.13 *RC* Circuits The charging or discharging of a capacitor in a dc series circuit (resistance R, capacitance C) does not occur instantaneously. The charge on a capacitor builds up gradually, as described by Equation 20.20, where q is the charge on the capacitor at time t and q_0 is the equilibrium value of the charge. The time constant τ of the circuit is given by Equation 20.21. The discharging of a capacitor through a resistor is described by Equation 20.22, where q_0 is the charge on the capacitor at time $t = 0$ s.

$$q = q_0[1 - e^{-t/(RC)}] \quad (20.20)$$

$$\tau = RC \quad (20.21)$$

$$q = q_0 e^{-t/(RC)} \quad (20.22)$$

Focus on Concepts

Online

Additional questions are available for assignment in WileyPLUS.

Section 20.1 Electromotive Force and Current

1. In 2.0 s, 1.9×10^{19} electrons pass a certain point in a wire. What is the current I in the wire?

Section 20.2 Ohm's Law

2. Which one of the following graphs correctly represents Ohm's law, where V is the voltage and I is the current? **(a)** A **(b)** B **(c)** C **(d)** D

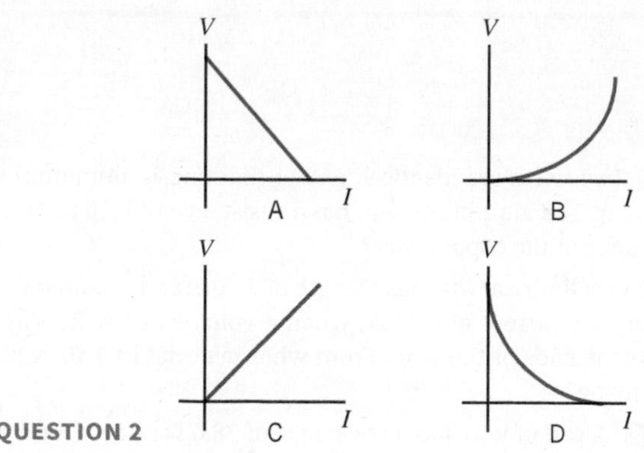

QUESTION 2

Section 20.3 Resistance and Resistivity

3. Two wires are made from the same material. One wire has a resistance of 0.10 Ω. The other wire is twice as long as the first wire and has a radius that is half as much. What is the resistance of the second wire? (a) 0.40 Ω (b) 0.20 Ω (c) 0.10 Ω (d) 0.050 Ω (e) 0.80 Ω

Section 20.4 Electric Power

4. A single resistor is connected across the terminals of a battery. Which one or more of the following changes in voltage and current leaves unchanged the electric power dissipated in the resistor?

(A) Doubling the voltage and reducing the current by a factor of two

(B) Doubling the voltage and increasing the resistance by a factor of four

(C) Doubling the current and reducing the resistance by a factor of four

(a) A, B, C (b) A, B (c) B, C (d) A (e) B

Section 20.5 Alternating Current

5. The average power dissipated in a 47-Ω resistor is 2.0 W. What is the peak value I_0 of the ac current in the resistor?

Section 20.6 Series Wiring

6. For the circuit shown in the drawing, what is the voltage V_1 across resistance R_1?

(a) $V_1 = \left(\frac{R_1}{R_2}\right)V$ (b) $V_1 = \left(\frac{R_2}{R_1}\right)V$ (c) $V_1 = V$ (d) $V_1 = \left(\frac{R_1}{R_1+R_2}\right)V$ (e) $V_1 = \left(\frac{R_1+R_2}{R_1}\right)V$

QUESTION 6

Section 20.7 Parallel Wiring

7. For the circuit shown in the drawing, what is the ratio of the current I_1 in resistance R_1 to the current I_2 in resistance R_2?

(a) $\frac{I_1}{I_2} = \frac{R_1}{R_2}$ (b) $\frac{I_1}{I_2} = \frac{R_2}{R_1+R_2}$ (c) $\frac{I_1}{I_2} = \frac{R_1}{R_1+R_2}$ (d) $\frac{I_1}{I_2} = 1$ (e) $\frac{I_1}{I_2} = \frac{R_2}{R_1}$

QUESTION 7

Section 20.8 Circuits Wired Partially in Series and Partially in Parallel

8. In the following three arrangements each resistor has the same resistance R. Rank the equivalent resistances of the arrangements in descending order (largest first). (a) A, B, C (b) B, A, C (c) B, C, A (d) A, C, B (e) C, B, A

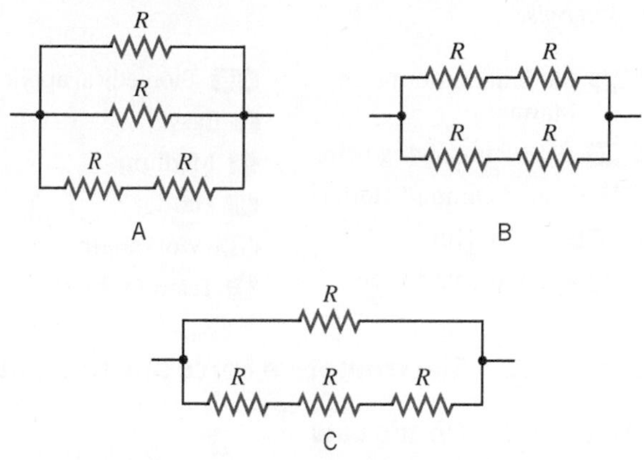

QUESTION 8

Section 20.9 Internal Resistance

9. A battery has an emf of V and an internal resistance of r. What resistance R, when connected across the terminals of this battery, will cause the terminal voltage of the battery to be $\frac{1}{2}V$? (a) $R = \frac{1}{2}r$ (b) $R = 2r$ (c) $R = 4r$ (d) $R = r$ (e) $R = \frac{1}{4}r$

Section 20.10 Kirchhoff's Rules

10. When applying Kirchhoff's rules, one of the essential steps is to mark each resistor with plus and minus signs to label how the potential changes from one end of the resistor to the other. The circuit in the drawing contains four resistors, each marked with the associated plus and minus signs. However, one resistor is marked incorrectly. Which one is it? (a) R_1 (b) R_2 (c) R_3 (d) R_4

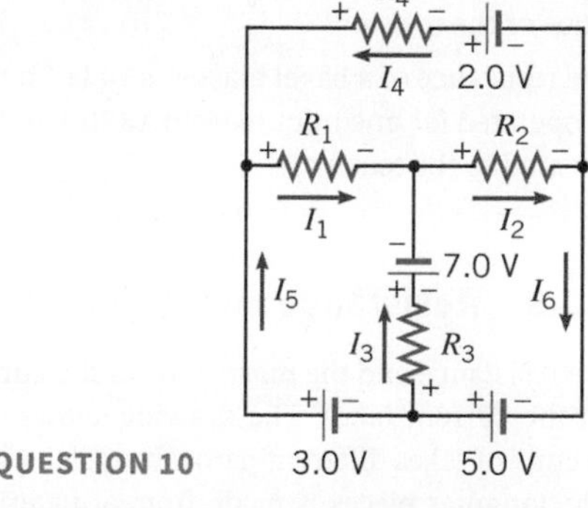

QUESTION 10

Section 20.12 Capacitors in Series and in Parallel

11. Three capacitors are identical, each having a capacitance C. Two of them are connected in series. Then, this series combination is connected in parallel with the third capacitor. What is the equivalent capacitance of the entire connection? (a) $\frac{1}{2}C$ (b) $\frac{1}{3}C$ (c) $3C$ (d) $\frac{2}{3}C$ (e) $\frac{3}{2}C$

Section 20.13 *RC* Circuits

12. The time constant of an *RC* circuit is [illegible]. How much time t is required for the capacitor (uncharged initially) to gain one-half of its full equilibrium charge?

Problems

Online

Additional questions are available for assignment in WileyPLUS.
***Note:** For problems that involve ac conditions, the current and voltage are rms values and the power is an average value, unless indicated otherwise.*

SSM Student Solutions Manual
MMH Problem-solving help
GO Guided Online Tutorial
V-HINT Video Hints
CHALK Chalkboard Videos
BIO Biomedical application
E Easy
M Medium
H Hard
WS Worksheet
T Team Problem

Section 20.1 Electromotive Force and Current, Section 20.2 Ohm's Law

1. E An especially violent lightning bolt has an average current of 1.26×10^3 A lasting 0.138 s. How much charge is delivered to the ground by the lightning bolt?

2. E SSM A battery charger is connected to a dead battery and delivers a current of 6.0 A for 5.0 hours, keeping the voltage across the battery terminals at 12 V in the process. How much energy is delivered to the battery?

3. E CHALK A coffee-maker contains a heating element that has a resistance of 14 Ω. This heating element is energized by a 120-V outlet. What is the current in the heating element?

4. M A car battery has a rating of 220 ampere · hours (A · h). This rating is one indication of the *total charge* that the battery can provide to a circuit before failing. **(a)** What is the total charge (in coulombs) that this battery can provide? **(b)** Determine the maximum current that the battery can provide for 38 minutes.

5. M A resistor is connected across the terminals of a 9.0-V battery, which delivers 1.1×10^5 J of energy to the resistor in six hours. What is the resistance of the resistor?

6. M GO The resistance of a bagel toaster is 14 Ω. To prepare a bagel, the toaster is operated for one minute from a 120-V outlet. How much energy is delivered to the toaster?

Section 20.3 Resistance and Resistivity

7. E GO The resistance and the magnitude of the current depend on the path that the current takes. The drawing shows three situations in which the current takes different paths through a piece of material. Each of the rectangular pieces is made from a material whose resistivity is $\rho = 1.50 \times 10^{-2}\ \Omega \cdot \text{m}$, and the unit of length in the drawing is $L_0 = 5.00$ cm. Each piece of material is connected to a 3.00-V battery. Find **(a)** the resistance and **(b)** the current in each case.

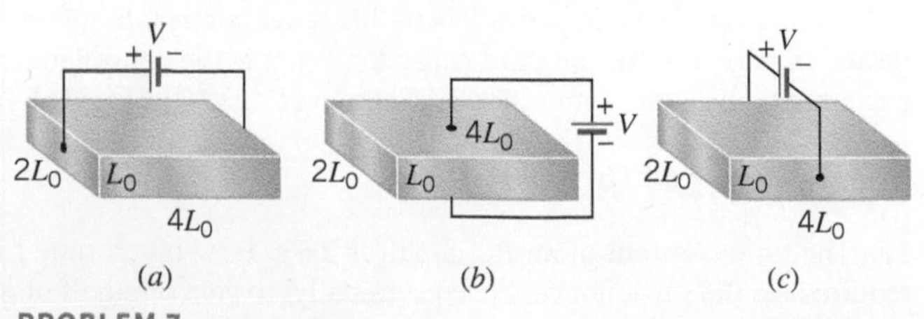

PROBLEM 7

8. E Two wires are identical, except that one is aluminum and one is copper. The aluminum wire has a resistance of 0.20 Ω. What is the resistance of the copper wire?

9. E A cylindrical wire has a length of 2.80 m and a radius of 1.03 mm. It carries a current of 1.35 A, when a voltage of 0.0320 V is applied across the ends of the wire. From what material in **Table 20.1** is the wire made?

10. E A coil of wire has a resistance of 38.0 Ω at 25 °C and 43.7 Ω at 55 °C. What is the temperature coefficient of resistivity?

11. E A large spool in an electrician's workshop has 75 m of insulation-coated wire coiled around it. When the electrician connects a battery to the ends of the spooled wire, the resulting current is 2.4 A. Some weeks later, after cutting off various lengths of wire for use in repairs, the electrician finds that the spooled wire carries a 3.1-A current when the same battery is connected to it. What is the length of wire remaining on the spool?

12. E High-voltage power lines are a familiar sight throughout the country. The aluminum wire used for some of these lines has a cross-sectional area of 4.9×10^{-4} m^2. What is the resistance of ten kilometers of this wire?

13. M The temperature coefficient of resistivity for the metal gold is 0.0034 (C°)$^{-1}$, and for tungsten it is 0.0045 (C°)$^{-1}$. The resistance of a gold wire increases by 7.0% due to an increase in temperature. For the same increase in temperature, what is the percentage increase in the resistance of a tungsten wire?

14. M GO A tungsten wire has a radius of 0.075 mm and is heated from 20.0 to 1320 °C. The temperature coefficient of resistivity is $\alpha = 4.5 \times 10^{-3}$ (C°)$^{-1}$. When 120 V is applied across the ends of the hot wire, a current of 1.5 A is produced. How long is the wire? Neglect any effects due to thermal expansion of the wire.

15. M V-HINT CHALK Two cylindrical rods, one copper and the other iron, are identical in lengths and cross-sectional areas. They are joined end to end to form one long rod. A 12-V battery is connected across the free ends of the copper–iron rod. What is the voltage between the ends of the copper rod?

Section 20.4 Electric Power

16. E An electric blanket is connected to a 120-V outlet and consumes 140 W of power. What is the resistance of the heater wire in the blanket?

17. E The heating element in an iron has a resistance of 24 Ω. The iron is plugged into a 120-V outlet. What is the power delivered to the iron?

18. E A blow-dryer and a vacuum cleaner each operate with a voltage of 120 V. The current rating of the blow-dryer is 11 A, and that of the vacuum cleaner is 4.0 A. Determine the power consumed by **(a)** the blow-dryer and **(b)** the vacuum cleaner. **(c)** Determine the ratio of the energy used by the blow-dryer in 15 minutes to the energy used by the vacuum cleaner in one-half hour.

19. E There are approximately 110 million households that use TVs in the United States. Each TV uses, on average, 75 W of power and is turned on for 6.0 hours a day. If electrical energy costs $0.12 per kWh, how much money is spent every day in keeping 110 million TVs turned on?

20. [E] An MP3 player operates with a voltage of 3.7 V, and is using 0.095 W of power. Find the current being supplied by the player's battery.

21. [E] [SSM] In doing a load of clothes, a clothes dryer uses 16 A of current at 240 V for 45 min. A personal computer, in contrast, uses 2.7 A of current at 120 V. With the energy used by the clothes dryer, how long (in hours) could you use this computer to "surf" the Internet?

22. [M] [V-HINT] An electric heater used to boil small amounts of water consists of a 15-Ω coil that is immersed directly in the water. It operates from a 120-V socket. How much time is required for this heater to raise the temperature of 0.50 kg of water from 13 °C to the normal boiling point?

23. [M] The rear window of a van is coated with a layer of ice at 0 °C. The density of ice is 917 kg/m^3. The driver of the van turns on the rear-window defroster, which operates at 12 V and 23 A. The defroster directly heats an area of 0.52 m^2 of the rear window. What is the maximum thickness of ice coating this area that the defroster can melt in 3.0 minutes?

24. [M] [GO] [CHALK] A piece of Nichrome wire has a radius of 6.5×10^{-4} m. It is used in a laboratory to make a heater that uses 4.00×10^2 W of power when connected to a voltage source of 120 V. Ignoring the effect of temperature on resistance, estimate the necessary length of wire.

25. [M] [SSM] Tungsten has a temperature coefficient of resistivity of 0.0045 (C°)$^{-1}$. A tungsten wire is connected to a source of constant voltage via a switch. At the instant the switch is closed, the temperature of the wire is 28 °C, and the initial power delivered to the wire is P_0. At what wire temperature will the power that is delivered to the wire be decreased to $\frac{1}{2}P_0$?

Section 20.5 Alternating Current

26. [E] According to Equation 20.7, an ac voltage V is given as a function of time t by $V = V_0 \sin 2\pi ft$, where V_0 is the peak voltage and f is the frequency (in hertz). For a frequency of 60.0 Hz, what is the smallest value of the time at which the voltage equals one-half of the peak value?

27. [E] The rms current in a copy machine is 6.50 A, and the resistance of the machine is 18.6 Ω. What are **(a)** the average power and **(b)** the peak power delivered to the machine?

28. [E] [GO] The rms current in a 47-Ω resistor is 0.50 A. What is the peak value of the voltage across this resistor?

29. [E] A 550-W space heater is designed for operation in Germany, where household electrical outlets supply 230 V (rms) service. What is the power output of the heater when plugged into a 120-V (rms) electrical outlet in a house in the United States? Ignore the effects of temperature on the heater's resistance.

30. [E] [V-HINT] Review Conceptual Example 7 as an aid in solving this problem. A portable electric heater uses 18 A of current. The manufacturer recommends that an extension cord attached to the heater receive no more than 2.0 W of power per meter of length. What is the smallest radius of copper wire that can be used in the extension cord? (*Note:* An extension cord contains two wires.)

31. [E] [SSM] The average power used by a stereo speaker is 55 W. Assuming that the speaker can be treated as a 4.0-Ω resistance, find the peak value of the ac voltage applied to the speaker.

32. [M] [GO] [CHALK] The *recovery time* of a hot water heater is the time required to heat all the water in the unit to the desired temperature. Suppose that a 52-gal ($1.00 \text{ gal} = 3.79 \times 10^{-3}\ \text{m}^3$) unit starts with cold water at 11 °C and delivers hot water at 53 °C. The unit is electric and utilizes a resistance heater (120 V ac, 3.0 Ω) to heat the water. Assuming that no heat is lost to the environment, determine the recovery time (in hours) of the unit.

33. [M] [SSM] A light bulb is connected to a 120.0-V wall socket. The current in the bulb depends on the time t according to the relation $I = (0.707\ \text{A}) \sin [(314\ \text{Hz})t]$. **(a)** What is the frequency f of the alternating current? **(b)** Determine the resistance of the bulb's filament. **(c)** What is the average power delivered to the light bulb?

Section 20.6 Series Wiring

34. [E] [SSM] Three resistors, 25, 45, and 75 Ω, are connected in series, and a 0.51-A current passes through them. What are **(a)** the equivalent resistance and **(b)** the potential difference across the three resistors?

35. [E] [GO] A 60.0-W lamp is placed in series with a resistor and a 120.0-V source. If the voltage across the lamp is 25 V, what is the resistance R of the resistor?

36. [E] [SSM] The current in a series circuit is 15.0 A. When an additional 8.00-Ω resistor is inserted in series, the current drops to 12.0 A. What is the resistance in the original circuit?

37. [E] [V-HINT] Multiple-Concept Example 9 discusses the physics principles used in this problem. Three resistors, 2.0, 4.0, and 6.0 Ω, are connected in series across a 24-V battery. Find the power delivered to each resistor.

38. [E] The current in a 47-Ω resistor is 0.12 A. This resistor is in series with a 28-Ω resistor, and the series combination is connected across a battery. What is the battery voltage?

39. [M] [V-HINT] Multiple-Concept Example 9 reviews the concepts that are important to this problem. A light bulb is wired in series with a 144-Ω resistor, and they are connected across a 120.0-V source. The power delivered to the light bulb is 23.4 W. What is the resistance of the light bulb? Note that there are two possible answers.

40. [M] [SSM] [MMH] [CHALK] Three resistors are connected in series across a battery. The value of each resistance and its maximum power rating are as follows: 2.0 Ω and 4.0 W, 12.0 Ω and 10.0 W, and 3.0 Ω and 5.0 W. **(a)** What is the greatest voltage that the battery can have without one of the resistors burning up? **(b)** How much power does the battery deliver to the circuit in (a)?

41. [M] [V-HINT] One heater uses 340 W of power when connected by itself to a battery. Another heater uses 240 W of power when connected by itself to the same battery. How much total power do the heaters use when they are both connected in series across the battery?

42. [H] Two resistances, R_1 and R_2, are connected in series across a 12-V battery. The current increases by 0.20 A when R_2 is removed, leaving R_1 connected across the battery. However, the current increases by just 0.10 A when R_1 is removed, leaving R_2 connected across the battery. Find **(a)** R_1 and **(b)** R_2.

Section 20.7 Parallel Wiring

43. [E] A coffee-maker (14 Ω) and a toaster (19 Ω) are connected in parallel to the same 120-V outlet in a kitchen. How much total power is supplied to the two appliances when both are turned on?

44. [E] For the three-way bulb (50 W, 100 W, 150 W) discussed in Conceptual Example 11, find the resistance of each of the two filaments. Assume that the wattage ratings are not limited by significant figures, and ignore any heating effects on the resistances.

45. [E] [GO] The drawing shows three different resistors in two different circuits. The battery has a voltage of $V = 24.0$ V, and the resistors have values of $R_1 = 50.0\ \Omega$, $R_2 = 25.0\ \Omega$, and $R_3 = 10.0\ \Omega$. **(a)** For the circuit on the left, determine the current through and the voltage across each resistor. **(b)** Repeat part (a) for the circuit on the right.

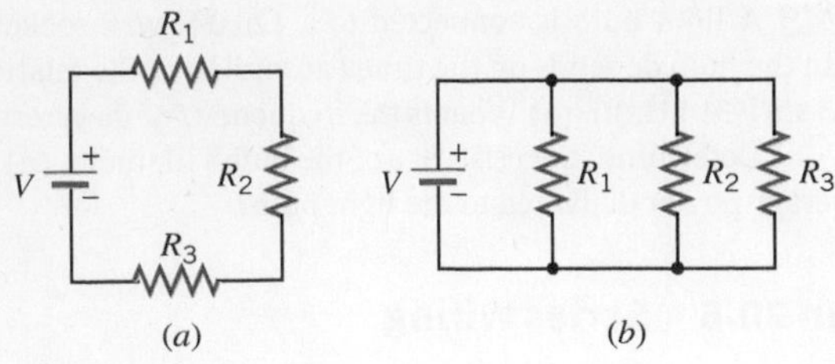

PROBLEM 45

46. **E SSM** The drawing shows a circuit that contains a battery, two resistors, and a switch. What is the equivalent resistance of the circuit when the switch is **(a)** open and **(b)** closed? What is the total power delivered to the resistors when the switch is **(c)** open and **(d)** closed?

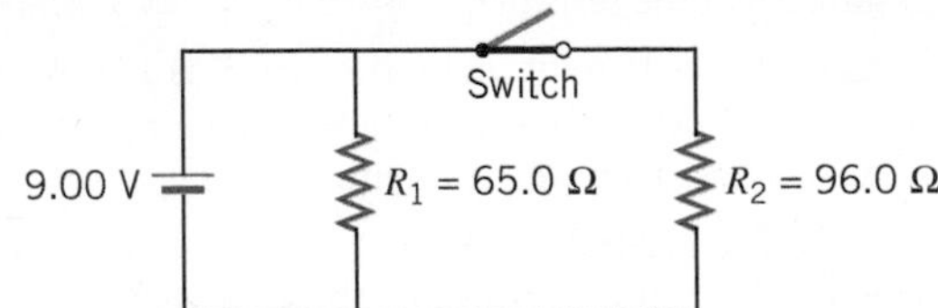

PROBLEM 46

47. **E** A 16-Ω loudspeaker, an 8.0-Ω loudspeaker, and a 4.0-Ω loudspeaker are connected in parallel across the terminals of an amplifier. Determine the equivalent resistance of the three speakers, assuming that they all behave as resistors.

48. **E SSM** Two resistors, 42.0 and 64.0 Ω, are connected in parallel. The current through the 64.0-Ω resistor is 3.00 A. **(a)** Determine the current in the other resistor. **(b)** What is the total power supplied to the two resistors?

49. **E V-HINT** Two identical resistors are connected in parallel across a 25-V battery, which supplies them with a total power of 9.6 W. While the battery is still connected, one of the resistors is heated so that its resistance doubles. The resistance of the other resistor remains unchanged. Find **(a)** the initial resistance of each resistor and **(b)** the total power delivered to the resistors after one resistor has been heated.

50. **E MMH** A coffee cup heater and a lamp are connected in parallel to the same 120-V outlet. Together, they use a total of 111 W of power. The resistance of the heater is 4.0×10^2 Ω. Find the resistance of the lamp.

51. **M CHALK** Two resistors have resistances R_1 and R_2. When the resistors are connected in series to a 12.0-V battery, the current from the battery is 2.00 A. When the resistors are connected in parallel to the battery, the total current from the battery is 9.00 A. Determine R_1 and R_2.

52. **M GO SSM** A cylindrical aluminum pipe of length 1.50 m has an inner radius of 2.00×10^{-3} m and an outer radius of 3.00×10^{-3} m. The interior of the pipe is completely filled with copper. What is the resistance of this unit? (*Hint:* Imagine that the pipe is connected between the terminals of a battery and decide whether the aluminum and copper parts of the pipe are in series or in parallel.)

53. **M GO** The drawing shows two circuits, and the same battery is used in each. The two resistances R_A in circuit A are the same, and the two resistances R_B in circuit B are the same. Knowing that the same total power is delivered in each circuit, find the ratio R_B/R_A for the circuits.

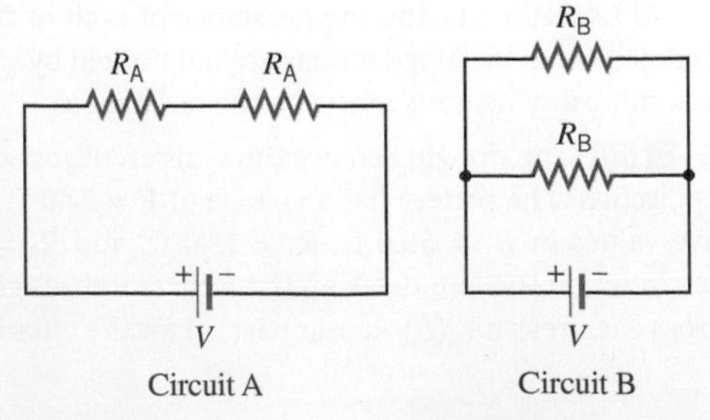

PROBLEM 53

Section 20.8 Circuits Wired Partially in Series and Partially in Parallel

54. **E** A 60.0-Ω resistor is connected in parallel with a 120.0-Ω resistor. This parallel group is connected in series with a 20.0-Ω resistor. The total combination is connected across a 15.0-V battery. Find **(a)** the current and **(b)** the power delivered to the 120.0-Ω resistor.

55. **E SSM** A 14-Ω coffee maker and a 16-Ω frying pan are connected in series across a 120-V source of voltage. A 23-Ω bread maker is also connected across the 120-V source and is in parallel with the series combination. Find the total current supplied by the source of voltage.

56. **E GO** Find the equivalent resistance between points *A* and *B* in the drawing.

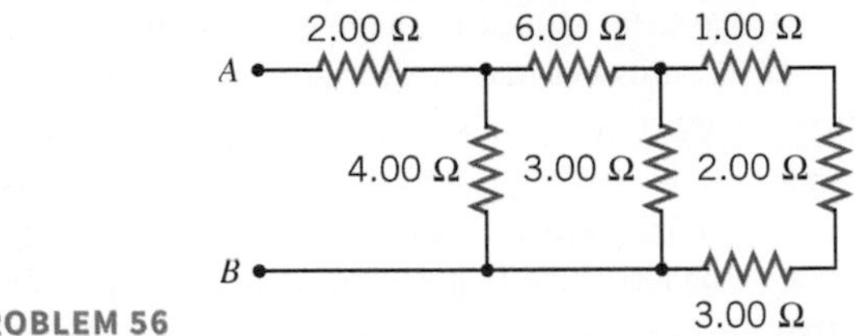

PROBLEM 56

57. **E CHALK SSM** Determine the equivalent resistance between the points *A* and *B* for the group of resistors in the drawing.

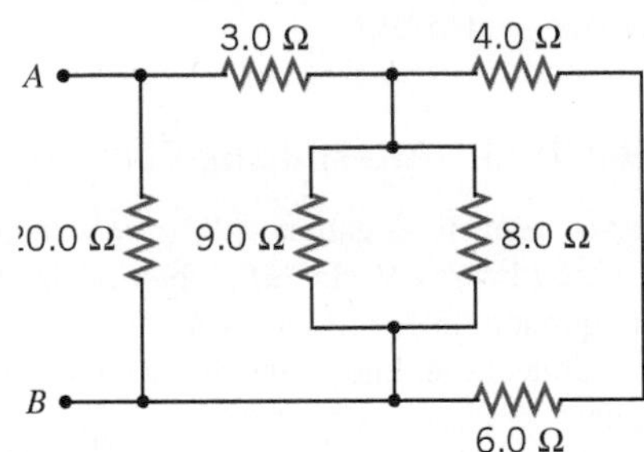

PROBLEM 57

58. **E GO** The circuit in the drawing contains three identical resistors. Each resistor has a value of 10.0 Ω. Determine the equivalent resistance between the points *a* and *b*, *b* and *c*, and *a* and *c*.

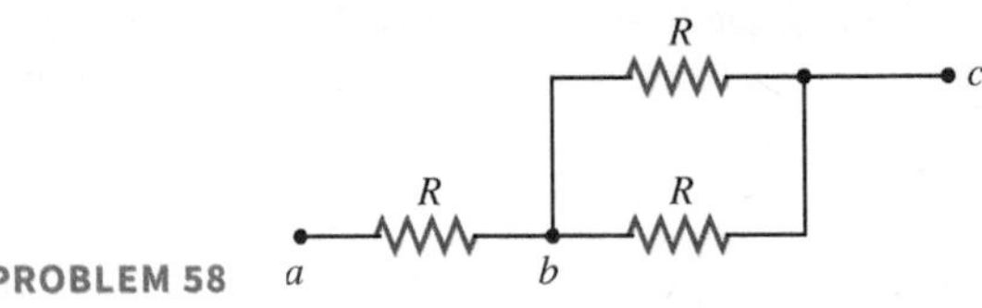

PROBLEM 58

59. **E GO** Find the equivalent resistance between the points *A* and *B* in the drawing.

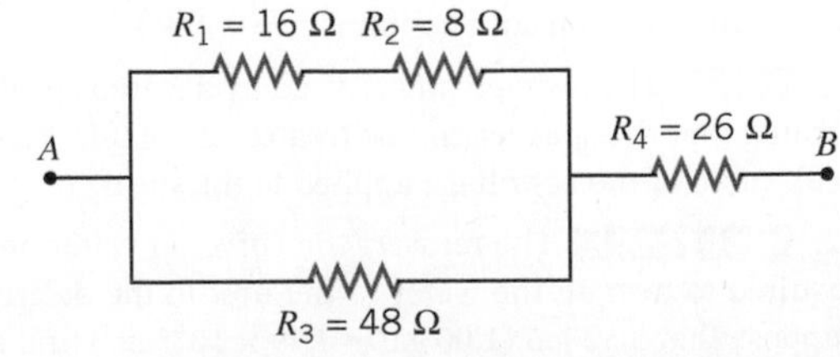

PROBLEM 59

60. **M GO** Each resistor in the three circuits in the drawing has the same resistance *R*, and the batteries have the same voltage *V*. The values for *R* and *V* are 9.0 Ω and 6.0 V, respectively. Determine the total power delivered by the battery in each of the three circuits.

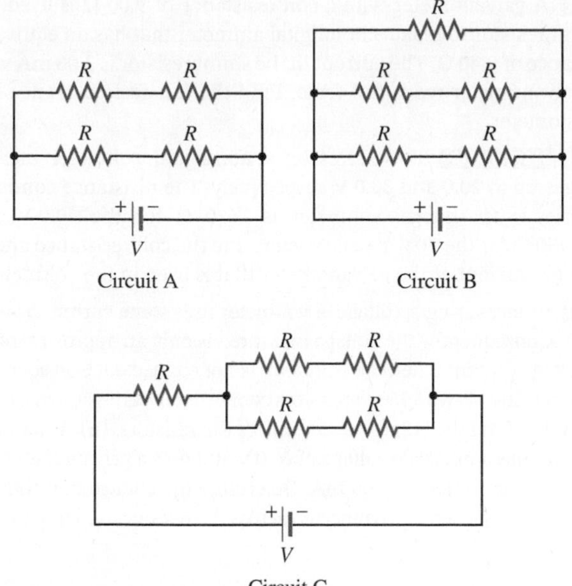

PROBLEM 60

61. **M** Eight different values of resistance can be obtained by connecting together three resistors (1.00, 2.00, and 3.00 Ω) in all possible ways. What are the values?

62. **M V-HINT MMH** Determine the power supplied to each of the resistors in the drawing.

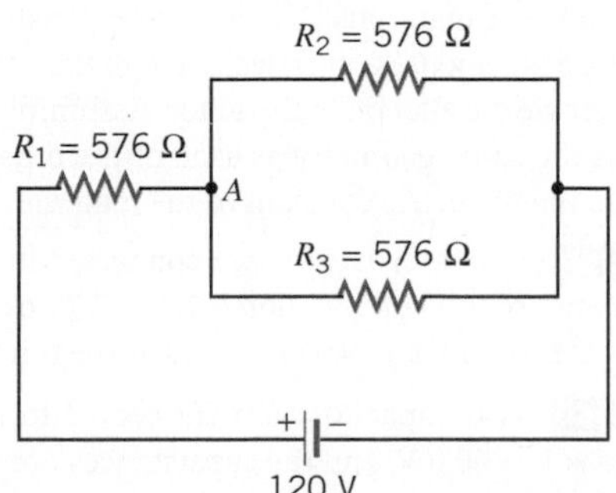

PROBLEM 62

63. **M SSM** The circuit in the drawing contains five identical resistors. The 45-V battery delivers 58 W of power to the circuit. What is the resistance R of each resistor?

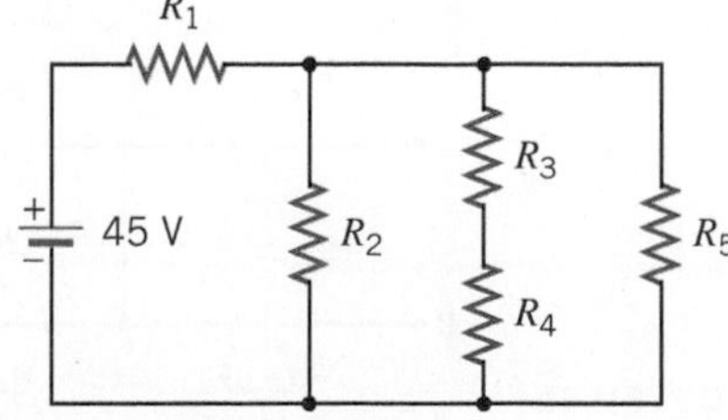

PROBLEM 63

Section 20.9 Internal Resistance

64. **E** A 1.40-Ω resistor is connected across a 9.00-V battery. The voltage between the terminals of the battery is observed to be only 8.30 V. Find the internal resistance of the battery.

65. **E** When a light bulb is connected across the terminals of a battery, the battery delivers 24 W of power to the bulb. A voltage of 11.8 V exists between the terminals of the battery, which has an internal resistance of 0.10 Ω. What is the emf of the battery?

66. **E** A battery has an internal resistance of 0.012 Ω and an emf of 9.00 V. What is the maximum current that can be drawn from the battery without the terminal voltage dropping below 8.90 V?

67. **M GO CHALK** A battery delivering a current of 55.0 A to a circuit has a terminal voltage of 23.4 V. The electric power being dissipated by the internal resistance of the battery is 34.0 W. Find the emf of the battery.

Section 20.10 Kirchhoff's Rules

68. **E SSM** Consider the circuit in the drawing. Determine **(a)** the magnitude of the current in the circuit and **(b)** the magnitude of the voltage between the points labeled A and B. **(c)** State which point, A or B, is at the higher potential.

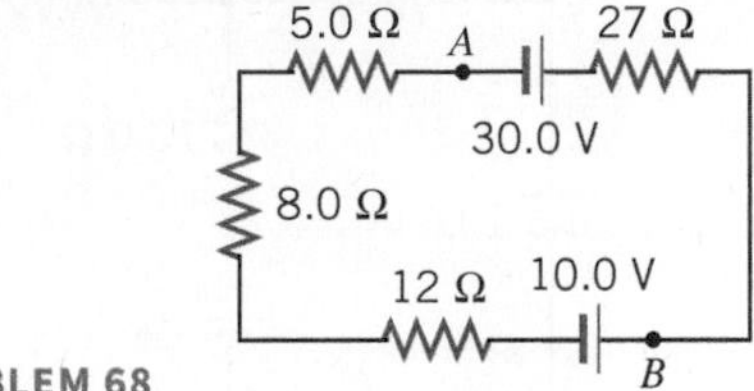

PROBLEM 68

69. **E GO** The drawing shows a portion of a larger circuit. Current flows left to right in each resistor. What is the current in the resistor R?

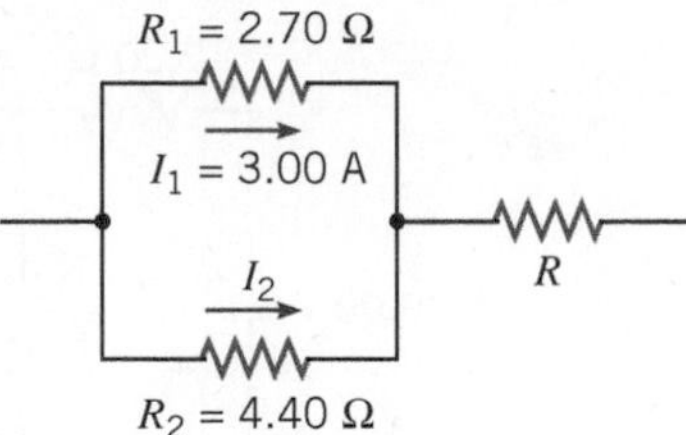

PROBLEM 69

70. **E MMH** Find the magnitude and the direction of the current in the 2.0-Ω resistor in the drawing.

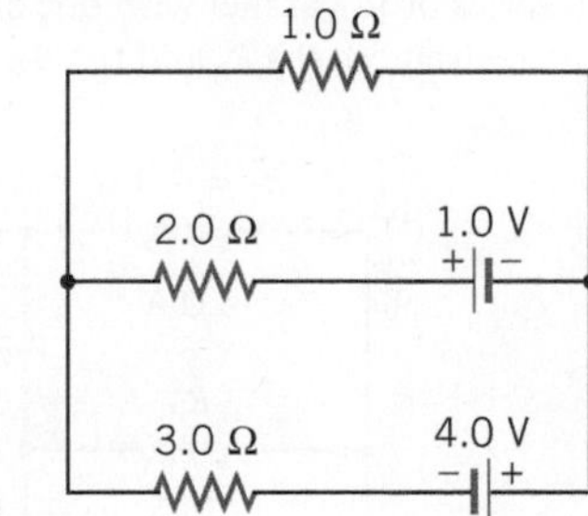

PROBLEM 70

71. **E GO** Using Kirchhoff's loop rule, find the value of the current I in part c of the drawing, where R = 5.0 Ω. (*Note:* Parts a and b of the drawing are used in the online tutorial help that is provided for this problem in the *WileyPLUS* homework management program.)

(a) (b) (c)

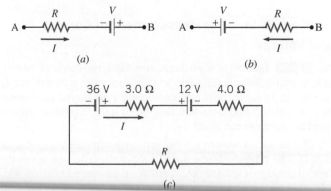

PROBLEM 71

72. **E** Determine the current (both magnitude and direction) in the 8.0- and 2.0-Ω resistors in the drawing.

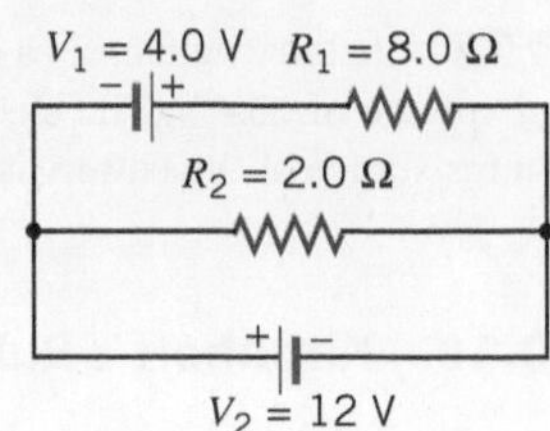

PROBLEM 72

73. M V-HINT Determine the voltage across the 5.0-Ω resistor in the drawing. Which end of the resistor is at the higher potential?

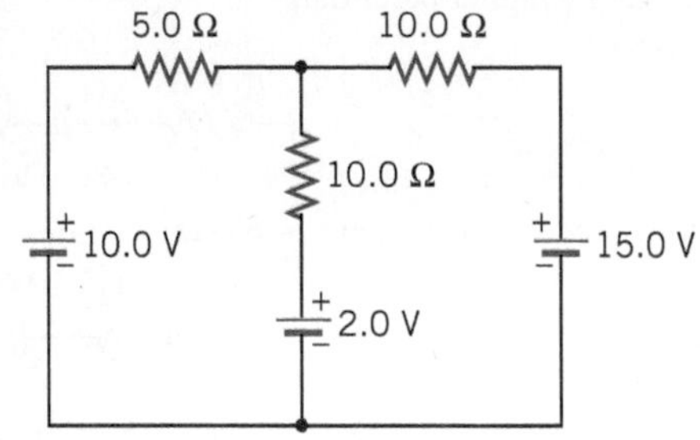

PROBLEM 73

74. M CHALK SSM MMH Find the current in the 4.00-Ω resistor in the drawing. Specify the direction of the current.

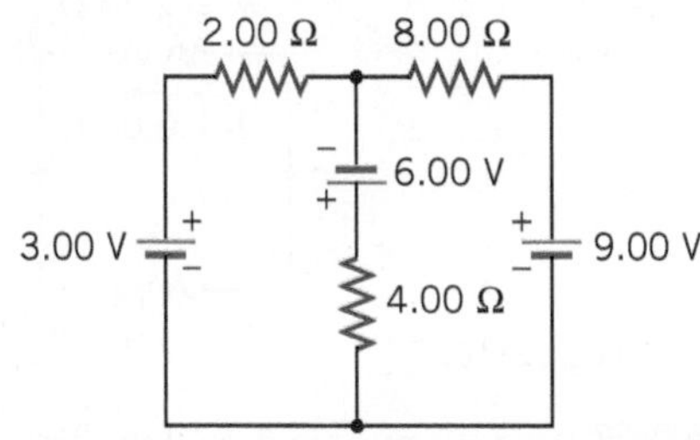

PROBLEM 74

75. H None of the resistors in the circuit shown in the drawing is connected in series or in parallel with one another. Find (a) the current I_5 and the resistances (b) R_2 and (c) R_3.

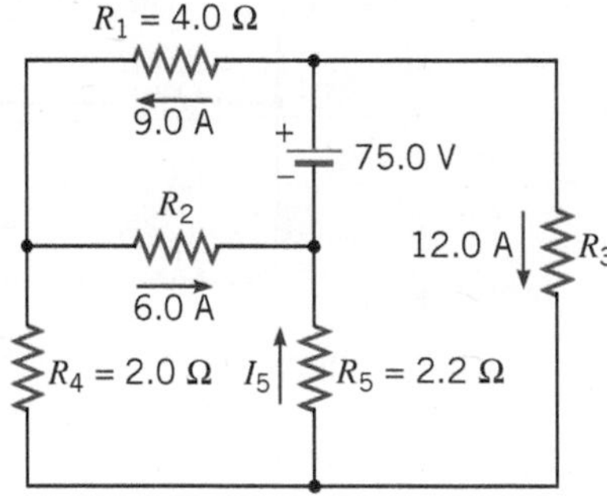

PROBLEM 75

Section 20.11 The Measurement of Current and Voltage

76. E SSM The coil of a galvanometer has a resistance of 20.0 Ω, and its meter deflects full scale when a current of 6.20 mA passes through it. To make the galvanometer into a nondigital ammeter, a 24.8-mΩ shunt resistor is added to it. What is the maximum current that this ammeter can read?

77. E The coil of wire in a galvanometer has a resistance of R_C = 60.0 Ω. The galvanometer exhibits a full-scale deflection when the current through it is 0.400 mA. A resistor is connected in series with this combination so as to produce a nondigital voltmeter. The voltmeter is to have a full-scale deflection when it measures a potential difference of 10.0 V. What is the resistance of this resistor?

78. E A galvanometer with a coil resistance of 9.00 Ω is used with a shunt resistor to make a nondigital ammeter that has an equivalent resistance of 0.40 Ω. The current in the shunt resistor is 3.00 mA when the galvanometer reads full scale. Find the full-scale current of the galvanometer.

79. M SSM CHALK Two scales on a nondigital voltmeter measure voltages up to 20.0 and 30.0 V, respectively. The resistance connected in series with the galvanometer is 1680 Ω for the 20.0-V scale and 2930 Ω for the 30.0-V scale. Determine the coil resistance and the full-scale current of the galvanometer that is used in the voltmeter.

80. H In measuring a voltage, a voltmeter uses some current from the circuit. Consequently, the voltage measured is only an approximation to the voltage present when the voltmeter is not connected. Consider a circuit consisting of two 1550-Ω resistors connected in series across a 60.0-V battery. (a) Find the voltage across one of the resistors. (b) A nondigital voltmeter has a full-scale voltage of 60.0 V and uses a galvanometer with a full-scale deflection of 5.00 mA. Determine the voltage that this voltmeter registers when it is connected across the resistor used in part (a).

Section 20.12 Capacitors in Series and in Parallel

81. E Two capacitors are connected in parallel across the terminals of a battery. One has a capacitance of 2.0 μF and the other a capacitance of 4.0 μF. These two capacitors together store 5.4×10^{-5} C of charge. What is the voltage of the battery?

82. E Three parallel plate capacitors are connected in series. These capacitors have identical geometries. However, they are filled with three different materials. The dielectric constants of these materials are 3.30, 5.40, and 6.70. It is desired to replace this series combination with a single parallel plate capacitor. Assuming that this single capacitor has the same geometry as each of the other three capacitors, determine the dielectric constant of the material with which it is filled.

83. E SSM Three capacitors are connected in series. The equivalent capacitance of this combination is 3.00 μF. Two of the individual capacitances are 6.00 μF and 9.00 μF. What is the third capacitance (in μF)?

84. E GO Two capacitors are connected to a battery. The battery voltage is V = 60.0 V, and the capacitances are C_1 = 2.00 μF and C_2 = 4.00 μF. Determine the total energy stored by the two capacitors when they are wired (a) in parallel and (b) in series.

85. E Determine the equivalent capacitance between A and B for the group of capacitors in the drawing.

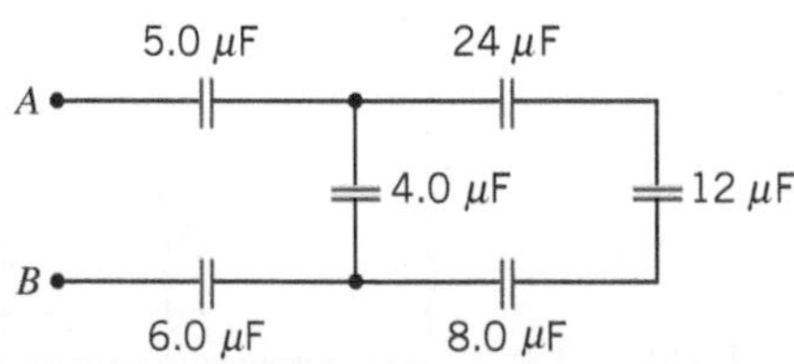

PROBLEM 85

86. E V-HINT A 2.00-μF and a 4.00-μF capacitor are connected to a 60.0-V battery. What is the total charge supplied to the capacitors when they are wired (a) in parallel and (b) in series with each other?

87. E Suppose that two capacitors (C_1 and C_2) are connected in series. Show that the sum of the energies stored in these capacitors is equal to the energy stored in the equivalent capacitor. [*Hint:* The energy stored in a capacitor can be expressed as $q^2/(2C)$.]

88. M A 3.00-μF and a 5.00-μF capacitor are connected in series across a 30.0-V battery. A 7.00-μF capacitor is then connected in parallel across the 3.00-μF capacitor. Determine the voltage across the 7.00-μF capacitor.

89. M CHALK SSM A 7.0-μF and a 3.0-μF capacitor are connected in series across a 24-V battery. What voltage is required to charge a parallel combination of the two capacitors to the same total energy?

90. H The drawing shows two capacitors that are fully charged ($C_1 = 2.00\ \mu$F, $q_1 = 6.00\ \mu$C; $C_2 = 8.00\ \mu$F, $q_2 = 12.0\ \mu$C). The switch is closed, and charge flows until equilibrium is reestablished (i.e., until both capacitors have the same voltage across their plates). Find the resulting voltage across either capacitor.

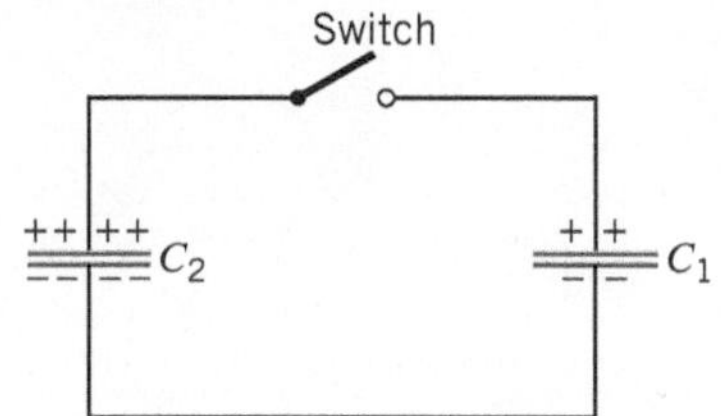

PROBLEM 90

Section 20.13 *RC* Circuits

91. E A circuit contains a resistor in series with a capacitor, the series combination being connected across the terminals of a battery, as in Figure 20.38a. The time constant for charging the capacitor is 1.5 s when the resistor has a resistance of $2.0 \times 10^4\ \Omega$. What would the time constant be if the resistance had a value of $5.2 \times 10^4\ \Omega$?

92. E V-HINT The circuit in the drawing contains two resistors and two capacitors that are connected to a battery via a switch. When the switch is closed, the capacitors begin to charge up. What is the time constant for the charging process?

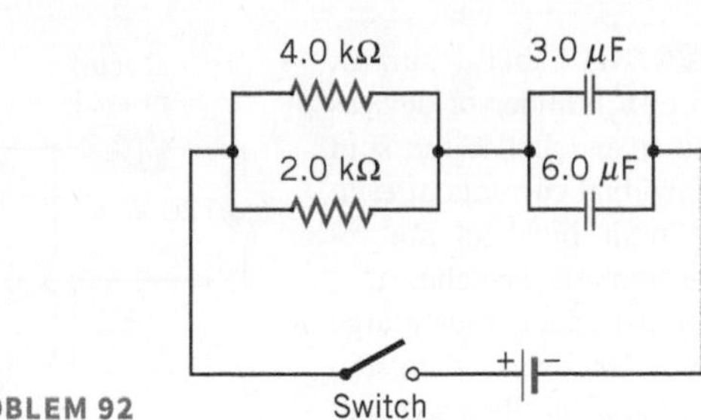

PROBLEM 92

93. M MMH How many time constants must elapse before a capacitor in a series *RC* circuit is charged to 80.0% of its equilibrium charge?

94. M MMH CHALK Four identical capacitors are connected with a resistor in two different ways. When they are connected as in part *a* of the drawing, the time constant to charge up this circuit is 0.72 s. What is the time constant when they are connected with the same resistor, as in part *b*?

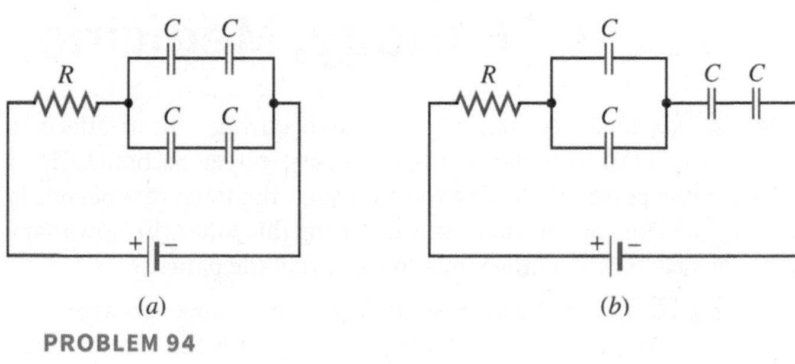

PROBLEM 94

Additional Problems

Online

95. E GO Each of the four circuits in the drawing consists of a single resistor whose resistance is either *R* or 2*R*, and a single battery whose voltage is either *V* or 2*V*. The unit of voltage in each circuit is *V* = 12.0 V and the unit of resistance is *R* = 6.00 Ω. Determine (a) the power supplied to each resistor and (b) the current delivered to each resistor.

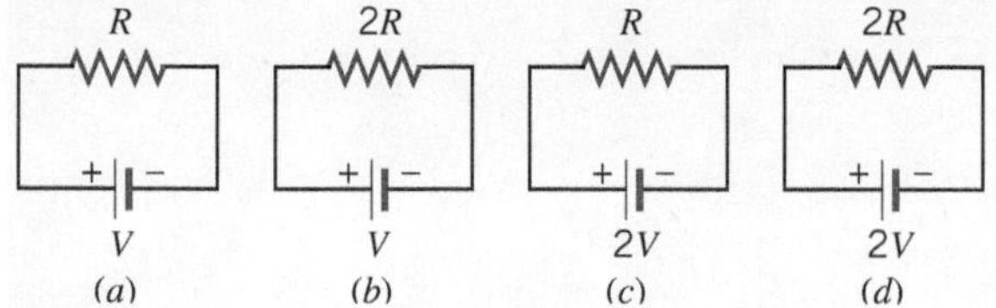

PROBLEM 95

96. E SSM In the Arctic, electric socks are useful. A pair of socks uses a 9.0-V battery pack for each sock. A current of 0.11 A is drawn from each battery pack by wire woven into the socks. Find the resistance of the wire in one sock.

97. E In Section 12.3 it was mentioned that temperatures are often measured with electrical resistance thermometers made of platinum wire. Suppose that the resistance of a platinum resistance thermometer is 125 Ω when its temperature is 20.0 °C. The wire is then immersed in boiling chlorine, and the resistance drops to 99.6 Ω. The temperature coefficient of resistivity of platinum is $\alpha = 3.72 \times 10^{-3}\ (\mathrm{C}^\circ)^{-1}$. What is the temperature of the boiling chlorine?

98. E The circuit in the drawing shows two resistors, a capacitor, and a battery. When the capacitor is fully charged, what is the magnitude *q* of the charge on one of its plates?

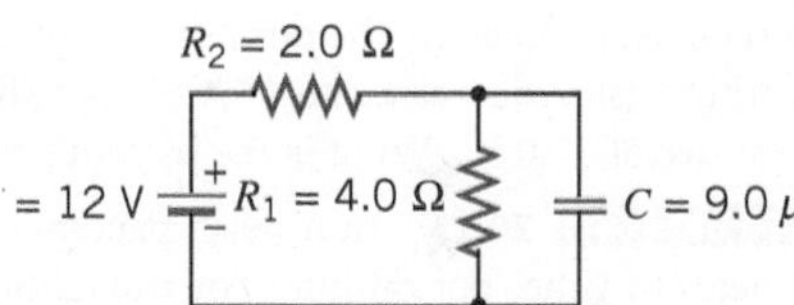

PROBLEM 98

99. E An 86-Ω resistor and a 67-Ω resistor are connected in series across a battery. The voltage across the 86-Ω resistor is 27 V. What is the voltage across the 67-Ω resistor?

100. M SSM The current in the 8.00-Ω resistor in the drawing is 0.500 A. Find the current in (a) the 20.0-Ω resistor and in (b) the 9.00-Ω resistor.

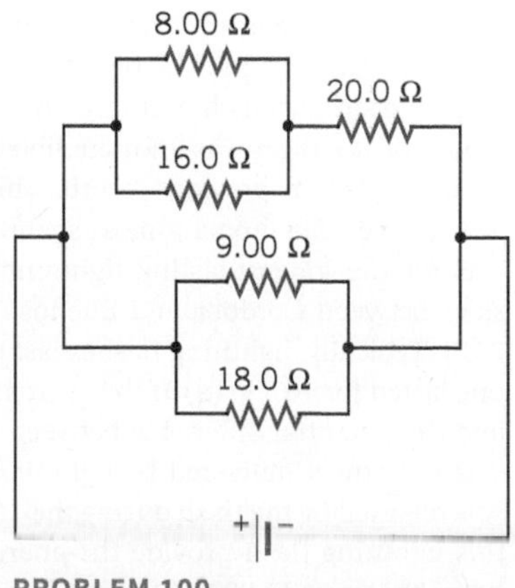

PROBLEM 100

101. M SSM An extension cord is used with an electric weed trimmer that has a resistance of 15.0 Ω. The extension cord is made of copper wire that has a cross-sectional area of $1.3 \times 10^{-6}\ \mathrm{m}^2$. The combined length of the two wires in the extension cord is 52 m. (a) Determine the resistance of the extension cord. (b) The extension cord is plugged into a 120-V socket. What voltage is applied to the trimmer itself?

102. M GO The total current delivered to a number of devices connected in parallel is the sum of the individual currents in each device. Circuit breakers are resettable automatic switches that protect against a dangerously large total current by "opening" to stop the current at a specified safe value. A 1650-W toaster, a 1090-W iron, and a 1250-W microwave oven are turned on in a kitchen. As the drawing shows, they are all connected through a 20-A circuit breaker (which has negligible resistance) to an ac voltage of 120 V. (a) Find the equivalent resistance of the three devices. (b) Obtain the total current delivered by the source and determine whether the breaker will "open" to prevent an accident.

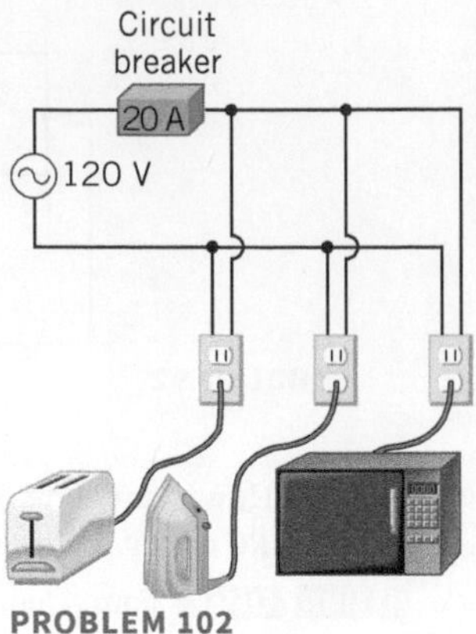

PROBLEM 102

103. M V-HINT The filament in an incandescent light bulb is made from tungsten. The light bulb is plugged into a 120-V outlet and draws a current of 1.24 A. If the radius of the tungsten wire is 0.0030 mm, how long must the wire be?

Physics in Biology, Medicine, and Sports

104. E BIO 20.2 A defibrillator is used during a heart attack to restore the heart to its normal beating pattern (see Section 19.5). A defibrillator passes 18 A of current through the torso of a person in 2.0 ms. (a) How much charge moves during this time? (b) How many electrons pass through the wires connected to the patient?

105. E BIO GO 20.2 Suppose that the resistance between the walls of a biological cell is $5.0 \times 10^9\ \Omega$. (a) What is the current when the potential difference between the walls is 75 mV? (b) If the current is composed of Na^+ ions ($q = +e$), how many such ions flow in 0.50 s?

106. H BIO 20.3 A digital thermometer employs a thermistor as the temperature-ensing element. A thermistor is a kind of semiconductor and has a large negative temperature coefficient of resistivity α. Suppose that $\alpha = -0.060\ (\text{C}°)^{-1}$ for the thermistor in a digital thermometer used to measure the temperature of a patient. The resistance of the thermistor decreases to 85% of its value at the normal body temperature of 37.0 °C. What is the patient's temperature?

107. E BIO SSM 20.13 In a heart pacemaker, a pulse is delivered to the heart 81 times per minute. The capacitor that controls this pulsing rate discharges through a resistance of $1.8 \times 10^6\ \Omega$. One pulse is delivered every time the fully charged capacitor loses 63.2% of its original charge. What is the capacitance of the capacitor?

108. M BIO 20.4 The World Meteorological Organization uses a system of satellites to continuously monitor electrical activity in the earth's atmosphere. Very recently, they observed two world-record lightning events. The first record was for the longest (in length) lightning strike, which occurred on October 31, 2018, over Brazil. The length of the flash was an incredible 709 km (441 miles)! This more than doubled the previous record, which was recorded in 2007, where a flash over Oklahoma almost spanned the entire state. The second was for the longest-lasting lightning strike, which occurred in the skies between Cordoba and Buenos Aires in Argentina on March 4, 2019. Typically, lightning flashes last just for a brief moment, but this one lasted for 16.7 s! (a) If the current in this strike was 2.00×10^5 A, and the potential difference between its ends was 1.20×10^5 V, what was the power delivered by lightning flash? (b) How much energy was released by the flash during the 16.7 s? (c) How many years could this lightning flash provide the energy requirements for an average home that uses 11 000 kWh/year?

109. M BIO 20.12 Electrogenic fish, like electric eels, have the ability to create fairly large voltages that can be used for defense or to stun their prey (see the photo). The electric eel has three organs along the length of its body that allow it to produce electricity. The organs contain special cells, called *electrocytes,* that can generate a voltage of 0.15 V/cell. When aligned, the cells resemble a stack of capacitors connected in series that allow a current of ions to flow through them. Eels are capable of producing a discharge at 860 V with a current of 1.0 A. (a) How much electrical power is delivered by the eel during a discharge? Use the information above. (b) How many cells would have to be aligned to produce a discharge voltage of 860 V? (c) What is the ratio of the capacitance of a single electrocyte to that of the equivalent capacitance of the entire stack of cells?

PROBLEM 109

110. E BIO 20.2 and 20.14 Human body tissue has a fairly low resistance, as the electrolytic fluids in our tissues are good conductors of electricity. However, the electrical resistance of human skin will largely determine how much current will flow through the body. Skin resistance will vary greatly, depending on the condition of the skin. Thick, dry skin has a much higher resistance than thin skin that is wet from, say, perspiration. Under dry conditions, the resistance of human skin can be 100 kΩ. Under wet conditions, this value can drop below 1 kΩ. What is the minimum skin resistance a person can have while receiving a shock from a 120-V outlet, if the current they receive cannot exceed 2.0×10^2 mA?

111. M BIO 20.1 In 50.0 ms, a current of 3.70 nA flows through a cell membrane wall as the result of the motion of sodium ions out

of the cell. How many moles of sodium pass through the wall during this time?

112. M BIO 20.4 After open-heart surgery, the heart may need to be "kickstarted" or have its normal sinus rhythm restored. This is often done with an internal defibrillator that consists of two conducting metal paddle electrodes that are placed directly on opposite sides of the heart (see the photos). Usually an electric shock with an energy as small as 5.0 J is sufficient to eliminate v-fib, or ventricular fibrillation, in which the ventricular chambers of the heart quiver, as opposed to pumping blood. (a) If the capacitor used in the internal defibrillator has a value of 45 μF, what is the charging voltage? (b) If the energy stored in the capacitor is sent to the patient's heart in 210 ms, what is the effective electrical resistance of the heart?

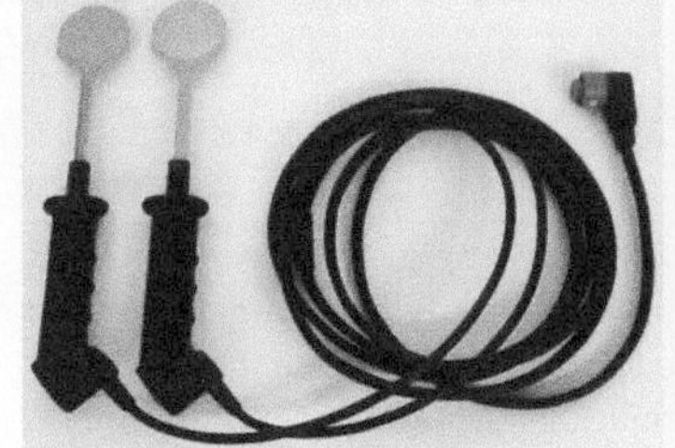

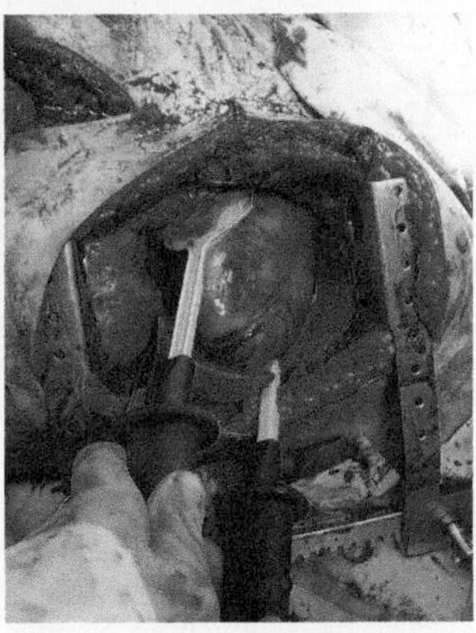

PROBLEM 112

Concepts and Calculations Problems

Online

Series and parallel wiring are two common ways in which devices, such as light bulbs, can be connected in a circuit. Problem 113 reviews the concepts of voltage, resistance, and power in the context of these two types of circuits. Kirchhoff's junction rule and loop rule are important tools for analyzing the currents and voltages in complex circuits. The rules are easy to use, once some of the subtleties are understood. Problem 114 explores these subtleties in a two-loop circuit.

113. M CHALK SSM A circuit contains a 48-V battery and a single light bulb whose resistance is 240 Ω. A second, identical, light bulb can be wired either in series or in parallel with the first one (see the figure). *Concepts:* (i) How is the power P that is delivered to a light bulb related to the bulb's resistance R and the voltage V across it? (ii) When there is only one bulb in the circuit, what is the voltage across it? (iii) The more power delivered to a bulb, the brighter it is. When two bulbs are wired in series, does the brightness of each bulb increase, decrease, or remain the same relative to the brightness of the bulb in the single-bulb circuit? (iv) When two bulbs are wired in parallel, does the brightness of each bulb increase, decrease, or remain the same relative to the brightness of the bulb in the single-bulb circuit? *Calculations:* Determine the power delivered to a single bulb when the circuit contains (a) only one bulb, (b) two bulbs in series and (c) two bulbs in parallel. Assume that the battery has no internal resistance.

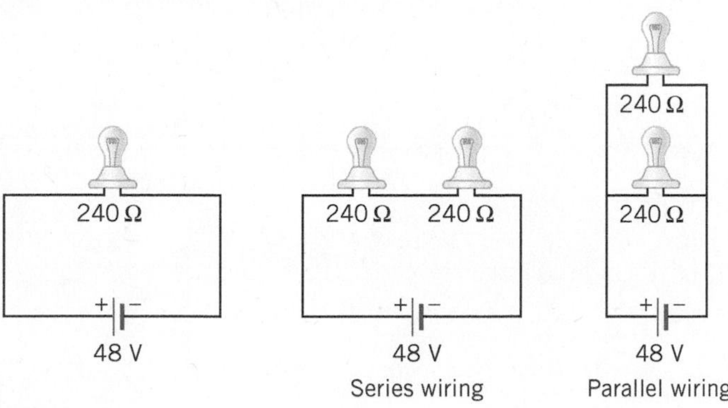

PROBLEM 113

114. M CHALK For this problem concerning Kirchhoff's junction rule and loop rule, refer to the figure. *Concepts:* (i) Notice that there are two loops, labeled 1 and 2 in this circuit. Does it matter that there is no battery in loop 1, but only two resistors? Explain. (ii) The currents through the three resistors are labeled as I_1, I_2, and I_3. Does it matter which direction, left-to-right or right-to-left, has been chosen for each circuit? (iii) When we place + and − signs on the ends of each resistor, does it matter which side is + and which is −? (iv) When we evaluate the potential drops and rises around a closed loop, does it matter which direction, clockwise or counterclockwise, is chosen for the evaluation? *Calculations:* Use Kirchhoff's junction and loop rules to determine the currents through the three resistors.

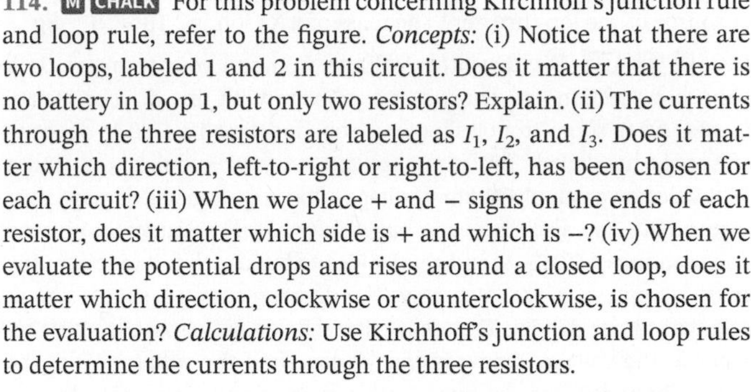

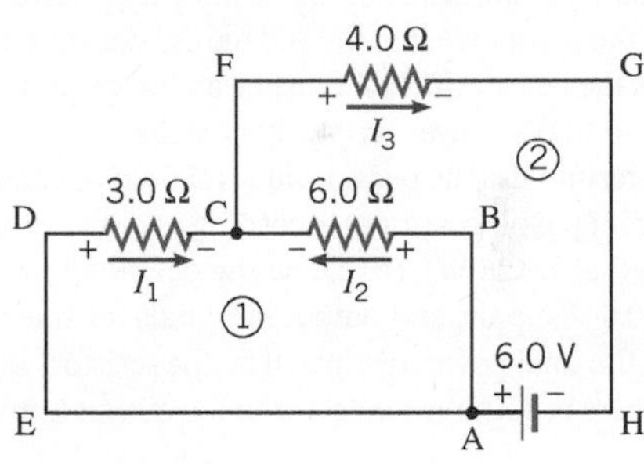

PROBLEM 114

Team Problems

Online

115. E T WS **Ion Thrusters III: Current and Power.** You and your team are designing a new propulsion engine that ejects xenon (Xe) ions in one direction, causing a spacecraft to accelerate in the opposite direction, in accordance with the conservation of linear momentum and the impulse-momentum theorem. This is exactly what happens when a rocket engine expels high velocity particles when burning fuel, except that, in the case of the ion drive, the ions are accelerated with an electric field. In this process, Xe gas atoms are ionized so that they each acquire a net charge of $+e$. They are then subjected to an electric field, which accelerates and ejects the ions from the engine. In the thruster you are designing, the Xe atoms accelerate through a potential difference of 2.06 kV between a grounded electrode and a negatively charged grid, through which the ions are ejected at a speed of 55.0 km/s relative to the engine. **(a)** If Xe ions are ejected from the thruster at a rate of 9.05×10^{-6} kg/s, what is the magnitude of the electric current that passes through the grid, that is, ejected from the thruster? **(b)** What is the electrical power in watts supplied by the power source of the ion thruster? The mass of a Xe ion is 2.18×10^{-25} kg.

116. M T WS **Dividing the Voltage.** You and your team are on a science expedition and encounter a problem with one of your instruments. After some troubleshooting, you find that its 5.00 V dc power supply has failed. You rummage around your supplies and find a 10.0-V dc power supply, but that voltage is too large and will damage your instrument. You also discover a bag of fifty 22.0 Ω resistors that have a power rating of 0.250 W, and get the idea to build a circuit to reduce the 10.0 V supply to a 5.00 V supply. The drawing shows a simple circuit called a voltage divider. The idea is that the voltage across the lower resistor in the voltage divider is equal to one-half the voltage (V) of the source, provided that the two resisters in the circuit have equal values. **(a)** Determine the resistance R in the voltage divider shown in the drawing that utilizes the resistors (i.e., R must be an integer number of 22.0 Ω resistors) and the 10.0-V power supply that results in a dc voltage of 5.00 V at the output terminals. The resistor values should be chosen to provide the largest current possible without exceeding one-half the power rating of the resistors (i.e., 0.125 W). **(b)** Draw the circuit and explicitly show the resistors, 10.0-V supply, and output terminals for the 5.00 V supply. **(c)** Determine the total power dissipated by the resistors in your voltage divider when there is nothing connected to the output terminals.

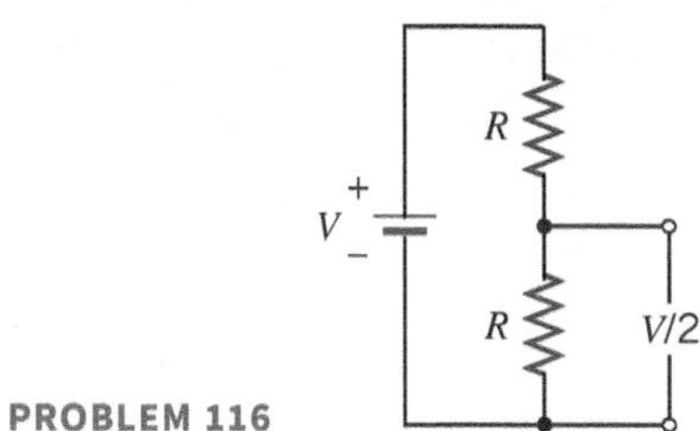

PROBLEM 116

117. M T **Fixing a Radio.** You and your team are stranded on a tropical island that hosts a deserted military base. In your efforts to get rescued, you find a radio. However, when you power it up some lights turn on, but the radio does not transmit or receive. You open up the back cover and quickly identify the problem: There is a burnt resistor on the circuit board. You have located some basic electronics tools and supplies, but the charred resistor is unidentifiable, so you do not know what to use for its replacement. There is a crude schematic on the inside of the cover of the radio, but it only gives the value of the current through the part of the circuit with the burnt resistor. You sketch the circuit with all the elements you can identify (see the drawing). **(a)** Determine the value of the damaged resistor. **(b)** The next problem is that the resistors that you have found are all 47.0 Ω. The available space will only allow for six or fewer resistors. Find a configuration of six or fewer 47.0-Ω resistors that has an equivalent resistance within 10% of that calculated in part (a).

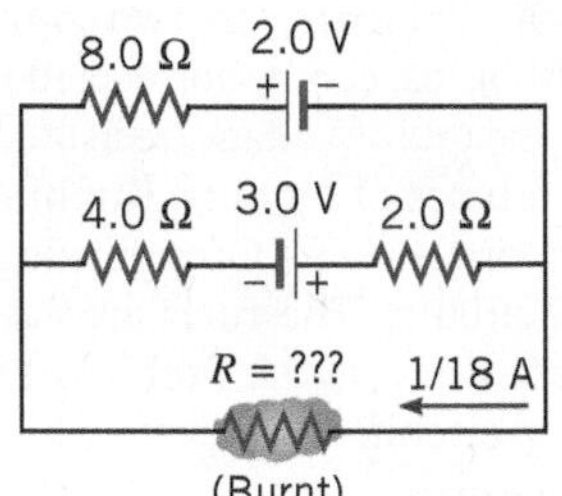

PROBLEM 117

118. M T **A Resistive Heater.** You and your team are designing a small tube heater that consists of a small ceramic tube wrapped with a special heater wire composed of Nichrome. When you run an electrical current through the wire, the wire (and therefore the tube) heats up through resistive heating. Nichrome is an alloy composed of 80% nickel and 20% chromium, and has a resistivity of $\rho = 1.25 \times 10^{-6}\ \Omega \cdot \text{m}$. **(a)** What is the resistance per centimeter of 28-gauge Nichrome wire? ("28 gauge" means the wire has a diameter $D = 0.320$ mm.) **(b)** You wrap the tube with 45.0 cm of the Nichrome wire, and you want the power output of your heater to be 120 W. What current is needed? **(c)** What voltage will provide the current calculated in (b)?

CHAPTER 21

Active regions on the surface of the sun correspond to areas of intense magnetic fields. The looping magnetic field lines above this area are illuminated by the motion of charged particles. The moving charges experience a force in the magnetic field and spiral along them. In this chapter, we will study magnetic fields and the forces they apply to charged particles. The sizes of these magnetic structures can be enormous, as the earth, drawn to scale at the lower right, demonstrates.

Solar Dynamics Observatory, NASA

Magnetic Forces and Magnetic Fields

LEARNING OBJECTIVES

After reading this module, you should be able to...

21.1 Define magnetic field.

21.2 Calculate the magnetic force on a moving charge in a magnetic field.

21.3 Analyze the motion of a charged particle in a magnetic field.

21.4 Describe how the masses of ions are determined using a mass spectrometer.

21.5 Calculate the magnetic force on a current in a magnetic field.

21.6 Calculate the torque on a current-carrying coil.

21.7 Calculate magnetic fields produced by currents.

21.8 Apply Ampère's law to calculate the magnetic field due to a steady current.

21.9 Describe magnetic materials.

21.1 Magnetic Fields

Permanent magnets have long been used in navigational compasses. As **Figure 21.1** illustrates, the compass needle is a permanent magnet supported so it can rotate freely in a plane. When the compass is placed on a horizontal surface, the needle rotates until one end points approximately to the north. The end of the needle that points north is labeled the **north magnetic pole**; the opposite end is the **south magnetic pole**.

Magnets can exert forces on each other. **Figure 21.2** shows that the magnetic forces between north and south poles have the property that

like poles repel each other, and unlike poles attract.

This behavior is similar to that of like and unlike electric charges. However, there is a significant difference between magnetic poles and electric charges. It is possible to separate positive from negative electric charges and produce isolated charges of either kind. In contrast, no one has

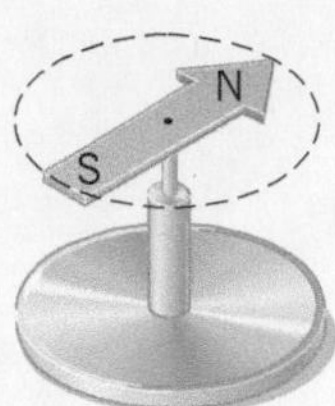

FIGURE 21.1 The needle of a compass is a permanent magnet that has a north magnetic pole (N) at one end and a south magnetic pole (S) at the other.

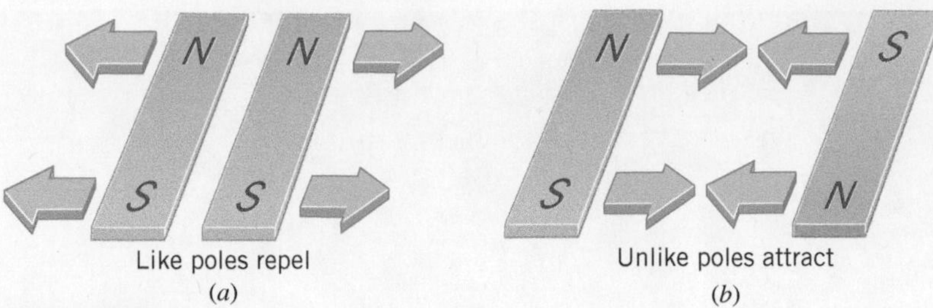

FIGURE 21.2 Bar magnets have a north magnetic pole at one end and a south magnetic pole at the other end. (*a*) Like poles repel each other, and (*b*) unlike poles attract.

found a magnetic monopole (an isolated north or south pole). Any attempt to separate north and south poles by cutting a bar magnet in half fails, because each piece becomes a smaller magnet with its own north and south poles.

Surrounding a magnet, there is a **magnetic field**. The magnetic field is analogous to the electric field that exists in the space around electric charges. Like the electric field, the magnetic field has both a magnitude and a direction. We postpone a discussion of the magnitude until Section 21.2, concentrating our attention here on the direction. ***The direction of the magnetic field at any point in space is the direction indicated by the north pole of a small compass needle placed at that point.*** In **Figure 21.3** the compass needle is symbolized by an arrow, with the head of the arrow representing the north pole. The drawing shows how compasses can be used to map out the magnetic field in the space around a bar magnet. Since like poles repel and unlike poles attract, the needle of each compass becomes aligned relative to the magnet in the manner shown in the picture. The compass needles provide a visual picture of the magnetic field that the bar magnet creates.

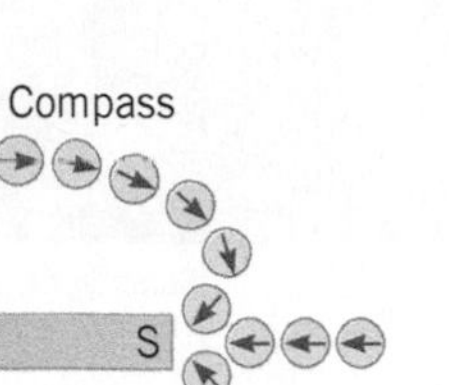

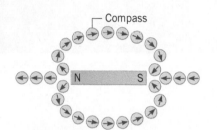

FIGURE 21.3 At any location in the vicinity of a magnet, the north pole (the arrowhead in this drawing) of a small compass needle points in the direction of the magnetic field at that location.

To help visualize the electric field, we introduced electric field lines in Section 18.7. In a similar fashion, it is possible to draw magnetic field lines, and **Figure 21.4*a*** illustrates some of the lines around a bar magnet. The lines appear to originate from the north pole and end on the south pole; they do not start or stop in midspace. A visual image of the magnetic field lines can be created by sprinkling finely ground iron filings on a piece of paper that covers the magnet. Iron filings in a magnetic field behave like tiny compasses and align themselves along the field lines, as the photograph in **Figure 21.4*b*** shows.

As is the case with electric field lines, the magnetic field at any point is tangent to the magnetic field line at that point. Furthermore, the strength of the field is proportional to the number of lines per unit area that passes through a surface oriented perpendicular to the lines. Thus, the magnetic field is stronger in regions where the field lines are relatively close together and weaker where they are relatively far apart. For instance, in **Figure 21.4*a*** the lines are closest together near the north and south poles, reflecting the fact that the strength of the field is greatest in these regions. Away from the poles, the magnetic field becomes weaker. Notice in part *c* of the drawing that the field lines in

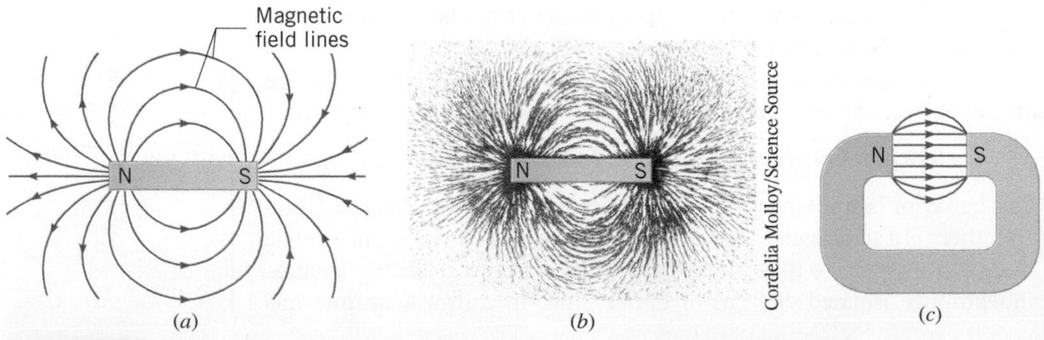

FIGURE 21.4 (*a*) The magnetic field lines and (*b*) the pattern of iron filings (black, curved regions) in the vicinity of a bar magnet. (*c*) The magnetic field lines in the gap of a horseshoe magnet.

the gap between the poles of the horseshoe magnet are nearly parallel and equally spaced, indicating that the magnetic field there is approximately constant.

Although the north pole of a compass needle points northward, it does not point exactly at the north geographic pole. The north geographic pole is that point where the earth's axis of rotation crosses the surface in the northern hemisphere (see **Figure 21.5**). Measurements of the magnetic field surrounding the earth show that the earth behaves magnetically almost as if it were a bar magnet.* As the drawing illustrates, the orientation of this fictitious bar magnet defines a magnetic axis for the earth. The location at which the magnetic axis crosses the surface in the northern hemisphere is known as the north magnetic pole. It is so named because it is the location toward which the north end of a compass needle points. Since unlike poles attract, the south pole of the earth's fictitious bar magnet lies beneath the north magnetic pole, as **Figure 21.5** indicates.

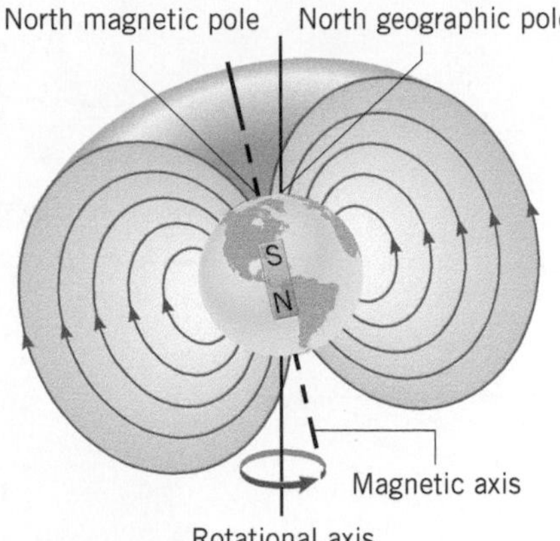

FIGURE 21.5 The earth behaves magnetically almost as if a bar magnet were located near its center. The axis of this fictitious bar magnet does not coincide with the earth's rotational axis; the two axes are currently about 9.4° apart.

The north magnetic pole does not coincide with the north geographic pole but, instead, lies at a latitude of nearly 80°, just northwest of Ellef Ringnes Island in extreme northern Canada. It is interesting to note that the position of the north magnetic pole is not fixed, but moves over the years. Pointing as it does at the north magnetic pole, a compass needle deviates from the north geographic pole. The angle that a compass needle deviates is called the *angle of declination.* For New York City, the present angle of declination is about 13° west, meaning that a compass needle points 13° west of geographic north.

Figure 21.5 shows that the earth's magnetic field lines are not parallel to the surface at all points. For instance, near the north magnetic pole the field lines are almost perpendicular to the surface of the earth. The angle that the magnetic field makes with respect to the surface at any point is known as the *angle of dip.*

BIO THE PHYSICS OF . . . navigation in animals. Some animals can sense the earth's magnetic field and use it for navigational purposes. Until recently (2004), the only examples of this ability were in vertebrates, or animals that have a backbone, such as migratory birds. Now, however, researchers have found that the spiny lobster (see **Figure 21.6**), which is an invertebrate, can also use the earth's magnetic field to navigate and can determine its geographic location in a way similar to that of a person using the Global Positioning System (see Section 5.5). This ability may be related to the presence in the lobsters of the mineral magnetite, a magnetic material used for compass needles.

D.P. Wilson/FLPA/Science Source

FIGURE 21.6 Spiny lobsters use the earth's magnetic field to navigate and determine their geographic position.

21.2 The Force That a Magnetic Field Exerts on a Moving Charge

When a charge is placed in an electric field, it experiences an electric force, as Section 18.6 discusses. When a charge is placed in a magnetic field, it also experiences a force, provided that certain conditions are met, as we will see. The **magnetic force**, like all the forces we have studied (e.g., the gravitational, elastic, and electric forces), may contribute to the net force that causes an object to accelerate. Thus, when present, the magnetic force must be included in Newton's second law.

The following two conditions must be met for a charge to experience a magnetic force when placed in a magnetic field:

1. The charge must be moving, because no magnetic force acts on a stationary charge.
2. The velocity of the moving charge must have a component that is perpendicular to the direction of the magnetic field.

To examine the second condition, consider **Figure 21.7**, which shows a positive test charge $+q_0$ moving with a velocity $\vec{\mathbf{v}}$ through a magnetic field $\vec{\mathbf{B}}$. The field is produced

*At present it is not known with certainty what causes the earth's magnetic field. The magnetic field seems to originate from electric currents that in turn arise from electric charges circulating within the liquid outer region of the earth's core. Section 21.7 discusses how a current produces a magnetic field.

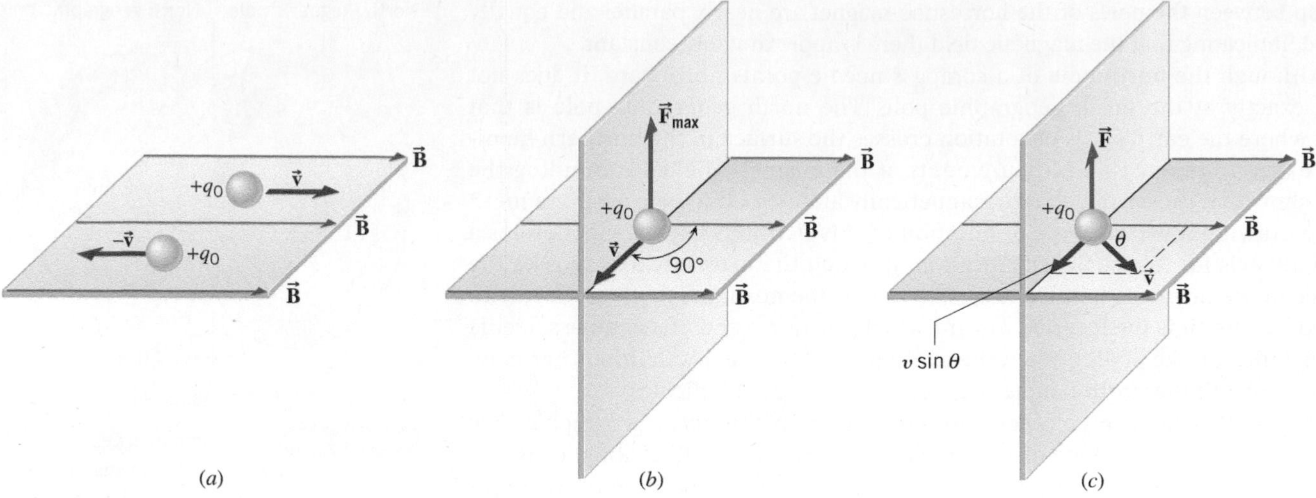

FIGURE 21.7 (*a*) No magnetic force acts on a charge moving with a velocity $\vec{v}$ that is parallel or antiparallel to a magnetic field $\vec{B}$. (*b*) The charge experiences a maximum force $\vec{F}_{max}$ when the charge moves perpendicular to the field. (*c*) If the charge travels at an angle θ with respect to $\vec{B}$, only the velocity component perpendicular to $\vec{B}$ gives rise to a magnetic force $\vec{F}$, which is smaller than $\vec{F}_{max}$. This component is $v \sin\theta$.

by magnets not shown in the drawing and is assumed to be constant in both magnitude and direction. If the charge moves *parallel or antiparallel* to the field, as in Figure 21.7*a*, the charge experiences *no magnetic force*. If, however, the charge moves *perpendicular* to the field, as in Figure 21.7*b*, the charge experiences the *maximum possible force* $\vec{F}_{max}$. In general, if a charge moves at an angle θ* with respect to the field (see Figure 21.7*c*), only the velocity component $v \sin\theta$, which is perpendicular to the field, gives rise to a magnetic force. This force $\vec{F}$ is smaller than the maximum possible force. The component of the velocity that is parallel to the magnetic field yields no force.

Figure 21.7 shows that the direction of the magnetic force $\vec{F}$ is perpendicular to both the velocity $\vec{v}$ and the magnetic field $\vec{B}$; in other words, $\vec{F}$ is perpendicular to the plane defined by $\vec{v}$ and $\vec{B}$. As an aid in remembering the direction of the force, it is convenient to use *Right-Hand Rule No. 1 (RHR-1)*, as Animated Figure 21.8 illustrates:

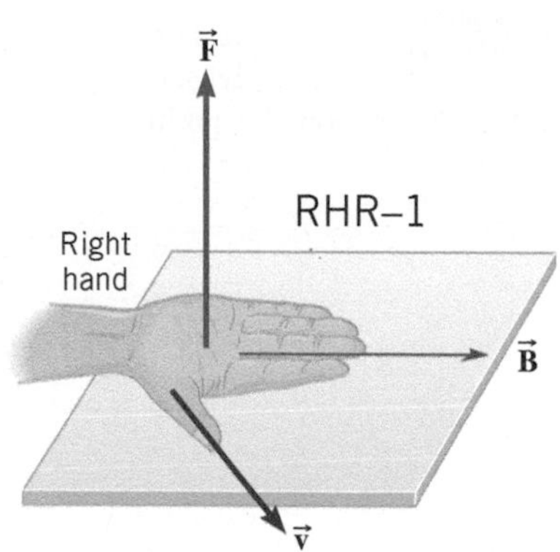

ANIMATED FIGURE 21.8 Right-Hand Rule No. 1 is illustrated. When the right hand is oriented so the fingers point along the magnetic field $\vec{B}$ and the thumb points along the velocity $\vec{v}$ of a positively charged particle, the palm faces in the direction of the magnetic force $\vec{F}$ applied to the particle.

> ***Right-Hand Rule No. 1:* The RHR to Find the Direction of the Magnetic Force That Acts on Moving Charges and Currents (Section 21.5) in a Magnetic Field.** Extend the right hand so the fingers point along the direction of the magnetic field $\vec{B}$ and the thumb points along the velocity $\vec{v}$ of the charge. The palm of the hand then faces in the direction of the magnetic force $\vec{F}$ that acts on a positive charge.

It is as if the open palm of the right hand pushes on the positive charge in the direction of the magnetic force.

Problem-Solving Insight **If the moving charge is negative instead of positive, the direction of the magnetic force is opposite to that predicted by RHR-1.**

Thus, there is an easy method for finding the force on a moving negative charge. First, assume that the charge is positive and use RHR-1 to find the direction of the force. Then, reverse this direction to find the direction of the force acting on the negative charge.

We will now use what we know about the magnetic force to define the magnetic field, in a procedure that is analogous to that used in Section 18.6 to define the electric field. Recall that the electric field at any point in space is the force per unit charge that acts on a test charge q_0 placed at that point. In other words, to determine the electric field $\vec{E}$, we divide the electrostatic force $\vec{F}$ by the charge q_0: $\vec{E} = \vec{F}/q_0$. In the magnetic case, however, the test charge is moving, and the force depends not only on the charge q_0, but also on the velocity component $v \sin\theta$ that is perpendicular to the magnetic field. Therefore, to determine the magnitude of the magnetic field, we divide the magnitude of the magnetic force by the magnitude $|q_0|$ of the charge and also by $v \sin\theta$, according to the following definition:

*The angle θ between the velocity of the charge and the magnetic field is chosen so that it lies in the range $0 \le \theta \le 180°$.

DEFINITION OF THE MAGNETIC FIELD

The magnitude B of the magnetic field at any point in space is defined as

$$B = \frac{F}{|q_0|(v \sin \theta)} \tag{21.1}$$

where F is the magnitude of the magnetic force on a test charge, $|q_0|$ is the magnitude of the test charge, and v is the magnitude of the charge's velocity, which makes an angle θ ($0 \le \theta \le 180°$) with the direction of the magnetic field. The magnetic field $\vec{B}$ is a vector, and its direction can be determined by using a small compass needle.

SI Unit of Magnetic Field: $\dfrac{\text{newton} \cdot \text{second}}{\text{coulomb} \cdot \text{meter}} = 1$ **tesla (T)**

The unit of magnetic field strength that follows from Equation 21.1 is the N · s/(C · m). This unit is called the *tesla* (T), a tribute to the Croatian-born American engineer Nikola Tesla (1856–1943). Thus, one tesla is the strength of the magnetic field in which a unit test charge, traveling perpendicular to the magnetic field at a speed of one meter per second, experiences a force of one newton. Because a coulomb per second is an ampere (1 C/s = 1 A, see Section 20.1), the tesla is often written as 1 T = 1 N/(A · m).

In many situations the magnetic field has a value that is considerably less than one tesla. For example, the strength of the magnetic field near the earth's surface is approximately 10^{-4} T. In such circumstances, a magnetic field unit called the *gauss* (G) is sometimes used. Although not an SI unit, the gauss is a convenient size for many applications involving magnetic fields. The relation between the gauss and the tesla is

$$1 \text{ gauss} = 10^{-4} \text{ tesla}$$

Example 1 deals with the magnetic force exerted on a moving proton and on a moving electron.

EXAMPLE 1 | Magnetic Forces on Charged Particles

A proton in a particle accelerator has a speed of 5.0×10^6 m/s. The proton encounters a magnetic field whose magnitude is 0.40 T and whose direction makes an angle of $\theta = 30.0°$ with respect to the proton's velocity (see **Figure 21.7c**). Find the magnitude and direction of **(a)** the magnetic force on the proton and **(b)** the acceleration of the proton. **(c)** What would be the force and acceleration if the particle were an electron instead of a proton?

Reasoning For both the proton and the electron, the magnitude of the magnetic force is given by Equation 21.1. The magnetic forces that act on these particles have opposite directions, however, because the charges have opposite signs. In either case, the acceleration is given by Newton's second law, which applies to the magnetic force just as it does to any force. In using the second law, we must take into account the fact that the masses of the proton and the electron are different.

Solution **(a)** The positive charge on a proton is 1.60×10^{-19} C, and according to Equation 21.1, the magnitude of the magnetic force is $F = |q_0|vB \sin \theta$. Therefore,

$$F = (1.60 \times 10^{-19}\text{ C})(5.0 \times 10^6\text{ m/s})(0.40\text{ T})(\sin 30.0°)$$

$$= \boxed{1.6 \times 10^{-13}\text{ N}}$$

The direction of the magnetic force is given by RHR-1 and is ***upward*** in **Figure 21.7c**, with the magnetic field pointing to the right.

(b) The magnitude a of the proton's acceleration follows directly from Newton's second law as the magnitude of the net force divided by the mass m_p of the proton. Since the only force acting on the proton is the magnetic force F, it is the net force. Thus,

$$a = \frac{F}{m_p} = \frac{1.6 \times 10^{-13}\text{ N}}{1.67 \times 10^{-27}\text{ kg}} = \boxed{9.6 \times 10^{13}\text{ m/s}^2} \tag{4.1}$$

The direction of the acceleration is the same as the direction of the net force (the magnetic force).

(c) The magnitude of the magnetic force on the electron is the same as that on the proton, since both have the same velocity and charge magnitude. However, the direction of the force on the electron is opposite to that on the proton, or ***downward*** in **Figure 21.7c**, since the electron charge is negative. Furthermore, the electron has a smaller mass m_e and, therefore, experiences a significantly greater acceleration:

$$a = \frac{F}{m_e} = \frac{1.6 \times 10^{-13}\text{ N}}{9.11 \times 10^{-31}\text{ kg}} = \boxed{1.8 \times 10^{17}\text{ m/s}^2}$$

The direction of this acceleration is downward in **Figure 21.7c**.

Check Your Understanding

(*The answers are given at the end of the book.*)

1. Suppose that you accidentally use your left hand, instead of your right hand, to determine the direction of the magnetic force that acts on a positive charge moving in a magnetic field. Do you get the correct answer? **(a)** Yes, because either hand can be used **(b)** No, because the direction you get will be perpendicular to the correct direction **(c)** No, because the direction you get will be opposite to the correct direction

2. Two particles, having the same charge but different velocities, are moving in a constant magnetic field (see **CYU Figure 21.1**, where the velocity vectors are drawn to scale). Which particle, if either, experiences the greater magnetic force? **(a)** Particle 1 experiences the greater force, because it is moving perpendicular to the magnetic field. **(b)** Particle 2 experiences the greater force, because it has the greater speed. **(c)** Particle 2 experiences the greater force, because a component of its velocity is parallel to the magnetic field. **(d)** Both particles experience the same magnetic force, because the component of each velocity that is perpendicular to the magnetic field is the same.

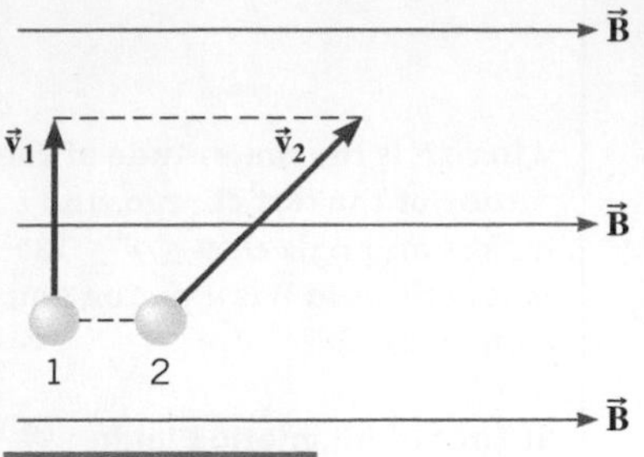

CYU FIGURE 21.1

3. A charged particle, passing through a certain region of space, has a velocity whose magnitude and direction remain constant. **(a)** If it is known that the external magnetic field is zero everywhere in this region, can you conclude that the external electric field is also zero? **(b)** If it is known that the external electric field is zero everywhere, can you conclude that the external magnetic field is also zero?

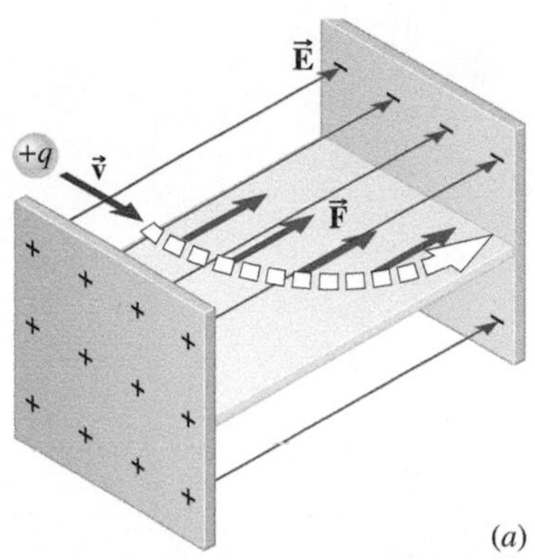

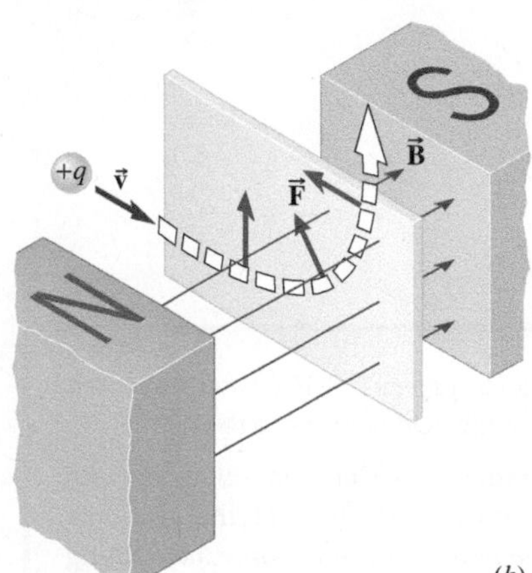

FIGURE 21.9 (*a*) The electric force $\vec{F}$ that acts on a positive charge is parallel to the electric field $\vec{E}$. (*b*) The magnetic force $\vec{F}$ is perpendicular to both the magnetic field $\vec{B}$ and the velocity $\vec{v}$.

21.3 The Motion of a Charged Particle in a Magnetic Field

Comparing Particle Motion in Electric and Magnetic Fields

The motion of a charged particle in an electric field is noticeably different from the motion in a magnetic field. For example, **Figure 21.9*a*** shows a positive charge moving between the plates of a parallel plate capacitor. Initially, the charge is moving perpendicular to the direction of the electric field. Since the direction of the electric force on a positive charge is in the same direction as the electric field, the particle is deflected sideways. Part *b* of the drawing shows the same particle traveling initially at right angles to a magnetic field. An application of RHR-1 shows that when the charge enters the field, the charge is deflected upward (not sideways) by the magnetic force. As the charge moves upward, the direction of the magnetic force changes, always remaining perpendicular to both the magnetic field and the velocity. Conceptual Example 2 focuses on the difference in how electric and magnetic fields apply forces to a moving charge.

CONCEPTUAL EXAMPLE 2 | The Physics of a Velocity Selector

A velocity selector is a device for measuring the velocity of a charged particle. The device operates by applying electric and magnetic forces to the particle in such a way that these forces balance. **Figure 21.10*a*** shows a particle with a positive charge $+q$ and a velocity $\vec{v}$ that is perpendicular to a constant magnetic field* $\vec{B}$. **Figure 21.10*b*** illustrates a velocity selector, which is a

*In many instances it is convenient to orient the magnetic field $\vec{B}$ so its direction is perpendicular to the page. In these cases it is customary to use a dot to symbolize the magnetic field pointing *out of the page* (toward the reader); this dot symbolizes the tip of the arrow representing the $\vec{B}$ vector. A region in which a magnetic field is directed *into the page* is drawn as a series of crosses that indicate the tail feathers of the arrows representing the $\vec{B}$ vectors. Therefore, regions in which a magnetic field is directed out of the page or into the page are drawn as shown below:

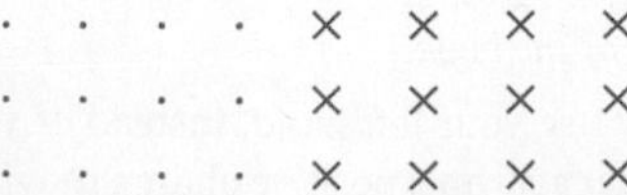

cylindrical tube that is located within the magnetic field. Inside the tube there is a parallel plate capacitor that produces an electric field $\vec{\mathbf{E}}$ (not shown) perpendicular to the magnetic field. The charged particle enters the left end of the tube, moving perpendicular to the magnetic field. If the strengths of $\vec{\mathbf{E}}$ and $\vec{\mathbf{B}}$ are adjusted properly, the electric and magnetic forces acting on the particle will cancel each other. With no net force acting on the particle, the velocity remains unchanged, according to Newton's first law. As a result, the particle moves in a straight line at a constant speed and exits at the right end of the tube. The magnitude of the velocity that is "selected" can be determined from a knowledge of the strengths of the electric and magnetic fields. Particles with velocities different from the one "selected" are deflected and do not exit at the right end of the tube.

How should the electric field $\vec{\mathbf{E}}$ be directed so that the force it applies to the particle can balance the magnetic force produced by $\vec{\mathbf{B}}$? The electric field should be directed: **(a)** in the same direction as the magnetic field; **(b)** in a direction opposite to that of the magnetic field; **(c)** from the upper plate of the parallel plate capacitor toward the lower plate; **(d)** from the lower plate of the parallel plate capacitor toward the upper plate.

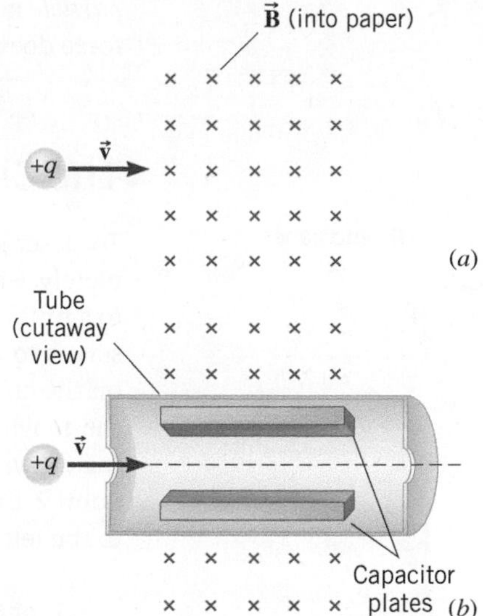

FIGURE 21.10 (*a*) A particle with a positive charge q and velocity $\vec{\mathbf{v}}$ moves perpendicularly into a magnetic field $\vec{\mathbf{B}}$. (*b*) A velocity selector is a tube in which an electric field (not shown) is perpendicular to a magnetic field, and the field magnitudes are adjusted so that the electric and magnetic forces acting on the particle cancel each other.

Reasoning If the electric and magnetic forces are to cancel each other, they must have opposite directions. The direction of the magnetic force can be found by applying Right-Hand Rule No. 1 (RHR-1) to the moving charged particle. This rule reveals that the magnetic force acting on the positively charged particle in **Figure 21.10*a*** is directed upward, toward the top of the page when the particle enters the field region. Since the particle is positively charged, the direction of the electric force is the same as the direction of the electric field produced by the capacitor plates.

Answers (a), (b), and (d) are incorrect. The electric force on the positively charged particle is in the direction of the electric field. Thus, the electric force would point perpendicularly into the page in answer (a), perpendicularly out of the page in answer (b), and upward toward the top of the page in answer (d). Since the magnetic force points upward toward the top of the page when the particle enters the field region, these electric forces do not have the proper direction to cancel the magnetic force.

Answer (c) is correct. Since the magnetic force is directed upward when the particle enters the field region, the electric force must be directed downward. The force applied to a positive charge by an electric field has the same direction as the field itself, so the electric field must point downward, from the upper plate of the capacitor toward the lower plate. As a result, the upper plate must be positively charged.

Related Homework: Problems 23, 27

We have seen that a charged particle traveling in a magnetic field experiences a magnetic force that is always perpendicular to the field. In contrast, the force applied by an electric field is always parallel (or antiparallel) to the field direction. Because of the difference in the way that electric and magnetic fields exert forces, the work done on a charged particle by each field is different, as we now discuss.

The Work Done on a Charged Particle Moving Through Electric and Magnetic Fields

In **Figure 21.9*a*** an electric field applies a force to a positively charged particle, and, consequently, the path of the particle bends in the direction of the force. Because there is a component of the particle's displacement in the direction of the electric force, the force does work on the particle, according to Equation 6.1. This work increases the kinetic energy and, hence, the speed of the particle, in accord with the work–energy theorem (see Section 6.2). In contrast, the magnetic force in **Figure 21.9*b*** always acts in a direction that is perpendicular to the motion of the charge. Consequently, the displacement of the moving charge never has a component in the direction of the magnetic force. As a result, *the magnetic force cannot do work and change the kinetic energy of the charged* *particle* in **Figure 21.9b**. Thus, the speed of the particle *does not* change, although the force does alter the direction of the motion.

The Circular Trajectory

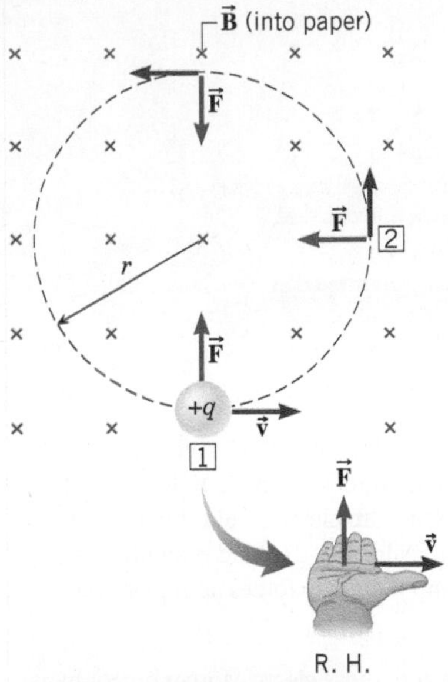

To describe the motion of a charged particle in a constant magnetic field more completely, let's discuss the special case in which the velocity of the particle is perpendicular to a uniform magnetic field. As **Interactive Figure 21.11** illustrates, the magnetic force serves to move the particle in a circular path. To understand why, consider two points on the circumference labeled 1 and 2. When the positively charged particle is at point 1, the magnetic force $\vec{F}$ is perpendicular to the velocity $\vec{v}$ and points directly upward in the drawing. This force causes the trajectory to bend upward. When the particle reaches point 2, the magnetic force still remains perpendicular to the velocity but is now directed to the left in the drawing.

INTERACTIVE FIGURE 21.11 A positively charged particle is moving perpendicular to a constant magnetic field. The magnetic force $\vec{F}$ causes the particle to move on a circular path (R.H. = right hand).

Problem-Solving Insight The magnetic force always remains perpendicular to the velocity and is directed toward the center of the circular path.

To find the radius of the path in **Interactive Figure 21.11**, we use the concept of centripetal force from Section 5.3. The centripetal force is the net force, directed toward the center of the circle, that is needed to keep a particle moving along a circular path. The magnitude F_c of this force depends on the speed v and mass m of the particle, as well as the radius r of the circle:

$$F_c = \frac{mv^2}{r} \tag{5.3}$$

In the present situation, the magnetic force furnishes the centripetal force. Being perpendicular to the velocity, the magnetic force does no work in keeping the charge $+q$ on the circular path. According to Equation 21.1, the magnitude of the magnetic force is given by $|q|vB \sin 90°$, so $|q|vB = mv^2/r$ or

$$r = \frac{mv}{|q|B} \tag{21.2}$$

This result shows that the radius of the circle is inversely proportional to the magnitude of the magnetic field, with stronger fields producing "tighter" circular paths. Example 3 illustrates an application of Equation 21.2.

EXAMPLE 3 | The Motion of a Proton

A proton is released from rest at point A, which is located next to the positive plate of a parallel plate capacitor (see **Figure 21.12**). The proton then accelerates toward the negative plate, leaving the capacitor at point B through a small hole in the plate. The electric potential of the positive plate is 2100 V greater than that of the negative plate, so $V_A - V_B = 2100$ V. Once outside the capacitor, the proton travels at a constant velocity until it enters a region of constant magnetic field of magnitude 0.10 T. The velocity is perpendicular to the magnetic field, which is directed out of the page in **Figure 21.12**. Find **(a)** the speed v_B of the proton when it leaves the negative plate of the capacitor, and **(b)** the radius r of the circular path on which the proton moves in the magnetic field.

Reasoning The only force that acts on the proton (charge = $+e$) while it is between the capacitor plates is the conservative electric force. Thus, we can use the conservation of energy to find the speed of the proton when it leaves the negative plate. The total energy of the proton is the sum of its kinetic energy, $\frac{1}{2}mv^2$, and its electric potential energy, EPE. Following Example 4 in Chapter 19, we set the total energy at point B equal to the total energy at point A:

$$\underbrace{\tfrac{1}{2}mv_B^2 + \text{EPE}_B}_{\text{Total energy at } B} = \underbrace{\tfrac{1}{2}mv_A^2 + \text{EPE}_A}_{\text{Total energy at } A}$$

We note that $v_A = 0$ m/s, since the proton starts from rest, and use Equation 19.4 to set $\text{EPE}_A - \text{EPE}_B = e(V_A - V_B)$. Then the conservation

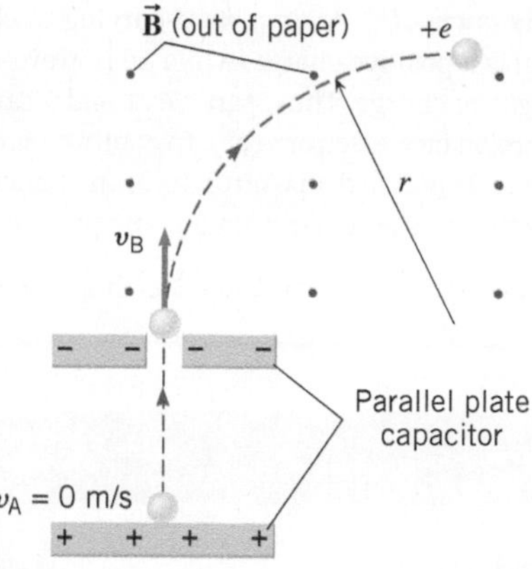

FIGURE 21.12 A proton, starting from rest at the positive plate of the capacitor, accelerates toward the negative plate. After leaving the capacitor, the proton enters a magnetic field, where it moves on a circular path of radius r.

of energy reduces to $\frac{1}{2}mv_B^2 = e(V_A - V_B)$. Solving for v_B gives $v_B = \sqrt{2e(V_A - V_B)/m}$. The proton enters the magnetic field with this speed and moves on a circular path with a radius that is given by Equation 21.2.

Solution **(a)** The speed of the proton is

$$v_B = \sqrt{\frac{2e(V_A - V_B)}{m}} = \sqrt{\frac{2(1.60 \times 10^{-19}\ \text{C})(2100\ \text{V})}{1.67 \times 10^{-27}\ \text{kg}}}$$

$$= \boxed{6.3 \times 10^5\ \text{m/s}}$$

(b) When the proton moves in the magnetic field, the radius of the circular path is

$$r = \frac{mv_B}{eB} = \frac{(1.67 \times 10^{-27}\ \text{kg})(6.3 \times 10^5\ \text{m/s})}{(1.60 \times 10^{-19}\ \text{C})(0.10\ \text{T})}$$

$$= \boxed{6.6 \times 10^{-2}\ \text{m}} \qquad \textbf{(21.2)}$$

One of the important and exciting areas in physics today is the study of elementary particles, which are the basic building blocks from which all matter is constructed. Important information about an elementary particle can be obtained from its motion in a magnetic field, with the aid of a device known as a bubble chamber. A bubble chamber contains a superheated liquid such as hydrogen, which will boil and form bubbles readily. When an electrically charged particle passes through the chamber, a thin track of bubbles is left in its wake. This track can be photographed to show how a magnetic field affects the particle motion. Conceptual Example 4 illustrates how physicists deduce information from such photographs.

CONCEPTUAL EXAMPLE 4 | Particle Tracks in a Bubble Chamber

Figure 21.13*a* shows the bubble-chamber tracks resulting from an event that begins at point A. At this point a gamma ray (emitted by certain radioactive substances), traveling in from the left, spontaneously transforms into two charged particles. There is no track from the gamma ray itself. These particles move away from point A, producing the two spiral tracks. A third charged particle is knocked out of a hydrogen atom and moves forward, producing the long track with the slight upward curvature. Each of the three particles has the same mass and carries a charge of the same magnitude. A uniform magnetic field is directed out of the paper toward you. What is the sign (+ or −) of the charge carried by each particle?

	Particle 1	Particle 2	Particle 3
(a)	−	−	+
(b)	−	+	−
(c)	+	−	−
(d)	+	−	+

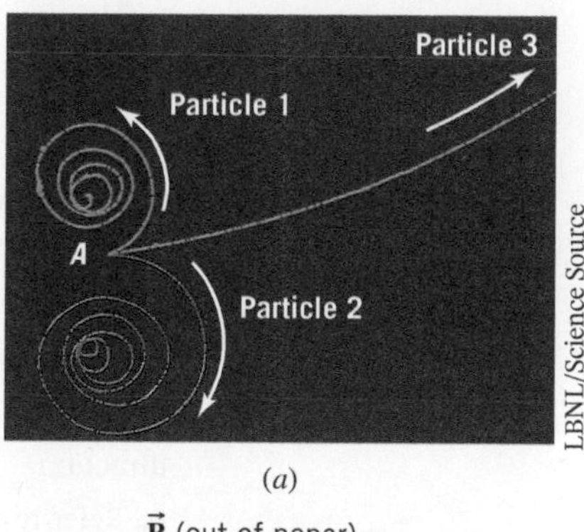

(*a*)

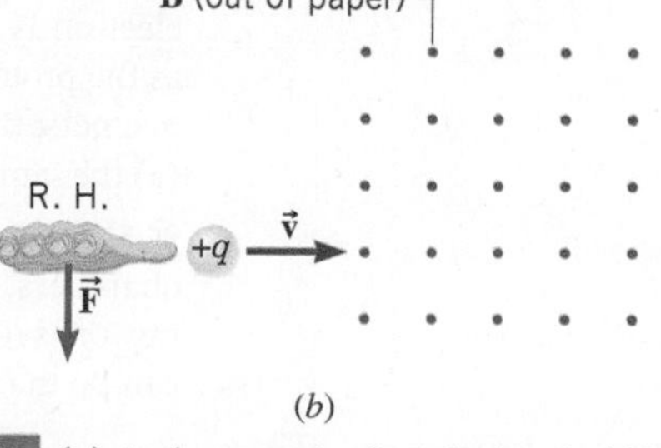

(*b*)

FIGURE 21.13 (*a*) A photograph of tracks in a bubble chamber. A magnetic field is directed perpendicularly out of the paper. At point A, a gamma ray (not visible) spontaneously transforms into two charged particles, and a third charged particle is knocked out of a hydrogen atom in the chamber. (*b*) In accord with RHR-1, the magnetic field applies a downward force to a positively charged particle that moves to the right.

Reasoning Figure 21.13*b* shows a positively charged particle traveling with a velocity $\vec{v}$ that is perpendicular to a magnetic field. The field is directed out of the paper, just like it is in part *a* of the drawing. RHR-1 indicates that the magnetic force points downward. This magnetic force provides the centripetal force that causes a particle to move on a circular path (see Section 5.3).

The centripetal force is directed toward the center of the circular path. Thus, in **Figure 21.13a** a positive charge would move on a downward-curving track, and a negative charge would move on an upward-curving track.

Answers (a), (c), and (d) are incorrect. Since particles 1 and 3 move on upward-curving tracks, they are negatively charged, not positively charged.

Answer (b) is correct. A downward-curving track in the photograph indicates a positive charge, while an upward-curving track indicates a negative charge. Thus, particles 1 and 3 carry a negative charge. They are, in fact, electrons (e^-). In contrast, particle 2 carries a positive charge. It is called a positron (e^+), an elementary particle that has the same mass as an electron but an opposite charge.

Related Homework: Check Your Understanding 5, 6, Problem 25

Check Your Understanding

(The answers are given at the end of the book.)

4. Suppose that the positive charge in **Figure 21.9a** were launched from the negative plate toward the positive plate, in a direction *opposite* to the electric field $\vec{\mathbf{E}}$. A sufficiently strong electric field would prevent the charge from striking the positive plate. Suppose that the positive charge in **Figure 21.9b** were launched from the south pole toward the north pole, in a direction *opposite* to the magnetic field $\vec{\mathbf{B}}$. Would a sufficiently strong magnetic field prevent the charge from reaching the north pole? **(a)** Yes **(b)** No, because a magnetic field cannot exert a force on a charged particle that is moving antiparallel to the field **(c)** No, because the magnetic force would cause the charge to move faster as it moved toward the north pole

5. Review Conceptual Example 4 as background for this question. Three particles move through a constant magnetic field and follow the paths shown in **CYU Figure 21.2**. Determine whether each particle is positively (+) charged, negatively (−) charged, or neutral.

	Particle 1	Particle 2	Particle 3
(a)	neutral	+	neutral
(b)	−	neutral	+
(c)	−	−	−
(d)	+	neutral	−
(e)	+	+	+

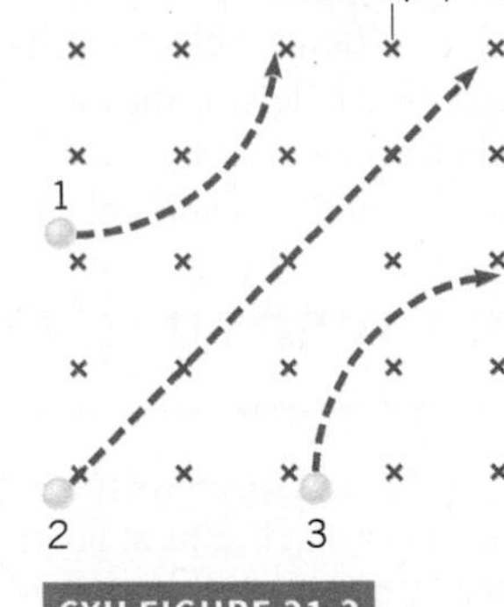

CYU FIGURE 21.2

6. Suppose that the three particles in **Figure 21.13a** have identical charge magnitudes and masses. Which particle has the greatest speed? Refer to Conceptual Example 4 as needed.

7. A positive charge moves along a circular path under the influence of a magnetic field. The magnetic field is perpendicular to the plane of the circle, as in **Interactive Figure 21.11**. If the velocity of the particle is reversed at some point along the path, will the particle retrace its path? **(a)** Yes **(b)** No, because it will move around a *different* circle in a counterclockwise direction

8. Refer to **Interactive Figure 21.11**. Assume that the particle in the picture is a proton. If an electron is projected at point 1 with the same velocity $\vec{\mathbf{v}}$, it will not follow exactly the same path as the proton, unless the magnetic field is adjusted in the following manner: the magnitude of the magnetic field must be ___________, and the direction of the magnetic field must be ___________. **(a)** the same, reversed **(b)** increased, the same **(c)** reduced, reversed

9. **CYU Figure 21.3** shows a top view of four interconnected chambers. A *negative* charge is fired into chamber 1. By turning on separate magnetic fields in each chamber, the charge can be made to exit from chamber 4, as shown. How should the

	Chamber 1	Chamber 2	Chamber 3	Chamber 4
(a)	out of	into	out of	into
(b)	into	out of	out of	into
(c)	out of	into	into	out of
(d)	into	out of	into	out of

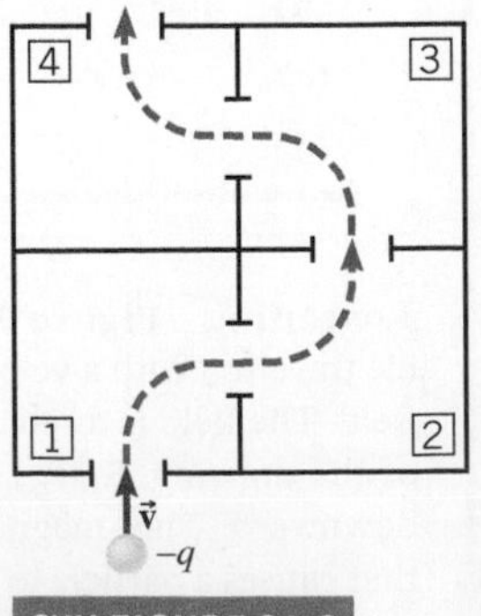

CYU FIGURE 21.3

magnetic field in each chamber be directed: out of the page or into the page?

10. CYU **Figure 21.4** shows a particle carrying a positive charge $+q$ at the coordinate origin, as well as a target located in the third quadrant. A uniform magnetic field is directed perpendicularly into the plane of the paper. The charge can be projected in the plane of the paper only, along the positive or negative x or y axis. There are four possible directions ($+x$, $-x$, $+y$, $-y$) for the initial velocity of the particle. The particle can be made to hit the target for only two of the four directions. Which two directions are they? **(a)** $+y$, $-y$ **(b)** $-y$, $+x$ **(c)** $-x$, $+y$ **(d)** $+x$, $-x$

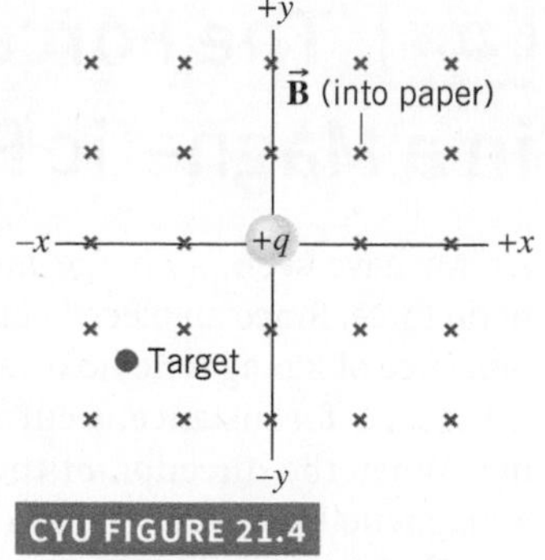

CYU FIGURE 21.4

21.4 The Mass Spectrometer

Physicists use mass spectrometers for determining the relative masses and abundances of isotopes.* Chemists use these instruments to help identify unknown molecules produced in chemical reactions. Mass spectrometers are also used during surgery, where they give the anesthesiologist information on the gases, including the anesthetic, in the patient's lungs.

THE PHYSICS OF . . . a mass spectrometer. In the type of mass spectrometer illustrated in **Figure 21.14**, the atoms or molecules are first vaporized and then ionized by the ion source. The ionization process removes one electron from the particle, leaving it with a net positive charge of $+e$. The positive ions are then accelerated through the potential difference V, which is applied between the ion source and the metal plate. With a speed v, the ions pass through a hole in the plate and enter a region of constant magnetic field $\vec{\mathbf{B}}$, where they are deflected in semicircular paths. Only those ions following a path with the proper radius r strike the detector, which records the number of ions arriving per second.

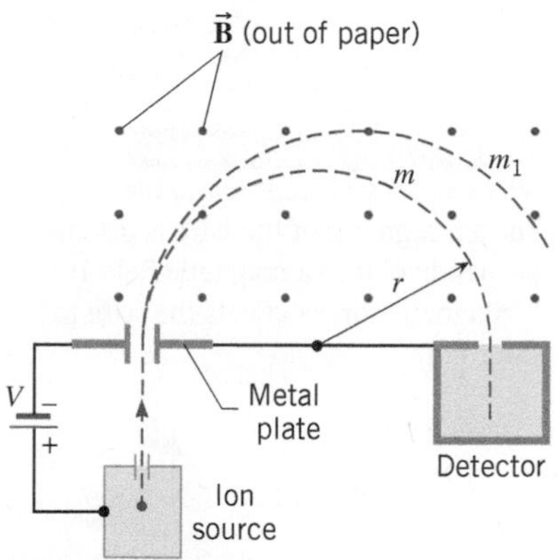

FIGURE 21.14 In this mass spectrometer the dashed lines are the paths traveled by ions of different masses. Ions with mass m follow the path of radius r and enter the detector. Ions with the larger mass m_1 follow the outer path and miss the detector.

The mass m of the detected ions can be expressed in terms of r, B, and v by recalling that the radius of the path followed by a particle of charge $+e$ is $r = mv/(eB)$ (Equation 21.2). In addition, the Reasoning section in Example 3 shows that the ion speed v can be expressed in terms of the potential difference V as $v = \sqrt{2eV/m}$. This expression for the ion speed is the same as that used in Example 3, except that, for convenience, we have replaced the potential difference, $V_A - V_B$, by the symbol V. Eliminating v from these two equations algebraically and solving for the mass gives

$$m = \left(\frac{er^2}{2V}\right)B^2$$

This result shows that the mass of each ion reaching the detector is proportional to B^2. Experimentally changing the value of B and keeping the term in the parentheses constant will allow ions of different masses to enter the detector. A plot of the detector output as a function of B^2 then gives an indication of what masses are present and the abundance of each mass.

Figure 21.15 shows a record obtained by a mass spectrometer for naturally occurring neon gas. The results show that the element neon has three isotopes whose atomic mass numbers are 20, 21, and 22. These isotopes occur because neon atoms exist with different numbers of neutrons in the nucleus. Notice that the isotopes have different abundances, with neon-20 being the most abundant.

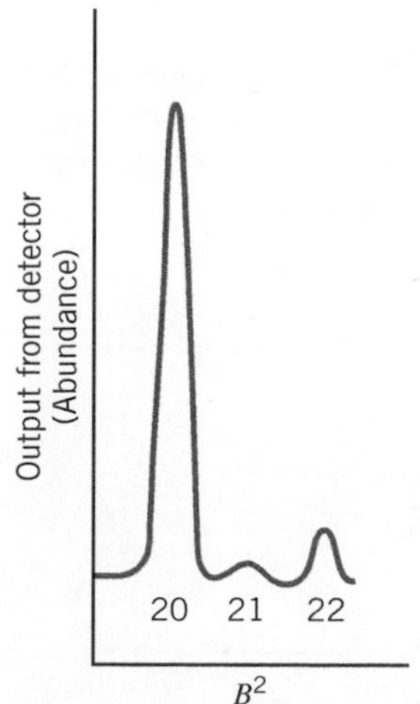

FIGURE 21.15 The mass spectrum (not to scale) of naturally occurring neon, showing three isotopes whose atomic mass numbers are 20, 21, and 22. The larger the peak, the more abundant the isotope.

*Isotopes are atoms that have the same atomic number but different atomic masses due to the presence of different numbers of neutrons in the nucleus. They are discussed in Section 31.1.

21.5 The Force on a Current in a Magnetic Field

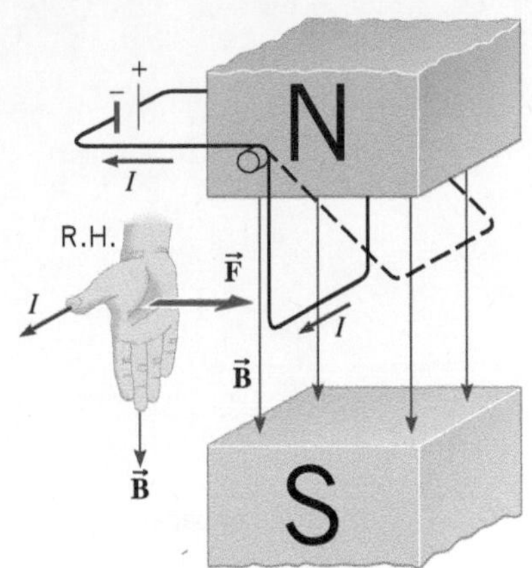

INTERACTIVE FIGURE 21.16 The wire carries a current I, and the bottom segment of the wire is oriented perpendicular to a magnetic field $\vec{B}$. A magnetic force deflects the wire to the right.

As we have seen, a charge moving through a magnetic field can experience a magnetic force. Since an electric current is a collection of moving charges, a current in the presence of a magnetic field can also experience a magnetic force. In **Interactive Figure 21.16**, for instance, a current-carrying wire is placed between the poles of a magnet. When the direction of the current I is as shown, the moving charges experience a magnetic force that pushes the wire to the right in the drawing. The direction of the force is determined in the usual manner by using RHR-1, with the minor modification that the direction of the velocity of a positive charge is replaced by the direction of the conventional current I. If the current in the drawing were reversed by switching the leads to the battery, the direction of the force would be reversed, and the wire would be pushed to the left.

Problem-Solving Insight Whenever the current in a wire reverses direction, the force exerted on the wire by a given magnetic field also reverses direction.

When a charge moves through a magnetic field, the magnitude of the force that acts on the charge is $F = |q|vB \sin\theta$ (Equation 21.1). With the aid of **Figure 21.17**, this expression can be put into a form that is more suitable for use with an electric current. The drawing shows a wire of length L that carries a current I. The wire is oriented at an angle θ with respect to a magnetic field $\vec{B}$. This picture is similar to **Figure 21.7c**, except that now the charges move in a wire. The magnetic force exerted on this length of wire is the net force acting on the total amount of charge moving in the wire. Suppose that an amount of conventional positive charge Δq travels the length of the wire in a time interval Δt. The magnitude of the magnetic force on this amount of charge is given by Equation 21.1 as $F = (\Delta q)vB \sin\theta$. Multiplying and dividing the right side of this equation by Δt, we find that

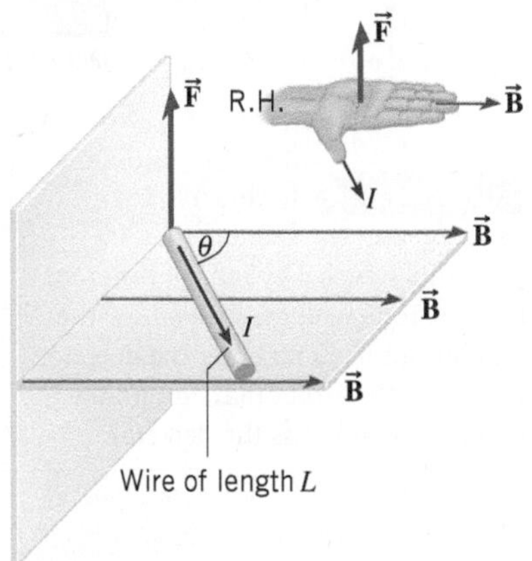

FIGURE 21.17 The current I in the wire, oriented at an angle θ with respect to a magnetic field $\vec{B}$, is acted upon by a magnetic force $\vec{F}$.

$$F = \underbrace{\left(\frac{\Delta q}{\Delta t}\right)}_{I}\underbrace{(v\,\Delta t)}_{L}B \sin\theta$$

The term $\Delta q/\Delta t$ is the current I in the wire (see Equation 20.1), and the term $v\,\Delta t$ is the length L of the wire. With these two substitutions, the expression for the magnetic force exerted on a current-carrying wire becomes

Magnetic force on a current-carrying wire of length L

$$F = ILB \sin\theta \qquad \textbf{(21.3)}$$

As in the case of a single charge traveling in a magnetic field, the magnetic force on a current-carrying wire is a maximum when the wire is oriented perpendicular to the field ($\theta = 90°$) and vanishes when the current is parallel or antiparallel to the field ($\theta = 0°$ or 180°). The direction of the magnetic force is given by RHR-1, as **Figure 21.17** indicates.

THE PHYSICS OF . . . a loudspeaker. Most loudspeakers operate on the principle that a magnetic field exerts a force on a current-carrying wire. **Figure 21.18*a*** shows a speaker design that consists of three basic parts: a cone, a voice coil, and a permanent magnet. The cone is mounted so it can vibrate back and forth. When vibrating, it pushes and pulls on the air in front of it, thereby creating sound waves. Attached to the apex of the cone is the voice coil, which is a hollow cylinder around which coils of wire are wound. The voice coil is slipped over one of the poles of the stationary permanent magnet (the north pole in the drawing) and can move freely. The two ends of the voice-coil wire are connected to the speaker terminals on the back panel of a receiver. The receiver acts as an ac generator, sending an alternating current to the voice coil. The alternating current interacts with the magnetic field to generate an alternating force that pushes and pulls on the voice coil and the attached cone. To see how the

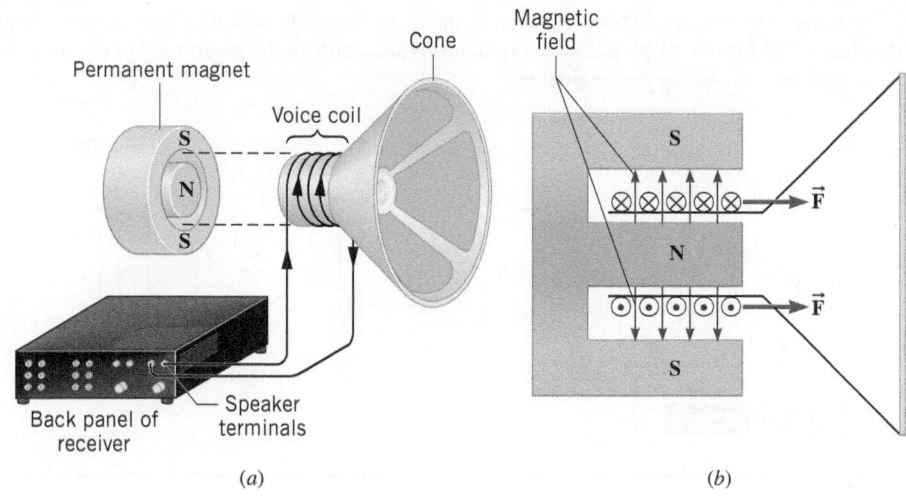

FIGURE 21.18 (*a*) An "exploded" view of one type of speaker design, which shows a cone, a voice coil, and a permanent magnet. (*b*) Because of the current in the voice coil (shown as ⊗ and ⊙), the magnetic field causes a force $\vec{F}$ to be exerted on the voice coil and cone.

magnetic force arises, consider **Figure 21.18*b***, which is a cross-sectional view of the voice coil and the magnet. In the cross-sectional view, the current is directed into the page in the upper half of the voice coil (⊗⊗⊗) and out of the page in the lower half (⊙⊙⊙). In both cases the magnetic field is perpendicular to the current, so the maximum possible force is exerted on the wire. An application of RHR-1 to both the upper and lower halves of the voice coil shows that the magnetic force $\vec{F}$ in the drawing is directed to the right, causing the cone to accelerate in that direction. One-half of a cycle later when the current is reversed, the direction of the magnetic force is also reversed, and the cone accelerates to the left. If, for example, the alternating current from the receiver has a frequency of 1000 Hz, the alternating magnetic force causes the cone to vibrate back and forth at the same frequency, and a 1000-Hz sound wave is produced. Thus, it is the magnetic force on a current-carrying wire that is responsible for converting an electrical signal into a sound wave. In Example 5 a typical force and acceleration in a loudspeaker are determined.

EXAMPLE 5 | The Force and Acceleration in a Loudspeaker

The voice coil of a speaker has a diameter of $d = 0.025$ m, contains 55 turns of wire, and is placed in a 0.10-T magnetic field. The current in the voice coil is 2.0 A. **(a)** Determine the magnetic force that acts on the coil and cone. **(b)** The voice coil and cone have a combined mass of 0.020 kg. Find their acceleration.

Reasoning The magnetic force that acts on the current-carrying voice coil is given by Equation 21.3 as $F = ILB \sin\theta$. The effective length L of the wire in the voice coil is very nearly the number of turns N times the circumference (πd) of one turn: $L = N\pi d$. The acceleration of the voice coil and cone is given by Newton's second law as the magnetic force divided by the combined mass.

Solution **(a)** Since the magnetic field acts perpendicular to all parts of the wire, $\theta = 90°$ and the force on the voice coil is

$$F = ILB\sin\theta = I(N\pi d)B\sin\theta = (2.0\text{ A})[55\pi(0.025\text{ m})](0.10\text{ T})\sin 90° = \boxed{0.86\text{ N}} \qquad (21.3)$$

(b) According to Newton's second law, the acceleration of the voice coil and cone is

$$a = \frac{F}{m} = \frac{0.86\text{ N}}{0.020\text{ kg}} = \boxed{43\text{ m/s}^2} \qquad (4.1)$$

This acceleration is more than four times the acceleration due to gravity.

Check Your Understanding

(*The answers are given at the end of the book.*)

11. Refer to Interactive Figure 21.16. (a) What happens to the direction of the magnetic force if the current is reversed? **(b)** What happens to the direction of the force if *both* the current *and* the magnetic poles are reversed?

12. The same current-carrying wire is placed in the same magnetic field $\vec{B}$ in four different orientations (see **CYU Figure 21.5**). Rank the orientations according to the magnitude of the magnetic force exerted on the wire, largest to smallest.

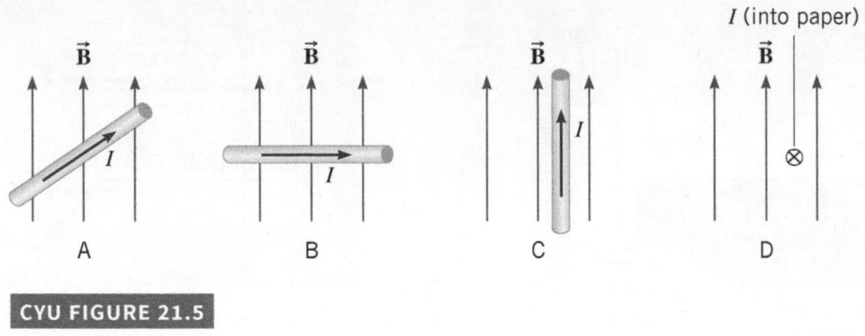

CYU FIGURE 21.5

21.6 The Torque on a Current-Carrying Coil

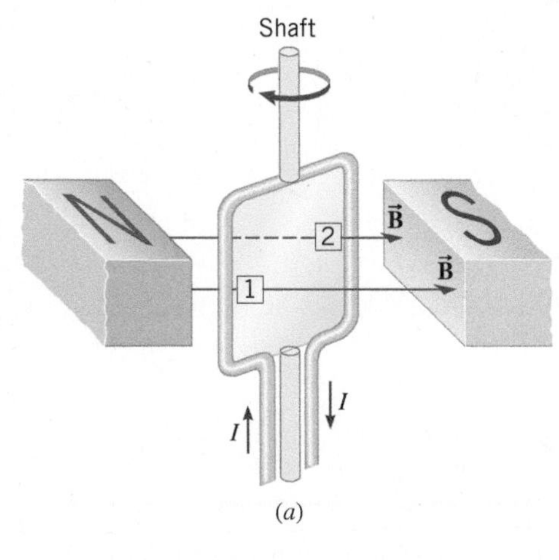

(a)

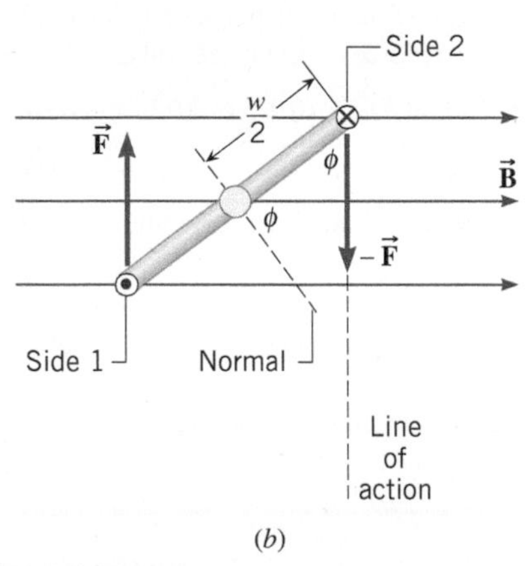

(b)

FIGURE 21.19 (a) A current-carrying loop of wire, which can rotate about a vertical shaft, is situated in a magnetic field. (b) A top view of the loop. The current in side 1 is directed out of the page (⊙), while the current in side 2 is directed into the page (⊗). The current in side 1 experiences a force $\vec{F}$ that is opposite to the force $-\vec{F}$ exerted on side 2. The two forces produce a clockwise torque about the shaft.

We have seen that a current-carrying wire can experience a force when placed in a magnetic field. If a loop of wire is suspended properly in a magnetic field, the magnetic force produces a torque that tends to rotate the loop. This torque is responsible for the operation of a widely used type of electric motor.

Figure 21.19*a* shows a rectangular loop of wire attached to a vertical shaft. The shaft is mounted so that it is free to rotate in a uniform magnetic field. When there is a current in the loop, the loop rotates because magnetic forces act on the vertical sides, labeled 1 and 2 in the drawing. Part *b* shows a top view of the loop and the magnetic forces $\vec{F}$ and $-\vec{F}$ that act on the two sides. These two forces have the same magnitude, but an application of RHR-1 shows that they point in opposite directions, so the loop experiences no net force. The loop does, however, experience a net torque that tends to rotate it in a clockwise fashion about the vertical shaft. **Figure 21.20*a*** shows that the torque is maximum when the normal to the plane of the loop is perpendicular to the field. In contrast, part *b* shows that the torque is zero when the normal is parallel to the field. ***When a current-carrying loop is placed in a magnetic field, the loop tends to rotate such that its normal becomes aligned with the magnetic field.*** In this respect, a current loop behaves like a magnet (e.g., a compass needle) suspended in a magnetic field, since a magnet also rotates to align itself with the magnetic field.

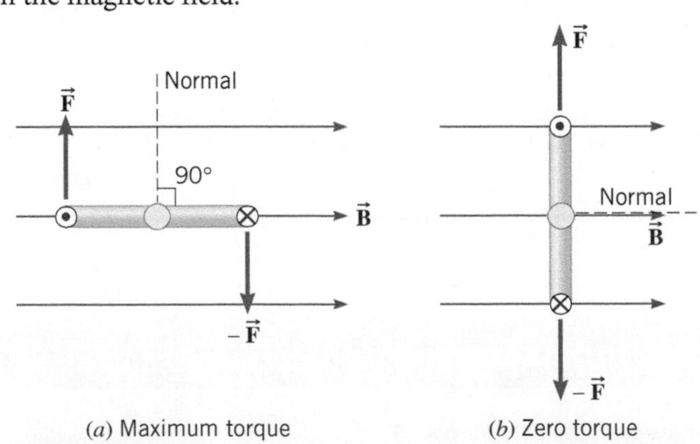

FIGURE 21.20 (a) Maximum torque occurs when the normal to the plane of the loop is perpendicular to the magnetic field. (b) The torque is zero when the normal is parallel to the field.

It is possible to determine the magnitude of the torque on the loop in **Figure 21.19a**. From Equation 21.3 the magnetic force on each vertical side has a magnitude of $F = ILB \sin 90°$, where L is the length of side 1 or side 2, and $\theta = 90°$ because the current I always remains perpendicular to the magnetic field as the loop rotates. As Section 9.1 discusses, the torque produced by a force is the product of the magnitude of the force and the lever arm. In **Figure 21.19b** the lever arm ℓ is the perpendicular distance from the line of action of the force to the shaft. This distance is given by $\ell = (w/2) \sin \phi$, where w is the width of the loop, and ϕ is the angle between the normal to the plane of the loop and the direction of the magnetic field. The net torque is the sum of the torques on the two sides, so

$$\text{Net torque} = \tau = ILB(\tfrac{1}{2}w \sin \phi) + ILB(\tfrac{1}{2}w \sin \phi) = IAB \sin \phi$$

In this result the product Lw has been replaced by the area A of the loop. If the wire is wrapped so as to form a coil containing N loops, each of area A, the force on each side is N times larger, and the torque becomes proportionally greater:

$$\tau = \underbrace{NIA}_{\text{Magnetic moment}}(B \sin \phi) \quad \textbf{(21.4)}$$

Equation 21.4 has been derived for a rectangular coil, but it is valid for any shape of flat coil, such as a circular coil. The torque depends on the geometric properties of the coil and the current in it through the quantity NIA. This quantity is known as the **magnetic moment** of the coil, and its unit is ampere · meter^2 ($A \cdot m^2$). The greater the magnetic moment of a current-carrying coil, the greater is the torque that the coil experiences when placed in a magnetic field. Example 6 discusses the torque that a magnetic field applies to such a coil.

Math Skills To understand why the lever arm for the torque calculation is $\ell = (w/2) \sin \phi$, it is first necessary to know why the two angles labeled ϕ in **Figure 21.19b** are equal. Referring to **Figure 21.21a** and noting that the normal is perpendicular to the plane of the loop, we can see that $\alpha + \phi = 90°$ or $\alpha = 90° - \phi$. Furthermore, the force $-\vec{\mathbf{F}}$ is perpendicular to the magnetic field $\vec{\mathbf{B}}$, so the triangle including angles α and β is a right triangle. Therefore, it follows that $\alpha + \beta = 90°$. Solving this equation gives $\beta = 90° - \alpha$, and substituting $\alpha = 90° - \phi$ reveals that

$$\beta = 90° - \alpha = 90° - \underbrace{(90° - \phi)}_{\alpha} = \phi$$

Recognizing that the two angles labeled ϕ in **Figure 21.19b** are equal, we can now understand why the lever arm is $\ell = (w/2)\sin \phi$ for the force $-\vec{\mathbf{F}}$. According to the definition given in Equation 1.1, the sine function is $\sin \phi = \frac{h_o}{h}$, where h_o is the length of the side of a right triangle that is opposite the angle ϕ and h is the length of the hypotenuse (see **Figure 21.21b**). By comparing **Figure 21.21b** with **Figure 21.21c**, we can see that

$$\sin \phi = \frac{h_o}{h} = \frac{\ell}{w/2} \quad \text{or} \quad \ell = \frac{w}{2} \sin \phi$$

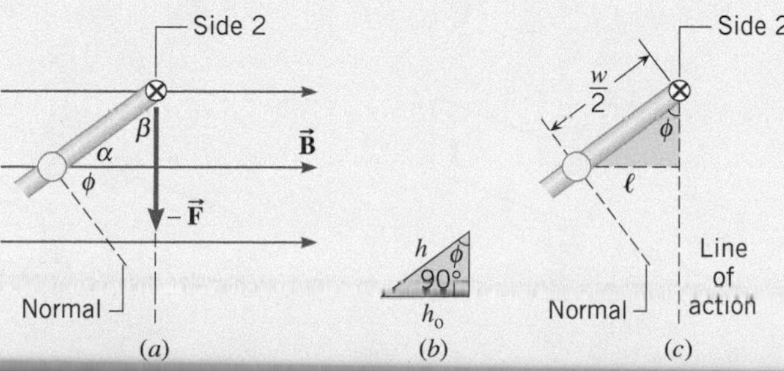

FIGURE 21.21 Math Skills drawing.

EXAMPLE 6 | The Torque Exerted on a Current-Carrying Coil

A coil of wire has an area of 2.0×10^{-4} m^2, consists of 100 loops or turns, and contains a current of 0.045 A. The coil is placed in a uniform magnetic field of magnitude 0.15 T. **(a)** Determine the magnetic moment of the coil. **(b)** Find the maximum torque that the magnetic field can exert on the coil.

Reasoning and Solution **(a)** The magnetic moment of the coil is

$$\text{Magnetic moment} = NIA = (100)(0.045\ \text{A})(2.0 \times 10^{-4}\ \text{m}^2) = \boxed{9.0 \times 10^{-4}\ \text{A} \cdot \text{m}^2}$$

(b) According to Equation 21.4, the torque is the product of the magnetic moment NIA and $B \sin \phi$. However, the maximum torque occurs when $\phi = 90°$, so

$$\tau = \underbrace{(NIA)}_{\text{Magnetic moment}}(B \sin 90°) = (9.0 \times 10^{-4}\ \text{A} \cdot \text{m}^2)(0.15\ \text{T}) = \boxed{1.4 \times 10^{-4}\ \text{N} \cdot \text{m}}$$

THE PHYSICS OF . . . a direct-current electric motor. The electric motor is found in many devices, such as CD players, automobiles, washing machines, and air conditioners. **Figure 21.22** shows that a direct-current (dc) motor consists of a coil of wire placed in a magnetic field and free to rotate about a vertical shaft. The coil of wire contains many turns and is wrapped around an iron cylinder that rotates with the coil, although these features have been omitted to simplify the drawing. The coil and iron cylinder assembly is known as the armature. Each end of the wire coil is attached to a metallic half-ring. Rubbing against each of the half-rings is a graphite contact called a brush. While the half-rings rotate with the coil, the graphite brushes remain stationary. The two half-rings and the associated brushes are referred to as a split-ring commutator (see below).

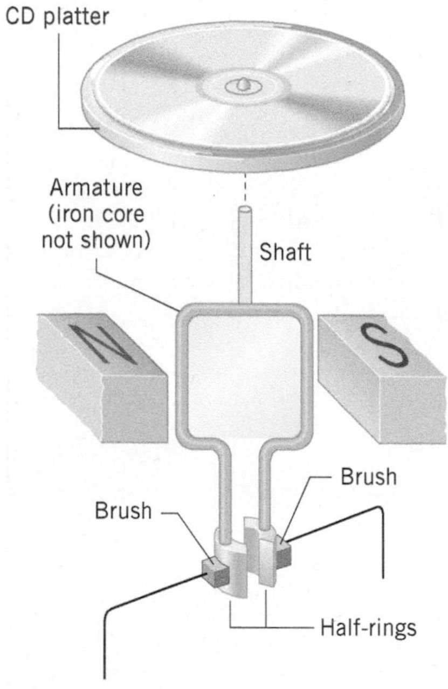

FIGURE 21.22 The basic components of a dc motor. A CD platter is shown as it might be attached to the motor.

The operation of a motor can be understood by considering **Figure 21.23**. In part *a* the current from the battery enters the coil through the left brush and half-ring, goes around the coil, and then leaves through the right half-ring and brush. Consistent with RHR-1, the directions of the magnetic forces $\vec{F}$ and $-\vec{F}$ on the two sides of the coil are as shown in the drawing. These forces produce the torque that turns the coil. Eventually, the coil reaches the position shown in part *b* of the drawing. In this position the half-rings momentarily lose electrical contact with the brushes, so that there is no current in the coil and no applied torque. However, like any moving object, the rotating coil does not stop immediately, for its inertia carries it onward. When the half-rings reestablish contact with the brushes, there again is a current in the coil, and a magnetic torque again rotates the coil in the same direction. The split-ring commutator ensures that the current is always in the proper direction to yield a torque that produces a continuous rotation of the coil.

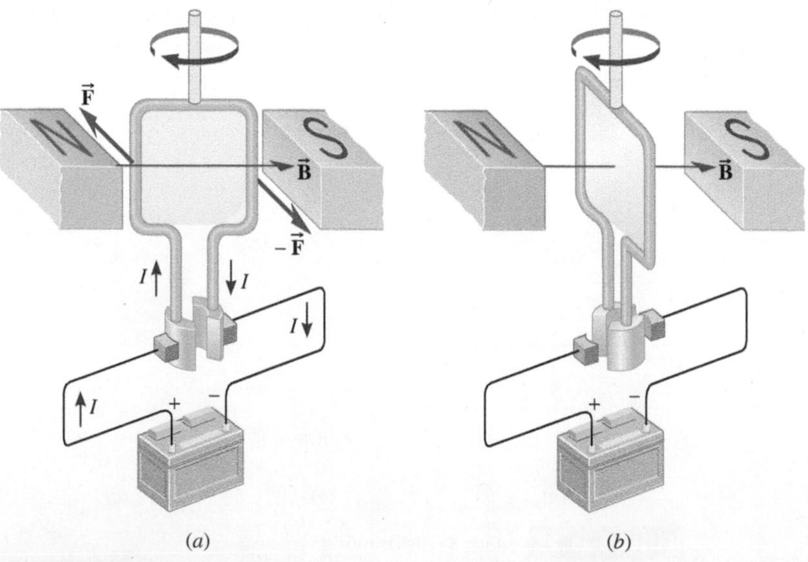

FIGURE 21.23 (*a*) When a current exists in the coil, the coil experiences a torque. (*b*) Because of its inertia, the coil continues to rotate when there is no current.

21.7 Magnetic Fields Produced by Currents

We have seen that a current-carrying wire can experience a magnetic force when placed in a magnetic field that is produced by an external source, such as a permanent magnet. ***A current-carrying wire also produces a magnetic field of its own***, as we will see in this section. Hans Christian Oersted (1777–1851) first discovered this effect in 1820 when he observed that a current-carrying wire influences the orientation of a nearby compass needle. The compass needle aligns itself with the net magnetic field produced by the current and the magnetic field of the earth. Oersted's discovery, which linked the motion of electric charges with the creation of a magnetic field, marked the beginning of an important discipline called **electromagnetism**.

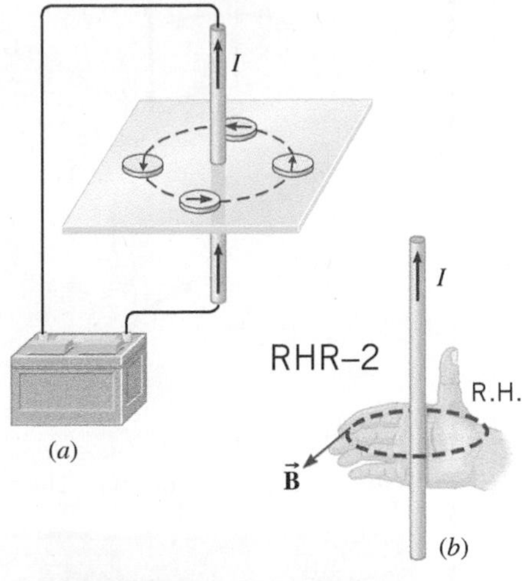

FIGURE 21.24 (*a*) A very long, straight, current-carrying wire produces magnetic field lines that are circular about the wire, as indicated by the compass needles. (*b*) With the thumb of the right hand (R.H.) along the current *I*, the curled fingers point in the direction of the magnetic field, according to RHR-2.

A Long, Straight Wire

Figure 21.24*a* illustrates Oersted's discovery with a very long, straight wire. When a current is present, the compass needles point in a circular pattern about the wire. The pattern indicates that the magnetic field lines produced by the current are circles centered on the wire. If the direction of the current is reversed, the needles also reverse their directions, indicating that the direction of the magnetic field has reversed. The direction of the field can be obtained by using *Right-Hand Rule No. 2 (RHR-2),* as part *b* of the drawing indicates:

> ***Right-Hand Rule No. 2:*** **The RHR for Determining the Direction of the Magnetic Field Near Long, Straight Wires.** Curl the fingers of the right hand into the shape of a half-circle. Point the thumb in the direction of the conventional current *I*, and the tips of the fingers will point in the direction of the magnetic field $\vec{\mathbf{B}}$.

Experimentally, it is found that the magnitude *B* of the magnetic field produced by an infinitely long, straight wire is directly proportional to the current *I* and inversely proportional to the radial distance *r* from the wire: $B \propto I/r$. As usual, this proportionality is converted into an equation by introducing a proportionality constant, which, in this instance, is written as $\mu_0/(2\pi)$. Thus, the magnitude of the magnetic field is

Infinitely long, straight wire
$$B = \frac{\mu_0 I}{2\pi r} \quad \textbf{(21.5)}$$

The constant μ_0 is referred to as the **permeability of free space**, and its value is $\mu_0 = 4\pi \times 10^{-7}\ \mathrm{T \cdot m/A}$. The magnetic field becomes stronger nearer the wire, where *r* is smaller. Therefore, the field lines near the wire are closer together than those located farther away, where the field is weaker. **Figure 21.25** shows the pattern of field lines.

The magnetic field that surrounds a current-carrying wire can exert a force on a moving charge, as the next example illustrates.

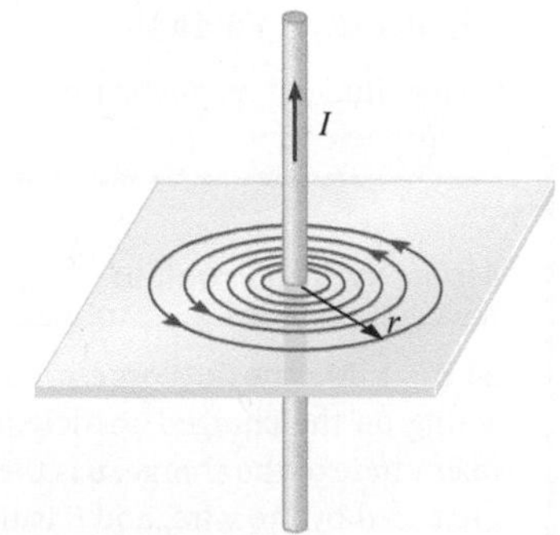

FIGURE 21.25 The magnetic field becomes stronger as the radial distance *r* decreases, so the field lines are closer together near the wire.

Analyzing Multiple-Concept Problems

EXAMPLE 7 | A Current Exerts a Magnetic Force on a Moving Charge

Figure 21.26 shows a very long, straight wire carrying a current of 3.0 A. A particle has a charge of $+6.5 \times 10^{-6}$ C and is moving parallel to the wire at a distance of 0.050 m. The speed of the particle is 280 m/s. Determine the magnitude and direction of the magnetic force exerted on the charged particle by the current in the wire.

Reasoning The current generates a magnetic field in the space around the wire. The charged particle moves in the presence of this field and, therefore, can experience a magnetic force. The magnitude of this force is given by Equation 21.1, and the direction can be determined by applying RHR-1 (see Section 21.2). Note in **Figure 21.26** that the magnetic field $\vec{\mathbf{B}}$ produced by the current lies in a plane that is perpendicular to both the wire and the velocity $\vec{\mathbf{v}}$ of the particle. Thus, the angle between $\vec{\mathbf{B}}$ and $\vec{\mathbf{v}}$ is $\theta = 90.0°$.

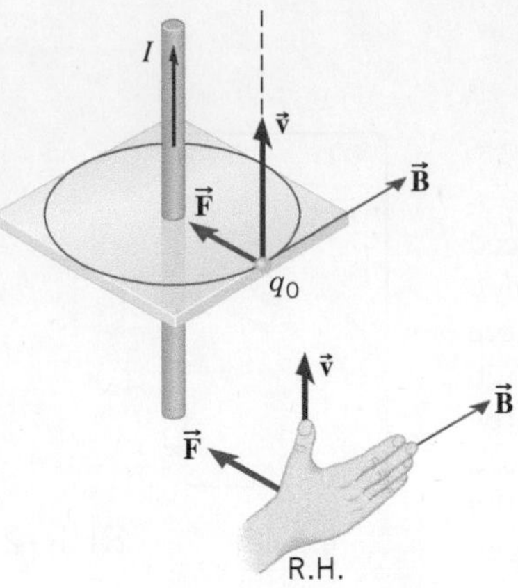

FIGURE 21.26 The positive charge q_0 moves with a velocity $\vec{v}$ and experiences a magnetic force $\vec{F}$ because of the magnetic field $\vec{B}$ produced by the current in the wire.

Knowns and Unknowns The following list summarizes the data that are given:

Description	Symbol	Value	Comment
Explicit Data			
Current in wire	I	3.0 A	
Electric charge of particle	q_0	$+6.5 \times 10^{-6}$ C	
Distance of particle from wire	r	0.050 m	Particle moves parallel to wire; see **Figure 21.26**.
Speed of particle	v	280 m/s	
Implicit Data			
Directional angle of particle velocity with respect to magnetic field	θ	90.0°	Particle moves parallel to wire; see *Reasoning*.
Unknown Variable			
Magnitude of magnetic force exerted on particle	F	?	

Modeling the Problem

STEP 1 Magnetic Force on the Particle The magnitude F of the magnetic force acting on the charged particle is given at the right by Equation 21.1, where $|q_0|$ is the magnitude of the charge, v is the particle speed, B is the magnitude of the magnetic field produced by the wire, and θ is the angle between the particle velocity and the magnetic field. Values for $|q_0|$, v, and θ are given. The value of B, however, is unknown, and we determine it in Step 2.

$$F = |q_0| vB \sin\theta \qquad \textbf{(21.1)}$$

?

STEP 2 Magnetic Field Produced by the Wire The magnitude B of the magnetic field produced by a current I in an infinitely long, straight wire is given by Equation 21.5:

$$\boxed{B = \frac{\mu_0 I}{2\pi r}} \qquad \textbf{(21.5)}$$

where μ_0 is the permeability of free space and r is the distance from the wire. This expression can be substituted into Equation 21.1, as shown at the right.

$$F = |q_0| vB \sin\theta \qquad \textbf{(21.1)}$$

$$\boxed{B = \frac{\mu_0 I}{2\pi r}} \qquad \textbf{(21.5)}$$

Solution Combining the results of each step algebraically, we find that

STEP 1 STEP 2

$$F = |q_0| vB \sin\theta = |q_0| v\left(\frac{\mu_0 I}{2\pi r}\right) \sin\theta$$

The magnitude of the magnetic force on the charged particle is

$$F = |q_0|v\left(\frac{\mu_0 I}{2\pi r}\right)\sin\theta$$

$$= (6.5\times10^{-6}\ \text{C})(280\ \text{m/s})\frac{(4\pi\times10^{-7}\ \text{T}\cdot\text{m/A})(3.0\ \text{A})}{2\pi(0.050\ \text{m})}\sin 90.0° = \boxed{2.2\times10^{-8}\ \text{N}}$$

The direction of the magnetic force is predicted by RHR-1 and, as shown in **Figure 21.26**, is radially inward toward the wire.

Related Homework: Problem 66

We have now seen that an electric current can create a magnetic field of its own. Earlier, we have also seen that an electric current can experience a force created by another magnetic field. Therefore, the magnetic field that one current creates can exert a force on another nearby current. Conceptual Example 8 deals with this magnetic interaction between currents.

CONCEPTUAL EXAMPLE 8 | The Net Force That a Current-Carrying Wire Exerts on a Current-Carrying Coil

Figure 21.27 shows a very long, straight wire carrying a current I_1 and a rectangular coil carrying a current I_2. The wire and the coil lie in the same plane, with the wire parallel to the long sides of the rectangle. The coil is **(a)** attracted to the wire, **(b)** repelled from the wire, **(c)** neither attracted to nor repelled from the wire.

Reasoning The current in the straight wire exerts a force on each of the four sides of the coil. The net force acting on the coil is the vector sum of these four forces. To determine whether the net force is attractive, repulsive, or neither, we need to consider the directions and magnitudes of the individual forces. For each side of the rectangular coil we will first use RHR-2 to determine the direction of the magnetic field produced by the long, straight wire. Then, we will employ RHR-1 to find the direction of the magnetic force exerted on each side.

It should be noted that the magnetic forces that act on the two short sides of the rectangular coil play no role. To see why, consider a small segment of each of the short sides, located at the same distance from the straight wire, as indicated by the dashed line in **Figure 21.27**. Each of these segments experiences the same magnetic field from the current I_1. RHR-2 shows that this field is directed downward into the plane of the paper, so that it is perpendicular to the current I_2 in each segment. However, the directions of I_2 in the segments are opposite. As a result, RHR-1 reveals that the magnetic field from the straight wire applies a magnetic force to one segment that is opposite to the magnetic force applied to the other segment. Thus, the forces on the two short sides of the coil cancel.

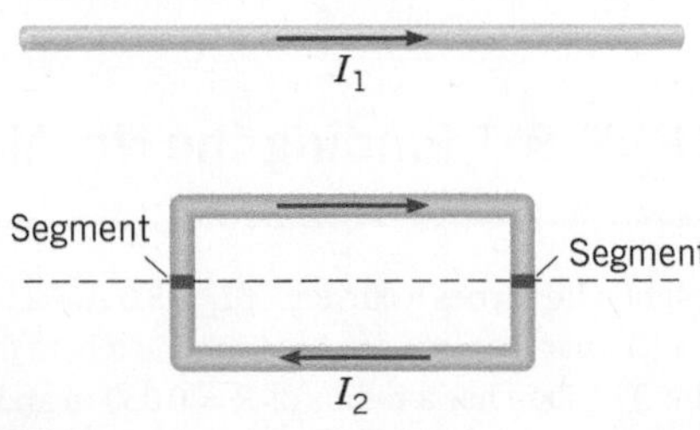

FIGURE 21.27 A very long, straight wire carries a current I_1, and a rectangular coil carries a current I_2. The dashed line is parallel to the wire and locates a small segment on each short side of the coil.

Answers (b) and (c) are incorrect. There is indeed a net magnetic force that acts on the rectangular coil, but it is not a force that repels the coil from the straight wire. To see why, consider the following explanation.

Answer (a) is correct. In the long side of the coil near the wire, the current I_2 has the same direction as the current I_1, and two such currents *attract* each other. In the other long side of the coil the current I_2 has a direction opposite to that of I_1, and they *repel* one another. However, the attractive force is stronger than the repulsive force because the magnetic field produced by the current I_1 is stronger at shorter distances than at greater distances. Consequently, the rectangular coil is attracted to the long, straight wire.

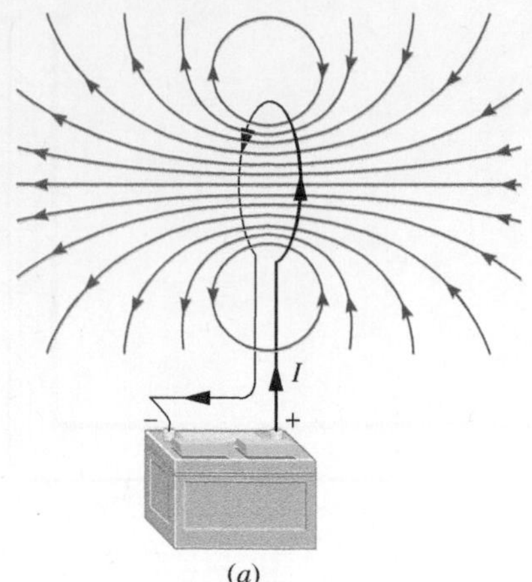

(a)

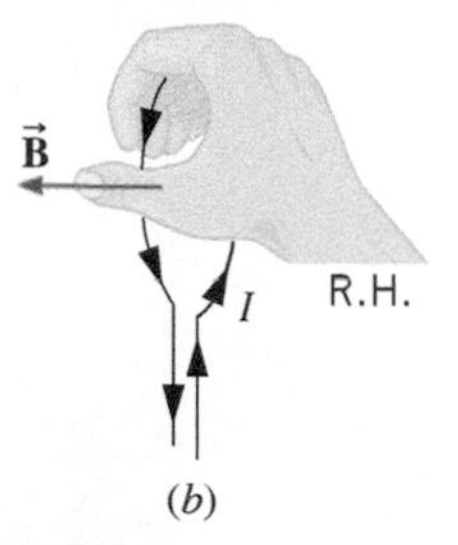

(b)

FIGURE 21.28 (a) The magnetic field lines in the vicinity of a current-carrying circular loop. (b) The direction of the magnetic field at the center of the loop is given by RHR-3.

A Loop of Wire

If a current-carrying wire is bent into a circular loop, the magnetic field lines around the loop have the pattern shown in Figure 21.28a. At the *center* of a loop of radius R, the magnetic field is perpendicular to the plane of the loop and has the value $B = \mu_0 I/(2R)$, where I is the current in the loop. Often, the loop consists of N turns of wire that are wound sufficiently close together that they form a flat coil with a single radius. In this case, the magnetic fields of the individual turns add together to give a net field that is N times greater than the field of a single loop. For such a coil the magnetic field at the center is

Center of a circular loop
$$B = N\frac{\mu_0 I}{2R} \tag{21.6}$$

The direction of the field through the loop can be obtained by using *Right-Hand Rule No. 3 (RHR-3)*, as demonstrated in part (b) of the figure:

> ***Right-Hand Rule No. 3*: The RHR Used to Determine the Direction of the Magnetic Field at the Center of a Loop or Solenoid.** Curl the fingers of the right hand along the direction of the conventional current I around the loop or solenoid, and the thumb will point in the direction of the magnetic field $\vec{\mathbf{B}}$ at the center of the loop or solenoid.

A solenoid, which we will describe later in this section, is just a stack of loops.

We can apply RHR-3 to determine the direction of the magnetic field for the single loop in Figure 21.28. If the fingers of the right hand curl along the direction of the current in the loop, as in Figure 21.28b, then the thumb indicates the direction the magnetic field points at the center of the loop. Here it points from right to left.

Example 9 shows how the magnetic fields produced by the current in a loop of wire and the current in a long, straight wire combine to form a net magnetic field.

EXAMPLE 9 | Finding the Net Magnetic Field

A long, straight wire carries a current of $I_1 = 8.0$ A. As Figure 21.29a illustrates, a circular loop of wire lies immediately to the right of the straight wire. The loop has a radius of $R = 0.030$ m and carries a current of $I_2 = 2.0$ A. Assuming that the thickness of each wire is negligible, find the magnitude and direction of the net magnetic field at the center C of the loop.

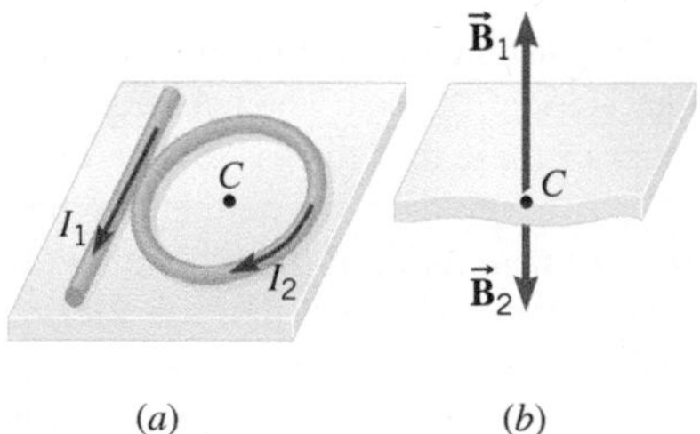

FIGURE 21.29 (a) A long, straight wire carrying a current I_1 lies next to a circular loop that carries a current I_2. (b) The magnetic fields at the center C of the loop produced by the straight wire ($\vec{\mathbf{B}}_1$) and the loop ($\vec{\mathbf{B}}_2$).

Reasoning The net magnetic field at the point C is the sum of two contributions: (1) the field $\vec{\mathbf{B}}_1$ produced by the long, straight wire according to Equation 21.5 [$B_1 = \mu_0 I_1/(2\pi r)$], and (2) the field $\vec{\mathbf{B}}_2$ produced by the circular loop according to Equation 21.6 with $N = 1$ [$B_2 = \mu_0 I_2/(2R)$]. An application of RHR-2 shows that at point C the field $\vec{\mathbf{B}}_1$ is directed upward, perpendicular to the plane containing the straight wire and the loop (see part *b* of the drawing). Applying RHR-3 to the circular loop carrying a current I_2 shows that the magnetic field $\vec{\mathbf{B}}_2$ is directed downward, opposite to the direction of $\vec{\mathbf{B}}_1$.

> **Problem-Solving Insight Do not confuse the formula for the magnetic field produced at the center of a circular loop with that of a very long, straight wire. The formulas are similar, differing only by a factor of π in the denominator.**

Solution If we choose the upward direction in Figure 21.29b as positive, the net magnetic field at point C is

$$B = \underbrace{\frac{\mu_0 I_1}{2\pi r}}_{\substack{\text{Long,}\\ \text{straight wire}}} - \underbrace{\frac{\mu_0 I_2}{2R}}_{\substack{\text{Center of a}\\ \text{circular loop}}} = \frac{\mu_0}{2}\left(\frac{I_1}{\pi r} - \frac{I_2}{R}\right)$$

$$B = \frac{(4\pi \times 10^{-7}\ \text{T}\cdot\text{m/A})}{2}\left[\frac{8.0\ \text{A}}{\pi(0.030\ \text{m})} - \frac{2.0\ \text{A}}{0.030\ \text{m}}\right] = \boxed{1.1 \times 10^{-5}\ \text{T}}$$

The net field is positive, so it is directed upward, perpendicular to the plane.

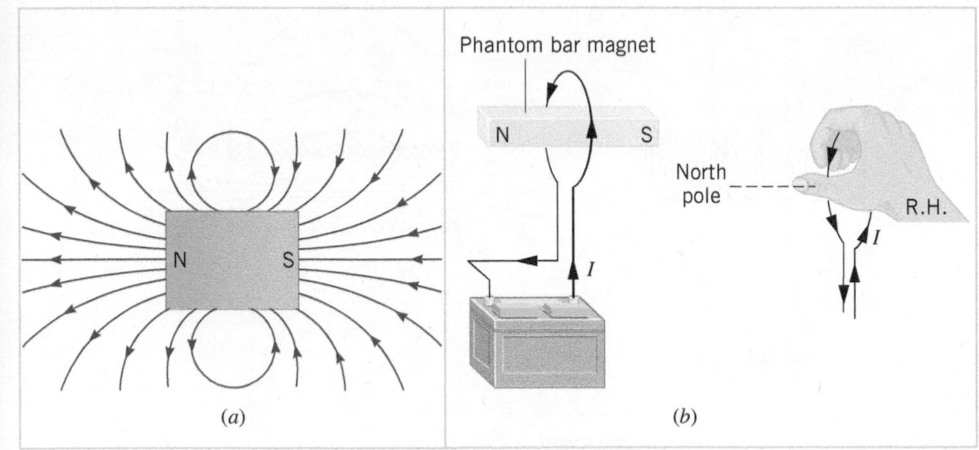

FIGURE 21.30 (*a*) The field lines around the bar magnet resemble those around the loop in **Figure 21.28*a***. (*b*) The current loop can be imagined to be a "phantom" bar magnet with a north pole and a south pole.

A comparison of the magnetic field lines around the current loop in **Figure 21.28*a*** with those in the vicinity of the short bar magnet in **Figure 21.30*a*** shows that the two patterns are similar. Not only are the patterns similar, but the loop itself behaves as a bar magnet with a "north pole" on one side and a "south pole" on the other side. To emphasize that the loop may be imagined to be a bar magnet, **Figure 21.30*b*** includes a "phantom" bar magnet at the center of the loop. The side of the loop that acts like a north pole can be determined with the aid of RHR-3: curl the fingers of the right hand along the direction of the current in the loop. The thumb not only points in the direction of B at the center of the loop, but it also points toward the north pole of the loop.

Because a current-carrying loop acts like a bar magnet, two adjacent loops can be either attracted to or repelled from each other, depending on the relative directions of the currents. **Figure 21.31** includes a "phantom" magnet for each loop and shows that the loops are attracted to each other when the currents are in the same direction and repelled from each other when the currents are in opposite directions.

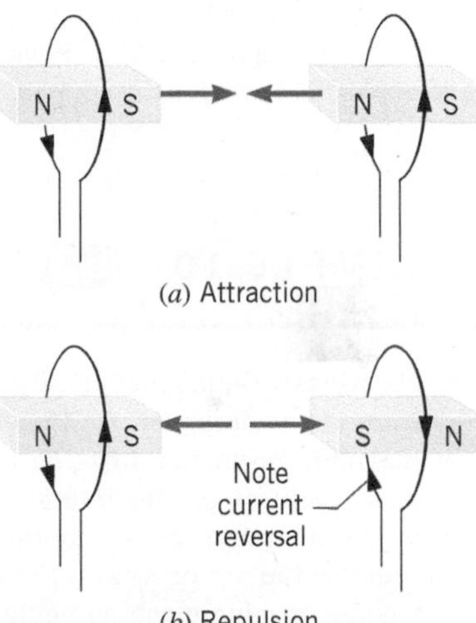

FIGURE 21.31 (*a*) The two current loops attract each other if the directions of the currents are the same and (*b*) repel each other if the directions are opposite. The "phantom" magnets help explain the attraction and repulsion.

A Solenoid

A solenoid is a long coil of wire in the shape of a helix (see **Figure 21.32**). If the wire is wound so the turns are packed close to each other and the solenoid is long compared to its diameter, the magnetic field lines have the appearance shown in the drawing (**Figure 21.32*b***). Notice that the field inside the solenoid and away from its ends is nearly constant in magnitude and directed parallel to the axis. The direction of the field inside the solenoid is given by RHR-3 (**Figure 21.32*c***), just as it is for a circular current loop. The magnitude of the magnetic field in the interior of a long solenoid is

Interior of a long solenoid

$$B = \mu_0 n I \qquad (21.7)$$

where n is the number of turns per unit length of the solenoid and I is the current. If, for example, the solenoid contains 100 turns and has a length of 0.05 m, the number of turns per unit length is $n = (100 \text{ turns})/(0.05 \text{ m}) = 2000 \text{ turns/m}$. The magnetic field outside the solenoid is not constant and is much weaker than the interior field. In fact, the magnetic field outside is nearly zero if the length of the solenoid is much greater than its diameter.

As with a single loop of wire, a solenoid can also be imagined to be a bar magnet, for the solenoid is just an array of connected current loops. And, as with a circular current loop, the location of the north pole can be determined with RHR-3. **Figure 21.32*b*** shows that the left end of the solenoid acts as a north pole, and the right end behaves as a south pole. Solenoids are often referred to as *electromagnets,* and they have several advantages over permanent magnets. For one thing, the strength of the magnetic field can be altered by changing the current and/or the number of turns per unit length. Furthermore, the north and south poles of an electromagnet can be readily switched by reversing the current.

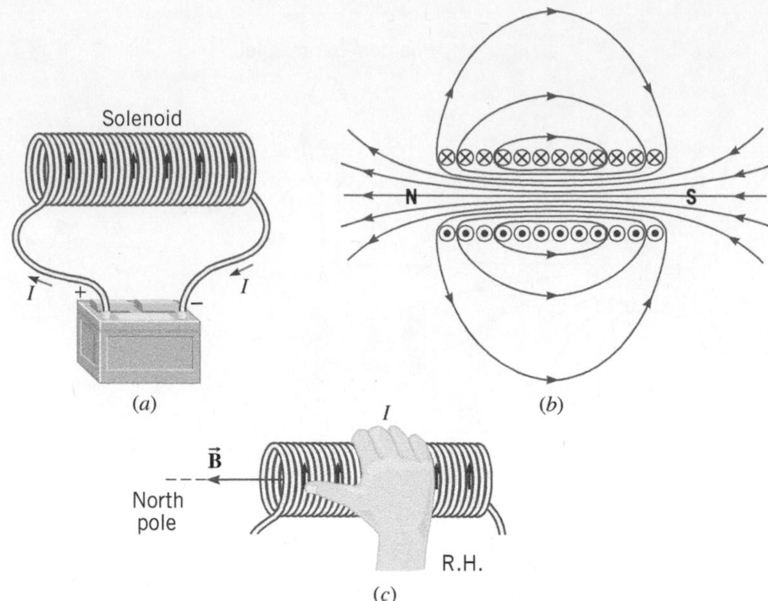

FIGURE 21.32 (*a*) A solenoid and (*b*) a cross-sectional view of it, showing the magnetic field lines and north and south poles. (*c*) Application of RHR-3 to the solenoid. The fingers of the right hand curl around the solenoid in the direction of the current flow, and the thumb points in the direction of the magnetic field everywhere inside the solenoid and toward the north pole.

EXAMPLE 10 | BIO The Physics of . . . Magnetic Resonance Imaging (MRI)

Applications of the magnetic field produced by a current-carrying solenoid are widespread. An important medical application is in magnetic resonance imaging (MRI). With this technique, detailed pictures of the internal parts of the body can be obtained in a noninvasive way that involves none of the risks inherent in the use of X-rays. **Figure 21.33** shows a magnetic resonance imaging machine being used. The circular opening visible in the photograph provides access to the interior of a solenoid, which is typically made from superconducting wire. The superconducting wire facilitates the use of a large current to produce a strong magnetic field. In the presence of this field, the nuclei of certain atoms can be made to behave as tiny radio transmitters and emit radio waves similar to those used by FM stations. The hydrogen atom, which is so prevalent in the human body, can be made to behave in this fashion. The strength of the magnetic field determines where a given collection of hydrogen atoms will "broadcast" on an imaginary FM dial. With a magnetic field that has a slightly different strength at different places, it is possible to associate the location on this imaginary FM dial with a physical location within the body. Computer processing of these locations produces the magnetic resonance image. When hydrogen atoms are used in this way, the image is essentially a map showing their distribution within the body. Remarkably detailed images can now be obtained, such as the one in **Figure 21.34**. They provide doctors with a powerful diagnostic tool that complements those from X-ray and other techniques. Surgeons can now perform operations

Javier Larrea/agefotostock/Alamy Stock Photo

FIGURE 21.33 A magnetic resonance imaging (MRI) machine being used. The circular opening in which the patient is positioned provides access to the interior of a solenoid. MRI scans of the knee are shown on the monitor.

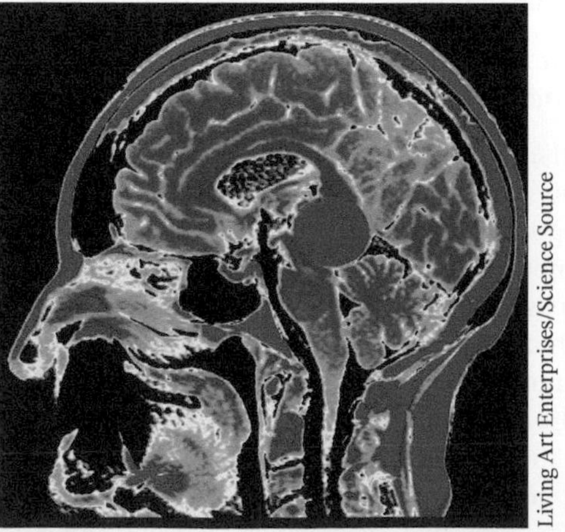

Living Art Enterprises/Science Source

FIGURE 21.34 Magnetic resonance imaging provides one way to diagnose brain disorders. This color-enhanced MRI scan shows the brain of a patient with a large pineal region meningioma (see the red circular region), which is a usually benign, slow-growing tumor.

more accurately by stepping inside specially designed MRI scanners along with the patient and seeing live images of the area into which they are cutting.

The diameter of an MRI magnet has to be fairly large, so that, if need be, the patient's entire body can fit inside. The machine shown in **Figure 21.33** is an open-bore magnet, meaning the solenoid is open at both ends. Older machines would often use a closed-bore configuration, in which the solenoid was closed at one end. This resulted in many patients feeling claustrophobic. One open-bore MRI magnet with a length of 1.20 m produces a magnetic field of 2.50 T at its center. The coils of the magnet are comprised of a superconducting wire with a total length of 3.75×10^4 m carrying a current of 125 A. **(a)** What is the number of turns in the solenoid? **(b)** What is the diameter of the solenoid?

Reasoning The magnetic field at the center of the solenoid can be determined by Equation 21.7, which we can solve for the number of turns in the solenoid. The total length of wire used to wind the solenoid will be equal to the circumference of each turn multiplied by the total number of turns. From this, we can calculate the diameter of the solenoid.

Solution **(a)** Equation 21.7 gives the magnetic field at the center of the solenoid: $B = \mu_0 I n$. Here, n is the number of turns per unit length of the solenoid, which we can write as N/ℓ, where ℓ is the solenoid's length. Making this substitution and then solving for N, we have

$$B = \frac{\mu_0 I N}{\ell} \Rightarrow N = \frac{B\ell}{\mu_0 I} = \frac{(2.50\ \text{T})(1.20\ \text{m})}{(4\pi \times 10^{-7}\ \text{T} \cdot \text{m/A})(125\ \text{A})}$$

$$= \boxed{1.91 \times 10^4\ \text{turns}}$$

(b) The total length of wire in the solenoid (L_w) will be equal to the circumference of one turn (πd) multiplied by the total number of turns N, we have

$$L_w = \pi d N \Rightarrow d = \frac{L_w}{\pi N} = \frac{3.75 \times 10^4\ \text{m}}{\pi(1.91 \times 10^4\ \text{turns})} = 0.625\ \text{m} = \boxed{62.5\ \text{cm}}$$

THE PHYSICS OF . . . old television screens and computer display monitors. Some older television sets and some computer display monitors use electromagnets (solenoids) to produce images by exerting magnetic forces on moving electrons. An evacuated glass tube, called a cathode-ray tube (CRT), contains an electron gun that sends a narrow beam of high-speed electrons toward the screen of the tube, as **Figure 21.35*a*** illustrates. The inner surface of the screen is covered with a phosphor coating, and when the electrons strike it, they generate a spot of visible light. This spot is called a pixel (a contraction of "picture element").

To create a black-and-white picture, the electron beam is scanned rapidly from left to right across the screen. As the beam makes each horizontal scan, the number of electrons per second striking the screen is changed by electronics controlling the electron gun, making the scan line brighter in some places and darker in others. When the beam reaches the right side of the screen, it is turned off and returned to the left side slightly below where it started (see **Figure 21.35*b***). The beam is then scanned across the next line, and so on. In these older TV sets, a complete picture consists of 525 scan lines (or 625 in Europe) and is formed in $\frac{1}{30}$ of a second. High-definition TV sets that utilized this older technology had about 1100 scan lines, giving a much sharper, more detailed picture.

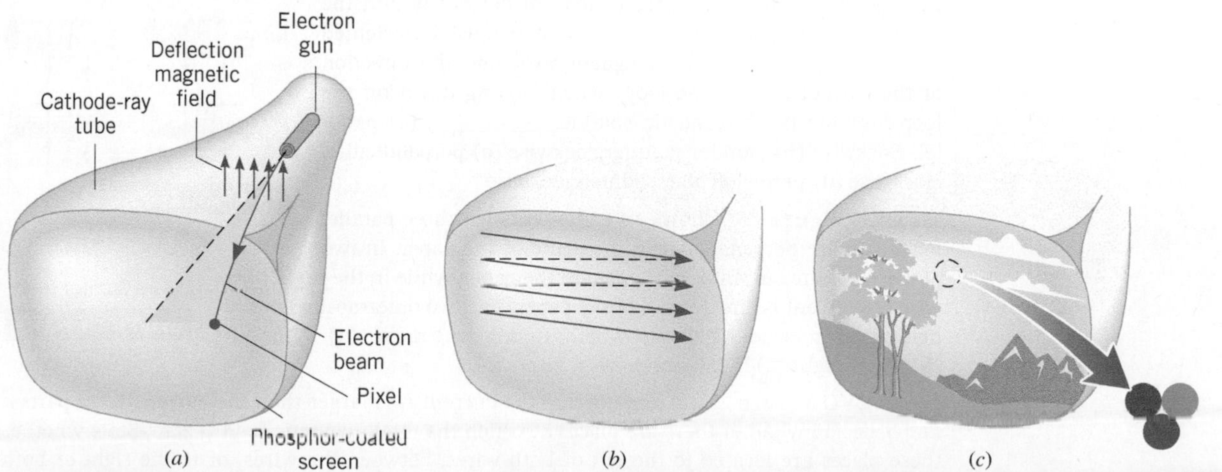

FIGURE 21.35 (*a*) A cathode-ray tube contains an electron gun, a magnetic field for deflecting the electron beam, and a phosphor-coated screen. A CRT color TV actually uses three guns, although only one is shown here for clarity. (*b*) The image is formed by scanning the electron beam across the screen. (*c*) The red, green, and blue phosphors of a color TV.

The electron beam is deflected by a pair of electromagnets placed around the neck of the tube, between the electron gun and the screen. One electromagnet is responsible for producing the horizontal deflection of the beam and the other for the vertical deflection. For clarity, **Figure 21.35*a*** shows the net magnetic field generated by the electromagnets at one instant, and not the electromagnets themselves. The electric current in the electromagnets produces a net magnetic field that exerts a force on the moving electrons, causing their trajectories to bend and reach different points on the screen. Changing the current changes the field, so the electrons can be deflected to any point on the screen.

A color TV operates with three electron guns instead of one. Furthermore, the single phosphor of a black-and-white TV is replaced by a large number of three-dot clusters of phosphors that glow red, green, and blue when struck by an electron beam, as indicated in **Figure 21.35*c***. Each red, green, and blue color in a cluster is produced when electrons from one of the three guns strike the corresponding phosphor dot. The three dots are so close together that, from a normal viewing distance, they cannot be separately distinguished. Red, green, and blue are primary colors, so virtually all other colors can be created by varying the intensities of the three beams focused on a cluster.

Check Your Understanding

(The answers are given at the end of the book.)

13. **CYU Figure 21.6** shows a conducting wire wound into a helical shape. The helix acts like a spring and expands back toward its original shape after the coils are squeezed together and released. The bottom end of the wire just barely touches the mercury (a good electrical conductor) in the cup. After the switch is closed, current in the circuit causes the light bulb to glow. Does the bulb **(a)** repeatedly turn on and off like a turn signal on a car, **(b)** glow continually, or **(c)** glow briefly and then go out?

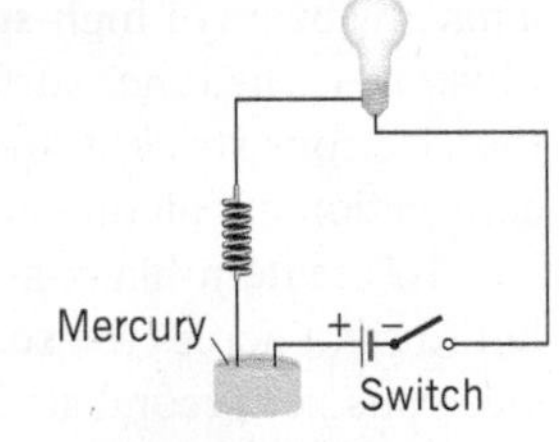

CYU FIGURE 21.6

14. For each electromagnet at the left in **CYU Figure 21.7**, will it be attracted to or repelled from the permanent magnet immediately to its right?

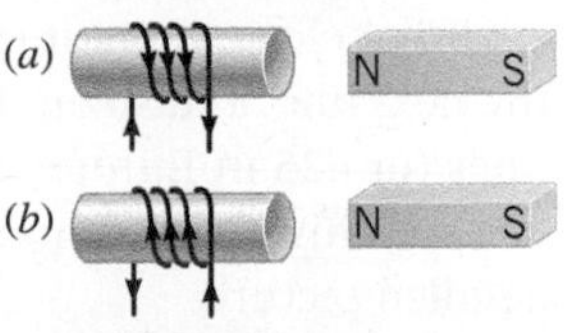

CYU FIGURE 21.7

15. For each electromagnet at the left in **CYU Figure 21.8**, will it be attracted to or repelled from the electromagnet immediately to its right?

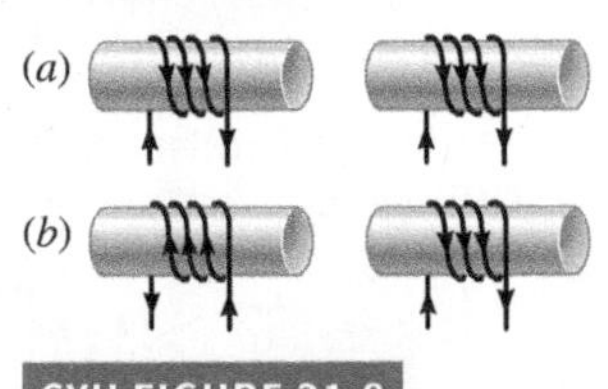

CYU FIGURE 21.8

16. Refer to **Figure 21.5**. If the earth's magnetism is assumed to originate from a large circular loop of current within the earth, then the plane of the current loop must be oriented __________ to the earth's magnetic axis, and the direction of the current around the loop (when looking down on the loop from the north magnetic pole) is __________. **(a)** parallel, clockwise **(b)** parallel, counterclockwise **(c)** perpendicular, clockwise **(d)** perpendicular, counterclockwise

17. **CYU Figure 21.9** shows an end-on view of three parallel wires that are perpendicular to the plane of the paper. In two of the wires the current is directed into the paper, while in the remaining wire the current is directed out of the paper. The two outermost wires are held rigidly in place. Which way will the middle wire move? **(a)** To the left **(b)** To the right **(c)** It will not move at all.

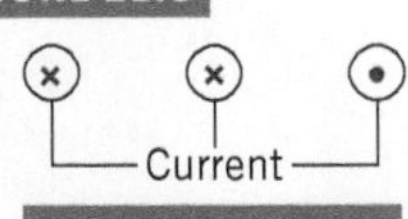

CYU FIGURE 21.9

18. In **CYU Figure 21.10**, assume that the current I_1 is larger than the current I_2. In parts *a* and *b*, decide whether there are places at which the total magnetic field is zero. State whether these places are located to the left of both wires, between the wires, or to the right of both wires.

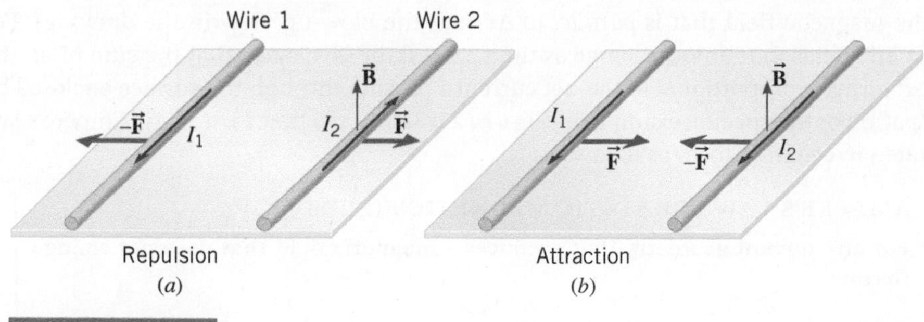

CYU FIGURE 21.10

19. Each of the four drawings **CYU Figure 21.11** shows the same three concentric loops of wire. The currents in the loops have the same magnitude I and have the directions shown. Rank the magnitude of the net magnetic field produced at the center of each of the four drawings, largest to smallest.

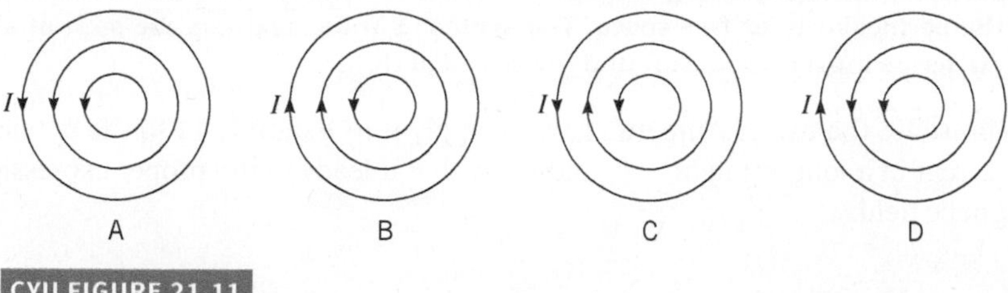

CYU FIGURE 21.11

20. There are four wires viewed end-on in **CYU Figure 21.12**. They are long, straight, and perpendicular to the plane of the paper. Their cross sections lie at the corners of a square. The magnitude of the current in each wire is the same. What must be the direction of the current in each wire (into or out of the page), so that when *any* single current is turned off, the total magnetic field at P (the center of the square) is directed toward a corner of the square?

	Wire 1	Wire 2	Wire 3	Wire 4
(a)	out of	out of	into	out of
(b)	into	out of	into	into
(c)	out of	into	out of	out of
(d)	into	into	into	into

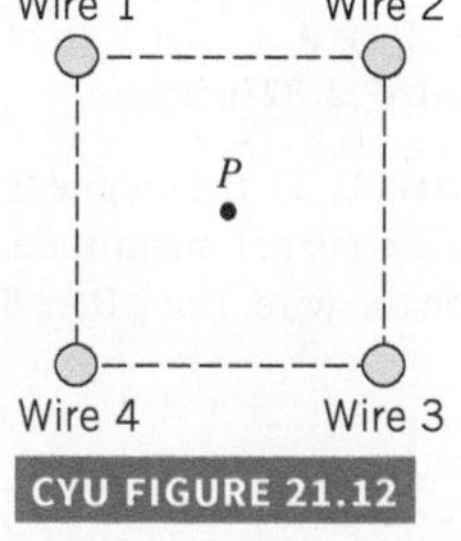

CYU FIGURE 21.12

21.8 Ampère's Law

We have seen that an electric current creates a magnetic field. However, the magnitude and direction of the field at any point in space depends on the specific geometry of the wire carrying the current. For instance, distinctly different magnetic fields surround a long, straight wire, a circular loop of wire, and a solenoid. Although different, each of these fields can be obtained from a general law known as **Ampère's law**, which is valid for a wire of any geometrical shape. Ampère's law specifies the relationship between an electric current and the magnetic field that it creates.

To see how Ampère's law is stated, consider **Figure 21.36**, which shows two wires carrying currents I_1 and I_2. In general, there may be any number of currents. Around the wires we construct an arbitrarily shaped but closed path. This path encloses a surface and is constructed from a large number of short segments, each of length $\Delta\ell$. Ampère's law deals with the product of $\Delta\ell$ and $B_{\parallel}$ for each segment, where $B_{\parallel}$ is the component

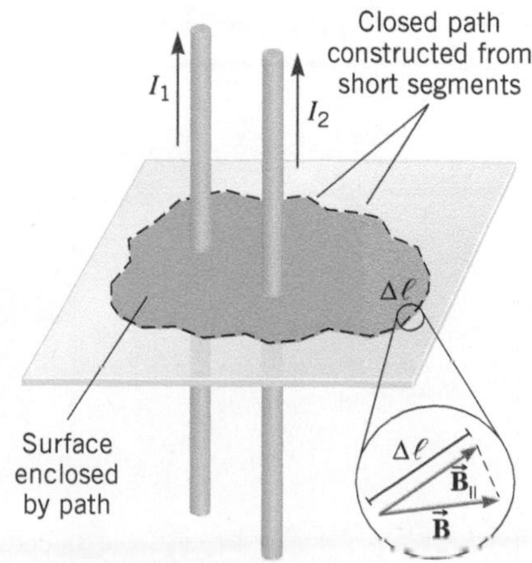

FIGURE 21.36 This setup is used in the text to explain Ampère's law.

of the magnetic field that is *parallel* to $\Delta\ell$ (see the blow-up view in the drawing). For magnetic fields that do not change as time passes, the law states that the sum of all the $B_{\parallel}\,\Delta\ell$ terms is proportional to the net current I passing through the surface enclosed by the path. For the specific example in **Figure 21.36**, we see that $I = I_1 + I_2$. Ampère's law is stated in equation form as follows:

AMPÈRE'S LAW FOR STATIC MAGNETIC FIELDS

For any current geometry that produces a magnetic field that does not change in time,

$$\Sigma B_{\parallel}\,\Delta\ell = \mu_0 I \qquad (21.8)$$

where $\Delta\ell$ is a small segment of length along a closed path of arbitrary shape around the current, $B_{\parallel}$ is the component of the magnetic field parallel to $\Delta\ell$, I is the net current passing through the surface bounded by the path, and μ_0 is the permeability of free space. The symbol Σ indicates that the sum of all $B_{\parallel}\,\Delta\ell$ terms must be taken around the closed path.

To illustrate the use of Ampère's law, we apply it in Example 11 to the special case of the current in a long, straight wire and show that it leads to the proper expression for the magnetic field.

EXAMPLE 11 | An Infinitely Long, Straight, Current-Carrying Wire

Use Ampère's law to obtain the magnetic field produced by the current in an infinitely long, straight wire.

Reasoning **Figure 21.24a** shows that compass needles point in a circular pattern around the wire, so we know that the magnetic field lines are circular. Therefore, it is convenient to use a circular path of radius r when applying Ampère's law, as **Figure 21.37** indicates.

Solution Along the circular path in **Figure 21.37**, the magnetic field is everywhere parallel to $\Delta\ell$ and has a constant magnitude, since each point is at the same distance from the wire. Thus, $B_{\parallel} = B$ and, according to Ampère's law, we have

$$\Sigma B_{\parallel}\,\Delta\ell = B(\Sigma\,\Delta\ell) = \mu_0 I$$

However, $\Sigma\Delta\ell$ is just the circumference $2\pi r$ of the circle, so Ampère's law reduces to

$$B(\Sigma\,\Delta\ell) = B(2\pi r) = \mu_0 I$$

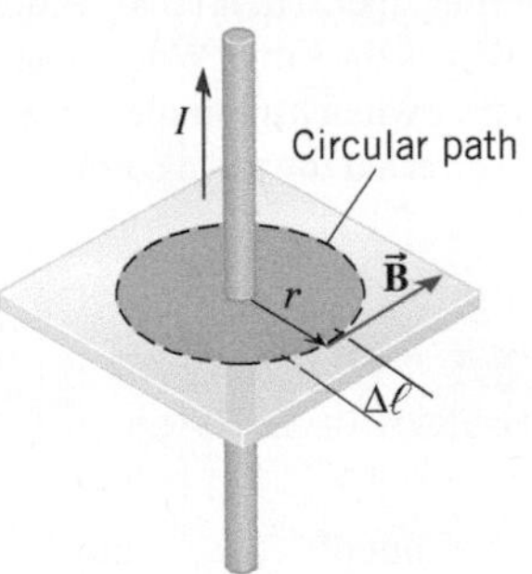

FIGURE 21.37 Example 11 uses Ampère's law to find the magnetic field in the vicinity of this long, straight, current-carrying wire.

Dividing both sides by $2\pi r$ shows that $\boxed{B = \mu_0 I/(2\pi r)}$, as given earlier in Equation 21.5.

21.9 Magnetic Materials

Ferromagnetism

The similarity between the magnetic field lines in the neighborhood of a bar magnet and those around a current loop suggests that the magnetism in each case arises from a common cause. The field that surrounds the loop is created by the charges moving in the wire. The magnetic field around a bar magnet is also due to the motion of charges, but the motion is not that of a bulk current through the magnetic material. Instead, the motion responsible for the magnetism is that of the electrons within the atoms of the material.

The magnetism produced by electrons within an atom can arise from two motions. First, each electron orbiting the nucleus behaves like an atomic-sized loop of current that generates a small magnetic field, similar to the field created by the current loop in **Figure 21.28a**. Second, each electron possesses a spin that also gives rise to a magnetic field. The net magnetic field created by the electrons within an atom is due to the combined fields created by their orbital and spin motions.

In most substances the magnetism produced at the atomic level tends to cancel out, with the result that the substance is nonmagnetic overall. However, there are some materials, known as **ferromagnetic materials**, in which the cancellation does not occur for groups of approximately 10^{16}–10^{19} neighboring atoms, because they have electron spins that are naturally aligned parallel to each other. This alignment results from a special type of quantum mechanical* interaction between the spins. The result of the interaction is a small but highly magnetized region of about 0.01 to 0.1 mm in size, depending on the nature of the material; this region is called a **magnetic domain**. Each domain behaves as a small magnet with its own north and south poles. Common ferromagnetic materials are iron, nickel, cobalt, chromium dioxide, and alnico (an *al*uminum–*ni*ckel–*co*balt alloy).

Induced Magnetism

Often magnetic domains in ferromagnetic materials are arranged randomly, as **Figure 21.38*a*** illustrates for a piece of iron. In such a situation, the magnetic fields of the domains cancel each other, so the iron displays little, if any, overall magnetism. However, an unmagnetized piece of iron can be magnetized by placing it in an external magnetic field provided by a permanent magnet or an electromagnet. The external magnetic field penetrates the unmagnetized iron and *induces* (or brings about) a state of magnetism in the iron by causing two effects. Those domains whose magnetism is parallel or nearly parallel to the external magnetic field grow in size at the expense of other domains that are not so oriented. In **Figure 21.38*b*** the growing domains are colored gold. In addition, the magnetic alignment of some domains may rotate and become more oriented in the direction of the external field. The resulting preferred alignment of the domains gives the iron an overall magnetism, so the iron behaves like a magnet with associated north and south poles. In some types of ferromagnetic materials, such as the chromium dioxide used in magnetic recording tapes, the domains remain aligned for the most part when the external magnetic field is removed, and the material thus becomes permanently magnetized.

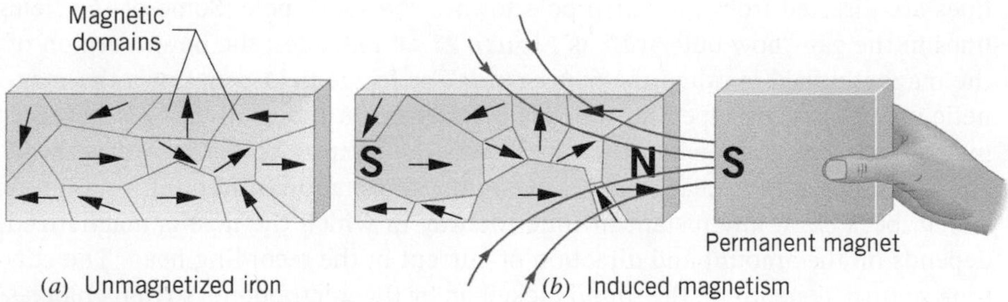

FIGURE 21.38 (*a*) Each magnetic domain is a highly magnetized region that behaves like a small magnet (represented by an arrow whose head indicates a north pole). Unmagnetized iron consists of many domains that are randomly aligned. The size of each domain is exaggerated for clarity. (*b*) The external magnetic field of the permanent magnet causes those domains that are parallel or nearly parallel to the field to grow in size (shown in gold color).

The magnetism induced in a ferromagnetic material can be surprisingly large, even in the presence of a weak external field. For instance, it is not unusual for the induced magnetic field to be a hundred to a thousand times stronger than the external field that causes the alignment. For this reason, high-field electromagnets are constructed by wrapping the current-carrying wire around a solid core made from iron or other ferromagnetic material.

Induced magnetism explains why a permanent magnet sticks to a refrigerator door and why an electromagnet can pick up scrap iron at a junkyard. Notice in **Figure 21.38*b*** that there is a north pole at the end of the iron that is closest to the south pole of the

*The branch of physics called quantum mechanics is mentioned in Section 29.5. A detailed discussion of it is beyond the scope of this book.

FIGURE 21.39 Magnetic fingerprint powder has been used to reveal fingerprints on this neoprene glove. The powder consists of tiny iron flakes coated with an organic material that enables them to stick to the greasy residue in the print. Because of induced magnetism, a permanent magnet can be used to remove any excess powder that would obscure the evidence. This eliminates the need for brushing and possible damage to the print.

permanent magnet. The net result is that the two opposite poles give rise to an attraction between the iron and the permanent magnet. Conversely, the north pole of the permanent magnet would also attract the piece of iron by inducing a south pole in the nearest side of the iron. In nonferromagnetic materials, such as aluminum and copper, the formation of magnetic domains does not occur, so magnetism cannot be induced into these substances. Consequently, magnets do not stick to aluminum cans.

THE PHYSICS OF . . . detecting fingerprints. The attraction that induced magnetism creates between a permanent magnet and a ferromagnetic material is used in crime-scene investigations, where powder is dusted onto surfaces to make fingerprints visible. Magnetic fingerprint powder allows investigators to recover evidence from surfaces like the neoprene glove in **Figure 21.39**, which are very difficult to examine without damaging the prints in the process. Brushing excess conventional powder away ruins the delicate ridges of the pattern that allow the print to be identified reliably. Magnetic fingerprint powder, however, consists of tiny iron flakes coated with an organic material that allows them to stick to the greasy residue in the print. A permanent magnet eliminates the need for brushing away excess powder by creating induced magnetism in the iron and pulling away the powder not stuck directly to the greasy residue.

Magnetic Tape Recording

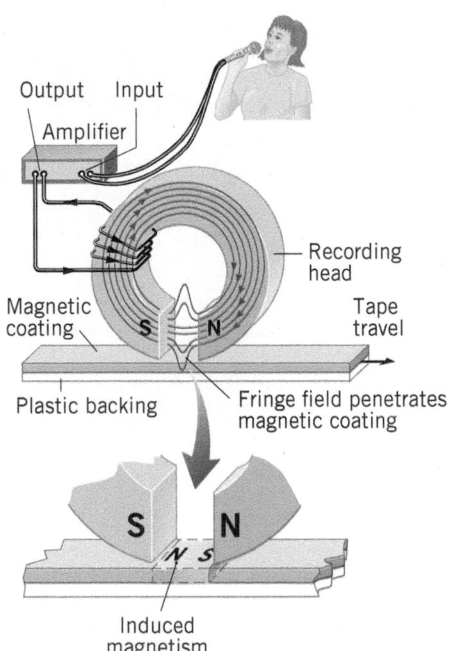

FIGURE 21.40 The magnetic fringe field of the recording head penetrates the magnetic coating on the tape and magnetizes it.

THE PHYSICS OF . . . magnetic tape recording. The process of magnetic tape recording uses induced magnetism, as **Figure 21.40** illustrates. The weak electrical signal from a microphone is routed to an amplifier where it is amplified. The current from the output of the amplifier is then sent to the recording head, which is a coil of wire wrapped around an iron core. The iron core has the approximate shape of a horseshoe with a small gap between the two ends. The ferromagnetic iron substantially enhances the magnetic field produced by the current in the coil.

When there is a current in the coil, the recording head becomes an electromagnet with a north pole at one end and a south pole at the other end. The magnetic field lines pass through the iron core and cross the gap. Within the gap, the lines are directed from the north pole toward the south pole. Some of the field lines in the gap "bow outward," as **Figure 21.40** indicates; the bowed region of the magnetic field is called the *fringe field.* The fringe field penetrates the magnetic coating on the tape and induces magnetism in the coating. This induced magnetism is retained when the tape leaves the vicinity of the recording head and, thus, provides a means for storing audio information. Audio information is stored, because at any instant in time the way in which the tape is magnetized depends on the amount and direction of current in the recording head. The current, in turn, depends on the sound picked up by the microphone, so that changes in the sound that occur from moment to moment are preserved as changes in the tape's induced magnetism.

Maglev Trains

THE PHYSICS OF . . . a magnetically levitated train. A magnetically levitated train—maglev, for short—uses forces that arise from induced magnetism to levitate or float above a guideway. Since it rides a few centimeters above the guideway, a maglev does not need wheels. Freed from friction due to the guideway, the train can achieve significantly greater speeds than do conventional trains. For example, the Transrapid maglev in **Figure 21.41** has achieved speeds of 110 m/s (250 mph).

Figure 21.41*a* shows that the Transrapid maglev achieves levitation with electromagnets mounted on arms that extend around and under the guideway. When a current is sent to an electromagnet, the resulting magnetic field creates induced magnetism in a rail mounted in the guideway. The upward attractive force from the induced magnetism is balanced by the weight of the train, so the train moves without touching the rail or the guideway.

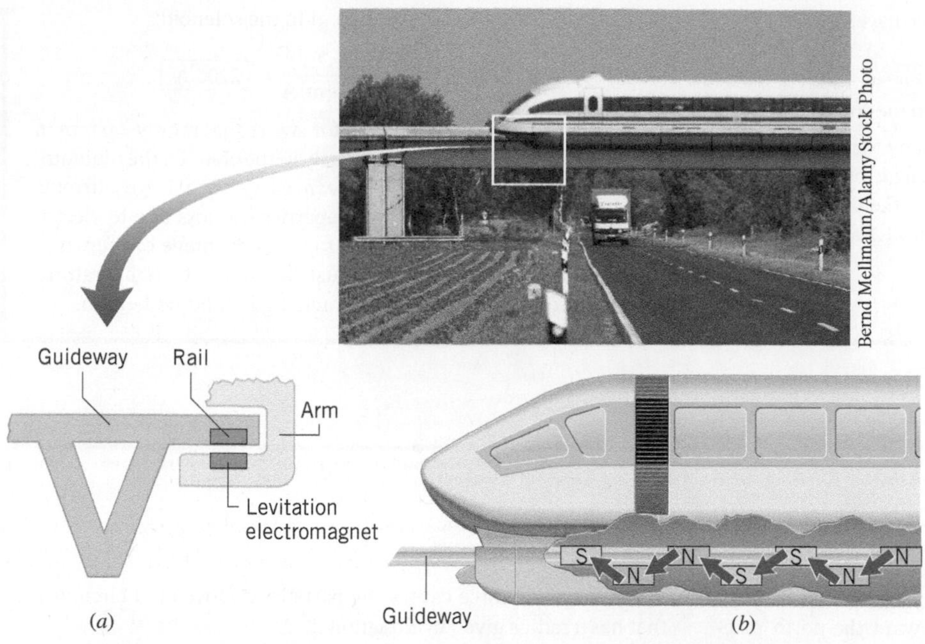

FIGURE 21.41 (*a*) The Transrapid maglev (a German train) has achieved speeds of 110 m/s (250 mph). The levitation electromagnets are drawn up toward the rail in the guideway, levitating the train. (*b*) The magnetic propulsion system.

Magnetic levitation only lifts the train and does not move it forward. **Figure 21.41*b*** illustrates how magnetic propulsion is achieved. In addition to the levitation electromagnets, propulsion electromagnets are also placed along the guideway. By controlling the direction of the currents in the train and guideway electromagnets, it is possible to create an unlike pole in the guideway just ahead of each electromagnet on the train and a like pole just behind. Each electromagnet on the train is thus both pulled and pushed forward by electromagnets in the guideway. By adjusting the timing of the like and unlike poles in the guideway, the speed of the train can be adjusted. Reversing the poles in the guideway electromagnets serves to brake the train.

Check Your Understanding

(The answers are given at the end of the book.)

21. In a TV commercial that advertises a soda pop, a strong electromagnet picks up a delivery truck carrying cans of the soft drink. The picture switches to the interior of the truck, where cans are seen to fly upward and stick to the roof just beneath the electromagnet. Are these cans made entirely of aluminum?

22. Suppose that you have two bars. Bar 1 is a permanent magnet and bar 2 is not a magnet, but is made from a ferromagnetic material like iron. A third bar (bar 3), which is known to be a permanent magnet, is brought close to one end of bar 1 and then to one end of bar 2. Which one of the following statements is true? **(a)** Bars 1 and 3 will either be attracted to or repelled from each other, while bars 2 and 3 will always be repelled from each other. **(b)** Bars 1 and 3 will either be attracted to or repelled from each other, while bars 2 and 3 will always be attracted to each other. **(c)** Bars 1 and 3 will always be repelled from each other, while bars 2 and 3 will either be attracted to or repelled from each other. **(d)** Bars 1 and 3 will always be attracted to each other, while bars 2 and 3 will either be attracted to or repelled from each other.

EXAMPLE 12 | BIO The Physics of MRI Machines—The Superconducting Magnet

As discussed in Example 10, the medical diagnostic technique of magnetic resonance imaging relies on a large solenoid electromagnet to produce strong magnetic fields. Assume the magnet is wound with a single layer of 2.0-mm diameter wire that is insulated. If the highest magnetic field produced by the magnet is 1.7 T, what is the maximum current in the magnet?

Reasoning Since the magnetic field in the MRI machine is produced by a solenoid, we can apply Equation 21.7 to calculate the current in the magnet. To do this, we will need the value of n (the number of turns per unit length). This can be determined from the length of the magnet and the thickness of the wire.

Solution Beginning with Equation 21.7, we have:

$$B = \mu_0 nI = \mu_0\left(\frac{N}{\ell}\right)I,$$

where $n = \left(\frac{N}{\ell}\right)$ is the number of turns (N) in the solenoid divided by its length (ℓ). N is determined by how many turns of wire in a single layer can fit side-by-side along the length of the solenoid. Therefore, $N = \ell/d$, where we let d represent the diameter of the wire. Making this substitution into Equation 21.7 above, we have the following:

$$B = \mu_0\left(\frac{\ell/d}{\ell}\right)I = \frac{\mu_0 I}{d}$$

We can now solve this for the current in the solenoid:

$$I = \frac{dB}{\mu_0} = \frac{(0.0020\ \text{m})(1.7\ \text{T})}{4\pi \times 10^{-7}\ \text{T}\cdot\text{m/A}} = \boxed{2700\ \text{A}}$$

This is an incredibly large current for a wire that is only 2.0 mm in diameter! As we mentioned previously in the chapter, the niobium-based wire in the MRI magnet can carry such large currents because of its superconducting properties (no loss due to electrical resistance). However, in order to take advantage of the wire's superconducting properties, it must be cooled to temperatures close to absolute zero (0 K) by submersing it in liquid helium.

Concept Summary

21.1 Magnetic Fields A magnet has a north pole and a south pole. The north pole is the end that points toward the north magnetic pole of the earth when the magnet is freely suspended. Like magnetic poles repel each other, and unlike poles attract each other.

A magnetic field exists in the space around a magnet. The magnetic field is a vector whose direction at any point is the direction indicated by the north pole of a small compass needle placed at that point. As an aid in visualizing the magnetic field, magnetic field lines are drawn in the vicinity of a magnet. The lines appear to originate from the north pole and end on the south pole. The magnetic field at any point in space is tangent to the magnetic field line at the point. Furthermore, the strength of the magnetic field is proportional to the number of lines per unit area that passes through a surface oriented perpendicular to the lines.

21.2 The Force That a Magnetic Field Exerts on a Moving Charge The direction of the magnetic force acting on a charge moving with a velocity $\vec{\mathbf{v}}$ in a magnetic field $\vec{\mathbf{B}}$ is perpendicular to both $\vec{\mathbf{v}}$ and $\vec{\mathbf{B}}$. For a positive charge the direction can be determined with the aid of Right-Hand Rule No. 1 (see below). The magnetic force on a moving negative charge is opposite to the force on a moving positive charge.

Right-Hand Rule No. 1: Extend the right hand so the fingers point along the direction of the magnetic field $\vec{\mathbf{B}}$ and the thumb points along the velocity $\vec{\mathbf{v}}$ of the charge. The palm of the hand then faces in the direction of the magnetic force $\vec{\mathbf{F}}$ that acts on a positive charge.

The magnitude B of the magnetic field at any point in space is defined according to Equation 21.1, where F is the magnitude of the magnetic force on a test charge, $|q_0|$ is the magnitude of the test charge, and v is the magnitude of the charge's velocity, which makes an angle θ with the direction of the magnetic field. The SI unit for the magnetic field is the tesla (T). Another, smaller unit for the magnetic field is the gauss; 1 gauss $= 10^{-4}$ tesla. The gauss is not an SI unit.

$$B = \frac{F}{|q_0|v \sin\theta} \qquad \textbf{(21.1)}$$

21.3 The Motion of a Charged Particle in a Magnetic Field When a charged particle moves in a region that contains both magnetic and electric fields, the net force on the particle is the vector sum of the magnetic and electric forces.

A magnetic force does no work on a charged particle moving as in **Figure 21.9b**, because the direction of the force is always perpendicular to the motion of the particle. Being unable to do work, the magnetic force cannot change the kinetic energy, and hence the speed, of the particle; however, the magnetic force does change the direction in which the particle moves.

When a particle of charge q (magnitude $= |q|$) and mass m moves with speed v perpendicular to a uniform magnetic field of magnitude B, the magnetic force causes the particle to move on a circular path that has a radius given by Equation 21.2.

$$r = \frac{mv}{|q|B} \qquad \textbf{(21.2)}$$

21.4 The Mass Spectrometer The mass spectrometer is an instrument for measuring the abundance of ionized atoms or molecules that have different masses. The atoms or molecules are ionized (charge $= +e$), accelerated to a speed v by a potential difference V, and sent into a uniform magnetic field of magnitude B. The magnetic field causes the particles (each with a mass m) to move on a circular path of radius r. The relation between m and B is given by Equation 1.

$$m = \left(\frac{er^2}{2V}\right)B^2 \qquad \textbf{(1)}$$

21.5 The Force on a Current in a Magnetic Field An electric current, being composed of moving charges, can experience a magnetic force when placed in a magnetic field of magnitude B. For a straight wire that has a length L and carries a current I, the magnetic force has a magnitude that is given by Equation 21.3, where θ is the angle between the directions of the current and the magnetic field. The direction of the force is perpendicular to both the current and the magnetic field and is given by Right-Hand Rule No. 1.

$$F = ILB \sin\theta \qquad \textbf{(21.3)}$$

21.6 The Torque on a Current-Carrying Coil Magnetic forces can exert a torque on a current-carrying loop of wire and thus cause the loop to rotate. When a current I exists in a coil of wire with N turns, each of area A, in the presence of a magnetic field of magnitude B, the coil experiences a net torque that has a magnitude given by Equation 21.4, where ϕ is the angle between the direction of the magnetic field and the normal to the plane of the coil. The quantity NIA is known as the magnetic moment of the coil.

$$\tau = NIAB \sin\phi \qquad \textbf{(21.4)}$$

21.7 Magnetic Fields Produced by Currents An electric current produces a magnetic field, with different current geometries giving rise to different field patterns. For an infinitely long, straight wire, the magnetic field lines are circles centered on the wire, and their direction is given by Right-Hand Rule No. 2 (see below). The magnitude of the magnetic field at a radial distance r from the wire is given by Equation 21.5, where I is the current in the wire and μ_0 is a constant known as the permeability of free space ($\mu_0 = 4\pi \times 10^{-7}\ \text{T}\cdot\text{m/A}$).

$$B = \frac{\mu_0 I}{2\pi r} \qquad \textbf{(21.5)}$$

Right-Hand Rule No. 2: Curl the fingers of the right hand into the shape of a half-circle. Point the thumb in the direction of the conventional current I, and the tips of the fingers will point in the direction of the magnetic field $\vec{\mathbf{B}}$.

The magnitude of the magnetic field at the center of a flat circular loop consisting of N turns, each of radius R and carrying a current I, is given by Equation 21.6.

$$B = N\frac{\mu_0 I}{2R} \tag{21.6}$$

The loop has associated with it a north pole on one side and a south pole on the other side. The side of the loop that behaves like a north pole can be predicted by using Right-Hand Rule No. 3: Curl the fingers of the right hand along the direction of the conventional current I around the loop or solenoid, and the thumb will point in the direction of the magnetic field at the center of the loop, or everywhere inside the solenoid, and toward their respective north poles.

A solenoid is a coil of wire wound in the shape of a helix. Inside a long solenoid the magnetic field is nearly constant and has a magnitude that is given by Equation 21.7, where n is the number of turns per unit length of the solenoid and I is the current in the wire. One end of the solenoid behaves like a north pole, and the other end like a south pole. The end that is the north pole can be predicted by using Right-Hand Rule No. 3.

$$B = \mu_0 n I \tag{21.7}$$

21.8 Ampère's Law Ampère's law specifies the relationship between a current and its associated magnetic field. For any current geometry that produces a magnetic field that does not change in time, Ampère's law is given by Equation 21.8, where $\Delta\ell$ is a small segment of length along a closed path of arbitrary shape around the current, $B_{\parallel}$ is the component of the magnetic field parallel to $\Delta\ell$, I is the net current passing through the surface bounded by the path, and μ_0 is the permeability of free space. The symbol Σ indicates that the sum of all $B_{\parallel}\,\Delta\ell$ terms must be taken around the closed path.

$$\Sigma B_{\parallel}\,\Delta\ell = \mu_0 I \tag{21.8}$$

21.9 Magnetic Materials Ferromagnetic materials, such as iron, are made up of tiny regions called magnetic domains, each of which behaves as a small magnet. In an unmagnetized ferromagnetic material, the domains are randomly aligned. In a permanent magnet, many of the domains are aligned, and a high degree of magnetism results. An unmagnetized ferromagnetic material can be induced into becoming magnetized by placing it in an external magnetic field.

Focus on Concepts

Online

Additional questions are available for assignment in WileyPLUS.

Section 21.2 The Force That a Magnetic Field Exerts on a Moving Charge

1. At a location near the equator, the earth's magnetic field is horizontal and points north. An electron is moving vertically upward from the ground. What is the direction of the magnetic force that acts on the electron? (a) North (b) East (c) South (d) West (e) The magnetic force is zero.

2. The drawing shows four situations in which a positively charged particle is moving with a velocity $\vec{\mathbf{v}}$ through a magnetic field $\vec{\mathbf{B}}$. In each case, the magnetic field is directed out of the page toward you, and the velocity is directed to the right. In only one of these drawings is the magnetic force $\vec{\mathbf{F}}$ physically reasonable. Which drawing is it? (a) 1 (b) 2 (c) 3 (d) 4

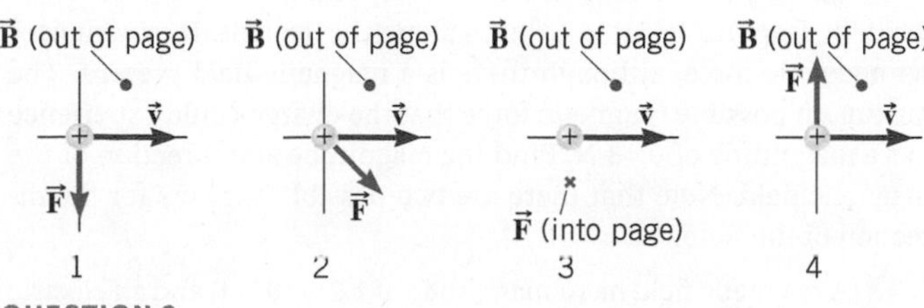

QUESTION 2

Section 21.3 The Motion of a Charged Particle in a Magnetic Field

3. Three particles are moving perpendicular to a uniform magnetic field and travel on circular paths (see the drawing). The particles have the same mass and speed. List the particles in order of their charge magnitude, largest to smallest. (a) 3, 2, 1 (b) 3, 1, 2 (c) 2, 3, 1 (d) 1, 3, 2 (e) 1, 2, 3

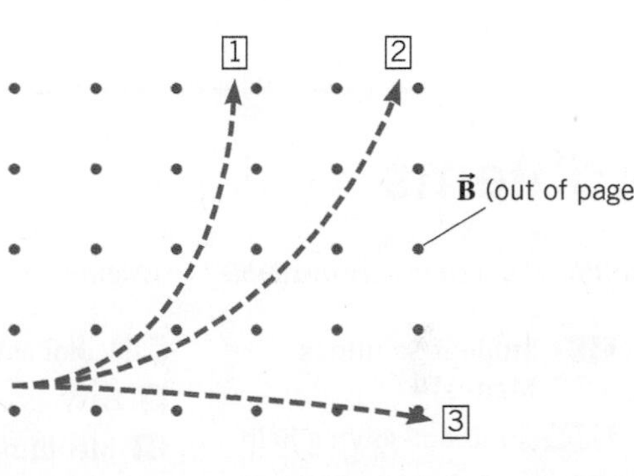

QUESTION 3

4. The drawing shows the circular paths of an electron and a proton. These particles have the same charge magnitudes, but the proton is more massive. They travel at the same speed in a uniform magnetic field $\vec{\mathbf{B}}$, which is directed into the page everywhere. Which particle follows the larger circle, and does it travel clockwise or counterclockwise?

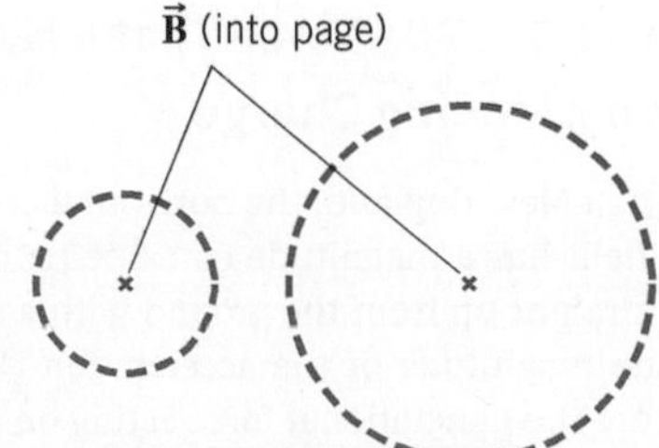

QUESTION 4

	Particle	Direction of Travel
(a)	electron	clockwise
(b)	electron	counterclockwise
(c)	proton	clockwise
(d)	proton	counterclockwise

Section 21.5 The Force on a Current in a Magnetic Field

5. Four views of a horseshoe magnet and a current-carrying wire are shown in the drawing. The wire is perpendicular to the page, and the current is directed out of the page toward you. In which one or more of these situations does the magnetic force on the current point due north? (a) 1 and 2 (b) 3 and 4 (c) 2 (d) 3 (e) 1

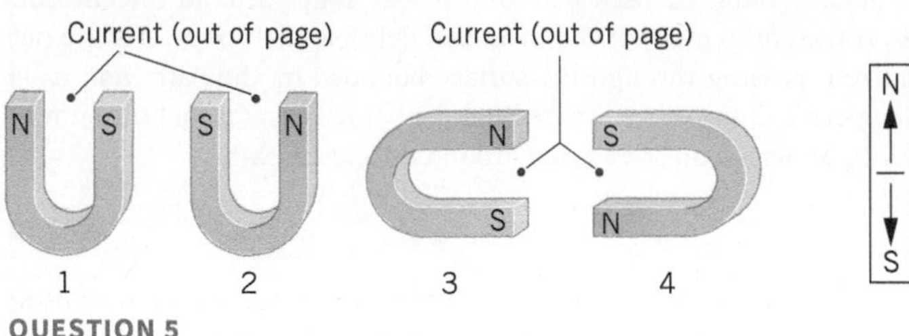

QUESTION 5

Section 21.6 The Torque on a Current-Carrying Coil

6. A square, current-carrying loop is placed in a uniform magnetic field $\vec{\mathbf{B}}$ with the plane of the loop parallel to the magnetic field (see the drawing). The dashed line is the axis of rotation. The magnetic field exerts _______. (a) a net force and a net torque on the loop (b) a net force, but not a net torque, on the loop (c) a net torque, but not a net force, on the loop (d) neither a net force nor a net torque on the loop

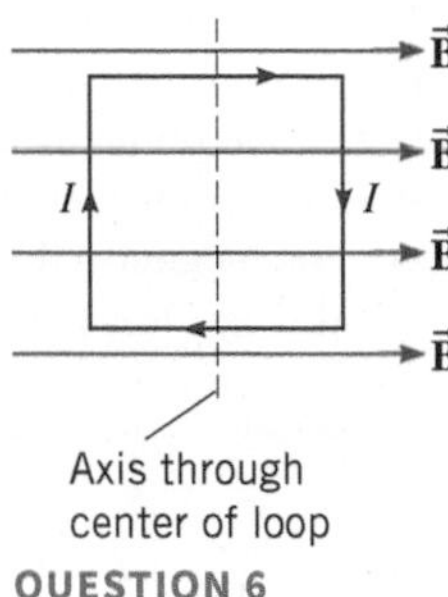

QUESTION 6

Section 21.7 Magnetic Fields Produced by Currents

7. The drawing shows four situations in which two very long wires are carrying the same current, although the directions of the currents may be different. The point P in the drawings is equidistant from each wire. Which one (or more) of these situations gives rise to a zero net magnetic field at P? (a) 2 and 4 (b) Only 1 (c) Only 2 (d) 2 and 3 (e) 3 and 4

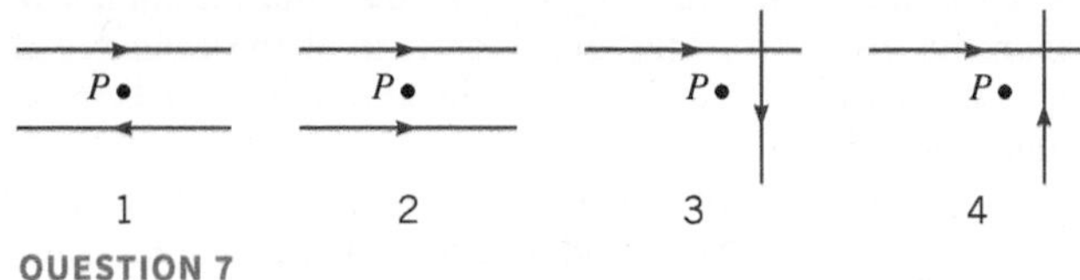

QUESTION 7

8. Three long, straight wires are carrying currents that have the same magnitude. In C the current is opposite to the current in A and B. The wires are equally spaced. Each wire experiences a net force due to the other two wires. Which wire experiences a net force with the greatest magnitude? (a) A (b) B (c) C (d) All three wires experience a net force that has the same magnitude.

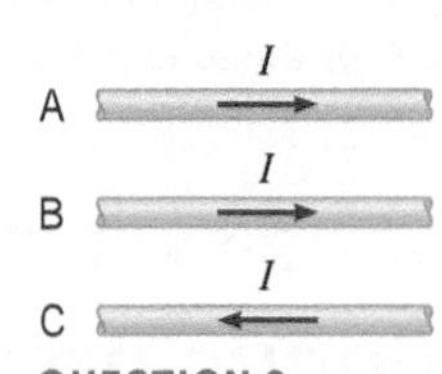

QUESTION 8

Problems

Online

Additional questions are available for assignment in WileyPLUS.

- SSM Student Solutions Manual
- MMH Problem-solving help
- GO Guided Online Tutorial
- V-HINT Video Hints
- CHALK Chalkboard Videos
- BIO Biomedical application
- E Easy
- M Medium
- H Hard
- WS Worksheet
- T Team Problem

Section 21.1 Magnetic Fields,

Section 21.2 The Force That a Magnetic Field Exerts on a Moving Charge

1. E SSM In New England, the horizontal component of the earth's magnetic field has a magnitude of 1.6×10^{-5} T. An electron is shot vertically straight up from the ground with a speed of 2.1×10^6 m/s. What is the magnitude of the acceleration caused by the magnetic force? Ignore the gravitational force acting on the electron.

2. E (a) A proton, traveling with a velocity of 4.5×10^6 m/s due east, experiences a magnetic force that has a maximum magnitude of 8.0×10^{-14} N and a direction of due south. What are the magnitude and direction of the magnetic field causing the force? (b) Repeat part (a) assuming the proton is replaced by an electron.

3. E SSM At a certain location, the horizontal component of the earth's magnetic field is 2.5×10^{-5} T, due north. A proton moves eastward with just the right speed for the magnetic force on it to balance its weight. Find the speed of the proton.

4. E A charge of -8.3 μC is traveling at a speed of 7.4×10^6 m/s in a region of space where there is a magnetic field. The angle between the velocity of the charge and the field is 52°. A force of magnitude 5.4×10^{-3} N acts on the charge. What is the magnitude of the magnetic field?

5. E When a charged particle moves at an angle of 25° with respect to a magnetic field, it experiences a magnetic force of magnitude F. At what angle (less than 90°) with respect to this field will this particle, moving at the same speed, experience a magnetic force of magnitude $2F$?

6. E GO A particle that has an 8.2 μC charge moves with a velocity of magnitude 5.0×10^5 m/s along the $+x$ axis. It experiences no magnetic force, although there is a magnetic field present. The maximum possible magnetic force that the charge could experience has a magnitude of 0.48 N. Find the magnitude and direction of the magnetic field. Note that there are two possible answers for the direction of the field.

7. E A magnetic field has a magnitude of 1.2×10^{-3} T, and an electric field has a magnitude of 4.6×10^3 N/C. Both fields point in the same direction. A positive 1.8 μC charge moves at a speed of 3.1×10^6 m/s in a direction that is perpendicular to both fields. Determine the magnitude of the net force that acts on the charge.

8. E GO Two charged particles move in the same direction with respect to the same magnetic field. Particle 1 travels three times faster than particle 2. However, each particle experiences a magnetic force of the same magnitude. Find the ratio $|q_1|/|q_2|$ of the magnitudes of the charges.

9. **M** **V-HINT** The drawing shows a parallel plate capacitor that is moving with a speed of 32 m/s through a 3.6-T magnetic field. The velocity $\vec{v}$ is perpendicular to the magnetic field. The electric field within the capacitor has a value of 170 N/C, and each plate has an area of 7.5×10^{-4} m^2. What is the magnetic force (magnitude and direction) exerted on the positive plate of the capacitor?

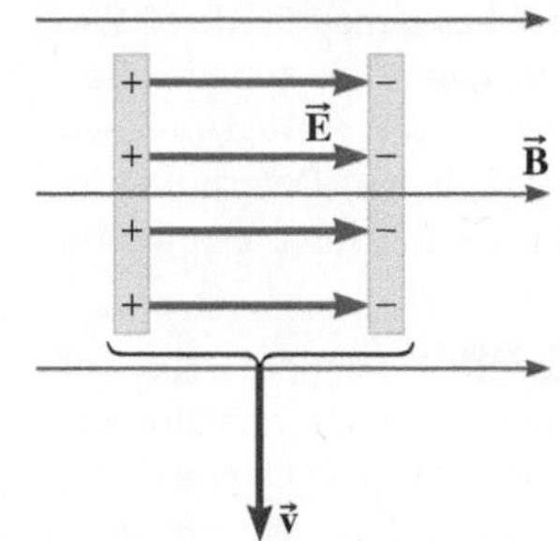

PROBLEM 9

10. **M** **V-HINT** One component of a magnetic field has a magnitude of 0.048 T and points along the $+x$ axis, while the other component has a magnitude of 0.065 T and points along the $-y$ axis. A particle carrying a charge of $+2.0 \times 10^{-5}$ C is moving along the $+z$ axis at a speed of 4.2×10^3 m/s. (a) Find the magnitude of the net magnetic force that acts on the particle. (b) Determine the angle that the net force makes with respect to the $+x$ axis.

11. **M** **SSM** The electrons in the beam of a television tube have a kinetic energy of 2.40×10^{-15} J. Initially, the electrons move horizontally from west to east. The vertical component of the earth's magnetic field points down, toward the surface of the earth, and has a magnitude of 2.00×10^{-5} T. **(a)** In what direction are the electrons deflected by this field component? **(b)** What is the acceleration of an electron in part (a)?

Section 21.3 The Motion of a Charged Particle in a Magnetic Field,

Section 21.4 The Mass Spectrometer

12. **E** An ionized helium atom has a mass of 6.6×10^{-27} kg and a speed of 4.4×10^5 m/s. It moves perpendicular to a 0.75-T magnetic field on a circular path that has a 0.012-m radius. Determine whether the charge of the ionized atom is $+e$ or $+2e$.

13. **E** **GO** A charged particle with a charge-to-mass ratio of $|q|/m = 5.7 \times 10^8$ C/kg travels on a circular path that is perpendicular to a magnetic field whose magnitude is 0.72 T. How much time does it take for the particle to complete one revolution?

14. **E** A charged particle enters a uniform magnetic field and follows the circular path shown in the drawing. (a) Is the particle positively or negatively charged? Why? (b) The particle's speed is 140 m/s, the magnitude of the magnetic field is 0.48 T, and the radius of the path is 960 m. Determine the mass of the particle, given that its charge has a magnitude of 8.2×10^{-4} C.

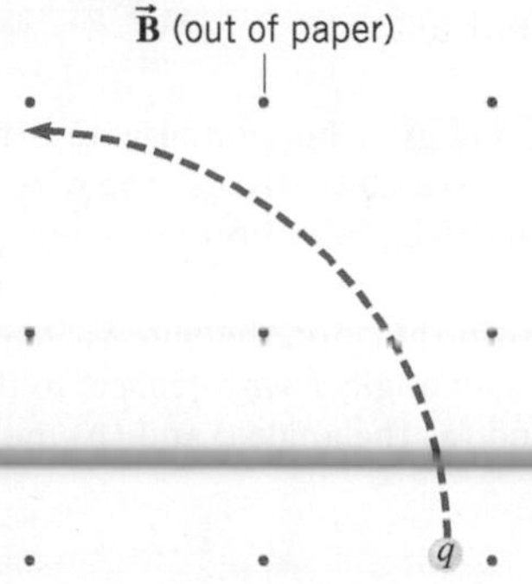

PROBLEM 14

15. **E** **GO** A proton is projected perpendicularly into a magnetic field that has a magnitude of 0.50 T. The field is then adjusted so that an electron will follow a circular path of the same radius when it is projected perpendicularly into the field with the same velocity that the proton had. What is the magnitude of the field used for the electron?

16. **E** **SSM** When beryllium-7 ions ($m = 11.65 \times 10^{-27}$ kg) pass through a mass spectrometer, a uniform magnetic field of 0.283 T curves their path directly to the center of the detector (see **Figure 21.14**). For the same accelerating potential difference, what magnetic field should be used to send beryllium-10 ions ($m = 16.63 \times 10^{-27}$ kg) to the same location in the detector? Both types of ions are singly ionized ($q = +e$).

17. **E** **GO** Suppose that an ion source in a mass spectrometer produces *doubly* ionized gold ions (Au^{2+}), each with a mass of 3.27×10^{-25} kg. The ions are accelerated from rest through a potential difference of 1.00 kV. Then, a 0.500-T magnetic field causes the ions to follow a circular path. Determine the radius of the path.

18. **E** An α-particle has a charge of $+2e$ and a mass of 6.64×10^{-27} kg. It is accelerated from rest through a potential difference that has a value of 1.20×10^6 V and then enters a uniform magnetic field whose magnitude is 2.20 T. The α-particle moves perpendicular to the magnetic field at all times. What is (a) the speed of the α-particle, (b) the magnitude of the magnetic force on it, and (c) the radius of its circular path?

19. **E** **GO** Particle 1 and particle 2 have masses of $m_1 = 2.3 \times 10^{-8}$ kg and $m_2 = 5.9 \times 10^{-8}$ kg, but they carry the same charge q. The two particles accelerate from rest through the same electric potential difference V and enter the same magnetic field, which has a magnitude B. The particles travel perpendicular to the magnetic field on circular paths. The radius of the circular path for particle 1 is $r_1 = 12$ cm. What is the radius (in cm) of the circular path for particle 2?

20. **E** Two of the isotopes of carbon, carbon-12 and carbon-13, have masses of 19.93×10^{-27} kg and 21.59×10^{-27} kg, respectively. These two isotopes are singly ionized ($+e$), each given a speed of 6.667×10^5 m/s. The ions then enter the bending region of a mass spectrometer where the magnetic field is 0.8500 T. Determine the spatial separation between the two isotopes after they have traveled through a half-circle.

21. **M** **V-HINT** The ion source in a mass spectrometer produces both singly and doubly ionized species, X^+ and X^{2+}. The difference in mass between these species is too small to be detected. Both species are accelerated through the same electric potential difference, and both experience the same magnetic field, which causes them to move on circular paths. The radius of the path for the species X^+ is r_1, while the radius for species X^{2+} is r_2. Find the ratio r_1/r_2 of the radii.

22. **M** **SSM** **MMH** A proton with a speed of 3.5×10^6 m/s is shot into a region between two plates that are separated by a distance of 0.23 m. As the drawing shows, a magnetic field exists between the plates, and it is perpendicular to the velocity of the proton. What must be the magnitude of the magnetic field so the proton just misses colliding with the opposite plate?

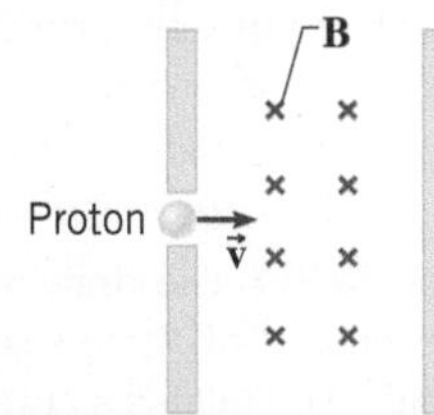

PROBLEM 22

23. **M** **V-HINT** Review Conceptual Example 2 as an aid in understanding this problem. A velocity selector has an electric field of magnitude 2470 N/C, directed vertically upward, and a horizontal magnetic field that is directed south. Charged particles, traveling east at a speed of 6.50×10^3 m/s, enter the velocity selector and are able to pass completely through without being deflected. When a different particle with an electric charge of $+4.00 \times 10^{-12}$ C enters the velocity selector traveling east, the net force (due to the electric and magnetic fields) acting on it is 1.90×10^{-9} N, pointing directly upward. What is the speed of this particle?

24. M SSM A particle of mass 6.0×10^{-8} kg and charge $+7.2\ \mu C$ is traveling due east. It enters perpendicularly a magnetic field whose magnitude is 3.0 T. After entering the field, the particle completes one-half of a circle and exits the field traveling due west. How much time does the particle spend traveling in the magnetic field?

25. M V-HINT Conceptual Example 4 provides background pertinent to this problem. An electron has a kinetic energy of 2.0×10^{-17} J. It moves on a circular path that is perpendicular to a uniform magnetic field of magnitude 5.3×10^{-5} T. Determine the radius of the path.

26. M CHALK A positively charged particle of mass 7.2×10^{-8} kg is traveling due east with a speed of 85 m/s and enters a 0.31-T uniform magnetic field. The particle moves through one-quarter of a circle in a time of 2.2×10^{-3} s, at which time it leaves the field heading due south. All during the motion the particle moves perpendicular to the magnetic field. (a) What is the magnitude of the magnetic force acting on the particle? (b) Determine the magnitude of its charge.

27. M Review Conceptual Example 2 as background for this problem. A charged particle moves through a velocity selector at a constant speed in a straight line. The electric field of the velocity selector is 3.80×10^{3} N/C, while the magnetic field is 0.360 T. When the electric field is turned off, the charged particle travels on a circular path whose radius is 4.30 cm. Find the charge-to-mass ratio of the particle.

Section 21.5 The Force on a Current in a Magnetic Field

28. E At New York City, the earth's magnetic field has a vertical component of 5.2×10^{-5} T that points downward (perpendicular to the ground) and a horizontal component of 1.8×10^{-5} T that points toward geographic north (parallel to the ground). What are the magnitude and direction of the magnetic force on a 6.0-m long, straight wire that carries a current of 28 A perpendicularly into the ground?

29. E SSM A 45-m length of wire is stretched horizontally between two vertical posts. The wire carries a current of 75 A and experiences a magnetic force of 0.15 N. Find the magnitude of the earth's magnetic field at the location of the wire, assuming the field makes an angle of 60.0° with respect to the wire.

30. E A straight wire in a magnetic field experiences a force of 0.030 N when the current in the wire is 2.7 A. The current in the wire is changed, and the wire experiences a force of 0.047 N as a result. What is the new current?

31. E A horizontal wire of length 0.53 m, carrying a current of 7.5 A, is placed in a uniform external magnetic field. When the wire is horizontal, it experiences no magnetic force. When the wire is tilted upward at an angle of 19°, it experiences a magnetic force of 4.4×10^{-3} N. Determine the magnitude of the external magnetic field.

32. E GO The drawing shows a wire composed of three segments, AB, BC, and CD. There is a current of $I = 2.8$ A in the wire. There is also a magnetic field $\vec{B}$ (magnitude = 0.26 T) that is the same everywhere and points in the direction of the $+z$ axis. The lengths of the wire segments are $L_{AB} = 1.1$ m, $L_{BC} = 0.55$ m, and $L_{CD} = 0.55$ m. Find the magnitude of the force that acts on each segment.

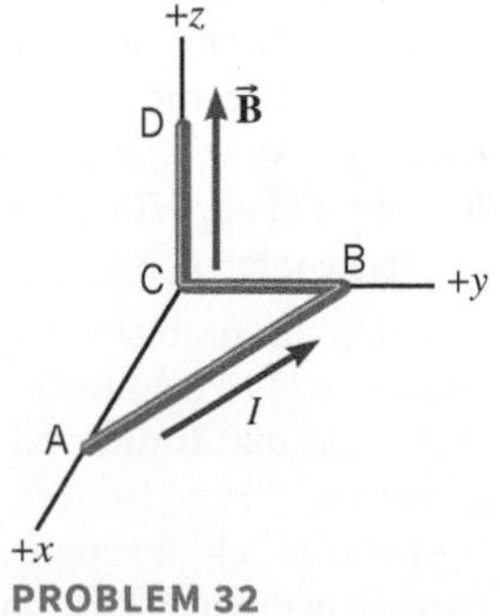

PROBLEM 32

33. E SSM A wire carries a current of 0.66 A. This wire makes an angle of 58° with respect to a magnetic field of magnitude 4.7×10^{-5} T. The wire experiences a magnetic force of magnitude 7.1×10^{-5} N. What is the length of the wire?

34. E A loop of wire has the shape of a right triangle (see the drawing) and carries a current of $I = 4.70$ A. A uniform magnetic field is directed parallel to side AB and has a magnitude of 1.80 T. (a) Find the magnitude and direction of the magnetic force exerted on each side of the triangle. (b) Determine the magnitude of the net force exerted on the triangle.

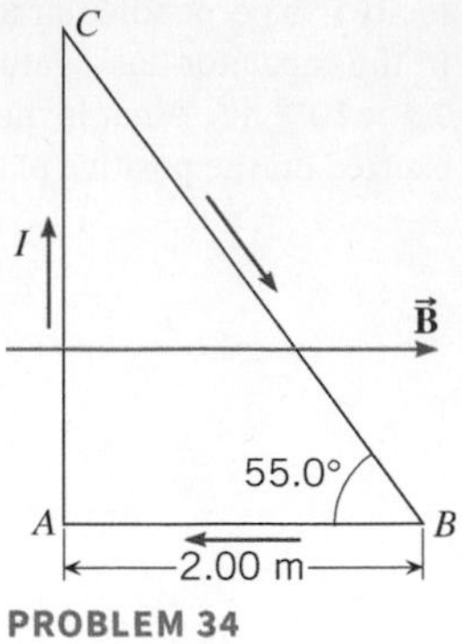

PROBLEM 34

35. M CHALK GO A copper rod of length 0.85 m is lying on a frictionless table (see the drawing). Each end of the rod is attached to a fixed wire by an unstretched spring that has a spring constant of $k = 75$ N/m. A magnetic field with a strength of 0.16 T is oriented perpendicular to the surface of the table. (a) What must be the direction of the current in the copper rod that causes the springs to stretch? (b) If the current is 12 A, by how much does each spring stretch?

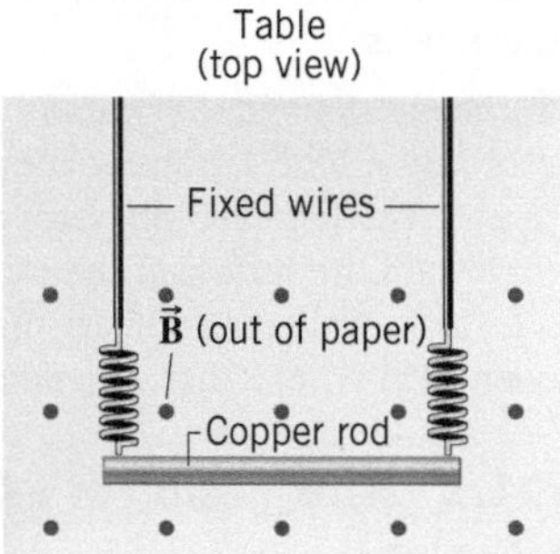

PROBLEM 35

36. M SSM MMH The drawing shows a thin, uniform rod that has a length of 0.45 m and a mass of 0.094 kg. This rod lies in the plane of the paper and is attached to the floor by a hinge at point P. A uniform magnetic field of 0.36 T is directed perpendicularly into the plane of the paper. There is a current $I = 4.1$ A in the rod, which does not rotate clockwise or counterclockwise. Find the angle θ. (*Hint:* The magnetic force may be taken to act at the center of gravity.)

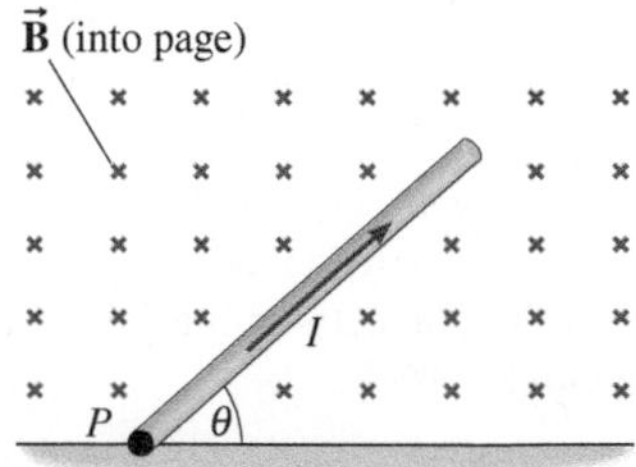

PROBLEM 36

37. M CHALK GO A horizontal wire is hung from the ceiling of a room by two massless strings. The wire has a length of 0.20 m and a mass of 0.080 kg. A uniform magnetic field of magnitude 0.070 T is directed from the ceiling to the floor. When a current of $I = 42$ A exists in the wire, the wire swings upward and, at equilibrium, makes an angle ϕ with respect to the vertical, as the drawing shows. Find (a) the angle ϕ and (b) the tension in each of the two strings.

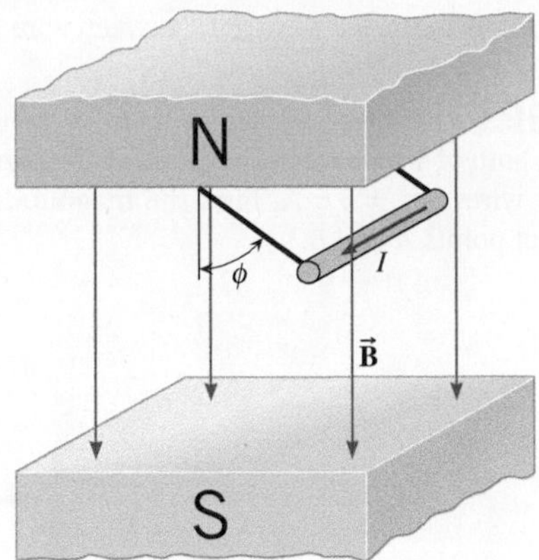

PROBLEM 37

38. H The two conducting rails in the drawing are tilted upward so they each make an angle of 30.0° with respect to the ground. The vertical magnetic field has a magnitude of 0.050 T. The 0.20-kg aluminum rod (length = 1.6 m) slides *without friction* down the rails at a constant velocity. How much current flows through the rod?

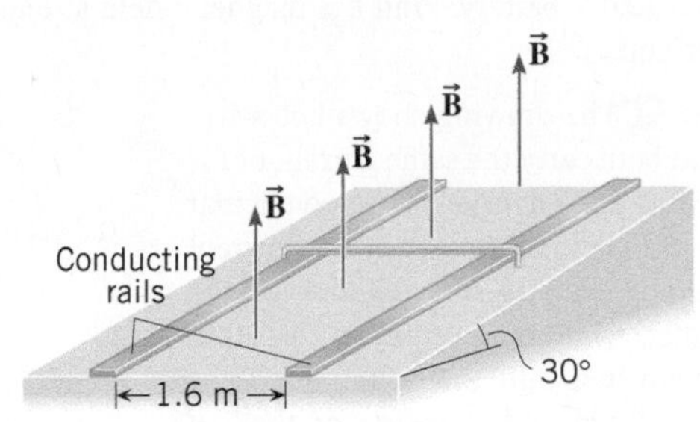

PROBLEM 38

Section 21.6 The Torque on a Current-Carrying Coil

39. E Two coils have the same number of circular turns and carry the same current. Each rotates in a magnetic field as in **Figure 21.19**. Coil 1 has a radius of 5.0 cm and rotates in a 0.18-T field. Coil 2 rotates in a 0.42-T field. Each coil experiences the same maximum torque. What is the radius (in cm) of coil 2?

40. E The 1200-turn coil in a dc motor has an area per turn of 1.1×10^{-2} m^2. The design for the motor specifies that the magnitude of the maximum torque is 5.8 N · m when the coil is placed in a 0.20-T magnetic field. What is the current in the coil?

41. E Two circular coils of current-carrying wire have the same magnetic moment. The first coil has a radius of 0.088 m, has 140 turns, and carries a current of 4.2 A. The second coil has 170 turns and carries a current of 9.5 A. What is the radius of the second coil?

42. E A wire has a length of 7.00×10^{-2} m and is used to make a circular coil of one turn. There is a current of 4.30 A in the wire. In the presence of a 2.50-T magnetic field, what is the maximum torque that this coil can experience?

43. E GO The coil of wire in the drawing is a right triangle and is free to rotate about an axis that is attached along side *AC*. The current in the loop is $I = 4.70$ A, and the magnetic field (parallel to the plane of the loop and side *AB*) is $B = 1.80$ T. **(a)** What is the magnetic moment of the loop, and **(b)** what is the magnitude of the net torque exerted on the loop by the magnetic field?

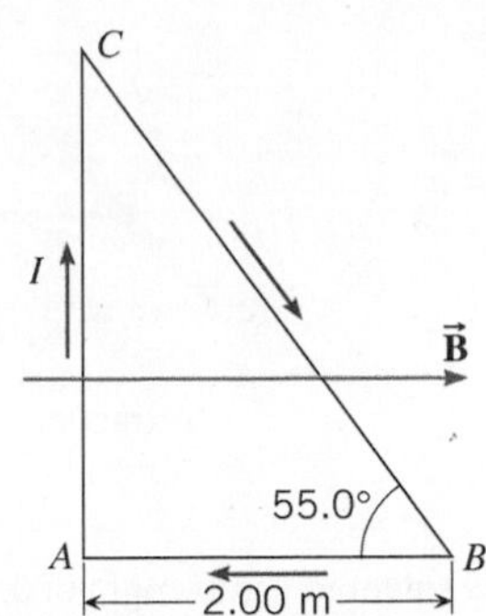

PROBLEM 43

44. E GO You have a wire of length $L = 1.00$ m from which to make the square coil of a dc motor. The current in the coil is $I = 1.7$ A, and the magnetic field of the motor has a magnitude of $B = 0.34$ T. Find the maximum torque exerted on the coil when the wire is used to make a single-turn square coil and a two-turn square coil.

45. E SSM The rectangular loop in the drawing consists of 75 turns and carries a current of $I = 4.4$ A. A 1.8-T magnetic field is directed along the $+y$ axis. The loop is free to rotate about the z axis. **(a)** Determine the magnitude of the net torque exerted on the loop and **(b)** state whether the 35° angle will increase or decrease.

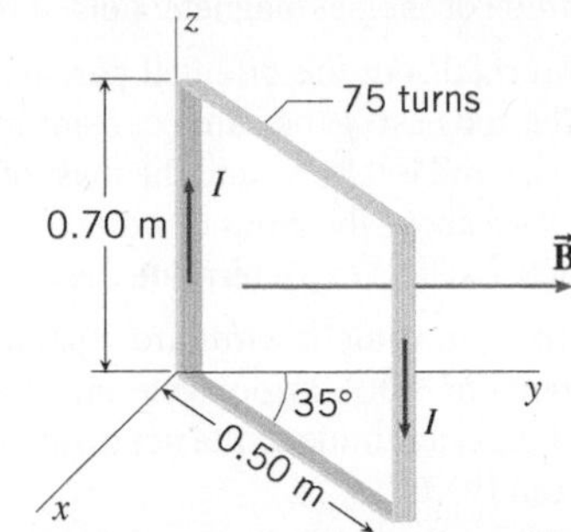

PROBLEM 45

46. M A square coil and a rectangular coil are each made from the same length of wire. Each contains a single turn. The long sides of the rectangle are twice as long as the short sides. Find the ratio $\tau_{square}/\tau_{rectangle}$ of the maximum torques that these coils experience in the same magnetic field when they contain the same current.

Section 21.7 Magnetic Fields Produced by Currents

47. E SSM Suppose in the figure that $I_1 = I_2 = 25$ A and that the separation between the wires is 0.016 m. By applying an external magnetic field (created by a source other than the wires) it is possible to cancel the mutual repulsion of the wires. This external field must point along the vertical direction. **(a)** Does the external field point up or down? Explain. **(b)** What is the magnitude of the external field?

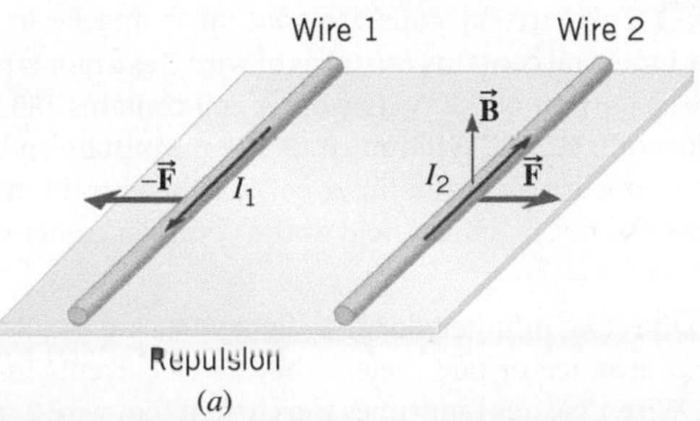

PROBLEM 47

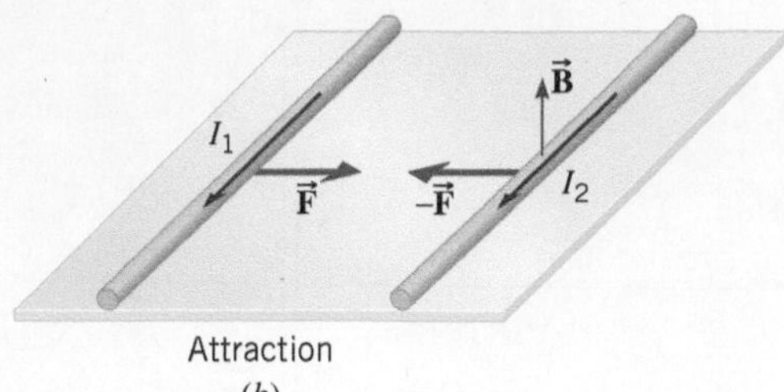

PROBLEM 47

48. E A long solenoid has a length of 0.65 m and contains 1400 turns of wire. There is a current of 4.7 A in the wire. What is the magnitude of the magnetic field within the solenoid?

49. E GO A long solenoid has 1400 turns per meter of length, and it carries a current of 3.5 A. A small circular coil of wire is placed inside the solenoid with the normal to the coil oriented at an angle of 90.0° with respect to the axis of the solenoid. The coil consists of 50 turns, has an area of 1.2×10^{-3} m^2, and carries a current of 0.50 A. Find the torque exerted on the coil.

50. E SSM Two circular loops of wire, each containing a single turn, have the same radius of 4.0 cm and a common center. The planes of the loops are perpendicular. Each carries a current of 1.7 A. What is the magnitude of the net magnetic field at the common center?

51. E Two rigid rods are oriented parallel to each other and to the ground. The rods carry the same current in the same direction. The length of each rod is 0.85 m, and the mass of each is 0.073 kg. One rod is held in place above the ground, while the other floats beneath it at a distance of 8.2×10^{-3} m. Determine the current in the rods.

52. E Two long, straight wires are separated by 0.120 m. The wires carry currents of 8.0 A in opposite directions, as the drawing indicates. Find the magnitude of the net magnetic field at the points labeled (a) A and (b) B.

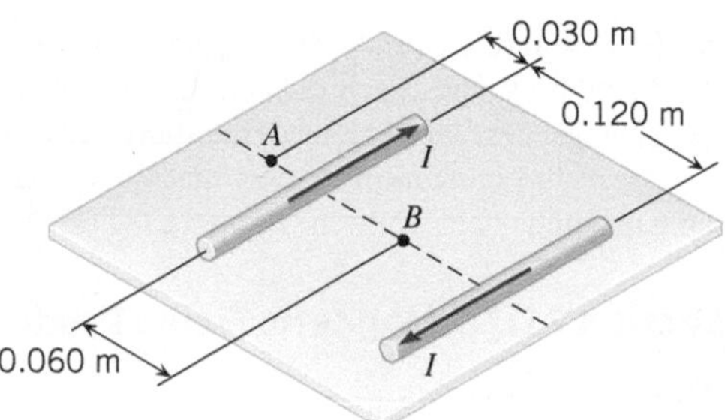

PROBLEM 52

53. E GO A long, straight wire carrying a current of 305 A is placed in a uniform magnetic field that has a magnitude of 7.00×10^{-3} T. The wire is perpendicular to the field. Find a point in space where the net magnetic field is zero. Locate this point by specifying its perpendicular distance from the wire.

54. M MMH Two circular coils are concentric and lie in the same plane. The inner coil contains 140 turns of wire, has a radius of 0.015 m, and carries a current of 7.2 A. The outer coil contains 180 turns and has a radius of 0.023 m. What must be the magnitude and direction (relative to the current in the inner coil) of the current in the outer coil, so that the net magnetic field at the common center of the two coils is zero?

55. M V-HINT Two infinitely long, straight wires are parallel and separated by a distance of one meter. They carry currents in the same direction. Wire 1 carries four times the current that wire 2 carries. On a line drawn perpendicular to both wires, locate the spot (relative to wire 1) where the net magnetic field is zero. Assume that wire 1 lies to the left of wire 2 and note that there are three regions to consider on this line: to the left of wire 1, between wire 1 and wire 2, and to the right of wire 2.

56. M GO The drawing shows two perpendicular, long, straight wires, both of which lie in the plane of the paper. The current in each of the wires is $I = 5.6$ A. Find the magnitudes of the net magnetic fields at points A and B.

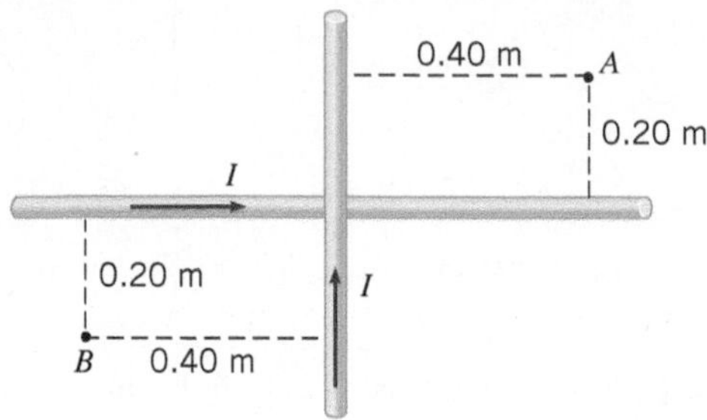

PROBLEM 56

57. M SSM A piece of copper wire has a resistance per unit length of 5.90×10^{-3} Ω/m. The wire is wound into a thin, flat coil of many turns that has a radius of 0.140 m. The ends of the wire are connected to a 12.0-V battery. Find the magnetic field strength at the center of the coil.

58. H The drawing shows two wires that both carry the same current of $I = 85.0$ A and are oriented perpendicular to the plane of the paper. The current in one wire is directed out of the paper, while the current in the other is directed into the paper. Find the magnitude and direction of the net magnetic field at point P.

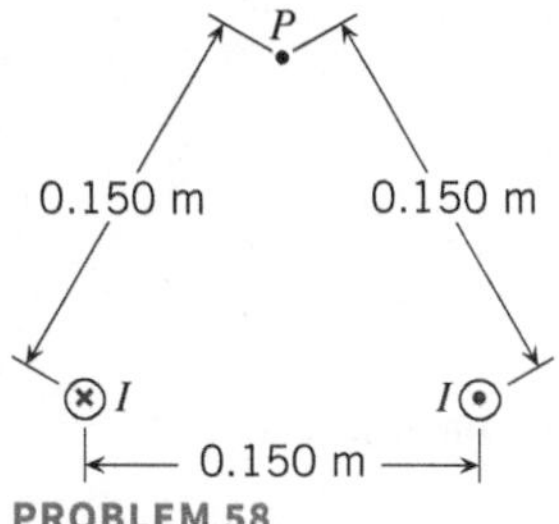

PROBLEM 58

59. H The drawing shows two long, straight wires that are suspended from a ceiling. The mass per unit length of each wire is 0.050 kg/m. Each of the four strings suspending the wires has a length of 1.2 m. When the wires carry identical currents in opposite directions, the angle between the strings holding the two wires is 15°. What is the current in each wire?

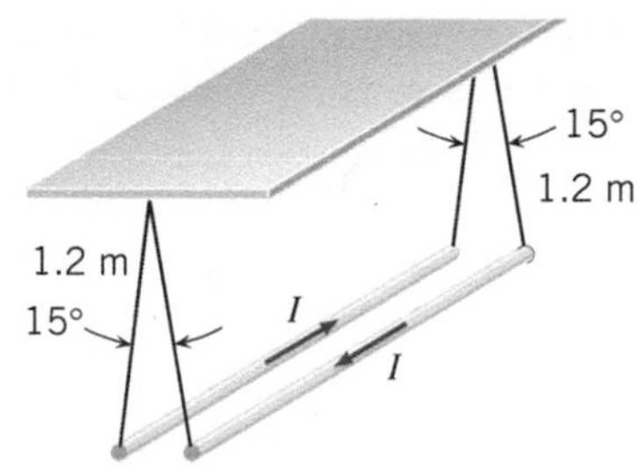

PROBLEM 59

Section 21.8 Ampère's Law

60. E Suppose that a uniform magnetic field is everywhere perpendicular to this page. The field points directly upward toward you. A circular path is drawn on the page. Use Ampère's law to show that there can be no net current passing through the circular surface.

61. E V-HINT The wire in Figure 21.37 carries a current of 12 A. Suppose that a second long, straight wire is placed right next to this wire. The current in the second wire is 28 A. Use Ampère's law to find the magnitude of the magnetic field at a distance of $r = 0.72$ m from the wires when the currents are (a) in the same direction and (b) in opposite directions.

Additional Problems

Online

62. **E** In a certain region, the earth's magnetic field has a magnitude of 5.4×10^{-5} T and is directed north at an angle of 58° below the horizontal. An electrically charged bullet is fired north and 11° above the horizontal, with a speed of 670 m/s. The magnetic force on the bullet is 2.8×10^{-10} N, directed due east. Determine the bullet's electric charge, including its algebraic sign (+ or −).

63. **E SSM** An electron is moving through a magnetic field whose magnitude is 8.70×10^{-4} T. The electron experiences only a magnetic force and has an acceleration of magnitude 3.50×10^{14} m/s². At a certain instant, it has a speed of 6.80×10^{6} m/s. Determine the angle θ (less than 90°) between the electron's velocity and the magnetic field.

64. **E GO** A very long, straight wire carries a current of 0.12 A. This wire is tangent to a single-turn, circular wire loop that also carries a current. The directions of the currents are such that the net magnetic field at the center of the loop is zero. Both wires are insulated and have diameters that can be neglected. How much current is there in the loop?

65. **E SSM** The maximum torque experienced by a coil in a 0.75-T magnetic field is 8.4×10^{-4} N · m. The coil is circular and consists of only one turn. The current in the coil is 3.7 A. What is the length of the wire from which the coil is made?

66. **E** Multiple-Concept Example 7 discusses how problems like this one can be solved. A +6.00 μC charge is moving with a speed of 7.50×10^4 m/s parallel to a very long, straight wire. The wire is 5.00 cm from the charge and carries a current of 67.0 A in a direction opposite to that of the moving charge. Find the magnitude and direction of the force on the charge.

67. **E V-HINT** The x, y, and z components of a magnetic field are B_x = 0.10 T, B_y = 0.15 T, and B_z = 0.17 T. A 25-cm wire is oriented along the z axis and carries a current of 4.3 A. What is the magnitude of the magnetic force that acts on this wire?

68. **E** In a lightning bolt, a large amount of charge flows during a time of 1.8×10^{-3} s. Assume that the bolt can be treated as a long, straight line of current. At a perpendicular distance of 27 m from the bolt, a magnetic field of 8.0×10^{-5} T is measured. How much charge has flowed during the lightning bolt? Ignore the earth's magnetic field.

69. **E** A charge is moving perpendicular to a magnetic field and experiences a force whose magnitude is 2.7×10^{-3} N. If this same charge were to move at the same speed and the angle between its velocity and the same magnetic field were 38°, what would be the magnitude of the magnetic force that the charge would experience?

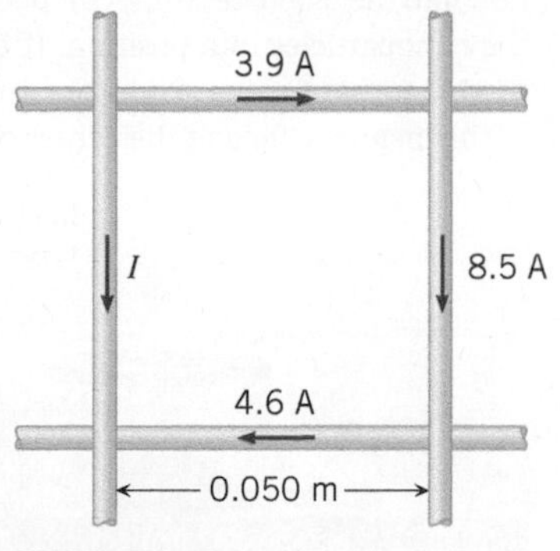

PROBLEM 70

70. **E GO** The drawing shows four insulated wires overlapping one another, forming a square with 0.050-m sides. All four wires are much longer than the sides of the square. The net magnetic field at the center of the square is 61 μT. Calculate the current I.

71. **M GO** A particle has a charge of q = +5.60 μC and is located at the coordinate origin. As the drawing shows, an electric field of E_x = +245 N/C exists along the +x axis. A magnetic field also exists, and its x and y components are B_x = +1.80 T and B_y = +1.40 T. Calculate the force (magnitude and direction) exerted on the particle by each of the three fields when it is **(a)** stationary, **(b)** moving along the +x axis at a speed of 375 m/s, and **(c)** moving along the +z axis at a speed of 375 m/s.

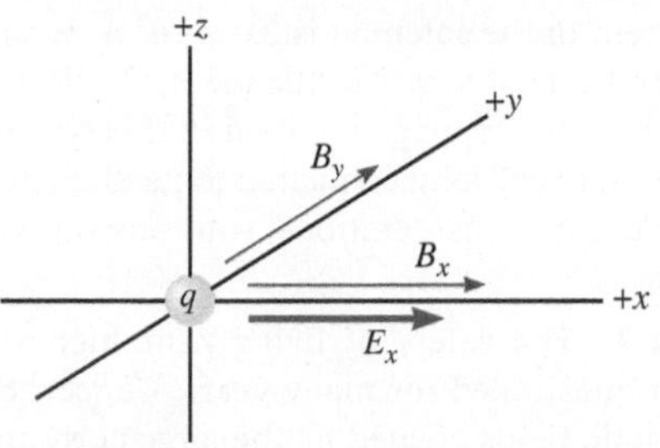

PROBLEM 71

72. **M GO** The figure shows two parallel, straight wires that are very long. The wires are separated by a distance of 0.065 m and carry currents of I_1 = 15 A and I_2 = 7.0 A. Find the magnitude and direction of the force that the magnetic field of wire 1 applies to a 1.5-m section of wire 2 when the currents have **(a)** opposite and **(b)** the same directions.

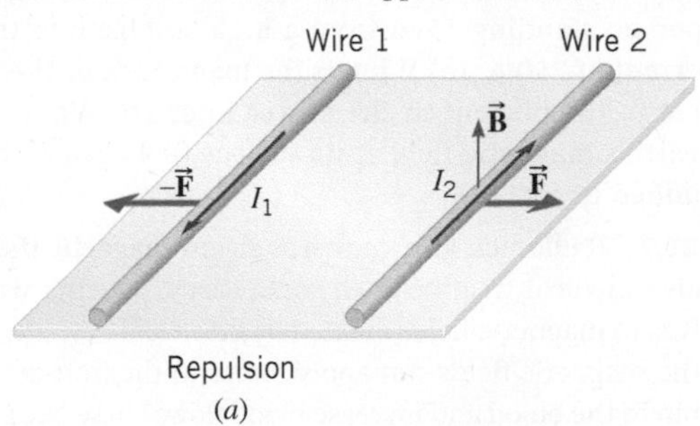

(a)

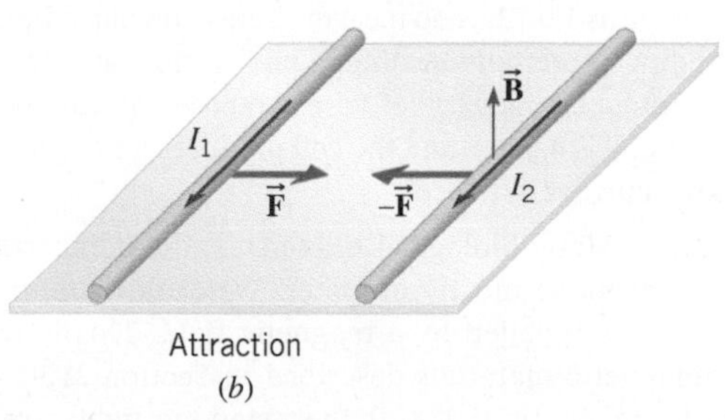

(b)

PROBLEM 72

Physics in Biology, Medicine, and Sports

73. **E BIO SSM 21.4** In the operating room, anesthesiologists use mass spectrometers to monitor the respiratory gases of patients undergoing surgery. One gas that is often monitored is the anesthetic isoflurane (molecular mass = 3.06×10^{-25} kg). In a spectrometer, a singly ionized molecule of isoflurane (charge = $+e$) moves at a speed of 7.2×10^3 m/s on a circular path that has a radius of 0.10 m. What is the magnitude of the magnetic field that the spectrometer uses?

74. **E BIO SSM 21.7** The magnetic field produced by the solenoid in a magnetic resonance imaging (MRI) system designed for measurements on whole human bodies has a field strength of 7.0 T, and the current in the solenoid is 2.0×10^2 A. What is the number of turns per meter of length of the solenoid? Note that the solenoid used to produce the magnetic field in this type of system has a length that is not very long compared to its diameter. Because of this and other design considerations, your answer will be only an approximation.

75. **M BIO 21.7** The safety of living near high voltage power lines has been questioned for many years. Concerned groups believe the magnetic fields created by the large currents in the lines can lead to an increased risk of cancer and other health problems. However, multiple studies since 1995 by the American Physical Society and the National Academy of Sciences (National Research Council) have concluded that the relationship between higher rates of cancer and the proximity to high voltage power lines is not scientifically substantiated. Current research suggests that exposure to these fields does not represent a human health hazard. Consider a person standing 55 m from a high voltage line that is carrying a current of 250 A. **(a)** What is the magnitude of the magnetic field due to the current in the power line? **(b)** What is the ratio of the earth's magnetic field at its surface (0.45 gauss) to the field you obtained in part (a)?

76. **M BIO 21.7** Radionics, also known as electromagnetic therapy, is an alternative medical treatment. In some cases, patients will expose themselves to magnetic fields created by electrical devices. They believe that the magnetic fields can apply forces to the iron-containing hemoglobin in the blood and increase blood flow. These claims are unproven, and no health benefits have ever been established. In fact, even a field as large as 1.0 T has no measured effect on blood hemoglobin. In an attempt to promote healing, a professional athlete inserts a broken wrist into a circular coil of wire composed of 5200 turns. If the radius of the coil is 4.5 cm, and the coil produces a 1.0-T magnetic field, what is the current in the coil?

77. **M BIO 21.7** Many biological organisms, including you and your authors, are made mostly of water. Water is a *diamagnetic* substance, which is repelled by a magnetic field. This is in contrast to ferromagnetic materials described in Section 21.9, which are attracted by magnetic fields. If the magnetic fields are large enough, and the mass of the object is small, the repulsive force can cause the object to levitate (see the photo). The magnetic field at the center of the electromagnet that is levitating the frog is 16 T. These powerful magnets use a stack of thin conducting plates instead of wire windings to form a single spiraling conducting path, as shown in the photo, effectively creating a solenoid. If the magnet contains 3.2 plates/mm of length, what is the current flowing through the plates?

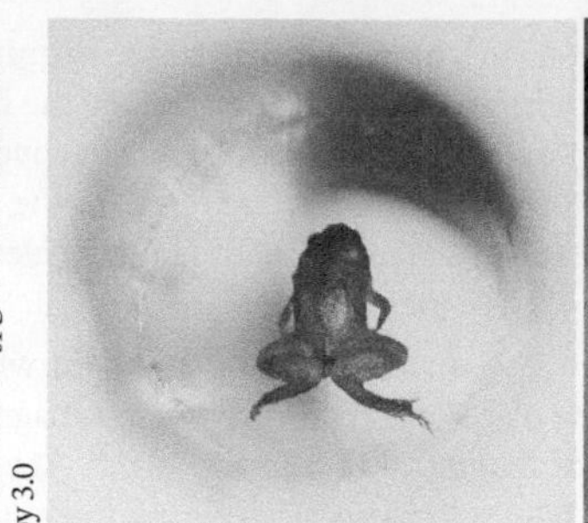

PROBLEM 77

78. **M BIO 21.2** One technique of biological medical imaging that does not require the high magnetic fields utilized in MRI is *magneto-acoustic (MA) imaging*. The sample is placed in a uniform magnetic field, and a low frequency ac current is passed through the tissues of the body, perpendicular to the magnetic field. The current will experience a force due to the field. Since the current is switching direction, the magnetic force switches direction too. Different tissues have different conductivities, and this results in different forces at the boundaries between different types of tissue. This creates ultrasonic waves that can be detected with millimeter resolution. Consider a 0.10-mA current of sodium ions moving as an electrolyte through the extracellular fluid within the tissue. If the current is oriented perpendicular to a 0.15-T magnetic field, what is the magnitude of the force that acts on a 2.0-mm length of the current?

79. **M BIO 21.7** Magnetic drug delivery is one option to target a drug onto a very localized part of the body. For example, if a blood clot, or thrombus, is detected inside a blood vessel, it should be removed in an effort to maintain normal blood flow through the vessel. Novel cancer therapies have also used magnetic drug delivery to treat tumors. Here, nanometer-sized particles (or nanoparticles) of iron oxide, a ferromagnetic material, are coated with functionalized drug components that can be precisely guided to specific areas of the body using magnetic fields. Once in position, the nanoparticles are exposed to a high frequency oscillating magnetic field that heats them, which releases the drug from their surface (see the figure). After injection into the bloodstream, a coil placed above the skin is used to guide the nanoparticles into position. If the 2.00-cm-diameter coil consists of 3750 turns of wire carrying a current of 5.00 A, what is the strength of the magnetic field at the center of the coil?

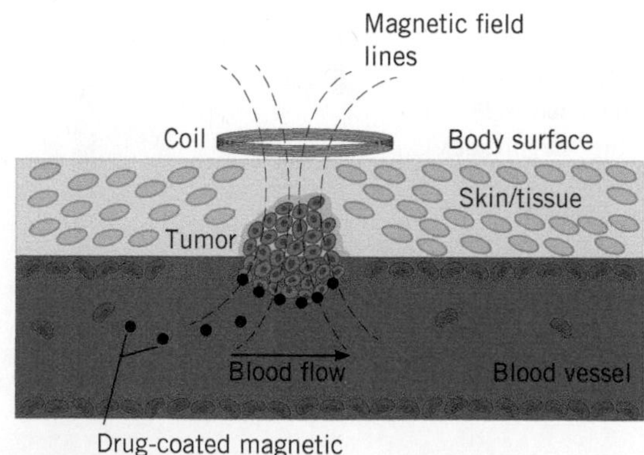

PROBLEM 79

Concepts and Calculations Problems

Online

Both magnetic and electric fields can apply forces to an electric charge. However, there are distinct differences in the way the two types of fields apply their forces. Problem 80 serves to review how the two types of fields behave. A number of forces can act on a rigid object, and some of them can produce torques, as Chapter 9 discusses. If, as a result of the forces and torques, the object has no acceleration of any kind, it is in equilibrium. Problem 81 illustrates how the magnetic force can be one of the forces keeping an object in equilibrium.

80. M CHALK SSM The figure shows a particle that carries a charge of $q_0 = -2.80 \times 10^{-6}$ C. It is moving along the $+y$ axis at a speed of $v = 4.8 \times 10^6$ m/s. A magnetic field $\vec{\mathbf{B}}$ of magnitude 3.35×10^{-5} T is directed along the $+z$ axis, and an electric field $\vec{\mathbf{E}}$ of magnitude 123 N/C points along the $-x$ axis. *Concepts:* (i) What forces make up the net force acting on the particle? (ii) How do you determine the direction of the magnetic force acting on the negative charge? (iii) How do you determine the direction of the electric force acting on the negative charge? (iv) Does the fact that the charge is moving affect the values of the magnetic and electric forces? *Calculations:* Determine the magnitude and direction of the net force that acts on the particle.

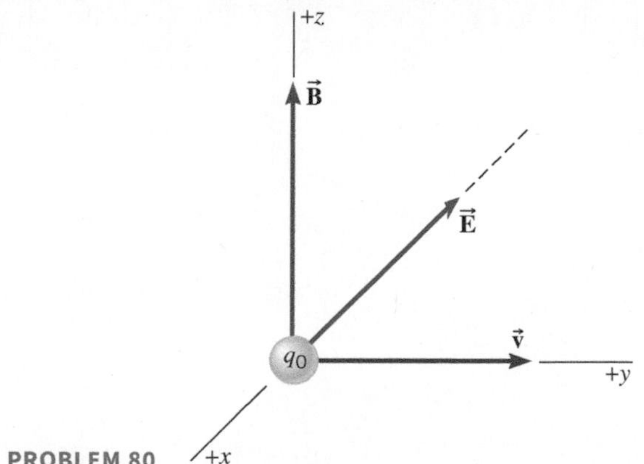

PROBLEM 80

81. M CHALK A 125-turn rectangular coil of wire is hung from one arm of a balance, as the figure shows. With the magnetic field $\vec{\mathbf{B}}$ turned off, an object of mass M is added to the pan on the other arm to balance the weight of the coil. *Concepts:* (i) In a balanced, or equilibrium, condition the device has no angular acceleration. What does this imply about the net torque acting on the device? (ii) What is torque? (iii) In calculating the torques acting on an object in equilibrium, where do you locate the axis of rotation? *Calculations:* When a constant 0.200-T magnetic field is turned on and there is a current of 8.50 A in the coil, how much additional mass m must be added to regain the balance?

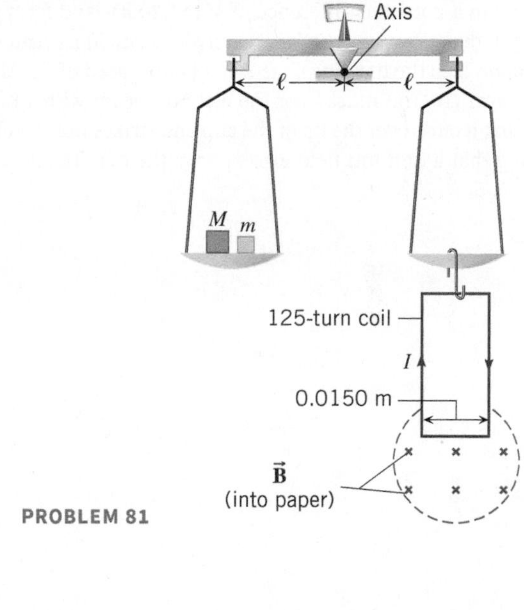

PROBLEM 81

Team Problems

Online

82. E T WS **A Crude Magnetometer.** You and your team have been tasked with estimating the magnetic field between the poles of a giant electromagnet. All you have to use to make this estimate is a power source that provides dc current, a piece of stiff copper wire that is 6.5 cm long, light wires to make connections, and a plastic spring that has an unstrained length of 3.5 cm and a spring constant of $k = 110$ N/m. The poles of the magnet are arranged so that the uniform field between them is horizontal. You attach one end of the spring to a support so that it hangs vertically between the poles of the magnet. You then attach the stiff copper wire to the lower end of the spring so that the wire is horizontal and perpendicular to the field. Finally, you connect the light, flexible wires to each end of the stiff wire and let them hang vertically. You run a current of 2.5 A through the wire. (a) With the magnet turned off, the spring stretches 0.25 cm vertically downward from its equilibrium position. What is the combined weight of the wires? (b) You turn on the magnet and measure the total (stretched) length of the spring to be 5.2 cm. What is the magnitude of the magnetic field?

83. E T WS **Building an Adjustable Electromagnet.** You and your team are tasked with designing a source to provide a uniform magnetic field for an experiment. The field inside a solenoid is uniform, provided that the length of the solenoid is much longer than its diameter. One useful characteristic of an electromagnet of this type is that the field inside can be adjusted anywhere between 0 T and some maximum value by controlling the current. In addition, the direction of the field, which is always pointed along the axis, can be reversed by reversing the current. The solenoid you are designing must be 3.60 cm in diameter and 25.0 cm long, and must generate a maximum magnetic field of magnitude $B = 0.150$ T. You intend to wind a cylinder of the given dimensions with wire that can safely pass a maximum current of 6.50 A. (a) How many windings should the solenoid have in order that the magnetic field at the center of the solenoid is 0.150 T when the current is 6.50 A? (b) What total length of wire is required? (c) What current should you pass through the coil to generate a smaller magnetic field of magnitude 3.50×10^{-2} T at its center, but directed antiparallel to the field generated in part (a)? Assume the current was positive in part (a).

84. M T **An Electron Evaporator.** Electron beams are sometimes used to melt and evaporate metals in order to deposit thin metallic films on surfaces (similar to gold plating). One method is to put the material to be evaporated (called the "target") into a small tungsten cup (a crucible that has a very high melting point) and direct a beam of electrons at the target. Your team has been given the task of designing an electron-beam evaporator. The crucible is a cylinder, 2.0 cm in diameter and 1.5 cm in height, and contains a small target of pure nickel (Ni). The electrons are accelerated through a potential difference of $V = 1.20$ kV, and form a beam that originates below the crucible, exactly 3.70 cm off its center, in the $+x$ direction (see the drawing). (a) What is the speed of the electrons in the beam? (b) You must steer the electron beam with a magnetic field so that it curls over the lip of the cup and strikes the nickel target. Assuming that a uniform field exists above the cup (the field is zero below), what must be the radius of the beam's circular path? (c) In what direction should the field point if the beam initially approaches the cup from the $-y$ axis? (d) What must be the magnitude of the uniform magnetic field?

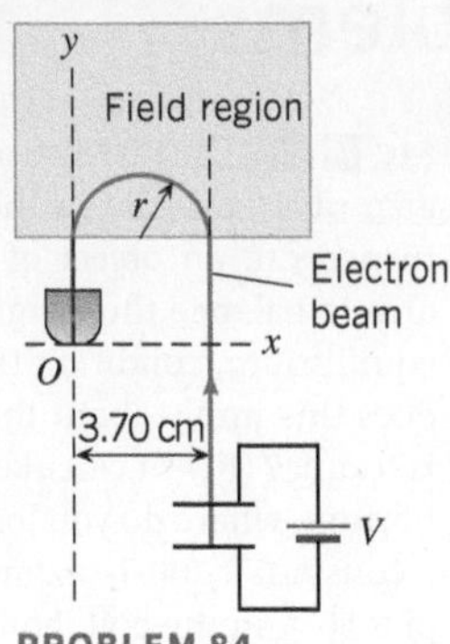

PROBLEM 84

85. M T **An Isotope Separator.** Hydrogen has three isotopes ^{1}H ($m_1 = m_p$), ^{2}H ($m_2 \cong 2m_p$), and ^{3}H ($m_s \cong 3m_p$), where m_p is the mass of a proton (1.67×10^{-27} kg). You and your team are tasked with constructing an isotope separator that will separate a gas of mixed hydrogen isotopes. The gas first passes through a device that atomizes it (i.e., makes sure the atoms are separate, and do not form H_2 molecules), and then ionizes the atoms (strips off their only electron) so that they have a net charge of $+e$. Next, the atoms (now positive ions) are accelerated between the plates of a parallel-plate capacitor with a voltage of 2.50 kV across it, and emerge through a hole in one of the plates as a beam. (a) What are the speeds of the three isotopes when they emerge from the capacitor plates? (b) The accelerated ions enter a region of uniform magnetic field oriented perpendicular to the velocity vector of the ion beam. What should be the magnitude of the magnetic field if you want the largest diameter of the three ion paths to be 20.0 cm? (c) If you are collecting the atoms after completing a half-circle, one collector should be located 20.0 cm from the point where the beam enters the magnetic field. Where should the other two be located?

CHAPTER 22

At the end of 2019, there were 1.5 million electric cars on the roadways in the United States, and 5.6 million worldwide. As their popularity continues to grow, so too will the unique ways to charge them. Here, an electric car is parked over a charging pad that is located on the floor. A similar pad is mounted on the bottom of the car, so that the two pads are close to each other. Embedded in each pad is a flat coil of wire with many turns. An ac current passes through the charging coil, which creates a changing magnetic field. This field passes through the coil on the car, which creates a current in that coil that is used to charge the car's battery. The transmission of electrical power between the two coils occurs wirelessly through the process of electromagnetic induction, which is the topic of this chapter.

Electromagnetic Induction

LEARNING OBJECTIVES

After reading this module, you should be able to...

22.1 Predict when an induced current will flow.

22.2 Solve motional emf problems.

22.3 Calculate magnetic flux.

22.4 Solve problems using Faraday's law of induction.

22.5 Predict the direction of an induced current using Lenz's law.

22.6 Describe how sound is reproduced via induction.

22.7 Solve problems involving generators.

22.8 Define mutual induction and self-inductance.

22.9 Solve problems involving transformers.

22.1 Induced Emf and Induced Current

There are a number of ways a magnetic field can be used to generate an electric current, and **Interactive Figure 22.1** illustrates one of them. This drawing shows a bar magnet and a helical coil of wire to which an ammeter is connected. When there is no *relative* motion between the magnet and the coil, as in part *a* of the drawing, the ammeter reads zero, indicating that no current exists. However, when the magnet moves toward the coil, as in part *b*, a current *I* appears. As the magnet approaches, the magnetic field $\vec{\mathbf{B}}$ that it creates at the location of the coil becomes stronger and stronger, and it is this *changing* field that produces the current. When the magnet moves away from the coil, as in part *c*, a

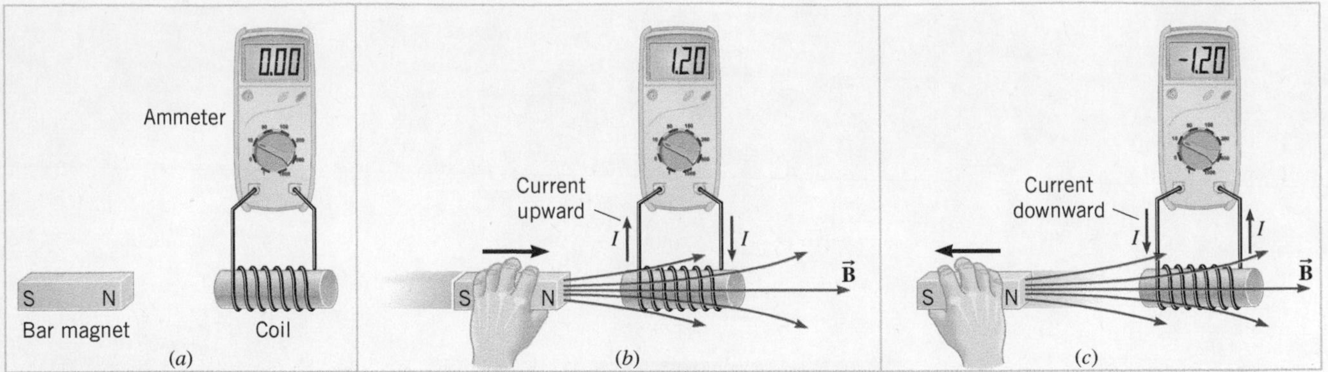

INTERACTIVE FIGURE 22.1 (*a*) When there is no relative motion between the coil of wire and the bar magnet, there is no current in the coil. (*b*) A current is created in the coil when the magnet moves toward the coil. (*c*) A current also exists when the magnet moves away from the coil, but the direction of the current is opposite to that in (*b*).

current is also produced, but with a reversed direction. Now the magnetic field at the coil becomes weaker as the magnet moves away. Once again it is the *changing* field that generates the current.

A current would also be created in **Interactive Figure 22.1** if the magnet were held stationary and the coil were moved, because the magnetic field at the coil would be changing as the coil approached or receded from the magnet. Only relative motion between the magnet and the coil is needed to generate a current; it does not matter which one moves.

The current in the coil is called an **induced current** because it is brought about (or "induced") by a changing magnetic field. Since a source of emf (electromotive force) is always needed to produce a current, the coil itself behaves as if it were a source of emf. This emf is known as an **induced emf**. Thus, a changing magnetic field induces an emf in the coil, and the emf leads to an induced current.

THE PHYSICS OF . . . an automobile cruise control. Induced emf and induced current are frequently used in the cruise controls found in many cars. **Interactive Figure 22.2** illustrates how a cruise control operates. Usually two magnets are mounted on opposite sides of the vehicle's drive shaft, with a stationary sensing coil positioned nearby. As the shaft turns, the magnets pass by the coil and cause an induced emf and current to appear in it. A microprocessor (the "brain" of a computer) counts the pulses of current and, with the aid of its internal clock and a knowledge of the shaft's radius, determines the rotational speed of the drive shaft. The rotational speed, in turn, is related to the car's speed. Thus, once the driver sets the desired cruising speed with the speed control switch (mounted near the steering wheel), the microprocessor can compare it with the measured speed. To the extent that the selected cruising speed and the measured

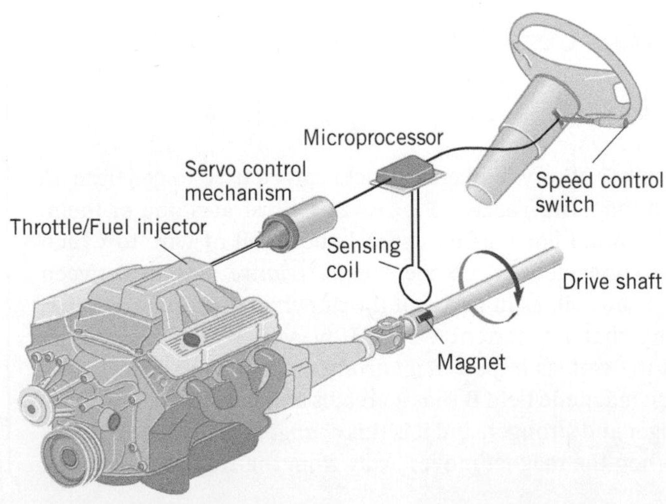

INTERACTIVE FIGURE 22.2 Induced emf lies at the heart of an automobile cruise control, in which an emf is induced in a sensing coil by magnets attached to the rotating drive shaft.

speed differ, a signal is sent to a servo, or control, mechanism, which causes the throttle/fuel injector to send more or less fuel to the engine. The car speeds up or slows down accordingly, until the desired cruising speed is reached.

Figure 22.3 shows another way to induce an emf and a current in a coil. An emf can be induced by *changing the area* of a coil in a constant magnetic field. Here the shape of the coil is being distorted to reduce the area. As long as the area is changing, an induced emf and current exist; they vanish when the area is no longer changing. If the distorted coil is returned to its original shape, thereby increasing the area, an oppositely directed current is generated while the area is changing.

In each of the previous examples, both an emf and a current are induced in the coil because the coil is part of a complete, or closed, circuit. If the circuit were open—perhaps because of an open switch—there would be no induced current. However, an emf would still be induced in the coil, whether the current exists or not.

Changing a magnetic field and changing the area of a coil are methods that can be used to create an induced emf. The phenomenon of producing an induced emf with the aid of a magnetic field is called **electromagnetic induction**. The next section discusses yet another method by which an induced emf can be created.

FIGURE 22.3 While the area of the coil is changing, an induced emf and current are generated.

Check Your Understanding

(The answer is given at the end of the book.)

1. Suppose that the coil and the magnet in **Interactive Figure 22.1*a*** were each moving with the same velocity relative to the earth. Would there be an induced current in the coil?

22.2 Motional Emf

The Emf Induced in a Moving Conductor

When a conducting rod moves through a constant magnetic field, an emf is induced in the rod. This special case of electromagnetic induction arises as a result of the magnetic force that acts on a moving charge (see Section 21.2). Consider the metal rod of length L moving to the right in **Animated Figure 22.4*a***. The velocity $\vec{\mathbf{v}}$ of the rod is constant and is perpendicular to a uniform magnetic field $\vec{\mathbf{B}}$. Each charge q within the rod also moves with a velocity $\vec{\mathbf{v}}$ and experiences a magnetic force of magnitude $F = |q|vB$, according to Equation 21.1. By using RHR-1, it can be seen that the mobile, free electrons are driven to the bottom of the rod, leaving behind an equal amount of positive charge at the top. (Remember to reverse the direction of the force that RHR-1 predicts, since the electrons have a negative charge. See Section 21.2.) The positive and negative charges accumulate until the attractive electric force that they exert on each other becomes equal in magnitude to the magnetic force. When the two forces balance, equilibrium is reached and no further charge separation occurs.

The separated charges on the ends of the moving conductor give rise to an induced emf, called a **motional emf** because it originates from the motion of charges through a magnetic field. The emf exists as long as the rod moves. If the rod is brought to a halt, the magnetic force vanishes, with the result that the attractive electric force reunites the positive and negative charges and the emf disappears. The emf of the moving rod is analogous to the emf between the terminals of a battery. However, the emf of a battery is produced by chemical reactions, whereas the motional emf is created by the agent that moves the rod through the magnetic field (like the hand in **Animated Figure 22.4*b***).

The fact that the electric and magnetic forces balance at equilibrium in **Animated Figure 22.4*a*** can be used to determine the magnitude of the motional emf $\mathcal{E}$. According to Equation 18.2, the magnitude of the electric force acting on the positive charge q at the top of the rod is Eq, where E is the magnitude of the electric field due to the separated

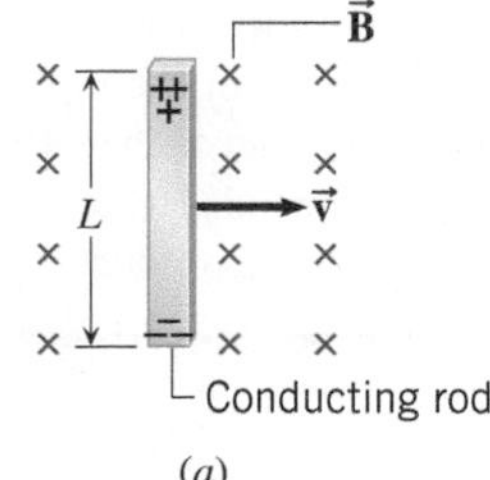

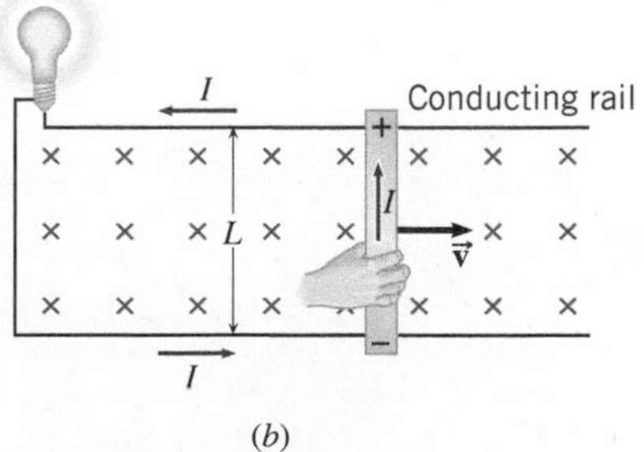

ANIMATED FIGURE 22.4 (*a*) When a conducting rod moves at right angles to a constant magnetic field, the magnetic force causes opposite charges to appear at the ends of the rod, giving rise to an induced emf. (*b*) The induced emf causes an induced current I to appear in the circuit.

charges. And according to Equation 19.7a (without the minus sign), the electric field magnitude is given by the voltage between the ends of the rod (the emf $\mathscr{E}$) divided by the length L of the rod. Thus, the electric force is $Eq = (\mathscr{E}/L)q$. Since we are dealing now with a positive charge, the magnetic force is $B|q|(v \sin 90°) = Bqv$, according to Equation 21.1, because the charge q moves perpendicular to the magnetic field. Since these two forces balance, it follows that $(\mathscr{E}/L)q = Bqv$. The emf, then, is

Motional emf when $\vec{v}$, $\vec{B}$, and L are mutually perpendicular

$$\mathscr{E} = vBL \tag{22.1}$$

As expected, $\mathscr{E} = 0$ V when $v = 0$ m/s, because no motional emf is developed in a stationary rod. Greater speeds and stronger magnetic fields lead to greater emfs for a given length L. As with batteries, $\mathscr{E}$ is expressed in volts. In **Animated Figure 22.4b** the rod is sliding on conducting rails that form part of a closed circuit, and L is the length of the rod between the rails. Due to the emf, electrons flow in a clockwise direction around the circuit. Positive charge would flow in the direction opposite to the electron flow, so the conventional current I is drawn counterclockwise in the picture. Example 1 illustrates how to determine the electrical energy that the motional emf delivers to a device such as the light bulb in the drawing.

EXAMPLE 1 | Operating a Light Bulb with Motional Emf

Suppose that the rod in **Animated Figure 22.4b** is moving at a speed of 5.0 m/s in a direction perpendicular to a 0.80-T magnetic field. The rod has a length of 1.6 m and a negligible electrical resistance. The rails also have negligible resistance. The light bulb, however, has a resistance of 96 Ω. Find **(a)** the emf produced by the rod, **(b)** the current induced in the circuit, **(c)** the electric power delivered to the bulb, and **(d)** the energy used by the bulb in 60.0 s.

Reasoning The moving rod acts like an imaginary battery and supplies a motional emf of vBL to the circuit. The induced current can be determined from Ohm's law as the motional emf divided by the resistance of the bulb. The electric power delivered to the bulb is the product of the induced current and the potential difference across the bulb (which, in this case, is the motional emf). The energy used is the product of the power and the time.

Solution **(a)** The motional emf is given by Equation 22.1 as

$$\mathscr{E} = vBL = (5.0 \text{ m/s})(0.80 \text{ T})(1.6 \text{ m}) = \boxed{6.4 \text{ V}}$$

(b) According to Ohm's law, the induced current is equal to the motional emf divided by the resistance of the circuit:

$$I = \frac{\mathscr{E}}{R} = \frac{6.4 \text{ V}}{96\ \Omega} = \boxed{0.067 \text{ A}} \tag{20.2}$$

(c) The electric power P delivered to the light bulb is the product of the current I and the potential difference across the bulb:

$$P = I\mathscr{E} = (0.067 \text{ A})(6.4 \text{ V}) = \boxed{0.43 \text{ W}} \tag{20.6a}$$

(d) Since power is energy per unit time, the energy E used in 60.0 s is the product of the power and the time:

$$E = Pt = (0.43 \text{ W})(60.0 \text{ s}) = \boxed{26 \text{ J}} \tag{6.10b}$$

Motional Emf and Electrical Energy

Motional emf arises because a magnetic force acts on the charges in a conductor that is moving through a magnetic field. Whenever this emf causes a current, a second magnetic force enters the picture. In **Animated Figure 22.4b**, for instance, the second force arises because the current I in the rod is perpendicular to the magnetic field. The current, and hence the rod, experiences a magnetic force $\vec{F}$ whose magnitude is given by Equation 21.3 as $F = ILB \sin 90°$. Using the values of I, L, and B given in Example 1, we see that $F = (0.067 \text{ A})(1.6 \text{ m})(0.80 \text{ T}) = 0.086$ N. The direction of $\vec{F}$ is specified by RHR-1 and is *opposite* to the velocity $\vec{v}$ of the rod, and thus points to the left (see **Figure 22.5**). By

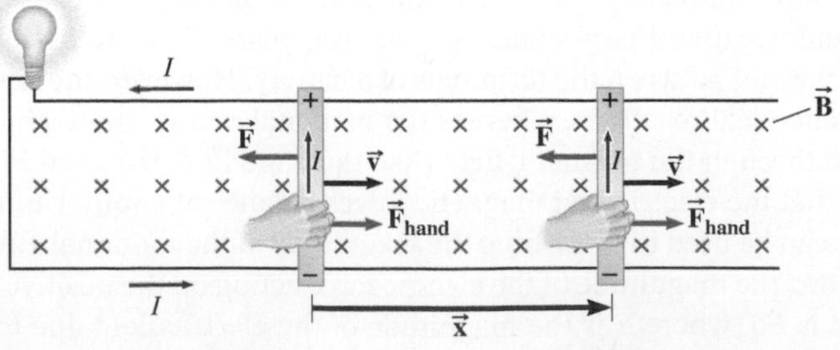

FIGURE 22.5 A magnetic force $\vec{F}$ is exerted on the current I in the moving rod and is directed opposite to the rod's velocity $\vec{v}$. Since the force $\vec{F}_{hand}$ counterbalances the magnetic force $\vec{F}$, the rod moves to the right at a constant velocity.

itself, $\vec{\mathbf{F}}$ would *slow down* the rod, and here lies the crux of the matter. To keep the rod moving to the right with a constant velocity, a counterbalancing force must be applied to the rod by an external agent, such as the hand in the picture. This force is labeled $\vec{\mathbf{F}}_{\text{hand}}$ in the drawing. The counterbalancing force must have a magnitude of 0.086 N and must be directed opposite to the magnetic force $\vec{\mathbf{F}}$. If the counterbalancing force were removed, the rod would decelerate under the influence of $\vec{\mathbf{F}}$ and eventually come to rest. During the deceleration, the motional emf would decrease and the light bulb would eventually go out.

We can now answer an important question—Who or what provides the 26 J of electrical energy that the light bulb in Example 1 uses in 60.0 seconds? The provider is the external agent that applies the 0.086-N counterbalancing force needed to keep the rod moving. This agent does work, and Example 2 shows that the work done is equal to the electrical energy used by the bulb.

Analyzing Multiple-Concept Problems

EXAMPLE 2 | The Work Needed to Keep the Light Bulb Burning

As we saw in Example 1, an induced current of 0.067 A exists in the circuit due to the moving rod. As **Figure 22.5** shows, the hand provides a force $\vec{\mathbf{F}}_{\text{hand}}$ that keeps the rod moving to the right. Determine the work done by this force in a time of 60.0 s. Assume, as in Example 1, that the magnetic field has a magnitude of 0.80 T and that the rod has a length of 1.6 m and moves at a constant speed of 5.0 m/s.

Reasoning According to the discussion in Section 6.1, the work done by the hand is equal to the product of (1) the magnitude F_{hand} of the force exerted by the hand, (2) the magnitude x of the rod's displacement, and (3) the cosine of the angle θ' between the force and the displacement. Since the rod moves to the right at a constant speed, it has no acceleration and is, therefore, in equilibrium (see Section 4.11). Thus, the force exerted by the hand must be equal in magnitude, but opposite in direction, to the magnetic force $\vec{\mathbf{F}}$ exerted on the rod, since they are the only two forces acting on the rod along the direction of the motion. We will determine the magnitude F of the magnetic force by using Equation 21.3, and this will enable us to find F_{hand}. Since the rod moves at a constant speed, the magnitude x of its displacement is the product of its speed and the time of travel.

Knowns and Unknowns The data for this problem are:

Description	Symbol	Value
Current	I	0.067 A
Length of rod	L	1.6 m
Speed of rod	v	5.0 m/s
Time during which rod moves	t	60.0 s
Magnitude of magnetic field	B	0.80 T
Unknown Variable		
Work done by hand	W	?

Modeling the Problem

STEP 1 Work The work done by the hand in **Figure 22.5** is given by $W = F_{\text{hand}} x \cos\theta'$ (Equation 6.1). In this equation, F_{hand} is the magnitude of the force that the hand exerts on the rod, x is the magnitude of the rod's displacement, and θ' is the angle between the force and the displacement. Since the force and displacement point in the same direction, $\theta' = 0°$, so

$$W = F_{\text{hand}} x \cos 0°$$

Two forces act on the rod; the force $\vec{\mathbf{F}}_{\text{hand}}$, which points to the right, and the magnetic force $\vec{\mathbf{F}}$, which points to the left. Since the rod moves at a constant velocity, the magnitudes of these two forces are equal, so that $F_{\text{hand}} = F$. Substituting this relation into the expression for the work gives Equation 1 at the right. At this point, neither F nor x is known, but they will be evaluated in Steps 2 and 3, respectively.

$$W = Fx \cos 0° \quad (1)$$

? ?

STEP 2 Magnetic Force Exerted on a Current-Carrying Rod We have seen in Section 21.5 that a rod of length L that carries a current I in a magnetic field of magnitude B experiences a magnetic force of magnitude F. The magnitude of the force is given by $F = ILB \sin\theta$ (Equation 21.3), where θ is the angle between the direction of the current and the magnetic field. In this case, the current and magnetic field are perpendicular to each other (see Figure 22.5), so $\theta = 90°$. Thus, the magnitude of the magnetic force is

$$\boxed{F = ILB \sin 90°}$$

The quantities, I, L, and B are known, and we substitute this expression into Equation 1 at the right.

$$W = Fx \cos 0° \quad (1)$$

$F = ILB \sin 90°$; ?

STEP 3 Kinematics Since the rod is moving at a constant speed, the distance x it travels is the product of its speed v and the time t:

$$\boxed{x = vt}$$

The variables v and t are known. We can also substitute this relation into Equation 1, as shown in the right column.

$$W = Fx \cos 0° \quad (1)$$

$F = ILB \sin 90°$; $x = vt$

Solution Algebraically combining the results of the three steps, we have

$$W = Fx \cos 0° \overset{\text{STEP 1}}{=} (ILB \sin 90°)x \cos 0° \overset{\text{STEP 2}}{=} (ILB \sin 90°)(vt) \cos 0°$$

(STEP 1, STEP 2, STEP 2)

The work done by the force of the hand is

$$W = (ILB \sin 90°)(vt) \cos 0°$$
$$= (0.067\ \text{A})(1.6\ \text{m})(0.80\ \text{T})(\sin 90°)(5.0\ \text{m/s})(60.0\ \text{s})(\cos 0°) = \boxed{26\ \text{J}}$$

The 26 J of work done on the rod by the hand is mechanical energy and is the same as the 26 J of energy consumed by the light bulb (see Example 1). Hence, the moving rod and the magnetic force convert mechanical energy into electrical energy, much as a battery converts chemical energy into electrical energy.

Related Homework: Problem 7

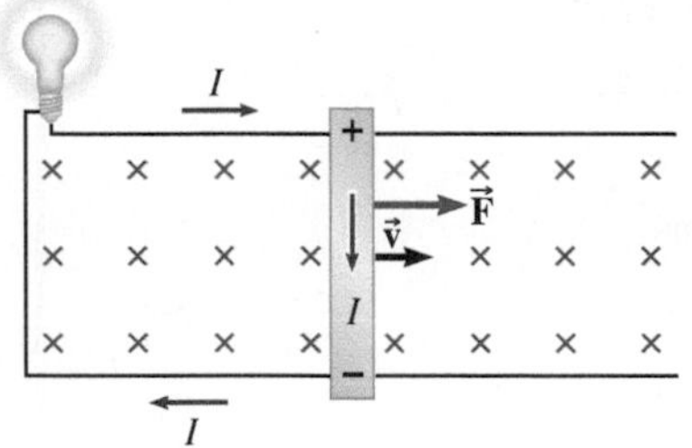

FIGURE 22.6 The current cannot be directed clockwise in this circuit, because the magnetic force $\vec{F}$ exerted on the rod would then be in the same direction as the velocity $\vec{v}$. The rod would accelerate to the right and create energy on its own, violating the principle of conservation of energy.

It is important to realize that the direction of the current in Figure 22.5 is consistent with the principle of conservation of energy. Consider what would happen if the direction of the current were reversed, as in Figure 22.6. With the direction of the current reversed, the direction of the magnetic force $\vec{F}$ would also be reversed and would point in the direction of the velocity $\vec{v}$ of the rod. As a result, the force would cause the rod to accelerate rather than decelerate. The rod would accelerate without the need for an external force (like that provided by the hand in Figure 22.5) and would create a motional emf that supplies energy to the light bulb. Thus, this hypothetical generator would produce energy out of nothing, since there is no external agent. Such a device cannot exist because it violates the principle of conservation of energy, which states that energy cannot be created or destroyed, but can only be converted from one form to another. Therefore, the current cannot be directed clockwise around the circuit, as in Figure 22.6. In situations such as the one in Examples 1 and 2, when a motional emf leads to an induced current, a magnetic force always appears that opposes the motion, in accord with the principle of conservation of energy. Conceptual Example 3 deals further with the important issue of energy conservation.

CONCEPTUAL EXAMPLE 3 | Conservation of Energy

Figure 22.7*a* illustrates a conducting rod that is free to slide down between two vertical copper tracks. There is no kinetic friction between the rod and the tracks, although the rod maintains electrical contact with the tracks during its fall. A constant magnetic field $\vec{B}$ is directed perpendicular to the motion of the rod, as the drawing shows. Because there is no friction, the only force that acts on the rod is its weight $\vec{W}$, so the rod falls with an acceleration equal to the acceleration due to gravity, which has a magnitude of

$g = 9.8\ m/s^2$. Suppose that a resistance R is connected between the tops of the tracks, as in part *b* of the drawing. Is the magnitude of the acceleration with which the rod now falls **(a)** equal to g, **(b)** greater than g, or **(c)** less than g?

Reasoning As the rod falls perpendicular to the magnetic field, a motional emf is induced between its ends. This emf is induced whether or not the resistance R is attached between the tracks. However, when R is present, a complete circuit is formed, and the emf produces an induced current I that is perpendicular to the field. The direction of this current is such that the rod experiences an upward magnetic force $\vec{\mathbf{F}}$, opposite to the weight of the rod (see part *b* of the drawing and use RHR-1). The net force acting on the rod is $\vec{\mathbf{W}} + \vec{\mathbf{F}}$, which is *less* than the weight, since $\vec{\mathbf{F}}$ points upward and the weight $\vec{\mathbf{W}}$ points downward. In accord with Newton's second law of motion, the downward acceleration is proportional to the net force.

Answers (a) and (b) are incorrect. Since the net downward force on the rod in **Figure 22.7*b*** is less than the rod's weight and since the downward acceleration is proportional to the net force, the rod cannot have an acceleration whose magnitude is equal to or greater than g.

Answer (c) is correct. Since the net downward force on the rod when R is present is less than the rod's weight and since the downward acceleration is proportional to the net force, the downward acceleration has a magnitude less than g. Thus, the rod gains speed as it falls but does so less rapidly than if R were not present. As the speed of the rod in **Figure 22.7*b*** increases during the descent, the magnetic force increases, until the time comes when its magnitude equals the magnitude of the rod's weight. When this occurs, the net force and the rod's acceleration will be zero. From this moment on, the rod will fall at a constant velocity. In any event, the rod always has a smaller speed than does a freely falling rod (i.e., R is absent) at the same place. The speed is smaller because only part of the gravitational potential energy (GPE) is being converted into kinetic energy (KE) as the rod falls, with part also being dissipated as heat in the resistance R. In fact, when the rod eventually attains a constant velocity, none of the GPE is converted into KE and all of it is dissipated as heat.

Related Homework: Problem 9

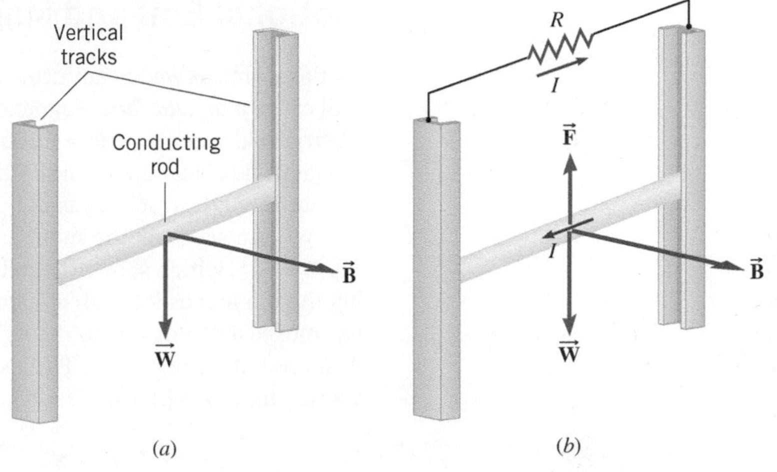

FIGURE 22.7 (*a*) There is no kinetic friction between the falling rod and the tracks, so the only force acting on the rod is its weight $\vec{\mathbf{W}}$. (*b*) When an induced current I exists in the circuit, a magnetic force $\vec{\mathbf{F}}$ also acts on the rod.

Check Your Understanding

(*The answers are given at the end of the book.*)

2. Consider the induced emf being generated in **Animated Figure 22.4**. Suppose that the length of the rod is reduced by a factor of four. For the induced emf to be the same, what should be done? **(a)** Without changing the speed of the rod, increase the strength of the magnetic field by a factor of four. **(b)** Without changing the magnetic field, increase the speed of the rod by a factor of four. **(c)** Increase both the speed of the rod and the strength of the magnetic field by a factor of two. **(d)** All of the previous three methods may be used.

3. In the discussion concerning **Figure 22.5**, we saw that a force of 0.086 N from an external agent was required to keep the rod moving at a constant velocity. Suppose that friction is absent and that the light bulb is suddenly removed from its socket while the rod is moving. How much force does the external agent then need to apply to the rod to keep it moving at a constant velocity? **(a)** 0 N **(b)** Greater than 0 N but less than 0.086 N **(c)** More than 0.086 N **(d)** 0.086 N

4. Eddy currents are electric currents that can arise in a piece of metal when it moves through a region where the magnetic field is not the same everywhere. **CYU Figure 22.1** shows, for example, a metal sheet moving to the right at a velocity $\vec{\mathbf{v}}$ and a magnetic field $\vec{\mathbf{B}}$ that is directed perpendicular to the sheet. At the instant represented, the field only extends over the left half of the sheet. An

Metal sheet

$\vec{\mathbf{v}}$

$\vec{\mathbf{B}}$ (into paper)

Eddy current

CYU FIGURE 22.1

emf is induced that leads to the eddy current indicated. Such eddy currents cause the velocity of the moving sheet to decrease and are used in various devices as a brake to damp out unwanted motion. Does the eddy current in the drawing circulate **(a)** counterclockwise or **(b)** clockwise?

22.3 Magnetic Flux

Motional Emf and Magnetic Flux

Motional emf, as well as any other type of induced emf, can be described in terms of a concept called *magnetic flux.* Magnetic flux is analogous to electric flux, which deals with the electric field and the surface through which it passes (see Section 18.9 and Figure 18.33). Magnetic flux is defined in a similar way by bringing together the magnetic field and the surface through which it passes.

We can see how the motional emf is related to the magnetic flux with the aid of **Figure 22.8**, which shows the rod used to derive Equation 22.1 ($\mathscr{E} = vBL$). In this drawing the rod moves through a magnetic field beginning at time $t = 0$ s. In part *a* the rod has moved a distance x_0 to the right at time t_0, whereas in part *b* it has moved a greater distance x at a later time t. The speed v of the rod is the distance traveled divided by the elapsed time: $v = (x - x_0)/(t - t_0)$. Substituting this expression for v into $\mathscr{E} = vBL$ gives

$$\mathscr{E} = \left(\frac{x - x_0}{t - t_0}\right)BL = \left(\frac{xL - x_0 L}{t - t_0}\right)B$$

As the drawing indicates, the term $x_0 L$ is the area A_0 swept out by the rod in moving a distance x_0, and xL is the area A swept out in moving a distance x. Thus, the emf becomes

$$\mathscr{E} = \left(\frac{A - A_0}{t - t_0}\right)B = \frac{(BA) - (BA)_0}{t - t_0}$$

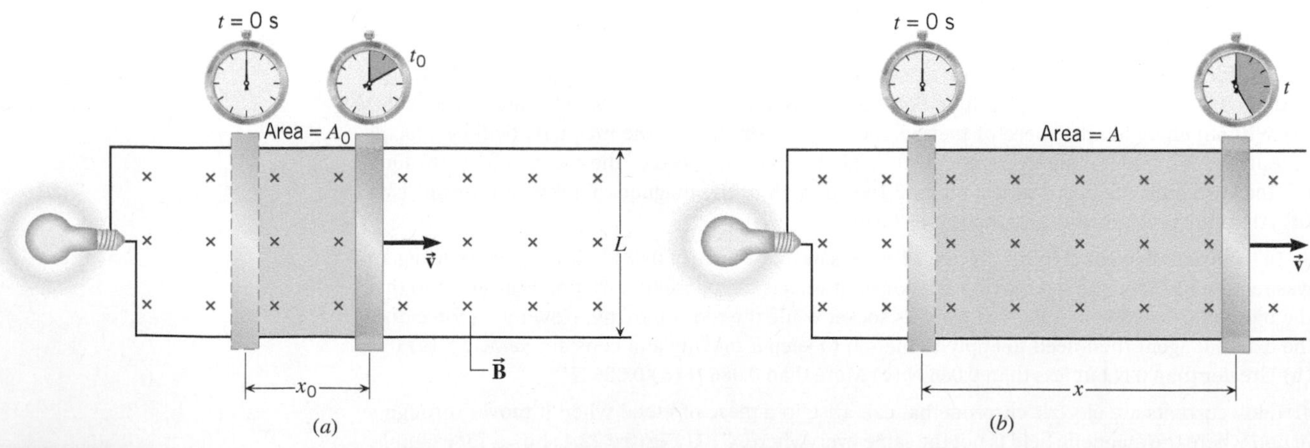

FIGURE 22.8 (*a*) In a time t_0, the moving rod sweeps out an area $A_0 = x_0 L$. (*b*) The area swept out in a time t is $A = xL$. In both parts of the figure the areas are shaded in color.

The product *BA* of the magnetic field strength and the area appears in the numerator of this expression. This product is called **magnetic flux** and is represented by the symbol Φ (Greek capital letter phi); thus $\Phi = BA$. The magnitude of the induced emf is the *change* in flux $\Delta\Phi = \Phi - \Phi_0$ divided by the time interval $\Delta t = t - t_0$ during which the change occurs:

$$\mathscr{E} = \frac{\Phi - \Phi_0}{t - t_0} = \frac{\Delta\Phi}{\Delta t}$$

In other words, the induced emf equals the time rate of change of the magnetic flux.

You will almost always see the previous equation written with a minus sign—namely, $\mathscr{E} = -\Delta\Phi/\Delta t$. The minus sign is introduced for the following reason: The direction of the current induced in the circuit is such that the magnetic force $\vec{\mathbf{F}}$ acts on the rod to *oppose* its motion, thereby tending to slow down the rod (see **Figure 22.5**). The minus sign ensures that the polarity of the induced emf sends the induced current in the proper direction so as to give rise to this opposing magnetic force.* This issue of the polarity of the induced emf will be discussed further in Section 22.5.

The advantage of writing the induced emf as $\mathscr{E} = -\Delta\Phi/\Delta t$ is that this relation is far more general than our present discussion suggests. In Section 22.4 we will see that $\mathscr{E} = -\Delta\Phi/\Delta t$ can be applied to *all possible ways of generating induced emfs.*

A General Expression for Magnetic Flux

In **Figure 22.8** the direction of the magnetic field $\vec{\mathbf{B}}$ is perpendicular to the surface swept out by the moving rod. In general, however, $\vec{\mathbf{B}}$ may not be perpendicular to the surface. For instance, in **Figure 22.9** the direction perpendicular to the surface is indicated by the normal to the surface, but the magnetic field is inclined at an angle ϕ with respect to this direction. In such a case the flux is computed using only the component of the field perpendicular to the surface, $B \cos \phi$. The general expression for magnetic flux is

$$\Phi = (B \cos \phi)A = BA \cos \phi \tag{22.2}$$

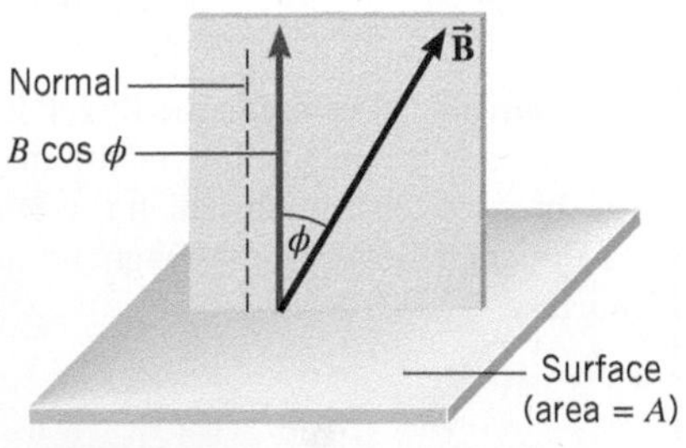

FIGURE 22.9 When computing the magnetic flux, the component of the magnetic field that is perpendicular to the surface must be used; this component is $B \cos \phi$.

Math Skills Equation 22.2 uses the component of the magnetic field perpendicular to the surface in defining the magnetic flux Φ. In order to determine this component of the field, we use the cosine function. According to the definition given in Equation 1.1, the cosine function is $\cos \phi = \frac{h_a}{h}$, where h_a is the length of the side of a right triangle that is adjacent to the angle ϕ and h is the length of the hypotenuse (see **Figure 22.10*a***). **Figure 22.10*b*** shows the field $\vec{\mathbf{B}}$ (magnitude *B*) and its components B_x and B_y. The field $\vec{\mathbf{B}}$ is oriented at an angle ϕ with respect to the normal, and the component that we seek is B_y, the one that points along the direction of the normal. By comparing **Figures 22.10*a*** and ***b***, we can see that

$$\cos \phi = \frac{h_a}{h} = \frac{B_y}{B} \quad \text{or} \quad B_y = B \cos \phi$$

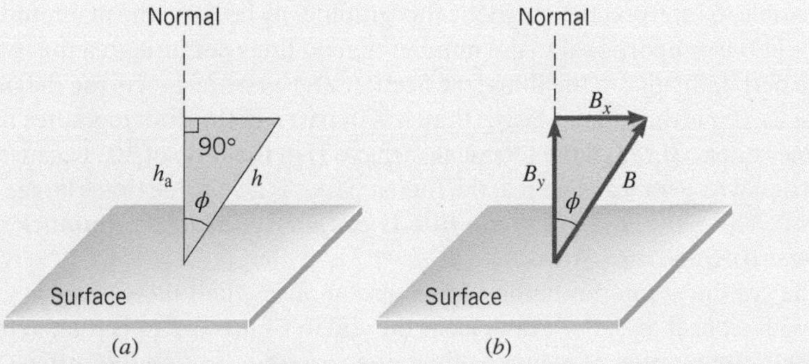

Normal 90° h_a h ϕ Surface (*a*)

Normal B_x B_y B ϕ Surface (*b*)

FIGURE 22.10 Math Skills drawing.

*A detailed mathematical discussion of why the minus sign arises is beyond the scope of this book.

If either the magnitude B of the magnetic field or the angle ϕ is not constant over the surface (i.e., they are not the same at each point on the surface), an average value for the product $B \cos \phi$ must be used to compute the flux. Equation 22.2 shows that the unit of magnetic flux is the tesla · meter² ($T \cdot m^2$). This unit is called a *weber* (Wb), after the German physicist Wilhelm Weber (1804–1891): $1 \text{ Wb} = 1 \text{ T} \cdot \text{m}^2$. Example 4 illustrates how to determine the magnetic flux for three different orientations of the surface of a coil relative to the magnetic field.

EXAMPLE 4 | Magnetic Flux

A rectangular coil of wire is situated in a constant magnetic field whose magnitude is 0.50 T. The coil has an area of 2.0 m². Determine the magnetic flux for the following three orientations, $\phi = 0°$, 60.0°, and 90.0°, shown in Figure 22.11.

Problem-Solving Insight The magnetic flux Φ is determined by more than just the magnitude B of the magnetic field and the area A. It also depends on the angle ϕ (see Figure 22.9 and Equation 22.2).

Reasoning The magnetic flux Φ is defined as $\Phi = BA \cos \phi$, where B is the magnitude of the magnetic field, A is the area of the surface through which the magnetic field passes, and ϕ is the angle between the magnetic field and the normal to the surface.

Solution The magnetic flux for the three cases is:

$\boldsymbol{\phi = 0^{\circ}}$ $\quad \Phi = (0.50 \text{ T})(2.0 \text{ m}^2) \cos 0° = \boxed{1.0 \text{ Wb}}$

$\boldsymbol{\phi = 60.0^{\circ}}$ $\quad \Phi = (0.50 \text{ T})(2.0 \text{ m}^2) \cos 60.0° = \boxed{0.50 \text{ Wb}}$

$\boldsymbol{\phi = 90.0^{\circ}}$ $\quad \Phi = (0.50 \text{ T})(2.0 \text{ m}^2) \cos 90.0° = \boxed{0 \text{ Wb}}$

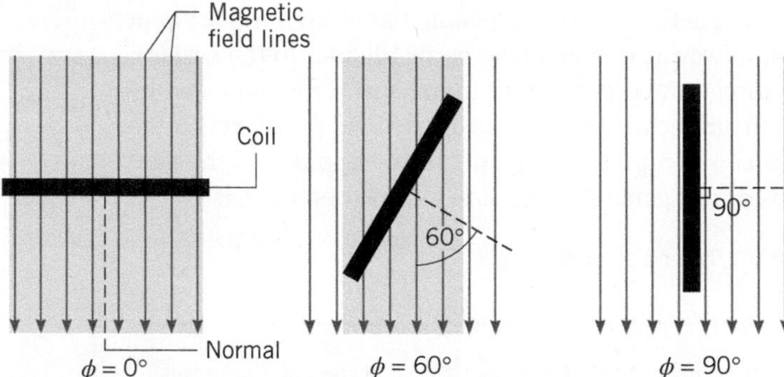

FIGURE 22.11 Three orientations of a rectangular coil (drawn as an edge view) relative to the magnetic field lines. The magnetic field lines that pass through the coil are those in the regions shaded in blue.

Graphical Interpretation of Magnetic Flux

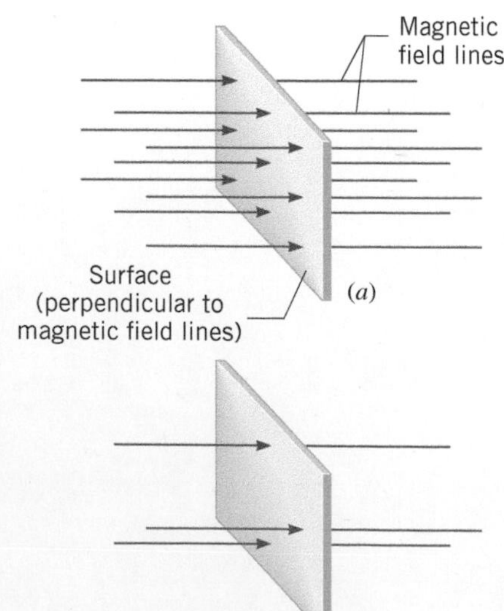

It is possible to interpret the magnetic flux graphically because the magnitude of the magnetic field $\vec{\mathbf{B}}$ is proportional to the number of field lines per unit area that pass through a surface perpendicular to the lines (see Section 21.1). For instance, the magnitude of $\vec{\mathbf{B}}$ in Figure 22.12*a* is three times larger than it is in part *b* of the drawing, since the number of field lines drawn through the identical surfaces is in the ratio of 3:1. Because Φ is directly proportional to B for a given area, the flux in part *a* is also three times larger than the flux in part *b*. Therefore, ***the magnetic flux is proportional to the number of field lines that pass through a surface***.

The graphical interpretation of flux also applies when the surface is oriented at an angle with respect to $\vec{\mathbf{B}}$. For example, as the coil in Figure 22.11 is rotated from $\phi = 0°$ to 60° to 90°, the number of magnetic field lines passing through the surface (see the field lines in the regions shaded in blue) changes in the ratio of 8 : 4 : 0 or 2 : 1 : 0. The results of Example 4 show that the flux in the three orientations changes by the same ratio.

FIGURE 22.12 The magnitude of the magnetic field in (*a*) is three times as great as that in (*b*) because the number of magnetic field lines crossing the surfaces is in the ratio of 3 : 1.

Because the magnetic flux is proportional to the number of field lines passing through a surface, we often use phrases such as "the flux that passes through a surface bounded by a loop of wire."

Check Your Understanding

(*The answers are given at the end of the book.*)

5. A magnetic field has the same direction and the same magnitude B everywhere. A circular area A is bounded by a loop of wire. Which of the following statements is true concerning the magnitude of the magnetic flux that passes through this area? **(a)** It is zero. **(b)** It is BA. **(c)** Its maximum possible value is BA. **(d)** Its minimum possible value is BA.

6. Suppose that a magnetic field is constant everywhere on a flat 1.0-m^2 surface and that the magnetic flux through this surface is 2.0 Wb. From these data, which one of the following pieces of information can be determined about the magnetic field? **(a)** The magnitude of the field **(b)** The magnitude of the component of the field that is perpendicular to the surface **(c)** The magnitude of the component of the field that is parallel to the surface

22.4 Faraday's Law of Electromagnetic Induction

Two scientists are given credit for the discovery of electromagnetic induction: the Englishman Michael Faraday (1791–1867) and the American Joseph Henry (1797–1878). Since Faraday investigated electromagnetic induction in more detail and published his findings first, the law that describes the phenomenon bears his name.

Faraday discovered that whenever there is a *change in magnetic flux* through a loop of wire, an emf is induced in the loop. In this context, the word "change" refers to a change as time passes. A magnetic flux that is constant in time creates no emf. Faraday's law of electromagnetic induction is expressed by bringing together the idea of magnetic flux and the time interval during which it changes. In fact, Faraday found that the magnitude of the induced emf is equal to the time rate of change of the magnetic flux. This is consistent with the relation we obtained in Section 22.3 for the specific case of motional emf: $\mathcal{E} = -\Delta\Phi/\Delta t$.

Often the magnetic flux passes through a coil of wire containing more than one loop (or turn). If the coil consists of N loops, and if the same flux passes through each loop, it is found experimentally that the total induced emf is N times that induced in a single loop. An analogous situation occurs in a flashlight when two 1.5-V batteries are stacked in series on top of one another to give a total emf of 3.0 volts. For the general case of N loops, the total induced emf is described by Faraday's law of electromagnetic induction in the following manner:

FARADAY'S LAW OF ELECTROMAGNETIC INDUCTION

The average emf $\mathcal{E}$ induced in a coil of N loops is

$$\mathcal{E} = -N\left(\frac{\Phi - \Phi_0}{t - t_0}\right) = -N\frac{\Delta\Phi}{\Delta t} \quad (22.3)$$

where $\Delta\Phi$ is the change in magnetic flux through one loop and Δt is the time interval during which the change occurs. The term $\Delta\Phi/\Delta t$ is the average time rate of change of the flux that passes through one loop.

SI Unit of Induced Emf: volt (V)

Faraday's law states that an emf is generated if the magnetic flux changes for any reason. Since the flux is given by Equation 22.2 as $\Phi = BA\cos\phi$, it depends on the three factors, B, A, and ϕ, any of which may change. Example 5 considers a change in B.

EXAMPLE 5 | The Emf Induced by a Changing Magnetic Field

A coil of wire consists of 20 turns, or loops, each with an area of 1.5×10^{-3} m^2. A magnetic field is perpendicular to the surface of each loop at all times, so that $\phi = \phi_0 = 0°$. At time $t_0 = 0$ s, the magnitude of the field at the location of the coil is $B_0 = 0.050$ T. At a later time $t = 0.10$ s, the magnitude of the field at the coil has increased to $B = 0.060$ T. **(a)** Find the average emf induced in the coil during this time. **(b)** What would be the value of the average induced emf if the magnitude of the magnetic field decreased from 0.060 T to 0.050 T in 0.10 s?

Reasoning To find the induced emf, we use Faraday's law of electromagnetic induction (Equation 22.3), combining it with the definition of magnetic flux from Equation 22.2. We note that only the magnitude of the magnetic field changes in time. All other factors remain constant.

> **Problem-Solving Insight** **The change in any quantity is the final value minus the initial value: e.g., the change in flux is $\Delta\Phi = \Phi - \Phi_0$ and the change in time is $\Delta t = t - t_0$.**

Solution **(a)** Since $\phi = \phi_0$, the induced emf is

$$\mathscr{E} = -N\left(\frac{\Phi - \Phi_0}{t - t_0}\right) = -N\left(\frac{BA\cos\phi - B_0 A\cos\phi}{t - t_0}\right)$$

$$= -NA\cos\phi\left(\frac{B - B_0}{t - t_0}\right)$$

$$\mathscr{E} = -(20)(1.5 \times 10^{-3}\ \text{m}^2)(\cos 0°)\left(\frac{0.060\ \text{T} - 0.050\ \text{T}}{0.10\ \text{s} - 0\ \text{s}}\right)$$

$$= \boxed{-3.0 \times 10^{-3}\ \text{V}}$$

(b) The calculation here is similar to that in part (a), except the initial and final values of B are interchanged. This interchange reverses the sign of the emf, so $\boxed{\mathscr{E} = +3.0 \times 10^{-3}\ \text{V}}$. Because the algebraic sign or polarity of the emf is reversed, the direction of the induced current would be opposite to that in part (a).

The next example demonstrates that an emf can be created when a coil is rotated in a magnetic field.

EXAMPLE 6 | The Emf Induced in a Rotating Coil

A flat coil of wire has an area of 0.020 m^2 and consists of 50 turns. At $t_0 = 0$ s the coil is oriented so the normal to its surface has the same direction ($\phi_0 = 0°$) as a constant magnetic field of magnitude 0.18 T. The coil is then rotated through an angle of $\phi = 30.0°$ in a time of 0.10 s (see **Figure 22.11**). **(a)** Determine the average induced emf. **(b)** What would be the induced emf if the coil were returned to its initial orientation in the same time of 0.10 s?

Reasoning As in Example 5 we can determine the induced emf by using Faraday's law of electromagnetic induction, along with the definition of magnetic flux. In the present case, however, only ϕ (the angle between the normal to the surface of the coil and the magnetic field) changes in time. All other factors remain constant.

Solution **(a)** Faraday's law yields

$$\mathscr{E} = -N\left(\frac{\Phi - \Phi_0}{t - t_0}\right) = -N\left(\frac{BA\cos\phi - BA\cos\phi_0}{t - t_0}\right)$$

$$= -NBA\left(\frac{\cos\phi - \cos\phi_0}{t - t_0}\right)$$

$$\mathscr{E} = -(50)(0.18\ \text{T})(0.020\ \text{m}^2)\left(\frac{\cos 30.0° - \cos 0°}{0.10\ \text{s} - 0\ \text{s}}\right) = \boxed{+0.24\ \text{V}}$$

(b) When the coil is rotated back to its initial orientation in a time of 0.10 s, the initial and final values of ϕ are interchanged. As a result, the induced emf has the same magnitude but opposite polarity, so $\boxed{\mathscr{E} = -0.24\ \text{V}}$.

THE PHYSICS OF . . . a ground fault interrupter. One application of Faraday's law that is found in the home is a safety device called a ground fault interrupter. This device protects against electrical shock from an appliance, such as a clothes dryer. It plugs directly into a wall socket, as in **Figure 22.13** or, in new home construction, replaces the socket entirely. The interrupter consists of a circuit breaker that can be triggered to stop the current to the dryer, depending on whether an induced voltage appears across a sensing coil. This coil is wrapped around an iron ring, through which the current-carrying wires pass. In the drawing, the current going to the dryer is shown in red, and the returning current is shown in green. Each of the currents creates a magnetic field that encircles the corresponding wire, according to RHR-2 (see Section 21.7). However, the field lines have opposite directions since the currents have opposite directions. As the drawing shows, the iron ring guides the field lines through the sensing coil. Since the current is ac, the fields from the red and green current are changing, but the red and green field lines always have opposite directions and the opposing fields cancel at all times. As a result, the net flux through the coil remains zero, and no induced emf appears in the coil. Thus,

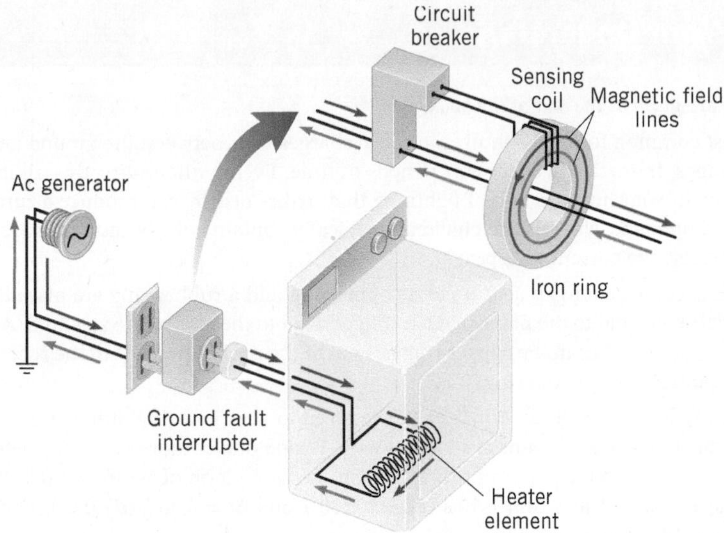

FIGURE 22.13 The clothes dryer is connected to the wall socket through a ground fault interrupter. The dryer is operating normally.

when the dryer operates normally, the circuit breaker is not triggered and does not shut down the current. The picture changes if the dryer malfunctions, as when a wire inside the unit breaks and accidentally contacts the metal case. When someone touches the case, some of the current begins to pass through the person's body and into the ground, returning to the ac generator *without using the return wire that passes through the ground fault interrupter*. Under this condition, the net magnetic field through the sensing coil is no longer zero and changes with time, since the current is ac. The changing flux causes an induced voltage to appear in the sensing coil, which triggers the circuit breaker to stop the current. Ground fault interrupters work very fast (in less than a millisecond) and turn off the current before it reaches a dangerous level.

Conceptual Example 7 discusses another application of electromagnetic induction —namely, how a stove can cook food without getting hot.

CONCEPTUAL EXAMPLE 7 | The Physics of an Induction Stove

Figure 22.14 shows two pots of water that were placed on an induction stove at the same time. There are two interesting features in this drawing. First, the stove itself is cool to the touch. Second, the water in the ferromagnetic metal pot is boiling while the water in the glass pot is not. How can such a "cool" stove boil water, and why isn't the water in the glass pot boiling?

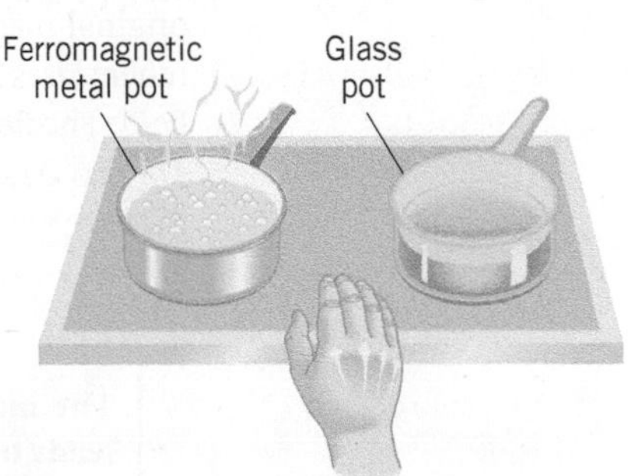

FIGURE 22.14 The water in the ferromagnetic metal pot is boiling—yet the water in the glass pot is not boiling, and the stove top is cool to the touch. The stove operates in this way by using electromagnetic induction.

Reasoning and Solution The key to this puzzle is related to the fact that one pot is made from a ferromagnetic metal and one from glass. We know that metals are good conductors, while glass is an insulator. Perhaps the stove causes electricity to flow directly in the metal pot. This is exactly what happens. The stove is called an *induction stove* because it operates by using electromagnetic induction. Just beneath the cooking surface is a metal coil that carries an ac current (frequency about 25 kHz). This current produces an alternating magnetic field that extends outward to the location of the metal pot. As the changing field crosses the pot's bottom surface, an emf is induced in it. Because the pot is metallic, an induced current is generated by the induced emf. The metal has a finite resistance to the induced current, however, and heats up as energy is dissipated in this resistance. The fact that the metal is ferromagnetic is important. Ferromagnetic materials contain magnetic domains (see Section 21.9), and the boundaries between them move extremely rapidly in response to the external magnetic field, thus enhancing the induction effect. A normal aluminum cooking pot, in contrast, is not ferromagnetic, so this enhancement is absent and such cookware is not used with induction stoves. An emf is also induced in the glass pot and the cooking surface of the stove. However, these materials are insulators, so very little induced current exists within them. Thus, they do not heat up very much and remain cool to the touch.

Check Your Understanding

(The answers are given at the end of the book.)

7. In the most common form of lightning, electric charges flow between the ground and a cloud. The flow changes dramatically over short periods of time. Even without directly striking an electrical appliance in your house, a bolt of lightning that strikes nearby can produce a current in the circuits of the appliance. Note that such circuits typically contain coils or loops of wire. Why can the lightning cause the current to appear?

8. A solenoid is connected to an ac source. A copper ring and a rubber ring are placed inside the solenoid, with the normal to the plane of each ring parallel to the axis of the solenoid. An induced emf appears __________. **(a)** in the copper ring but not in the rubber ring **(b)** in the rubber ring but not in the copper ring **(c)** in both rings

9. A magnetic field of magnitude $B = 0.20$ T is reduced to zero in a time interval of $\Delta t = 0.10$ s, thereby creating an induced current in a loop of wire. Which one or more of the following choices would cause the same induced current to appear in the same loop of wire? **(a)** $B = 0.40$ T and $\Delta t = 0.20$ s **(b)** $B = 0.30$ T and $\Delta t = 0.10$ s **(c)** $B = 0.30$ T and $\Delta t = 0.30$ s **(d)** $B = 0.10$ T and $\Delta t = 0.050$ s **(e)** $B = 0.50$ T and $\Delta t = 0.40$ s

10. A coil is placed in a magnetic field, and the normal to the plane of the coil remains parallel to the field. Which one of the following options causes the magnitude of the average emf induced in the coil to be as large as possible? **(a)** The magnitude of the field is small, and its rate of change is large. **(b)** The magnitude of the field is large, and its rate of change is small. **(c)** The magnitude of the field is large, and it does not change.

22.5 Lenz's Law

An induced emf drives current around a circuit just as the emf of a battery does. With a battery, conventional current is directed out of the positive terminal, through the attached device, and into the negative terminal. The same is true for an induced emf, although the locations of the positive and negative terminals are generally not as obvious. Therefore, a method is needed for determining the polarity or algebraic sign of the induced emf, so the terminals can be identified. As we discuss this method, it will be helpful to keep in mind that the net magnetic field penetrating a coil of wire results from two contributions. One is the original magnetic field that produces the changing flux that leads to the induced emf. The other arises because of the induced current, which, like any current, creates its own magnetic field. The field created by the induced current is called the **induced magnetic field**.

To determine the polarity of the induced emf, we will use a method based on a discovery made by the Russian physicist Heinrich Lenz (1804–1865). This discovery is known as **Lenz's law**.

> LENZ'S LAW
>
> **The induced emf resulting from a changing magnetic flux has a polarity that leads to an induced current whose direction is such that the induced magnetic field opposes the original flux change.**

Lenz's law is best illustrated with examples. Each will be worked out according to the following reasoning strategy:

> REASONING STRATEGY **Determining the Polarity of the Induced Emf**
>
> 1. **Determine whether the magnetic flux that penetrates a coil is increasing or decreasing.**
> 2. **Find what the direction of the induced magnetic field must be so that it can *oppose the change in flux* by adding to or subtracting from the original field.**
> 3. **Having found the direction of the induced magnetic field, use RHR-3 (see Section 21.7) to determine the direction of the induced current. Then the polarity of the induced emf can be assigned because conventional current is directed out of the positive terminal, through the external circuit, and into the negative terminal.**

CONCEPTUAL EXAMPLE 8 | The Emf Produced by a Moving Magnet

Figure 22.15*a* shows a permanent magnet approaching a loop of wire. The external circuit attached to the loop consists of the resistance *R*, which could represent the filament in a light bulb, for instance. In **Figure 22.15*a***, what is the polarity of the induced emf? In other words, **(a)** is point *A* positive and point *B* negative or **(b)** is point *A* negative and point *B* positive?

Reasoning We will apply Lenz's law, the essence of which is that the change in magnetic flux must be opposed by the induced magnetic field. The flux through the loop is increasing, since the magnitude of the magnetic field at the loop is increasing as the magnet approaches. To oppose the increase in the flux, the direction of the induced magnetic field must be opposite to the field of the bar magnet. Thus, since the field of the bar magnet passes through the loop from left to right in part *a* of the drawing, the induced field must pass through the loop from right to left. An induced current creates this induced field, and from the direction of this current we will be able to decide the polarity of the induced emf.

Answer (b) is incorrect. If point *A* were negative and point *B* were positive, as in **Figure 22.15*b***, the induced current in the loop would be as shown in that part of the drawing, because conventional current exits from the positive terminal and returns through the external circuit (the resistance *R*) to the negative terminal. Application of RHR-3 reveals that this induced current would lead to an induced field that passes through the loop from left to right, not right to left as needed to oppose the flux change.

Answer (a) is correct. **Figure 22.15*c*** shows the situation with point *A* positive and point *B* negative and the induced current that results. An application of RHR-3 reveals that the induced field produced by this current indeed passes through the loop from right to left, as needed. We conclude, therefore, that the polarity shown in **Figure 22.15*c*** is correct.

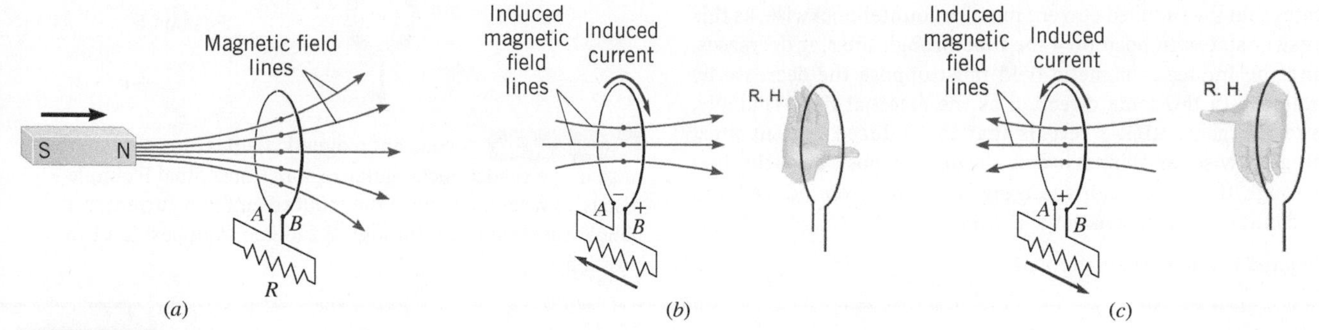

FIGURE 22.15 (*a*) As the magnet moves to the right, the magnetic flux through the loop increases. The external circuit attached to the loop has a resistance *R*. (*b*) One possibility and (*c*) another possibility for the direction of the induced current. See Conceptual Example 8.

In Conceptual Example 8 the direction of the induced magnetic field is opposite to the direction of the external field of the bar magnet.

Problem-Solving Insight The induced field is not always opposite to the external field, however, because Lenz's law requires only that it must oppose the *change* in the flux that generates the emf.

Conceptual Example 9 illustrates this point.

CONCEPTUAL EXAMPLE 9 | The Emf Produced by a Moving Copper Ring

In **Figure 22.16** there is a constant magnetic field in a rectangular region of space. This field is directed perpendicularly into the page. Outside this region there is no magnetic field. A copper ring slides through the region, from position 1 to position 5. Since the field is zero outside the rectangular region, no flux passes through the ring in positions 1 and 5, there is no change in the flux through the ring, and there is no induced emf or current in the ring. Which one of the following options correctly describes the induced current in the ring as it passes through positions 2, 3, and 4? **(a)** I_2 is clockwise, I_3 is counterclockwise, I_4 is counterclockwise. **(b)** I_2 is counterclockwise, I_3 is clockwise, I_4 is clockwise. **(c)** I_2 is clockwise, $I_3 = 0$ A, I_4 is counterclockwise. **(d)** I_2 is counterclockwise, $I_3 = 0$ A, I_4 is clockwise.

Reasoning Lenz's law will guide us. It requires that the induced magnetic field oppose the change in flux. Sometimes this means that the induced field is opposite to the external magnetic field, as in Example 8. However, we will see that the induced field sometimes has the same direction as the external field in order to oppose the flux change.

Answers (a) and (b) are incorrect. Both of these answers specify that there is an induced current I_3 in the ring as it passes through position 3, contrary to fact. The external field within the rectangular region certainly produces a flux through the ring. (See **Figure 22.16**.) However, the external field is constant, so the flux through the ring does not change as the ring moves. In order for an induced emf to exist and to cause an induced current, the flux must change.

Answer (c) is incorrect. As the ring moves out of the field region in position 4, the flux through the ring decreases, so there is an induced emf and an induced current. Lenz's law requires that the induced current must lead to an induced magnetic field that opposes this flux decrease. To oppose the decrease, the induced field must point in the same direction as the external field. To create an induced field pointing into the page, the induced current I_4 must be clockwise (use RHR-3), not counterclockwise as this answer specifies.

Answer (d) is correct. In position 2 the flux increases and, according to Lenz's law, the induced current must create an induced magnetic field that opposes the increase. To oppose the increase the induced field must point opposite to the external field and, therefore, must point out of the page. RHR-3 indicates that the induced current must be counterclockwise, as this answer states. In position 4 the flux through the ring decreases, and the induced magnetic field must oppose the decrease by pointing in the same direction as the external field—namely, into the page. RHR-3 reveals that the induced current must be clockwise, as this answer indicates. In position 3 the flux through the loop is not changing, so there is no induced emf and current, as this answer specifies.

Related Homework: Problem 62

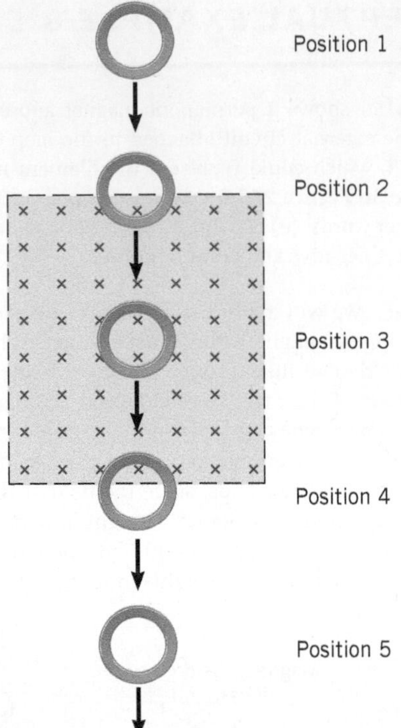

FIGURE 22.16 A constant magnetic field is directed into the page in the shaded rectangular region. Conceptual Example 9 discusses what happens to the induced emf and current in a copper ring that slides through the region from position 1 to position 5.

Lenz's law should not be thought of as an independent law, because it is a consequence of the law of conservation of energy. The connection between energy conservation and induced emf has already been discussed in Section 22.2 for the specific case of motional emf. However, the connection is valid for any type of induced emf. In fact, the polarity of the induced emf, as specified by Lenz's law, ensures that energy is conserved.

Check Your Understanding

(The answers are given at the end of the book.)

11. In **Figure 22.3** a coil of wire is being stretched. What would be the direction of the induced current if the direction of the external magnetic field in the figure were reversed? **(a)** Clockwise **(b)** Counterclockwise

12. A circular loop of wire is lying flat on a horizontal table, and you are looking down at it. An external magnetic field has a constant direction that is perpendicular to the table, and there is an induced clockwise current in the loop. Is the external field directed upward toward you or downward away from you, and is its magnitude increasing or decreasing? Note that there are two possible answers.

13. When the switch in **CYU Figure 22.2** is closed, the current in the coil increases to its equilibrium value. While the current is increasing there is an induced current in the metal ring. The ring is free to move. What happens to the ring? **(a)** It does not move. **(b)** It jumps downward. **(c)** It jumps upward.

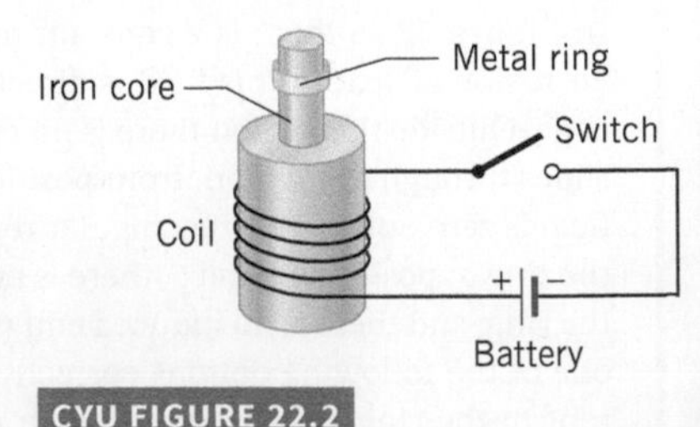

CYU FIGURE 22.2

14. A conducting rod is free to slide along a pair of conducting rails, in a region where a uniform and constant

(in time) magnetic field is directed into the plane of the paper, as **CYU Figure 22.3** illustrates. Initially the rod is at rest. There is no friction between the rails and the rod. What happens to the rod after the switch is closed? If any induced emf develops, be sure to account for its effect. **(a)** The rod accelerates to the right, its velocity increasing without limit. **(b)** The rod does not move. **(c)** The rod accelerates to the right for a while and then slows down and comes to a halt. **(d)** The rod accelerates to the right and eventually reaches a constant velocity at which it continues to move.

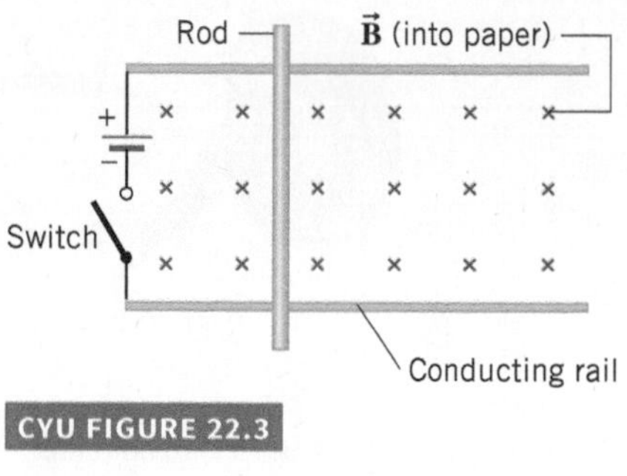

CYU FIGURE 22.3

22.6 *Applications of Electromagnetic Induction to the Reproduction of Sound

THE PHYSICS OF . . . the electric guitar pickup. Electromagnetic induction plays an important role in the technology used for the reproduction of sound. Virtually all electric guitars, for example, use electromagnetic pickups in which an induced emf is generated in a coil of wire by a vibrating string. Each pickup is located below the strings, as **Figure 22.17** illustrates, and each is sensitive to different harmonics that the strings produce. Each string is made from a magnetizable metal, and the pickup consists of a coil of wire within which a permanent magnet is located. The magnetic field of the magnet penetrates the guitar string, causing it to become magnetized with north and south poles. When the magnetized string is plucked, it oscillates, thereby changing the magnetic flux that passes through the coil. The changing flux induces an emf in the coil, and the polarity of this emf alternates with the vibratory motion of the string. A string vibrating at 440 Hz, for example, induces a 440-Hz ac emf in the coil. This signal, after being amplified, is sent to loudspeakers, which produce a 440-Hz sound wave (concert A).

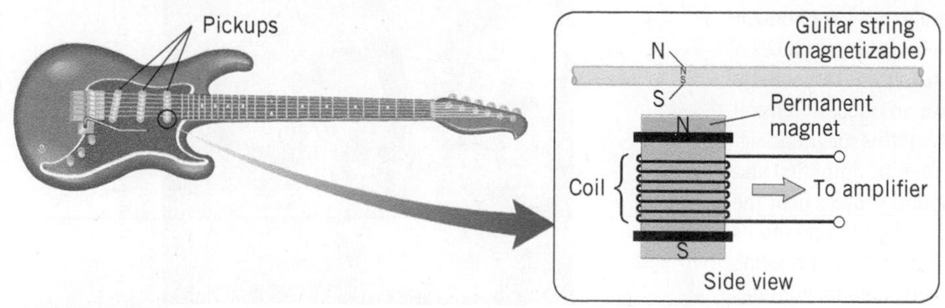

FIGURE 22.17 When the string of an electric guitar vibrates, an emf is induced in the coil of the pickup. The two ends of the coil are connected to the input of an amplifier.

THE PHYSICS OF . . . a tape-deck playback head. The playback head of a tape deck uses a moving tape to generate an emf in a coil of wire. **Figure 22.18** shows a section of magnetized tape in which a series of "tape magnets" have been created in the magnetic layer of the tape during the recording process (see Section 21.9). The tape moves beneath the playback head, which consists of a coil of wire wrapped around an iron core. The iron core has the approximate shape of a horseshoe with a small gap between the two ends. Some of the field lines of the tape magnet under the gap are routed through the highly magnetizable iron core, and hence through the coil, as they proceed from the north pole to the south pole. Consequently, the flux through the coil changes as the tape moves past the gap. The change in flux leads to an ac emf, which is amplified and sent to the speakers, which reproduce the original sound.

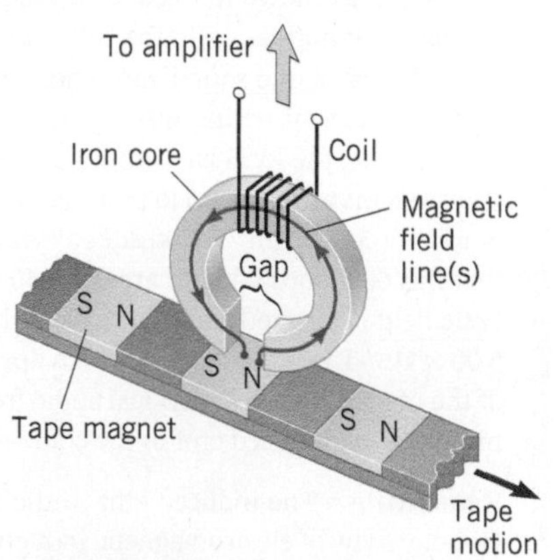

FIGURE 22.18 The playback head of a tape deck. As each tape magnet goes by the gap, some magnetic field lines pass through the core and coil. The changing flux in the coil creates an induced emf. The gap width has been exaggerated.

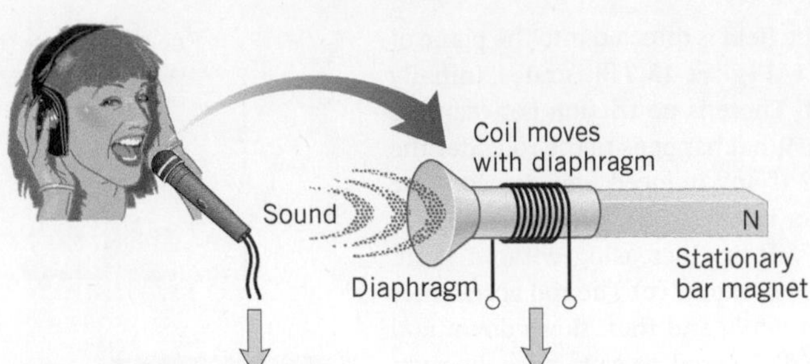

FIGURE 22.19 A moving-coil microphone.

THE PHYSICS OF . . . microphones. There are a number of types of microphones, and Figure 22.19 illustrates the one known as a moving-coil microphone. When a sound wave strikes the diaphragm of the microphone, the diaphragm vibrates back and forth, and the attached coil moves along with it. Nearby is a stationary magnet. As the coil alternately approaches and recedes from the magnet, the flux through the coil changes. Consequently, an ac emf is induced in the coil. This electrical signal is sent to an amplifier and then to the speakers. In a moving-magnet microphone, the magnet is attached to the diaphragm and moves relative to a stationary coil.

Example 10 describes a useful application of electromagnetic induction, called a hearing loop, that can provide hearing alternatives for those who have difficulties in places that have a lot of background noise.

EXAMPLE 10 | BIO A Hearing Loop

People that use hearing aids often encounter problems in places that have a lot of background noise, such as classrooms, theaters, and auditoriums. This occurs because, in its normal setting, a hearing aid directly detects sound waves, including those emanating from nearby sources, obscuring those from the desired source located some distance away. One solution to this problem is an audio induction loop, or a "hearing loop." The idea is that a cable is placed around a designated area, such as an auditorium, and then connected to a microphone or other electronic sound amplification source. The cable produces a magnetic field in the looped space that oscillates at the same audio frequencies as the source. Hearing aids that have an onboard telecoil or "T-coil," as shown in Figure 22.20, can pick up this magnetic signal via the induced emf in the coil, which can then be amplified and converted back into sound. Since none of the sound sources near the listener (e.g., whispering spectators) are picked up by the presenter's microphone, the local background noise is eliminated. Such hearing loop systems are required to produce peak magnetic field magnitudes of at least 5.00×10^{-7} T. Consider a cylindrical hearing induction coil that has 600 loops with an area of $1.50\ \text{mm}^2$. Suppose that the magnetic field is directed along the axis of the coil and ramps from 0 T to 5.00×10^{-7} T in 1.30×10^{-4} s, which approximates the rate of change of the magnetic field at a typical audio frequency. Determine the magnitude of the induced emf in the coil.

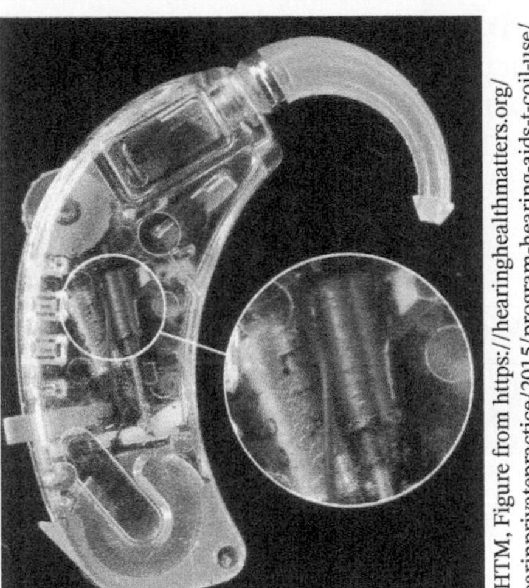

HHTM, Figure from https://hearinghealthmatters.org/hearinprivatepractice/2015/program-hearing-aids-t-coil-use/

FIGURE 22.20 The telecoil, or "T-coil," inside a hearing aid.

Reasoning The induced emf in the coil can be determined from Faraday's law of electromagnetic induction (Equation 22.3). Since the magnetic field is directed along the axis of the coil, $\cos\phi = 1$ in Equation 22.2 for the magnetic flux through the coil, and $\Phi = BA$. Since the area is constant, the time rate of change of the magnetic flux will be determined solely by the changing magnetic field magnitude.

Solution The time rate of change of the magnetic flux is given by $\Delta\Phi/\Delta t$, where $\Phi = BA$, since $\cos\phi = 1$ in this case. Since the area A does not change with time, the change in magnetic flux with time can be written as

$$\frac{\Delta\Phi}{\Delta t} = \frac{\Delta(BA)}{\Delta t} = A\left(\frac{\Delta B}{\Delta t}\right)$$

The magnitude of the emf is given by Equation 22.3. With $N = 600$, we have

$$|\mathscr{E}| = N\left|\frac{\Delta\Phi}{\Delta t}\right| = NA\left|\frac{\Delta B}{\Delta t}\right| = (600)(1.50 \times 10^{-6}\ \text{m}^2)\left(\frac{5.00 \times 10^{-7}\ \text{T}}{1.30 \times 10^{-4}\ \text{s}}\right)$$

$$= \boxed{3.46 \times 10^{-6}\ \text{V}}$$

Related homework: Problem 73

Check Your Understanding

(The answer is given at the end of the book.)

15. The string of an electric guitar vibrates in a standing wave pattern that consists of nodes and antinodes. (Section 17.5 discusses standing waves.) Where should an electromagnetic pickup be located in the standing wave pattern to produce a maximum emf? **(a)** At a node **(b)** At an antinode

22.7 The Electric Generator

How a Generator Produces an Emf

Electric generators, such as those in **Figure 22.21**, produce virtually all of the world's electrical energy. A generator produces electrical energy from mechanical work, which is just the opposite of what a motor does. In a motor, an *input* electric current causes a coil to rotate, thereby doing mechanical work on any object attached to the shaft of the motor. In a generator, the shaft is rotated by some mechanical means, such as an engine or a turbine, and an emf is induced in a coil. If the generator is connected to an external circuit, an electric current is the *output* of the generator.

THE PHYSICS OF . . . an electric generator. In its simplest form, an ac generator consists of a coil of wire that is rotated in a uniform magnetic field, as **Figure 22.22*a*** indicates. Although not shown in the picture, the wire is usually wound around an iron core. As in an electric motor, the coil/core combination is called the *armature.* Each end of the wire forming the coil is connected to the external circuit by means of a metal ring that rotates with the coil. Each ring slides against a stationary carbon brush, to which the external circuit (the lamp in the drawing) is connected.

To see how current is produced by the generator, consider the two vertical sides of the coil in **Figure 22.22*b***. Since each is moving in a magnetic field $\vec{\mathbf{B}}$, the magnetic force exerted on the charges in the wire causes them

FIGURE 22.21 Electric generators such as these supply electric power by producing an induced emf according to Faraday's law of electromagnetic induction.

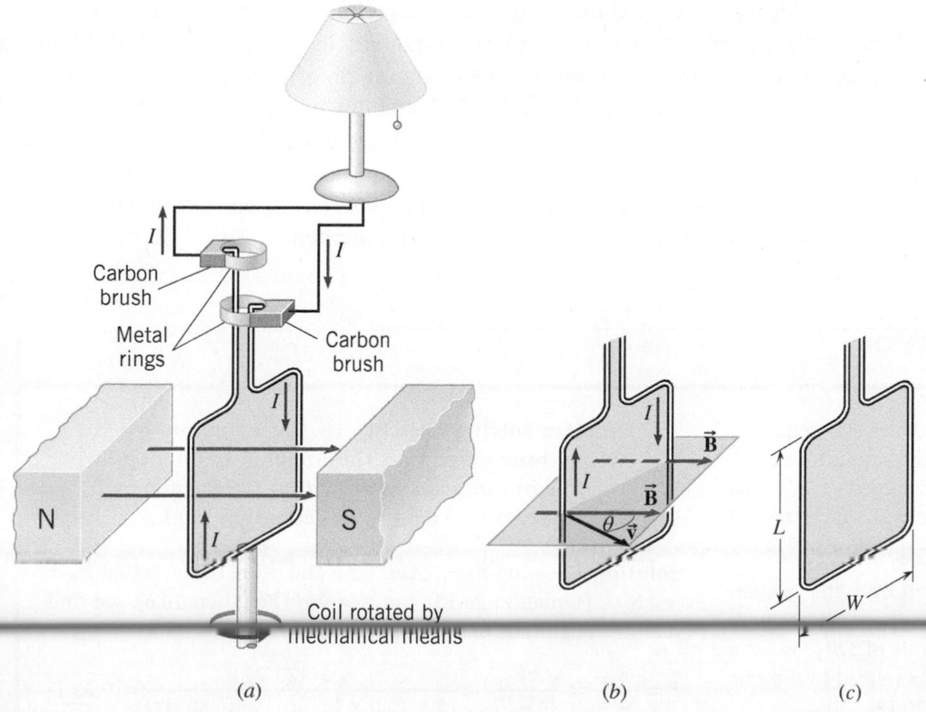

FIGURE 22.22 (*a*) This electric generator consists of a coil (only one loop is shown) of wire that is rotated in a magnetic field $\vec{\mathbf{B}}$ by some mechanical means. (*b*) The current I arises because of the magnetic force exerted on the charges in the moving wire. (*c*) The dimensions of the coil.

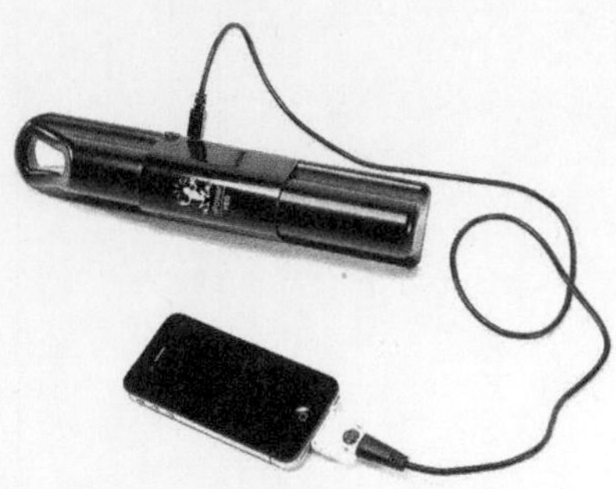

This personal energy generator (PEG) is a small device that can fit into a backpack. It uses Faraday's law of electromagnetic induction to convert some of the kinetic energy of your normal movements into electric energy, which keeps a battery in the device fully charged. The PEG serves as a back-up power source for your mobile phone or other handheld electronic equipment.

to flow, thus creating a current. With the aid of RHR-1 (fingers of extended right hand point along $\vec{B}$, thumb along the velocity $\vec{v}$, palm pushes in the direction of the force on a positive charge), it can be seen that the direction of the current is from bottom to top in the left side and from top to bottom in the right side. Thus, charge flows around the loop. The upper and lower segments of the loop are also moving. However, these segments can be ignored because the magnetic force on the charges within them points toward the sides of the wire and not along the length.

The magnitude of the motional emf developed in a conductor moving through a magnetic field is given by Equation 22.1. To apply this expression to the left side of the coil, whose length is L (see **Figure 22.22c**), we need to use the velocity component $v_\perp$ that is perpendicular to $\vec{B}$. Letting θ be the angle between $\vec{v}$ and $\vec{B}$ (see **Figure 22.22b**), it follows that $v_\perp = v \sin\theta$, and, with the aid of Equation 22.1, the emf can be written as

$$\mathscr{E} = BLv_\perp = BLv \sin\theta$$

The emf induced in the right side has the same magnitude as that in the left side. Since the emfs from both sides drive current in the same direction around the loop, the emf for the complete loop is $\mathscr{E} = 2BLv \sin\theta$. If the coil consists of N loops, the net emf is N times as great as that of one loop, so

$$\mathscr{E} = N(2BLv \sin\theta)$$

It is convenient to express the variables v and θ in terms of the angular speed ω at which the coil rotates. Equation 8.2 shows that the angle θ is the product of the angular speed and the time, $\theta = \omega t$, if it is assumed that $\theta = 0$ rad when $t = 0$ s. Furthermore, any point on each vertical side moves on a circular path of radius $r = W/2$, where W is the width of the coil (see **Figure 22.22c**). Thus, the tangential speed v of each side is related to the angular speed ω via Equation 8.9 as $v = r\omega = (W/2)\omega$. Substituting these expressions for θ and v in the previous equation for $\mathscr{E}$, and recognizing that the product LW is the area A of the coil, we can write the induced emf as

Emf induced in a rotating planar coil

$$\mathscr{E} = NAB\omega \sin\omega t = \mathscr{E}_0 \sin\omega t \quad \text{where } \omega = 2\pi f \qquad \textbf{(22.4)}$$

In this result, the angular speed ω is in radians per second and is related to the frequency f [in cycles per second or hertz (Hz)] according to $\omega = 2\pi f$ (Equation 10.6).

Although Equation 22.4 was derived for a rectangular coil, the result is valid for any planar shape of area A (e.g., circular) and shows that the emf varies sinusoidally with time. The peak, or maximum, emf $\mathscr{E}_0$ occurs when $\sin\omega t = 1$ and has the value $\mathscr{E}_0 = NAB\omega$. **Figure 22.23** shows a plot of Equation 22.4 and reveals that the emf changes polarity as the coil rotates. This changing polarity is exactly the same as that discussed for an ac voltage in Section 20.5 and illustrated in **Figure 20.10**. If the external circuit connected to the generator is a closed circuit, an alternating current results that changes direction at the same frequency f as the emf changes polarity. Therefore, this electric generator is also called an *alternating current (ac) generator*. The next two examples show how Equation 22.4 is applied.

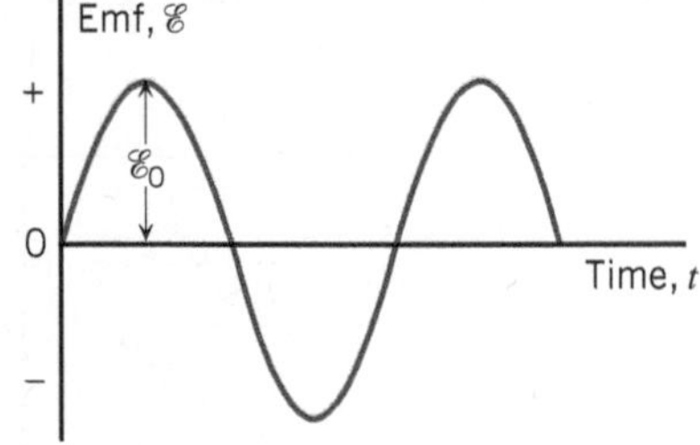

FIGURE 22.23 An ac generator produces this alternating emf $\mathscr{E}$ according to $\mathscr{E} = \mathscr{E}_0 \sin\omega t$.

EXAMPLE 11 | An Ac Generator

In **Figure 22.22** the coil of the ac generator rotates at a frequency of $f = 60.0$ Hz and develops an emf of 120 V (rms; see Section 20.5). The coil has an area of $A = 3.0 \times 10^{-3}$ m^2 and consists of $N = 500$ turns. Find the magnitude of the magnetic field in which the coil rotates.

Reasoning The magnetic field can be found from the relation $\mathscr{E}_0 = NAB\omega$. However, in using this equation we must remember that $\mathscr{E}_0$ is the peak emf, whereas the given value of 120 V is not a peak value but an rms value. The peak emf is related to the rms emf by $\mathscr{E}_0 = \sqrt{2}\mathscr{E}_{rms}$, according to Equation 20.13.

Problem-Solving Insight In the equation $\mathscr{E}_0 = NAB\omega$, remember that the angular frequency ω must be in rad/s and is related to the frequency f (in Hz) according to $\omega = 2\pi f$ (Equation 10.6).

Solution Solving $\mathscr{E}_0 = NAB\omega$ for B and using the facts that $\mathscr{E}_0 = \sqrt{2}\mathscr{E}_{rms}$ (Equation 20.13) and $\omega = 2\pi f$ (Equation 10.6), we find that the magnitude of the magnetic field is

$$B = \frac{\mathscr{E}_0}{NA\omega} = \frac{\sqrt{2}\mathscr{E}_{rms}}{NA2\pi f} = \frac{\sqrt{2}(120\text{ V})}{(500)(3.0\times10^{-3}\text{ m}^2)2\pi(60.0\text{ Hz})} = \boxed{0.30\text{ T}}$$

Analyzing Multiple-Concept Problems

EXAMPLE 12 | The Physics of a Bike Generator

A bicyclist is traveling at night, and a generator mounted on the bike powers a headlight. A small rubber wheel on the shaft of the generator presses against the bike tire and turns the coil of the generator at an angular speed that is 44 times as great as the angular speed of the tire itself. The tire has a radius of 0.33 m. The coil consists of 75 turns, has an area of 2.6×10^{-3} m^2, and rotates in a 0.10-T magnetic field. When the peak emf being generated is 6.0 V, what is the linear speed of the bike?

Reasoning Since the tires are rolling, the linear speed v of the bike is related to the angular speed ω_{tire} of its tires by $v = r\omega_{tire}$ (see Section 8.6), where r is the radius of a tire. We are given that the angular speed ω_{coil} of the coil is 44 times as great as that of the tire. Thus, $\omega_{tire} = \frac{1}{44}\omega_{coil}$ and the linear speed of the bike can be related to the angular speed of the coil. Furthermore, according to the discussion in this section on electric generators, the angular speed of the coil is related (see Equation 22.4) to the peak emf developed by the rotating coil, the number of turns in the coil, the area of the coil, and the magnetic field, all of which are known.

Knowns and Unknowns The data for this problem are:

Description	Symbol	Value	Comment
Radius of tire	r	0.33 m	
Number of turns in coil	N	75	
Angular speed of coil	ω_{coil}	$44\omega_{tire}$	Angular speed of coil is 44 times as great as that of tire.
Area of coil	A	2.6×10^{-3} m^2	
Magnitude of magnetic field	B	0.10 T	
Peak emf produced by generator	$\mathscr{E}_0$	6.0 V	
Unknown Variable			
Linear speed of bike	v	?	

Modeling the Problem

STEP 1 Rolling Motion When a tire rolls without slipping on the ground, the linear speed v of the tire (the speed at which its axle is moving forward) is related to the angular speed ω_{tire} of the tire about the axle. This relationship is given by

$$v = r\omega_{tire} \tag{8.12}$$

where r is the radius of the tire. We are given that the angular speed ω_{coil} of the coil is 44 times as great as the angular speed of the tire, so $\omega_{coil} = 44\omega_{tire}$. Solving this equation for ω_{tire} and substituting the result into $v = r\omega_{tire}$, we obtain Equation 1 at the right. The radius of the tire is known, but the angular speed of the coil is not; we will evaluate it in Step 2.

$$v = r(\tfrac{1}{44})\omega_{coil} \tag{1}$$

?

STEP 2 Peak Emf Induced in a Rotating Planar Coil A generator produces an emf when a coil of wire rotates in a magnetic field. According to Equation 22.4, the peak emf $\mathscr{E}_0$ is given by $\mathscr{E}_0 = NAB\omega_{coil}$, where N is the number of turns in the coil, A is the area of the coil, B is the magnitude of the magnetic field, and ω_{coil} is the angular speed of the rotating coil. Solving this relation for ω_{coil} gives

$$\boxed{\omega_{coil} = \frac{\mathscr{E}_0}{NAB}}$$

Note from the data table that all the variables on the right side of this equation are known. We substitute this result into Equation 1, as indicated at the right.

$$v = r(\tfrac{1}{44})\omega_{coil} \tag{1}$$

$$\omega_{coil} = \frac{\mathscr{E}_0}{NAB}$$

Solution Algebraically combining the results of the two steps, we have

$$v = \overbrace{r\left(\tfrac{1}{44}\right)\omega_{\text{coil}}}^{\text{STEP 1}} = \overbrace{r\left(\tfrac{1}{44}\right)\left(\frac{\mathscr{E}}{NAB}\right)}^{\text{STEP 2}}$$

The linear speed of the bicycle is

$$v = r\left(\tfrac{1}{44}\right)\left(\frac{\mathscr{E}_0}{NAB}\right) = (0.33\text{ m})\left(\tfrac{1}{44}\right)\left[\frac{6.0\text{ V}}{(75)(2.6\times10^{-3}\text{ m}^2)(0.10\text{ T})}\right] = \boxed{2.3\text{ m/s}}$$

Related Homework: Problem 42

The Electrical Energy Delivered by a Generator and the Countertorque

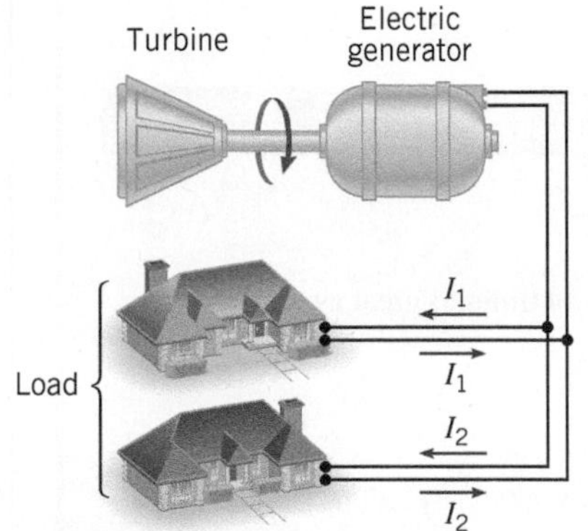

FIGURE 22.24 The generator supplies a total current of $I = I_1 + I_2$ to the load.

Some power-generating stations burn fossil fuel (coal, gas, or oil) to heat water and produce pressurized steam for turning the blades of a turbine whose shaft is linked to the generator. Others use nuclear fuel or falling water as a source of energy. As the turbine rotates, the generator coil also rotates and mechanical work is transformed into electrical energy.

The devices to which the generator supplies electricity are known collectively as the "load," because they place a burden or load on the generator by taking electrical energy from it. If all the devices are switched off, the generator runs under a no-load condition, because there is no current in the external circuit and the generator does not supply electrical energy. Then, work needs to be done on the turbine only to overcome friction and other mechanical losses within the generator itself, and fuel consumption is at a minimum.

Figure 22.24 illustrates a situation in which a load is connected to a generator. Because there is now a current $I = I_1 + I_2$ in the coil of the generator and the coil is situated in a magnetic field, the current experiences a magnetic force $\vec{\mathbf{F}}$. Figure 22.25 shows the magnetic force acting on the left side of the coil, with the direction of $\vec{\mathbf{F}}$ given by RHR-1. A force of equal magnitude but opposite direction also acts on the right side of the coil, although this force is not shown in the drawing. The magnetic force $\vec{\mathbf{F}}$ gives rise to a *countertorque* that opposes the rotational motion. The greater the current drawn from the generator, the greater the countertorque, and the harder it is for the turbine to turn the coil. To compensate for this countertorque and keep the coil rotating at a constant angular speed, work must be done by the turbine, which means that more fuel must be burned. This is another example of the law of conservation of energy, since the electrical energy consumed by the load must ultimately come from the energy source used to drive the turbine.

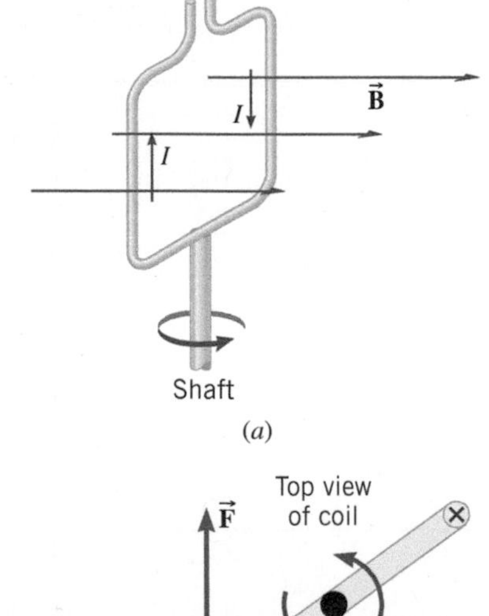

FIGURE 22.25 (*a*) A current I exists in the rotating coil of a generator. (*b*) A top view of the coil, showing the magnetic force $\vec{\mathbf{F}}$ exerted on the left side of the coil.

The Back Emf Generated by an Electric Motor

A generator converts mechanical work into electrical energy; in contrast, an electric motor converts electrical energy into mechanical work. Both devices are similar and consist of a coil of wire that rotates in a magnetic field. In fact, as the armature of a motor rotates, the magnetic flux passing through the coil changes and an emf is induced in the coil. Thus, when a motor is operating, two sources of emf are present: (1) the applied emf V that provides current to drive the motor (e.g., from a 120-V outlet), and (2) the emf $\mathscr{E}$ induced by the generator-like action of the rotating coil. The circuit diagram in Figure 22.26 shows these two emfs.

Consistent with Lenz's law, the induced emf $\mathscr{E}$ acts to oppose the applied emf V and is called the **back emf** or the **counter emf** of the motor. The greater the speed of the motor, the greater is the flux change through the coil, and the greater is the back emf. Because V and $\mathscr{E}$ have opposite polarities, the net emf in the circuit is $V - \mathscr{E}$. In Figure 22.26, R is the

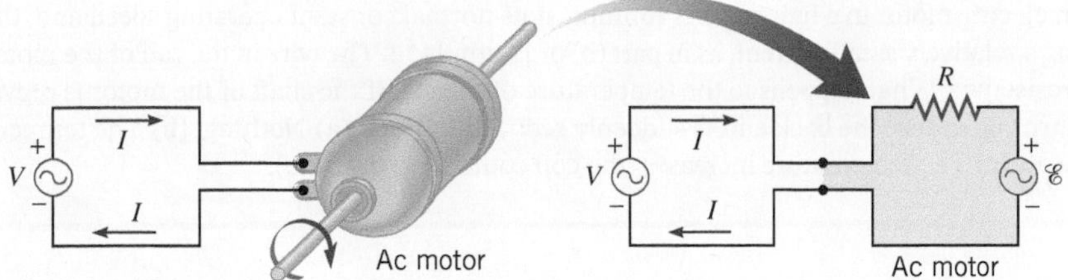

FIGURE 22.26 The applied emf V supplies the current I to drive the motor. The circuit on the right shows V along with the electrical equivalent of the motor, including the resistance R of its coil and the back emf $\mathscr{E}$.

resistance of the wire in the coil, and the current I drawn by the motor is determined from Ohm's law as the net emf divided by the resistance:

$$I = \frac{V - \mathscr{E}}{R} \tag{22.5}$$

The next example uses this result to illustrate that the current in a motor depends on both the applied emf V and the back emf $\mathscr{E}$.

EXAMPLE 13 | The Physics of Operating a Motor

The coil of an ac motor has a resistance of $R = 4.1\ \Omega$. The motor is plugged into an outlet where $V = 120.0$ volts (rms), and the coil develops a back emf of $\mathscr{E} = 118.0$ volts (rms) when rotating at normal constant speed. The motor is turning a wheel. Find **(a)** the current when the motor first starts up and **(b)** the current when the motor is operating at normal speed.

Reasoning Once normal operating speed is attained, the motor need only work to compensate for frictional losses. But in bringing the wheel up to speed from rest, the motor must also do work to increase the wheel's rotational kinetic energy. Thus, bringing the wheel up to speed requires more work, and hence more current, than maintaining the normal operating speed. We expect our answers to parts (a) and (b) to reflect this fact.

Problem-Solving Insight The current in an electric motor depends on both the applied emf V and any back emf $\mathscr{E}$ developed because the coil of the motor is rotating.

Solution (a) When the motor just starts up, the coil is not rotating, so there is no back emf induced in the coil and $\mathscr{E} = 0$ V. The start-up current drawn by the motor is

$$I = \frac{V - \mathscr{E}}{R} = \frac{120\ \text{V} - 0\ \text{V}}{4.1\ \Omega} = \boxed{29\ \text{A}} \tag{22.5}$$

(b) At normal speed, the motor develops a back emf of $\mathscr{E} = 118.0$ volts, so the current is

$$I = \frac{V - \mathscr{E}}{R} = \frac{120.0\ \text{V} - 118.0\ \text{V}}{4.1\ \Omega} = \boxed{0.49\ \text{A}}$$

Example 13 illustrates that when a motor is just starting, there is little back emf, and, consequently, a relatively large current exists in the coil. As the motor speeds up, the back emf increases until it reaches a maximum value when the motor is rotating at normal speed. The back emf becomes almost equal to the applied emf, and the current is reduced to a relatively small value, which is sufficient to provide the torque on the coil needed to overcome frictional and other losses in the motor and to drive the load (e.g., a fan).

Check Your Understanding

(*The answers are given at the end of the book.*)

16. In a car, the generator-like action of the alternator occurs while the engine is running and keeps the battery fully charged. The headlights would discharge an old and failing battery quickly if it were not for the alternator. Why does the engine of a parked car run more quietly with the headlights off than with them on when the battery is in bad shape?

17. You have a fixed length of wire and need to design a generator that will produce the greatest peak emf for a given frequency and magnetic field strength. You should use **(a)** a one-turn square coil, **(b)** a two-turn square coil, **(c)** either a one- or a two-turn square coil because both give the same peak emf for a given frequency and magnetic field strength.

18. An electric motor in a hair dryer is running at its normal constant operating speed and, thus, is drawing a relatively small current, as in part (b) of Example 13. The wire in the coil of the motor has some resistance. What happens to the temperature of the coil if the shaft of the motor is prevented from turning, so that the back emf is suddenly reduced to zero? **(a)** Nothing. **(b)** The temperature decreases. **(c)** The temperature increases (the coil could even burn up).

22.8 Mutual Inductance and Self-Inductance

Mutual Inductance

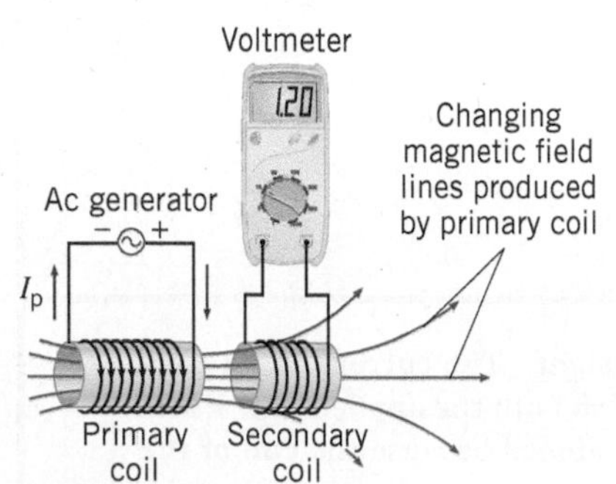

FIGURE 22.27 An alternating current I_p in the primary coil creates an alternating magnetic field. This changing field induces an emf in the secondary coil.

We have seen that an emf can be induced in a coil by keeping the coil stationary and moving a magnet nearby, or by moving the coil near a stationary magnet. **Figure 22.27** illustrates another important method of inducing an emf. Here, two coils of wire, the *primary coil* and the *secondary coil*, are placed close to each other. The primary coil is the one connected to an ac generator, which sends an alternating current I_p through it. The secondary coil is not attached to a generator, although a voltmeter is connected across it to register any induced emf.

The current-carrying primary coil is an electromagnet and creates a magnetic field in the surrounding region. If the two coils are close to each other, a significant fraction of this magnetic field penetrates the secondary coil and produces a magnetic flux. The flux is changing, since the current in the primary coil and its associated magnetic field are changing. Because of the change in flux, an emf is induced in the secondary coil.

The effect in which a changing current in one circuit induces an emf in another circuit is called **mutual induction**. According to Faraday's law of electromagnetic induction, the average emf $\mathscr{E}_s$ induced in the secondary coil is proportional to the change in flux $\Delta\Phi_s$ passing through it. However, $\Delta\Phi_s$ is produced by the change in current ΔI_p in the primary coil. Therefore, it is convenient to recast Faraday's law into a form that relates $\mathscr{E}_s$ to ΔI_p. To see how this recasting is accomplished, note that the net magnetic flux passing through the secondary coil is $N_s\Phi_s$, where N_s is the number of loops in the secondary coil and Φ_s is the flux through one loop (assumed to be the same for all loops). The net flux is proportional to the magnetic field, which, in turn, is proportional to the current I_p in the primary coil. Thus, we can write $N_s\Phi_s \propto I_p$. This proportionality can be converted into an equation in the usual manner by introducing a proportionality constant M, known as the **mutual inductance**:

$$N_s\Phi_s = MI_p \quad \text{or} \quad M = \frac{N_s\Phi_s}{I_p} \tag{22.6}$$

Substituting this equation into Faraday's law, we find that

$$\mathscr{E}_s = -N_s\frac{\Delta\Phi_s}{\Delta t} = -\frac{\Delta(N_s\Phi_s)}{\Delta t} = -\frac{\Delta(MI_p)}{\Delta t} = -M\frac{\Delta I_p}{\Delta t}$$

Emf due to mutual induction

$$\mathscr{E}_s = -M\frac{\Delta I_p}{\Delta t} \tag{22.7}$$

Writing Faraday's law in this manner makes it clear that the average emf $\mathscr{E}_s$ induced in the secondary coil is due to the change in the current ΔI_p in the primary coil.

Equation 22.7 shows that the measurement unit for the mutual inductance M is V·s/A, which is called a henry (H) in honor of Joseph Henry: 1 V·s/A = 1 H. The mutual inductance depends on the geometry of the coils and the nature of any ferromagnetic core material that is present. Although M can be calculated for some highly symmetrical arrangements, it is usually measured experimentally. In most situations, values of M are less than 1 H and are often on the order of millihenries ($1\text{ mH} = 1 \times 10^{-3}\text{ H}$) or microhenries ($1\ \mu\text{H} = 1 \times 10^{-6}\text{ H}$).

BIO THE PHYSICS OF . . . transcranial magnetic stimulation (TMS). A new technique that shows promise for the treatment of psychiatric disorders such as depression is based on mutual induction. This technique is called transcranial magnetic stimulation (TMS) and is a type of indirect and gentler electric shock therapy. In traditional electric shock therapy, electric current is delivered directly through the skull and penetrates the brain, disrupting its electrical circuitry and in the process alleviating the symptoms of the psychiatric disorder. The treatment is not gentle and requires an anesthetic, because relatively large electric currents must be used to penetrate the skull. In contrast, TMS produces its electric current by using a time-varying magnetic field. A primary coil is positioned over the part of the brain to be treated (see **Figure 22.28**), and a time-varying current is applied to this coil. The arrangement is analogous to that in **Figure 22.27**, except that the brain and the electrically conductive pathways within it take the place of the secondary coil. The magnetic field produced by the primary coil penetrates the brain and, since the field is changing in time, it induces an emf in the brain. This induced emf causes an electric current to flow in the conductive brain tissue, with therapeutic results similar to those of conventional electric shock treatment. The current delivered to the brain, however, is much smaller than the current in the conventional treatment, so that patients receive TMS treatments without anesthetic and without severe after-effects such as headaches and memory loss. TMS remains in the experimental stage, however, and the optimal protocol for applying the technique has not yet been determined.

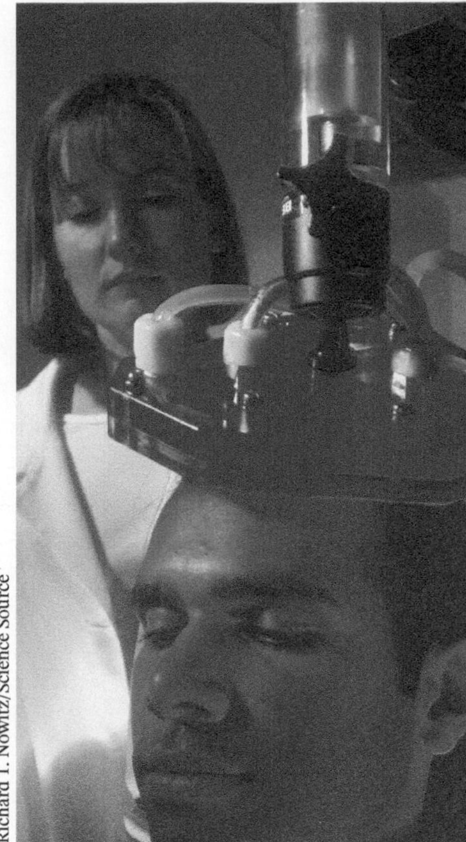

Richard T. Nowitz/Science Source

FIGURE 22.28 In the technique of transcranial magnetic stimulation (TMS), a time-varying electric current is applied to a primary coil, which is positioned over a region of the brain, as this photograph illustrates. The time-varying magnetic field produced by the coil penetrates the brain and creates an induced emf within it. This induced emf leads to an induced current that disrupts the electric circuits of the brain, thereby relieving some of the symptoms of psychiatric disorders such as depression.

Self-Inductance

In all the examples of induced emfs presented so far, the magnetic field has been produced by an external source, such as a permanent magnet or an electromagnet. However, the magnetic field need not arise from an external source. An emf can be induced in a current-carrying coil by a change in the magnetic field that the current *itself* produces. For instance, **Figure 22.29** shows a coil connected to an ac generator. The alternating current creates an alternating magnetic field that, in turn, creates a changing flux through the coil. The change in flux induces an emf in the coil, in accord with Faraday's law. The effect in which a changing current in a circuit induces an emf in the same circuit is referred to as **self-induction**.

When dealing with self-induction, as with mutual induction, it is customary to recast Faraday's law into a form in which the induced emf is proportional to the change in current in the coil rather than to the change in flux. If Φ is the magnetic flux that passes through one turn of the coil, then $N\Phi$ is the net flux through a coil of N turns. Since Φ is proportional to the magnetic field, and the magnetic field is proportional to the current I, it follows that $N\Phi \propto I$. By inserting a constant L, called the **self-inductance**, or simply the **inductance**, of the coil, we can convert this proportionality into Equation 22.8:

$$N\Phi = LI \quad \text{or} \quad L = \frac{N\Phi}{I} \tag{22.8}$$

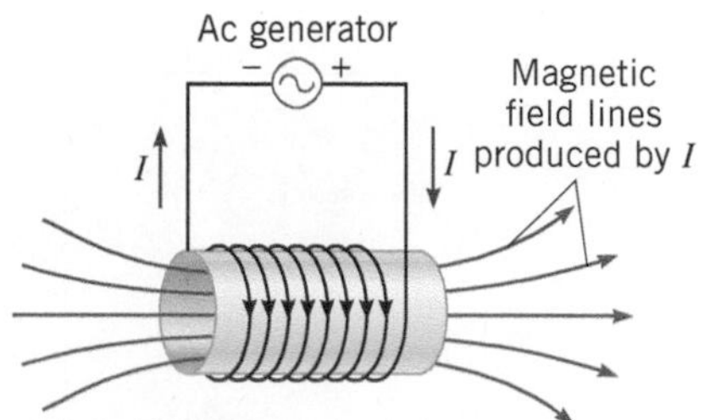

FIGURE 22.29 The alternating current in the coil generates an alternating magnetic field that induces an emf in the coil.

Faraday's law of induction now gives the average induced emf as

$$\mathscr{E} = -N\frac{\Delta\Phi}{\Delta t} = -\frac{\Delta(N\Phi)}{\Delta t} = -\frac{\Delta(LI)}{\Delta t} = -L\frac{\Delta I}{\Delta t}$$

Emf due to self-induction

$$\mathscr{E} = -L\frac{\Delta I}{\Delta t} \tag{22.9}$$

Like mutual inductance, L is measured in henries. The magnitude of L depends on the geometry of the coil and on the core material. Wrapping the coil around a ferromagnetic (iron) core substantially increases the magnetic flux—and therefore the inductance—relative to that for an air core. Because of their self-inductance, coils are known as **inductors** and are widely used in electronics. Inductors come in all sizes, typically in the range between millihenries and microhenries.

The Energy Stored in an Inductor

An inductor, like a capacitor, can store energy. This stored energy arises because a generator does work to establish a current in an inductor. Suppose that an inductor is connected to a generator whose terminal voltage can be varied continuously from zero to some final value. As the voltage is increased, the current I in the circuit rises continuously from zero to its final value. While the current is rising, an induced emf $\mathscr{E} = -L(\Delta I/\Delta t)$ appears across the inductor. Conforming to Lenz's law, the polarity of the induced emf $\mathscr{E}$ is opposite to the polarity of the generator voltage, so as to oppose the increase in the current. Thus, the generator must do work to push the charges through the inductor against this induced emf. The increment of work ΔW done by the generator in moving a small amount of charge ΔQ through the inductor is $\Delta W = -(\Delta Q)\mathscr{E} = -(\Delta Q)[-L(\Delta I/\Delta t)]$, according to Equation 19.4. Since $\Delta Q/\Delta t$ is the current I, the work done is

$$\Delta W = LI(\Delta I)$$

In this expression ΔW represents the work done by the generator to increase the current in the inductor by an amount ΔI. To determine the total work W done while the current is changed from zero to its final value, all the small increments ΔW must be added together. The result is $W = \frac{1}{2}LI^2$, where I represents the final current in the inductor. This work is stored as energy in the inductor, so that

Energy stored in an inductor

$$\text{Energy} = \frac{1}{2}LI^2 \tag{22.10}$$

It is possible to regard the energy in an inductor as being stored in its magnetic field. For the special case of a long solenoid (see Problem 68), the self-inductance is $L = \mu_0 n^2 A\ell$, where n is the number of turns per unit length, A is the cross-sectional area, and ℓ is the length of the solenoid. As a result, the energy stored in a long solenoid is

$$\text{Energy} = \frac{1}{2}LI^2 = \frac{1}{2}\mu_0 n^2 A\ell I^2$$

Since $B = \mu_0 nI$ at the interior of a long solenoid (Equation 21.7), this energy can be expressed as

$$\text{Energy} = \frac{1}{2\mu_0}B^2 A\ell$$

The term $A\ell$ is the volume inside the solenoid where the magnetic field exists, so the energy per unit volume, or **energy density**, is

$$\text{Energy density} = \frac{\text{Energy}}{\text{Volume}} = \frac{1}{2\mu_0}B^2 \tag{22.11}$$

Although this result was obtained for the special case of a long solenoid, it is quite general and is valid for any point where a magnetic field exists in air or vacuum or in a nonmagnetic material. Thus, energy can be stored in a magnetic field, just as it can in an electric field.

22.9 Transformers

One of the most important applications of mutual induction and self-induction takes place in a transformer. A **transformer** is a device that is used to increase or decrease an ac voltage. For instance, whenever a cordless device (e.g., a cell phone) is plugged into a wall receptacle to recharge the batteries, a transformer plays a role in reducing the 120-V ac voltage to a much smaller value. Typically, between 3 and 9 V are needed to energize batteries. In another example, a picture tube in a television set needs about 15 000 V to accelerate the electron beam, and a transformer is used to obtain this high voltage from the 120 V at a wall socket.

THE PHYSICS OF . . . transformers. **Figure 22.30** shows a drawing of a transformer. The transformer consists of an iron core on which two coils are wound: a primary coil with N_p turns and a secondary coil with N_s turns. The primary coil is the one connected to the ac generator. For the moment, suppose that the switch in the secondary circuit is open, so there is no current in this circuit. The alternating current in the primary coil establishes a changing magnetic field in the iron core. Because iron is easily magnetized, it greatly enhances the magnetic field relative to that in an air core and guides the field lines to the secondary coil. In a well-designed core, nearly all the magnetic flux Φ that passes through each turn of the primary also goes through each turn of the secondary. Since the magnetic field is changing, the flux through the primary and secondary coils is also changing, and consequently an emf is induced in both coils. In the secondary coil the induced emf $\mathscr{E}_s$ arises from mutual induction and is given by Faraday's law as

$$\mathscr{E}_s = -N_s \frac{\Delta\Phi}{\Delta t}$$

In the primary coil the induced emf $\mathscr{E}_p$ is due to self-induction and is specified by Faraday's law as

$$\mathscr{E}_p = -N_p \frac{\Delta\Phi}{\Delta t}$$

The term $\Delta\Phi/\Delta t$ is the same in both of these equations, since the same flux penetrates each turn of both coils. Dividing the two equations shows that

$$\frac{\mathscr{E}_s}{\mathscr{E}_p} = \frac{N_s}{N_p}$$

In a high-quality transformer the resistances of the coils are negligible, so the magnitudes of the emfs, $\mathscr{E}_s$ and $\mathscr{E}_p$, are nearly equal to the terminal voltages, V_s and V_p, across the coils (see Section 20.9 for a discussion of terminal voltage). The relation $\mathscr{E}_s/\mathscr{E}_p = N_s/N_p$ is called the **transformer equation** and is usually written in terms of the terminal voltages:

Transformer equation

$$\frac{V_s}{V_p} = \frac{N_s}{N_p} \tag{22.12}$$

According to the transformer equation, if N_s is greater than N_p, the secondary (output) voltage is greater than the primary (input) voltage. In this case we have a *step-up*

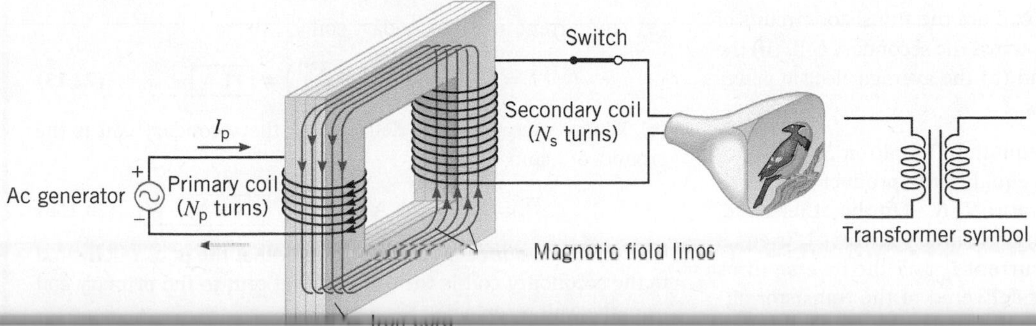

FIGURE 22.30 A transformer consists of a primary coil and a secondary coil, both wound on an iron core. The changing magnetic flux produced by the current in the primary coil induces an emf in the secondary coil. At the far right is the symbol for a transformer.

Sukpaiboonwat/Shutterstock.com

Power distribution stations use high-voltage transformers similar to this one to step up or step down voltages.

transformer. On the other hand, if N_s is less than N_p, the secondary voltage is less than the primary voltage, and we have a *step-down* transformer. The ratio N_s/N_p is referred to as the *turns ratio* of the transformer. A turns ratio of 8/1 (often written as 8 : 1) means, for example, that the secondary coil has eight times more turns than the primary coil. Conversely, a turns ratio of 1:8 implies that the secondary has one-eighth as many turns as the primary.

A transformer operates with ac electricity and not with dc. A steady direct current in the primary coil produces a flux that does not change in time, and thus no emf is induced in the secondary coil. The ease with which transformers can change voltages from one value to another is a principal reason why ac is preferred over dc.

With the switch in the secondary circuit of **Figure 22.30** closed, a current I_s exists in the circuit and electrical energy is fed to the TV tube. This energy comes from the ac generator connected to the primary coil. Although the secondary voltage V_s may be larger or smaller than the primary voltage V_p, energy is not being created or destroyed by the transformer. Energy conservation requires that the energy delivered to the secondary coil must be the same as the energy delivered to the primary coil, provided no energy is dissipated in heating these coils or is otherwise lost. In a well-designed transformer, less than 1% of the input energy is lost in the form of heat. Noting that power is energy per unit time, and assuming 100% energy transfer, the average power $\overline{P}_p$ delivered to the primary coil is equal to the average power $\overline{P}_s$ delivered to the secondary coil: $\overline{P}_p = \overline{P}_s$. However, $\overline{P} = I_{rms}V_{rms}$ (Equation 20.15a), so $I_pV_p = I_sV_s$, or

$$\frac{I_s}{I_p} = \frac{V_p}{V_s} = \frac{N_p}{N_s} \qquad \textbf{(22.13)}$$

Observe that V_s/V_p is equal to the turns ratio N_s/N_p, while I_s/I_p is equal to the inverse turns ratio N_p/N_s. Note that, for clarity, we have dropped the rms subscripts in Equation 22.13, and we will continue with this truncated notation going forward with the understanding that the currents and voltages represent their rms values.

Problem-Solving Insight **Consequently, a transformer that steps up the voltage simultaneously steps down the current, and a transformer that steps down the voltage steps up the current.**

However, the power is neither stepped up nor stepped down, since $\overline{P}_p = \overline{P}_s$. Example 14 emphasizes this fact.

EXAMPLE 14 | A Step-Down Transformer

A step-down transformer inside a stereo receiver has 330 turns in the primary coil and 25 turns in the secondary coil. The plug connects the primary coil to a 120-V wall socket, and there is a current of 0.83 A in the primary coil while the receiver is turned on. Connected to the secondary coil are the transistor circuits of the receiver. Find **(a)** the voltage across the secondary coil, **(b)** the current in the secondary coil, and **(c)** the average electric power delivered to the transistor circuits.

Reasoning The transformer equation, Equation 22.12, states that the secondary voltage V_s is equal to the product of the primary voltage V_p and the turns ratio N_s/N_p. On the other hand, Equation 22.13 indicates that the secondary current I_s is equal to the product of the primary current I_p and the inverse turns ratio N_p/N_s. The average power delivered to the transistor circuits is the product of the secondary current and the secondary voltage.

Solution **(a)** The voltage across the secondary coil can be found from the transformer equation:

$$V_s = V_p\frac{N_s}{N_p} = (120\text{ V})\left(\frac{25}{330}\right) = \boxed{9.1\text{ V}} \qquad \textbf{(22.12)}$$

(b) The current in the secondary coil is

$$I_s = I_p\frac{N_p}{N_s} = (0.83\text{ A})\left(\frac{330}{25}\right) = \boxed{11\text{ A}} \qquad \textbf{(22.13)}$$

(c) The average power P_s delivered to the secondary coil is the product of I_s and V_s:

$$\overline{P}_s = I_sV_s = (11\text{ A})(9.1\text{ V}) = \boxed{1.0 \times 10^2\text{ W}} \qquad \textbf{(20.15a)}$$

As a check on our calculation, we verify that the power delivered to the secondary coil is the same as that sent to the primary coil from the wall receptacle:

$$\overline{P}_p = I_pV_p = (0.83\text{ A})(120\text{ V}) = 1.0 \times 10^2\text{ W}$$

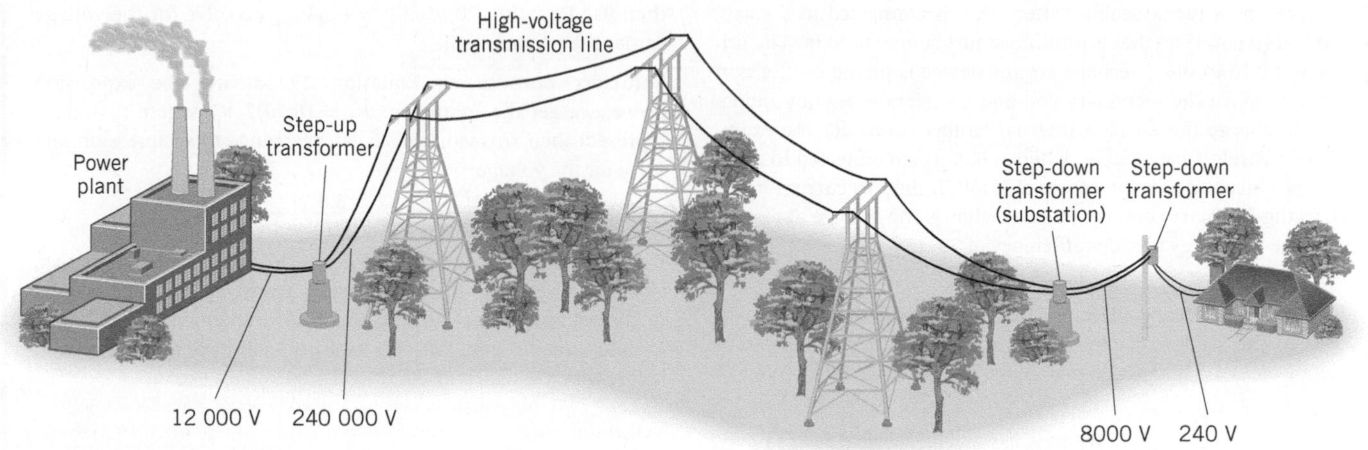

INTERACTIVE FIGURE 22.31 Transformers play a key role in the transmission of electric power.

Transformers play an important role in the transmission of power between electrical generating plants and the communities they serve. Whenever electricity is transmitted, there is always some loss of power in the transmission lines themselves due to resistive heating. Since the resistance of the wires is proportional to their length, the longer the wires the greater is the power loss. The resistance of the wires is also inversely proportional to their cross-sectional area, so thicker wires are used to help minimize the power loss. To reduce the loss further, power companies use transformers that step up the voltage to high levels while reducing the current. A smaller current means less power loss, since $P = I^2R$, where R is the resistance of the transmission wires (see Problem 58). **Interactive Figure 22.31** shows one possible way of transmitting power. The power plant produces a voltage of 12 000 V. This voltage is then raised to 240 000 V by a 20:1 step-up transformer. The high-voltage power is sent over the long-distance transmission line. Upon arrival at the city, the voltage is reduced to about 8000 V at a substation using a 1:30 step-down transformer. However, before any domestic use, the voltage is further reduced to 240 V (or possibly 120 V) by another step-down transformer that is often mounted on a utility pole. The power is then distributed to consumers.

Check Your Understanding

(*The answers are given at the end of the book.*)

19. A transformer changes the 120-V voltage at a wall socket to 12 000 V. The current delivered by the wall socket is **(a)** stepped up by a factor of 100, **(b)** stepped down by a factor of 100, **(c)** neither stepped up nor stepped down.

20. A transformer that stepped up the voltage and the current simultaneously would **(a)** produce less power at the secondary coil than was supplied at the primary coil, **(b)** produce more power at the secondary coil than was supplied at the primary coil, **(c)** produce the same amount of power at the secondary coil that was supplied at the primary coil, **(d)** violate the law of conservation of energy. Choose one or more.

EXAMPLE 15 | BIO The Physics of Wireless Charging—Implantable Medical Devices

Because of electromagnetic induction, energy can be transferred between two coils that are not in contact, as long as the changing magnetic flux created by one coil passes through the second one. This arrangement can be used to charge batteries connected to small medical devices that are implanted inside the body for many different applications (**Figure 22.32**). The implantable device is powered by a rechargeable battery that is connected to a small, flat coil (secondary) that is positioned just below the skin. The primary coil from the external charging device is placed on the skin directly above the secondary coil and transfers energy by induction to charge the device's internal battery. Consider the example of a wireless pacemaker, where a battery is connected to a coil that produces an average power of 3.0 W. If the rms current in the charging (primary) coil is 500 mA, what is the voltage across it? Assume an energy transfer efficiency of 70%.

Reasoning The problem tells us that the average power delivered to the secondary coil by the primary occurs with an efficiency of 70%. This means that $\overline{P}_s = (70\%)\overline{P}_p = (0.70)\overline{P}_p$. We can then use Equation 20.15a ($\overline{P} = I_{rms}V_{rms}$) to solve for the voltage across the primary coil.

Solution Substituting Equation 20.15a into the expression above, we get the following: $\overline{P}_s = (0.70)I_pV_p$, where I_p and V_p represent their rms values. We can rearrange this expression and solve for the voltage:

$$V_p = \frac{\overline{P}_s}{(0.70)I_p} = \frac{3.0 \text{ W}}{(0.70)(0.50 \text{ mA})} = \boxed{8.6 \text{ V}}$$

Implantable devices offer the advantage of lower infection risk, since no wires have to exit the body. However, they take longer to charge, and further advances in their implementation, as well as other areas of concern, like security, will have to be improved before they are widely used.

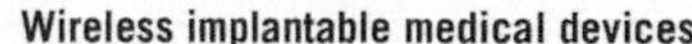

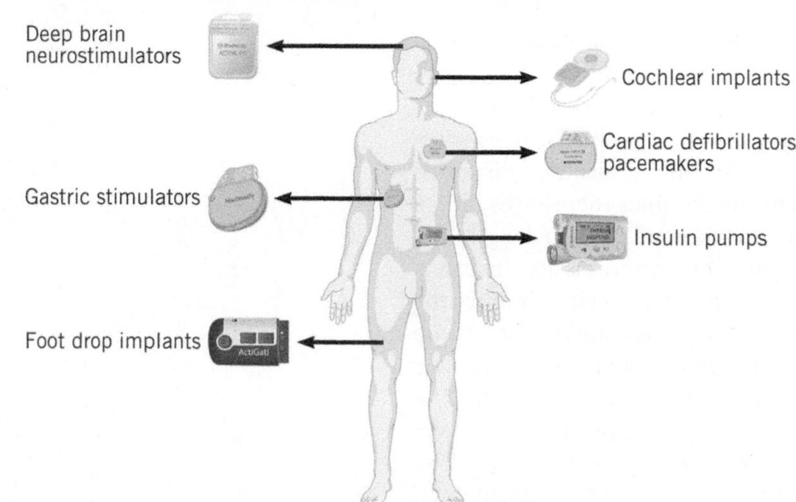

FIGURE 22.32 Examples of medical devices that could be implanted inside the body and charged inductively.

Concept Summary

22.1 Induced Emf and Induced Current Electromagnetic induction is the phenomenon in which an emf is induced in a piece of wire or a coil of wire with the aid of a magnetic field. The emf is called an induced emf, and any current that results from the emf is called an induced current.

22.2 Motional Emf An emf $\mathscr{E}$ is induced in a conducting rod of length L when the rod moves with a speed v in a magnetic field of magnitude B, according to Equation 22.1, which applies when the velocity of the rod, the length of the rod, and the magnetic field are mutually perpendicular. When the motional emf is used to operate an electrical device, such as a light bulb, the energy delivered to the device originates in the work done to move the rod, and the law of conservation of energy applies.

$$\mathscr{E} = vBL \tag{22.1}$$

22.3 Magnetic Flux The magnetic flux Φ that passes through a surface is given by Equation 22.2, where B is the magnitude of the magnetic field, A is the area of the surface, and ϕ is the angle between the field and the normal to the surface. The magnetic flux is proportional to the number of magnetic field lines that pass through the surface.

$$\Phi = BA\cos\phi \tag{22.2}$$

22.4 Faraday's Law of Electromagnetic Induction Faraday's law of electromagnetic induction states that the average emf $\mathscr{E}$ induced in a coil of N loops is given by Equation 22.3, where $\Delta\Phi$ is the change in magnetic flux through one loop and Δt is the time interval during which the change occurs. Motional emf is a special case of induced emf.

$$\mathscr{E} = -N\left(\frac{\Phi - \Phi_0}{t - t_0}\right) = -N\frac{\Delta\Phi}{\Delta t} \tag{22.3}$$

22.5 Lenz's Law Lenz's law provides a way to determine the polarity of an induced emf. Lenz's law is stated as follows: The induced emf resulting from a changing magnetic flux has a polarity that leads to an induced current whose direction is such that the induced magnetic field opposes the original flux change. This statement is a consequence of the law of conservation of energy.

22.7 The Electric Generator In its simplest form, an electric generator consists of a coil of N loops that rotates in a uniform magnetic field $\vec{\mathbf{B}}$. The emf produced by this generator is given by Equation 22.4, where A is the area of the coil, ω is the angular speed (in rad/s) of the coil, and $\mathscr{E}_0 = NAB\omega$ is the peak emf. The angular speed in rad/s is related to the frequency f in cycles/s, or Hz, according to $\omega = 2\pi f$.

$$\mathscr{E} = NAB\omega\sin\omega t = \mathscr{E}_0\sin\omega t \tag{22.4}$$

When an electric motor is running, it exhibits a generator-like behavior by producing an induced emf, called the back emf. The current I needed to keep the motor running at a constant speed is given by Equation 22.5, where V is the emf applied to the motor by an external source, $\mathscr{E}$ is the back emf, and R is the resistance of the motor coil.

$$I = \frac{V - \mathscr{E}}{R} \tag{22.5}$$

22.8 Mutual Inductance and Self-Inductance Mutual induction is the effect in which a changing current in the primary coil induces an emf in the secondary coil. The average emf $\mathscr{E}_s$ induced in the secondary coil by a change in current ΔI_p in the primary coil is given by Equation 22.7, where Δt is the time interval during which the change occurs. The constant M is the mutual inductance between the two coils and is measured in henries (H).

$$\mathscr{E}_s = -M\frac{\Delta I_p}{\Delta t} \tag{22.7}$$

Self-induction is the effect in which a change in current ΔI in a coil induces an average emf $\mathscr{E}$ in the same coil, according to Equation 22.9. The constant L is the self-inductance, or inductance, of the coil and is measured in henries.

$$\mathscr{E} = -L\frac{\Delta I}{\Delta t} \tag{22.9}$$

To establish a current I in an inductor, work must be done by an external agent. This work is stored as energy in the inductor, the amount of energy being given by Equation 22.10. The energy stored in an inductor can be regarded as being stored in its magnetic field. At any point in air or vacuum or in a nonmagnetic material where a magnetic field $\vec{\mathbf{B}}$ exists, the energy density, or the energy stored per unit volume, is given by Equation 22.11.

$$\text{Energy} = \tfrac{1}{2}LI^2 \tag{22.10}$$

$$\text{Energy density} = \frac{1}{2\mu_0}B^2 \tag{22.11}$$

22.9 Transformers A transformer consists of a primary coil of N_p turns and a secondary coil of N_s turns. If the resistances of the coils are negligible, the voltage V_p across the primary coil and the voltage V_s across the secondary coil are related according to the transformer equation, which is Equation 22.12, where the ratio N_s/N_p is called the turns ratio of the transformer. A transformer functions with ac electricity, not with dc. If the transformer is 100% efficient in transferring power from the primary to the secondary coil, the ratio of the secondary current I_s to the primary current I_p is given by Equation 22.13.

$$\frac{V_s}{V_p} = \frac{N_s}{N_p} \tag{22.12}$$

$$\frac{I_s}{I_p} = \frac{N_p}{N_s} \tag{22.13}$$

Focus on Concepts

Online

Additional questions are available for assignment in WileyPLUS.

Section 22.2 Motional Emf

1. You have three light bulbs; bulb A has a resistance of 240 Ω, bulb B has a resistance of 192 Ω, and bulb C has a resistance of 144 Ω, Each of these bulbs is used for the same amount of time in a setup like the one in the drawing. In each case the speed of the rod and the magnetic field strength are the same. Rank the setups in descending order, according to how much work the hand in the drawing must do (largest amount of work first). **(a)** A, B, C **(b)** A, C, B **(c)** B, C, A **(d)** B, A, C **(e)** C, B, A

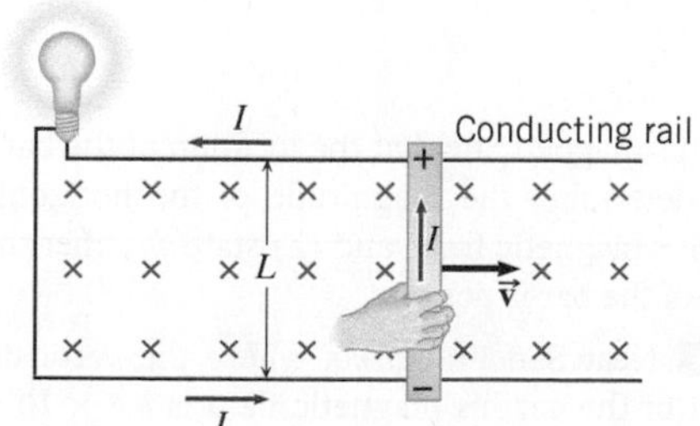

QUESTION 1

Section 22.3 Magnetic Flux

2. The drawing shows a cube. The dashed lines in the drawing are perpendicular to faces 1, 2, and 3 of the cube. Magnetic fields are oriented with respect to these faces as shown, and each of the three fields $\vec{\mathbf{B}}_1$, $\vec{\mathbf{B}}_2$, and $\vec{\mathbf{B}}_3$ has the same magnitude. Note that $\vec{\mathbf{B}}_2$ is parallel to face 2 of the cube. Rank the magnetic fluxes that pass through the faces 1, 2, and 3 of the cube in decreasing order (largest first). **(a)** Φ_1, Φ_2, Φ_3 **(b)** Φ_1, Φ_3, Φ_2 **(c)** Φ_2, Φ_1, Φ_3 **(d)** Φ_2, Φ_3, Φ_1 **(e)** Φ_3, Φ_2, Φ_1

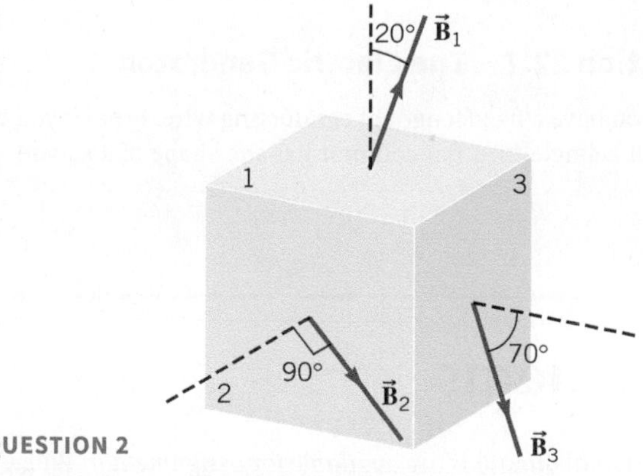

QUESTION 2

Section 22.4 Faraday's Law of Electromagnetic Induction

3. The drawing shows three flat coils, one square and two rectangular, that are each being pushed into a region where there is a uniform magnetic field directed into the page. Outside of this region the magnetic field is zero. In each case the magnetic field within the region has the same magnitude, and the coil is being pushed at the same velocity $\vec{\mathbf{v}}$. Each coil begins with one side just at the edge of the field region. Consider the magnitude of the average emf induced as each coil is pushed from the starting position shown in the drawing until the coil is just completely within the field region. Rank the magnitudes of the average emfs in descending order (largest first). **(a)** $\mathscr{E}_A, \mathscr{E}_B, \mathscr{E}_C$ **(b)** $\mathscr{E}_A, \mathscr{E}_C, \mathscr{E}_B$ **(c)** $\mathscr{E}_B$, $\mathscr{E}_A$ and $\mathscr{E}_C$ (a tie) **(d)** $\mathscr{E}_C$, $\mathscr{E}_A$ and $\mathscr{E}_B$ (a tie)

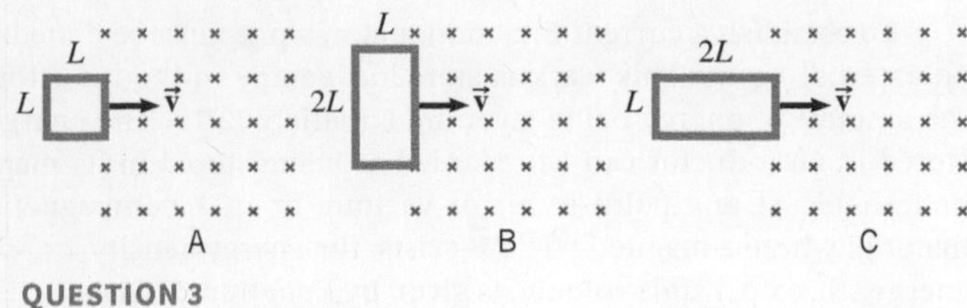

QUESTION 3

4. A long, vertical, straight wire carries a current I. The wire is perpendicular to the plane of a circular metal loop and passes through the center of the loop (see the drawing). The loop is allowed to fall and maintains its orientation with respect to the straight wire while doing so. In what direction does the current induced in the loop flow? (a) There is no induced current. (b) It is flowing around the loop from A to B to C to A. (c) It is flowing around the loop from C to B to A to C.

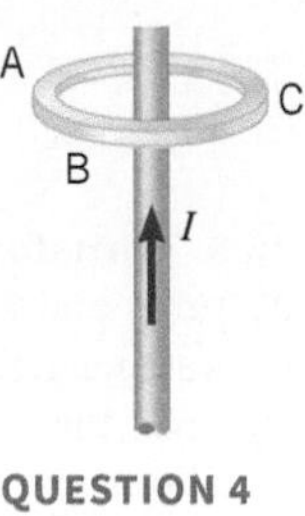

QUESTION 4

Section 22.5 Lenz's Law

5. The drawing shows a top view of two circular coils of conducting wire lying on a flat surface. The centers of the coils coincide. In the larger coil there are a switch and a battery. The smaller coil contains no switch and no battery. Describe the induced current that appears in the smaller coil when the switch in the larger coil is closed. (a) It flows counterclockwise forever after the switch is closed. (b) It flows clockwise forever after the switch is closed. (c) It flows counterclockwise, but only for a short period just after the switch is closed. (d) It flows clockwise, but only for a short period just after the switch is closed.

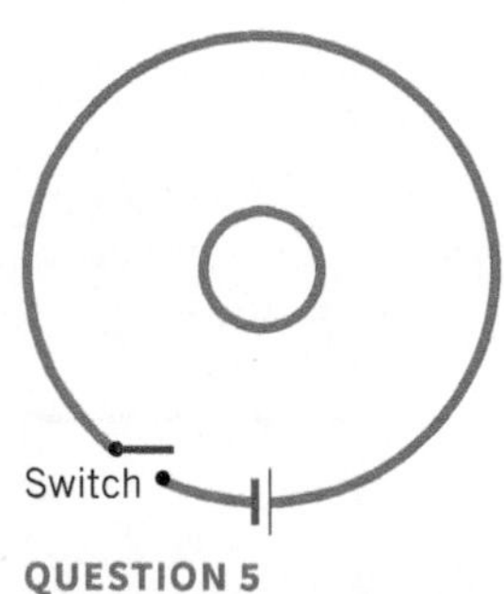

QUESTION 5

Section 22.7 The Electric Generator

6. You have a fixed length of conducting wire. From it you can construct a single-turn flat coil that has the shape of a square, a circle, or a rectangle with the long side twice the length of the short side. Each can be used with the same magnetic field to produce a generator that operates at the same frequency. Rank the peak emfs $\mathscr{E}_0$ of the three generators in descending order (largest first). (a) $\mathscr{E}_{0,\,\text{square}}$, $\mathscr{E}_{0,\,\text{circle}}$, $\mathscr{E}_{0,\,\text{rectangle}}$ (b) $\mathscr{E}_{0,\,\text{circle}}$, $\mathscr{E}_{0,\,\text{square}}$, $\mathscr{E}_{0,\,\text{rectangle}}$ (c) $\mathscr{E}_{0,\,\text{square}}$, $\mathscr{E}_{0,\,\text{rectangle}}$, $\mathscr{E}_{0,\,\text{circle}}$ (d) $\mathscr{E}_{0,\,\text{rectangle}}$, $\mathscr{E}_{0,\,\text{square}}$, $\mathscr{E}_{0,\,\text{circle}}$ (e) $\mathscr{E}_{0,\,\text{rectangle}}$, $\mathscr{E}_{0,\,\text{circle}}$, $\mathscr{E}_{0,\,\text{square}}$

7. An electric motor is plugged into a standard wall socket and is running at normal speed. Suddenly, some dirt prevents the shaft of the motor from turning quite so rapidly. What happens to the back emf of the motor, and what happens to the current that the motor draws from the wall socket? (a) The back emf increases, and the current drawn from the socket decreases. (b) The back emf increases, and the current drawn from the socket increases. (c) The back emf decreases, and the current drawn from the socket decreases. (d) The back emf decreases, and the current drawn from the socket increases.

Section 22.8 Mutual Inductance and Self-Inductance

8. Inductor 1 stores the same amount of energy as inductor 2, although it has only one-half the inductance of inductor 2. What is the ratio I_1/I_2 of the currents in the two inductors? (a) 2.000 (b) 1.414 (c) 4.000 (d) 0.500 (e) 0.250

Section 22.9 Transformers

9. The primary coil of a step-up transformer is connected across the terminals of a standard wall socket, and resistor 1 with a resistance R_1 is connected across the secondary coil. The current in the resistor is then measured. Next, resistor 2 with a resistance R_2 is connected directly across the terminals of the wall socket (without the transformer). The current in this resistor is also measured and found to be the same as the current in resistor 1. How does the resistance R_2 compare to the resistance R_1? (a) The resistance R_2 is less than the resistance R_1. (b) The resistance R_2 is greater than the resistance R_1. (c) The resistance R_2 is the same as the resistance R_1.

Problems

Online

Additional questions are available for assignment in WileyPLUS.

- SSM Student Solutions Manual
- MMH Problem-solving help
- GO Guided Online Tutorial
- V-HINT Video Hints
- CHALK Chalkboard Videos
- BIO Biomedical application
- E Easy
- M Medium
- H Hard
- WS Worksheet
- T Team Problem

Section 22.2 Motional Emf

1. E A 0.80-m aluminum bar is held with its length parallel to the east–west direction and dropped from a bridge. Just before the bar hits the river below, its speed is 22 m/s, and the emf induced across its length is 6.5×10^{-4} V. Assuming the horizontal component of the earth's magnetic field at the location of the bar points directly north, (a) determine the magnitude of the horizontal component of the earth's magnetic field, and (b) state whether the east end or the west end of the bar is positive.

2. E Near San Francisco, where the vertically downward component of the earth's magnetic field is 4.8×10^{-5} T, a car is traveling forward at 25 m/s. The width of the car is 2.0 m. (a) Find the emf induced between the two sides of the car. (b) Which side of the car is positive—the driver's side or the passenger's side?

3. E In 1996, NASA performed an experiment called the Tethered Satellite experiment. In this experiment a 2.0×10^4-m length of wire was let out by the space shuttle *Atlantis* to generate a motional emf. The shuttle had an orbital speed of 7.6×10^3 m/s, and the magnitude of the earth's magnetic field at the location of the wire was 5.1×10^{-5} T. If the wire had moved perpendicular to the earth's magnetic field, what would have been the motional emf generated between the ends of the wire?

4. E SSM The drawing shows three identical rods (A, B, and C) moving in different planes. A constant magnetic field of magnitude 0.45 T is directed along the $+y$ axis. The length of each rod is $L = 1.3$ m, and the rods each have the same speed, $v_A = v_B = v_C = 2.7$ m/s. For each rod, find the magnitude of the motional emf, and indicate which end (1 or 2) of the rod is positive.

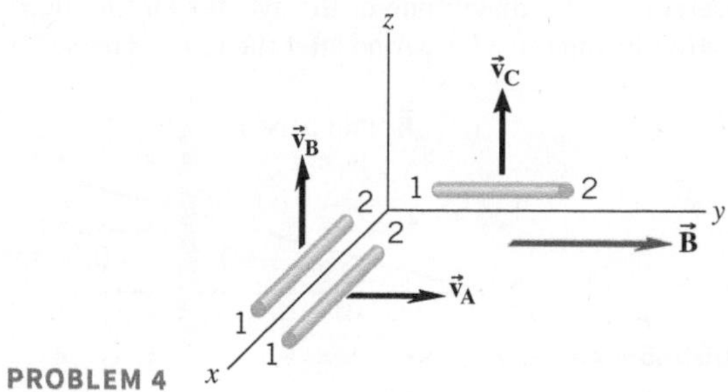

PROBLEM 4

5. E GO Two circuits contain an emf produced by a moving metal rod, like that shown in **Animated Figure 22.4b**. The speed of the rod is the same in each circuit, but the bulb in circuit 1 has one-half the resistance of the bulb in circuit 2. The circuits are otherwise identical. The resistance of the light bulb in circuit 1 is 55 Ω, and that in circuit 2 is 110 Ω. Determine (a) the ratio $\mathcal{E}_1/\mathcal{E}_2$ of the emfs and (b) the ratio I_1/I_2 of the currents in the circuits. (c) If the speed of the rod in circuit 1 were twice that in circuit 2, what would be the ratio P_1/P_2 of the powers in the circuits?

6. M GO Refer to the drawing (**CYU Figure 22.3**) that accompanies Check Your Understanding Question 14. Suppose that the voltage of the battery in the circuit is 3.0 V, the magnitude of the magnetic field (directed perpendicularly into the plane of the paper) is 0.60 T, and the length of the rod between the rails is 0.20 m. Assuming that the rails are very long and have negligible resistance, find the maximum speed attained by the rod after the switch is closed.

7. M V-HINT Multiple-Concept Example 2 discusses the concepts that are used in this problem. Suppose that the magnetic field in **Figure 22.5** has a magnitude of 1.2 T, the rod has a length of 0.90 m, and the hand keeps the rod moving to the right at a constant speed of 3.5 m/s. If the current in the circuit is 0.040 A, what is the average power being delivered to the circuit by the hand?

8. M CHALK SSM Suppose that the light bulb in **Animated Figure 22.4b** is a 60.0-W bulb with a resistance of 240 Ω. The magnetic field has a magnitude of 0.40 T, and the length of the rod is 0.60 m. The only resistance in the circuit is that due to the bulb. What is the shortest distance along the rails that the rod would have to slide for the bulb to remain lit for one-half second?

9. H Review Conceptual Example 3 and **Figure 22.7b**. A conducting rod slides down between two frictionless vertical copper tracks at a constant speed of 4.0 m/s perpendicular to a 0.50-T magnetic field. The resistance of the rod and tracks is negligible. The rod maintains electrical contact with the tracks at all times and has a length of 1.3 m. A 0.75-Ω resistor is attached between the tops of the tracks. (a) What is the mass of the rod? (b) Find the change in the gravitational potential energy that occurs in a time of 0.20 s. (c) Find the electrical energy dissipated in the resistor in 0.20 s.

Section 22.3 Magnetic Flux

For problems in this set, assume that the magnetic flux is a positive quantity.

10. E SSM The drawing shows two surfaces that have the same area. A uniform magnetic field $\vec{B}$ fills the space occupied by these surfaces, and it is oriented parallel to the yz plane as shown. Find the ratio Φ_{xz}/Φ_{xy} of the magnetic fluxes that pass through the surfaces.

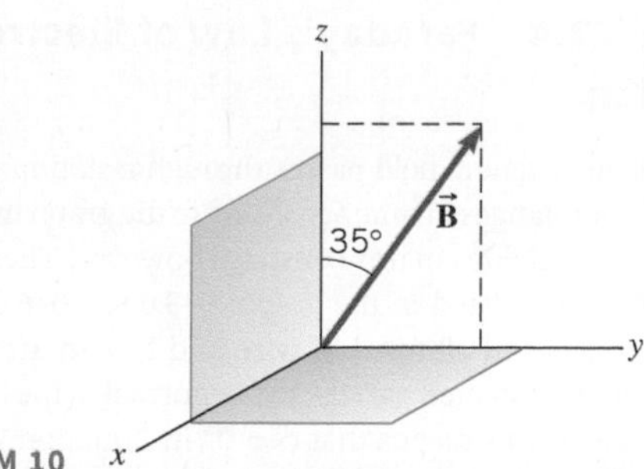

PROBLEM 10

11. E Two flat surfaces are exposed to a uniform, horizontal magnetic field of magnitude 0.47 T. When viewed edge-on, the first surface is tilted at an angle of 12° from the horizontal, and a net magnetic flux of 8.4×10^{-3} Wb passes through it. The same net magnetic flux passes through the second surface. (a) Determine the area of the first surface. (b) Find the smallest possible value for the area of the second surface.

12. E SSM A standard door into a house rotates about a vertical axis through one side, as defined by the door's hinges. A uniform magnetic field is parallel to the ground and perpendicular to this axis. Through what angle must the door rotate so that the magnetic flux that passes through it decreases from its maximum value to one-third of its maximum value?

13. E GO A loop of wire has the shape shown in the drawing. The top part of the wire is bent into a semicircle of radius $r = 0.20$ m. The normal to the plane of the loop is parallel to a constant magnetic field ($\phi = 0°$) of magnitude 0.75 T. What is the change $\Delta\Phi$ in the magnetic flux that passes through the loop when, starting with the position shown in the drawing, the semicircle is rotated through half a revolution?

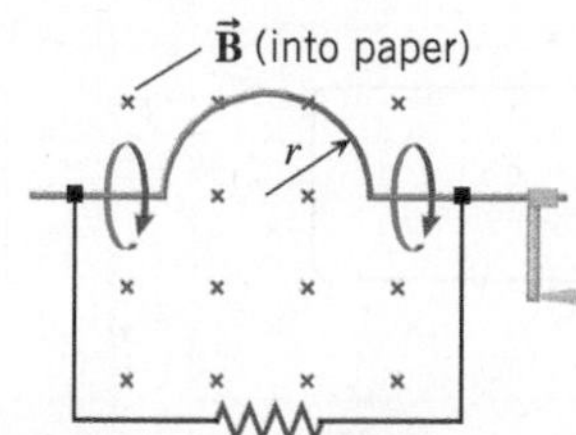

PROBLEM 13

14. E A magnetic field has a magnitude of 0.078 T and is uniform over a circular surface whose radius is 0.10 m. The field is oriented at an angle of $\phi = 25°$ with respect to the normal to the surface. What is the magnetic flux through the surface?

15. M GO CHALK A square loop of wire consisting of a single turn is perpendicular to a uniform magnetic field. The square loop is then re-formed into a circular loop, which also consists of a single turn and is also perpendicular to the same magnetic field. The magnetic flux that passes through the square loop is 7.0×10^{-3} Wb. What is the flux that passes through the circular loop?

16. M V-HINT A five-sided object, whose dimensions are shown in the drawing, is placed in a uniform magnetic field. The magnetic field has a magnitude of 0.25 T and points along the positive y direction. Determine the magnetic flux through each of the five sides.

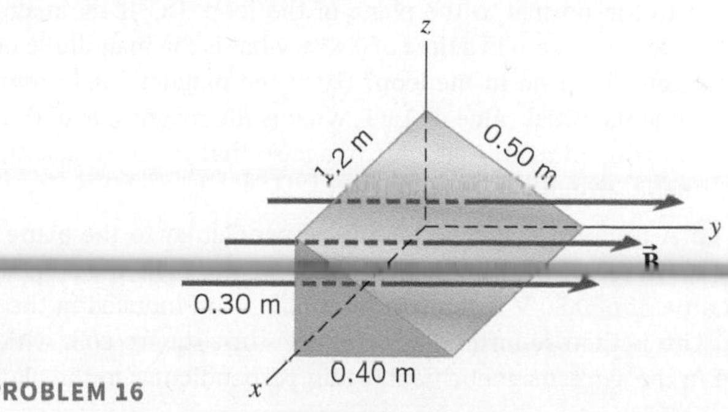

PROBLEM 16

Section 22.4 Faraday's Law of Electromagnetic Induction

17. **E** **GO** A magnetic field passes through a stationary wire loop, and its magnitude changes in time according to the graph in the drawing. The direction of the field remains constant, however. There are three equal time intervals indicated in the graph: 0–3.0 s, 3.0–6.0 s, and 6.0–9.0 s. The loop consists of 50 turns of wire and has an area of 0.15 m^2. The magnetic field is oriented parallel to the normal to the loop. For purposes of this problem, this means that $\phi = 0°$ in Equation 22.2. **(a)** For each interval, determine the induced emf. **(b)** The wire has a resistance of 0.50 Ω. Determine the induced current for the first and third intervals.

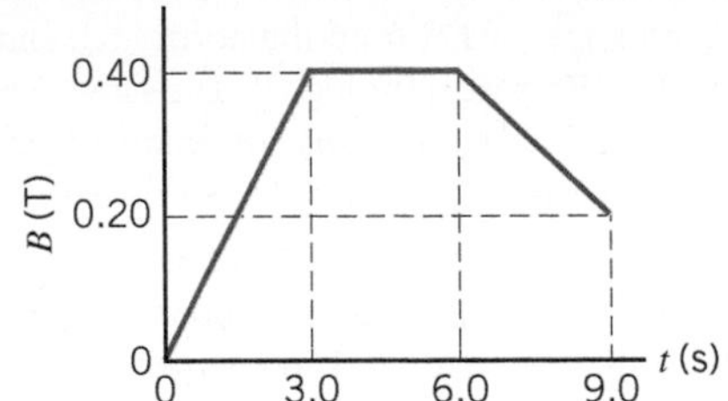

PROBLEM 17

18. **E** A rectangular loop of wire with sides 0.20 and 0.35 m lies in a plane perpendicular to a constant magnetic field (see part *a* of the drawing). The magnetic field has a magnitude of 0.65 T and is directed parallel to the normal of the loop's surface. In a time of 0.18 s, one-half of the loop is then folded back onto the other half, as indicated in part *b* of the drawing. Determine the magnitude of the average emf induced in the loop.

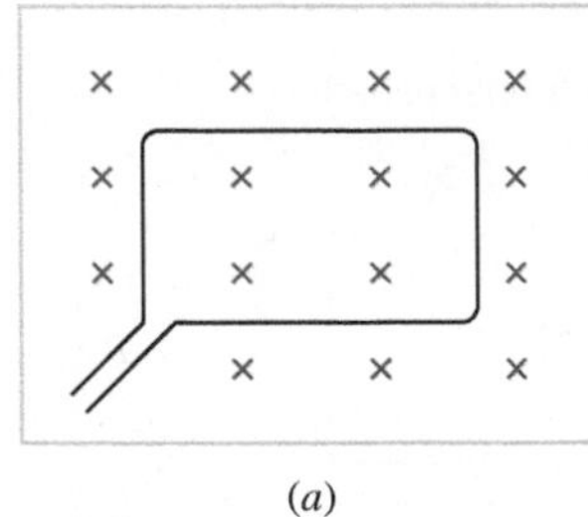

(*a*)

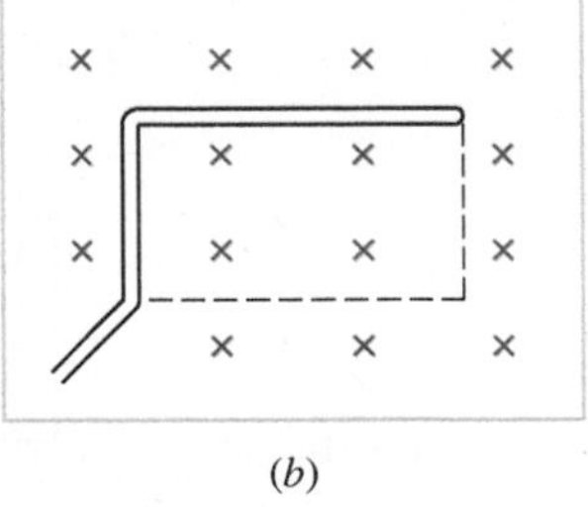

(*b*)

PROBLEM 18

19. **E** **SSM** A circular coil (950 turns, radius = 0.060 m) is rotating in a uniform magnetic field. At $t = 0$ s, the normal to the coil is perpendicular to the magnetic field. At $t = 0.010$ s, the normal makes an angle of $\phi = 45°$ with the field because the coil has made one-eighth of a revolution. An average emf of magnitude 0.065 V is induced in the coil. Find the magnitude of the magnetic field at the location of the coil.

20. **E** The magnetic flux that passes through one turn of a 12-turn coil of wire changes to 4.0 from 9.0 Wb in a time of 0.050 s. The average induced current in the coil is 230 A. What is the resistance of the wire?

21. **E** **MMH** A constant magnetic field passes through a single rectangular loop whose dimensions are 0.35 m × 0.55 m. The magnetic field has a magnitude of 2.1 T and is inclined at an angle of 65° with respect to the normal to the plane of the loop. **(a)** If the magnetic field decreases to zero in a time of 0.45 s, what is the magnitude of the average emf induced in the loop? **(b)** If the magnetic field remains constant at its initial value of 2.1 T, what is the magnitude of the rate $\Delta A/\Delta t$ at which the area should change so that the average emf has the same magnitude as in part (a)?

22. **E** A uniform magnetic field is perpendicular to the plane of a single-turn circular coil. The magnitude of the field is changing, so that an emf of 0.80 V and a current of 3.2 A are induced in the coil. The wire is then re-formed into a single-turn square coil, which is used in the same magnetic field (again perpendicular to the plane of the coil and with a magnitude changing at the same rate). What emf and current are induced in the square coil?

23. **M** **MMH** A copper rod is sliding on two conducting rails that make an angle of 19° with respect to each other, as in the drawing. The rod is moving to the right with a constant speed of 0.60 m/s. A 0.38-T uniform magnetic field is perpendicular to the plane of the paper. Determine the magnitude of the average emf induced in the triangle *ABC* during the 6.0-s period after the rod has passed point *A*.

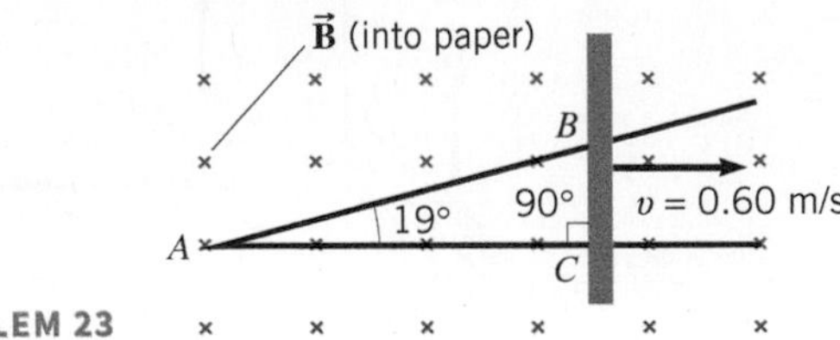

PROBLEM 23

24. **M** **CHALK** **GO** A flat coil of wire has an area *A*, *N* turns, and a resistance *R*. It is situated in a magnetic field, such that the normal to the coil is parallel to the magnetic field. The coil is then rotated through an angle of 90°, so that the normal becomes perpendicular to the magnetic field. The coil has an area of 1.5×10^{-3} m^2, 50 turns, and a resistance of 140 Ω. During the time while it is rotating, a charge of 8.5×10^{-5} C flows in the coil. What is the magnitude of the magnetic field?

25. **E** **SSM** A magnetic field is passing through a loop of wire whose area is 0.018 m^2. The direction of the magnetic field is parallel to the normal to the loop, and the magnitude of the field is increasing at the rate of 0.20 T/s. **(a)** Determine the magnitude of the emf induced in the loop. **(b)** Suppose that the area of the loop can be enlarged or shrunk. If the magnetic field is increasing as in part (a), at what rate (in m^2/s) should the area be changed at the instant when $B = 1.8$ T if the induced emf is to be zero? Explain whether the area is to be enlarged or shrunk.

26. **M** **V-HINT** A flat circular coil with 105 turns, a radius of 4.00×10^{-2} m, and a resistance of 0.480 Ω is exposed to an external magnetic field that is directed perpendicular to the plane of the coil. The magnitude of the external magnetic field is changing at a rate of $\Delta B/\Delta t = 0.783$ T/s, thereby inducing a current in the coil. Find the magnitude of the magnetic field at the center of the coil that is produced by the induced current.

27. **M** **GO** The drawing shows a coil of copper wire that consists of two semicircles joined by straight sections of wire. In part *a* the coil is lying flat on a horizontal surface. The dashed line also lies in the plane of the horizontal surface. Starting from the orientation in part *a*, the smaller semicircle rotates at an angular frequency ω about the dashed line, until its plane becomes perpendicular to the horizontal surface, as shown in part *b*. A uniform magnetic field is constant in time and is directed upward, perpendicular to the horizontal surface. The field completely fills the region occupied by the coil in either part of the drawing. The magnitude of the magnetic field is 0.35 T. The resistance of the coil is 0.025 Ω, and the smaller semicircle has a radius of 0.20 m. The angular frequency at which the small semicircle rotates is 1.5 rad/s. Determine the average current *I*, if any, induced in the coil as the coil changes shape from that in part *a* of the drawing to that in part *b*. Be sure to include an explicit plus or minus sign along with your answer.

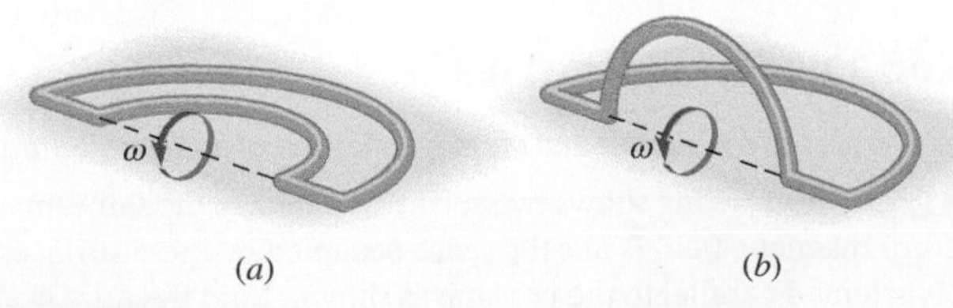

(*a*) (*b*)

PROBLEM 27

28. **M GO** A conducting coil of 1850 turns is connected to a galvanometer, and the total resistance of the circuit is 45.0 Ω. The area of each turn is 4.70×10^{-4} m². This coil is moved from a region where the magnetic field is zero into a region where it is nonzero, the normal to the coil being kept parallel to the magnetic field. The amount of charge that is induced to flow around the circuit is measured to be 8.87×10^{-3} C. Find the magnitude of the magnetic field.

Section 22.5 Lenz's Law

29. **E** Starting from the position indicated in the drawing, the semicircular piece of wire rotates through half a revolution in the direction shown. Which end of the resistor is positive—the left or the right end? Explain your reasoning.

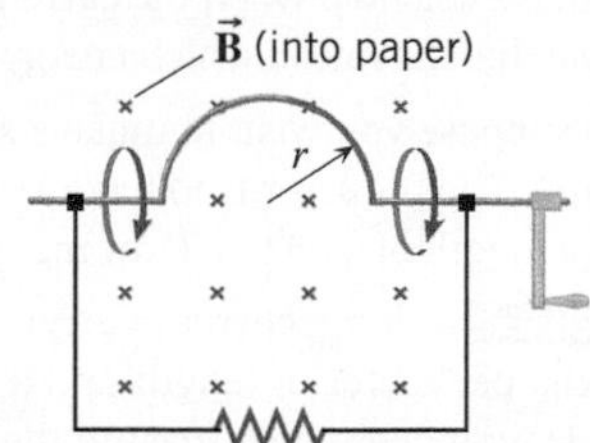

PROBLEM 29

30. **E SSM** The plane of a flat, circular loop of wire is horizontal. An external magnetic field is directed perpendicular to the plane of the loop. The magnitude of the external magnetic field is increasing with time. Because of this increasing magnetic field, an induced current is flowing clockwise in the loop, as viewed from above. What is the direction of the external magnetic field? Justify your conclusion.

31. **E GO** The drawing shows a straight wire carrying a current I. Above the wire is a rectangular loop that contains a resistor R. If the current I is decreasing in time, what is the direction of the induced current through the resistor R—left-to-right or right-to-left?

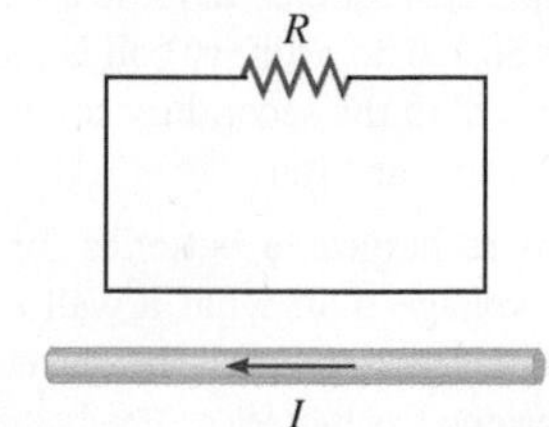

PROBLEM 31

32. **E SSM** The drawing depicts a copper loop lying flat on a table (not shown) and connected to a battery via a closed switch. The current I in the loop generates the magnetic field lines shown in the drawing. The switch is then opened and the current goes to zero. There are also two smaller conducting loops A and B lying flat on the table, but not connected to batteries. Determine the direction of the induced current in **(a)** loop A and **(b)** loop B. Specify the direction of each induced current to be clockwise or counterclockwise when viewed from above the table. Provide a reason for each answer.

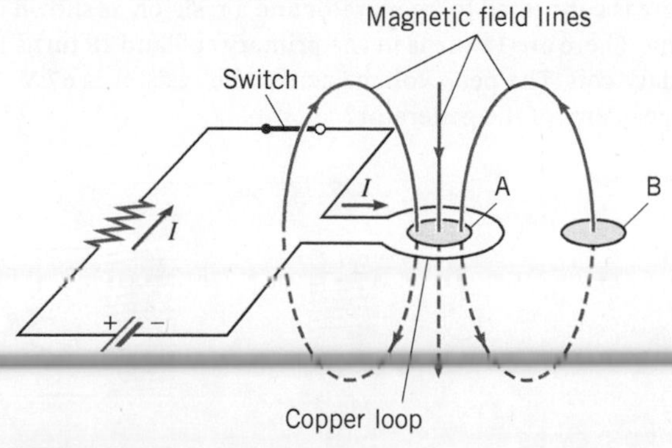

PROBLEM 32

33. **E** The drawing shows that a uniform magnetic field is directed perpendicularly into the plane of the paper and fills the entire region to the left of the y axis. There is no magnetic field to the right of the y axis. A rigid right triangle ABC is made of copper wire. The triangle rotates counterclockwise about the origin at point C. What is the direction (clockwise or counterclockwise) of the induced current when the triangle is crossing **(a)** the $+y$ axis, **(b)** the $-x$ axis, **(c)** the $-y$ axis, and **(d)** the $+x$ axis? For each case, justify your answer.

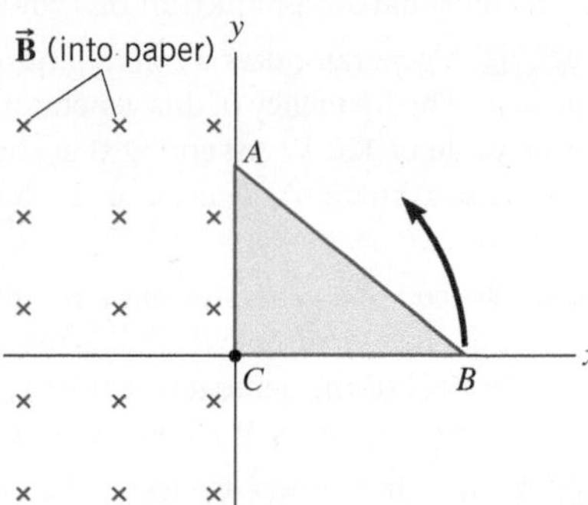

PROBLEM 33

34. **M MMH** A circular loop of wire rests on a table. A long, straight wire lies on this loop, directly over its center, as the drawing illustrates. The current I in the straight wire is decreasing. In what direction is the induced current, if any, in the loop? Give your reasoning.

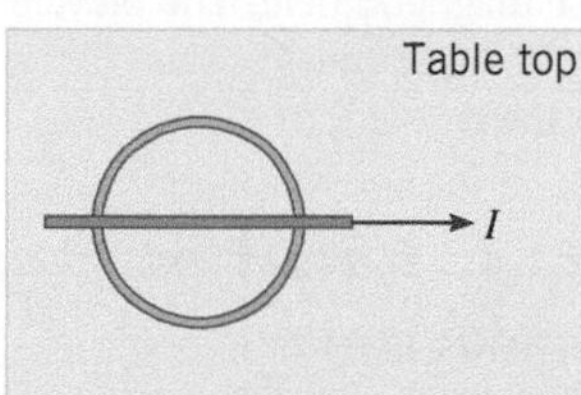

PROBLEM 34

35. **M V-HINT** The drawing shows a bar magnet falling through a metal ring. In part *a* the ring is solid all the way around, but in part *b* it has been cut through. **(a)** Explain why the motion of the magnet in part *a* is retarded when the magnet is above the ring and below the ring as well. Draw any induced currents that appear in the ring. **(b)** Explain why the motion of the magnet is unaffected by the ring in part *b*.

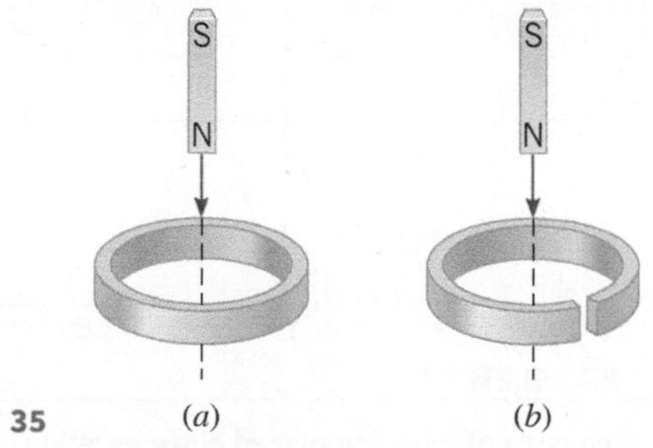

PROBLEM 35 (*a*) (*b*)

Section 22.7 The Electric Generator

36. **E** A 120.0-V motor draws a current of 7.00 A when running at normal speed. The resistance of the armature wire is 0.720 Ω. **(a)** Determine the back emf generated by the motor. **(b)** What is the current at the instant when the motor is just turned on and has not begun to rotate? **(c)** What series resistance must be added to limit the starting current to 15.0 A?

37. **E SSM** A generator has a square coil consisting of 248 turns. The coil rotates at 79.1 rad/s in a 0.170-T magnetic field. The peak output of the generator is 75.0 V. What is the length of one side of the coil?

38. E You need to design a 60.0-Hz ac generator that has a maximum emf of 5500 V. The generator is to contain a 150-turn coil that has an area per turn of 0.85 m^2. What should be the magnitude of the magnetic field in which the coil rotates?

39. E MMH The maximum strength of the earth's magnetic field is about 6.9×10^{-5} T near the south magnetic pole. In principle, this field could be used with a rotating coil to generate 60.0-Hz ac electricity. What is the minimum number of turns (area per turn = 0.022 m^2) that the coil must have to produce an rms voltage of 120 V?

40. M V-HINT A generator uses a coil that has 100 turns and a 0.50-T magnetic field. The frequency of this generator is 60.0 Hz, and its emf has an rms value of 120 V. Assuming that each turn of the coil is a square (an approximation), determine the length of the wire from which the coil is made.

41. M GO The coil of a generator has a radius of 0.14 m. When this coil is unwound, the wire from which it is made has a length of 5.7 m. The magnetic field of the generator is 0.20 T, and the coil rotates at an angular speed of 25 rad/s. What is the peak emf of this generator?

42. M SSM Consult Multiple-Concept Example 12 for background material relating to this problem. A small rubber wheel on the shaft of a bicycle generator presses against the bike tire and turns the coil of the generator at an angular speed that is 38 times as great as the angular speed of the tire itself. Each tire has a radius of 0.300 m. The coil consists of 125 turns, has an area of 3.86×10^{-3} m^2, and rotates in a 0.0900-T magnetic field. The bicycle starts from rest and has an acceleration of +0.550 m/s^2. What is the peak emf produced by the generator at the end of 5.10 s?

Section 22.8 Mutual Inductance and Self-Inductance

43. E SSM The earth's magnetic field, like any magnetic field, stores energy. The maximum strength of the earth's field is about 7.0×10^{-5} T. Find the maximum magnetic energy stored in the space above a city if the space occupies an area of 5.0×10^8 m^2 and has a height of 1500 m.

44. E The current through a 3.2-mH inductor varies with time according to the graph shown in the drawing. What is the average induced emf during the time intervals (a) 0–2.0 ms, (b) 2.0–5.0 ms, and (c) 5.0–9.0 ms?

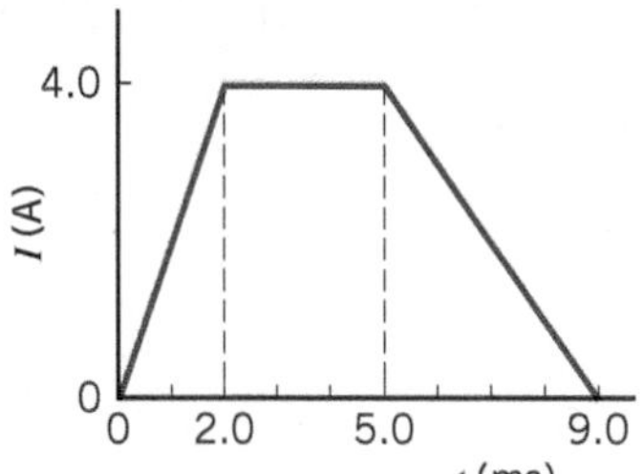

PROBLEM 44

45. E Two coils of wire are placed close together. Initially, a current of 2.5 A exists in one of the coils, but there is no current in the other. The current is then switched off in a time of 3.7×10^{-2} s. During this time, the average emf induced in the other coil is 1.7 V. What is the mutual inductance of the two-coil system?

46. E GO During a 72-ms interval, a change in the current in a primary coil occurs. This change leads to the appearance of a 6.0-mA current in a nearby secondary coil. The secondary coil is part of a circuit in which the resistance is 12 Ω. The mutual inductance between the two coils is 3.2 mH. What is the change in the primary current?

47. E Mutual induction can be used as the basis for a metal detector. A typical setup uses two large coils that are parallel to each other and have a common axis. Because of mutual induction, the ac generator connected to the primary coil causes an emf of 0.46 V to be induced in the secondary coil. When someone without metal objects walks through the coils, the mutual inductance and, thus, the induced emf do not change much. But when a person carrying a handgun walks through, the mutual inductance increases. The change in emf can be used to trigger an alarm. If the mutual inductance increases by a factor of three, find the new value of the induced emf.

48. E GO A constant current of $I = 15$ exists in a solenoid whose inductance is $L = 3.1$ H. The current is then reduced to zero in a certain amount of time. (a) If the current goes from 15 to 0 A in a time of 75 ms, what is the emf induced in the solenoid? (b) How much electrical energy is stored in the solenoid? (c) At what rate must the electrical energy be removed from the solenoid when the current is reduced to 0 A in a time of 75 ms? Note that the rate at which energy is removed is the power.

49. E SSM Suppose you wish to make a solenoid whose self-inductance is 1.4 mH. The inductor is to have a cross-sectional area of 1.2×10^{-3} m^2 and a length of 0.052 m. How many turns of wire are needed?

50. M GO CHALK A long, current-carrying solenoid with an air core has 1750 turns per meter of length and a radius of 0.0180 m. A coil of 125 turns is wrapped tightly around the outside of the solenoid, so it has virtually the same radius as the solenoid. What is the mutual inductance of this system?

Section 22.9 Transformers

51. E The battery charger for an MP3 player contains a step-down transformer with a turns ratio of 1:32, so that the voltage of 120 V available at a wall socket can be used to charge the battery pack or operate the player. What voltage does the secondary coil of the transformer provide?

52. E MMH The secondary coil of a step-up transformer provides the voltage that operates an electrostatic air filter. The turns ratio of the transformer is 50:1. The primary coil is plugged into a standard 120-V outlet. The current in the secondary coil is 1.7×10^{-3} A. Find the power consumed by the air filter.

53. E GO The rechargeable batteries for a laptop computer need a much smaller voltage than what a wall socket provides. Therefore, a transformer is plugged into the wall socket and produces the necessary voltage for charging the batteries. The batteries are rated at 9.0 V, and a current of 225 mA is used to charge them. The wall socket provides a voltage of 120 V. (a) Determine the turns ratio of the transformer. (b) What is the current coming from the wall socket? (c) Find the average power delivered by the wall socket and the average power sent to the batteries.

54. E The resistances of the primary and secondary coils of a transformer are 56 and 14 Ω, respectively. Both coils are made from lengths of the same copper wire. The circular turns of each coil have the same diameter. Find the turns ratio N_s/N_p.

55. E A transformer consisting of two coils wrapped around an iron core is connected to a generator and a resistor, as shown in the drawing. There are 11 turns in the primary coil and 18 turns in the secondary coil. The peak voltage across the resistor is 67 V. What is the peak emf of the generator?

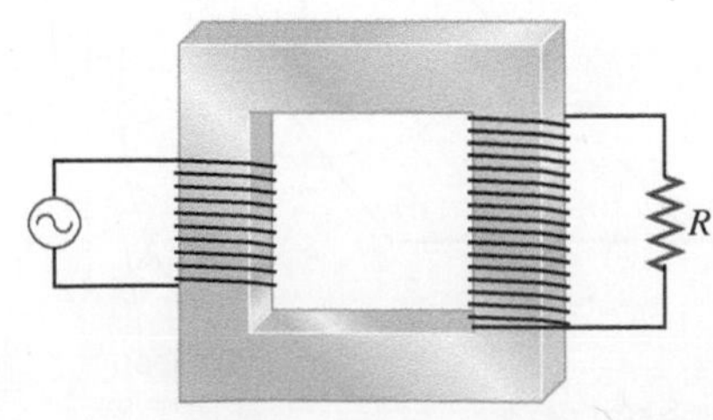

PROBLEM 55

56. E SSM A step-down transformer (turns ratio = 1:8) is used with an electric train to reduce the voltage from the wall receptacle to a value needed to operate the train. When the train is running, the current in the secondary coil is 1.6 A. What is the current in the primary coil?

57. M GO In a television set the power needed to operate the picture tube comes from the secondary of a transformer. The primary of the transformer is connected to a 120-V receptacle on a wall. The picture tube of the television set uses 91 W, and there is 5.5 mA of current in the secondary coil of the transformer to which the tube is connected. Find the turns ratio N_s/N_p of the transformer.

58. M SSM CHALK A generating station is producing 1.2×10^6 W of power that is to be sent to a small town located 7.0 km away. Each of the two wires that comprise the transmission line has a resistance per kilometer of 5.0×10^{-2} Ω/km. **(a)** Find the power used to heat the wires if the power is transmitted at 1200 V. **(b)** A 100:1 step-up transformer is used to raise the voltage before the power is transmitted. How much power is now used to heat the wires?

59. H A generator is connected across the primary coil (N_p turns) of a transformer, while a resistance R_2 is connected across the secondary coil (N_s turns). This circuit is equivalent to a circuit in which a single resistance R_1 is connected directly across the generator, without the transformer. Show that $R_1 = (N_p/N_s)^2R_2$, by starting with Ohm's law as applied to the secondary coil.

Additional Problems

Online

60. E In each of two coils the rate of change of the magnetic flux in a single loop is the same. The emf induced in coil 1, which has 184 loops, is 2.82 V. The emf induced in coil 2 is 4.23 V. How many loops does coil 2 have?

61. E GO A planar coil of wire has a single turn. The normal to this coil is parallel to a uniform and constant (in time) magnetic field of 1.7 T. An emf that has a magnitude of 2.6 V is induced in this coil because the coil's area A is shrinking. What is the magnitude of $\Delta A/\Delta t$, which is the rate (in m²/s) at which the area changes?

62. E CHALK SSM Review Conceptual Example 9 as an aid in understanding this problem. A long, straight wire lies on a table and carries a current I. As the drawing shows, a small circular loop of wire is pushed across the top of the table from position 1 to position 2. Determine the direction of the induced current, clockwise or counterclockwise, as the loop moves past **(a)** position 1 and **(b)** position 2. Justify your answers.

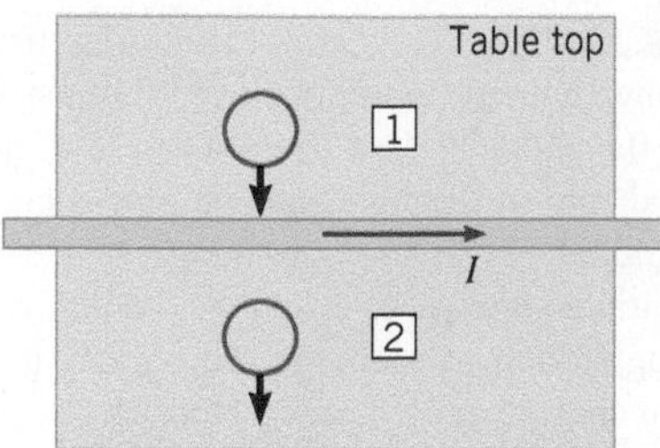

PROBLEM 62

63. E In some places, insect "zappers," with their blue lights, are a familiar sight on a summer's night. These devices use a high voltage to electrocute insects. One such device uses an ac voltage of 4320 V, which is obtained from a standard 120.0-V outlet by means of a transformer. If the primary coil has 21 turns, how many turns are in the secondary coil?

64. M GO At its normal operating speed, an electric fan motor draws only 15.0% of the current it draws when it just begins to turn the fan blade. The fan is plugged into a 120.0-V socket. What back emf does the motor generate at its normal operating speed?

65. M V-HINT Parts a and b of the drawing show the same uniform and constant (in time) magnetic field $\vec{\mathbf{B}}$ directed perpendicularly into the paper over a rectangular region. Outside this region, there is no field. Also shown is a rectangular coil (one turn), which lies in the plane of the paper. In part a the long side of the coil (length = L) is just at the edge of the field region, while in part b the short side (width = W) is just at the edge. It is known that $L/W = 3.0$. In both parts of the drawing the coil is pushed into the field with the same velocity $\vec{\mathbf{v}}$ until it is completely within the field region. The magnitude of the average emf induced in the coil in part a is 0.15 V. What is its magnitude in part b?

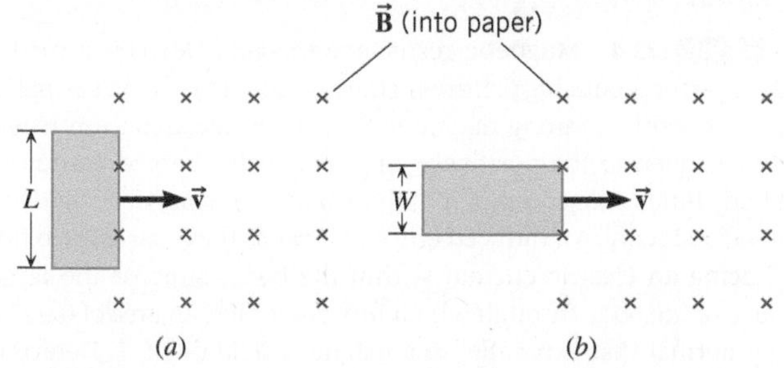

PROBLEM 65

66. M V-HINT Indicate the direction of the electric field between the plates of the parallel plate capacitor shown in the drawing if the magnetic field is decreasing in time. Give your reasoning.

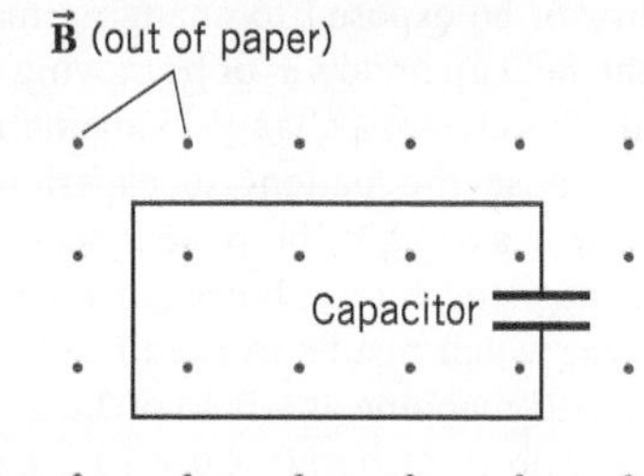

PROBLEM 66

67. M SSM A piece of copper wire is formed into a single circular loop of radius 12 cm. A magnetic field is oriented parallel to the normal to the loop, and it increases from 0 to 0.60 T in a time of 0.45 s. The wire has a resistance per unit length of 3.3×10^{-2} Ω/m. What is the average electrical energy dissipated in the resistance of the wire?

68. M GO SSM A long solenoid of length 8.0×10^{-2} m and cross-sectional area 5.0×10^{-5} m² contains 6500 turns per meter of length. Determine the emf induced in the solenoid when the current in the solenoid changes from 0 to 1.5 A during the time interval from 0 to 0.20 s.

Physics in Biology, Medicine, and Sports

69. **E** **BIO** **22.2** The drawing shows a type of flow meter that can be used to measure the speed of blood in situations when a blood vessel is sufficiently exposed (e.g., during surgery). Blood is conductive enough that it can be treated as a moving conductor. When it flows perpendicularly with respect to a magnetic field, as in the drawing, electrodes can be used to measure the small voltage that develops across the vessel. Suppose that the speed of the blood is 0.30 m/s and the diameter of the vessel is 5.6 mm. In a 0.60-T magnetic field what is the magnitude of the voltage that is measured with the electrodes in the drawing?

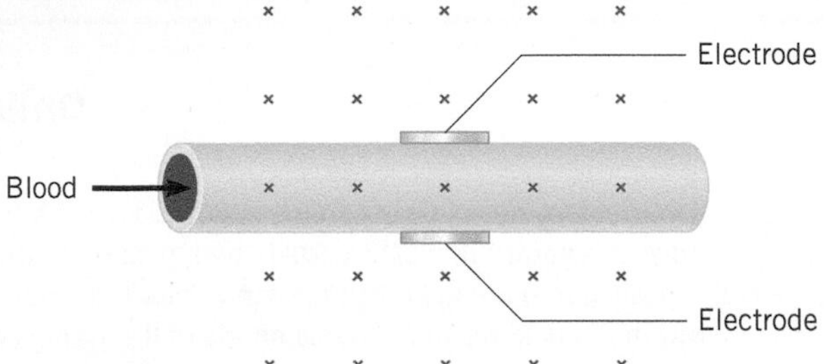

PROBLEM 69

70. **E** **BIO** **22.4** Magnetic resonance imaging (MRI) is a medical technique for producing pictures of the interior of the body. The patient is placed within a strong magnetic field. One safety concern is what would happen to the positively and negatively charged particles in the body fluids if an equipment failure caused the magnetic field to be shut off suddenly. An induced emf could cause these particles to flow, producing an electric current within the body. Suppose the largest surface of the body through which flux passes has an area of 0.032 m^2 and a normal that is parallel to a magnetic field of 1.5 T. Determine the smallest time period during which the field can be allowed to vanish if the magnitude of the average induced emf is to be kept less than 0.010 V.

71. **M** **BIO** **22.4** A typical MRI machine creates a constant magnetic field of 3.0 T. However, some patients receiving a CAT scan on their brain can be exposed to changing magnetic fields, either by ramping the field up or down, or by moving their head quickly while in the field. This creates a change in magnetic flux through the head, which can cause the patient to experience *magnetophosphenes*. These are flashes of light the patient sees in their visual field. The change in magnetic flux induces currents to flow in the retina or visual cortex, which results in the illusion of light. Consider a patient in an MRI machine at a field of 3.0 T. The field is switched off and suddenly drops to zero in 4.0×10^{-4} s, creating a single loop of current enclosing an area of the brain that is 5.0×10^{-3} m^2. What is the magnitude of the induced emf in the brain? Assume the changing magnetic field passes perpendicularly through the surface of the current loop.

72. **M** **BIO** **22.4** Many athletes and cool tech lovers wear a smartwatch, like the Apple Watch, to monitor their fitness and sleep patterns, receive emails, and make phone calls. The watch has an internal battery that is charged wirelessly by induction. The watch is placed faceup onto a charging base (see the photo). There is a charging coil (the primary) in the base and a receiving coil (the secondary) just inside the back faceplate of the watch. An alternating current is sent through the base coil, which creates a changing magnetic field. This produces a change in magnetic flux through the watch's coil, which induces an emf in that coil, thereby charging the watch's battery. Assume the secondary coil has 10 turns with an average area of 1.77 cm^2. If the emf induced in the coil is 3.80 V, what is the rate of change of the magnetic field within the charging coil?

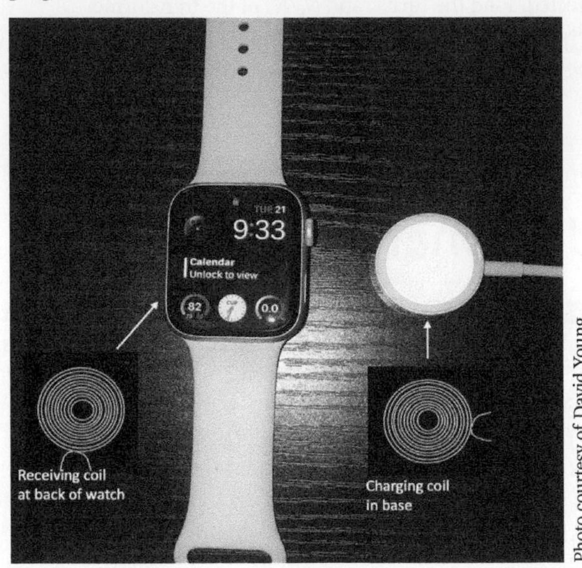

Photo courtesy of David Young

PROBLEM 72

73. **M** **BIO** **22.4** Consider again the hearing aid equipped with the T-coil, as described in Example 10. When the loop system in the room is initially turned on, a surge of current around the loop induces an emf in the T-coil of 7.2×10^{-6} V. What is the value of the peak magnetic field in the room, if it is reached after 6.5 ms?

74. **E** **BIO** **22.9** Tesla Motors manufactures fully electric vehicles. They have gained popularity over the last several years, primarily due to the Model 3 (see the photo), and electric cars will play an essential role in the fight against global climate change. The long-range model of this car is powered by a 350-V lithium-ion battery pack with an energy of 80.5 kWh. When connected to an at-home charging station to recharge the battery pack, the transformer inside the charger only draws 80% of the current allowed by the circuit breaker that protects the circuit that powers the charger. If the homeowner is using a charger that is connected to a 220-V outlet with a 40-A circuit breaker, what is the average electrical power delivered to the Model 3 during charging?

Jteder/Pixabay

PROBLEM 74

75. M BIO 22.2 As a prelude to this problem, you should review Problem 75 in Chapter 21. Some researchers have noticed that cows and wild deer will align themselves in a north–south direction while resting or grazing (see the photo). This suggests they may sense the direction of the earth's magnetic field and align with it. One possible explanation for this ability is to coordinate movements over a large herd. For example, roe deer, when startled, tend to run north to south. However, if they are close enough to high voltage power lines, they do not exhibit this behavior. (a) How close would a deer have to be to a power line carrying 250 A, so that the magnetic field created by the line was equal to the earth's magnetic field (0.45 gauss)? (b) What would be the magnetic flux passing through the deer's body at this position? Assume the magnetic field passes perpendicularly through a 0.70-m^2 cross section of the deer.

PROBLEM 75

Olavfin, Image taken from https://commons.wikimedia.org/wiki/File:R%C3%A5dyr_1.jpg

Concepts and Calculations Problems

Online

In this chapter we have seen that there are three ways to create an induced emf in a coil: by changing the magnitude of a magnetic field, by changing the direction of the field relative to the coil, and by changing the area of the coil. Problem 76 explores the third method and, in the process, provides a review of Faraday's law of electromagnetic induction. Problem 77 explores the characteristics of the emf produced in a generator that utilizes a coil rotating in a fixed magnetic field.

76. M CHALK A circular coil of radius 0.11 m contains a single turn and is located in a constant magnetic field of magnitude 0.27 T. The magnetic field has the same direction as the normal to the plane of the coil. The radius increases to 0.30 m in a time of 0.080 s. *Concepts:* (i) Why is there an emf induced in the coil? (ii) Does the magnitude of the induced emf depend on whether the area is increasing or decreasing? Explain. (iii) What determines the amount of current induced in the coil? (iv) If the coil is cut so it is no longer one continuous piece, are there an induced emf and an induced current? Explain. *Calculations:* (a) Determine the magnitude of the emf induced in the coil. (b) The coil has a resistance of 0.70 Ω. Find the magnitude of the induced current.

77. M CHALK SSM The graph in the figure shows the emf produced by a generator as a function of time t. The coil for the generator has an area of $A = 0.15\ m^2$ and consists of $N = 10$ turns. The coil rotates in a field of magnitude 0.27 T. *Concepts:* (i) Can the period of the rotating coil be determined from the graph? (ii) The emf produced by a generator depends on its angular frequency. How is the angular frequency of the coil related to its period? (iii) Starting at $t = 0$ s, how much time is required for the generator to produce its peak emf? Express the answer in terms of the period T of the motion (e.g., $t = (0.1)\ T$). (iv) How often does the polarity of the emf change in one cycle? *Calculations:* (a) Determine the period of the motion. (b) What is the angular frequency of the rotating coil? (c) Find the value of the emf when $t = \frac{1}{4}T$, where T denotes the period of the coil motion. (d) What is the emf when $t = 0.025$ s?

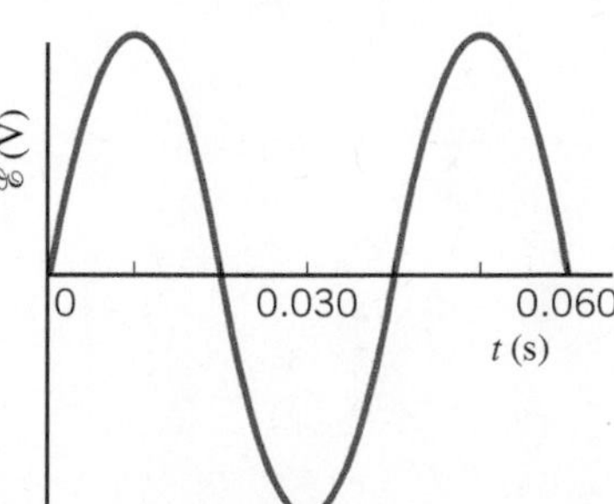

PROBLEM 77

Team Problems

Online

78. M T WS **A Magnetic Fan Light.** You and your team are designing a device in which a square loop of wire of side length $L = 3.30$ cm is glued to the end of a fan blade. Into one of the sides of the wire loop you insert a light-emitting diode (LED) so that whatever emf is induced in the loop appears across the LED. The idea is to orient the fan so that the end of the fan blade (and the square circuit) passes between the poles of a magnet, and therefore through a region of uniform, horizontally oriented magnetic field that has a square cross section of side length 5.00 cm. You align the fan so that the loop area is perpendicular to the field and so that the leading edge of the square loop is nearly horizontal when it enters the region of magnetic field. In order for the LED to turn on, an emf of at least 1.80 V must be induced in the loop. (a) If the center of the loop is 30.0 cm from the rotation axis of the fan, and the rate of rotation of the fan is 1.00×10^3 rpm, estimate the minimum magnitude of the magnetic field needed to turn on the LED. (b) For how long is the LED illuminated as it passes into the field region, that is, for what period of time is the required emf maintained? Note that LEDs illuminate only when the emf across them has the correct polarity (i.e., they will not illuminate with the opposite polarity). You can assume that the required emf polarity is satisfied as the loop enters the field.

79. E T WS **Passing by a Magnetar.** A type of neutron star called a *magnetar* is believed to have an extremely large magnetic field on the order of 10^{11} T, about a hundred million times greater than the strongest magnetic fields humans have been able to produce on earth. You and your team have been assigned to evaluate some of the risks that a probe might encounter while passing by such an object, espe-

cially regarding its electronics. Suppose that the probe passes through a region of space near a magnetar where the magnetic field is nearly uniform and has a constant magnitude of 275 T. The translational speed of the craft is 65.0 km/s relative to the star, and rotates about its axis at a rate of 1.75 rev/s. (a) The probe has a long, straight antenna (i.e., a conductor) that is 2.80 m in length. Neglecting the rotation of the probe, what is the maximum possible induced emf between the two ends of the antenna as a result of it moving through the magnetic field? (b) What is the maximum emf induced in an electrical circuit loop of area 1.33×10^{-4} m^2 due to the rotation of the craft? (c) There are other conducting "loops" that are a part of the structure of the craft, and not a part of the functional electronics. What is the maximum emf induced in a loop of area 1.50 m^2 due to the spinning of the probe? (d) Based on the results of (a) through (c), should the probe take a more distant path around the magnetar? *Note:* any induced emf that exceeds 1000 V is problematic for structures, and anything above a few volts will damage sensitive circuits.

80. M T **Reconfiguring a Transformer.** You and your team are exploring an abandoned science facility on the coast of western Antarctica when a large storm hits, and it is clear that you will be stranded there for a few days. You and the others search for supplies and find a generator and a tank of fuel. Having electrical power would allow you to keep your communication devices operational and make your stay more comfortable. A team member gets the generator running, but there is a complication: The output of the generator is 50 Hz at 440 VAC (rms), and your devices require 60 Hz, 110 VAC (rms). The electrical power in many European countries runs on 240 V at 50 Hz, so a few of your team members have converters. However, 440 V is still too high to use them. You search and eventually find a large transformer that, according to a worn tag on its case, is designed to step down from 5000 V to 880 V. The tag also indicates that its primary coil has 1500 turns, but you cannot read the number of turns in the secondary coil. (a) How many turns should its secondary coil have? (b) It will be a difficult job, but you can change the number of primary turns by cutting some of them out. How many turns should you leave on the primary coil so that, with the primary connected to 440 V, the secondary outputs 240 V (so that you can use the 240 V to 110 V converters)? (c) You find that the current at the source (i.e., that connected to the primary) is limited to a maximum of 20.0 A. What is the maximum current limit through the secondary coil? (d) What is the maximum average power available at the secondary coil?

81. M T **A Generator Bike.** You and your team are designing a generator using a stationary bike to rotate a coil in a uniform magnetic field. The gearing is set up so that the coil rotates 60 times for one complete rotation of the bike pedals. Therefore, one revolution of the pedals per second results in a 60-Hz alternating current in the coil. The circular coil has 350 turns and a diameter of 15.0 cm, and its axis of rotation is along its diameter. (a) If a uniform magnetic field is oriented perpendicular to the coil's axis of rotation and has a magnitude of $B = 0.225$ T, what is the peak emf produced by the generator bike? (b) What is the rms emf? (c) To what magnitude should you reduce the field if you want the rms emf to be 110 VAC? (d) Instead of reducing the field, you could use a step-down transformer to reduce the rms emf to 110 VAC. What should be the ratio of primary to secondary turns of the transformer coils?

CHAPTER 23

Thousands of lights dot the spectacular Chicago skyline at night. Lighting control, and the delivery of electrical energy to cities in general, would not be possible without the aid of alternating current (ac) circuits. Ac circuits lie at the heart of most modern technologies, and we will study them in this chapter.

Gian Lorenzo Ferretti Photography/ E+/Getty Images

Alternating Current Circuits

LEARNING OBJECTIVES

After reading this module, you should be able to...

23.1 Calculate capacitive reactance.

23.2 Calculate inductive reactance.

23.3 Calculate impedance in an RCL circuit.

23.4 Calculate the resonance frequency of an RCL circuit.

23.5 Describe how semiconductor devices operate.

23.1 Capacitors and Capacitive Reactance

Our experience with capacitors so far has been in dc circuits. As we have seen in Section 20.13, charge flows in a dc circuit only for the brief period after the battery voltage is applied across the capacitor. In other words, charge flows only while the capacitor is charging up. After the capacitor becomes fully charged, no more charge leaves the battery. However, suppose that the battery connections to the fully charged capacitor were suddenly reversed. Then charge would flow again, but in the reverse direction, until the battery recharges the capacitor according to the new connections. In an ac circuit what happens is similar. The polarity of the voltage applied to the capacitor continually switches back and forth, and, in response, charges flow first one way around the circuit and then the other way. This flow of charge, surging back and forth, constitutes an alternating current. Thus, charge flows continuously in an ac circuit containing a capacitor.

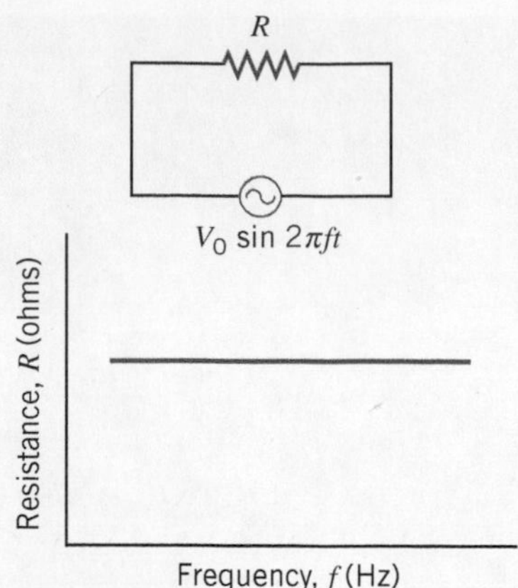

FIGURE 23.1 The resistance in a purely resistive circuit has the same value at all frequencies. The maximum emf of the generator is V_0.

To help set the stage for the present discussion, recall from Section 20.5 that the rms voltage V_{rms} across the resistor in a purely resistive ac circuit is related to the rms current I_{rms} by $V_{rms} = I_{rms}R$ (Equation 20.14). The resistance R has the same value for any frequency of the ac voltage or current. **Figure 23.1** emphasizes this fact by showing that a graph of resistance versus frequency is a horizontal straight line.

For the rms voltage across a capacitor the following expression applies, which is analogous to $V_{rms} = I_{rms}R$:

$$V_{rms} = I_{rms}X_C \tag{23.1}$$

The term X_C appears in place of the resistance R and is called the **capacitive reactance.** The capacitive reactance, like resistance, is measured in *ohms* and determines how much rms current exists in a capacitor in response to a given rms voltage across the capacitor. It is found experimentally that the capacitive reactance X_C is inversely proportional to both the frequency f and the capacitance C, according to the following equation:

$$X_C = \frac{1}{2\pi fC} \tag{23.2}$$

For a fixed value of the capacitance C, **Figure 23.2** gives a plot of X_C versus frequency, according to Equation 23.2. A comparison of this drawing with **Figure 23.1** reveals that a capacitor and a resistor behave differently. As the frequency becomes very large, **Figure 23.2** shows that X_C approaches zero, signifying that a capacitor offers only a negligibly small opposition to the alternating current. In contrast, in the limit of zero frequency (i.e., direct current), X_C becomes infinitely large, and a capacitor provides so much opposition to the motion of charges that there is no current.

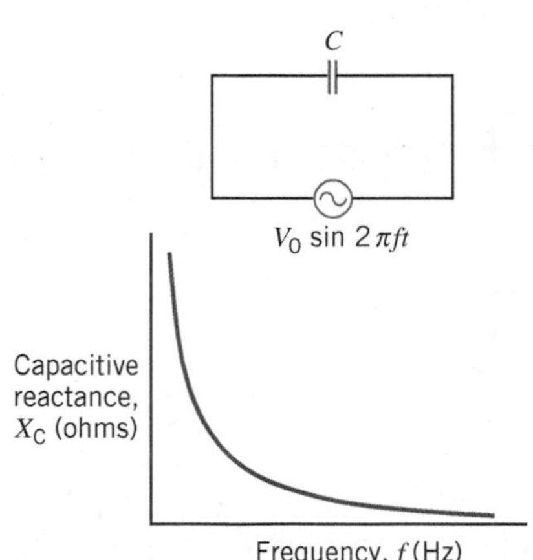

FIGURE 23.2 The capacitive reactance X_C is inversely proportional to the frequency f according to $X_C = 1/(2\pi fC)$.

Example 1 illustrates the use of Equation 23.2 and also demonstrates how frequency and capacitance determine the amount of current in an ac circuit.

EXAMPLE 1 | A Capacitor in an Ac Circuit

For the circuit in **Figure 23.2**, the capacitance of the capacitor is 1.50 μF, and the rms voltage of the generator is 25.0 V. What is the rms current in the circuit when the frequency of the generator is **(a)** 1.00×10^2 Hz and **(b)** 5.00×10^3 Hz?

Reasoning The current can be found from $I_{rms} = V_{rms}/X_C$, once the capacitive reactance X_C is determined. The values for the capacitive reactance will reflect the fact that the capacitor provides more opposition to the current when the frequency is smaller.

Problem-Solving Insight The capacitive reactance X_C is inversely proportional to the frequency f of the voltage, so if the frequency increases by a factor of 50, as in this example, the capacitive reactance decreases by a factor of 50.

Solution (a) At a frequency of 1.00×10^2 Hz, we find

$$X_C = \frac{1}{2\pi fC} = \frac{1}{2\pi(1.00 \times 10^2\ \text{Hz})(1.50 \times 10^{-6}\ \text{F})} = 1060\ \Omega \quad \textbf{(23.2)}$$

$$I_{\text{rms}} = \frac{V_{\text{rms}}}{X_C} = \frac{25.0\ \text{V}}{1060\ \Omega} = \boxed{0.0236\ \text{A}} \quad \textbf{(23.1)}$$

(b) When the frequency is 5.00×10^3 Hz, the calculations are similar:

$$X_C = \frac{1}{2\pi fC} = \frac{1}{2\pi(5.00 \times 10^3\ \text{Hz})(1.50 \times 10^{-6}\ \text{F})} = 21.2\ \Omega \quad \textbf{(23.2)}$$

$$I_{\text{rms}} = \frac{V_{\text{rms}}}{X_C} = \frac{25.0\ \text{V}}{21.2\ \Omega} = \boxed{1.18\ \text{A}} \quad \textbf{(23.1)}$$

We now consider the behavior of the instantaneous (not rms) voltage and current. For comparison, **Figure 23.3** shows graphs of voltage and current versus time in a resistive circuit. These graphs indicate that when only resistance is present, the voltage and current are proportional to each other at every moment. For example, when the voltage increases from A to B on the graph, the current follows along in step, increasing from A' to B' during the same time interval. Likewise, when the voltage decreases from B to C, the current decreases from B' to C'. For this reason, the current in a resistance R is said to be **in phase** with the voltage across the resistance.

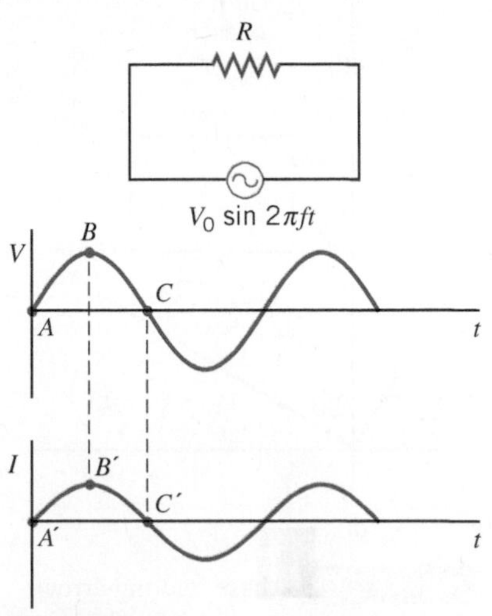

FIGURE 23.3 The instantaneous voltage V and current I in a purely resistive circuit are *in phase,* which means that they increase and decrease in step with one another.

For a capacitor, this in-phase relation between instantaneous voltage and current does *not* exist. **Interactive Figure 23.4** shows graphs of the ac voltage and current versus time for a circuit that contains only a capacitor. As the voltage increases from A to B, the charge on the capacitor increases and reaches its full value at B. The current, however, is not the same thing as the charge. The current is the rate of flow of charge and has a maximum positive value at the start of the charging process at A'. It is a maximum because there is no charge on the capacitor at the start and hence no capacitor voltage to oppose the generator voltage. When the capacitor is fully charged at B, the capacitor voltage has a magnitude equal to that of the generator and completely opposes the generator voltage. The result is that the current decreases to zero at B'. While the capacitor voltage decreases from B to C, the charges flow out of the capacitor in a direction opposite to that of the charging current, as indicated by the negative current from B' to C'. Thus, voltage and current are not in phase but are, in fact, one-quarter wave cycle out of step, or out of phase. More specifically, assuming that the voltage fluctuates as $V_0 \sin 2\pi ft$, the current varies as $I_0 \sin(2\pi ft + \pi/2) = I_0 \cos 2\pi ft$. Since $\pi/2$ radians correspond to 90° and since the current reaches its maximum value *before* the voltage does, it is said that ***the current in a capacitor leads the voltage across the capacitor by a phase angle of 90°.***

The fact that the current and voltage for a capacitor are 90° out of phase has an important consequence from the point of view of electric power, since power is the product of current and voltage. For the time interval between points A and B (or A' and B') in **Interactive Figure 23.4**, both current and voltage are positive. Therefore,

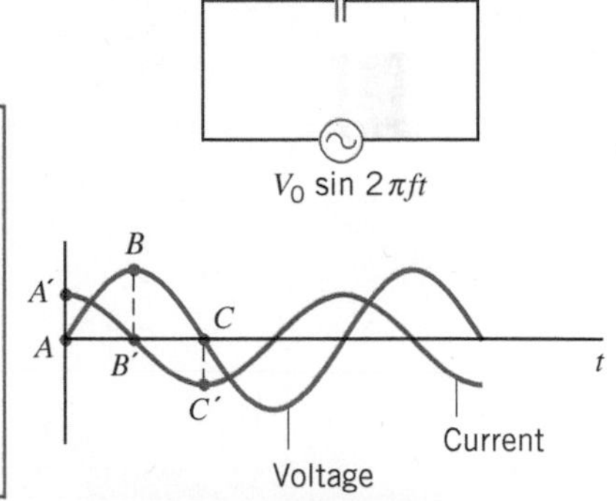

INTERACTIVE FIGURE 23.4 In a circuit containing only a capacitor, the instantaneous voltage and current are not in phase. Instead, the current *leads* the voltage by one-quarter of a cycle or by a phase angle of 90°.

Math Skills Why does $I_0 \sin\left(2\pi ft + \frac{\pi}{2}\right) = I_0 \cos 2\pi ft$? To see why, consider the trigonometric identity $\sin(\alpha + \beta) = \sin\alpha\cos\beta + \cos\alpha\sin\beta$ (see Appendix E.2, Other Trigonometric Identities, Equation 6). Using $\alpha = 2\pi ft$ and $\beta = \frac{\pi}{2}$ in this identity, we find that

$$I_0 \sin\left(2\pi f + \frac{\pi}{2}\right) = I_0 \sin 2\pi ft \underbrace{\cos\frac{\pi}{2}}_{0} + I_0 \cos 2\pi ft \underbrace{\sin\frac{\pi}{2}}_{1}$$

But $\frac{\pi}{2}$ radians corresponds to 90°, so that $\cos\frac{\pi}{2} = 0$ and $\sin\frac{\pi}{2} = 1$. With these two substitutions, we have

$$I_0 \sin\left(2\pi ft + \frac{\pi}{2}\right) = I_0 \cos 2\pi ft$$

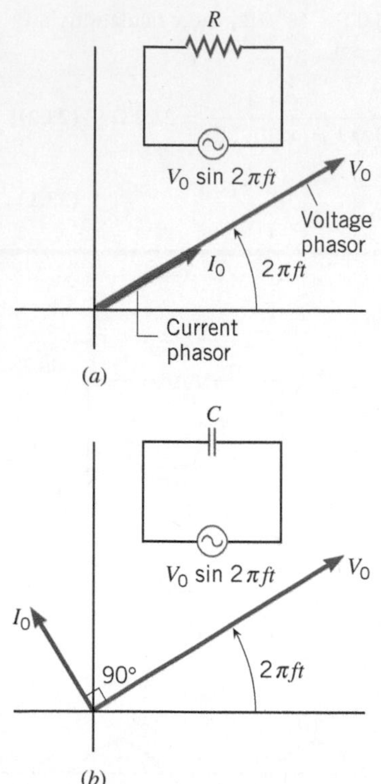

FIGURE 23.5 These rotating-arrow models (or phasor models) represent the voltage and the current in ac circuits that contain (*a*) only a resistor and (*b*) only a capacitor.

the instantaneous power is also positive, meaning that the generator is delivering energy to the capacitor. However, during the period between B and C (or B' and C'), the current is negative while the voltage remains positive, and the power, which is the product of the two, is negative. During this period, the capacitor is returning energy to the generator. Thus, the power alternates between positive and negative values for equal periods of time. In other words, the capacitor alternately absorbs and releases energy.

Problem-Solving Insight Consequently, the average power (and, hence, the average energy) used by a capacitor in an ac circuit is zero.

It will prove useful later on to use a model for the voltage and current when analyzing ac circuits. In this model, voltage and current are represented by rotating arrows, often called **phasors**, whose lengths correspond to the maximum voltage V_0 and maximum current I_0, as **Figure 23.5** indicates. These phasors rotate counterclockwise at a frequency f. For a resistor, the phasors are co-linear as they rotate (see part *a* of the drawing) because voltage and current are in phase. For a capacitor (see part *b*), the phasors remain perpendicular while rotating because the phase angle between the current and the voltage is 90°. Since current leads voltage for a capacitor, the current phasor is ahead of the voltage phasor in the direction of rotation.

Note from the two circuit drawings in **Figure 23.5** that the instantaneous voltage across the resistor or the capacitor is $V_0 \sin 2\pi ft$. We can find this instantaneous value of the voltage directly from the phasor diagram. Imagine that the voltage phasor V_0 in this diagram represents the hypotenuse of a right triangle. Then, the *vertical component* of the phasor would be $V_0 \sin 2\pi ft$. In a similar manner, the instantaneous current can be found as the vertical component of the current phasor.

Check Your Understanding

(*The answer is given at the end of the book.*)

1. One circuit contains only an ac generator and a resistor, and the rms current in this circuit is I_R. Another circuit contains only an ac generator and a capacitor, and the rms current in this circuit is I_C. The maximum, or peak, voltage of the generator is the same in both circuits and does not change. If the frequency of each generator is tripled, by what factor does the ratio I_R/I_C change? Specify whether the change is an increase or a decrease.

23.2 Inductors and Inductive Reactance

As Section 22.8 discusses, an inductor is usually a coil of wire, and the basis of its operation is Faraday's law of electromagnetic induction. According to Faraday's law, an inductor develops a voltage that opposes a change in the current. This voltage V is given by $V = -L(\Delta I/\Delta t)$ (see Equation 22.9*), where $\Delta I/\Delta t$ is the rate at which the current changes and L is the inductance of the inductor. In an ac circuit the current is always changing, and Faraday's law can be used to show that the rms voltage across an inductor is

$$V_{rms} = I_{rms} X_L \tag{23.3}$$

Equation 23.3 is analogous to $V_{rms} = I_{rms}R$, with the term X_L appearing in place of the resistance R and being called the **inductive reactance**. The inductive reactance is measured in ohms and determines how much rms current exists in an inductor for a given

*When an inductor is used in a circuit, the notation is simplified if we designate the potential difference across the inductor as the voltage V, rather than the emf $\mathcal{E}$.

rms voltage across the inductor. It is found experimentally that the inductive reactance X_L is directly proportional to the frequency f and the inductance L, as indicated by the following equation:

$$X_L = 2\pi f L \quad (23.4)$$

This relation indicates that the larger the inductance, the larger is the inductive reactance. Note that the inductive reactance is directly proportional to the frequency ($X_L \propto f$), whereas the capacitive reactance is inversely proportional to the frequency ($X_C \propto 1/f$).

Figure 23.6 shows a graph of the inductive reactance versus frequency for a fixed value of the inductance, according to Equation 23.4. As the frequency becomes very large, X_L also becomes very large. In such a situation, an inductor provides a large opposition to the alternating current. In the limit of zero frequency (i.e., direct current), X_L becomes zero, indicating that an inductor does not oppose direct current at all. The next example demonstrates the effect of inductive reactance on the current in an ac circuit.

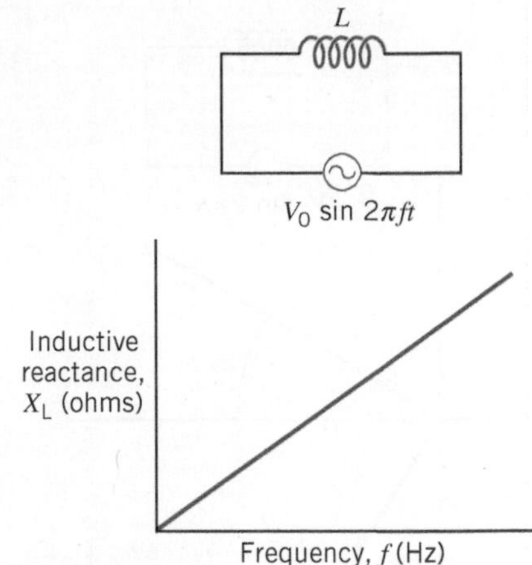

FIGURE 23.6 In an ac circuit the inductive reactance X_L is directly proportional to the frequency f, according to $X_L = 2\pi f L$.

EXAMPLE 2 | An Inductor in an Ac Circuit

The circuit in **Figure 23.6** contains a 3.60-mH inductor. The rms voltage of the generator is 25.0 V. Find the rms current in the circuit when the generator frequency is **(a)** 1.00×10^2 Hz and **(b)** 5.00×10^3 Hz.

Reasoning The current can be calculated from $I_{rms} = V_{rms}/X_L$, provided the inductive reactance is obtained first. The inductor offers more opposition to the changing current when the frequency is larger, and the values for the inductive reactance will reflect this fact.

Problem-Solving Insight The inductive reactance X_L is directly proportional to the frequency f of the voltage. If the frequency increases by a factor of 50, as here, the inductive reactance also increases by a factor of 50.

Solution **(a)** At a frequency of 1.00×10^2 Hz, we find

$$X_L = 2\pi f L = 2\pi(1.00 \times 10^2\,\text{Hz})(3.60 \times 10^{-3}\,\text{H}) = 2.26\,\Omega \quad (23.4)$$

$$I_{rms} = \frac{V_{rms}}{X_L} = \frac{25.0\,\text{V}}{2.26\,\Omega} = \boxed{11.1\,\text{A}} \quad (23.3)$$

(b) The calculation is similar when the frequency is 5.00×10^3 Hz:

$$X_L = 2\pi f L = 2\pi(5.00 \times 10^3\,\text{Hz})(3.60 \times 10^{-3}\,\text{H}) = 113\,\Omega \quad (23.4)$$

$$I_{rms} = \frac{V_{rms}}{X_L} = \frac{25.0\,\text{V}}{113\,\Omega} = \boxed{0.221\,\text{A}} \quad (23.3)$$

Due to its inductive reactance, an inductor affects the amount of current in an ac circuit. The inductor also influences the current in another way, as **Interactive Figure 23.7** shows. This figure displays graphs of voltage and current versus time for a circuit containing only an inductor. At a maximum or minimum on the current graph, the current does not change much with time, so the voltage generated by the inductor to oppose a change in the current is zero. At the points on the current graph where the current is zero, the graph is at its steepest, and the current has the largest rate of increase or decrease. Correspondingly, the voltage generated by the inductor to oppose a change in the current has the largest positive or negative value. Thus, current and voltage are not in phase but are one-quarter of a wave cycle out of phase. If the voltage varies as $V_0 \sin 2\pi ft$, the current fluctuates as $I_0 \sin(2\pi ft - \pi/2) = -I_0 \cos 2\pi ft$. The current reaches its maximum *after* the voltage does, and it is said that ***the current in an inductor lags behind the voltage by a phase angle of 90°*** ($\pi/2$ radians). In a purely capacitive circuit, in contrast, the current leads the voltage by 90° (see **Interactive Figure 23.4**).

In an inductor the 90° phase difference between current and voltage leads to the same result for average power that it does in a capacitor. An inductor alternately absorbs and releases energy for equal periods of time.

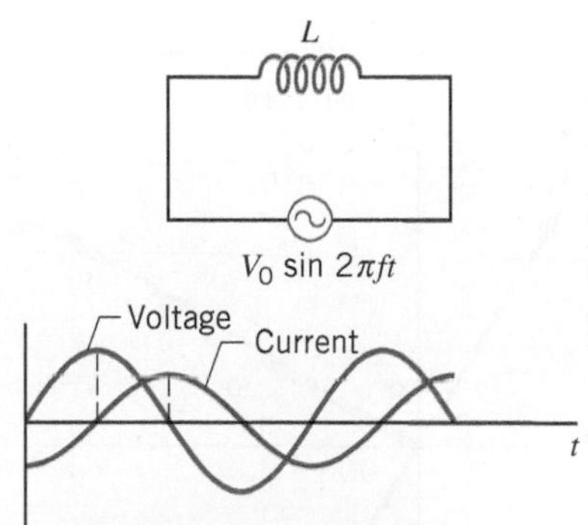

INTERACTIVE FIGURE 23.7 The instantaneous voltage and current in a circuit containing only an inductor are not in phase. The current *lags behind* the voltage by one-quarter of a cycle or by a phase angle of 90°.

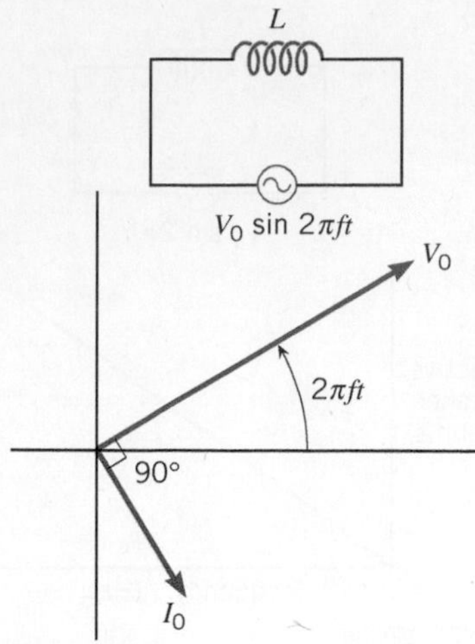

FIGURE 23.8 This phasor model represents the voltage and current in a circuit that contains only an inductor.

Problem-Solving Insight Thus, the average power (and, hence, the average energy) used by an inductor in an ac circuit is zero.

As an alternative to the graphs in **Interactive Figure 23.7**, **Figure 23.8** uses phasors to describe the instantaneous voltage and current in a circuit containing only an inductor. The voltage and current phasors remain perpendicular as they rotate, because there is a 90° phase angle between them. The current phasor lags behind the voltage phasor, relative to the direction of rotation, in contrast to the equivalent picture in **Figure 23.5*b*** for a capacitor. Once again, the instantaneous values are given by the vertical components of the phasors.

Check Your Understanding

(*The answer is given at the end of the book.*)

2. CYU Figure 23.1 shows three ac circuits: one contains a resistor, one a capacitor, and one an inductor. The frequency of each ac generator is reduced to one-half its initial value. Which circuit experiences **(a)** the *greatest increase* in current and **(b)** the *least change* in current?

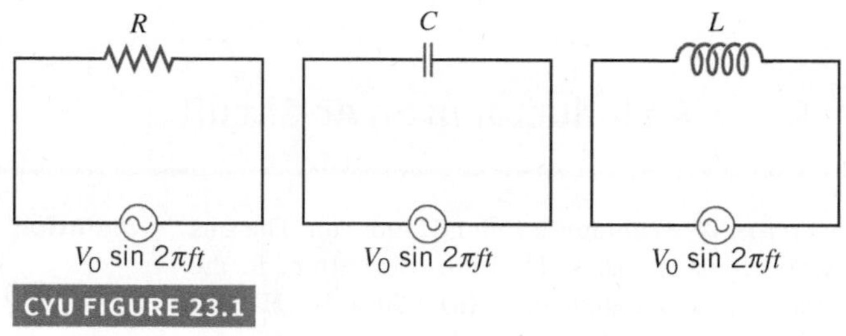

CYU FIGURE 23.1

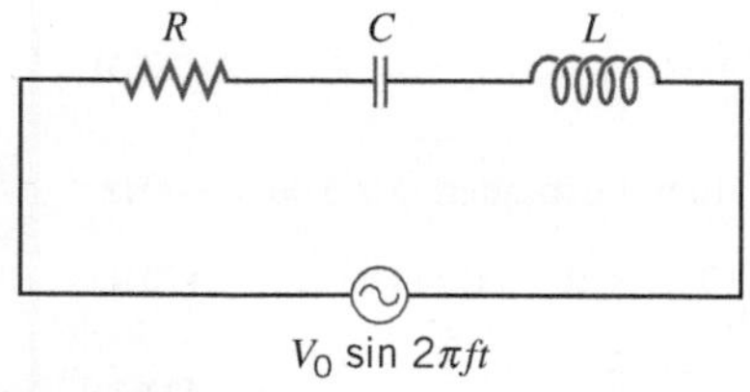

FIGURE 23.9 A series RCL circuit contains a resistor, a capacitor, and an inductor.

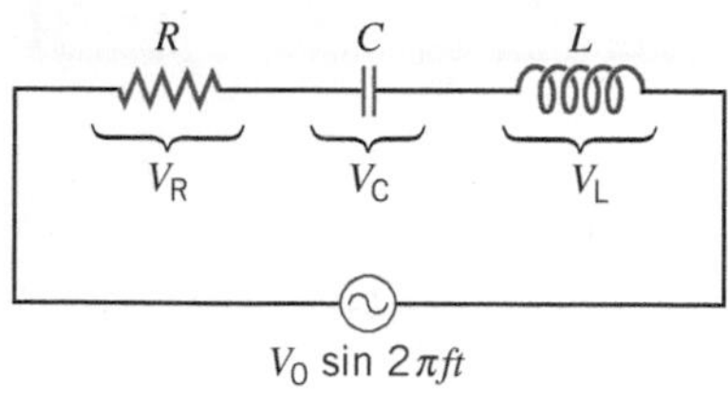

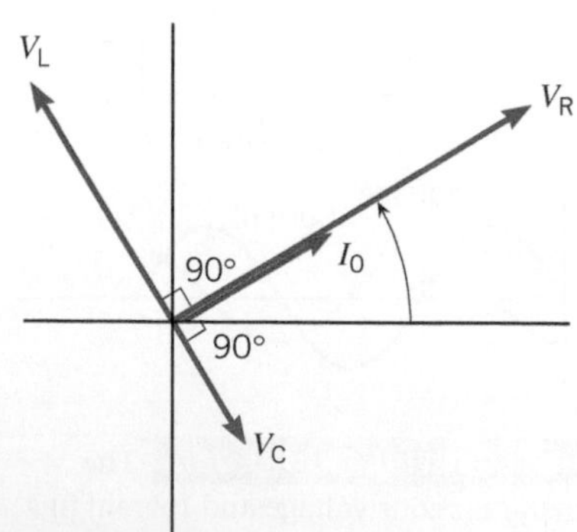

FIGURE 23.10 The three voltage phasors (V_R, V_C, and V_L) and the current phasor (I_0) for a series RCL circuit.

23.3 Circuits Containing Resistance, Capacitance, and Inductance

Capacitors and inductors can be combined along with resistors in a single circuit. The simplest combination contains a resistor, a capacitor, and an inductor in series, as **Figure 23.9** shows. In a series RCL circuit the total opposition to the flow of charge is called the **impedance** of the circuit and comes partially from (1) the resistance R, (2) the capacitive reactance X_C, and (3) the inductive reactance X_L. It is tempting to follow the analogy of a series combination of resistors and calculate the impedance by simply adding together R, X_C, and X_L. However, such a procedure is not correct. Instead, the phasors shown in **Figure 23.10** must be used. The lengths of the voltage phasors in this drawing represent the maximum voltages V_R, V_C, and V_L across the resistor, the capacitor, and the inductor, respectively. The current is the same for each device, since the circuit is wired in series. The length of the current phasor represents the maximum current I_0. Notice that the drawing shows the current phasor to be (1) in phase with the voltage phasor for the resistor, (2) ahead of the voltage phasor for the capacitor by 90°, and (3) behind the voltage phasor for the inductor by 90°. These three facts are consistent with our earlier discussion in Sections 23.1 and 23.2.

The basis for dealing with the voltage phasors in **Figure 23.10** is Kirchhoff's loop rule. In an ac circuit this rule applies to the *instantaneous* voltages across each circuit component and the generator. Therefore, it is necessary to take into account the fact that these voltages do not have the same phase; that is, the phasors V_R, V_C, and V_L point in different directions in the drawing. Kirchhoff's loop rule indicates that the phasors add together to give the total voltage V_0 that is supplied to the circuit by the generator. The addition, however, must be like a vector addition, to take into account the different directions of the phasors. Since V_L and V_C point in opposite directions, they combine to give a resultant phasor of $V_L - V_C$, as **Figure 23.11** shows. In this drawing the resultant $V_L - V_C$ is perpendicular to V_R and may be combined with it to give the total voltage V_0. Using the Pythagorean theorem, we find

$$V_0^2 = V_R^2 + (V_L - V_C)^2$$

In this equation each of the symbols stands for a maximum voltage and when divided by $\sqrt{2}$ gives the corresponding rms voltage. Therefore, it is possible to divide both sides of the equation by $(\sqrt{2})^2$ and obtain a result for $V_{rms} = V_0/\sqrt{2}$. This result has exactly the same form as that above, but involves the rms voltages $V_{R\text{-}rms}$, $V_{C\text{-}rms}$, and $V_{L\text{-}rms}$. However, to avoid such awkward symbols, we simply interpret V_R, V_C, and V_L as rms quantities in the following expression:

$$V_{rms}^2 = V_R^2 + (V_L - V_C)^2 \tag{23.5}$$

The last step in determining the impedance of the circuit is to remember that $V_R = I_{rms}R$, $V_C = I_{rms}X_C$, and $V_L = I_{rms}X_L$. With these substitutions Equation 23.5 can be written as

$$V_{rms} = I_{rms}\sqrt{R^2 + (X_L - X_C)^2}$$

Therefore, for the entire RCL circuit, it follows that

$$V_{rms} = I_{rms}Z \tag{23.6}$$

where the impedance Z of the circuit is defined as

Series RCL combination

$$Z = \sqrt{R^2 + (X_L - X_C)^2} \tag{23.7}$$

The impedance of the circuit, like R, X_C, and X_L, is measured in ohms. In Equation 23.7, $X_L = 2\pi fL$ and $X_C = 1/(2\pi fC)$.

The phase angle between the current in and the voltage across a series RCL combination is the angle ϕ between the current phasor I_0 and the voltage phasor V_0 in **Figure 23.11**. According to **Figure 23.11**, the tangent of this angle is

$$\tan\phi = \frac{V_L - V_C}{V_R} = \frac{I_{rms}X_L - I_{rms}X_C}{I_{rms}R}$$

Series RCL combination

$$\tan\phi = \frac{X_L - X_C}{R} \tag{23.8}$$

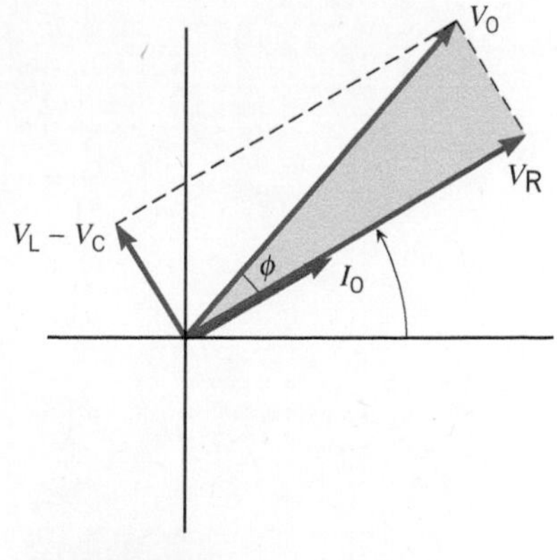

FIGURE 23.11 This simplified version of **Figure 23.10** results when the phasors V_L and V_C, which point in opposite directions, are combined to give a resultant of $V_L - V_C$.

Math Skills To see why $\tan\phi = \frac{V_L - V_C}{V_R}$, remember that the tangent function is defined as $\tan\phi = \frac{h_o}{h_a}$ (Equation 1.3). As shown in **Figure 23.12*a***, h_o is the length of the side of a right triangle opposite the angle ϕ, and h_a is the length of the side adjacent to the angle ϕ. **Figure 23.11** shows the angle ϕ as the phase angle between the current phasor I_0 and the voltage phasor V_0 and is reproduced in **Figure 23.12*b***. The corresponding right triangles are shaded in **Figure 23.12**, and a comparison reveals that $h_o = V_L - V_C$ and $h_a = V_R$. It follows that

$$\tan\phi = \frac{h_o}{h_a} = \frac{V_L - V_C}{V_R}$$

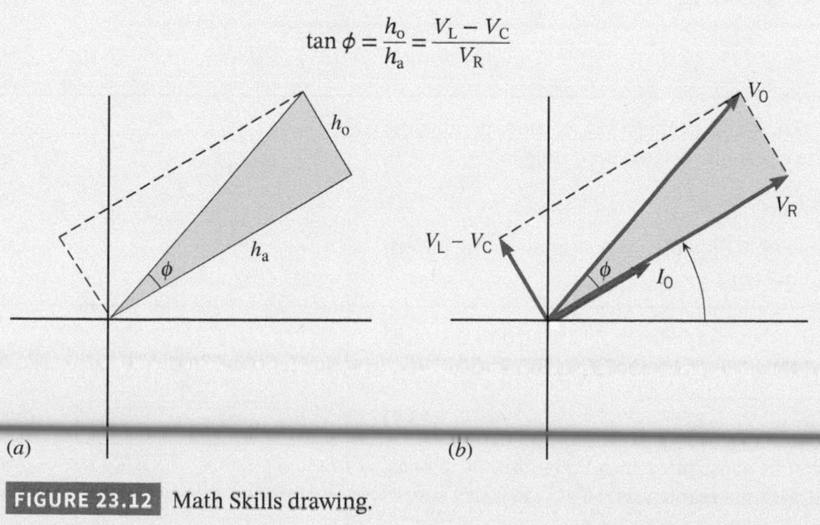

FIGURE 23.12 Math Skills drawing.

The phase angle ϕ is important because it has a major effect on the average power $\overline{P}$ delivered to the circuit. Remember that, on the average, only the resistance consumes power; that is, $\overline{P} = I_{\text{rms}}^2 R$ (Equation 20.15b). According to **Figure 23.11**, $\cos\phi = V_R/V_0 = (I_{\text{rms}} R)/(I_{\text{rms}} Z) = R/Z$, so that $R = Z\cos\phi$. Therefore,

$$\overline{P} = I_{\text{rms}}^2 Z \cos\phi = I_{\text{rms}}(I_{\text{rms}} Z)\cos\phi$$

$$\overline{P} = I_{\text{rms}} V_{\text{rms}} \cos\phi \qquad \textbf{(23.9)}$$

where $V_{\text{rms}} = I_{\text{rms}} Z$ is the rms voltage of the generator, according to Equation 23.6. The term $\cos\phi$ is called the **power factor** of the circuit. As a check on the validity of Equation 23.9, note that if no resistance is present, $R = 0\ \Omega$, and $\cos\phi = R/Z = 0$. Consequently, $\overline{P} = I_{\text{rms}} V_{\text{rms}} \cos\phi = 0$, a result that is expected since, on the average, neither a capacitor nor an inductor consumes energy. Conversely, if only resistance is present, $Z = \sqrt{R^2 + (X_L - X_C)^2} = R$, and $\cos\phi = R/Z = 1$. In this case, $\overline{P} = I_{\text{rms}} V_{\text{rms}} \cos\phi = I_{\text{rms}} V_{\text{rms}}$, which is the expression for the average power delivered to a resistor. Examples 3 and 4 deal with the current, voltages, and power for a series RCL circuit.

Analyzing Multiple-Concept Problems

EXAMPLE 3 | Current in a Series RCL Circuit

A series RCL circuit contains a 148-Ω resistor, a 1.50-μF capacitor, and a 35.7-mH inductor. The generator has a frequency of 512 Hz and an rms voltage of 35.0 V. Determine the rms current in the circuit.

Reasoning The rms current in the circuit is equal to the rms voltage of the generator divided by the impedance of the circuit, according to Equation 23.6. The impedance of the circuit can be found from the resistance of the resistor and the reactances of the capacitor and the inductor via Equation 23.7. The reactances of the capacitor and the inductor can be found with the aid of Equations 23.2 and 23.4.

Knowns and Unknowns The following table summarizes the data that we have:

Description	Symbol	Value	Comment
Resistance of resistor	R	148 Ω	
Capacitance of capacitor	C	1.50 μF	$1\ \mu\text{F} = 10^{-6}\ \text{F}$
Inductance of inductor	L	35.7 mH	$1\ \text{mH} = 10^{-3}\ \text{H}$
Frequency of generator	f	512 Hz	
Rms voltage of generator	V_{rms}	35.0 V	
Unknown Variable			
Rms current in circuit	I_{rms}	?	

Modeling the Problem

STEP 1 Current According to Equation 23.6, the rms voltage V_{rms} of the generator, the rms current I_{rms}, and the impedance Z of the circuit are related according to

$$V_{\text{rms}} = I_{\text{rms}} Z$$

Solving for the current gives Equation 1 at the right, in which V_{rms} is known. The impedance is unknown, but it can be dealt with as in Step 2.

$$I_{\text{rms}} = \frac{V_{\text{rms}}}{Z} \qquad \textbf{(1)}$$

?

STEP 2 Impedance For a series RCL circuit, the impedance Z is related to the resistance R, the inductive reactance X_L, and the capacitive reactance X_C, according to

$$\boxed{Z = \sqrt{R^2 + (X_L - X_C)^2}} \qquad \textbf{(23.7)}$$

As indicated at the right, this expression can be substituted into Equation 1. The resistance is given, and we turn to Step 3 to deal with the reactances.

$$I_{\text{rms}} = \frac{V_{\text{rms}}}{Z} \qquad \textbf{(1)}$$

$$\boxed{Z = \sqrt{R^2 + (X_L - X_C)^2}} \qquad \textbf{(23.7)}$$

?

STEP 3 Inductive and Capacitive Reactances The inductive reactance X_L and the capacitive reactance X_C are given, respectively, by Equation 23.4 and Equation 23.2 as

$$X_L = 2\pi f L \quad \text{and} \quad X_C = \frac{1}{2\pi f C}$$

where L is the inductance, C is the capacitance, and f is the frequency. Using these two expressions, we find that

$$\boxed{X_L - X_C = 2\pi f L - \frac{1}{2\pi f C}}$$

This result can now be substituted into Equation 23.7, as shown at the right.

$$I_{rms} = \frac{V_{rms}}{Z} \quad \textbf{(1)}$$

$$\boxed{Z = \sqrt{R^2 + (X_L - X_C)^2}} \quad \textbf{(23.7)}$$

$$\boxed{X_L - X_C = 2\pi f L - \frac{1}{2\pi f C}}$$

Solution Combining the results of each step algebraically, we find that

$$\underset{\text{STEP 1}}{I_{rms} = \frac{V_{rms}}{Z}} = \underset{\text{STEP 2}}{\frac{V_{rms}}{\sqrt{R^2 + (X_L - X_C)^2}}} = \underset{\text{STEP 2}}{\frac{V_{rms}}{\sqrt{R^2 + \left(2\pi f L - \dfrac{1}{2\pi f C}\right)^2}}}$$

The rms current in the circuit is

$$I_{rms} = \frac{V_{rms}}{\sqrt{R^2 + \left(2\pi f L - \dfrac{1}{2\pi f C}\right)^2}}$$

$$= \frac{35.0\text{ V}}{\sqrt{(148\ \Omega)^2 + \left[2\pi(512\text{ Hz})(35.7 \times 10^{-3}\text{ H}) - \dfrac{1}{2\pi(512\text{ Hz})(1.50 \times 10^{-6}\text{ F})}\right]^2}}$$

$$= \boxed{0.201\text{A}}$$

Related Homework: Problem 18

EXAMPLE 4 | Voltages and Power in a Series RCL Circuit

For the series RCL circuit discussed in Example 3, the resistance, capacitance, and inductance are $R = 148\ \Omega$, $C = 1.50\ \mu\text{F}$, and $L = 35.7$ mH, respectively. The generator has a frequency of 512 Hz and an rms voltage of 35.0 V. In Example 3, it is found that the rms current in the circuit is $I_{rms} = 0.201$ A. Find **(a)** the rms voltage across each circuit element and **(b)** the average electric power delivered by the generator.

Reasoning The rms voltages across each circuit element can be determined from $V_R = I_{rms}R$, $V_C = I_{rms}X_C$, and $V_L = I_{rms}X_L$. In these expressions, the rms current is known. The resistance R is given, and the capacitive reactance X_C and inductive reactance X_L can be determined from Equations 23.2 and 23.4. The average power delivered to the circuit by the generator is specified by Equation 23.9.

Solution **(a)** The individual reactances are

$$X_C = \frac{1}{2\pi f C} = \frac{1}{2\pi(512\text{ Hz})(1.50 \times 10^{-6}\text{ F})} = 207\ \Omega \quad \textbf{(23.2)}$$

$$X_L = 2\pi f L = 2\pi(512\text{ Hz})(35.7 \times 10^{-3}\text{ H}) = 115\ \Omega \quad \textbf{(23.4)}$$

The rms voltages across each circuit element are

$$V_R = I_{rms} R = (0.201\text{ A})(148\ \Omega) = \boxed{29.7\text{ V}} \quad \textbf{(20.14)}$$

$$V_C = I_{rms} X_C = (0.201\text{ A})(207\ \Omega) = \boxed{41.6\text{ V}} \quad \textbf{(23.1)}$$

$$V_L = I_{rms} X_L = (0.201\text{ A})(115\ \Omega) = \boxed{23.1\text{ V}} \quad \textbf{(23.3)}$$

Observe that these three voltages do not add up to give the generator's rms voltage of 35.0 V. Instead, the rms voltages satisfy Equation 23.5. It is the sum of the *instantaneous* voltages across R, C, and L that equals the generator's *instantaneous* voltage, according to Kirchhoff's loop rule. *The rms voltages do not satisfy the loop rule.*

Math Skills The sum of the three voltages calculated in Example 4 is

$$V_R + V_C + V_L = 29.7\text{ V} + 41.6\text{ V} + 23.1\text{ V} = 94.4\text{ V}$$

Since the rms voltage of the generator is 35.0 V in Example 4, the voltages V_R, V_C, and V_L (respectively across the resistor, the capacitor, and the inductor) in a series RCL circuit clearly do not add up to give the rms voltage of the generator. However, they do satisfy $V_{rms}^2 = V_R^2 + (V_L - V_C)^2$ (Equation 23.5). You can use this fact to check the correctness of your calculations in problems such as that in the example. For instance, using $V_R = 29.7$ V, $V_C = 41.6$ V, and $V_L = 23.1$ V in Equation 23.5, we find

$$V_{rms}^2 = V_R^2 + (V_L - V_C)^2 = (29.7\text{ V})^2 + (23.1\text{ V} - 41.6\text{ V})^2$$

$$V_{rms} = \sqrt{(29.7\text{ V})^2 + (23.1\text{ V} - 41.6\text{ V})^2} = 35.0\text{ V}$$

This result confirms the correctness of the calculations in Example 4, since the rms voltage of the generator is, in fact, 35.0 V.

(b) The average power delivered by the generator is $\overline{P} = I_{rms}V_{rms}\cos\phi$ (Equation 23.9). Therefore, a value for the phase angle ϕ is needed and can be obtained from Equation 23.8 as follows:

$$\tan\phi = \frac{X_L - X_C}{R} \quad \text{or} \quad \phi = \tan^{-1}\left(\frac{X_L - X_C}{R}\right)$$

$$= \tan^{-1}\left(\frac{115\ \Omega - 207\ \Omega}{148\ \Omega}\right) = -32°$$

The phase angle is negative since the circuit is more capacitive than inductive (X_C is greater than X_L), and the current leads the voltage. The average power delivered by the generator is

$$\overline{P} = I_{rms}V_{rms}\cos\phi = (0.201\ \text{A})(35.0\ \text{V})\cos(-32°) = \boxed{6.0\ \text{W}}$$

This amount of power is delivered only to the resistor, since neither the capacitor nor the inductor uses power, on average.

In addition to the series RCL circuit, there are many different ways to connect resistors, capacitors, and inductors. In analyzing these additional possibilities, it helps to keep in mind the behavior of capacitors and inductors at the extreme limits of the frequency. When the frequency approaches zero (i.e., dc conditions), the reactance of a capacitor becomes so large that no charge can flow through the capacitor. It is as if the capacitor were cut out of the circuit, leaving an open gap in the connecting wire. In the limit of zero frequency the reactance of an inductor is vanishingly small. The inductor offers no opposition to a dc current and behaves as if it were replaced with a wire of zero resistance. In the limit of very large frequency, the behaviors of a capacitor and an inductor are reversed. The capacitor has a very small reactance and offers little opposition to the current, as if it were replaced by a wire with zero resistance. In contrast, the inductor has a very large reactance when the frequency is very large. The inductor offers so much opposition to the current that it might as well be cut out of the circuit, leaving an open gap in the connecting wire. Conceptual Example 5 illustrates how to gain insight into more complicated circuits using the limiting behaviors of capacitors and inductors.

CONCEPTUAL EXAMPLE 5 | The Limiting Behavior of Capacitors and Inductors

Figure 23.13*a* shows two circuits. The rms voltage of the generator is the same in each case. The values of the resistance *R*, the capacitance *C*, and the inductance *L* in these circuits are the same. The frequency of the ac generator is very nearly zero. In which circuit does the generator supply more rms current, **(a)** circuit I or **(b)** circuit II?

Reasoning According to Equation 23.6, the rms current is given by $I_{rms} = V_{rms}/Z$. However, the impedance *Z* cannot be obtained from Equation 23.7, since the circuits in **Figure 23.13*a*** are not series RCL circuits. Since V_{rms} is the same in each case, the greater current is delivered to the circuit with the smaller impedance *Z*. In the limit of very small frequencies, the capacitors have very large impedances and, thus, allow very little current to flow through them. In essence, the capacitors behave as if they were cut out of the circuit, leaving gaps in the connecting wires. On the other hand, in the limit of very small frequencies, the inductors have very small impedances and behave as if they were replaced by wires with zero resistance. **Figure 23.13*b*** shows the circuits as they would appear according to these changes.

Answer (a) is incorrect. According to **Figure 23.13*b***, circuit I behaves as if it contained only two identical resistors wired in series, with a total impedance of $Z = R + R = 2R$. In contrast, circuit II behaves as if it contained two identical resistors wired in parallel, in which case the total impedance is given by $1/Z = 1/R + 1/R = 2/R$, or $Z = R/2$. Circuit I contains the greater impedance, so the generator supplies less, not more, rms current to that circuit.

Answer (b) is correct. The rms current I_{rms} in a circuit is given by $I_{rms} = V_{rms}/Z$. Since V_{rms} is the same for both circuits and circuit II has the smaller impedance [see the explanation for why Answer (a) is incorrect], its generator supplies the greater rms current.

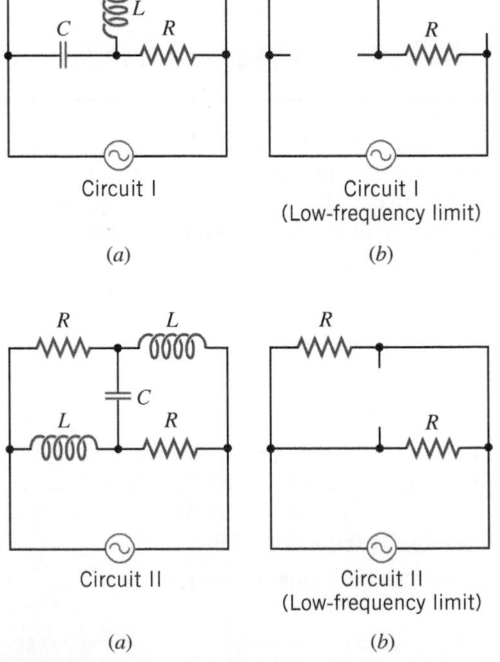

FIGURE 23.13 (*a*) These circuits are discussed in the limit of very small or low frequency in Conceptual Example 5. (*b*) For a frequency very near zero, the circuits in part *a* behave as if they were as shown here.

Related Homework: Check Your Understanding 7, 8, Problems 38, 55

BIO THE PHYSICS OF . . . body-fat scales. The impedance of an ac circuit contains important information about the resistance, capacitance, and inductance in the circuit. As an example of a very complex circuit, consider the human body. It contains muscle, which is a relatively good conductor of electricity due to its high water content, and also fat, which is a relatively poor conductor due to its low water content. The impedance that the body offers to ac electricity is referred to as bioelectrical impedance and is largely determined by resistance and capacitive reactance. Capacitance enters the picture because cell membranes can act like tiny capacitors. Bioelectric impedance analysis provides the basis for the determination of body-fat percentage by the body-fat scales (see **Figure 23.14**) that are widely available for home use. When you stand barefoot on such a scale, electrodes beneath your feet send a weak ac current (approximately 800 μA, 50 kHz) through your lower body in order to measure your body's impedance. The scale also measures your weight. A built-in computer combines the impedance and weight with information you provide about height, age, and sex to determine the percentage of fat in your body to an accuracy of about 5%. For men (age 20 to 39) a percentage of 8 to 19% is considered average, whereas the corresponding range of values for women of similar ages is 21 to 33%.

Electrodes for balls of feet

Electrodes for heels

FIGURE 23.14 Bathroom scales are now widely available that can provide estimates of your body-fat percentage. When you stand barefoot on the scale, electrodes beneath your feet send a small ac current through your lower body that allows the body's electrical impedance to be measured. This impedance is correlated with the percentage of fat in the body.

BIO THE PHYSICS OF . . . transcutaneous electrical nerve stimulation (TENS). Weak ac electricity with a much lower frequency than that used to measure bioelectrical impedance is used in a technique called transcutaneous electrical nerve stimulation (TENS). TENS is the most commonly used form of electroanalgesia in pain-management situations and, in its conventional form, uses an ac frequency between 40 and 150 Hz. Ac current is passed between two electrodes attached to the body and inhibits the transmission of pain-related nerve impulses. The technique is thought to work by affecting the "gates" in a nerve cell that control the passage of sodium ions into and out of the cell (see Section 19.6). **Figure 23.15** shows TENS being applied during assessment of pain control following suspected damage to the radial nerve in the forearm.

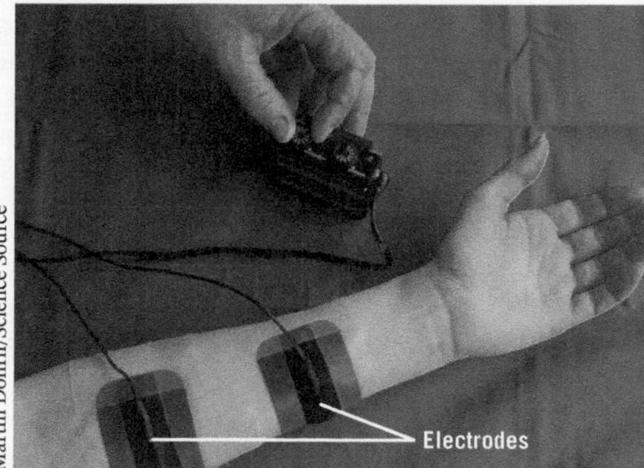

Martin Dohrn/Science Source

FIGURE 23.15 Transcutaneous electrical nerve stimulation (TENS) is shown here being applied to the forearm, in an assessment of pain control following suspected damage to the radial nerve.

Check Your Understanding

(The answers are given at the end of the book.)

3. A long wire of finite resistance is connected to an ac generator. The wire is then wound into a coil of many loops and reconnected to the generator. Is the current in the circuit with the coil greater than, less than, or the same as the current in the circuit with the uncoiled wire?

4. A light bulb and a parallel plate capacitor (containing a dielectric material between the plates) are connected in series to the 60-Hz ac voltage present at a wall outlet. When the dielectric material is removed from the space between the plates, does the brightness of the bulb increase, decrease, or remain the same?

5. An air-core inductor is connected in series with a light bulb, and this circuit is plugged into an ac electrical outlet. When a piece of iron is inserted inside the inductor, does the brightness of the bulb increase, decrease, or remain the same?

6. Consider the circuit in **Figure 23.9**. With the capacitor and the inductor present, a certain amount of current is in the circuit. Then the capacitor and the inductor are removed, so that only the resistor remains connected to the generator. Is it possible that, under a certain condition, the current in the simplified circuit has the same rms value as in the original circuit? **(a)** No **(b)** Yes, when $X_L = R$ **(c)** Yes, when $X_C = R$ **(d)** Yes, when $X_C = X_L$

7. Review Conceptual Example 5 as an aid in understanding this question. An inductor and a capacitor are connected in parallel across the terminals of an ac generator. Does the current from the generator decrease, remain the same, or increase as the frequency becomes **(a)** very large and **(b)** very small?

8. Review Conceptual Example 5 as an aid in understanding this question. For which of the two circuits discussed there does the generator deliver more current when the frequency is very large? **(a)** Circuit I **(b)** Circuit II

23.4 Resonance in Electric Circuits

The behavior of the current and voltage in a series RCL circuit can give rise to a condition of **resonance**. Resonance occurs when the frequency of a vibrating force exactly matches a natural (resonant) frequency of the object to which the force is applied. When resonance occurs, the force can transmit a large amount of energy to the object, leading to a large-amplitude vibratory motion. We have already encountered several instances of resonance. For example, resonance can occur when a vibrating force acts on an object of mass m that is attached to a spring whose spring constant is k (Section 10.6). In this case there is one natural frequency f_0, whose value is $f_0 = [1/(2\pi)]\sqrt{k/m}$. Resonance also occurs when standing waves are set up on a string (Section 17.5) or in a tube of air (Section 17.6). The string and tube of air have many natural frequencies, one for each allowed standing wave. As we will now see, a condition of resonance can also be established in a series RCL circuit. In this case there is only one natural frequency, and the vibrating force is provided by the oscillating electric field that is related to the voltage of the generator.

Animated Figure 23.16 helps us understand why a resonant frequency exists for an ac circuit. This drawing presents an analogy between the electrical case (ignoring resistance) and the mechanical case of an object attached to a horizontal spring (ignoring friction). Part *a* shows a fully stretched spring that has just been released, and the initial speed v of the object is zero. All the energy is stored in the form of elastic potential energy. When the object begins to move, it gradually loses potential energy and gains kinetic energy. In part *b*, the object moves with speed v_{max} and maximum kinetic energy through the position where the spring is unstretched (zero potential energy). Because of its inertia, the moving object coasts through this position and eventually comes to a halt in part *c* when the spring is fully compressed and all kinetic energy has been converted back into elastic potential energy. Part *d* of the picture is like part *b*, except that the direction of motion is reversed. The resonant frequency f_0 of the object on the spring is the natural frequency at which the object vibrates and is given as $f_0 = [1/(2\pi)]\sqrt{k/m}$ according to Equations 10.6 and 10.11. In this expression, m is the mass of the object, and k is the spring constant.

In the electrical case, **Animated Figure 23.16*a*** begins with a fully charged capacitor that has just been connected to an inductor. At this instant the energy is stored in the electric field between the capacitor plates. As the capacitor discharges, the electric field $\vec{\mathbf{E}}$ between the plates decreases, while a magnetic field $\vec{\mathbf{B}}$ builds up around the inductor because of the increasing current in the circuit. The maximum current and the maximum magnetic field exist at the instant when the capacitor is

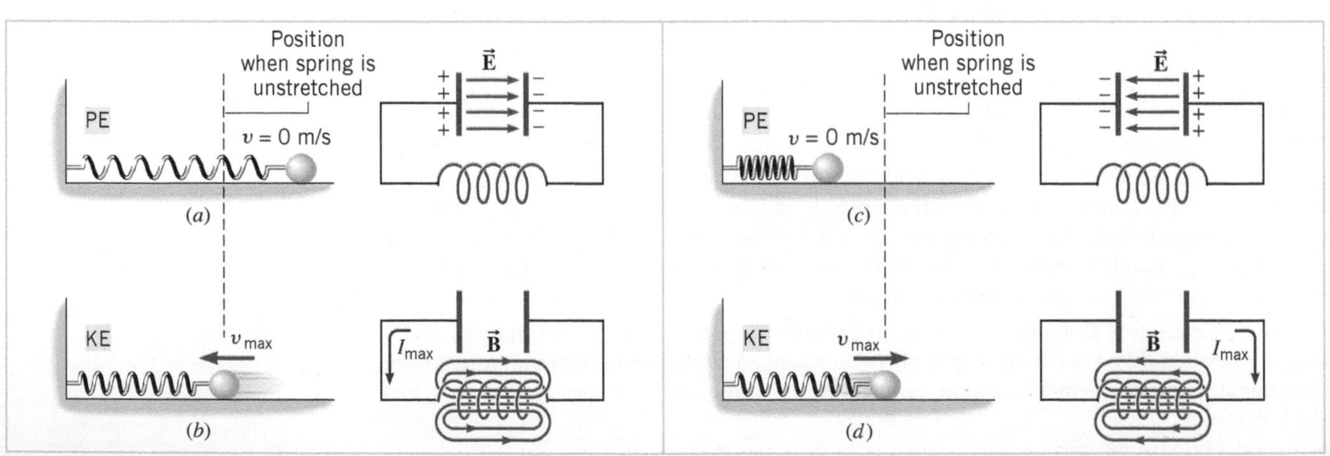

ANIMATED FIGURE 23.16 The oscillation of an object on a spring is analogous to the oscillation of the electric and magnetic fields that occur, respectively, in a capacitor and in an inductor. (PE, potential energy; KE, kinetic energy.)

completely discharged, as in part *b* of the figure. Energy is now stored entirely in the magnetic field of the inductor. The voltage induced in the inductor keeps the charges flowing until the capacitor again becomes fully charged, but now with reversed polarity, as in part *c*. Once again, the energy is stored in the electric field between the plates, and no energy resides in the magnetic field of the inductor. Part *d* of the cycle repeats part *b*, but with reversed directions of current and magnetic field. Thus, an ac circuit can have a resonant frequency because there is a natural tendency for energy to shuttle back and forth between the electric field of the capacitor and the magnetic field of the inductor.

To determine the resonant frequency at which energy shuttles back and forth between the capacitor and the inductor, we note that the current in a series RCL circuit is $I_{rms} = V_{rms}/Z$ (Equation 23.6). In this expression Z is the impedance of the circuit and is given by $Z = \sqrt{R^2 + (X_L - X_C)^2}$ (Equation 23.7). As **Figure 23.17** illustrates, the rms current is a maximum when the impedance is a minimum, assuming a given generator voltage. The minimum impedance of $Z = R$ occurs when the frequency is f_0, such that $X_L = X_C$ or $2\pi f_0 L = 1/(2\pi f_0 C)$. This result can be solved for f_0, which is the resonant frequency:

$$f_0 = \frac{1}{2\pi\sqrt{LC}} \tag{23.10}$$

The resonant frequency is determined by the inductance and the capacitance, but not the resistance.

The effect of resistance on electrical resonance is to make the "sharpness" of the circuit response less pronounced, as **Figure 23.18** indicates. When the resistance is small, the current–frequency graph falls off suddenly on either side of the maximum current. When the resistance is large, the falloff is more gradual, and there is less current at the maximum.

Example 6 demonstrates the use of Equation 23.10 to calculate the resonant frequency of an MRI magnet.

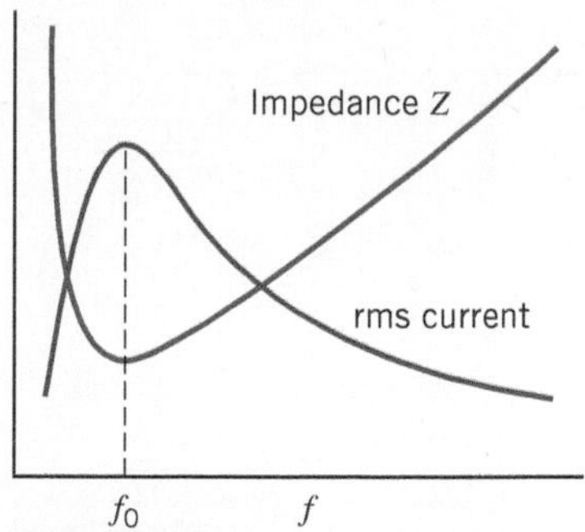

FIGURE 23.17 In a series RCL circuit the impedance is a minimum, and the current is a maximum, when the frequency f equals the resonant frequency f_0 of the circuit.

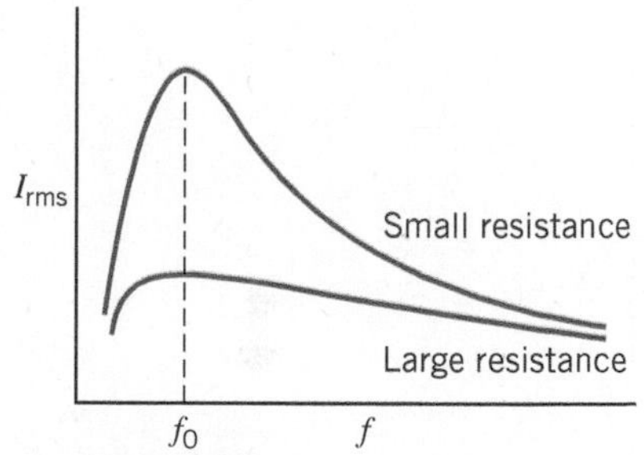

FIGURE 23.18 The effect of resistance on the current in a series RCL circuit.

EXAMPLE 6 | BIO The Physics of Nuclear Magnetic Resonance—NMR

In Chapter 21 we described the medical diagnostic technique of magnetic resonance imaging, or MRI. This technique is based on the more fundamental principle of nuclear magnetic resonance, or NMR. Here, nuclei in the substance under study will absorb and re-emit radio waves when placed in a strong magnetic field. They do this by rotating at a frequency that is typically hundreds of MHz. A pickup coil detects the AC voltage created by the rotating nuclei, and different nuclei will produce different characteristic frequencies. Consider an NMR spectrometer that is configured with a pickup coil designed to detect one particular resonant frequency. This occurs because the coil is part of an *RLC* circuit with a capacitance of 9.4×10^{-16} F. If the inductance of the pickup coil is 7.5 mH, what is the resonant frequency of the coil?

Reasoning In an *RLC* circuit, the resonance frequency depends on the value of the circuit's capacitance and inductance, as given by Equation 23.10. We can use this relationship to calculate the frequency.

Solution Applying Equation 23.10, we have:

$$f_0 = \frac{1}{2\pi\sqrt{LC}} = \frac{1}{2\pi\sqrt{(0.0075\text{ H})(9.4 \times 10^{-16}\text{ F})}} = 6.0 \times 10^7\text{ Hz}$$

$$= \boxed{60\text{ MHz}}$$

Check Your Understanding

(The answers are given at the end of the book.)

9. The resistance in a series RCL circuit is doubled. **(a)** Does the resonant frequency increase, decrease, or remain the same? **(b)** Does the maximum current in the circuit increase, decrease, or remain the same?

10. In a series RCL circuit at resonance, does the current lead or lag behind the voltage across the generator, or is the current in phase with the voltage?

11. Is it possible for two series RCL circuits to have the same resonant frequencies and yet have **(a)** different R values and **(b)** different C and L values?

12. Suppose the generator connected to a series RCL circuit has a frequency that is greater than the resonant frequency of the circuit. Suppose, in addition, that it is necessary to match the resonant frequency of the circuit to the frequency of the generator. To accomplish this, should you add a second capacitor **(a)** in series or **(b)** in parallel with the one already present?

23.5 Semiconductor Devices

Semiconductor devices such as diodes and transistors are widely used in modern electronics, and **Figure 23.19** illustrates one application. The drawing shows an audio system in which small ac voltages (originating in a compact disc player, an FM tuner, or a cassette deck) are amplified so they can drive the speaker(s). The electric circuits that accomplish the amplification do so with the aid of a dc voltage provided by the power supply. In portable units the power supply is simply a battery. In nonportable units, however, the power supply is a separate electric circuit containing diodes, along with other elements. As we will see, the diodes convert the 60-Hz ac voltage present at a wall outlet into the dc voltage needed by the amplifier, which, in turn, performs its job of amplification with the aid of transistors.

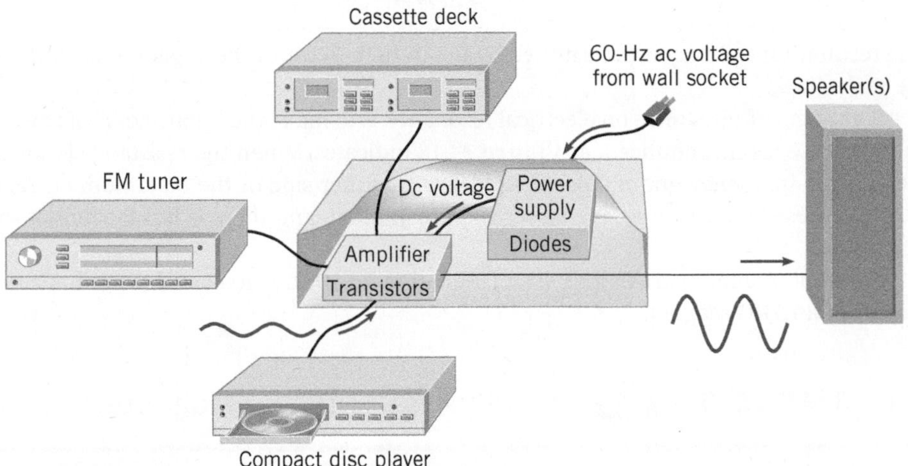

FIGURE 23.19 In a typical audio system, diodes are used in the power supply to create a dc voltage from the ac voltage present at the wall socket. This dc voltage is necessary so the transistors in the amplifier can perform their task of enlarging the small ac voltage (in blue) originating in the compact disc player, or other device.

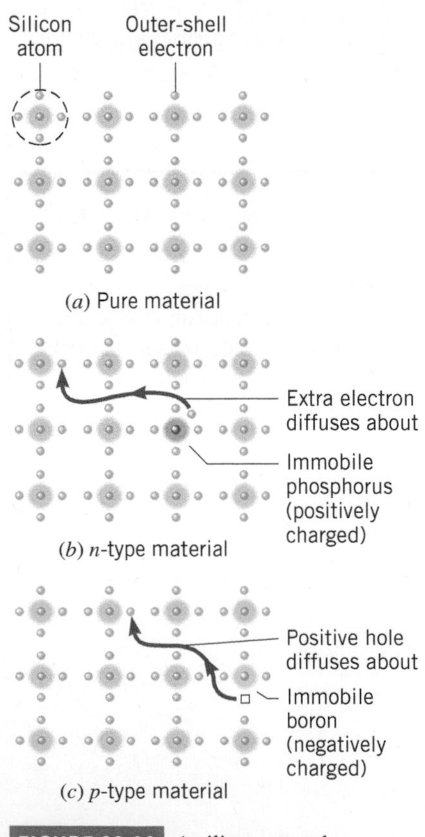

FIGURE 23.20 A silicon crystal that is (*a*) undoped, or pure, (*b*) doped with phosphorus to produce an *n*-type material, and (*c*) doped with boron to produce a *p*-type material.

n-Type and *p*-Type Semiconductors

The materials used in diodes and transistors are semiconductors, such as silicon and germanium. However, they are not pure materials because small amounts of "impurity" atoms (about one part in a million) have been added to them to change their conductive properties. For instance, **Figure 23.20*a*** shows an array of atoms that symbolizes the crystal structure in pure silicon. Each silicon atom has four outer-shell* electrons, and each electron participates with electrons from neighboring atoms in forming the bonds that hold the crystal together. Since they participate in forming bonds, these electrons generally do not move throughout the crystal. Consequently, pure silicon and germanium are not good conductors of electricity. It is possible, however, to increase their conductivities by adding tiny amounts of impurity atoms, such as phosphorus or arsenic, whose atoms have five outer-shell electrons. For example, when a phosphorus atom replaces a silicon atom in the crystal, only four of the five outer-shell electrons of phosphorus fit into the crystal structure. The extra fifth electron does not fit in and is relatively free to diffuse throughout the crystal, as part *b* of the drawing suggests. A semiconductor containing small amounts of phosphorus can therefore be envisioned as containing immobile, positively charged phosphorus atoms and a pool of electrons that are free to wander throughout the material. These mobile electrons allow the semiconductor to conduct electricity.

*Section 30.6 discusses the electronic structure of the atom in terms of "shells."

The process of adding impurity atoms is called *doping*. A semiconductor doped with an impurity that contributes mobile electrons is called an ***n*-type semiconductor**, since the mobile charge carriers have a **n**egative charge. Note that an *n*-type semiconductor is overall electrically neutral, since it contains equal numbers of positive and negative charges.

It is also possible to dope a silicon crystal with an impurity whose atoms have only three outer-shell electrons (e.g., boron or gallium). Because of the missing fourth electron, there is a "hole" in the lattice structure at the boron atom, as **Figure 23.20c** illustrates. An electron from a neighboring silicon atom can move into this hole, in which event the region around the boron atom, having acquired the electron, becomes negatively charged. Of course, when a nearby electron does move, it leaves behind a hole. This hole is positively charged, since it results from the removal of an electron from the vicinity of a neutral silicon atom. The vast majority of atoms in the lattice are silicon, so the hole is almost always next to another silicon atom. Consequently, an electron from one of these adjacent atoms can move into the hole, with the result that the hole moves to yet another location. In this fashion, a positively charged hole can wander through the crystal. This type of semiconductor can, therefore, be viewed as containing immobile, negatively charged boron atoms and an equal number of positively charged, mobile holes. Because of the mobile holes, the semiconductor can conduct electricity. In this case the charge carriers are positive. A semiconductor doped with an impurity that introduces mobile **p**ositive holes is called a ***p*-type semiconductor**.

The Semiconductor Diode

THE PHYSICS OF . . . a semiconductor diode. A ***p-n* junction diode** is a device that is formed from a *p*-type semiconductor and an *n*-type semiconductor. The *p-n* junction between the two materials is of fundamental importance to the operation of diodes and transistors. **Figure 23.21** shows separate *p*-type and *n*-type semiconductors, each electrically neutral. **Figure 23.22a** shows them joined together to form a diode. Mobile electrons from the *n*-type semiconductor and mobile holes from the *p*-type semiconductor flow across the junction and combine. This process leaves the *n*-type material with a positive charge layer and the *p*-type material with a negative charge layer, as part *b* of the drawing indicates. The positive and negative charge layers on the two sides of the junction set up an electric field $\vec{E}$, much like the field in a parallel plate capacitor. This electric field tends to prevent any further movement of charge across the junction, and all charge flow quickly stops.

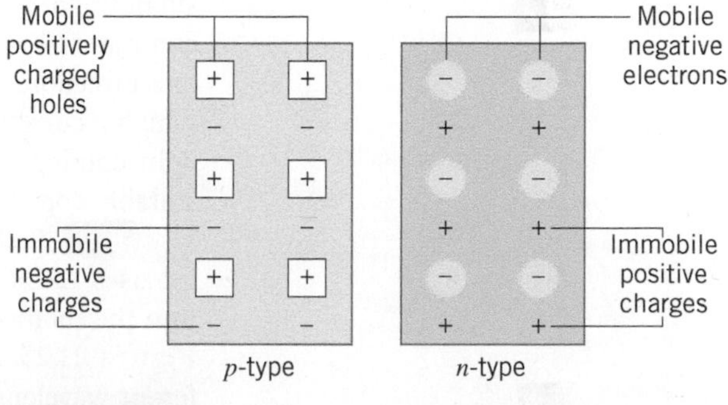

FIGURE 23.21 A *p*-type semiconductor and an *n*-type semiconductor.

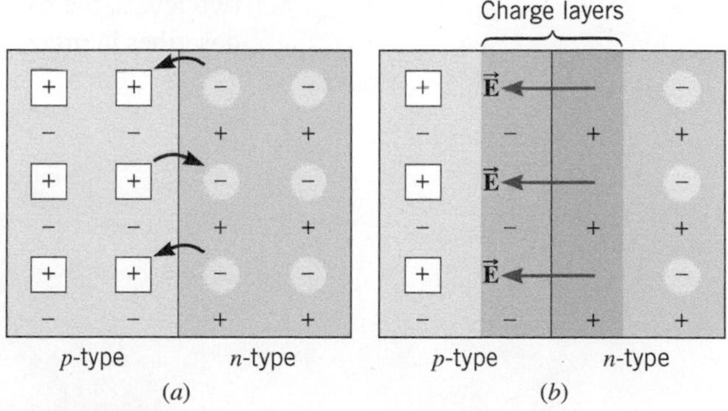

FIGURE 23.22 At the junction between *n* and *p* materials, (*a*) mobile electrons and holes combine and (*b*) create positive and negative charge layers. The electric field produced by the charge layers is $\vec{E}$.

Suppose now that a battery is connected across the *p-n* junction, as in **Figure 23.23a**, where the negative terminal of the battery is attached to the *n*-material, and the positive terminal is attached to the *p*-material. In this situation the junction is said to be in a condition of **forward bias**, and, as a result, there is a current in the circuit. The negative terminal of the battery repels the mobile electrons in the *n*-type material, and they move toward the junction. Likewise, the positive terminal repels the positive holes in the *p*-type material, and they also move toward the junction. At the junction the electrons fill the holes. In the meantime, the negative terminal provides a fresh supply of electrons to the *n*-material, and the positive terminal pulls off electrons from the *p*-material, forming new holes in the process. Consequently, a continual flow of charge, and hence a current, is maintained.

In **Figure 23.23b** the battery polarity has been reversed, and the *p-n* junction is in a condition known as **reverse bias**. The battery forces electrons in the *n*-material and holes in the *p*-material away from the junction. As a result, the potential across the junction builds up until it opposes the battery potential, and very little current can be sustained through the diode. The diode, then, is a unidirectional device, for it allows current to pass in only one direction.

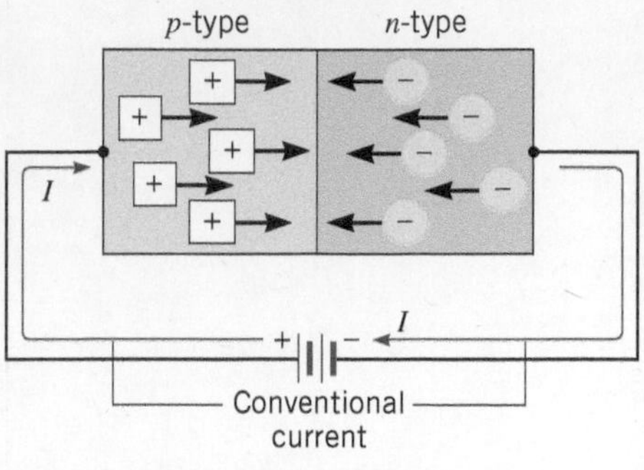

(*a*) Forward bias

(*b*) Reverse bias

FIGURE 23.23 There is an appreciable current through the diode when the diode is forward-biased. (*b*) Under a reverse-bias condition, there is almost no current through the diode.

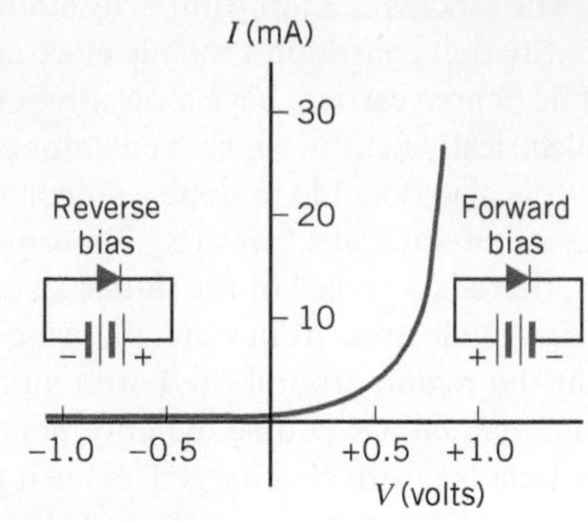

FIGURE 23.24 The current–voltage characteristics of a typical *p-n* junction diode.

The graph in **Figure 23.24** shows the dependence of the current on the magnitude and polarity of the voltage applied across a *p-n* junction diode. The exact values of the current depend on the nature of the semiconductor and the extent of the doping. Also shown in the drawing is the symbol used for a diode (—▶|—). The direction of the arrowhead in the symbol indicates the direction of the conventional current in the diode under a forward-bias condition. In a forward-bias condition, the side of the symbol that contains the arrowhead has a positive potential relative to the other side.

THE PHYSICS OF . . . light-emitting diodes (LEDs). A special kind of diode is called an **LED**, which stands for light-emitting diode. You can see LEDs in the form of small bright red, green, or yellow lights that appear on most electronic devices, such as computers, TV sets, and audio systems. These diodes, like others, carry current in only one direction. Imagine a forward-biased diode, like that shown in **Figure 23.23*a***, in which a current exists. An LED emits light whenever electrons and holes combine, the light coming from the *p-n* junction. Commercial LEDs are often made from gallium, suitably doped with arsenic and phosphorus atoms.

BIO **THE PHYSICS OF . . . a fetal oxygen monitor.** A fetal oxygen monitor uses LEDs to measure the level of oxygen in a fetus's blood. A sensor is inserted into the mother's uterus and positioned against the cheek of the fetus, as indicated in **Figure 23.25**. Two LEDs are located within the sensor, and each shines light of a different wavelength (or color) into the fetal tissue. The light is reflected by the oxygen-carrying red blood cells and is detected by an adjacent photodetector. Light from one of the LEDs is used to measure the level of oxyhemoglobin in the blood, and light from the other LED is used to measure the level of deoxyhemoglobin. From a comparison of these two levels, the oxygen saturation in the blood is determined. Example 3 in Chapter 24 describes in more detail the operation of an adult version of the oxygen monitor.

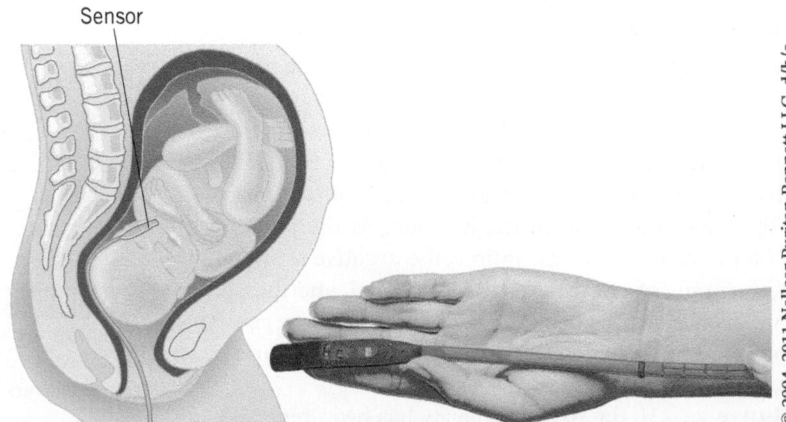

FIGURE 23.25 A fetal oxygen monitor uses a sensor that contains LEDs to measure the level of oxygen in the blood of an unborn child.

THE PHYSICS OF . . . rectifier circuits. Because diodes are unidirectional devices, they are commonly used in rectifier circuits, which convert an ac voltage into a dc voltage. For instance, **Figure 23.26** shows a circuit in which charges flow through the resistance R only while the ac generator biases the diode in the forward direction. Since

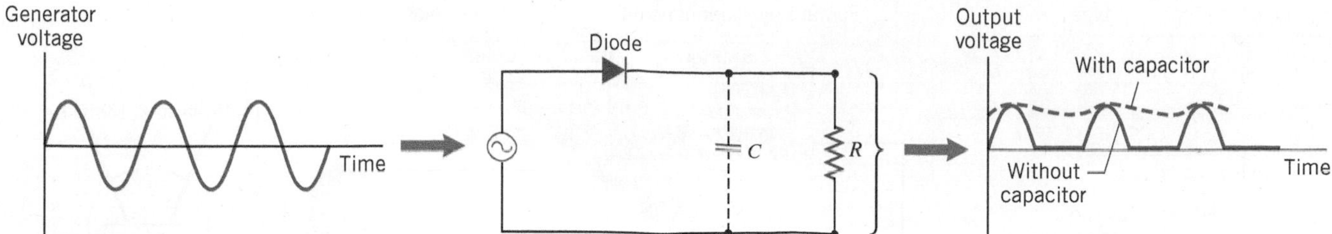

FIGURE 23.26 A half-wave rectifier circuit, together with a capacitor and a transformer (not shown), constitutes a dc power supply because the rectifier converts an ac voltage into a dc voltage.

current occurs only during one-half of every generator voltage cycle, the circuit is called a *half-wave rectifier*. A plot of the output voltage across the resistor reveals that only the positive halves of each cycle are present. If a capacitor is added in parallel with the resistor, as indicated in **Figure 23.26**, the capacitor charges up and keeps the voltage from dropping to zero between each positive half-cycle. It is also possible to construct *full-wave rectifier circuits*, in which both halves of every cycle of the generator voltage drive current through the load resistor in the same direction (see Check Your Understanding Question 13).

When a circuit such as the one in **Figure 23.26** includes a capacitor and also a transformer to establish the desired voltage level, the circuit is called a *power supply*. In the audio system in **Figure 23.19**, the power supply receives the 60-Hz ac voltage from a wall socket and produces a dc output voltage that is used for the transistors within the amplifier. Power supplies using diodes are also found in virtually all electronic appliances, such as televisions and microwave ovens.

Solar Cells

THE PHYSICS OF . . . solar cells. Solar cells use *p-n* junctions to convert sunlight directly into electricity, as **Figure 23.27** illustrates. The solar cell in this drawing consists of a *p*-type semiconductor surrounding an *n*-type semiconductor. As discussed earlier, charge layers form at the junction between the two types of semiconductors, leading to an electric field $\vec{\mathbf{E}}$ pointing from the *n*-type toward the *p*-type layer. The outer covering of *p*-type material is so thin that sunlight penetrates into the charge layers and ionizes some of the atoms there. In the process of ionization, the energy of the sunlight causes a negative electron to be ejected from the atom, leaving behind a positive hole. As the drawing indicates, the electric field in the charge layers causes the electron and the hole to move away from the junction. The electron moves into the *n*-type material, and the hole moves into the *p*-type material. As a result, the sunlight causes the solar cell to develop negative and positive terminals, much like the terminals of a battery. The current that a single solar cell can provide is small, so applications of solar cells often use many of them mounted to form large panels, as **Figure 23.28** illustrates.

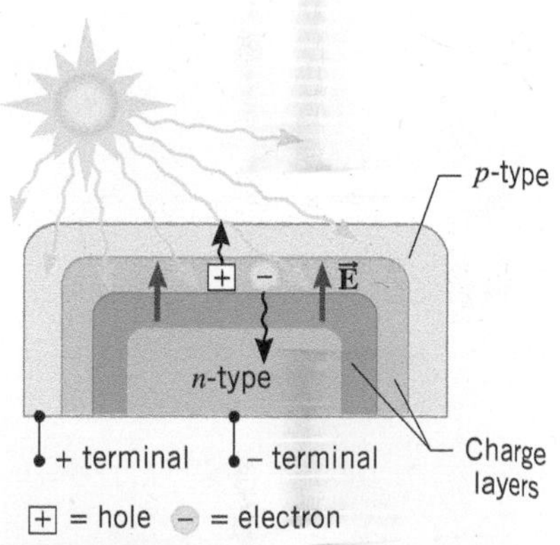

FIGURE 23.27 A solar cell formed from a *p-n* junction. When sunlight strikes it, the solar cell acts like a battery, with + and − terminals.

Transistors

THE PHYSICS OF . . . transistors. A number of different kinds of transistors are in use today. One type is the bipolar-junction transistor, which consists of two *p-n* junctions formed by three layers of doped semiconductors. As **Figure 23.29** indicates, there are *pnp* and *npn* transistors. In either case, the middle region is made very thin compared to the outer regions.

A transistor is useful because it can be used in circuits that amplify a smaller voltage into a larger one. A transistor plays the same kind of role in an amplifier circuit that a valve does when it controls the flow of water through a pipe. A small change in the valve setting produces a large change in the amount of water per second that flows through the pipe. Similarly, a small change in the voltage input to a transistor produces a large change in the output from the transistor.

CHRISTIAN CHARISIUS/Reuters/Landov LLC

FIGURE 23.28 This boat, the *PlanetSolar*, is powered by the electrical energy generated by approximately 500 m^2 of solar cells that cover its top surface. It was built by the shipyard of Knierim Yachtbau in Kiel, Germany.

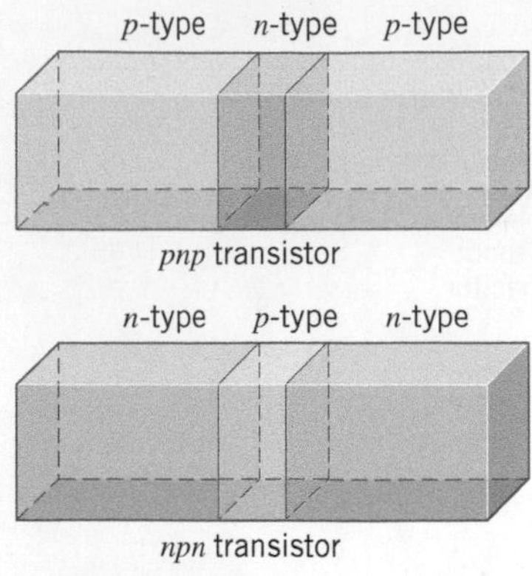

FIGURE 23.29 There are two kinds of bipolar-junction transistors, *pnp* and *npn*.

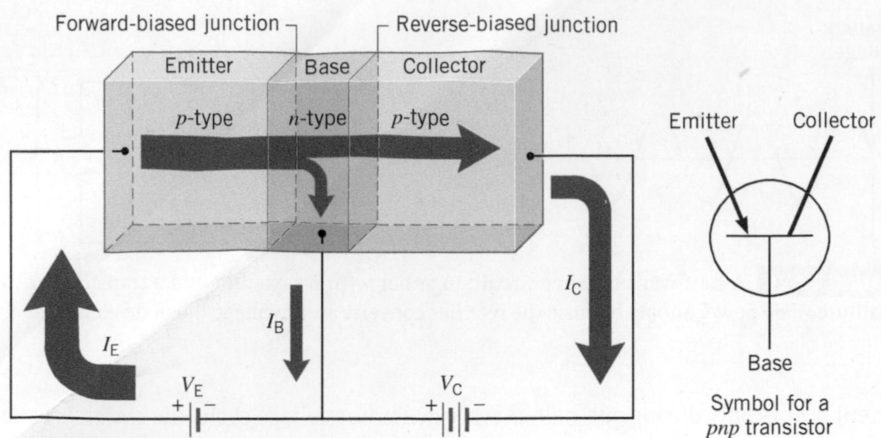

FIGURE 23.30 A *pnp* transistor, along with its bias voltages V_E and V_C. On the symbol for the *pnp* transistor at the right, the emitter is marked with an arrow that denotes the direction of conventional current through the emitter.

Figure 23.30 shows a *pnp* transistor connected to two batteries, labeled V_E and V_C. The voltages V_E and V_C are applied in such a way that the *p-n* junction on the left has a forward bias, while the *p-n* junction on the right has a reverse bias. Moreover, the voltage V_C is usually much larger than V_E for a reason to be discussed shortly. The drawing also shows the standard symbol and nomenclature for the three sections of the *pnp* transistor—namely, the *emitter,* the *base,* and the *collector.* The arrow in the symbol points in the direction of the conventional current through the emitter.

The positive terminal of V_E pushes the mobile positive holes in the *p*-type material of the emitter toward the emitter/base junction. Since this junction has a forward bias, the holes enter the base region readily. Once in the base region, the holes come under the strong influence of V_C and are attracted to its negative terminal. Since the base is so thin (about 10^{-6} m or so), approximately 98% of the holes are drawn through the base and into the collector. The remaining 2% of the holes combine with free electrons in the base region, thereby giving rise to a small base current I_B. As the drawing shows, the moving holes in the emitter and collector constitute currents that are labeled I_E and I_C, respectively. From Kirchhoff's junction rule it follows that $I_C = I_E - I_B$.

Because the base current I_B is small, the collector current is determined primarily by current from the emitter ($I_C = I_E - I_B \approx I_E$). This means that a change in I_E will cause a change in I_C of nearly the same amount. Furthermore, a substantial change in I_E can be caused by only a small change in the forward-bias voltage V_E. To see that this is the case, look at **Figure 23.24** and notice how steep the current–voltage curve is for a *p-n* junction: small changes in the forward-bias voltage give rise to large changes in the current.

With the help of **Interactive Figure 23.31** we can now appreciate what was meant by the earlier statement that a small change in the voltage input to a transistor leads to a large change in the output. This picture shows an ac generator connected in series with the battery V_E and a resistance R connected in series with the collector. The generator voltage could originate from many sources, such as an electric guitar pickup or a compact disc player, while the resistance R could represent a loudspeaker. The generator introduces small voltage changes in the forward bias across the emitter/base junction and, thus, causes large corresponding changes in the current leaving the collector and passing through the resistance R. As a result, the output voltage across R is an enlarged or amplified version of the input voltage of the generator. The operation of an *npn* transistor is similar to that of a *pnp* transistor. The main difference is that the bias voltages and current directions are reversed.

It is important to realize that the increased power available at the output of a transistor amplifier does *not* come from the transistor itself. Rather, it comes from the power provided by the voltage source V_C. The transistor, acting like an automatic valve, merely allows the small signals from the input generator to control the power taken from the source V_C and delivered to the resistance R.

Today it is possible to combine arrays of billions of transistors, diodes, resistors, and capacitors on a tiny chip of silicon that usually measures less than a centimeter on a side. These arrays are called *integrated circuits* (ICs) and can be designed to perform almost any

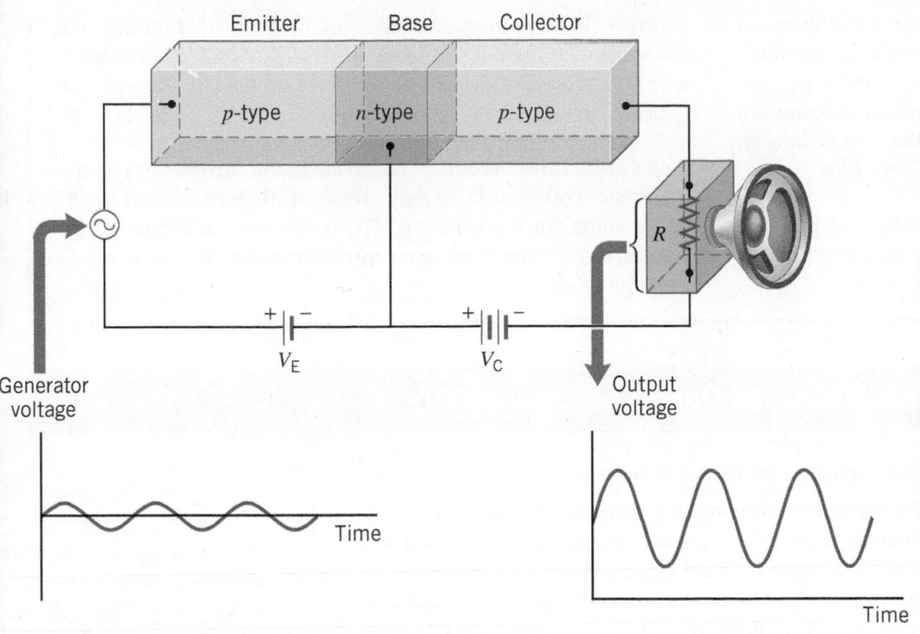

INTERACTIVE FIGURE 23.31 The basic *pnp* transistor amplifier in this drawing amplifies a small generator voltage to produce an enlarged voltage across the resistance *R*.

Adam Hart-Davis/Science Source

FIGURE 23.32 Integrated circuit (IC) chips are manufactured on wafers of semiconductor material. Shown here is one wafer containing many chips.

desired electronic function. Integrated circuits composed of semiconductors, such as the type in **Figure 23.32**, have revolutionized the electronics industry and lie at the heart of computers, cellular phones, digital watches, and programmable appliances. The following example describes just one of the many unique applications of semiconductors.

CONCEPTUAL EXAMPLE 7 | BIO The Physics of Health-Monitoring Smart Rings

As mentioned earlier, the impurity atoms that are chemically doped into semiconductor materials alter their conductivities, and it is useful to consider the qualitative behavior of their electrical resistivity as a function of temperature. In Section 20.3, we mentioned that the resistivity of a metal decreases with decreasing temperature, whereas for semiconductors, the reverse is true. Materials for which the resistivity decreases with increasing temperature are referred to as NTCs, or ***negative temperature coefficient*** materials. Semiconductors become better conductors as their temperature increases. A comprehensive understanding of why this occurs requires the band theory of solids, which is beyond the scope of this book. However, it is sufficient to know that the number of mobile charge carriers increases exponentially with increasing temperature. These carriers participate in conduction, thereby lowering the resistivity.

The resistivity in many semiconductors is extremely sensitive to temperature, much more so than in metals, which makes them particularly well-suited for sensor applications. These highly temperature-sensitive resistors are also called *thermistors*. One application where thermistors are routinely utilized is in wearable health-monitoring devices. Smartwatches, smartbracelets, and smartrings use sensors to monitor in real time a person's movements, sleep patterns, and vital signs, such as heart rate, blood pressure, blood-oxygen level, and many others (see, for example, the Oura ring in **Figure 23.33**).

Thermistor sensors measure a person's temperature, and the current technology can detect temperature changes as small as 0.1 °C. Since the ring can detect an increase in a person's temperature, it can indicate the onset of a fever, which is often associated with infectious disease. In fact, during the COVID-19 pandemic in 2020, thousands of medical workers wore the smart rings to study their effectiveness at predicting the infection. Players in the NBA also wore the rings to monitor their wellness while finishing the 2019–2020 season at the Walt Disney World Resort in Orlando, Florida.

©2020 Oura, https://ouraring.com/meet-oura

FIGURE 23.33 Smart rings, like the Oura shown here, use different sensors to monitor a person's movement and vital signs. A finely calibrated thermistor measures body temperature, which, in some cases, may provide early detection of fever and possible infection.

The Oura is powered by a 3.7-V lithium-ion battery, and the current through the thermistor is monitored by the ring's circuitry. Which of the following changes in the thermistor's current might indicate the onset of fever? **(a)** The current through the thermistor remains constant. **(b)** The current through the thermistor increases. **(c)** The current through the thermistor decreases.

Reasoning Given that the voltage output from the lithium-ion battery is a constant, a change in the thermistor's current must correspond to a change in its resistance. Ohm's law can be used to determine whether or not the resistance increases or decreases, and this change in resistance will depend on the change in temperature. An increase in temperature may indicate a fever.

Answer (a) is incorrect. If the current through the thermistor remains constant, then the thermistor's resistance would also remain constant, indicating no change in temperature.

Answer (b) is correct. An increase in the current through the thermistor corresponds to a decrease in its resistance, in accordance with Ohm's law. For an NTC thermistor, a decrease in its resistance indicates an increase in its temperature, which may be due to fever.

Answer (c) is incorrect. A decrease in the current through the thermistor corresponds to an increase in its resistance, in accordance with Ohm's law. For an NTC thermistor, an increase in its resistance indicates a decrease in its temperature, which would suggest no fever.

Check Your Understanding

(The answer is given at the end of the book.)

13. CYU Figure 23.2 shows a full-wave rectifier circuit, in which the direction of the current through the load resistor R is the same for both positive and negative halves of the generator's voltage cycle. What is the direction of the current through the resistor (left to right, or right to left) when **(a)** the top of the generator is positive and the bottom is negative and **(b)** the top of the generator is negative and the bottom is positive?

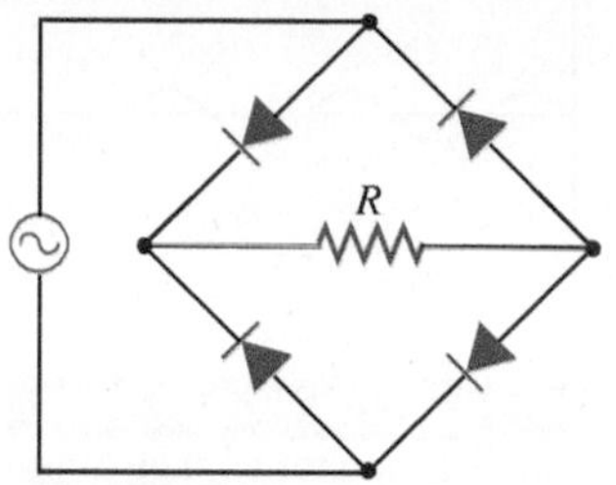

CYU FIGURE 23.2

Concept Summary

23.1 Capacitors and Capacitive Reactance In an ac circuit the rms voltage V_{rms} across a capacitor is related to the rms current I_{rms} by Equation 23.1, where X_C is the capacitive reactance. The capacitive reactance is measured in ohms (Ω) and is given by Equation 23.2, where f is the frequency and C is the capacitance. The ac current in a capacitor leads the voltage across the capacitor by a phase angle of 90° or $\pi/2$ radians. As a result, a capacitor consumes no power, on average.

$$V_{rms} = I_{rms} X_C \quad (23.1)$$

The phasor model is useful for analyzing the voltage and current in an ac circuit. In this model, the voltage and current are represented by rotating arrows, called phasors. The length of the voltage phasor represents the maximum voltage V_0, and the length of the current phasor represents the maximum current I_0. The phasors rotate in a counterclockwise direction at a frequency f. Since the current leads the voltage by 90° in a capacitor, the current phasor is ahead of the voltage phasor by 90° in the direction of rotation. The instantaneous values of the voltage and current are equal to the vertical components of the corresponding phasors.

$$X_C = \frac{1}{2\pi f C} \quad (23.2)$$

23.2 Inductors and Inductive Reactance In an ac circuit the rms voltage V_{rms} across an inductor is related to the rms current I_{rms} by Equation 23.3, where X_L is the inductive reactance. The inductive reactance is measured in ohms (Ω) and is given by Equation 23.4, where f is the frequency and L is the inductance. The ac current in an inductor lags behind the voltage across the inductor by a phase angle of 90° or $\pi/2$ radians. Consequently, an inductor, like a capacitor, consumes no power, on average.

$$V_{rms} = I_{rms} X_L \quad (23.3)$$

$$X_L = 2\pi f L \quad (23.4)$$

The voltage and current phasors in a circuit containing only an inductor also rotate in a counterclockwise direction at a frequency f. However, since the current lags the voltage by 90° in an inductor, the current phasor is behind the voltage phasor by 90° in the direction of rotation. The instantaneous values of the voltage and current are equal to the vertical components of the corresponding phasors.

23.3 Circuits Containing Resistance, Capacitance, and Inductance When a resistor, a capacitor, and an inductor are connected in series, the rms voltage across the combination is related to the rms current according to Equation 23.6, where Z is the impedance of the combination. The impedance is measured in ohms (Ω) and is given by Equation 23.7, where R is the resistance, and X_L and X_C are, respectively, the inductive and capacitive reactances.

$$V_{rms} = I_{rms} Z \quad (23.6)$$

$$Z = \sqrt{R^2 + (X_L - X_C)^2} \quad (23.7)$$

The tangent of the phase angle ϕ between current and voltage in a series RCL circuit is given by Equation 23.8.

$$\tan \phi = \frac{X_L - X_C}{R} \quad (23.8)$$

Only the resistor in the RCL combination consumes power, on average. The average power $\overline{P}$ consumed in the circuit is given by Equation 23.9, where $\cos \phi$ is called the power factor of the circuit.

$$\overline{P} = I_{rms} V_{rms} \cos \phi \quad (23.9)$$

23.4 Resonance in Electric Circuits A series RCL circuit has a resonant frequency f_0 that is given by Equation 23.10, where L is the inductance and C is the capacitance. At resonance, the impedance of the circuit has a minimum value equal to the resistance R, and the rms current has a maximum value.

$$f_0 = \frac{1}{2\pi\sqrt{LC}} \quad (23.10)$$

23.5 Semiconductor Devices In an n-type semiconductor, mobile negative electrons carry the current. An n-type material is produced by doping a semiconductor such as silicon with a small amount of impurity atoms such as phosphorus. In a p-type semiconductor, mobile positive "holes" in the crystal structure carry the current. A p-type material is produced by doping a semiconductor with a small amount of impurity atoms such as boron. These two types of semiconductors are used in p-n junction diodes, light-emitting diodes, solar cells, and *pnp* and *npn* bipolar junction transistors.

Focus on Concepts

Online

Additional questions are available for assignment in WileyPLUS.

Section 23.1 Capacitors and Capacitive Reactance

1. A circuit contains an ac generator and a resistor. What happens to the average power dissipated in the resistor when the frequency is doubled and the rms voltage is tripled? **(a)** Nothing happens, because the average power does not depend on either the frequency or the rms voltage. **(b)** The average power doubles because it is proportional to the frequency. **(c)** The average power triples because it is proportional to the rms voltage. **(d)** The average power increases by a factor of $3^2 = 9$ because it is proportional to the square of the rms voltage. **(e)** The average power increases by a factor of $2 \times 3 = 6$ because it is proportional to the product of the frequency and the rms voltage.

Section 23.2 Inductors and Inductive Reactance

2. What happens to the capacitive reactance X_C and the inductive reactance X_L if the frequency of the ac voltage is doubled? **(a)** X_C increases by a factor of 2, and X_L decreases by a factor of 2. **(b)** X_C and X_L both increase by a factor of 2. **(c)** X_C and X_L do not change. **(d)** X_C and X_L both decrease by a factor of 2. **(e)** X_C decreases by a factor of 2, and X_L increases by a factor of 2.

3. Each of the four phasor diagrams represents a different circuit. V_0 and I_0 represent, respectively, the maximum voltage of the generator and the current in the circuit. Which circuit contains only a resistor? **(a)** A **(b)** B **(c)** C **(d)** D

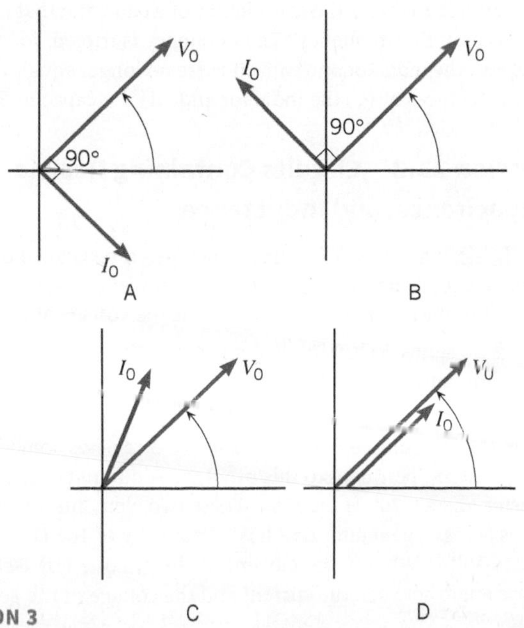

QUESTION 3

Section 23.3 Circuits Containing Resistance, Capacitance, and Inductance

4. The table shows the rms voltage V_C across the capacitor and the rms voltage V_L across the inductor for three series RCL circuits. In which circuit does the rms voltage across the entire RCL combination lead the current through the combination? **(a)** Circuit 1 **(b)** Circuit 2 **(c)** Circuit 3 **(d)** The total rms voltage across the RCL combination does not lead the current in any of the circuits.

Circuit	V_C	V_L
1	50 V	100 V
2	100 V	50 V
3	50 V	50 V

5. A capacitor and an inductor are connected to an ac generator in two ways: in series and in parallel (see the drawing). At low frequencies, which circuit has the greater current? **(a)** The series circuit, because the impedance of the circuit is small due to the small reactances of both the inductor and the capacitor. **(b)** The series circuit, because the impedance of the circuit is large due to the large reactances of both the inductor and the capacitor. **(c)** The parallel circuit, because the impedance of the circuit is small due to the large reactance of the inductor. **(d)** The parallel circuit, because the impedance of the circuit is large due to the large reactance of the inductor. **(e)** The parallel circuit, because the impedance of the circuit is small due to the small reactance of the inductor.

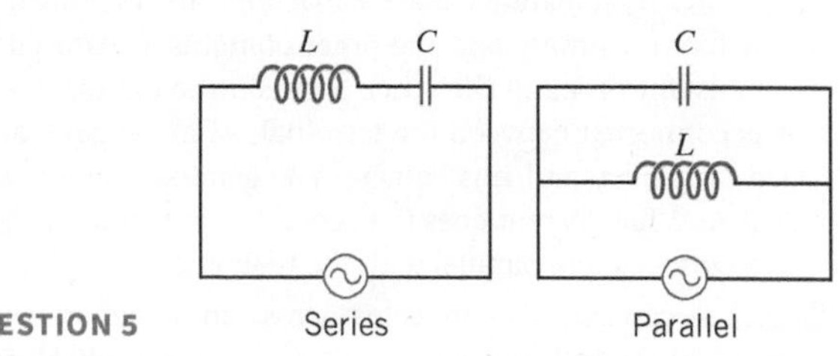

QUESTION 5

Section 23.4 Resonance in Electric Circuits

6. In an RCL circuit a second capacitor is added in parallel to the capacitor already present. Does the resonant frequency of the circuit increase, decrease, or remain the same? **(a)** The resonant frequency

increases, because it depends inversely on the square root of the capacitance, and the equivalent capacitance decreases when a second capacitor is added in parallel. **(b)** The resonant frequency increases, because it is directly proportional to the capacitance, and the equivalent capacitance increases when a second capacitor is added in parallel. **(c)** The resonant frequency decreases, because it is directly proportional to the capacitance, and the equivalent capacitance decreases when a second capacitor is added in parallel. **(d)** The resonant frequency decreases, because it depends inversely on the square root of the capacitance, and the equivalent capacitance increases when a second capacitor is added in parallel. **(e)** The resonant frequency remains the same.

Problems

Online

Additional questions are available for assignment in WileyPLUS.

Note: *For problems in this set, the ac current and voltage are rms values, and the power is an average value, unless indicated otherwise.*

- SSM Student Solutions Manual
- MMH Problem-solving help
- GO Guided Online Tutorial
- V-HINT Video Hints
- CHALK Chalkboard Videos
- BIO Biomedical application
- E Easy
- M Medium
- H Hard
- WS Worksheet
- T Team Problem

Section 23.1 Capacitors and Capacitive Reactance

1. E SSM A 63.0-μF capacitor is connected to a generator operating at a low frequency. The rms voltage of the generator is 4.00 V and is constant. A fuse in series with the capacitor has negligible resistance and will burn out when the rms current reaches 15.0 A. As the generator frequency is increased, at what frequency will the fuse burn out?

2. E GO Two identical capacitors are connected in parallel to an ac generator that has a frequency of 610 Hz and produces a voltage of 24 V. The current in the circuit is 0.16 A. What is the capacitance of each capacitor?

3. E The reactance of a capacitor is 68 Ω when the ac frequency is 460 Hz. What is the reactance when the frequency is 870 Hz?

4. E GO A capacitor is connected to an ac generator that has a frequency of 3.4 kHz and produces a voltage of 2.0 V. The current in the capacitor is 35 mA. When the same capacitor is connected to a second ac generator that has a frequency of 5.0 kHz, the current in the capacitor is 85 mA. What voltage does the second generator produce?

5. E SSM A capacitor is connected across the terminals of an ac generator that has a frequency of 440 Hz and supplies a voltage of 24 V. When a second capacitor is connected in parallel with the first one, the current from the generator increases by 0.18 A. Find the capacitance of the second capacitor.

6. M CHALK GO Two parallel plate capacitors are identical, except that one of them is empty and the other contains a material with a dielectric constant of 4.2 in the space between the plates. The empty capacitor is connected between the terminals of an ac generator that has a fixed frequency and rms voltage. The generator delivers a current of 0.22 A. What current does the generator deliver after the other capacitor is connected in parallel with the first one?

7. M V-HINT A capacitor is connected across an ac generator whose frequency is 750 Hz and whose *peak* output voltage is 140 V. The rms current in the circuit is 3.0 A. **(a)** What is the capacitance of the capacitor? **(b)** What is the magnitude of the *maximum* charge on one plate of the capacitor?

8. H A capacitor (capacitance C_1) is connected across the terminals of an ac generator. Without changing the voltage or frequency of the generator, a second capacitor (capacitance C_2) is added in series with the first one. As a result, the current delivered by the generator decreases by a factor of three. Suppose that the second capacitor had been added in parallel with the first one, instead of in series. By what factor would the current delivered by the generator have increased?

Section 23.2 Inductors and Inductive Reactance

9. E SSM An 8.2-mH inductor is connected to an ac generator (10.0 V rms, 620 Hz). Determine the *peak value* of the current supplied by the generator.

10. E An inductor has an inductance of 0.080 H. The voltage across this inductor is 55 V and has a frequency of 650 Hz. What is the current in the inductor?

11. E An inductor is to be connected to the terminals of a generator (rms voltage = 15.0 V) so that the resulting rms current will be 0.610 A. Determine the required inductive reactance.

12. E GO An ac generator has a frequency of 7.5 kHz and a voltage of 39 V. When an inductor is connected between the terminals of this generator, the current in the inductor is 42 mA. What is the inductance of the inductor?

13. E V-HINT A 40.0-μF capacitor is connected across a 60.0-Hz generator. An inductor is then connected in parallel with the capacitor. What is the value of the inductance if the rms currents in the inductor and capacitor are equal?

14. E GO An ac generator has a frequency of 2.2 kHz and a voltage of 240 V. An inductance $L_1 = 6.0$ mH is connected across its terminals. Then a second inductance $L_2 = 9.0$ mH is connected in parallel with L_1. Find the current that the generator delivers to L_1 and to the parallel combination.

15. M SSM CHALK A 30.0-mH inductor has a reactance of 2.10 kΩ. **(a)** What is the frequency of the ac current that passes through the inductor? **(b)** What is the capacitance of a capacitor that has the same reactance at this frequency? The frequency is tripled, so that the reactances of the inductor and capacitor are no longer equal. What are the new reactances of **(c)** the inductor and **(d)** the capacitor?

Section 23.3 Circuits Containing Resistance, Capacitance, and Inductance

16. E SSM A series RCL circuit includes a resistance of 275 Ω, an inductive reactance of 648 Ω, and a capacitive reactance of 415 Ω. The current in the circuit is 0.233 A. What is the voltage of the generator?

17. E A series RCL circuit contains a 47.0-Ω resistor, a 2.00-μF capacitor, and a 4.00-mH inductor. When the frequency is 2550 Hz, what is the power factor of the circuit?

18. E SSM Multiple-Concept Example 3 reviews some of the basic ideas that are pertinent to this problem. A circuit consists of a 215-Ω resistor and a 0.200-H inductor. These two elements are connected in series across a generator that has a frequency of 106 Hz and a voltage of 234 V. **(a)** What is the current in the circuit? **(b)** Determine the phase angle between the current and the voltage of the generator.

19. **E GO** An ac series circuit has an impedance of 192 Ω, and the phase angle between the current and the voltage of the generator is $\phi = -75°$. The circuit contains a resistor and either a capacitor or an inductor. Find the resistance R and the capacitive reactance X_C or the inductive reactance X_L, whichever is appropriate.

20. **E** When only a resistor is connected across the terminals of an ac generator (112 V) that has a fixed frequency, there is a current of 0.500 A in the resistor. When only an inductor is connected across the terminals of this same generator, there is a current of 0.400 A in the inductor. When both the resistor and the inductor are connected in series between the terminals of this generator, what are **(a)** the impedance of the series combination and **(b)** the phase angle between the current and the voltage of the generator?

21. **E GO** A 2700-Ω resistor and a 1.1-μF capacitor are connected in series across a generator (60.0 Hz, 120 V). Determine the power delivered to the circuit.

22. **M CHALK GO** Part *a* of the drawing shows a resistor and a charged capacitor wired in series. When the switch is closed, the capacitor discharges as charge moves from one plate to the other. Part *b* shows the amount q of charge remaining on each plate of the capacitor as a function of time. In part *c* of the drawing, the switch has been removed and an ac generator has been inserted into the circuit. The circuit elements in the drawing have the following values: $R = 18$ Ω, $V_{rms} = 24$ V for the generator, and $f = 380$ Hz for the generator. The time constant for the circuit in part *a* is $\tau = 3.0 \times 10^{-4}$ s. What is the rms current in the circuit in part *c*?

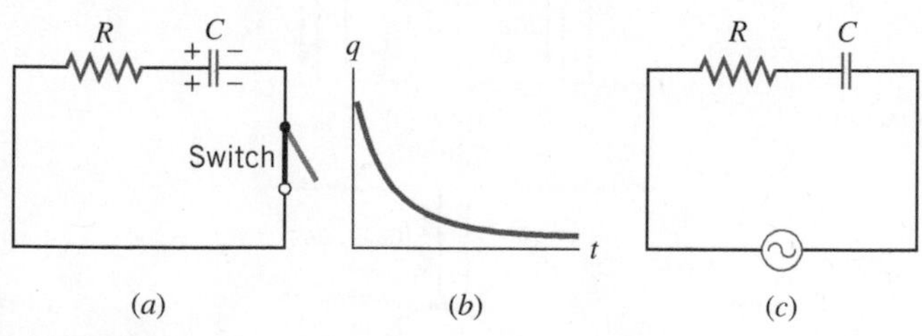

PROBLEM 22

23. **M SSM** A circuit consists of an 85-Ω resistor in series with a 4.0-μF capacitor, and the two are connected between the terminals of an ac generator. The voltage of the generator is fixed. At what frequency is the current in the circuit one-half the value that exists when the frequency is very large?

24. **M V-HINT** An 84.0-mH inductor and a 5.80-μF capacitor are connected in series with a generator whose frequency is 375 Hz. The rms voltage across the capacitor is 2.20 V. Determine the rms voltage across the inductor.

Section 23.4 Resonance in Electric Circuits

25. **E** A *tank circuit* in a radio transmitter is a series RCL circuit connected to an antenna. The antenna broadcasts radio signals at the resonant frequency of the tank circuit. Suppose that a certain tank circuit in a shortwave radio transmitter has a fixed capacitance of 1.8×10^{-11} F and a variable inductance. If the antenna is intended to broadcast radio signals ranging in frequency from 4.0 MHz to 9.0 MHz, find the **(a)** minimum and **(b)** maximum inductance of the tank circuit.

26. **E SSM** A series RCL circuit has a resonant frequency of 690 kHz. If the value of the capacitance is 2.0×10^{-9} F, what is the value of the inductance?

27. **E** The power dissipated in a series RCL circuit is 65.0 W, and the current is 0.530 A. The circuit is at resonance. Determine the voltage of the generator.

28. **E SSM** A 10.0-Ω resistor, a 12.0-μF capacitor, and a 17.0-mH inductor are connected in series with a 155-V generator. **(a)** At what frequency is the current a maximum? **(b)** What is the maximum value of the rms current?

29. **E GO** The capacitance in a series RCL circuit is $C_1 = 2.60$ μF, and the corresponding resonant frequency is $f_{01} = 7.30$ kHz. The generator frequency is 5.60 kHz. What is the value of the capacitance C_2 that should be added to the circuit so that the circuit will have a resonant frequency that matches the generator frequency? Note that you must decide whether C_2 is added in series or in parallel with C_1.

30. **E** A series RCL circuit is at resonance and contains a variable resistor that is set to 175 Ω. The power delivered to the circuit is 2.6 W. Assuming that the voltage remains constant, how much power is delivered when the variable resistor is set to 562 Ω?

31. **E GO** The resonant frequency of an RCL circuit is 1.3 kHz, and the value of the inductance is 7.0 mH. What is the resonant frequency (in kHz) when the value of the inductance is 1.5 mH?

32. **M** A series RCL circuit has a resonant frequency of 1500 Hz. When operating at a frequency other than 1500 Hz, the circuit has a capacitive reactance of 5.0 Ω and an inductive reactance of 30.0 Ω. What are the values of **(a)** L and **(b)** C?

33. **M GO** In a series RCL circuit the generator is set to a frequency that is *not* the resonant frequency. This nonresonant frequency is such that the ratio of the inductive reactance to the capacitive reactance of the circuit is observed to be 5.36. The resonant frequency is 225 Hz. What is the frequency of the generator?

34. **M CHALK GO** A charged capacitor and an inductor are connected as shown in the drawing (this circuit is the same as that in **Animated Figure 23.16a**). There is no resistance in the circuit. As Section 23.4 discusses, the electrical energy initially present in the charged capacitor then oscillates back and forth between the inductor and the capacitor. The initial charge on the capacitor has a magnitude of $q = 2.90$ μC. The capacitance is $C = 3.60$ μF, and the inductance is $L = 75.0$ mH. **(a)** What is the electrical energy stored initially by the charged capacitor? **(b)** Find the maximum current in the inductor.

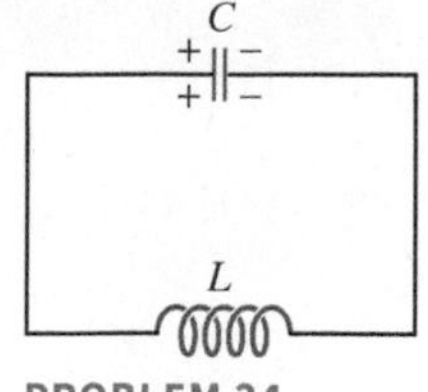

PROBLEM 34

35. **M V-HINT** In the absence of a nearby metal object, the two inductances (L_A and L_B) in a heterodyne metal detector are the same, and the resonant frequencies of the two oscillator circuits have the same value of 630.0 kHz. When the search coil (inductor B) is brought near a buried metal object, a beat frequency of 7.30 kHz is heard. By what percentage does the buried object increase the inductance of the search coil?

Additional Problems

Online

36. **E GO** A circuit consists of a resistor in series with an inductor and an ac generator that supplies a voltage of 115 V. The inductive reactance is 52.0 Ω, and the current in the circuit is 1.75 A. Find the average power delivered to the circuit.

37. **E** In a series circuit, a generator (1350 Hz, 15.0 V) is connected to a 16.0-Ω resistor, a 4.10-μF capacitor, and a 5.30-mH inductor. Find the voltage across each circuit element.

38. E Review Conceptual Example 5 and Figure 23.13. Find the ratio of the current in circuit I to the current in circuit II in the *high-frequency limit* for the same generator voltage.

39. E SSM At what frequency (in Hz) are the reactances of a 52-mH inductor and a 76-μF capacitor equal?

40. E GO The resistor in a series RCL circuit has a resistance of 92 Ω, while the voltage of the generator is 3.0 V. At resonance, what is the average power delivered to the circuit?

41. E Two ac generators supply the same voltage. However, the first generator has a frequency of 1.5 kHz, and the second has a frequency of 6.0 kHz. When an inductor is connected across the terminals of the first generator, the current delivered is 0.30 A. How much current is delivered when this inductor is connected across the terminals of the second generator?

42. M V-HINT A series circuit contains only a resistor and an inductor. The voltage V of the generator is fixed. If $R = 16\ \Omega$ and $L = 4.0$ mH, find the frequency at which the current is one-half its value at zero frequency.

43. M SSM A series RCL circuit contains a 5.10-μF capacitor and a generator whose voltage is 11.0 V. At a resonant frequency of 1.30 kHz the power delivered to the circuit is 25.0 W. Find the values of (a) the inductance and (b) the resistance. (c) Calculate the power factor when the generator frequency is 2.31 kHz.

44. M GO SSM Part *a* of the figure shows a heterodyne metal detector being used. As part *b* of the figure illustrates, this device utilizes two capacitor/inductor oscillator circuits, A and B. Each produces its own resonant frequency, $f_{0A} = 1/[2\pi(L_A C)^{1/2}]$ and $f_{0B} = 1/[2\pi(L_B C)^{1/2}]$. Any difference between these frequencies is detected through earphones as a beat frequency $|f_{0B} - f_{0A}|$. In the absence of any nearby metal object, the inductances L_A and L_B are identical. When inductor B (the search coil) comes near a piece of metal, the inductance L_B increases, the corresponding oscillator frequency f_{0B} decreases, and a beat frequency is heard. Suppose that initially each inductor is adjusted so that $L_B = L_A$, and each oscillator has a resonant frequency of 855.5 kHz. Assuming that the inductance of search coil B increases by 1.000% due to a nearby piece of metal, determine the beat frequency heard through the earphones.

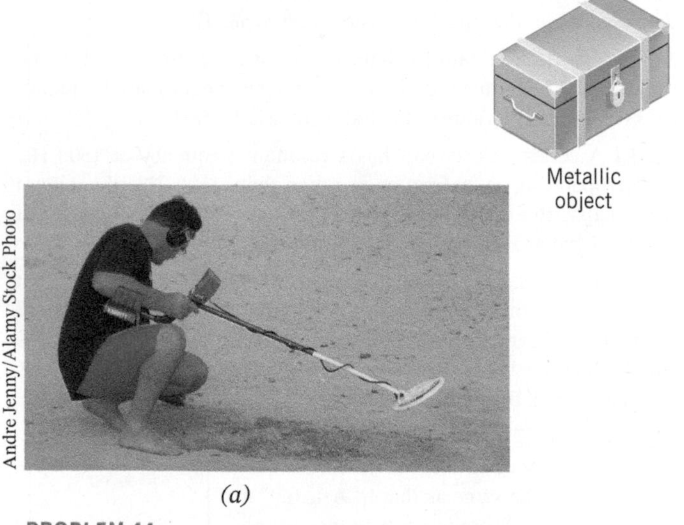

(*a*)

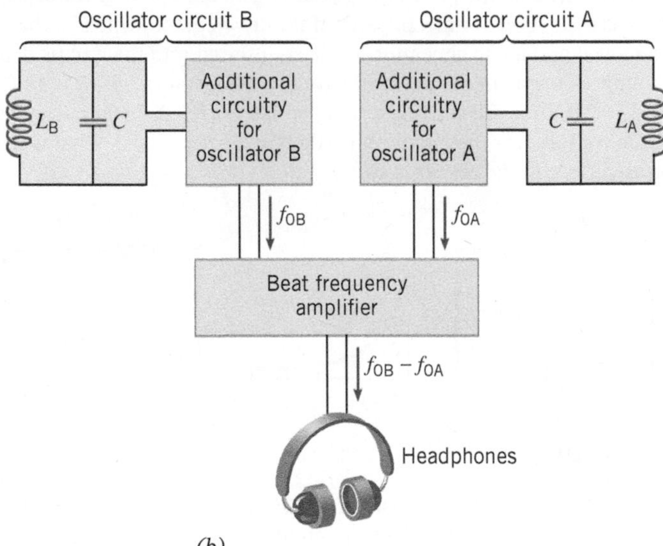

(*b*)

PROBLEM 44

Physics in Biology, Medicine, and Sports

45. M BIO V-HINT 23.3 In one measurement of the body's bioelectric impedance, values of $Z = 4.50 \times 10^2\ \Omega$ and $\phi = -9.80°$ are obtained for the total impedance and the phase angle, respectively. These values assume that the body's resistance R is in series with its capacitance C and that there is no inductance L. Determine the body's resistance and capacitive reactance.

46. M BIO 23.2 A high-field superconducting magnet in an MRI machine has an inductance of 60.0 H. If this magnet were connected to a generator with an rms voltage of 2.50×10^2 V, what would be the maximum current in the magnet, if the frequency of the generator is 17.0 Hz?

47. M BIO 23.3 In Chapter 20, we analyzed a cardiac pacemaker that uses a 4.1×10^{-7}-F capacitor and a 1.8-MΩ resistor connected in series. When an rms voltage of 3.7 V is supplied to this circuit, the rms current is 1.7×10^{-6} A. What is the (a) impedance and (b) the capacitive reactance of this circuit? (c) What is the frequency of the rms voltage?

48. E BIO 23.2 As medical technologies advance, there is a greater need for remote monitoring of patient vital signs and the equipment that provides their care. This has led to an increase in high frequency communication signals from systems such as Bluetooth and WiFi. These high frequency signals can potentially interfere with sensitive medical equipment. One common approach to combating high frequency noise in the power input to sensitive devices is to use something called a *choke*. Chokes come in multiple types, but one of the simplest is just an inductor. Consider an inductor that is placed in series with a power supply for a medical monitor. The inductor is designed to filter out high frequency noise. An inductor with an inductance of 5.5×10^{-3} H is used to filter noise in a power line with a frequency of 150 kHz. (a) What is the inductive reactance? (b) What is the I_{rms} current in the circuit, if the power cord is plugged into a standard wall outlet with $V_{rms} = 120$ V?

49. M BIO 23.4 The game-day coordinator at NFL football games communicates wirelessly with officials on the field using a frequency of about 467 MHz. **(a)** If the detection circuit in the communication equipment uses a 2.75-pF capacitor, what inductance would be required for a resonant frequency of 467 MHz? **(b)** If during one game, the resonant frequency had to be changed to 457 MHz, and assuming the capacitance in the detection circuit could be varied, what would be the new capacitance for this lower frequency?

50. M BIO 23.4 The ionosphere is the ionized part of the upper layer of the earth's atmosphere. The air molecules there are ionized by solar radiation. This layer of the atmosphere is a fairly good conductor, and radio waves are often "bounced" off the bottom of the ionosphere back toward the earth, in a process called skip or skywave propagation. Due to these properties, the space between the surface of the earth and the bottom of the ionosphere acts like a closed waveguide that will exhibit resonance for very low frequencies. Resonance excitations in the cavity are caused by lightning strikes, which hit the earth about 50 to 100 times a second. These low atmospheric resonance frequencies are known as *Schumann resonances*, named after the physicist Winfried Otto Schumann, who first calculated them in 1952. There are several Schumann frequencies that occur in the low frequency background, which ranges from 3 to 60 Hz. The highest intensity resonance mode (called the fundamental) occurs at 7.83 Hz. Other Schumann resonances occur at 14.3, 20.8, 27.3, and 33.8 Hz. Around 20 years ago, research showed that the majority of biological systems exhibit electrical activity in the same range as the Schumann resonances. New theoretical work suggests that primordial cells synchronized their electrical activity to the natural atmospheric resonances, and this evolutionary process is evident today across a gigantic scale—from single-celled organisms to the electrical activity in our brains. The surface of the earth and the bottom of the ionosphere act like a giant capacitor, and we can model one path that stretches between them as a circuit consisting of a resistor and capacitor connected in series. Assume the atmosphere has an average resistance of $2.00 \times 10^2\ \Omega$, and the capacitance is equal to 0.070 F. **(a)** What is the capacitive reactance of the circuit when it is active at the fundamental Schumann resonance frequency? **(b)** What is the impedance of this circuit? **(c)** What is the rms value of the current in the circuit, if the maximum voltage is 3.50×10^5 V?

Concepts and Calculations Problems

Online

A capacitor is one of the important elements found in ac circuits. As we have seen in this chapter, the capacitance of a capacitor influences the amount of current in a circuit. The capacitance, in turn, is determined by the geometry of the capacitor and the material that fills the space between its plates, as Section 19.5 discusses. When capacitors are connected together, the equivalent capacitance depends on the nature of the connection—for example, whether it is a series or parallel connection. Problem 51 provides a review of these issues concerning capacitors. In ac circuits that contain capacitance, inductance, and resistance, it is only the resistance that, on the average, consumes power. The average power delivered to a capacitor or an inductor is zero. However, the presence of a capacitor and/or an inductor does influence the rms current in the circuit. When the current changes for any reason, the power in the resistor also changes, as Problem 52 illustrates.

51. M CHALK Two parallel plate capacitors are filled with the same dielectric material and have the same plate area. However, the plate separation of capacitor 1 is twice that of capacitor 2. When capacitor 1 is connected across the terminals of an ac generator, the generator delivers an rms current of 0.60 A. *Concepts:* (i) Which of the two capacitors has the greater capacitance? (ii) Is the equivalent capacitance of the parallel combination (C_P) greater or smaller than the capacitance of capacitor 1? (iii) Is the capacitive reactance of C_P greater or smaller than for C_1? (iv) When both capacitors are connected in parallel across the terminals of the generator, is the current from the generator greater or smaller than when capacitor 1 is connected alone? *Calculations:* What is the current delivered by the generator when both capacitors are connected in parallel across the terminals?

52. M CHALK SSM An ac generator has a frequency of 1200 Hz and a constant rms voltage. When a 470-Ω resistor is connected between the terminals of the generator, an average power of 0.25 W is consumed by the resistor. Then, a 0.080-H inductor is connected in series with the resistor, and the combination is connected between the generator terminals. *Concepts:* (i) In which case does the generator deliver a greater rms current? (ii) In which case is the greater average power consumed by the circuit? *Calculations:* What is the average power consumed in the inductor–resistor series circuit?

Team Problems

Online

53. E T WS **Power in Extreme Frequency Limits.** You and your team have been assigned to find a power supply for the circuit in the drawing, which can be used to supply a dc voltage at 15.0 V, or a high frequency ac signal with a root-mean-square (rms) voltage of 15.0 V. The components in the circuit have the following values: $R = 4.60\ \Omega$, $C = 20.0$ nF, and $L = 22.0$ mH. Your task is to estimate the peak wattage (i.e., power) required of the power supply for **(a)** the dc and **(b)** the high frequency signals. Conceptual Example 5 will provide insight into this problem.

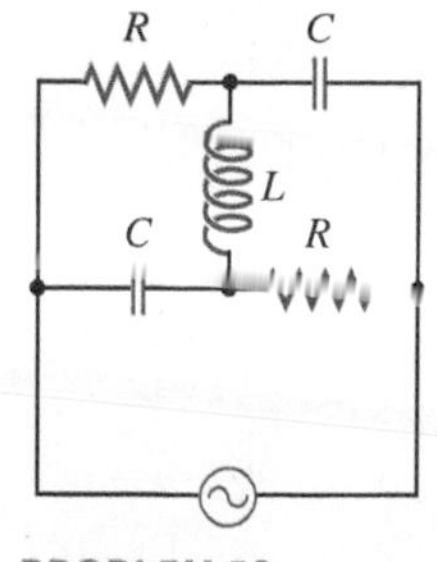

PROBLEM 53

54 M T WS **Extending the Range.** You and your team are tasked with repairing part of a communications circuit that consists of an inductor and a resistor in series, as shown in the drawing. The ac power source supplies a sinusoidal root-mean-square voltage of 30.0 V at an adjustable frequency between 0 Hz and 775 kHz. The problem was that the resistor in the circuit burned out as the ac source

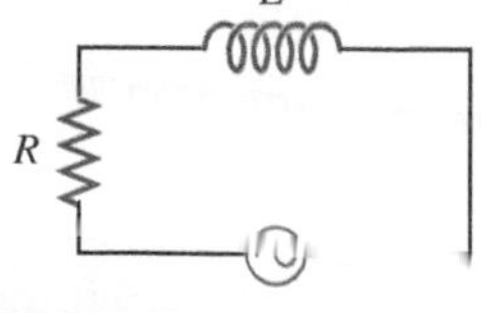

PROBLEM 54

frequency was being reduced in order to produce a signal at some low frequency. The resistor had a resistance of $R = 50\ \Omega$, and an instantaneous, or peak, power rating of 0.50 W, meaning that it would burn up if forced to dissipate power above its power rating. The inductor has an inductance of $L = 3.2$ mH. **(a)** What was the minimum frequency (in Hz) that this circuit was originally designed to accommodate without burning the resistor? **(b)** Your supplies are very limited but you were able to find a few other resistors including two 25-Ω resistors (each rated for a peak power of 0.25 W), a 50-Ω (0.50 W) resistor, and two 100-Ω resistors (0.50 W). Assume all resistor values are known to two significant figures. Find a way to replace the burned resistor with a configuration of the resistors you have available that will have the same effective resistance, but will extend the low frequency limit of the circuit (i.e., so that the circuit can go to lower frequencies without burning a resistor). What is the new low frequency limit?

55. M T **A Communications Jammer.** Radio jamming is the intentional disruption or interference of radio communications by overwhelming the intended receivers of the signal with random noise. You and your team have been tasked with jamming a specific radio signal at 720 kHz. You have access to a high-powered transmitter, but the part of its circuitry that tunes the broadcast frequency, called the tank circuit, has been damaged. A tank circuit is a series RCL circuit whose resonance frequency determines the frequency broadcasted by the antenna. At your disposal are two 220-Ω resistors, one variable capacitor that ranges from 2.0 to 6.0 nF, and four inductors with the following values: $L_1 = 5.0 \times 10^{-6}$ H, $L_2 = 7.2 \times 10^{-6}$ H, $L_3 = 6.5 \times 10^{-5}$ H, and $L_4 = 5.4 \times 10^{-6}$ H. **(a)** If you set your variable capacitor at the center of its range, what must be the value of the inductance of your RCL circuit so that it resonates at 720 kHz? **(b)** How should you configure the available inductors to give you the needed equivalent inductance? (Hint: the rules for adding inductors in series and parallel are the same as for resistors.) **(c)** With the inductance set as calculated in (a), what resonant frequency range does the variable capacitor provide? **(d)** The two resistors can be configured to give different equivalent resistance values. How should you configure the resistors in the RCL circuit in order to maximize the current at the resonant frequency? (Refer to Section 23.4.)

56. M T **A Reconfigurable RCL Circuit.** A series RCL circuit is composed of a resistor ($R = 220.0\ \Omega$), two identical capacitors ($C = 3.50$ nF) connected in series, and two identical inductors ($L = 5.50 \times 10^{-5}$ H) connected in series. You and your team need to determine: **(a)** the resonant frequency of this configuration. **(b)** What are all of the other possible resonant frequencies that can be attained by reconfiguring the capacitors and inductors (while using all of the components and keeping the proper series RCL order)? **(c)** If you were to design a circuit using only one of the given inductors and one adjustable capacitor, what would the range of the variable capacitor need to be in order to cover all of the resonant frequencies found in (a) and (b)?

CHAPTER **24**

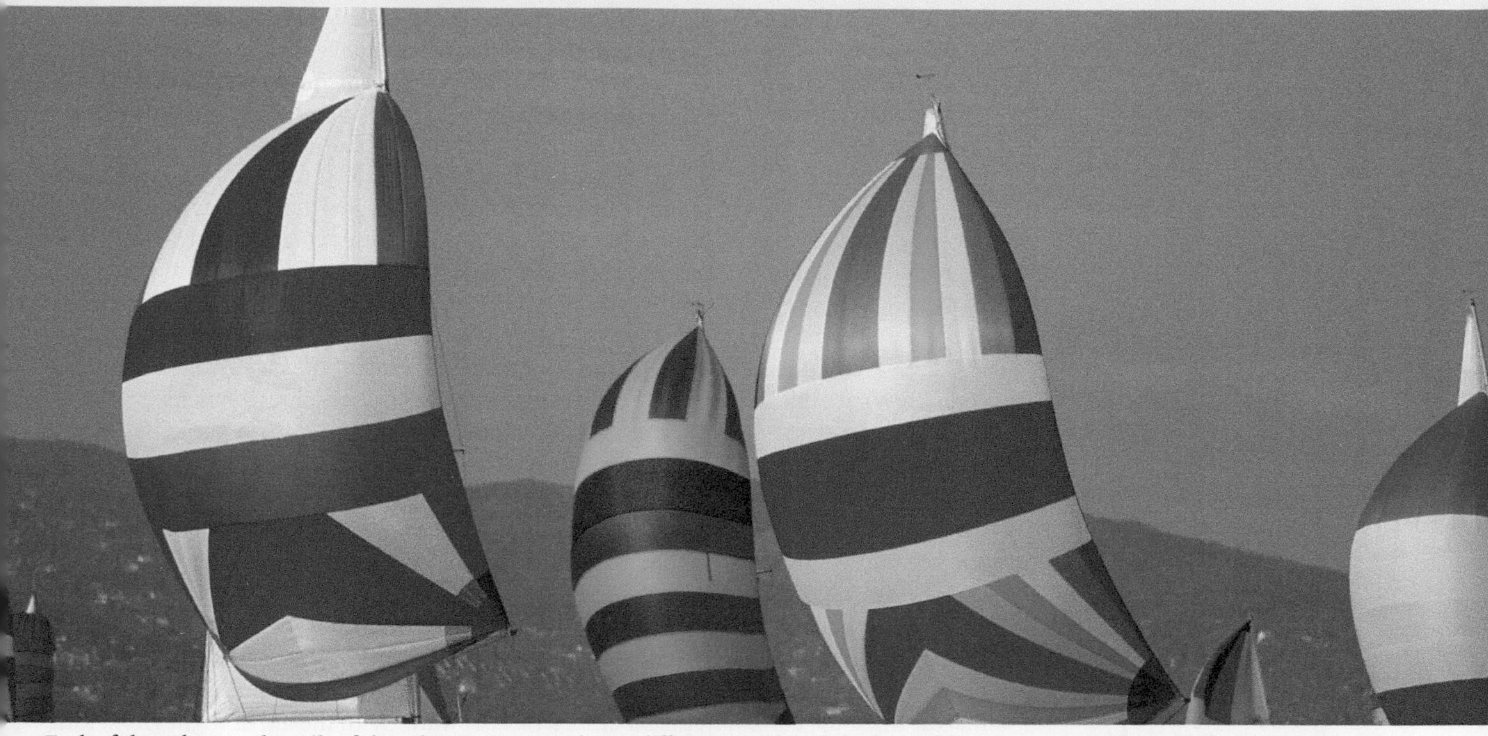

Each of the colors on the sails of these boats corresponds to a different wavelength in the visible region of the spectrum of electromagnetic waves. As we will see in this chapter, however, the visible wavelengths comprise only a small part of the total electromagnetic spectrum.

Andrew Woodley/Alamy Stock Photo

Electromagnetic Waves

LEARNING OBJECTIVES

After reading this module, you should be able to...

24.1 Describe the nature of electromagnetic waves.

24.2 Calculate speed, frequency, and wavelength for electromagnetic waves.

24.3 Relate the speed of light to electromagnetic quantities.

24.4 Calculate energy, power, and intensity for electromagnetic waves.

24.5 Solve problems involving the Doppler effect for electromagnetic waves.

24.6 Solve polarization problems using Malus' law.

24.1 The Nature of Electromagnetic Waves

In Section 13.3 we saw that energy is transported to us from the sun via a class of waves known as electromagnetic waves. This class includes the familiar visible, ultraviolet, and infrared waves. In Sections 18.6, 21.1, and 21.2 we studied the concepts of electric and magnetic fields. It was the great Scottish physicist James Clerk Maxwell (1831–1879) who showed that these two fields fluctuating together can form a propagating **electromagnetic wave**. We will now bring together our knowledge of electric and magnetic fields in order to understand this important type of wave.

Animated Figure 24.1 illustrates one way to create an electromagnetic wave. The setup consists of two straight metal wires that are connected to the terminals of an ac generator and serve as an antenna. The potential difference between the terminals changes sinusoidally with time t and has a period T. Part a shows the instant $t = 0$ s, when there is no charge at the ends of either wire. Since there is no charge, there is

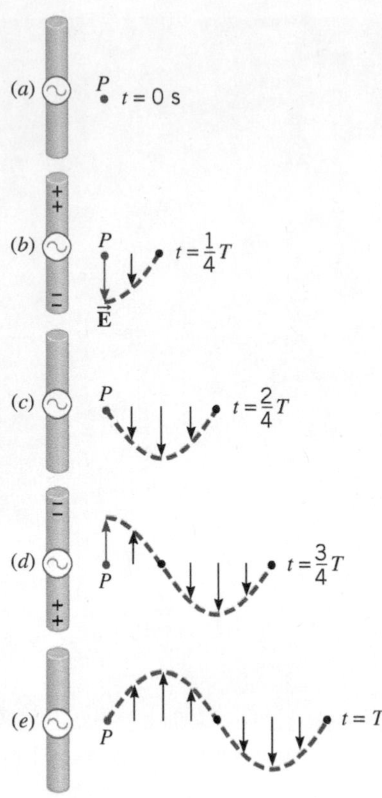

ANIMATED FIGURE 24.1 In each part of the drawing, the red arrow represents the electric field $\vec{E}$ produced at point P by the oscillating charges on the antenna at the indicated time. The black arrows represent the electric fields created at earlier times. For simplicity, only the fields propagating to the right are shown.

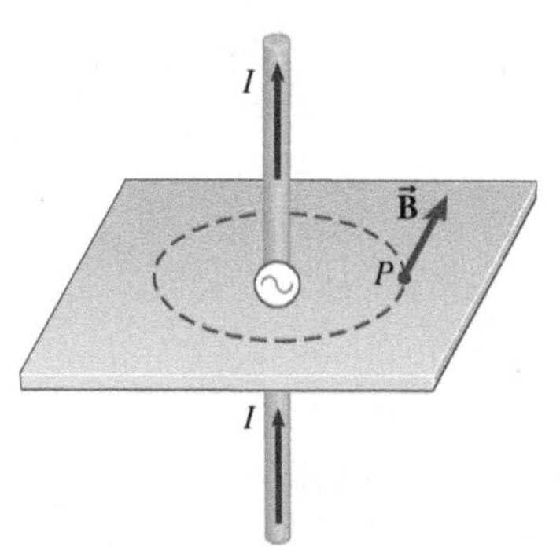

FIGURE 24.2 The oscillating current I in the antenna wires creates a magnetic field $\vec{B}$ at point P that is tangent to a circle centered on the wires. The field is directed as shown when the current is upward and is directed in the opposite direction when the current is downward.

no electric field at the point P just to the right of the antenna. As time passes, the top wire becomes positively charged and the bottom wire negatively charged. One-quarter of a cycle later ($t = \frac{1}{4}T$), the charges have attained their maximum values, as part b of the drawing indicates. The corresponding electric field $\vec{E}$ at point P is represented by the red arrow and has increased to its maximum strength in the downward direction.* Part b also shows that the electric field created at earlier times (see the black arrow in the picture) has not disappeared but has moved to the right. Here lies the crux of the matter: At distant points, the electric field of the charges is not felt immediately. Instead, the field is created first near the wires and then, like the effect of a pebble dropped into a pond, moves outward as a wave in all directions. Only the field moving to the right is shown in the picture for the sake of clarity.

Parts c–e of **Animated Figure 24.1** show the creation of the electric field at point P (red arrow) at later times during the generator cycle. In each part, the fields produced earlier in the sequence (black arrows) continue propagating toward the right. Part d shows the charges on the wires when the polarity of the generator has reversed, so the top wire is negative and the bottom wire is positive. As a result, the electric field at P has reversed its direction and points upward. In part e of the sequence, a complete sine wave has been drawn through the tips of the electric field vectors to emphasize that the field changes sinusoidally.

Along with the electric field in **Animated Figure 24.1**, a magnetic field $\vec{B}$ is also created, because the charges flowing in the antenna constitute an electric current, which produces a magnetic field. **Figure 24.2** illustrates the field direction at point P at the instant when the current in the antenna wire is upward. With the aid of Right-Hand Rule No. 2 (thumb of the right hand points along the current I, fingers curl in the direction of $\vec{B}$), the magnetic field at P can be seen to point into the page. As the oscillating current changes, the magnetic field changes accordingly. The magnetic fields created at earlier times propagate outward as a wave, just as the electric fields do.

Notice that the magnetic field in **Figure 24.2** is perpendicular to the page, whereas the electric field in **Animated Figure 24.1** lies in the plane of the page. Thus, the electric and magnetic fields created by the antenna are mutually perpendicular and remain so as they move outward. Moreover, both fields are perpendicular to the direction of travel. These perpendicular electric and magnetic fields, moving together, constitute an electromagnetic wave.

The electric and magnetic fields in **Animated Figure 24.1** and **Figure 24.2** decrease to zero rapidly with increasing distance from the antenna. Therefore, they exist mainly near the antenna and together are called the **near field**. Electric and magnetic fields do form a wave at large distances from the antenna, however. These fields arise from an effect that is different from the one that produces the near field and are referred to as the **radiation field**. Faraday's law of induction provides part of the basis for the radiation field. As Section 22.4 discusses, this law describes the emf or potential difference produced by a changing magnetic field. And, as Section 19.4 explains, a potential difference can be related to an electric field. Thus, a changing magnetic field produces an electric field. Maxwell predicted that the reverse effect also occurs—namely, that a changing electric field produces a magnetic field. The radiation field arises because the changing magnetic field creates an electric field that fluctuates in time and the changing electric field creates the magnetic field.

Figure 24.3 shows the electromagnetic wave of the radiation field far from the antenna. The picture shows only the part of the wave traveling along the $+x$ axis. The parts traveling in the other directions have been omitted for clarity. It should be clear from the drawing that ***an electromagnetic wave is a transverse wave*** because the electric and magnetic fields are both perpendicular to the direction in which the wave travels. Moreover, an electromagnetic wave, unlike a wave on a string or a sound wave, does not require a medium in which to propagate. ***Electromagnetic waves can travel through a vacuum or a material substance***, since electric and magnetic fields can exist in either one.

Electromagnetic waves can be produced in situations that do not involve a wire antenna. In general, any electric charge that is accelerating emits an electromagnetic wave, whether the charge is inside a wire or not. In an alternating current, an electron

*The direction of the electric field can be obtained by imagining a positive test charge at P and determining the direction in which it would be pushed because of the charges on the wires.

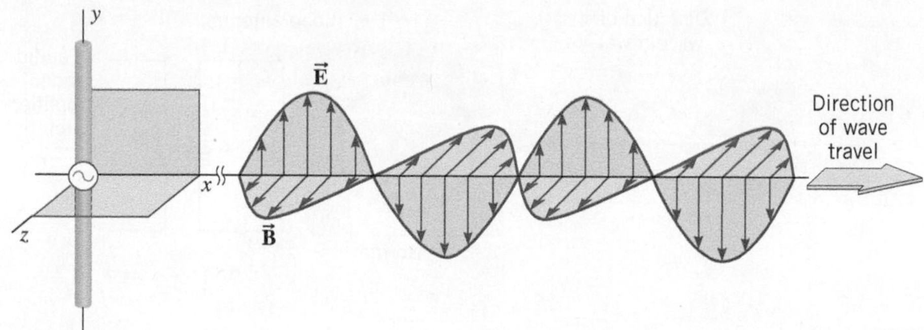

FIGURE 24.3 This picture shows the wave of the radiation field far from the antenna. Observe that $\vec{E}$ and $\vec{B}$ are perpendicular to each other, and both are perpendicular to the direction of travel.

oscillates in simple harmonic motion along the length of the wire and is one example of an accelerating charge.

All electromagnetic waves move through a vacuum at the same speed, and the symbol c is used to denote its value. This speed is called the ***speed of light in a vacuum*** and is $c = 3.00 \times 10^8$ m/s. In air, electromagnetic waves travel at nearly the same speed as they do in a vacuum, but, in general, they move through a substance such as glass at a speed that is substantially less than c.

The frequency of an electromagnetic wave is determined by the oscillation frequency of the electric charges at the source of the wave. In **Figures 24.1–24.3** the wave frequency would equal the frequency of the ac generator. Suppose, for example, that the antenna is broadcasting electromagnetic waves known as radio waves. The frequencies of AM radio waves lie between 545 and 1605 kHz, these numbers corresponding to the limits of the AM broadcast band on the radio dial. In contrast, frequencies of FM radio waves lie between 88 and 108 MHz on the dial. Television channels 2–6, on the other hand, utilize electromagnetic waves with frequencies between 54 and 88 MHz, and channels 7–13 use frequencies between 174 and 216 MHz.

THE PHYSICS OF . . . radio and television reception. Radio and television reception involves a process that is the reverse of that outlined earlier for the creation of electromagnetic waves. When broadcasted waves reach a receiving antenna, they interact with the electric charges in the antenna wires. Either the electric field or the magnetic field of the waves can be used. To take full advantage of the electric field, the wires of the receiving antenna must be parallel to the electric field, as **Figure 24.4** indicates. The electric field acts on the electrons in the wire, forcing them to oscillate back and forth along the length of the wire. Thus, an ac current exists in the antenna and the circuit connected to it. The variable-capacitor C (variable-capacitor symbol) and the inductor L in the circuit provide one way to select the frequency of the desired electromagnetic wave. By adjusting the value of the capacitance, it is possible to adjust the corresponding resonant frequency f_0 of the circuit [$f_0 = 1/(2\pi\sqrt{LC})$, Equation 23.10] to match the frequency of the wave. Under the condition of resonance, there will be a maximum oscillating current in the inductor. Because of mutual inductance, this current creates a maximum voltage in the second coil in the drawing, and this voltage can then be amplified and processed by the remaining radio or television circuitry.

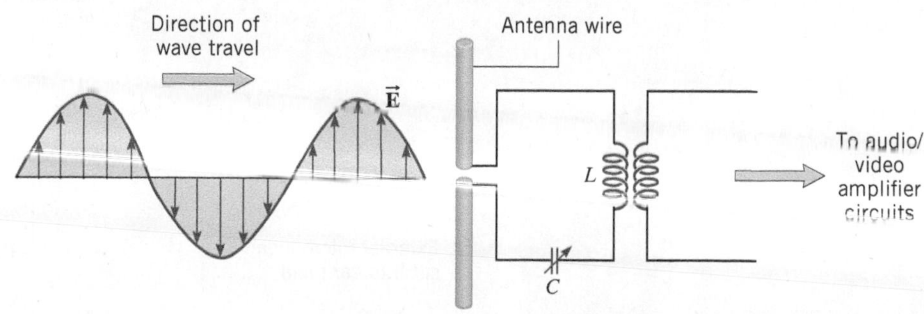

FIGURE 24.4 A radio wave can be detected with a receiving antenna wire that is parallel to the electric field of the wave. The magnetic field of the radio wave has been omitted for simplicity.

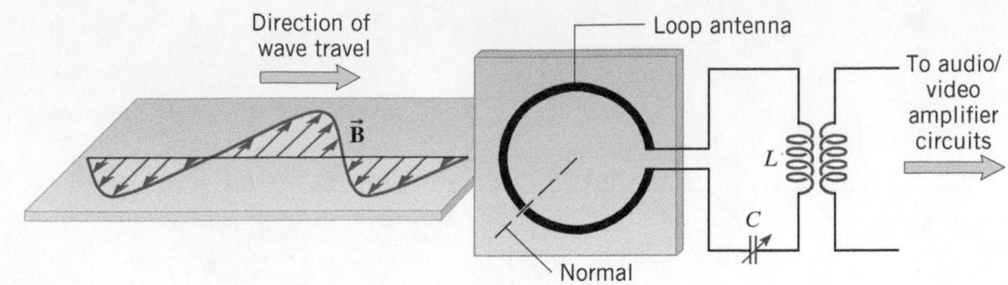

FIGURE 24.5 With a receiving antenna in the form of a loop, the magnetic field of a broadcasted radio wave can be detected. The normal to the plane of the loop should be parallel to the magnetic field for best reception. For clarity, the electric field of the radio wave has been omitted.

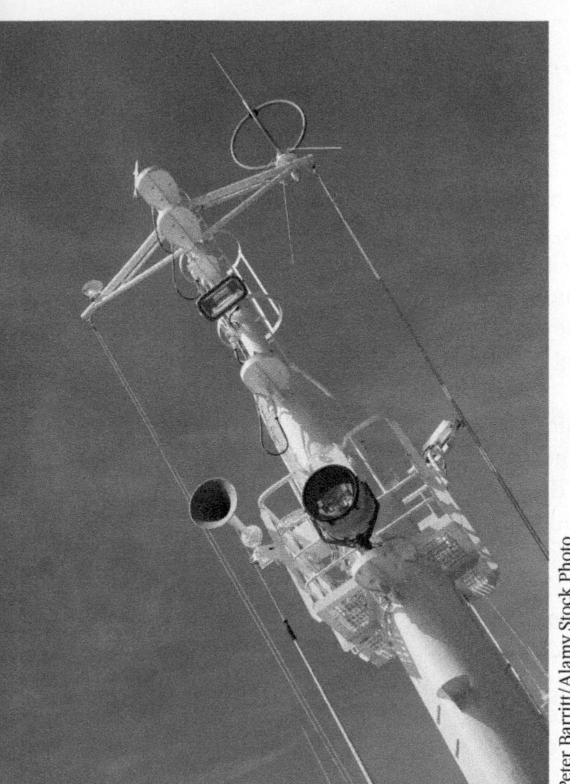

Peter Barritt/Alamy Stock Photo

FIGURE 24.6 This ship's mast indicates that both straight and loop antennas are being used to communicate with other vessels and on-shore stations.

To detect the magnetic field of a broadcasted radio wave, a receiving antenna in the form of a loop can be used, as Figure 24.5 shows. For best reception, the normal to the plane of the wire loop must be oriented parallel to the magnetic field. Then, as the wave sweeps by, the magnetic field penetrates the loop, and the changing magnetic flux induces a voltage and a current in the loop, in accord with Faraday's law. Once again, the resonant frequency of a capacitor/inductor combination can be adjusted to match the frequency of the desired electromagnetic wave. Both straight wire and loop antennas can be seen on the ship's mast in Figure 24.6.

BIO **THE PHYSICS OF . . . cochlear implants.** Cochlear implants use the broadcasting and receiving of radio waves to provide assistance to hearing-impaired people who have auditory nerves that are at least partially intact. These implants utilize radio waves to bypass the damaged part of the hearing mechanism and access the auditory nerve directly, as Interactive Figure 24.7 illustrates. An external microphone (often set into an ear mold) detects sound waves and sends a corresponding electrical signal to a speech processor small enough to be carried in a pocket. The speech processor encodes these signals into a radio wave, which is broadcast from an external transmitter coil placed over the site of a miniature receiver (and its receiving antenna) that has been surgically inserted beneath the skin. The receiver acts much like a radio. It detects the broadcasted wave and from the encoded audio information produces electrical signals that represent the sound wave. These signals are sent along a wire to electrodes that are implanted in the cochlea of the inner ear. The electrodes stimulate the auditory nerves that feed directly between structures within the cochlea and the brain. To the extent that the nerves are intact, a person can learn to recognize sounds.

BIO **THE PHYSICS OF . . . wireless capsule endoscopy.** The broadcasting and receiving of radio waves are also now being used in the practice of endoscopy. In this medical diagnostic technique a device called an endoscope is

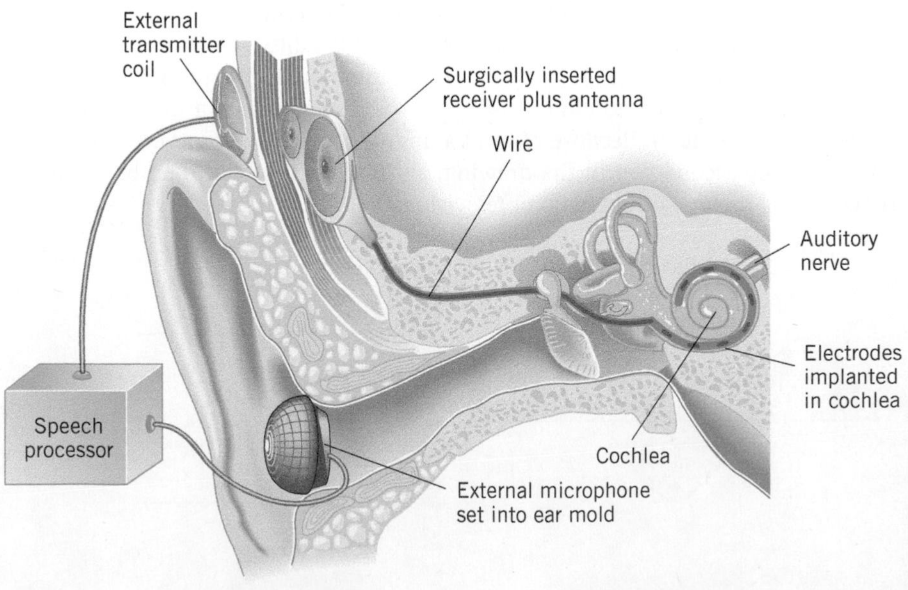

INTERACTIVE FIGURE 24.7 Hearing-impaired people can sometimes recover part of their hearing with the help of a cochlear implant. Broadcasting and receiving electromagnetic waves play central roles in this device.

used to peer inside the body. For example, to examine the interior of the colon for signs of cancer, a conventional endoscope (known as a colonoscope) is inserted through the rectum. (See Section 26.3.) The wireless capsule endoscope shown in **Figure 24.8** bypasses this invasive procedure completely. With a size of about 11×26 mm, this capsule can be swallowed and carried through the gastrointestinal tract by the involuntary contractions of the walls of the intestines (peristalsis). The capsule is self-contained and uses no external wires. A marvel of miniaturization, it contains a radio transmitter and its associated antenna, batteries, a white-light-emitting diode (see Section 23.5) for illumination, and an optical system to capture the digital images. As the capsule moves through the intestine, the transmitter broadcasts the images to an array of small receiving antennas attached to the patient's body. These receiving antennas also are used to determine the position of the capsule within the body. The radio waves that are used lie in the ultrahigh-frequency, or UHF, band, from 3×10^8 to 3×10^9 Hz.

Radio waves are only one part of the broad spectrum of electromagnetic waves that has been discovered. The next section discusses the entire spectrum.

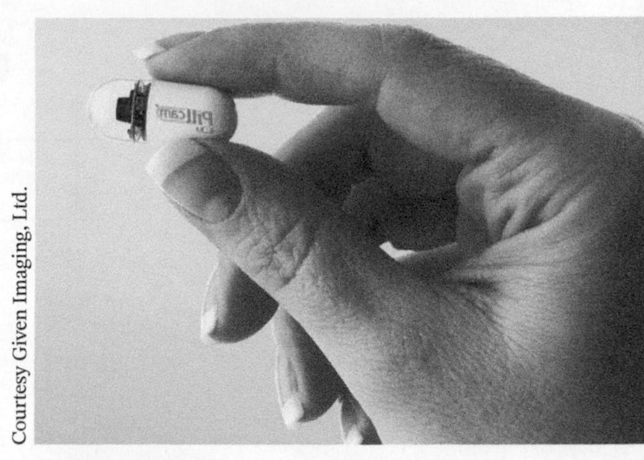

Courtesy Given Imaging, Ltd.

FIGURE 24.8 This wireless capsule endoscope is designed to be swallowed. As it passes through a patient's intestines, it broadcasts video images of the interior of the intestines.

Check Your Understanding

(The answers are given at the end of the book.)

1. Refer to **Animated Figure 24.1**. Between the times indicated in parts *c* and *d* in the drawing, what is the direction of the magnetic field at the point *P* for the electromagnetic wave being generated? Is it directed **(a)** upward along the length of the wire, **(b)** downward along the length of the wire, **(c)** into the plane of the paper, or **(d)** out of the plane of the paper?

2. A transmitting antenna is located at the origin of an *x, y, z* axis system and broadcasts an electromagnetic wave whose electric field oscillates along the *y* axis. The wave travels along the $+x$ axis. Three possible wire loops are available for use with an LC-tuned circuit to detect this wave: **(a)** a loop that lies in the *xy* plane, **(b)** a loop that lies in the *xz* plane, and **(c)** a loop that lies in the *yz* plane. Which one of the loops will detect the wave?

3. Why does the peak value of the emf induced in a loop antenna (see **Figure 24.5**) depend on the frequency of the electromagnetic wave?

24.2 The Electromagnetic Spectrum

An electromagnetic wave, like any periodic wave, has frequency f and wavelength λ that are related to the speed v of the wave by $v = f\lambda$ (Equation 16.1). For electromagnetic waves traveling through a vacuum or, to a good approximation, through air, the speed is $v = c$, so $c = f\lambda$.

As **Figure 24.9** shows, electromagnetic waves exist with an enormous range of frequencies, from values less than 10^4 Hz to greater than 10^{24} Hz. Since all these waves travel through a vacuum at the same speed of $c = 3.00 \times 10^8$ m/s, Equation 16.1 can be used to find the correspondingly wide range of wavelengths that the picture also displays. The ordered series of electromagnetic wave frequencies or wavelengths in **Figure 24.9** is called the **electromagnetic spectrum**. Historically, regions of the spectrum have been given names such as radio waves and infrared waves. Although the boundary between adjacent regions is shown as a sharp line in the drawing, the boundary is not so well defined in practice, and the regions often overlap.

Beginning on the left in **Figure 24.9**, we find radio waves. Lower-frequency radio waves are generally produced by electrical oscillator circuits, while higher-frequency radio waves (called microwaves) are usually generated using electron tubes called klystrons. Infrared radiation, sometimes loosely called heat waves, originates with the vibration and rotation of molecules within a material. Visible light is emitted by hot objects, such as the sun, a burning log, or the filament of an incandescent light bulb, when the temperature is high enough to excite the electrons within an atom. Ultraviolet frequencies can be

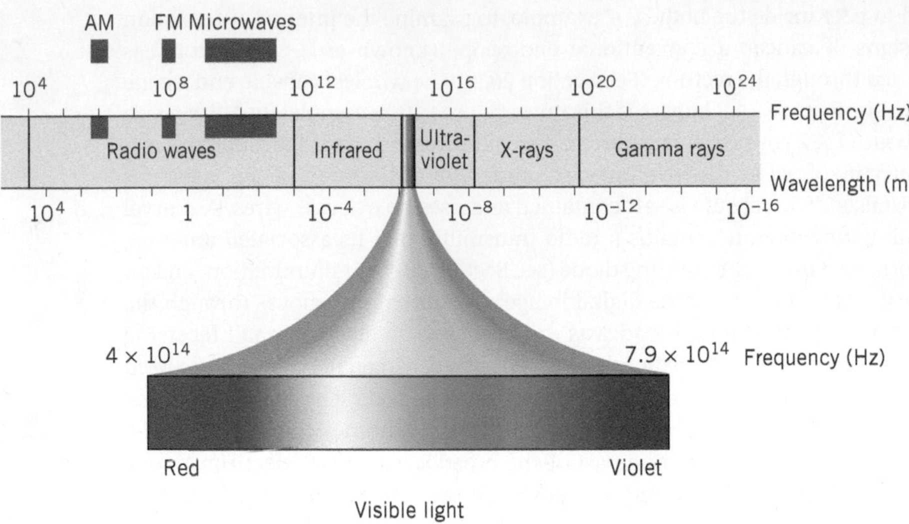

FIGURE 24.9 The electromagnetic spectrum.

produced from the discharge of an electric arc. X-rays are produced by the sudden deceleration of high-speed electrons. And, finally, gamma rays are radiation from nuclear decay.

THE PHYSICS OF . . . astronomy and the electromagnetic spectrum. Astronomers use the different regions of the electromagnetic spectrum to gather information about distant celestial objects. **Interactive Figure 24.10**, for example, shows four views of the Crab Nebula, each in a different region of the spectrum. The Crab Nebula is located 6.0×10^{16} km away from the earth and is the remnant of a star that underwent a supernova explosion in 1054 AD.

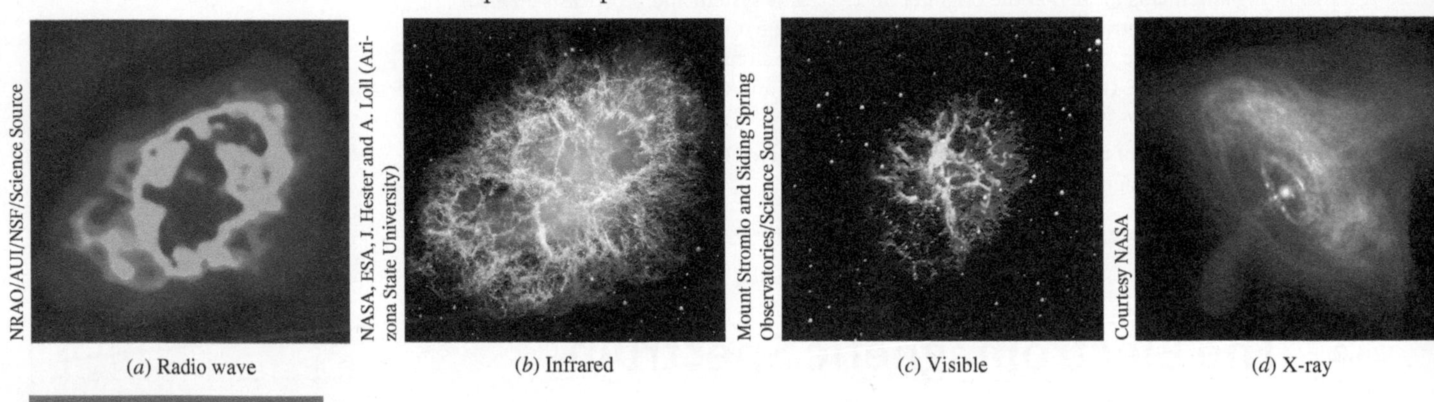

INTERACTIVE FIGURE 24.10 Four views of the Crab Nebula. Each view is in a different region of the electromagnetic spectrum, as indicated.

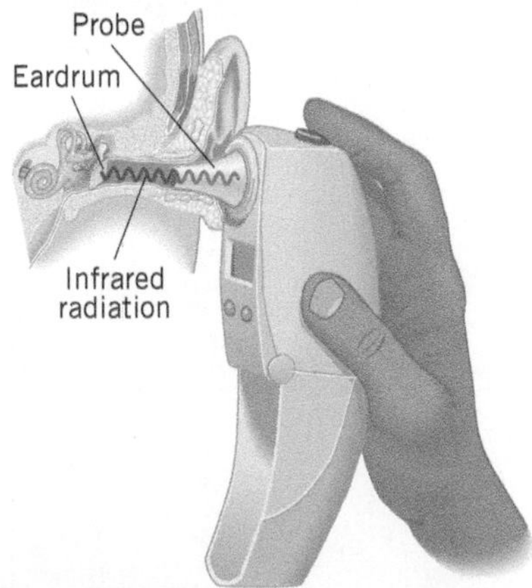

FIGURE 24.11 A pyroelectric thermometer measures body temperature by determining the amount of infrared radiation emitted by the eardrum and surrounding tissue.

BIO THE PHYSICS OF . . . a pyroelectric ear thermometer. The human body, like any object, radiates infrared radiation, and the amount emitted depends on the temperature of the body. Although infrared radiation cannot be seen by the human eye, it can be detected by sensors. An ear thermometer, like the pyroelectric thermometer shown in **Figure 24.11**, measures the body's temperature by determining the amount of infrared radiation that emanates from the eardrum and surrounding tissue. The ear is one of the best places for this measurement because it is close to the hypothalamus, an area at the bottom of the brain that controls body temperature. The ear is also not cooled or warmed by eating, drinking, or breathing. When the probe of the thermometer is inserted into the ear canal, infrared radiation travels down the barrel of the probe and strikes the sensor. The absorption of infrared radiation warms the sensor, and, as a result, its electrical conductivity changes. The change in electrical conductivity is measured by an electronic circuit. The output from the circuit is sent to a microprocessor, which calculates the body temperature and displays the result on a digital readout.

Of all the frequency ranges in the electromagnetic spectrum, the most familiar is that of visible light, although it is the most narrow (see **Figure 24.9**). Only waves with frequencies between about 4.0×10^{14} Hz and 7.9×10^{14} Hz are perceived by the human eye as visible light. Usually visible light is discussed in terms of wavelengths (in vacuum) rather than frequencies. As Example 1 indicates, the wavelengths of visible light are extremely small and, therefore, are normally expressed in *nanometers* (nm); 1 nm = 10^{-9} m. An obsolete (non-SI) unit still occasionally used for wavelengths is the *angstrom* (Å); 1 Å = 10^{-10} m.

EXAMPLE 1 | The Wavelengths of Visible Light

Find the range in wavelengths (in vacuum) for visible light in the frequency range between 4.0×10^{14} Hz (red light) and 7.9×10^{14} Hz (violet light). Express the answers in nanometers.

Reasoning According to Equation 16.1, the wavelength (in vacuum) λ of a light wave is equal to the speed of light c in a vacuum divided by the frequency f of the wave, $\lambda = c/f$.

Solution The wavelength corresponding to a frequency of 4.0×10^{14} Hz is

$$\lambda = \frac{c}{f} = \frac{3.00 \times 10^8 \text{ m/s}}{4.0 \times 10^{14} \text{ Hz}} = 7.5 \times 10^{-7} \text{ m}$$

Since 1 nm = 10^{-9} m, it follows that

$$\lambda = (7.5 \times 10^{-7} \cancel{\text{m}})\left(\frac{1 \text{ nm}}{10^{-9} \cancel{\text{m}}}\right) = \boxed{750 \text{ nm}}$$

The calculation for a frequency of 7.9×10^{14} Hz is similar:

$$\lambda = \frac{c}{f} = \frac{3.00 \times 10^8 \text{ m/s}}{7.9 \times 10^{14} \text{ Hz}} = 3.8 \times 10^{-7} \text{ m} \quad \text{or} \quad \lambda = \boxed{380 \text{ nm}}$$

The eye/brain recognizes light of different wavelengths as different colors. A wavelength of 750 nm (in vacuum) is approximately the longest wavelength of red light, whereas 380 nm (in vacuum) is approximately the shortest wavelength of violet light. Between these limits are found the other familiar colors, as **Figure 24.9** indicates.

The association between color and wavelength in the visible part of the electromagnetic spectrum is well known. The wavelength also plays a central role in governing the behavior and use of electromagnetic waves in all regions of the spectrum. For instance, Conceptual Example 2 considers the influence of the wavelength on diffraction. Example 3 describes an important medical device that simultaneously employs two wavelengths of light in order to estimate the oxygen saturation in the bloodstream.

CONCEPTUAL EXAMPLE 2 | The Physics of AM and FM Radio Reception

As we have discussed in Section 17.3, diffraction is the ability of a wave to bend around an obstacle or around the edges of an opening. Based on that discussion, which type of radio wave would you expect to bend more readily around an obstacle such as a building, **(a)** an AM radio wave or **(b)** an FM radio wave?

Reasoning Section 17.3 points out that, other things being equal, sound waves exhibit diffraction to a greater extent when the wavelength is longer than when it is shorter. Based on this information, we expect that longer-wavelength electromagnetic waves will bend more readily around obstacles than will shorter-wavelength waves.

Answer (b) is incorrect. **Figure 24.9** shows that FM radio waves have considerably shorter wavelengths than do AM waves. Therefore, FM radio waves exhibit less diffraction than AM waves do and bend less readily around obstacles.

Answer (a) is correct. Since AM radio waves have greater wavelengths than FM waves do (see **Figure 24.9**), they exhibit greater diffraction and bend more readily around obstacles than FM waves do.

EXAMPLE 3 | BIO Pulse Oximetry

The amount of oxygen carried by hemoglobin in the blood is a critical parameter that is precisely regulated by the human body. *Oxygen saturation* is the fraction of the total hemoglobin that is carrying bound oxygen and should have a value between 95% and 100%. Below 90% is considered to be low and is called *hypoxemia*, and below 80% may result in organ failure. A noninvasive device that is used to measure oxygen saturation, called a *pulse oximeter*, clips onto a fleshy semitransparent part of the body, such as a fingertip (see **Figure 24.12**). The clip uses light-emitting diodes (LEDs) to pass light of two wavelengths, usually 6.60×10^{-7} m and 9.40×10^{-7} m (in air), through the blood-carrying tissue, and monitors how much light of each wavelength gets absorbed on the way through. Since oxygenated and deoxygenated hemoglobin absorb light of those wavelengths differently, the percentage of oxygenated hemoglobin can be determined. **(a)** What are the frequencies of light emitted by the LEDs? **(b)** In what regions of the electromagnetic spectrum do the two frequencies fall? Can one see with the naked eye both types of light being emitted by the device? (Caution: Never look directly into LED devices.) Refer to **Figure 24.9** to answer this question.

Reasoning (a) According to Equation 16.1, the frequency of a light wave can be calculated from its wavelength using $v = c = f\lambda$, where c is the speed of light in vacuum. **(b)** Comparing the resulting frequencies (or wavelengths) to the electromagnetic spectrum depicted in **Figure 24.9**, we can identify the regions of the electromagnetic spectrum in which the two light frequencies fall, and can then determine whether they are visible to the naked eye.

Solution (a) From Equation 16.1, for the light of wavelength $\lambda_1 = 6.60 \times 10^{-7}$ m, we have

$$f_1 = \frac{c}{\lambda_1} = \frac{3.00 \times 10^8 \text{ m/s}}{6.60 \times 10^{-7} \text{ m}} = \boxed{4.55 \times 10^{14} \text{ Hz}}$$

Similarly, for the light of wavelength $\lambda_2 = 9.40 \times 10^{-7}$ m, we have

$$f_2 = \frac{c}{\lambda_2} = \frac{3.00 \times 10^8 \text{ m/s}}{9.40 \times 10^{-7} \text{ m}} = \boxed{3.19 \times 10^{14} \text{ Hz}}$$

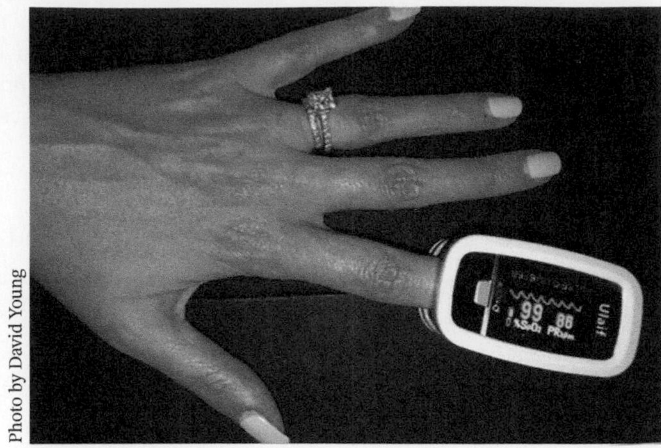

Photo by David Young

FIGURE 24.12 The sale of fingertip pulse oximeters that are designed for at-home use, like the one shown here, soared during the COVID-19 pandemic in 2020. One of the more dangerous symptoms of COVID-19 is respiratory pneumonia, which attacks a person's lungs and inhibits the blood oxygenation process. Careful monitoring of a person's blood oxygen saturation is critical in determining the level of care that is needed.

(b) Referring to **Figure 24.9**, the light with $f_1 = 4.55 \times 10^{14}$ Hz falls into the visible (red) part of the electromagnetic spectrum, whereas the light of $f_2 = 3.19 \times 10^{14}$ Hz falls into the infrared. Thus, the light of frequency f_1 can be seen with naked eye, and that with frequency f_2 cannot.

The picture of light as a wave is supported by experiments that will be discussed in Chapter 27. However, there are also experiments indicating that light can behave as if it were composed of discrete particles rather than waves. These experiments will be discussed in Chapter 29. Wave theories and particle theories of light have been around for hundreds of years, and it is now widely accepted that light, as well as other electromagnetic radiation, exhibits a dual nature. Either wave-like or particle-like behavior can be observed, depending on the kind of experiment being performed.

24.3 The Speed of Light

At a speed of 3.00×10^8 m/s, light travels from the earth to the moon in a little over a second, so the time required for light to travel between two places on earth is very short. Therefore, the earliest attempts at measuring the speed of light had only limited success. One of the first accurate measurements employed a rotating mirror, and **Figure 24.13** shows a simplified version of the setup. It was used first by the French scientist Jean Foucault (1819–1868) and later in a more refined version by the American physicist Albert Michelson (1852–1931). If the angular speed of the rotating eight-sided mirror in **Figure 24.13** is adjusted correctly, light reflected from one side travels to the fixed mirror, reflects, and can be detected after reflecting from another side that has rotated into place at just the right time. The minimum angular speed must be such that one side of the mirror rotates one-eighth of a revolution during the time it takes for the light to make the round trip between the mirrors. For one of his experiments, Michelson placed his fixed mirror and rotating mirror on Mt. San Antonio and Mt. Wilson in California, a distance of 35 km apart. From a value of the minimum angular speed in such experiments, he obtained the value of $c = (2.997\,96 \pm 0.000\,04) \times 10^8$ m/s in 1926.

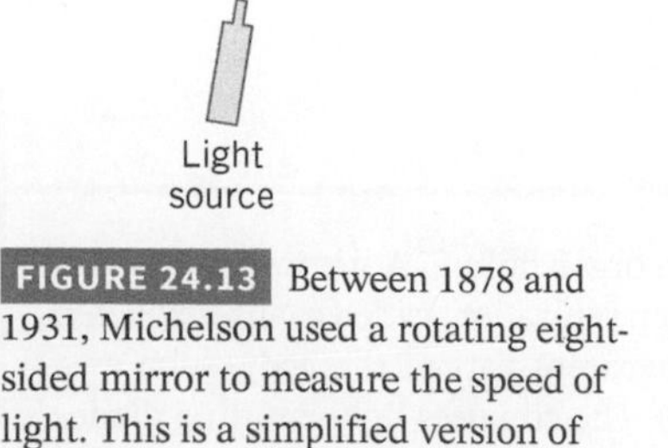

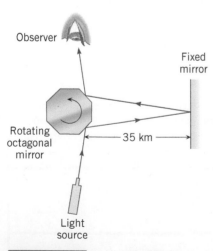

FIGURE 24.13 Between 1878 and 1931, Michelson used a rotating eight-sided mirror to measure the speed of light. This is a simplified version of the setup.

Today, the speed of light has been determined with such high accuracy that it is used to define the meter. As discussed in Section 1.2, the speed of light is now *defined* to be

Speed of light in a vacuum

$$c = 299\,792\,458 \text{ m/s}$$

However, a value of 3.00×10^8 m/s is adequate for most calculations. The second is defined in terms of a cesium clock, and the meter is then defined as the distance light travels in a vacuum during a time of 1/(299 792 458) s. Although the speed of light in a vacuum is large, it is finite, so it takes a finite amount of time for light to travel from one place to another. The travel time is especially long for light traveling between astronomical objects, as Conceptual Example 4 discusses.

CONCEPTUAL EXAMPLE 4 | Looking Back in Time

A supernova is a violent explosion that occurs at the death of certain stars. For a few days after the explosion, the intensity of the emitted light can become a billion times greater than that of our own sun. After several years, however, the intensity usually returns to zero. Supernovae are relatively rare events in the universe; only six have been observed in our galaxy within the past 400 years. A supernova that occurred in a neighboring galaxy, approximately 1.66×10^{21} m away, was recorded in 1987. **Figure 24.14** shows a photograph of the sky just a few hours after the explosion. Astronomers say that viewing an event like the supernova is like looking back in time. Which one of the following statements correctly describes what we see when we view such events? **(a)** The nearer the event is to the earth, the further back in time we are looking. **(b)** The farther the event is from the earth, the further back in time we are looking.

Reasoning The light from the supernova traveled to earth at a speed of $c = 3.00 \times 10^8$ m/s. The time t required for the light to travel the distance d between the event and the earth is $t = d/c$ and is proportional to the distance.

Answer (a) is incorrect. Since the time required for the light to travel the distance between the event and the earth is proportional to the distance, the light from near-earth events reaches us sooner rather than later. Therefore, the nearer the event is to the earth, the less into the past it allows us to see, contrary to what this answer implies.

Answer (b) is correct. The travel time for light from the supernova is

$$t = \frac{d}{c} = \frac{1.66 \times 10^{21}\ \text{m}}{3.00 \times 10^8\ \text{m/s}} = 5.53 \times 10^{12}\ \text{s}$$

FIGURE 24.14 True color image of the 1987 supernova (bright spot at the lower right). The larger cloud-like object near the middle left is the Tarantula nebula, whose light also takes approximately 175 000 years to reach the earth.

Celestial Image Co./Science Source

This corresponds to 175 000 years, so when astronomers saw the explosion in 1987, they were actually seeing the light that left the supernova 175 000 years earlier. In other words, they were looking back in time. Greater values for the distance d mean greater values for the time t.

Related Homework: Problem 12

In 1865, Maxwell determined theoretically that electromagnetic waves propagate through a vacuum at a speed given by

$$c = \frac{1}{\sqrt{\varepsilon_0 \mu_0}} \tag{24.1}$$

where $\varepsilon_0 = 8.85 \times 10^{-12}\ \text{C}^2/(\text{N} \cdot \text{m}^2)$ is the (electric) permittivity of free space and $\mu_0 = 4\pi \times 10^{-7}\ \text{T} \cdot \text{m/A}$ is the (magnetic) permeability of free space. Originally ε_0 was introduced in Section 18.5 as an alternative way of writing the proportionality constant k in Coulomb's law $[k = 1/(4\pi\varepsilon_0)]$ and, hence, plays a basic role in determining the strengths of the electric fields created by point charges. The role of μ_0 is similar for magnetic fields; it was introduced in Section 21.7 as part of a proportionality constant in the expression for the magnetic field created by the current in a long, straight wire. Substituting the values for ε_0 and μ_0 into Equation 24.1 shows that

$$c = \frac{1}{\sqrt{[8.85 \times 10^{-12}\ \text{C}^2/(\text{N} \cdot \text{m}^2)](4\pi \times 10^{-7}\ \text{T} \cdot \text{m/A})}} = 3.00 \times 10^8\ \text{m/s}$$

The experimental and theoretical values for c agree. Maxwell's success in predicting c provided a basis for inferring that light behaves as a wave consisting of oscillating electric and magnetic fields.

Check Your Understanding

(The answer is given at the end of the book.)

4. The frequency of electromagnetic wave A is twice that of electromagnetic wave B. For these two waves, what is the ratio λ_A/λ_B of the wavelengths in a vacuum? **(a)** $\lambda_A/\lambda_B = 2$, because wave A has twice the speed that wave B has. **(b)** $\lambda_A/\lambda_B = 2$, because wave A has one-half the speed that wave B has. **(c)** $\lambda_A/\lambda_B = \frac{1}{2}$, because wave A has one-half the speed that wave B has. **(d)** $\lambda_A/\lambda_B = \frac{1}{2}$, because wave A has twice the speed that wave B has. **(e)** $\lambda_A/\lambda_B = \frac{1}{2}$, because both waves have the same speed.

24.4 The Energy Carried by Electromagnetic Waves

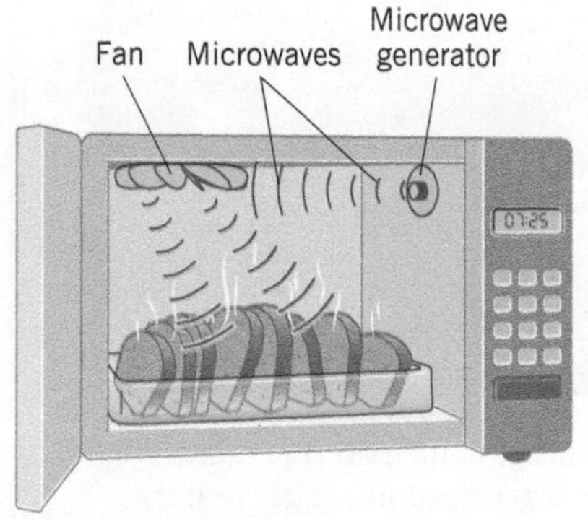

FIGURE 24.15 A microwave oven. The rotating fan blades reflect the microwaves to all parts of the oven.

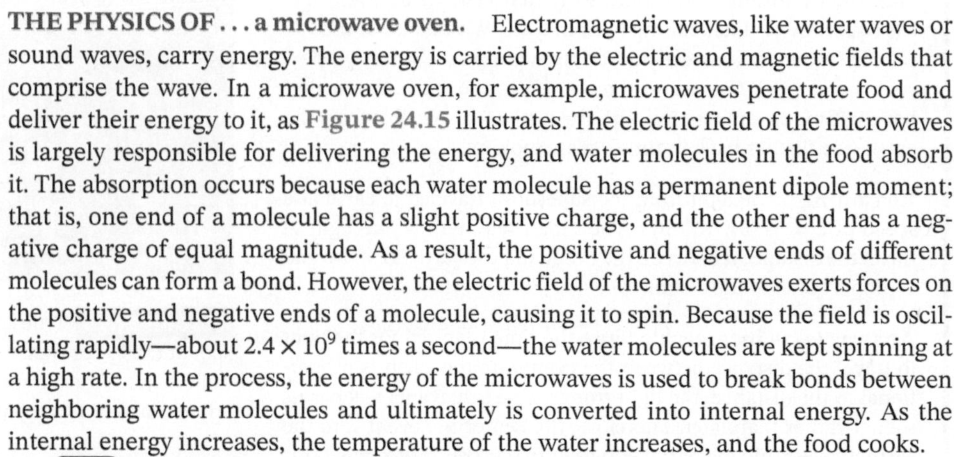

THE PHYSICS OF . . . a microwave oven. Electromagnetic waves, like water waves or sound waves, carry energy. The energy is carried by the electric and magnetic fields that comprise the wave. In a microwave oven, for example, microwaves penetrate food and deliver their energy to it, as **Figure 24.15** illustrates. The electric field of the microwaves is largely responsible for delivering the energy, and water molecules in the food absorb it. The absorption occurs because each water molecule has a permanent dipole moment; that is, one end of a molecule has a slight positive charge, and the other end has a negative charge of equal magnitude. As a result, the positive and negative ends of different molecules can form a bond. However, the electric field of the microwaves exerts forces on the positive and negative ends of a molecule, causing it to spin. Because the field is oscillating rapidly—about 2.4×10^9 times a second—the water molecules are kept spinning at a high rate. In the process, the energy of the microwaves is used to break bonds between neighboring water molecules and ultimately is converted into internal energy. As the internal energy increases, the temperature of the water increases, and the food cooks.

BIO **THE PHYSICS OF . . . the greenhouse effect.** As discussed in Section 15.7, the energy carried by electromagnetic waves in the infrared and visible regions of the spectrum plays the key role in the greenhouse effect that is a contributing factor to global warming. The infrared waves from the sun are largely prevented from reaching the earth's surface by carbon dioxide and water in the atmosphere, which reflect them back into space. The visible waves do reach the earth's surface, however, and the energy they carry heats the earth. Heat also flows to the surface from the interior of the earth. The heated surface in turn radiates infrared waves outward, which, if they could, would carry their energy into space. However, the atmospheric carbon dioxide and water reflect these infrared waves back toward the earth, just as they reflect the infrared waves from the sun. Thus, their energy is trapped, and the earth becomes warmer, like plants in a greenhouse. In a greenhouse, however, energy is trapped mainly for a different reason—namely, the lack of effective convection currents to carry warm air past the cold glass walls.

A measure of the energy stored in the electric field $\vec{\mathbf{E}}$ of an electromagnetic wave, such as a microwave, is provided by the electric energy density. As we saw in Section 19.5, this density is the electric energy per unit volume of space in which the electric field exists:

$$\text{Electric energy density} = \frac{\text{Electric energy}}{\text{Volume}} = \frac{1}{2}\kappa\varepsilon_0 E^2 = \frac{1}{2}\varepsilon_0 E^2 \quad \textbf{(19.12)}$$

In this equation, the dielectric constant κ has been set equal to unity, since we are dealing with an electric field in a vacuum (or in air). From Section 22.8, the analogous expression for the magnetic energy density is

$$\text{Magnetic energy density} = \frac{\text{Magnetic energy}}{\text{Volume}} = \frac{1}{2\mu_0}B^2 \quad \textbf{(22.11)}$$

The **total energy density** u of an electromagnetic wave in a vacuum is the sum of these two energy densities:

$$u = \frac{\text{Total energy}}{\text{Volume}} = \frac{1}{2}\varepsilon_0 E^2 + \frac{1}{2\mu_0}B^2 \tag{24.2a}$$

In an electromagnetic wave propagating through a vacuum or air, the electric field and the magnetic field carry equal amounts of energy per unit volume of space. Since $\frac{1}{2}\varepsilon_0 E^2 = \frac{1}{2}(B^2/\mu_0)$, it is possible to rewrite Equation 24.2a for the total energy density in two additional, but equivalent, forms:

$$u = \varepsilon_0 E^2 \tag{24.2b}$$

$$u = \frac{1}{\mu_0}B^2 \tag{24.2c}$$

The fact that the two energy densities are equal implies that the electric and magnetic fields are related. To see how, we set the electric energy density equal to the magnetic energy density and obtain

$$\frac{1}{2}\varepsilon_0 E^2 = \frac{1}{2\mu_0}B^2 \quad \text{or} \quad E^2 = \frac{1}{\varepsilon_0\mu_0}B^2$$

However, according to Equation 24.1, $c = 1/\sqrt{\varepsilon_0\mu_0}$, so it follows that $E^2 = c^2B^2$. Taking the square root of both sides of this result shows that the relation between the magnitudes of the electric and magnetic fields in an electromagnetic wave is

$$E = cB \tag{24.3}$$

In an electromagnetic wave, the electric and magnetic fields fluctuate sinusoidally in time, so Equations 24.2a–c give the energy density of the wave at any instant in time. If an average value $\overline{u}$ for the total energy density is desired, average values are needed for E^2 and B^2. In Section 20.5 we faced a similar situation for alternating currents and voltages and introduced rms (root mean square) quantities. Using an analogous procedure here, we find that the rms values for the electric and magnetic fields, E_{rms} and B_{rms}, are related to the maximum values of these fields, E_0 and B_0, by

$$E_{\text{rms}} = \frac{1}{\sqrt{2}}E_0 \quad \text{and} \quad B_{\text{rms}} = \frac{1}{\sqrt{2}}B_0$$

Equations 24.2a–c can now be interpreted as giving the average energy density $\overline{u}$, provided the symbols E and B are interpreted to mean the rms values given above. The average energy density of the sunlight reaching the earth is determined in the next example.

EXAMPLE 5 | The Average Energy Density of Sunlight

Sunlight enters the top of the earth's atmosphere with an electric field whose rms value is $E_{\text{rms}} = 720$ N/C. Find **(a)** the average total energy density of this electromagnetic wave and **(b)** the rms value of the sunlight's magnetic field.

Reasoning The average total energy density $\overline{u}$ of the sunlight can be obtained with the aid of Equation 24.2b, provided the rms value is used for the electric field. Since the magnitudes of the magnetic and electric fields are related according to Equation 24.3, the rms value of the magnetic field is $B_{\text{rms}} = E_{\text{rms}}/c$.

Solution **(a)** According to Equation 24.2b, the average total energy density is

$$\overline{u} = \varepsilon_0 E_{\text{rms}}^2 = [8.85 \times 10^{-12}\ \text{C}^2/(\text{N}\cdot\text{m}^2)](720\ \text{N/C})^2 = \boxed{4.6 \times 10^{-6}\ \text{J/m}^3}$$

(b) Using Equation 24.3, we find that the rms magnetic field is

$$B_{\text{rms}} = \frac{E_{\text{rms}}}{c} = \frac{720\ \text{N/C}}{3.0 \times 10^8\ \text{m/s}} = \boxed{2.4 \times 10^{-6}\ \text{T}}$$

As an electromagnetic wave moves through space, it carries energy from one region to another. This energy transport is characterized by the **intensity** of the wave. We have encountered the concept of intensity before, in connection with sound waves in Section 16.7. The sound intensity is the sound power that passes perpendicularly through a surface divided by the area of the surface. The intensity of an electromagnetic wave is defined similarly. For an electromagnetic wave, the intensity is the electromagnetic power divided by the area of the surface.

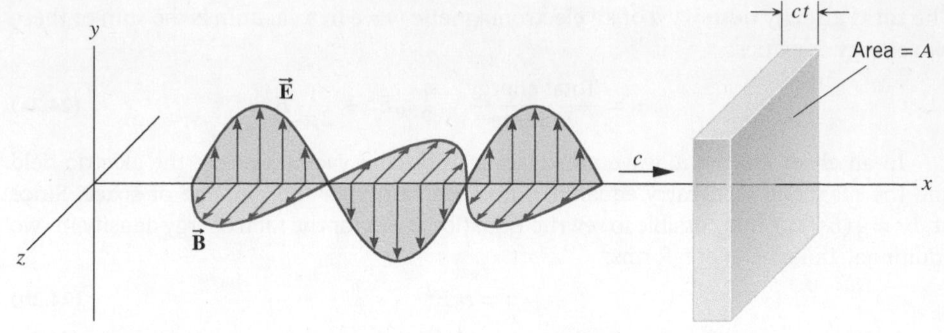

FIGURE 24.16 In a time t, an electromagnetic wave moves a distance ct along the x axis and passes through a surface of area A.

Using this definition of intensity, we can show that the electromagnetic intensity S is related to the energy density u. According to Equation 16.8 the intensity is the power P that passes perpendicularly through a surface divided by the area A of that surface, or $S = P/A$. Furthermore, the power is cqual to the total energy passing through the surface divided by the elapsed time t (Equation 6.10b), so that $P =$ (Total energy)/t. Combining these two relations gives

$$S = \frac{P}{A} = \frac{\text{Total energy}}{tA}$$

Problem-Solving Insight The concepts of power and intensity are similar, but they are not the same. Intensity is the power that passes perpendicularly through a surface divided by the area of the surface.

Now, consider **Figure 24.16**, which shows an electromagnetic wave traveling in a vacuum along the x axis. In a time t the wave travels the distance ct, passing through the surface of area A. Consequently, the volume of space through which the wave passes is ctA. The total (electric and magnetic) energy in this volume is

$$\text{Total energy} = (\text{Total energy density}) \times \text{Volume} = u(ctA)$$

Using this result in the expression for the intensity, we obtain

$$S = \frac{\text{Total energy}}{tA} = \frac{uctA}{tA} = cu \tag{24.4}$$

Thus, the intensity and the energy density are related by the speed of light, c. If the average total energy density $(\overline{u})$ is used in Equation 24.4, then the average intensity $(\overline{S})$ can be determined, as illustrated in Example 6. Substituting Equations 24.2a–c, one at a time, into Equation 24.4 shows that the intensity of an electromagnetic wave depends on the electric and magnetic fields according to the following equivalent relations:

$$S = cu = \frac{1}{2}c\varepsilon_0 E^2 + \frac{c}{2\mu_0}B^2 \tag{24.5a}$$

$$S = c\varepsilon_0 E^2 \tag{24.5b}$$

$$S = \frac{c}{\mu_0}B^2 \tag{24.5c}$$

If the rms values for the electric and magnetic fields are used in Equations 24.5a–c, the intensity becomes an average intensity, $\overline{S}$, as Example 7 illustrates.

EXAMPLE 6 | BIO The Physics of the Laser Scalpel

A laser scalpel is a tool used for general surgery, which may include cutting or vaporizing biological tissues using the energy from coherent laser light. This surgical technique is very precise and can minimize bleeding, swelling, and general discomfort commonly associated with traditional surgical techniques. A gas laser using CO_2 (carbon dioxide) is the highest power continuous-operation laser that is currently available. It operates at wavelengths of 9.4–10.6 μm, which correspond to frequencies that are readily absorbed by water molecules, and therefore soft tissues. One such laser is being used to remove a precancerous mole from a person's arm. The average power of the laser is 7.5 W, and it creates a circular beam (spot size) with a diameter of 0.25 mm. What is the average total energy density contained in the beam? How does this value compare to the average total energy density of sunlight?

Reasoning The average total energy density ($\overline{u}$) of the laser light in the beam is related to the average intensity ($\overline{S}$) by Equation 24.4. We can then use Equation 16.8 to write the average intensity in terms of the average power of the laser.

Solution According to Equation 24.4, the average total energy density is directly related to the average intensity: $\overline{S} = c\overline{u}$. Using Equation 16.8, we write the average intensity in terms of the average power, $\overline{S} = \frac{\overline{P}}{A}$, and then substitute this into the equation above: $\frac{\overline{P}}{A} = c\overline{u}$. Rearranging, we solve for the average total energy density: $\overline{u} = \frac{\overline{P}}{cA}$. Since the beam is circular, the cross-sectional area will be equal to $A = \pi r^2 = \pi\left(\frac{d}{2}\right)^2 = \frac{\pi d^2}{4}$, where d is the diameter of the circular spot size. Substituting this expression for the area into the average total energy density equation, we get our final result:

$$\overline{u} = \frac{\overline{P}}{cA} = \frac{4\overline{P}}{c\pi d^2} = \frac{4(7.5\ \text{W})}{\pi(3.0\times10^8\ \text{m/s})(0.25\times10^{-3}\ \text{m})^2} = \boxed{0.51\ \text{J/m 3}}$$

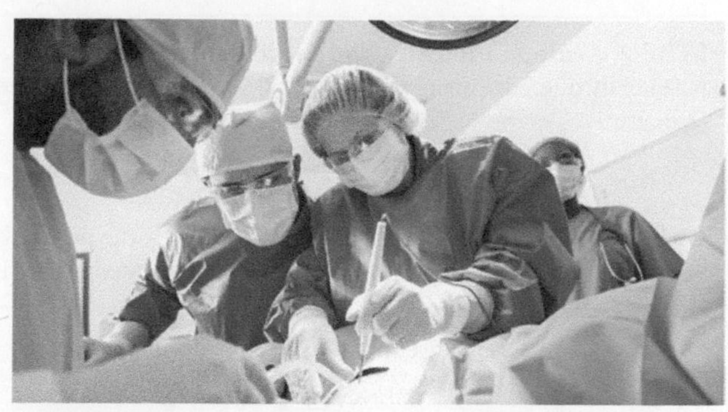
SDI Productions/E+/Getty Images

FIGURE 24.17 Surgeon using a laser scalpel to operate on the heart.

From our result in Example 5 above, we see the laser has an average total energy density that is approximately 100 000 times greater than that of sunlight!

Analyzing Multiple-Concept Problems

EXAMPLE 7 | Power and Intensity

Figure 24.18 shows a tiny source that is emitting light uniformly in all directions. At a distance of 2.50 m from the source, the rms electric field strength of the light is 19.0 N/C. Assuming that the light does not reflect from anything in the environment, determine the average power of the light emitted by the source.

Reasoning Recall from Section 16.7 that the power crossing a surface perpendicularly is equal to the intensity at the surface times the area of the surface (see Equation 16.8). Since the source emits light uniformly in all directions, the light intensity is the same at all points on the imaginary spherical surface in Figure 24.18. Moreover, the light crosses this surface perpendicularly. Equation 24.5b relates the average light intensity at the surface to its rms electric field strength (which is known), and the area of the surface can be found from a knowledge of its radius.

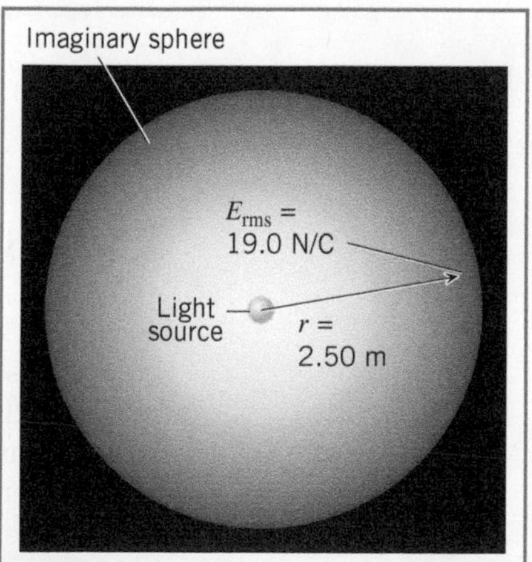

FIGURE 24.18 At a distance of 2.50 m from the light source, the rms electric field of the light has a value of 19.0 N/C.

Knowns and Unknowns The following data are available:

Description	Symbol	Value
Rms electric field strength 2.50 m from light source	E_{rms}	19.0 N/C
Distance from light source	r	2.50 m
Unknown Variable		
Average power emitted by light source	$\overline{P}$	?

Modelling the Problem

STEP 1 Average Intensity According to the discussion in Section 16.7, the average power $\overline{P}$ that passes perpendicularly through the imaginary spherical surface is equal to the average intensity $\overline{S}$ times the area A of the surface, or $\overline{P} = \overline{S}A$. The area of a spherical surface is $A = 4\pi r^2$, where r is the radius of the sphere. Thus, the average power can be written as in Equation 1 at the right. The radius is known, but the average light intensity is not, so we turn to Step 2 to evaluate it.

$$\overline{P} = \overline{S}(4\pi r^2) \quad (1)$$

?

STEP 2 Average Intensity and Electric Field The average intensity $\overline{S}$ of the light passing through the imaginary spherical surface is related to the known rms electric field strength E_{rms} at the surface by Equation 24.5b:

$$\boxed{\overline{S} = c\varepsilon_0 E_{\text{rms}}^2}$$

where c is the speed of light in a vacuum and ε_0 is the permittivity of free space. We can substitute this expression into Equation 1, as indicated at the right.

$$\overline{P} = \overline{S}(4\pi r^2) \quad (1)$$

$$\uparrow$$

$$\boxed{\overline{S} = c\varepsilon_0 E_{\text{rms}}^2}$$

Solution Algebraically combining the results of each step, we have

STEP 1 STEP 2

$$\overline{P} = \overline{S}(4\pi r^2) = c\varepsilon_0\, E_{\text{rms}}^2(4\pi r^2)$$

The average power emitted by the light source is

$$\overline{P} = c\varepsilon_0 E_{\text{rms}}^2(4\pi r^2)$$
$$= (3.00 \times 10^8\ \text{m/s})[8.85 \times 10^{-12}\ \text{C}^2/(\text{N} \cdot \text{m}^2)](19.0\ \text{N/C})^2\, 4\pi(2.50\ \text{m})^2 = \boxed{75.3\ \text{W}}$$

Related Homework: Problems 24, 25

Check Your Understanding

(The answers are given at the end of the book.)

5. If both the electric and magnetic fields of an electromagnetic wave double in magnitude, how does the intensity of the wave change? The intensity **(a)** decreases by a factor of four **(b)** decreases by a factor of two **(c)** increases by a factor of two **(d)** increases by a factor of four **(e)** increases by a factor of eight.

6. Suppose that the electric field of an electromagnetic wave decreases in magnitude. Does the magnitude of the magnetic field **(a)** increase, **(b)** decrease, or **(c)** remain the same?

24.5 The Doppler Effect and Electromagnetic Waves

Section 16.9 presents a discussion of the Doppler effect that sound waves exhibit when either the source of a sound wave, the observer of the wave, or both are moving with respect to the medium of propagation (e.g., air). This effect is one in which the observed sound frequency is greater or smaller than the frequency emitted by the source. A different Doppler effect arises when the source moves than when the observer moves.

Electromagnetic waves also can exhibit a Doppler effect, but it differs from that for sound waves for two reasons. First, sound waves require a medium such as air in which to propagate. In the Doppler effect for sound, it is the motion (of the source, the observer, and the waves themselves) relative to this medium that is important. In the Doppler effect for electromagnetic waves, motion relative to a medium plays no role, because the waves do not require a medium in which to propagate. They can travel in a vacuum. Second, in the equations for the Doppler effect in Section 16.9, the speed of sound plays an important role, and it depends on the reference frame relative to which it is measured. For example, the speed of sound with respect to moving air is different than it is with respect to stationary air. As we will see in Section 28.2, electromagnetic waves behave in a different way. The speed at which they travel has the same value, whether it is measured relative to a stationary observer or relative to one moving at a constant velocity. For these two reasons, the same Doppler effect arises for electromagnetic waves when either the source or the observer of the waves moves; only the relative motion of the source and the observer with respect to one another is important.

When electromagnetic waves and the source and the observer of the waves all travel along the same line in a vacuum (or in air, to a good degree of approximation), the single equation that specifies the Doppler effect is

$$f_o = f_s\left(1 \pm \frac{v_{rel}}{c}\right) \quad \text{if } v_{rel} \ll c \tag{24.6}$$

In this expression, f_o is the observed frequency, and f_s is the frequency emitted by the source. The symbol v_{rel} stands for the speed of the source and the observer relative to one another, and c is the speed of light in a vacuum. Equation 24.6 applies only if v_{rel} is very small compared to c—that is, if $v_{rel} \ll c$. Since v_{rel} is the **relative speed** of the source and the observer, it is like any **speed** and has no algebraic sign associated with it to denote the direction. The direction of the relative motion is taken into account by choosing the plus or minus sign in Equation 24.6. ***The plus sign is used when the source and the observer come together, and the minus sign is used when they move apart.***

For instance, suppose that the source and the observer are both traveling due east, the source at a speed of 28 m/s with respect to the ground and the observer at a speed of 22 m/s with respect to the ground. Neither of these speeds with respect to the ground is used for the symbol v_{rel} in Equation 24.6. Instead, the value for v_{rel} is $|28 \text{ m/s} - 22 \text{ m/s}| = 6$ m/s. If the faster source is behind the slower observer, the source and the observer come together because the source is catching up. Therefore, the plus sign is chosen in Equation 24.6. On the other hand, if the slower observer is behind the faster source, the source and the observer move apart because the source is pulling away. In this case, the minus sign is chosen in Equation 24.6.

Math Skills The choice of the plus or minus sign in Equation 24.6 is critical. Without the right choice, the equation cannot be used successfully to solve problems. Whether the plus or the minus sign is used depends on the nature of the individual problem, and the following two lists outline the possibilities.

PLUS SIGN (SOURCE AND OBSERVER COME TOGETHER)

(1) The source is catching up with the observer.
(2) The observer is catching up with the source.
(3) The source and the observer both move toward one another.

MINUS SIGN (SOURCE AND OBSERVER MOVE APART)

(1) The source is pulling away from the observer.
(2) The observer is pulling away from the source.
(3) The source and the observer both move away from one another.

THE PHYSICS OF . . . astronomy and the Doppler effect. The Doppler effect of electromagnetic waves provides a powerful tool for astronomers. For instance, Example 11 in Chapter 5 discusses how astronomers have identified a supermassive black hole at the center of galaxy M87 by using the Hubble Space Telescope. They focused the telescope on regions to either side of the center of the galaxy (see Figure 5.15). From the light emitted by these two regions, they were able to use the Doppler effect to determine that one side is moving away from the earth, while the other side is moving toward the earth. In other words, the galaxy is rotating. The speeds of recession and approach enabled astronomers to determine the rotational speed of the galaxy, and Example 11 in Chapter 5 shows how the value for this speed leads to the identification of the black hole. Astronomers routinely study the Doppler effect of the light that reaches the earth from distant parts of the universe. From such studies, they have determined the speeds at which distant light-emitting objects are receding from the earth.

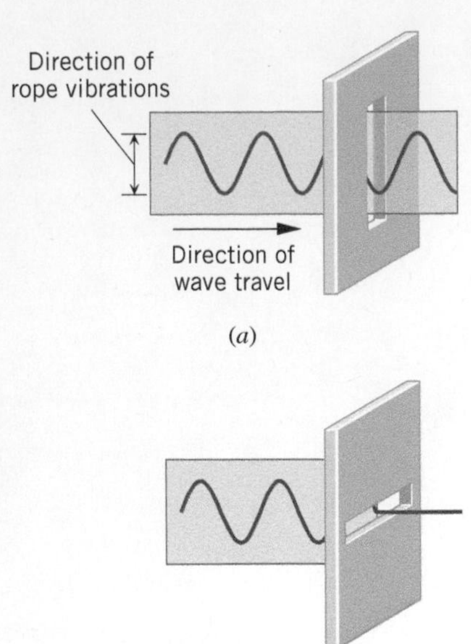

FIGURE 24.19 A transverse wave is linearly polarized when its vibrations always occur along one direction. (*a*) A linearly polarized wave on a rope can pass through a slit that is parallel to the direction of the rope vibrations, but (*b*) cannot pass through a slit that is perpendicular to the vibrations.

Check Your Understanding

(*The answers are given at the end of the book.*)

7. An astronomer measures the Doppler change in frequency for the light reaching the earth from a distant star. From this measurement, can the astronomer tell whether the star is moving away from the earth or the earth is moving away from the star?

8. **CYU Figure 24.1** shows three situations—A, B, and C—in which an observer and a source of electromagnetic waves are moving along the same line. In each case the source emits a wave of the same frequency. The arrows in each situation denote velocity vectors relative to the ground and have magnitudes of either v or $2v$. Rank the magnitudes of the frequencies of the observed waves in descending order (largest first).

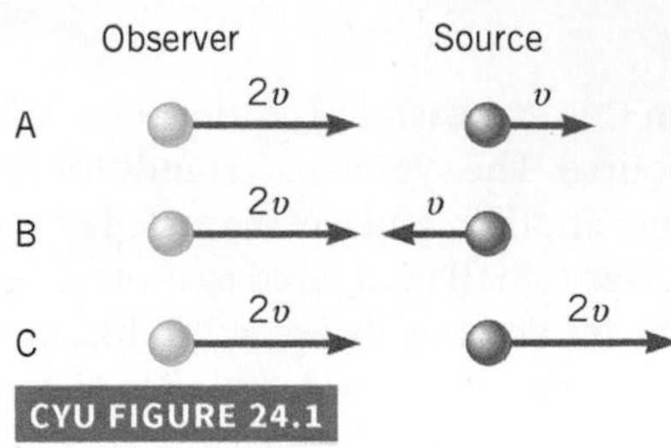

CYU FIGURE 24.1

24.6 Polarization

Polarized Electromagnetic Waves

One of the essential features of electromagnetic waves is that they are transverse waves, and because of this feature they can be polarized. **Figure 24.19** illustrates the idea of polarization by showing a transverse wave as it travels along a rope toward a slit. The wave is said to be **linearly polarized**, which means that its vibrations always occur along one direction. This direction is called the direction of polarization. In part *a* of the picture, the direction of polarization is vertical, parallel to the slit. Consequently, the wave passes through easily. However, when the slit is turned perpendicular to the direction of polarization, as in part *b*, the wave cannot pass, because the slit prevents the rope from oscillating. For longitudinal waves, such as sound waves, the notion of polarization has no meaning. In a longitudinal wave the direction of vibration is along the direction of travel, and the orientation of the slit would have no effect on the wave.

In an electromagnetic wave such as the one in **Figure 24.3**, the electric field oscillates along the y axis. Similarly, the magnetic field oscillates along the z axis. Therefore, the wave is linearly polarized, with the direction of polarization taken arbitrarily to be the direction along which the electric field oscillates. If the wave is a radio wave generated by a straight-wire antenna, the direction of polarization is determined by the orientation of the antenna. In comparison, the visible light given off by an incandescent light bulb consists of electromagnetic waves that are completely unpolarized. In this case the waves are emitted by a large number of atoms in the hot filament of the bulb. When an electron in an atom oscillates, the atom behaves as a miniature antenna that broadcasts light for brief periods of time, about 10^{-8} seconds. However, the directions of these atomic antennas change randomly as a result of collisions. Unpolarized light, then, consists of many individual waves, emitted in short bursts by many "atomic antennas," each with its own direction of polarization. **Figure 24.20** compares polarized and unpolarized light. In the unpolarized case, the arrows shown around the direction of wave travel symbolize the random directions of polarization of the individual waves that comprise the light.

Linearly polarized light can be produced from unpolarized light with the aid of certain materials. One commercially available material goes under the name of Polaroid. Such materials allow only the component of the electric field along one direction to pass through, while absorbing the field component perpendicular to this direction. As **Figure 24.21** indicates, the direction of polarization that a polarizing material allows through is called the **transmission axis**. No matter how this axis is oriented, the average intensity of the transmitted polarized light is one-half the average intensity of the incident unpolarized light. The reason for this is that the unpolarized light contains all polarization directions to an equal extent. Moreover, the electric field for each direction

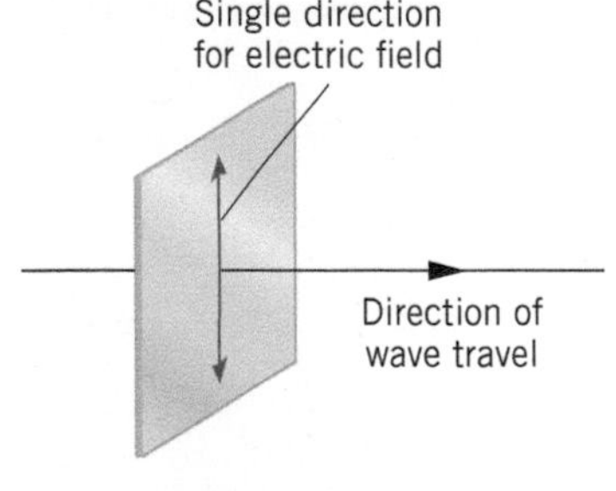

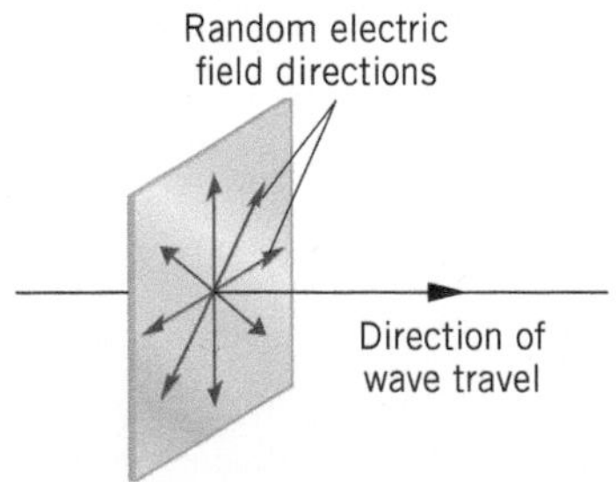

FIGURE 24.20 (*a*) In polarized light, the electric field of the electromagnetic wave fluctuates along a single direction. (*b*) Unpolarized light consists of short bursts of electromagnetic waves emitted by many different atoms. The electric field directions of these bursts are perpendicular to the direction of wave travel but are distributed randomly about it.

can be resolved into components perpendicular and parallel to the transmission axis, with the result that the average components perpendicular and parallel to the axis are equal. As a result, the polarizing material absorbs as much of the electric (and magnetic) field strength as it transmits.

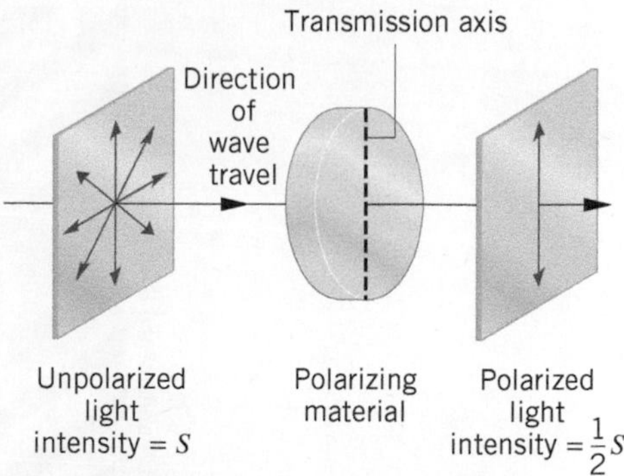

FIGURE 24.21 With the aid of a piece of polarizing material, polarized light may be produced from unpolarized light. The transmission axis of the material is the direction of polarization of the light that passes through the material.

Malus' Law

Once polarized light has been produced with a piece of polarizing material, it is possible to use a second piece to change the polarization direction and simultaneously adjust the intensity of the light. **Figure 24.22** shows how. As in this picture, the first piece of polarizing material is called the **polarizer** and the second piece is referred to as the **analyzer**. The transmission axis of the analyzer is oriented at an angle θ relative to the transmission axis of the polarizer. If the electric field strength of the polarized light incident on the analyzer is E, the field strength passing through is the component parallel to the transmission axis, or $E \cos \theta$. According to Equation 24.5b, the intensity is proportional to the square of the electric field strength. Consequently, the average intensity of polarized light passing through the analyzer is proportional to $\cos^2 \theta$. Thus, both the polarization direction and the intensity of the light can be adjusted by rotating the transmission axis of the analyzer relative to that of the polarizer. The average intensity $\overline{S}$ of the light leaving the analyzer, then, is

Malus' law
$$\overline{S} = \overline{S}_0 \cos^2 \theta \tag{24.7}$$

where $\overline{S}_0$ is the average intensity of the light entering the analyzer. Equation 24.7 is sometimes called **Malus' law**, for it was discovered by the French engineer Étienne Louis Malus (1775–1812). Example 8 illustrates the use of Malus' law.

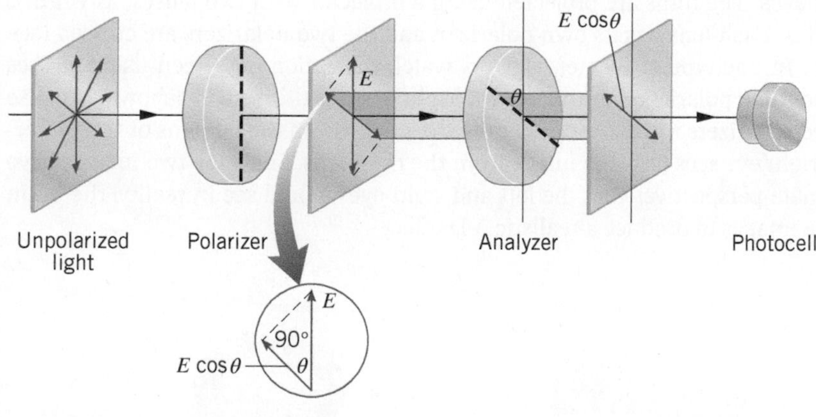

FIGURE 24.22 Two sheets of polarizing material, called the polarizer and the analyzer, may be used to adjust the polarization direction and intensity of the light reaching the photocell. This can be done by changing the angle θ between the transmission axes of the polarizer and analyzer.

EXAMPLE 8 | Using Polarizers and Analyzers

What value of θ should be used in **Figure 24.22**, so that the average intensity of the polarized light reaching the photocell will be one-tenth the average intensity of the unpolarized light?

Reasoning Both the polarizer and the analyzer reduce the intensity of the light. The polarizer reduces the intensity by a factor of one-half, as discussed earlier. Therefore, if the average intensity of the unpolarized light is $\overline{I}$, the average intensity of the polarized light leaving the polarizer and striking the analyzer is $\overline{S}_0 = \overline{I}/2$. The angle θ must now be selected so that the average intensity of the light leaving the analyzer will be $\overline{S} = \overline{I}/10$. Malus' law provides the solution.

Problem-Solving Insight Remember that when unpolarized light strikes a polarizer, only one-half of the incident light is transmitted, the other half being absorbed by the polarizer.

Solution Using $\overline{S}_0 = \overline{I}/2$ and $\overline{S} = \overline{I}/10$ in Malus' law, we find

$$\frac{1}{10}\overline{I} = \frac{1}{2}\overline{I} \cos^2 \theta$$

$$\frac{1}{5} = \cos^2 \theta \quad \text{or} \quad \theta = \cos^{-1}\left(\frac{1}{\sqrt{5}}\right) = \boxed{63.4^\circ}$$

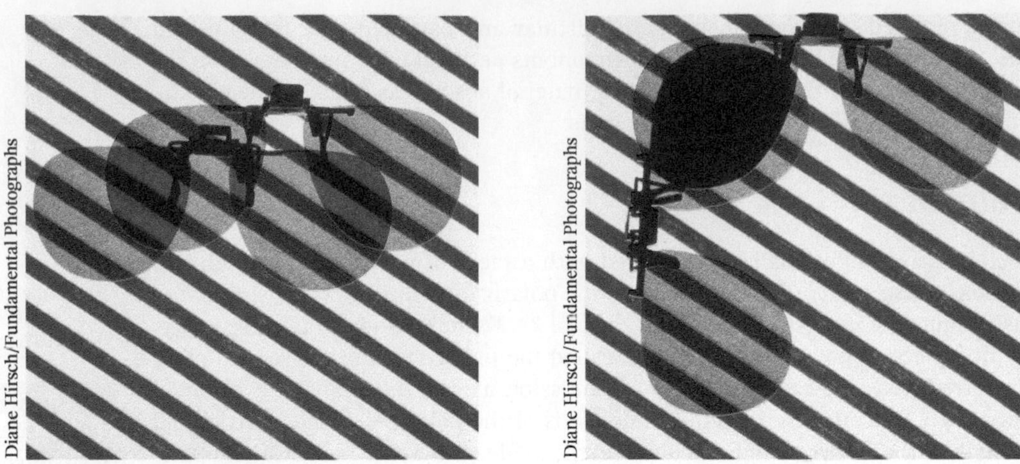

INTERACTIVE FIGURE 24.23 When Polaroid sunglasses are uncrossed (left side), the transmitted light is dimmed due to the extra thickness of tinted plastic. However, when they are crossed (right side), the intensity of the transmitted light is reduced to zero because of the effects of polarization.

When $\theta = 90°$ in **Figure 24.22**, the polarizer and analyzer are said to be **crossed**, and no light is transmitted by the polarizer/analyzer combination. As an illustration of this effect, **Interactive Figure 24.23** shows two pairs of Polaroid sunglasses in uncrossed and crossed configurations.

THE PHYSICS OF . . . IMAX 3-D films. An exciting application of crossed polarizers is used in viewing IMAX 3-D movies. These movies are recorded on two separate rolls of film, using a camera that provides images from the two different perspectives that correspond to what is observed by human eyes and allow us to see in three dimensions. The camera has two apertures or openings located at roughly the spacing between our eyes. The films are projected using a projector with two lenses, as **Figure 24.24** indicates. Each lens has its own polarizer, and the two polarizers are crossed (see the drawing). In one type of theater, viewers watch the action on-screen using glasses with corresponding polarizers for the left and right eyes, as the drawing shows. Because of the crossed polarizers the left eye sees only the image from the left lens of the projector, and the right eye sees only the image from the right lens. Since the two images have the approximate perspectives that the left and right eyes would see in reality, the brain combines the images to produce a realistic 3-D effect.

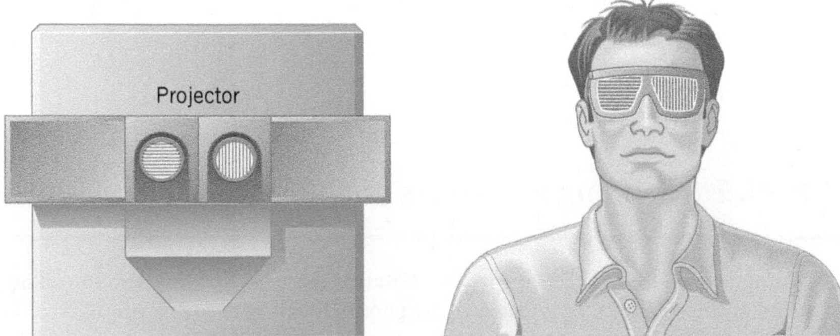

FIGURE 24.24 In an IMAX 3-D film, two separate rolls of film are projected using a projector with two lenses, each with its own polarizer. The two polarizers are crossed. Viewers watch the action on-screen through glasses that have corresponding crossed polarizers for each eye. The result is a 3-D moving picture, as the text discusses.

Conceptual Example 9 illustrates an interesting result that occurs when a piece of polarizing material is inserted between a crossed polarizer and analyzer.

CONCEPTUAL EXAMPLE 9 | How Can a Crossed Polarizer and Analyzer Transmit Light?

As explained earlier, no light reaches the photocell in Figure 24.22 when the polarizer and the analyzer are crossed. Suppose that a third piece of polarizing material is inserted between the polarizer and analyzer, as in Figure 24.25*a*. With the insert in place, will light reach the photocell when **(a)** $\theta = 0°$, **(b)** $\theta = 90°$, or **(c)** θ is between 0 and 90°?

Reasoning If any light is to pass through the analyzer, it must have an electric field component parallel to the transmission axis of the analyzer. Thus, without the insert in Figure 24.25*a*, no light reaches the photocell, because the analyzer and polarizer are crossed, which means that the electric field of the light reaching the analyzer has no component parallel to the analyzer's transmission axis. We need to consider, then, whether the presence of the insert leads to an electric field component parallel to the analyzer's transmission axis.

Answers (a) and (b) are incorrect. With $\theta = 0°$, the polarizer and the insert have parallel transmission axes, so the light leaving the polarizer passes through the insert unaffected. It reaches the analyzer with its electric field perpendicular to the analyzer's transmission axis and is, thus, prevented from reaching the photocell. With $\theta = 90°$, the polarizer and the insert are crossed, so no light leaves the insert to reach the analyzer and the photocell.

Answer (c) is correct. Parts *b* and *c* of Figure 24.25 show that, with the insert present, the light reaching the analyzer has an electric field component that is parallel to the analyzer's transmission axis when θ is between 0 and 90°. In part *b* the electric field E of the light leaving the polarizer makes an angle θ with respect to the transmission axis of the insert and has a component $E \cos \theta$ with respect to that axis. This component passes through the insert. In part *c* the field $(E \cos \theta)$ incident on the analyzer has a component parallel to the transmission axis of the analyzer—namely, $(E \cos \theta) \sin \theta$. This component passes through the analyzer and reaches the photocell.

Related Homework: Problem 39

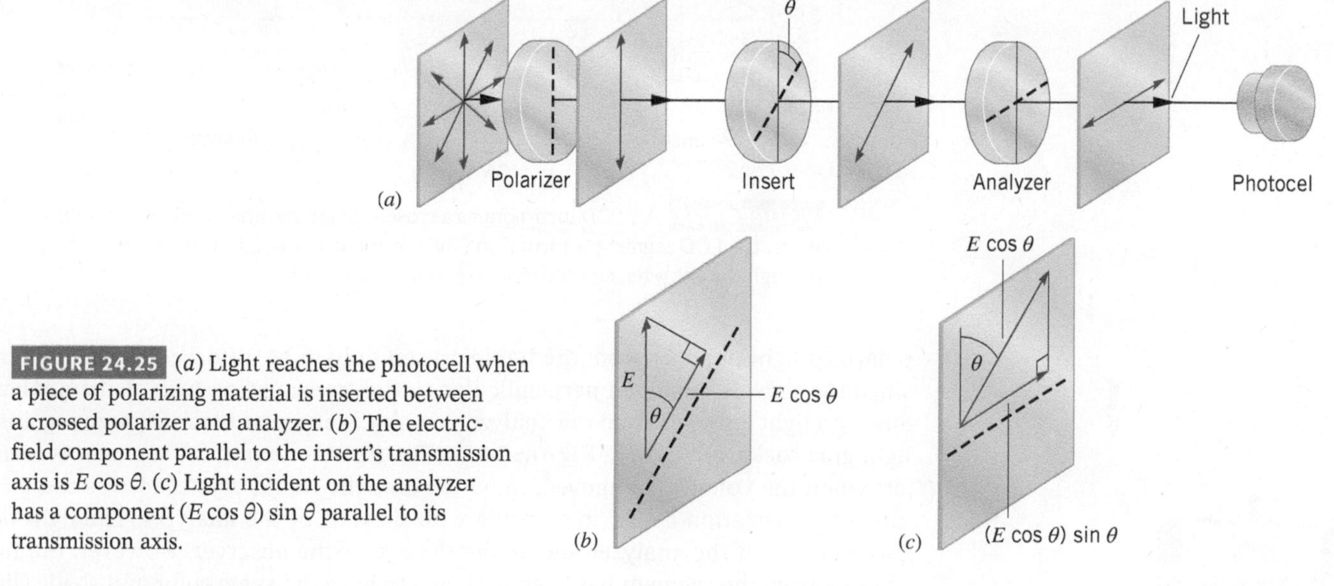

FIGURE 24.25 (*a*) Light reaches the photocell when a piece of polarizing material is inserted between a crossed polarizer and analyzer. (*b*) The electric-field component parallel to the insert's transmission axis is $E \cos \theta$. (*c*) Light incident on the analyzer has a component $(E \cos \theta) \sin \theta$ parallel to its transmission axis.

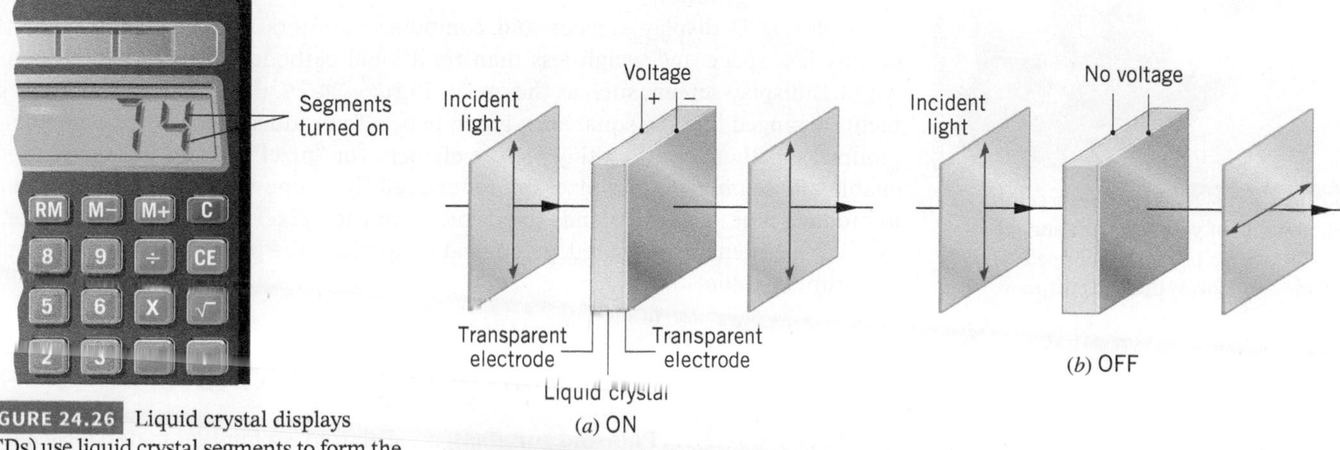

FIGURE 24.26 Liquid crystal displays (LCDs) use liquid crystal segments to form the numbers.

FIGURE 24.27 A liquid crystal in its (*a*) "on" state and (*b*) "off" state.

THE PHYSICS OF . . . a liquid crystal display (LCD). An application of a crossed polarizer/analyzer combination occurs in one kind of liquid crystal display (LCD). LCDs are widely used in pocket calculators and cell phones. The display usually consists of blackened numbers and letters set against a light gray background. As **Figure 24.26** indicates, each number or letter is formed from a combination of liquid crystal segments that have been turned on and appear black. The liquid crystal part of an LCD segment consists of the liquid crystal material sandwiched between two transparent electrodes, as in **Figure 24.27**. When a voltage is applied between the electrodes, the liquid crystal is said to be "on." Part *a* of the picture shows that linearly polarized incident light passes through the "on" material without having its direction of polarization affected. When the voltage is removed, as in part *b*, the liquid crystal is said to be "off" and now rotates the direction of polarization by 90°. A complete LCD segment also includes a crossed polarizer/analyzer combination, as **Figure 24.28** illustrates. The polarizer, analyzer, electrodes, and liquid crystal material are packaged as a single unit. The polarizer produces polarized light from incident unpolarized light. With the display segment turned on, as in **Figure 24.28**, the

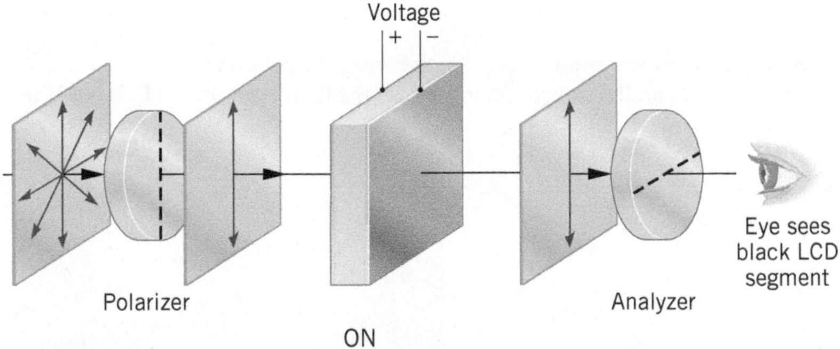

FIGURE 24.28 An LCD incorporates a crossed polarizer/analyzer combination. When the LCD segment is turned on (voltage applied), no light is transmitted through the analyzer, and the observer sees a black segment.

polarized light emerges from the liquid crystal only to be absorbed by the analyzer, since the light is polarized perpendicular to the transmission axis of the analyzer. Since no light emerges from the analyzer, an observer sees a black segment against a light gray background, as in **Figure 24.26**. On the other hand, the segment is turned off when the voltage is removed, in which case the liquid crystal rotates the direction of polarization by 90° to coincide with the axis of the analyzer. The light now passes through the analyzer and enters the eye of the observer. However, the light coming from the segment has been designed to have the same color and shade (light gray) as the background of the display, so the segment becomes indistinguishable from the background.

FIGURE 24.29 Want to see a color photograph of yourself? Just snap a "selfie" with a properly equipped cell phone and look at the LCD display.

Color LCD display screens and computer monitors are popular because they occupy less space and weigh less than traditional cathode-ray tube (CRT) units do. An LCD display screen, such as the one in **Figure 24.29**, uses thousands of LCD segments arranged like the squares on graph paper. To produce color, three segments are grouped together to form a tiny picture element (or "pixel"). Color filters are used to enable one segment in the pixel to produce red light, one to produce green, and one to produce blue. The eye blends the colors from each pixel into a composite color. By varying the intensity of the red, green, and blue colors, the pixel can generate an entire spectrum of colors.

The Occurrence of Polarized Light in Nature

THE PHYSICS OF . . . Polaroid sunglasses. Polaroid is a familiar material because of its widespread use in sunglasses. Such sunglasses are designed so that the transmission

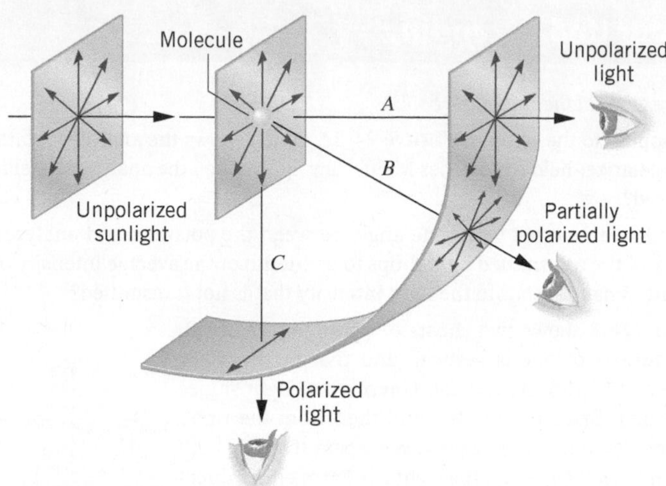

FIGURE 24.30 In the process of being scattered from atmospheric molecules, unpolarized light from the sun becomes partially polarized.

axis of the Polaroid is oriented vertically when the glasses are worn in the usual fashion. Thus, the glasses prevent any light that is polarized horizontally from reaching the eye. Light from the sun is unpolarized, but a considerable amount of horizontally polarized sunlight originates by reflection from horizontal surfaces such as that of a lake. Section 26.4 discusses this effect. Polaroid sunglasses reduce glare by preventing the horizontally polarized reflected light from reaching the eyes.

Polarized sunlight also originates from the scattering of light by molecules in the atmosphere. **Figure 24.30** shows light being scattered by a single atmospheric molecule. The electric fields in the unpolarized sunlight cause the electrons in the molecule to vibrate perpendicular to the direction in which the light is traveling. The electrons, in turn, reradiate the electromagnetic waves in different directions, as the drawing illustrates. The light radiated straight ahead in direction A is unpolarized, just like the incident light; but light radiated perpendicular to the incident light in direction C is polarized. Light radiated in the intermediate direction B is partially polarized.

BIO THE PHYSICS OF . . . butterflies and polarized light. Researchers have discovered that at least one butterfly species uses polarized light to attract members of the opposite sex. The butterfly species *Heliconius* has patterns on its wings that cause light reflected from them to be polarized. This polarized light, invisible to the human eye but visible to other butterflies, is attractive to potential mates. When males were shown the female wings with their polarized light patterns, they swarmed toward the wings. When the males were shown the wings through a filter that blocked out the polarization effects, they largely ignored the wings. **Figure 24.31** shows the polarized light reflected from the wings of the *Heliconius cydno* butterfly. The left wing is shown as it normally appears. The light reflected from the white pattern is highly polarized. The right wing is shown as it appears when viewed through a polarizing filter whose transmission axis is crossed with respect to the direction in which the reflected light is polarized. The white patterns in the right wing are black when viewed through the filter, a clear indication that the light reflected from them is indeed polarized. There is also experimental evidence that some bird species use polarized light as a navigational aid.

Courtesy Alison Sweeney, Duke University. Image from *Nature* 423: 31–32 (May 1, 2003). Reproduced with permission.

FIGURE 24.31 A *Heliconius cydno* butterfly. The left wing is shown as it appears normally, and the right wing as it appears when viewed through a polarizing filter. The light reflected from the white patterns is polarized. These patterns in the right wing are black because the transmission axis of the filter is crossed with respect to the polarization direction of the reflected light.

Check Your Understanding

(The answers are given at the end of the book.)

9. Malus' law applies to the setup in Figure 24.22, which shows the analyzer rotated through an angle θ and the polarizer held fixed. Does Malus' law apply when the analyzer is held fixed and the polarizer is rotated?

10. In Example 8, we saw that when the angle between the polarizer and analyzer is 63.4°, the average intensity of the transmitted light drops to one-tenth of the average intensity of the incident unpolarized light. What happens to the light intensity that is not transmitted?

11. CYU Figure 24.2 shows two sheets of polarizing material. The transmission axis of one is vertical, and that of the other makes an angle of 45° with the vertical. Unpolarized light shines on this arrangement first from the left and then from the right. From which direction does at least some light pass through both sheets? **(a)** From the left **(b)** From the right **(c)** From either direction **(d)** From neither direction. What is the answer when the light is horizontally polarized? What is the answer when the light is vertically polarized?

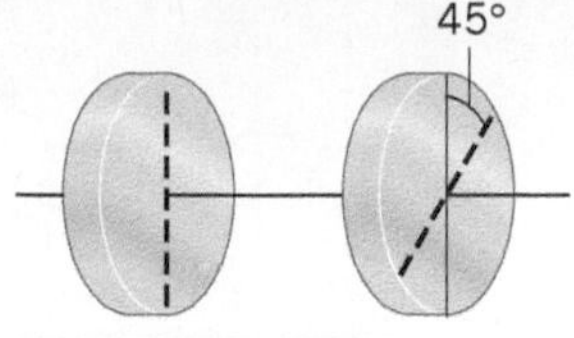

CYU FIGURE 24.2

12. You are sitting upright on the beach near a lake on a sunny day, wearing Polaroid sunglasses. When you lie down on your side, facing the lake, the sunglasses don't work as well as they do while you are sitting upright. Why not?

Concept Summary

24.1 The Nature of Electromagnetic Waves An electromagnetic wave consists of mutually perpendicular and oscillating electric and magnetic fields. The wave is a transverse wave, since the fields are perpendicular to the direction in which the wave travels. Electromagnetic waves can travel through a vacuum or a material substance. All electromagnetic waves travel through a vacuum at the same speed, which is known as the speed of light c ($c = 3.00 \times 10^8$ m/s).

24.2 The Electromagnetic Spectrum The frequency f and wavelength λ of an electromagnetic wave in a vacuum are related to its speed c through the relation $c = f\lambda$.

The series of electromagnetic waves, arranged in order of their frequencies or wavelengths, is called the electromagnetic spectrum. In increasing order of frequency (decreasing order of wavelength), the spectrum includes radio waves, infrared radiation, visible light, ultraviolet radiation, X-rays, and gamma rays. Visible light has frequencies between about 4.0×10^{14} and 7.9×10^{14} Hz. The human eye and brain perceive different frequencies or wavelengths as different colors.

24.3 The Speed of Light James Clerk Maxwell showed that the speed of light in a vacuum is given by Equation 24.1, where ε_0 is the (electric) permittivity of free space and μ_0 is the (magnetic) permeability of free space.

$$c = \frac{1}{\sqrt{\varepsilon_0 \mu_0}} \tag{24.1}$$

24.4 The Energy Carried by Electromagnetic Waves The total energy density u of an electromagnetic wave is the total energy per unit volume of the wave and, in a vacuum, is given by Equation 24.2a, where E and B, respectively, are the magnitudes of the electric and magnetic fields of the wave. Since the electric and magnetic parts of the total energy density are equal, Equations 24.2b and 24.2c are equivalent to Equation 24.2a. In a vacuum, E and B are related according to Equation 24.3.

$$u = \frac{1}{2}\varepsilon_0 E^2 + \frac{1}{2\mu_0}B^2 \tag{24.2a}$$

$$u = \varepsilon_0 E^2 \tag{24.2b}$$

$$u = \frac{1}{\mu_0}B^2 \tag{24.2c}$$

$$E = cB \tag{24.3}$$

Equations 24.2a–c can be used to determine the average total energy density, if the rms average values E_{rms} and B_{rms} are used in place of the symbols E and B. The rms values are related to the peak values E_0 and B_0 in the usual way, as shown in Equations 1 and 2.

The intensity of an electromagnetic wave is the power that the wave carries perpendicularly through a surface divided by the area of the surface. In a vacuum, the intensity S is related to the total energy density u according to Equation 24.4.

$$E_{rms} = \frac{1}{\sqrt{2}}E_0 \quad (1) \qquad B_{rms} = \frac{1}{\sqrt{2}}B_0 \quad (2)$$

$$S = cu \tag{24.4}$$

24.5 The Doppler Effect and Electromagnetic Waves When electromagnetic waves and the source and observer of the waves all travel along the same line in a vacuum, the Doppler effect is given by Equation 24.6, where f_o and f_s are, respectively, the observed and emitted wave frequencies and v_{rel} is the relative speed of the source and the observer. The plus sign is used when the source and the observer come together, and the minus sign is used when they move apart.

$$f_o = f_s\left(1 \pm \frac{v_{rel}}{c}\right) \quad \text{if } v_{rel} \ll c \tag{24.6}$$

24.6 Polarization A linearly polarized electromagnetic wave is one in which the oscillation of the electric field occurs only along one direction, which is taken to be the direction of polarization. The magnetic field also oscillates along only one direction, which is perpendicular to the electric field direction. In an unpolarized wave such as the light from an incandescent bulb, the direction of polarization does not remain fixed, but fluctuates randomly in time.

Polarizing materials allow only the component of the wave's electric field along one direction (and the associated magnetic field component) to pass through them. The preferred transmission direction for the electric field is called the transmission axis of the material.

When unpolarized light is incident on a piece of polarizing material, the transmitted polarized light has an average intensity that is one-half the average intensity of the incident light.

When two pieces of polarizing material are used one after the other, the first is called the polarizer, and the second is referred to as the analyzer. If the average intensity of polarized light falling on the analyzer is $\overline{S}_0$, the average intensity $\overline{S}$ of the light leaving the analyzer is given by Malus' law, as shown in Equation 24.7, where θ is the angle between the transmission axes of the polarizer and analyzer. When $\theta = 90°$, the polarizer and the analyzer are said to be "crossed," and no light passes through the analyzer.

$$\overline{S} = \overline{S}_0 \cos^2 \theta \qquad \textbf{(24.7)}$$

Focus on Concepts

Online

Additional questions are available for assignment in WileyPLUS.

Section 24.1 The Nature of Electromagnetic Waves

1. The drawing shows an x, y, z coordinate system. A circular loop of wire lies in the z, x plane and, when used with an LC-tuned circuit, detects an electromagnetic wave. Which one of the following statements is correct? **(a)** The wave travels along the x axis, and its electric field oscillates along the y axis. **(b)** The wave travels along the z axis, and its electric field oscillates along the x axis. **(c)** The wave travels along the z axis, and its electric field oscillates along the y axis. **(d)** The wave travels along the y axis, and its electric field oscillates along the x axis. **(e)** The wave travels along the y axis, and its electric field oscillates along the z axis.

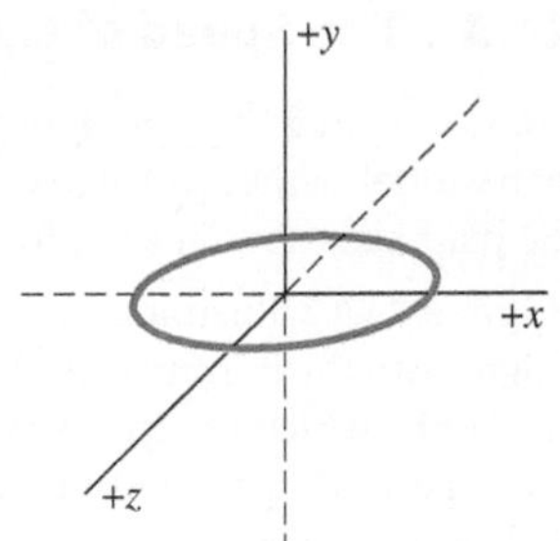

QUESTION 1

Section 24.2 The Electromagnetic Spectrum

2. An electromagnetic wave travels in a vacuum. The wavelength of the wave is tripled. How is this accomplished? **(a)** By tripling the frequency of the wave **(b)** By tripling the speed of the wave **(c)** By reducing the frequency of the wave by a factor of three **(d)** By reducing the speed of the wave by a factor of three **(e)** By tripling the magnitudes of the electric and magnetic fields that comprise the wave

Section 24.4 The Energy Carried by Electromagnetic Waves

3. An electromagnetic wave is traveling in a vacuum. The magnitudes of the electric and magnetic fields of the wave are ______, and the electric and magnetic energies carried by the wave are ______. **(a)** equal, proportional (but not equal) to each other **(b)** proportional (but not equal) to each other, equal **(c)** equal, equal **(d)** proportional (but not equal) to each other, unequal

Section 24.5 The Doppler Effect and Electromagnetic Waves

4. The drawing shows four situations—A, B, C, and D—in which an observer and a source of electromagnetic waves can move along the same line. In each case the source emits a wave of the same frequency, and in each case only the source or the observer is moving. The arrow in each situation denotes the velocity vector, which has the same magnitude in each situation. When there is no arrow, the observer or the source is stationary. Rank the frequencies of the observed electromagnetic waves in descending order (largest first) according to magnitude. **(a)** A and B (a tie), C and D (a tie) **(b)** C and D (a tie), A and B (a tie) **(c)** A and D (a tie), B and C (a tie) **(d)** B and D (a tie), A and C (a tie) **(e)** B and C (a tie), A and D (a tie)

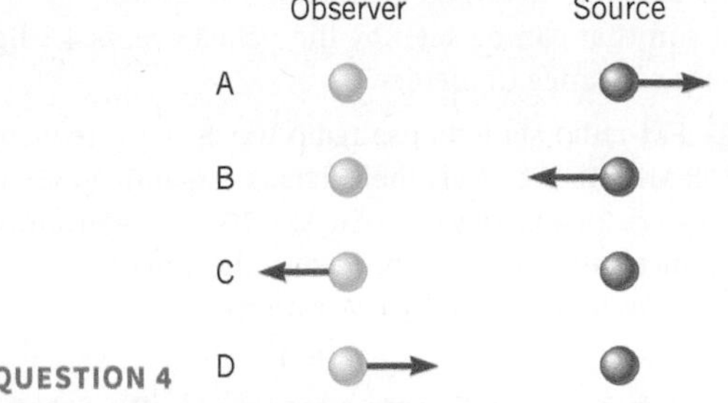

QUESTION 4

Section 24.6 Polarization

5. The drawing shows two sheets of polarizing material. Polarizer 1 has its transmission axis aligned vertically, and polarizer 2 has its transmission axis aligned at an angle of 45° with respect to the vertical. Light that is completely polarized along the vertical direction is incident either from the far left or from the far right. In either case, the average intensity of the incident light is the same. Which one of the following statements is true concerning the average intensity of the light that is transmitted by the pair of sheets? **(a)** When the light is incident from either the left or the right, the transmitted intensity is one-half the incident intensity. **(b)** When the light is incident from either the left or the right, the transmitted intensity is one-fourth the incident intensity. **(c)** When the light is incident from the left, the transmitted intensity is one-half the incident intensity; when the light

is incident from the right, the transmitted intensity is zero. **(d)** When the light is incident from the left, the transmitted intensity is one-fourth the incident intensity; when the light is incident from the right, the transmitted intensity is one-half the incident intensity. **(e)** When the light is incident from the left, the transmitted intensity is one-half the incident intensity; when the light is incident from the right, the transmitted intensity is one-fourth the incident intensity.

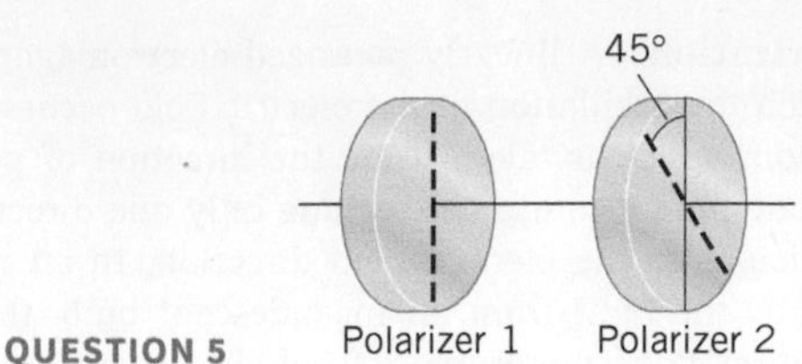

QUESTION 5

Problems

Online

Additional questions are available for assignment in WileyPLUS.

- SSM Student Solutions Manual
- MMH Problem-solving help
- GO Guided Online Tutorial
- V-HINT Video Hints
- CHALK Chalkboard Videos
- BIO Biomedical application
- E Easy
- M Medium
- H Hard
- WS Worksheet
- T Team Problem

Section 24.1 The Nature of Electromagnetic Waves

1. E The team monitoring a space probe exploring the outer solar system finds that radio transmissions from the probe take 2.53 hours to reach earth. How distant (in meters) is the probe?

2. E **(a)** Neil A. Armstrong was the first person to walk on the moon. The distance between the earth and the moon is 3.85×10^8 m. Find the time it took for his voice to reach the earth via radio waves. **(b)** Someday a person will walk on Mars, which is 5.6×10^{10} m from the earth at the point of closest approach. Determine the minimum time that will be required for a message from Mars to reach the earth via radio waves.

3. E SSM CHALK In astronomy, distances are often expressed in light-years. One light-year is the distance traveled by light in one year. The distance to Alpha Centauri, the closest star other than our own sun that can be seen by the naked eye, is 4.3 light-years. Express this distance in meters.

4. E GO FM radio stations use radio waves with frequencies from 88.0 to 108 MHz to broadcast their signals. Assuming that the inductance in **Figure 24.4** has a value of 6.00×10^{-7} H, determine the range of capacitance values that are needed so the antenna can pick up all the radio waves broadcasted by FM stations.

Section 24.2 The Electromagnetic Spectrum

5. E A truck driver is broadcasting at a frequency of 26.965 MHz with a CB (citizen's band) radio. Determine the wavelength of the electromagnetic wave being used. The speed of light is $c = 2.9979 \times 10^8$ m/s.

6. E In a dentist's office an X-ray of a tooth is taken using X-rays that have a frequency of 6.05×10^{18} Hz. What is the wavelength in vacuum of these X-rays?

7. E SSM In a certain UHF radio wave, the shortest distance between positions at which the electric and magnetic fields are zero is 0.34 m. Determine the frequency of this UHF radio wave.

8. E FM radio waves have frequencies between 88.0 and 108.0 MHz. Determine the range of wavelengths for these waves.

9. E GO A certain type of laser emits light that has a frequency of 5.2×10^{14} Hz. The light, however, occurs as a series of short pulses, each lasting for a time of 2.7×10^{-11} s. **(a)** How many wavelengths are there in one pulse? **(b)** The light enters a pool of water. The frequency of the light remains the same, but the speed of the light slows down to 2.3×10^8 m/s. How many wavelengths are there now in one pulse?

10. E Two radio waves are used in the operation of a cellular telephone. To receive a call, the phone detects the wave emitted at one frequency by the transmitter station or base unit. To send your message to the base unit, your phone emits its own wave at a different frequency. The difference between these two frequencies is fixed for all channels of cell phone operation. Suppose that the wavelength of the wave emitted by the base unit is 0.34339 m and the wavelength of the wave emitted by the phone is 0.36205 m. Using a value of 2.9979×10^8 m/s for the speed of light, determine the difference between the two frequencies used in the operation of a cell phone.

11. M V-HINT CHALK A positively charged object with a mass of 0.115 kg oscillates at the end of a spring, generating ELF (extremely low frequency) radio waves that have a wavelength of 4.80×10^7 m. The frequency of these radio waves is the same as the frequency at which the object oscillates. What is the spring constant of the spring?

Section 24.3 The Speed of Light

12. E Review Conceptual Example 4 for information pertinent to this problem. When we look at the star Polaris (the North Star), we are seeing it as it was 680 years ago. How far away from us (in meters) is Polaris?

13. E GO **Figure 24.13** illustrates Michelson's setup for measuring the speed of light with the mirrors placed on Mt. San Antonio and Mt. Wilson in California, which are 35 km apart. Using a value of 3.00×10^8 m/s for the speed of light, find the minimum angular speed (in rev/s) for the rotating mirror.

14. E SSM Two astronauts are 1.5 m apart in their spaceship. One speaks to the other. The conversation is transmitted to earth via electromagnetic waves. The time it takes for sound waves to travel at 343 m/s through the air between the astronauts equals the time it takes for the electromagnetic waves to travel to the earth. How far away from the earth is the spaceship?

15. E V-HINT A laptop computer communicates with a router wirelessly, by means of radio signals. The router is connected by cable directly to the Internet. The laptop is 8.1 m from the router, and is downloading text and images from the Internet at an average rate of 260 Mbps, or 260 megabits per second. (A *bit*, or *binary digit*, is the smallest unit of digital information.) On average, how many bits are downloaded to the laptop in the time it takes the wireless signal to travel from the router to the laptop?

16. E A lidar (laser radar) gun is an alternative to the standard radar gun that uses the Doppler effect to catch speeders. A lidar gun uses an infrared laser and emits a precisely timed series of pulses of infrared electromagnetic waves. The time for each pulse to travel to the speeding vehicle and return to the gun is measured. In one situation a lidar gun in a stationary police car observes a difference of 1.27×10^{-7} s in round-trip travel times for two pulses that are emitted 0.450 s apart. Assuming that the speeding vehicle is approaching the police car essentially head-on, determine the speed of the vehicle.

17. M GO A politician holds a press conference that is televised live. The sound picked up by the microphone of a TV news network is broadcast via electromagnetic waves and heard by a television viewer. This viewer is seated 2.3 m from his television set. A reporter at the press conference is located 4.1 m from the politician, and the sound of the words travels directly from the celebrity's mouth, through the air, and into the reporter's ears. The reporter hears the words *exactly at the same instant* that the television viewer hears them. Using a value of 343 m/s for the speed of sound, determine the maximum distance between the television set and the politician. Ignore the small distance between the politician and the microphone. In addition, assume that the only delay between what the microphone picks up and the sound being emitted by the television set is that due to the travel time of the electromagnetic waves used by the network.

18. M GO CHALK A mirror faces a cliff located some distance away. Mounted on the cliff is a second mirror, directly opposite the first mirror and facing toward it. A gun is fired very close to the first mirror. The speed of sound is 343 m/s. How many times does the flash of the gunshot travel the round-trip distance between the mirrors before the echo of the gunshot is heard?

Section 24.4 The Energy Carried by Electromagnetic Waves

19. E A laser emits a narrow beam of light. The radius of the beam is 1.0×10^{-3} m, and the power is 1.2×10^{-3} W. What is the intensity of the laser beam?

20. E An industrial laser is used to burn a hole through a piece of metal. The average intensity of the light is $\overline{S} = 1.23 \times 10^{9}$ W/m^2. What is the rms value of (a) the electric field and (b) the magnetic field in the electromagnetic wave emitted by the laser?

21. E The maximum strength of the magnetic field in an electromagnetic wave is 3.3×10^{-6} T. What is the maximum strength of the wave's electric field?

22. E SSM The microwave radiation left over from the Big Bang explosion of the universe has an average energy density of 4×10^{-14} J/m^3. What is the rms value of the electric field of this radiation?

23. E GO On a cloudless day, the sunlight that reaches the surface of the earth has an intensity of about 1.0×10^{3} W/m^2. What is the electromagnetic energy contained in 5.5 m^3 of space just above the earth's surface?

24. E GO Consult Multiple-Concept Example 7 to review the concepts on which this problem depends. A light bulb emits light uniformly in all directions. The average emitted power is 150.0 W. At a distance of 5.00 m from the bulb, determine (a) the average intensity of the light, (b) the rms value of the electric field, and (c) the peak value of the electric field.

25. E Multiple-Concept Example 7 provides some pertinent background for this problem. The mean distance between earth and the sun is 1.50×10^{11} m. The average intensity of solar radiation incident on the upper atmosphere of the earth is 1390 W/m^2. Assuming that the sun emits radiation uniformly in all directions, determine the total power radiated by the sun.

26. E GO A stationary particle of charge $q = 2.6 \times 10^{-8}$ C is placed in a laser beam (an electromagnetic wave) whose intensity is 2.5×10^{3} W/m^2. Determine the magnitudes of the (a) electric and (b) magnetic forces exerted on the charge. If the charge is moving at a speed of 3.7×10^{4} m/s perpendicular to the magnetic field of the electromagnetic wave, find the magnitudes of the (c) electric and (d) magnetic forces exerted on the particle.

27. M The power radiated by the sun is 3.9×10^{26} W. The earth orbits the sun in a nearly circular orbit of radius 1.5×10^{11} m. The earth's axis of rotation is tilted by 27° relative to the plane of the orbit (see the drawing), so sunlight does not strike the equator perpendicularly. What power strikes a 0.75-m^2 patch of flat land at the equator at point Q?

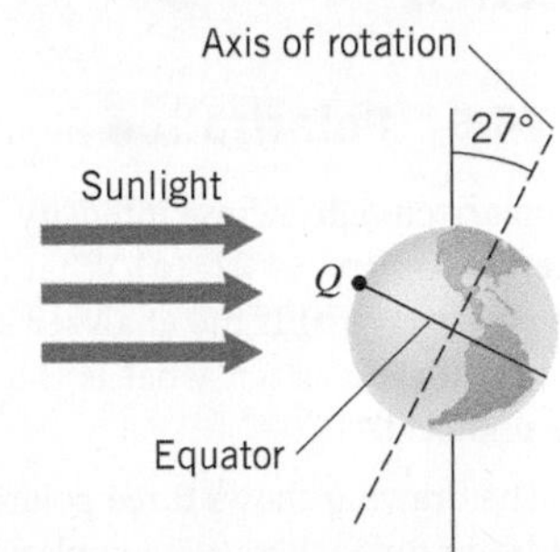

PROBLEM 27

28. M V-HINT An electromagnetic wave strikes a 1.30-cm^2 section of wall perpendicularly. The rms value of the wave's magnetic field is determined to be 6.80×10^{-4} T. How long does it take for the wave to deliver 1850 J of energy to the wall?

Section 24.5 The Doppler Effect and Electromagnetic Waves

29. E MMH A distant galaxy emits light that has a wavelength of 434.1 nm. On earth, the wavelength of this light is measured to be 438.6 nm. (a) Decide whether this galaxy is approaching or receding from the earth. Give your reasoning. (b) Find the speed of the galaxy relative to the earth.

30. E GO A speeder is pulling directly away and increasing his distance from a police car that is moving at 25 m/s with respect to the ground. The radar gun in the police car emits an electromagnetic wave with a frequency of 7.0×10^{9} Hz. The wave reflects from the speeder's car and returns to the police car, where its frequency is measured to be 320 Hz less than the emitted frequency. Find the speeder's speed with respect to the ground.

31. M CHALK SSM A distant galaxy is simultaneously rotating and receding from the earth. As the drawing shows, the galactic center is receding from the earth at a relative speed of $u_G = 1.6 \times 10^{6}$ m/s. Relative to the center, the tangential speed is $v_T = 0.4 \times 10^{6}$ m/s for locations A and B, which are equidistant from the center. When the frequencies of the light coming from regions A and B are measured on earth, they are not the same and each is different from the emitted frequency of 6.200×10^{14} Hz. Find the measured frequency for the light from (a) region A and (b) region B.

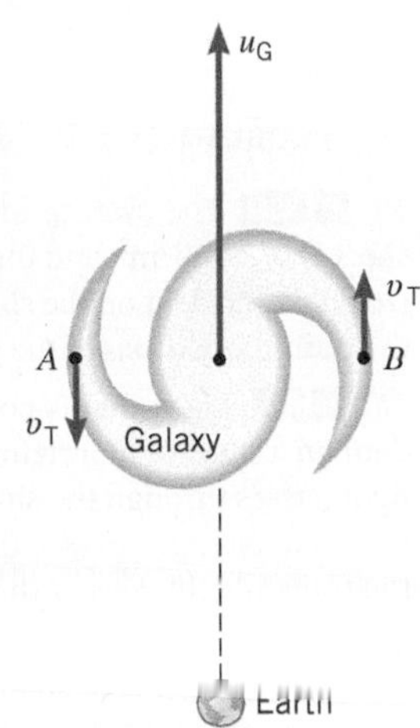

PROBLEM 31

32. M GO The drawing shows three situations— A, B, and C—in which an observer and a source of electromagnetic waves

are moving along the same line. In each case the source emits a wave that has a frequency of 4.57×10^{14} Hz. The arrows in each situation denote velocity vectors of the observer and source relative to the ground and have the magnitudes indicated (v or $2v$), where the speed v is 1.50×10^{6} m/s. Calculate the observed frequency in each of the three cases.

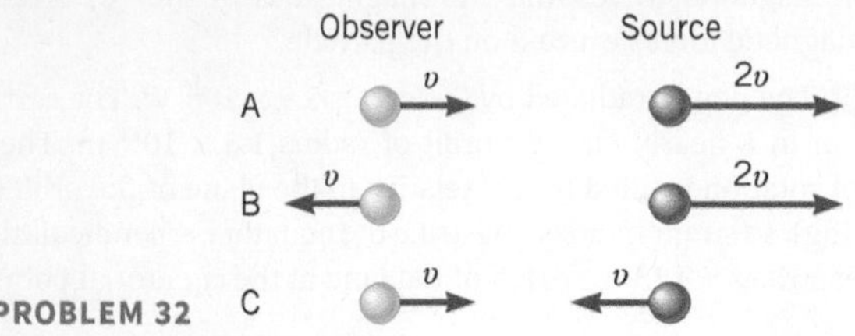

PROBLEM 32

Section 24.6 Polarization

33. **E** Unpolarized light whose intensity is 1.10 W/m^2 is incident on the polarizer in **Figure 24.22**. **(a)** What is the intensity of the light leaving the polarizer? **(b)** If the analyzer is set at an angle of $\theta = 75°$ with respect to the polarizer, what is the intensity of the light that reaches the photocell?

34. **E GO** The drawing shows three polarizer/analyzer pairs. The incident light beam for each pair is unpolarized and has the same average intensity of 48 W/m^2. Find the average intensity of the transmitted beam for each of the three cases (A, B, and C) shown in the drawing.

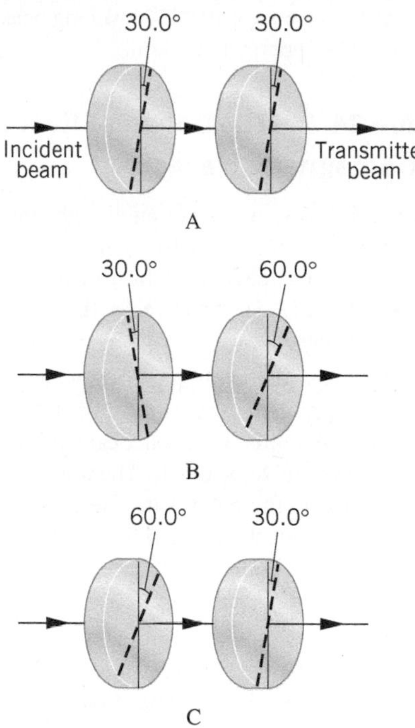

PROBLEM 34

35. **E SSM** The average intensity of light emerging from a polarizing sheet is 0.764 W/m^2, and the average intensity of the horizontally polarized light incident on the sheet is 0.883 W/m^2. Determine the angle that the transmission axis of the polarizing sheet makes with the horizontal.

36. **E GO** Light that is polarized along the vertical direction is incident on a sheet of polarizing material. Only 94% of the intensity of the light passes through the sheet and strikes a second sheet of polarizing material. No light passes through the second sheet. What angle does the transmission axis of the second sheet make with the vertical?

37. **E GO** The drawing shows light incident on a polarizer whose transmission axis is parallel to the z axis. The polarizer is rotated clockwise through an angle α. The average intensity of the incident light is 7.0 W/m^2. Determine the average intensity of the transmitted light for each of the six cases shown in the table.

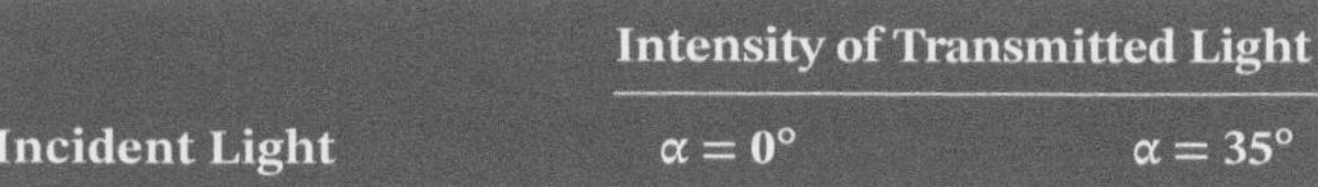

	Intensity of Transmitted Light	
Incident Light	$\alpha = 0°$	$\alpha = 35°$
(a) Unpolarized		
(b) Polarized parallel to z axis		
(c) Polarized parallel to y axis		

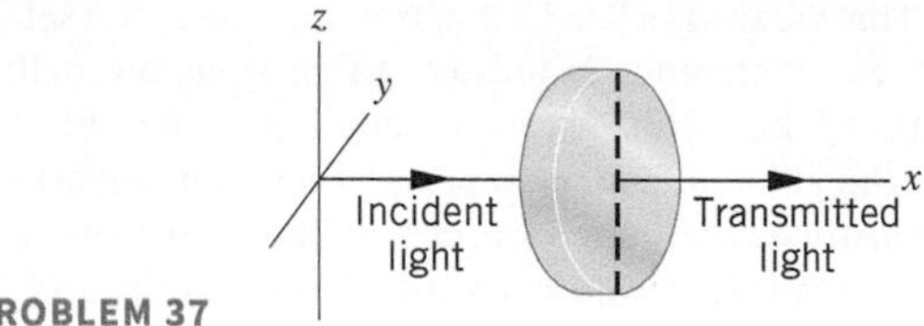

PROBLEM 37

38. **E MMH** For each of the three sheets of polarizing material shown in the drawing, the orientation of the transmission axis is labeled relative to the vertical. The incident beam of light is unpolarized and has an intensity of 1260.0 W/m^2. What is the intensity of the beam transmitted through the three sheets when $\theta_1 = 19.0°$, $\theta_2 = 55.0°$, and $\theta_3 = 100.0°$?

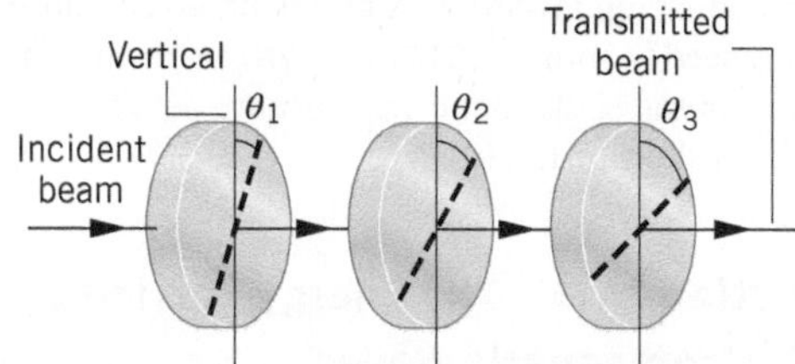

PROBLEM 38

39. **M V-HINT** Before attempting this problem, review Conceptual Example 9. The intensity of the light that reaches the photocell in **Figure 24.25*a*** is 110 W/m^2, when $\theta = 23°$. What would be the intensity reaching the photocell if the *analyzer* were removed from the setup, everything else remaining the same?

40. **M SSM** More than one analyzer can be used in a setup like the one in **Figure 24.22**, each analyzer following the previous one. Suppose that the transmission axis of the first analyzer is rotated 27° relative to the transmission axis of the polarizer, and that the transmission axis of each additional analyzer is rotated 27° relative to the transmission axis of the previous one. What is the minimum number of analyzers needed for the light reaching the photocell to have an intensity that is reduced by at least a factor of 100 relative to the intensity of the light striking the first analyzer?

41. **M CHALK GO** The drawing shows four sheets of polarizing material, each with its transmission axis oriented differently. Light that is polarized in the vertical direction is incident from the left and has an average intensity of 27 W/m^2. Determine the average intensity of the light that emerges on the right in the drawing when sheet A alone is removed, when sheet B alone is removed, when sheet C alone is removed, and when sheet D alone is removed.

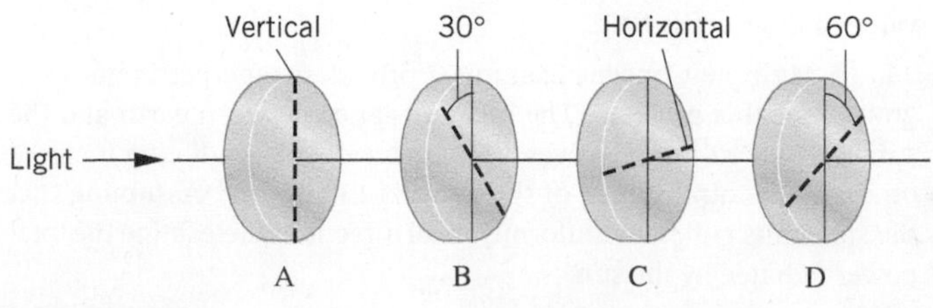

PROBLEM 41

Additional Problems

Online

42. E Obtain the wavelengths in vacuum for **(a)** blue light whose frequency is 6.34×10^{14} Hz, and **(b)** orange light whose frequency is 4.95×10^{14} Hz. Express your answers in nanometers (1 nm $= 10^{-9}$ m).

43. E GO The magnitude of the electric field of an electromagnetic wave increases from 315 to 945 N/C. **(a)** Determine the wave intensities for the two values of the electric field. **(b)** What is the magnitude of the magnetic field associated with each electric field? **(c)** Determine the wave intensity for each value of the magnetic field.

44. E SSM A future space station in orbit about the earth is being powered by an electromagnetic beam from the earth. The beam has a cross-sectional area of 135 m^2 and transmits an average power of 1.20×10^4 W. What are the rms values of the **(a)** electric and **(b)** magnetic fields?

45. E V-HINT Suppose that a police car is moving to the right at 27 m/s, while a speeder is coming up from behind at a speed of 39 m/s, both speeds being with respect to the ground. Assume that the electromagnetic wave emitted by the police car's radar gun has a frequency of 8.0×10^9 Hz. Find the difference between the frequency of the wave that returns to the police car after reflecting from the speeder's car and the original frequency emitted by the police car.

46. M CHALK GO The electromagnetic wave that delivers a cellular phone call to a car has a magnetic field with an rms value of 1.5×10^{-10} T. The wave passes perpendicularly through an open window, the area of which is 0.20 m^2. How much energy does this wave carry through the window during a 45-s phone call?

47. M V-HINT A beam of polarized light with an average intensity of 15 W/m^2 is sent through a polarizer. The transmission axis makes an angle of 25° with respect to the direction of polarization. Determine the rms value of the electric field of the transmitted beam.

48. M An argon–ion laser produces a cylindrical beam of light whose average power is 0.750 W. How much energy is contained in a 2.50-m length of the beam?

49. M V-HINT What fraction of the power radiated by the sun is intercepted by the planet Mercury? The radius of Mercury is 2.44×10^6 m, and its mean distance from the sun is 5.79×10^{10} m. Assume that the sun radiates uniformly in all directions.

50. M SSM MMH The drawing shows an edge-on view of the solar panels on a communications satellite. The dashed line specifies the normal to the panels. Sunlight strikes the panels at an angle θ with respect to the normal. If the solar power impinging on the panels is 2600 W when $\theta = 65°$, what is it when $\theta = 25°$?

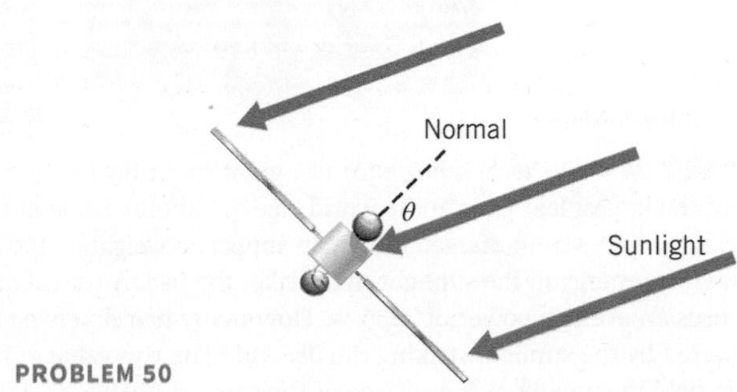

PROBLEM 50

51. M GO SSM Police use radar guns and the Doppler effect to catch speeders. The figure illustrates a moving car approaching a stationary police car. A radar gun emits an electromagnetic wave that reflects from the oncoming car. The reflected wave returns to the police car with a frequency (measured by on-board equipment) that is different from the emitted frequency. One such radar gun emits a wave whose frequency is 8.0×10^9 Hz. When the speed of the car is 39 m/s and the approach is essentially head-on, what is the *difference* between the frequency of the wave returning to the police car and that emitted by the radar gun?

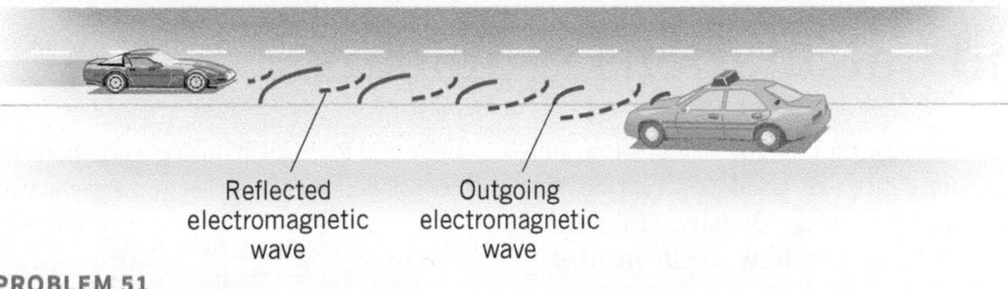

PROBLEM 51

Physics in Biology, Medicine, and Sports

52. E BIO **24.2** Magnetic resonance imaging, or MRI (see Section 21.7), and positron emission tomography, or PET scanning (see Section 32.6), are two medical diagnostic techniques. Both employ electromagnetic waves. For these waves, find the ratio of the MRI wavelength (frequency $= 6.38 \times 10^7$ Hz) to the PET scanning wavelength (frequency $= 1.23 \times 10^{20}$ Hz).

53. M BIO **24.4** A heat lamp emits infrared radiation whose rms electric field is $E_{rms} = 2800$ N/C. **(a)** What is the average intensity of the radiation? **(b)** The radiation is focused on a person's leg over a circular area of radius 4.0 cm. What is the average power delivered to the leg? **(c)** The portion of the leg being irradiated has a mass of 0.28 kg and a specific heat capacity of 3500 J/(kg · C°). How long does it take to raise its temperature by 2.0 C°? Assume that there is no other heat transfer into or out of the portion of the leg being heated.

54. E BIO **24.2** The human eye is most sensitive to light with a frequency of about 5.5×10^{14} Hz, which is in the yellow-green region of the electromagnetic spectrum. How many wavelengths of this light can fit across the width of your thumb, a distance of about 2.0 cm?

55. E BIO 24.2 X-rays are an important component of dental treatment and in evaluating oral health. They can help identify cavities, tooth decay, and impacted teeth. The photo shows a bitewing X-ray, in which the patient bites down on a plastic tab that holds the X-ray film or sensor in place. Here, the X-ray reveals decay inside a tooth that would otherwise go undetected. X-rays are produced in a special device called an X-ray tube (see Section 30.7). (a) What is the speed of the X-rays when they are produced by the tube? (b) If the wavelength of the X-rays is 0.0230 nm, what is their frequency?

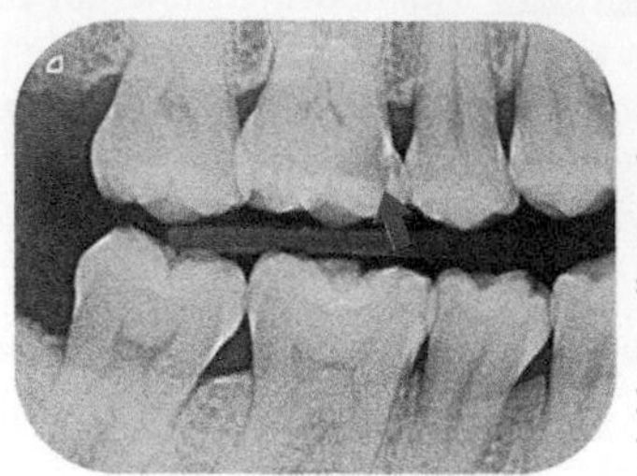

Twinkle Family Dentalcare Pte Ltd. https://www.twinkledental.com.sg/xrays/

PROBLEM 55

56. M BIO 24.4 *Victoria amazonica* is a giant water lily that grows some of the largest leaf pads in the world (see the photo). Some of the pads can be over 3 m in diameter and can support a weight of 100 lb. Consider the energy in the sunlight that strikes the pad. A typical hair dryer uses an average power of 1250 W. How many hair dryers could be powered by the sunlight striking the lily pad? The rms value of the electric field in sunlight at the surface of the earth is 720.0 N/C. Treat the lily pad as a circle with a diameter of 3.20 m.

Stevebidmead/Pixabay

PROBLEM 56

57. M BIO 24.5 Radar guns have many applications in sports. They are used to measure the speed of baseballs, bowling balls, and serves in tennis, as well as the running speed of athletes themselves. A Major League Baseball scout is evaluating a Minor League pitcher by standing behind home plate and measuring the speed of his pitches with a radar gun. The gun emits electromagnetic waves in the microwave part of the EM spectrum at a frequency of 24.25 GHz. Waves that reflect off of a pitched baseball return to the gun with a frequency that has increased by 7.388 kHz. What is the speed of the pitch in mph?

58. E BIO 24.2 Frequencies in the range of 30 to 300 MHz in the radio wave part of the electromagnetic spectrum correspond to the VHF (very high frequency) bands. These frequencies are heavily used by the U.S. Coast Guard (USCG), and the international distress frequency has been chosen to be channel 16, with a frequency of 156.8 MHz. The USCG, as well as most coastal stations, maintains a constant monitoring of this frequency. Cell phones are not a sufficient substitute for a good handheld VHF radio. Cell phone service is unavailable at about 1 mile offshore, whereas VHF radios can transmit and receive signals from ship to ship and ship to shore at distances of more than 20 miles. The VHF signals that are used for marine communications are also vertically polarized, so the transmitting and receiving antennas should be linear dipoles oriented vertically. The simplest form is the quarterwave antenna, in which the length of the antenna is equal to onequarter of the wavelength of the signal. What would be the length of a quarter-wave antenna for the USCG's distress frequency?

59. M BIO 24.2 The human heart creates the body's largest electromagnetic field. During normal beating, an electrocardiogram measures a voltage of 2.7 mV between two electrical contact pads on the chest that are separated by 25 cm. This creates an electromagnetic wave with a frequency of 1.2 Hz. (a) What is the wavelength of this wave? (b) What is the maximum value of the electric field in this wave? (c) What is the maximum value of the magnetic field? (d) What is the average total energy density in the wave?

Concepts and Calculations Problems

Online

One of the central ideas of this chapter is that electromagnetic waves carry energy. Two concepts are used to describe this energy—the wave's intensity and its energy density. Problem 60 reviews this important idea. We have seen how the intensities of completely polarized or completely unpolarized light beams can change when they pass through a polarizer. But what about light that is partially polarized or partially unpolarized? Can the concepts that we discussed in Section 24.6 be applied to such light? The answer is "yes," and Problem 61 illustrates how.

60. M CHALK The figure shows the popular dish antenna that receives digital TV signals from a satellite. The average intensity of the electromagnetic wave that carries a particular TV program is $\overline{S} = 7.5 \times 10^{-14}\ \text{W/m}^2$, and the circular aperture of the antenna has a radius of $r = 15$ cm. *Concepts:* (i) How is the average power passing through the circular aperture of the antenna related to the average intensity of the TV signal? (ii) How much energy does the antenna receive in a time t? (iii) What is the average energy density, or average energy per unit volume, of the electromagnetic wave? *Calculations:* (a) Determine the electromagnetic energy delivered to the dish during a one-hour program. (b) What is the average energy density of the electromagnetic wave?

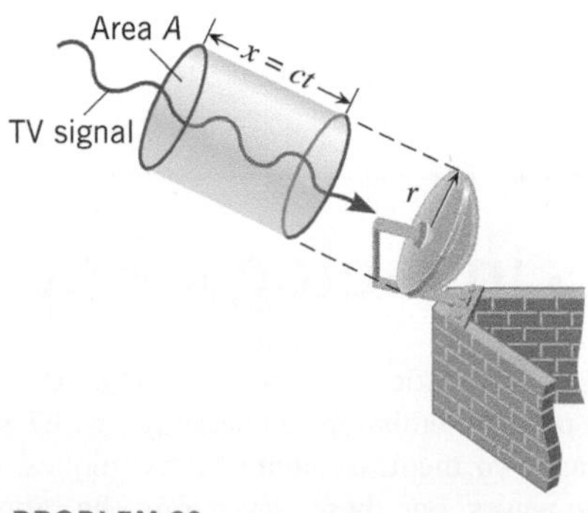

PROBLEM 60

61. M CHALK SSM The light beam in the figure passes through a polarizer whose transmission axis makes an angle ϕ with the vertical. The beam is partially polarized and partially unpolarized, and the

average intensity $\overline{S}_0$ of the incident light is the sum of the average intensity $\overline{S}_{0,\text{ polar}}$ of the polarized light and the average intensity $\overline{S}_{0,\text{ unpolar}}$ of the unpolarized light; $\overline{S}_0 = \overline{S}_{0,\text{ polar}} + \overline{S}_{0,\text{ unpolar}}$. The intensity $\overline{S}$ of the transmitted light is also the sum of two parts: $\overline{S} = \overline{S}_{\text{polar}} + \overline{S}_{\text{unpolar}}$. As the polarizer is rotated clockwise, the intensity of the transmitted light has a minimum value of $\overline{S} = 2.0$ W/m^2 when $\phi = 20.0°$ and has a maximum value of $\overline{S} = 8.0$ W/m^2 when the angle is $\phi = \phi_{\text{max}}$. *Concepts:* (i) How is $\overline{S}_{\text{unpolar}}$ related to $\overline{S}_{0,\text{ unpolar}}$? (ii) How is $\overline{S}_{\text{polar}}$ related to $\overline{S}_{0,\text{ polar}}$? (iii) The minimum transmitted intensity is 2.0 W/m^2. Why isn't it 0 W/m^2? *Calculations:* **(a)** What is the intensity $\overline{S}_{0,\text{ unpolar}}$ of the incident light that is unpolarized? **(b)** What is the intensity $\overline{S}_{0,\text{ polar}}$ of the incident light that is polarized?

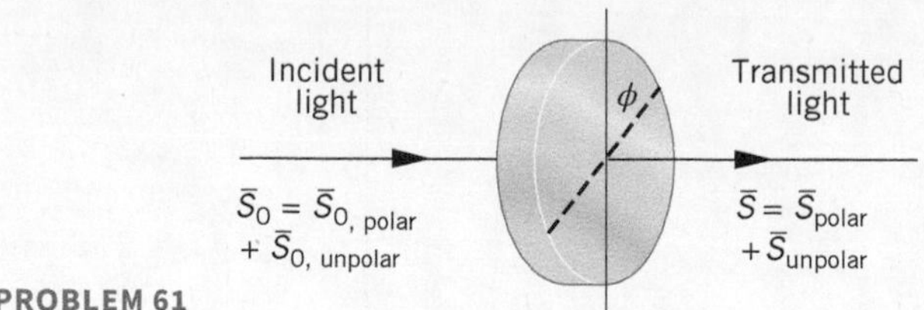

PROBLEM 61

Team Problems

Online

62. E T WS **Laser Power Transmission.** You and your team are assigned to use a new power delivery system to charge the batteries of a spacecraft. The new power station consists of a large solar array that orbits the earth and harnesses solar energy to power a large laser. The laser beam has an average intensity of 2.44×10^6 W/m^2 and a circular cross section with a diameter of 20.5 cm. The power receiver on your ship that converts the light into electrical energy accepts the entire beam, but only has an efficiency of 34.0% (i.e., only 34.0% of the incoming power gets converted to energy stored in the batteries). The maximum total energy storage capacity of the batteries is 5.04×10^7 J. **(a)** What is the average energy density of the beam? **(b)** Assuming that the batteries of your craft are completely depleted, determine how long it will take to charge them.

63. E T WS **A Space Beacon.** You and your team are on a spacecraft far from earth (outside the solar system), and your navigation system has malfunctioned due to an encounter with a cloud of ionized gas. You have unknowingly been going off course for weeks. In addition, the ionized gas damaged your communications systems, so not only do you not know exactly where you are, but you cannot send out a distress signal for help. Fortunately, there is a high power radio beacon orbiting earth that can guide you. However, the only detection equipment on the ship that still works measures the root-mean-square electric field of incoming radio waves. The detector system is highly directional, meaning it has to be aimed directly at the source. The average output power of the beacon is 125 kW, and it radiates isotropically—that is, uniformly in all directions. If the detection device on the ship measures the magnitude of the rms electric field to be 4.84×10^{-9} N/C, how far is your ship from earth?

64. M T **A Turbo-shaft Helicopter.** You and your team are tasked with covertly acquiring information about a new helicopter being developed by an adversary. You are given a handheld device that has many functions, one of which is to send out a broad pulse of electromagnetic waves at a frequency of $f_0 = 9.00 \times 10^9$ Hz, and then collect the scattered light that returns from whatever it hits. The device then analyzes the scattered light and returns a graph of maximum *frequency shift* Δf as a function of collection angle θ (see the drawing). The device also has a built-in rangefinder that measures the distance to an object. As the helicopter engine powers up and the helicopter just starts to lift off the ground, you point the device at it and pull the trigger. It sends out a pulse and, a few seconds later, it gives you (i) the distance to the helicopter ($d = 82.7$ m), and (ii) a graph of the frequency shift Δf versus collection angle θ, as shown in the figure. **(a)** Assuming that light detected by your device has scattered from the moving rotor of the helicopter, what is the diameter of the rotor? **(b)** What is the speed of the tips of the rotor? **(c)** What is the rotational speed (in rpm) of the rotor?

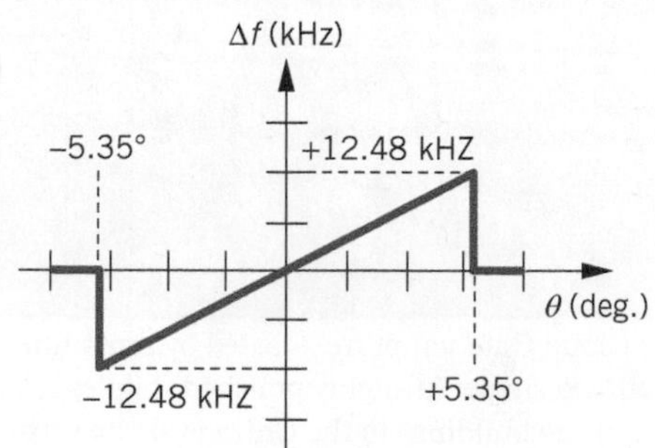

PROBLEM 64

65. M T **A Magneto-optic Device.** The *Faraday effect* is a magneto-optic phenomenon where the polarization of linearly polarized light rotates when the light passes through a magnetized material. You and your team are tasked with analyzing the results of an experiment designed to measure how much a certain thin magnetic film rotates the polarization of the light that passes through it. As the drawing shows, an unpolarized laser beam of average intensity $\overline{S}_0 = 1.00 \times 10^3$ W/m^2 passes through a vertical polarizer. It then passes through a thin magnetic film that can be in one of three states: (i) unmagnetized, (ii) magnetized to the right (when looking from the laser), and (iii) magnetized to the left. It is assumed that the film does not absorb any of the light. The light then passes through another polarizer (called the analyzer) whose transmission axis is rotated 7.50° clockwise from the horizontal (when looking from the laser). **(a)** What is the intensity of the beam after it passes through the first polarizer? **(b)** In the case that the film is unmagnetized, the polarization will not rotate when it passes through the film. What should be the light intensity measured at the detector? **(c)** When the film is magnetized to the left, the intensity measured by the detector is $I_L = 12.2$ W/m^2, and when magnetized to the right the intensity is $I_R = 5.46$ W/m^2. In each case, determine the direction (clockwise or counterclockwise when looking from the laser) and the magnitude of the angle through which the magnetic film rotates the polarization of the light from the vertical.

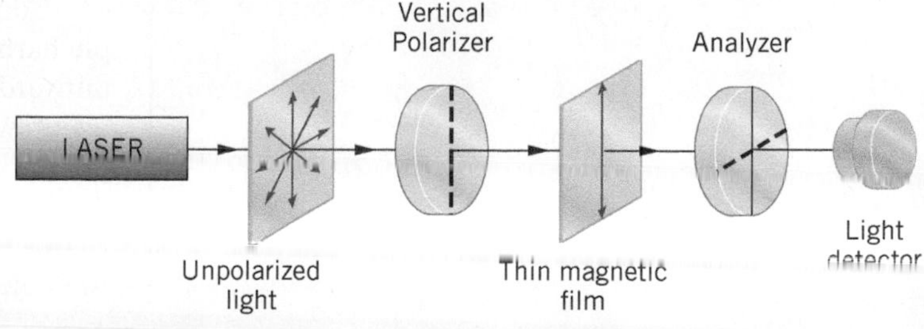

PROBLEM 65

CHAPTER **25**

The Cloud Gate sculpture, created by artist Anish Kapoor, is affectionately known as "The Bean," due to its curved shape. Its highly polished stainless steel surface reflects the light around it, forming an image here of the buildings in the Chicago skyline. The reflection of light and the properties of the images formed by this and other kinds of mirrors will be the focus of this chapter.

rlobes/Pixabay

LEARNING OBJECTIVES

After reading this module, you should be able to...

25.1 Relate wave fronts and rays.

25.2 Apply the law of reflection to plane mirrors.

25.3 Describe image formation by a plane mirror.

25.4 Calculate the focal length of a spherical mirror.

25.5 Perform ray tracing for spherical mirrors.

25.6 Use the mirror and magnification equations to solve problems.

The Reflection of Light: Mirrors

25.1 Wave Fronts and Rays

Mirrors are usually close at hand. It is difficult, for example, to put on makeup, shave, or drive a car without them. We see images in mirrors because some of the light that strikes them is reflected into our eyes. To discuss reflection, it is necessary to introduce the concepts of a wave front and a ray of light, and we can do so by taking advantage of the familiar topic of sound waves (see Chapter 16). Both sound and light are kinds of waves. Sound is a pressure wave, whereas light is electromagnetic in nature. However, the ideas of a wave front and a ray apply to both.

Consider a small spherical object whose surface is pulsating in simple harmonic motion. A sound wave is emitted that moves spherically outward from the object at a constant speed. To represent this wave, we draw surfaces through all points of the wave that are in the same phase of motion. These surfaces of constant phase are called **wave fronts**. **Figure 25.1** shows a hemispherical view of the wave fronts. In this view they appear as concentric spherical shells about the vibrating object. If the wave fronts are drawn through the condensations, or crests, of the sound wave, as they are in the picture, the distance between adjacent wave fronts equals the wavelength λ. The radial lines pointing outward from the source and perpendicular to the wave fronts are called **rays**. The rays point in the direction of the velocity of the wave.

Figure 25.2a shows small sections of two adjacent spherical wave fronts. At large distances from the source, the wave fronts become less and less curved and approach the shape of flat surfaces, as in part *b* of the drawing. Waves whose wave fronts are flat surfaces (i.e., planes) are known as **plane waves** and are important in understanding the properties of mirrors and lenses. Since rays are perpendicular to the wave fronts, the rays for a plane wave are parallel to each other.

The concepts of wave fronts and rays can also be used to describe light waves. For light waves, the ray concept is particularly convenient when showing the path taken by the light. We will make frequent use of light rays, which can be regarded essentially as narrow beams of light much like those that lasers produce.

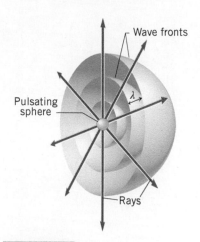

FIGURE 25.1 A hemispherical view of a sound wave emitted by a pulsating sphere. The wave fronts are drawn through the condensations of the wave, so the distance between two successive wave fronts is the wavelength λ. The rays are perpendicular to the wave fronts and point in the direction of the velocity of the wave.

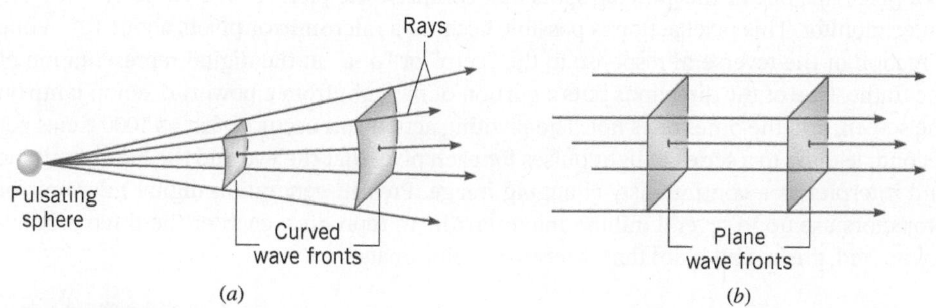

FIGURE 25.2 (*a*) Portions of two spherical wave fronts are shown. The rays are perpendicular to the wave fronts and diverge. (*b*) For a plane wave, the wave fronts are flat surfaces, and the rays are parallel to each other.

25.2 The Reflection of Light

Most objects reflect a certain portion of the light falling on them. Suppose that a ray of light is incident on a flat, shiny surface, such as the mirror in **Figure 25.3**. As the drawing shows, the *angle of incidence* θ_i is the angle that the incident ray makes with respect to the normal, which is a line drawn perpendicular to the surface at the point of incidence. The *angle of reflection* θ_r is the angle that the reflected ray makes with the normal. The **law of reflection** describes the behavior of the incident and reflected rays.

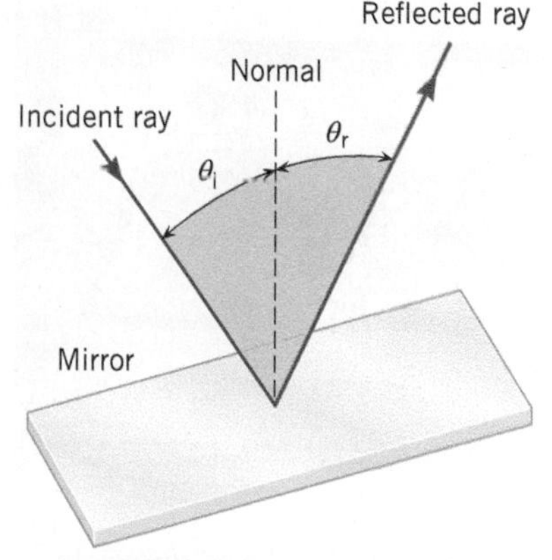

FIGURE 25.3 The angle of reflection θ_r equals the angle of incidence θ_i. These angles are measured with respect to the normal, which is a line drawn perpendicular to the surface of the mirror at the point of incidence.

LAW OF REFLECTION

The incident ray, the reflected ray, and the normal to the surface all lie in the same plane, and the angle of reflection θ_r equals the angle of incidence θ_i:

$$\theta_r = \theta_i$$

When parallel light rays strike a smooth, plane surface, such as the ones in **Figure 25.4a**, the reflected rays are parallel to each other. This type of reflection is one example of what is known as *specular reflection* and is important in determining the properties of mirrors. Most

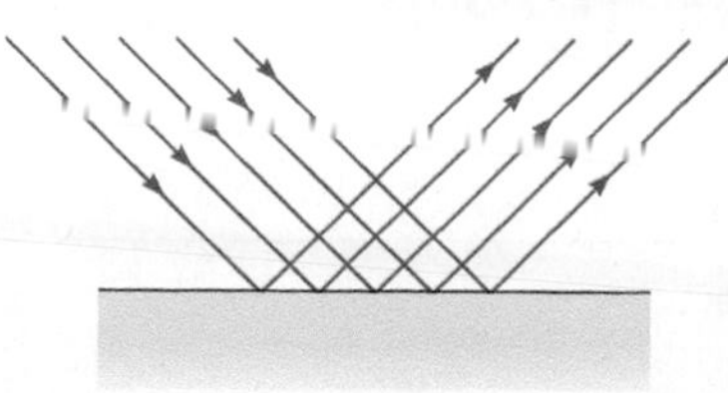

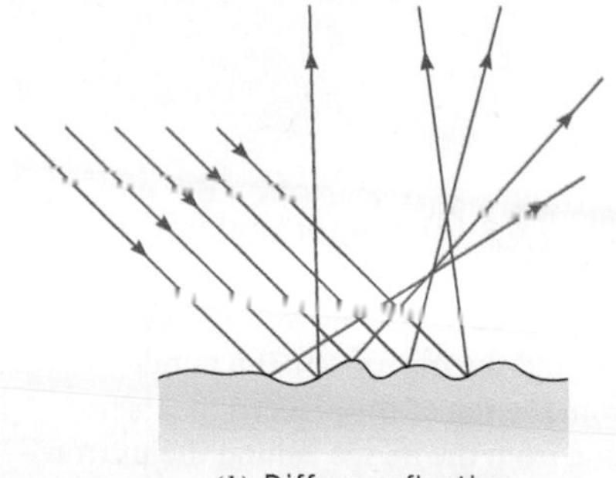

FIGURE 25.4 (*a*) The drawing shows specular reflection from a polished plane surface, such as a mirror. The reflected rays are parallel to each other. (*b*) A rough surface reflects the light rays in all directions; this type of reflection is known as diffuse reflection.

surfaces, however, are not perfectly smooth, because they contain irregularities the sizes of which are equal to or greater than the wavelength of the light. The law of reflection applies to each ray, but the irregular surface reflects the light rays in various directions, as **Figure 25.4b** suggests. This type of reflection is known as *diffuse reflection*. Common surfaces that give rise to diffuse reflection are most papers, wood, nonpolished metals, and walls covered with a "flat" (nongloss) paint.

THE PHYSICS OF . . . digital movie projectors and micromirrors. A revolution in digital technology is occurring in the movie industry, where digital techniques are now being used to produce films. Until recently, films have been viewed primarily by using projectors that shine light directly through a strip of film containing the images. Now, however, projectors are available that allow a movie produced using digital techniques to be viewed completely without film by using a digital representation (zeros and ones) of the images. One form of digital projector depends on the law of reflection and tiny mirrors called micromirrors, each about the size of one-fourth the diameter of a human hair. Each micromirror creates a tiny portion of an individual movie frame on the screen and serves as a pixel, like one of the glowing spots that comprise the picture on a TV screen or computer monitor. This pixel action is possible because a micromirror pivots about 10° in one direction or the reverse in response to the "zero" or "one" in the digital representation of the frame. One of the directions puts a portion of the light from a powerful xenon lamp on the screen, and the other does not. The pivoting action can occur as fast as 1000 times per second, leading to a series of light pulses for each pixel that the eye and the brain combine and interpret as a continuously changing image. Present-generation digital micromirror projectors use up to several million micromirrors to reproduce each of the three primary colors (red, green, and blue) that comprise a color image.

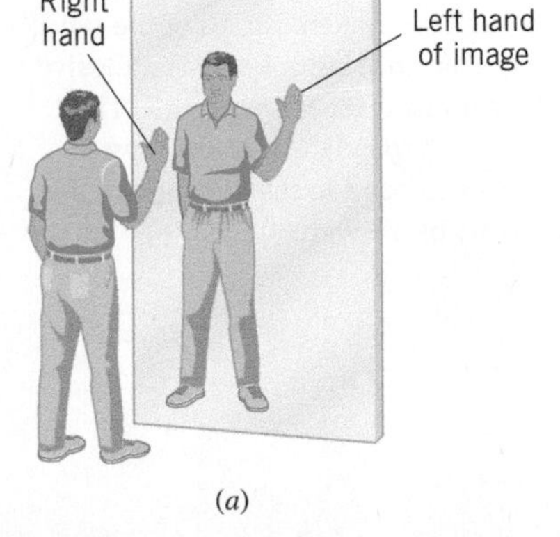

(*a*)

Dennis MacDonald/Age Fotostock

(*b*)

FIGURE 25.5 (*a*) The person's right hand becomes the image's left hand in a plane mirror. (*b*) Many emergency vehicles are reverse-lettered so the lettering appears normal when viewed in the rearview mirror of a car.

25.3 The Formation of Images by a Plane Mirror

When you look into a plane (flat) mirror, you see an image of yourself that has three properties:

1. The image is upright.
2. The image is the same size as you are.
3. The image is located as far behind the mirror as you are in front of it.

As **Figure 25.5a** illustrates, the image of yourself in the mirror is also reversed right to left and left to right. If you wave your *right* hand, it is the *left* hand of the image that waves back. Similarly, letters and words held up to a mirror are reversed. Ambulances and other emergency vehicles are often lettered in reverse, as in **Figure 25.5b**, so that the letters will appear normal when seen in the rearview mirror of a car.

To illustrate why an image appears to originate from behind a plane mirror, **Figure 25.6a** shows a light ray leaving the top of an object. This ray reflects from the mirror (angle of reflection equals angle of incidence) and enters the eye. To the eye, it appears that the ray originates from behind the mirror, somewhere back along the dashed line. Actually, rays going in all directions leave each point on the object, but only a small bundle of such rays is intercepted by the

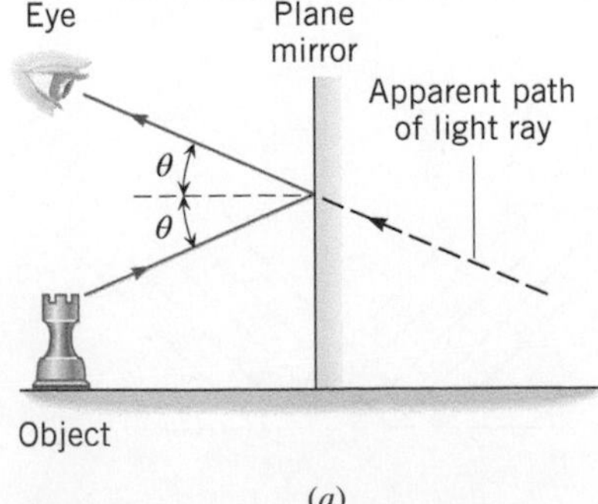

(*a*)

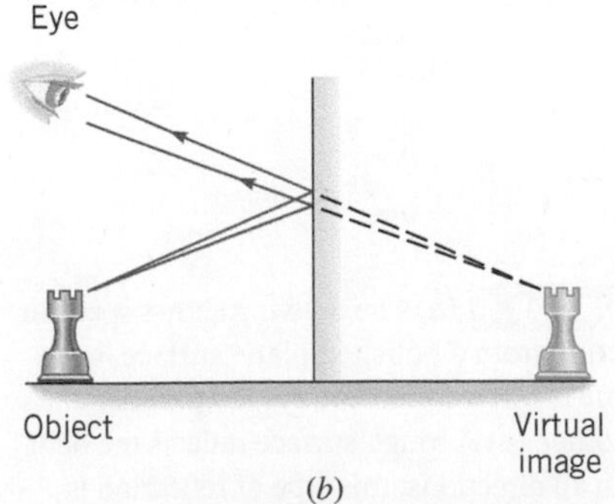

(*b*)

FIGURE 25.6 (*a*) A ray of light from the top of the chess piece reflects from the mirror. To the eye, the ray seems to come from behind the mirror. (*b*) The bundle of rays from the top of the object appears to originate from the image behind the mirror.

eye. Part *b* of the figure shows a bundle of two rays leaving the top of the object. All the rays that leave a given point on the object, no matter what angle θ they have when they strike the mirror, appear to originate from a corresponding point on the image behind the mirror (see the dashed lines in part *b*). For each point on the object, there is a single corresponding point on the image, and it is this fact that makes the image in a plane mirror a sharp and undistorted one.

Although rays of light *seem* to come from the image, it is evident from **Figure 25.6b** that they do not originate from behind the plane mirror where the image appears to be. Because none of the light rays actually emanate from the image, it is called a **virtual image**. In this text the parts of the light rays that appear to come from a virtual image are represented by dashed lines. *Curved* mirrors, on the other hand, can produce images from which all the light rays actually do emanate. Such images are known as **real images** and are discussed later.

With the aid of the law of reflection, it is possible to show that the image is located as far behind a plane mirror as the object is in front of it. In **Figure 25.7** the object distance is d_o and the image distance is d_i. A ray of light leaves the base of the object, strikes the mirror at an angle of incidence θ, and is reflected at the same angle. To the eye, this ray appears to come from the base of the image. For the angles β_1 and β_2 in the drawing it follows that $\theta + \beta_1 = 90°$ and $\alpha + \beta_2 = 90°$. But the angle α is equal to the angle of reflection θ, since the two are opposite angles formed by intersecting lines. Therefore, $\beta_1 = \beta_2$. As a result, triangles *ABC* and *DBC* are identical (congruent) because they share a common side *BC* and have equal angles ($\beta_1 = \beta_2$) at the top and equal angles (90°) at the base. Thus, the magnitude of the object distance d_o equals the magnitude of the image distance d_i.

By starting with a light ray from the top of the object, rather than the bottom, we can use the same line of reasoning to show also that the height of the image equals the height of the object.

Conceptual Examples 1 and 2 discuss some interesting features of plane mirrors.

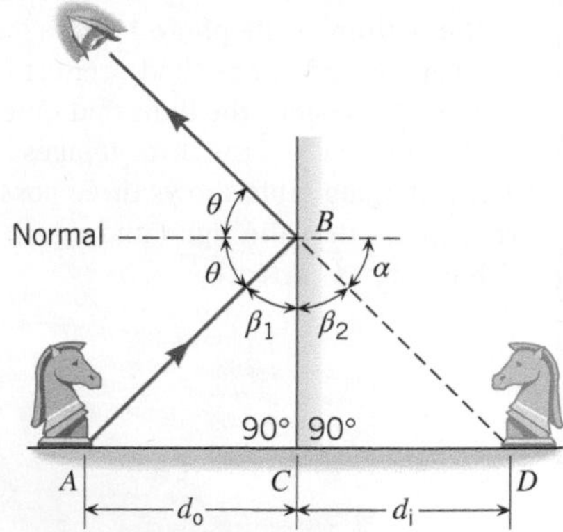

FIGURE 25.7 This drawing illustrates the geometry used with a plane mirror to show that the image distance d_i equals the object distance d_o.

CONCEPTUAL EXAMPLE 1 | Full-Length Versus Half-Length Mirrors

In **Figure 25.8** a woman is standing in front of a plane mirror. Is the minimum mirror height that is necessary for her to see her full image **(a)** equal to her height, or **(b)** equal to one-half her height?

Reasoning The woman sees her image because light emanating from her body is reflected by the mirror (labeled *ABCD* in **Figure 25.8**) and enters her eyes. Consider, for example, a ray of light from her foot *F*. This ray strikes the mirror at *B* and enters her eyes at *E*. According to the law of reflection, the angles of incidence and reflection are both θ. This law will allow us to deduce how the height of the mirror is related to her own height.

Answer (a) is incorrect. The mirror in **Figure 25.8** is the same height as the woman. Any light from her foot that strikes the mirror below *B* is reflected toward a point on her body that is below her eyes. Since light striking the mirror below *B* does not enter her eyes, the part of the mirror between *B* and *A* may be removed. Thus, the necessary minimum height of the mirror is not equal to the woman's height.

Answer (b) is correct. As discussed above, the section *AB* of the mirror is not necessary in order for the woman to see her full image. The section *BC* of the mirror that produces the image is one-half the woman's height between *F* and *E*. This follows because the right triangles *FBM* and *EBM* are identical. They are identical because they share a common side *BM* and have two angles, θ and 90°, that are the same. The blowup in **Figure 25.8** illustrates a similar line of reasoning, starting with a ray from the woman's head at *H*. This ray is reflected from the mirror at *P* and enters her eyes. The top mirror section *PD* may be removed without disturbing this reflection. The necessary section *CP* is one-half the woman's height between her head at *H* and her eyes at *E*. We find, then, that only the sections *BC* and *CP* are needed for the woman to see her full length. The height of section *BC* plus section *CP* is exactly one-half the woman's height. The conclusions here are valid regardless of how far the person stands from the mirror. Thus, to view one's full length in a mirror, only a half-length mirror is needed.

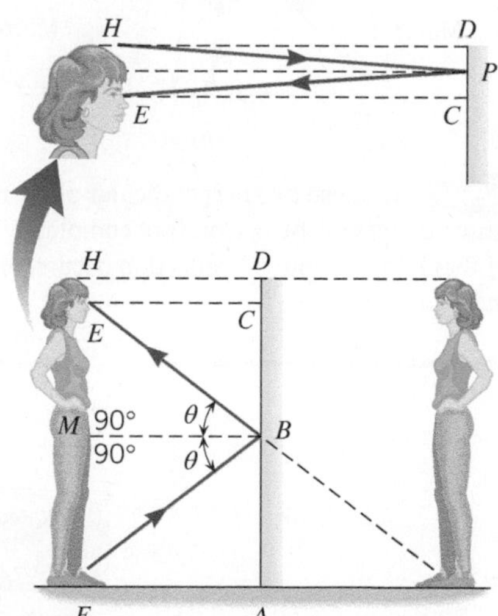

FIGURE 25.8 A woman stands in front of a plane mirror and sees her full image.

Related Homework: Problem 34

CONCEPTUAL EXAMPLE 2 | Multiple Reflections

A person is sitting in front of two mirrors that intersect at a 90° angle. As **Figure 25.9a** illustrates, the person sees three images of herself. (The person herself is only partially visible at the bottom of the photo.) These images arise because rays of light emanate from her body, reflect from the mirrors, and enter her eyes. Consider the light that enters her eyes and appears to come from each of the three images identified in **Figure 25.9b**. The following table shows three possibilities for the number of reflections that the light undergoes before entering her eyes. Which one is correct?

(*a*)

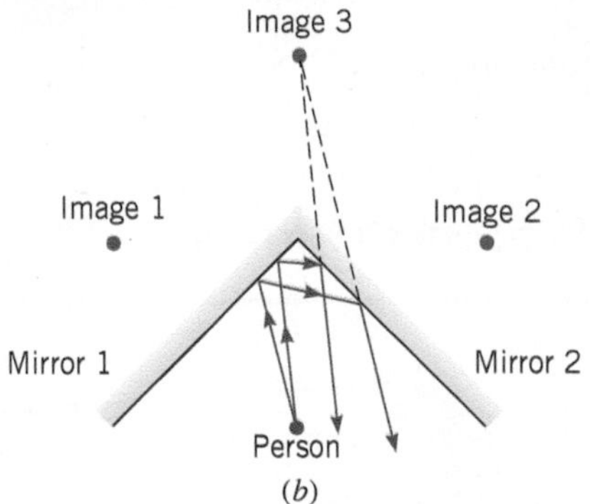

(*b*)

FIGURE 25.9 (*a*) These two perpendicular plane mirrors produce three images of the person (not completely visible) sitting in front of them. (*b*) A "double" reflection occurs, one from each mirror, and produces Image 3.

	Number of Reflections		
Possibility	**Image 1**	**Image 2**	**Image 3**
(a)	2	2	3
(b)	3	3	3
(c)	1	1	2

Reasoning Images of the woman are formed when light emanating from her body enters her eyes after being reflected by one, or both, mirrors. For each reflection, the angle of the light reflected from a mirror is equal to the angle of the light incident on the mirror (law of reflection). We will see that there are three ways that light can reach her eyes from the two mirrors.

Answers (a) and (b) are incorrect. **Figure 25.9b** represents a top view of the person in front of the two mirrors. It is a straightforward matter to understand two of the images that she sees. These are the images that are normally seen when one sits in front of a mirror. Sitting in front of mirror 1, she sees image 1, which is located as far behind that mirror as she is in front of it. She also sees image 2 behind mirror 2, at a distance that matches her distance in front of that mirror. Each of these images arises from light emanating from her body and reflecting only once from a single mirror. Therefore, each ray of light does not reflect two or three times before entering her eyes.

Answer (c) is correct. As discussed above, images 1 and 2 arise, respectively, from *single* reflections from mirrors 1 and 2. The third image arises when light undergoes two reflections in sequence, first from one mirror and then from the other. When such a double reflection occurs, an additional image becomes possible. **Figure 25.9b** shows two rays of light that strike mirror 1. Each one, according to the law of reflection, has an angle of reflection that equals the angle of incidence. The rays then strike mirror 2, where they again are reflected according to the law of reflection. When the outgoing rays are extended backward (see the dashed lines in the drawing), they intersect and appear to originate from image 3. Thus, the third image arises when an incident ray of light is reflected twice, once from each mirror, before entering her eyes.

Check Your Understanding

(*The answers are given at the end of the book.*)

1. **CYU Figure 25.1** shows a light ray undergoing multiple reflections from a mirrored corridor. The walls of the corridor are either parallel or perpendicular to one another. If the initial angle of incidence is 35°, what is the angle of reflection when the ray makes its last reflection?

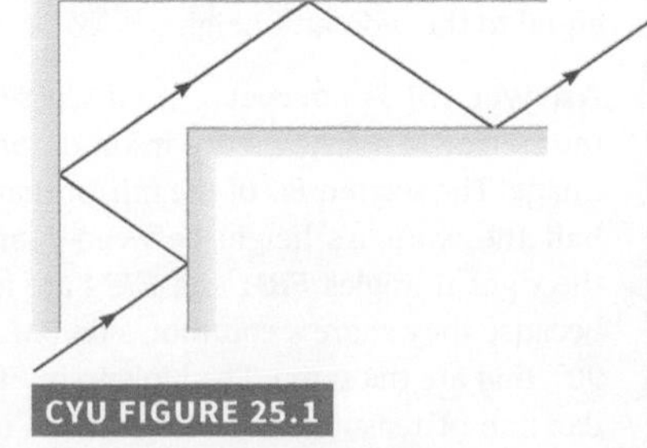

CYU FIGURE 25.1

2. A sign painted on a store window is reversed when viewed from inside the store. If a person inside the store views the reversed sign in a plane mirror, does the sign appear as it would when viewed from outside the store? (Try it by writing some letters on a transparent sheet of paper and then holding the back side of the paper up to a mirror.)

3. If a clock is held in front of a mirror, its image is reversed left to right. From the point of view of a person looking into the mirror, does the image of the second hand rotate in the reverse (counterclockwise) direction?

25.4 Spherical Mirrors

The most common type of curved mirror is a spherical mirror. As **Interactive Figure 25.10** shows, a spherical mirror has the shape of a section from the surface of a hollow sphere. If the inside surface of the mirror is polished, it is a **concave mirror**. If the outside surface is polished, it is a **convex mirror**. The drawing shows both types of mirrors, with a light ray reflecting from the polished surface. The law of reflection applies, just as it does for a plane mirror. For either type of spherical mirror, the normal is drawn perpendicular to the mirror at the point of incidence. For each type, the center of curvature is located at point *C*, and the radius of curvature is *R*. The **principal axis** of the mirror is a straight line drawn through *C* and the midpoint of the mirror.

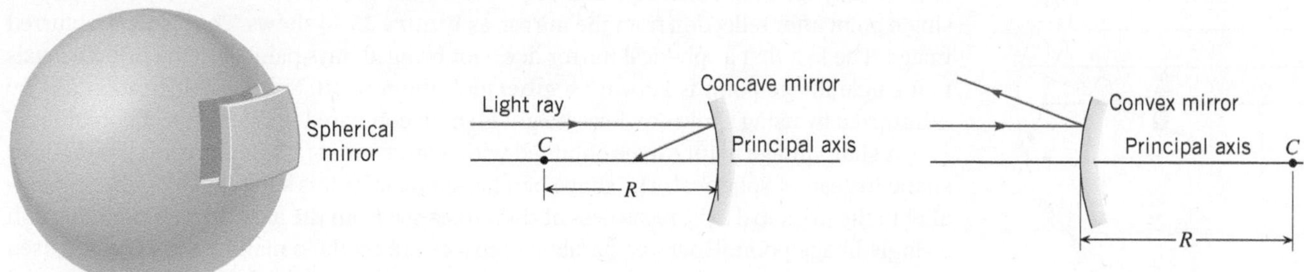

INTERACTIVE FIGURE 25.10 A spherical mirror has the shape of a segment of a spherical surface. The center of curvature is point *C* and the radius is *R*. For a concave mirror, the reflecting surface is the inner one; for a convex mirror it is the outer one.

Figure 25.11 shows a tree in front of a concave mirror. A point on this tree lies on the principal axis of the mirror and is beyond the center of curvature *C*. Light rays emanate from this point and reflect from the mirror, consistent with the law of reflection. If the rays are near the principal axis, they cross it at a common point after reflection. This point is called the *image point*. The rays continue to diverge from the image point as if there were an object there. Since light rays actually come from the image point, the image is a real image.

If the tree in **Figure 25.11** is infinitely far from the mirror, the rays are parallel to each other and to the principal axis as they approach the mirror. **Figure 25.12** shows rays near and parallel to the principal axis, as they reflect from the mirror and pass

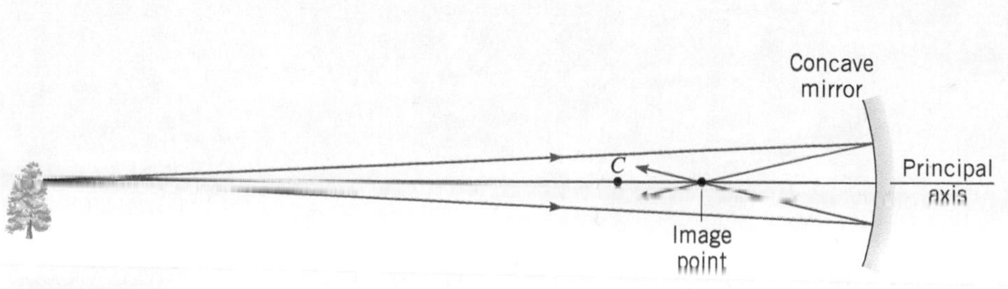

FIGURE 25.11 A point on the tree lies on the principal axis of the concave mirror. Rays from this point that are near the principal axis are reflected from the mirror and cross the axis at the image point.

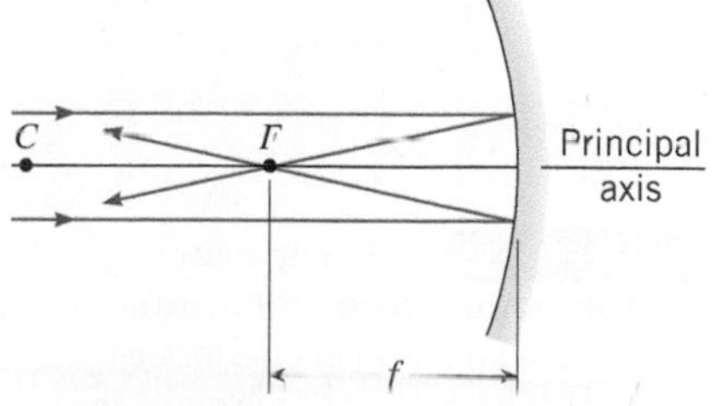

FIGURE 25.12 Light rays near and parallel to the principal axis are reflected from a concave mirror and converge at the focal point *F*. The focal length *f* is the distance between *F* and the mirror.

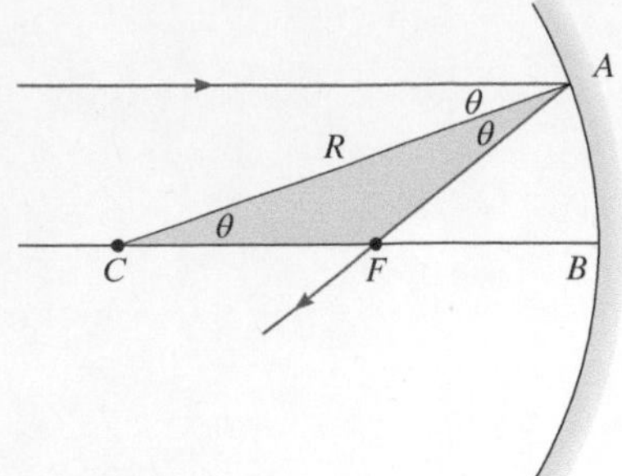

FIGURE 25.13 This drawing is used to show that the focal point F of a concave mirror is halfway between the center of curvature C and the mirror at point B.

through an image point. In this special case the image point is referred to as the **focal point** F of the mirror. Therefore, an object infinitely far away on the principal axis gives rise to an image at the focal point of the mirror. The distance between the focal point and the middle of the mirror is the **focal length** f of the mirror.

We can show that the focal point F lies halfway between the center of curvature C and the middle of a concave mirror. In **Figure 25.13**, a light ray parallel to the principal axis strikes the mirror at point A. The line CA is the radius of the mirror and, therefore, is the normal to the spherical surface at the point of incidence. The ray reflects from the mirror, and the angle of reflection θ equals the angle of incidence. Furthermore, the angle ACF is also θ because the radial line CA is a transversal of two parallel lines. Since two of its angles are equal, the colored triangle CAF is an isosceles triangle; thus, sides CF and FA are equal. However, when the incoming ray lies close to the principal axis, the angle of incidence θ is small, and the distance FA does not differ appreciably from the distance FB. Therefore, in the limit that θ is small, $CF = FA = FB$, and so the focal point F lies halfway between the center of curvature and the mirror. In other words, the focal length f is one-half of the radius R:

Focal length of a concave mirror

$$f = \tfrac{1}{2}R \qquad \textbf{(25.1)}$$

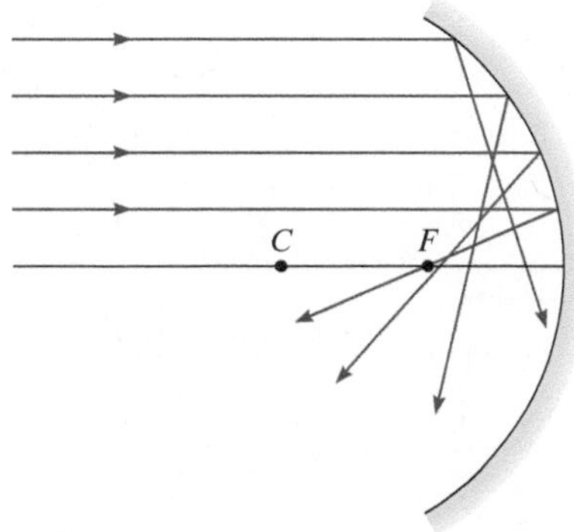

FIGURE 25.14 Rays that are farthest from the principal axis have the greatest angle of incidence and miss the focal point F after reflection from the mirror.

Rays that lie close to the principal axis are known as **paraxial rays**,* and Equation 25.1 is valid only for such rays. Rays that are far from the principal axis do not converge to a single point after reflection from the mirror, as **Figure 25.14** shows. The result is a blurred image. The fact that a spherical mirror does not bring all rays parallel to the principal axis to a single image point is known as **spherical aberration**. Spherical aberration can be minimized by using a mirror whose height is small compared to the radius of curvature.

A sharp image point can be obtained with a large mirror, if the mirror is parabolic in shape instead of spherical. The shape of a parabolic mirror is such that all light rays parallel to the principal axis, regardless of their distance from the axis, are reflected through a single image point. However, parabolic mirrors are costly to manufacture and are used where the sharpest images are required, as in research-quality telescopes.

THE PHYSICS OF . . . capturing solar energy with mirrors. Parabolic mirrors are also used in one method of collecting solar energy for commercial purposes. **Figure 25.15** shows a long row of concave parabolic mirrors that reflect the sun's rays to the focal point. Located at the focal point and running the length of the row is an

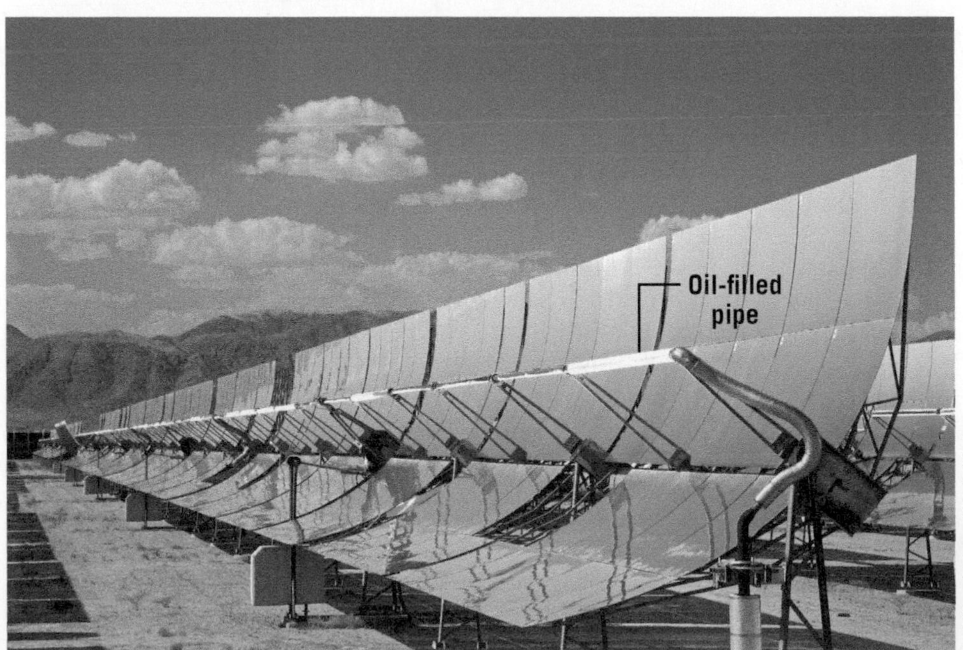

Jim West / Alamy Stock Photo

FIGURE 25.15 This long row of parabolic mirrors focuses the sun's rays to heat an oil-filled pipe located at the focal point of each mirror. Many such rows are used by a solar-thermal electric plant in the Mojave Desert.

*Paraxial rays are close to the principal axis but not necessarily parallel to it.

oil-filled pipe. The focused rays of the sun heat the oil. In a solar-thermal electric plant, the heat from many such rows is used to generate steam. The steam, in turn, drives a turbine connected to an electric generator.

THE PHYSICS OF . . . automobile headlights. Another application of parabolic mirrors is in automobile headlights. Here, however, the situation is reversed from the operation of a solar collector. In a headlight, a high-intensity light source is placed at the focal point of the mirror, and light emerges parallel to the principal axis.

A convex mirror also has a focal point, and **Figure 25.16** illustrates its meaning. In this picture, parallel rays are incident on a convex mirror. Clearly, the rays diverge after being reflected. If the incident parallel rays are paraxial, the reflected rays seem to come from a single point F behind the mirror. This point is the focal point of the convex mirror, and its distance from the midpoint of the mirror is the **focal length** f. The focal length of a convex mirror is also one-half of the radius of curvature, just as it is for a concave mirror. However, we assign the focal length of a convex mirror a negative value because it will be convenient later on:

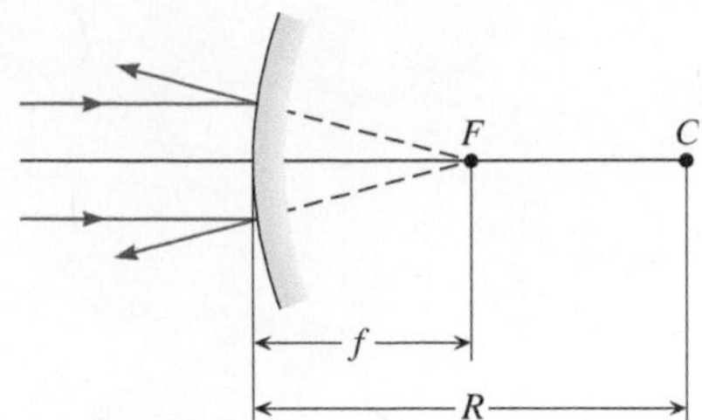

FIGURE 25.16 When paraxial light rays that are parallel to the principal axis strike a convex mirror, the reflected rays appear to originate from the focal point F. The radius of curvature is R and the focal length is f.

Focal length of a convex mirror

$$f = -\tfrac{1}{2}R \tag{25.2}$$

Spherical aberration is a problem with convex mirrors, just as it is with concave mirrors. Rays that emanate from a single point on an object but are far from the principal axis do not appear to originate from a single image point after reflection from the mirror. As with a concave mirror, the result is a blurred image.

Check Your Understanding

(*The answers are given at the end of the book.*)

4. A section of the surface of a hollow sphere has a radius of curvature of 0.60 m, and both the inside and outside surfaces have a mirror-like polish. What are the focal lengths of the inside and outside surfaces?

5. **Photo 25.1** shows an experimental device at Sandia National Laboratories in New Mexico. This device is a mirror that focuses sunlight to heat sodium to a boil, which then heats helium gas in an engine. The engine does the work of driving a generator to produce electricity. The sodium unit and the engine are labeled in the photo. **(a)** What kind of mirror, concave or convex, is being used? **(b)** Where is the sodium unit located relative to the mirror? Express your answer in terms of the focal length of the mirror.

PHOTO 25.1

6. Refer to **Figure 25.14** and the related discussion about spherical aberration. To bring the top ray closer to the focal point F after reflection, describe how you would change the shape of the mirror. Would you open it up to produce a more gently curving shape or bring the top and bottom edges closer to the principal axis?

25.5 The Formation of Images by Spherical Mirrors

As we have seen, some of the light rays emitted from an object in front of a mirror strike the mirror, reflect from it, and form an image. We can analyze the image produced by either concave or convex mirrors by using a graphical method called **ray tracing**. This method is based on the law of reflection and the notion that a spherical mirror has a center of curvature C and a focal point F. Ray tracing enables us to find the location of the image, as well as its size, by taking advantage of the following fact: paraxial rays leave from a point on the object and intersect, or appear to intersect, at a corresponding point on the image after reflection.

Concave Mirrors

Three specific paraxial rays are especially convenient to use in the ray-tracing method. Interactive Figure 25.17 shows an object in front of a concave mirror, and these three rays leave from a point on the top of the object. The rays are labeled 1, 2, and 3, and when tracing their paths, we use the following reasoning strategy.

REASONING STRATEGY Ray Tracing for a Concave Mirror

Ray 1. **This ray is initially parallel to the principal axis and, therefore, passes through the focal point F after reflection from the mirror.**

Ray 2. **This ray initially passes through the focal point F and is reflected parallel to the principal axis. Ray 2 is analogous to ray 1 except that the reflected, rather than the incident, ray is parallel to the principal axis.**

Ray 3. **This ray travels along a line that passes through the center of curvature C and follows a radius of the spherical mirror; as a result, the ray strikes the mirror perpendicularly and reflects back on itself.**

If rays 1, 2, and 3 are superimposed on a scale drawing, they converge at a point on the top of the image, as can be seen in Animated Figure 25.18*a*.* Although three rays have been used here to locate the image, only two are really needed; the third ray is usually drawn to serve as a check. In a similar fashion, rays from all other points on the object locate corresponding points on the image, and the mirror forms a complete image of the object. If you were to place your eye as shown in the drawing, you would see an image that is *larger* and *inverted* relative to the object. The image is *real* because the light rays actually pass through the image.

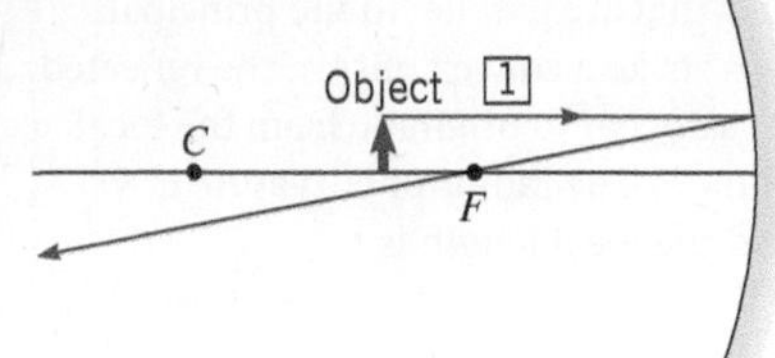

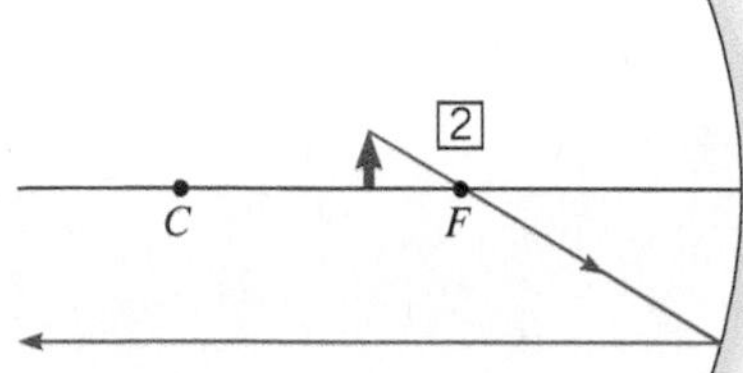

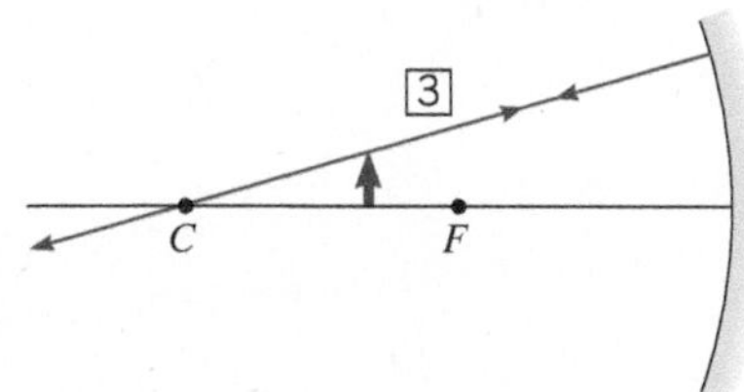

INTERACTIVE FIGURE 25.17 The rays labeled 1, 2, and 3 are useful in locating the image of an object placed in front of a concave spherical mirror. The object is represented as a vertical arrow.

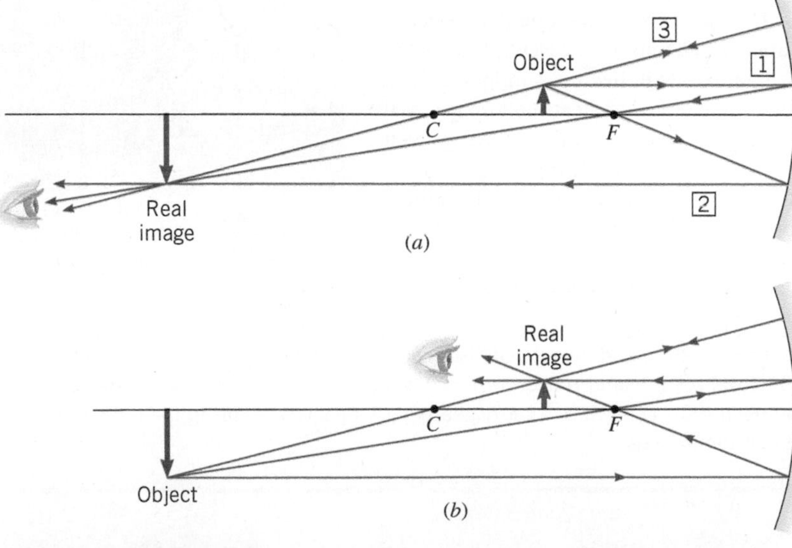

ANIMATED FIGURE 25.18 (*a*) When an object is placed between the focal point F and the center of curvature C of a concave mirror, a real image is formed. The image is enlarged and inverted relative to the object. (*b*) When the object is located beyond the center of curvature C, a real image is created that is reduced in size and inverted relative to the object.

*In the drawings that follow, we assume that the rays are paraxial, although the distance between the rays and the principal axis is often exaggerated for clarity.

If the locations of the object and image in **Animated Figure 25.18a** are interchanged, the situation in part *b* of the drawing results. The three rays in part *b* are the same as those in part *a*, except that the directions are reversed. These drawings illustrate the **principle of reversibility**, which states the following:

> **Problem-Solving Insight** **If the direction of a light ray is reversed, the light retraces its original path.**

This principle is quite general and is not restricted to reflection from mirrors. The image is *real,* and it is *smaller* and *inverted* relative to the object.

When the object is placed between the focal point *F* and a concave mirror, as in **Figure 25.19a**, three rays can again be drawn to find the image. Now, however, ray 2 does not go through the focal point on its way to the mirror, since the object is closer to the mirror than the focal point is. When projected backward, though, ray 2 appears to come from the focal point. Therefore, after reflection, ray 2 is directed parallel to the principal axis. In this case the three reflected rays diverge from each other and do not converge to a common point. However, when projected behind the mirror, the three rays appear to come from a common point; thus, a *virtual* image is formed. This virtual image is *larger* than the object and *upright.*

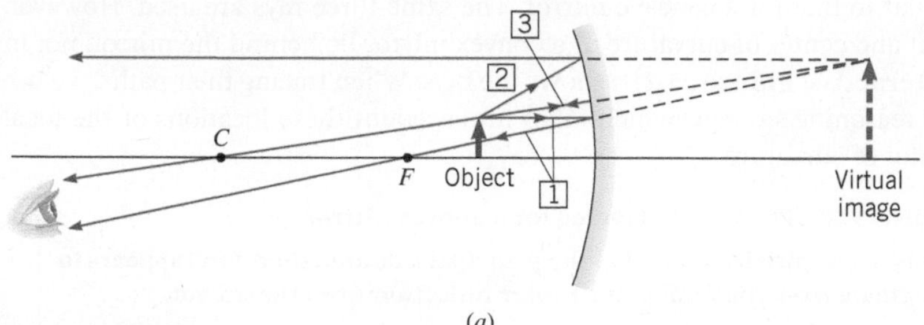

(*a*)

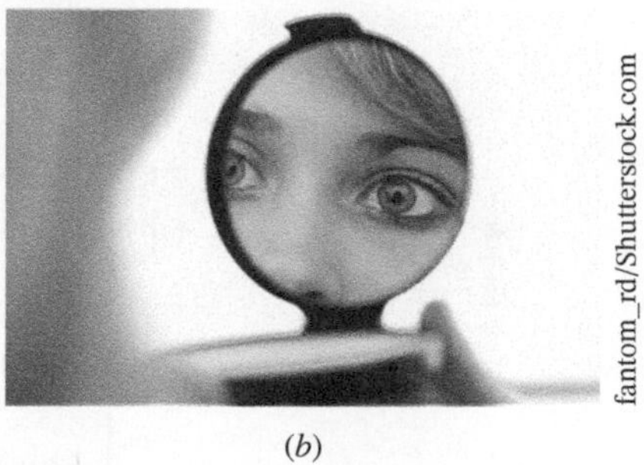

(*b*)

FIGURE 25.19 (*a*) When an object is located between a concave mirror and its focal point *F*, an enlarged, upright, and virtual image is produced. (*b*) A makeup mirror (or shaving mirror) is a concave mirror that functions in exactly this fashion, as this photograph shows.

THE PHYSICS OF . . . makeup and shaving mirrors. Makeup and shaving mirrors are concave mirrors. When you place your face between the mirror and its focal point, you see an enlarged virtual image of yourself, as **Figure 25.19b** shows.

THE PHYSICS OF . . . a head-up display for automobiles. Concave mirrors are also used in one method for displaying the speed of a car. The method presents a digital readout (e.g., "51 km/h") that the driver sees when looking directly through the windshield, as in **Figure 25.20a**. The advantage of the method, which is called a head-up display (HUD), is that the driver does not need to take his or her eyes off the

(*a*)

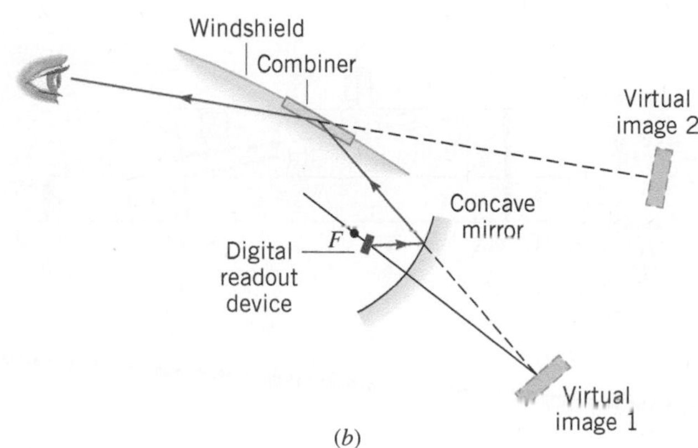

(*b*)

FIGURE 25.20 (*a*) A head-up display (HUD) presents the driver with a digital readout of the car's speed in the field of view seen through the windshield. (*b*) One version of a HUD uses a concave mirror. (See the text for explanation.)

road to monitor the speed. **Figure 25.20*b*** shows how one type of HUD works. Below the windshield is a readout device that displays the speed in digital form. This device is located between a concave mirror and its focal point. The arrangement is similar to the one in **Figure 25.19*a*** and produces a virtual, upright, and enlarged image of the speed readout (see virtual image 1 in **Figure 25.20*b***). Light rays that appear to come from this image strike the windshield at a place where a so-called "combiner" is located. The purpose of the combiner is to combine the digital readout information with the field of view that the driver sees through the windshield. The combiner is virtually undetectable by the driver because it allows all colors except one to pass through it unaffected. The one exception is the color produced by the digital readout device. For this color, the combiner behaves as a plane mirror and reflects the light that appears to originate from image 1. Thus, the combiner produces image 2, which is what the driver sees. The location of image 2 is out above the front bumper. The driver can then read the speed with eyes focused just as they are to see the road.

Convex Mirrors

The ray-tracing procedure for determining the location and size of an image in a convex mirror is similar to that for a concave mirror. The same three rays are used. However, the focal point and center of curvature of a convex mirror lie behind the mirror, not in front of it. **Interactive Figure 25.21*a*** shows the rays. When tracing their paths, we use the following reasoning strategy, which takes into account these locations of the focal point and center of curvature.

REASONING STRATEGY **Ray Tracing for a Convex Mirror**

Ray 1. **This ray is initially parallel to the principal axis and, therefore, appears to originate from the focal point *F* after reflection from the mirror.**

Ray 2. **This ray heads toward *F*, emerging parallel to the principal axis after reflection. Ray 2 is analogous to ray 1, except that the reflected, rather than the incident, ray is parallel to the principal axis.**

Ray 3. **This ray travels toward the center of curvature *C*; as a result, the ray strikes the mirror perpendicularly and reflects back on itself.**

The three rays in **Interactive Figure 25.21*a*** appear to come from a point on a *virtual* image that is behind the mirror. The virtual image is *diminished in size* and *upright,* relative to the object. A convex mirror *always* forms a virtual image of the object, no matter where in front of the mirror the object is placed. **Interactive Figure 25.21*b*** shows an example of such an image.

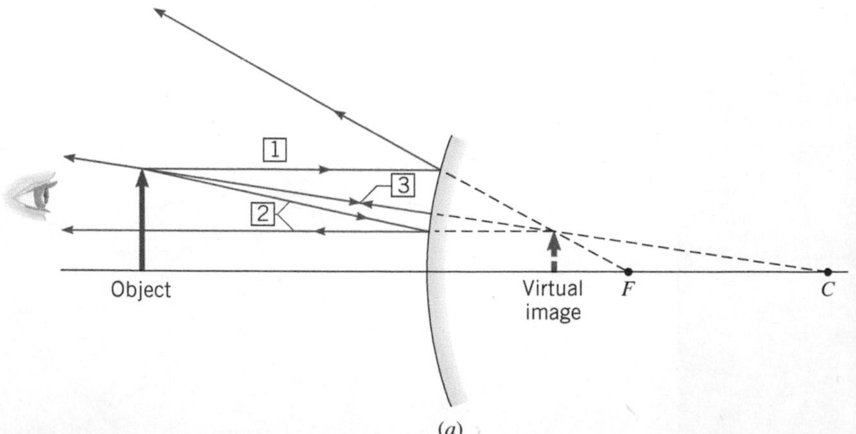

(*a*)

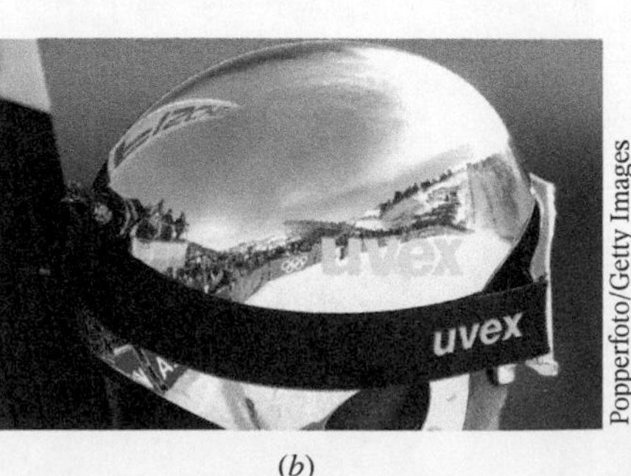

Popperfoto/Getty Images

(*b*)

INTERACTIVE FIGURE 25.21 (*a*) An object placed in front of a convex mirror always produces a virtual image behind the mirror; the image is reduced in size and is upright. (*b*) The skier's chromed helmet acts as a convex mirror and produces an image of the spectators who are watching the event.

THE PHYSICS OF . . . passenger-side automobile mirrors. Because of its shape, a convex mirror gives a wider field of view than do other types of mirrors. A mirror with a wide field of view is needed to give a driver a good rear view. Thus, the outside mirror on the passenger side is often a convex mirror. Printed on such a mirror is usually the warning "OBJECTS IN MIRROR ARE CLOSER THAN THEY APPEAR." The reason for the warning is that, as in **Interactive Figure 25.21*a***, the virtual image is reduced in size and therefore looks smaller, just as a distant object would appear in a plane mirror. An unwary driver, thinking that the side-view mirror is a plane mirror, might incorrectly deduce from the small size of the image that the car behind is far enough away to ignore. Because of their wide field of view, convex mirrors are also used in stores for security purposes.

Check Your Understanding

(The answers are given at the end of the book.)

7. Is it possible to use a convex mirror to produce an image that is larger than the object? Explain.

8. **(a)** When you look at the back side of a shiny teaspoon held at arm's length, do you see yourself upright or upside down? **(b)** When you look at the other side of the spoon, do you see yourself upright or upside down? Assume in both cases that the distance between you and the spoon is greater than the focal length of the spoon.

9. **(a)** Can the image formed by a concave mirror ever be projected directly onto a screen without the help of other mirrors or lenses? If so, specify where the object should be placed relative to the mirror. **(b)** Repeat part (a) assuming that the mirror is convex.

10. Suppose that you stand in front of a spherical mirror (concave or convex). Is it possible for your image to be **(a)** real and upright **(b)** virtual and inverted?

11. An object is placed between the focal point and the center of curvature of a concave mirror. The object is then moved closer to the mirror, but still remains between the focal point and the center of curvature. Do the magnitudes of **(a)** the image distance and **(b)** the image height become larger or smaller?

12. When you see the image of yourself formed by a mirror, it is because (1) light rays actually coming from a real image enter your eyes or (2) light rays appearing to come from a virtual image enter your eyes. If light rays from the image do not enter your eyes, you do not see yourself. Are there any places on the principal axis where you cannot see yourself when you are standing in front of a mirror that is **(a)** convex **(b)** concave? If so, where are these places? Assume that you have only the one mirror to use.

25.6 The Mirror Equation and the Magnification Equation

Ray diagrams drawn to scale are useful for determining the location and size of the image formed by a mirror. However, for an accurate description of the image, a more analytical technique is needed, so we will derive two equations, known as the **mirror equation** and the **magnification equation**. These equations are based on the law of reflection and provide relationships between:

f = the focal length of the mirror
d_o = the object distance, which is the distance between the object and the mirror
d_i = the image distance, which is the distance between the image and the mirror
m = the magnification of the mirror, which is the ratio of the height of the image to the height of the object.

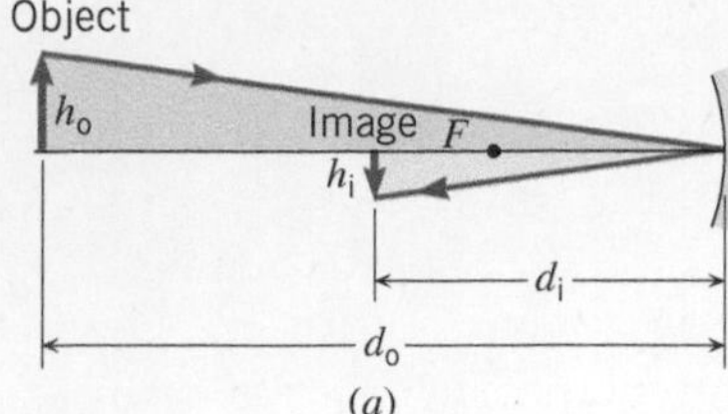

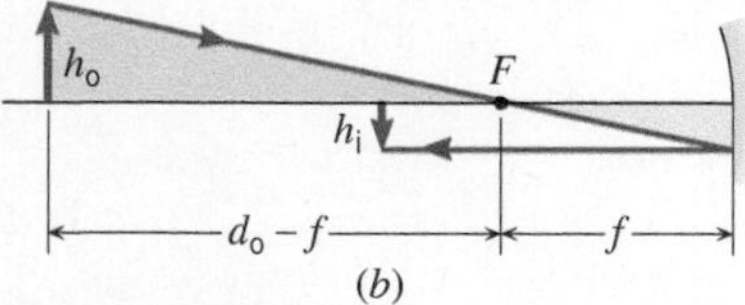

FIGURE 25.22 These diagrams are used to derive the mirror equation and the magnification equation. (*a*) The two colored triangles are similar triangles. (*b*) If the ray is close to the principal axis, the two colored regions are almost similar triangles.

Concave Mirrors

We begin our derivation of the mirror equation by referring to **Figure 25.22*a***, which shows a ray leaving the top of the object and striking a concave mirror at the point where the principal axis intersects the mirror. Since the principal axis is perpendicular to the mirror, it is also the normal at this point of incidence. Therefore, the ray reflects at an equal angle and passes through the image. The two colored triangles are similar triangles because they have equal angles, so

$$\frac{h_o}{-h_i} = \frac{d_o}{d_i}$$

where h_o is the height of the object and h_i is the height of the image. The minus sign appears on the left in this equation because the image is inverted in **Figure 25.22*a***. In part *b* another ray leaves the top of the object, this one passing through the focal point *F*, reflecting parallel to the principal axis, and then passing through the image. Provided the ray remains close to the axis, the two colored areas can be considered to be similar triangles, with the result that

$$\frac{h_o}{-h_i} = \frac{d_o - f}{f}$$

Setting the two equations above equal to each other yields $d_o/d_i = (d_o - f)/f$. Rearranging this result gives the **mirror equation**:

Mirror equation

$$\frac{1}{d_o} + \frac{1}{d_i} = \frac{1}{f} \tag{25.3}$$

We have derived this equation for a real image formed in front of a concave mirror. In this case, the image distance is a positive quantity, as are the object distance and the focal length. However, we have seen in the last section that a concave mirror can also form a virtual image, if the object is located between the focal point and the mirror. Equation 25.3 can also be applied to such a situation, provided that we adopt the convention that d_i is negative for an image behind the mirror, as it is for a virtual image.

In deriving the magnification equation, we remember that the **magnification *m*** of a mirror is the ratio of the image height to the object height: $m = h_i/h_o$. If the image height is less than the object height, the magnitude of *m* is less than one, and if the image is larger than the object, the magnitude of *m* is greater than one. We have already shown that $h_o/(-h_i) = d_o/d_i$, so it follows that

Magnification equation

$$m = \frac{\text{Image height, } h_i}{\text{Object height, } h_o} = -\frac{d_i}{d_o} \tag{25.4}$$

As Examples 3 and 4 show, the value of *m* is negative if the image is inverted and positive if the image is upright.

EXAMPLE 3 | A Real Image Formed by a Concave Mirror

A 2.0-cm-high object is placed 7.10 cm from a concave mirror whose radius of curvature is 10.20 cm. Find **(a)** the location of the image and **(b)** its size.

Reasoning For a concave mirror, Equation 25.1 gives the focal length as $f = \frac{1}{2}R$. Therefore, the focal length is $f = \frac{1}{2}(10.20 \text{ cm}) = 5.10 \text{ cm}$, and the object is located between the focal point *F* and the center of curvature *C* of the mirror, as in **Animated Figure 25.18*a***. Based on this figure, we expect that the image is real and that, relative to the object, it is farther away from the mirror, inverted, and larger.

Problem-Solving Insight **According to the mirror equation, the image distance d_i has a reciprocal given by $d_i^{-1} = f^{-1} - d_o^{-1}$. After combining the reciprocals f^{-1} and d_o^{-1}, do not forget to take the reciprocal of the result to find d_i.**

Solution **(a)** With $d_o = 7.10$ cm and $f = 5.10$ cm, the mirror equation (Equation 25.3) can be used to find the image distance:

$$\frac{1}{d_i} = \frac{1}{f} - \frac{1}{d_o} = \frac{1}{5.10 \text{ cm}} - \frac{1}{7.10 \text{ cm}} = 0.055 \text{ cm}^{-1} \quad \text{or} \quad \boxed{d_i = 18 \text{ cm}}$$

In this calculation, f and d_o are positive numbers, indicating that the focal point and the object are in front of the mirror. The positive answer for d_i means that the image is also in front of the mirror, and the reflected rays actually pass through the image, as **Animated Figure 25.18*a*** shows. In other words, the positive value for d_i indicates that the image is a real image.

(b) According to the magnification equation (Equation 25.4), the image height h_i is related to the object height h_o and the magnification m by $h_i = mh_o$, where $m = -d_i/d_o$. Thus, we find that

$$h_i = -\left(\frac{d_i}{d_o}\right)h_o = -\left(\frac{18\ \text{cm}}{7.10\ \text{cm}}\right)(2.0\ \text{cm}) = \boxed{-5.1\ \text{cm}}$$

The image is larger than the object, and the negative value for h_i indicates that the image is inverted with respect to the object, as in **Animated Figure 25.18*a***.

EXAMPLE 4 | A Virtual Image Formed by a Concave Mirror

An object is placed 6.00 cm in front of a concave mirror that has a 10.0-cm focal length. **(a)** Determine the location of the image. **(b)** The object is 1.2 cm high. Find the image height.

Reasoning The object is located between the focal point and the mirror, as in **Figure 25.19*a***. The setup is analogous to a person using a makeup or shaving mirror. Therefore, we expect that the image is virtual and that, relative to the object, it is upright and larger.

Solution **(a)** Using the mirror equation with $d_o = 6.00$ cm and $f = 10.0$ cm, we have

$$\frac{1}{d_i} = \frac{1}{f} - \frac{1}{d_o} = \frac{1}{10.0\ \text{cm}} - \frac{1}{6.00\ \text{cm}}$$

$$= -0.067\ \text{cm}^{-1} \quad \text{or} \quad \boxed{d_i = -15\ \text{cm}}$$

The answer for d_i is negative, indicating that the image is *behind* the mirror. Thus, as expected, the image is virtual.

(b) The image height h_i can be found from the magnification equation, which indicates that $h_i = mh_o$, where h_o is the object height and $m = -d_i/d_o$. It follows, then, that

$$h_i = -\left(\frac{d_i}{d_o}\right)h_o = -\left(\frac{-15\ \text{cm}}{6.00\ \text{cm}}\right)(1.2\ \text{cm}) = \boxed{3.0\ \text{cm}}$$

The image is larger than the object, and the positive value for h_i indicates that the image is upright (see **Figure 25.19*a***).

There are many applications of concave mirrors. As the following example shows, the James Webb Space Telescope will use one of the largest.

EXAMPLE 5 | BIO The Physics of Searching for Extraterrestrial Life

The James Webb Space Telescope (JWST) will be the largest and most powerful telescope ever launched into space. The successor to the Hubble Space Telescope (see Examples 10 and 11, Chapter 5), the JWST will expand on Hubble's discoveries and further improve our fundamental understanding of the universe. Both Hubble and the JWST are similar in how they operate. They each have a large primary mirror that reflects the incoming light off a smaller secondary mirror, which then directs the light back through the center of the primary mirror and focuses it at a point behind the mirror. This focal point coincides with the location of a suite of scientific instruments that analyzes the light (see **Figure 25.23*a***). While the two telescopes operate in a similar fashion, the JWST is specifically designed to view the universe using longer wavelength infrared light. This makes it particularly well-suited for analyzing the atmospheres of new *exoplanets*, which are planets that exist outside the solar system. (At the time of this writing, researchers have discovered over 4395 exoplanets in 3242 star systems.) Molecules in the atmosphere of an exoplanet that might be conducive to life, such as oxygen, water vapor, and methane, have a large number of spectral features, or signatures, when viewed in the infrared, which makes the JWST the perfect instrument for this type of research. Even though the telescope will utilize wavelengths of light outside of the visible spectrum, the law of reflection still governs how the light will reflect from the surface of its mirrors. The primary mirror of the JWST is composed of an array of 18 hexagonal mirrors that are each 1.3 m in width, making the overall diameter of the primary mirror 6.5 m. This is over 2 ½ times larger in diameter than Hubble's primary mirror (see **Figure 25.23*b***). At a temperature of 30 K, which will be close to the operating temperature of the telescope in deep space, the radius of curvature of the primary mirror is 15.880 m. At what distance from the primary mirror should the secondary mirror be placed, if light from a faraway object is to be focused at that location?

Reasoning The primary mirror of the telescope can be treated as a concave spherical mirror. Since the radius of curvature of the mirror is known, we can calculate its focal length using Equation 25.1. Applying the mirror equation (Equation 25.3) shows that, for an object located at infinity, the image distance will correspond to the focal length.

Solution Applying Equation 25.1, we find the focal length of the primary mirror to be

$$f = \frac{R}{2} = \frac{15.880\ \text{m}}{2} = 7.9400\ \text{m}$$

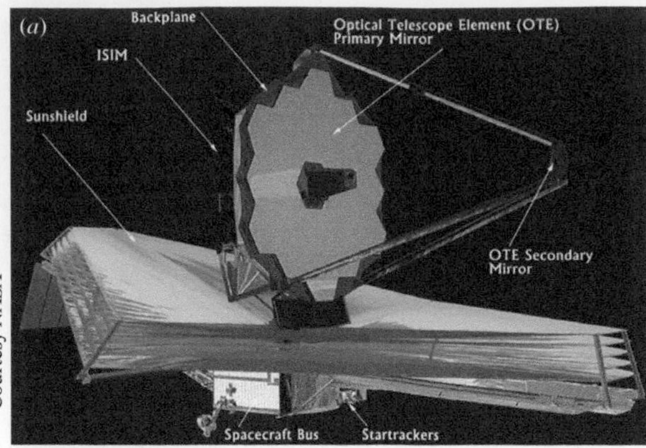

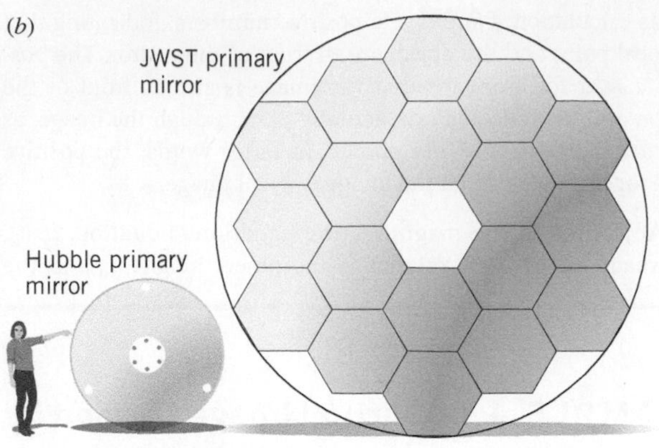

FIGURE 25.23 (*a*) The geometry of the James Webb Space Telescope (JWST). The large primary mirror consists of an array of 18 individual hexagonal mirrors coated with beryllium and gold. Light from the primary mirror reflects off a secondary mirror and back toward analysis equipment at the center of the primary. (*b*) A relative comparison of the size of the primary mirrors in the Hubble Space Telescope and the JWST.

For a faraway object, like a distant galaxy, the object distance d_o will equal infinity. Applying Equation 25.3 and solving for d_i, we find

$$\frac{1}{d_i} = \frac{1}{f} - \frac{1}{d_o} = \frac{1}{f} - \frac{1}{\infty} = \frac{1}{f} - 0 \Rightarrow d_i = f = \boxed{7.9400 \text{ m}}$$

The actual position of the secondary mirror is about 7.32 m from the primary mirror, somewhat closer than the calculated result. Thus, the focal point of the light from the primary is located behind the secondary. The secondary intercepts the light from the primary and reflects it back toward the instruments that reside behind the center of the primary mirror.

Convex Mirrors

The mirror equation and the magnification equation can also be used with convex mirrors, provided the focal length f is taken to be a *negative number,* as indicated explicitly in Equation 25.2. One way to remember this is to recall that the focal point of a convex mirror lies *behind* the mirror. Example 6 deals with a convex mirror.

EXAMPLE 6 | A Virtual Image Formed by a Convex Mirror

A convex mirror is used to reflect light from an object placed 66 cm in front of the mirror. The focal length of the mirror is $f = -46$ cm (note the minus sign). Find **(a)** the location of the image and **(b)** the magnification.

Reasoning We have seen that a convex mirror always forms a virtual image, as in **Interactive Figure 25.21*a***, where the image is upright and smaller than the object. These characteristics should also be indicated by the results of our analysis here.

> **Problem-Solving Insight When using the mirror equation, it is useful to construct a ray diagram to guide your thinking and to check your calculation.**

Solution **(a)** With $d_o = 66$ cm and $f = -46$ cm, the mirror equation gives

$$\frac{1}{d_i} = \frac{1}{f} - \frac{1}{d_o} = \frac{1}{-46 \text{ cm}} - \frac{1}{66 \text{ cm}} = -0.037 \text{ cm}^{-1} \quad \text{or} \quad \boxed{d_i = -27 \text{ cm}}$$

The negative sign for d_i indicates that the image is behind the mirror and, therefore, is a virtual image.

(b) According to the magnification equation, the magnification is

$$m = -\frac{d_i}{d_o} = -\frac{(-27 \text{ cm})}{66 \text{ cm}} = \boxed{0.41}$$

The image is smaller (m is less than one) and upright (m is positive) with respect to the object.

Convex mirrors, like plane (flat) mirrors, always produce virtual images behind the mirror. However, the virtual image in a convex mirror is closer to the mirror than it would be if the mirror were planar, as Example 7 illustrates.

EXAMPLE 7 | A Convex Versus a Plane Mirror

An object is placed 9.00 cm in front of a mirror. The image is 3.00 cm closer to the mirror when the mirror is convex than when it is planar (see **Figure 25.24**). Find the focal length of the convex mirror.

Reasoning For a plane mirror, the image and the object are the same distance on either side of the mirror. Thus, the image would be 9.00 cm behind a plane mirror. If the image in a convex mirror is 3.00 cm closer than this, the image must be located 6.00 cm behind the convex mirror. In other words, when the object distance is $d_o = 9.00$ cm, the image distance for the convex mirror is $d_i = -6.00$ cm (negative because the image is virtual). The mirror equation can be used to find the focal length of the mirror.

Solution According to the mirror equation, the reciprocal of the focal length is

$$\frac{1}{f} = \frac{1}{d_o} + \frac{1}{d_i} = \frac{1}{9.00\text{ cm}} + \frac{1}{-6.00\text{ cm}} = -0.056\text{ cm}^{-1} \quad \text{or} \quad \boxed{f = -18\text{ cm}}$$

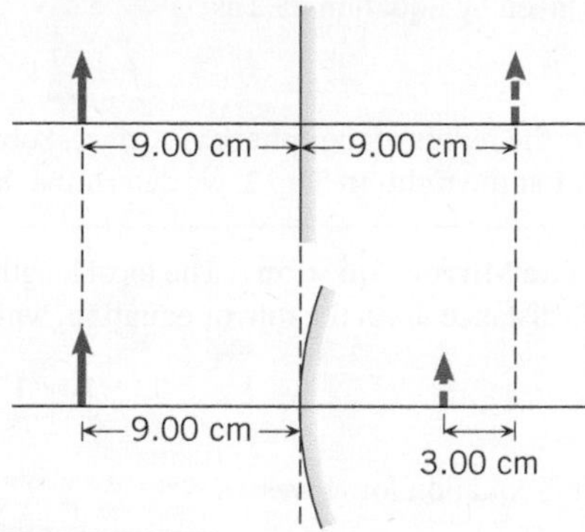

FIGURE 25.24 The object distance (9.00 cm) is the same for the plane mirror (top part of drawing) as for the convex mirror (bottom part of drawing). However, as discussed in Example 7, the image formed by the convex mirror is 3.00 cm closer to the mirror.

Contact lenses are worn to correct vision problems. Optometrists take advantage of the mirror equation and the magnification equation in providing lenses that fit the patient's eyes properly, as the next example illustrates.

Analyzing Multiple-Concept Problems

EXAMPLE 8 | BIO The Physics of Keratometers

A contact lens rests against the cornea of the eye. **Figure 25.25** shows an optometrist using a keratometer to measure the radius of curvature of the cornea, thereby ensuring that the prescribed lenses fit accurately. In the keratometer, light from an illuminated object reflects from the corneal surface, which acts like a convex mirror and forms an upright virtual image that is smaller than the object (see **Interactive Figure 25.21*a***). With the object placed 9.0 cm in front of the cornea, the magnification of the corneal surface is measured to be 0.046. Determine the radius of the cornea.

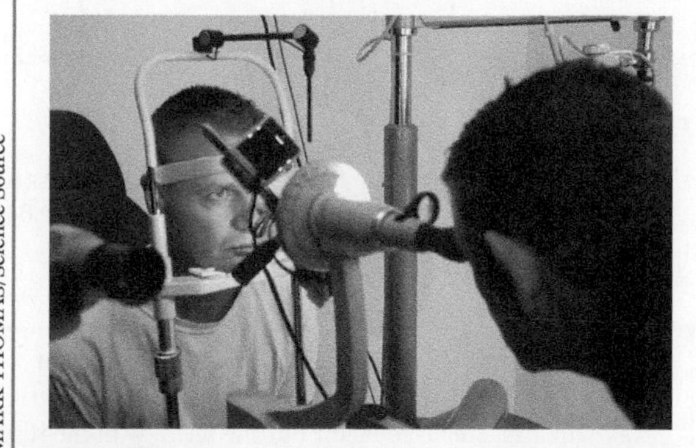

MARK THOMAS/Science Source

FIGURE 25.25 An optometrist is using a keratometer to measure the radius of curvature of the cornea of the eye, which is the surface against which a contact lens rests.

Reasoning The radius of a convex mirror can be determined from the mirror's focal length, since the two are related. The focal length is related to the distances of the object and its image from the mirror via the mirror equation. The magnification of the mirror is also related to the object and image distances according to the magnification equation. By using the mirror equation and the magnification equation, we will be able to determine the focal length and, hence, the radius.

Knowns and Unknowns The following table summarizes the available data:

Description	Symbol	Value	Comment
Object distance	d_o	9.0 cm	Distance of object from cornea
Magnification of corneal surface	m	0.046	Cornea acts like a convex mirror and forms a virtual image.
Unknown Variable			
Radius of cornea	R	?	

Modeling the Problem

STEP 1 Relation Between Radius and Focal Length The focal length f of a convex mirror is given by Equation 25.2 as

$$f = -\tfrac{1}{2}R$$

where R is the radius of the spherical surface. Solving this expression for the radius gives Equation 1 at the right. In Step 2, we determine the unknown focal length.

$$R = -2f \quad (1)$$

$$\boxed{?}$$

STEP 2 The Mirror Equation The focal length is related to the object distance d_o and the image distance d_i via the mirror equation, which specifies that

$$\frac{1}{f} = \frac{1}{d_o} + \frac{1}{d_i} \quad (25.3)$$

Solving this equation for f gives

$$\boxed{f = \left(\frac{1}{d_o} + \frac{1}{d_i}\right)^{-1}}$$

which can be substituted into Equation 1 as shown in the right column. A value for d_o is given in the data table, and we turn to Step 3 to determine a value for d_i.

$$R = -2f \quad (1)$$

$$\boxed{f = \left(\frac{1}{d_o} + \frac{1}{d_i}\right)^{-1}} \quad (2)$$

$$\boxed{?}$$

STEP 3 The Magnification Equation According to the magnification equation, the magnification m is given by

$$m = -\frac{d_i}{d_o} \quad (25.4)$$

Solving for d_i, we obtain

$$\boxed{d_i = -md_o}$$

and can substitute this result into Equation 2, as shown at the right.

$$R = -2f \quad (1)$$

$$\boxed{f = \left(\frac{1}{d_o} + \frac{1}{d_i}\right)^{-1}} \quad (2)$$

$$\boxed{d_i = -md_o}$$

Solution Combining the results of each step algebraically, we find that

$$\underset{\text{STEP 1}}{R = -2f} \underset{\text{STEP 2}}{=} -2\left(\frac{1}{d_o} + \frac{1}{d_i}\right)^{-1} \underset{\text{STEP 3}}{=} -2\left[\frac{1}{d_o} + \frac{1}{(-md_o)}\right]^{-1}$$

Thus, the radius is

$$R = -2\left[\frac{1}{d_o} + \frac{1}{(-md_o)}\right]^{-1} = \frac{2d_o m}{1-m} = \frac{2(9.0\text{ cm})(0.046)}{1-0.046} = \boxed{0.87\text{ cm}}$$

Related Homework: Problem 29

Math Skills To show that the radius R is $R = \frac{2d_o m}{1-m}$, we proceed in the following way. The first step is to factor out the term $\frac{1}{d_o}$ in the result for R:

$$R = -2\left[\frac{1}{d_o} + \frac{1}{(-md_o)}\right]^{-1} = -2\left[\frac{1}{d_o}\left(1 - \frac{1}{m}\right)\right]^{-1}$$

Rearranging the term within the brackets is the next step:

$$R = -2\left[\frac{1}{d_o}\left(1 - \frac{1}{m}\right)\right]^{-1} = -2\left[\frac{1}{d_o}\left(\frac{m}{m} - \frac{1}{m}\right)\right]^{-1} = -2\left(\frac{m-1}{d_o m}\right)^{-1}$$

Finally, taking the reciprocal of the term within the parentheses shows that

$$R = -2\left(\frac{m-1}{d_o m}\right)^{-1} = -2\left(\frac{d_o m}{m-1}\right) = \frac{2d_o m}{1-m}$$

The following Reasoning Strategy summarizes the sign conventions that are used with the mirror equation and the magnification equation. These conventions apply to both concave and convex mirrors.

REASONING STRATEGY Summary of Sign Conventions for Spherical Mirrors

Focal length

f is + for a concave mirror.

f is − for a convex mirror.

Object distance

d_o is + if the object is in front of the mirror (real object).

d_o is − if the object is behind the mirror (virtual object).*

Image distance

d_i is + if the image is in front of the mirror (real image).

d_i is − if the image is behind the mirror (virtual image).

Magnification

m is + for an image that is upright with respect to the object.

m is − for an image that is inverted with respect to the object.

Check Your Understanding

(*The answers are given at the end of the book.*)

13. An object is placed in front of a spherical mirror, and the magnification of the system is $m = -6$. What does this number tell you about the image? (Select one or more of the following choices.) **(a)** The image is larger than the object. **(b)** The image is smaller than the object. **(c)** The image is upright relative to the object. **(d)** The image is inverted relative to the object. **(e)** The image is a real image. **(f)** The image is a virtual image.

14. Plane mirrors and convex mirrors form virtual images. With a plane mirror, the image may be infinitely far behind the mirror, depending on where the object is located in front of the mirror. For an object in front of a single convex mirror, what is the greatest distance behind the mirror at which the image can be found?

EXAMPLE 9 | BIO The Physics of a Head Mirror

A head mirror is a simple medical diagnostic tool that is used to illuminate a patient's ear, throat, and nasal passages. It consists of a concave spherical mirror with a small hole drilled in the center (**Figure 25.26**). The physician places the mirror over one

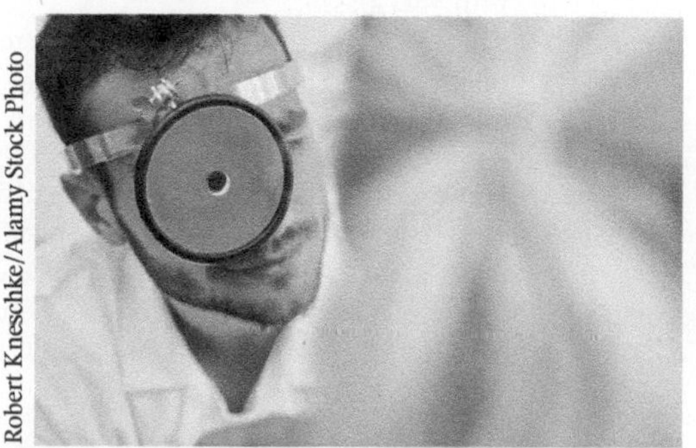
Robert Kneschke/Alamy Stock Photo

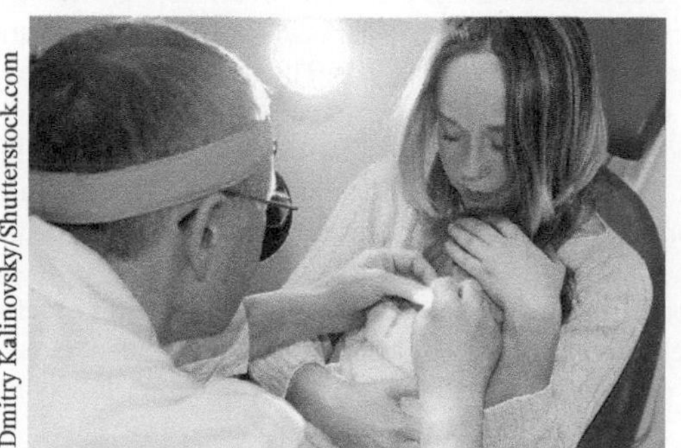
Dmitry Kalinovsky/Shutterstock.com

FIGURE 25.26 On the left is the patient's view of a physician wearing a head mirror. The image on the right shows the bright light reflected by the mirror onto the patient under examination. Notice the lamp adjacent to the patient's head above her right shoulder.

*Sometimes optical systems use two (or more) mirrors, and the image formed by the first mirror serves as the object for the second mirror. Occasionally, such an object falls *behind* the second mirror. In this case the object distance is negative, and the object is said to be a virtual object.

eye and looks through the hole in the center. Light from a bright lamp that is placed near the patient's head reflects off the mirror along the line of sight of the doctor and illuminates the area on the patient that is under examination. During a routine physical, a patient sitting 0.50 m away from the physician sees the reflection of her nose in the head mirror. If the focal length of the mirror is 25 cm, **(a)** where is the image located, and **(b)** what is its magnification?

Reasoning The head mirror is a small concave spherical mirror. The mirror equation (Equation 25.3) can be used to calculate the location of the image, and then Equation 25.4 is used to calculate the magnification.

Solution **(a)** Rearranging the mirror equation, we have: $\frac{1}{d_i} = \frac{1}{f} - \frac{1}{d_o}$. Plugging in the values in the problem, we find:

$$\frac{1}{d_i} = \frac{1}{0.25\text{ m}} - \frac{1}{0.50\text{ m}} \Rightarrow d_i = \boxed{0.50\text{ m}}$$

(b) The magnification is determined from Equation 25.4:

$$m = \frac{-d_i}{d_o} = -\frac{0.50\text{ m}}{0.50\text{ m}} = \boxed{-1}$$

Since the absolute value of the magnification is 1, the image is the same size as the object, and the negative sign tells us the image is real and inverted with respect to the object.

Related Homework: Problem 44

Concept Summary

25.1 Wave Fronts and Rays Wave fronts are surfaces on which all points of a wave are in the same phase of motion. Waves whose wave fronts are flat surfaces are known as plane waves. Rays are lines that are perpendicular to the wave fronts and point in the direction of the velocity of the wave.

25.2 The Reflection of Light When light reflects from a smooth surface, the reflected light obeys the law of reflection: The incident ray, the reflected ray, and the normal to the surface all lie in the same plane, and the angle of reflection θ_r equals the angle of incidence θ_i ($\theta_r = \theta_i$).

25.3 The Formation of Images by a Plane Mirror A virtual image is one from which all the rays of light do not actually come, but only appear to do so. A real image is one from which all the rays of light actually do emanate.

A plane mirror forms an upright, virtual image that is located as far behind the mirror as the object is in front of it. In addition, the heights of the image and the object are equal.

25.4 Spherical Mirrors A spherical mirror has the shape of a section from the surface of a hollow sphere. If the inside surface of the mirror is polished, it is a concave mirror. If the outside surface is polished, it is a convex mirror.

The principal axis of a mirror is a straight line drawn through the center of curvature and the middle of the mirror's surface. Rays that are close to the principal axis are known as paraxial rays. Paraxial rays are not necessarily parallel to the principal axis. The radius of curvature R of a mirror is the distance from the center of curvature to the mirror.

The focal point of a concave spherical mirror is a point on the principal axis, in front of the mirror. Incident paraxial rays that are parallel to the principal axis converge to the focal point after being reflected from the concave mirror.

The focal point of a convex spherical mirror is a point on the principal axis, behind the mirror. For a convex mirror, incident paraxial rays that are parallel to the principal axis diverge after reflecting from the mirror. These rays seem to originate from the focal point.

The fact that a spherical mirror does not bring all rays parallel to the principal axis to a single image point after reflection is known as spherical aberration.

The focal length f indicates the distance along the principal axis between the focal point and the mirror. The focal length and the radius of curvature R are related by Equations 25.1 and 25.2.

$$f = \tfrac{1}{2}R \quad \text{(Concave mirror)} \qquad \textbf{(25.1)}$$

$$f = -\tfrac{1}{2}R \quad \text{(Convex mirror)} \qquad \textbf{(25.2)}$$

25.5 The Formation of Images by Spherical Mirrors The image produced by a mirror can be located by a graphical method known as ray tracing.

For a concave mirror, the following paraxial rays are useful for ray tracing (see **Interactive Figure 25.17**):

Ray 1. This ray leaves the object traveling parallel to the principal axis. The ray reflects from the mirror and passes through the focal point.

Ray 2. This ray leaves the object and passes through the focal point. The ray reflects from the mirror and travels parallel to the principal axis.

Ray 3. This ray leaves the object and travels along a line that passes through the center of curvature. The ray strikes the mirror perpendicularly and reflects back on itself.

For a convex mirror, the following paraxial rays are useful for ray tracing (see **Interactive Figure 25.21*a***):

Ray 1. This ray leaves the object traveling parallel to the principal axis. After reflection from the mirror, the ray appears to originate from the focal point of the mirror.

Ray 2. This ray leaves the object and heads toward the focal point. After reflection, the ray travels parallel to the principal axis.

Ray 3. This ray leaves the object and travels toward the center of curvature. The ray strikes the mirror perpendicularly and reflects back on itself.

25.6 The Mirror Equation and the Magnification Equation The mirror equation (Equation 25.3) specifies the relation between the object distance d_o, the image distance d_i, and the focal length f of the mirror. The mirror equation can be used with either concave or convex mirrors.

$$\frac{1}{d_o} + \frac{1}{d_i} = \frac{1}{f} \qquad \textbf{(25.3)}$$

The magnification m of a mirror is the ratio of the image height h_i to the object height h_o: $m = h_i/h_o$. The magnification is also related to d_i and d_o by the magnification equation (Equation 25.4). The algebraic sign conventions for the variables appearing in these equations are summarized in the Reasoning Strategy at the end of Section 25.6.

$$m = -\frac{d_i}{d_o} \qquad \textbf{(25.4)}$$

Focus on Concepts

Online

Additional questions are available for assignment in WileyPLUS.

Section 25.1 Wave Fronts and Rays

1. A ray is ______. (a) always parallel to other rays (b) parallel to the velocity of the wave (c) perpendicular to the velocity of the wave (d) parallel to the wave fronts

Section 25.2 The Reflection of Light

2. The drawing shows a top view of an object located to the right of a mirror. A single ray of light is shown leaving the object. After reflection from the mirror, through which location, A, B, C, or D, does the ray pass? (a) A (b) B (c) C (d) D

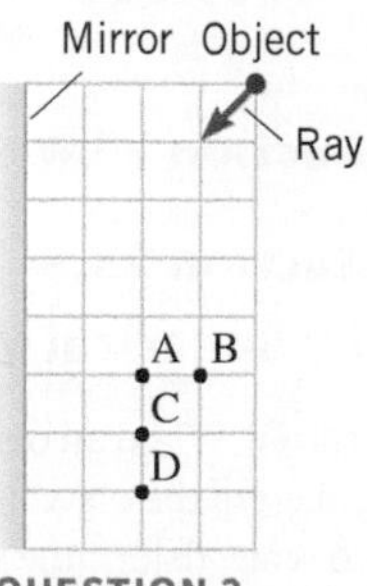

QUESTION 2

Section 25.3 The Formation of Images by a Plane Mirror

3. A friend is standing 2 m in front of a plane mirror. You are standing 3 m directly behind your friend. What is the distance between you and the image of your friend? (a) 2 m (b) 3 m (c) 5 m (d) 7 m (e) 10 m

4. You hold the words **TOP DOG** in front of a plane mirror. What does the image of these words look like? (a) **ꟼOT ꓷOG** (b) **ꓷOꓷ ꟼOT** (c) **TOP DOG** (d) **DOG TOP** (e) **ꓕOꓒ DOG**

Section 25.4 Spherical Mirrors

5. Rays of light coming from the sun (a very distant object) are near and parallel to the principal axis of a concave mirror. After reflecting from the mirror, where will the rays cross each other at a single point? The rays ______. (a) will not cross each other after reflecting from a concave mirror (b) will cross at the point where the principal axis intersects the mirror (c) will cross at the center of curvature (d) will cross at the focal point (e) will cross at a point beyond the center of curvature

Section 25.5 The Formation of Images by Spherical Mirrors

6. Which one of the following statements concerning spherical mirrors is correct? (a) Only a convex mirror can produce an enlarged image. (b) Both concave and convex mirrors can produce an enlarged image. (c) Only a concave mirror can produce an enlarged image, provided the object distance is less than the radius of curvature. (d) Only a concave mirror can produce an enlarged image, provided the object distance is greater than the radius of curvature.

7. Suppose that you hold up a small convex mirror in front of your face. Which answer describes the image of your face? (a) Virtual, inverted (b) Virtual, upright (c) Virtual, enlarged (d) Real, inverted (e) Real, reduced in size

Section 25.6 The Mirror Equation and the Magnification Equation

8. An object is placed at a known distance in front of a mirror whose focal length is also known. You apply the mirror equation and find that the image distance is a negative number. This result tells you that ______. (a) the image is larger than the object (b) the image is smaller than the object (c) the image is inverted relative to the object (d) the image is real (e) the image is virtual

9. An object is situated at a known distance in front of a convex mirror whose focal length is also known. A friend of yours does a calculation that shows that the magnification is −2. After some thought, you conclude correctly that ______. (a) your friend's answer is correct (b) the magnification should be +2 (c) the magnification should be $+\frac{1}{2}$ (d) the magnification should be $-\frac{1}{2}$

Problems

Online

Additional questions are available for assignment in WileyPLUS.

- SSM Student Solutions Manual
- MMH Problem-solving help
- GO Guided Online Tutorial
- V-HINT Video Hints
- CHALK Chalkboard Videos
- BIO Biomedical application
- E Easy
- M Medium
- H Hard
- WS Worksheet
- T Team Problem

Section 25.2 The Reflection of Light, Section 25.3 The Formation of Images by a Plane Mirror

1. E On the $+y$ axis a laser is located at $y = +3.0$ cm. The coordinates of a small target are $x = +9.0$ cm and $y = +6.0$ cm. The $+x$ axis represents the edge-on view of a plane mirror. At what point on the $+x$ axis should the laser be aimed in order for the laser light to hit the target after reflection?

2. E SSM CHALK You are trying to photograph a bird sitting on a tree branch, but a tall hedge is blocking your view. However, as the drawing shows, a plane mirror reflects light from the bird into your camera. For what distance must you set the focus of the camera lens in order to snap a sharp picture of the bird's image?

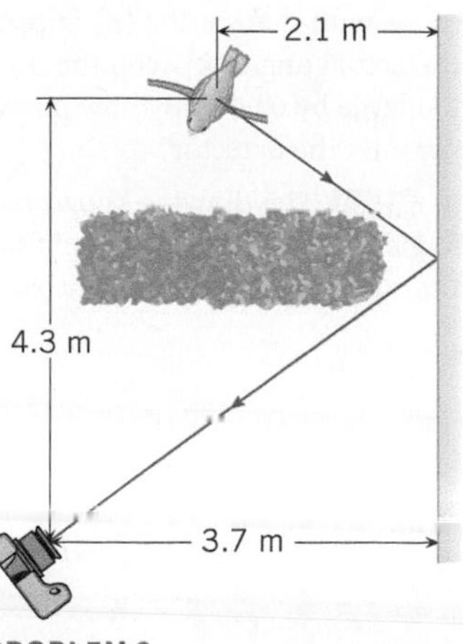

PROBLEM 2

3. E GO Suppose that you are walking perpendicularly with a velocity of +0.90 m/s toward a stationary plane mirror. What is the velocity of your image relative to you? The direction in which you walk is the positive direction.

4. E SSM Two plane mirrors are separated by 120°, as the drawing illustrates. If a ray strikes mirror M_1 at a 65° angle of incidence, at what angle θ does it leave mirror M_2?

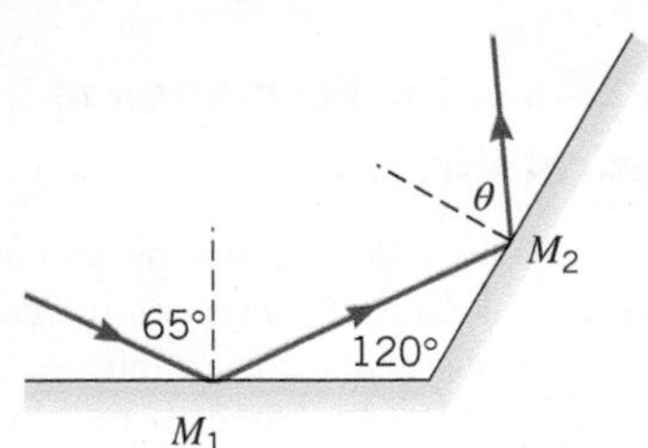

PROBLEM 4

5. E CHALK The drawing shows a laser beam shining on a plane mirror that is perpendicular to the floor. The beam's angle of incidence is 33.0°. The beam emerges from the laser at a point that is 1.10 m from the mirror and 1.80 m above the floor. After reflection, how far from the base of the mirror does the beam strike the floor?

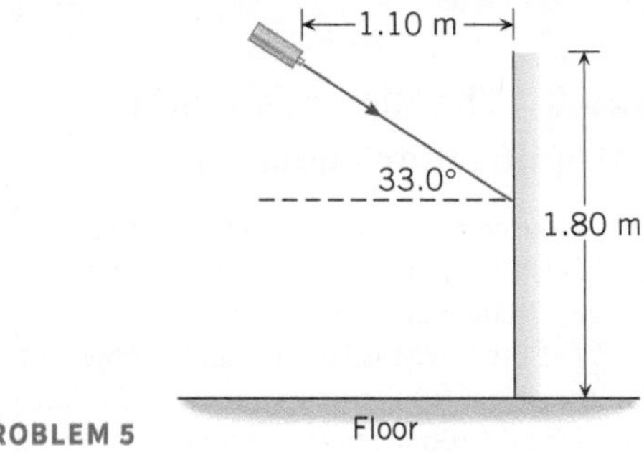

PROBLEM 5

6. M GO A small mirror is attached to a vertical wall, and it hangs a distance of 1.80 m above the floor. The mirror is facing due east, and a ray of sunlight strikes the mirror early in the morning and then again later in the morning. The incident and reflected rays lie in a plane that is perpendicular to both the wall and the floor. Early in the morning, the reflected ray strikes the floor at a distance of 3.86 m from the base of the wall. Later on in the morning, the ray is observed to strike the floor at a distance of 1.26 m from the wall. The earth rotates at a rate of 15.0° per hour. How much time (in hours) has elapsed between the two observations?

7. M V-HINT In an experiment designed to measure the speed of light, a laser is aimed at a mirror that is 50.0 km due north. A detector is placed 117 m due east of the laser. The mirror is to be aligned so that light from the laser reflects into the detector. (a) When properly aligned, what angle should the normal to the surface of the mirror make with due south? (b) Suppose the mirror is misaligned, so that the actual angle between the normal to the surface and due south is too large by 0.004°. By how many meters (due east) will the reflected ray miss the detector?

8. M GO The drawing shows two plane mirrors that intersect at an angle of 50°. An incident light ray reflects from one mirror and then the other. What is the angle θ between the incident and outgoing rays?

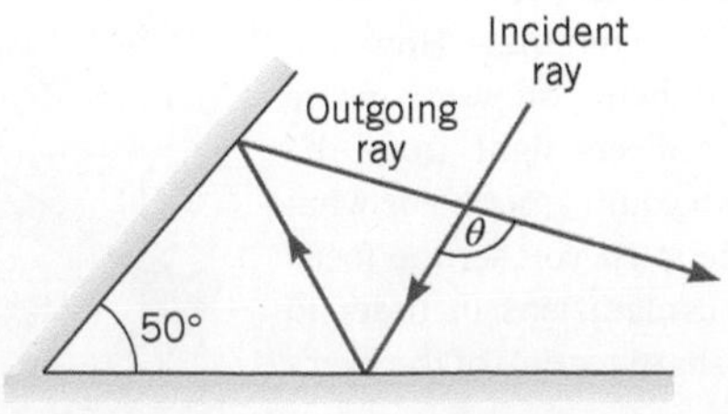

PROBLEM 8

9. M Two plane mirrors are facing each other. They are parallel, 3.00 cm apart, and 17.0 cm in length, as the drawing indicates. A laser beam is directed at the top mirror from the left edge of the bottom mirror. What is the smallest angle of incidence with respect to the top mirror, such that the laser beam (a) hits only one of the mirrors and (b) hits each mirror only once?

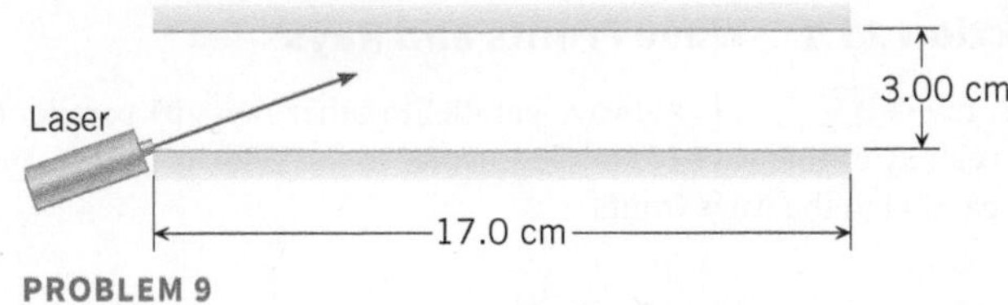

PROBLEM 9

Section 25.4 Spherical Mirrors,

Section 25.5 The Formation of Images by Spherical Mirrors

10. E When an object is located very far away from a convex mirror, the image of the object is 18 cm behind the mirror. Using a ray diagram drawn to scale, determine where the image is located when the object is placed 9.0 cm in front of the mirror. Note that the mirror must be drawn to scale also. In your drawing, assume that the height of the object is 3.0 cm.

11. E CHALK The image of a very distant car is located 12 cm behind a convex mirror. (a) What is the radius of curvature of the mirror? (b) Draw a ray diagram to scale showing this situation.

12. E SSM An object is placed 11 cm in front of a concave mirror whose focal length is 18 cm. The object is 3.0 cm tall. Using a ray diagram drawn to scale, measure (a) the location and (b) the height of the image. The mirror must be drawn to scale.

13. E MMH A 2.0-cm-high object is situated 15.0 cm in front of a concave mirror that has a radius of curvature of 10.0 cm. Using a ray diagram drawn to scale, measure (a) the location and (b) the height of the image. The mirror must be drawn to scale.

14. E MMH A convex mirror has a focal length of −40.0 cm. A 12.0-cm-tall object is located 40.0 cm in front of this mirror. Using a ray diagram drawn to scale, determine the (a) location and (b) size of the image. Note that the mirror must be drawn to scale.

15. M CHALK A plane mirror and a concave mirror ($f = 8.0$ cm) are facing each other and are separated by a distance of 20.0 cm. An object is placed between the mirrors and is 10.0 cm from each mirror. Consider the light from the object that reflects first from the plane mirror and then from the concave mirror. Using a ray diagram drawn to scale, find the location of the image that this light produces in the concave mirror. Specify this distance relative to the concave mirror.

Section 25.6 The Mirror Equation and the Magnification Equation

16. E SSM The image produced by a concave mirror is located 26 cm in front of the mirror. The focal length of the mirror is 12 cm. How far in front of the mirror is the object located?

17. E The image behind a convex mirror (radius of curvature = 68 cm) is located 22 cm from the mirror. (a) Where is the object located and (b) what is the magnification of the mirror? Determine whether the image is (c) upright or inverted and (d) larger or smaller than the object.

18. E MMH A concave mirror (R = 56.0 cm) is used to project a transparent slide onto a wall. The slide is located at a distance of 31.0 cm from the mirror, and a small flashlight shines light

through the slide and onto the mirror. The setup is similar to that in **Animated Figure 25.18*a***. (a) How far from the wall should the mirror be located? (b) The height of the object on the slide is 0.95 cm. What is the height of the image? (c) How should the slide be oriented, so that the picture on the wall looks normal?

19. E GO A small statue has a height of 3.5 cm and is placed in front of a concave mirror. The image of the statue is inverted, 1.5 cm tall, and located 13 cm in front of the mirror. Find the focal length of the mirror.

20. E GO SSM A mirror produces an image that is located 34.0 cm behind the mirror when the object is located 7.50 cm in front of the mirror. What is the focal length of the mirror, and is the mirror concave or convex?

21. E GO A concave mirror ($f = 45$ cm) produces an image whose distance from the mirror is one-third the object distance. Determine (a) the object distance and (b) the (positive) image distance.

22. E The outside mirror on the passenger side of a car is convex and has a focal length of −7.0 m. Relative to this mirror, a truck traveling in the rear has an object distance of 11 m. Find (a) the image distance of the truck and (b) the magnification of the mirror.

23. E GO A convex mirror has a focal length of −27.0 cm. Find the magnification produced by the mirror when the object distance is 9.0 cm and 18.0 cm.

24. E SSM When viewed in a spherical mirror, the image of a setting sun is a virtual image. The image lies 12.0 cm behind the mirror. (a) Is the mirror concave or convex? Why? (b) What is the radius of curvature of the mirror?

25. E V-HINT A concave mirror has a focal length of 12 cm. This mirror forms an image located 36 cm in front of the mirror. What is the magnification of the mirror?

26. M V-HINT An object is located 14.0 cm in front of a convex mirror, the image being 7.00 cm behind the mirror. A second object, twice as tall as the first one, is placed in front of the mirror, but at a different location. The image of this second object has the same height as the other image. How far in front of the mirror is the second object located?

27. M GO A tall tree is growing across a river from you. You would like to know the distance between yourself and the tree, as well as its height, but are unable to make the measurements directly. However, by using a mirror to form an image of the tree and then measuring the image distance and the image height, you can calculate the distance to the tree as well as its height. Suppose that this mirror produces an image of the sun, and the image is located 0.9000 m from the mirror. The same mirror is then used to produce an image of the tree. The image of the tree is 0.9100 m from the mirror. (a) How far away is the tree? (b) The image height of the tree has a magnitude of 0.12 m. How tall is the tree?

28. M A spherical mirror is polished on both sides. When the concave side is used as a mirror, the magnification is +2.0. What is the magnification when the convex side is used as a mirror, the object remaining the same distance from the mirror?

29. M GO Consult Multiple-Concept Example 8 to see a model for solving this type of problem. A concave makeup mirror is designed so the virtual image it produces is twice the size of the object when the distance between the object and the mirror is 14 cm. What is the radius of curvature of the mirror?

30. H A concave mirror has a focal length of 30.0 cm. The distance between an object and its image is 45.0 cm. Find the object and image distances, assuming that (a) the object lies beyond the center of curvature and (b) the object lies between the focal point and the mirror.

Additional Problems

Online

31. E An object is placed in front of a convex mirror. Draw the convex mirror (radius of curvature = 15 cm) to scale, and place the object 25 cm in front of it. Make the object height 4 cm. Using a ray diagram, locate the image and measure its height. Now move the object closer to the mirror, so the object distance is 5 cm. Again, locate its image using a ray diagram. As the object moves closer to the mirror, (a) does the magnitude of the image distance become larger or smaller, and (b) does the magnitude of the image height become larger or smaller? (c) What is the ratio of the image height when the object distance is 5 cm to its height when the object distance is 25 cm? Give your answer to one significant figure.

32. E SSM An object that is 25 cm in front of a convex mirror has an image located 17 cm behind the mirror. How far behind the mirror is the image located when the object is 19 cm in front of the mirror?

33. E GO A concave mirror has a focal length of 42 cm. The image formed by this mirror is 97 cm in front of the mirror. What is the object distance?

34. E Review Conceptual Example 1 before attempting this problem. A person whose eyes are 1.70 m above the floor stands in front of a plane mirror. The top of her head is 0.12 m above her eyes. (a) What is the height of the shortest mirror in which she can see her entire image? (b) How far above the floor should the bottom edge of the mirror be placed?

35. E V-HINT A drop of water on a countertop reflects light from a flower held 3.0 cm directly above it. The flower's diameter is 2.0 cm, and the diameter of the flower's image is 0.10 cm. What is the focal length of the water drop, assuming that it may be treated as a convex spherical mirror?

36. M GO You walk at an angle of $\theta = 50.0°$ toward a plane mirror, as in the drawing. Your walking velocity has a magnitude of 0.90 m/s. What is the velocity of your image relative to you (magnitude and direction)?

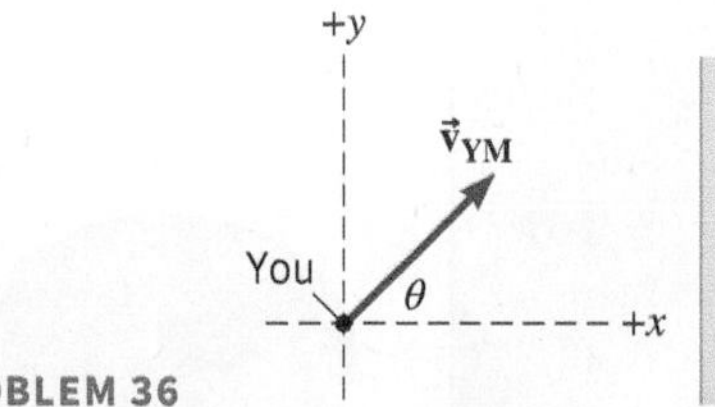

PROBLEM 36

37. M CHALK GO A candle is placed 15.0 cm in front of a convex mirror. When the convex mirror is replaced with a plane mirror, the image moves 7.0 cm farther away from the mirror. Find the focal length of the convex mirror.

38. H A man holds a double-sided spherical mirror so that he is looking directly into its convex surface, 45 cm from his face. The magnification of the image of his face is +0.20. What will be the image distance when he reverses the mirror (looking into its concave surface), maintaining the same distance between the mirror and his face? Be sure to include the algebraic sign (+ or −) with your answer.

Physics in Biology, Medicine, and Sports

39. **M BIO GO** **25.6** A dentist's mirror is placed 2.0 cm from a tooth. The *enlarged* image is located 5.6 cm behind the mirror. **(a)** What kind of mirror (plane, concave, or convex) is being used? **(b)** Determine the focal length of the mirror. **(c)** What is the magnification? **(d)** How is the image oriented relative to the object?

40. **E BIO** **25.2** At major golfing events, spectators surround the greens in an attempt to see their favorite players. A handy device that can help them see over the crowd is a periscope (see the photo). Inside the periscope there are two plane mirrors separated by the length of the periscope and oriented such that the normal to their surface makes an angle of 45° with the horizontal (see the drawing). Consider a horizontal light ray that enters the periscope. **(a)** If $\theta_1 = 45°$, what is the value of θ_2? **(b)** If it takes light 1.0 ns to travel the distance between the two mirrors, what is this distance?

41. **E BIO** **25.2** A laser range finder is a device that quickly and accurately determines the distance between it and an object. The devices are frequently used in hunting and in golf to measure the distance to a wild animal or to a flag on the green, respectively. The range finder works by emitting laser light toward the target. Some of that light reflects from the target and returns, where it is detected by a sensor in the range finder (see the photo). An accurate time circuit in the range finder measures the time that elapses between the emission and detection of the laser light. In the figure shown here, the range finder is used to determine the distance to mirror 2. Because of a tree line, a direct measurement of the distance to mirror 2 is not possible. Two other mirrors are used to reflect the light around the tree line. Laser light from the range finder is reflected by the three mirrors and returns to the finder in 3.0 μs. The light strikes mirror 1 at an angle of 70° with respect to the mirror's normal. **(a)** What is the angle θ_1? **(b)** What is the angle θ_2? **(c)** What is the distance between mirrors 1 and 2? **(d)** What is the distance between the range finder and mirror 2?

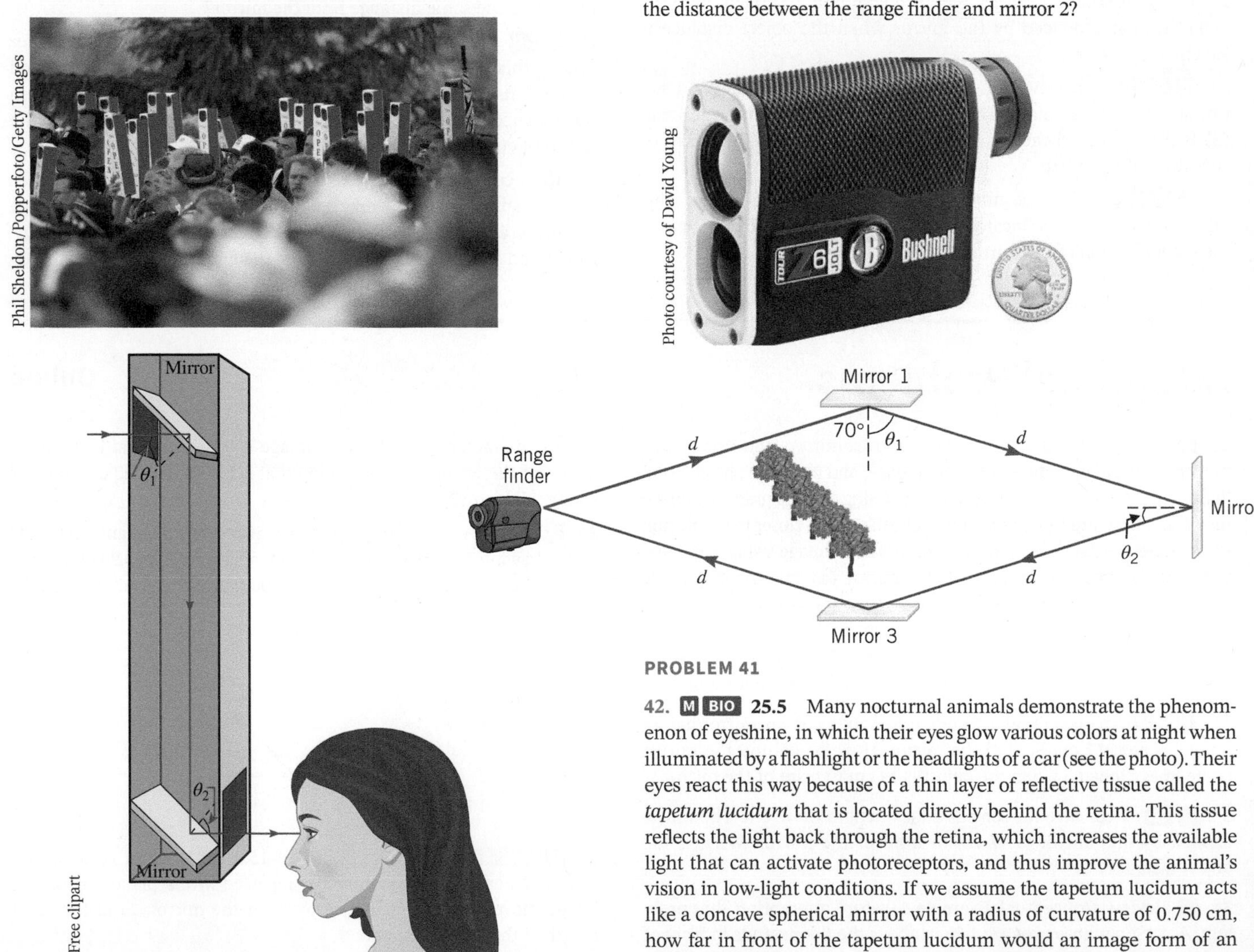

PROBLEM 40

PROBLEM 41

42. **M BIO** **25.5** Many nocturnal animals demonstrate the phenomenon of eyeshine, in which their eyes glow various colors at night when illuminated by a flashlight or the headlights of a car (see the photo). Their eyes react this way because of a thin layer of reflective tissue called the *tapetum lucidum* that is located directly behind the retina. This tissue reflects the light back through the retina, which increases the available light that can activate photoreceptors, and thus improve the animal's vision in low-light conditions. If we assume the tapetum lucidum acts like a concave spherical mirror with a radius of curvature of 0.750 cm, how far in front of the tapetum lucidum would an image form of an object located 30.0 cm away? Neglect the effects of the other structures of the eye, such as the cornea and lens.

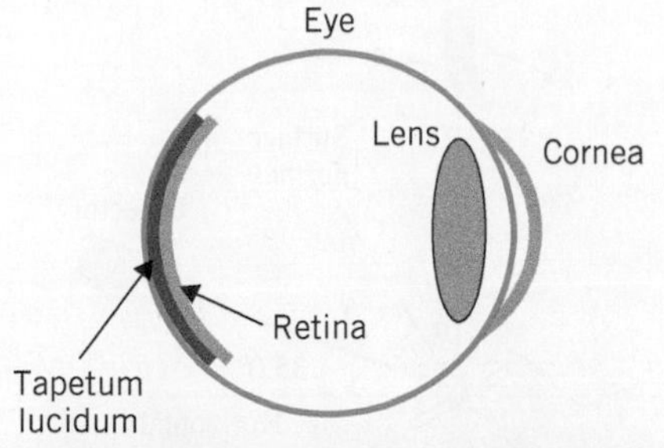

In response to a camera flash, the eyes of a cat named Einstein glow orange-red due to the reflection of light from the tapetum lucidum located behind the retina.

PROBLEM 42

43. H BIO **25.3** Currently, nine nonhuman species of animals pass the *mirror self-recognition test* (MSR), which means they demonstrate the ability of self-recognition when they look at their reflection. Some of the animals on this list include the great apes, Asian elephants, bottlenose dolphins, and orca whales. In the figure, an Asian elephant is standing 3.5 m from a vertical wall. Given the dimensions shown in the drawing, what should be the minimum length of the mirror (L), such that the elephant can see the entire height of its body—from the top of its head to the bottom of its feet?

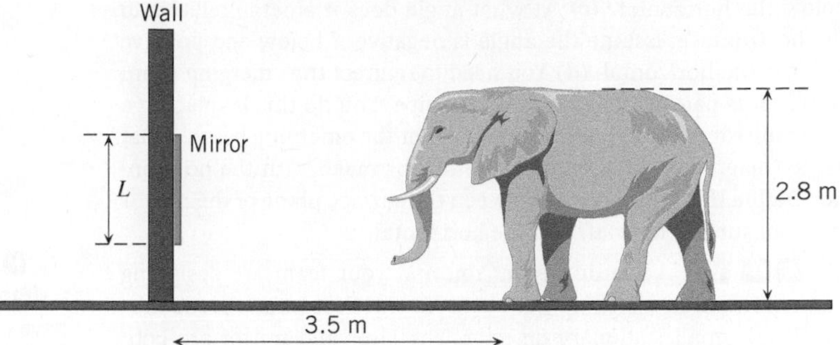

PROBLEM 43

44. E BIO **25.3** Mirror laryngoscopy is a medical diagnostic technique that traditionally uses both a head mirror (see Example 9) and a laryngeal mirror. The laryngeal mirror can come in many sizes. Size 0 corresponds to a circular plane mirror with a diameter of 10.0 mm. The length of the mirror's handle (20.3 cm) allows the physical examination of the larynx, a procedure that is often used to identify early stages of malignancy in swollen lymph nodes. A physician using a laryngeal mirror to examine a patient discovers a granularcell tumor on the patient's larynx. The approximately spherical tumor is 5.2 mm in diameter and is located 1.2 cm in front of the mirror. **(a)** What kind of image is formed by the laryngeal mirror—real or virtual? **(b)** Where is the image located? **(c)** What is the magnification of the image?

Concepts and Calculations Problems

Online

Relative to the object in front of a spherical mirror, the image can differ in a number of respects. The image can be real (in front of the mirror) or virtual (behind the mirror). It can be larger or smaller than the object, and it can be upright or inverted. As you solve problems dealing with spherical mirrors, keep these image characteristics in mind. They can help guide you to the correct answer, as Problems 45 and 46 illustrate.

45. M CHALK An object is located 7.0 cm in front of a mirror. The virtual image is located 4.5 cm away from the mirror and is smaller than the object. *Concepts:* (i) Based solely on the fact that the image is virtual, is the mirror concave or convex, or is either type possible? (ii) The image is smaller than the object, as well as virtual. Do these characteristics together indicate a concave or convex mirror, or is either type possible? (iii) Is the focal length positive or negative? Explain. *Calculation:* Find the focal length of the mirror.

46. M CHALK SSM The radius of curvature of a mirror is 24 cm. A diamond ring is placed in front of this mirror. The image is twice the size of the ring. *Concepts:* (i) Is the mirror concave or convex, or is either type possible? (ii) How many places are there in front of a concave mirror where the ring can be placed and produce an image twice the size of the object? (iii) What are the possible values for the magnification of the image of the ring? *Calculation:* Find the object distance of the ring.

Team Problems

Online

47. E T WS **A Light Pipe.** You and your team have been tasked with safely guiding a laser beam across a lab through a straight, horizontal pipe with a square cross section and mirrored internal surfaces. You are unable to aim the beam directly down the center, so it will have to reflect multiple times from the internal surfaces. You are, however, able to align the laser so that the beam is centered

between the left and right sides, and reflects only from the top and bottom surfaces. The pipe is 3.50 meters in length and has a square cross section with a side length of 3.75 cm. If the laser beam enters at the lower edge of the pipe at 10.0° angle above the horizontal, **(a)** how many reflections does the beam undergo before emerging from the other side? **(b)** Does the beam emerge at an angle above or below the horizontal? **(c)** At what angle does it emerge relative to the horizontal? Assume the angle is negative if below and positive if above the horizontal. **(d)** You need to redirect the emerging beam so that it is parallel to the axis of the pipe. You do this by placing a plane mirror at the exit of the pipe to steer the emerging beam. What angle (magnitude only) should the mirror make with the horizontal? Define this angle as that between the surface plane of the mirror (i.e., not surface normal) and the horizontal.

48. E T WS **A Goniometer.** You and your team are designing an instrument in which a laser beam must reflect from a plane mirror into a small stationary detector. The laser and mirror are both mounted to a mechanical instrument called a *goniometer* that allows for the precise independent angular positioning of each. The laser is mounted on an arm that can rotate like the hand of a clock. As shown in the drawing, the arm is initially positioned so that the laser points vertically downward, toward the axis of rotation. In this geometry, no matter what the angular position of the arm, the laser beam will always pass though the axis of rotation. Located exactly on the axis of rotation is the top surface of a plane mirror that can rotate independently of the laser arm. The idea is that the laser reflects from the mirror and toward the detector, which is located to the right at an angle of 35.0° above an imaginary horizontal line that passes through the axis of rotation. **(a)** With the laser arm in the vertical position (i.e., the laser is aimed vertically downward), at what angle relative to the horizontal should the surface normal of the mirror be directed so that the reflected beam will strike the detector? Be sure to indicate whether the angle is clockwise or counterclockwise. **(b)** If the laser arm rotates counterclockwise at a rate of 2.00° per minute, what should the angular velocity of the mirror (in degrees/min) be so that the reflected laser beam always strikes the detector?

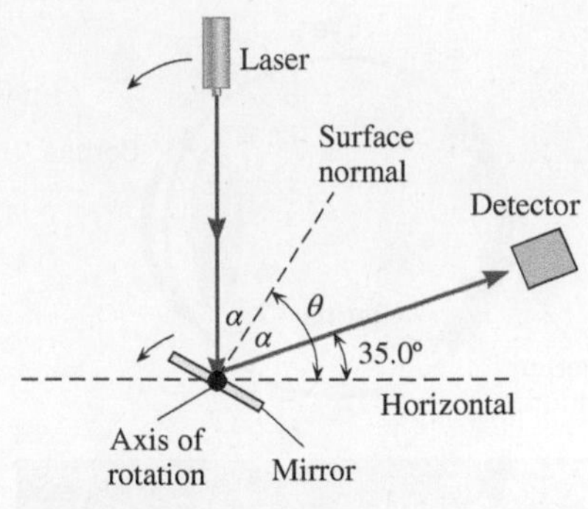

PROBLEM 48

49. M T **An Unidentified Flying Object.** You and your team spot a strange object hovering over the open ocean in broad daylight. You have nothing to reference in order to estimate its distance, height, or size, but you get the idea to use a concave mirror from a signal light. You disassemble the light, extract the mirror, and point it at the sun. You measure the distance to the sun's image from the center of the mirror to be 3.405 m. You then point the mirror at the object and find that its image forms 3.443 m from the center of the mirror. The angle between the line of sight to the object and the horizontal is 28.50°. **(a)** What is the distance between you and the object? **(b)** What is the height of the object above the ground? **(c)** If the width of the object's image is 11.50 cm, what is the width of the object?

50. M T **Radius of Curvature of a Partial Sphere.** You and your team need to estimate the radius of curvature of a panel that had been removed from some unknown object. You were told that the object was spherical and, by the size of the panel (about one square meter in area) and its slight curvature, you estimate that the spherical object from which it came had been quite large. You notice that the outer surface has a mirrored metallic finish (like a convex mirror) and you get an idea. You find a wrench that is 21.0 cm long, and hold it at a distance of 10.0 m from the middle of the mirrored surface. The virtual image of the wrench is 14.5 cm long. Determine the radius of the sphere from which the panel came.

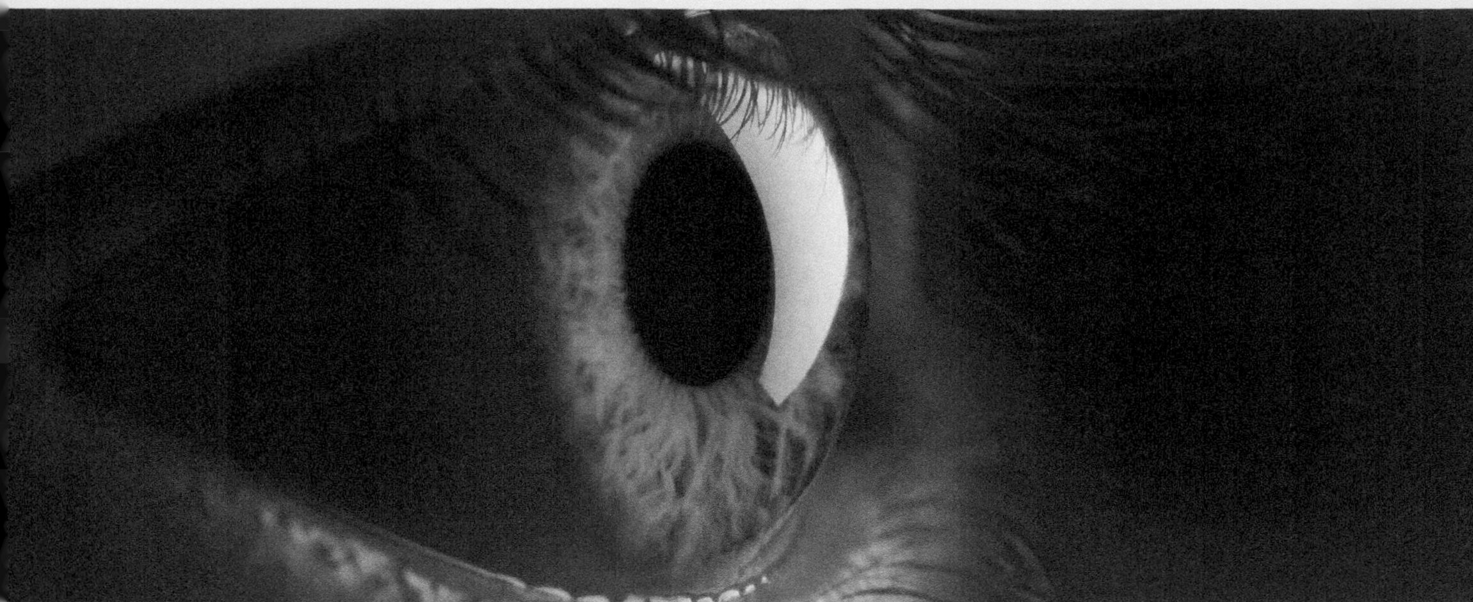

When light moves from one medium into another, its direction of travel changes, and this change in direction is called refraction. The human eye is an incredible optical instrument and, as we will see, refraction plays a major role in the way it works to produce clear vision.

Design Pics/Don Hammond/Getty Images

The Refraction of Light: Lenses and Optical Instruments

LEARNING OBJECTIVES

After reading this module, you should be able to . . .

26.1 Define the index of refraction.

26.2 Use Snell's law of refraction to solve problems.

26.3 Analyze total internal reflection problems.

26.4 Define Brewster's angle.

26.5 Analyze examples involving dispersion of light.

26.6 Trace rays passing through converging and diverging lenses.

26.7 Trace rays from objects through lenses to form images.

26.8 Apply the lens and magnification equations to solve problems.

26.9 Solve problems involving lenses in combination.

26.10 Apply ray tracing and the lens and magnification equations to the human eye.

26.11 Calculate angular size and angular magnification.

26.12 Apply optical principles to the compound microscope.

26.13 Apply optical principles to the telescope.

26.14 Describe spherical and chromatic lens aberrations.

26.1 The Index of Refraction

As Section 24.3 discusses, light travels through a vacuum at a speed of $c = 3.00 \times 10^8$ m/s. It can also travel through many materials, such as air, water, and glass. Atoms in the material absorb, reemit, and scatter the light, however. Therefore, light travels through the material at a speed that is less than *c*, the actual speed depending on the nature of the material. In general, we will see that the change in speed as a ray of light goes from one material to another causes the ray to deviate from its incident direction. This change in direction is called **refraction**. To describe the extent to which the speed of light in a material medium differs from that in a vacuum, we use a parameter called the **index of refraction** (or **refractive index**). The index of refraction is an important parameter because it appears in Snell's law of refraction, which will be discussed in the next section. This law is the basis of all the phenomena discussed in this chapter.

DEFINITION OF THE INDEX OF REFRACTION

The index of refraction n of a material is the ratio of the speed c of light in a vacuum to the speed v of light in the material:

$$n = \frac{\text{Speed of light in a vacuum}}{\text{Speed of light in the material}} = \frac{c}{v} \tag{26.1}$$

TABLE 26.1 Index of Refraction[a] for Various Substances

Substance	Index of Refraction, n
Solids at 20 °C	
Diamond	2.419
Glass, crown	1.523
Ice (0 °C)	1.309
Sodium chloride	1.544
Quartz	
Crystalline	1.544
Fused	1.458
Liquids at 20 °C	
Benzene	1.501
Carbon disulfide	1.632
Carbon tetrachloride	1.461
Ethyl alcohol	1.362
Water	1.333
Gases at 0 °C, 1 atm	
Air	1.000293
Carbon dioxide	1.00045
Oxygen, O_2	1.000271
Hydrogen, H_2	1.000139

[a] Measured with light whose wavelength in a vacuum is 589 nm.

Table 26.1 lists the refractive indices for some common substances. The values of n are greater than unity because the speed of light in a material medium is less than it is in a vacuum. For example, the index of refraction for diamond is $n = 2.419$, so the speed of light in diamond is $v = c/n = (3.00 \times 10^8 \text{ m/s})/2.419 = 1.24 \times 10^8 \text{ m/s}$. In contrast, the index of refraction for air (and also for other gases) is so close to unity that $n_{air} = 1$ for most purposes. The index of refraction depends slightly on the wavelength of the light, and the values in Table 26.1 correspond to a wavelength of $\lambda = 589$ nm in a vacuum.

26.2 Snell's Law and the Refraction of Light

Snell's Law

When light strikes the interface between two transparent materials, such as air and water, the light generally divides into two parts, as Figure 26.1a illustrates. Part of the light is reflected, with the angle of reflection equaling the angle of incidence. The remainder is transmitted across the interface. If the incident ray does not strike the interface at normal incidence, the transmitted ray has a different direction than the incident ray. When a ray enters the second material and changes direction, it is said to be refracted and behaves in one of the following two ways:

1. When light travels from a medium where the refractive index is smaller into a medium where it is larger, the refracted ray is bent toward the normal, as in Figure 26.1*a*.

2. When light travels from a medium where the refractive index is larger into a medium where it is smaller, the refracted ray is bent away from the normal, as in Figure 26.1*b*.

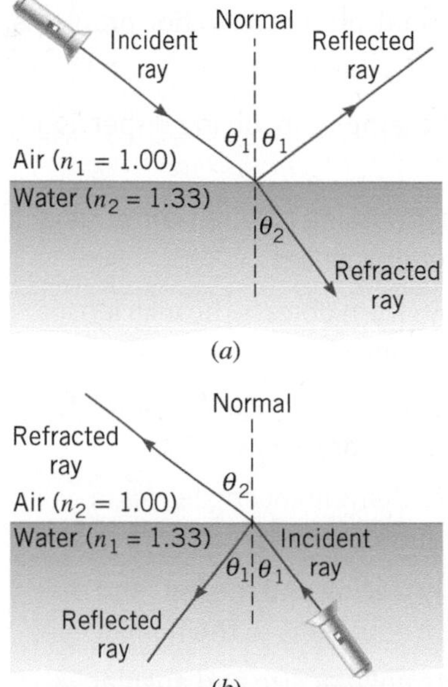

FIGURE 26.1 (*a*) When a ray of light is directed from air into water, part of the light is reflected at the interface and the remainder is refracted into the water. The refracted ray is bent *toward* the normal ($\theta_2 < \theta_1$). (*b*) When a ray of light is directed from water into air, the refracted ray in air is bent *away* from the normal ($\theta_2 > \theta_1$).

These two possibilities illustrate that both the incident and refracted rays obey the principle of reversibility. Thus, the directions of the rays in part *a* of the drawing can be reversed to give the situation depicted in part *b*. In part *b* the reflected ray lies in the water rather than in the air.

In both parts of Figure 26.1 the angles of incidence, refraction, and reflection are measured relative to the normal. Note that the index of refraction of air is labeled n_1 in part *a*, whereas it is n_2 in part b, because *we label all variables associated with the incident (and reflected) ray with subscript 1 and all variables associated with the refracted ray with subscript 2.*

The angle of refraction θ_2 depends on the angle of incidence θ_1 and on the indices of refraction, n_2 and n_1, of the two media. The relation between these quantities is known as **Snell's law of refraction**, after the Dutch mathematician Willebrord Snell (1591–1626), who discovered it experimentally. At the end of this section is a proof of Snell's law.

SNELL'S LAW OF REFRACTION

When light travels from a material with refractive index n_1 into a material with refractive index n_2, the refracted ray, the incident ray, and the normal to the interface between the materials all lie in the same plane. The angle of refraction θ_2 is related to the angle of incidence θ_1 by

$$n_1 \sin \theta_1 = n_2 \sin \theta_2 \tag{26.2}$$

Example 1 illustrates the use of Snell's law.

EXAMPLE 1 | Determining the Angle of Refraction

A light ray strikes an air/water surface at an angle of 46° with respect to the normal. The refractive index for water is 1.33. Find the angle of refraction when the direction of the ray is **(a)** from air to water and **(b)** from water to air.

Reasoning Snell's law of refraction applies to both part (a) and part (b). However, in part (a) the incident ray is in air, whereas in part (b) it is in water. We keep track of this difference by always labeling the variables associated with the incident ray with a subscript 1 and the variables associated with the refracted ray with a subscript 2.

Problem-Solving Insight The angle of incidence θ_1 and the angle of refraction θ_2 that appear in Snell's law are measured with respect to the normal to the surface, and not with respect to the surface itself.

Solution **(a)** The incident ray is in air, so $\theta_1 = 46°$ and $n_1 = 1.00$. The refracted ray is in water, so $n_2 = 1.33$. Snell's law can be used to find the angle of refraction θ_2:

$$\sin\theta_2 = \frac{n_1 \sin\theta_1}{n_2} = \frac{(1.00)\sin 46°}{1.33} = 0.54 \qquad \textbf{(26.2)}$$

$$\theta_2 = \sin^{-1}(0.54) = \boxed{33°}$$

Since θ_2 is less than θ_1, the refracted ray is bent *toward* the normal, as **Figure 26.1*a*** shows.

(b) With the incident ray in water, we find that

$$\sin\theta_2 = \frac{n_1 \sin\theta_1}{n_2} = \frac{(1.33)\sin 46°}{1.00} = 0.96$$

$$\theta_2 = \sin^{-1}(0.96) = \boxed{74°}$$

Since θ_2 is greater than θ_1, the refracted ray is bent *away* from the normal, as in **Figure 26.1*b***.

We have seen that reflection and refraction of light waves occur simultaneously at the interface between two transparent materials. It is important to keep in mind that light waves are composed of electric and magnetic fields, which carry energy. The principle of conservation of energy (see Chapter 6) indicates that the energy reflected plus the energy refracted must add up to equal the energy carried by the incident light, provided that none of the energy is absorbed by the materials. The percentage of incident energy that appears as reflected versus refracted light depends on the angle of incidence and the refractive indices of the materials on either side of the interface. For instance, when light travels from air toward water at perpendicular incidence, most of its energy is refracted and little is reflected. But when the angle of incidence is nearly 90° and the light barely grazes the water surface, most of its energy is reflected, with only a small amount refracted into the water. On a rainy night, you probably have experienced the annoying glare that results when light from an oncoming car just grazes the wet road. Under such conditions, most of the light energy reflects into your eyes.

THE PHYSICS OF . . . rearview mirrors. The simultaneous reflection and refraction of light have applications in a number of devices. For instance, interior rearview mirrors in cars often have adjustment levers. One position of the lever is for day driving, while the other is for night driving and reduces glare from the headlights of the car behind. As **Interactive Figure 26.2*a*** indicates, this kind of mirror is a glass wedge with a back side that is silvered and highly reflecting. Part *b* of the picture shows the day setting. Light from the car behind follows path *ABCD* in reaching the driver's eye. At points *A* and *C*, where the light strikes the air–glass surface, there are both reflected and refracted rays. The reflected rays are drawn as thin lines, the thinness denoting that

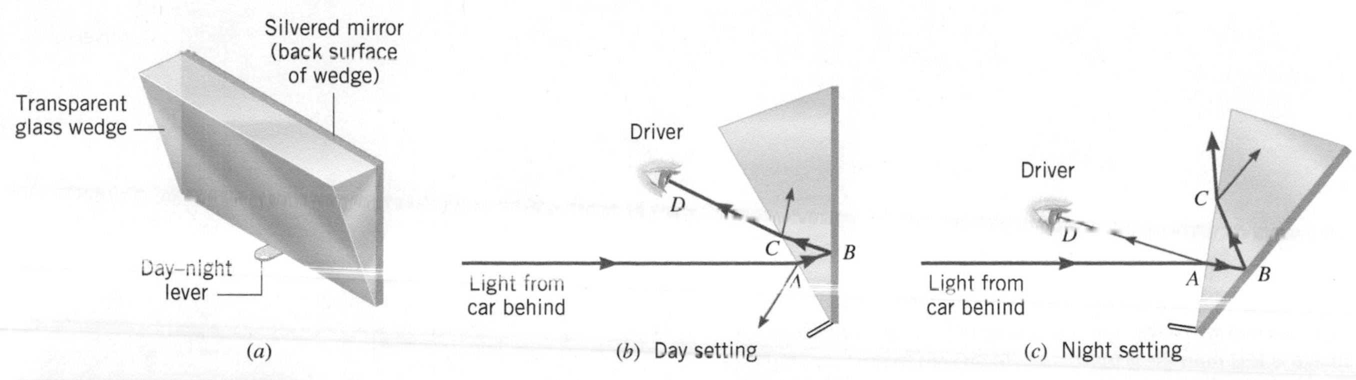

INTERACTIVE FIGURE 26.2 A car's interior rearview mirror with a day–night adjustment lever.

only a small percentage (about 10%) of the light during the day is reflected at *A* and *C*. The weak reflected rays at *A* and *C* do not reach the driver's eye. In contrast, almost all the light reaching the silvered back surface at *B* is reflected toward the driver. Since most of the light follows path *ABCD*, the driver sees a bright image of the car behind. During the night, the adjustment lever can be used to rotate the top of the mirror away from the driver (see part *c* of the drawing). Now, most of the light from the headlights behind follows path *ABC* and does not reach the driver. Only the light that is weakly reflected from the front surface along path *AD* is seen, and the result is significantly less glare.

Apparent Depth

One interesting consequence of refraction is that an object lying under water appears to be closer to the surface than it actually is. Example 2 sets the stage for explaining why, by showing what must be done to shine a light on such an object.

EXAMPLE 2 | Finding a Sunken Chest

A searchlight on a yacht is being used at night to illuminate a sunken chest, as in **Figure 26.3**. At what angle of incidence θ_1 should the light be aimed?

Reasoning The angle of incidence θ_1 must be such that, after refraction, the light strikes the chest. The angle of incidence can be obtained from Snell's law, once the angle of refraction θ_2 is determined. This angle can be found using the data in **Figure 26.3** and trigonometry. The light travels from a region of lower into a region of higher refractive index, so the light is bent toward the normal and we expect θ_1 to be greater than θ_2.

Problem-Solving Insight Remember that the refractive indices are written as n_1 for the medium in which the incident light travels and n_2 for the medium in which the refracted light travels.

Solution From the data in the drawing it follows that $\tan\theta_2 = (2.0\text{ m})/(3.3\text{ m})$, so $\theta_2 = 31°$. With $n_1 = 1.00$ for air and $n_2 = 1.33$ for water, Snell's law gives

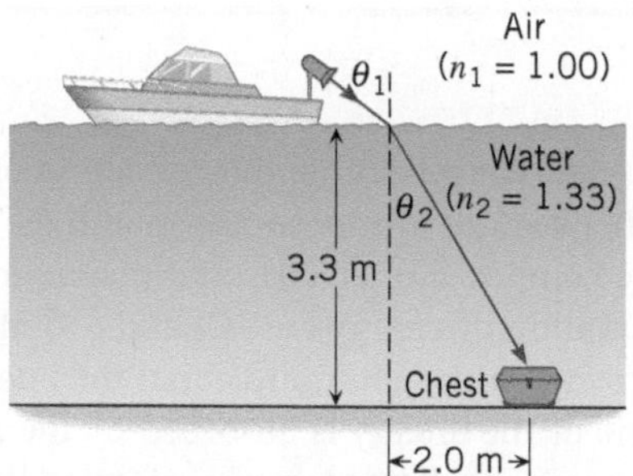

FIGURE 26.3 The beam from the searchlight is refracted when it enters the water.

$$\sin\theta_1 = \frac{n_2\sin\theta_2}{n_1} = \frac{(1.33)\sin 31°}{1.00} = 0.69$$

$$\theta_1 = \sin^{-1}(0.69) = \boxed{44°}$$

As expected, θ_1 is greater than θ_2.

When the sunken chest in Example 2 is viewed from the boat (**Figure 26.4*a***), light rays from the chest pass upward through the water, refract away from the normal when they enter the air, and then travel to the observer. This picture is similar to **Figure 26.3**, except that the direction of the rays is reversed and the searchlight is replaced by an observer. When the rays entering the air are extended back into the water (see the dashed lines), they indicate that the observer sees a virtual image of the chest at an *apparent depth* that is less than the actual depth. The image is virtual because light rays do not actually pass through

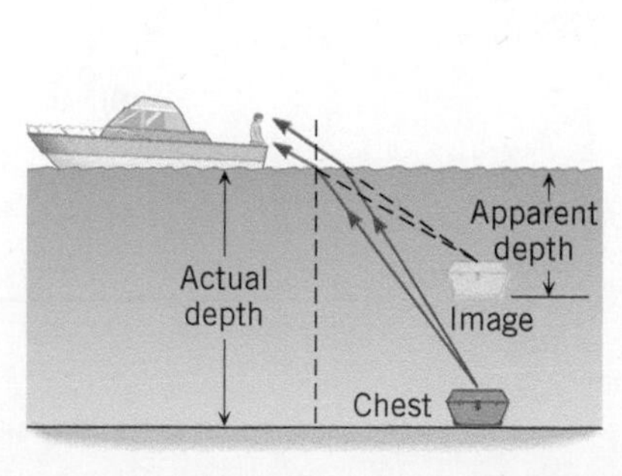

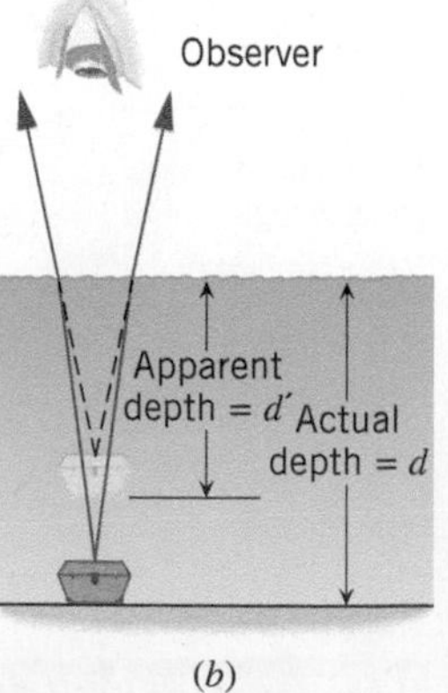

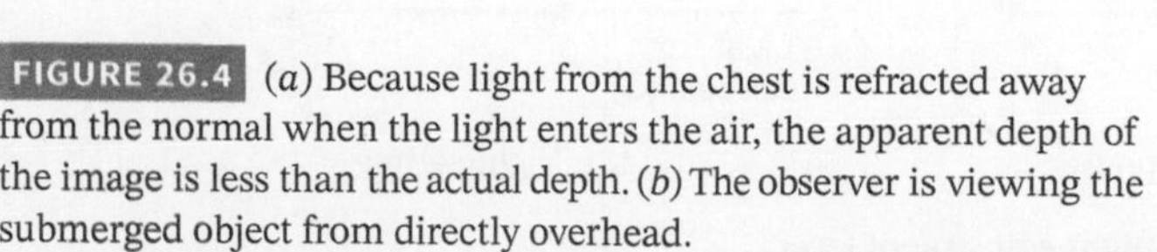
FIGURE 26.4 (*a*) Because light from the chest is refracted away from the normal when the light enters the air, the apparent depth of the image is less than the actual depth. (*b*) The observer is viewing the submerged object from directly overhead.

it. For the situation shown in **Figure 26.4a**, it is difficult to determine the apparent depth. A much simpler case is shown in part *b*, where the observer is *directly above* the submerged object, and the apparent depth d' is related to the actual depth d by

Apparent depth, observer directly above object

$$d' = d\left(\frac{n_2}{n_1}\right) \quad \textbf{(26.3)}$$

In this result, n_1 is the refractive index of the medium associated with the incident ray (the medium in which the object is located), and n_2 refers to the medium associated with the refracted ray (the medium in which the observer is situated). The proof of Equation 26.3 is the focus of Problem 23 at the end of the chapter. Example 3 illustrates that the effect of apparent depth is quite noticeable in water.

EXAMPLE 3 | The Apparent Depth of a Swimming Pool

A swimmer is treading water (with her head above the water) at the surface of a pool 3.00 m deep. She sees a coin on the bottom directly below. How deep does the coin appear to be?

Reasoning Equation 26.3 may be used to find the apparent depth, provided we remember that the light rays travel from the coin to the swimmer. Therefore, the incident ray is coming from the coin under the water ($n_1 = 1.33$), while the refracted ray is in the air ($n_2 = 1.00$).

Solution The apparent depth d' of the coin is

$$d' = d\left(\frac{n_2}{n_1}\right) = (3.00 \text{ m})\left(\frac{1.00}{1.33}\right) = \boxed{2.26 \text{ m}} \quad \textbf{(26.3)}$$

In Example 3, a person sees a coin on the bottom of a pool at an apparent depth that is less than the actual depth. Conceptual Example 4 considers the reverse situation—namely, a person looking from under the water at a coin in the air.

CONCEPTUAL EXAMPLE 4 | On the Inside Looking Out

A swimmer is under water and looking up at the surface. Someone holds a coin in the air, directly above the swimmer's eyes. To the swimmer, the coin appears to be at a certain height above the water. Is the apparent height of the coin **(a)** greater than, **(b)** less than, or **(c)** the same as its actual height?

Reasoning **Figure 26.5** shows two rays of light leaving a point P on the coin. When the rays enter the water, they are refracted toward the normal because water has a larger index of refraction than air has. By extending the refracted rays backward (see the dashed lines in the drawing), we find that they appear to originate from a point P' on a virtual image, which is what the swimmer sees.

Answers (b) and (c) are incorrect. These answers are incorrect because the point P' in **Figure 26.5** is located at a height that is greater than, not less than or the same as, the actual height of the coin.

Answer (a) is correct. The point P' in **Figure 26.5** is on a virtual image that is located at an apparent height d' that is greater than the actual height d. Equation 26.3 [$d' = d(n_2/n_1)$] reveals the same result, because n_1 represents the medium (air) associated with the incident ray and n_2 represents the medium (water) associated with the refracted ray. Since n_2 for water is greater than n_1 for air, the ratio n_2/n_1 is greater than one and d' is larger than d. This situation is the opposite of that in **Figure 26.4b**, where an object beneath the water appears to a person above the water to be closer to the surface than it actually is.

Related Homework: Problems 19, 98

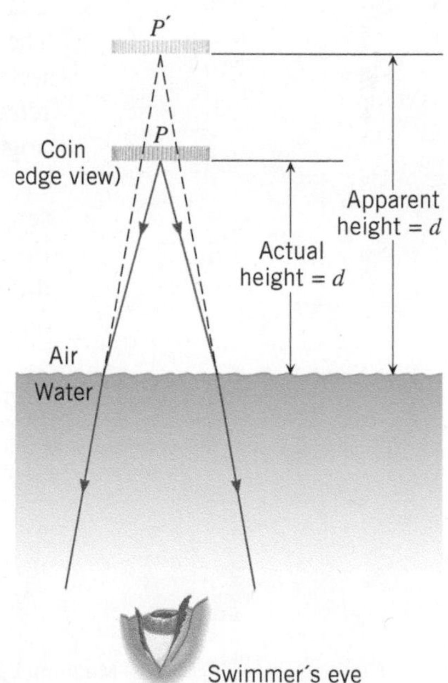

FIGURE 26.5 Rays from point P on a coin in the air above the water refract toward the normal as they enter the water. An underwater swimmer perceives the rays as originating from a point P' that is farther above the surface than the actual point P.

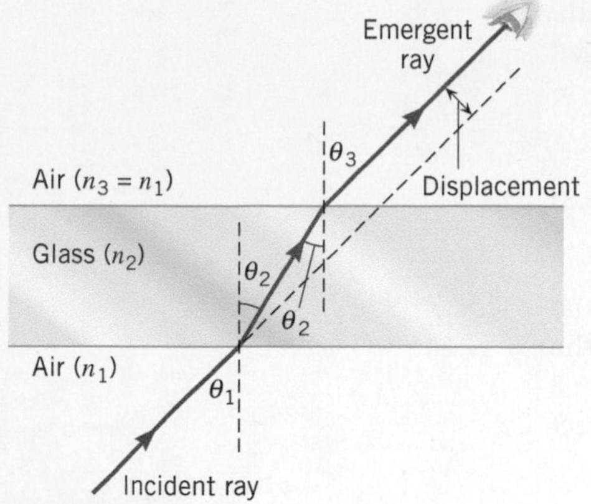

FIGURE 26.6 When a ray of light passes through a pane of glass that has parallel surfaces and is surrounded by air, the emergent ray is parallel to the incident ray ($\theta_3 = \theta_1$) but is displaced from it.

The Displacement of Light by a Transparent Slab of Material

A windowpane is an example of a transparent slab of material. It consists of a plate of glass with parallel surfaces. When a ray of light passes through the glass, the emergent ray is parallel to the incident ray but displaced from it, as **Figure 26.6** shows. This result can be verified by applying Snell's law to each of the two glass surfaces, with the result that $n_1 \sin \theta_1 = n_2 \sin \theta_2 = n_3 \sin \theta_3$. Since air surrounds the glass, $n_1 = n_3$, and it follows that $\sin \theta_1 = \sin \theta_3$. Therefore, $\theta_1 = \theta_3$, and the emergent and incident rays are parallel. However, as the drawing shows, the emergent ray is displaced laterally relative to the incident ray. The extent of the displacement depends on the angle of incidence, the thickness of the slab, and the refractive index of the slab.

Derivation of Snell's Law

Snell's law can be derived by considering what happens to the wave fronts when the light passes from one medium into another. **Figure 26.7*a*** shows light propagating from medium 1, where the speed is relatively large, into medium 2, where the speed is smaller; therefore, n_1 is less than n_2. The plane wave fronts in this picture are drawn perpendicular to the incident and refracted rays. Since the part of each wave front that penetrates medium 2 slows down, the wave fronts in medium 2 are rotated clockwise relative to those in medium 1. Correspondingly, the refracted ray in medium 2 is bent toward the normal, as the drawing shows.

Although the incident and refracted waves have different speeds, *they have the same frequency f.* The fact that the frequency does not change can be understood in terms of the atomic mechanism underlying the generation of the refracted wave. When the electromagnetic wave strikes the surface, the oscillating electric field forces the electrons in the molecules of medium 2 to oscillate at the same frequency as the wave. The accelerating electrons behave like atomic antennas that radiate "extra" electromagnetic waves, which combine with the original wave. The net electromagnetic wave within medium 2 is a superposition of the original wave plus the extra radiated waves, and it is this superposition that constitutes the refracted wave. Since the extra waves are radiated at the same frequency as the original wave, the refracted wave also has the same frequency as the original wave.

The distance between successive wave fronts in **Figure 26.7*a*** has been chosen to be the wavelength λ. Since the frequencies are the same in both media but the speeds are different, it follows from Equation 16.1 that the wavelengths are different: $\lambda_1 = v_1/f$ and $\lambda_2 = v_2/f$. Since v_1 is assumed to be larger than v_2, λ_1 is larger than λ_2, and the wave fronts are farther apart in medium 1.

Figure 26.7*b* shows an enlarged view of the incident and refracted wave fronts at the surface. The angles θ_1 and θ_2 within the colored right triangles are,

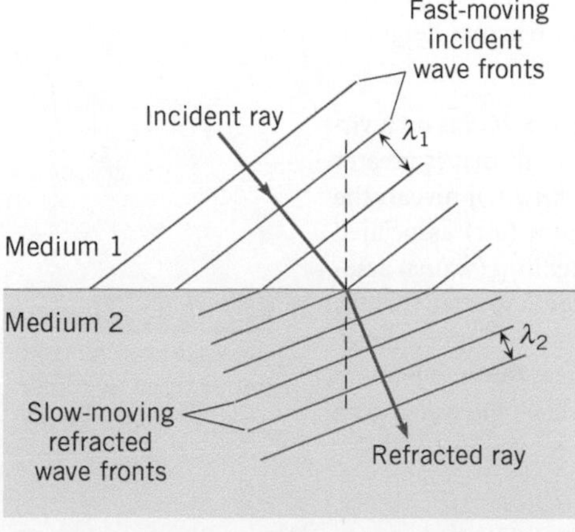

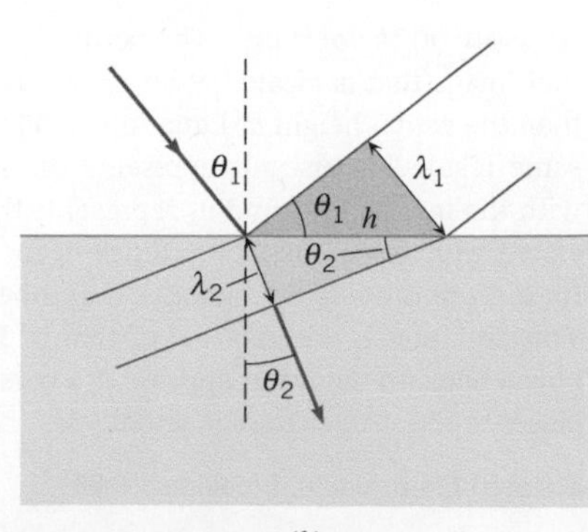

FIGURE 26.7 (*a*) The wave fronts are refracted as the light passes from medium 1 into medium 2. (*b*) An enlarged view of the incident and refracted wave fronts at the surface.

respectively, the angles of incidence and refraction. In addition, the triangles share the same hypotenuse h. Therefore,

$$\sin \theta_1 = \frac{\lambda_1}{h} = \frac{v_1/f}{h} = \frac{v_1}{hf}$$

and

$$\sin \theta_2 = \frac{\lambda_2}{h} = \frac{v_2/f}{h} = \frac{v_2}{hf}$$

Combining these two equations into a single equation by eliminating the common term hf gives

$$\frac{\sin \theta_1}{v_1} = \frac{\sin \theta_2}{v_2}$$

By multiplying each side of this result by c, the speed of light in a vacuum, and recognizing that the ratio c/v is the index of refraction n, we arrive at Snell's law of refraction: $n_1 \sin \theta_1 = n_2 \sin \theta_2$.

Check Your Understanding

(*The answers are given at the end of the book.*)

1. Two slabs with parallel faces are made from different types of glass. A ray of light travels through air and enters each slab at the same angle of incidence, as **CYU Figure 26.1** shows. Which slab has the greater index of refraction?

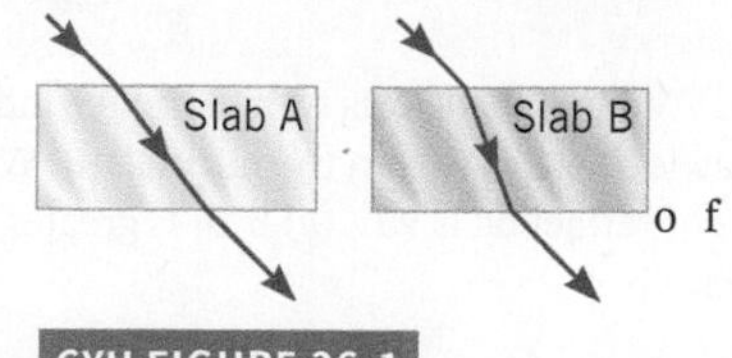

CYU FIGURE 26.1

2. **CYU Figure 26.2** shows three layers of liquids, A, B, and C, each with a different index of refraction. Light begins in liquid A, passes into B, and eventually into C, as the ray of light in the drawing shows. The dashed lines denote the normals to the interfaces between the layers. Which liquid has the smallest index of refraction?

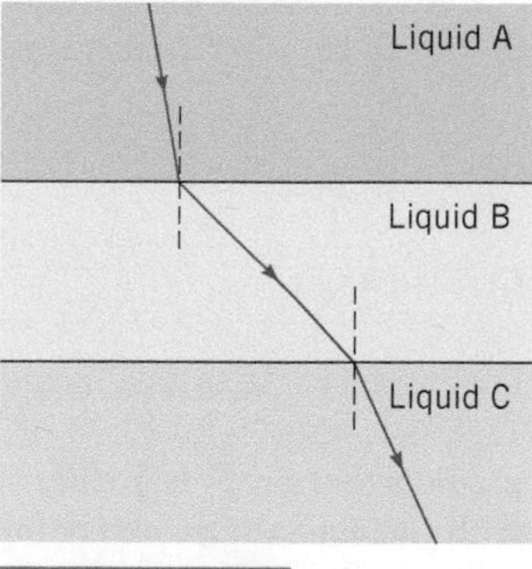

CYU FIGURE 26.2

3. Light traveling through air is incident on a flat piece of glass at a 35° angle of incidence and enters the glass at an angle of refraction θ_{glass}. Suppose that a layer of water is added on top of the glass. Then the light travels through air and is incident on the water at the 35° angle of incidence. Does the light enter the glass at the same angle of refraction θ_{glass} as it did when the water was not present?

4. Two identical containers, one filled with water ($n = 1.33$) and the other with benzene ($n = 1.50$), are viewed from directly above. Which container (if either) appears to have a greater depth of fluid?

5. When an observer peers over the edge of a deep, empty, metal bowl on a kitchen table, he does not see the entire bottom surface. Therefore, a small object lying on the bottom is hidden from view, but the object can be seen when the bowl is filled with liquid A. When the bowl is filled with liquid B, however, the object remains hidden from view. Which liquid has the greater index of refraction?

6. A man is fishing from a dock, using a bow and arrow. To strike a fish that he sees beneath the water, should he aim **(a)** somewhat above the fish, **(b)** directly at the fish, or **(c)** somewhat below the fish?

7. A man is fishing from a dock. He is using a laser gun that emits an intense beam of light. To strike a fish that he sees beneath the water, should he aim **(a)** somewhat above the fish, **(b)** directly at the fish, or **(c)** somewhat below the fish?

8. Two rays of light converge to a point on a screen. A thick plate of glass with parallel surfaces is placed in the path of this converging light, with the parallel surfaces parallel to the screen. Will the point of convergence **(a)** move away from the glass plate, **(b)** move toward the glass plate, or **(c)** remain on the screen?

26.3 Total Internal Reflection

When light passes from a medium of larger refractive index into one of smaller refractive index—for example, from water to air—the refracted ray bends *away* from the normal, as in **Animated Figure 26.8*a***. As the angle of incidence increases, the angle of refraction also increases. When the angle of incidence reaches a certain value, called the **critical angle** θ_c, the angle of refraction is 90°. Then the refracted ray points along the surface; **Animated Figure 26.8*b*** illustrates what happens at the critical angle. When the angle of incidence exceeds the critical angle, as in **Animated Figure 26.8*c***, there is no refracted light. All the incident light is reflected back into the medium from which it came, a phenomenon called **total internal reflection**. Total internal reflection occurs only when light travels from a higher-index medium toward a lower-index medium. It does not occur when light propagates in the reverse direction—for example, from air to water.

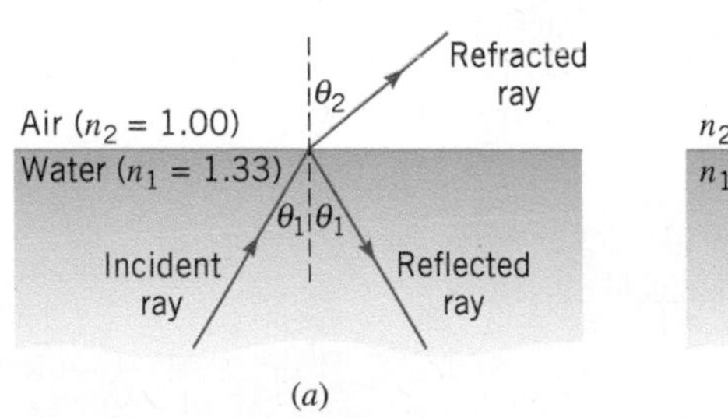

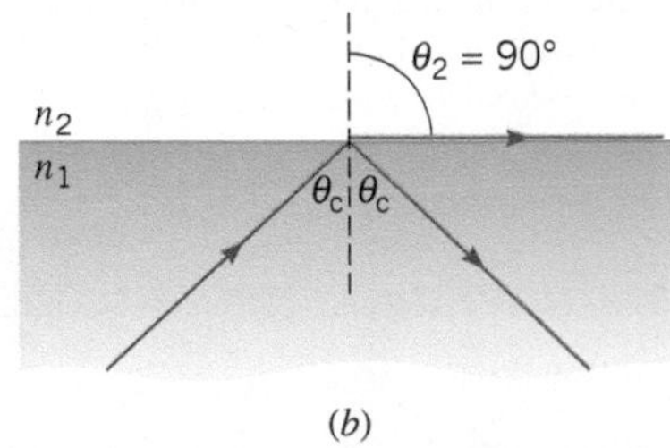

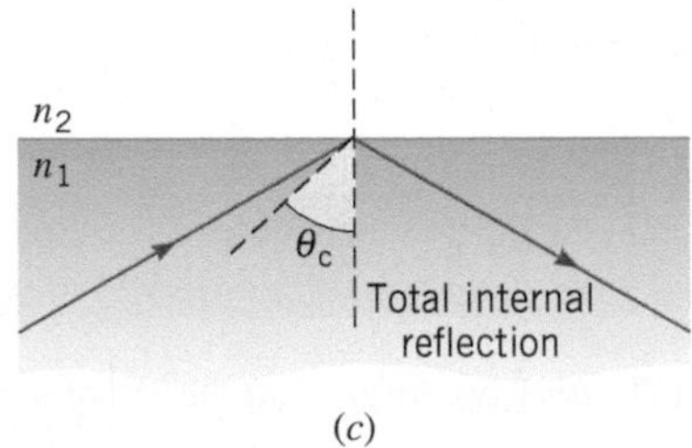

ANIMATED FIGURE 26.8 (*a*) When light travels from a higher-index medium (water) into a lower-index medium (air), the refracted ray is bent away from the normal. (*b*) When the angle of incidence is equal to the critical angle θ_c, the angle of refraction is 90°. (*c*) If θ_1 is greater than θ_c, there is no refracted ray, and total internal reflection occurs.

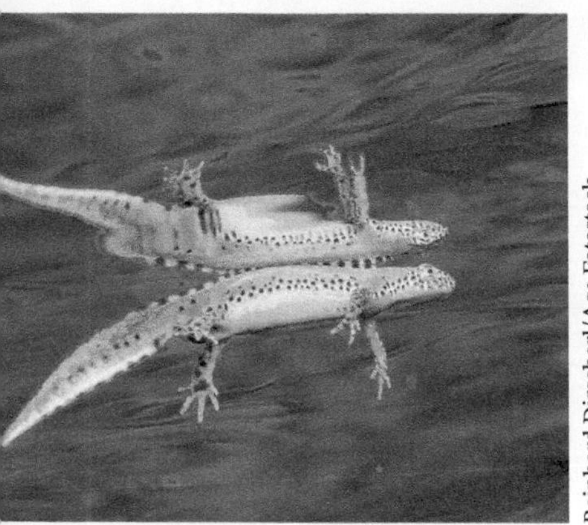

Reinhard Dirscherl/Age Fotostock

FIGURE 26.9 This underwater photograph shows a salamander with its snout near the surface of the water. Some of the light from its body strikes the air–water interface at angles greater than the critical angle and is reflected. Thus, the interface acts like a mirror and forms the image in the upper part of the photograph.

An expression for the critical angle θ_c can be obtained from Snell's law by setting $\theta_1 = \theta_c$ and $\theta_2 = 90°$ (see **Animated Figure 26.8*b***):

$$\sin\theta_c = \frac{n_2 \sin 90°}{n_1}$$

Critical angle

$$\sin\theta_c = \frac{n_2}{n_1} \quad (n_1 > n_2) \qquad \textbf{(26.4)}$$

For instance, the critical angle for light traveling from water (n_1 = 1.33) to air (n_2 = 1.00) is $\theta_c = \sin^{-1}(1.00/1.33) = 48.8°$. For incident angles greater than 48.8°, Snell's law predicts that $\sin\theta_2$ is greater than unity, a value that is not possible. Thus, light rays with incident angles exceeding 48.8° yield no refracted light, and the light is totally reflected back into the water, as **Animated Figure 26.8*c*** indicates. Then, the air–water interface acts like a mirror. **Figure 26.9**, for example, shows the mirror-like ability of the interface to form a reflected image of a salamander with its snout near the surface of the water. Light from the salamander's body that strikes the surface at angles exceeding the critical angle is reflected to form the image in the upper part of the photograph.

The next example illustrates how the critical angle changes when the indices of refraction change.

EXAMPLE 5 | Total Internal Reflection

A beam of light is propagating through diamond (n_1 = 2.42) and strikes a diamond–air interface at an angle of incidence of 28°. **(a)** Will part of the beam enter the air (n_2 = 1.00) or will the beam be totally reflected at the interface? **(b)** Repeat part (a), assuming that the diamond is surrounded by water (n_2 = 1.33) instead of air.

Reasoning Total internal reflection occurs only when the beam of light has an angle of incidence that is greater than the critical angle θ_c. The critical angle is different in parts (a) and (b), since it depends on the ratio n_2/n_1 of the refractive indices of the incident (n_1) and refracting (n_2) media.

Solution **(a)** The critical angle θ_c for total internal reflection at the diamond–air interface is given by Equation 26.4 as

$$\theta_c = \sin^{-1}\left(\frac{n_2}{n_1}\right) = \sin^{-1}\left(\frac{1.00}{2.42}\right) = 24.4°$$

Because the angle of incidence of 28° is greater than the critical angle, there is no refraction, and the light is totally reflected back into the diamond.

(b) If water, rather than air, surrounds the diamond, the critical angle for total internal reflection becomes larger:

$$\theta_c = \sin^{-1}\left(\frac{n_2}{n_1}\right) = \sin^{-1}\left(\frac{1.33}{2.42}\right) = 33.3°$$

Now a beam of light that has an angle of incidence of 28° (less than the critical angle of 33.3°) at the diamond–water interface is refracted into the water.

The critical angle plays an important role in why a diamond sparkles, as Conceptual Example 6 discusses.

CONCEPTUAL EXAMPLE 6 | The Physics of Why a Diamond Sparkles

A diamond gemstone is famous for its sparkle in air because the light coming from it glitters as the diamond is moved about. The sparkle is related to the total internal reflection of light that occurs within the diamond. What happens to the sparkle when the diamond is placed under water? **(a)** Nothing happens, for the water has no effect on total internal reflection. **(b)** The water reduces the sparkle markedly by making the total internal reflection less likely to occur.

Reasoning When a diamond is held in a certain way in air, the intensity of the light coming from within it is greatly enhanced. **Figure 26.10** helps to explain that this enhancement or sparkle is related to total internal reflection. Part *a* of the drawing shows a ray of light striking a lower facet of the diamond at an angle of incidence that exceeds the critical angle for a diamond–air interface. As a result, this ray undergoes total internal reflection back into the diamond and eventually exits the top surface. Since diamond has a relatively small critical angle in air, many of the rays striking a lower facet behave in this fashion and create the diamond's sparkle. Part (a) of Example 5 shows that the critical angle is 24.4° and is so small because the index of refraction of diamond ($n = 2.42$) is large compared to that of air ($n = 1.00$).

Answer (a) is incorrect. The water does indeed have an effect on the total internal reflection that occurs. This is because the critical angle depends on the index of refraction of the water as well as that of the diamond (see Equation 26.4).

Answer (b) is correct. **Figure 26.10*b*** illustrates what happens to the same ray of light within the diamond when the diamond is surrounded by water. Because water has a larger index of refraction than air does, the critical angle for the diamond–water interface is no longer 24.4° but increases to 33.3°, as part (b) of Example 5 shows. Therefore, this particular ray is no longer totally internally reflected. As **Figure 26.10*b*** indicates, only some of the light is now reflected back into the diamond, while the remainder escapes into the water. Consequently, less light exits from the top of the diamond, causing it to lose much of its sparkle.

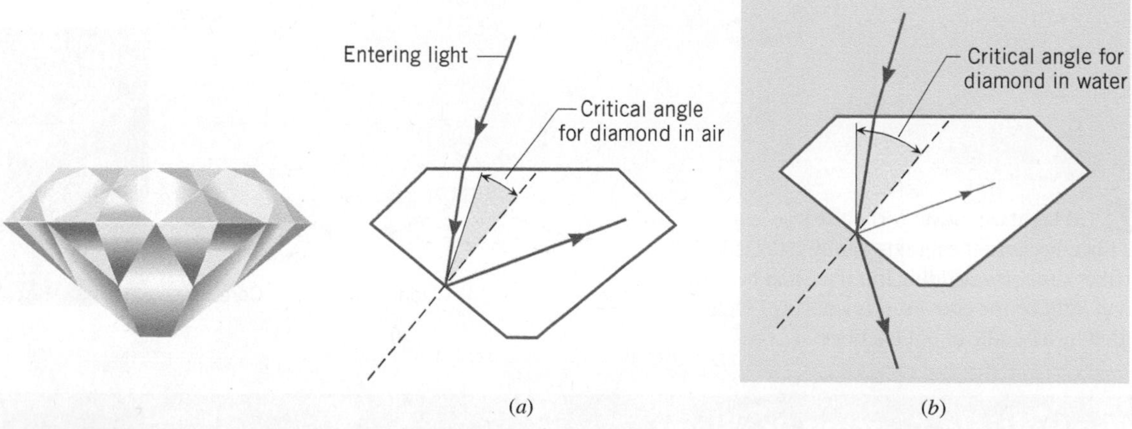

FIGURE 26.10 *(a)* Near the bottom of the diamond, light is totally internally reflected, because the incident angle exceeds the critical angle for diamond and air. *(b)* When the diamond is in water, the same light is partially reflected and partially refracted, since the incident angle is less than the critical angle for diamond and water.

Many optical instruments, such as binoculars, periscopes, and telescopes, use glass prisms and total internal reflection to turn a beam of light through 90° or 180°. **Figure 26.11*a*** shows a light ray entering a 45°–45°–90° glass prism ($n_1 = 1.5$) and striking the hypotenuse of the prism at an angle of incidence of $\theta_1 = 45°$. Equation 26.4 shows that the critical angle for a glass–air interface is $\theta_c = \sin^{-1}(n_2/n_1) = \sin^{-1}(1.0/1.5) = 42°$. Since the angle of incidence is greater than the critical angle, the light is totally reflected at the hypotenuse and is directed vertically upward in the drawing, having been turned through an angle of 90°. **Figure 26.11*b*** shows how the same prism can turn the beam through 180° when total

FIGURE 26.11 Total internal reflection at a glass–air interface can be used to turn a ray of light through an angle of (*a*) 90° or (*b*) 180°. (*c*) Two prisms, each reflecting the light twice by total internal reflection, are sometimes used in binoculars to produce a lateral displacement of a light ray.

internal reflection occurs twice. Prisms can also be used in tandem to produce a lateral displacement of a light ray, while leaving its initial direction unaltered. **Figure 26.11c** illustrates such an application in binoculars.

THE PHYSICS OF . . . fiber optics. An important application of total internal reflection occurs in fiber optics, where hair-thin threads of glass or plastic, called optical fibers, "pipe" light from one place to another. **Figure 26.12*a*** shows that an optical fiber consists of a cylindrical inner *core* that carries the light and an outer concentric shell, the *cladding*. The core is made from transparent glass or plastic that has a relatively high index of refraction. The cladding is also made of glass, but of a type that has a relatively low index of refraction. Light enters one end of the core, strikes the core/cladding interface at an angle of incidence greater than the critical angle, and, therefore, is reflected back into the core. Light thus travels inside the optical fiber along a zigzag path. In a well-designed fiber, little light is lost as a result of absorption by the core, so light can travel many kilometers before its intensity diminishes appreciably. Optical fibers are often bundled together to produce cables. Because the fibers themselves are so thin, the cables are relatively small and flexible and can fit into places inaccessible to larger metal wires. Example 7 deals with the light entering and traveling in an optical fiber.

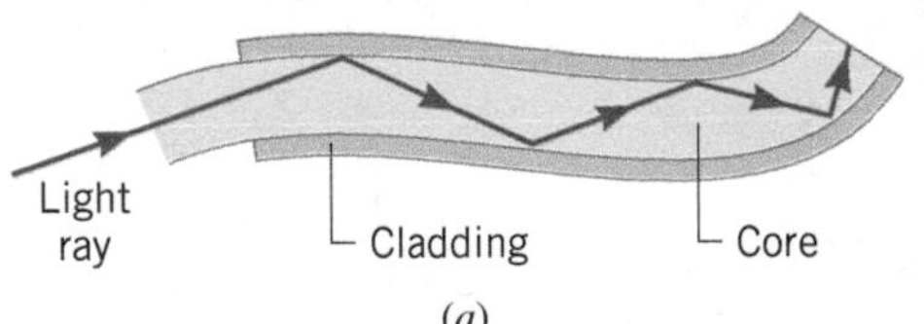

George Doyle/Getty Images

FIGURE 26.12 (*a*) Light can travel with little loss in a curved optical fiber, because the light is totally reflected whenever it strikes the core–cladding interface and because the absorption of light by the core itself is small. (*b*) Light being transmitted by a bundle of optical fibers.

Analyzing Multiple-Concept Problems

EXAMPLE 7 | An Optical Fiber

Figure 26.13 shows an optical fiber that consists of a core made of flint glass ($n_{\text{flint}} = 1.667$) surrounded by a cladding made of crown glass ($n_{\text{crown}} = 1.523$). A ray of light in air enters the fiber at an angle θ_1 with respect to the normal. What is θ_1 if this light also strikes the core-cladding interface at an angle that just barely exceeds the critical angle?

Reasoning The angle of incidence θ_1 is related to the angle of refraction θ_2 (see **Figure 26.13**) by Snell's law, where θ_2 is part of the right triangle in the drawing. The critical angle θ_c for the core-cladding interface is also part of the same right triangle, so that $\theta_2 = 90° - \theta_c$. The critical angle can be determined from a knowledge of the indices of refraction of the core and the cladding.

Knowns and Unknowns The data used in this problem are:

Description	Symbol	Value	Comment
Index of refraction of core material (flint glass)	n_{flint}	1.667	
Index of refraction of cladding material (crown glass)	n_{crown}	1.523	
Index of refraction of air	n_{air}	1.000	See **Table 26.1**.
Unknown Variable			
Angle of incidence of light ray entering optical fiber	θ_1	?	

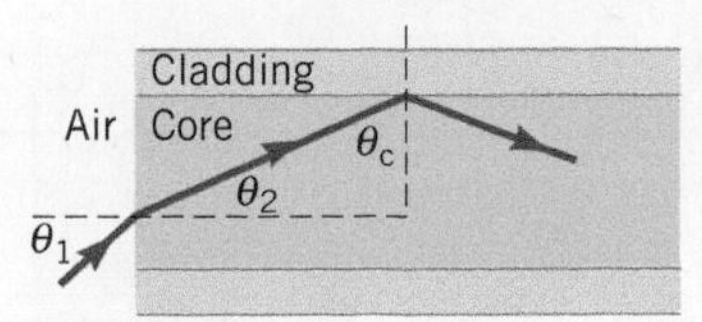

FIGURE 26.13 A ray of light enters the left end of an optical fiber and strikes the core–cladding interface at an angle that just barely exceeds the critical angle θ_c.

Modeling the Problem

STEP 1 Snell's Law of Refraction The ray of light, initially traveling in air, strikes the left end of the optical fiber at an angle of incidence labeled θ_1 in **Figure 26.13**. When the light enters the flint-glass core, its angle of refraction is θ_2. Snell's law of refraction gives the relation between these angles as

$$n_{\text{air}} \sin \theta_1 = n_{\text{flint}} \sin \theta_2 \qquad \textbf{(26.2)}$$

Solving this equation for θ_1 yields Equation 1 at the right. Values for n_{air} and n_{flint} are known. The angle θ_2 will be evaluated in the next step.

$$\theta_1 = \sin^{-1}\left(\frac{n_{\text{flint}} \sin \theta_2}{n_{\text{air}}}\right) \qquad \textbf{(1)}$$

?

STEP 2 The Critical Angle We know that the light ray inside the core strikes the core–cladding interface at an angle that just barely exceeds the critical angle θ_c. When the angle of incidence exceeds the critical angle, all the light is reflected back into the core. From the right triangle in **Figure 26.13**, we see that the critical angle is related to θ_2 by $\theta_2 = 90° - \theta_c$. The critical angle depends on the indices of refraction of the core (flint glass) and cladding (crown glass) according to Equation 26.4:

$$\sin \theta_c = \frac{n_{\text{crown}}}{n_{\text{flint}}} \quad \text{or} \quad \theta_c = \sin^{-1}\left(\frac{n_{\text{crown}}}{n_{\text{flint}}}\right)$$

Substituting this expression for θ_c into $\theta_2 = 90° - \theta_c$ gives

$$\boxed{\theta_2 = 90° - \sin^{-1}\left(\frac{n_{\text{crown}}}{n_{\text{flint}}}\right)}$$

This result for θ_2 can be substituted into Equation 1, as indicated at the right.

$$\theta_1 = \sin^{-1}\left(\frac{n_{\text{flint}} \sin \theta_2}{n_{\text{air}}}\right) \qquad \textbf{(1)}$$

$$\boxed{\theta_2 = 90° - \sin^{-1}\left(\frac{n_{\text{crown}}}{n_{\text{flint}}}\right)} \qquad \textbf{(2)}$$

Solution Combining the results of Steps 1 and 2 algebraically to produce a single equation gives a rather cumbersome result. Hence, we follow the simpler procedure of evaluating Equation 2 numerically and then substituting the result into Equation 1:

$$\theta_2 = 90° - \sin^{-1}\left(\frac{n_{\text{crown}}}{n_{\text{flint}}}\right) = 90° - \underbrace{\sin^{-1}\left(\frac{1.523}{1.667}\right)}_{66.01°} = 23.99° \qquad \textbf{(2)}$$

$$\theta_1 = \sin^{-1}\left(\frac{n_{\text{flint}} \sin \theta_2}{n_{\text{air}}}\right) = \sin^{-1}\left(\frac{1.667 \sin 23.99°}{1.000}\right) = \boxed{42.67°}$$

Related Homework: Problem 32

Optical fiber cables are the medium of choice for high-quality telecommunications because the cables are relatively immune to external electrical interference and because a light beam can carry information through optical fibers just as electricity carries

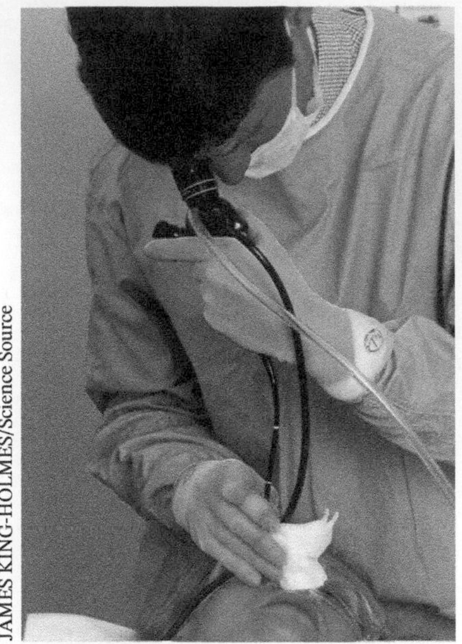
JAMES KING-HOLMES/Science Source

FIGURE 26.14 A doctor is using a bronchoscope to examine the lungs of a patient who has a history of asthma and allergies.

information through copper wires. The information-carrying capacity of light, however, is thousands of times greater than that of electricity. A laser beam traveling through a single optical fiber can carry tens of thousands of telephone conversations and several TV programs simultaneously.

BIO **THE PHYSICS OF . . . endoscopy.** In the field of medicine, optical fiber cables have had extraordinary impact. In the practice of endoscopy, for instance, a device called an endoscope is used to peer inside the body. Figure 26.14 shows a bronchoscope being used, which is a kind of endoscope that is inserted through the nose or mouth, down the bronchial tubes, and into the lungs. It consists of two optical fiber cables. One provides light to illuminate interior body parts, while the other sends back an image for viewing. A bronchoscope greatly simplifies the diagnosis of pulmonary disease. Tissue samples can even be collected with some bronchoscopes. A colonoscope is another kind of endoscope, and its design is similar to that of the bronchoscope. It is inserted through the rectum and used to examine the interior of the colon (see Figure 26.15). The colonoscope currently offers the best hope for diagnosing colon cancer in its early stages, when it can be treated.

BIO **THE PHYSICS OF . . . arthroscopic surgery.** The use of optical fibers has also revolutionized surgical techniques. In arthroscopic surgery, a small surgical instrument, several millimeters in diameter, is mounted at the end of an optical fiber cable. The surgeon can insert the instrument and cable into a joint, such as the knee, with only a tiny incision and minimal damage to the surrounding tissue (see Figure 26.16). Consequently, recovery from the procedure is relatively rapid compared to recovery from traditional surgical techniques.

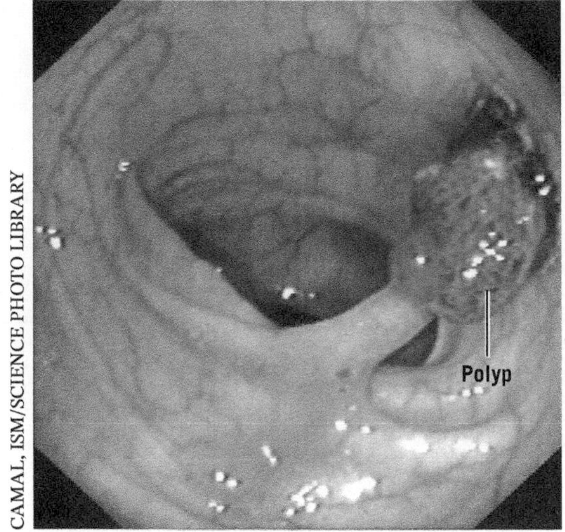

CAMAL, ISM/SCIENCE PHOTO LIBRARY

FIGURE 26.15 A colonoscope revealed this benign (noncancerous) polyp attached to the wall of the colon (large intestine). Polyps that can turn cancerous or grow large enough to obstruct the colon are removed surgically.

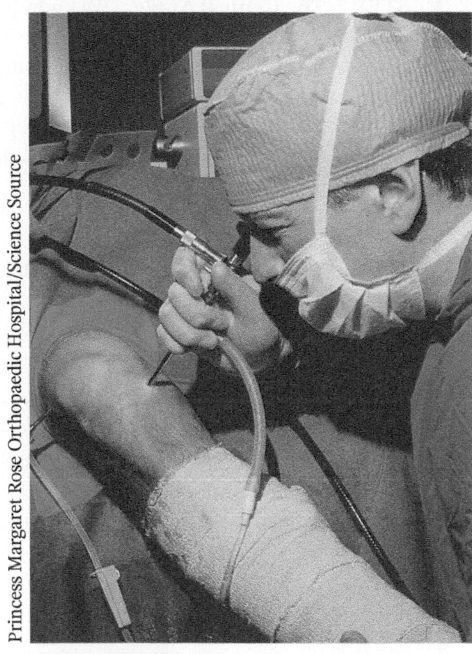
Princess Margaret Rose Orthopaedic Hospital/Science Source

FIGURE 26.16 Optical fibers have made arthroscopic surgery possible, such as the repair of the damaged knee shown here.

Check Your Understanding

(The answers are given at the end of the book.)

9. CYU Figure 26.3 shows a 30°–60°–90° prism and two light rays, A and B, that both strike the prism perpendicularly. The prism is surrounded by an unknown liquid, which is the same in both parts of the drawing. When ray A reaches the hypotenuse in the drawing, it is totally internally reflected. Which one of the following statements applies to ray B when it reaches the hypotenuse?

(a) It may or may not be totally internally reflected, depending on what the surrounding liquid is. **(b)** It is not totally internally reflected, no matter what the surrounding liquid is. **(c)** It is totally internally reflected, no matter what the surrounding liquid is.

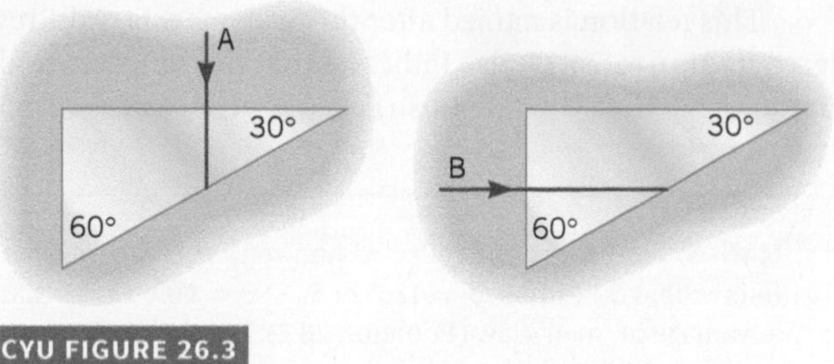

CYU FIGURE 26.3

10. A shallow swimming pool has a constant depth. A point source of light is located in the middle of the bottom of this pool and emits light in all directions. However, no light exits the surface of the water except through a relatively small circular area that is centered on and directly above the light source. Why does the light exit the water through such a limited area?

11. Refer to **Figure 26.6**. Note that the ray within the glass slab is traveling from a medium with a larger refractive index toward a medium with a smaller refractive index. Is it possible, for θ_1 less than 90°, that the ray within the glass will experience total internal reflection at the glass–air interface?

26.4 Polarization and the Reflection and Refraction of Light

For incident angles other than 0°, unpolarized light becomes partially polarized in reflecting from a nonmetallic surface, such as water. To demonstrate this fact, rotate a pair of Polaroid sunglasses in the sunlight reflected from a lake. You will see that the light intensity transmitted through the glasses is a minimum when the glasses are oriented as they are normally worn. Since the transmission axis of the glasses is aligned vertically, it follows that the light reflected from the lake is partially polarized in the horizontal direction.

There is one special angle of incidence at which the reflected light is completely polarized parallel to the surface, the refracted ray being only partially polarized. This angle is called the **Brewster angle** θ_B. **Figure 26.17** summarizes what happens when unpolarized light strikes a nonmetallic surface at the Brewster angle. The value of θ_B is given by **Brewster's law**, in which n_1 and n_2 are, respectively, the refractive indices of the materials in which the incident and refracted rays propagate:

Brewster's law

$$\tan \theta_B = \frac{n_2}{n_1} \tag{26.5}$$

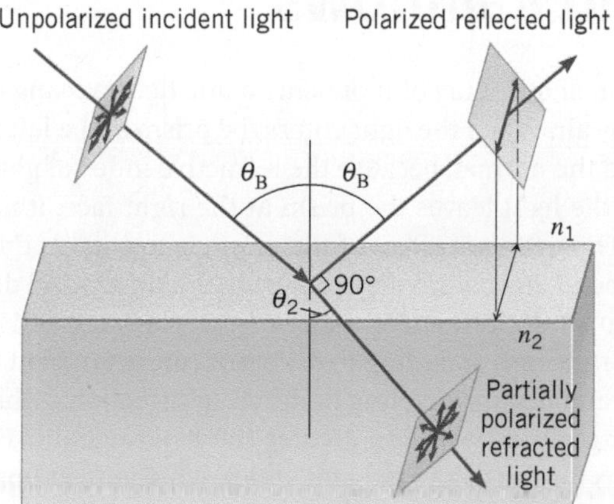

FIGURE 26.17 When unpolarized light is incident on a nonmetallic surface at the Brewster angle θ_B, the reflected light is 100% polarized in a direction parallel to the surface. The angle between the reflected and refracted rays is 90°.

This relation is named after the Scotsman David Brewster (1781–1868), who discovered it. **Figure 26.17** also indicates that the reflected and refracted rays are perpendicular to each other when light strikes the surface at the Brewster angle.

Math Skills Since the reflected and refracted rays are perpendicular in **Figure 26.17**, it follows that $\theta_B + 90° + \theta_2 = 180°$ or $\theta_B + \theta_2 = 90°$. To see that this is indeed the case, we take advantage of Snell's law (Equation 26.2). For an incident angle $\theta_1 = \theta_B$, this law is

$$\sin\theta_B = \frac{n_2 \sin\theta_2}{n_1}$$

In addition, Brewster's law states that $\tan\theta_B = \frac{n_2}{n_1}$ (Equation 26.5), and $\tan\theta_B = \frac{\sin\theta_B}{\cos\theta_B}$ (see Appendix E.2, Other Trigonometric Identities, Equation 4). Therefore, we can substitute $\frac{\sin\theta_B}{\cos\theta_B} = \frac{n_2}{n_1}$ into Snell's law and obtain

$$\cancel{\sin\theta_B} = \left(\frac{n_2}{n_1}\right)\sin\theta_2 = \frac{\cancel{\sin\theta_B}}{\cos\theta_B}\sin\theta_2 \quad \text{or} \quad \cos\theta_B = \sin\theta_2$$

This result is what we are looking for, because $\sin\theta_2 = \cos(90° - \theta_2)$ (see Appendix E.2, Other Trigonometric Identities, Equation 7). Thus, we have

$$\cos\theta_B = \sin\theta_2 = \cos(90° - \theta_2) \quad \text{or} \quad \theta_B = 90° - \theta_2 \quad \text{or} \quad \theta_B + \theta_2 = 90°$$

Check Your Understanding

(The answer is given at the end of the book.)

12. You are sitting by the shore of a lake on a sunny and windless day. When are your Polaroid sunglasses most effective in reducing the glare of the sunlight reflected from the lake surface? When the angle of incidence of the sunlight on the lake is ______. **(a)** almost 90° because the sun is low in the sky **(b)** 0° because the sun is directly overhead **(c)** somewhere between 90° and 0°

26.5 The Dispersion of Light: Prisms and Rainbows

Figure 26.18*a* shows a ray of monochromatic light passing through a glass prism surrounded by air. When the light enters the prism at the left face, the refracted ray is bent toward the normal, because the refractive index of glass is greater than that of air. When the light leaves the prism at the right face, it is refracted away from the normal. Thus, the net effect of the prism is to change the direction of the ray, causing it to bend downward upon entering the prism, and downward again upon leaving. Because the refractive index of the glass depends on wavelength (see **Table 26.2**), rays corresponding to different colors are bent by different amounts by the prism and depart traveling in different directions. The greater the index of refraction for a given color, the greater the bending, and **Figure 26.18*b*** shows the refractions for the colors red and violet, which are at opposite ends of the visible spectrum. If a beam of sunlight, which contains all colors, is sent through the prism, the sunlight is separated into a spectrum of colors, as **Figure 26.18*c*** shows. The spreading of light into its color components is called **dispersion**.

In **Figure 26.18*a*** the ray of light is refracted twice by a glass prism surrounded by air. Conceptual Example 8 explores what happens to the light when the prism is surrounded by materials other than air.

TABLE 26.2 Indices of Refraction *n* of Crown Glass at Various Wavelengths

Color[a]	Vacuum Wavelength (nm)	Index of Refraction, *n*
Red	660	1.520
Orange	610	1.522
Yellow	580	1.523
Green	550	1.526
Blue	470	1.531
Violet	410	1.538

[a]Approximate

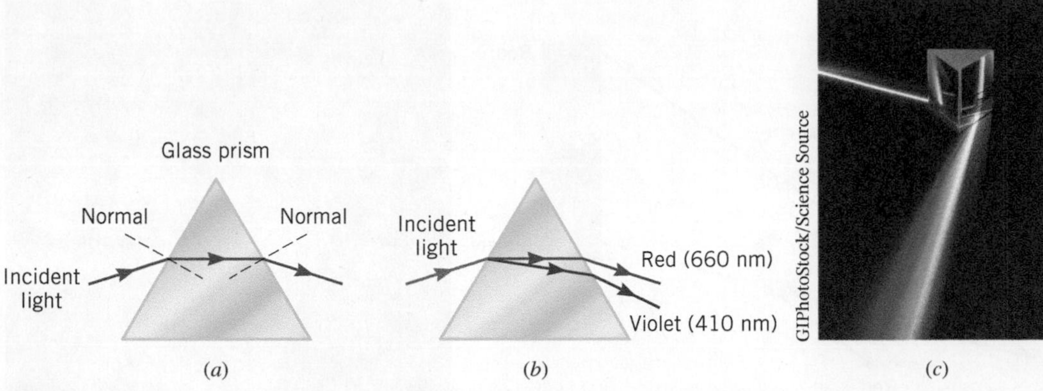

FIGURE 26.18 (*a*) A ray of light is refracted as it passes through a prism. The prism is surrounded by air. (*b*) Two different colors are refracted by different amounts. For clarity, the amount of refraction has been exaggerated. (*c*) Sunlight is dispersed into its color components by this prism.

CONCEPTUAL EXAMPLE 8 | The Refraction of Light Depends on Two Refractive Indices

In **Figure 26.18*a*** the glass prism is surrounded by air and bends the ray of light downward. It is also possible for the prism to bend the ray upward, as in **Figure 26.19*a***, or to not bend the ray at all, as in part *b* of the drawing. How can the situations illustrated in **Figure 26.19** arise?

Reasoning and Solution Snell's law of refraction includes the refractive indices of *both* materials on either side of an interface. With this in mind, we note that the ray bends upward, or away from the normal, as it enters the prism in **Figure 26.19*a***. A ray bends away from the normal when it travels from a medium with a larger refractive index into a medium with a smaller refractive index. When the ray leaves the prism, it again bends upward, which is toward the normal at the point of exit. A ray bends toward the normal when traveling from a smaller toward a larger refractive index. Thus, ***the situation in Figure 26.19a could arise if the prism were immersed in a fluid, such as carbon disulfide, that has a larger refractive index than does glass*** (see **Table 26.1**).

We have seen in **Figures 26.18*a*** and **26.19*a*** that a glass prism can bend a ray of light either downward or upward, depending on whether the surrounding fluid has a smaller or larger index of refraction than the glass. It is logical to conclude, then, that ***a prism will not bend a ray at all, neither up nor down, if the surrounding fluid has the same index of refraction as the glass***—a condition known as *index matching*. This is exactly what is happening in **Figure 26.19*b***, where the ray proceeds straight through the prism as if the prism were not even there. If the index of refraction of the surrounding fluid equals that of the glass prism, then $n_1 = n_2$, and Snell's law ($n_1 \sin \theta_1 = n_2 \sin \theta_2$) reduces to $\sin \theta_1 = \sin \theta_2$. Therefore, the angle of refraction equals the angle of incidence, and no bending of the light occurs.

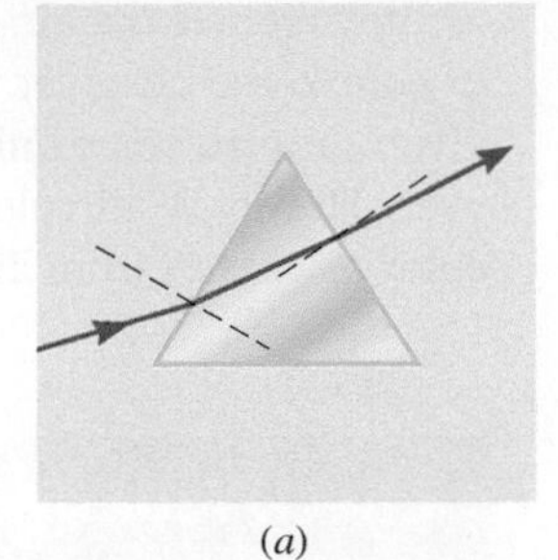

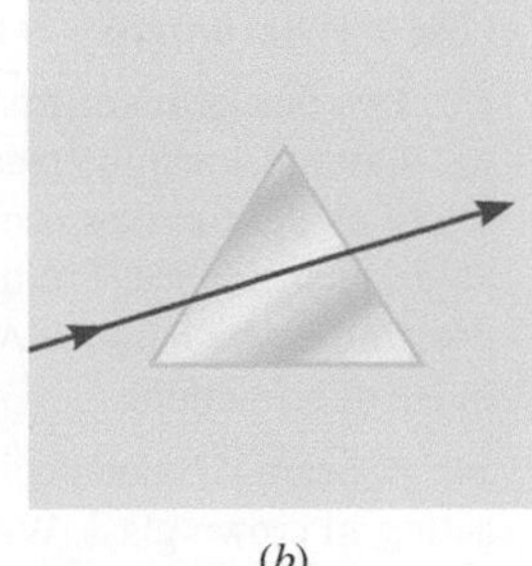

FIGURE 26.19 A ray of light passes through identical prisms, each surrounded by a different fluid. The ray of light is (*a*) refracted upward and (*b*) not refracted at all.

Related Homework: Check Your Understanding 16

THE PHYSICS OF . . . rainbows. Another example of dispersion occurs in rainbows, in which refraction by water droplets gives rise to the colors. You can often see a rainbow just as a storm is leaving, if you look at the departing rain with the sun at your back. When light from the sun enters a spherical raindrop, as in **Figure 26.20**, light of each color is refracted or bent by an amount that depends on the refractive index of water for that wavelength. After

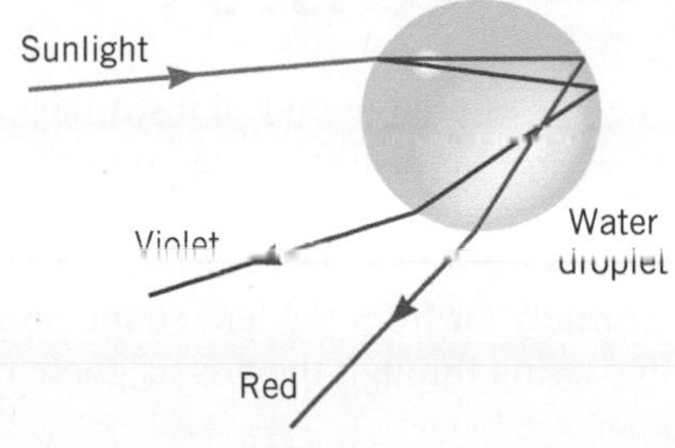

FIGURE 26.20 When sunlight emerges from a water droplet, the light is dispersed into its constituent colors, of which only two are shown.

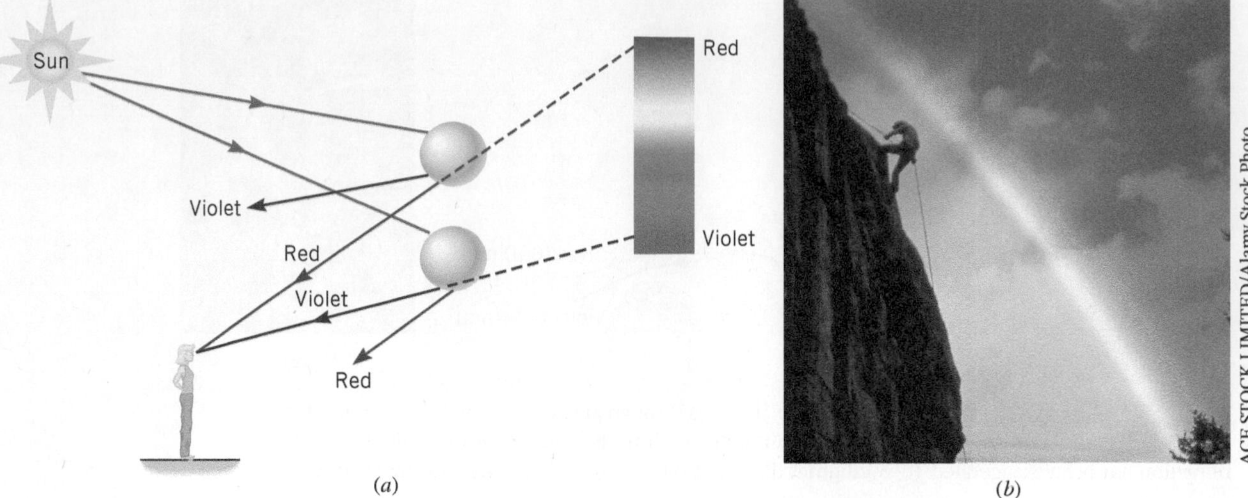

FIGURE 26.21 (*a*) The different colors seen in a rainbow originate from water droplets at different angles of elevation. (*b*) A rock climber beneath a rainbow.

reflection from the back surface of the droplet, the different colors are again refracted as they reenter the air. Although each droplet disperses the light into its full spectrum of colors, the observer in **Figure 26.21a** sees only one color of light coming from any given droplet, since only one color travels in the right direction to reach the observer's eyes. However, all colors are visible in a rainbow (see **Figure 26.21b**) because each color originates from different droplets at different angles of elevation.

Check Your Understanding

(*The answers are given at the end of the book.*)

13. Two blocks, made from the same transparent material, are immersed in different liquids. A ray of light strikes each block at the same angle of incidence. From **CYU Figure 26.4**, determine which liquid, A or B, has the greater index of refraction.

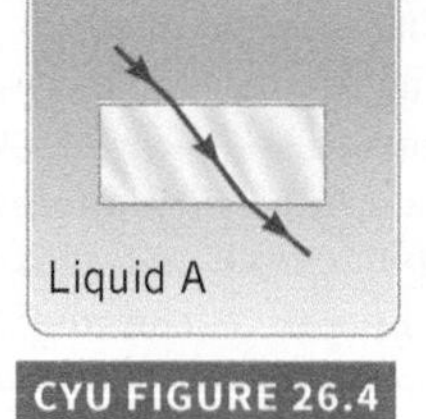

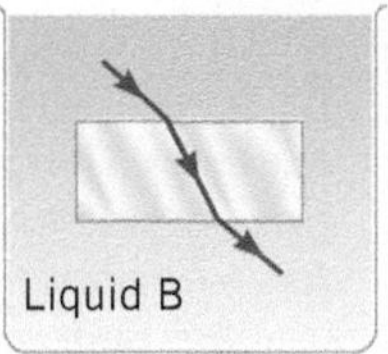

CYU FIGURE 26.4

14. A beam of violet-colored light is propagating in crown glass. When the light reaches the boundary between the glass and the surrounding air, the beam is totally reflected back into the glass. What happens if the light is red and has the same angle of incidence θ_1 at the glass–air interface as does the violet-colored light? **(a)** Depending on the value for θ_1, red light may not be totally reflected, and some of it may be refracted into the air. **(b)** No matter what the value for θ_1, the red light behaves exactly the same as the violet-colored light. (*Hint:* Refer to **Table 26.2** and review Section 26.3.)

26.6 Lenses

The lenses used in optical instruments, such as eyeglasses, cameras, and telescopes, are made from transparent materials that refract light. They refract the light in such a way that an image of the source of the light is formed. **Figure 26.22a** shows a crude lens formed from two glass prisms. Suppose that an object centered on the principal axis is infinitely far from the lens so the rays from the object are parallel to the principal axis. In passing through the prisms, these rays are bent toward the axis because of refraction.

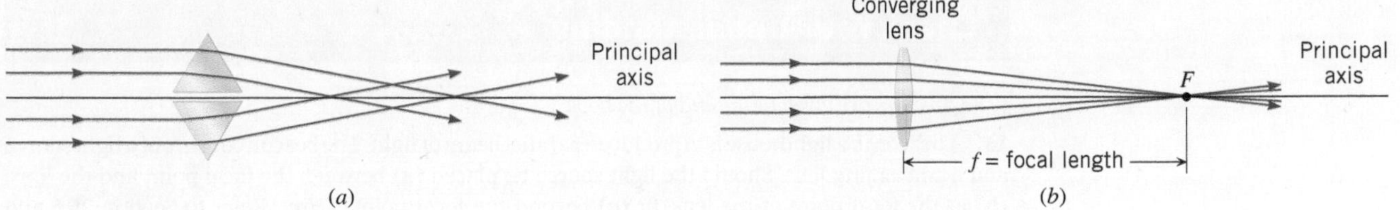

FIGURE 26.22 (*a*) These two prisms cause rays of light that are parallel to the principal axis to change direction and cross the axis at different points. (*b*) With a converging lens, paraxial rays that are parallel to the principal axis converge to the focal point *F* after passing through the lens.

Unfortunately, the rays do not all cross the axis at the same place, and, therefore, such a crude lens gives rise to a blurred image of the object.

A better lens can be constructed from a single piece of transparent material with properly curved surfaces, often spherical, as in **Figure 26.22*b***. With this improved lens, rays that are near the principal axis (paraxial rays) and parallel to it converge to a single point on the axis after emerging from the lens. This point is called the **focal point** *F* of the lens. Thus, an object located infinitely far away on the principal axis leads to an image at the focal point of the lens. The distance between the focal point and the lens is the **focal length** *f*. In what follows, we assume the lens is so thin compared to *f* that it makes no difference whether *f* is measured between the focal point and either surface of the lens or the center of the lens. The type of lens in **Figure 26.22*b*** is known as a **converging lens** because it causes incident parallel rays to converge at the focal point.

Another type of lens found in optical instruments is a **diverging lens**, which causes incident parallel rays to diverge after exiting the lens. Two prisms can also be used to form a crude diverging lens, as in **Figure 26.23*a***. In a properly designed diverging lens, such as the one in part *b* of the picture, paraxial rays that are parallel to the principal axis appear to originate from a single point on the axis after passing through the lens. This point is the focal point *F*, and its distance *f* from the lens is the focal length. Again, we assume that the lens is thin compared to the focal length.

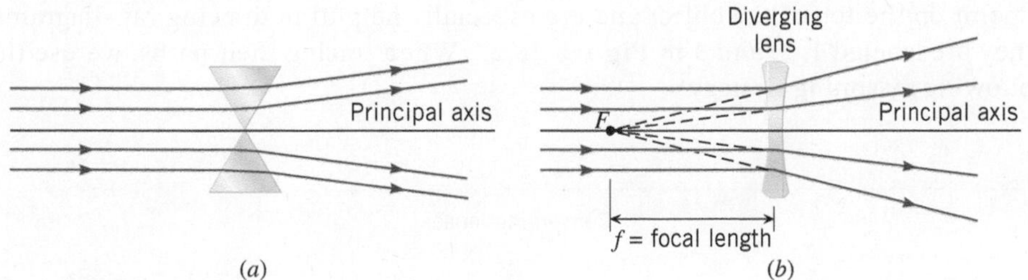

FIGURE 26.23 (*a*) These two prisms cause parallel rays to diverge. (*b*) With a diverging lens, paraxial rays that are parallel to the principal axis appear to originate from the focal point *F* after passing through the lens.

Converging and diverging lenses come in a variety of shapes, as **Figure 26.24** illustrates. Observe that converging lenses are thicker at the center than at the edges, whereas diverging lenses are thinner at the center.

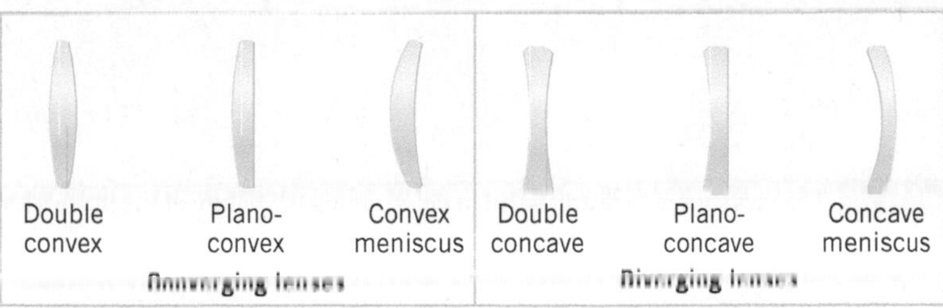

FIGURE 26.24 Converging and diverging lenses come in a variety of shapes.

Check Your Understanding

(The answers are given at the end of the book.)

15. A beacon in a lighthouse is to produce a parallel beam of light. The beacon consists of a light source and a converging lens. Should the light source be placed **(a)** between the focal point and the lens, **(b)** at the focal point of the lens, or **(c)** beyond the focal point? (*Hint:* Refer to Section 25.5 and review the principle of reversibility.)

16. Review Conceptual Example 8 as an aid in answering this question. Is it possible for a lens to behave as a converging lens when surrounded by air but to behave as a diverging lens when surrounded by another medium?

26.7 The Formation of Images by Lenses

Ray Diagrams and Ray Tracing

Each point on an object emits light rays in all directions, and when some of these rays pass through a lens, they form an image. As with mirrors, the ray-tracing method can be used to determine the location and size of the image. Lenses differ from mirrors, however, in that light can pass through a lens from left to right or from right to left. Therefore, when constructing ray diagrams, begin by locating a focal point F on *each side of the lens;* each point lies on the principal axis at the same distance f from the lens. The lens is assumed to be thin, in that its thickness is small compared with the focal length and the distances of the object and the image from the lens. For convenience, it is also assumed that the object is located to the left of the lens and is oriented perpendicular to the principal axis. There are three paraxial rays that leave a point on the top of the object and are especially helpful in drawing ray diagrams. They are labeled 1, 2, and 3 in **Figure 26.25**. When tracing their paths, we use the following reasoning strategy.

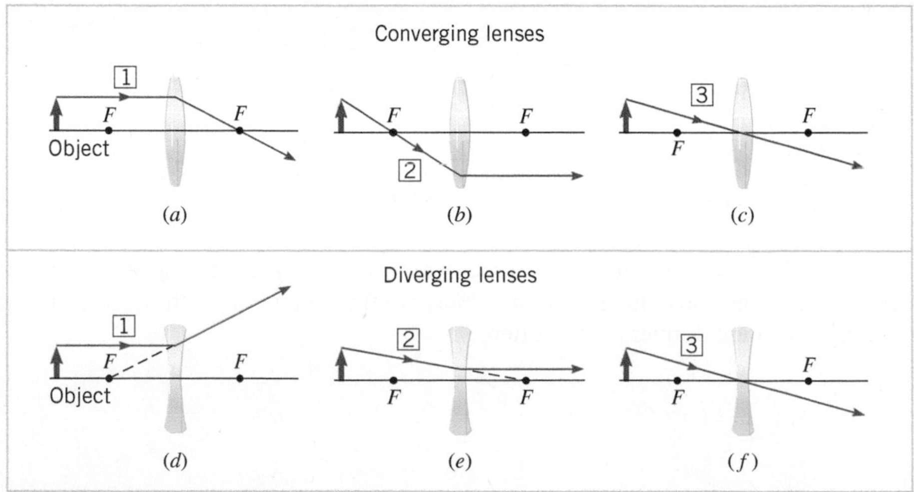

FIGURE 26.25 The rays shown here are useful in determining the nature of the images formed by converging and diverging lenses.

REASONING STRATEGY Ray Tracing for Converging and Diverging Lenses

Converging Lens	Diverging Lens
Ray 1	
This ray initially travels parallel to the principal axis. In passing through a converging lens, the ray is refracted toward the axis and travels through the focal point on the right side of the lens, as Figure 26.25*a* shows.	This ray initially travels parallel to the principal axis. In passing through a diverging lens, the ray is refracted away from the axis, and *appears* to have originated from the focal point on the left of the lens. The dashed line in Figure 26.25*d* represents the apparent path of the ray.
Ray 2	
This ray first passes through the focal point on the left and then is refracted by the lens in such a way that it leaves traveling parallel to the axis, as in Figure 26.25*b*.	This ray leaves the object and moves toward the focal point on the right of the lens. Before reaching the focal point, however, the ray is refracted by the lens so as to exit parallel to the axis. See Figure 26.25*e*, where the dashed line indicates the ray's path in the absence of the lens.
Ray 3*	
This ray travels directly through the center of the thin lens without any appreciable bending, as in Figure 26.25*c*.	This ray travels directly through the center of the thin lens without any appreciable bending, as in Figure 26.25*f*.

*Ray 3 does not bend as it proceeds through the lens because the left and right surfaces of each type of lens are nearly parallel at the center. Thus, in either case, the lens behaves as a transparent slab. As Figure 26.6 shows, the rays incident on and exiting from a slab travel in the same direction with only a lateral displacement. If the lens is sufficiently thin, the displacement is negligibly small.

Image Formation by a Converging Lens

Figure 26.26*a* illustrates the formation of a real image by a converging lens. Here the object is located at a distance from the lens that is greater than twice the focal length (beyond the point labeled 2*F*). To locate the image, any two of the three special rays can be drawn from the tip of the object, although all three are shown in the drawing. The point on the right side of the lens where these rays intersect locates the tip of the image. The ray diagram indicates that the image is real, inverted, and smaller than the object.

THE PHYSICS OF . . . a camera. This optical arrangement is similar to that used in a camera, where an image sensor* or a piece of film records the image (see **Figure 26.26*b***).

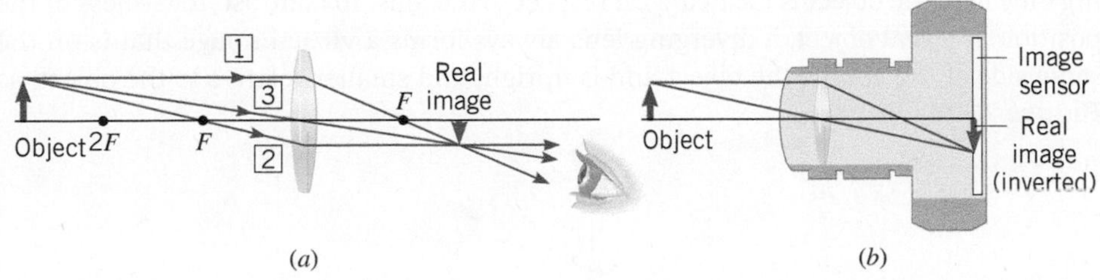

FIGURE 26.26 (*a*) When the object is placed to the left of the point labeled 2*F*, a real, inverted, and smaller image is formed. (*b*) The arrangement in part *a* is like that used in a camera.

*One type of image sensor used in today's digital cameras utilizes a charge-coupled device (CCD). See Section 29.3 for a discussion of CCDs.

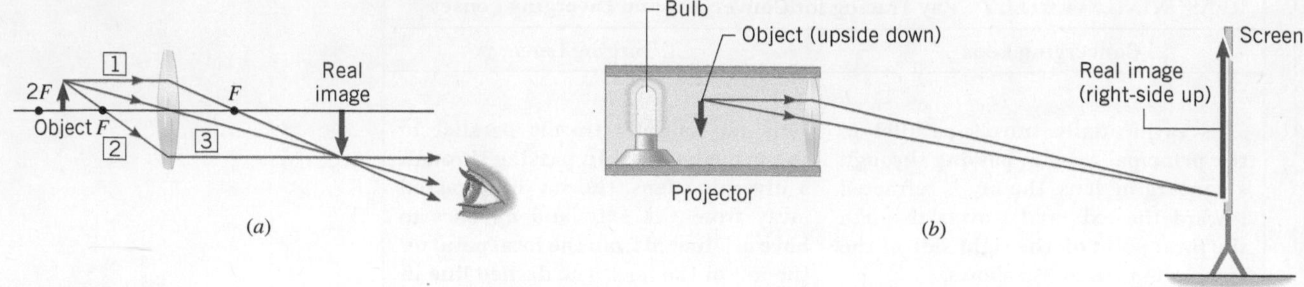

FIGURE 26.27 (*a*) When the object is placed between 2*F* and *F*, the image is real, inverted, and larger than the object. (*b*) This arrangement is found in projectors.

When the object is placed between 2*F* and *F*, as in **Figure 26.27*a***, the image is still real and inverted; however, the image is now larger than the object.

THE PHYSICS OF . . . a slide or film projector. This optical system is used in a slide or film projector in which a small piece of film is the object and the enlarged image falls on a screen. However, to obtain an image that is right-side up, the film must be placed in the projector upside down.

When the object is located between the focal point and the lens, as in **Figure 26.28**, the rays diverge after leaving the lens. To a person viewing the diverging rays, they appear to come from an image behind (to the left of) the lens. Because none of the rays actually come from the image, it is a virtual image. The ray diagram shows that the virtual image is upright and enlarged.

THE PHYSICS OF . . . a magnifying glass. A magnifying glass uses this arrangement, as can be seen in part *b* of the drawing.

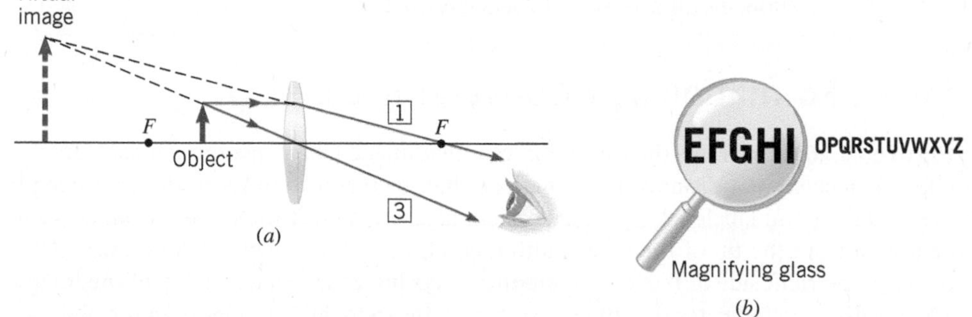

FIGURE 26.28 (*a*) When an object is placed between the focal point *F* of a converging lens and the lens, an upright, enlarged, and virtual image is created. (*b*) Such an image is seen when looking through a magnifying glass.

Image Formation by a Diverging Lens

We have seen that a converging lens can form a real image or a virtual image, depending on where the object is located with respect to the lens. In contrast, regardless of the position of a real object, a diverging lens always forms a virtual image that is on the same side of the lens as the object and is upright and smaller relative to the object, as **Figure 26.29** illustrates.

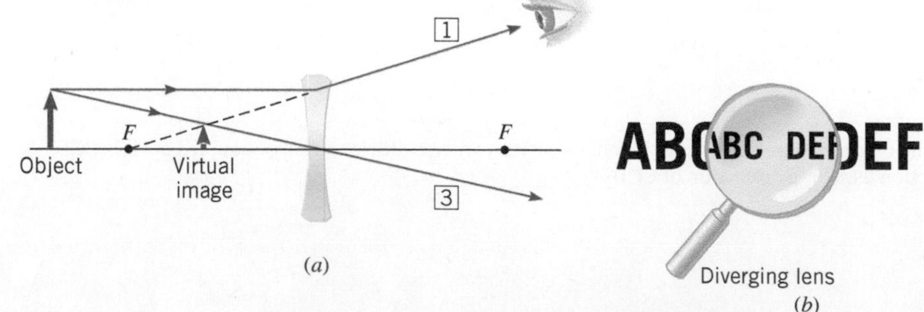

FIGURE 26.29 (*a*) A diverging lens always forms a virtual image of a real object. The image is upright and smaller relative to the object. (*b*) The image seen through a diverging lens.

Check Your Understanding

(*The answer is given at the end of the book.*)

17. A converging lens is used to produce a real image, as in **Figure 26.27a**. A piece of black tape is then placed over the upper half of the lens. Which one of the following statements is true concerning the image that results with the tape in place? **(a)** The image is of the entire object, although its brightness is reduced since fewer rays produce it. **(b)** The image is of the object's lower half only, but its brightness is not reduced. **(c)** The image is of the object's upper half only, but its brightness is not reduced.

26.8 The Thin-Lens Equation and the Magnification Equation

When an object is placed in front of a spherical mirror, we can determine the location, size, and nature of its image by using the technique of ray tracing or the mirror and magnification equations. Both options are based on the law of reflection. The mirror and magnification equations relate the distances d_o and d_i of the object and image from the mirror to the focal length f and magnification m. For an object placed in front of a lens, Snell's law of refraction leads to the technique of ray tracing and to equations that are identical to the mirror and magnification equations. Thus, mirrors work because of the reflection of light, whereas lenses work because of the refraction of light, a distinction between the two devices that is important to keep in mind.

The equations that result from applying Snell's law to lenses are referred to as the thin-lens equation and the magnification equation:

Thin-lens equation

$$\frac{1}{d_o}+\frac{1}{d_i}=\frac{1}{f} \tag{26.6}$$

Math Skills The thin-lens equation $\left(\frac{1}{d_o}+\frac{1}{d_i}=\frac{1}{f}\right)$ is sometimes thought to imply that $d_o+d_i=f$. To emphasize that the focal length f does not equal the object distance d_o plus the image distance d_i, we can solve the thin lens equation for f. First, we multiply the left side of the thin-lens equation by 1 in the form of $\frac{d_o d_i}{d_o d_i}$:

$$\underbrace{\left(\frac{d_o d_i}{d_o d_i}\right)}_{1}\left(\frac{1}{d_o}+\frac{1}{d_i}\right)=\frac{1}{f} \quad \text{or} \quad \left(\frac{1}{d_o d_i}\right)\left(\frac{d_o d_i}{d_o}+\frac{d_o d_i}{d_i}\right)=\frac{1}{f}$$

Simplifying this result gives

$$\left(\frac{1}{d_o d_i}\right)\left(\frac{\cancel{d_o} d_i}{\cancel{d_o}}+\frac{d_o \cancel{d_i}}{\cancel{d_i}}\right)=\frac{1}{f} \quad \text{or} \quad \frac{d_i+d_o}{d_o d_i}=\frac{1}{f}$$

Taking the reciprocal of both sides of the simplified result shows that

$$\left(\frac{d_i+d_o}{d_o d_i}\right)^{-1}=\left(\frac{1}{f}\right)^{-1} \quad \text{or} \quad \frac{d_o d_i}{d_i+d_o}=f$$

Clearly, it is not true that $d_o+d_i=f$. Do not make this mistake when solving problems.

Magnification equation

$$m=\frac{\text{Image height}}{\text{Object height}}=\frac{h_i}{h_o}=-\frac{d_i}{d_o} \tag{26.7}$$

Figure 26.30 defines the symbols in these expressions with the aid of a thin converging lens, but the expressions also apply to a diverging lens, if it is thin. The derivations of these equations are presented at the end of this section.

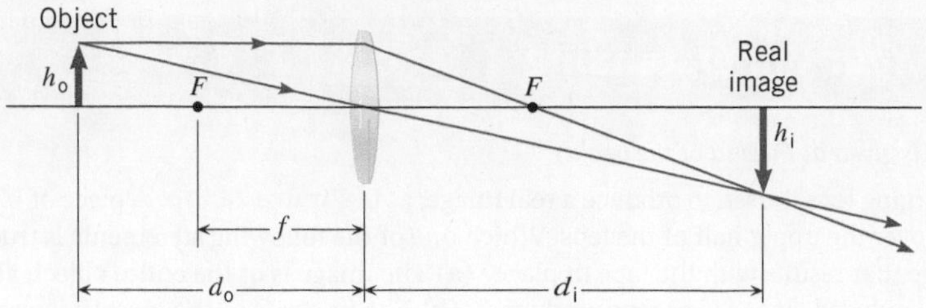

FIGURE 26.30 The drawing shows the focal length f, the object distance d_o, and the image distance d_i for a converging lens. The object and image heights are, respectively, h_o and h_i.

Certain sign conventions accompany the use of the thin-lens and magnification equations, and the conventions are similar to those used with mirrors in Section 25.6. The issue of real versus virtual images, however, is slightly different with lenses than with mirrors. With a mirror, a real image is formed on the *same side* of the mirror as the object (see Figure 25.18), in which case the image distance d_i is a positive number. With a lens, a positive value for d_i also means the image is real. However, starting with an actual object, a real image is formed on the side of the lens *opposite to* the object (see **Figure 26.30**). The **sign conventions** listed in the following Reasoning Strategy apply to light rays traveling from left to right from a real object.

REASONING STRATEGY **Summary of Sign Conventions for Lenses**

Focal length

f is + for a converging lens.

f is − for a diverging lens.

Object distance

d_o is + if the object is to the left of the lens (real object), as is usual.

d_o is − if the object is to the right of the lens (virtual object).*

Image distance

d_i is + for an image (real) formed to the right of the lens by a real object.

d_i is − for an image (virtual) formed to the left of the lens by a real object.

Magnification

m is + for an image that is upright with respect to the object.

m is − for an image that is inverted with respect to the object.

***This situation arises in systems containing more than one lens, where the image formed by the first lens becomes the object for the second lens. In such a case, the object of the second lens may lie to the right of that lens, in which event d_o is assigned a negative value and the object is called a virtual object.**

Examples 9 and 10 illustrate the use of the thin-lens and magnification equations.

EXAMPLE 9 | The Real Image Formed by a Camera Lens

A person 1.70 m tall is standing 2.50 m in front of a digital camera. The camera uses a converging lens whose focal length is 0.0500 m. **(a)** Find the image distance (the distance between the lens and the image sensor) and determine whether the image is real or virtual. **(b)** Find the magnification and the height of the image on the image sensor.

Reasoning This optical arrangement is similar to that in **Figure 26.26*a***, where the object distance is greater than twice the focal length of the lens. Therefore, we expect the image to be real, inverted, and smaller than the object.

Solution **(a)** To find the image distance d_i we use the thin-lens equation with $d_o = 2.50$ m and $f = 0.0500$ m:

$$\frac{1}{d_i} = \frac{1}{f} - \frac{1}{d_o} = \frac{1}{0.0500 \text{ m}} - \frac{1}{2.50 \text{ m}} = 19.6 \text{ m}^{-1} \quad \text{or} \quad \boxed{d_i = 0.0510 \text{ m}}$$

Since the image distance is a positive number, a real image is formed on the image sensor.

(b) The magnification follows from the magnification equation:

$$m = -\frac{d_i}{d_o} = -\frac{0.0510 \text{ m}}{2.50 \text{ m}} = \boxed{-0.0204}$$

The image is 0.0204 times as large as the object, and it is inverted since m is negative. Since the object height is $h_o = 1.70$ m, the image height is

$$h_i = mh_o = (-0.0204)(1.70 \text{ m}) = \boxed{-0.0347 \text{ m}}$$

EXAMPLE 10 | The Virtual Image Formed by a Diverging Lens

An object is placed 7.10 cm to the left of a diverging lens whose focal length is $f = -5.08$ cm (a diverging lens has a negative focal length). **(a)** Find the image distance and determine whether the image is real or virtual. **(b)** Obtain the magnification.

Reasoning This situation is similar to that in **Figure 26.29*a***. The ray diagram shows that the image is virtual, erect, and smaller than the object.

Problem-Solving Insight In the thin-lens equation, the reciprocal of the image distance d_i is given by $d_i^{-1} = f^{-1} - d_o^{-1}$, where f is the focal length and d_o is the object distance. After combining the reciprocals f^{-1} and d_o^{-1}, do not forget to take the reciprocal of the result to find d_i.

Solution (a) The thin-lens equation can be used to find the image distance d_i:

$$\frac{1}{d_i} = \frac{1}{f} - \frac{1}{d_o} = \frac{1}{-5.08 \text{ cm}} - \frac{1}{7.10 \text{ cm}}$$

$$= -0.338 \text{ cm}^{-1} \quad \text{or} \quad \boxed{d_i = -2.96 \text{ cm}}$$

The image distance is negative, indicating that the image is $\boxed{\text{virtual}}$ and located to the left of the lens.

(b) Since d_i and d_o are known, the magnification equation shows that

$$m = -\frac{d_i}{d_o} = -\frac{-2.96 \text{ cm}}{7.10 \text{ cm}} = \boxed{0.417}$$

The image is upright (m is positive) and smaller ($m < 1$) than the object.

The thin-lens and magnification equations can be derived by considering rays 1 and 3 in **Figure 26.31*a***. Ray 1 is shown separately in part *b* of the drawing, where the angle θ is the same in each of the two colored triangles. Thus, $\tan\theta$ is the same for each triangle:

$$\tan\theta = \frac{h_o}{f} = \frac{-h_i}{d_i - f}$$

A minus sign has been inserted in the numerator of the ratio $h_i/(d_i - f)$ for the following reason. The angle θ in **Figure 26.31*b*** is assumed to be positive. Since the image is inverted relative to the object, the image height h_i is a negative number. The insertion of the minus sign ensures that the term $-h_i/(d_i - f)$, and hence $\tan\theta$, is a positive quantity.

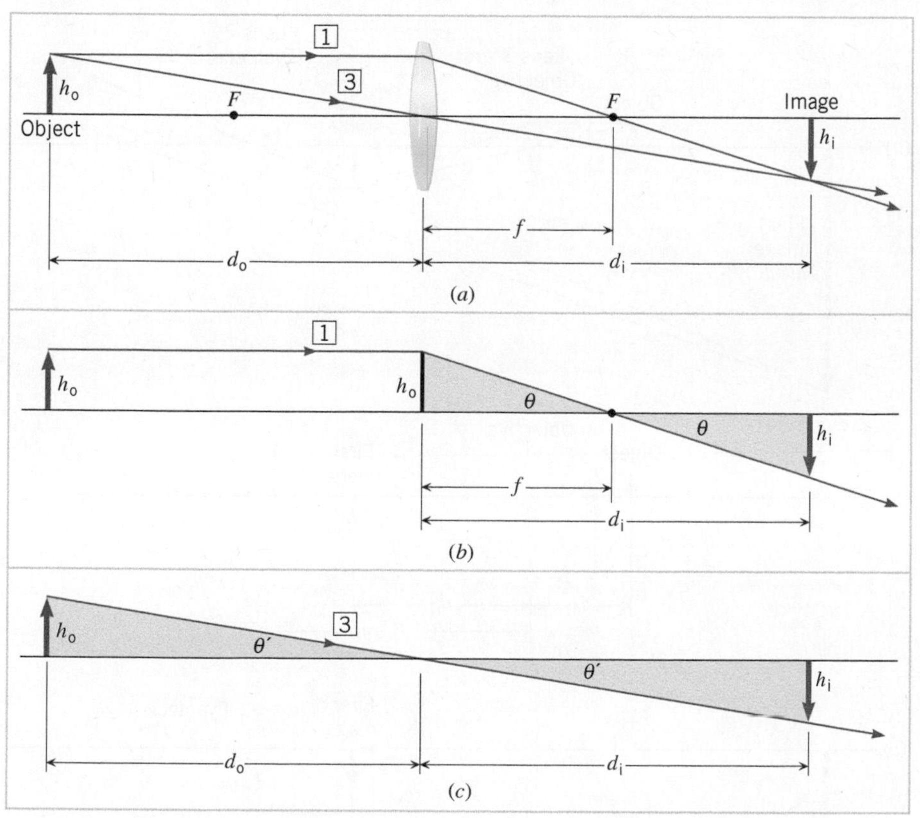

FIGURE 26.31 These ray diagrams are used for deriving the thin-lens and magnification equations.

Ray 3 is shown separately in **Figure 26.31*c***, where the two angles labeled θ' are the same. Therefore,

$$\tan\theta' = \frac{h_o}{d_o} = \frac{-h_i}{d_i}$$

A minus sign has been inserted in the numerator of the term h_i/d_i for the same reason that a minus sign was inserted earlier—namely, to ensure that $\tan\theta'$ is a positive quantity. The first equation gives $h_i/h_o = -(d_i - f)/f$, while the second equation yields

$h_i/h_o = -d_i/d_o$. Equating these two expressions for h_i/h_o and rearranging the result produces the thin-lens equation, $1/d_o + 1/d_i = 1/f$. The magnification equation follows directly from the equation $h_i/h_o = -d_i/d_o$, if we recognize that h_i/h_o is the magnification m of the lens.

Check Your Understanding

(The answers are given at the end of the book.)

18. A spherical mirror and a lens are immersed in water. Compared to the way they work in air, which one do you expect will be more affected by the water?

19. An object is located at a distance d_o in front of a lens. The lens has a focal length f and produces an upright image that is twice as tall as the object. What kind of lens is it, and what is the object distance? Express your answer as a fraction or multiple of the focal length.

20. In an old movie a photographic film negative is introduced as evidence in a trial. The negative shows an image of a house that no longer exists. The verdict depends on knowing exactly how far above the ground a window ledge was (the object height h_o). The distance between the ground and the ledge on the negative (the image height h_i) can be measured. What additional information is needed to calculate h_o? **(a)** Nothing else is needed. **(b)** Just the object distance d_o, which is the distance between the house and the camera lens. **(c)** Just the focal length f of the lens. **(d)** Both d_o and f are needed.

26.9 Lenses in Combination

Many optical instruments, such as microscopes and telescopes, use a number of lenses together to produce an image. Among other things, a multiple-lens system can produce an image that is magnified more than is possible with a single lens. For instance, **Figure 26.32*a***

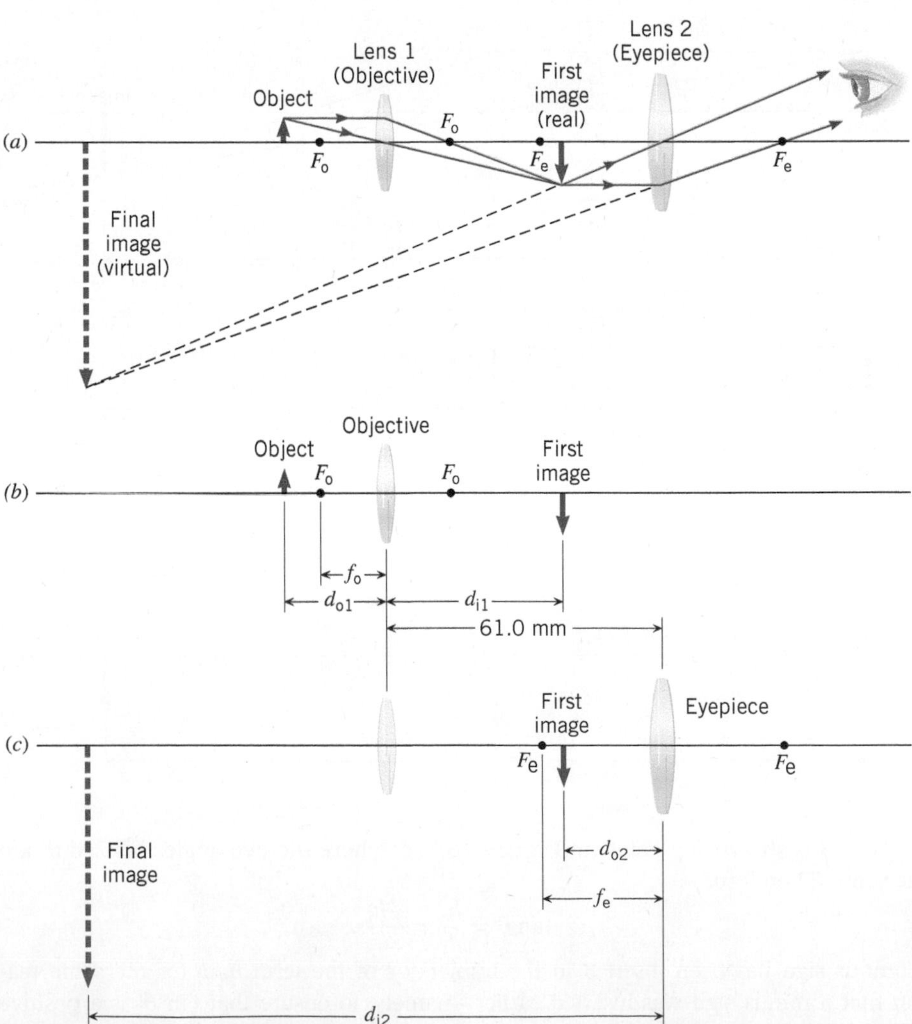

FIGURE 26.32 *(a)* This two-lens system can be used as a compound microscope to produce a virtual, enlarged, and inverted final image. *(b)* The objective forms the first image and *(c)* the eyepiece forms the final image.

shows a two-lens system used in a microscope. The first lens, the lens closest to the object, is referred to as the *objective*. The second lens is known as the *eyepiece* (or *ocular*). The object is placed just outside the focal point F_o of the objective. The image formed by the objective—called the "first image" in the drawing—is real, inverted, and enlarged compared to the object. This first image then serves as the object for the eyepiece. Since the first image falls between the eyepiece and its focal point F_e, the eyepiece forms an enlarged, virtual, final image, which is what the observer sees.

The location of the final image in a multiple-lens system can be determined by applying the thin-lens equation to each lens separately. The key point to remember in such situations is the following:

Problem-Solving Insight The image produced by one lens serves as the object for the next lens.

The next example illustrates this point.

EXAMPLE 11 | A Microscope—Two Lenses in Combination

The objective and eyepiece of the compound microscope in **Figure 26.32** are both converging lenses and have focal lengths of $f_o = 15.0$ mm and $f_e = 25.5$ mm. A distance of 61.0 mm separates the lenses. The microscope is being used to examine an object placed $d_{o1} = 24.1$ mm in front of the objective. Find the final image distance relative to the eyepiece.

Reasoning The thin-lens equation can be used to locate the final image produced by the eyepiece. We know the focal length of the eyepiece, but to determine the final image distance from the thin-lens equation we also need to know the object distance, which is not given. To obtain this distance, we recall that the image produced by one lens (the objective) is the object for the next lens (the eyepiece). We can use the thin-lens equation to locate the image produced by the objective, since the focal length and the object distance for this lens are given. The location of this image relative to the eyepiece will tell us the object distance for the eyepiece.

Solution The final image distance relative to the eyepiece is d_{i2}, and we can determine it by using the thin-lens equation:

$$\frac{1}{d_{i2}} = \frac{1}{f_e} - \frac{1}{d_{o2}}$$

The focal length f_e of the eyepiece is known, but to obtain a value for the object distance d_{o2} we must locate the first image produced by the objective. The first image distance d_{i1} (see **Figure 26.32*b***) can be determined using the thin-lens equation with $d_{o1} = 24.1$ mm and $f_o = 15.0$ mm.

$$\frac{1}{d_{i1}} = \frac{1}{f_o} - \frac{1}{d_{o1}} = \frac{1}{15.0\text{ mm}} - \frac{1}{24.1\text{ mm}}$$

$$= 0.0252\text{ mm}^{-1} \quad \text{or} \quad d_{i1} = 39.7\text{ mm}$$

The first image now becomes the object for the eyepiece, as indicated in **Figure 26.32*c***. Since the distance between the lenses is 61.0 mm, the object distance for the eyepiece is $d_{o2} = 61.0\text{ mm} - d_{i1} = 61.0\text{ mm} - 39.7\text{ mm} = 21.3$ mm. Noting that the focal length of the eyepiece is $f_e = 25.5$ mm, we can find the final image distance with the thin-lens equation:

$$\frac{1}{d_{i2}} = \frac{1}{f_e} - \frac{1}{d_{o2}} = \frac{1}{25.5\text{ mm}} - \frac{1}{21.3\text{ mm}}$$

$$= -0.0077\text{ mm}^{-1} \quad \text{or} \quad \boxed{d_{i2} = -130\text{ mm}}$$

The fact that d_{i2} is negative indicates that the final image is virtual. It lies to the left of the eyepiece, as the drawing shows.

Problem-Solving Insight The overall magnification m of a two-lens system is the product of the magnifications m_1 and m_2 of the individual lenses, or $m = m_1 \times m_2$.

Suppose, for example, that the image of lens 1 is magnified by a factor of 5 relative to the original object, so that $m_1 = 5$. As we know, the image of lens 1 serves as the object for lens 2. Suppose, in addition, that lens 2 magnifies this object further by a factor of 8, so that $m_2 = 8$. The final image of the two-lens system, then, would be $5 \times 8 = 40$ times as large as the original object. In other words, the overall magnification is $m = m_1 \times m_2$, and this calculation can be extended to any number of lenses.

26.10 The Human Eye

Anatomy

Without doubt, the human eye is the most remarkable of all optical devices. **Interactive Figure 26.33** shows some of its main anatomical features. The eyeball is approximately spherical with a diameter of about 25 mm. Light enters the eye through a transparent membrane (the *cornea*). This membrane covers a clear liquid region (the *aqueous*

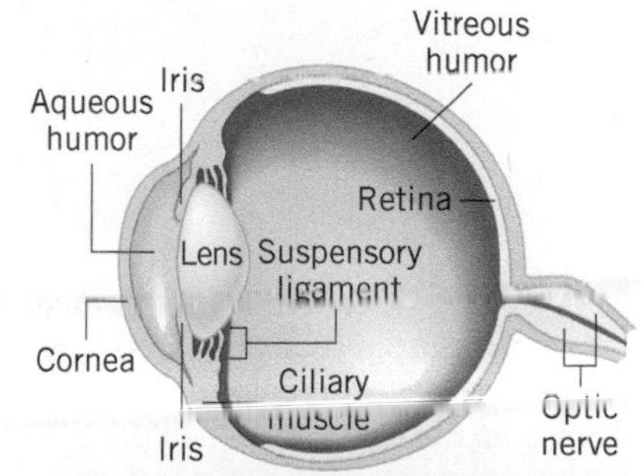

INTERACTIVE FIGURE 26.33 A cross-sectional view of the human eye.

humor), behind which are a diaphragm (the *iris*), the *lens*, a region filled with a jelly-like substance (the *vitreous humor*), and, finally, the *retina*. The retina is the light-sensitive part of the eye, consisting of millions of structures called *rods* and *cones*. When stimulated by light, these structures send electrical impulses via the *optic nerve* to the brain, which interprets the image detected by the retina.

The iris is the colored portion of the eye and controls the amount of light reaching the retina. The iris acts as a controller because it is a muscular diaphragm with a variable opening at its center, through which the light passes. The opening is called the *pupil*. The diameter of the pupil varies from about 2 to 7 mm, decreasing in bright light and increasing (dilating) in dim light.

Of prime importance to the operation of the eye is the fact that the lens is flexible, and its shape can be altered by the action of the *ciliary muscle*. The lens is connected to the ciliary muscle by the *suspensory ligaments* (see the drawing). We will see shortly how the shape-changing ability of the lens affects the focusing ability of the eye.

Optics

BIO **THE PHYSICS OF . . . the human eye.** Optically, the eye and the camera are similar; both have a lens system and a diaphragm with a variable opening or aperture at its center. Moreover, the retina of the eye and the image sensor in a camera serve similar functions, for both record the image formed by the lens system. In the eye, the image formed on the retina is real, inverted, and smaller than the object, just as it is in a camera. Although the image on the retina is inverted, it is interpreted by the brain as being right-side up.

For clear vision, the eye must refract the incoming light rays, so as to form a sharp image on the retina. In reaching the retina, the light travels through five different media, each with a different index of refraction n: air (n = 1.00), the cornea (n = 1.38), the aqueous humor (n = 1.33), the lens (n = 1.40, on the average), and the vitreous humor (n = 1.34). Each time light passes from one medium into another, it is refracted at the boundary. The greatest amount of refraction, about 70% or so, occurs at the air/cornea boundary. According to Snell's law, the large refraction at this interface occurs primarily because the refractive index of air (n = 1.00) is so different from that of the cornea (n = 1.38). The refraction at all the other boundaries is relatively small because the indices of refraction on either side of these boundaries are nearly equal. The lens itself contributes only about 20–25% of the total refraction, since the surrounding aqueous and vitreous humors have indices of refraction that are nearly the same as that of the lens.

Even though the lens contributes only a quarter of the total refraction or less, its function is an important one. The eye has a fixed image distance; that is, the distance between the lens and the retina is constant. Therefore, the only way that objects located at different distances can produce sharp images on the retina is for the focal length of the lens to be adjustable. It is the ciliary muscle that adjusts the focal length. When the eye looks at a very distant object, the ciliary muscle is not tensed. The lens has its least curvature and, consequently, its longest focal length. Under this condition the eye is said to be "fully relaxed," and the rays form a sharp image on the retina, as in **Figure 26.34*a***. When the object moves closer to the eye, the ciliary muscle

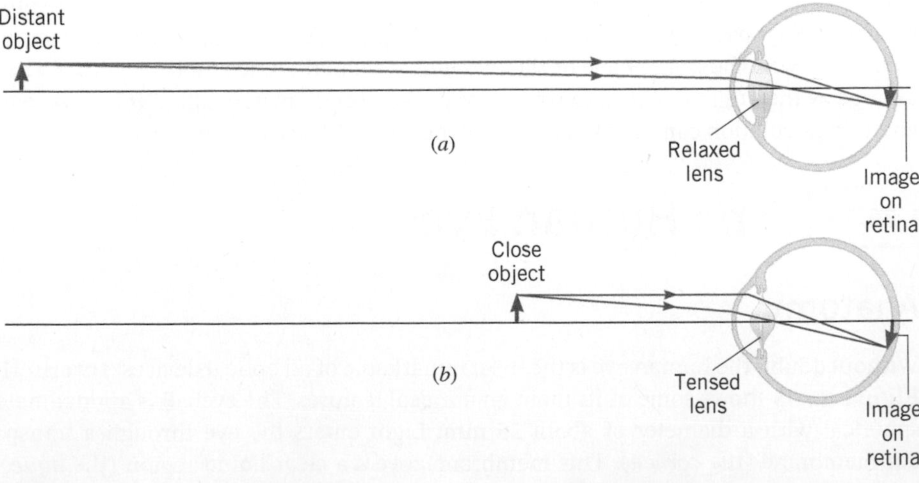

FIGURE 26.34 (*a*) When fully relaxed, the lens of the eye has its longest focal length, and an image of a very distant object is formed on the retina. (*b*) When the ciliary muscle is tensed, the lens has a shorter focal length. Consequently, an image of a closer object is formed on the retina.

tenses automatically, thereby increasing the curvature of the lens, shortening the focal length, and permitting a sharp image to form again on the retina (**Figure 26.34*b***). When a sharp image of an object is formed on the retina, we say the eye is "focused" on the object. The process in which the lens changes its focal length to focus on objects at different distances is called **accommodation**.

When you hold a book too close, the print is blurred because the lens cannot adjust enough to bring the book into focus. The point nearest the eye at which an object can be placed and still produce a sharp image on the retina is called the **near point** of the eye. The ciliary muscle is fully tensed when an object is placed at the near point. For people in their early twenties with normal vision, the near point is located about 25 cm from the eye. It increases to about 50 cm at age 40 and to roughly 500 cm at age 60. Since most reading material is held at a distance of 25–45 cm from the eye, older adults typically need eyeglasses to overcome the loss of accommodation. The **far point** of the eye is the location of the farthest object on which the fully relaxed eye can focus. A person with normal eyesight can see objects very far away, such as the planets and stars, and thus has a far point located nearly at infinity.

Nearsightedness

BIO THE PHYSICS OF . . . nearsightedness. A person who is **nearsighted (myopic)** can focus on nearby objects but cannot clearly see objects far away. For such a person, the far point of the eye is not at infinity and may even be as close to the eye as three or four meters. When a nearsighted eye tries to focus on a distant object, the eye is fully relaxed, like a normal eye. However, the nearsighted eye has a focal length that is shorter than it should be, so rays from the distant object form a sharp image in front of the retina, as **Figure 26.35*a*** shows, and blurred vision results.

The nearsighted eye can be corrected with glasses or contacts that use *diverging* lenses, as **Figure 26.35*b*** suggests. The rays from the object diverge after leaving the eyeglass lens. Therefore, when they are subsequently refracted toward the principal axis by the eye, a sharp image is formed farther back and falls on the retina. Since the relaxed (but nearsighted) eye can focus on an object at the eye's far point—but not on objects farther away—the diverging lens is designed to transform a very distant object into an image located at the

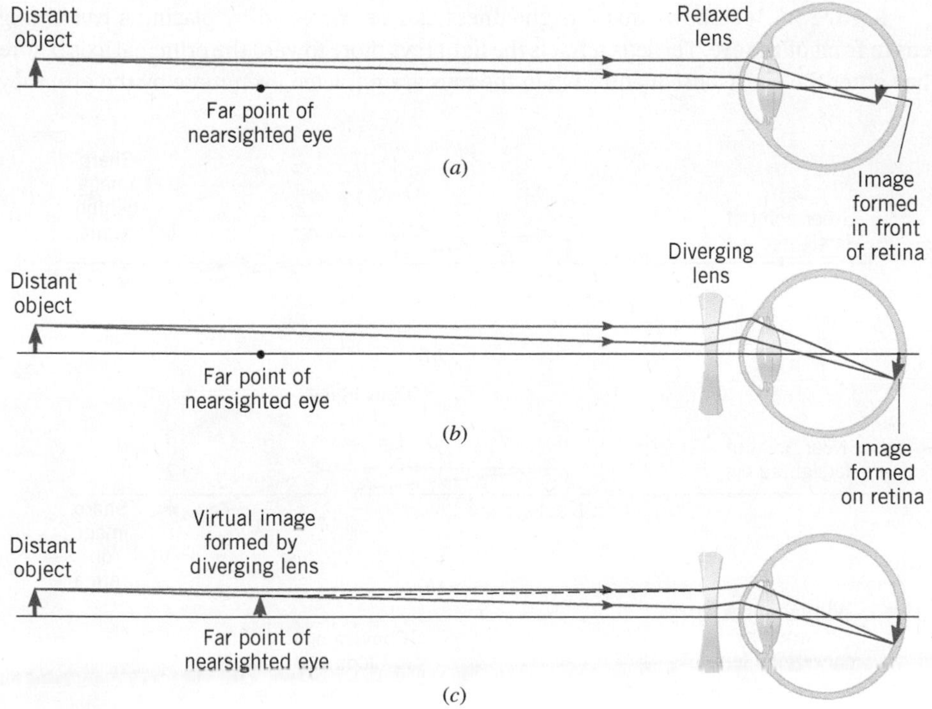

FIGURE 26.35 (*a*) When a nearsighted person views a distant object, the image is formed in front of the retina. The result is blurred vision. (*b*) With a diverging lens in front of the eye, the image is moved onto the retina and clear vision results. (*c*) The diverging lens is designed to form a virtual image at the far point of the nearsighted eye.

far point. **Figure 26.35c** shows this transformation, and the next example illustrates how to determine the focal length of the diverging lens that accomplishes it.

EXAMPLE 12 | BIO Eyeglasses for the Nearsighted Person

A nearsighted person has a far point located only 521 cm from the eye. Assuming that eyeglasses are to be worn 2 cm in front of the eye, find the focal length needed for the diverging lenses of the glasses so the person can see distant objects.

Reasoning In **Figure 26.35c** the far point is 521 cm away from the eye. Since the glasses are worn 2 cm from the eye, the far point is 519 cm to the left of the diverging lens. The image distance, then, is −519 cm, the negative sign indicating that the image is a virtual image formed to the left of the lens. The object is assumed to be infinitely far from the diverging lens. The thin-lens equation can be used to find the focal length of the eyeglasses. We expect the focal length to be negative, since the lens is a diverging lens.

Problem-Solving Insight Eyeglasses are worn about 2 cm from the eyes. Be sure, if necessary, to take this 2 cm into account when determining the object and image distances (d_o and d_i) that are used in the thin-lens equation.

Solution With $d_i = -519$ cm and $d_o = \infty$, the focal length can be found as follows:

$$\frac{1}{f} = \frac{1}{d_o} + \frac{1}{d_i} = \frac{1}{\infty} + \frac{1}{-519 \text{ cm}} \quad \text{or} \quad \boxed{f = -519 \text{ cm}} \qquad \textbf{(26.6)}$$

The value for f is negative, as expected for a diverging lens.

Farsightedness

BIO THE PHYSICS OF . . . farsightedness. A **farsighted (hyperopic)** person can usually see distant objects clearly, but cannot focus on those nearby. Whereas the near point of a young and normal eye is located about 25 cm from the eye, the near point of a farsighted eye may be considerably farther away than that, perhaps as far as several hundred centimeters. When a farsighted eye tries to focus on a book held closer than the near point, it accommodates and shortens its focal length as much as it can. However, even at its shortest, the focal length is longer than it should be. Therefore, the light rays from the book would form a sharp image behind the retina if they could do so, as **Figure 26.36a** suggests. In reality, no light passes through the retina, but a blurred image does form on it.

Figure 26.36b shows that farsightedness can be corrected by placing a *converging* lens in front of the eye. The lens refracts the light rays more toward the principal axis before they enter the eye. Consequently, when the rays are refracted even more by the eye, they

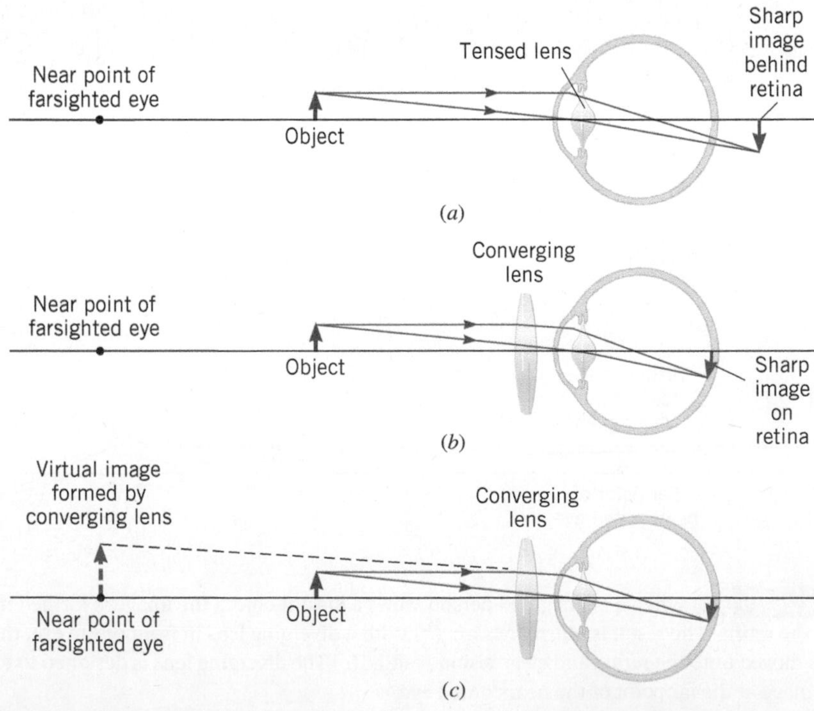

FIGURE 26.36 (*a*) When a farsighted person views an object located between the near point and the eye, a sharp image would be formed behind the retina if light could pass through it. Only a blurred image forms on the retina. (*b*) With a converging lens in front of the eye, the sharp image is moved onto the retina and clear vision results. (*c*) The converging lens is designed to form a virtual image at the near point of the farsighted eye.

converge to form an image on the retina. Part *c* of the figure illustrates what the eye sees when it looks through the converging lens. The lens is designed so that the eye perceives the light to be coming from a virtual image located at the near point. Example 13 shows how the focal length of the converging lens is determined to correct for farsightedness.

EXAMPLE 13 | BIO Contact Lenses for the Farsighted Person

A farsighted person has a near point located 210.0 cm from the eyes. Obtain the focal length of the converging lenses in a pair of contacts that can be used to read a book held 25.0 cm from the eyes.

Reasoning A contact lens is placed directly against the eye. Thus, the object distance, which is the distance from the book to the lens, is 25.0 cm. The lens forms an image of the book at the near point of the eye, so the image distance is −210.0 cm. The minus sign indicates that the image is a virtual image formed to the left of the lens, as in **Figure 26.36c**. The focal length can be obtained from the thin-lens equation.

Solution With $d_o = 25.0$ cm and $d_i = -210.0$ cm, the focal length can be determined from the thin-lens equation as follows:

$$\frac{1}{f} = \frac{1}{d_o} + \frac{1}{d_i} = \frac{1}{25.0\ \text{cm}} + \frac{1}{-210.0\ \text{cm}} = 0.0352\ \text{cm}^{-1} \text{ or } \boxed{f = 28.4\ \text{cm}}$$

BIO THE PHYSICS OF . . . Correcting color vision deficiency. The most common cause of color vision deficiency in humans is an inherited condition affecting one or more of the three sets of color-sensing cells, called cones, in the *fovea centralis*, located at the back of the eye. A common problem is the inability to distinguish red and green, a condition that can sometimes be diagnosed with a simple test (see **Figure 26.37**). The three types of cones cover three overlapping wavelength ranges in the visible spectrum. The cones that are most sensitive to the red part of spectrum span a range of about 500 to 675 nm with a peak sensitivity at about 570 nm. Those most sensitive to the green portion of the spectrum sense wavelengths from about 450 to 600 nm with a peak at about 535 nm, and those that best detect blue wavelengths are most sensitive in the 400 to 500 nm range with a peak at 445 nm. These three ranges overlap, but the most significant overlap in sensitivities is between those of the red and greens cones. The ranges can vary and, in some people, the overlap between those of the red and green cones is too drastic, meaning that light that is supposed to be identified as "green" is instead is detected as "red" and vice versa, making those wavelengths indistinguishable.

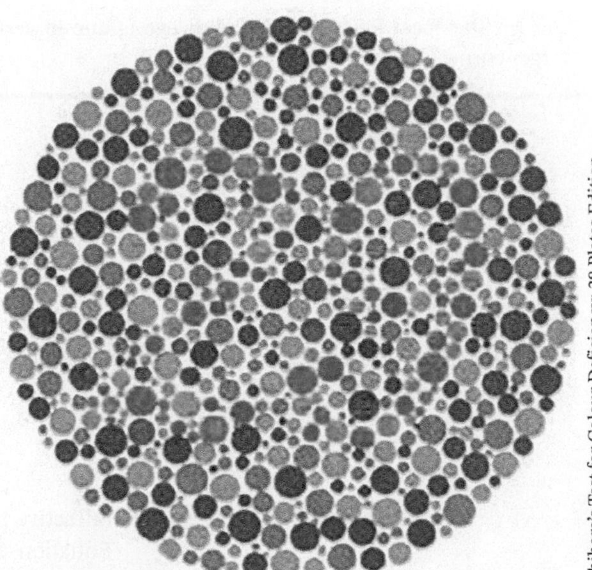

Ishihara's Test for Colour Deficiency: 38 Plates Edition. 2006-2018 by Colblindor. https://www.color-blindness.com/ishiharas-test-for-colour-deficiency-38-plates-edition/#prettyPhoto

FIGURE 26.37 Someone with a red-green color vision deficiency might not be able to see the number 29 in the figure, which is called an Ishihara plate. In some cases, optical filters may provide a means to help distinguish these colors.

One way to lessen the overlap is to remove some of the light that is entering the eye, specifically the light at the wavelengths in which the red and green cones most strongly overlap, which is typically at about 565 nm (i.e., in the yellow-orange part of the spectrum). This is done with something called a "notch" filter, which absorbs light only in a certain wavelength interval, and allows all other wavelengths to pass. Since filtering attenuates the light, it is often recommended that glasses utilizing such filters not be used at night. Notch filters can be constructed from optical materials that naturally absorb in a given wavelength range, or from multilayer thin films that exhibit interference effects, a topic that will be discussed in Chapter 27.

Example 14 explains why babies can only see nearby objects.

EXAMPLE 14 | BIO Through a Baby's Eyes—Infant Vision

Healthy newborn babies have the ability to see shapes of objects and colors, but they cannot see objects more than 8–15 inches away (**Figure 26.38**). They naturally prefer rounded objects with contrast (light and dark shades), so it is not surprising they often fixate on the faces of their parents. They also have diminished visual acuity, or the ability of their eyes to resolve objects in fine detail. This is in large part due to the smaller size of a baby's eye compared to an adult. The cornea-to-retina distance in a baby's eye can be 6–9 mm shorter than in an adult, which results in smaller retinal images. The newborn's pupil also grows with age. Assume that an infant and an adult are both staring at a teddy bear that is located 35 cm from their eyes. If the lens-to-retina distance in the adult and infant is 25 mm and 16 mm, respectively, what is the ratio of the effective focal lengths of their eyes (f_A/f_I)?

Reasoning We can apply the thin-lens equation (Equation 26.6) twice—once for the adult, and once for the infant—to calculate the focal lengths of the eyes and then take their ratio.

Solution Applying Equation 26.6 to the adult, we have:

$$\frac{1}{f_A} = \frac{1}{d_o} + \frac{1}{d_{iA}} = \frac{1}{35\text{ cm}} + \frac{1}{2.5\text{ cm}} \Rightarrow f_A = 2.33\text{ cm}.$$

Repeating this calculation for the infant, we find:

$$\frac{1}{f_I} = \frac{1}{d_o} + \frac{1}{d_{iI}} = \frac{1}{35\text{ cm}} + \frac{1}{1.6\text{ cm}} \Rightarrow f_I = 1.53\text{ cm}.$$

We can now calculate the ratio of the two focal lengths:

$$\left(\frac{f_A}{f_I}\right) = \frac{2.33\text{ cm}}{1.53\text{ cm}} = \boxed{1.52}$$

The greater focal length in the adult eye is due in part to its larger cornea.

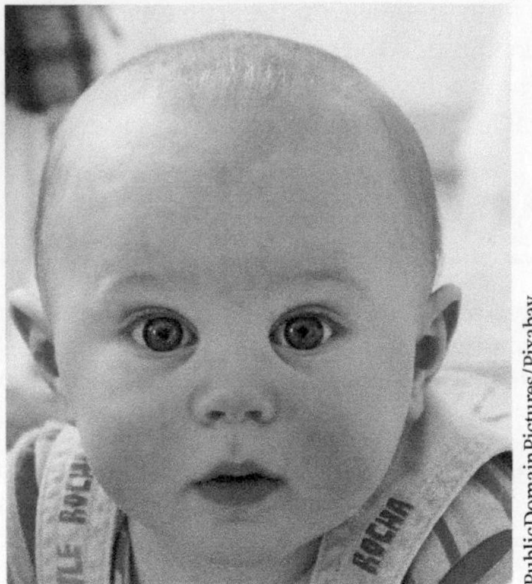

FIGURE 26.38 The smaller structures of an infant's eyes result in less visual acuity as compared to an adult with normal vision. However, the infant's vision improves dramatically in just a few months.

The Refractive Power of a Lens—The Diopter

The extent to which rays of light are refracted by a lens depends on its focal length. However, optometrists who prescribe correctional lenses and opticians who make the lenses do not specify the focal length directly in prescriptions. Instead, they use the concept of **refractive power** to describe the extent to which a lens refracts light:

$$\begin{array}{c}\text{Refractive power}\\ \text{of a lens (in diopters)}\end{array} = \frac{1}{f\,(\text{in meters})} \tag{26.8}$$

The refractive power is measured in units of *diopters*. One diopter is 1 m^{-1}.

Equation 26.8 shows that a converging lens has a refractive power of 1 diopter if it focuses parallel light rays to a focal point 1 m beyond the lens. If a lens refracts parallel rays even more and converges them to a focal point only 0.25 m beyond the lens, the lens has four times more refractive power, or 4 diopters. Since a converging lens has a positive focal length and a diverging lens has a negative focal length, the refractive power of a converging lens is positive and that of a diverging lens is negative. Thus, the eyeglasses in Example 12 would be described in an optometrist's prescription in the following way: Refractive power = 1/(−5.19 m) = −0.193 diopters. The contact lenses in Example 13 would be described in a similar fashion: Refractive power = 1/(0.284 m) = 3.52 diopters.

Check Your Understanding

(*The answers are given at the end of the book.*)

21. Two people who wear glasses are camping. One is nearsighted, and the other is farsighted. Whose glasses may be useful in starting a fire by concentrating the sun's rays into a small region at the focal point of the lens used in the glasses?

22. Suppose that a person with a near point of 26 cm is standing in front of a plane mirror. How close can he stand to the mirror and still see himself in focus?

23. BIO To a swimmer under water, objects look blurred. However, goggles that keep the water away from the eyes allow the swimmer to see objects in sharp focus. Why?

24. When glasses use diverging lenses to correct for nearsightedness or converging lenses to correct for farsightedness, the eyes of the person wearing the glasses lie between the lenses and their focal points. When you look at the eyes of this person, they do not appear to have their normal size. Which one of the following describes what you see? **(a)** The converging lenses make the eyes appear smaller, and the diverging lenses make the eyes appear larger. **(b)** The converging lenses make the eyes appear larger, and the diverging lenses make the eyes appear smaller. **(c)** Both types of lenses make the eyes appear larger. **(d)** Both types of lenses make the eyes appear smaller.

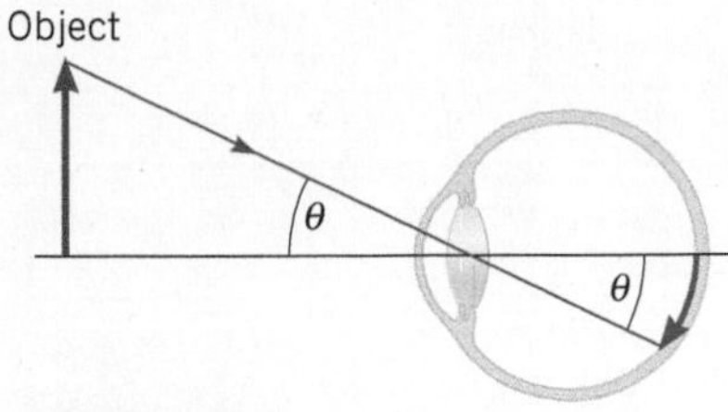

FIGURE 26.39 The angle θ is the angular size of both the image and the object.

26.11 Angular Magnification and the Magnifying Glass

If you hold a penny at arm's length, the penny looks larger than the moon. The reason is that the penny, being so close, forms a larger image on the retina of the eye than does the more distant moon. The brain interprets the larger image of the penny as arising from a larger object. The size of the image on the retina determines how large an object appears to be. However, the size of the image on the retina is difficult to measure. Alternatively, the angle θ subtended by the image can be used as an indication of the image size. **Figure 26.39** shows this alternative, which has the advantage that θ is also the angle subtended by the object and, hence, can be measured more easily. The angle θ is called the **angular size** of both the image and the object. The larger the angular size, the larger the image on the retina, and the larger the object appears to be.

According to Equation 8.1, the angle θ (measured in radians) is the length of the circular arc that is subtended by the angle divided by the radius of the arc, as **Figure 26.40*a*** indicates. Part *b* of the drawing shows the situation for an object of height h_o viewed at a distance d_o from the eye. When θ is small, h_o is approximately equal to the arc length and d_o is nearly equal to the radius, so that

$$\theta \text{ (in radians)} = \text{Angular size} \approx \frac{h_o}{d_o}$$

This approximation is good to within one percent for angles of 9° or smaller. In the next example the angular size of a penny is compared with that of the moon.

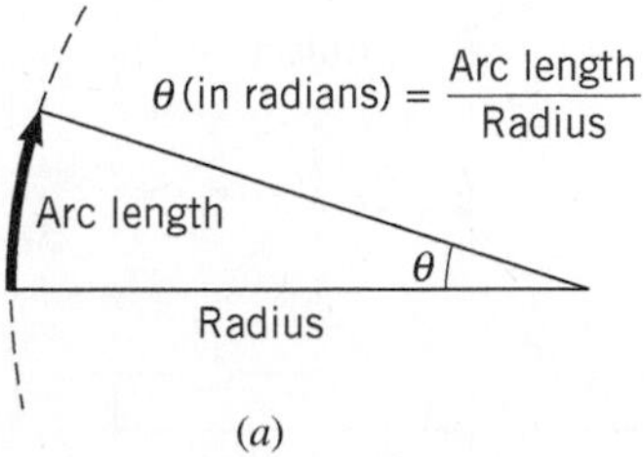

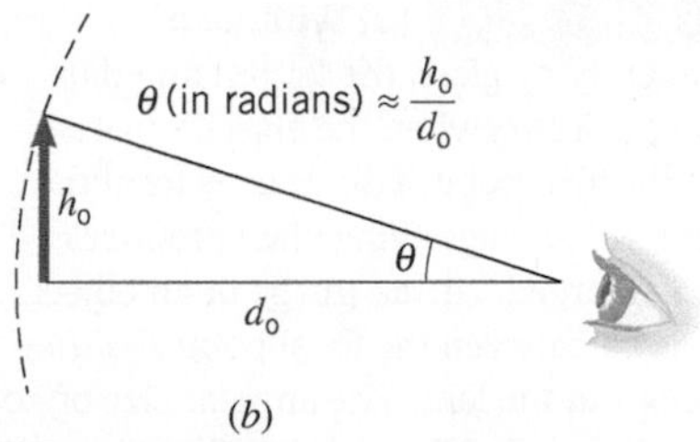

FIGURE 26.40 (*a*) The angle θ, measured in radians, is the arc length divided by the radius. (*b*) For small angles (less than 9°), θ in radians is approximately equal to h_o/d_o, where h_o and d_o are the object height and distance.

EXAMPLE 15 | A Penny and the Moon

Compare the angular size of a penny (diameter = h_o = 1.9 cm) held at arm's length (d_o = 71 cm) with the angular size of the moon (diameter = $h_o = 3.5 \times 10^6$ m, and $d_o = 3.9 \times 10^8$ m).

Reasoning The angular size θ of an object is given approximately by its height h_o divided by its distance d_o from the eye, $\theta \approx h_o/d_o$, provided that the angle involved is less than roughly 9°; this approximation applies here. The "heights" of the penny and the moon are their diameters.

Solution The angular sizes of the penny and moon are

Penny $$\theta \approx \frac{h_o}{d_o} = \frac{1.9 \text{ cm}}{71 \text{ cm}} = \boxed{0.027 \text{ rad } (1.5°)}$$

Moon $$\theta \approx \frac{h_o}{d_o} = \frac{3.5 \times 10^6 \text{ m}}{3.9 \times 10^8 \text{ m}} = \boxed{0.0090 \text{ rad } (0.52°)}$$

The penny thus appears to be about three times as large as the moon.

An optical instrument, such as a magnifying glass, allows us to view small or distant objects because it produces a larger image on the retina than would be possible otherwise. In other words, an optical instrument magnifies the angular size of the object. The **angular magnification** (or **magnifying power**) M is the angular size θ' of the final image produced by the instrument divided by a reference angular size θ. The reference angular size is the angular size of the object when seen without the instrument.

Angular magnification

$$M = \frac{\text{Angular size of final image produced by optical instrument}}{\text{Reference angular size of object seen without optical instrument}} = \frac{\theta'}{\theta} \tag{26.9}$$

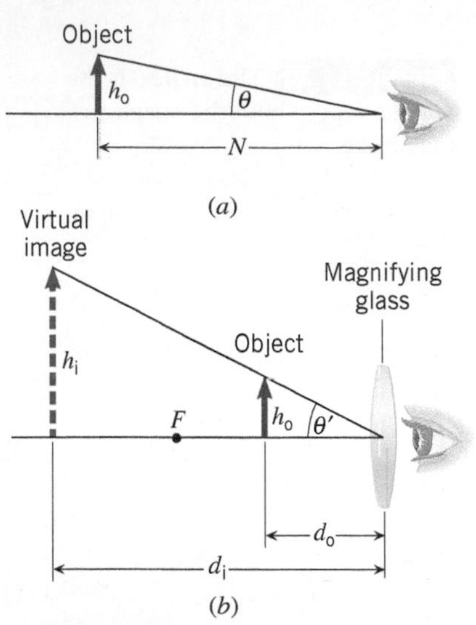

FIGURE 26.41 (*a*) Without a magnifying glass, the largest angular size θ occurs when the object is placed at the near point, a distance *N* from the eye. (*b*) A magnifying glass produces an enlarged, virtual image of an object placed between the focal point *F* of the lens and the lens. The angular size of both the image and the object is θ'.

A magnifying glass is the simplest device that provides angular magnification. In this case, the reference angular size θ is chosen to be the angular size of the object when placed at the near point of the eye and seen without the magnifying glass. Since an object cannot be brought closer than the near point and still produce a sharp image on the retina, θ represents the largest angular size obtainable without the magnifying glass. **Figure 26.41*a*** indicates that the reference angular size is $\theta \approx h_o/N$, where N is the distance from the eye to the near point. To compute θ', recall from Section 26.7 and **Figure 26.28** that a magnifying glass is usually a single converging lens, with the object located between the focal point of the lens and the lens. In this situation, **Figure 26.41*b*** indicates that the lens produces a virtual image that is enlarged and upright with respect to the object. Assuming that the eye is next to the magnifying glass, the angular size θ' seen by the eye is $\theta' \approx h_o/d_o$, where d_o is the object distance. The angular magnification is

$$M = \frac{\theta'}{\theta} \approx \frac{h_o/d_o}{h_o/N} = \frac{N}{d_o}$$

According to the thin-lens equation, d_o is related to the image distance d_i and the focal length f of the lens by

$$\frac{1}{d_o} = \frac{1}{f} - \frac{1}{d_i}$$

Substituting this expression for $1/d_o$ into the previous expression for M leads to the following result:

Angular magnification of a magnifying glass

$$M = \frac{\theta'}{\theta} \approx \left(\frac{1}{f} - \frac{1}{d_i}\right)N \tag{26.10}$$

Two special cases of this result are of interest, depending on whether the image is located as close to the eye as possible or as far away as possible. To be seen clearly, the closest the image can be relative to the eye is at the near point, or $d_i = -N$. The minus sign indicates that the image lies to the left of the lens and is virtual. In this event, Equation 26.10 becomes $M \approx (N/f) + 1$. The farthest the image can be from the eye is at infinity ($d_i = -\infty$); this occurs when the object is placed at the focal point of the lens. When the image is at infinity, Equation 26.10 simplifies to $M \approx N/f$. Clearly, the angular magnification is greater when the image is at the near point of the eye rather than at infinity. In either case, however, the greatest magnification is achieved by using a magnifying glass with the shortest possible focal length. Example 16 illustrates how to determine the angular magnification of a magnifying glass that is used in these two ways.

EXAMPLE 16 | Examining a Diamond with a Magnifying Glass

A jeweler, whose near point is 40.0 cm from his eye and whose far point is at infinity, is using a small magnifying glass (called a loupe) to examine a diamond. The lens of the magnifying glass has a focal length of 5.00 cm, and the image of the gem is −185 cm from the lens. The image distance is negative because the image is virtual and is formed on the same side of the lens as the object.

(a) Determine the angular magnification of the magnifying glass. **(b)** Where should the image be located so the jeweler's eye is fully relaxed and has the least strain? What is the angular magnification under this "least strain" condition?

Reasoning The angular magnification of the magnifying glass can be determined from Equation 26.10. In part (a) the image distance is −185 cm. In part (b) the ciliary muscle of the jeweler's eye is fully relaxed, so the image must be located infinitely far from the eye, at its far point, as Section 26.10 discusses.

Solution (a) With $f = 5.00$ cm, $d_i = -185$ cm, and $N = 40.0$ cm, the angular magnification is

$$M = \left(\frac{1}{f} - \frac{1}{d_i}\right) N = \left(\frac{1}{5.00\text{ cm}} - \frac{1}{-185\text{ cm}}\right)(40.0\text{ cm}) = \boxed{8.22}$$

(b) With $f = 5.00$ cm, $d_i = -\infty$, and $N = 40.0$ cm, the angular magnification is

$$M = \left(\frac{1}{f} - \frac{1}{d_i}\right) N = \left(\frac{1}{5.00\text{ cm}} - \frac{1}{-\infty}\right)(40.0\text{ cm}) = \boxed{8.00}$$

Jewelers often prefer to minimize eyestrain when viewing objects, even though it means a slight reduction in angular magnification.

Check Your Understanding

(The answers are given at the end of the book.)

25. A bird-watcher sees the following three raptors in the air at the distances indicated: a kestrel (wing span = 0.58 m at a distance of 21 m), a bald eagle (wing span = 2.29 m at a distance of 95 m), and a red-tailed hawk (wing span = 1.27 m at a distance of 41 m). Rank the raptors in descending order (largest first) according to the angular size seen by the bird-watcher.

26. Who benefits more from using a magnifying glass, a person whose near point is located at a distance away from the eyes of **(a)** 75 cm or **(b)** 25 cm?

27. A person who has a near point of 25.0 cm is looking with unaided eyes at an object that is located at the near point. The object has an angular size of 0.012 rad. Then, holding a magnifying glass ($f = 10.0$ cm) next to her eye, she views the image of this object, the image being located at her near point. What is the angular size of the image?

26.12 The Compound Microscope

THE PHYSICS OF . . . the compound microscope. To increase the angular magnification beyond that possible with a magnifying glass, an additional converging lens can be included to "premagnify" the object before the magnifying glass comes into play. The result is an optical instrument known as the **compound microscope** (**Figure 26.42**). As discussed in Section 26.9, the magnifying glass is called the eyepiece, and the additional lens is called the objective.

The angular magnification of the compound microscope is $M = \theta'/\theta$ (Equation 26.9), where θ' is the angular size of the final image and θ is the reference angular size. As with the magnifying glass in **Figure 26.41**, the reference angular size is determined by the height h_o of the object when the object is located at the near point of the unaided eye: $\theta \approx h_o/N$, where N is the distance between the eye and the near point. Assuming that the object is placed just outside the focal point F_o of the objective (see **Figure 26.32*a***) and that the final image is very far from the eyepiece (i.e., near infinity; see **Figure 26.32*c***), it can be shown that

Angular magnification of a compound microscope

$$M \approx -\frac{(L - f_e)N}{f_o f_e} \qquad (L > f_o + f_e) \tag{26.11}$$

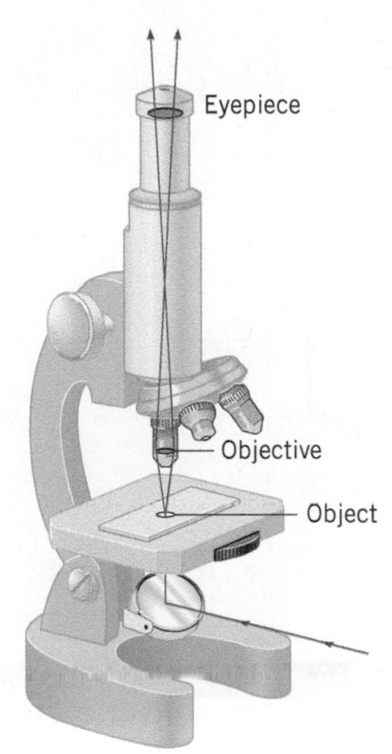

FIGURE 26.42 A compound microscope.

In Equation 26.11, f_o and f_e are, respectively, the focal lengths of the objective and the eyepiece. The angular magnification is greatest when f_o and f_e are as small as possible (since they are in the denominator in Equation 26.11) and when the distance L between the lenses is as large as possible. Furthermore, L must be greater than the sum of f_o and f_e for this equation to be valid. Example 17 deals with the angular magnification of a compound microscope.

EXAMPLE 17 | The Angular Magnification of a Compound Microscope

The focal length of the objective of a compound microscope is $f_o = 0.40$ cm, and the focal length of the eyepiece is $f_e = 3.0$ cm. The two lenses are separated by a distance of $L = 20.0$ cm. A person with a near point distance of $N = 25$ cm is using the microscope. **(a)** Determine the angular magnification of the microscope. **(b)** Compare the answer in part (a) with the largest angular magnification obtainable by using the eyepiece alone as a magnifying glass.

Reasoning The angular magnification of the compound microscope can be obtained directly from Equation 26.11, since all the variables are known. When the eyepiece is used alone as a magnifying glass, as in **Figure 26.41b**, the largest angular magnification occurs when the image seen through the eyepiece is as close as possible to the eye. The image in this case is at the near point, and according to Equation 26.10, the angular magnification is $M \approx (N/f_e) + 1$.

Solution **(a)** The angular magnification of the compound microscope is

$$M \approx -\frac{(L - f_e)N}{f_o f_e} = -\frac{(20.0 \text{ cm} - 3.0 \text{ cm})(25 \text{ cm})}{(0.40 \text{ cm})(3.0 \text{ cm})} = \boxed{-350}$$

The minus sign indicates that the final image is inverted relative to the initial object.

(b) The maximum angular magnification of the eyepiece by itself is

$$M \approx \frac{N}{f_e} + 1 = \frac{25 \text{ cm}}{3.0 \text{ cm}} + 1 = \boxed{9.3}$$

The effect of the objective is to increase the angular magnification of the compound microscope by a factor of $350/9.3 = 38$ compared to the angular magnification of a magnifying glass.

26.13 The Telescope

THE PHYSICS OF . . . the telescope. A telescope is an instrument for magnifying distant objects, such as stars and planets. Like a microscope, a telescope consists of an objective and an eyepiece (also called the ocular). When the objective is a lens, as is the case in this section, the telescope is referred to as a *refracting* telescope, since lenses utilize the refraction of light.*

Usually the object being viewed is far away, so the light rays entering the telescope are nearly parallel, and the "first image" is formed just beyond the focal point F_o of the objective, as **Figure 26.43a** illustrates. The first image is real and inverted. Unlike the first image in the compound microscope, however, this image is *smaller* than the object. If, as in part *b* of the drawing, the telescope is constructed so the first image lies just inside the focal point F_e of the eyepiece, the eyepiece acts like a magnifying glass. It forms a final image that is greatly enlarged, virtual, and located near infinity. This final image can then be viewed with a fully relaxed eye.

The angular magnification M of a telescope, like that of a magnifying glass or a microscope, is the angular size θ' subtended by the final image of the telescope divided by the reference angular size θ of the object. For an astronomical object, such as a planet, it is convenient to use as a reference the angular size of the object seen in the sky with the unaided eye. Since the object is far away, the angular size seen by the unaided eye is nearly the same as the angle θ subtended at the objective of the telescope in **Figure 26.43a**. Moreover, θ is also the angle subtended by the first image, so

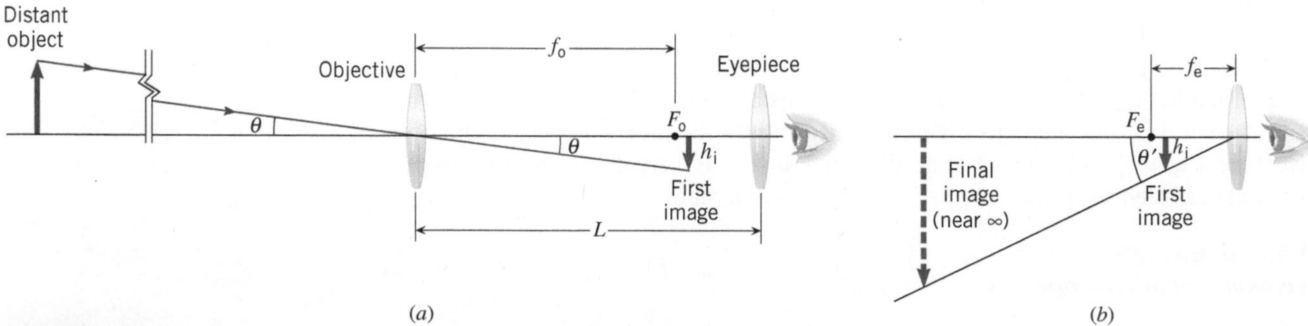

FIGURE 26.43 (*a*) An astronomical telescope is used to view distant objects. (Note the "break" in the principal axis, between the object and the objective.) The objective produces a real, inverted first image. (*b*) The eyepiece magnifies the first image to produce the final image near infinity.

*Another type of telescope utilizes a mirror instead of a lens for the objective and is called a *reflecting* telescope.

$\theta \approx -h_i/f_o$, where h_i is the height of the first image and f_o is the focal length of the objective. A minus sign has been inserted into this equation because the first image is inverted relative to the object and the image height h_i is a negative number. The insertion of the minus sign ensures that the term $-h_i/f_o$, and hence θ, is a positive quantity. To obtain an expression for θ', we refer to **Figure 26.43*b*** and note that the first image is located very near the focal point F_e of the eyepiece, which has a focal length f_e. Therefore, $\theta' \approx h_i/f_e$. The angular magnification of the telescope is approximately

Angular magnification of an astronomical telescope

$$M = \frac{\theta'}{\theta} \approx \frac{h_i/f_e}{-h_i/f_o} \approx -\frac{f_o}{f_e} \qquad \textbf{(26.12)}$$

The angular magnification is determined by the ratio of the focal length of the objective to the focal length of the eyepiece. For large angular magnifications, the objective should have a long focal length and the eyepiece a short one. Some of the design features of a telescope are the topic of the next example.

EXAMPLE 18 | The Angular Magnification of an Astronomical Telescope

A telescope similar to that in **Figure 26.44** has the following specifications: $f_o = 985$ mm and $f_e = 5.00$ mm. From these data, find **(a)** the angular magnification and **(b)** the approximate length of this telescope.

Reasoning The angular magnification of the telescope follows directly from Equation 26.12, since the focal lengths of the objective and eyepiece are known. We can find the length of the telescope by noting that it is approximately equal to the distance L between the objective and eyepiece. **Figure 26.43*a*** shows that the first image is located just beyond the focal point F_o of the objective. **Figure 26.43*b*** shows that the first image is also just to the right of the focal point F_e of the eyepiece. These two focal points are, therefore, very close together, so the distance L is approximately the sum of the two focal lengths: $L \approx f_o + f_e$.

Solution **(a)** The angular magnification is approximately

$$M \approx -\frac{f_o}{f_e} = -\frac{985 \text{ mm}}{5.00 \text{ mm}} = \boxed{-197} \qquad \textbf{(26.12)}$$

(b) The approximate length of the telescope is

$$L \approx f_o + f_e = 985 \text{ mm} + 5.00 \text{ mm} = \boxed{990 \text{ mm}}$$

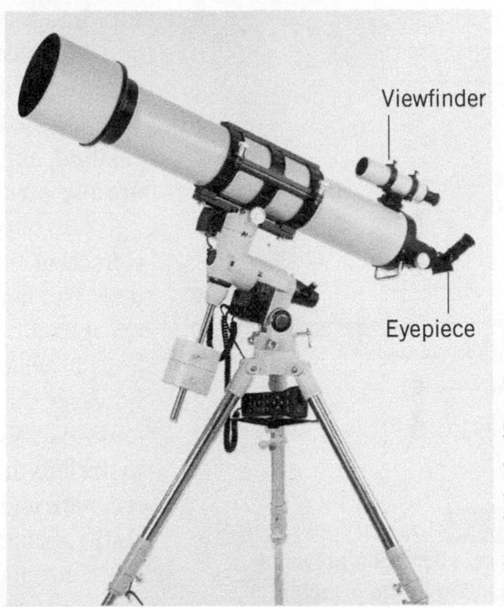

FIGURE 26.44 An astronomical telescope typically includes a viewfinder, which is a separate small telescope with low magnification and serves as an aid in locating the object. Once the object has been found, the viewer uses the eyepiece to obtain the full magnification of the telescope.

Check Your Understanding

(The answers are given at the end of the book.)

28. In the construction of a telescope, one of two lenses is to be used as the objective and one as the eyepiece. The focal lengths of the lenses are **(a)** 3 cm and **(b)** 45 cm. Which lens should be used as the objective?

29. Two refracting telescopes have identical eyepieces, although one telescope is twice as long as the other. Which telescope has the greater angular magnification?

30. A well-designed optical instrument is composed of two converging lenses separated by 14 cm. The focal lengths of the lenses are 0.60 and 4.5 cm. Is the instrument a microscope or a telescope?

31. It is often thought that virtual images are somehow less important than real images. To show that this is not true, identify which of the following instruments normally produce final images that are virtual: **(a)** a projector, **(b)** a camera, **(c)** a magnifying glass, **(d)** eyeglasses, **(e)** a compound microscope, and **(f)** an astronomical telescope.

26.14 Lens Aberrations

Rather than forming a sharp image, a single lens typically forms an image that is slightly out of focus. This lack of sharpness arises because the rays originating from a single point on the object are not focused to a single point on the image. As a result, each point on the image becomes a small blur. The lack of point-to-point correspondence between object and image is called an *aberration*.

One common type of aberration is **spherical aberration**, and it occurs with converging and diverging lenses made with spherical surfaces. **Figure 26.45*a*** shows how spherical aberration arises with a converging lens. Ideally, all rays traveling parallel to the principal axis are refracted so they cross the axis at the same point after passing through the lens. However, rays far from the principal axis are refracted more by the lens than are those closer in. Consequently, the outer rays cross the axis closer to the lens than do the inner rays, so a lens with spherical aberration does not have a unique focal point. Instead, as the drawing suggests, there is a location along the principal axis where the light converges to the smallest cross-sectional area. This area is circular and is known as the **circle of least confusion**. The circle of least confusion is where the most satisfactory image can be formed by the lens.

Spherical aberration can be reduced substantially by using a variable-aperture diaphragm to allow only those rays close to the principal axis to pass through the lens. **Figure 26.45*b*** indicates that a reasonably sharp focal point can be achieved by this method, although less light now passes through the lens. Lenses with parabolic surfaces are also used to reduce this type of aberration, but they are difficult and expensive to make.

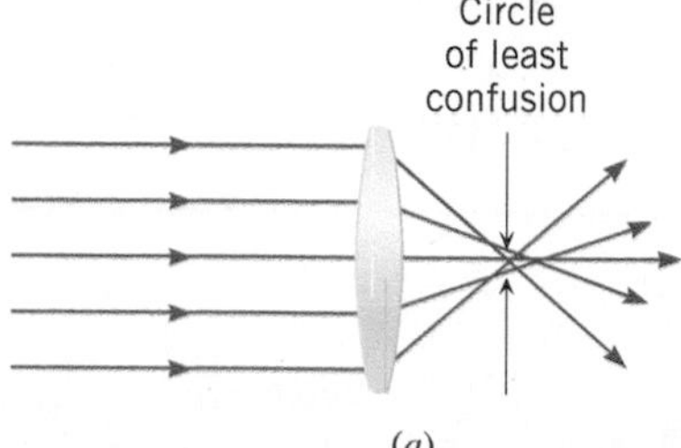

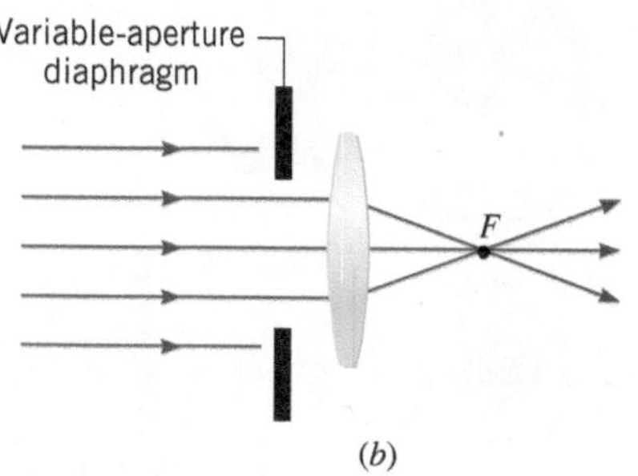

FIGURE 26.45 (*a*) In a converging lens, spherical aberration prevents light rays parallel to the principal axis from converging to a common point. (*b*) Spherical aberration can be reduced by allowing only rays near the principal axis to pass through the lens. The refracted rays now converge more nearly to a single focal point *F*.

Chromatic aberration also causes blurred images. It arises because the index of refraction of the material from which the lens is made varies with wavelength. Section 26.5 discusses how this variation leads to the phenomenon of dispersion, in which different colors refract by different amounts. **Figure 26.46*a*** shows sunlight incident on a converging lens, in which the light spreads into its color spectrum because of dispersion. For clarity, however, the picture shows only the colors at the opposite ends of the visible spectrum—red and violet. Violet is refracted more than red, so the violet ray crosses the principal axis closer to the lens than does the red ray. Thus, the focal length of the lens is shorter for violet than for red, with intermediate values of the focal length corresponding to the colors in between. As a result of chromatic aberration, an undesirable color fringe surrounds the image.

Chromatic aberration can be greatly reduced by using a compound lens, such as the combination of a converging lens and a diverging lens shown in **Figure 26.46*b***. Each lens is made from a different type of glass. With this lens combination the red and violet rays almost come to a common focus and, thus, chromatic aberration is reduced. A lens combination designed to reduce chromatic aberration is called an *achromatic lens* (from the Greek "achromatos," meaning "without color"). All high-quality cameras use achromatic lenses.

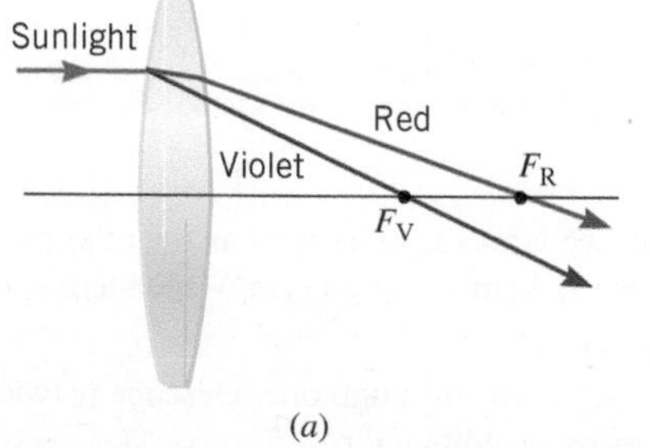

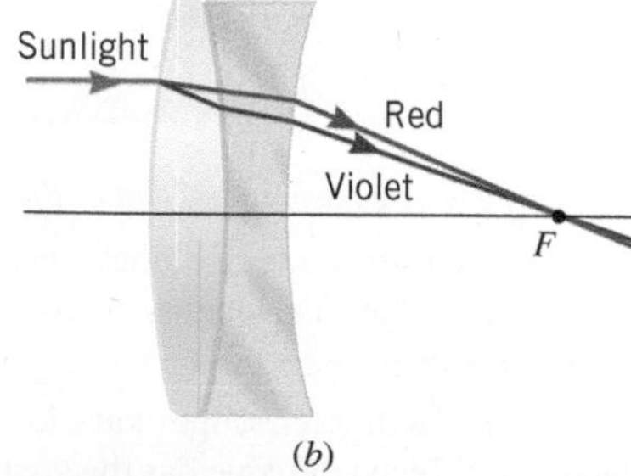

FIGURE 26.46 (*a*) Chromatic aberration arises when different colors are focused at different points along the principal axis: F_V = focal point for violet light, F_R = focal point for red light. (*b*) A converging and a diverging lens in tandem can be designed to bring different colors more nearly to the same focal point *F*.

Check Your Understanding

(*The answer is given at the end of the book.*)

32. Why does chromatic aberration occur in lenses but not in mirrors?

Concept Summary

26.1 The Index of Refraction The change in speed as a ray of light goes from one material to another causes the ray to deviate from its incident direction. This change in direction is called refraction. The index of refraction n of a material is the ratio of the speed c of light in a vacuum to the speed v of light in the material, as shown in Equation 26.1. The values for n are greater than unity, because the speed of light in a material medium is less than it is in a vacuum.

$$n = \frac{c}{v} \qquad \textbf{(26.1)}$$

26.2 Snell's Law and the Refraction of Light The refraction that occurs at the interface between two materials obeys Snell's law of refraction. This law states that (1) the refracted ray, the incident ray, and the normal to the interface all lie in the same plane, and (2) the angle of refraction θ_2 is related to the angle of incidence θ_1 according to Equation 26.2, where n_1 and n_2 are the indices of refraction of the incident and refracting media, respectively. The angles are measured relative to the normal.

$$n_1 \sin \theta_1 = n_2 \sin \theta_2 \qquad \textbf{(26.2)}$$

Because of refraction, a submerged object has an apparent depth that is different from its actual depth. If the observer is directly above (or below) the object, the apparent depth (or height) d' is related to the actual depth (or height) d according to Equation 26.3, where n_1 and n_2 are the refractive indices of the materials (the media) in which the object and the observer, respectively, are located.

$$d' = d\left(\frac{n_2}{n_1}\right) \qquad \textbf{(26.3)}$$

26.3 Total Internal Reflection When light passes from a material with a larger refractive index n_1 into a material with a smaller refractive index n_2, the refracted ray is bent away from the normal. If the incident ray is at the critical angle θ_c, the angle of refraction is 90°. The critical angle is determined from Snell's law and is given by Equation 26.4. When the angle of incidence exceeds the critical angle, all the incident light is reflected back into the material from which it came, a phenomenon known as total internal reflection.

$$\sin \theta_c = \frac{n_2}{n_1} \quad (n_1 > n_2) \qquad \textbf{(26.4)}$$

26.4 Polarization and the Reflection and Refraction of Light When light is incident on a nonmetallic surface at the Brewster angle θ_B, the reflected light is completely polarized parallel to the surface. The Brewster angle is given by Equation 26.5, where n_1 and n_2 are the refractive indices of the incident and refracting media, respectively. When light is incident at the Brewster angle, the reflected and refracted rays are perpendicular to each other.

$$\tan \theta_B = \frac{n_2}{n_1} \qquad \textbf{(26.5)}$$

26.5 The Dispersion of Light: Prisms and Rainbows A glass prism can spread a beam of sunlight into a spectrum of colors because the index of refraction of the glass depends on the wavelength of the light. Thus, a prism bends the refracted rays corresponding to different colors by different amounts. The spreading of light into its color components is known as dispersion. The dispersion of light by water droplets in the air leads to the formation of rainbows. A prism will not bend a light ray at all, neither up nor down, if the surrounding fluid has the same refractive index as the material from which the prism is made, a condition known as index matching.

26.6 Lenses 26.7 The Formation of Images by Lenses Converging lenses and diverging lenses depend on the phenomenon of refraction in forming an image. With a converging lens, paraxial rays that are parallel to the principal axis are focused to a point on the axis by the lens. This point is called the focal point of the lens, and its distance from the lens is the focal length f. Paraxial light rays that are parallel to the principal axis of a diverging lens appear to originate from its focal point after passing through the lens. The distance of this point from the lens is the focal length f. The image produced by a converging or a diverging lens can be located via a technique known as ray tracing, which utilizes the three rays outlined in the Reasoning Strategy given in Section 26.7.

The nature of the image formed by a converging lens depends on where the object is situated relative to the lens. When the object is located at a distance from the lens that is greater than twice the focal length, the image is real, inverted, and smaller than the object. When the object is located at a distance from the lens that is between the focal length and twice the focal length, the image is real, inverted, and larger than the object. When the object is located between the focal point and the lens, the image is virtual, upright, and larger than the object.

Regardless of the position of a real object, a diverging lens always produces an image that is virtual, upright, and smaller than the object.

26.8 The Thin-Lens Equation and the Magnification Equation The thin-lens equation (Equation 26.6) can be used with either converging or diverging lenses that are thin, and it relates the object distance d_o, the image distance d_i, and the focal length f of the lens.

$$\frac{1}{d_o} + \frac{1}{d_i} = \frac{1}{f} \qquad \textbf{(26.6)}$$

The magnification m of a lens is the ratio of the image height h_i to the object height h_o and is also related to d_o and d_i by the magnification equation (Equation 26.7).

$$m = \frac{h_i}{h_o} = -\frac{d_i}{d_o} \qquad \textbf{(26.7)}$$

The algebraic sign conventions for the variables appearing in the thin-lens and magnification equations are summarized in the Reasoning Strategy given in Section 26.8.

26.9 Lenses in Combination When two or more lenses are used in combination, the image produced by one lens serves as the object for the next lens.

26.10 The Human Eye In the human eye, a real, inverted image is formed on a light-sensitive surface, called the retina. Accommodation is the process by which the focal length of the eye is automatically adjusted, so that objects at different distances produce sharp images on the retina. The near point of the eye is the point nearest the eye at which an object can be placed and still have a sharp image produced on the retina. The far point of the eye is the location of the farthest object on which the fully relaxed eye can focus. For a young and normal eye, the near point is located 25 cm from the eye, and the far point is located at infinity.

A nearsighted (myopic) eye is one that can focus on nearby objects, but not on distant objects. Nearsightedness can be corrected with eyeglasses or contacts made from diverging lenses. A farsighted (hyperopic) eye can see distant objects clearly, but not objects close up. Farsightedness can be corrected with converging lenses.

The refractive power of a lens is measured in diopters and is given by Equation 26.8, where f is the focal length of the lens and must be expressed in meters. A converging lens has a positive refractive power, and a diverging lens has a negative refractive power.

$$\text{Refractive power (in diopters)} = \frac{1}{f\text{ (in meters)}} \quad \textbf{(26.8)}$$

26.11 Angular Magnification and the Magnifying Glass The angular size of an object is the angle that it subtends at the eye of the viewer. For small angles, the angular size θ in radians is given approximately by Equation 1, where h_o is the height of the object and d_o is the object distance. The angular magnification M of an optical instrument is the angular size θ' of the final image produced by the instrument divided by the reference angular size θ of the object, which is that seen without the instrument (see Equation 26.9).

$$\theta \text{ (in radians)} \approx \frac{h_o}{d_o} \quad \textbf{(1)}$$

$$M = \frac{\theta'}{\theta} \quad \textbf{(26.9)}$$

A magnifying glass is usually a single converging lens that forms an enlarged, upright, and virtual image of an object placed at the focal point of the lens or between the focal point and the lens. For a magnifying glass held close to the eye, the angular magnification M is given approximately by Equation 26.10, where f is the focal length of the lens, d_i is the image distance, and N is the distance of the viewer's near point from the eye.

$$M \approx \left(\frac{1}{f} - \frac{1}{d_i}\right)N \quad \textbf{(26.10)}$$

26.12 The Compound Microscope A compound microscope usually consists of two lenses, an objective and an eyepiece. The final image is enlarged, inverted, and virtual. The angular magnification M of such a microscope is given approximately by Equation 26.11, where f_o and f_e are, respectively, the focal lengths of the objective and eyepiece, L is the distance between the two lenses, and N is the distance of the viewer's near point from his or her eye.

$$M \approx -\frac{(L - f_e)N}{f_o f_e} \quad (L > f_o + f_e) \quad \textbf{(26.11)}$$

26.13 The Telescope An astronomical telescope magnifies distant objects with the aid of an objective and an eyepiece, and it produces a final image that is enlarged, inverted, and virtual. The angular magnification M of a telescope is given approximately by Equation 26.12, where f_o and f_e are, respectively, the focal lengths of the objective and the eyepiece.

$$M \approx -\frac{f_o}{f_e} \quad \textbf{(26.12)}$$

26.14 Lens Aberrations Lens aberrations limit the formation of perfectly focused or sharp images by optical instruments. Spherical aberration occurs because rays that pass through the outer edge of a lens with spherical surfaces are not focused at the same point as rays that pass through near the center of the lens. Chromatic aberration arises because a lens focuses different colors at slightly different points.

Focus on Concepts

Online

Additional questions are available for assignment in WileyPLUS.

Section 26.2 Snell's Law and the Refraction of Light

1. The drawings show two examples in which a ray of light is refracted at the interface between two liquids. In each example the incident ray is in liquid A and strikes the interface at the same angle of incidence. In one case the ray is refracted into liquid B, and in the other it is refracted into liquid C. The dashed lines denote the normals to the interfaces. Rank the indices of refraction of the three liquids in descending order (largest first). (a) n_A, n_B, n_C (b) n_A, n_C, n_B (c) n_C, n_A, n_B (d) n_B, n_A, n_C (e) n_C, n_B, n_A

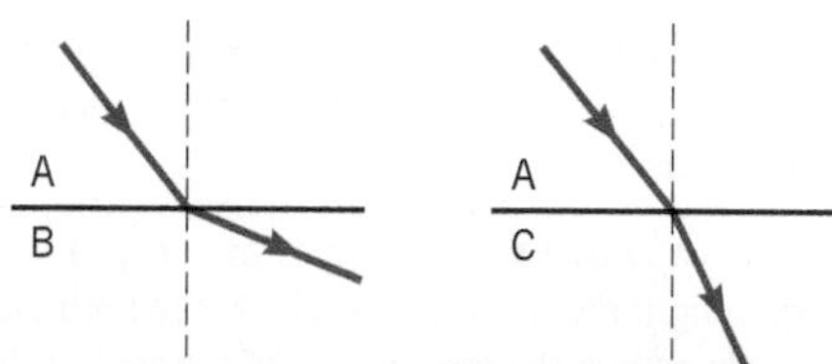

QUESTION 1

2. A coin is resting on the bottom of an empty container. The container is then filled to the brim three times, each time with a different liquid. An observer (in air) is directly above the coin and looks down at it. With liquid A in the container, the apparent depth of the coin is 7 cm, with liquid B it is 6 cm, and with liquid C it is 5 cm. Rank the indices of refraction of the liquids in descending order (largest first). (a) n_A, n_B, n_C (b) n_A, n_C, n_B (c) n_C, n_A, n_B (d) n_C, n_B, n_A (e) n_B, n_A, n_C

Section 26.3 Total Internal Reflection

3. The refractive index of material A is greater than the refractive index of material B. A ray of light is incident on the interface between these two materials in a number of ways, as the drawings illustrate. The dashed lines denote the normals to the interfaces. Which one of the drawings shows a situation that is *not* possible? (a) Drawing 1 (b) Drawing 2 (c) Drawing 3 (d) Drawing 4

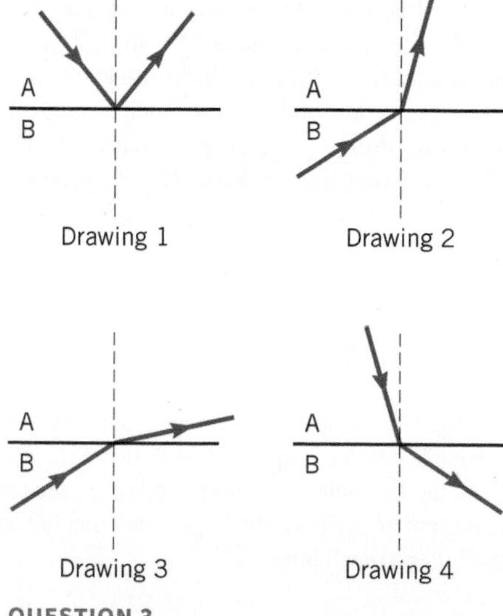

QUESTION 3

4. The drawing shows a rectangular block of glass ($n = 1.52$) surrounded by air. A ray of light starts out within the glass and travels toward point A, where some or all of it is reflected toward point B. At which points does some of the light escape the glass? (a) Only at point A (b) Only at point B (c) At both points A and B (d) At neither point A nor point B

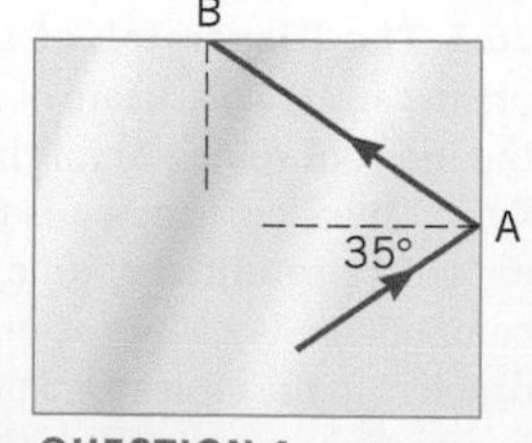

QUESTION 4

Section 26.4 Polarization and the Reflection and Refraction of Light

5. A diamond (n = 2.42) is lying on a table. At what angle of incidence θ is the light that is reflected from one of the facets of the diamond completely polarized?

Section 26.5 The Dispersion of Light: Prisms and Rainbows

6. The indices of refraction for red, green, and violet light in glass are $n_{red} = 1.520$, $n_{green} = 1.526$, and $n_{violet} = 1.538$. When a ray of light passes through a transparent slab of glass, the emergent ray is parallel to the incident ray, but can be displaced relative to it. For light passing through a glass slab that is surrounded by air, which color is displaced the most? (a) All colors are displaced equally. (b) Red (c) Green (d) Violet

Section 26.7 The Formation of Images by Lenses

7. An object is situated to the left of a lens. A ray of light from the object is close to and parallel to the principal axis of the lens. The ray passes through the lens. Which one of the following statements is true? (a) The ray crosses the principal axis at a distance from the lens equal to twice the focal length, no matter whether the lens is converging or diverging. (b) The ray passes through the lens without changing direction, no matter whether the lens is converging or diverging. (c) The ray passes through a focal point of the lens, no matter whether the lens is converging or diverging. (d) The ray passes through a focal point of the lens only if the lens is a diverging lens. (e) The ray passes through a focal point of the lens only if the lens is a converging lens.

8. What type of single lens produces a virtual image that is inverted with respect to the object? (a) Both a converging and a diverging lens can produce such an image. (b) Neither a converging nor a diverging lens produces such an image. (c) A converging lens (d) A diverging lens

Section 26.9 Lenses in Combination

9. Two converging lenses have the same focal length of 5.00 cm. They have a common principal axis and are separated by 21.0 cm. An object is located 10.0 cm to the left of the left-hand lens. What is the image distance (relative to the lens on the right) of the final image produced by this two-lens system?

Section 26.10 The Human Eye

10. Here are a number of statements concerning the refractive power of lenses.

A. A positive refractive power means that a lens always creates an image that is larger than the object.
B. Two lenses with the same refractive power have the same focal lengths.
C. A lens with a positive refractive power is a converging lens, whereas a lens with a negative refractive power is a diverging lens.
D. Two lenses with different refractive powers can have the same focal length.
E. The fact that lens A has twice the refractive power of lens B means that the focal length of lens A is twice that of lens B.

Which of these statements are false? (a) A, B, C (b) C, D, E (c) A, D, E (d) B, C, E (e) B, C, D

Section 26.11 Angular Magnification and the Magnifying Glass

11. The table lists the angular sizes in radians and distances from the eye for three objects, A, B, and C. In each case the angular size is small.

Object	Angular Size (in Radians)	Distance of Object from Eye
A	θ	d_o
B	2θ	$2d_o$
C	θ	$2d_o$

Rank the heights of these objects in descending order (largest first). (a) B, C, A (b) B, A, C (c) A, B, C (d) A, C, B (e) C, A, B

Section 26.13 The Telescope

12. An astronomical telescope has an angular magnification of −125 when used properly. What would the angular magnification M be if the objective and the eyepiece were interchanged?

Section 26.14 Lens Aberrations

13. Which one of the five choices below best completes the following statement? The fact that the refractive index depends on the wavelength of light is the cause of __________. (a) dispersion (b) chromatic aberration (c) spherical aberration (d) dispersion and chromatic aberration (e) spherical aberration and chromatic aberration

Problems

Online

Additional questions are available for assignment in WileyPLUS.
***Note:** Unless specified otherwise, use the values given in Table 26.1 for the refractive indices.*

SSM Student Solutions Manual
MMH Problem-solving help
GO Guided Online Tutorial
V-HINT Video Hints
CHALK Chalkboard Videos
BIO Biomedical application

E Easy
M Medium
H Hard
WS Worksheet
T Team Problem

Section 26.1 The Index of Refraction

1. E SSM A plate glass window (n = 1.5) has a thickness of 4.0 × 10^{-3} m. How long does it take light to pass perpendicularly through the plate?

2. E In an ultra-low temperature experiment, a collection of sodium atoms enter a special state called a *Bose-Einstein condensate* in which the index of refraction is 1.57 × 10^7. What is the speed of light in this condensate?

3. E The refractive indices of materials A and B have a ratio of n_A/n_B = 1.33. The speed of light in material A is 1.25 × 10^8 m/s. What is the speed of light in material B?

4. E GO The frequency of a light wave is the same when the light travels in ethyl alcohol or in carbon disulfide. Find the ratio of the wavelength of the light in ethyl alcohol to that in carbon disulfide.

5. E SSM Light travels at a speed of 2.201×10^8 m/s in a certain substance. What substance from Table 26.1 could this be? For the speed of light in a vacuum use 2.998×10^8 m/s; show your calculations.

6. E Light has a wavelength of 340.0 nm and a frequency of 5.403×10^{14} Hz when traveling through a certain substance. What substance from Table 26.1 could this be? Show your calculations.

7. M MMH CHALK In a certain time, light travels 6.20 km in a vacuum. During the same time, light travels only 3.40 km in a liquid. What is the refractive index of the liquid?

8. M V-HINT A flat sheet of ice has a thickness of 2.0 cm. It is on top of a flat sheet of crystalline quartz that has a thickness of 1.1 cm. Light strikes the ice perpendicularly and travels through it and then through the quartz. In the time it takes the light to travel through the two sheets, how far (in centimeters) would it have traveled in a vacuum?

Section 26.2 Snell's Law and the Refraction of Light

9. E GO The drawing shows four different situations in which a light ray is traveling from one medium into another. In some of the cases, the refraction is not shown correctly. For cases (*a*), (*b*), and (*c*), the angle of incidence is 55°; for case (*d*), the angle of incidence is 0°. Determine the angle of refraction in each case. If the drawing shows the refraction incorrectly, explain why it is incorrect.

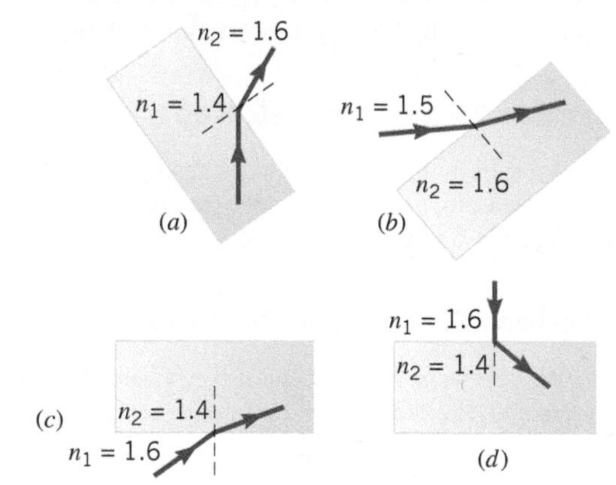

PROBLEM 9

10. E A layer of oil ($n = 1.45$) floats on an unknown liquid. A ray of light originates in the oil and passes into the unknown liquid. The angle of incidence is 64.0°, and the angle of refraction is 53.0°. What is the index of refraction of the unknown liquid?

11. E SSM A ray of light impinges from air onto a block of ice ($n = 1.309$) at a 60.0° angle of incidence. Assuming that this angle remains the same, find the difference $\theta_{2,\text{ice}} - \theta_{2,\text{water}}$ in the angles of refraction when the ice turns to water ($n = 1.333$).

12. E A narrow beam of light from a laser travels through air ($n = 1.00$) and strikes point A on the surface of the water ($n = 1.33$) in a lake. The angle of incidence is 55°. The depth of the lake is 3.0 m. On the flat lake-bottom is point B, directly below point A. **(a)** If refraction did not occur, how far away from point B would the laser beam strike the lake-bottom? **(b)** Considering refraction, how far away from point B would the laser beam strike the lake-bottom?

13. E SSM The drawing shows a coin resting on the bottom of a beaker filled with an unknown liquid. A ray of light from the coin travels to the surface of the liquid and is refracted as it enters into the air. A person sees the ray as it skims just above the surface of the liquid. How fast is the light traveling in the liquid?

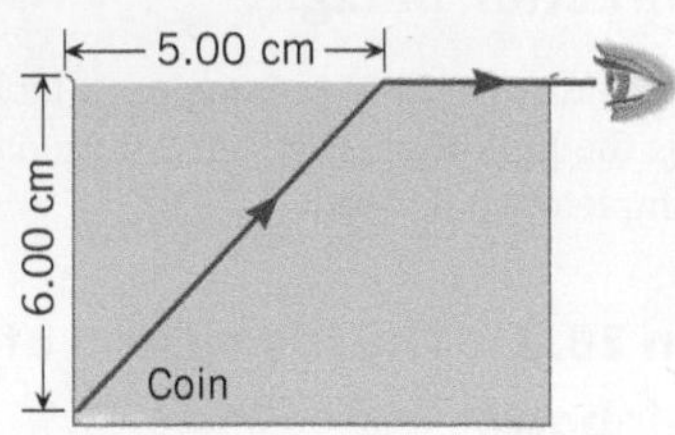

PROBLEM 13

14. E Amber ($n = 1.546$) is a transparent brown-yellow fossil resin. An insect, trapped and preserved within the amber, appears to be 2.5 cm beneath the surface when viewed directly from above. How far below the surface is the insect actually located?

15. E SSM A beam of light is traveling in air and strikes a material. The angles of incidence and refraction are 63.0° and 47.0°, respectively. Obtain the speed of light in the material.

16. E GO The drawing shows a ray of light traveling through three materials whose surfaces are parallel to each other. The refracted rays (but not the reflected rays) are shown as the light passes through each material. A ray of light strikes the *a*–*b* interface at a 50.0° angle of incidence. The index of refraction of material *a* is $n_a = 1.20$. The angles of refraction in materials *b* and *c* are, respectively, 45.0° and 56.7°. Find the indices of refraction in these two media.

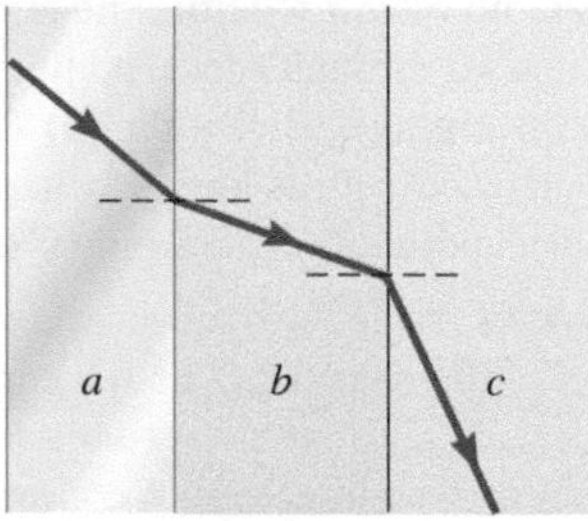

PROBLEM 16

17. E GO Light in a vacuum is incident on a transparent glass slab. The angle of incidence is 35.0°. The slab is then immersed in a pool of liquid. When the angle of incidence for the light striking the slab is 20.3°, the angle of refraction for the light entering the slab is the same as when the slab was in a vacuum. What is the index of refraction of the liquid?

18. M V-HINT CHALK A stone held just beneath the surface of a swimming pool is released and sinks to the bottom at a constant speed of 0.48 m/s. What is the apparent speed of the stone, as viewed from directly above by an observer who is in air?

19. M SSM Review Conceptual Example 4 as background for this problem. A man in a boat is looking straight down at a fish in the water directly beneath him. The fish is looking straight up at the man. They are equidistant from the air–water interface. To the man, the fish appears to be 2.0 m beneath his eyes. To the fish, how far above its eyes does the man appear to be?

20. M V-HINT The drawing shows a rectangular block of glass ($n = 1.52$) surrounded by liquid carbon disulfide ($n = 1.63$). A ray of light is incident on the glass at point A with a 30.0° angle of incidence. At what angle of refraction does the ray leave the glass at point B?

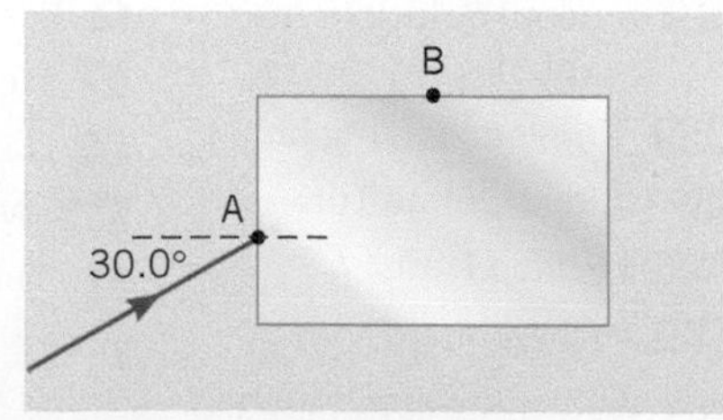

PROBLEM 20

21. **M** **SSM** In **Figure 26.6**, suppose that the angle of incidence is $\theta_1 = 30.0°$, the thickness of the glass pane is 6.00 mm, and the refractive index of the glass is $n_2 = 1.52$. Find the amount (in mm) by which the emergent ray is displaced relative to the incident ray.

22. **M** **GO** The back wall of a home aquarium is a mirror that is a distance of 40.0 cm away from the front wall. The walls of the tank are negligibly thin. A fish, swimming midway between the front and back walls, is being viewed by a person looking through the front wall. The index of refraction of air is $n_{air} = 1.000$ and that of water is $n_{water} = 1.333$. **(a)** Calculate the apparent distance between the fish and the front wall. **(b)** Calculate the apparent distance between the image of the fish and the front wall.

23. **M** Refer to **Figure 26.4b** and assume the observer is nearly above the submerged object. For this situation, derive the expression for the apparent depth: $d' = d(n_2/n_1)$, Equation 26.3. (*Hint:* Use Snell's law of refraction and the fact that the angles of incidence and refraction are small, so $\tan\theta \approx \sin\theta$.)

24. **H** A small logo is embedded in a thick block of crown glass ($n = 1.52$), 3.20 cm beneath the top surface of the glass. The block is put under water, so there is 1.50 cm of water above the top surface of the block. The logo is viewed from directly above by an observer in air. How far beneath the top surface of the water does the logo appear to be?

Section 26.3 Total Internal Reflection

25. **E** For the liquids in **Table 26.1**, determine the smallest critical angle for light that originates in one of them and travels toward the air–liquid interface.

26. **E** **SSM** A glass is half-full of water, with a layer of vegetable oil ($n = 1.47$) floating on top. A ray of light traveling downward through the oil is incident on the water at an angle of 71.4°. Determine the critical angle for the oil–water interface and decide whether the ray will penetrate into the water.

27. **E** A point source of light is submerged 2.2 m below the surface of a lake and emits rays in all directions. On the surface of the lake, directly above the source, the area illuminated is a circle. What is the maximum radius that this circle could have?

28. **E** **MMH** A ray of light is traveling in glass and strikes a glass–liquid interface. The angle of incidence is 58.0°, and the index of refraction of glass is $n = 1.50$. **(a)** What must be the index of refraction of the liquid so that the direction of the light entering the liquid is not changed? **(b)** What is the largest index of refraction that the liquid can have, so that none of the light is transmitted into the liquid and all of it is reflected back into the glass?

29. **E** **GO** The drawing shows three layers of different materials, with air above and below the layers. The interfaces between the layers are parallel. The index of refraction of each layer is given in the drawing. Identical rays of light are sent into the layers, and light zigzags through each layer, reflecting from the top and bottom surfaces. The index of refraction for air is $n_{air} = 1.00$. For each layer, the ray of light has an angle of incidence of 75.0°. For the cases in which total internal refection is possible from either the top or bottom surface of a layer, determine the amount by which the angle of incidence exceeds the critical angle.

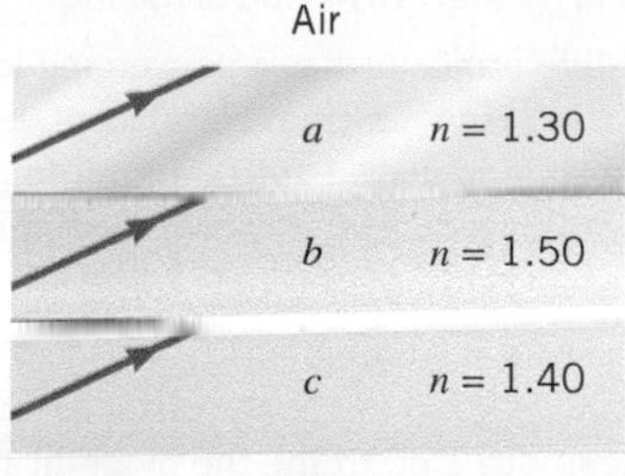

PROBLEM 29

30. **E** The drawing shows a crown glass slab with a rectangular cross section. As illustrated, a laser beam strikes the upper surface at an angle of 60.0°. After reflecting from the upper surface, the beam reflects from the side and bottom surfaces. **(a)** If the glass is surrounded by air, determine where part of the beam first exits the glass, at point *A*, *B*, or *C*. **(b)** Repeat part (a), assuming that the glass is surrounded by water instead of air.

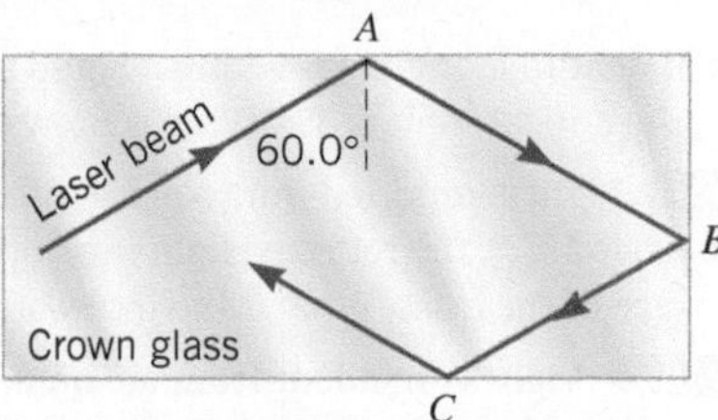

PROBLEM 30

31. **E** **GO** The drawing shows three materials, *a*, *b*, and *c*. A ray of light strikes the *a–b* interface at an angle that just barely exceeds its critical angle of 40.0°. The reflected ray then strikes the *a–c* interface at an angle of incidence that just barely exceeds its critical angle (which is not 40.0°). The index of refraction of material *a* is $n_a = 1.80$. Find the indices of refraction for the two other materials.

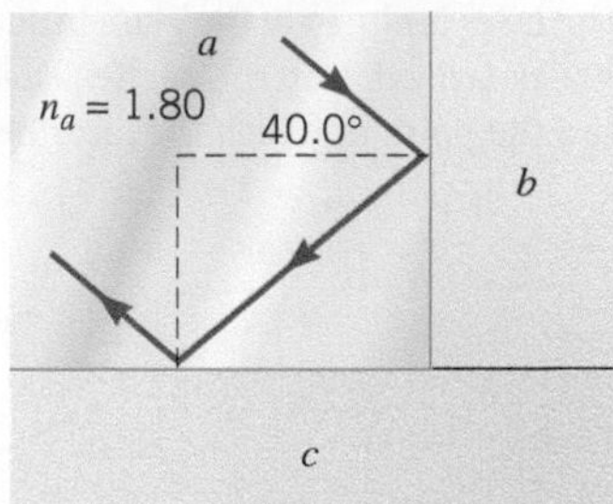

PROBLEM 31

32. **M** **CHALK** **SSM** **MMH** Multiple-Concept Example 7 provides helpful background for this problem. The drawing shows a crystalline quartz slab with a rectangular cross section. A ray of light strikes the slab at an incident angle of $\theta_1 = 34°$, enters the quartz, and travels to point *P*. This slab is surrounded by a fluid with a refractive index *n*. What is the maximum value of *n* for which total internal reflection occurs at point *P*?

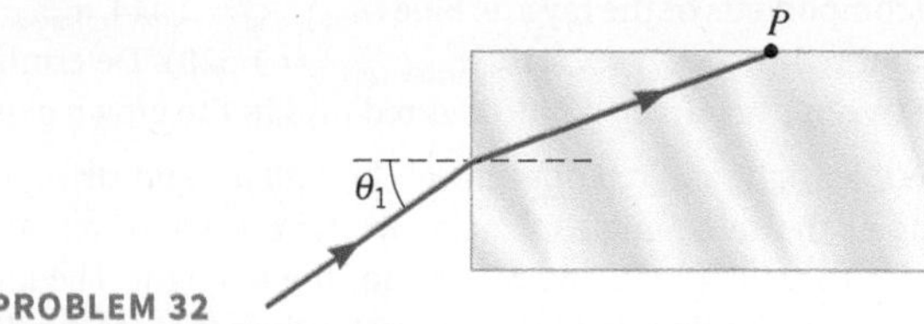

PROBLEM 32

33. **M** **V-HINT** The drawing shows a ray of light traveling from point *A* to point *B*, a distance of 4.60 m in a material that has an index of refraction n_1. At point *B*, the light encounters a different substance whose index of refraction is $n_2 = 1.63$. The light strikes the interface at the critical angle of $\theta_c = 48.1°$. How much time does it take for the light to travel from *A* to *B*?

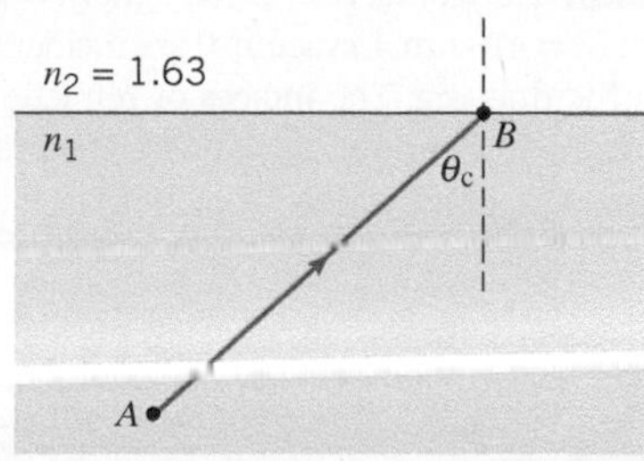

PROBLEM 33

34. **M** **GO** A layer of liquid *B* floats on liquid *A*. A ray of light begins in liquid *A* and undergoes total internal reflection at the interface

between the liquids when the angle of incidence exceeds 36.5°. When liquid B is replaced with liquid C, total internal reflection occurs for angles of incidence greater than 47.0°. Find the ratio n_B/n_C of the refractive indices of liquids B and C.

Section 26.4 Polarization and the Reflection and Refraction of Light

35. E For light that originates within a liquid and strikes the liquid–air interface, the critical angle is 39°. What is Brewster's angle for this light?

36. E SSM Light is reflected from a glass coffee table. When the angle of incidence is 56.7°, the reflected light is completely polarized parallel to the surface of the glass. What is the index of refraction of the glass?

37. E V-HINT Light is incident from air onto the surface of a liquid. The angle of incidence is 53.0°, and the angle of refraction is 34.0°. At what angle of incidence would the reflected light be 100% polarized?

38. E SSM CHALK When light strikes the surface between two materials from above, the Brewster angle is 65.0°. What is the Brewster angle when the light encounters the same surface from below?

39. E A laser is mounted in air, at a distance of 0.476 m above the edge of a large, horizontal pane of crown glass, as shown in the drawing. The laser is aimed at the glass in such a way that the reflected beam is 100% polarized. Determine the distance d between the edge of the pane and the point at which the laser beam is reflected.

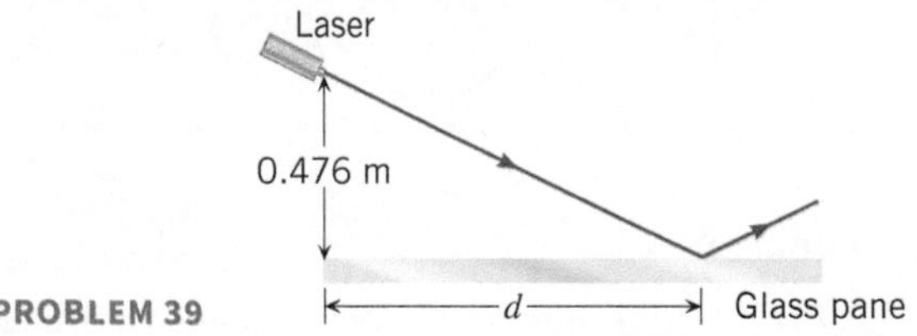

PROBLEM 39

Section 26.5 The Dispersion of Light: Prisms and Rainbows

40. E A ray of sunlight is passing from diamond into crown glass; the angle of incidence is 35.00°. The indices of refraction for the blue and red components of the ray are: blue ($n_{\text{diamond}} = 2.444$, $n_{\text{crown glass}} = 1.531$), and red ($n_{\text{diamond}} = 2.410$, $n_{\text{crown glass}} = 1.520$). Determine the angle between the refracted blue and red rays in the crown glass.

41. E Violet light and red light travel through air and strike a block of plastic at the same angle of incidence. The angle of refraction is 30.400° for the violet light and 31.200° for the red light. The index of refraction for violet light in plastic is greater than that for red light by 0.0400. Delaying any rounding off of calculations until the very end, find the index of refraction for violet light in plastic.

42. E SSM A beam of sunlight encounters a plate of crown glass at a 45.00° angle of incidence. Using the data in Table 26.2, find the angle between the violet ray and the red ray in the glass.

43. E CHALK Horizontal rays of red light ($\lambda = 660$ nm, in vacuum) and violet light ($\lambda = 410$ nm, in vacuum) are incident on the flint-glass prism shown in the drawing. The indices of refraction for the red and violet light are $n_{\text{red}} = 1.662$ and $n_{\text{violet}} = 1.698$. The prism is surrounded by air. What is the angle of refraction for each ray as it emerges from the prism?

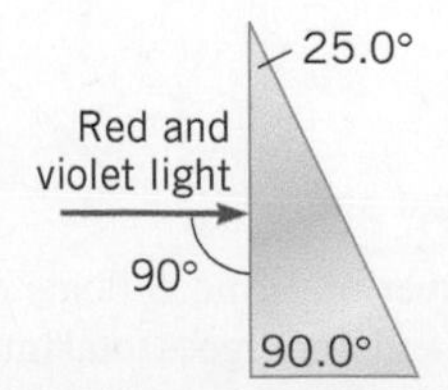

PROBLEM 43

44. M SSM This problem relates to Figure 26.18, which illustrates the dispersion of light by a prism. The prism is made from glass, and its cross section is an equilateral triangle. The indices of refraction for the red and violet light are 1.662 and 1.698, respectively. The angle of incidence for both the red and the violet light is 60.0°. Find the angles of refraction at which the red and violet rays emerge into the air from the prism.

45. M V-HINT The drawing shows a horizontal ray of white light incident perpendicularly on the vertical face of a prism (crown glass). The light enters the prism, and part of it undergoes refraction at the slanted face and emerges into the surrounding material. The rest of it is totally internally reflected and exits through the horizontal base of the prism. The colors of light that emerge from the slanted face may be chosen by altering the index of refraction of the material surrounding the prism. Find the required index of refraction of the surrounding material so that (a) only red light and (b) all colors except violet emerge from the slanted face. (See Table 26.2.)

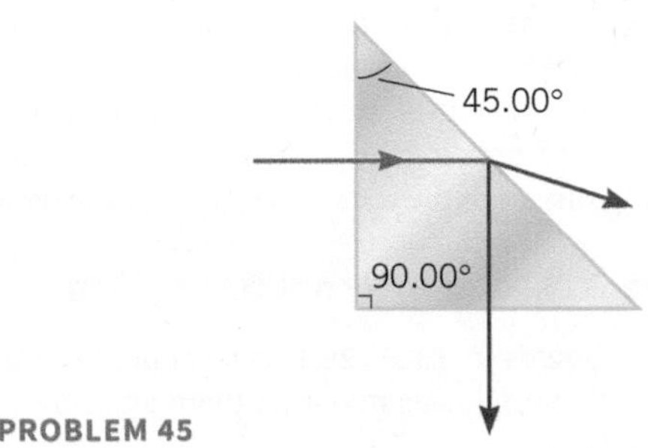

PROBLEM 45

Section 26.6 Lenses, Section 26.7 The Formation of Images by Lenses, Section 26.8 The Thin-Lens Equation and the Magnification Equation

(*Note:* When drawing ray diagrams, be sure that the object height h_o is much smaller than the focal length f of the lens or mirror.)

46. E SSM CHALK An object is located 9.0 cm in front of a converging lens ($f = 6.0$ cm). Using an accurately drawn ray diagram, determine where the image is located.

47. E The owner of a van installs a rear-window lens that has a focal length of −0.300 m. When the owner looks out through the lens at a person standing directly behind the van, the person appears to be just 0.240 m from the back of the van, and appears to be 0.34 m tall. (a) How far from the van is the person actually standing, and (b) how tall is the person?

48. E MMH A camera is supplied with two interchangeable lenses, whose focal lengths are 35.0 and 150.0 mm. A woman whose height is 1.60 m stands 9.00 m in front of the camera. What is the height (including sign) of her image on the image sensor, as produced by (a) the 35.0-mm lens and (b) the 150.0-mm lens?

49. E When a diverging lens is held 13.0 cm above a line of print, as in Figure 26.29, the image is 5.0 cm beneath the lens. (a) Is the image real or virtual? (b) What is the focal length of the lens?

50. E A slide projector has a converging lens whose focal length is 105.00 mm. (a) How far (in meters) from the lens must the screen be located if a slide is placed 108.00 mm from the lens? (b) If the slide measures 24.0 mm × 36.0 mm, what are the dimensions (in mm) of its image?

51. E (a) For a diverging lens ($f = -20.0$ cm), construct a ray diagram to scale and find the image distance for an object that is 20.0 cm from the lens. (b) Determine the magnification of the lens from the diagram.

52. **E GO** A tourist takes a picture of a mountain 14 km away using a camera that has a lens with a focal length of 50.0 mm. She then takes a second picture when she is only 5.0 km away. What is the ratio of the height of the mountain's image on the camera's image sensor for the second picture to its height on the image sensor for the first picture?

53. **E GO** An object is placed to the left of a lens, and a real image is formed to the right of the lens. The image is inverted relative to the object and is one-half the size of the object. The distance between the object and the image is 90.0 cm. **(a)** How far from the lens is the object? **(b)** What is the focal length of the lens?

54. **E MMH** A converging lens has a focal length of 88.00 cm. An object 13.00 cm tall is located 155.0 cm in front of this lens. **(a)** What is the image distance? **(b)** Is the image real or virtual? **(c)** What is the image height? Be sure to include the proper algebraic sign.

55. **M CHALK MMH** The distance between an object and its image formed by a diverging lens is 49.0 cm. The focal length of the lens is −233.0 cm. Find **(a)** the image distance and **(b)** the object distance.

56. **M SSM** The moon's diameter is 3.48×10^6 m, and its mean distance from the earth is 3.85×10^8 m. The moon is being photographed from the earth by a camera whose lens has a focal length of 50.0 mm. **(a)** Find the diameter of the moon's image on the slide film. **(b)** When the slide is projected onto a screen that is 15.0 m from the lens of the projector (f = 110.0 mm), what is the diameter of the moon's image on the screen?

57. **M** When a converging lens is used in a camera (as in **Figure 26.26*b***), the film must be at a distance of 0.210 m from the lens to record an image of an object that is 4.00 m from the lens. The same lens and film are used in a projector (see **Figure 26.27*b***), with the screen 0.500 m from the lens. How far from the projector lens should the film be placed?

58. **M SSM** An object is 18 cm in front of a diverging lens that has a focal length of −12 cm. How far in front of the lens should the object be placed so that the size of its image is reduced by a factor of 2.0?

59. **M V-HINT** An object is placed in front of a converging lens in such a position that the lens (f = 12.0 cm) creates a real image located 21.0 cm from the lens. Then, with the object remaining in place, the lens is replaced with another converging lens (f = 16.0 cm). A new, real image is formed. What is the image distance of this new image?

60. **H** A converging lens (f = 25.0 cm) is used to project an image of an object onto a screen. The object and the screen are 125 cm apart, and between them the lens can be placed at either of two locations. Find the two object distances.

Section 26.9 Lenses in Combination

61. **E GO** Two identical diverging lenses are separated by 16 cm. The focal length of each lens is −8.0 cm. An object is located 4.0 cm to the left of the lens that is on the left. Determine the final image distance relative to the lens on the right.

62. **E GO** Two systems are formed from a converging lens and a diverging lens, as shown in parts *a* and *b* of the drawing. (The point labeled "$F_{converging}$" is the focal point of the converging lens.) An object is placed inside the focal point of lens 1 at a distance of 10.00 cm to the left of lens 1. The focal lengths of the converging and diverging lenses are 15.00 and −20.0 cm, respectively. The distance between the lenses is 50.0 cm. Determine the final image distance for each system, measured with respect to lens 2.

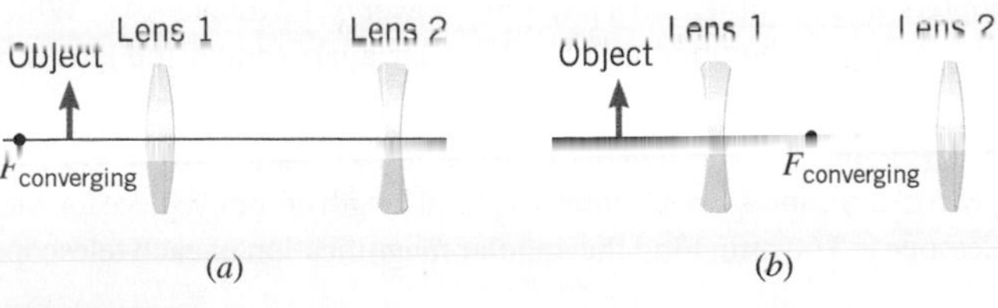

PROBLEM 62

63. **E CHALK** Two converging lenses are separated by 24.00 cm. The focal length of each lens is 12.00 cm. An object is placed 36.00 cm to the left of the lens that is on the left. Determine the final image distance relative to the lens on the right.

64. **E GO** A converging lens (f_1 = 24.0 cm) is located 56.0 cm to the left of a diverging lens (f_2 = −28.0 cm). An object is placed to the left of the converging lens, and the final image produced by the two-lens combination lies 20.7 cm to the left of the diverging lens. How far is the object from the converging lens?

65. **E SSM** A converging lens (f = 12.0 cm) is located 30.0 cm to the left of a diverging lens (f = −6.00 cm). A postage stamp is placed 36.0 cm to the left of the converging lens. **(a)** Locate the final image of the stamp relative to the diverging lens. **(b)** Find the overall magnification. **(c)** Is the final image real or virtual? With respect to the original object, is the final image **(d)** upright or inverted, and is it **(e)** larger or smaller?

66. **E MMH** A diverging lens (f = −10.0 cm) is located 20.0 cm to the left of a converging lens (f = 30.0 cm). A 3.00-cm-tall object stands to the left of the diverging lens, exactly at its focal point. **(a)** Determine the distance of the final image relative to the converging lens. **(b)** What is the height of the final image (including the proper algebraic sign)?

67. **M SSM** An object is placed 20.0 cm to the left of a diverging lens (f = −8.00 cm). A concave mirror (f = 12.0 cm) is placed 30.0 cm to the right of the lens. **(a)** Find the final image distance, measured relative to the mirror. **(b)** Is the final image real or virtual? **(c)** Is the final image upright or inverted with respect to the original object?

68. **M** Two converging lenses (f_1 = 9.00 cm and f_2 = 6.00 cm) are separated by 18.0 cm. The lens on the left has the longer focal length. An object stands 12.0 cm to the left of the left-hand lens in the combination. **(a)** Locate the final image relative to the lens on the right. **(b)** Obtain the overall magnification. **(c)** Is the final image real or virtual? With respect to the original object, **(d)** is the final image upright or inverted and **(e)** is it larger or smaller?

69. **M V-HINT** Visitors at a science museum are invited to sit in a chair to the right of a full-length diverging lens (f_1 = −3.00 m) and observe a friend sitting in a second chair, 2.00 m to the left of the lens. The visitor then presses a button and a converging lens (f_2 = +4.00 m) rises from the floor to a position 1.60 m to the right of the diverging lens, allowing the visitor to view the friend through both lenses at once. Find **(a)** the magnification of the friend when viewed through the diverging lens only and **(b)** the overall magnification of the friend when viewed through both lenses. Be sure to include the algebraic signs (+ or −) with your answers.

Section 26.10 The Human Eye

70. **E BIO** A student is reading material written on a blackboard. Her contact lenses have a refractive power of 57.50 diopters; the lens-to-retina distance is 1.750 cm. **(a)** How far (in meters) is the blackboard from her eyes? **(b)** If the material written on the blackboard is 5.00 cm high, what is the size of the image on her retina?

71. **E BIO** A nearsighted person cannot read a sign that is more than 5.2 m from his eyes. To deal with this problem, he wears contact lenses that do not correct his vision completely, but do allow him to read signs located up to distances of 12.0 m from his eyes. What is the focal length of the contacts?

72. **E BIO GO** A woman can read the large print in a newspaper only when it is at a distance of 65 cm or more from her eyes. **(a)** Is she nearsighted (myopic) or farsighted (hyperopic), and what kind of lens is used in her glasses to correct her eyesight? **(b)** What should be the refractive power (in diopters) of her glasses (worn

2.0 cm from the eyes), so she can read the newspaper at a distance of 25 cm from her eyes?

73. E BIO SSM Your friend has a near point of 138 cm, and she wears contact lenses that have a focal length of 35.1 cm. How close can she hold a magazine and still read it clearly?

74. E BIO GO A farsighted woman breaks her current eyeglasses and is using an old pair whose refractive power is 1.660 diopters. Since these eyeglasses do not completely correct her vision, she must hold a newspaper 42.00 cm from her eyes in order to read it. She wears the eyeglasses 2.00 cm from her eyes. How far is her near point from her eyes?

75. E BIO SSM CHALK A person has far points of 5.0 m from the right eye and 6.5 m from the left eye. Write a prescription for the refractive power of each corrective contact lens.

76. M BIO V-HINT A farsighted man uses eyeglasses with a refractive power of 3.80 diopters. Wearing the glasses 0.025 m from his eyes, he is able to read books held no closer than 0.280 m from his eyes. He would like a prescription for contact lenses to serve the same purpose. What is the correct contact lens prescription, in diopters?

77. H BIO The far point of a nearsighted person is 6.0 m from her eyes, and she wears contacts that enable her to see distant objects clearly. A tree is 18.0 m away and 2.0 m high. (a) When she looks through the contacts at the tree, what is its image distance? (b) How high is the image formed by the contacts?

78. H BIO The contacts worn by a farsighted person allow her to see objects clearly that are as close as 25.0 cm, even though her uncorrected near point is 79.0 cm from her eyes. When she is looking at a poster, the contacts form an image of the poster at a distance of 217 cm from her eyes. (a) How far away is the poster actually located? (b) If the poster is 0.350 m tall, how tall is the image formed by the contacts?

Section 26.11 Angular Magnification and the Magnifying Glass

79. E SSM A jeweler whose near point is 72 cm from his eye uses a magnifying glass as in Figure 26.41*b* to examine a watch. The watch is held 4.0 cm from the magnifying glass. Find the angular magnification of the magnifying glass.

80. E A spectator, seated in the left-field stands, is watching a baseball player who is 1.9 m tall and is 75 m away. On a TV screen, located 3.0 m from a person watching the game at home, the image of this same player is 0.12 m tall. Find the angular size of the player as seen by (a) the spectator watching the game live and (b) the TV viewer. (c) To whom does the player appear to be larger?

81. E An engraver uses a magnifying glass ($f = 9.50$ cm) to examine some work, as in Figure 26.41*b*. The image he sees is located 25.0 cm from his eye, which is his near point. (a) What is the distance between the work and the magnifying glass? (b) What is the angular magnification of the magnifying glass?

82. E GO CHALK The near point of a naked eye is 32 cm. When an object is placed at the near point and viewed by the naked eye, it has an angular size of 0.060 rad. A magnifying glass has a focal length of 16 cm, and is held next to the eye. The enlarged image that is seen is located 64 cm from the magnifying glass. Determine the angular size of the image.

83. E V-HINT An object has an angular size of 0.0150 rad when placed at the near point (21.0 cm) of an eye. When the eye views this object using a magnifying glass, the largest possible angular size of the image is 0.0380 rad. What is the focal length of the magnifying glass?

84. H SSM A farsighted person can read printing as close as 25.0 cm when she wears contacts that have a focal length of 45.4 cm. One day, she forgets her contacts and uses a magnifying glass, as in Figure 26.41*b*. Its maximum angular magnification is 7.50 for a young person with a normal near point of 25.0 cm. What is the maximum angular magnification that the magnifying glass can provide for her?

Section 26.12 The Compound Microscope

85. E A forensic pathologist is viewing heart muscle cells with a microscope that has two selectable objectives with refracting powers of 100 and 300 diopters. When he uses the 100-diopter objective, the image of a cell subtends an angle of 3×10^{-3} rad with the eye. What angle is subtended when he uses the 300-diopter objective?

86. E SSM A compound microscope has a barrel whose length is 16.0 cm and an eyepiece whose focal length is 1.4 cm. The viewer has a near point located 25 cm from his eyes. What focal length must the objective have so that the angular magnification of the microscope will be −320?

87. E CHALK The distance between the lenses in a compound microscope is 18 cm. The focal length of the objective is 1.5 cm. If the microscope is to provide an angular magnification of −83 when used by a person with a normal near point (25 cm from the eye), what must be the focal length of the eyepiece?

88. E GO The near point of a naked eye is 25 cm. When placed at the near point and viewed by the naked eye, a tiny object would have an angular size of 5.2×10^{-5} rad. When viewed through a compound microscope, however, it has an angular size of -8.8×10^{-3} rad. (The minus sign indicates that the image produced by the microscope is inverted.) The objective of the microscope has a focal length of 2.6 cm, and the distance between the objective and the eyepiece is 16 cm. Find the focal length of the eyepiece.

89. M In a compound microscope, the focal length of the objective is 3.50 cm and that of the eyepiece is 6.50 cm. The distance between the lenses is 26.0 cm. (a) What is the angular magnification of the microscope if the person using it has a near point of 35.0 cm? (b) If, as usual, the first image lies just inside the focal point of the eyepiece (see Figure 26.32), how far is the object from the objective? (c) What is the magnification (not the angular magnification) of the objective?

Section 26.13 The Telescope

90. E An astronomical telescope has an angular magnification of −132. Its objective has a refractive power of 1.50 diopters. What is the refractive power of its eyepiece?

91. E SSM Mars subtends an angle of 8.0×10^{-5} rad at the unaided eye. An astronomical telescope has an eyepiece with a focal length of 0.032 m. When Mars is viewed using this telescope, it subtends an angle of 2.8×10^{-3} rad. Find the focal length of the telescope's objective lens.

92. E GO A telescope has an objective with a refractive power of 1.25 diopters and an eyepiece with a refractive power of 250 diopters. What is the angular magnification of the telescope?

93. E An amateur astronomer decides to build a telescope from a discarded pair of eyeglasses. One of the lenses has a refractive power of 11 diopters, and the other has a refractive power of 1.3 diopters. (a) Which lens should be the objective? (b) How far apart should the lenses be separated? (c) What is the angular magnification of the telescope?

94. M GO CHALK The lengths of three telescopes are $L_A = 455$ mm, $L_B = 615$ mm, and $L_C = 824$ mm. The focal length of the eyepiece for each telescope is 3.00 mm. Find the angular magnification of each telescope.

95. **H** An astronomical telescope is being used to examine a relatively close object that is only 114.00 m away from the objective of the telescope. The objective and eyepiece have focal lengths of 1.500 and 0.070 m, respectively. Noting that the expression $M \approx -f_o/f_e$ is no longer applicable because the object is so close, use the thin-lens and magnification equations to find the angular magnification of this telescope. (*Hint:* See **Figure 26.43** and note that the focal points F_o and F_e are so close together that the distance between them may be ignored.)

Additional Problems

Online

96. **E SSM** An object is located 30.0 cm to the left of a converging lens whose focal length is 50.0 cm. (a) Draw a ray diagram to scale and from it determine the image distance and the magnification. (b) Use the thin-lens and magnification equations to verify your answers to part (a).

97. **E SSM MMH** A glass block ($n = 1.56$) is immersed in a liquid. A ray of light within the glass hits a glass–liquid surface at a 75.0° angle of incidence. Some of the light enters the liquid. What is the smallest possible refractive index for the liquid?

98. **E** As an aid in understanding this problem, refer to Conceptual Example 4. A swimmer, who is looking up from under the water, sees a diving board directly above at an apparent height of 4.0 m above the water. What is the actual height of the diving board above the water?

99. **E** A camper is trying to start a fire by focusing sunlight onto a piece of paper. The diameter of the sun is 1.39×10^9 m, and its mean distance from the earth is 1.50×10^{11} m. The camper is using a converging lens whose focal length is 10.0 cm. (a) What is the area of the sun's image on the paper? (b) If 0.530 W of sunlight passes through the lens, what is the intensity of the sunlight at the paper?

100. **E GO** Red light ($n = 1.520$) and violet light ($n = 1.538$) traveling in air are incident on a slab of crown glass. Both colors enter the glass at the same angle of refraction. The red light has an angle of incidence of 30.00°. What is the angle of incidence of the violet light?

101. **E** A converging lens ($f = 12.0$ cm) is held 8.00 cm in front of a newspaper that has a print size with a height of 2.00 mm. Find (a) the image distance (in cm) and (b) the height (in mm) of the magnified print.

102. **E GO** To focus a camera on objects at different distances, the converging lens is moved toward or away from the image sensor, so a sharp image always falls on the sensor. A camera with a telephoto lens ($f = 200.0$ mm) is to be focused on an object located first at a distance of 3.5 m and then at 50.0 m. Over what distance must the lens be movable?

103. **E SSM** An office copier uses a lens to place an image of a document onto a rotating drum. The copy is made from this image. (a) What kind of lens is used, converging or diverging? If the document and its copy are to have the same size, but are inverted with respect to one another, (b) how far from the document is the lens located and (c) how far from the lens is the image located? Express your answers in terms of the focal length f of the lens.

104. **M V-HINT** An object is in front of a converging lens ($f = 0.30$ m). The magnification of the lens is $m = 4.0$. (a) Relative to the lens, in what direction should the object be moved so that the magnification changes to $m = -4.0$? (b) Through what distance should the object be moved?

105. **H SSM** The angular magnification of a telescope is 32 800 times as large when you look through the correct end of the telescope as when you look through the wrong end. What is the angular magnification of the telescope?

Physics in Biology, Medicine, and Sports

106. **M BIO SSM 26.10** At age forty, a man requires contact lenses ($f = 65.0$ cm) to read a book held 25.0 cm from his eyes. At age forty-five, while wearing these contacts he must now hold a book 29.0 cm from his eyes. (a) By what distance has his near point *changed?* (b) What focal-length lenses does he require at age forty-five to read a book at 25.0 cm?

107. **H BIO 26.10** Bill is farsighted and has a near point located 125 cm from his eyes. Anne is also farsighted, but her near point is 75.0 cm from her eyes. Both have glasses that correct their vision to a normal near point (25.0 cm from the eyes), and both wear the glasses 2.0 cm from the eyes. Relative to the eyes, what is the closest object that can be seen clearly (a) by Anne when she wears Bill's glasses and (b) by Bill when he wears Anne's glasses?

108. **M BIO 26.11** Dermatologists will often use the technique of *dermatoscopy,* in which they examine skin irregularities using a device called a *dermatoscope.* The dermatoscope consists of a light source and a magnifying lens (see the photo). The light source is capable of producing both unpolarized and polarized light, which allows for the examination of surface skin lesions and abnormalities in deeper skin structures, respectively. This technique allows dermatologists to distinguish benign skin lesions from those that are malignant (or cancerous), indicating the presence of melanoma. Consider a farsighted dermatologist who has a near point located at 37.0 cm from her eyes. She is using a dermatoscope with an angular magnification of 10.0 to examine a suspicious looking mole on a patient's back. (a) What is the distance between the mole and the lens in the dermatoscope? (b) What is the focal length of the dermatoscope's lens, if the image of the mole forms at a distance of −14.2 cm from the lens?

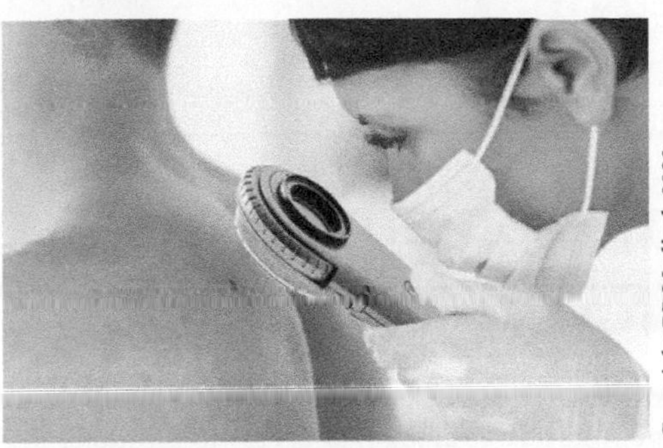

PROBLEM 108

109. M BIO **26.10** A farsighted person has a near point of 72.0 cm. What should be the refractive power of their glasses, if they are able to see objects clearly at 25.0 cm? Assume their glasses are located 2.00 cm in front of their eyes.

110. M BIO **26.10** A person with abnormal vision dives into a swimming pool, and while underwater, can see distant objects clearly. **(a)** Is this person nearsighted or farsighted? This same person visits an optometrist who prescribes corrective contact lenses with a refractive power of −12.0 diopters. **(b)** How far from the person's eyes is their far point?

111. M BIO **26.11** A technician in a pathology laboratory is using a compound microscope to examine tissue samples. The focal length of the objective of the microscope is 0.35 cm, and the focal length of the eyepiece is 2.5 cm. The two lenses are separated by 27 cm. One day, the technician forgot to wear his contact lenses, which have a prescription of +1.5 diopters and allow him to see objects clearly at 25 cm. When he looks into the microscope without wearing his contacts, what is the approximate angular magnification of the scope?

112. M BIO **26.10** A person with myopic vision has a right eye with a far point of 4.0 m and a left eye with a far point of 5.0 m. Write a prescription for their corrective glasses that will sit 2.7 cm in front of their eyes. **(a)** What will be the refractive power of the right lens of their glasses, and **(b)** what will be the refractive power of the left lens?

Concepts and Calculations Problems

Online

One important phenomenon discussed in this chapter is how a ray of light is refracted when it goes from one medium into another. Problem 113 reviews some of the important aspects of refraction, including Snell's law, the concept of a critical angle, and the notion of index matching. One of the most important applications of refraction is in lenses, the behaviors of which are governed by the thin-lens and magnification equations. Problem 114 reviews how these equations are applied to a two-lens system, along with the all-important sign conventions that must be followed.

113. M CHALK A ray of light is incident on a glass–water interface at the critical angle θ_c as the figure illustrates. The reflected light then passes through a liquid (immiscible with water) and into air. The indices of refraction for the four substances are given in the drawing. *Concepts:* (i) What determines the critical angle when the ray strikes the glass–water interface? (ii) When the light is incident at the glass–water interface at the critical angle, what is the angle of refraction, and how is the angle θ_1 related to the critical angle? (iii) When the reflected ray strikes the glass–liquid interface, how is the angle of refraction θ_3 related to the angle of incidence θ_2? Note that the two materials have the same indices of refraction. (iv) When the ray passes from the liquid into the air, is the ray refracted? Explain. *Calculation:* Determine the angle of refraction θ_5 for the ray as it passes into the air.

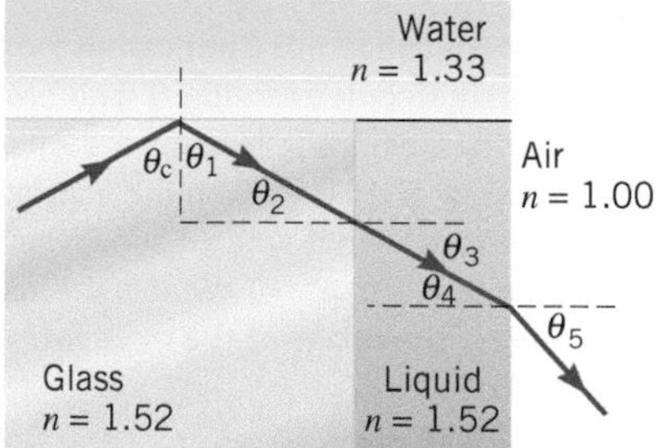

PROBLEM 113

114. M CHALK SSM In the figure, a converging lens (f_1 = +20.0 cm) and a diverging lens (f_2 = −15.0 cm) are separated by a distance of 10.0 cm. An object with a height of h_{o1} = 5.00 mm is placed at a distance of d_{o1} = 45.0 cm to the left of the first (converging) lens. *Concepts:* (i) Is the image produced by the first (converging) lens real or virtual? (ii) As far as the second lens is concerned, what role does the image produced by the first lens play? (iii) Note in the figure that the image produced by the first lens is called the "first image," and it falls to the right of the second lens. This image acts as the object for the second lens. Normally, however, an object would lie to the left of the lens. How do we take into account that this object lies to the right of the diverging lens? (iv) How do we find the location of the image produced by the second lens when its object is a virtual object? *Calculation:* What are **(a)** the image distance d_{i1} and **(b)** the height h_{i1} of the image produced by the first lens? **(c)** What is the object distance for the second (diverging) lens? Find **(d)** the image distance d_{i2} and **(e)** the height h_{i2} of the image produced by the second lens.

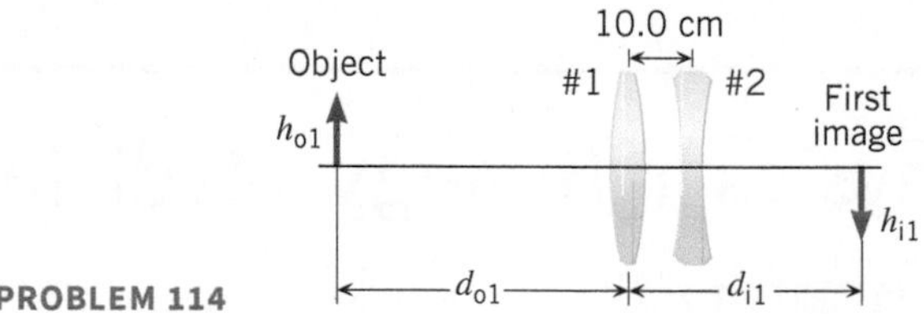

PROBLEM 114

Team Problems

Online

115. M T WS **Laser Ablation.** You and your team have been assigned to set up a powerful laser to be used in a process called laser ablation. Laser ablation is a process in which a powerful, focused laser beam delivers a large amount of energy to the atoms of a surface, called the target, in such a short time that the atoms are expelled from the surface before the target material even has time to melt. Such a process can be used to shape the cornea in laser eye surgery (i.e., Lasik surgery), or to coat nearby surfaces with the ablated material to make novel coatings, such as those used in semiconductor devices. Suppose the laser emits pulses of ultraviolet light with a circular cross section of diameter 1.70 cm, each with a duration of 25.0 nanoseconds and a total energy of 650.0 mJ. (a) What is the energy density of each pulse? (b) What is the intensity of the light striking the surface during each pulse? (c) In order for the laser to properly ablate the target, the intensity must be 7.00×10^{11} W/m^2. What should be the magnification of a lens system that focuses the laser on the target so that this intensity is achieved? (d) An optical system processes the incoming beam so that the laser source can be treated as an object located in front of a converging spherical lens. If the lens has a focal length of 30.0 cm, determine the object distance so that the focused beam (i.e., the image) has an intensity of 7.00×10^{11} W/m^2. (e) With the object distance found in part (d), how far from the lens should the target be located, that is, what is the image distance?

116. **E T WS An Underwater Transparent Reflector.** You and your team have been assigned to optically engineer a transparent optical device that can reverse a laser beam under water. The idea is that the laser beam travels in water ($n_W = 1.33$) and passes through one surface of a triangular prism with an angle of incidence equal to 90.0°. It then undergoes a reflection from each of the two other internal surfaces of the prism with equal angles of incidence, and then exits through the surface from which it entered in a direction antiparallel to the incoming beam. The exiting beam has the same intensity as the incoming beam (i.e., all of the incident light is reflected) (a) What is the angle between the two internal reflecting surfaces of the prism? (b) What should be the minimum index of refraction (n_G) of the glass?

117. **M T An Optical Spectrometer.** You and your team are tasked with characterizing an equilateral, triangular prism to be used in an optical spectrometer. An optical spectrometer contains a dispersive element (in this case, the prism) that separates an incoming beam of light into its constituent wavelengths (or colors), and a photocell that measures the intensity of each color (wavelength) by measuring the intensity of the dispersed light as a function of angle. The index of refraction of the prism material you are using depends on the wavelength as

$$n(\lambda) = -(1.080 \times 10^{-4}\ \text{nm}^{-1})\ \lambda\ (\text{nm}) + 1.586$$

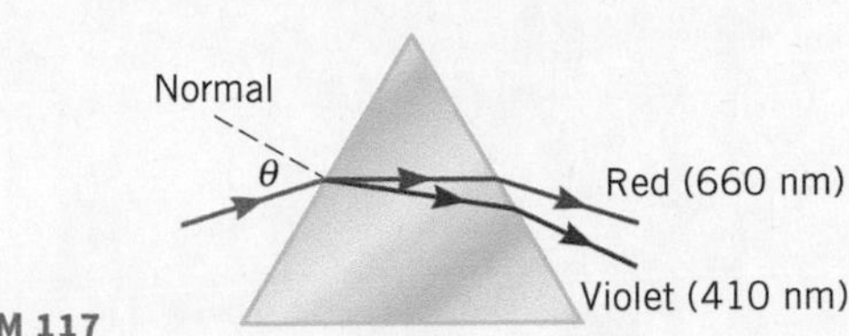

PROBLEM 117

An incident beam of white light impinges on the surface (see the drawing) at an angle of 45.00 degrees below the normal. The prism is in air ($n = 1.000$). Relative to the prism base (i.e., the horizontal in the drawing), at what angles do (a) red light ($\lambda = 660.0$ nm) and (b) violet light ($\lambda = 410.0$ nm) emerge on the opposite side? (c) What is the angular range of the full spectrum, from red to violet?

118. **M T Emergency Replacement Glasses.** One of your team members lost her glasses in a river. She is nearsighted and cannot see long distances without them. You are on an expedition to map a remote area in southern Argentina, and her long-distance vision is crucial to her role in the group. You can request a new pair of glasses for her that will be delivered with the next airdrop of supplies, but she does not know her lens prescription. (a) Does she need a converging or diverging lens? (b) You do a simple eye test and estimate her far point to be 623 cm from her eyes. Assuming she will wear her glasses 2.00 cm in front of her eyes, what should be the focal length of her new lenses? (c) What should be the refractive power of her new lenses (in diopters)?

CHAPTER 27

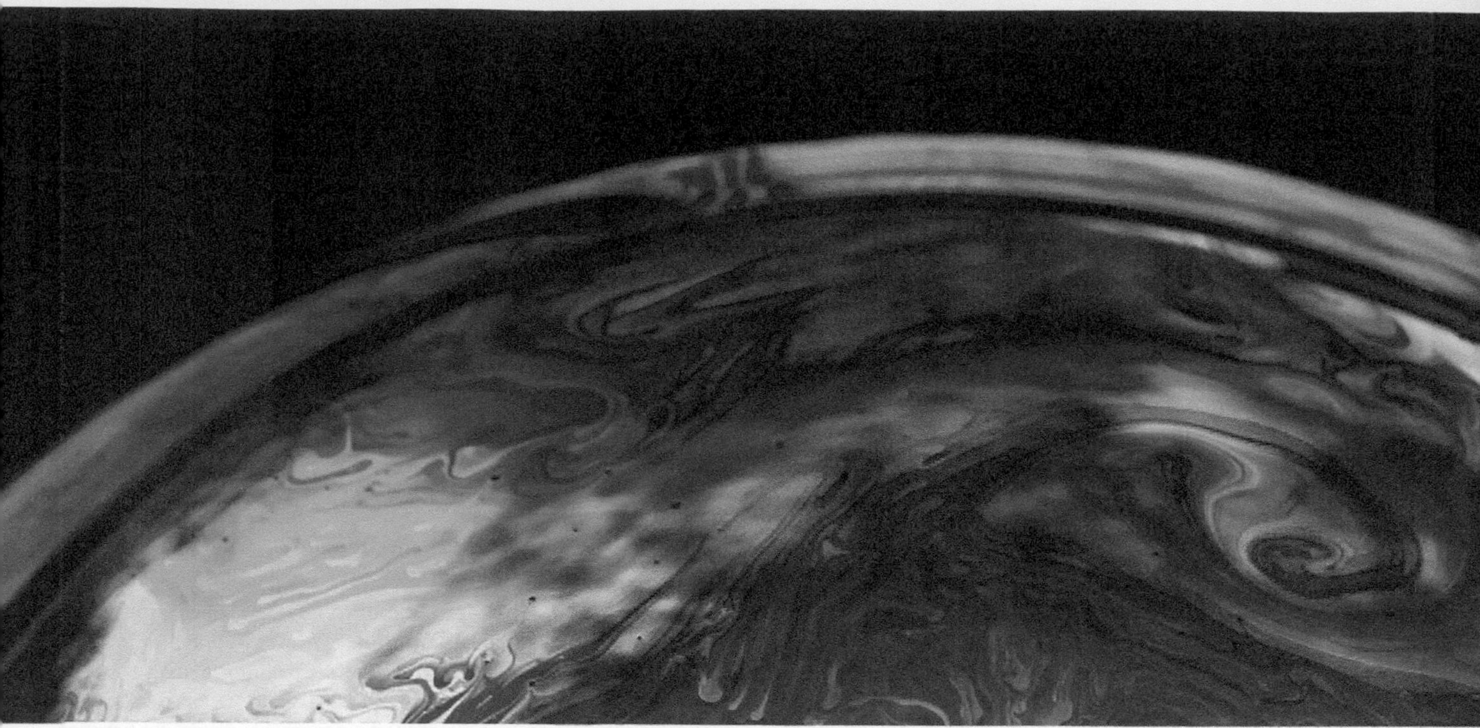

A rainbow of colors is observed in a bubble, which is a thin film of soap surrounded by air on both sides. But unlike a real rainbow's colors, which are due to dispersion, the colors in the bubble are created by wave interference. Interference occurs when two or more waves exist simultaneously at the same place. Here, some of the light reaching the eye of the observer has reflected from the top surface of the bubble, while some of the light has reflected from the inner surface. The thickness of the bubble, which varies across its surface, affects how the light waves interfere, and thus determines the colors we see. Interference effects, which we will study in this chapter, provide direct evidence of the wave nature of light.

Squirrel_photos/Pixabay

LEARNING OBJECTIVES

After reading this module, you should be able to...

27.1 Apply the principle of linear superposition to light waves.

27.2 Analyze double-slit interference.

27.3 Analyze thin-film interference.

27.4 Understand the operation of the Michelson interferometer.

27.5 Analyze single-slit diffraction.

27.6 Determine the resolving power of lenses.

27.7 Apply interference principles to the diffraction grating.

27.8 Describe X-ray diffraction in crystals.

Interference and the Wave Nature of Light

27.1 The Principle of Linear Superposition

Chapter 17 examines what happens when several sound waves are present at the same place at the same time. The pressure disturbance that results is governed by the ***principle of linear superposition, which states that the resultant disturbance is the sum of the disturbances from the individual waves.*** Light is also a wave, an electromagnetic wave, and it too obeys the superposition principle. When two or more light waves pass through a given point, their electric fields combine

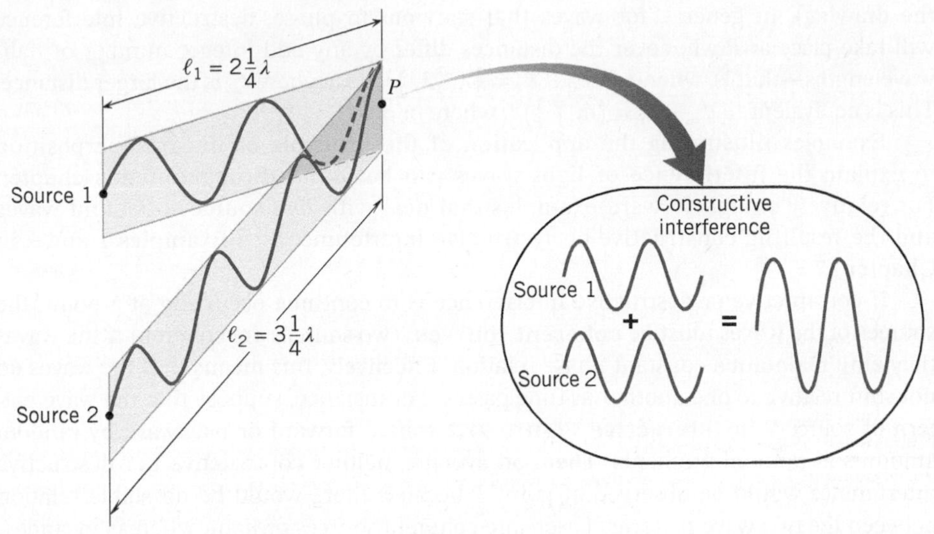

INTERACTIVE FIGURE 27.1 The waves emitted by source 1 and source 2 start out in phase and arrive at point P in phase, leading to constructive interference at that point.

according to the principle of linear superposition and produce a resultant electric field. According to Equation 24.5b, the square of the electric field strength is proportional to the intensity of the light, which, in turn, is related to its brightness. Thus, interference can and does alter the brightness of light, just as it affects the loudness of sound.

Interactive Figure 27.1 illustrates what happens when two identical waves (same wavelength λ and same amplitude) arrive at the point P in phase—that is, crest-to-crest and trough-to-trough. According to the principle of linear superposition, the waves reinforce each other and **constructive interference** occurs. The resulting total wave at P has an amplitude that is twice the amplitude of either individual wave, and, in the case of light waves, the brightness at P is greater than that due to either wave alone. The waves start out in phase and are in phase at P because the distances ℓ_1 and ℓ_2 between this spot and the sources of the waves differ by one wavelength λ. In **Interactive Figure 27.1**, these distances are $\ell_1 = 2\frac{1}{4}$ wavelengths and $\ell_2 = 3\frac{1}{4}$ wavelengths. In general, when the waves start out in phase, constructive interference will result at P whenever the distances are the same or differ by any integer number of wavelengths—in other words, assuming that ℓ_2 is the larger distance, whenever $\ell_2 - \ell_1 = m\lambda$, where $m = 0, 1, 2, 3, \ldots$.

Interactive Figure 27.2 shows what occurs when two identical waves arrive at the point P out of phase with one another, or crest-to-trough. Now the waves mutually cancel, according to the principle of linear superposition, and **destructive interference** results. With light waves this would mean that there is no brightness. The waves begin with the same phase but are out of phase at P because the distances through which they travel in reaching this spot differ by one-half of a wavelength ($\ell_1 = 2\frac{3}{4}\lambda$ and $\ell_2 = 3\frac{1}{4}\lambda$ in

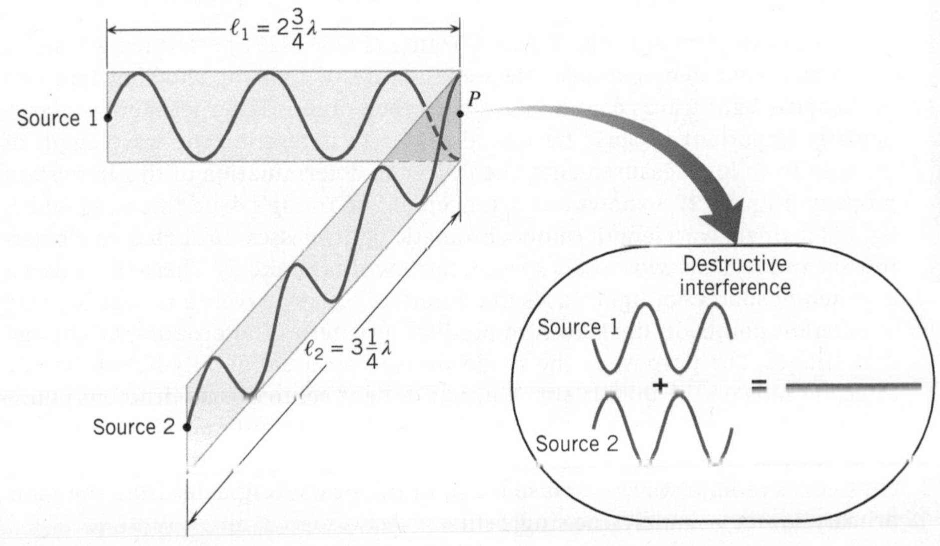

INTERACTIVE FIGURE 27.2 The waves emitted by the two sources have the same phase to begin with, but they arrive at point P out of phase. As a result, destructive interference occurs at P.

the drawing). In general, for waves that start out in phase, destructive interference will take place at P whenever the distances differ by any odd integer number of half-wavelengths—that is, whenever $\ell_2 - \ell_1 = \frac{1}{2}\lambda, \frac{3}{2}\lambda, \frac{5}{2}\lambda, \ldots$, where ℓ_2 is the larger distance. This is equivalent to $\ell_2 - \ell_1 = \left(m + \frac{1}{2}\right)\lambda$, where $m = 0, 1, 2, 3, \ldots$.

Examples illustrating the application of the principle of linear superposition to explain the interference of light waves can be found throughout this chapter. For relatively straightforward examples that deal with two sources of sound waves and the resulting constructive or destructive interference, see Examples 1 and 2 in Chapter 17.

If constructive or destructive interference is to continue occurring at a point, the sources of the waves must be **coherent sources**. Two sources are coherent if the waves they emit maintain a constant phase relation. Effectively, this means that the waves do not shift relative to one another as time passes. For instance, suppose that the wave pattern of source 1 in **Interactive Figure 27.2** shifted forward or backward by random amounts at random moments. Then, on average, neither constructive nor destructive interference would be observed at point P because there would be no stable relation between the two wave patterns. Lasers are coherent sources of light, whereas incandescent light bulbs and fluorescent lamps are incoherent sources.

Check Your Understanding

(The answers are given at the end of the book.)

1. Two separate coherent sources produce waves whose wavelengths are 0.10 m. The two waves spread out and overlap at a certain point. Does constructive or destructive interference occur at this point when **(a)** one wave travels 3.20 m and the other travels 3.00 m, **(b)** one wave travels 3.20 m and the other travels 3.05 m, **(c)** one wave travels 3.20 m and the other travels 2.95 m?

2. Suppose that a radio station broadcasts simultaneously from two transmitting antennas at *two different locations*. Compared with only one transmitting antenna, the reception with two transmitting antennas **(a)** is always better **(b)** is always worse **(c)** can be either better or worse, depending on the distance traveled by each wave.

3. Two sources of waves are in phase and produce identical waves. These sources are mounted at the corners of a square and broadcast waves uniformly in all directions. At the center of the square, will the waves always produce constructive interference no matter which two corners of the square are occupied by the sources?

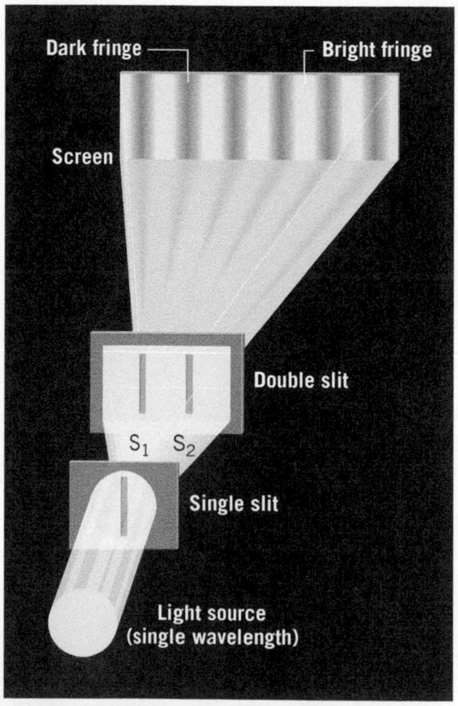

FIGURE 27.3 In Young's double-slit experiment, two slits S_1 and S_2 act as coherent sources of light. Light waves from these slits interfere constructively and destructively on the screen to produce, respectively, the bright and dark fringes. The slit widths and the distance between the slits have been exaggerated for clarity.

27.2 Young's Double-Slit Experiment

In 1801 the English scientist Thomas Young (1773–1829) performed a historic experiment that demonstrated the wave nature of light by showing that two overlapping light waves can interfere with each other. His experiment was particularly important because he was also able to determine the wavelength of the light from his measurements, the first such determination of this important property. **Figure 27.3** shows one arrangement of Young's experiment, in which light of a single wavelength (monochromatic light) passes through a single narrow slit and falls on two closely spaced, narrow slits S_1 and S_2. These two slits act as coherent sources of light waves that interfere constructively and destructively at different points on the screen to produce a pattern of alternating bright and dark fringes. The purpose of the single slit is to ensure that only light from one direction falls on the double slit. Without it, light coming from different points on the light source would strike the double slit from different directions and cause the pattern on the screen to be washed out. The slits S_1 and S_2 act as coherent sources of light waves because the light from each originates from the same primary source—namely, the single slit.

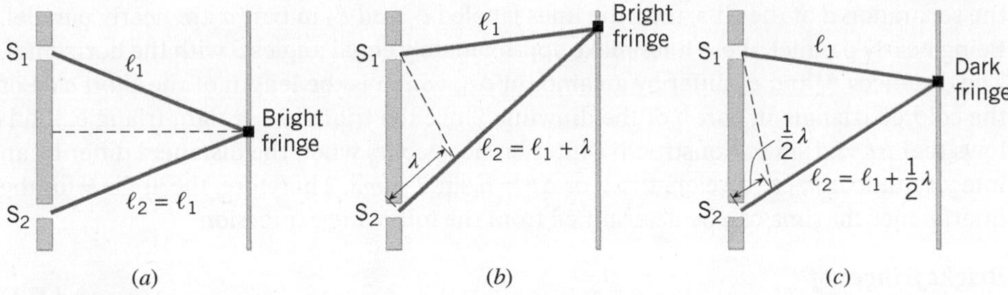

FIGURE 27.4 The waves that originate from slits S_1 and S_2 interfere constructively (parts *a* and *b*) or destructively (part *c*) on the screen, depending on the difference in distances between the slits and the screen. Note that the slit widths and the distance between the slits have been exaggerated for clarity.

To help explain the origin of the bright and dark fringes, **Figure 27.4** presents three top views of the double slit and the screen. Part *a* illustrates how a bright fringe arises directly opposite the midpoint between the two slits. In this part of the drawing the waves (identical) from each slit travel to the midpoint on the screen. At this location, the distances ℓ_1 and ℓ_2 to the slits are equal, each containing the same number of wavelengths. Therefore, constructive interference results, leading to the bright fringe. Part *b* indicates that constructive interference produces another bright fringe on one side of the midpoint when the distance ℓ_2 is larger than ℓ_1 by exactly one wavelength. A bright fringe also occurs symmetrically on the other side of the midpoint when the distance ℓ_1 exceeds ℓ_2 by one wavelength; for clarity, however, this additional bright fringe is not shown. Constructive interference produces other bright fringes (also not shown) on both sides of the middle wherever the difference between ℓ_1 and ℓ_2 is an integer number of wavelengths: λ, 2λ, 3λ, and so on. Part *c* shows how the first dark fringe arises. Here the distance ℓ_2 is larger than ℓ_1 by exactly one-half a wavelength, so the waves interfere destructively, giving rise to the dark fringe. Destructive interference creates other dark fringes on both sides of the center wherever the difference between ℓ_1 and ℓ_2 equals an odd integer number of half-wavelengths: $\frac{1}{2}\lambda$, $\frac{3}{2}\lambda$, $\frac{5}{2}\lambda$, and so on.

The brightness of the fringes in Young's experiment varies, as the fringe pattern in **Figure 27.5** shows. Below the fringe pattern is a graph to suggest the way in which the intensity varies for the fringes. The central fringe is labeled with a zero, and the other bright fringes are numbered in ascending order on either side of the center. It can be seen that the central fringe has the greatest intensity. To either side of the center, the intensities of the other fringes decrease symmetrically in a way that depends on how small the slit widths are relative to the wavelength of the light.

The position of the fringes observed on the screen in Young's experiment can be calculated with the aid of **Figure 27.6**. If the screen is located far away compared with

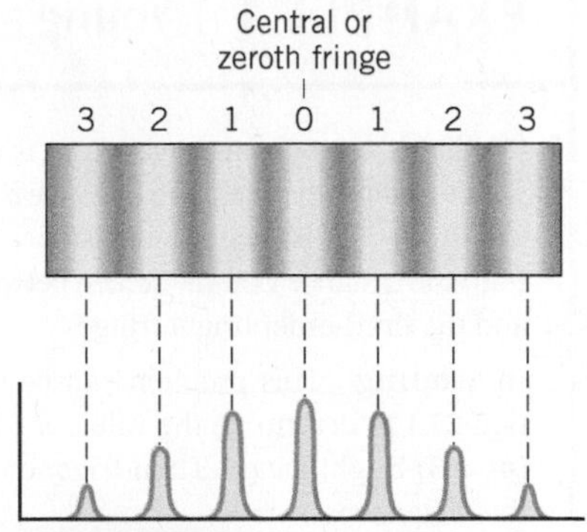

FIGURE 27.5 The results of Young's double-slit experiment, showing the pattern of the bright and dark fringes formed on the screen and a graph of the light intensity. The central or zeroth fringe is the brightest fringe (greatest intensity).

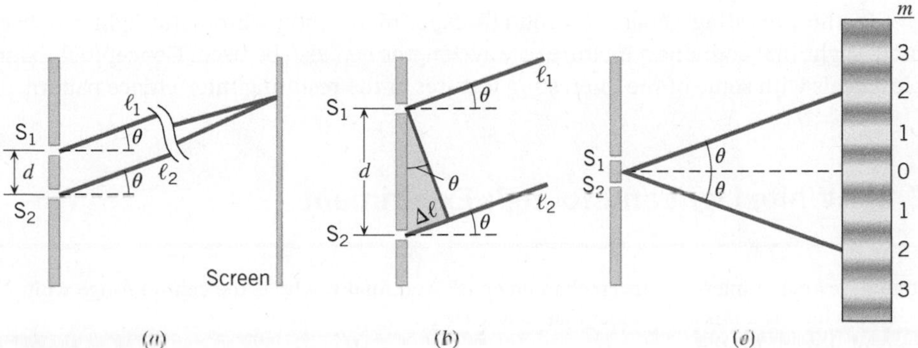

FIGURE 27.6 (*a*) Rays from slits S_1 and S_2, which make approximately the same angle θ with the horizontal, strike a distant screen at the same spot. (*b*) The difference in the path lengths of the two rays is $\Delta\ell = d \sin\theta$. (*c*) The angle θ is the angle at which a bright fringe ($m = 2$, here) occurs on either side of the central bright fringe ($m = 0$).

the separation d of the slits, then the lines labeled ℓ_1 and ℓ_2 in part a are nearly parallel. Being nearly parallel, these lines make approximately equal angles θ with the horizontal. The distances ℓ_1 and ℓ_2 differ by an amount $\Delta\ell$, which is the length of the short side of the colored triangle in part b of the drawing. Since the triangle is a right triangle, it follows that $\Delta\ell = d \sin\theta$. Constructive interference occurs when the distances differ by an integer number m of wavelengths λ, or $\Delta\ell = d\sin\theta = m\lambda$. Therefore, the angle θ for the interference maxima can be determined from the following expression:

Bright fringes of a double slit

$$\sin\theta = m\frac{\lambda}{d} \quad m = 0, 1, 2, 3, \ldots \tag{27.1}$$

The value of m specifies the *order* of the fringe. Thus, $m = 2$ identifies the "second-order" bright fringe. Part c of **Figure 27.6** stresses that the angle θ given by Equation 27.1 locates bright fringes on either side of the midpoint between the slits. A similar line of reasoning leads to the conclusion that the dark fringes, which lie between the bright fringes, are located according to

Dark fringes of a double slit

$$\sin\theta = \left(m + \frac{1}{2}\right)\frac{\lambda}{d} \quad m = 0, 1, 2, 3, \ldots \tag{27.2}$$

Example 1 illustrates how to determine the distance of a higher-order bright fringe from the central bright fringe with the aid of Equation 27.1.

EXAMPLE 1 | Young's Double-Slit Experiment

Red light ($\lambda = 664$ nm in vacuum) is used in Young's experiment with the slits separated by a distance $d = 1.20 \times 10^{-4}$ m. The screen in **Figure 27.7** is located at a distance of $L = 2.75$ m from the slits. Find the distance y on the screen between the central bright fringe and the third-order bright fringe.

Reasoning This problem can be solved by first using Equation 27.1 to determine the value of θ that locates the third-order ($m = 3$) bright fringe. Then trigonometry can be used to obtain the distance y.

Solution According to Equation 27.1, we find

$$\theta = \sin^{-1}\left(\frac{m\lambda}{d}\right) = \sin^{-1}\left[\frac{3(664 \times 10^{-9}\ \text{m})}{1.20 \times 10^{-4}\ \text{m}}\right] = 0.951°$$

According to **Figure 27.7**, the distance y can be calculated from $\tan\theta = y/L$:

$$y = L\tan\theta = (2.75\ \text{m})\tan 0.951° = \boxed{0.0456\ \text{m}}$$

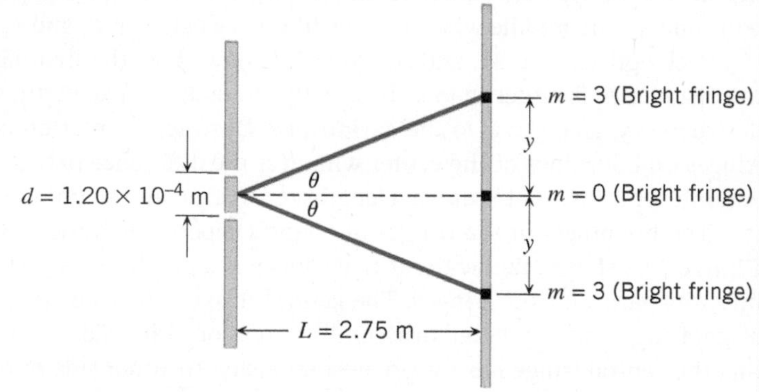

FIGURE 27.7 The third-order bright fringe ($m = 3$) is observed on the screen at a distance y from the central bright fringe ($m = 0$).

In the preceding version of Young's experiment, monochromatic light has been used. Light that contains a mixture of wavelengths can also be used. Conceptual Example 2 deals with some of the interesting features of the resulting interference pattern.

CONCEPTUAL EXAMPLE 2 | White Light and Young's Experiment

Figure 27.8 shows a photograph that illustrates the kind of interference fringes that can result when white light, which is a mixture of all colors, is used in Young's experiment. Except for the central fringe, which is white, the bright fringes are a rainbow of colors. Why does Young's experiment separate white light into its constituent colors? In any group of colored fringes, such as the two singled out in **Figure 27.8**, why is red farther out from the central fringe than green is? And finally, why is the central fringe white rather than colored?

Reasoning and Solution To understand how the color separation arises, we need to remember that each color corresponds to a different wavelength λ and that constructive and destructive interference depend on the wavelength. According to Equation 27.1

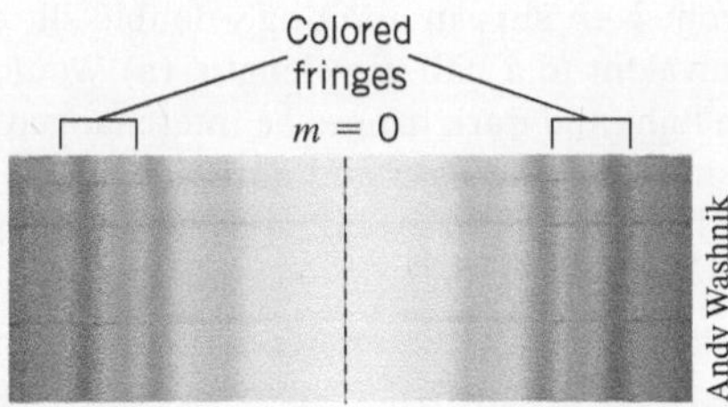

FIGURE 27.8 This photograph shows the results observed on the screen in one version of Young's experiment in which white light (a mixture of all colors) is used.

($\sin\theta = m\lambda/d$), there is a different angle that locates a bright fringe for each value of λ, and thus for each color. These different angles lead to the separation of colors on the observation screen. In fact, on either side of the central fringe, there is one group of colored fringes for $m = 1$ and another for each additional value of m.

Now, consider what it means that, within any single group of colored fringes, red is farther out from the central fringe than green is. It means that, in the equation $\sin\theta = m\lambda/d$, red light has a larger angle θ than green light does. Does this make sense? Yes, because red has the larger wavelength (see Table 26.2, where $\lambda_{red} = 660$ nm and $\lambda_{green} = 550$ nm).

In **Figure 27.8**, the central fringe is distinguished from all the other colored fringes by being white. In Equation 27.1, the central fringe is different from the other fringes because it is the only one for which $m = 0$. In Equation 27.1, a value of $m = 0$ means that $\sin\theta = m\lambda/d = 0$, which reveals that $\theta = 0°$, no matter what the wavelength λ is. In other words, all wavelengths have a zeroth-order bright fringe located at the same place on the screen, so that all colors strike the screen there and mix together to produce the white central fringe.

Related Homework: Problem 46

Historically, Young's experiment provided strong evidence that light has a wave-like character. If light behaved only as a stream of "tiny particles," as others believed at the time,* then the two slits would deliver the light energy into only two bright fringes located directly opposite the slits on the screen. Instead, Young's experiment shows that wave interference redistributes the energy from the two slits into many bright fringes.

Check Your Understanding

(The answers are given at the end of the book.)

4. Replace the slits S_1 and S_2 in **Figure 27.3** with identical in-phase loudspeakers and use the same ac electrical signal to drive them. The two sound waves produced will then be identical, and you will have the audio equivalent of Young's double-slit experiment. In terms of loudness and softness, what would you hear as you walk along the screen, starting from the center and going to either end? **(a)** Loud, then soft, then loud, then soft, etc., with the loud sounds *decreasing in intensity* as you walk away from the center **(b)** Loud, then soft, then loud, then soft, etc., with the loud sounds *increasing in intensity* as you walk away from the center **(c)** Soft, then loud, then soft, then loud, etc., with the loud sounds *decreasing in intensity* as you walk away from the center **(d)** Soft, then loud, then soft, then loud, etc., with the loud sounds *increasing in intensity* as you walk away from the center

5. **CYU Figure 27.1** shows two double slits that have slit separations of d_1 and d_2. Light whose wavelength is either λ_1 or λ_2 passes through the slits. For comparison, the wavelengths are also illustrated in the drawing. For which combination of slit separation and wavelength would the pattern of bright and dark fringes on the observation screen be **(a)** the most spread out and **(b)** the least spread out?

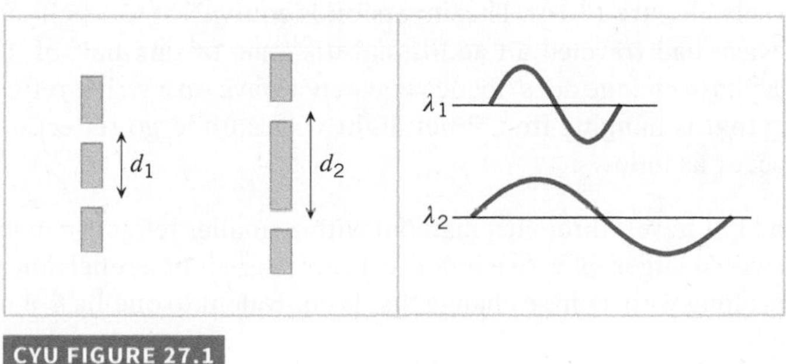

CYU FIGURE 27.1

*It is now known that the particle, or corpuscular, theory of light, which Isaac Newton promoted, does indeed explain some experiments that the wave theory cannot explain. Today, light is regarded as having both particle and wave characteristics. Chapter 29 discusses this dual nature of light.

6. Suppose the light waves coming from *both* slits in a Young's double-slit experiment had their phases shifted by an amount equivalent to a half-wavelength. **(a)** Would the pattern be the same or would the positions of the light and dark fringes be interchanged? **(b)** Would the pattern be the same or would the positions of the light and dark fringes be interchanged if the light coming from *only one* of the slits had its phase shifted by an amount equivalent to a half-wavelength?

7. In Young's double-slit experiment, is it possible to see interference fringes when the wavelength of the light is greater than the distance between the slits?

27.3 Thin-Film Interference

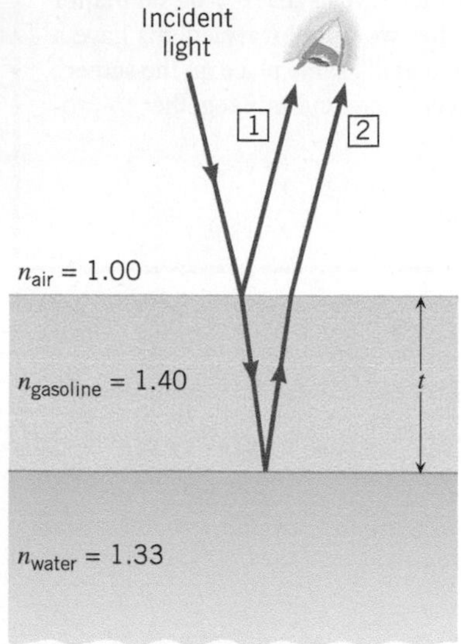

INTERACTIVE FIGURE 27.9 Because of reflection and refraction, two light waves, represented by rays 1 and 2, enter the eye when light shines on a thin film of gasoline floating on a thick layer of water.

Young's double-slit experiment is one example of interference between light waves. Interference also occurs in more common circumstances. For instance, **Interactive Figure 27.9** shows a thin film such as gasoline floating on water. To begin with, let us assume that the film has a constant thickness. Consider what happens when monochromatic light (a single wavelength) strikes the film nearly perpendicularly. At the top surface of the film reflection occurs and produces the light wave represented by ray 1. However, refraction also occurs, and some light enters the film. Part of this light reflects from the bottom surface of the film and passes back up through the film, eventually reentering the air. Thus, a second light wave, which is represented by ray 2, also exists. Moreover, this wave, having traversed the film twice, has traveled farther than wave 1. Because of the extra travel distance, there can be interference between the two waves. If constructive interference occurs, an observer whose eyes detect the superposition of waves 1 and 2 would see a uniformly bright film. If destructive interference occurs, an observer would see a uniformly dark film.

In **Interactive Figure 27.9** the difference in path lengths between waves 1 and 2 occurs inside the thin film. Therefore, we note the following:

> **Problem-Solving Insight The wavelength that is important for thin-film interference is the wavelength within the film, not the wavelength in vacuum.**

The wavelength within the film can be calculated from the wavelength in vacuum by using the index of refraction n_{film} for the film. With the aid of Equations 26.1 and 16.1, it can be shown that $n_{\text{film}} = c/v = (c/f)/(v/f) = \lambda_{\text{vacuum}}/\lambda_{\text{film}}$. In other words,

$$\lambda_{\text{film}} = \frac{\lambda_{\text{vacuum}}}{n_{\text{film}}} \tag{27.3}$$

In explaining the interference that can occur in **Interactive Figure 27.9**, we need to add one other important part to the story. Whenever waves reflect at a boundary, it is possible for them to change phase. **Animated Figure 27.10**, for example, shows that a wave on a string is inverted when it reflects from the end that is tied to a wall (see also Figure 17.16). This inversion is equivalent to a half-cycle of the wave, as if the wave had traveled an additional distance of one-half of a wavelength. In contrast, a phase change does not occur when a wave on a string reflects from the end of a string that is hanging free. When light waves undergo reflection, similar phase changes occur as follows:

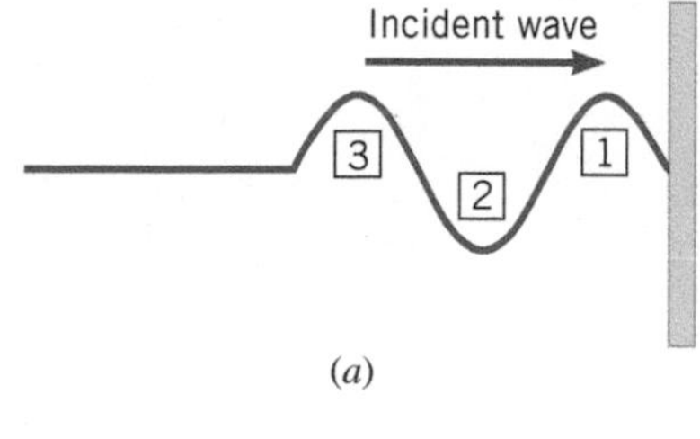

(a)

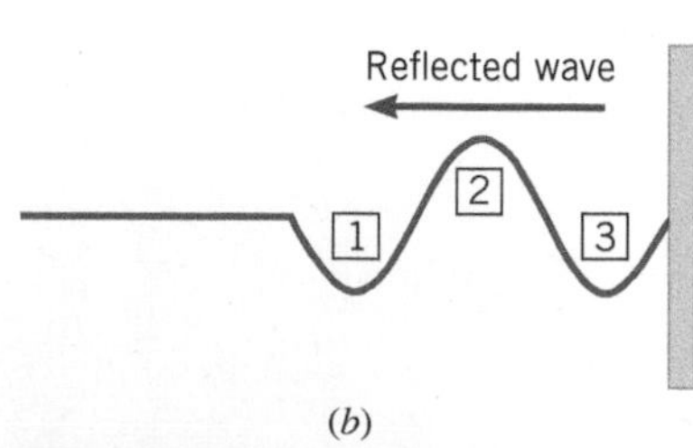

(b)

ANIMATED FIGURE 27.10 When a wave on a string reflects from a wall, the wave undergoes a phase change. Thus, an upward-pointing half-cycle of the wave becomes, after reflection, a downward-pointing half-cycle, and vice versa, as the numbered labels in the drawing indicate.

1. When light travels through a material with a smaller refractive index toward a material with a larger refractive index (e.g., air to gasoline), reflection at the boundary occurs along with a phase change that is equivalent to one-half of a wavelength in the film.
2. When light travels from a larger toward a smaller refractive index, there is no phase change upon reflection at the boundary.

The next example indicates how the phase change that can accompany reflection is taken into account when dealing with thin-film interference.

EXAMPLE 3 | A Colored Thin Film of Gasoline

A thin film of gasoline floats on a puddle of water. Sunlight falls almost perpendicularly on the film and reflects into your eyes. Although sunlight is white since it contains all colors, the film looks yellow because destructive interference eliminates the color of blue ($\lambda_{\text{vacuum}} = 469$ nm) from the reflected light. The refractive indices of the blue light in gasoline and in water are 1.40 and 1.33, respectively. Determine the minimum nonzero thickness t of the film.

Reasoning To solve this problem, we must express the condition for destructive interference in terms of the film thickness t and the wavelength λ_{film} in the gasoline film. We must also take into account any phase changes that occur upon reflection.

In **Interactive Figure 27.9**, the phase change for wave 1 is equivalent to one-half of a wavelength, since this light travels from a smaller refractive index ($n_{\text{air}} = 1.00$) toward a larger refractive index ($n_{\text{gasoline}} = 1.40$). In contrast, there is no phase change when wave 2 reflects from the bottom surface of the film, since this light travels from a material with a larger refractive index ($n_{\text{gasoline}} = 1.40$) toward a material with a smaller one ($n_{\text{water}} = 1.33$). The net phase change between waves 1 and 2 due to reflection is, thus, equivalent to one-half of a wavelength, $\frac{1}{2}\lambda_{\text{film}}$. This half-wavelength must be combined with the extra travel distance for wave 2, to determine the condition for destructive interference. For destructive interference, the combined total must be an odd integer number of half-wavelengths. Since wave 2 travels back and forth through the film and since light strikes the film nearly perpendicularly, the extra travel distance is twice the film thickness, or $2t$. Thus, the condition for destructive interference is

$$\underbrace{2t}_{\substack{\text{Extra distance}\\ \text{traveled by}\\ \text{wave 2}}} + \underbrace{\tfrac{1}{2}\lambda_{\text{film}}}_{\substack{\text{Half-wavelength}\\ \text{net phase change}\\ \text{due to reflection}}} = \underbrace{\tfrac{1}{2}\lambda_{\text{film}}, \tfrac{3}{2}\lambda_{\text{film}}, \tfrac{5}{2}\lambda_{\text{film}}, \ldots}_{\substack{\text{Condition for destructive}\\ \text{interference}}}$$

or

$$2t + \tfrac{1}{2}\lambda_{\text{film}} = \left(m + \tfrac{1}{2}\right)\lambda_{\text{film}} \qquad m = 0, 1, 2, 3, \ldots$$

After subtracting the term $\frac{1}{2}\lambda_{\text{film}}$ from both sides of this equation, we can solve for the thickness t of the film that yields destructive interference:

$$t = \frac{m\lambda_{\text{film}}}{2} \qquad m = 0, 1, 2, 3, \ldots$$

Problem-Solving Insight When analyzing thin-film interference effects, remember to use the wavelength of the light in the film (λ_{film}) instead of the wavelength in a vacuum (λ_{vacuum}).

Solution In order to calculate t, we need to know the wavelength of the blue light in the film. Equation 27.3, with $n_{\text{film}} = 1.40$, gives this wavelength as

$$\lambda_{\text{film}} = \frac{\lambda_{\text{vacuum}}}{n_{\text{film}}} = \frac{469 \text{ nm}}{1.40} = 335 \text{ nm}$$

With this value for λ_{film} and $m = 1$, our result for t gives the minimum nonzero film thickness for which the blue color is missing in the reflected light as follows:

$$t = \frac{m\lambda_{\text{film}}}{2} = \frac{(1)(335 \text{ nm})}{2} = \boxed{168 \text{ nm}}$$

In Example 3 a half-wavelength net phase change occurs due to the reflections at the upper and lower surfaces of the thin film. Depending on the refractive indices of the materials above and below the film, it is also possible that these reflections yield a zero net phase change.

The thin film in Example 3 has the same yellow color everywhere. In nature, such a uniformly colored thin film would be unusual; the next example is more realistic.

CONCEPTUAL EXAMPLE 4 | Multicolored Thin Films

Under natural conditions, thin films, like gasoline on water or like the soap bubble in **Figure 27.11**, have a multicolored appearance that often changes while you are watching them. Why are such films multicolored, and what can be inferred from the fact that the colors change in time?

Reasoning and Solution In Example 3 we have seen that a thin film can appear yellow if destructive interference removes blue light from the reflected sunlight. The thickness of the film is the key. If the thickness were different, so that destructive interference removed green light from the reflected sunlight, the film would appear magenta. Constructive interference can also cause certain colors to appear brighter than others in the reflected light and thereby give the film a colored appearance. The colors that are enhanced by constructive interference, like those removed by destructive interference, depend on the thickness of the film. Thus, we conclude the following:

Problem-Solving Insight The different colors in a thin film of gasoline on water or in a soap bubble arise because the thickness is different in different places on the film. Moreover, the fact that the colors change as you watch them indicates that the thickness is changing.

A number of factors can cause the thickness to change, including air currents, temperature fluctuations, and the pull of gravity, which tends to make a vertical film sag, leading to thicker regions at the bottom than at the top.

Related Homework: Problem 17

FIGURE 27.11 This fantastic soap bubble is multicolored when viewed in sunlight because of the effects of thin-film interference.

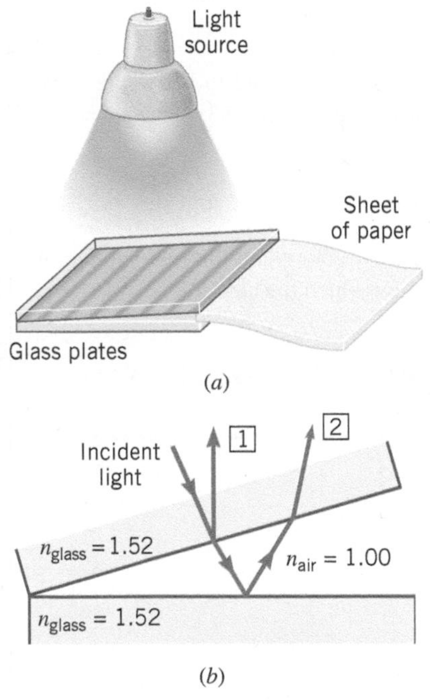

FIGURE 27.12 (*a*) The wedge of air formed between two flat glass plates causes an interference pattern of alternating dark and bright fringes to appear in reflected light. (*b*) A side view of the glass plates and the air wedge.

The colors that you see when sunlight is reflected from a thin film also depend on your viewing angle. At an oblique angle, the light corresponding to ray 2 in **Interactive Figure 27.9**, for instance, would travel a greater distance within the film than it does at nearly perpendicular incidence. The greater distance would lead to destructive interference for a different wavelength.

THE PHYSICS OF . . . nonreflecting lens coatings. Thin-film interference can be beneficial in optical instruments. For example, some cameras contain six or more lenses. Reflections from all the lens surfaces can reduce considerably the amount of light directly reaching the film. In addition, multiple reflections from the lenses often reach the film indirectly and degrade the quality of the image. To minimize such unwanted reflections, high-quality lenses are often covered with a thin nonreflective coating of magnesium fluoride ($n = 1.38$). The thickness of the coating is usually chosen to ensure that destructive interference eliminates the reflection of green light, which is in the middle of the visible spectrum. It should be pointed out that the absence of any reflected light does not mean that it has been destroyed by the nonreflective coating. Rather, the "missing" light has been transmitted into the coating and the lens.

Another interesting illustration of thin-film interference is the air wedge. As **Figure 27.12*a*** shows, an air wedge is formed when two flat plates of glass are separated along one side, perhaps by a thin sheet of paper. The thickness of this film of air varies between zero, where the plates touch, and the thickness of the paper. When monochromatic light reflects from this arrangement, alternate bright and dark fringes are formed by constructive and destructive interference, as the drawing indicates and Example 5 discusses.

EXAMPLE 5 | An Air Wedge

(a) Assuming that green light ($\lambda_{\text{vacuum}} = 552$ nm) strikes the glass plates nearly perpendicularly in **Figure 27.12**, determine the number of bright fringes that occur between the place where the plates touch and the edge of the sheet of paper (thickness $= 4.10 \times 10^{-5}$ m). **(b)** Explain why there is a dark fringe where the plates touch.

Reasoning A bright fringe occurs wherever there is constructive interference, as determined by any phase changes due to reflection and by the thickness of the air wedge. There is no phase change upon reflection for wave 1, since this light travels from a larger (glass) toward a smaller (air) refractive index. In contrast, there is a half-wavelength phase change for wave 2, since the ordering of the refractive indices is reversed at the lower air/glass boundary where reflection occurs. The net phase change due to reflection for waves 1 and 2, then, is equivalent to a half-wavelength. Now we combine any extra distance traveled by ray 2 with this half-wavelength and determine the condition for the constructive interference that creates the bright fringes. Constructive interference occurs whenever the *combination* yields an integer number of wavelengths. At nearly perpendicular incidence, the extra travel distance for wave 2 is approximately twice the thickness t of the wedge at any point, so the condition for constructive interference is

$$\underbrace{2t}_{\substack{\text{Extra distance}\\\text{traveled by}\\\text{wave 2}}} + \underbrace{\tfrac{1}{2}\lambda_{\text{film}}}_{\substack{\text{Half-wavelength}\\\text{net phase change}\\\text{due to reflection}}} = \underbrace{\lambda_{\text{film}}, 2\lambda_{\text{film}}, 3\lambda_{\text{film}}, \ldots}_{\substack{\text{Condition for}\\\text{constructive interference}}}$$

Subtracting the term $\frac{1}{2}\lambda_{\text{film}}$ from the left-hand side of this equation and from each term on the right-hand side yields

$$2t = \underbrace{\tfrac{1}{2}\lambda_{\text{film}}, \tfrac{3}{2}\lambda_{\text{film}}, \tfrac{5}{2}\lambda_{\text{film}}, \ldots}_{(m+\frac{1}{2})\lambda_{\text{film}} \quad m = 0, 1, 2, 3, \ldots}$$

Therefore,

$$t = \frac{\left(m + \frac{1}{2}\right)\lambda_{\text{film}}}{2} \qquad m = 0, 1, 2, 3, \ldots$$

In this expression, note that the "film" is a film of air. Since the refractive index of air is nearly one, λ_{film} is virtually the same as that in a vacuum, so $\lambda_{\text{film}} = 552$ nm.

Solution **(a)** When t equals the thickness of the paper holding the plates apart, the corresponding value of m can be obtained from the equation above:

$$m = \frac{2t}{\lambda_{\text{film}}} - \frac{1}{2} = \frac{2(4.10 \times 10^{-5}\ \text{m})}{552 \times 10^{-9}\ \text{m}} - \frac{1}{2} = 148$$

Since the first bright fringe occurs when $m = 0$, the number of bright fringes is $m + 1 = \boxed{149}$.

(b) Where the plates touch, there is a dark fringe because of destructive interference between the light waves represented by rays 1 and 2. Destructive interference occurs because the thickness of the wedge is zero here and the only difference between the rays is the half-wavelength phase change due to reflection from the lower plate.

Another type of air wedge can also be used to determine the degree to which the surface of a lens or mirror is spherical. When an accurate spherical surface is put in contact with an optically flat plate, as in **Figure 27.13*a***, the circular interference fringes shown in part *b* of the figure can be observed. The circular fringes are called *Newton's rings.* They arise in the same way that the straight fringes arise in **Figure 27.12*a***.

Problem-Solving Insight **Reflected light will experience a phase change only if the light travels from a material with a smaller refractive index toward a material with a larger refractive index. Be sure to take such a phase change into account when analyzing thin-film interference.**

As a final example of thin-film interference, we consider the phenomenon of structural coloration in the animal world in Example 6.

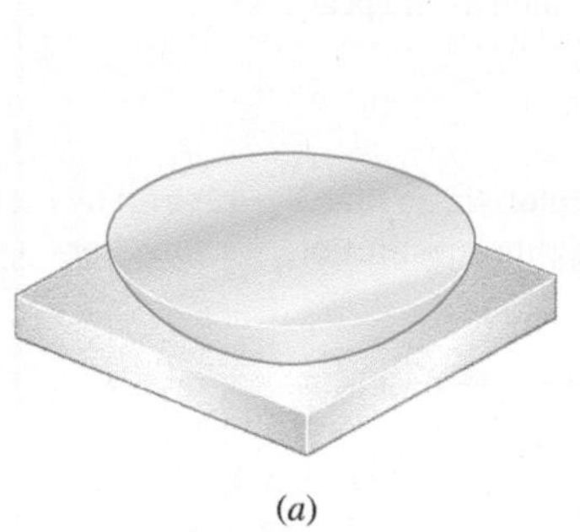
(*a*)

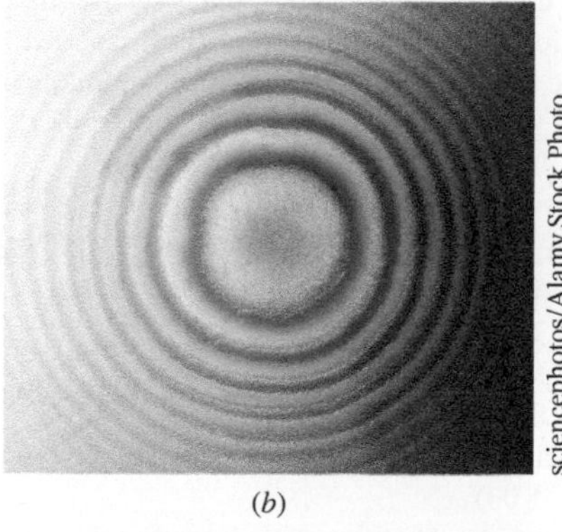
(*b*)

sciencephotos/Alamy Stock Photo

FIGURE 27.13 (*a*) The air wedge between a convex spherical glass surface and an optically flat plate leads to (*b*) a pattern of circular interference fringes that are known as Newton's rings.

EXAMPLE 6 | BIO The Physics of Structural Coloration

As its name suggests, the blue morpho butterfly (**Figure 27.14*a***) has beautiful metallic blue wings. The color of its wings is not due to pigmentation–they are actually brown–but rather due to an interference effect. An electron microscope image of its wings reveals multiple layers of horizontal rows of shingle-like scales. Along the vertical direction of these rows are Christmas tree-like nanostructures composed of chitin–a long-chain polymer structurally similar to cellulose (**Figure 27.14*b***). When light is incident on the wings, it undergoes multiple reflections from the many layers in these structures. We can model the reflections as a thin-film interference problem with the geometry shown in **Figure 27.14*c***. Along the vertical direction of the tree-like structures, there are alternating layers (or films) of chitin and air. The thickness of the chitin layers is 40.0 nm, and the thickness of the air layers is 115 nm. The thickness of the chitin layers is too small to produce constructive interference in the visible spectrum. However, the thickness of the air layers is large enough, so we treat the problem as an air film surrounded by chitin on both sides. Which wavelengths of visible light in vacuum are strongly reflected from the butterfly wings due to constructive interference? The indices of refraction of air and chitin are 1.00 and 1.56, respectively.

Reasoning We can solve this problem by applying the thin-film interference condition for constructive interference, where we consider any phase shifts that occur upon reflection from the

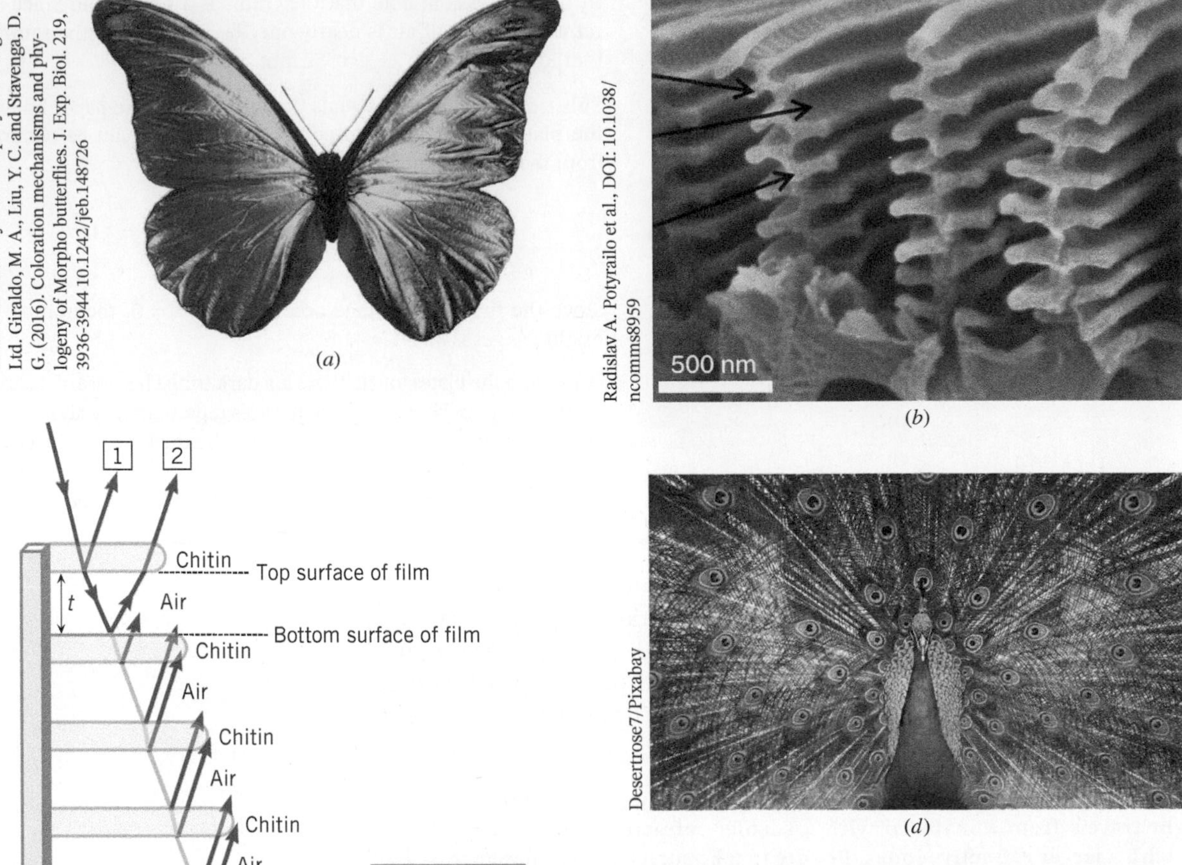

FIGURE 27.14 (*a*) The iridescent blue color of the morpho butterfly's wings is not due to pigmentation, but rather the interference effect of structural coloration. (*b*) Electron microscope image of the tree-like nanostructures of chitin that form the horizontal rows of scales in the butterfly's wings. (*c*) A sketch showing two waves that reflect from the top and bottom surface of the air film ($n = 1.00$) that is surrounded on both sides by chitin ($n = 1.56$). Some of the additional multiple reflections are also shown. (*d*) Structural coloration is also responsible for the beautiful iridescent plumage on many birds, such as the peacock.

top (wave 1) and bottom surface (wave 2) of the air film. In **Figure 27.14c**, wave 1 reflects from the top surface of the air film. It is traveling from a larger index of refraction ($n_{\text{chitin}} = 1.56$) toward a smaller index of refraction ($n_{\text{air}} = 1.00$). Therefore, wave 1 undergoes no phase shift upon reflection. Wave 2 reflects from the bottom surface of the air film. In this case, wave 2 is traveling from a smaller index of refraction ($n_{\text{air}} = 1.00$) toward a larger index ($n_{\text{chitin}} = 1.56$). Therefore, wave 2 undergoes a $\frac{1}{2}\lambda_{\text{film}}$ phase shift upon reflection. Here, the optical path length difference is just twice the film thickness ($2t$), we have one phase shift, and the condition for constructive interference is

$$\underbrace{2t}_{\substack{\text{Extra distance}\\ \text{traveled by}\\ \text{wave 2}}} + \underbrace{\tfrac{1}{2}\lambda_{\text{film}}}_{\substack{\text{Half-wavelength}\\ \text{net phase change}\\ \text{due to reflection}}} = \underbrace{\lambda_{\text{film}}, 2\lambda_{\text{film}}, 3\lambda_{\text{film}}, \ldots}_{\substack{\text{Condition for}\\ \text{constructive interference}}}$$

Subtracting the $\frac{1}{2}\lambda_{\text{film}}$ phase shift from the left-hand side of this equation and from each term on the right-hand side yields

$$2t = \left(m + \tfrac{1}{2}\right)\lambda_{\text{film}},\ m = 0, 1, 2, 3, \ldots$$

As in Example 5, the "film" is a film of air, so that λ_{film} is virtually the same as λ_{vacuum}. After making this substitution, we solve for λ_{vacuum} in terms of m:

$$\lambda_{\text{vacuum}} = \frac{2t}{\left(m + \frac{1}{2}\right)} =,\ m = 0, 1, 2, 3, \ldots \Rightarrow \lambda_{\text{vacuum}} = \frac{2(115\text{ nm})}{\left(m + \frac{1}{2}\right)},\quad m = 0, 1, 2, 3, \ldots$$

Solution Using the first three values of m (0, 1, and 2) returns the following results for λ_{vacuum}: 460 nm, 153 nm, and 92.0 nm, respectively. The 460 nm result for $m = 0$ is the only one that resides in the visible spectrum. The other two satisfy the constructive interference condition for wavelengths of light smaller than visible light, as do all the larger values of m. As expected, 460 nm is located in the blue region of the visible spectrum. While we only considered the interference from the air film closest to the surface of the wing, light shining on the butterfly undergoes multiple reflections from the many structural components inside its wings. This, along with the size of the air gap, effectively enhances the blue light, while averaging out the other colors of visible light. When they change their viewing angle, observers will notice subtle variations in the blue color, which is referred to as *iridescence*. Examples of iridescence are also seen in shellfish, beetles, and bird feathers–most notably, the peacock (**Figure 27.14d**).

Check Your Understanding

(*The answers are given at the end of the book.*)

8. A camera lens is covered with a nonreflective coating that eliminates the reflection of perpendicularly incident green light. Recalling Snell's law of refraction (see Section 26.2), would you expect the reflected green light to be eliminated if it were incident on the nonreflective coating at an angle of 45° rather than perpendicularly? **(a)** No, because the distance traveled by the light in the film is less than twice the film thickness. **(b)** No, because the distance traveled by the light in the film is greater than twice the film thickness. **(c)** Yes, the green light will still be eliminated.

9. Two pieces of the same glass are covered with thin films of different materials. In reflected sunlight, however, the films have different colors. Why? **(a)** The films could have the same thickness, but different refractive indices. **(b)** The films could have different thicknesses, but the same refractive indices. **(c)** Both of the preceding answers could be correct.

10. A transparent coating is deposited on a glass plate and has a refractive index that is *larger than that of the glass.* For a certain wavelength within the coating, the thickness of the coating is a quarter-wavelength. Does the coating enhance or reduce the reflection of the light corresponding to this wavelength?

11. The drawings in **CYU Figure 27.2** show three situations—A, B, and C—in which light reflects almost perpendicularly from the top and bottom surfaces of a thin film, with the indices of refraction as shown. **(a)** For which situation(s) will there be a net phase shift (due to reflection) between waves 1 and 2 that is equivalent to either zero wavelengths or one wavelength (λ_{film}), where λ_{film} is the wavelength of the light in the film? **(b)** For which situation(s) will the film appear dark when the thickness of the film is equal to $\frac{1}{2}\lambda_{film}$?

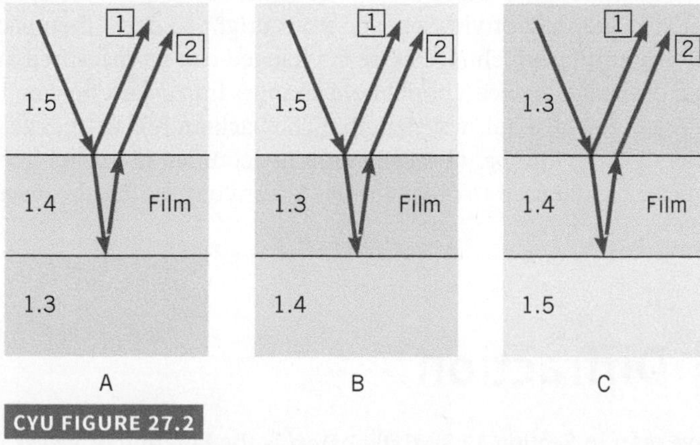

CYU FIGURE 27.2

12. When sunlight reflects from a thin film of soapy water (air on both sides), the film appears multicolored, in part because destructive interference removes different wavelengths from the light reflected at different places, depending on the thickness of the film. What happens as the film becomes thinner and thinner? **(a)** Nothing happens, and the film remains multicolored. **(b)** The film looks brighter and brighter in reflected light, appearing totally white just before it breaks. **(c)** The film looks darker and darker in reflected light, appearing black just before it breaks.

13. Two thin films are floating on water ($n = 1.33$). The films have refractive indices of $n_1 = 1.20$ and $n_2 = 1.45$. Suppose that the thickness of each film approaches zero. In reflected light, film 1 will look____ and film 2 will look____. **(a)** bright, bright **(b)** bright, dark **(c)** dark, bright **(d)** dark, dark

27.4 The Michelson Interferometer

THE PHYSICS OF . . . the Michelson interferometer. An interferometer is an apparatus that can be used to measure the wavelength of light by utilizing interference between two light waves. One particularly famous interferometer was developed by Albert A. Michelson (1852–1931). The Michelson interferometer uses reflection to set up conditions where two light waves interfere. **Figure 27.15** presents a schematic drawing of the instrument. Waves emitted by the monochromatic light source strike a

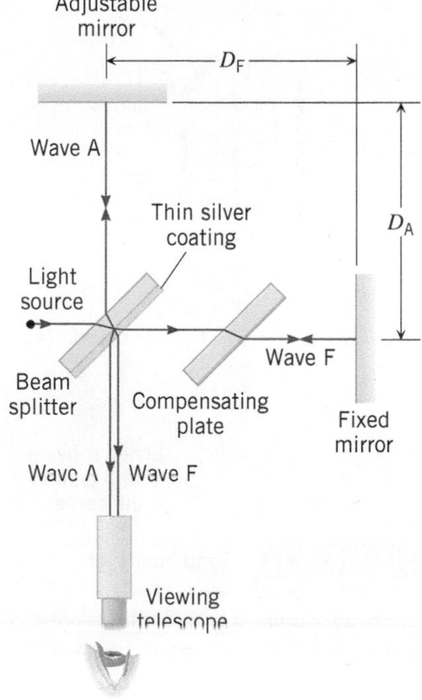

FIGURE 27.15 A schematic drawing of a Michelson interferometer.

beam splitter, so called because it splits the beam of light into two parts. The beam splitter is a glass plate, the far side of which is coated with a thin layer of silver that reflects part of the beam upward as wave A in the drawing. The coating is so thin, however, that it also allows the remainder of the beam to pass directly through as wave F. Wave A strikes an adjustable mirror and reflects back on itself. It again crosses the beam splitter and then enters the viewing telescope. Wave F strikes a fixed mirror and returns, to be partly reflected into the viewing telescope by the beam splitter. Note that wave A passes through the glass plate of the beam splitter three times in reaching the viewing scope, while wave F passes through it only once. The compensating plate in the path of wave F has the same thickness as the beam splitter plate and ensures that wave F also passes three times through the same thickness of glass on the way to the viewing scope. Thus, an observer viewing the superposition of waves A and F through the telescope sees constructive or destructive interference, depending only on the difference in path lengths D_A and D_F traveled by the two waves.

Now suppose that the mirrors are perpendicular to each other, the beam splitter makes a 45° angle with each, and the distances D_A and D_F are equal. Waves A and F travel the same distance, and the field of view in the telescope is uniformly bright due to constructive interference. However, if the adjustable mirror is moved away from the telescope by a distance of $\frac{1}{4}\lambda$, wave A travels back and forth by an amount that is twice this value, leading to an extra distance of $\frac{1}{2}\lambda$. Then, the waves are out of phase when they reach the viewing scope, destructive interference occurs, and the viewer sees a dark field. If the adjustable mirror is moved farther, full brightness returns as soon as the waves are in phase and interfere constructively. The in-phase condition occurs when wave A travels a total extra distance of λ relative to wave F. Thus, as the mirror is continuously moved, the viewer sees the field of view change from bright to dark, then back to bright, and so on. The amount by which D_A has been changed can be measured and related to the wavelength of the light, since a bright field changes into a dark field and back again each time D_A is changed by a half-wavelength. (The back-and-forth change in distance is λ.) If a sufficiently large number of wavelengths are counted in this manner, the Michelson interferometer can be used to obtain a very accurate value for the wavelength from the measured changes in D_A.

27.5 Diffraction

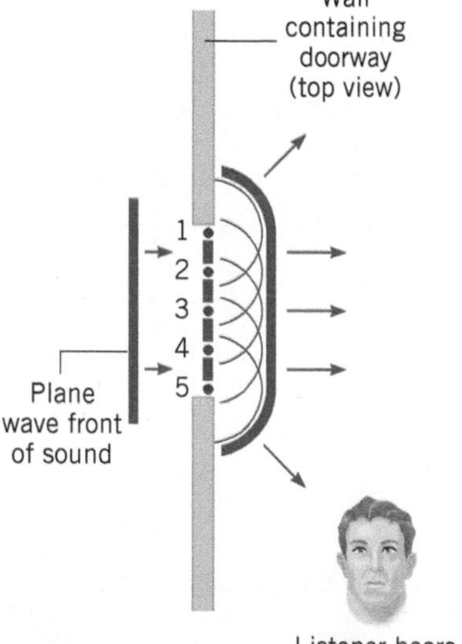

FIGURE 27.16 Sound bends, or diffracts, around the edges of a doorway, so even a person who is not standing directly in front of the opening can hear the sound. The five red points within the doorway act as sources and emit the five Huygens wavelets shown in red.

As we have seen in Section 17.3, **diffraction** is the bending of waves around obstacles or the edges of an opening. In **Figure 27.16**, sound waves are leaving a room through an open doorway and bend, or diffract, around the edges of the opening. Therefore, a listener hears the sound even when he is around the corner from the doorway.

Diffraction is an interference effect, and the Dutch scientist Christian Huygens (1629–1695) developed a principle that is useful in explaining why diffraction arises. **Huygens' principle** describes how a wave front that exists at one instant gives rise to the wave front that exists later on. This principle states that:

> ***Every point on a wave front acts as a source of tiny wavelets that move forward with the same speed as the wave; the wave front at a later instant is the surface that is tangent to the wavelets.***

We begin by using Huygens' principle to explain the diffraction of sound waves in **Figure 27.16**. The drawing shows the top view of a plane wave front of sound approaching a doorway and identifies five points on the wave front just as it is leaving the opening. According to Huygens' principle, each of these points acts as a source of wavelets, which are shown as red circular arcs at some moment after they are emitted. The tangent to the wavelets from points 2, 3, and 4 indicates that in front of the doorway the wave front is flat and moving straight ahead. At the edges, however, points 1 and 5 are the last points that produce wavelets. Huygens' principle suggests that in conforming to the curved shape of the wavelets near the edges, the new wave front

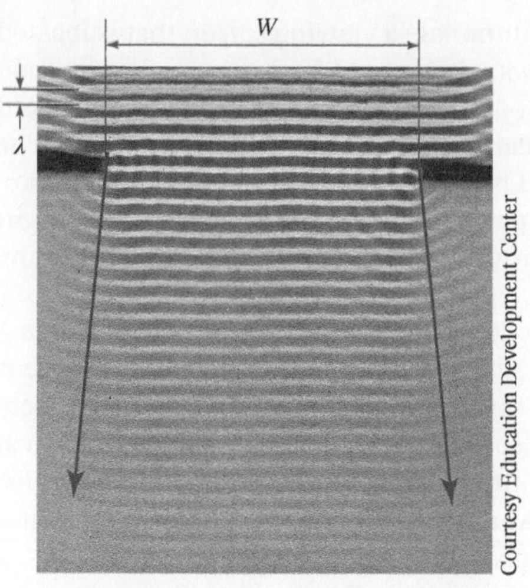

(*a*) Smaller value for λ/W, less diffraction

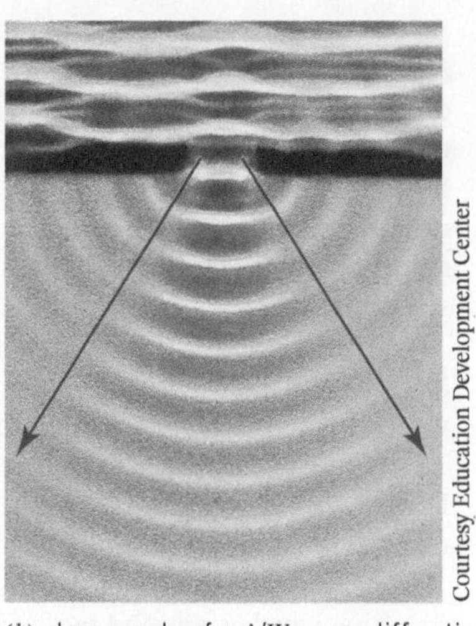

(*b*) Larger value for λ/W, more diffraction

FIGURE 27.17 These photographs show water waves (horizontal lines) approaching an opening whose width W is greater in (*a*) than in (*b*). In addition, the wavelength λ of the waves is smaller in (*a*) than in (*b*). Therefore, the ratio λ/W increases from (*a*) to (*b*) and so does the extent of the diffraction, as the red arrows indicate.

moves into regions that it would not reach otherwise. The sound wave, then, bends or diffracts around the edges of the doorway.

Huygens' principle applies not just to sound waves, but to all kinds of waves. For instance, light has a wave-like nature and, consequently, exhibits diffraction. Therefore, you may ask, "Since I can hear around the edges of a doorway, why can't I also see around them?" As a matter of fact, light waves do bend around the edges of a doorway. However, the degree of bending is extremely small, so the diffraction of light is not enough to allow you to see around the corner.

As we will learn, the extent to which a wave bends around the edges of an opening is determined by the ratio λ/W, where λ is the wavelength of the wave and W is the width of the opening. The photographs in **Figure 27.17** illustrate the effect of this ratio on the diffraction of water waves. The degree to which the waves are diffracted or bent is indicated by the two red arrows in each photograph. In part *a*, the ratio λ/W is small because the wavelength (as indicated by the distance between the wave fronts) is small relative to the width of the opening. The wave fronts move through the opening with little bending or diffraction into the regions around the edges. In part *b*, the wavelength is larger and the width of the opening is smaller. As a result, the ratio λ/W is larger, and the wave fronts bend more into the regions around the edges of the opening.

Based on the pictures in **Figure 27.17**, we expect that light waves of wavelength λ will bend or diffract appreciably when they pass through an opening whose width W is small enough to make the ratio λ/W sufficiently large. This is indeed the case, as **Figure 27.18** illustrates. In this picture, it is assumed that parallel rays (or plane wave

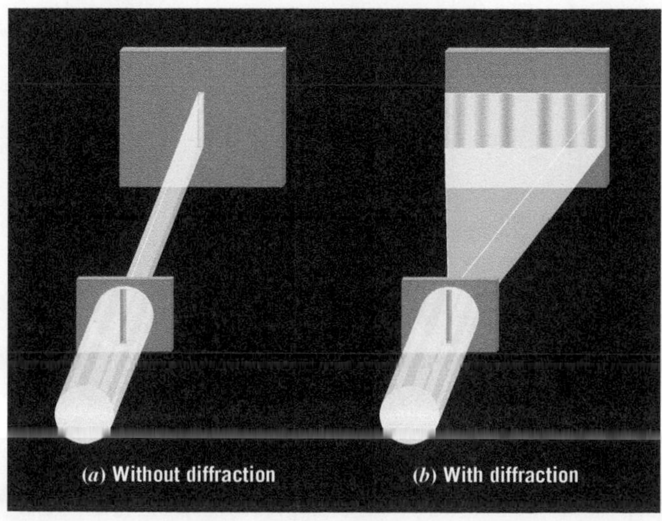

FIGURE 27.18 (*a*) If light were to pass through a very narrow slit *without* being diffracted, only the region on the screen directly opposite the slit would be illuminated. (*b*) Diffraction causes the light to bend around the edges of the slit into regions it would not otherwise reach, forming a pattern of alternating bright and dark fringes on the screen. The slit width has been exaggerated for clarity.

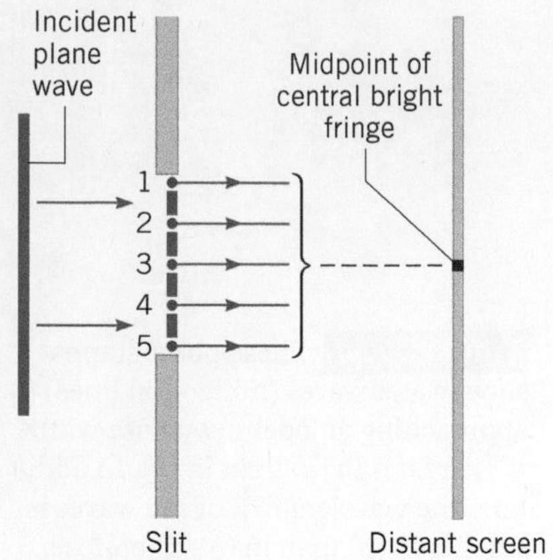

FIGURE 27.19 A plane wave front is incident on a single slit. This top view of the slit shows five sources of Huygens wavelets. The wavelets travel toward the midpoint of the central bright fringe on the screen, as the red rays indicate. The screen is very far from the slit.

fronts) of light fall on a very narrow slit and illuminate a viewing screen that is located far from the slit. Part *a* of the drawing shows what would happen if light were *not* diffracted: it would pass through the slit without bending around the edges and would produce an image of the slit on the screen. Part *b* shows what actually happens. The light diffracts around the edges of the slit and brightens regions on the screen that are not directly opposite the slit. The diffraction pattern on the screen consists of a bright central band, accompanied by a series of narrower faint fringes that are parallel to the slit itself.

To help explain how the pattern of diffraction fringes arises, **Figure 27.19** shows a top view of a plane wave front approaching the slit and singles out five sources of Huygens wavelets. Consider how the light from these five sources reaches the midpoint on the screen. To simplify things, the screen is assumed to be so far from the slit that the rays from each Huygens source are nearly parallel.* Then, all the wavelets travel virtually the same distance to the midpoint, arriving there in phase. As a result, constructive interference creates a bright central fringe on the screen, directly opposite the slit.

The wavelets emitted by the Huygens sources in the slit can also interfere destructively on the screen, as **Figure 27.20** illustrates. Part *a* shows light rays directed from each source toward the first dark fringe. The angle θ gives the position of this dark fringe relative to the line between the midpoint of the slit and the midpoint of the central bright fringe. Since the screen is very far from the slit, the rays from each Huygens source are nearly parallel and are oriented at nearly the same angle θ, as in part *b* of the drawing. The wavelet from source 1 travels the shortest distance to the screen, while the wavelet from source 5 travels the farthest. Destructive interference creates the first dark fringe when the extra distance traveled by the wavelet from source 5 is exactly one wavelength, as the colored right triangle in the drawing indicates. Under this condition, the extra distance traveled by the wavelet from source 3 at the center of the slit is exactly one-half of a wavelength. Therefore, wavelets from sources 1 and 3 in **Figure 27.20*b*** are exactly out of phase and interfere destructively when they reach the screen. Similarly, a wavelet that originates slightly below source 1 cancels a wavelet that originates the same distance below source 3. Thus, each wavelet from the upper half of the slit cancels a corresponding wavelet from the lower half, and no light reaches the screen. As can be seen from the

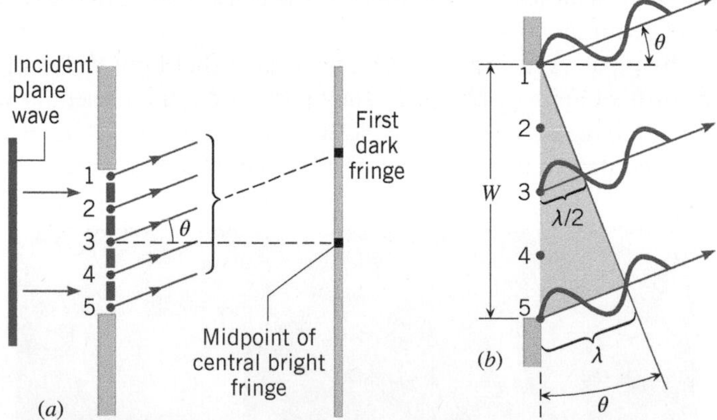

FIGURE 27.20 These drawings pertain to single-slit diffraction and show how destructive interference leads to the first dark fringe on either side of the central bright fringe. For clarity, only one of the dark fringes is shown. The screen is very far from the slit.

*When the rays are parallel, the diffraction is called Fraunhofer diffraction in tribute to the German optician Joseph von Fraunhofer (1787–1826). When the rays are not parallel, the diffraction is referred to as Fresnel diffraction, named for the French physicist Augustin Jean Fresnel (1788–1827).

colored right triangle, the angle θ locating the first dark fringe is given by $\sin\theta = \lambda/W$, where W is the width of the slit.

Math Skills To understand why the colored right triangle in **Figure 27.20b** implies that $\sin\theta = \lambda/W$, it is necessary to see why there are two angles labeled θ in the figure. To help explain these angles, we show a simplified version of **Figure 27.20b** in **Figure 27.21**, where we label the angle in the colored triangle as α. We do this so that we can show that the angle α is, in fact, the same as the angle θ between the red ray at the top of the slit and the horizontal dashed line. Since the slit is oriented vertically, we know that the line AC in **Figure 27.21** is perpendicular to the horizontal dashed line, with the result that the angles α and β form a right angle:

$$\alpha + \beta = 90^\circ \quad \text{or} \quad \alpha = 90^\circ - \beta$$

In addition, we know that the line AB in **Figure 27.21** has been drawn perpendicular to the red rays, which means that the angles θ and β also form a right angle:

$$\theta + \beta = 90^\circ \quad \text{or} \quad \beta = 90^\circ - \theta$$

Substituting this result for β into the result for α reveals that

$$\alpha = 90^\circ - \beta = 90^\circ - (90^\circ - \theta) = \theta$$

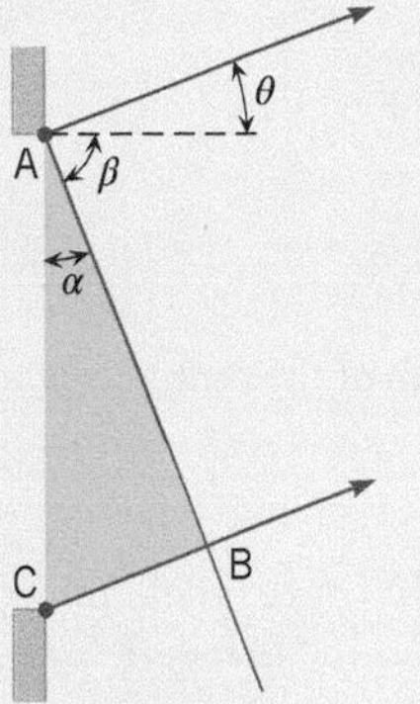

FIGURE 27.21 Math Skills drawing.

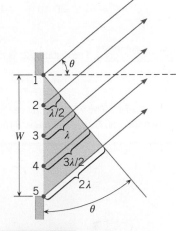

FIGURE 27.22 In a single-slit diffraction pattern, multiple dark fringes occur on either side of the central bright fringe. This drawing shows how destructive interference creates the second dark fringe on a very distant screen.

Figure 27.22 shows the condition that leads to destructive interference at the second dark fringe on either side of the midpoint on the screen. In reaching the screen, the light from source 5 now travels a distance of two wavelengths farther than the light from source 1. Under this condition, the wavelet from source 5 travels one wavelength farther than the wavelet from source 3, and the wavelet from source 3 travels one wavelength farther than the wavelet from source 1. Therefore, each half of the slit can be treated as the entire slit was in the previous paragraph; all the wavelets from the top half interfere destructively with each other, and all the wavelets from the bottom half do likewise. As a result, no light from either half reaches the screen, and another dark fringe occurs. The colored triangle in the drawing shows that this second dark fringe occurs when $\sin\theta = 2\lambda/W$. Similar arguments hold for the third- and higher-order dark fringes, with the general result being

Dark fringes for single-slit diffraction

$$\sin\theta = m\frac{\lambda}{W} \qquad m = 1, 2, 3, \ldots \tag{27.4}$$

Between each pair of dark fringes there is a bright fringe due to constructive interference. The brightness of the fringes is related to the light intensity, just as loudness is related to sound intensity. The intensity of the light at any location on the screen is the amount of light energy per second per unit area that strikes the screen there. **Figure 27.23** gives a graph of the light intensity, along with the single-slit diffraction pattern. The central bright fringe, which is approximately twice as wide as the other bright fringes, has by far the greatest intensity.

The width of the central fringe provides some indication of the extent of the diffraction, as Example 7 illustrates.

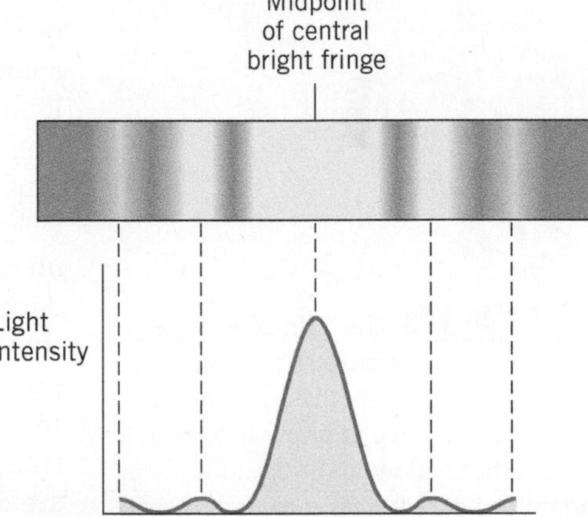

FIGURE 27.23 A single-slit diffraction pattern, with a bright and wide central fringe. The higher-order bright fringes are much less intense than the central fringe, as the graph indicates.

EXAMPLE 7 | Single-Slit Diffraction

Light passes through a slit and shines on a flat screen that is located $L = 0.40$ m away (see **Figure 27.24**). The wavelength of the light in a vacuum is $\lambda = 410$ nm. The distance between the midpoint of the central bright fringe and the first dark fringe is y. Determine the width $2y$ of the central bright fringe when the width of the slit is **(a)** $W = 5.0 \times 10^{-6}$ m and **(b)** $W = 2.5 \times 10^{-6}$ m.

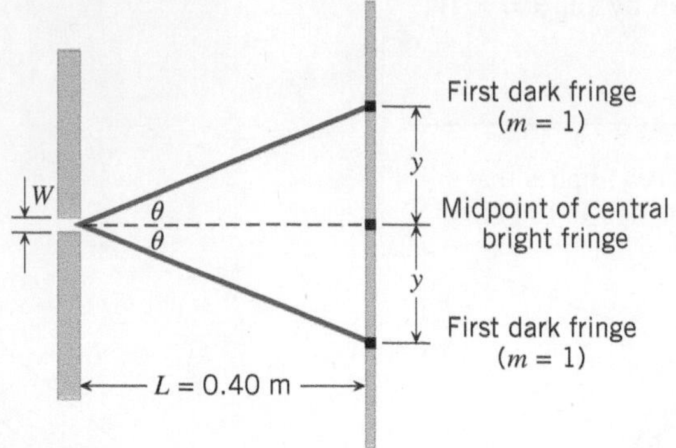

FIGURE 27.24 The distance $2y$ is the width of the central bright fringe.

Reasoning The width of the central bright fringe is determined by two factors. One is the angle θ that locates the first dark fringe on either side of the midpoint. The other is the distance L between the screen and the slit. Larger values for θ and L lead to a wider central bright fringe. Larger values of θ mean greater diffraction and occur when the ratio λ/W is larger. Thus, we expect the width of the central bright fringe to be greater when the slit width W is smaller.

Solution **(a)** The angle θ in Equation 27.4 locates the first dark fringe when $m = 1$: $\sin\theta = (1)\lambda/W$. Therefore,

$$\theta = \sin^{-1}\left(\frac{\lambda}{W}\right) = \sin^{-1}\left(\frac{410 \times 10^{-9}\text{ m}}{5.0 \times 10^{-6}\text{ m}}\right) = 4.7^\circ$$

According to **Figure 27.24**, $\tan\theta = y/L$, so the width $2y$ of the central bright fringe is

$$2y = 2L\tan\theta = 2(0.40\text{ m})\tan 4.7^\circ = \boxed{0.066\text{ m}}$$

(b) Repeating the same calculations as in part (a) with $W = 2.5 \times 10^{-6}$ m reveals that $\boxed{2y = 0.13\text{ m}}$. As expected for a given wavelength, the width $2y$ of the central maximum in the diffraction pattern is greater when the width of the slit is smaller.

THE PHYSICS OF . . . producing computer chips using photolithography. In the production of computer chips, it is important to minimize the effects of diffraction. Such chips are very small and yet contain enormous numbers of electronic components, as **Figure 23.32** illustrates. Such miniaturization is achieved using the techniques of photolithography. The patterns on the chip are created first on a "mask," which is similar to a photographic slide. Light is then directed through the mask onto silicon wafers that have been coated with a photosensitive material. The light-activated parts of the coating can be removed chemically, to leave the ultrathin lines that correspond to the miniature patterns on the chip. As the light passes through the narrow slit-like patterns on the mask, the light spreads out due to diffraction. If excessive diffraction occurs, the light spreads out so much that sharp patterns are not formed on the photosensitive material coating the silicon wafer. Ultraminiaturization of the patterns requires the absolute minimum of diffraction, and currently this is achieved by using ultraviolet light, which has a wavelength shorter than that of visible light. The shorter the wavelength λ, the smaller the ratio λ/W, and the less the diffraction. The wavelengths of X-rays are much shorter than those of ultraviolet light and, thus, will reduce diffraction even more, allowing further miniaturization.

Another example of diffraction can be seen when light from a point source falls on an opaque disk, such as a coin (**Figure 27.25**). The effects of diffraction modify the dark shadow cast by the disk in several ways. First, the light waves diffracted around the circular edge of the disk interfere constructively at the center of the shadow to produce a small bright spot. There are also circular bright fringes in the shadow area. In addition, the boundary between the circular shadow and the lighted screen is not sharply defined but consists of concentric bright and dark fringes. The various fringes are analogous to those produced by a single slit and are due to interference between Huygens wavelets that originate from different points near the edge of the disk.

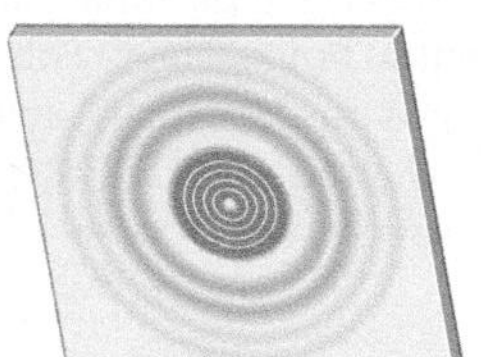

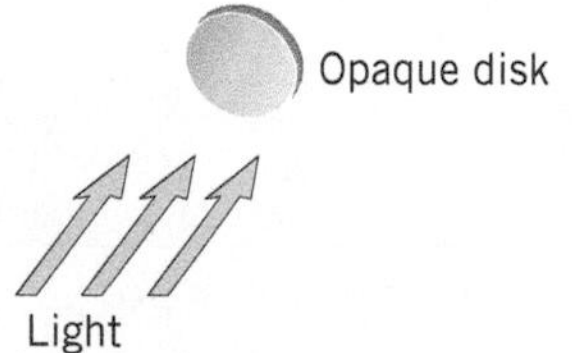

FIGURE 27.25 The diffraction pattern formed by an opaque disk consists of a small bright spot in the center of the dark shadow, circular bright fringes within the shadow, and concentric bright and dark fringes surrounding the shadow.

Check Your Understanding

(The answers are given at the end of the book.)

14. A diffraction pattern is produced on a viewing screen by using a single slit with blue light. Does the pattern broaden or contract (become narrower) **(a)** when the blue light is replaced by red light **(b)** when the slit width is increased?

15. A sound wave has a much greater wavelength than does a light wave. When the two waves pass through a doorway, which one, if either, diffracts to a greater extent? **(a)** The sound wave **(b)** The light wave **(c)** Both waves diffract by the same amount.

27.6 Resolving Power

Figure 27.26 shows three photographs of an automobile's headlights taken at progressively greater distances from the camera. In parts *a* and *b*, the two separate headlights can be seen clearly. In part *c*, however, the car is so far away that the headlights are barely distinguishable and appear almost as a single light. The **resolving power** of an optical instrument, such as a camera, is its ability to distinguish between two closely spaced objects. If a camera with a higher resolving power had taken these pictures, the photograph in part *c* would have shown two distinct and separate headlights. Any instrument used for viewing objects that are close together must have a high resolving power. This is true, for example, for a telescope used to view distant stars or for a microscope used to view tiny organisms. We will now see that diffraction occurs when light passes through the circular, or nearly circular, openings that admit light into cameras, telescopes, microscopes, and human eyes. The resulting diffraction pattern places a limit on the resolving power of these instruments.

Truax/The Image Finders

Truax/The Image Finders

Truax/The Image Finders

FIGURE 27.26 These automobile headlights were photographed at various distances from the camera, closest in part *a* and farthest in part *c*. In part *c*, the headlights are so far away that they are barely distinguishable.

Figure 27.27 shows the diffraction pattern created by a small circular opening when the viewing screen is far from the opening. The pattern consists of a central bright circular region, surrounded by alternating bright and dark circular fringes. These fringes are analogous to the rectangular fringes that a single slit produces. The angle θ in the picture locates the first circular dark fringe relative to the center of the central bright region and is given by

$$\sin\theta = 1.22\frac{\lambda}{D} \tag{27.5}$$

where λ is the wavelength of the light and D is the diameter of the opening. This expression is similar to Equation 27.4 for a slit ($\sin\theta = \lambda/W$, when $m = 1$) and is valid when the distance to the screen is much larger than the diameter D.

An optical instrument with the ability to resolve two closely spaced objects can produce images of them that can be identified separately. Think about the images on the image sensor when light from two widely separated point objects passes through the circular aperture of a camera. As **Figure 27.28** illustrates, each image is a circular diffraction pattern, but the two patterns do not overlap and are completely resolved. On the other hand, if the objects are sufficiently close together, the intensity patterns created by the diffraction overlap, as **Figure 27.29*a*** suggests. In fact, if the overlap is extensive, it may no longer be possible to distinguish the patterns separately. In such a case, the picture from a camera would show a single blurred object instead of two separate objects. In **Figure 27.29*b*** the diffraction patterns overlap, but not enough to prevent us from seeing that two objects are present. Ultimately, then, diffraction limits the ability of an optical instrument to produce distinguishable images of objects that are close together.

It is useful to have a criterion for judging whether two closely spaced objects will be resolved by an optical instrument. **Figure 27.29*a*** presents the **Rayleigh criterion** for resolution, first proposed by Lord Rayleigh (1842–1919):

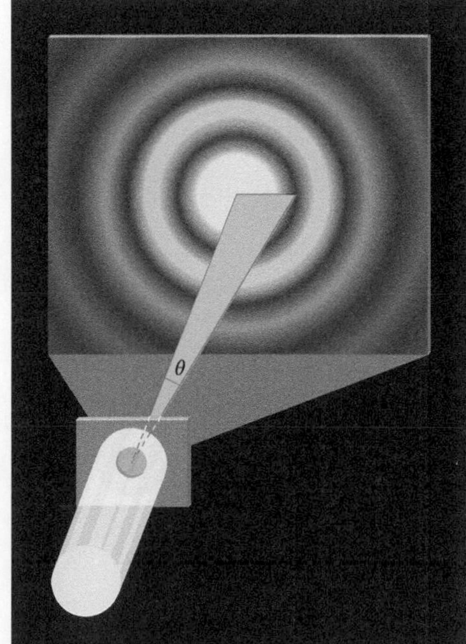

FIGURE 27.27 When light passes through a small circular opening, a circular diffraction pattern is formed on a screen. The angle θ locates the first dark fringe relative to the central bright region. The intensities of the bright fringes and the diameter of the opening have been exaggerated for clarity.

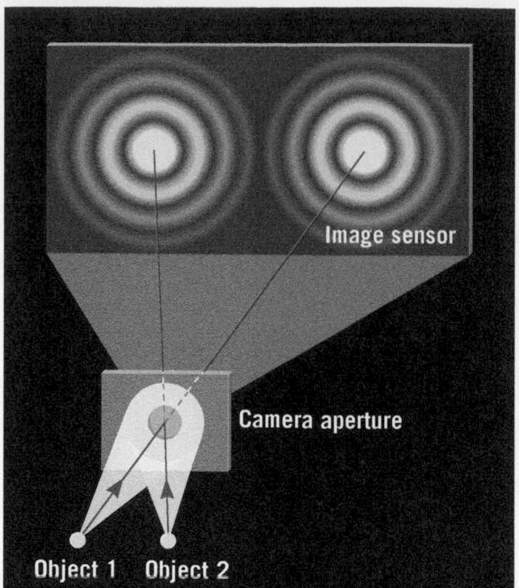

FIGURE 27.28 When light from two point objects passes through the circular aperture of a camera, two circular diffraction patterns are formed as images on the image sensor. The images here are completely separated or resolved because the objects are widely separated.

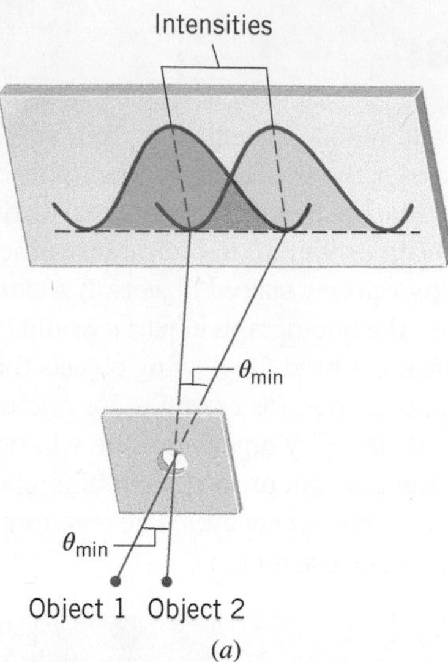

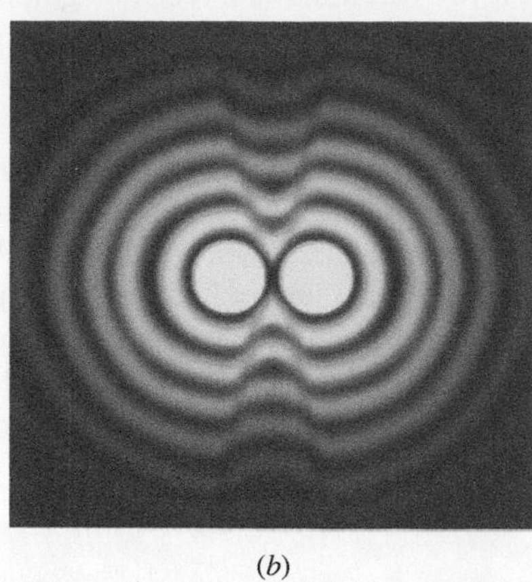

FIGURE 27.29 (*a*) According to the Rayleigh criterion, two point objects are just resolved when the first dark fringe (zero intensity) of one image falls on the central bright fringe (maximum intensity) of the other image. (*b*) This drawing shows two overlapping but still resolvable diffraction patterns.

Two point objects are just resolved when the first dark fringe in the diffraction pattern of one falls directly on the central bright fringe in the diffraction pattern of the other.

According to the Rayleigh criterion, the minimum angle θ_{min} between the two objects in **Figure 27.29a** is the angle given by Equation 27.5. If θ_{min} is small (less than about 9°) and is expressed in radians, $\sin\theta_{min} \approx \theta_{min}$. Then, Equation 27.5 becomes

$$\theta_{min} \approx 1.22\frac{\lambda}{D} \quad (\theta_{min} \text{ in radians}) \tag{27.6}$$

Math Skills To see why $\sin\theta_{min} \approx \theta_{min}$ when θ_{min} is small and is expressed in radians, we drop the subscript (for simplicity) and consider Equation 8.1, which defines an angle θ in radians. According to this definition, θ is the length s of the circular arc that the angle subtends divided by the radius r of the arc (see **Figure 27.30a**):

$$\theta = \frac{s}{r}$$

On the other hand, Equation 1.1 defines $\sin\theta$ as

$$\sin\theta = \frac{h_o}{h}$$

where h_o is the length of the side of a right triangle that is opposite the angle θ and h is the length of the hypotenuse (see **Figure 27.30b**). As θ becomes smaller and smaller in both parts of **Figure 27.30**, we can see that h_o also becomes smaller and approaches the arc length s, whereas h approaches the radius r. Therefore, for a small angle θ expressed in radians, we conclude that

$$\sin\theta = \frac{h_o}{h} \approx \frac{s}{r} = \theta \qquad \text{or} \qquad \sin\theta \approx \theta$$

Note that the symbol "≈" means "approximately equal to."

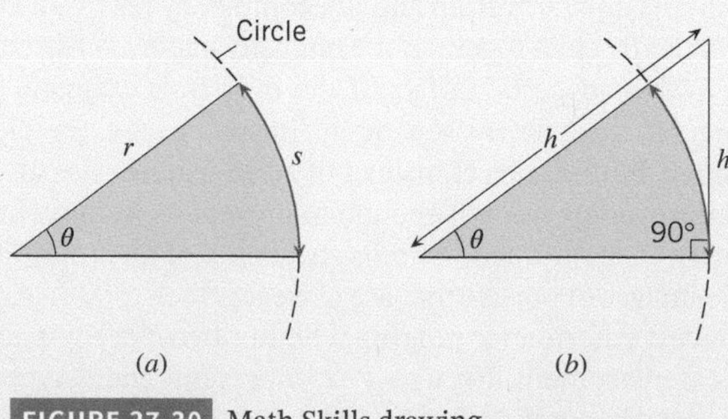

FIGURE 27.30 Math Skills drawing.

For a given wavelength λ and aperture diameter D, Equation 27.6 specifies the smallest angle that two point objects can subtend at the aperture and still be resolved. According to this equation, optical instruments designed to resolve closely spaced objects (small values of θ_{min}) must utilize the smallest possible wavelength and the largest possible aperture diameter. For example, when short-wavelength ultraviolet light is collected by its large 2.4-m-diameter mirror, the Hubble Space Telescope is capable of resolving two closely spaced stars that have an angular separation of about $\theta_{min} = 1 \times 10^{-7}$ rad. This angle is equivalent to resolving two objects only 1 cm apart when they are 1×10^5 m (about 62 miles) from the telescope. Example 8 deals with the resolving power of the human eye.

Analyzing Multiple-Concept Problems

EXAMPLE 8 | BIO The Physics of Comparing Human Eyes and Eagle Eyes

(a) A hang glider is flying at an altitude of 120 m. Green light (wavelength = 555 nm in vacuum) enters the pilot's eye through a pupil that has a diameter of 2.5 mm. Determine how far apart two point objects must be on the ground if the pilot is to have any hope of distinguishing between them (see **Figure 27.31**). **(b)** An eagle's eye has a pupil with a diameter of 6.2 mm. Repeat part (a) for an eagle flying at the same altitude as the glider.

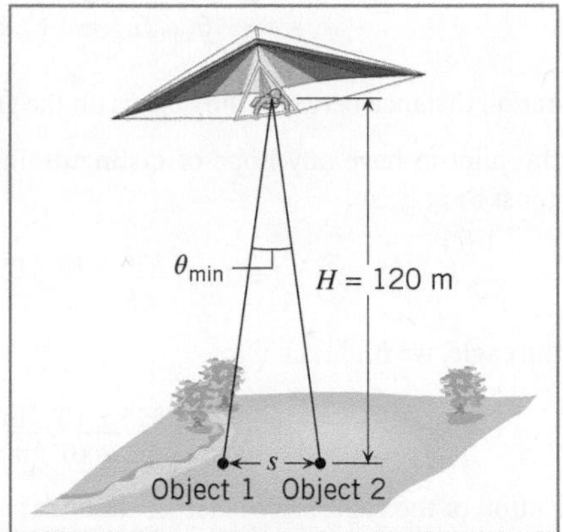

FIGURE 27.31 The Rayleigh criterion can be used to estimate the smallest distance s that can separate two objects on the ground, if a person on a hang glider is to be able to distinguish between them.

Reasoning A greater distance s of separation between the objects on the ground makes it easier for the eye of the observer (the pilot or the eagle; see **Figure 27.31**) to resolve them as separate objects. This is because the angle that the two objects subtend at the pupil of the eye is greater when the separation distance is greater. This angle must be at least as large as the angle θ_{min} specified by the Rayleigh criterion for resolution. In applying this criterion, we will use the concept of the radian to express the angle as an arc length (approximately the separation distance) divided by a radius (the altitude), as discussed in Section 8.1.

Knowns and Unknowns The following table summarizes the data that are given:

Description	Symbol	Value	Comment
Altitude	H	120 m	Same for pilot and eagle.
Wavelength of light in vacuum	λ	555 nm	1 nm = 10^{-9} m
Diameter of pupil of eye	D	2.5 mm or 6.2 mm	Smaller value is for pilot; larger value is for eagle.
Unknown Variables			
Separation between objects on ground	s	?	

Modeling the Problem

STEP 1 Radian Measure The angle that the two objects subtend at the pupil of the eye must be at least as large as the angle θ_{min} specified by the Rayleigh criterion for resolution. Using radian measure as discussed in Section 8.1, we refer to **Figure 27.31** and express this angle as

$$\theta_{min} \approx \frac{s}{H}$$

This is an approximate application of Equation 8.1, which states that an angle in radians is the arc length divided by the radius. Here, the arc length is approximately the separation distance s, assuming that the altitude H is much greater than s. Solving for s gives

$$s \approx \theta_{min} H \quad (1)$$

?

Equation 1 at the right. In this result, the altitude is known, and we proceed to Step 2 to evaluate the angle θ_{min}.

$$s \approx \theta_{min} H \quad (1)$$

STEP 2 The Rayleigh Criterion The Rayleigh criterion specifies the angle θ_{min} in radians as

$$\boxed{\theta_{min} \approx 1.22\frac{\lambda}{D}} \quad \textbf{(27.6)}$$

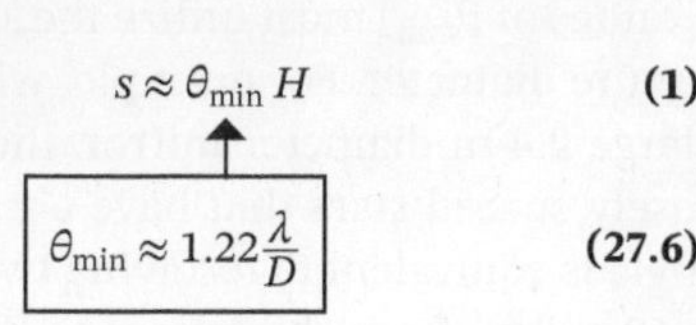

where λ is the wavelength of the light in vacuum* and D is the diameter of the pupil of the eye. The substitution of this expression into Equation 1 is shown at the right.

Problem-Solving Insight The minimum angle θ_{min} between two objects that are just resolved must be expressed in radians, not degrees, when using Equation 27.6 ($\theta_{min} \approx 1.22\,\lambda/D$).

Solution Combining the results of each step algebraically, we find that

$$s \approx \underbrace{\theta_{min} H}_{\text{STEP 1}} \approx \underbrace{\left(1.22\frac{\lambda}{D}\right)H}_{\text{STEP 2}}$$

The separation distance between the objects on the ground can now be obtained.

(a) For the pilot to have any hope of distinguishing between the objects, the separation distance must be at least

$$s \approx \left(1.22\frac{\lambda}{D}\right)H = 1.22\left(\frac{555 \times 10^{-9}\text{ m}}{2.5 \times 10^{-3}\text{ m}}\right)(120\text{ m}) = \boxed{0.033\text{ m}}$$

(b) For the eagle, we find that

$$s \approx \left(1.22\frac{\lambda}{D}\right)H = 1.22\left(\frac{555 \times 10^{-9}\text{ m}}{6.2 \times 10^{-3}\text{ m}}\right)(120\text{ m}) = \boxed{0.013\text{ m}}$$

Since the pupil of the eagle's eye is larger than that of a human eye, diffraction creates less of a limitation for the eagle; the two objects can be closer together and still be resolved by the eagle's eye.

Many optical instruments have a resolving power exceeding that of the human eye. The typical camera does, for instance. Conceptual Example 9 compares the abilities of the human eye and a camera to resolve two closely spaced objects.

CONCEPTUAL EXAMPLE 9 | Is What You See What You Get?

The French postimpressionist artist Georges Seurat developed a painting technique in which dots of color are placed close together on the canvas. From sufficiently far away the individual dots are not distinguishable, and the images in the picture take on a more normal appearance. **Figure 27.32** shows a person in a museum looking at one of Seurat's paintings. Suppose that the person stands close to the painting, then backs up until the dots just become indistinguishable to his eyes and takes a picture from this position. The light enters his eyes through pupils that have diameters of 2.5 mm and enters the digital camera through an aperture, or opening, with a diameter of 25 mm. He then goes home and prints an enlarged photograph of the painting. Can he see the individual dots in the photograph? **(a)** No, because if his eye cannot see the dots at the museum, the camera is also unable

*In applying the Rayleigh criterion, we use the given wavelength in vacuum because it is nearly identical to the wavelength in air. We use the wavelength in air or vacuum, even though the diffraction occurs within the eye, where the index of refraction is about $n_{eye} = 1.36$ and the wavelength is $\lambda_{eye} = \lambda_{vacuum}/n_{eye}$, according to Equation 27.3. The reason is that in entering the eye, the light is refracted according to Snell's law (Section 26.2), which also includes an effect due to the index of refraction. If the angle of incidence is small, the effect of n_{eye} in Snell's law cancels the effect of n_{eye} in Equation 27.3, to a good degree of approximation.

FIGURE 27.32 This person is about to take a photograph of a famous painting by Georges Seurat, who developed the technique of using tiny dots of color to construct his images. Conceptual Example 9 discusses what the person sees when the photograph is printed.

to record the individual dots. **(b)** Yes, because the camera gathers light through a much larger aperture than does the eye. **(c)** Yes, because, unlike the eye, a photograph taken by a camera is not limited by the effects of diffraction.

Reasoning To answer this question, we turn to Equation 27.6, which expresses the Rayleigh criterion for resolving two point objects (such as the dots)—namely, $\theta_{min} \approx 1.22\lambda/D$. Here θ_{min} is the minimum angle that exists between light rays from two adjacent dots as the rays pass through the aperture (see **Figure 27.29**), λ is the wavelength of the light, and D is the diameter of the aperture. A larger value of D implies a smaller value for θ_{min}, which, in turn, means that the instrument has a greater resolving power.

Answer (a) is incorrect. Diffraction limits the ability of any instrument to see two closely spaced objects as distinct. This ability depends on the diameter of the aperture through which the light enters the instrument. Since the eye and the camera have apertures with different diameters, the camera may record the individual dots in the painting even though the eye does not see them as distinct.

Answer (c) is incorrect. The effects of diffraction limit the resolving power of both the eye and the camera.

Answer (b) is correct. For the eye and the camera, the aperture diameters are $D_{eye} = 2.5$ mm and $D_{camera} = 25$ mm, so the diameter for the camera is ten times larger than that for the eye. Thus, at the distance at which the eye loses its ability to resolve the individual dots in the painting, the camera can still easily resolve them. As discussed in the footnote to Example 8, we can ignore the effect on the wavelength of the index of refraction of the material from which the eye is made.

Related Homework: Check Your Understanding Question 18, Problem 54

Conceptual Example 10 illustrates how diffraction affects the vision of animals that have evolved with pupils that are not circular.

CONCEPTUAL EXAMPLE 10 | BIO Predator or Prey? Pupils of Different Shapes

As discussed earlier in the chapter, pupils with different shapes will affect an animal's optical resolution along different directions. Some animals, like goats, and other large herbivores, have horizontal pupils, while others, like cats and crocodiles, have vertical pupils (**Figure 27.33**). Recent research suggests that animals that are ambush predators tend to have pupils with vertical slits, while plant-eating prey species have horizontal slits located near the sides of their heads. This is especially true for animals whose heads are close to the ground (i.e., grazing animals and small predators). Suppose a goat's eye has a horizontal, rectangular pupil that has a width-to-height ratio of 5:1. If the goat can optically resolve two-closely spaced objects separated by a distance s in the horizontal direction, how much closer would the objects have to be moved toward the goat for it to distinguish them in the vertical direction?

Reasoning We can apply Rayleigh's criterion for resolution (Equation 27.6) to the pupil in both directions.

Solution Beginning with Equation 27.6, we see that the minimum angle for resolution is inversely related to the pupil diameter:

$$\theta_{min} \propto \frac{1}{D}.$$

Using Equation 8.1, we can write the angle in terms of the separation distance (s) between the two objects and the distance from the objects (r):

$$\frac{s}{r} \propto \frac{1}{D}.$$

Rearranging this equation for r, we have: $r \propto sD$. Notice that the distance to the objects is directly proportional to the pupil diameter. Therefore, the objects would have to be moved five times closer for the goat to resolve them in the vertical direction. The goat's greater optical resolution in the horizontal direction allows it to detect predators along the ground at greater distances.

bones64/Pixabay

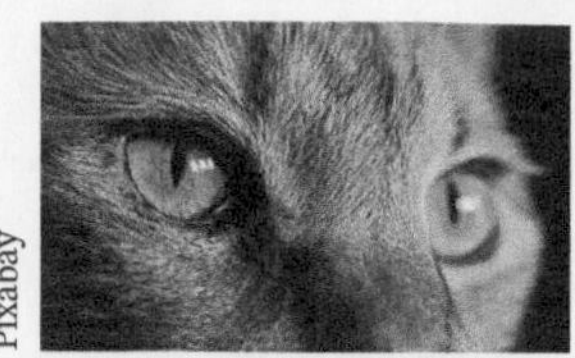
cocoparisienne/Pixabay

Felix Broennimann/Pixabay

FIGURE 27.33 The pupils of animals come in all shapes and sizes. Some are horizontal, such as in the goat, and others are vertical, such as in the domestic cat and crocodile. The shape of the pupil will affect the optical resolution of the eyes in different directions.

Check Your Understanding

(*The answers are given at the end of the book.*)

16. BIO Suppose that the pupil of your eye were elliptical instead of circular in shape, with the long axis of the ellipse oriented in the vertical direction. Would the resolving power of your eye be the same in the horizontal and vertical directions and, if not, in which direction would it be greater? The resolving power would **(a)** be the same in both directions **(b)** be greater in the horizontal direction **(c)** be greater in the vertical direction.

17. BIO Suppose that you were designing an eye and could select the size of the pupil and the wavelengths of the electromagnetic waves to which the eye is sensitive. As far as the limitation created by diffraction is concerned, rank the following design choices in order of decreasing resolving power (greatest first): **(a)** Large pupil and ultraviolet wavelengths **(b)** Small pupil and infrared wavelengths **(c)** Small pupil and ultraviolet wavelengths

18. BIO Review Conceptual Example 9 before answering this question. A person is viewing one of Seurat's paintings that consists of dots of color. She is so close to the painting that the dots are distinguishable. Without moving, she squints, thus reducing the size of the opening in her eyes. Does squinting make the painting take on a more normal appearance?

19. On many cameras one can select the *f*-number setting, or *f*-stop. The *f*-number gives the ratio of the focal length of the camera lens to the diameter of the aperture through which light enters the camera. If you want to resolve two closely spaced objects in a picture, should you use a small or a large *f*-number setting?

27.7 The Diffraction Grating

THE PHYSICS OF . . . a diffraction grating. Diffraction patterns of bright and dark fringes occur when monochromatic light passes through a single or double slit. Fringe patterns also result when light passes through more than two slits, and an arrangement consisting of a large number of parallel, closely spaced slits called a **diffraction grating** has proved very useful. Gratings with as many as 40 000 slits per centimeter can be made, depending on the production method. In one method a diamond-tipped cutting tool is used to inscribe closely spaced parallel lines on a glass plate, the spaces between the lines serving as the slits. In fact, the number of slits per centimeter is often quoted as the number of lines per centimeter.

Figure 27.34 illustrates how light travels to a distant viewing screen from each of five slits in a grating and forms the central bright fringe and the first-order bright fringes on either side. Higher-order bright fringes are also formed but are not shown in the drawing. Each bright fringe is located by an angle θ relative to the central fringe. These bright fringes are sometimes called the *principal fringes* or *principal maxima*, since they are places where the light intensity is a maximum. The term "principal" distinguishes them from other, much less bright, fringes that are referred to as *secondary fringes* or *secondary maxima*.

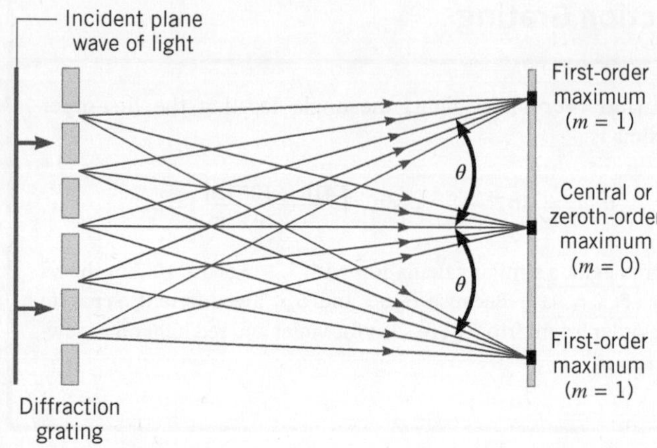

FIGURE 27.34 When light passes through a diffraction grating, a central bright fringe ($m = 0$) and higher-order bright fringes ($m = 1, 2, \ldots$) form when the light falls on a distant viewing screen.

Constructive interference creates the principal fringes. To show how, we assume the screen is far from the grating, so that the rays remain nearly parallel while the light travels toward the screen. In reaching the place on the screen where a first-order maximum is located, light from slit 2 travels a distance of one wavelength farther than light from slit 1, as in **Figure 27.35**. Similarly, light from slit 3 travels one wavelength

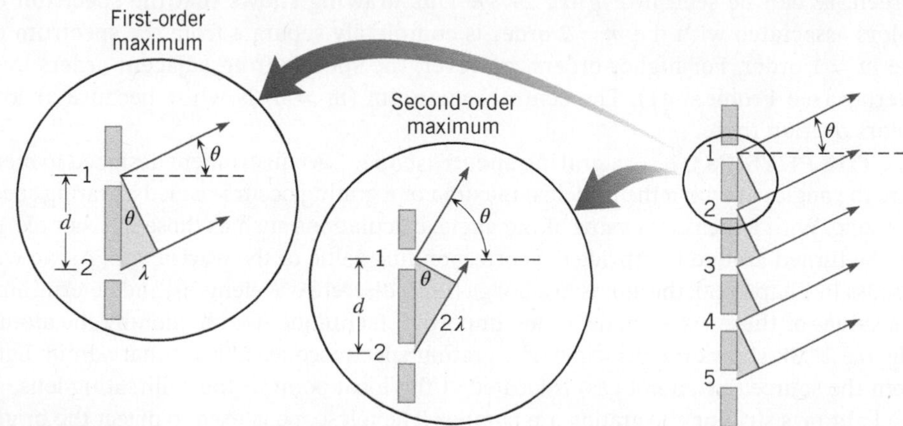

FIGURE 27.35 The conditions shown here lead to the first- and second-order intensity maxima in the diffraction pattern produced by the diffraction grating on the right.

farther than light from slit 2, and so forth, as emphasized by the four colored right triangles on the right-hand side of the drawing. For the first-order maximum, the blow-up view of slits 1 and 2 shows that constructive interference occurs when $\sin\theta = \lambda/d$, where d is the separation between slits. A second-order maximum forms when the extra distance traveled by light from adjacent slits is two wavelengths, so that $\sin\theta = 2\lambda/d$. The general result is

Principal maxima of a diffraction grating

$$\sin\theta = m\frac{\lambda}{d} \qquad m = 0, 1, 2, 3, \ldots \tag{27.7}$$

The separation d between the slits can be calculated from the number of slits per centimeter of grating; for instance, a grating with 2500 slits per centimeter has a slit separation of $d = (1/2500)\text{ cm} = 4.0 \times 10^{-4}\text{ cm}$. Equation 27.7 is identical to Equation 27.1 for the double slit. A grating, however, produces bright fringes that are much *narrower* or *sharper* than those from a double slit, as the intensity patterns in **Figure 27.36** reveal. Between the principal maxima of a diffraction grating there are secondary maxima with much smaller intensities. For a large number of slits, these secondary maxima are very small.

The next example illustrates the ability of a grating to separate the components in a mixture of colors.

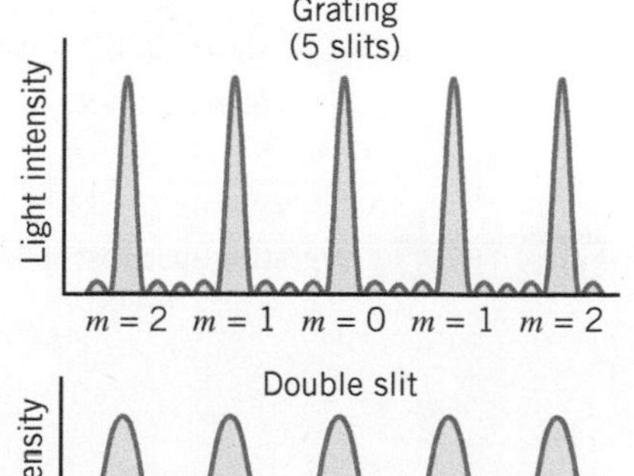

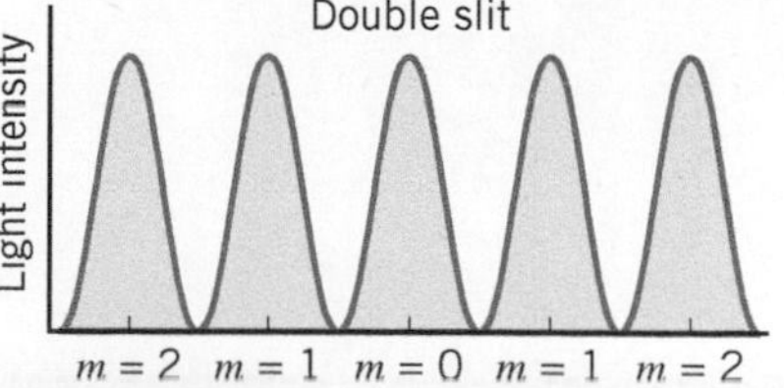

FIGURE 27.36 The bright fringes produced by a diffraction grating are much narrower than those produced by a double slit. Note the three small secondary bright fringes between the principal bright fringes of the grating.

EXAMPLE 11 | Separating Colors with a Diffraction Grating

A mixture of violet light ($\lambda = 410$ nm in vacuum) and red light ($\lambda = 660$ nm in vacuum) falls on a grating that contains 1.0×10^4 lines/cm. For each wavelength, find the angle θ that locates the first-order maxima.

Reasoning Before Equation 27.7 can be used here, a value for the separation d between the slits is needed: $d = 1/(1.0 \times 10^4 \text{ lines/cm}) = 1.0 \times 10^{-4}$ cm, or 1.0×10^{-6} m. For violet light, the angle θ_{violet} for the first-order maxima ($m = 1$) is given by $\sin\theta_{\text{violet}} = m\lambda_{\text{violet}}/d$, with an analogous equation applying for the red light.

Solution For violet light, the angle locating the first-order maxima is

$$\theta_{\text{violet}} = \sin^{-1}\frac{\lambda_{\text{violet}}}{d} = \sin^{-1}\left(\frac{410 \times 10^{-9}\text{ m}}{1.0 \times 10^{-6}\text{ m}}\right) = \boxed{24^\circ}$$

For red light, a similar calculation with $\lambda_{\text{red}} = 660 \times 10^{-9}$ m shows that $\boxed{\theta_{\text{red}} = 41^\circ}$. Because θ_{violet} and θ_{red} are different, separate first-order bright fringes are seen for violet and red light on a viewing screen.

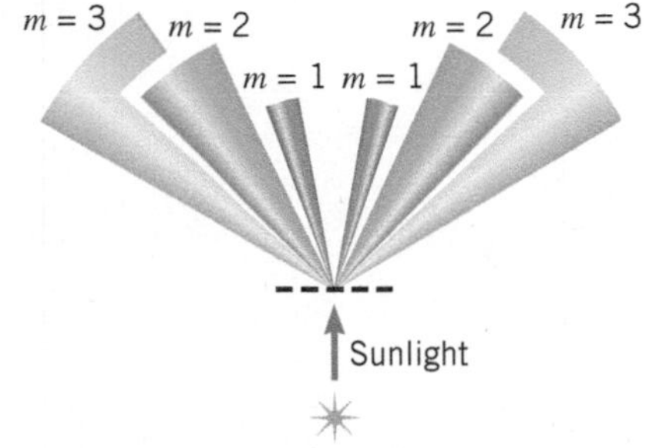

FIGURE 27.37 When sunlight falls on a diffraction grating, a rainbow of colors is produced at each principal maximum ($m = 1, 2, \ldots$). The central maximum ($m = 0$), however, is white but is not shown in the drawing.

If the light in Example 11 had been sunlight, the angles for the first-order maxima would cover all values in the range between 24° and 41°, since sunlight contains all colors or wavelengths between violet and red. Consequently, a rainbow-like dispersion of the colors would be observed to either side of the central fringe on a screen, as can be seen in **Figure 27.37**. This drawing shows that the spectrum of colors associated with the $m = 2$ order is completely separate from the spectrum of the $m = 1$ order. For higher orders, however, the spectra from adjacent orders may overlap (see Problem 41). The central maximum ($m = 0$) is white because all the colors overlap there.

THE PHYSICS OF . . . a grating spectroscope. An instrument designed to measure the angles at which the principal maxima of a grating occur is called a grating spectroscope. With a measured value of the angle, calculations such as those in Example 11 can be turned around to provide the corresponding value of the wavelength. As we will discuss in Chapter 30, the atoms in a hot gas emit discrete wavelengths, and determining the values of these wavelengths is one important technique used to identify the atoms. **Figure 27.38** shows the principle of a grating spectroscope. The slit that admits light from the source (e.g., a hot gas) is located at the focal point of the collimating lens, so the light rays striking the grating are parallel. The telescope is used to detect the bright fringes and, hence, to measure the angle θ.

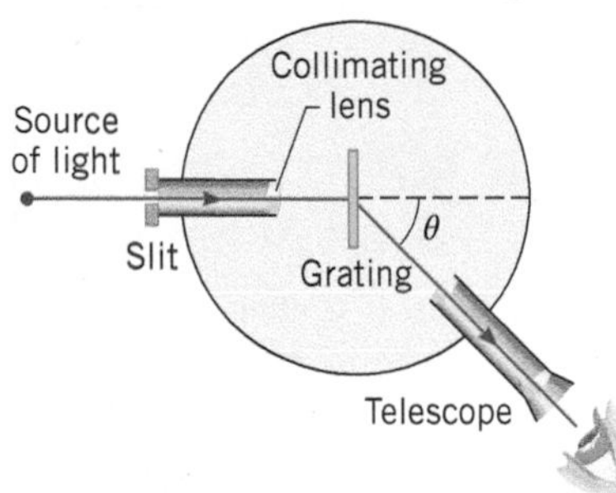

FIGURE 27.38 A grating spectroscope.

Check Your Understanding

(The answers are given at the end of the book.)

20. **CYU Figure 27.3** shows a top view of a diffraction grating and the mth-order principal maxima that are obtained with red and blue light. Red light has the longer wavelength. **(a)** Which principal maximum is associated with blue light, the one farther from or the one closer to the central maximum? **(b)** If the number of slits per centimeter in the grating were increased, would these two principal maxima move away from the central maximum, move toward the central maximum, or remain in the same place?

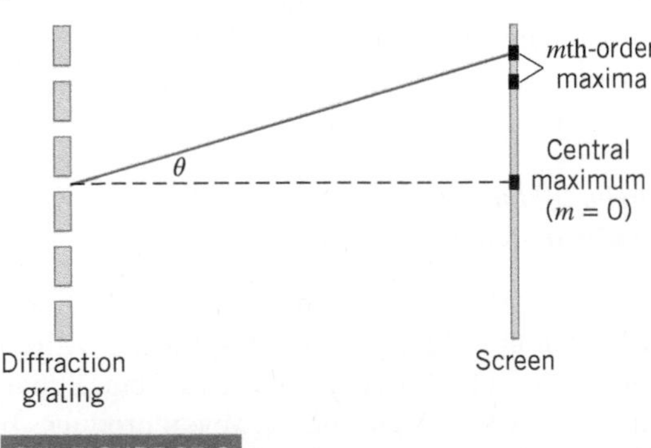

CYU FIGURE 27.3

21. What would happen to the distance between the bright fringes produced by a diffraction grating if the entire interference apparatus (light source, grating, and screen) were immersed in water?

27.8 X-Ray Diffraction

THE PHYSICS OF . . . X-ray diffraction. Not all diffraction gratings are commercially made. Nature also creates diffraction gratings, although these gratings do not look like an array of closely spaced slits. Instead, nature's gratings are the arrays of regularly spaced atoms that exist in crystalline solids. For example, **Figure 27.39** shows the structure of a crystal of ordinary salt (NaCl). Typically, the atoms in a crystalline solid are separated by distances of about 1.0×10^{-10} m, so we might expect a crystalline array of atoms to act like a grating with roughly this "slit" spacing for electromagnetic waves of the appropriate wavelength. Assuming that $\sin\theta = 0.5$ and that $m = 1$ in Equation 27.7, then $0.5 = \lambda/d$. A value of $d = 1.0 \times 10^{-10}$ m in this equation gives a wavelength of $\lambda = 0.5 \times 10^{-10}$ m. This wavelength is much shorter than that of visible light and falls in the X-ray region of the electromagnetic spectrum. (See Figure 24.9.)

A diffraction pattern does indeed result when X-rays are directed onto a crystalline material, as **Figure 27.40*a*** illustrates for a crystal of NaCl. The pattern consists of a complicated arrangement of spots because a crystal has a complex three-dimensional structure. It is from patterns such as this that the spacing between atoms and the nature of the crystal structure can be determined. X-ray diffraction has also been applied with great success toward understanding the structure of biologically important molecules, such as proteins and nucleic acids. One of the most famous results was the discovery in 1953 by James Watson and Francis Crick that the structure of the nucleic acid DNA is a double helix. X-ray diffraction patterns such as that in **Figure 27.40*b*** played the pivotal role in their research.

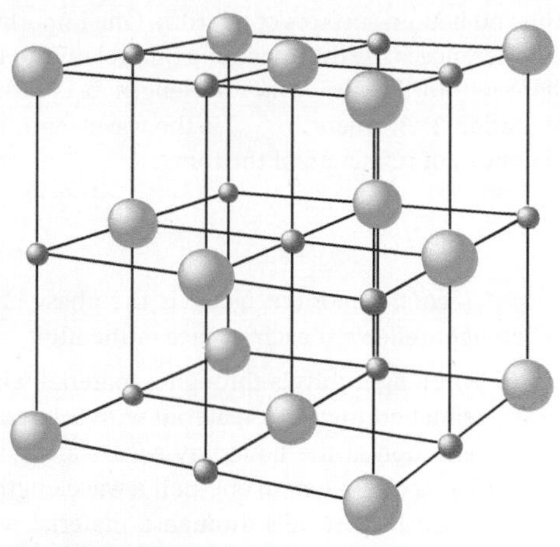

FIGURE 27.39 In this drawing of the crystalline structure of sodium chloride, the small red spheres represent positive sodium ions, and the large blue spheres represent negative chloride ions.

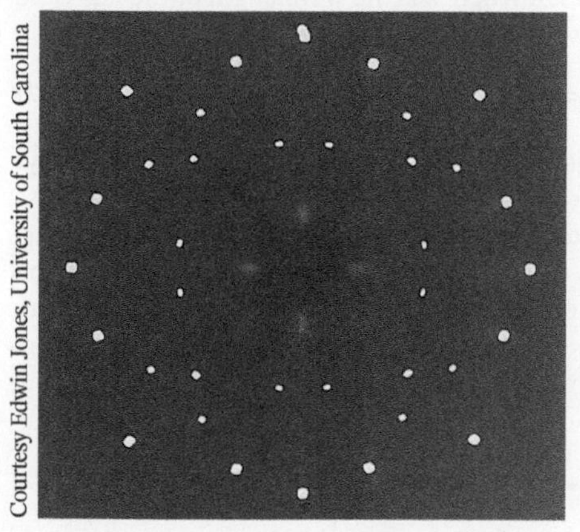

Courtesy Edwin Jones, University of South Carolina

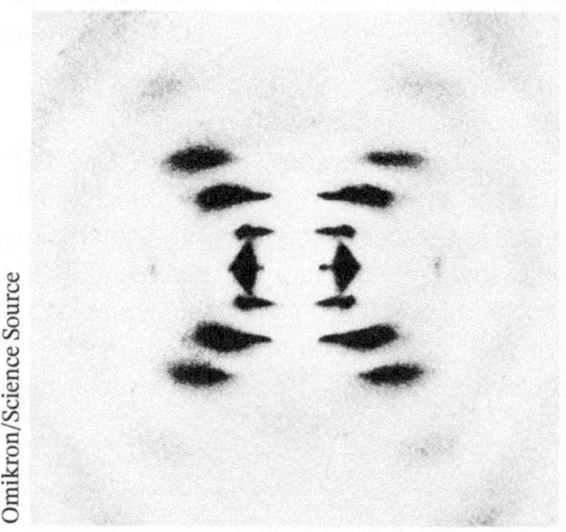

Omikron/Science Source

FIGURE 27.40 The X-ray diffraction patterns of (*a*) crystalline NaCl and (*b*) DNA. The image of DNA was obtained by Rosalind Franklin in 1953.

Concept Summary

27.1 The Principle of Linear Superposition The principle of linear superposition states that when two or more waves are present simultaneously in the same region of space, the resultant disturbance is the sum of the disturbances from the individual waves.

Constructive interference occurs at a point when two waves meet there crest-to-crest and trough-to-trough, thus reinforcing each other. When two waves that start out in phase and have traveled some distance meet at a point, constructive interference occurs whenever the travel distances are the same or differ by any integer number of wavelengths: $\ell_2 - \ell_1 = m\lambda$, where ℓ_1 and ℓ_2 are the distances traveled by the waves, and $m = 0, 1, 2, 3, \ldots$.

Destructive interference occurs at a point when two waves meet there crest-to-trough, thus mutually canceling each other. When two waves that start out in phase and have traveled some distance meet at a point, destructive interference occurs whenever the travel distances differ by any odd integer number of half-wavelengths: $\ell_2 - \ell_1 = (m + \frac{1}{2})\lambda$, where ℓ_1 and ℓ_2 are the distances traveled by the waves, and $m = 0, 1, 2, 3, \ldots$.

Two sources are coherent if the waves they emit maintain a constant phase relation. In other words, the waves do not shift relative to one another as time passes. If constructive and destructive interference are to be observed, coherent sources are necessary.

27.2 Young's Double-Slit Experiment In Young's double-slit experiment, light passes through a pair of closely spaced narrow slits and produces a pattern of alternating bright and dark fringes on a viewing screen. The fringes arise because of constructive and destructive interference. The angle θ that locates the mth-order bright fringe is given by Equation 27.1, where λ is the wavelength of the light, and d is the spacing between the slits. The angle that locates the mth dark fringe is given by Equation 27.2.

$$\sin\theta = \frac{m\lambda}{d} \qquad m = 0, 1, 2, 3, \ldots \qquad (27.1)$$

$$\sin\theta = \frac{(m + \frac{1}{2})\lambda}{d} \qquad m = 0, 1, 2, 3, \ldots \qquad (27.2)$$

27.3 Thin-Film Interference Constructive and destructive interference of light waves can occur with thin films of transparent materials. The interference occurs between light waves that reflect from the top and bottom surfaces of the film. One important factor in thin-film interference is the thickness of a film relative to the wavelength of the light within the film. The wavelength λ_{film} within a film is given by Equation 27.3, where λ_{vacuum} is the wavelength in a vacuum, and n is the index of refraction of the film.

$$\lambda_{\text{film}} = \frac{\lambda_{\text{vaccum}}}{n} \quad (27.3)$$

A second important factor is the phase change that can occur when light reflects at each surface of the film:

1. When light travels through a material with a smaller index of refraction toward a material with a larger index of refraction, reflection at the boundary occurs along with a phase change that is equivalent to one-half a wavelength in the film.
2. When light travels through a material with a larger index of refraction toward a material with a smaller index of refraction, there is no phase change upon reflection at the boundary.

27.4 The Michelson Interferometer An interferometer is an instrument that can be used to measure the wavelength of light by employing interference between two light waves. The Michelson interferometer splits the light into two beams. One beam travels to a fixed mirror, reflects from it, and returns. The other beam travels to a movable mirror, reflects from it, and returns. When the two returning beams are combined, interference is observed, the amount of which depends on the travel distances.

27.5 Diffraction Diffraction is the bending of waves around obstacles or around the edges of an opening. Diffraction is an interference effect that can be explained with the aid of Huygens' principle. This principle states that every point on a wave front acts as a source of tiny wavelets that move forward with the same speed as the wave; the wave front at a later instant is the surface that is tangent to the wavelets.

When light passes through a single narrow slit and falls on a viewing screen, a pattern of bright and dark fringes is formed because of the superposition of Huygens wavelets. The angle θ that specifies the mth dark fringe on either side of the central bright fringe is given by Equation 27.4, where λ is the wavelength of the light and W is the width of the slit.

$$\sin\theta = m\frac{\lambda}{W} \quad m = 1, 2, 3, \ldots \quad (27.4)$$

27.6 Resolving Power The resolving power of an optical instrument is the ability of the instrument to distinguish between two closely spaced objects. Resolving power is limited by the diffraction that occurs when light waves enter an instrument, often through a circular opening.

The Rayleigh criterion specifies that two point objects are just resolved when the first dark fringe in the diffraction pattern of one falls directly on the central bright fringe in the diffraction pattern of the other. According to this specification, the minimum angle (in radians) that two point objects can subtend at a circular aperture of diameter D and still be resolved as separate objects is given by Equation 27.6, where λ is the wavelength of the light.

$$\theta_{\text{min}} \approx 1.22\frac{\lambda}{D} \quad (\theta_{\text{min}} \text{ in radians}) \quad (27.6)$$

27.7 The Diffraction Grating A diffraction grating consists of a large number of parallel, closely spaced slits. When light passes through a diffraction grating and falls on a viewing screen, the light forms a pattern of bright and dark fringes. The bright fringes are referred to as principal maxima and are found at an angle θ that is specified by Equation 27.7, where λ is the wavelength of the light and d is the separation between two adjacent slits.

$$\sin\theta = m\frac{\lambda}{d} \quad m = 0, 1, 2, 3, \ldots \quad (27.7)$$

27.8 X-Ray Diffraction A diffraction pattern forms when X-rays are directed onto a crystalline material. The pattern arises because the regularly spaced atoms in a crystal act like a diffraction grating. Because the spacing is extremely small, on the order of 1×10^{-10} m, the wavelength of the electromagnetic waves must also be very small—hence, the use of X-rays. The crystal structure of a material can be determined from its X-ray diffraction pattern.

Focus on Concepts

Online

Additional questions are available for assignment in WileyPLUS.

Section 27.1 The Principle of Linear Superposition

1. The two loudspeakers in the drawing are producing identical sound waves. The waves spread out and overlap at the point P. What is the difference $\ell_2 - \ell_1$ in the two path lengths if point P is at the third sound intensity minimum from the central sound intensity maximum? Express this difference in terms of the wavelength λ of the sound. **(a)** $\frac{1}{2}\lambda$ **(b)** λ **(c)** $\frac{3}{2}\lambda$ **(d)** 3λ **(e)** $\frac{5}{2}\lambda$

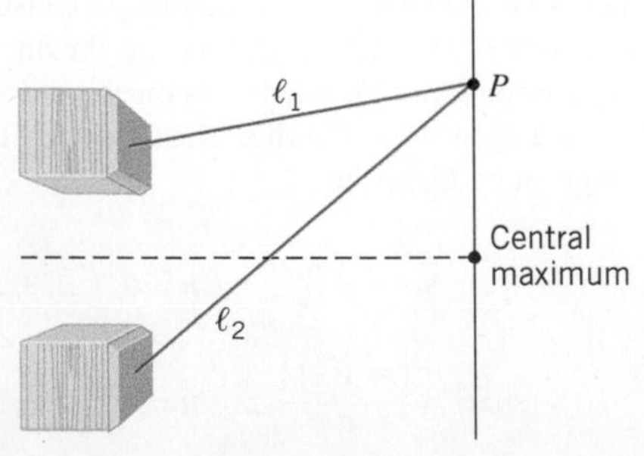

QUESTION 1

Section 27.2 Young's Double-Slit Experiment

2. In a certain Young's double-slit experiment, a diffraction pattern is formed on a distant screen, as the drawing shows. The angle that locates a given bright fringe is small, so that the approximation $\sin\theta \approx \theta$ is valid. Assuming that θ remains small, by what factor does it change if the wavelength λ is doubled and the slit separation d is doubled? **(a)** The angle does not change. **(b)** The angle increases by a factor of 2. **(c)** The angle increases by a factor of 4. **(d)** The angle decreases by a factor of 2. **(e)** The angle decreases by a factor of 4.

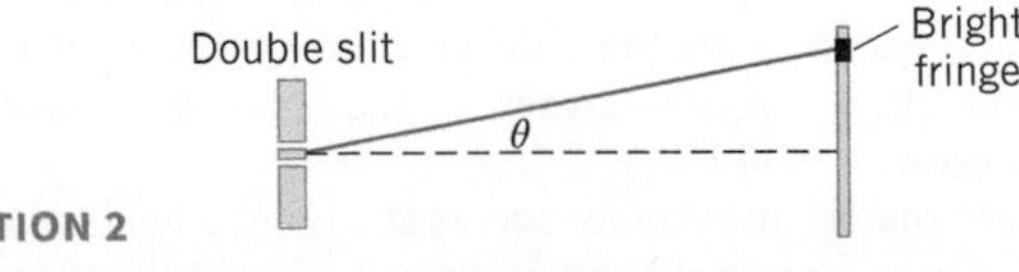

QUESTION 2

Section 27.3 Thin-Film Interference

3. Light of wavelength 600 nm in vacuum is incident nearly perpendicularly on a thin film whose index of refraction is 1.5. The light travels from the top surface of the film to the bottom surface, reflects from

the bottom surface, and returns to the top surface, as the drawing indicates. How far has the light traveled inside the film? Express your answer in terms of the wavelength λ_{film} of the light within the film. (a) $2\lambda_{\text{film}}$ (b) $3\lambda_{\text{film}}$ (c) $4\lambda_{\text{film}}$ (d) $6\lambda_{\text{film}}$ (e) $12\lambda_{\text{film}}$

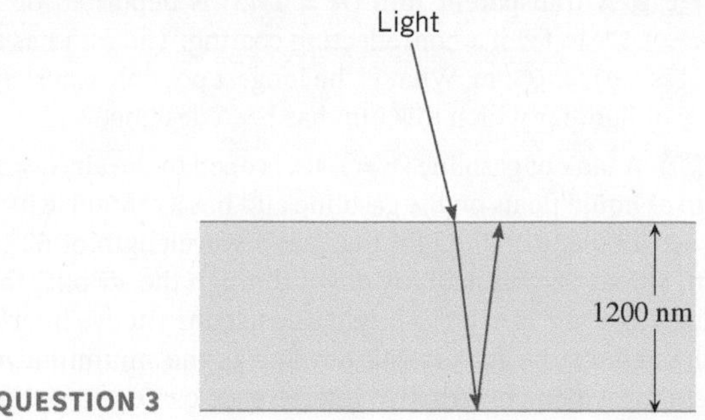

QUESTION 3

4. Light is incident perpendicularly on four transparent films of different thickness. The thickness of each film is given in the drawings in terms of the wavelength λ_{film} of the light within the film. The index of refraction of each film is 1.5, and each is surrounded by air. Which film (or films) will appear bright due to constructive interference when viewed from the top surface, upon which the light is incident? (a) 1, 2, 3, 4 (b) 2, 3 (c) 3 (d) 3, 4 (e) 4

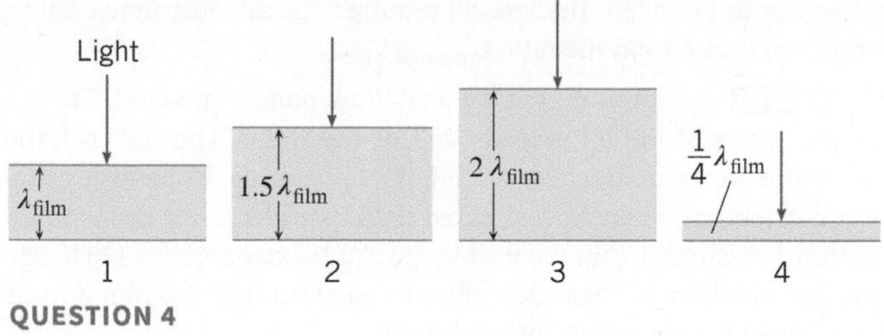

QUESTION 4

Section 27.5 Diffraction

5. Light passes through a single slit. If the width of the slit is reduced, what happens to the width of the central bright fringe? (a) The width of the central bright fringe does not change, because it depends only on the wavelength of the light and not on the width of the slit. (b) The central bright fringe becomes wider, because the angle that locates the first dark fringe on either side of the central bright fringe becomes smaller. (c) The central bright fringe becomes wider, because the angle that locates the first dark fringe on either side of the central bright fringe becomes larger. (d) The central bright fringe becomes narrower, because the angle that locates the first dark fringe on either side of the central bright fringe becomes larger. (e) The central bright fringe becomes narrower, because the angle that locates the first dark fringe on either side of the central bright fringe becomes smaller.

6. Light of wavelength λ passes through a single slit of width W and forms a diffraction pattern on a viewing screen. If this light is then replaced by light of wavelength 2λ, the original diffraction pattern is exactly reproduced if the width of the slit ______. (a) is changed to $\frac{1}{4}W$ (b) is changed to $\frac{1}{2}W$ (c) is changed to $2W$ (d) is changed to $4W$ (e) remains the same—no change is necessary

Section 27.6 Resolving Power

7. Suppose that you are using a microscope to view two closely spaced cells. For a given lens diameter, which color of light would you use to achieve the best possible resolving power? (a) Red (b) Yellow (c) Green (d) Blue (e) All the colors give the same resolving power.

Section 27.7 The Diffraction Grating

8. A diffraction grating is illuminated with yellow light. The diffraction pattern seen on a viewing screen consists of three yellow bright fringes, one at the central maximum ($\theta = 0°$) and one on either side of it at $\theta = \pm 50°$. Then the grating is simultaneously illuminated with red light. Where a red and a yellow fringe overlap, an orange fringe is produced. The new pattern consists of ______. (a) only red fringes at 0° and ±50° (b) only yellow fringes at 0° and ±50° (c) only orange fringes at 0° and ±50° (d) an orange fringe at 0°, yellow fringes at ±50°, and red fringes farther out (e) an orange fringe at 0°, yellow fringes at ±50°, and red fringes closer in

Problems

Online

Additional questions are available for assignment in WileyPLUS.

- SSM Student Solutions Manual
- MMH Problem-solving help
- GO Guided Online Tutorial
- V-HINT Video Hints
- CHALK Chalkboard Videos
- BIO Biomedical application
- E Easy
- M Medium
- H Hard
- WS Worksheet
- T Team Problem

Section 27.1 The Principle of Linear Superposition,

Section 27.2 Young's Double-Slit Experiment

1. E SSM In a Young's double-slit experiment, the wavelength of the light used is 520 nm (in vacuum), and the separation between the slits is 1.4×10^{-6} m. Determine the angle that locates (a) the dark fringe for which $m = 0$, (b) the bright fringe for which $m = 1$, (c) the dark fringe for which $m = 1$, and (d) the bright fringe for which $m = 2$.

2. E In a Young's double-slit experiment, the angle that locates the second dark fringe on either side of the central bright fringe is 5.4°. Find the ratio d/λ of the slit separation d to the wavelength λ of the light.

3. E CHALK Two in-phase sources of waves are separated by a distance of 4.00 m. These sources produce identical waves that have a wavelength of 5.00 m. On the line between them, there are two places at which the same type of interference occurs. (a) Is it constructive or destructive interference, and (b) where are the places located?

4. E The dark fringe for $m = 0$ in a Young's double-slit experiment is located at an angle of $\theta = 15°$. What is the angle that locates the dark fringe for $m = 1$?

5. E V-HINT In a Young's double-slit experiment, the seventh dark fringe is located 0.025 m to the side of the central bright fringe on a flat screen, which is 1.1 m away from the slits. The separation between the slits is 1.4×10^{-4} m. What is the wavelength of the light being used?

6. E GO MMH Two parallel slits are illuminated by light composed of two wavelengths. One wavelength is $\lambda_A = 645$ nm. The other wavelength is λ_B and is unknown. On a viewing screen, the light with wavelength $\lambda_A = 645$ nm produces its third-order bright fringe at the same place where the light with wavelength λ_B produces its fourth dark fringe. The fringes are counted relative to the central or zeroth-order bright fringe. What is the unknown wavelength?

7. E GO In a setup like that in **Figure 27.7**, a wavelength of 625 nm is used in a Young's double-slit experiment. The separation between the slits is $d = 1.4 \times 10^{-5}$ m. The total width of the screen is 0.20 m. In one version of the setup, the separation between the double slit and the screen is $L_A = 0.35$ m, whereas in another version it is $L_B = 0.50$ m. On one side of the central bright fringe, how many bright fringes lie on the screen in the two versions of the setup? Do not include the central bright fringe in your counting.

8. M V-HINT At most, how many bright fringes can be formed on either side of the central bright fringe when light of wavelength 625 nm falls on a double slit whose slit separation is 3.76×10^{-6} m?

9. M CHALK MMH In a Young's double-slit experiment the separation y between the second-order bright fringe and the central bright fringe on a flat screen is 0.0180 m when the light has a wavelength of 425 nm. Assume that the angles that locate the fringes on the screen are small enough so that $\sin\theta \approx \tan\theta$. Find the separation y when the light has a wavelength of 585 nm.

10. H In Young's experiment a mixture of orange light (611 nm) and blue light (471 nm) shines on the double slit. The centers of the first-order bright blue fringes lie at the outer edges of a screen that is located 0.500 m away from the slits. However, the first-order bright orange fringes fall off the screen. By how much and in which direction (toward or away from the slits) should the screen be moved so that the centers of the first-order bright orange fringes will just appear on the screen? It may be assumed that θ is small, so that $\sin\theta \approx \tan\theta$.

11. H SSM A sheet that is made of plastic ($n = 1.60$) covers *one slit* of a double slit (see the drawing). When the double slit is illuminated by monochromatic light ($\lambda_{vacuum} = 586$ nm), the center of the screen appears dark rather than bright. What is the minimum thickness of the plastic?

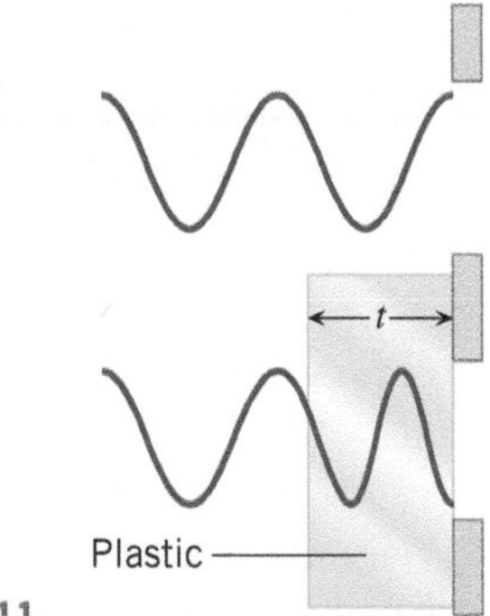

PROBLEM 11

Section 27.3 Thin-Film Interference

12. E You are standing in air and are looking at a flat piece of glass ($n = 1.52$) on which there is a layer of transparent plastic ($n = 1.61$). Light whose wavelength is 589 nm in vacuum is incident nearly perpendicularly on the coated glass and reflects into your eyes. The layer of plastic looks dark. Find the two smallest possible nonzero values for the thickness of the layer.

13. E SSM A nonreflective coating of magnesium fluoride ($n = 1.38$) covers the glass ($n = 1.52$) of a camera lens. Assuming that the coating prevents reflection of yellow-green light (wavelength in vacuum = 565 nm), determine the minimum nonzero thickness that the coating can have.

14. E GO When monochromatic light shines perpendicularly on a soap film ($n = 1.33$) with air on each side, the second smallest nonzero film thickness for which destructive interference of reflected light is observed is 296 nm. What is the vacuum wavelength of the light in nm?

15. E MMH A transparent film ($n = 1.43$) is deposited on a glass plate ($n = 1.52$) to form a nonreflecting coating. The film has a thickness that is 1.07×10^{-7} m. What is the longest possible wavelength (in vacuum) of light for which this film has been designed?

16. E GO A tank of gasoline ($n = 1.40$) is open to the air ($n = 1.00$). A thin film of liquid floats on the gasoline and has a refractive index that is between 1.00 and 1.40. Light that has a wavelength of 625 nm (in vacuum) shines perpendicularly down through the air onto this film, and in this light the film looks bright due to constructive interference. The thickness of the film is 242 nm and is the minimum nonzero thickness for which constructive interference can occur. What is the refractive index of the film?

17. E V-HINT Review Conceptual Example 4 before beginning this problem. A soap film with different thicknesses at different places has an unknown refractive index n and air on both sides. In reflected light it looks multicolored. One region looks yellow because destructive interference has removed blue ($\lambda_{vacuum} = 469$ nm) from the reflected light, while another looks magenta because destructive interference has removed green ($\lambda_{vacuum} = 555$ nm). In these regions the film has the minimum nonzero thickness t required for the destructive interference to occur. Find the ratio $t_{magenta}/t_{yellow}$.

18. M MMH A film of oil lies on a shallow puddle of water. The refractive index of the oil exceeds that of the water. The film has the minimum nonzero thickness such that it appears dark due to destructive interference when viewed in red light (wavelength = 640.0 nm in vacuum). Assuming that the visible spectrum extends from 380 to 750 nm, for which visible wavelength(s) in vacuum will the film appear bright due to constructive interference?

19. M CHALK SSM Orange light ($\lambda_{vacuum} = 611$ nm) shines on a soap film ($n = 1.33$) that has air on either side of it. The light strikes the film perpendicularly. What is the minimum thickness of the film for which constructive interference causes it to look bright in reflected light?

Section 27.5 Diffraction

20. E (a) As Section 17.3 discusses, high-frequency sound waves exhibit less diffraction than low-frequency sound waves do. However, even high-frequency sound waves exhibit much more diffraction under normal circumstances than do light waves that pass through the same opening. The highest frequency that a healthy ear can typically hear is 2.0×10^4 Hz. Assume that a sound wave with this frequency travels at 343 m/s and passes through a doorway that has a width of 0.91 m. Determine the angle that locates the first minimum to either side of the central maximum in the diffraction pattern for the sound. This minimum is equivalent to the first dark fringe in a single-slit diffraction pattern for light. (b) Suppose that yellow light (wavelength = 580 nm in vacuum) passes through a doorway and that the first dark fringe in its diffraction pattern is located at the angle determined in part (a). How wide would this hypothetical doorway have to be?

21. E A dark fringe in the diffraction pattern of a single slit is located at an angle of $\theta_A = 34°$. With the same light, the same dark fringe formed with another single slit is at an angle of $\theta_B = 56°$. Find the ratio W_A/W_B of the widths of the two slits.

22. E SSM A diffraction pattern forms when light passes through a single slit. The wavelength of the light is 675 nm. Determine the angle that locates the first dark fringe when the width of the slit is (a) 1.8×10^{-4} m and (b) 1.8×10^{-6} m.

23. E GO A slit has a width of $W_1 = 2.3 \times 10^{-6}$ m. When light with a wavelength of $\lambda_1 = 510$ nm passes through this slit, the width of the central bright fringe on a flat observation screen has a certain value. With the screen kept in the same place, this slit is replaced with a second slit (width W_2), and a wavelength of $\lambda_2 = 740$ nm is used. The width of the central bright fringe on the screen is observed to be unchanged. Find W_2.

24. E SSM Light that has a wavelength of 668 nm passes through a slit 6.73×10^{-6} m wide and falls on a screen that is 1.85 m away. What is the distance on the screen from the center of the central bright fringe to the third dark fringe on either side?

25. E GO Light shines through a single slit whose width is 5.6×10^{-4} m. A diffraction pattern is formed on a flat screen located 4.0 m away. The distance between the middle of the central bright fringe and the first dark fringe is 3.5 mm. What is the wavelength of the light?

26. M V-HINT Light waves with two different wavelengths, 632 nm and 474 nm, pass simultaneously through a single slit whose width is 7.15×10^{-5} m and strike a screen 1.20 m from the slit. Two diffraction patterns are formed on the screen. What is the distance (in cm) between the common center of the diffraction patterns and the first occurrence of a dark fringe from one pattern falling on top of a dark fringe from the other pattern?

27. M GO CHALK The central bright fringe in a single-slit diffraction pattern has a width that equals the distance between the screen and the slit. Find the ratio λ/W of the wavelength λ of the light to the width W of the slit.

28. M SSM How many dark fringes will be produced on either side of the central maximum if light ($\lambda = 651$ nm) is incident on a single slit that is 5.47×10^{-6} m wide?

29. H In a single-slit diffraction pattern, the central fringe is 450 times as wide as the slit. The screen is 18 000 times farther from the slit than the slit is wide. What is the ratio λ/W, where λ is the wavelength of the light shining through the slit and W is the width of the slit? Assume that the angle that locates a dark fringe on the screen is small, so that $\sin\theta \approx \tan\theta$.

Section 27.6 Resolving Power

30. E Two stars are 3.7×10^{11} m apart and are equally distant from the earth. A telescope has an objective lens with a diameter of 1.02 m and just detects these stars as separate objects. Assume that light of wavelength 550 nm is being observed. Also assume that diffraction effects, rather than atmospheric turbulence, limit the resolving power of the telescope. Find the maximum distance that these stars could be from the earth.

31. E GO An inkjet printer uses tiny dots of red, green, and blue ink to produce an image. Assume that the dot separation on the printed page is the same for all colors. At normal viewing distances, the eye does not resolve the individual dots, regardless of color, so that the image has a normal look. The wavelengths for red, green, and blue are $\lambda_{red} = 660$ nm, $\lambda_{green} = 550$ nm, and $\lambda_{blue} = 470$ nm. The diameter of the pupil through which light enters the eye is 2.0 mm. For a viewing distance of 0.40 m, what is the maximum allowable dot separation?

32. E V-HINT Astronomers have discovered a planetary system orbiting the star Upsilon Andromedae, which is at a distance of 4.2×10^{17} m from the earth. One planet is believed to be located at a distance of 1.2×10^{11} m from the star. Using visible light with a vacuum wavelength of 550 nm, what is the minimum necessary aperture diameter that a telescope must have so that it can resolve the planet and the star?

Section 27.7 The Diffraction Grating

33. E SSM A diffraction grating is 1.50 cm wide and contains 2400 lines. When used with light of a certain wavelength, a third-order maximum is formed at an angle of 18.0°. What is the wavelength (in nm)?

34. E The light shining on a diffraction grating has a wavelength of 495 nm (in vacuum). The grating produces a second-order bright fringe whose position is defined by an angle of 9.34°. How many lines per centimeter does the grating have?

35. E For a wavelength of 420 nm, a diffraction grating produces a bright fringe at an angle of 26°. For an unknown wavelength, the same grating produces a bright fringe at an angle of 41°. In both cases the bright fringes are of the same order m. What is the unknown wavelength?

36. E GO Two diffraction gratings, A and B, are located at the same distance from the observation screens. Light with the same wavelength λ is used for each. The separation between adjacent principal maxima for grating A is 2.7 cm, and for grating B it is 3.2 cm. Grating A has 2000 lines per meter. How many lines per meter does grating B have? (*Hint*: The diffraction angles are small enough that the approximation $\sin\theta \approx \tan\theta$ can be used.)

37. E SSM The wavelength of a laser beam is 780 nm. Suppose that a diffraction grating produces first-order maxima that are 1.2 mm apart at a distance of 3.0 mm from the grating. Estimate the spacing between the slits of the grating.

38. E The first-order principle maximum produced by a grating is located at an angle of $\theta = 18.0°$. What is the angle for the third-order maximum with the same light?

39. M V-HINT CHALK A diffraction grating has 2604 lines per centimeter, and it produces a principal maximum at $\theta = 30.0°$. The grating is used with light that contains all wavelengths between 410 and 660 nm. What is (are) the wavelength(s) of the incident light that could have produced this maximum?

40. M GO Light of wavelength 410 nm (in vacuum) is incident on a diffraction grating that has a slit separation of 1.2×10^{-5} m. The distance between the grating and the viewing screen is 0.15 m. A diffraction pattern is produced on the screen that consists of a central bright fringe and higher-order bright fringes (see the drawing). **(a)** Determine the distance y from the central bright fringe to the second-order bright fringe. (*Hint*: The diffraction angles are small enough that the approximation $\tan\theta \approx \sin\theta$ can be used.) **(b)** If the entire apparatus is submerged in water ($n_{water} = 1.33$), what is the distance y?

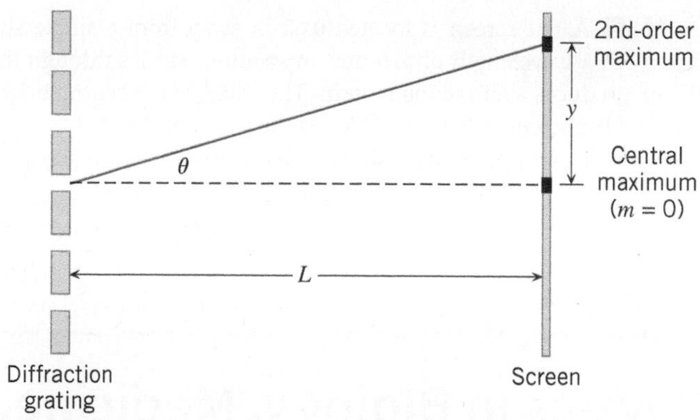

PROBLEM 40

41. M SSM Violet light (wavelength = 410 nm) and red light (wavelength = 660 nm) lie at opposite ends of the visible spectrum. **(a)** For each wavelength, find the angle θ that locates the first-order maximum produced by a grating with 3300 lines/cm. This grating converts a mixture of all colors between violet and red into a rainbow-like dispersion between the two angles. Repeat the calculation above for **(b)** the second-order maximum and **(c)** the third-order maximum. **(d)** From your results, decide whether there is an overlap between any of the "rainbows" and, if so, specify which orders overlap.

Additional Problems

Online

42. E GO A soap film ($n = 1.33$) is 465 nm thick and lies on a glass plate ($n = 1.52$). Sunlight, whose wavelengths (in vacuum) extend from 380 to 750 nm, travels through the air and strikes the film perpendicularly. For which wavelength(s) in this range does destructive interference cause the film to look dark in reflected light?

43. E SSM In a Young's double-slit experiment, two rays of monochromatic light emerge from the slits and meet at a point on a distant screen, as in **Figure 27.6a**. The point on the screen where these two rays meet is the eighth-order bright fringe. The difference in the distances that the two rays travel is 4.57×10^{-6} m. What is the wavelength (in nm) of the monochromatic light?

44. E GO Point A is the midpoint of one of the sides of a square. On the side opposite this spot, two in-phase loudspeakers are located at adjacent corners, as shown in the figure. Standing at point A you hear a loud sound because of constructive interference between the identical sound waves coming from the speakers. As you walk along the side of the square toward either empty corner, the loudness diminishes gradually to nothing and then increases again until you hear a maximally loud sound at the corner. If the length of each side of the square is 4.6 m, find the wavelength of the sound waves.

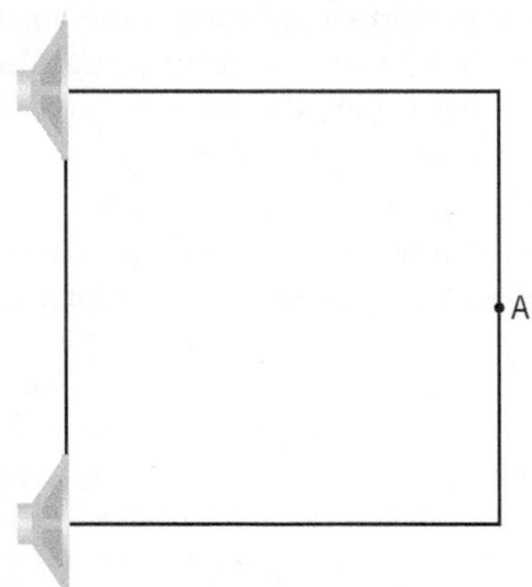

PROBLEM 44

45. E GO A flat screen is located 0.60 m away from a single slit. Light with a wavelength of 510 nm (in vacuum) shines through the slit and produces a diffraction pattern. The width of the central bright fringe on the screen is 0.050 m. What is the width of the slit?

46. M Review Conceptual Example 2 before attempting this problem. Two slits are 0.158 mm apart. A mixture of red light (wavelength = 665 nm) and yellow-green light (wavelength = 565 nm) falls on the slits. A flat observation screen is located 2.24 m away. What is the distance on the screen between the third-order red fringe and the third-order yellow-green fringe?

47. M GO A spotlight sends red light (wavelength = 694.3 nm) to the moon. At the surface of the moon, which is 3.77×10^8 m away, the light strikes a reflector left there by astronauts. The reflected light returns to the earth, where it is detected. When it leaves the spotlight, the circular beam of light has a diameter of about 0.20 m, and diffraction causes the beam to spread as the light travels to the moon. In effect, the first circular dark fringe in the diffraction pattern defines the size of the central bright spot on the moon. Determine the diameter (not the radius) of the central bright spot on the moon.

48. M In a single-slit diffraction pattern on a flat screen, the central bright fringe is 1.2 cm wide when the slit width is 3.2×10^{-5} m. When the slit is replaced by a second slit, the wavelength of the light and the distance to the screen remaining unchanged, the central bright fringe broadens to a width of 1.9 cm. What is the width of the second slit? It may be assumed that θ is so small that $\sin\theta \approx \tan\theta$.

49. M GO A beam of light is sent directly down onto a glass plate ($n = 1.5$) and a plastic plate ($n = 1.2$) that form a thin wedge of air (see the drawing). An observer looking down through the glass plate sees the fringe pattern shown in the lower part of the drawing, with the dark fringes at the ends A and B. The wavelength of the light is 520 nm. Using the fringe pattern shown in the drawing, determine the thickness of the air wedge at B.

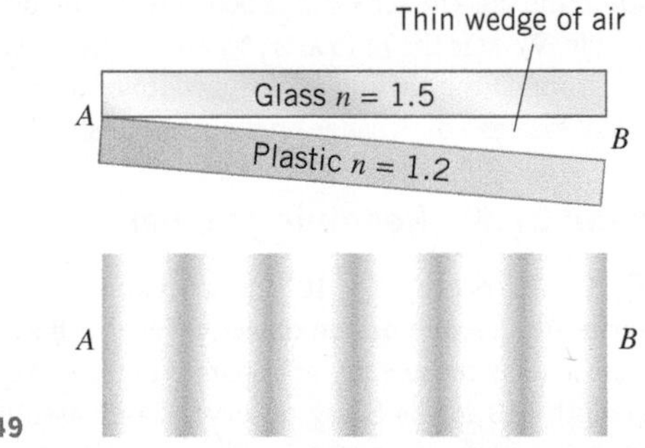

PROBLEM 49

50. H There are 5620 lines per centimeter in a grating that is used with light whose wavelength is 471 nm. A flat observation screen is located at a distance of 0.750 m from the grating. What is the minimum width that the screen must have so the *centers* of all the principal maxima formed on either side of the central maximum fall on the screen?

Physics in Biology, Medicine, and Sports

51. E BIO 27.6 It is claimed that some professional baseball players can see which way the ball is spinning as it travels toward home plate. One way to judge this claim is to estimate the distance at which a batter can first hope to resolve two points on opposite sides of a baseball, which has a diameter of 0.0738 m. **(a)** Estimate this distance, assuming that the pupil of the eye has a diameter of 2.0 mm and the wavelength of the light is 550 nm in vacuum. **(b)** Considering that the distance between the pitcher's mound and home plate is 18.4 m, can you rule out the claim based on your answer to part (a)?

52. E BIO CHALK SSM 27.6 Late one night on a highway, a car speeds by you and fades into the distance. Under these conditions the pupils of your eyes have diameters of about 7.0 mm. The taillights of this car are separated by a distance of 1.2 m and emit red light (wavelength = 660 nm in vacuum). How far away from you is this car when its taillights appear to merge into a single spot of light because of the effects of diffraction?

53. E BIO GO 27.6 A hunter who is a bit of a braggart claims that from a distance of 1.6 km he can selectively shoot either of two squirrels who are sitting ten centimeters apart on the same branch of a tree. What's more, he claims that he can do this without the aid of a telescopic sight on his rifle. **(a)** Determine the diameter of the pupils of his eyes that would be required for him to be able to resolve the squirrels

as separate objects. In this calculation use a wavelength of 498 nm (in vacuum) for the light. **(b)** State whether his claim is reasonable, and provide a reason for your answer. In evaluating his claim, consider that the human eye automatically adjusts the diameter of its pupil over a typical range of 2 to 8 mm, the larger values coming into play as the lighting becomes darker. Note also that under dark conditions, the eye is most sensitive to a wavelength of 498 nm.

54. E BIO **27.6** Review Conceptual Example 9 as background for this problem. In addition to the data given there, assume that the dots in the painting are separated by 1.5 mm and that the wavelength of the light is $\lambda_{\text{vacuum}} = 550$ nm. Find the distance at which the dots can just be resolved by **(a)** the eye and **(b)** the camera.

55. M BIO GO **27.6** The pupil of an eagle's eye has a diameter of 6.0 mm. Two field mice are separated by 0.010 m. From a distance of 176 m, the eagle sees them as one unresolved object and dives toward them at a speed of 17 m/s. Assume that the eagle's eye detects light that has a wavelength of 550 nm in vacuum. How much time passes until the eagle sees the mice as separate objects?

56. E BIO **27.6** A large group of football fans comes to the game with colored cards that spell out the name of their team when held upsimultaneously. Most of the cards are colored blue ($\lambda_{\text{vacuum}} = 480$ nm). When displayed, the average distance between neighboring cards is 5.0 cm. If the cards are to blur together into solid blocks of color when viewed by a spectator at the other end of the stadium (160 m away), what must be the maximum diameter (in mm) of the spectator's pupils?

57. M BIO **27.6** Under dark conditions, the maximum diameter of a human pupil is 7.0 mm, where an owl's pupil may be 8.5 mm. Assume a human can optically resolve two closely spaced objects at a distance r. **(a)** By what factor could the distance between the two objects be reduced and still have the owl optically resolve them at the same distance r? **(b)** If the distance between the two objects remains fixed, by what factor could r be increased and still have the owl optically resolve the two objects? In both *(a)* and *(b)*, assume the wavelength of the light remains constant.

58. M BIO **27.7** The bold colors observed in certain bird feathers are due to diffraction. They have microstructures on their surface that act like a diffraction grating. If a first-order maximum is observed at an angle of 25° relative to the surface normal for 530-nm light, what is the spacing between the microstructures?

59. M BIO **27.7** The human eye detects the wavelengths of light in the visible spectrum, which span the range from 380 to 740 nm. White light is incident on a diffraction grating with 3500 lines per cm. **(a)** Which wavelength in the visible spectrum occurs at the greatest angle for the $m = 2$ maximum, and **(b)** what is this angle?

60. M BIO **27.6** Review Example 5 from Chapter 25. If the James Webb Space Telescope were enclosed, such that it had a circular aperture with a diameter equal to the diameter of its primary mirror, calculate its optical resolution. More specifically, what is the smallest separation distance of two nickels that are optically resolved with 520-nm light, when viewed from a distance of 4.7×10^6 m, which is the distance between New York and San Francisco?

61. M BIO **27.3** Many outdoor athletes splurge on a high-end pair of sunglasses. The lenses in these glasses are not a single piece of glass, but are actually composed of multiple layers (see the photo). The antireflective coating at the back of the lens is there to reduce back glare, which is light that strikes the back of the lens and reflects toward the eyes. The antireflective coating consists of a transparent thin film with an index of refraction equal to 1.36. The film, with a thickness of 96.0 nm, is applied to the back of the glass lens, which has an index of refraction of 1.52. Which wavelength of light in the visible spectrum is not reflected from the coating, due to destructive interference?

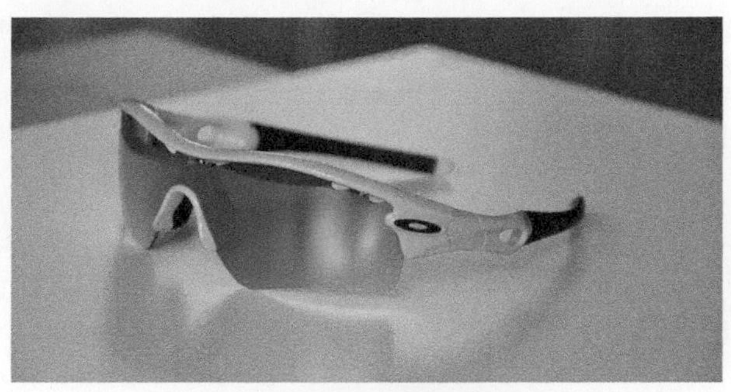

fielperson/Pixabay

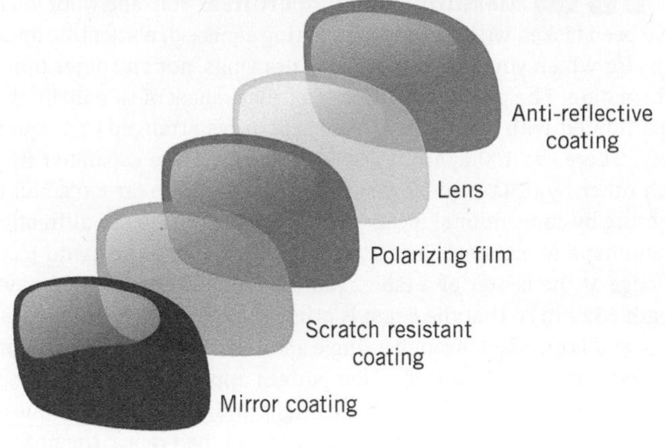

PROBLEM 61

Concepts and Calculations Problems

Online

The ability to exhibit interference effects is a fundamental characteristic of any kind of wave. Our understanding of these effects depends on the principle of linear superposition, which we first encountered in Chapter 17. Only by means of this principle can we understand the constructive and destructive interference of light waves that lie at the heart of every topic in this chapter. Problem 62 serves as a review of the essence of this principle. Problem 63 deals with thin-film interference and reviews the factors that must be considered in such cases.

62. M CHALK SSM A square is 3.5 m on a side, and point A is the midpoint of one of the sides. On the side opposite this spot, two in-phase loud-speakers are located at adjacent corners, as shown in the figure. Standing at point A, you hear a loud sound because constructive interference occurs between the identical sound waves coming from the speakers. As you walk along the side of the square toward either empty corner, the loudness diminishes gradually but does not entirely disappear until you reach either empty corner, where you hear no sound at all. Thus, at each empty corner destructive interference occurs. *Concepts:* (i) Why does constructive interference occur at point A? (ii) What is the general condition that leads to destructive interference? (iii) The condition that leads to destructive interference entails a number of possibilities. Which of them applies at either empty corner of the square? *Calculation:* Find the wavelength of the sound waves.

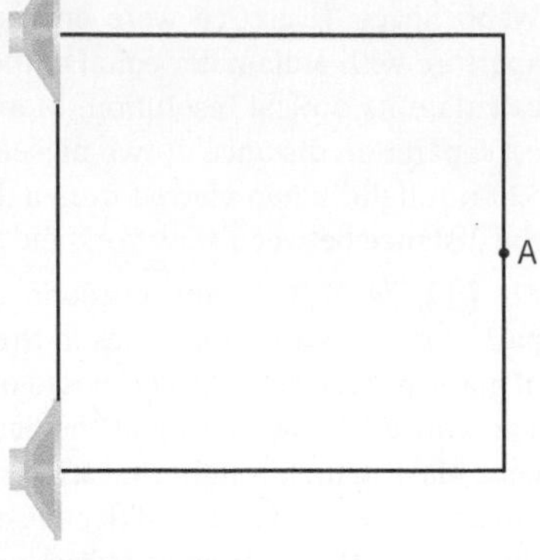

PROBLEM 62

63. M CHALK A soap film ($n = 1.33$) is 375 nm thick and coats a flat piece of glass ($n = 1.52$). Thus, air is on one side of the film and glass is on the other side, as the figure illustrates. Sunlight, whose wavelengths (in vacuum) extend from 380 to 750 nm, travels through air and strikes the film nearly perpendicularly. *Concepts:* (i) What, if any, phase change occurs when light, traveling in air, reflects from the air–film interface? (ii) What, if any, phase change occurs when light, traveling in the film, reflects from the film–glass interface? (iii) Is the wavelength of the light in the film greater than, smaller than, or equal to the wavelength in a vacuum? *Calculations:* For what wavelengths in the range of 380 to 750 nm does constructive interference cause the film to look bright in reflected light?

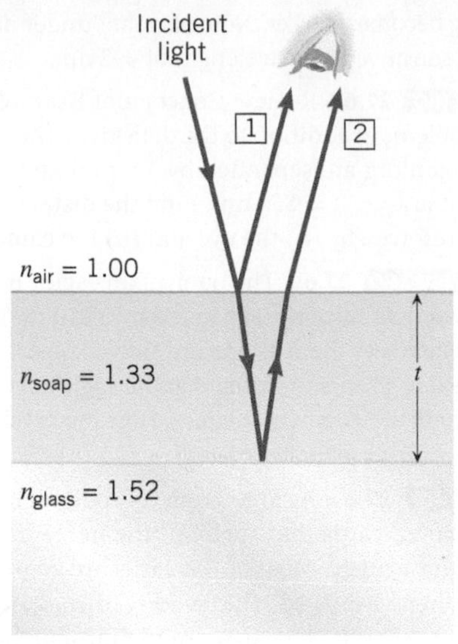

PROBLEM 63

Team Problems

Online

64. E T WS **Measuring small apertures.** You and your team have been tasked with reverse engineering a piece of a scientific apparatus for which you have no technical drawings, nor any other useful information. The piece in question is a square sheet of thin metal that is perforated with miniscule circular apertures arranged in a square array. There are 100 apertures in total and they are separated from each other by 0.500 cm. However, the apertures are far too small to measure by conventional means, so you get the idea to use diffraction phenomena to estimate their sizes. First, you secure the metal piece on edge at the center of a table. You then align a red laser of wavelength 632 nm so that the beam is oriented perpendicular the square piece and is directed through a single aperture. On a wall 11.6 m from the aperture, a circular diffraction pattern appears that consists of a central bright circle and concentric bright and dark rings. You measure the diameter of the smallest dark ring and then repeat the process for each aperture in the piece. You find that the 100 apertures have only three different sizes that correspond to smallest/innermost dark ring diameters of 2.75 cm, 4.10 cm, and 6.15 cm. What are the three different aperture diameters?

65. E T WS **Ranging With Diffraction.** You and your team need to estimate the distance between your spaceship and another spacecraft that has been damaged and is inoperable. On the front of the disabled ship are two small, bright red lights (wavelength $\lambda = 665$ nm) that you know are separated by 15.0 meters. Your own ship has bright blue lights ($\lambda = 475$ nm) on its leading edge that are separated by 12.0 m. The ships have identical cameras, each with a circular aperture of diameter 1.20 mm. Assume that you are heading directly toward the front of the damaged ship. **(a)** How far away from the damaged ship will your camera be able to just resolve its lights? **(b)** How far will you be away from the damaged ship when its camera can just resolve your lights?

66. M T **An Optical Monochromator.** You and your team are designing a device that inputs a beam of white light (i.e., a continuous spectrum of visible light spanning all wavelengths from 410 nm to 660 nm), and outputs a nearly monochromatic beam (i.e., a single color). Such a device is called an optical monochromator and is used in a wide variety of instruments and scientific experiments. In the instrument you are building, white light impinges upon the backside of a diffraction grating that has 1200 lines/cm. A movable rectangular aperture (a slit) is located on the opposite side of the grating, and can translate along a circular arc of radius 20.0 cm, the center of which is located at the grating. **(a)** At what angle relative to the normal of the grating should the center of the slit be located in order to pass green light ($\lambda = 550$ nm) from first order ($m = 1$) diffracted light? **(b)** How wide should the slit be so that the wavelengths passing through the slit fall in the range 540 nm $\leq \lambda \leq$ 560 nm?

67. M T **A Crude Thickness Monitor.** You and your team need to determine the thicknesses of two extremely thin sheets of plastic and do not have a measurement instrument capable of the job. You find a white light source with color filters (i.e., tinted glass windows that only let certain colors pass through) including red ($\lambda = 660.0$ nm), green ($\lambda = 550.0$ nm), and blue ($\lambda = 470.0$ nm). You locate two glass microscope slides, each of length 10.0 cm, stack one on top of the other, and tape their ends together on one side. On the very edge of the side opposite the tape, you sandwich a piece of plastic between the slides to form a wedge of air between the plates. **(a)** When you shine blue light perpendicular to the slides, 55.0 bright fringes form along the full length of the slide. How thick is the plastic? **(b)** When you replace the first sheet with the second sheet of plastic, the fringes that appear are too closely spaced to count. You switch to red light (a longer wavelength), and count 124 fringes. How thick is the plastic? **(c)** How many fringes would you expect in each case, (a) and (b), if you instead used green light?

CHAPTER 28

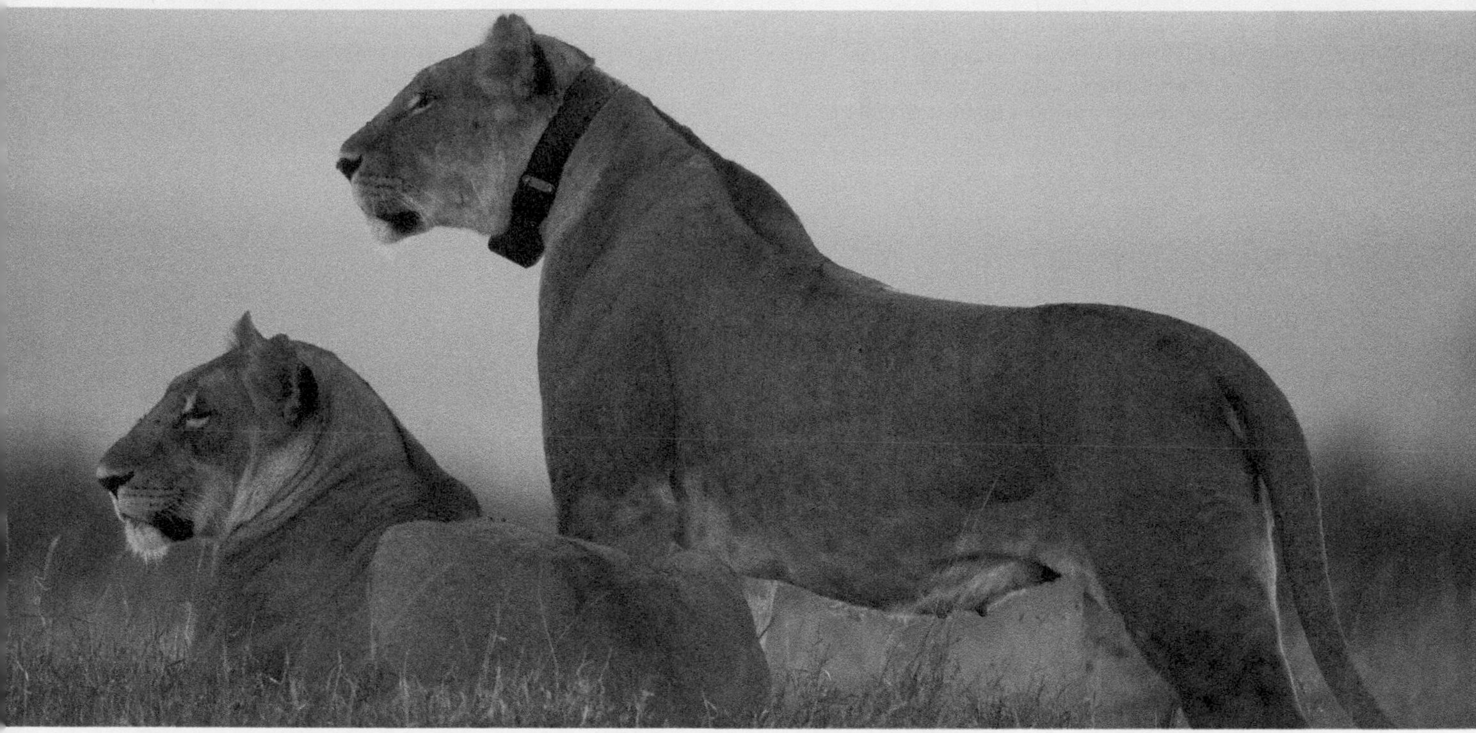

One of the lionesses is wearing a collar equipped with a Global Positioning System (GPS) tracking unit in the Ol Pejeta Conservancy in Kenya, Africa. The unit will allow researchers to track the lion's movements because GPS technology can locate objects on the earth with remarkable accuracy. The accuracy results, in part, because the system accounts for Einstein's theory of special relativity.

Special Relativity

LEARNING OBJECTIVES

After reading this module, you should be able to...

28.1 Define inertial reference frames.

28.2 List the postulates of special relativity.

28.3 Use time dilation to calculate time intervals in different frames of reference.

28.4 Use length contraction to calculate distances in different frames of reference.

28.5 Calculate the relativistic momentum of a high-speed particle.

28.6 Calculate the value of the various forms of energy a moving body possesses.

28.7 Calculate the relative velocity between relativistically moving bodies.

28.1 Events and Inertial Reference Frames

In the theory of special relativity, an **event**, such as the launching of the space shuttle in **Figure 28.1**, is a physical "happening" that occurs at a certain place and time. In this drawing two observers are watching the lift-off, one standing on the earth and one in an airplane that is flying at a constant velocity relative to the earth. To record the event, each observer uses a **reference frame** that consists of a set of *x, y, z* axes (called a *coordinate system*) and a clock. The coordinate systems are used to establish where the event occurs, and the clocks to specify when. Each observer is at rest relative to his own reference frame. However, the earth-based observer and the airborne observer are moving relative to each other and so, also, are their respective reference frames.

The theory of special relativity deals with a "special" kind of reference frame, called an **inertial reference frame**. As Section 4.2 discusses, an inertial reference frame is one in which **Newton's** law of inertia is valid. That is, if the net force acting on a body is zero, the body either remains at rest or moves at a **constant velocity**. In other words, the acceleration of such a body is zero when measured in an inertial

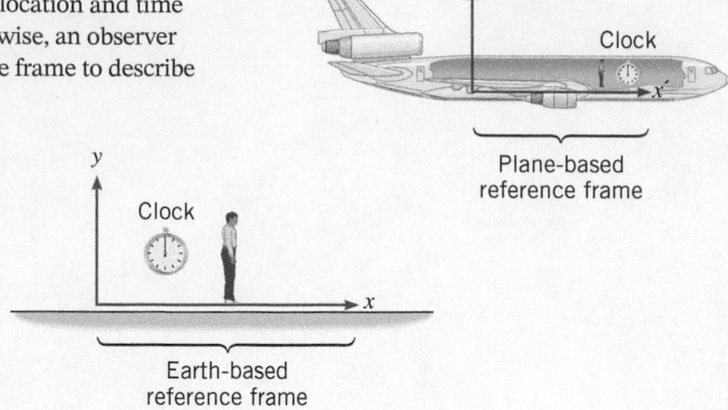

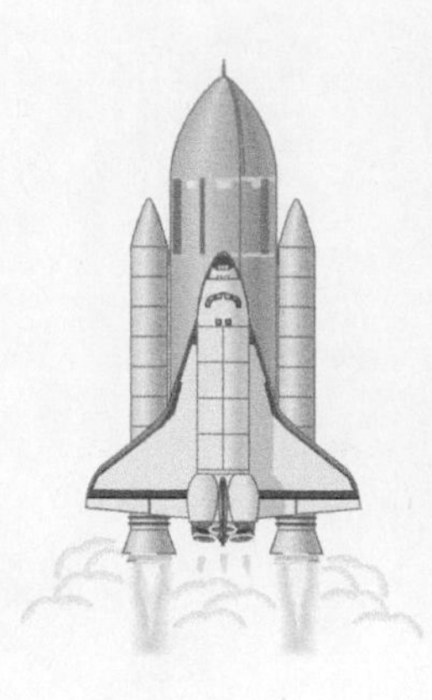

FIGURE 28.1 Using an earth-based reference frame, an observer standing on the earth records the location and time of an event (the space shuttle lift-off). Likewise, an observer in the airplane uses a plane-based reference frame to describe the event.

reference frame. Rotating and otherwise accelerating reference frames are not inertial reference frames. The earth-based reference frame in Figure 28.1 is not quite an inertial frame because it is subjected to centripetal accelerations as the earth spins on its axis and revolves around the sun. In most situations, however, the effects of these accelerations are small, and we can neglect them. To the extent that the earth-based reference frame is an inertial frame, so is the plane-based reference frame, because the plane moves at a constant velocity relative to the earth. The next section discusses why inertial reference frames are important in relativity.

28.2 The Postulates of Special Relativity

Einstein built his theory of special relativity on two fundamental assumptions or postulates about the way nature behaves.

THE POSTULATES OF SPECIAL RELATIVITY

1. ***The Relativity Postulate.*** **The laws of physics are the same in every inertial reference frame.**
2. ***The Speed-of-Light Postulate.*** **The speed of light in a vacuum, measured in any inertial reference frame, always has the same value of c, no matter how fast the source of light and the observer are moving relative to each other.**

It is not too difficult to accept the relativity postulate. For instance, in Figure 28.1 each observer, using his own inertial reference frame, can make measurements on the motion of the space shuttle. The relativity postulate asserts that both observers find their data to be consistent with Newton's laws of motion. Similarly, both observers find that the behavior of the electronics on the space shuttle is described by the laws of electromagnetism. According to the relativity postulate, ***any inertial reference frame is as good as any other for expressing the laws of physics.*** As far as inertial reference frames are concerned, nature does not play favorites.

Since the laws of physics are the same in all inertial reference frames, there is no experiment that can distinguish between an inertial frame that is at rest and one that is moving at a constant velocity. When you are seated on the aircraft in Figure 28.1, for instance, it is just as valid to say that you are at rest and the earth is moving as it is to say the converse. It is not possible to single out one particular inertial reference frame as being at "absolute rest." Consequently, it is meaningless to talk about the "absolute velocity" of an object—that is, its velocity measured relative to a reference frame at "absolute rest." Thus, the earth moves relative to the sun, which itself moves relative to the center of our galaxy. And the galaxy moves relative to other galaxies, and so on. According to Einstein, only the relative velocity between objects, not their absolute velocities, can be measured and is physically meaningful.

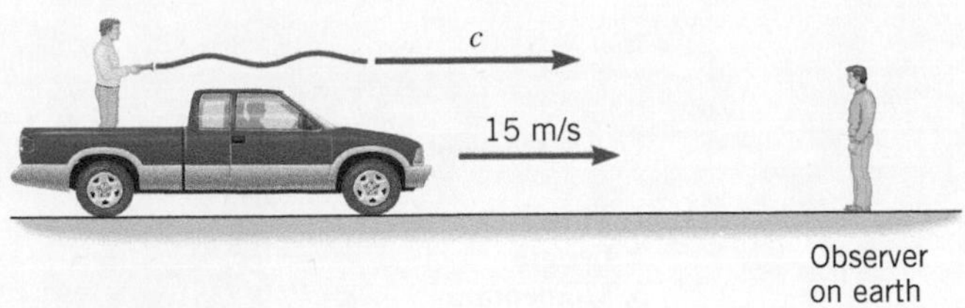

FIGURE 28.2 Both the person on the truck and the observer on the earth measure the speed of the light to be c, regardless of the speed of the truck.

Whereas the relativity postulate seems plausible, the speed-of-light postulate defies common sense. For instance, **Figure 28.2** illustrates a person standing on the bed of a truck that is moving at a constant speed of 15 m/s relative to the ground. Now, suppose that you are standing on the ground and the person on the truck shines a flashlight at you. The person on the truck observes the speed of light to be c. What do you measure for the speed of light? You might guess that the speed of light would be $c + 15$ m/s. However, this guess is inconsistent with the speed-of-light postulate, which states that all observers in inertial reference frames measure the speed of light to be c—nothing more, nothing less. Therefore, you must also measure the speed of light to be c, the same as that measured by the person on the truck. According to the speed-of-light postulate, the fact that the flashlight is moving has no influence whatsoever on the speed of the light approaching you. This property of light, although surprising, has been verified many times by experiment.

Since waves, such as water waves and sound waves, require a medium through which to propagate, it was natural for scientists before Einstein to assume that light did too. This hypothetical medium was called the *luminiferous ether* and was assumed to fill all of space. Furthermore, it was believed that light traveled at the speed c only when measured with respect to the ether. According to this view, an observer moving relative to the ether would measure a speed for light that was smaller or greater than c, depending on whether the observer moved with or against the light, respectively. During the years 1883–1887, however, the American scientists A. A. Michelson and E. W. Morley carried out a series of famous experiments whose results were not consistent with the ether theory. Their results indicated that the speed of light is indeed the same in all inertial reference frames and does not depend on the motion of the observer. These experiments, and others, led eventually to the demise of the ether theory and the acceptance of the theory of special relativity.

The remainder of this chapter reexamines, from the viewpoint of special relativity, a number of fundamental concepts that have been discussed in earlier chapters from the viewpoint of classical physics. These concepts are time, length, momentum, kinetic energy, and the addition of velocities. We will see that each is modified by special relativity in a way that depends on the speed v of a moving object relative to the speed c of light in a vacuum. **Figure 28.3** illustrates that when the object moves slowly [v is much smaller than c ($v \ll c$)], the modification is negligibly small, and the classical version of each concept provides an accurate description of reality. However, when the object moves so rapidly that v is an appreciable fraction of the speed of light [v is approximately equal to c ($v \approx c$)], the effects of special relativity must be considered. The gold panel in **Figure 28.3** lists the various equations that convey the modifications imposed by special relativity. Each of these equations will be discussed in later sections of this chapter.

It is important to realize that the modifications imposed by special relativity do not imply that the classical concepts of time, length, momentum, kinetic energy, and the addition of velocities, as developed by Newton and others, are wrong. They are just limited to speeds that are very small compared to the speed of light. In contrast, the relativistic view of the concepts applies to all speeds between zero and the speed of light.

E.O. Hoppe/Time & Life Pictures/Getty Images

Concepts

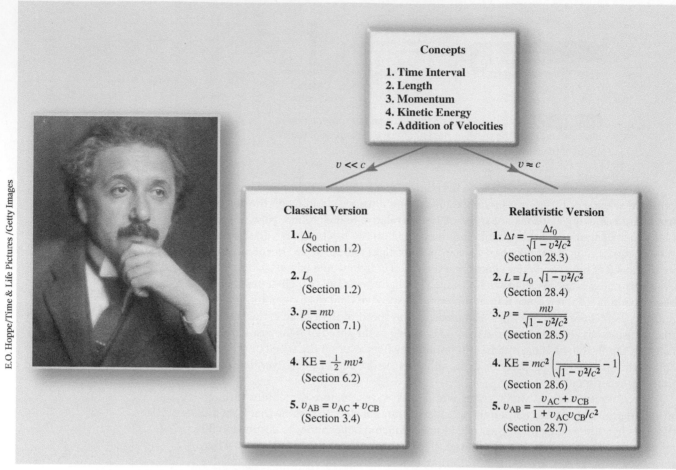

1. Time Interval
2. Length
3. Momentum
4. Kinetic Energy
5. Addition of Velocities

$v << c$ | $v \approx c$

Classical Version

1. Δt_0 (Section 1.2)
2. L_0 (Section 1.2)
3. $p = mv$ (Section 7.1)
4. $\text{KE} = \frac{1}{2}mv^2$ (Section 6.2)
5. $v_{AB} = v_{AC} + v_{CB}$ (Section 3.4)

Relativistic Version

1. $\Delta t = \dfrac{\Delta t_0}{\sqrt{1 - v^2/c^2}}$ (Section 28.3)
2. $L = L_0\sqrt{1 - v^2/c^2}$ (Section 28.4)
3. $p = \dfrac{mv}{\sqrt{1 - v^2/c^2}}$ (Section 28.5)
4. $\text{KE} = mc^2\left(\dfrac{1}{\sqrt{1 - v^2/c^2}} - 1\right)$ (Section 28.6)
5. $v_{AB} = \dfrac{v_{AC} + v_{CB}}{1 + v_{AC}v_{CB}/c^2}$ (Section 28.7)

FIGURE 28.3 Albert Einstein (1879–1955), the author of the theory of special relativity, is one of the most famous scientists of the twentieth century. This figure emphasizes that the ratio v/c of the speed v of a moving object to the speed c of light in a vacuum is what determines how great the effects of special relativity are.

28.3 The Relativity of Time: Time Dilation

Time Dilation

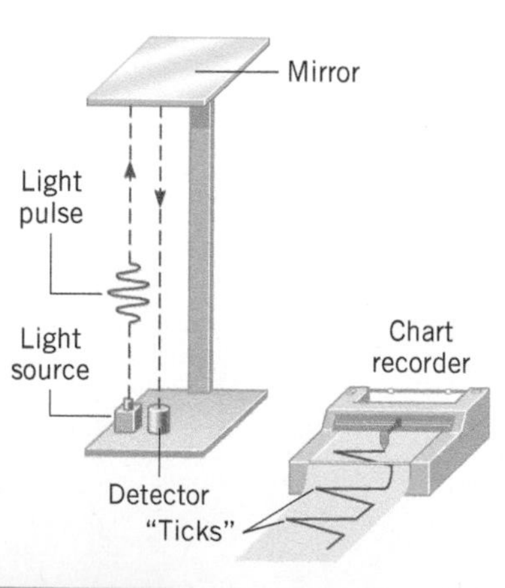

FIGURE 28.4 A light clock.

Common experience indicates that time passes just as quickly for a person standing on the ground as it does for an astronaut in a spacecraft. In contrast, special relativity reveals that the person on the ground measures time passing more slowly for the astronaut than for herself. We can see how this curious effect arises with the help of the clock illustrated in **Figure 28.4**, which uses a pulse of light to mark time. A short pulse of light is emitted by a light source, reflects from a mirror, and then strikes a detector that is situated next to the source. Each time a pulse reaches the detector, a "tick" registers on the chart recorder, another short pulse of light is emitted, and the cycle repeats. Thus, the time interval between successive "ticks" is marked by a beginning event (the firing of the light source) and an ending event (the pulse striking the detector). The source and detector are so close to each other that the two events can be considered to occur at the same location.

Suppose two identical clocks are built. One is kept on earth, and the other is placed aboard a spacecraft that travels at a constant velocity relative to the earth. The astronaut is at rest with respect to the clock on the spacecraft and, therefore, sees the light pulse move along the up/down path shown in **Animated Figure 28.5*a***. According to the astronaut, the time interval Δt_0 required for the light to follow this path is the distance $2D$ divided by the speed of light c; $\Delta t_0 = 2D/c$. To the astronaut, Δt_0 is the time interval

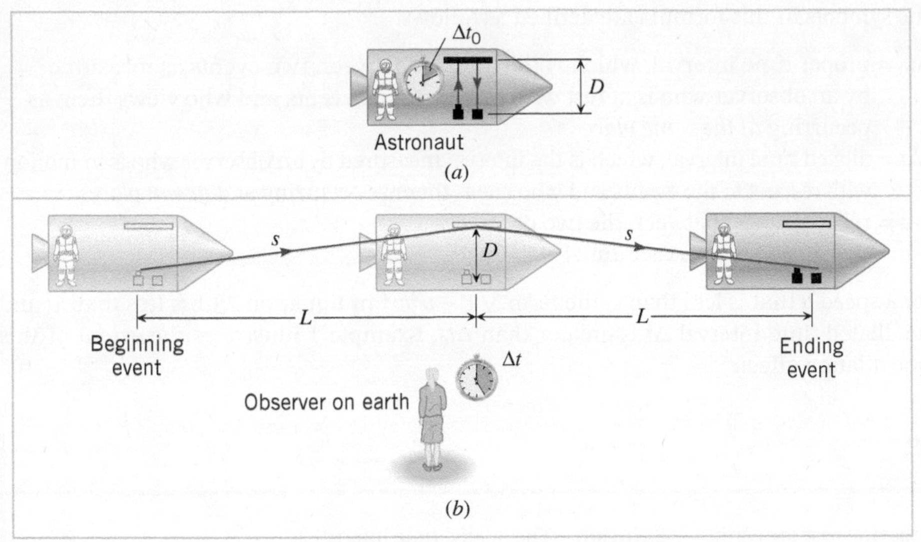

ANIMATED FIGURE 28.5 (*a*) The astronaut measures the time interval Δt_0 between successive "ticks" of his light clock. (*b*) An observer on earth watches the astronaut's clock and sees the light pulse travel a greater distance between "ticks" than it does in part *a*. Consequently, the earth-based observer measures a time interval Δt between "ticks" that is greater than Δt_0.

between the "ticks" of the spacecraft clock—that is, the time interval between the beginning and ending events of the clock. An earth-based observer, however, does *not* measure Δt_0 as the time interval between these two events. Since the spacecraft is moving, the earth-based observer sees the light pulse follow the diagonal path shown in red in part *b* of the drawing. This path is longer than the up/down path seen by the astronaut. But light travels at the *same speed* c for both observers, in accord with the speed-of-light postulate. Therefore, the earth-based observer measures a time interval Δt between the two events that is *greater* than the time interval Δt_0 measured by the astronaut. In other words, the earth-based observer, using her own earth-based clock to measure the performance of the astronaut's clock, finds that the astronaut's clock runs slowly. This result of special relativity is known as **time dilation**. (To *dilate* means to expand, and the time interval Δt is "expanded" relative to Δt_0.)

The time interval Δt that the earth-based observer measures in **Animated Figure 28.5b** can be determined as follows. While the light pulse travels from the source to the detector, the spacecraft moves a distance $2L = v\,\Delta t$ to the right, where v is the speed of the spacecraft relative to the earth. From the drawing it can be seen that the light pulse travels a total diagonal distance of $2s$ during the time interval Δt. Applying the Pythagorean theorem, we find that

$$2s = 2\sqrt{D^2 + L^2} = 2\sqrt{D^2 + \left(\frac{v\,\Delta t}{2}\right)^2}$$

But the distance $2s$ is also equal to the speed of light times the time interval Δt, so that $2s = c\,\Delta t$. Therefore,

$$c\,\Delta t = 2\sqrt{D^2 + \left(\frac{v\,\Delta t}{2}\right)^2}$$

Squaring this result and solving for Δt gives

$$\Delta t = \frac{2D}{c}\frac{1}{\sqrt{1 - \frac{v^2}{c^2}}}$$

However, $2D/c = \Delta t_0$, where Δt_0 is the time interval between successive "ticks" of the spacecraft's clock as measured by the astronaut. With this substitution, the equation for Δt can be expressed as

Time dilation

$$\Delta t = \frac{\Delta t_0}{\sqrt{1 - \frac{v^2}{c^2}}} \qquad \textbf{(28.1)}$$

The symbols in this formula are defined as follows:

Δt_0 = proper time interval, which is the interval between two events as measured by an observer who is at rest with respect to the events and who views them as occurring *at the same place*

Δt = dilated time interval, which is the interval measured by an observer who is in motion with respect to the events and who views them as occurring at *different places*

v = relative speed between the two observers

c = speed of light in a vacuum

For a speed v that is less than c, the term $\sqrt{1 - v^2/c^2}$ in Equation 28.1 is less than 1, and the dilated time interval Δt is greater than Δt_0. Example 1 illustrates the extent of this time dilation effect.

EXAMPLE 1 | Time Dilation

The spacecraft in **Animated Figure 28.5** is moving past the earth at a constant speed v that is 0.92 times the speed of light. Thus, $v = (0.92)(3.0 \times 10^8 \text{ m/s})$, which is often written as $v = 0.92c$. The astronaut measures the time interval between successive "ticks" of the spacecraft clock to be $\Delta t_0 = 1.0$ s. What is the time interval Δt that an earth observer measures between "ticks" of the astronaut's clock?

Reasoning Since the clock on the spacecraft is moving relative to the earth, the earth-based observer measures a greater time interval Δt between "ticks" than does the astronaut, who is at rest relative to the clock. The dilated time interval Δt can be determined from the time dilation relation, Equation 28.1.

Solution The dilated time interval is

$$\Delta t = \frac{\Delta t_0}{\sqrt{1 - \frac{v^2}{c^2}}} = \frac{1.0 \text{ s}}{\sqrt{1 - \left(\frac{0.92c}{c}\right)^2}} = \boxed{2.6 \text{ s}}$$

From the point of view of the earth-based observer, the astronaut is using a clock that is running slowly, because the earth-based observer measures a time between "ticks" that is longer (2.6 s) than what the astronaut measures (1.0 s).

THE PHYSICS OF . . . the Global Positioning System and special relativity. Present-day spacecrafts fly nowhere near as fast as the craft in Example 1. Yet circumstances exist in which time dilation can create appreciable errors if not accounted for. The Global Positioning System (GPS), for instance, uses highly accurate and stable atomic clocks on board each of 24 satellites orbiting the earth at speeds of about 4000 m/s. These clocks make it possible to measure the time it takes for electromagnetic waves to travel from a satellite to a ground-based GPS receiver. From the speed of light and the times measured for signals from three or more of the satellites, it is possible to locate the position of the receiver (see Section 5.5). The stability of the clocks must be better than one part in 10^{13} to ensure the positional accuracy demanded of the GPS. Using Equation 28.1 and the speed of the GPS satellites, we can calculate the difference between the dilated time interval and the proper time interval as a fraction of the proper time interval and compare the result to the stability of the GPS clocks:

$$\frac{\Delta t - \Delta t_0}{\Delta t_0} = \frac{1}{\sqrt{1 - v^2/c^2}} - 1 = \frac{1}{\sqrt{1 - (4000 \text{ m/s})^2/(3.00 \times 10^8 \text{ m/s})^2}} - 1$$

$$= \frac{1}{1.1 \times 10^{10}}$$

This result is approximately one thousand times greater than the GPS-clock stability of one part in 10^{13}. Thus, if not taken into account, time dilation would cause an error in the measured position of the earth-based GPS receiver roughly equivalent to the error caused by a thousand-fold degradation in the stability of the atomic clocks.

Proper Time Interval

In **Animated Figure 28.5** both the astronaut and the person standing on the earth are measuring the time interval between a beginning event (the firing of the light source) and an ending event (the light pulse striking the detector). For the astronaut, who is at rest with respect to the light clock, the two events occur at the same location. (Remember, we are assuming that the light source and detector are so close together that they are considered to be at the same place.) Being at rest with respect to a clock is the usual or "proper" situation, so the time interval Δt_0 measured by the astronaut is called the **proper time interval**. In general, the proper time interval Δt_0 between two events is the time interval measured by an observer who is at rest relative to the events and sees them at the *same location* in space. On the other hand, the earth-based observer does not see the two events occurring at the same location in space, since the spacecraft is in motion. The time interval Δt that the earth-based observer measures is, therefore, not a proper time interval in the sense that we have defined it.

To understand situations involving time dilation, it is essential to distinguish between Δt_0 and Δt. It is helpful if one first identifies the two events that define the time interval. These may be something other than the firing of a light source and the light pulse striking a detector. Then determine the reference frame in which the two events occur at the same place. An observer at rest in this reference frame measures the proper time interval Δt_0.

Space Travel

One of the intriguing aspects of time dilation occurs in conjunction with space travel. Since enormous distances are involved, travel to even the closest star outside our solar system would take a long time. However, as the following example shows, the travel time can be considerably less for the passengers than one might guess.

EXAMPLE 2 | The Physics of Space Travel and Special Relativity

Alpha Centauri, a nearby star in our galaxy, is 4.3 light-years away. This means that, as measured by a person on earth, it would take light 4.3 years to reach this star. If a rocket leaves for Alpha Centauri and travels at a speed of $v = 0.95c$ relative to the earth, by how much will the passengers have aged, according to their own clock, when they reach their destination? Assume that the earth and Alpha Centauri are stationary with respect to one another.

Reasoning The two events in this problem are the departure from earth and the arrival at Alpha Centauri. At departure, earth is just outside the spaceship. Upon arrival at the destination, Alpha Centauri is just outside. Therefore, relative to the passengers, the two events occur at the same place—namely, just outside the spaceship. Thus, the passengers measure the proper time interval Δt_0 on their clock, and it is this interval that we must find. For a person left behind on earth, the events occur at *different places*, so such a person measures the dilated time interval Δt rather than the proper time interval. To find Δt we note that the time to travel a given distance is inversely proportional to the speed. Since it takes 4.3 years to traverse the distance between earth and Alpha Centauri at the speed of light, it would take even longer at the slower speed of $v = 0.95c$. Thus, a person on earth measures the dilated time interval to be $\Delta t = (4.3 \text{ years})/0.95 = 4.5$ years. This value can be used with the time-dilation equation to find the proper time interval Δt_0.

Problem-Solving Insight In dealing with time dilation, decide which interval is the proper time interval as follows: (1) Identify the two events that define the interval. (2) Determine the reference frame in which the events occur at the same place; an observer at rest in this frame measures the proper time interval Δt_0.

Solution Using the time-dilation equation, we find that the proper time interval by which the passengers judge their own aging is

$$\Delta t_0 = \Delta t\sqrt{1 - \frac{v^2}{c^2}} = (4.5 \text{ years})\sqrt{1 - \left(\frac{0.95c}{c}\right)^2} = \boxed{1.4 \text{ years}}$$

Thus, the people aboard the rocket will have aged by only 1.4 years when they reach Alpha Centauri, and not the 4.5 years an earth-bound observer has calculated.

Verification of Time Dilation

A striking confirmation of time dilation was achieved in 1971 by an experiment carried out by J. C. Hafele and R. E. Keating.* They transported very precise cesium-beam atomic clocks around the world on commercial jets. Since the speed of a jet plane is considerably less than c, the time-dilation effect is extremely small. However, the atomic clocks were

*J. C. Hafele and R. E. Keating, "Around-the-World Atomic Clocks: Observed Relativistic Time Gains," *Science*, Vol. 177, July 14, 1972, p. 168.

accurate to about $\pm 10^{-9}$ s, so the effect could be measured. The clocks were in the air for 45 hours, and their times were compared to reference atomic clocks kept on earth. The experimental results revealed that, within experimental error, the readings on the clocks on the planes were different from those on earth by an amount that agreed with the prediction of relativity.

The behavior of subatomic particles called *muons* provides additional confirmation of time dilation. These particles are created high in the atmosphere, at altitudes of about 10 000 m. When at rest, muons exist only for about 2.2×10^{-6} s before disintegrating. With such a short lifetime, these particles could never make it down to the earth's surface, even traveling at nearly the speed of light. However, *a large number of muons do reach the earth.* The only way they can do so is to live longer because of time dilation, as Example 3 illustrates.

EXAMPLE 3 | The Lifetime of a Muon

The average lifetime of a muon at rest is 2.2×10^{-6} s. A muon created in the upper atmosphere, thousands of meters above sea level, travels toward the earth at a speed of $v = 0.998c$. Find, on the average, **(a)** how long a muon lives according to an observer on earth, and **(b)** how far the muon travels before disintegrating.

Reasoning The two events of interest are the generation and subsequent disintegration of the muon. When the muon is at rest, these events occur at the same place, so the muon's average (at rest) lifetime of 2.2×10^{-6} s is a proper time interval Δt_0. When the muon moves at a speed $v = 0.998c$ relative to the earth, an observer on the earth measures a dilated lifetime Δt that is given by Equation 28.1. The average distance x traveled by a muon, as measured by an earth observer, is equal to the muon's speed times the dilated time interval.

> **Problem-Solving Insight** **The proper time interval Δt_0 is always shorter than the dilated time interval Δt.**

Solution **(a)** The observer on earth measures a dilated lifetime. Using the time-dilation equation, we find that

$$\Delta t = \frac{\Delta t_0}{\sqrt{1-\frac{v^2}{c^2}}} = \frac{2.2 \times 10^{-6}\ \text{s}}{\sqrt{1-\left(\frac{0.998c}{c}\right)^2}} = \boxed{35 \times 10^{-6}\ \text{s}} \qquad \textbf{(28.1)}$$

(b) The distance traveled by the muon before it disintegrates is

$$x = v\,\Delta t = (0.998)(3.00 \times 10^8\ \text{m/s})(35 \times 10^{-6}\ \text{s}) = \boxed{1.0 \times 10^4\ \text{m}}$$

Thus, the dilated, or extended, lifetime provides sufficient time for the muon to reach the surface of the earth. If its lifetime were only 2.2×10^{-6} s, a muon would travel only 660 m before disintegrating and could never reach the earth.

Example 4 illustrates how a person subjected to time dilation not only perceives extended time relative to another reference frame, but is subjected to the additional aging (biological processes) that would occur in such a timespan.

EXAMPLE 4 | BIO The Physics of Space Travel Revisited

Relativistic time dilation affects not only the operation of clocks, but also all physiological and biological processes, including the rates of chemical reactions. This is why a person moving on a spaceship, for example, will age slower than one that stays at home on earth. Imagine an astronaut in a spaceship that is moving away from the earth at 98% the speed of light. Given that it takes the human body only 27 days to replace the entire outer layer of the skin, how many times has the person on earth replaced their skin, if 27 days pass on the spaceship?

Reasoning We recognize this to be an example of relativistic time dilation. We can apply Equation 28.1 to calculate the dilated time interval. Since the astronaut measures the time on his clock at the same location (inside the ship), he is measuring the proper time interval Δt_0. The ship is moving relative to the reference frame of the earth, so the earth-bound observer measures the dilated time interval.

Solution Beginning with Equation 28.1, we have:

$$\Delta t = \frac{\Delta t_0}{\sqrt{1-\frac{v^2}{c^2}}}.$$

Since we want to calculate the dilated time interval in days, we keep the proper time in those units as well. The speed of the ship is 98% the speed of light, which is $v = 0.98c$. We make these substitutions above and solve for Δt:

$$\Delta t = \frac{27\ \text{days}}{\sqrt{1-\frac{(0.98c)^2}{c^2}}} = \frac{27\ \text{days}}{\sqrt{1-(0.98)^2}} = \boxed{(5.0)(27\ \text{days})}$$

The length of time that passes on earth during the 27 days on the ship is five times longer. Therefore, the person on earth will have replaced their skin five times!

Check Your Understanding

(The answers are given at the end of the book.)

1. Which one of the following statements concerning the dilated time interval is false? **(a)** It is always greater than the proper time interval. **(b)** It depends on the relative speed between the observers who measure the proper and dilated time intervals. **(c)** It depends on the speed of light in a vacuum. **(d)** It is measured by an observer who is at rest with respect to the events that define the time interval.

2. A baseball player at home plate hits a pop fly straight up (the beginning event) that is caught by the catcher at home plate (the ending event). Which one or more of the following observers record(s) the proper time interval between the events? **(a)** A person sitting in the stands **(b)** A person watching the game on TV **(c)** The pitcher running in to cover the play

3. A playground carousel is a circular platform that can rotate about an axis perpendicular to the plane of the platform at its center. It behaves approximately as an inertial reference frame. An observer is looking down at this platform from an inertial reference frame directly above the rotational axis. Three clocks are attached to the platform. Clock A is attached directly to the axis. Clock B is attached to a point midway between the axis and the outer edge of the platform. Clock C is attached to the outer edge of the platform. Rank the clocks according to how slow the observer finds them to be running (slowest first).

28.4 The Relativity of Length: Length Contraction

Because of time dilation, observers moving at a constant velocity relative to each other measure different time intervals between two events. For instance, Example 2 in the previous section illustrates that a trip from earth to Alpha Centauri at a speed of $v = 0.95c$ takes 4.5 years according to a clock on earth, but only 1.4 years according to a clock in the rocket. These two times differ by the factor $\sqrt{1 - v^2/c^2}$. Since the times for the trip are different, one might ask whether the observers measure different distances between earth and Alpha Centauri. The answer, according to special relativity, is yes. After all, both the earth-based observer and the rocket passenger agree that the relative speed between the rocket and earth is $v = 0.95c$. Since speed is distance divided by time and the time is different for the two observers, it follows that the distances must also be different, if the relative speed is to be the same for both individuals. Thus, the earth observer determines the distance to Alpha Centauri to be $L_0 = v\,\Delta t = (0.95c)(4.5 \text{ years}) = 4.3$ light-years. On the other hand, a passenger aboard the rocket finds the distance is only $L = v\,\Delta t_0 = (0.95c)(1.4 \text{ years}) = 1.3$ light-years. The passenger, measuring the shorter time, also measures the shorter distance. This shortening of the distance between two points is one example of a phenomenon known as **length contraction.**

The relation between the distances measured by two observers in relative motion at a constant velocity can be obtained with the aid of **Interactive Figure 28.6.** Part *a* of the drawing shows the situation from the point of view of the earth-based observer. This person measures the time of the trip to be Δt, the distance to be L_0, and the relative speed of the rocket to be $v = L_0/\Delta t$. Part *b* of the drawing presents the point of view of the passenger, for whom the rocket is at rest, and the earth and Alpha Centauri appear to move by at a speed v. The passenger determines the distance of the trip to be L, the time to be Δt_0, and the relative speed to be $v = L/\Delta t_0$. Since the relative speed computed by the passenger equals that computed by the earth-based observer, it follows that $v = L/\Delta t_0 = L_0/\Delta t$. Using this result and the time-dilation equation, Equation 28.1, we obtain the following relation between L and L_0:

Length contraction

$$L = L_0\sqrt{1 - \frac{v^2}{c^2}} \qquad \textbf{(28.2)}$$

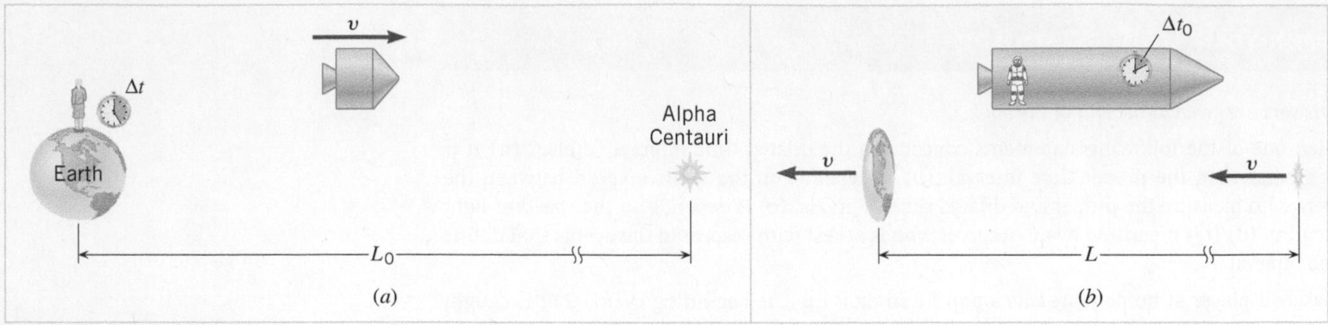

INTERACTIVE FIGURE 28.6 (*a*) As measured by an observer on the earth, the distance to Alpha Centauri is L_0, and the time required to make the trip is Δt. (*b*) According to the passenger on the spacecraft, the earth and Alpha Centauri move with speed v relative to the craft. The passenger measures the distance and time of the trip to be L and Δt_0, respectively, both quantities being less than those in part *a*.

The length L_0 is called the **proper length**; it is the length (or distance) between two points *as measured by an observer at rest with respect to them.* Since v is less than c, the term $\sqrt{1 - v^2/c^2}$ is less than 1, and L is less than L_0. It is important to note that this length contraction occurs only along the direction of the motion. Those dimensions that are perpendicular to the motion are not shortened, as the next example discusses.

EXAMPLE 5 | The Contraction of a Spacecraft

An astronaut, using a meter stick that is at rest relative to a cylindrical spacecraft, measures the length and diameter of the spacecraft to be 82 m and 21 m, respectively. The spacecraft moves with a constant speed of $v = 0.95c$ relative to the earth, as in **Interactive Figure 28.6**. What are the dimensions of the spacecraft, as measured by an observer on earth?

Reasoning The length of 82 m is a proper length L_0, since it is measured using a meter stick that is at rest relative to the spacecraft. The length L measured by the observer on earth can be determined from the length-contraction formula, Equation 28.2. On the other hand, the diameter of the spacecraft is perpendicular to the motion, so the earth observer does not measure any change in the diameter.

Problem-Solving Insight The proper length L_0 is always larger than the contracted length L.

Solution The length L of the spacecraft, as measured by the observer on earth, is

$$L = L_0\sqrt{1 - \frac{v^2}{c^2}} = (82\text{ m})\sqrt{1 - \left(\frac{0.95c}{c}\right)^2} = \boxed{26\text{ m}}$$

Both the astronaut and the observer on earth measure the same value for the diameter of the spacecraft: Diameter = $\boxed{21\text{ m}}$. **Interactive Figure 28.6*a*** shows the size of the spacecraft as measured by the earth observer, and part *b* shows the size as measured by the astronaut.

When dealing with relativistic effects we need to distinguish carefully between the criteria for the proper time interval and the proper length. The proper time interval Δt_0 between two events is the time interval measured by an observer who is at rest relative to the events and sees them occurring at the *same place.* All other moving inertial observers will measure a larger value for this time interval. The proper length L_0 of an object is the length measured by an observer who is *at rest* with respect to the object. All other moving inertial observers will measure a shorter value for this length. The observer who measures the proper time interval may not be the same one who measures the proper length. For instance, **Interactive Figure 28.6** shows that the astronaut measures the proper time interval Δt_0 for the trip between earth and Alpha Centauri, whereas the earth-based observer measures the proper length (or distance) L_0 for the trip.

It should be emphasized that the word "proper" in the phrases "proper time" and "proper length" does *not* mean that these quantities are the correct or preferred quantities in any absolute sense. If this were so, the observer measuring these quantities would be using a preferred reference frame for making the measurement, a situation that is prohibited by the relativity postulate. According to this postulate, there is no preferred inertial reference frame. When two observers are moving relative to each other at a constant velocity, each measures the other person's clock to run more slowly than his own, and each measures the other person's length, along that person's motion, to be contracted.

Check Your Understanding

(*The answers are given at the end of the book.*)

4. If the speed c of light in a vacuum were infinitely large instead of 3.0×10^8 m/s, would the effects of time dilation and length contraction be observable?

5. Suppose that you are standing at a railroad crossing, watching a train go by. Both you and a passenger in the train are looking at a clock on the train. Who measures the proper time interval, and who measures the proper length of a train car? **(a)** You measure the proper time interval, and the passenger measures the proper length. **(b)** You measure both the proper time interval and the proper length. **(c)** The passenger measures both the proper time interval and the proper length. **(d)** You measure the proper length, and the passenger measures the proper time interval.

6. Which one or more of the following quantities will two observers always measure to be the *same*, regardless of the relative velocity between the observers? **(a)** The time interval between two events **(b)** The length of an object **(c)** The speed of light in a vacuum **(d)** The relative speed between the observers

7. **CYU Figure 28.1** shows an object that has the shape of a square when it is at rest in inertial reference frame R. When the object moves relative to this reference frame, the object's velocity vector is in the plane of the square and is parallel to the diagonal AC. Since the speed of the motion is an appreciable fraction of the speed of light in a vacuum, noticeable length contraction occurs. Does an observer in reference frame R see the object as a square? (*Hint:* Consider what happens to each of the diagonals AC and BD.)

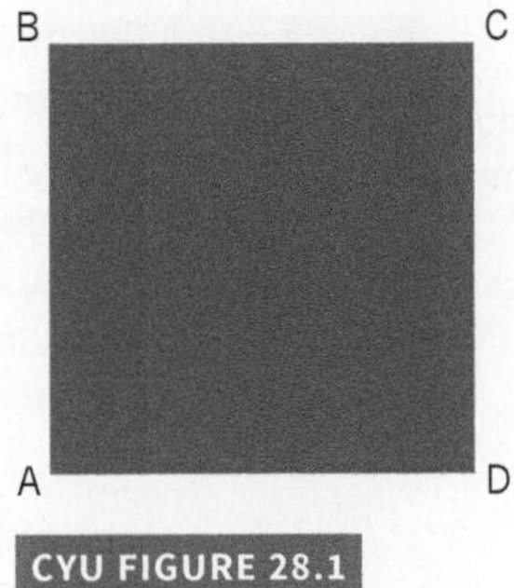

CYU FIGURE 28.1

28.5 Relativistic Momentum

Thus far we have discussed how time intervals and distances between two events are measured by observers moving at a constant velocity relative to each other. Special relativity also alters our ideas about momentum and energy.

Recall from Section 7.2 that when two or more objects interact, the principle of conservation of linear momentum applies if the system of objects is isolated. This principle states that the total linear momentum of an isolated system remains constant at all times. (An isolated system is one in which the sum of the external forces acting on the objects is zero.) The conservation of linear momentum is a law of physics and, in accord with the relativity postulate, is valid in all inertial reference frames. That is, when the total linear momentum is conserved in one inertial reference frame, it is conserved in all inertial reference frames.

As an example of momentum conservation, suppose that several people are watching two billiard balls collide on a frictionless pool table. One person is standing next to the pool table, and the other is moving past the table with a constant velocity. Since the two balls constitute an isolated system, the relativity postulate requires that both observers must find the total linear momentum of the two-ball system to be the same before, during, and after the collision. For this kind of situation, Section 7.1 defines the classical linear momentum $\vec{\mathbf{p}}$ of an object to be the product of its mass m and velocity $\vec{\mathbf{v}}$. As a result, the magnitude of the classical momentum is $p = mv$. As long as the speed of an object is considerably smaller than the speed of light, this definition is adequate. However, when the speed approaches the speed of light, an analysis of the collision shows that the total linear momentum is not conserved in all inertial reference frames if one defines linear momentum simply as the product of mass and velocity. In order to preserve the conservation of linear momentum, it is necessary to modify this definition.

The theory of special relativity reveals that the magnitude of the **relativistic momentum** must be defined as in Equation 28.3:

Magnitude of the relativistic momentum

$$p = \frac{mv}{\sqrt{1 - \frac{v^2}{c^2}}} \tag{28.3}$$

The total relativistic momentum of an isolated system is conserved in all inertial reference frames.

From Equation 28.3, we can see that the magnitudes of the relativistic and nonrelativistic momenta differ by the same factor of $\sqrt{1 - v^2/c^2}$ that occurs in the time-dilation and length-contraction equations. Since this factor is always less than 1 and occurs in the denominator in Equation 28.3, the relativistic momentum is always larger than the nonrelativistic momentum. To illustrate how the two quantities differ as the speed v increases, **Figure 28.7** shows a plot of the ratio of the momentum magnitudes (relativistic/nonrelativistic) as a function of v. According to Equation 28.3, this ratio is just $1/\sqrt{1 - v^2/c^2}$. The graph shows that for speeds attained by ordinary objects, such as cars and planes, the relativistic and nonrelativistic momenta are almost equal because their ratio is nearly 1. Thus, at speeds much less than the speed of light, either the nonrelativistic momentum or the relativistic momentum can be used to describe collisions. On the other hand, when the speed of the object becomes comparable to the speed of light, the relativistic momentum becomes significantly greater than the nonrelativistic momentum and must be used. Example 6 deals with the relativistic momentum of an electron traveling close to the speed of light.

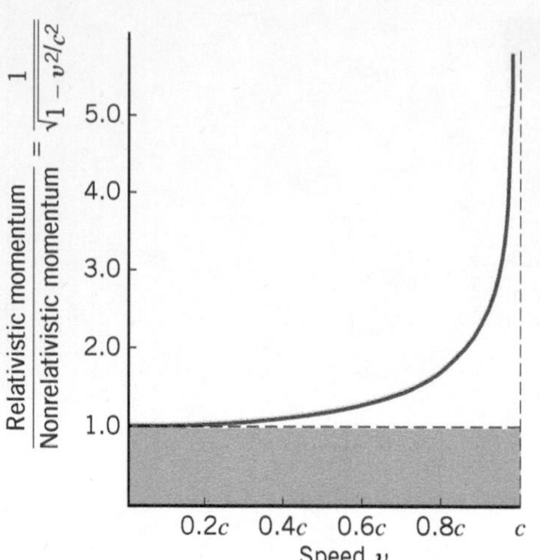

FIGURE 28.7 This graph shows how the ratio of the magnitude of the relativistic momentum to the magnitude of the nonrelativistic momentum increases as the speed of an object approaches the speed of light.

EXAMPLE 6 | The Relativistic Momentum of a High-Speed Electron

The particle accelerator at Stanford University (**Figure 28.8**) is 3 km long and accelerates electrons to a speed of 0.999 999 999 7c, which is very nearly equal to the speed of light. Find the magnitude of the relativistic momentum of an electron that emerges from the accelerator, and compare it with the nonrelativistic value.

Reasoning and Solution The magnitude of the electron's relativistic momentum can be obtained from Equation 28.3 if we recall that the mass of an electron is $m = 9.11 \times 10^{-31}$ kg:

$$p = \frac{mv}{\sqrt{1 - \frac{v^2}{c^2}}} = \frac{(9.11 \times 10^{-31}\text{ kg})(0.999\,999\,999\,7c)}{\sqrt{1 - \frac{(0.999\,999\,999\,7c)^2}{c^2}}}$$

$$= \boxed{1 \times 10^{-17}\text{ kg}\cdot\text{m/s}}$$

This value for the magnitude of the momentum agrees with the value measured experimentally. The relativistic momentum is greater than the nonrelativistic momentum by a factor of

$$\frac{1}{\sqrt{1 - \frac{v^2}{c^2}}} = \frac{1}{\sqrt{1 - \frac{(0.999\,999\,999\,7c)^2}{c^2}}} = \boxed{4 \times 10^4}$$

B.W.Marsh/Science Source

FIGURE 28.8 The Stanford 3-km linear accelerator accelerates electrons almost to the speed of light.

Math Skills Using your calculator to obtain the answers in Example 6 may not be possible due to the term $\sqrt{1-\frac{v^2}{c^2}}$ in the pertinent equations. The potential difficulty is that your calculator may not let you enter a number such as 0.999 999 999 7, because it cannot accept ten places after the decimal point. Since the value for v is nearly equal to c and $\frac{v}{c} \approx 1$, one way to deal with such a situation is to note that

$$\sqrt{1-\frac{v^2}{c^2}} = \sqrt{\left(1+\frac{v}{c}\right)\left(1-\frac{v}{c}\right)} \approx \sqrt{2\left(1-\frac{v}{c}\right)}$$

Thus, we find that

$$\frac{1}{\sqrt{1-\frac{v^2}{c^2}}} \approx \frac{1}{\sqrt{2\left(1-\frac{v}{c}\right)}}$$

$$= \frac{1}{\sqrt{2\left(1-\frac{0.999\,999\,999\,7\not{c}}{\not{c}}\right)}} = \frac{1}{\sqrt{2(3\times10^{-10})}}$$

$$\approx \boxed{4\times10^4}$$

In this answer the speed v is given to ten significant figures. However, the answer is given to only one significant figure! This is because the subtraction causes a loss of significant figures as follows:

$$1 - 0.999\,999\,999\,7 = 0.000\,000\,000\,3 = 3\times10^{-10}$$

The amount of time it would take for relativistic effects to become appreciable while accelerating at $g = 9.8 \text{ m/s}^2$ is explored in Example 7.

EXAMPLE 7 | BIO Artificial Gravity for Long Voyages

Weightlessness, or microgravity, can have detrimental effects on the human body. Short-term symptoms, including disorientation and nausea, can be overcome in a short time so that an astronaut can function over the course of a mission. Long-term effects, including muscle and bone atrophy, and the possible deterioration of cognitive ability, are highly undesirable, and need to be avoided in future long voyages. In Chapter 5, we encountered examples of how "artificial gravity" on a space station can be generated using rotational motion. However, in order for a spacecraft to make long voyages at high velocities relative to earth, linear accelerations are required. If the linear acceleration of a ship was $g = 9.8 \text{ m/s}^2$, a person standing on the "floor" of the ship would feel as if they were standing on earth, and would not experience any of the weightlessness-induced effects. It is interesting to know how long it would take a spacecraft accelerating at g to reach relativistic speeds. Figure 28.7 gives a graphical measure of how relativistic and nonrelativistic momenta differ, and we can estimate that relativistic effects become appreciable when $v \approx 0.30c$, that is, where the relativistic and nonrelativistic momenta differ by about 5%. Use nonrelativistic kinematics for constant acceleration ($a = g = 9.8 \text{ m/s}^2$) to estimate **(a)** the time it takes a ship to accelerate from rest to $v = 0.30c$, and **(b)** the distance it travels during that time.

Reasoning **(a)** Since the acceleration is constant, we can apply the kinematic equations, specifically Equation 2.4, with $a = g$ and the initial velocity set to 0 m/s, since the ship starts from rest. **(b)** With the time calculated in part (a), and the initial velocity again set to 0 m/s, we can apply Equation 2.8 to find the total distance traveled.

Solution **(a)** Using Equation 2.4 with the initial velocity v_0 set to 0 m/s, we have

$$v = \not{v_0} + at \quad\Rightarrow\quad t = \frac{v}{a} = \frac{(0.30)c}{a} = \frac{(0.30)(3.0\times10^8 \text{ m/s})}{9.8 \text{ m/s}^2}$$

$$= \boxed{9.2\times10^6 \text{ s}}$$

This is equal to about 106 days.

(b) We know the time from part (a) and that the initial velocity is 0 m/s. Using Equation 2.8 to find the distance traveled during this time, we have

$$x = \not{v_0}t + \tfrac{1}{2}at^2 = \tfrac{1}{2}(9.8 \text{ m/s}^2)(9.2\times10^6 \text{ s})^2 = \boxed{4.1\times10^{14} \text{ m}}$$

or 4.1×10^{11} km = 0.043 ly. For comparison, the radius of the solar system is about 4.5×10^9 km.

Although constant-acceleration (or constant-thrust) propulsion would be a quick way to travel, there are challenges that make it currently unfeasible. For instance, a large, constant thrust for an extended period of time would require a prohibitively large amount of fuel.

28.6 The Equivalence of Mass and Energy

The Total Energy of an Object

One of the most astonishing results of special relativity is that mass and energy are equivalent, in the sense that a gain or loss of mass can be regarded equally well as a gain or loss of energy. Consider, for example, an object of mass m traveling at a speed v. Einstein showed that the **total energy** E of the moving object is related to its mass and speed by the following relation:

Total energy of an object

$$E = \frac{mc^2}{\sqrt{1 - \frac{v^2}{c^2}}} \tag{28.4}$$

To gain some understanding of Equation 28.4, consider the special case in which the object is at rest. When $v = 0$ m/s, the total energy is called the **rest energy** E_0, and Equation 28.4 reduces to Einstein's now-famous equation:

Rest energy of an object

$$E_0 = mc^2 \tag{28.5}$$

The rest energy represents the energy equivalent of the mass of an object at rest. As Example 8 shows, even a small mass is equivalent to an enormous amount of energy.

Analyzing Multiple-Concept Problems

EXAMPLE 8 | The Energy Equivalent of a Golf Ball

A 0.046-kg golf ball is lying on the green, as **Figure 28.9** illustrates. If the rest energy of this ball were used to operate a 75-W light bulb, how long would the bulb remain lit?

Reasoning The average power delivered to the light bulb is 75 W, which means that it uses 75 J of energy per second. Therefore, the time that the bulb would remain lit is equal to the total energy used by the light bulb divided by the energy per second (i.e., the average power) delivered to it. This energy comes from the rest energy of the golf ball, which is equal to its mass times the speed of light squared.

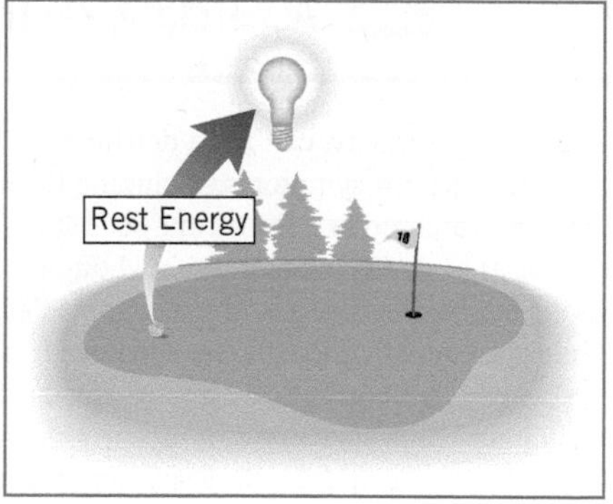

FIGURE 28.9 The rest energy of a golf ball is sufficient to keep a 75-W light bulb burning for an incredibly long time (see Example 8).

Knowns and Unknowns The data for this problem are:

Description	Symbol	Value
Mass of golf ball	m	0.046 kg
Average power delivered to light bulb	$\overline{P}$	75 W
Unknown Variable		
Time that light bulb would remain lit	t	?

Modeling the Problem

STEP 1 Power The average power $\overline{P}$ is equal to the energy delivered to the light bulb divided by the time t (see Section 6.7 and Equation 6.10b), or $\overline{P} = \text{Energy}/t$. In this case the energy comes from the rest energy E_0 of the golf ball, so $\overline{P} = E_0/t$. Solving for the time gives Equation 1 at the right. The average power is known, and the rest energy will be evaluated in Step 2.

$$t = \frac{E_0}{\overline{P}} \tag{1}$$

?

STEP 2 Rest Energy The rest energy E_0 is the total energy of the golf ball as it rests on the green. If the golf ball's mass is m, then its rest energy is

$$\boxed{E_0 = mc^2} \tag{28.5}$$

where c is the speed of light in a vacuum. Both m and c are known, so we substitute this expression for the rest energy into Equation 1, as indicated at the right.

$$t = \frac{E_0}{P} \tag{1}$$

$$\boxed{E_0 = mc^2} \tag{28.5}$$

Solution Algebraically combining the results of each step, we have

$$t \overset{\text{STEP 1}}{=} \frac{E_0}{P} \overset{\text{STEP 2}}{=} \frac{mc^2}{P}$$

The time that the light bulb would remain lit is

$$t = \frac{mc^2}{P} = \frac{(0.046\text{ kg})(3.0 \times 10^8\text{ m/s})^2}{75\text{ W}} = \boxed{5.5 \times 10^{13}\text{ s}}$$

Expressed in years (1 yr = 3.2×10^7 s), this time is equivalent to

$$(5.5 \times 10^{13}\text{ s})\left(\frac{1\text{ yr}}{3.2 \times 10^7\text{ s}}\right) = 1.7 \times 10^6\text{ yr or 1.7 million years!}$$

Related Homework: Problems 27, 28

When an object is accelerated from rest to a speed v, the object acquires kinetic energy in addition to its rest energy. The total energy E is the sum of the rest energy E_0 and the kinetic energy KE, or $E = E_0 + \text{KE}$. Therefore, the kinetic energy is the difference between the object's total energy and its rest energy. Using Equations 28.4 and 28.5, we can write the kinetic energy as

Kinetic energy of an object

$$\text{KE} = E - E_0 = mc^2\left(\frac{1}{\sqrt{1-\frac{v^2}{c^2}}} - 1\right) \tag{28.6}$$

This equation is the relativistically correct expression for the kinetic energy of an object of mass m moving at speed v.

Equation 28.6 looks nothing like the kinetic energy expression introduced in Section 6.2—namely, $\text{KE} = \frac{1}{2}mv^2$ (Equation 6.2). However, for speeds much less than the speed of light ($v \ll c$), the relativistic equation for the kinetic energy reduces to $\text{KE} = \frac{1}{2}mv^2$, as can be seen by using the binomial expansion* to represent the square root term in Equation 28.6:

$$\frac{1}{\sqrt{1-\frac{v^2}{c^2}}} = 1 + \frac{1}{2}\left(\frac{v^2}{c^2}\right) + \frac{3}{8}\left(\frac{v^2}{c^2}\right)^2 + \cdots$$

Suppose that v is much smaller than c—say, $v = 0.01c$. The second term in the expansion has the value $\frac{1}{2}(v^2/c^2) = 5.0 \times 10^{-5}$, while the third term has the much smaller value $\frac{3}{8}(v^2/c^2)^2 = 3.8 \times 10^{-9}$. The additional terms are smaller still, so if $v \ll c$, we can neglect the third and additional terms in comparison with the first and second terms. Substituting the first two terms into Equation 28.6 gives

$$\text{KE} \approx mc^2\left(1 + \frac{1}{2}\frac{v^2}{c^2} - 1\right) = \frac{1}{2}mv^2$$

which is the familiar form for the kinetic energy. However, Equation 28.6 gives the correct kinetic energy for all speeds and must be used for speeds near the speed of light, as in Example 9.

*The binomial expansion states that $(1 - x)^n = 1 - nx + n(n - 1)x^2/2 + \cdots$. In our case, $x = v^2/c^2$ and $n = -1/2$.

EXAMPLE 9 | A High-Speed Electron

An electron ($m = 9.109 \times 10^{-31}$ kg) is accelerated from rest to a speed of $v = 0.9995c$ in a particle accelerator. Determine the electron's **(a)** rest energy, **(b)** total energy, and **(c)** kinetic energy in millions of electron volts or MeV.

Reasoning and Solution **(a)** The electron's rest energy is

$$E_0 = mc^2 = (9.109 \times 10^{-31}\ \text{kg})(2.998 \times 10^8\ \text{m/s})^2$$

$$= 8.187 \times 10^{-14}\ \text{J} \qquad \textbf{(28.5)}$$

Since 1 eV = 1.602×10^{-19} J, the electron's rest energy is

$$(8.187 \times 10^{-14}\ \text{J})\left(\frac{1\ \text{eV}}{1.602 \times 10^{-19}\ \text{J}}\right) = \boxed{5.11 \times 10^5\ \text{eV or } 0.511\ \text{MeV}}$$

(b) The total energy of an electron traveling at a speed of $v = 0.9995c$ is

$$E = \frac{mc^2}{\sqrt{1 - \frac{v^2}{c^2}}} = \frac{(9.109 \times 10^{-31}\ \text{kg})(2.998 \times 10^8\ \text{m/s})^2}{\sqrt{1 - \left(\frac{0.9995c}{c}\right)^2}} \qquad \textbf{(28.4)}$$

$$= \boxed{2.59 \times 10^{-12}\ \text{J or } 16.2\ \text{MeV}}$$

(c) The kinetic energy is the difference between the total energy and the rest energy:

$$\text{KE} = E - E_0 = 2.59 \times 10^{-12}\ \text{J} - 8.2 \times 10^{-14}\ \text{J} \qquad \textbf{(28.6)}$$

$$= \boxed{2.51 \times 10^{-12}\ \text{J or } 15.7\ \text{MeV}}$$

For comparison, if the kinetic energy of the electron had been calculated from $\frac{1}{2}mv^2$, a value of only 0.255 MeV would have been obtained.

Since mass and energy are equivalent, any change in one is accompanied by a corresponding change in the other. For instance, life on earth is dependent on electromagnetic energy (light) from the sun. Because this energy is leaving the sun (see **Figure 28.10**), there is a decrease in the sun's mass. Example 10 illustrates how to determine this decrease.

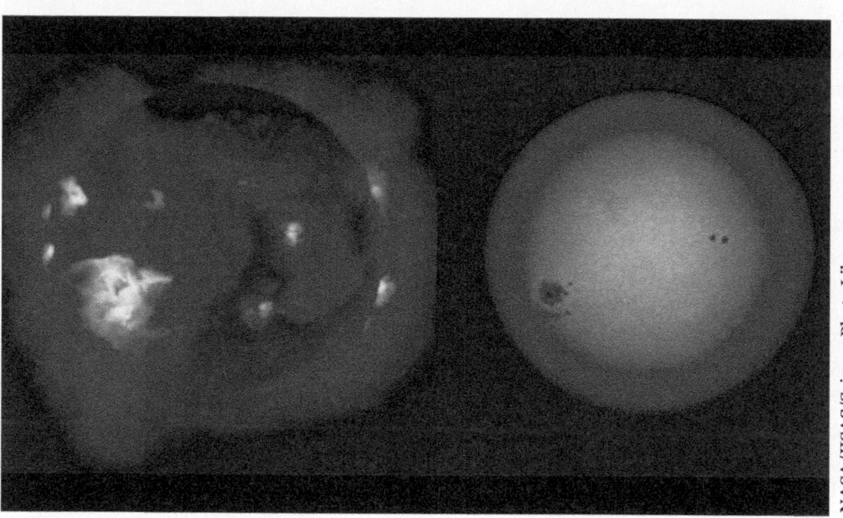

FIGURE 28.10 The sun emits electromagnetic energy over a broad portion of the electromagnetic spectrum. These photographs were obtained using that energy in the indicated regions of the spectrum.

EXAMPLE 10 | The Sun Is Losing Mass

The sun radiates electromagnetic energy at the rate of 3.92×10^{26} W. **(a)** What is the change in the sun's mass during each second that it is radiating energy? **(b)** The mass of the sun is 1.99×10^{30} kg. What fraction of the sun's mass is lost during a human lifetime of 75 years?

Reasoning Since 1 W = 1 J/s, the amount of electromagnetic energy radiated during each second is 3.92×10^{26} J. Thus, during each second, the sun's rest energy decreases by this amount. The change ΔE_0 in the sun's rest energy is related to the change Δm in its mass by $\Delta E_0 = (\Delta m)c^2$, according to Equation 28.5.

Solution **(a)** For each second that the sun radiates energy, the change in its mass is

$$\Delta m = \frac{\Delta E_0}{c^2} = \frac{3.92 \times 10^{26}\ \text{J}}{(3.00 \times 10^8\ \text{m/s})^2} = \boxed{4.36 \times 10^9\ \text{kg}}$$

Over 4 billion kilograms of mass are lost by the sun during each second.

(b) The amount of mass lost by the sun in 75 years is

$$\Delta m = (4.36 \times 10^9\ \text{kg/s})\left(\frac{3.16 \times 10^7\ \text{s}}{1\ \text{year}}\right)(75\ \text{years}) = 1.0 \times 10^{19}\ \text{kg}$$

Although this is an enormous amount of mass, it represents only a tiny fraction of the sun's total mass:

$$\frac{\Delta m}{m_{\text{sun}}} = \frac{1.0 \times 10^{19}\ \text{kg}}{1.99 \times 10^{30}\ \text{kg}} = \boxed{5.0 \times 10^{-12}}$$

Any change ΔE_0 in the rest energy of a system causes a change in the mass of the system according to $\Delta E_0 = (\Delta m)c^2$. It does not matter whether the change in energy is due to a change in electromagnetic energy, potential energy, thermal energy, or so on. Although any change in energy gives rise to a change in mass, in most instances the change in mass is too small to be detected. For instance, when 4186 J of heat is used to raise the temperature of 1 kg of water by 1 °C, the mass changes by only $\Delta m = \Delta E_0/c^2 = (4186 \text{ J})/(3.00 \times 10^8 \text{ m/s})^2 = 4.7 \times 10^{-14}$ kg. Conceptual Example 11 illustrates further how a change in the energy of an object leads to an equivalent change in its mass.

CONCEPTUAL EXAMPLE 11 | When Is a Massless Spring Not Massless?

Interactive Figure 28.11*a* shows a top view of a *massless* spring on a horizontal table. Initially the spring is unstrained. Then the spring is either stretched or compressed by an amount x from its unstrained length, as **Interactive Figure 28.11*b*** illustrates. What is the mass of the spring in **Interactive Figure 28.11*b***? **(a)** It is greater than zero by an amount that is larger when the spring is stretched. **(b)** It is greater than zero by an amount that is larger when the spring is compressed. **(c)** It is greater than zero by an amount that is the same when the spring is stretched or compressed. **(d)** It remains zero.

INTERACTIVE FIGURE 28.11 (*a*) This spring is unstrained and assumed to have no mass. (*b*) When the spring is either stretched or compressed by an amount x, it gains elastic potential energy and, hence, mass.

Reasoning When a spring is stretched or compressed, its elastic potential energy changes. As discussed in Section 10.3, the elastic potential energy of an ideal spring is equal to $\frac{1}{2}kx^2$, where k is the spring constant and x is the amount of stretch or compression. Consistent with the theory of special relativity, any change in the total energy of a system, including a change in the elastic potential energy, is equivalent to a change in the mass of the system.

Answers (a), (b), and (d) are incorrect. In being stretched or compressed by the same amount x, the spring acquires the same amount of elastic potential energy ($\frac{1}{2}kx^2$). Therefore, according to special relativity, the spring acquires the same mass regardless of whether it is stretched or compressed, so these answers must be incorrect.

Answer (c) is correct. The spring acquires elastic potential energy in being stretched or compressed. Special relativity indicates that this additional energy is equivalent to additional mass. Since the amount of stretch or compression is the same, the potential energy is the same in either case, and so is the additional mass.

It is also possible to transform matter itself into other forms of energy. For example, the positron (see Section 31.4) has the same mass as an electron but an opposite electrical charge. If these two particles of matter collide, they are completely annihilated, and a burst of high-energy electromagnetic waves is produced. Thus, matter is transformed into electromagnetic waves, the energy of the electromagnetic waves being equal to the total energies of the two colliding particles. The medical diagnostic technique known as positron emission tomography or PET scanning depends on the electromagnetic energy produced when a positron and an electron are annihilated (see Section 32.6).

The transformation of electromagnetic waves into matter also happens. In one experiment, an extremely high-energy electromagnetic wave, called a gamma ray (see Section 31.4), passes close to the nucleus of an atom. If the gamma ray has sufficient energy, it can create an electron and a positron. The gamma ray disappears, and the two particles of matter appear in its place. Except for picking up some momentum, the nearby nucleus remains unchanged. The process in which the gamma ray is transformed into the two particles is known as *pair production.*

The Relation Between Total Energy and Momentum

It is possible to derive a useful relation between the total relativistic energy E and the relativistic momentum p. We begin by rearranging Equation 28.3 for the momentum, to obtain

$$\frac{m}{\sqrt{1 - v^2/c^2}} = \frac{p}{v}$$

With this substitution, Equation 28.4 for the total energy becomes

$$E = \frac{mc^2}{\sqrt{1 - v^2/c^2}} = \frac{pc^2}{v} \quad \text{or} \quad \frac{v}{c} = \frac{pc}{E}$$

Using this expression to replace v/c in Equation 28.4 gives

$$E = \frac{mc^2}{\sqrt{1 - v^2/c^2}} = \frac{mc^2}{\sqrt{1 - p^2c^2/E^2}}$$

Math Skills To solve the equation $E = \frac{mc^2}{\sqrt{1 - p^2c^2/E^2}}$ for E, we begin by squaring both sides:

$$E^2 = \left(\frac{mc^2}{\sqrt{1 - p^2c^2/E^2}}\right)^2 = \frac{m^2c^4}{1 - p^2c^2/E^2} \tag{1}$$

The next step is to isolate the terms containing E^2 on the left side of Equation 1. To do this, we multiply each side of the equation by $1 - p^2c^2/E^2$:

$$E^2(1 - p^2c^2/E^2) = \frac{m^2c^4}{\cancel{1 - p^2c^2/E^2}}\cancel{(1 - p^2c^2/E^2)} \quad \text{or} \quad E^2\left(1 - \frac{p^2c^2}{E^2}\right) = m^2c^4 \tag{2}$$

Multiplying terms on the left side of Equation 2 gives

$$E^2 - p^2c^2 = m^2c^4 \tag{3}$$

Finally, to isolate E^2 on the left side of Equation 3, we add p^2c^2 to both sides of the equation:

$$E^2 - p^2c^2 + (p^2c^2) = (p^2c^2) + m^2c^4 \quad \text{or} \quad E^2 = p^2c^2 + m^2c^4$$

Solving this expression for E^2 reveals that

$$E^2 = p^2c^2 + m^2c^4 \tag{28.7}$$

The Speed of Light in a Vacuum Is the Ultimate Speed

One of the important consequences of the theory of special relativity is that objects with mass cannot reach the speed c of light in a vacuum. Thus, c represents the ultimate speed. To see that this speed limit is a consequence of special relativity, consider Equation 28.6, which gives the kinetic energy of an object moving at a speed v. As v approaches the speed of light c, the $\sqrt{1 - v^2/c^2}$ term in the denominator approaches zero. Hence, the kinetic energy becomes infinitely large. However, the work–energy theorem (Section 6.2) tells us that an infinite amount of work would have to be done to give the object an infinite kinetic energy. Since an infinite amount of work is not available, we are left with the conclusion that objects with mass cannot attain the speed of light c.

Check Your Understanding

(The answers are given at the end of the book.)

8. Consider the same cup of coffee sitting on the same table in the following four situations: **(a)** The coffee is hot (95 °C), and the table is at sea level. **(b)** The coffee is cold (60 °C), and the table is at sea level. **(c)** The coffee is hot (95 °C), and the table is on a mountain top. **(d)** The coffee is cold (60 °C), and the table is on a mountain top. In which situation does the cup of coffee have the greatest mass, and in which the smallest mass?

9. A system consists of two positive charges. Is the total mass of the system greater when the two charges are **(a)** separated by a finite distance or **(b)** infinitely far apart?

10. A parallel plate capacitor is initially uncharged. Then it is fully charged up by removing electrons from one plate and placing them on the other plate. Is the mass of the capacitor greater when it is **(a)** uncharged or **(b)** fully charged?

11. It takes work to accelerate a particle from rest to a given speed close to the speed of light in a vacuum. For which particle is less work required, **(a)** an electron or **(b)** a proton?

28.7 The Relativistic Addition of Velocities

The velocity of an object relative to an observer plays a central role in special relativity, and to determine this velocity, it is sometimes necessary to add two or more velocities together. We first encountered relative velocity in Section 3.4, so we will begin by reviewing some of the ideas presented there. **Figure 28.12** illustrates a truck moving at a constant velocity of $v_{TG} = +15$ m/s relative to an observer standing on the ground, where the plus sign denotes a direction to the right. Suppose someone on the truck throws a baseball toward the observer at a velocity of $v_{BT} = +8.0$ m/s relative to the truck. We might conclude that the observer on the ground would see the ball approaching at a velocity of $v_{BG} = v_{BT} + v_{TG} = 8.0\text{ m/s} + 15\text{ m/s} = +23\text{ m/s}$. These symbols are similar to those used in Section 3.4 and have the following meaning:

$v_{\boxed{BG}}$ = velocity of the **Baseball** relative to the **Ground** = +23 m/s

$v_{\boxed{BT}}$ = velocity of the **Baseball** relative to the **Truck** = +8.0 m/s

$v_{\boxed{TG}}$ = velocity of the **Truck** relative to the **Ground** = +15.0 m/s

Although the result for the velocity of the baseball relative to the ground ($v_{BG} = +23$ m/s) seems reasonable, careful measurements would show that it is not quite right. According to special relativity, the equation $v_{BG} = v_{BT} + v_{TG}$ is not valid for the following reason. If the velocity of the truck had a magnitude sufficiently close to the speed of light in a vacuum, the equation would predict that the observer on the earth could see the baseball moving faster than the speed of light. This is not possible, since no object with a finite mass can move faster than the speed of light in a vacuum.

For the case in which the truck and ball are moving along the same straight line, the theory of special relativity reveals that the velocities are related according to

$$v_{BG} = \frac{v_{BT} + v_{TG}}{1 + \frac{v_{BT}v_{TG}}{c^2}}$$

The subscripts in this equation have been chosen for the specific situation shown in **Figure 28.12**. For the general situation, the relative velocities are related by the following **velocity-addition formula**:

Velocity addition

$$v_{AB} = \frac{v_{AC} + v_{CB}}{1 + \frac{v_{AC}v_{CB}}{c^2}} \qquad (28.8)$$

where all the velocities are assumed to be constant and the symbols have the following meanings:

$v_{\boxed{AB}}$ = velocity of **object A** relative to **object B**

$v_{\boxed{AC}}$ = velocity of **object A** relative to **object C**

$v_{\boxed{CB}}$ = velocity of **object C** relative to **object B**

The ordering of the subscripts in Equation 28.8 follows the discussion in Section 3.4. For motion along a straight line, the velocities can have either positive or negative values,

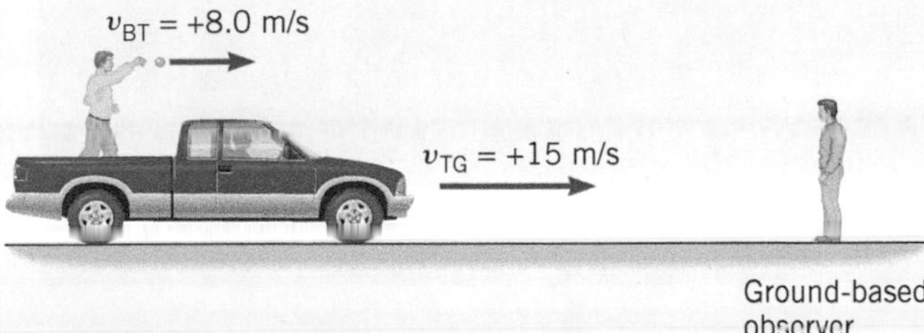

FIGURE 28.12 The truck is approaching the ground-based observer at a relative velocity of $v_{TG} = +15$ m/s. The velocity of the baseball relative to the truck is $v_{BT} = +8.0$ m/s.

depending on whether they are directed along the positive or negative direction. Furthermore, switching the order of the subscripts changes the sign of the velocity, so, for example, $v_{BA} = -v_{AB}$ (see Example 11 in Chapter 3).

Equation 28.8 differs from the nonrelativistic formula ($v_{AB} = v_{AC} + v_{CB}$) by the presence of the $v_{AC}\, v_{CB}/c^2$ term in the denominator. This term arises because of the effects of time dilation and length contraction that occur in special relativity. When v_{AC} and v_{CB} are small compared to c, the $v_{AC}\, v_{CB}/c^2$ term is small compared to 1, so the velocity-addition formula reduces to $v_{AB} \approx v_{AC} + v_{CB}$. However, when either v_{AC} or v_{CB} is comparable to c, the results can be quite different, as Example 12 illustrates.

EXAMPLE 12 | The Relativistic Addition of Velocities

Imagine a hypothetical situation in which the truck in Figure 28.12 is moving relative to the ground with a velocity of $v_{TG} = +0.8c$. A person riding on the truck throws a baseball at a velocity relative to the truck of $v_{TG} = +0.5c$. What is the velocity v_{BG} of the baseball relative to a person standing on the ground?

Reasoning The observer standing on the ground does *not* see the baseball approaching at $v_{BG} = 0.5c + 0.8c = 1.3c$. This cannot be, because the speed of the ball would then exceed the speed of light in a vacuum. The velocity-addition formula gives the correct velocity, which has a magnitude less than the speed of light.

Solution The ground-based observer sees the ball approaching with a velocity of

$$v_{BG} = \frac{v_{BT} + v_{TG}}{1 + \dfrac{v_{BT}v_{TG}}{c^2}} = \frac{0.5c + 0.8c}{1 + \dfrac{(0.5c)(0.8c)}{c^2}} = \boxed{0.93c} \quad \textbf{(28.8)}$$

Example 12 discusses how the speed of a baseball is viewed by observers in different inertial reference frames. The next example deals with a similar situation, except that the baseball is replaced by the light of a laser beam.

CONCEPTUAL EXAMPLE 13 | The Speed of a Laser Beam

Figure 28.13 shows an intergalactic cruiser approaching a hostile spacecraft. Both vehicles move at a constant velocity. The velocity of the cruiser relative to the spacecraft is $v_{CS} = +0.7c$, the direction to the right being the positive direction. The cruiser fires a beam of laser light at the hostile renegades. The velocity of the laser beam relative to the cruiser is $v_{LC} = +c$. Which one of the following statements correctly describes the velocity v_{LS} of the laser beam relative to the renegades' spacecraft and the velocity v at which the renegades see the laser beam move away from the cruiser? **(a)** $v_{LS} = +0.7c$ and $v = +c$ **(b)** $v_{LS} = +0.3c$ and $v = +c$ **(c)** $v_{LS} = +c$ and $v = +0.7c$ **(d)** $v_{LS} = +c$ and $v = +0.3c$

Reasoning Since both vehicles move at a constant velocity, each constitutes an inertial reference frame. According to the speed-of-light postulate, *all* observers in inertial reference frames measure the speed of light in a vacuum to be c.

Answers (a) and (b) are incorrect. Since the renegades' spacecraft constitutes an inertial reference frame, the velocity of the laser beam relative to it can only have a value of $v_{LS} = +c$, according to the speed-of-light postulate.

Answer (c) is incorrect. The velocity at which the renegades see the laser beam move away from the cruiser cannot be $v = +0.7c$, because they see the cruiser moving at a velocity of $+0.7c$ and the laser beam moving at a velocity of only $+c$ (not $+1.4c$).

Answer (d) is correct. The renegades see the cruiser approach them at a relative velocity of $v_{CS} = +0.7c$ and see the laser beam approach them at a relative velocity of $v_{LS} = +c$. Both these velocities are measured relative to the *same* inertial reference frame—namely, that of their own spacecraft. Therefore, the renegades see the laser beam move away from the cruiser at a velocity that is the difference between these two velocities, or $+c - (+0.7c) = +0.3c$. The velocity-addition formula, Equation 28.8, does not apply here because both velocities are measured relative to the *same* inertial reference frame. Equation 28.8 is used only when the velocities are measured relative to *different* inertial reference frames.

Related Homework: Problem 35

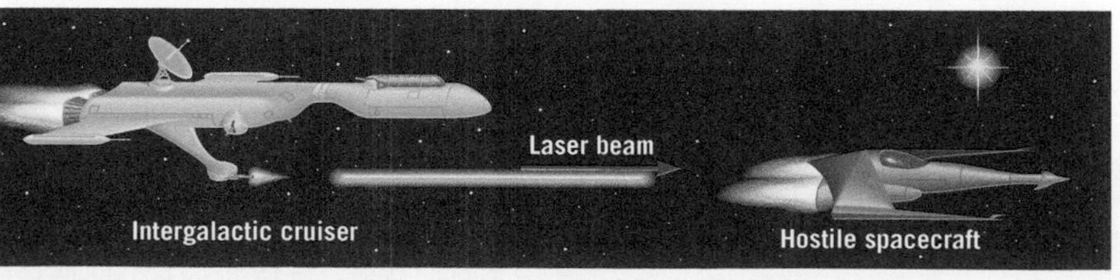

FIGURE 28.13 An intergalactic cruiser, closing in on a hostile spacecraft, fires a beam of laser light.

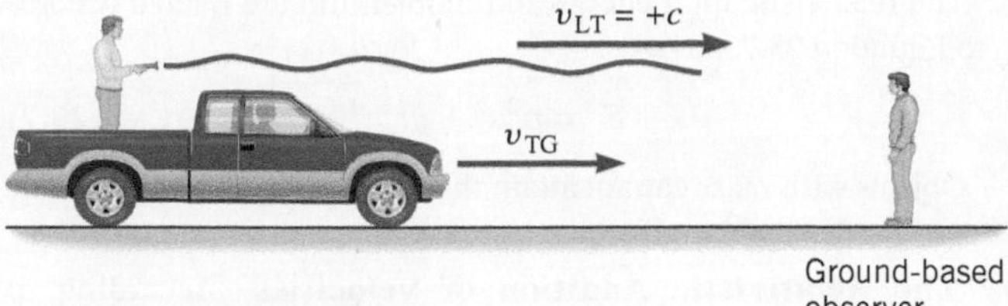

FIGURE 28.14 The speed of the light emitted by the flashlight is c relative to both the truck and the observer on the ground.

It is a straightforward matter to show that the velocity-addition formula is consistent with the speed-of-light postulate. Consider **Figure 28.14**, which shows a person riding on a truck and holding a flashlight. The velocity of the light relative to the person on the truck is $v_{LT} = +c$. The velocity v_{LG} of the light relative to the observer standing on the ground is given by the velocity-addition formula as

$$v_{LG} = \frac{v_{LT} + v_{TG}}{1 + \dfrac{v_{LT} v_{TG}}{c^2}} = \frac{c + v_{TG}}{1 + \dfrac{c v_{TG}}{c^2}} = \frac{(c + v_{TG})c}{(c + v_{TG})} = c$$

Thus, the velocity-addition formula indicates that the observer on the ground and the person on the truck both measure the speed of light to be c, independent of the relative velocity v_{TG} between them. This conclusion is completely consistent with the speed-of-light postulate.

Check Your Understanding

(The answer is given at the end of the book.)

12. Car A and car B are both traveling due east on a straight section of an interstate highway. The speed of car A relative to the ground is 10 m/s faster than the speed of car B relative to the ground. According to special relativity, is the speed of car A relative to car B **(a)** 10 m/s, **(b)** less than 10 m/s, or **(c)** greater than 10 m/s?

Concept Summary

28.1 Events and Inertial Reference Frames An event is a physical "happening" that occurs at a certain place and time. To record the event an observer uses a reference frame that consists of a coordinate system and a clock. Different observers may use different reference frames. The theory of special relativity deals with inertial reference frames. An inertial reference frame is one in which Newton's law of inertia is valid. Accelerating reference frames are not inertial reference frames.

28.2 The Postulates of Special Relativity The theory of special relativity is based on two postulates. The relativity postulate states that the laws of physics are the same in every inertial reference frame. The speed-of-light postulate says that the speed of light in a vacuum, measured in any inertial reference frame, always has the same value of c, no matter how fast the source of the light and the observer are moving relative to each other.

28.3 The Relativity of Time: Time Dilation The proper time interval Δt_0 between two events is the time interval measured by an observer who is at rest relative to the events and views them occurring at the same place. An observer who is in motion with respect to the events and who views them as occurring at different places measures a dilated time interval Δt. The dilated time interval is greater than the proper time interval, according to the time-dilation equation (Equation 28.1). In this expression, v is the relative speed between the observer who measures Δt_0 and the observer who measures Δt.

$$\Delta t = \frac{\Delta t_0}{\sqrt{1 - \dfrac{v^2}{c^2}}} \quad (28.1)$$

28.4 The Relativity of Length: Length Contraction The proper length L_0 between two points is the length measured by an observer who is at rest relative to the points. An observer moving with a relative speed v parallel to the line between the two points does not measure the proper length. Instead, such an observer measures a contracted length L given by the length-contraction formula (Equation 28.2). Length contraction occurs only along the direction of the motion. Those dimensions that are perpendicular to the motion are not shortened. The observer who measures the proper length may not be the observer who measures the proper time interval.

$$L = L_0\sqrt{1 - \frac{v^2}{c^2}} \quad (28.2)$$

28.5 Relativistic Momentum An object of mass m, moving with speed v, has a relativistic momentum whose magnitude p is given by Equation 28.3.

$$p = \frac{mv}{\sqrt{1 - \frac{v^2}{c^2}}} \quad (28.3)$$

28.6 The Equivalence of Mass and Energy Energy and mass are equivalent. The total energy E of an object of mass m, moving at speed v, is given by Equation 28.4. The rest energy E_0 is the total energy of an object at rest ($v = 0$ m/s), as given by Equation 28.5. An object's total energy is the sum of its rest energy and its kinetic energy KE, or $E = E_0 + \text{KE}$. Therefore, the kinetic energy is given by Equation 28.6.

$$E = \frac{mc^2}{\sqrt{1 - \frac{v^2}{c^2}}} \quad (28.4)$$

$$E_0 = mc^2 \quad (28.5)$$

$$\text{KE} = E - E_0 = mc^2\left(\frac{1}{\sqrt{1 - \frac{v^2}{c^2}}} - 1\right) \quad (28.6)$$

The relativistic total energy and momentum are related according to Equation 28.7.

$$E^2 = p^2c^2 + m^2c^4 \quad (28.7)$$

Objects with mass cannot attain the speed of light c, which is the ultimate speed for such objects.

28.7 The Relativistic Addition of Velocities According to special relativity, the velocity-addition formula specifies how the relative velocities of moving objects are related. For objects that move along the same straight line, this formula is given by Equation 28.8, where v_{AB} is the velocity of object A relative to object B, v_{AC} is the velocity of object A relative to object C, and v_{CB} is the velocity of object C relative to object B. The velocities can have positive or negative values, depending on whether they are directed along the positive or negative direction. Furthermore, switching the order of the subscripts changes the sign of the velocity, so that, for example, $v_{BA} = -v_{AB}$.

$$v_{AB} = \frac{v_{AC} + v_{CB}}{1 + \frac{v_{AC}v_{CB}}{c^2}} \quad (28.8)$$

Focus on Concepts

Online

Additional questions are available for assignment in WileyPLUS.

Section 28.1 Events and Inertial Reference Frames

1. Consider a person along with a frame of reference in each of the following situations. In which one or more of the following situations is the frame of reference an inertial frame of reference? (a) The person is oscillating in simple harmonic motion at the end of a bungee cord. (b) The person is in a car going around a circular curve at a constant speed. (c) The person is in a plane that is landing on an aircraft carrier. (d) The person is in the space shuttle during lift-off. (e) None of the above.

Section 28.3 The Relativity of Time: Time Dilation

2. On a highway there is a flashing light to mark the start of a section of the road where work is being done. Who measures the proper time between two flashes of light? (a) A worker standing still on the road (b) A driver in a car approaching at a constant velocity (c) Both the worker and the driver (d) Neither the worker nor the driver

Section 28.4 The Relativity of Length: Length Contraction

3. Two spacecrafts A and B are moving relative to each other at a constant velocity. Observers in spacecraft A see spacecraft B. Likewise, observers in spacecraft B see spacecraft A. Who sees the proper length of either spacecraft? (a) Observers in spacecraft A see the proper length of spacecraft B. (b) Observers in spacecraft B see the proper length of spacecraft A. (c) Observers in both spacecrafts see the proper length of the other spacecraft. (d) Observers in neither spacecraft see the proper length of the other spacecraft.

4. In a baseball game the batter hits the ball into center field and takes off for first base. The catcher can only stand and watch. Assume that the batter runs at a constant velocity. Who measures the proper time it takes for the runner to reach first base, and who measures the proper length between home plate and first base? (a) The catcher measures the proper time, and the runner measures the proper length. (b) The runner measures the proper time, and the catcher measures the proper length. (c) The catcher measures both the proper time and the proper length. (d) The runner measures both the proper time and the proper length.

5. To which one or more of the following situations do the time-dilation and length-contraction equations apply? (a) With respect to an inertial frame, two observers have different constant accelerations. (b) With respect to an inertial frame, two observers have the same constant acceleration. (c) With respect to an inertial frame, two observers are moving with different constant velocities. (d) With respect to an inertial frame, one observer has a constant velocity, and another observer has a constant acceleration. (e) All of the above.

Section 28.5 Relativistic Momentum

6. Which one of the following statements about linear momentum is true (p = magnitude of the momentum, m = mass, and v = speed)? (a) When the magnitude p of the momentum is defined as $p = \frac{mv}{\sqrt{1 - v^2/c^2}}$, the linear momentum of an isolated system is conserved only if the speeds of the various parts of the system are very high. (b) When the magnitude p of the momentum is defined as $p = mv$, the linear momentum of an isolated system is conserved only if the speeds of the various parts of the system are very high. (c) When the magnitude p of the momentum is defined as $p = \frac{mv}{\sqrt{1 - v^2/c^2}}$, the linear momentum of an isolated system is conserved no matter what the speeds of the various parts of the system are. (d) When the magnitude p of the momentum is defined as $p = mv$, the linear momentum of an isolated system is conserved no matter what the speeds of the various parts of the system are.

7. Which of the following two expressions for the magnitude p of the linear momentum can be used when the speed v of an object of mass m is very small compared to the speed of light c in a vacuum?

A. $p = mv$

B. $p = \dfrac{mv}{\sqrt{1 - \dfrac{v^2}{c^2}}}$

(a) Only A (b) Only B (c) Neither A nor B (d) Both A and B

Section 28.6 The Equivalence of Mass and Energy

8. Consider the following three possibilities for a glass of water at rest on a kitchen counter. The temperature of the water is 0 °C. Rank the mass of the water in descending order (largest first).

A. The water is half liquid and half ice.

B. The water is all liquid.

C. The water is all ice.

(a) C, A, B (b) B, A, C (c) A, C, B (d) B, C, A (e) C, B, A

9. An object has a kinetic energy KE and a potential energy PE. It also has a rest energy E_0. Which one of the following is the correct way to express the object's total energy E? (a) $E = \text{KE} + \text{PE}$ (b) $E = E_0 + \text{KE}$ (c) $E = E_0 + \text{KE} + \text{PE}$ (d) $E = E_0 + \text{KE} - \text{PE}$

10. The kinetic energy of an object of mass m is equal to its rest energy. What is the magnitude p of the object's momentum? (a) $p = \sqrt{3}\ mc$ (b) $p = 2mc$ (c) $p = 4mc$ (d) $p = \sqrt{2}mc$ (e) $p = 3mc$

Section 28.7 The Relativistic Addition of Velocities

11. Two spaceships are traveling in the same direction. With respect to an inertial frame of reference, spaceship A has a speed of 0.900*c*. With respect to the same inertial frame, spaceship B has a speed of 0.500*c*. Find the speed v_{AB} of spaceship A relative to spaceship B.

Problems

Online

Additional questions are available for assignment in WileyPLUS.

Note: *Before doing any calculations involving time dilation or length contraction, it is useful to identify which observer measures the proper time interval* Δt_0 *or the proper length* L_0.

- SSM Student Solutions Manual
- MMH Problem-solving help
- GO Guided Online Tutorial
- V-HINT Video Hints
- CHALK Chalkboard Videos
- BIO Biomedical application
- E Easy
- M Medium
- H Hard
- WS Worksheet
- T Team Problem

Section 28.3 The Relativity of Time: Time Dilation

1. E SSM A particle known as a pion lives for a short time before breaking apart into other particles. Suppose that a pion is moving at a speed of 0.990*c*, and an observer who is stationary in a laboratory measures the pion's lifetime to be 3.5×10^{-8} s. (a) What is the lifetime according to a hypothetical person who is riding along with the pion? (b) According to this hypothetical person, how far does the laboratory move before the pion breaks apart?

2. E A radar antenna is rotating and makes one revolution every 25 s, as measured on earth. However, instruments on a spaceship moving with respect to the earth at a speed v measure that the antenna makes one revolution every 42 s. What is the ratio v/c of the speed v to the speed c of light in a vacuum?

3. E SSM Suppose that you are planning a trip in which a spacecraft is to travel at a constant velocity for exactly six months, as measured by a clock on board the spacecraft, and then return home at the same speed. Upon your return, the people on earth will have advanced exactly one hundred years into the future. According to special relativity, how fast must you travel? Express your answer to five significant figures as a multiple of *c*—for example, 0.955 85*c*.

4. E CHALK GO Suppose that you are traveling on board a spacecraft that is moving with respect to the earth at a speed of 0.975*c*. You are breathing at a rate of 8.0 breaths per minute. As monitored on earth what is your breathing rate?

5. M GO MMH A 6.00-kg object oscillates back and forth at the end of a spring whose spring constant is 76.0 N/m. An observer is traveling at a speed of 1.90×10^8 m/s relative to the fixed end of the spring. What does this observer measure for the period of oscillation?

6. M V-HINT A spaceship travels at a constant speed from earth to a planet orbiting another star. When the spacecraft arrives, 12 years have elapsed on earth, and 9.2 years have elapsed on board the ship. How far away (in meters) is the planet, according to observers on earth?

7. H As observed on earth, a certain type of bacterium is known to double in number every 24.0 hours. Two cultures of these bacteria are prepared, each consisting initially of one bacterium. One culture is left on earth and the other placed on a rocket that travels at a speed of 0.866*c* relative to the earth. At a time when the earthbound culture has grown to 256 bacteria, how many bacteria are in the culture on the rocket, according to an earth-based observer?

Section 28.4 The Relativity of Length: Length Contraction

8. E Suppose the straight-line distance between New York and San Francisco is 4.1×10^6 m (neglecting the curvature of the earth). A UFO is flying between these two cities at a speed of 0.70*c* relative to the earth. What do the voyagers aboard the UFO measure for this distance?

9. E SSM How fast must a meter stick be moving if its length is observed to shrink to one-half of a meter?

10. E V-HINT MMH The distance from earth to the center of our galaxy is about 23 000 ly (1 ly = 1 light-year = 9.47×10^{15} m), as measured by an earth-based observer. A spaceship is to make this journey at a speed of 0.9990*c*. According to a clock on board the spaceship, how long will it take to make the trip? Express your answer in years (1 yr = 3.16×10^7 s).

11. E A tourist is walking at a speed of 1.3 m/s along a 9.0-km path that follows an old canal. If the speed of light in a vacuum were 3.0 m/s, how long would the path be, according to the tourist?

12. E GO A Martian leaves Mars in a spaceship that is heading to Venus. On the way, the spaceship passes earth with a speed $v = 0.80c$

relative to it. Assume that the three planets do not move relative to each other during the trip. The distance between Mars and Venus is 1.20×10^{11} m, as measured by a person on earth. **(a)** What does the Martian measure for the distance between Mars and Venus? **(b)** What is the time of the trip (in seconds) as measured by the Martian?

13. **E** Two spaceships A and B are exploring a new planet. Relative to this planet, spaceship A has a speed of 0.60*c*, and spaceship B has a speed of 0.80*c*. What is the ratio D_A/D_B of the values for the planet's diameter that each spaceship measures in a direction that is parallel to its motion?

14. **E** An unstable high-energy particle is created in the laboratory, and it moves at a speed of 0.990*c*. Relative to a stationary reference frame fixed to the laboratory, the particle travels a distance of 1.05×10^{-3} m before disintegrating. What are **(a)** the proper distance and **(b)** the distance measured by a hypothetical person traveling with the particle? Determine the particle's **(c)** proper lifetime and **(d)** its dilated lifetime.

15. **M** **CHALK** As the drawing shows, a carpenter on a space station has constructed a 30.0° ramp. A rocket moves past the space station with a relative speed of 0.730*c* in a direction parallel to side x_0. What does a person aboard the rocket measure for the angle of the ramp?

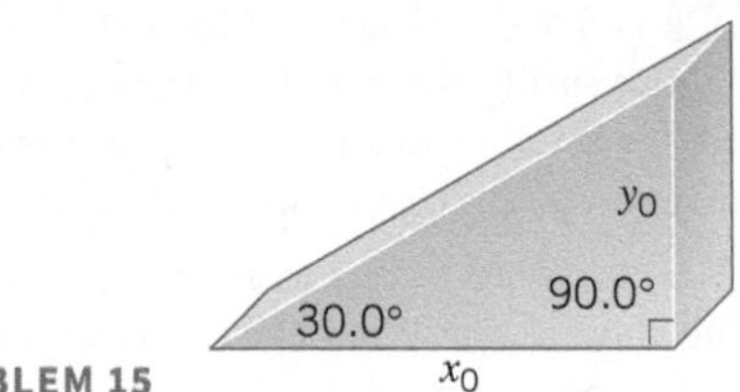

PROBLEM 15

Section 28.5 Relativistic Momentum

16. **E** At what speed is the magnitude of the relativistic momentum of a particle three times the magnitude of the nonrelativistic momentum?

17. **E** What is the magnitude of the relativistic momentum of a proton with a relativistic total energy of 2.7×10^{-10} J?

18. **E** **GO** A spacecraft has a nonrelativistic (or classical) momentum whose magnitude is 1.3×10^{13} kg · m/s. The spacecraft moves at such a speed that the pilot measures the proper time interval between two events to be one-half the dilated time interval. Find the relativistic momentum of the spacecraft.

19. **E** **SSM** A woman is 1.6 m tall and has a mass of 55 kg. She moves past an observer with the direction of the motion parallel to her height. The observer measures her relativistic momentum to have a magnitude of 2.0×10^{10} kg · m/s. What does the observer measure for her height?

20. **E** **GO** Three particles are listed in the table. The mass and speed of each particle are given as multiples of the variables *m* and *v*, which have the values $m = 1.20 \times 10^{-8}$ kg and $v = 0.200c$. The speed of light in a vacuum is $c = 3.00 \times 10^8$ m/s. Determine the momentum for each particle according to special relativity.

Particle	Mass	Speed
a	m	v
b	$\frac{1}{2}m$	$2v$
c	$\frac{1}{4}m$	$4v$

21. **M** **SSM** **CHALK** Starting from rest, two skaters push off against each other on smooth level ice, where friction is negligible. One is a woman and one is a man. The woman moves away with a velocity of +2.5 m/s relative to the ice. The mass of the woman is 54 kg, and the mass of the man is 88 kg. Assuming that the speed of light is 3.0 m/s, so that the relativistic momentum must be used, find the recoil velocity of the man relative to the ice. (*Hint:* This problem is similar to Example 7 in Chapter 7.)

Section 28.6 The Equivalence of Mass and Energy

22. **E** Radium is a radioactive element whose nucleus emits an α particle (a helium nucleus) with a kinetic energy of about 7.8×10^{-13} J (4.9 MeV). To what amount of mass is this energy equivalent?

23. **E** **SSM** How much work must be done on an electron to accelerate it from rest to a speed of 0.990*c*?

24. **E** Suppose that one gallon of gasoline produces 1.1×10^8 J of energy, and this energy is sufficient to operate a car for twenty miles. An aspirin tablet has a mass of 325 mg. If the aspirin could be converted completely into thermal energy, how many miles could the car go on a single tablet?

25. **E** **GO** Two kilograms of water are changed **(a)** from ice at 0 °C into liquid water at 0 °C and **(b)** from liquid water at 100 °C into steam at 100 °C. For each situation, determine the change in mass of the water.

26. **E** **CHALK** **SSM** Determine the ratio of the relativistic kinetic energy to the nonrelativistic kinetic energy ($\frac{1}{2}mv^2$) when a particle has a speed of **(a)** $1.00 \times 10^{-3}c$ and **(b)** 0.970*c*.

27. **E** Multiple-Concept Example 8 reviews the principles that play a role in this problem. A nuclear power reactor generates 3.0×10^9 W of power. In one year, what is the change in the mass of the nuclear fuel due to the energy being taken from the reactor?

28. **M** **V-HINT** Multiple-Concept Example 8 explores the approach taken in problems such as this one. Quasars are believed to be the nuclei of galaxies in the early stages of their formation. Suppose that a quasar radiates electromagnetic energy at the rate of 1.0×10^{41} W. At what rate (in kg/s) is the quasar losing mass as a result of this radiation?

29. **M** **GO** An electron is accelerated from rest through a potential difference that has a magnitude of 2.40×10^7 V. The mass of the electron is 9.11×10^{-31} kg, and the negative charge of the electron has a magnitude of 1.60×10^{-19} C. **(a)** What is the relativistic kinetic energy (in joules) of the electron? **(b)** What is the speed of the electron? Express your answer as a multiple of *c*, the speed of light in a vacuum.

Section 28.7 The Relativistic Addition of Velocities

30. **E** **GO** You are driving down a two-lane country road, and a truck in the opposite lane is traveling toward you. Suppose that the speed of light in a vacuum is $c = 65$ m/s. Determine the speed of the truck relative to you when **(a)** your speed is 25 m/s and the truck's speed is 35 m/s and **(b)** your speed is 5.0 m/s and the truck's speed is 55 m/s. The speeds given in parts (a) and (b) are relative to the ground.

31. **E** **SSM** A spacecraft approaching the earth launches an exploration vehicle. After the launch, an observer on earth sees the spacecraft approaching at a speed of 0.50*c* and the exploration vehicle approaching at a speed of 0.70*c*. What is the speed of the exploration vehicle relative to the spaceship?

32. **E** **GO** Spaceships of the future may be powered by ion-propulsion engines in which ions are ejected from the back of the ship to drive it forward. In one such engine the ions are to be ejected with a speed of 0.80*c* relative to the spaceship. The spaceship is traveling away from the earth at a speed of 0.70*c* relative to the earth. What is the velocity of the ions relative to the earth? Assume that the direction in which the spaceship is traveling is the positive direction, and be sure to assign the correct plus or minus signs to the velocities.

33. E The spaceship *Enterprise 1* is moving directly away from earth at a velocity that an earth-based observer measures to be $+0.65c$. A sister ship, *Enterprise 2*, is ahead of *Enterprise 1* and is also moving directly away from earth along the same line. The velocity of *Enterprise 2* relative to *Enterprise 1* is $+0.31c$. What is the velocity of *Enterprise 2*, as measured by the earth-based observer?

34. M V-HINT A person on earth notices a rocket approaching from the right at a speed of $0.75c$ and another rocket approaching from the left at $0.65c$. What is the relative speed between the two rockets, as measured by a passenger on one of them?

35. M SSM Refer to Conceptual Example 13 as an aid in solving this problem. An intergalactic cruiser has two types of guns: a photon cannon that fires a beam of laser light and an ion gun that shoots ions at a velocity of $0.950c$ relative to the cruiser. The cruiser closes in on an alien spacecraft at a velocity of $0.800c$ relative to this spacecraft. The captain fires both types of guns. At what velocity do the aliens see (a) the laser light and (b) the ions approach them? At what velocity do the aliens see (c) the laser light and (d) the ions move away from the cruiser?

36. M GO CHALK Two identical spaceships are under construction. The constructed length of each spaceship is 1.50 km. After being launched, spaceship A moves away from earth at a constant velocity (speed is $0.850c$) with respect to the earth. Spaceship B follows in the same direction at a different constant velocity (speed is $0.500c$) with respect to the earth. Determine the length that a passenger on one spaceship measures for the other spaceship.

Additional Problems

Online

37. E An electron and a positron have masses of 9.11×10^{-31} kg. They collide and both vanish, with only electromagnetic radiation appearing after the collision. If each particle is moving at a speed of $0.20c$ relative to the laboratory before the collision, determine the energy of the electromagnetic radiation.

38. E GO The speed of an ion in a particle accelerator is doubled from $0.460c$ to $0.920c$. The initial relativistic momentum of the ion is 5.08×10^{-17} kg · m/s. Determine (a) the mass and (b) the magnitude of the final relativistic momentum of the ion.

39. E GO A Klingon spacecraft has a speed of $0.75c$ with respect to the earth. The Klingons measure 37.0 h for the time interval between two events on the earth. What value for the time interval would they measure if their ship had a speed of $0.94c$ with respect to the earth?

40. M SSM The crew of a rocket that is moving away from the earth launches an escape pod, which they measure to be 45 m long. The pod is launched toward the earth with a speed of $0.55c$ relative to the rocket. After the launch, the rocket's speed relative to the earth is $0.75c$. What is the length of the escape pod as determined by an observer on earth?

41. M GO The table gives the total energy and the rest energy for three objects in terms of an energy increment ε. For each object, determine the speed as a multiple of the speed c of light in a vacuum.

Object	Total Energy (E)	Rest Energy (E_0)
A	2.00ε	ε
B	3.00ε	ε
C	3.00ε	2.00ε

42. H GO Twins who are 19.0 years of age leave the earth and travel to a distant planet 12.0 light-years away. Assume that the planet and earth are at rest with respect to each other. The twins depart at the same time on different spaceships. One twin travels at a speed of $0.900c$, and the other twin travels at $0.500c$. (a) According to the theory of special relativity, what is the difference between their ages when they meet again at the earliest possible time? (b) Which twin is older?

Physics in Biology, Medicine, and Sports

43. M BIO 28.4 Special relativity tells us that someone sprinting will shrink, or undergo length contraction, along the direction of motion relative to someone at rest. The world's fastest sprinters, like Usain Bolt, can reach speeds near 13.0 m/s. How fast would they have to run so that their length contraction was (a) equal to the width of a human hair (50.0 μm) and (b) equal to the width of an atom (1.00×10^{-10} m)? Assume the width of the sprinter's body at rest is 25.0 cm.

44. M BIO 28.4 A football player returns a kickoff the full length of the field for a 100-yard touchdown. How fast would he have to run so that his return appears to be only 99 yards to him?

45. M BIO 28.3 Protons and neutrons exist together in the nuclei of atoms and form the building blocks of all living things. However, if a neutron is removed from a nucleus, so that it exists as a free neutron, it is unstable. A free neutron at rest has a mean lifetime of 8.80×10^2 s, at which point, on average, it will decay into a proton, an electron, and another particle called an antineutrino. This process is known as β decay (see Chapter 31). If an observer measures the mean lifetime of a neutron to be twice as long (1760 s), how fast is the neutron moving relative to the observer?

46. M BIO 28.3 On earth, the greatest length of life occurs for females living in Hong Kong, whose average life expectancy is 88.2 years. If such a person lived their life on a spaceship moving at 20% the speed of light, what would be their mean lifetime, as measured by an observer at rest on earth?

47. M BIO 28.6 The African bush elephant is the largest terrestrial mammal. The tallest one ever recorded stood 4.21 m (13.8 ft) at the

shoulder and weighed 18 000 lb (8160 kg). How fast would a Ferrari sports car ($m = 1660$ kg) have to be moving, such that its kinetic energy is equal to the rest mass energy of the elephant? Use $c = 3.00 \times 10^8$ m/s for the speed of light.

Concepts and Calculations Problems

Online

There are many astonishing consequences of special relativity, two of which are time dilation and length contraction. Problem 48 reviews these important concepts in the context of a golf game in a hypothetical world where the speed of light is only a little faster than that of a golf cart. Other important consequences of special relativity are the equivalence of mass and energy, and the dependence of kinetic energy on the total energy and on the rest energy. Problem 49 serves as a review of the roles played by mass and energy in special relativity.

48. M CHALK Imagine playing golf in a world where the speed of light is only $c = 3.40$ m/s. Golfer A drives a ball down a flat horizontal fairway for a distance that he measures as 75.0 m. Golfer B, riding in a cart, happens to pass by just as the ball is hit (see the figure). Golfer A stands at the tee and watches while golfer B moves down the fairway toward the ball at a constant speed of 2.80 m/s. *Concepts:* (i) Who measures the proper length of the drive, and who measures the contracted length? (ii) Who measures the proper time interval, and who measures the dilated time interval? *Calculations:* **(a)** How far is the ball hit according to golfer B? **(b)** According to each golfer, how much time does it take golfer B to reach the ball?

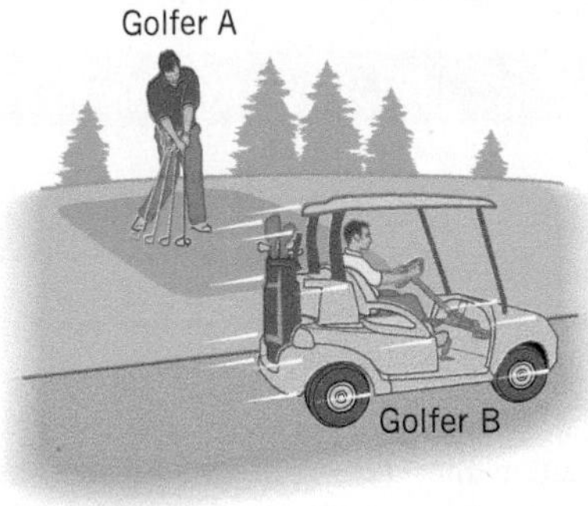

PROBLEM 48

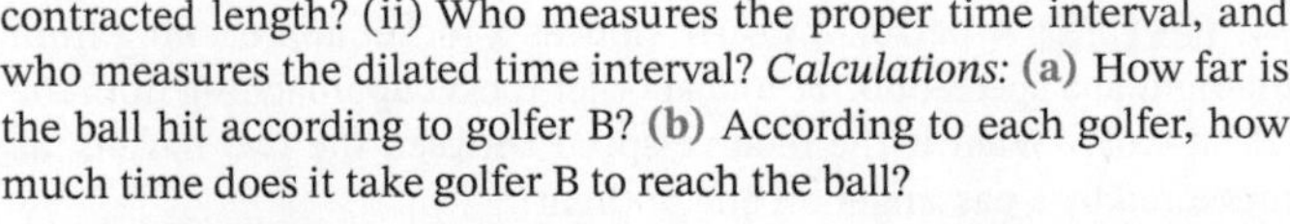

49. M CHALK SSM The rest energy E_0 and the total energy E of three particles, expressed in terms of a basic amount of energy $E' = 5.98 \times 10^{-10}$ J, are listed in the table. The speeds of these particles are large, in some cases approaching the speed of light. *Concepts:* (i) Given the rest energies specified in the table, what is the ranking (largest first) of the masses of the particles? (ii) Is the kinetic energy KE given by the expression $\text{KE} = \frac{1}{2}mv^2$, and what is the ranking (largest first) of the kinetic energies of the particles? *Calculations:* For each particle, determine its **(a)** mass and **(b)** kinetic energy.

Particle	Rest Energy	Total Energy
a	E'	$2E'$
b	E'	$4E'$
c	$5E'$	$6E'$

Team Problems

Online

50. E T WS **Space Smugglers.** You and your team are returning from a long voyage but have attracted a stalker along the way. Another ship that is transporting illegal contraband is following yours at a short distance in order that it might sneak into earth's orbit and eventually deliver its cargo. As you get closer to earth, you encounter a rotating beacon, the location of which is stationary in earth's frame of reference. It is known that, in the reference frame of earth, the beacon rotates with a frequency of 0.8030 Hz. **(a)** What is the velocity of your ship relative to the beacon, and therefore relative to earth, if the instruments on your ship measure the rotational frequency of the beacon to be 0.6080 Hz? **(b)** The smugglers' ship is traveling at exactly the same speed as yours, but is of a different type, and therefore likely to be of a different length, although you do not know whether it is longer or shorter. You decide to warn authorities on earth about the smugglers' ship, and want to give them enough information to identify it. You do this by informing them of the exact length of your ship in earth's frame of reference. Since the smugglers' ship will likely be of a different length, the authorities will be able to distinguish it from yours. If your ship is 175.8 meters in length in your frame of reference, what is its length in earth's frame?

51. M T WS **Special Relativity and the Doppler Effect.** You and your team are on a spacecraft that is approaching earth at a relativistic speed, but you are low on fuel and supplies. It is crucial that you start the deceleration process at exactly the right time. If you brake too early, you will spend too much time completing the trip home at the lower speed, and will run out of food. If you brake too late, you will pass by earth and will not have enough fuel to turn the ship around. To accurately determine when to brake, you need to have an accurate measurement of your current speed relative to earth. For this purpose, there is a beacon far from earth that emits light in all directions with a wavelength of 685 nm (in its frame of reference). The beacon is stationary relative to earth and your ship is heading directly toward it. The idea is that an incoming spacecraft will measure the wavelength of the light emitted from the beacon with onboard instruments to determine the Doppler shift, from which the relative velocity between the ship and the beacon can be calculated. There is one complication, however. Since your ship is moving at a relativistic speed relative to the source, you must apply a Doppler effect formula that takes relativistic effects (i.e., time dilation and length contraction) into account. Without taking relativistic effects into account, you would apply the nonrelativistic formula given by Equation 24.6:

$$f_o = f_s\left(1 \pm \frac{v_{rel}}{c}\right) \qquad (\text{if } v_{rel} \ll c)$$

where the plus sign is used when the source (f_s) and observer (f_o) come together, and the minus sign is used when they move apart. Accounting for relativistic effects, this equation must be modified as follows:

$$f_o = f_s\sqrt{\frac{1+\frac{v_{rel}}{c}}{1-\frac{v_{rel}}{c}}} \qquad (\text{relativistic Doppler equation})$$

where v_{rel} is positive when the source (f_s) and observer (f_o) come together, and negative when they move apart. **(a)** If your ship measures the beacon wavelength to be 395 nm, what is the ship's velocity relative to the beacon? **(b)** What speed would you have obtained using the nonrelativistic Doppler effect equation? **(c)** If you had used the nonrelativistic Doppler equation, would you have stopped too early, or too late? Explain.

52. M T **A Super Thruster.** You and your team are evaluating a new thruster that will be used to propel a 6.00×10^4 kg ship. The thruster has a power output of 2.88×10^6 W, calculated to initially accelerate the ship at 9.80 m/s^2 (safe for humans). **(a)** Assuming that all of the energy of the thruster is converted into kinetic energy of the ship, what is the velocity of the ship after 1 year? **(b)** After 1000 years? **(c)** Assuming the thruster converts mass directly into energy, how much mass would be converted into energy over the 1000-year trip?

53. M T **A Long Trip.** You and your team are on a journey to a solar system in another part of our galaxy. Your destination is 4.00×10^3 ly (1 ly = 1 light-year = 9.47×10^{15} m) from earth, as measured by an earth-based observer. **(a)** If the ship has a speed (relative to earth) of $0.9999c$, how long will the trip take according to the clocks on board the ship? Express your answer in years. **(b)** You have to pass by the sun on the way out of our solar system. According to the clocks on your ship, how much time (in minutes) will it take to get to the sun from earth? The sun-earth mean distance is 1.50×10^{11} m. **(c)** The sun will not look spherical as you pass by. What is the ratio of the shortest to the longest dimensions (i.e., the ratio of the semi-minor to the semi-major axes) of the sun as you see it?

CHAPTER 29

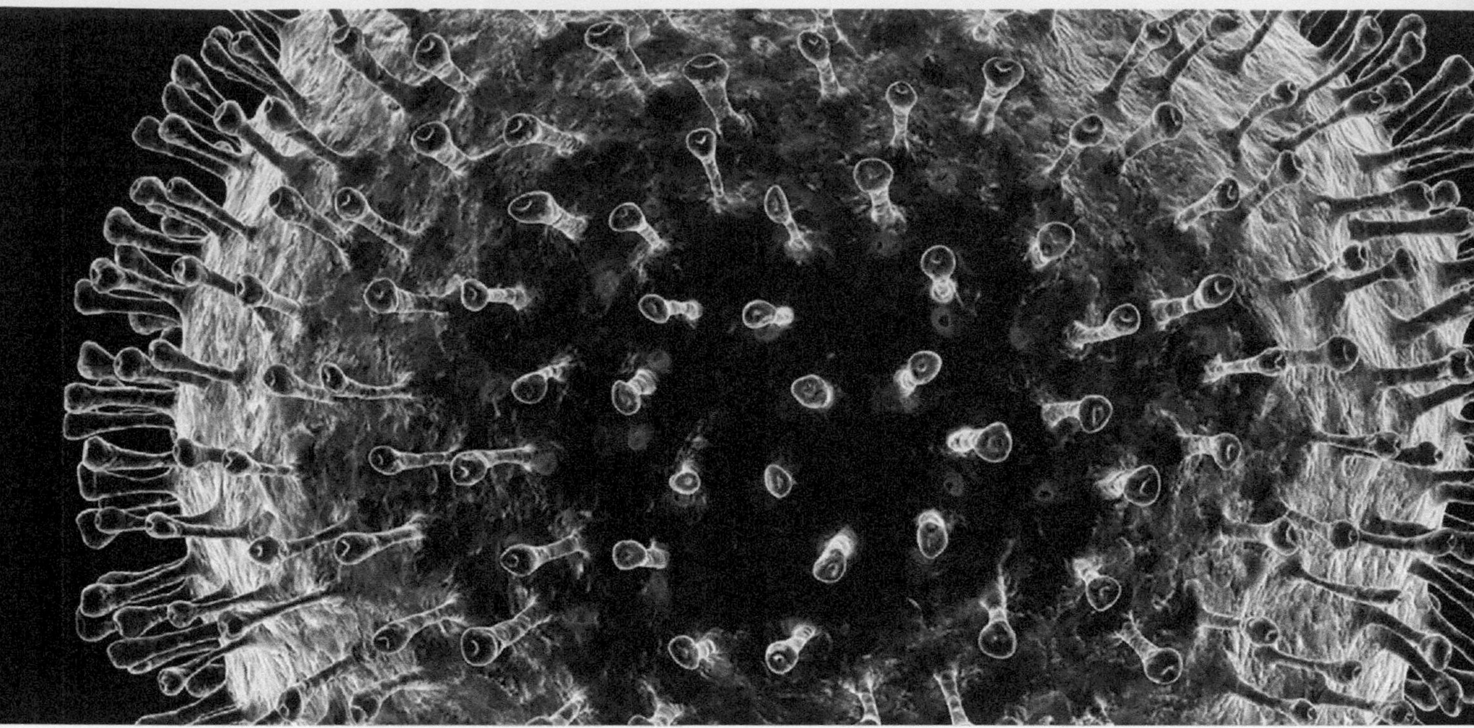

This photograph shows a highly magnified view of a coronavirus that was made with a scanning electron microscope (SEM). The diameter of the roughly spherical virus is approximately 120 nm. In the twentieth century, physicists were astonished when it was discovered that particles could behave like waves. In fact, we will see in this chapter that there is a wavelength associated with a moving particle such as an electron. The microscope used for the photograph takes advantage of the electron wavelength, which can be made much smaller than that of visible light. It is this small electron wavelength that is responsible for the exceptional resolution of the fine details in the photograph. The spike-like structures on the outer surface of the virus are composed of proteins that attach to the cell membrane of the host. A novel coronavirus (SARS-CoV-2) was responsible for an outbreak of coronavirus disease in 2019 (COVID-19) that infected millions of people worldwide.

PASIEKA/Getty Images

LEARNING OBJECTIVES

After reading this module, you should be able to...

29.1 Define wave–particle duality.

29.2 Explain the origin of Planck's constant from blackbody radiation.

29.3 Use photon energy to explain in detail the photoelectric effect.

29.4 Use photon momentum to explain in detail the Compton effect.

29.5 Solve problems involving the de Broglie wavelength of a particle.

29.6 Calculate quantum uncertainty using the Heisenberg uncertainty principle.

Particles and Waves

29.1 The Wave–Particle Duality

The ability to exhibit interference effects is an essential characteristic of waves. For instance, Section 27.2 discusses Young's famous experiment in which light passes through two closely spaced slits and produces a pattern of bright and dark fringes on a screen (see **Figure 27.3**). The fringe pattern is a direct indication that interference is occurring between the light waves coming from each slit.

One of the most incredible discoveries of twentieth-century physics is that particles can also behave like waves and exhibit interference effects. For instance, **Interactive Figure 29.1** shows a version of Young's experiment performed by directing *a beam of electrons* onto a double slit. In this experiment, the screen is like a television screen and glows wherever an electron strikes it. Part *a* of the drawing indicates the pattern that would be seen on the screen if each electron, behaving strictly as a particle, were to pass through one slit or the other and strike the screen. The pattern would consist of an image of each slit. Part *b* shows the pattern actually observed, which consists of bright and dark fringes, reminiscent of what is obtained when light waves pass through a double slit. The fringe pattern indicates that the electrons are exhibiting the interference effects associated with waves.

But how can electrons behave like waves in the experiment shown in **Interactive Figure 29.1*b***? And what kind of waves are they? The answers to these profound questions will be discussed later in this chapter. For the moment, we intend only to emphasize that the concept of an electron as a tiny discrete particle of matter does not account for the fact that the electron can behave as a wave in some circumstances. In other words, the electron exhibits a dual nature, with both particle-like characteristics and wave-like characteristics.

Here is another interesting question: If a particle can exhibit wave-like properties, can waves exhibit particle-like behavior? As the next three sections reveal, the answer is yes. In fact, experiments that demonstrated the particle-like behavior of waves were performed near the beginning of the twentieth century, before the experiments that demonstrated the wave-like properties of the electrons. Scientists now accept the **wave–particle duality** as an essential part of nature:

Waves can exhibit particle-like characteristics, and particles can exhibit wave-like characteristics.

Section 29.2 begins the remarkable story of the wave–particle duality by discussing the electromagnetic waves that are radiated by a perfect blackbody. It is appropriate to begin with blackbody radiation, because it provided the first link in the chain of experimental evidence leading to our present understanding of the wave–particle duality.

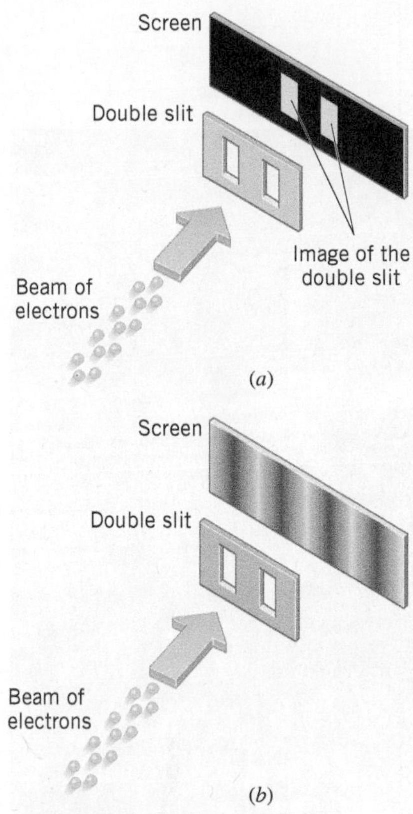

INTERACTIVE FIGURE 29.1 (*a*) If electrons behaved as discrete particles with no wave properties, they would pass through one or the other of the two slits and strike the screen, causing it to glow and produce exact images of the slits. (*b*) In reality, the screen reveals a pattern of bright and dark fringes, similar to the pattern produced when a beam of light is used and interference occurs between the light waves coming from each slit.

29.2 Blackbody Radiation and Planck's Constant

All bodies, no matter how hot or cold, continuously radiate electromagnetic waves. For instance, we see the glow of very hot objects because they emit electromagnetic waves in the visible region of the spectrum. Our sun, which has a surface temperature of about 6000 K, appears yellow, while the cooler star Betelgeuse has a red-orange appearance due to its lower surface temperature of 2900 K. However, at relatively low temperatures, cooler objects emit visible light waves only weakly and, as a result, do not appear to be glowing. Certainly the human body, at only 310 K, does not emit enough visible light to be seen in the dark with the unaided eye. But the body does emit electromagnetic waves in the infrared region of the spectrum, and these can be detected with infrared-sensitive devices.

At a given temperature, the intensities of the electromagnetic waves emitted by an object vary from wavelength to wavelength throughout the visible, infrared, and other regions of the spectrum. **Figure 29.2** illustrates how the intensity per unit wavelength depends on wavelength for a perfect blackbody emitter. As Section 13.3 discusses, a perfect blackbody at a constant temperature absorbs and reemits all the electromagnetic radiation that falls on it. The two curves in **Figure 29.2** show that at a higher temperature the maximum emitted intensity per unit wavelength increases and shifts to shorter wavelengths, toward the visible region of the spectrum. In accounting for the shape of these curves, the German physicist Max Planck (1858–1947) took the first step toward our present understanding of the wave–particle duality.

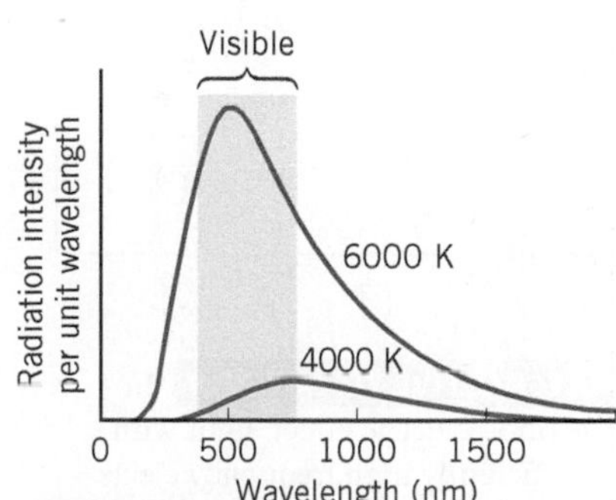

FIGURE 29.2 The electromagnetic radiation emitted by a perfect blackbody has an intensity per unit wavelength that varies from wavelength to wavelength, as each curve indicates. At the higher temperature, the intensity per unit wavelength is greater, and the maximum occurs at a shorter wavelength.

In 1900 Planck calculated the blackbody radiation curves, using a model that represents a blackbody as a large number of atomic oscillators, each of which emits and absorbs electromagnetic waves. To obtain agreement between the theoretical and experimental curves, Planck assumed that the energy E of an atomic oscillator could have only the discrete values of $E = 0, hf, 2hf, 3hf$, and so on. In other words, he assumed that

$$E = nhf \quad n = 0, 1, 2, 3, \ldots \tag{29.1}$$

where n is either zero or a positive integer, f is the frequency of vibration (in hertz), and h is a constant now called **Planck's constant**.* It has been determined experimentally that Planck's constant has a value of

$$h = 6.626\,068\,96 \times 10^{-34}\ \text{J} \cdot \text{s}$$

The radical feature of Planck's assumption was that the energy of an atomic oscillator could have only discrete values (hf, $2hf$, $3hf$, etc.), with energies in between these values being forbidden. Whenever the energy of a system can have only certain definite values, and nothing in between, the energy is said to be *quantized.* This quantization of the energy was unexpected on the basis of the traditional physics of the time. However, it was soon realized that energy quantization had wide-ranging and valid implications.

Conservation of energy requires that the energy carried off by the radiated electromagnetic waves must equal the energy lost by the atomic oscillators in Planck's model. Suppose, for example, that an oscillator with an energy of $3hf$ emits an electromagnetic wave. According to Equation 29.1, the next smallest allowed value for the energy of the oscillator is $2hf$. In such a case, the energy carried off by the electromagnetic wave would have the value of hf, equaling the amount of energy lost by the oscillator. Thus, Planck's model for blackbody radiation sets the stage for the idea that electromagnetic energy occurs as a collection of discrete amounts, or packets, of energy, with the energy of a packet being equal to hf. Einstein made the proposal that light consists of such energy packets.

John Lee/Masterfile

FIGURE 29.3 Although the spotlight beams in the photograph look like continuous beams of light, each is composed of discrete photons.

29.3 Photons and the Photoelectric Effect

We have seen in Chapter 24 that light is an electromagnetic wave and that such waves are continuous patterns of electric and magnetic fields. It is not unexpected, then, that light beams, such as those in the photograph in **Figure 29.3**, or those coming from flashlights, look like continuous beams. However, we must now discuss the surprising fact that visible light and all other types of electromagnetic waves are composed of discrete particle-like entities called **photons**. As discussed in Chapters 6 and 7, the total energy E and the linear momentum $\vec{\mathbf{p}}$ are fundamental concepts in physics that apply to moving particles such as electrons and protons. The total energy of a (nonrelativistic) particle is the sum of its kinetic energy (KE) and potential energy (PE), or $E = \text{KE} + \text{PE}$. The magnitude p of the particle's momentum is the product of its mass m and speed v, or $p = mv$. We will see that the ideas of energy and momentum also apply to photons. The defining equations for photon energy and photon momentum, however, are not the same as they are for particles such as electrons and protons.

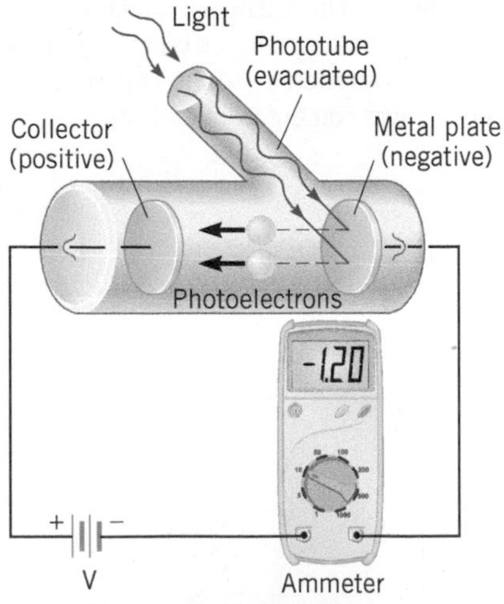

INTERACTIVE FIGURE 29.4 In the photoelectric effect, light with a sufficiently high frequency ejects electrons from a metal surface. These photoelectrons, as they are called, are drawn to the positive collector, thus producing a current.

Experimental evidence that light consists of photons comes from a phenomenon called the **photoelectric effect**, in which electrons are emitted from a metal surface when light shines on it. **Interactive Figure 29.4** illustrates the effect. The electrons are emitted if the light being used has a sufficiently high frequency. The ejected electrons move toward a positive electrode called the *collector* and cause a current to register on the ammeter. Because the electrons are ejected with the aid of light, they are called **photoelectrons**. As will be discussed shortly, a number of features of the photoelectric effect could not be explained solely with the ideas of classical physics.

In 1905 Einstein presented an explanation of the photoelectric effect that took advantage of Planck's work concerning blackbody radiation. It was primarily for his theory of the photoelectric effect that he was awarded the Nobel Prize in physics in 1921. In his photoelectric theory, Einstein proposed that light of frequency f could be regarded as a collection of discrete packets of energy (photons), each packet containing an amount of energy E given by

Energy of a photon $$E = hf \quad \textbf{(29.2)}$$

*It is now known that the energy of a harmonic oscillator is $E = \left(n + \frac{1}{2}\right)hf$; the extra term of $\frac{1}{2}$ is unimportant to the present discussion.

where h is Planck's constant. The light energy given off by a light bulb, for instance, is carried by photons. The brighter the bulb, the greater is the number of photons emitted per second. Example 1 estimates the number of photons emitted per second by a typical light bulb.

EXAMPLE 1 | Photons from a Light Bulb

In converting electrical energy into light energy, a sixty-watt incandescent light bulb operates at about 2.1% efficiency. Assuming that all the light is green light (vacuum wavelength = 555 nm), determine the number of photons per second given off by the bulb.

Reasoning The number of photons emitted per second can be found by dividing the amount of light energy emitted per second by the energy E of one photon. The energy of a single photon is $E = hf$, according to Equation 29.2. The frequency f of the photon is related to its wavelength λ by Equation 16.1 as $f = c/\lambda$.

Solution At an efficiency of 2.1%, the light energy emitted per second by a sixty-watt bulb is (0.021)(60.0 J/s) = 1.3 J/s. The energy of a single photon is

$$E = hf = \frac{hc}{\lambda} = \frac{(6.63 \times 10^{-34}\ \text{J}\cdot\text{s})(3.00 \times 10^{8}\ \text{m/s})}{555 \times 10^{-9}\ \text{m}} = 3.58 \times 10^{-19}\ \text{J}$$

Therefore,

$$\begin{array}{c}\text{Number of}\\ \text{photons emitted}\\ \text{per second}\end{array} = \frac{1.3\ \text{J/s}}{3.58 \times 10^{-19}\ \text{J/photon}} = \boxed{3.6 \times 10^{18}\ \text{photons/s}}$$

The solution to Example 1 shows that the energy of the photons in light increases as the wavelength of the light, λ, decreases. Example 2 discusses this wavelength dependence in the context of sunlight and its ability to damage our skin.

EXAMPLE 2 | BIO The Physics of The Solar Spectrum and Sunscreen

We are all aware of the importance of protecting our skin from prolonged exposure to the sun. **Figure 29.5** shows the solar radiation, or sunlight, that impacts the earth as a function of wavelength. The pink region represents the light that strikes the earth's upper atmosphere, while the blue region shows the reduced intensity that reaches the earth's surface. The several sharp dips in intensity in the solar spectrum at the earth's surface are due to atmospheric gases absorbing the light energy at those particular wavelengths. The sun's radiation spectrum is well approximated by an ideal blackbody with a temperature of 5778 K, which is the surface temperature of the sun (see the solid green line in **Figure 29.5**). The majority of the sun's radiation that reaches the earth spans a range of wavelengths from 200 to 3000 nm, although longer wavelength infrared radiation and shorter wavelength UV, and even X-rays, are also present, but they make up a very small fraction of the sun's total power output. At earth's upper atmosphere, the composition of sunlight is approximately 10% UV, 40% visible, and 50% in the infrared region of the electromagnetic spectrum. The UV radiation can be further separated into three types by wavelength range: UV-A (315 to 400 nm), UV-B (280 to 315 nm), and UV-C (100 to 280 nm).

Around 70% of the sun's UV light is absorbed by the atmosphere. The portion that reaches the earth's surface is represented by the blue striped region shown in **Figure 29.5**, and this is the region in which we are interested. The majority of the composition of the UV light that reaches the earth is UV-A (~95%). Almost all of the UV-C and much of the UV-B are absorbed by oxygen (O_2) and ozone (O_3) in the atmosphere. This is a good thing, since UV-C, also known as hard UV, has sufficient energy to severely damage DNA molecules. In fact, without the atmospheric absorption of UV-C light, DNA-based life on earth would not be possible. UV-B makes up ~5% of the UV light reaching the earth's surface, and it has longer wavelengths than UV-C, and thus less energy, but it also damages DNA in a similar way. It is this radiation that is responsible for damaging skin cells and producing sunburn. UV-A has longer wavelengths than UV-B and even less energy, and is often referred to as soft UV.

It was believed that UV-A was far less damaging to DNA, so artificial UV-A lamps were utilized in tanning beds at salons. However, research has shown that UV-A can also destroy DNA molecules, but by more indirect routes, such as in the creation of free radicals, which are highly reactive atoms or molecules with unpaired electrons that can cause cell damage, leading to premature aging of the skin and even cancer. Sunscreen provides protection from UV-A and UV-B. Historically, both titanium oxide and zinc oxide were used as sunscreen. After application, they would leave a white layer on the skin that reflects much of the UV light away, while also absorbing some of the energy harmlessly. Present-day sunscreens use a combination of inorganic and organic compounds that contain certain chemical bonds that absorb UV light and safely dissipate its energy as heat. The chemical structure of the compounds determines which wavelengths of UV light they absorb—A, B, or both. The chemical bonds will break down over time during exposure, which is why it is so important to frequently reapply sunscreen. This breaking of chemical bonds is the same mechanism that causes the damage in DNA. If it requires 4.28 eV to break the carbon–hydrogen bond in thymine—one of the base pairs of DNA—what is the longest wavelength of UV-B light that will produce this damage?

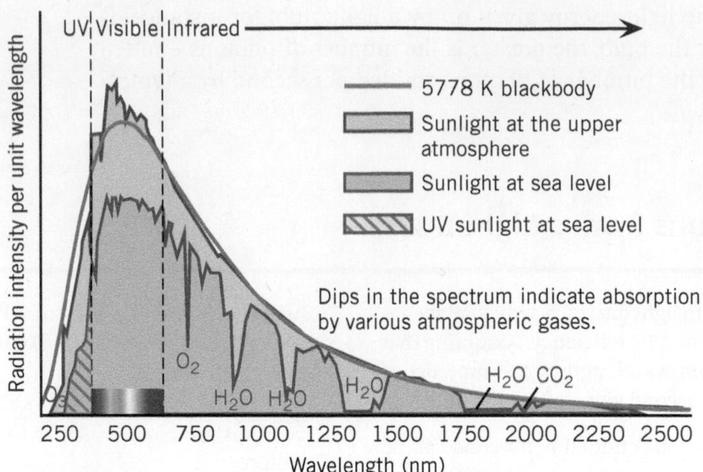

FIGURE 29.5 The solar radiation spectrum as a function of wavelength. The overall spectrum is similar to that produced by a blackbody with a surface temperature equal to that of the sun. The UV sunlight that reaches the earth's surface is responsible for causing sunburn.

Reasoning Given the energy required to break the bond, we can set this energy equal to the energy of the photons in the light with Equation 29.2. The frequency of the light is related to the wavelength by Equation 16.1 with $v = c$.

Solution Applying Equations 29.2 and 16.1, we have

$$E = hf = \frac{hc}{\lambda} \quad \text{or} \quad \lambda = \frac{hc}{E}$$

We can now calculate the wavelength, but we first convert the energy from eV to J:

$$E = (4.28\,\cancel{\text{eV}})\left(\frac{1.60 \times 10^{-19}\ \text{J}}{1\ \cancel{\text{eV}}}\right) = 6.85 \times 10^{-19}\ \text{J}$$

$$\lambda = \frac{hc}{E} = \frac{(6.63 \times 10^{-34}\ \text{J}\cdot\text{s})(3.00 \times 10^{8}\ \text{m/s})}{6.85 \times 10^{-19}\ \text{J}} = 2.90 \times 10^{-7}\ \text{m} = \boxed{290\ \text{nm}}$$

This wavelength is in the middle of the UV-B band.

Related Homework: Problem 45.

According to Einstein, when light shines on a metal, a photon can give up its energy to an electron in the metal. If the photon has enough energy to do the work of removing the electron from the metal, the electron can be ejected. The work required depends on how strongly the electron is held. For the *least strongly* held electrons, the necessary work has a minimum value W_0 and is called the **work function** of the metal. If a photon has energy in excess of the work needed to remove an electron, the excess appears as kinetic energy of the ejected electron. Thus, the least strongly held electrons are ejected with the maximum kinetic energy KE_{max}. Einstein applied the conservation-of-energy principle and proposed the following relation to describe the photoelectric effect:

$$\underbrace{hf}_{\text{Photon energy}} = \underbrace{KE_{max}}_{\text{Maximum kinetic energy of ejected electron}} + \underbrace{W_0}_{\text{Minimum work needed to eject electron}} \qquad (29.3)$$

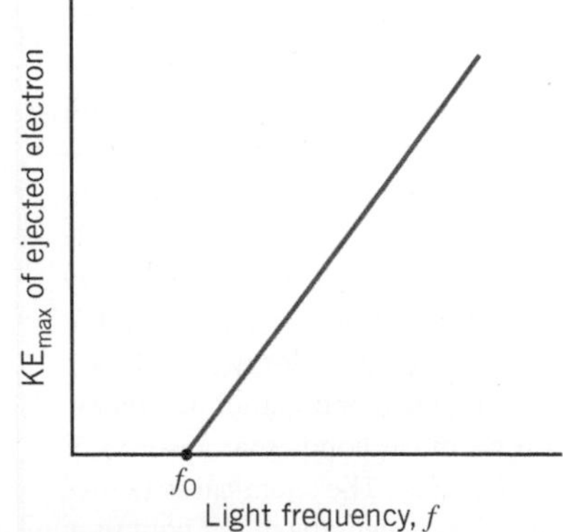

FIGURE 29.6 Photons can eject electrons from a metal when the light frequency is above a minimum value f_0. For frequencies above this value, ejected electrons have a maximum kinetic energy KE_{max} that is linearly related to the frequency, as the graph shows.

According to this equation, $KE_{max} = hf - W_0$, which is plotted in **Figure 29.6**, with KE_{max} along the y axis and f along the x axis. The graph is a straight line that crosses the x axis at $f = f_0$. At this frequency, the electron departs from the metal with no kinetic energy ($KE_{max} = 0$ J). According to Equation 29.3, when $KE_{max} = 0$ J the energy hf_0 of the incident photon is equal to the work function W_0 of the metal: $hf_0 = W_0$.

The photon concept provides an explanation for a number of features of the photoelectric experiment that are difficult to explain without photons. It is observed, for instance, that only light with a frequency above a certain minimum value f_0 will eject electrons. If the frequency is below this value, no electrons are ejected, regardless of how intense the light is. Example 3 determines the minimum frequency value for a silver surface.

EXAMPLE 3 | The Photoelectric Effect for a Silver Surface

The work function for a silver surface is $W_0 = 4.73$ eV. Find the minimum frequency that light must have to eject electrons from this surface.

Reasoning The minimum frequency f_0 is that frequency at which the photon energy equals the work function W_0 of the metal, so the electron is ejected with zero kinetic energy. Since $1 \text{ eV} = 1.60 \times 10^{-19}$ J, the work function expressed in joules is

$$W_0 = (4.73 \cancel{\text{eV}})\left(\frac{1.60 \times 10^{-19} \text{ J}}{1 \cancel{\text{eV}}}\right) = 7.57 \times 10^{-19} \text{ J}$$

Using Equation 29.3, we find

$$hf_0 = \underbrace{\text{KE}_{\text{max}}}_{= 0 \text{ J}} + W_0 \quad \text{or} \quad f_0 = \frac{W_0}{h}$$

Problem-Solving Insight **The work function of a metal is the minimum energy needed to eject an electron from the metal. An electron that has received this minimum energy has no kinetic energy once outside the metal.**

Solution The minimum frequency f_0 is

$$f_0 = \frac{W_0}{h} = \frac{7.57 \times 10^{-19} \text{ J}}{6.63 \times 10^{-34} \text{ J} \cdot \text{s}} = \boxed{1.14 \times 10^{15} \text{ Hz}}$$

Photons with frequencies less than f_0 do not have enough energy to eject electrons from a silver surface. Since $\lambda_0 = c/f_0$, the wavelength of this light is $\lambda_0 = 263$ nm, which is in the ultraviolet region of the electromagnetic spectrum.

In Example 3 the electrons are ejected with no kinetic energy, because the light shining on the silver surface has the minimum possible frequency that will eject them. When the frequency of the light exceeds this minimum value, the electrons that are ejected do have kinetic energy. The next example deals with such a situation.

Analyzing Multiple-Concept Problems

EXAMPLE 4 | The Maximum Speed of Ejected Photoelectrons

Light with a wavelength of 95 nm shines on a selenium surface, which has a work function of 5.9 eV. The ejected electrons have some kinetic energy. Determine the maximum speed with which electrons are ejected.

Reasoning The maximum speed of the ejected electrons is related to their maximum kinetic energy. Conservation of energy dictates that this maximum kinetic energy is related to the work function of the surface and the energy of the incident photons. The work function is given. The energy of the photons can be obtained from the frequency of the light, which is related to the wavelength.

Knowns and Unknowns We have the following data:

Description	Symbol	Value	Comment
Wavelength of light	λ	95 nm	$1 \text{ nm} = 10^{-9}$ m
Work function of selenium surface	W_0	5.9 eV	Will be converted to joules.
Unknown Variable			
Maximum speed of photoelectrons	v_{max}	?	

Modeling the Problem

STEP 1 Kinetic Energy and Speed The maximum kinetic energy KE_{max} of the ejected electrons is $\text{KE}_{\text{max}} = \frac{1}{2}mv_{\text{max}}^2$ (Equation 6.2), where m is the mass of an electron. Solving for the maximum speed v_{max} gives Equation 1 at the right. The mass of the electron is $m = 9.11 \times 10^{-31}$ kg (see inside of front cover of the textbook). The maximum kinetic energy is unknown, but we will evaluate it in Step 2.

$$v_{\text{max}} = \sqrt{\frac{2\text{KE}_{\text{max}}}{m}} \quad (1)$$

? (for KE_{max})

STEP 2 Conservation of Energy According to the principle of conservation of energy, as expressed by Equation 29.3, we have

$$\underbrace{hf}_{\substack{\text{Photon}\\\text{energy}}} = \underbrace{\text{KE}_{\text{max}}}_{\substack{\text{Maximum}\\\text{kinetic energy}\\\text{of ejected}\\\text{electron}}} + \underbrace{W_0}_{\substack{\text{Minimum work}\\\text{needed to eject}\\\text{electron}}}$$

where f is the frequency of the light. Solving for KE_{max} gives

$$\boxed{KE_{max} = hf - W_0} \quad (2)$$

which can be substituted into Equation 1, as shown at the right. In this expression the work function W_0 is known, and we will deal with the unknown frequency f in Step 3.

$$v_{max} = \sqrt{\frac{2KE_{max}}{m}} \quad (1)$$

$$\boxed{KE_{max} = hf - W_0} \quad (2)$$

$$\boxed{?}$$

STEP 3 Relationship Between Frequency and Wavelength The frequency and wavelength of the light are related to the speed of light c according to $f\lambda = c$ (Equation 16.1). Solving for the frequency gives

$$\boxed{f = \frac{c}{\lambda}}$$

which we substitute into Equation 2, as shown at the right.

$$v_{max} = \sqrt{\frac{2KE_{max}}{m}} \quad (1)$$

$$\boxed{KE_{max} = hf - W_0} \quad (2)$$

$$\boxed{f = \frac{c}{\lambda}}$$

Solution Combining the results of each step algebraically, we find that

STEP 1 STEP 2 STEP 3

$$v_{max} = \sqrt{\frac{2KE_{max}}{m}} = \sqrt{\frac{2(hf - W_0)}{m}} = \sqrt{\frac{2\left(h\frac{c}{\lambda} - W_0\right)}{m}}$$

Thus, the maximum speed of the photoelectrons is

$$v_{max} = \sqrt{\frac{2\left(h\frac{c}{\lambda} - W_0\right)}{m}}$$

$$v_{max} = \sqrt{\frac{2\left[(6.63 \times 10^{-34}\ \text{J}\cdot\text{s})\dfrac{(3.00 \times 10^{8}\ \text{m/s})}{(95 \times 10^{-9}\ \text{m})} - (5.9\ \text{eV})\dfrac{(1.60 \times 10^{-19}\ \text{J})}{(1\ \text{eV})}\right]}{9.11 \times 10^{-31}\ \text{kg}}}$$

$$= \boxed{1.6 \times 10^{6}\ \text{m/s}}$$

Note that in this calculation we have converted the value of the work function from electron volts to joules.

Related Homework: Problem 8

Another significant feature of the photoelectric effect is that the maximum kinetic energy of the ejected electrons remains the same when the intensity of the light increases, provided the light frequency remains the same. As the light intensity increases, more photons per second strike the metal, and consequently more electrons per second are ejected. However, since the frequency is the same for each photon, the energy of each photon is also the same. Thus, the ejected electrons always have the same maximum kinetic energy.

Whereas the photon model of light explains the photoelectric effect satisfactorily, the electromagnetic wave model of light does not. Certainly, it is possible to imagine that the electric field of an electromagnetic wave would cause electrons in the metal to oscillate and tear free from the surface when the amplitude of oscillation becomes large enough. However, were this the case, higher-intensity light would eject electrons with a greater maximum kinetic energy, a fact that experiment does not confirm. Moreover, in the electromagnetic wave model, a relatively long time would be required with low-intensity light before the electrons would build up a sufficiently large oscillation amplitude to tear free. Instead, experiment shows that even the weakest light intensity causes electrons to be ejected almost instantaneously, provided the frequency of the light is above the minimum value f_0. The failure of the electromagnetic wave model to explain the photoelectric effect does not mean that the wave model should be abandoned. However, we must recognize that the wave model does not account for all the characteristics of light. The photon model also makes an important contribution to our understanding of the way light behaves when it interacts with matter.

FIGURE 29.7 Digital cameras like this one use an array of charge-coupled devices instead of film to capture an image.

Because a photon has energy, the photon can eject an electron from a metal surface when it interacts with the electron. However, a photon is different from a normal particle. A normal particle has a mass and can travel at speeds up to, but not equal to, the speed of light. A photon, on the other hand, travels at the speed of light in a vacuum and does not exist as an object at rest. The energy of a photon is entirely kinetic in nature, because it has no rest energy and no mass. To show that a photon has no mass, we rewrite Equation 28.4 for the total energy E as

$$E\sqrt{1-\frac{v^2}{c^2}} = mc^2$$

The term $\sqrt{1-(v^2/c^2)}$ is zero because a photon travels at the speed of light, $v = c$. Since the energy E of the photon is finite, the left side of the equation above is zero. Thus, the right side must also be zero, so $m = 0$ kg and the photon has no mass.

THE PHYSICS OF . . . charge-coupled devices and digital cameras. One of the most exciting and useful applications of the photoelectric effect is the charge-coupled device (CCD). An array of these devices is used instead of film in digital cameras (see **Figure 29.7**) to capture images in the form of many small groups of electrons. CCD arrays are also used in digital camcorders and electronic scanners, and they provide the method of choice with which astronomers capture those spectacular images of the planets and the stars. For use with visible light, a CCD array consists of a sandwich of semiconducting silicon, insulating silicon dioxide, and a number of electrodes, as **Figure 29.8** shows. The array is divided into many small sections, or pixels, sixteen of which are shown in the drawing. Each pixel captures a small part of a picture. Digital cameras can have up to over 100 million pixels, depending on price. The greater the number of pixels, the better is the resolution of the photograph. The blow-up in **Figure 29.8** shows a single pixel. Incident photons of visible light strike the silicon and generate electrons via the photoelectric effect. The range of energies of the visible photons is such that approximately one electron is released when a photon interacts with a silicon atom. The electrons do not escape from the silicon, but are trapped within a pixel because of a positive voltage applied to the electrodes beneath the insulating layer. Thus, the number of electrons that are released and trapped is proportional to the number of photons striking the pixel. In this fashion, each pixel in the CCD array accumulates an accurate representation of the light intensity at that point on the image. Color information is provided using red, green, or blue filters or a system of prisms to separate the colors. Astronomers use CCD arrays not only in the visible region of the electromagnetic spectrum but in other regions as well.

In addition to trapping the photoelectrons, the electrodes beneath the pixels are used to read out the electron representation of the picture. By changing the positive voltages applied to the electrodes, it is possible to cause all of the electrons trapped in one row of pixels to be transferred to the adjacent row. In this fashion, for instance, row 1 in **Figure 29.8** is transferred into row 2, row 2 into row 3, and row 3 into the bottom row, which serves a special purpose. The bottom row functions as a horizontal shift register, from which the contents of each pixel can be shifted to the right, one at a time, and read into an analog signal processor.

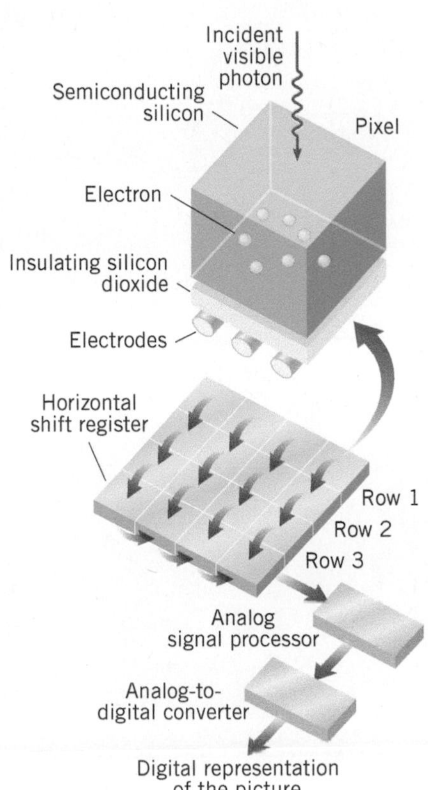

FIGURE 29.8 A CCD array can be used to capture photographic images using the photoelectric effect.

This processor senses the varying number of electrons in each pixel in the shift register as a kind of wave that has a fluctuating amplitude. After another shift in rows, the information in the next row is read out, and so forth. The output of the analog signal processor is sent to an analog-to-digital converter, which produces a digital representation of the image in terms of the zeros and ones that computers recognize.

THE PHYSICS OF . . . a safety feature of garage door openers. Another application of the photoelectric effect depends on the fact that the moving photoelectrons in **Interactive Figure 29.4** constitute a current—a current that changes as the intensity of the light changes. All automatic garage door openers have a safety feature that prevents the door from closing when it encounters an obstruction (person, vehicle, etc.). As **Figure 29.9** illustrates, a sending unit transmits an invisible (infrared) beam across the opening of the door. The beam is detected by a receiving unit that contains a photodiode. A photodiode is a type of *p-n* junction diode (see Section 23.5). When infrared photons strike the photodiode, electrons bound to the atoms absorb the photons and become liberated. These liberated, mobile electrons cause the current in the photodiode to increase. When a person walks through the beam, the light is momentarily blocked from reaching the receiving unit, and the current in the photodiode decreases. The change in current is sensed by electronic circuitry that immediately stops the downward motion of the door and then causes it to rise up.

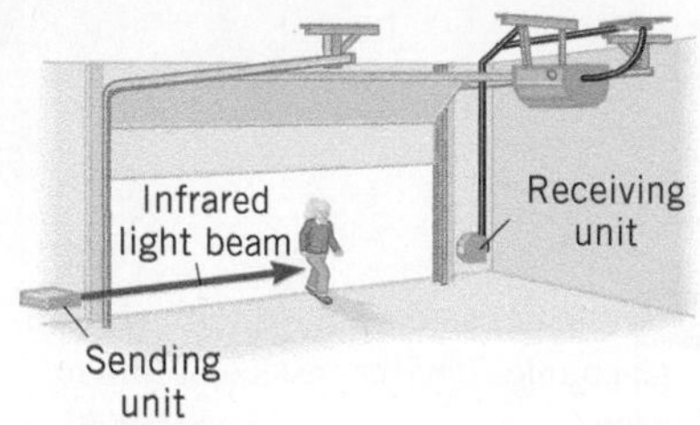

FIGURE 29.9 When an obstruction prevents the infrared light beam from reaching the receiving unit, the current in the receiving unit drops. This drop in current is detected by an electronic circuit that stops the downward movement of the garage door and then causes it to rise.

THE PHYSICS OF . . . photoevaporation and star formation. **Figure 29.10*a*** shows a portion of the Eagle Nebula, a giant star-forming region some 7000 light-years from earth. The photo was taken by the Hubble Space Telescope and reveals clouds of molecular gas and dust, in which there is dramatic evidence of the energy carried by photons. These clouds extend more than a light-year from base to tip and are the birthplace of stars. A star begins to form within a cloud when the gravitational force pulls together sufficient gas to create a high-density "ball." When the gaseous ball becomes sufficiently dense, thermonuclear fusion (see Section 32.5) occurs at its core, and the star begins to shine. The newly born stars are buried within the cloud and cannot be seen from earth. However, the process of photoevaporation allows astronomers to see many of the high-density regions where stars are being formed. Photoevaporation is the process in which high-energy, ultraviolet (UV) photons from hot stars outside the cloud heat it up, much like microwave photons heat food in a microwave oven. **Figure 29.10*a*** shows streamers of gas photoevaporating from the cloud as it is illuminated by stars located beyond the photograph's upper edge. As photoevaporation proceeds, globules of gas that are denser than their surroundings are exposed. The globules are known as *evaporating gaseous globules* (EGGs), and they are slightly larger than our solar system. The drawing in part *b* of **Figure 29.10** shows that the EGGs shade the gas and dust behind them from the UV photons, creating the many finger-like protrusions seen on the surface of the cloud. Astronomers believe that some of these EGGs contain young stars within them.

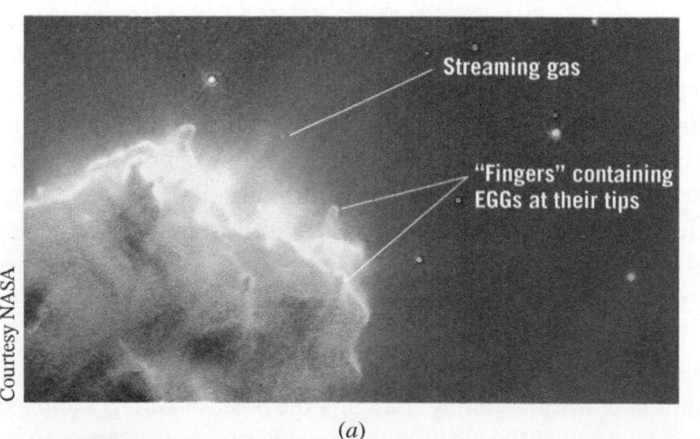

(*a*)

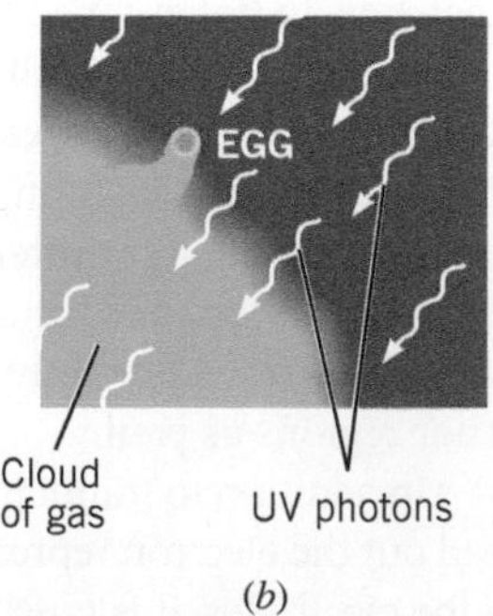

(*b*)

FIGURE 29.10 (*a*) Photoevaporation produces finger-like projections on the surface of the gas clouds in the Eagle Nebula. At the fingertips are high-density evaporating gaseous globules (EGGs). (*b*) This drawing illustrates the photoevaporation that is occurring in the photograph in part *a*.

Check Your Understanding

(The answers are given at the end of the book.)

1. The photons emitted by a source of light do *not* all have the same energy. Is the source monochromatic? (A monochromatic light source emits light that has a single wavelength.)

2. Which colored light bulb—red, orange, yellow, green, or violet—emits photons with **(a)** the least energy and **(b)** the greatest energy? (See Example 1 in Chapter 24.)

3. Does a photon emitted by a higher-wattage red light bulb have more energy than a photon emitted by a lower-wattage red bulb?

4. Radiation of a given wavelength causes electrons to be emitted from the surface of metal 1 but not from the surface of metal 2. Why could this be? **(a)** Metal 1 has a greater work function than metal 2 has. **(b)** Metal 1 has a smaller work function than metal 2 has. **(c)** The energy of a photon striking metal 1 is greater than the energy of a photon striking metal 2.

5. In the photoelectric effect, electrons are ejected from the surface of a metal when light shines on it. Which one or more of the following would lead to an increase in the maximum kinetic energy of the ejected electrons? **(a)** Increasing the frequency of the incident light **(b)** Increasing the number of photons per second striking the surface **(c)** Using photons whose frequency f_0 is less than W_0/h, where W_0 is the work function of the metal and h is Planck's constant **(d)** Selecting a metal that has a greater work function

6. In the photoelectric effect, suppose that the intensity of the light is increased, while the frequency of the light is kept constant. The frequency is greater than the minimum frequency f_0. State whether each of the following will increase, decrease, or remain the same: **(a)** The current in the phototube **(b)** The number of electrons emitted per second from the metal surface **(c)** The maximum kinetic energy that an electron could have **(d)** The maximum momentum that an electron could have

29.4 The Momentum of a Photon and the Compton Effect

Although Einstein presented his photon model for the photoelectric effect in 1905, it was not until 1923 that the photon concept began to achieve widespread acceptance. It was then that the American physicist Arthur H. Compton (1892–1962) used the photon model to explain his research on the scattering of X-rays by the electrons in graphite. X-rays are high-frequency electromagnetic waves and, like light, they are composed of photons.

Animated Figure 29.11 illustrates what happens when an X-ray photon strikes an electron in a piece of graphite. Like two billiard balls colliding on a pool table, the X-ray photon scatters in one direction after the collision, and the electron recoils in another direction. Compton observed that the scattered photon has a frequency f' that is smaller than the frequency f of the incident photon, indicating that the photon loses energy during the collision. In addition, he found that the difference between the two frequencies depends on the angle θ at which the scattered photon leaves the collision. The phenomenon in which an X-ray photon is scattered from an electron, with the scattered photon having a smaller frequency than the incident photon, is called the **Compton effect**.

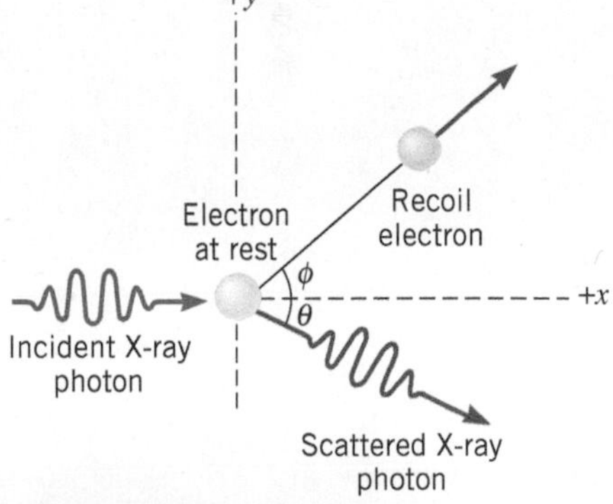

ANIMATED FIGURE 29.11 In an experiment performed by Arthur H. Compton, an X-ray photon collides with a stationary electron. The scattered photon and the recoil electron depart from the collision in different directions.

In Section 7.3, collisions between two objects are analyzed using the fact that the total kinetic energy and the total linear momentum of the objects are the same before and after an elastic collision. Similar analysis can be applied to the collision between a photon and an electron. The electron is assumed to be initially at rest and essentially free—that is, not bound to the atoms of the material. According to the principle of conservation of energy,

$$\underbrace{hf}_{\text{Energy of incident photon}} = \underbrace{hf'}_{\text{Energy of scattered photon}} + \underbrace{\text{KE}}_{\text{Kinetic energy of recoil electron}} \tag{29.4}$$

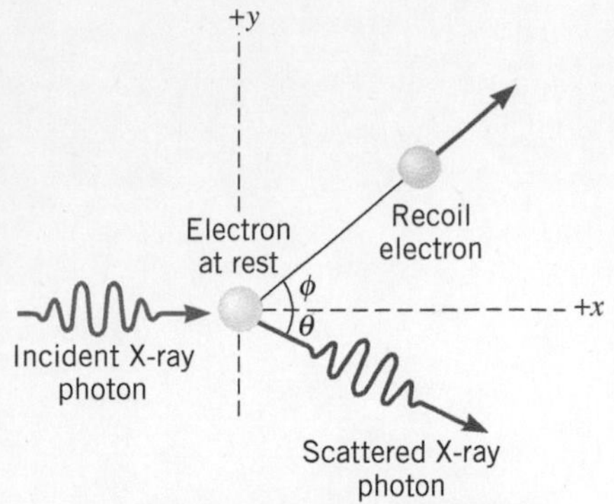

ANIMATED FIGURE 29.11 (REPEATED)
In an experiment performed by Arthur H. Compton, an X-ray photon collides with a stationary electron. The scattered photon and the recoil electron depart from the collision in different directions.

Math Skills Equation 29.5 is a relationship between the momentum $\vec{\mathbf{p}}$ of the incident photon, the momentum $\vec{\mathbf{p}}'$ of the scattered photon, and the momentum $\vec{\mathbf{p}}_{\text{electron}}$ of the recoil electron in **Animated Figure 29.11**. These momenta are vector quantities. Therefore, Equation 29.5 is equivalent to two equations; one is for the x components of the vectors and one for the y components of the vectors (see Sections 1.7 and 1.8). Using components with respect to the x, y axis in **Animated Figure 29.11**, we can see that the following two equations are equivalent to Equation 29.5:

$$\textbf{x component} \quad \underbrace{p}_{\text{Incident photon}} = \underbrace{p'\cos\theta}_{\text{Scattered photon}} + \underbrace{p_{\text{electron}}\cos\phi}_{\text{Recoil electron}}$$

$$\textbf{y component} \quad \underbrace{0}_{\text{Incident photon}} = \underbrace{-p'\sin\theta}_{\text{Scattered photon}} + \underbrace{p_{\text{electron}}\sin\phi}_{\text{Recoil electron}}$$

In these equations, the symbols p and p' denote the vector magnitudes of the momenta. Note that the y component of the momentum is zero for the incident photon, because that photon travels along the x axis in **Animated Figure 29.11**. Note also that the y component of the momentum is negative for the scattered photon, because the direction in which that photon travels is below the x axis.

where the relation $E = hf$ has been used for the photon energies. It follows, then, that $hf' = hf - \text{KE}$, which shows that the energy and corresponding frequency f' of the scattered photon are less than the energy and frequency of the incident photon, just as Compton observed. Since $\lambda' = c/f'$ (Equation 16.1), the wavelength of the scattered X-rays is larger than that of the incident X-rays.

For an initially stationary electron, conservation of total linear momentum requires that

$$\underbrace{\vec{\mathbf{p}}}_{\substack{\text{Momentum of}\\ \text{incident photon}}} = \underbrace{\vec{\mathbf{p}}'}_{\substack{\text{Momentum of}\\ \text{scattered photon}}} + \underbrace{\vec{\mathbf{p}}_{\text{electron}}}_{\substack{\text{Momentum}\\ \text{of recoil}\\ \text{electron}}} \tag{29.5}$$

To find an expression for the magnitude p of a photon's momentum, we use Equations 28.3 and 28.4. According to these equations, the momentum p and the total energy E of any particle are

$$p = \frac{mv}{\sqrt{1-(v^2/c^2)}} \tag{28.3}$$

$$E = \frac{mc^2}{\sqrt{1-(v^2/c^2)}} \tag{28.4}$$

Rearranging Equation 28.4 to show that $\dfrac{m}{\sqrt{1-(v^2/c^2)}} = \dfrac{E}{c^2}$ and substituting this result into Equation 28.3 gives

$$p = \frac{mv}{\sqrt{1-(v^2/c^2)}} = \frac{Ev}{c^2}$$

A photon travels at the speed of light, so that we have $v = c$, and the momentum of a photon is

$$p = \frac{Ev}{c^2} = \frac{Ec}{c^2} = \frac{E}{c}$$

This result only applies to a photon and does not apply to a particle with mass, because such a particle cannot travel at the speed of light. We also know that the energy of a photon is related to its frequency f according to $E = hf$ (Equation 29.2) and that the speed c of a photon is related to its frequency and wavelength λ according to $c = f\lambda$ (Equation 16.1). With these substitutions, our expression for the momentum of a photon becomes

$$p = \frac{E}{c} = \frac{hf}{f\lambda} = \frac{h}{\lambda} \tag{29.6}$$

Using Equations 29.4, 29.5, and 29.6, Compton showed that the difference between the wavelength λ' of the scattered photon and the wavelength λ of the incident photon is related to the scattering angle θ by

$$\lambda' - \lambda = \frac{h}{mc}(1 - \cos\theta) \tag{29.7}$$

In this equation m is the mass of the electron. The quantity $h/(mc)$ is called the **Compton wavelength of the electron** and has the value $h/(mc) = 2.43 \times 10^{-12}$ m. Since $\cos\theta$ varies between +1 and −1, the shift $\lambda' - \lambda$ in the wavelength can vary between zero and $2h/(mc)$, depending on the value of θ, a fact observed by Compton.

The photoelectric effect and the Compton effect provide compelling evidence that light can exhibit particle-like characteristics attributable to energy packets called photons. But what about the interference phenomena discussed in Chapter 27, such as Young's double-slit experiment and single-slit diffraction, which demonstrate that light behaves as a wave? Does light have two distinct personalities, in which it behaves like a stream of particles in some experiments and like a wave in others? The answer is yes, for physicists now believe that this wave–particle duality is an inherent property of light. Light is a far more interesting (and complex) phenomenon than just a stream of particles or an electromagnetic wave.

In the Compton effect the electron recoils because it gains some of the photon's momentum. In principle, then, the momentum that photons have can be used to make other objects move. Conceptual Example 5 considers a propulsion system for an interstellar spaceship that is based on the momentum of a photon.

CONCEPTUAL EXAMPLE 5 | The Physics of Solar Sails and Spaceship Propulsion

One propulsion method that is currently being studied for interstellar travel uses a large sail. The intent is that sunlight striking the sail will create a force that pushes the ship away from the sun (**Figure 29.12**), much as the wind propels a sailboat. To get the greatest possible force, the surface of the sail facing the sun **(a)** should be shiny like a mirror, **(b)** should be black, or **(c)** could be either shiny or black, since the same force will be created for either type of surface.

Reasoning In Conceptual Example 3 in Chapter 7, we found that hailstones striking the roof of a car exert a force on it because the collision changes their momentum. Photons also have momentum, so, like the hailstones, they can apply a force to the sail.

As in Chapter 7, we will be guided by the impulse–momentum theorem (Equation 7.4) in assessing the force. This theorem states that when a net force acts on an object, the impulse of the net force is equal to the change in momentum of the object. Greater impulses lead to greater forces for a given time interval. Thus, when a photon collides with the sail, the photon's momentum changes because of the force that the sail applies to the photon. Newton's action–reaction law (Section 4.5) indicates that the photon simultaneously applies a force of equal magnitude, but opposite direction, to the sail. It is this reaction force that propels the spaceship, and it will be greater when the momentum change experienced by the photon is greater. The surface of the sail facing the sun, then, should be such that it causes the largest possible momentum change for the impinging photons.

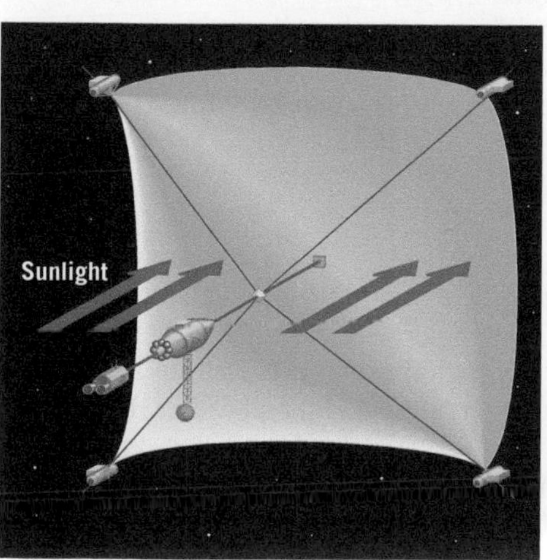

FIGURE 29.12 A solar sail provides the propulsion for this interstellar spaceship.

Answers (b) and (c) are incorrect. In Conceptual Example 3 in Chapter 7 we examined whether hailstones or raindrops exert the greater force when they strike the roof of a car. The difference is that hailstones, being hard objects, bounce off the roof, while raindrops splatter and do not bounce very much. We concluded that hailstones, because of the bounce, experience a greater change in momentum than raindrops do, so the roof exerts a greater force on the hailstones. By Newton's action–reaction law, the car roof, then, experiences a greater force from the hailstones than from the raindrops. We saw in Section 13.3 that radiation is reflected from a shiny mirror-like surface and is absorbed by a black surface. Therefore, by analogy with raindrops that stick to the car roof, the sail experiences a smaller force when the surface facing the sun is black.

Answer (a) is correct. The sun's radiation reflects from a shiny mirror-like surface and is absorbed by a black surface. Now, consider a photon that strikes the sail perpendicularly. When a photon reflects from a mirror-like surface, the photon's momentum changes from its value in the forward direction to a value of the same magnitude in the reverse direction. This is a greater change than the one that occurs when the photon is absorbed by a black surface. In the latter case, the momentum changes only from its value in the forward direction to a value of zero. Consequently, the surface of the sail facing the sun should be shiny in order to produce the greatest possible propulsion force. A shiny surface causes the photons to bounce like hailstones on the roof of a car and, in doing so, to apply a greater force to the sail.

Check Your Understanding

(*The answers are given at the end of the book.*)

7. In the Compton effect, an incident X-ray photon of wavelength λ is scattered by an electron, and the scattered photon has a wavelength of λ'. Suppose that the incident photon is scattered by a proton instead of an electron. For a given scattering angle θ, is the change $\lambda' - \lambda$ in the wavelength of the photon scattered by the proton greater than, less than, or the same as the wavelength of the photon scattered by the electron? (*Hint:* Use Equation 29.7 for a proton instead of an electron.)

8. In a Compton scattering experiment, an incident X-ray photon is traveling along the $+x$ direction. An electron, initially at rest, is struck by the photon and is accelerated straight ahead in the same direction as the incident X-ray photon. Which way does the scattered photon move? **(a)** Along the $+y$ direction **(b)** Along the $-y$ direction **(c)** Along the $-x$ direction (*Hint:* Use the principle of conservation of momentum to guide your reasoning.)

9. The speed of a particle is much less than the speed of light. Thus, when the particle's speed doubles, the particle's momentum doubles, and its kinetic energy becomes four times greater. However, when the momentum of a photon doubles, its energy becomes **(a)** two times greater **(b)** four times greater **(c)** one-half as much **(d)** one-fourth as much.

10. Review Conceptual Example 5 as background for this question. **CYU Figure 29.1** shows a device called a radiometer. The four regular panels are black on one side and shiny like a mirror on the other side. In bright light, the panel arrangement spins around in a direction from the black side of a panel toward the shiny side. Do photon collisions with both sides of the panels cause the observed spinning?

Charles D. Winters/Science Source

CYU FIGURE 29.1

29.5 The De Broglie Wavelength and the Wave Nature of Matter

As a graduate student in 1923, Louis de Broglie (1892–1987) made the astounding suggestion that since light waves could exhibit particle-like behavior, particles of matter should exhibit wave-like behavior. De Broglie proposed that all moving matter has a wavelength associated with it, just as a wave does. He made the explicit proposal that the wavelength λ of a particle is given by the same relation (Equation 29.6) that applies to a photon:

De Broglie wavelength

$$\lambda = \frac{h}{p} \tag{29.8}$$

where h is Planck's constant and p is the magnitude of the relativistic momentum of the particle. Today, λ is known as the **de Broglie wavelength** of the particle.

Confirmation of de Broglie's suggestion came in 1927 from the experiments of the American physicists Clinton J. Davisson (1881–1958) and Lester H. Germer (1896–1971) and, independently, those of the English physicist George P. Thomson (1892–1975). Davisson and Germer directed a beam of electrons onto a crystal of nickel and observed that the electrons exhibited a diffraction behavior, analogous to that seen when X-rays are diffracted by a crystal (see Section 27.9 for a discussion of X-ray diffraction). The wavelength of the electrons revealed by the diffraction pattern matched that predicted by de Broglie's hypothesis, $\lambda = h/p$. More recently, Young's double-slit experiment has

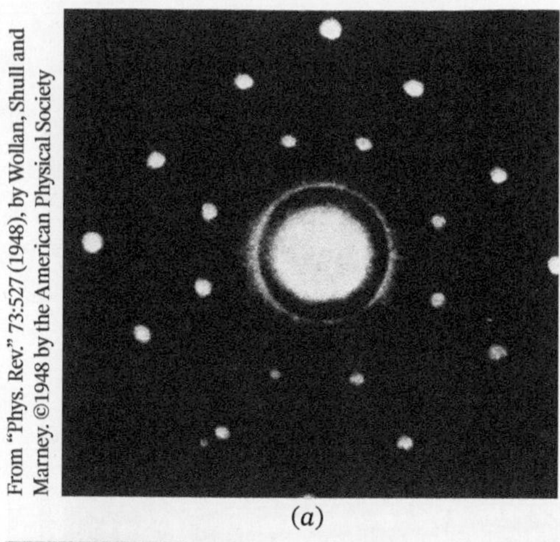

(a)

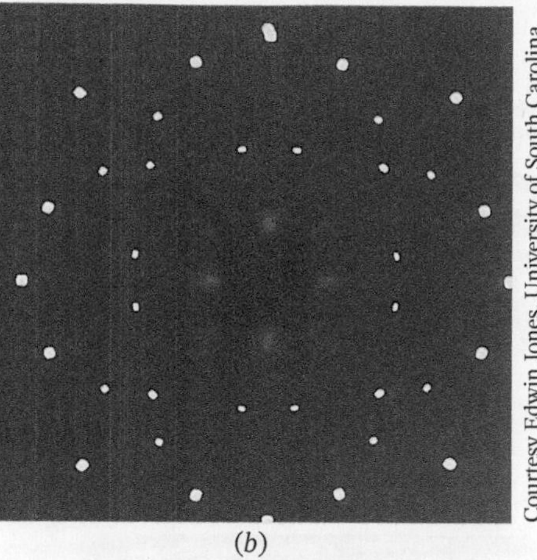

(b)

FIGURE 29.13 (*a*) The neutron diffraction pattern and (*b*) the X-ray diffraction pattern for a crystal of sodium chloride (NaCl).

been performed with electrons and reveals the effects of wave interference illustrated in **Interactive Figure 29.1**.

Particles other than electrons can also exhibit wave-like properties. For instance, neutrons are sometimes used in diffraction studies of crystal structure. **Figure 29.13** compares the neutron diffraction pattern and the X-ray diffraction pattern caused by a crystal of rock salt (NaCl).

Although all moving particles have a de Broglie wavelength, the effects of this wavelength are observable only for particles whose masses are very small, on the order of the mass of an electron or a neutron, for instance. Example 6 illustrates why.

EXAMPLE 6 | The De Broglie Wavelength of an Electron and of a Baseball

Determine the de Broglie wavelength for **(a)** an electron (mass = 9.1×10^{-31} kg) moving at a speed of 6.0×10^6 m/s and **(b)** a baseball (mass = 0.15 kg) moving at a speed of 13 m/s.

Reasoning In each case, the de Broglie wavelength is given by Equation 29.8 as Planck's constant divided by the magnitude of the momentum. Since the speeds are small compared to the speed of light, we can ignore relativistic effects and express the magnitude of the momentum as the product of the mass and the speed, as in Equation 7.2.

Solution (a) Since the magnitude p of the momentum is the product of the mass m of the particle and its speed v we have $p = mv$ (Equation 7.2). Using this expression in Equation 29.8 for the de Broglie wavelength, we obtain

$$\lambda = \frac{h}{p} = \frac{h}{mv} = \frac{6.63 \times 10^{-34}\ \text{J}\cdot\text{s}}{(9.1 \times 10^{-31}\ \text{kg})(6.0 \times 10^{6}\ \text{m/s})} = \boxed{1.2 \times 10^{-10}\ \text{m}}$$

A de Broglie wavelength of 1.2×10^{-10} m is about the size of the interatomic spacing in a solid, such as the nickel crystal used by Davisson and Germer, and, therefore, leads to the observed diffraction effects.

(b) A calculation similar to that in part (a) shows that the de Broglie wavelength of the baseball is $\boxed{\lambda = 3.3 \times 10^{-34}\ \text{m}}$. This wavelength is incredibly small, even by comparison with the size of an atom (10^{-10} m) or a nucleus (10^{-14} m). Thus, the ratio λ/W of this wavelength to the width W of an ordinary opening, such as a window, is so small that the diffraction of a baseball passing through the window cannot be observed.

The de Broglie equation for particle wavelength provides no hint as to what kind of wave is associated with a particle of matter. To gain some insight into the nature of this wave, we turn our attention to **Figure 29.14**. Part *a* shows the fringe pattern on the screen when electrons are used in a version of Young's double-slit experiment. The bright fringes occur in places where particle waves coming from each slit interfere constructively, while the dark fringes occur in places where the waves interfere destructively.

When an electron passes through the double-slit arrangement and strikes a spot on the screen, the screen glows at that spot, and parts *b*, *c*, and *d* of **Figure 29.14** illustrate how the spots accumulate in time. As more and more electrons strike the screen, the spots eventually form the fringe pattern that is evident in part *d*. Bright fringes occur where there is a high probability of electrons striking the screen, and dark fringes occur where there is a low probability. Here lies the key to understanding particle

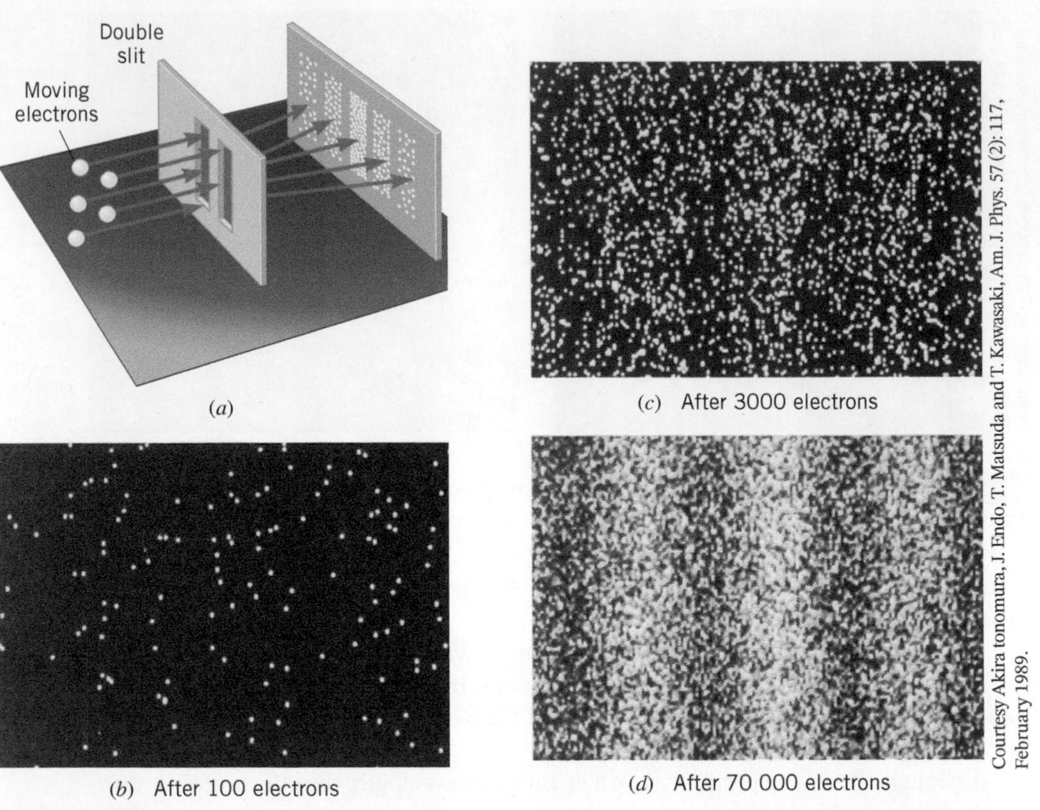

FIGURE 29.14 In this electron version of Young's double-slit experiment, the characteristic fringe pattern becomes recognizable only after a sufficient number of electrons have struck the screen.

waves. **Particle waves are waves of probability**, waves whose magnitude at a point in space gives an indication of the probability that the particle will be found at that point. At the place where the screen is located, the pattern of probabilities conveyed by the particle waves causes the fringe pattern to emerge. The fact that no fringe pattern is apparent in part *b* of the figure does not mean that there are no probability waves present; it just means that too few electrons have struck the screen for the pattern to be recognizable.

The pattern of probabilities that leads to the fringes in **Figure 29.14** is analogous to the pattern of light intensities that is responsible for the fringes in Young's original experiment with light waves (see Figure 27.3). Section 24.4 discusses the fact that the intensity of the light is proportional to either the square of the electric field strength or the square of the magnetic field strength of the wave. In an analogous fashion in the case of particle waves, the probability is proportional to the square of the magnitude ψ (Greek letter psi) of the wave. ψ is referred to as the **wave function** of the particle.

In 1925 the Austrian physicist Erwin Schrödinger (1887–1961) and the German physicist Werner Heisenberg (1901–1976) independently developed theoretical frameworks for determining the wave function. In so doing, they established a new branch of physics called **quantum mechanics**. The word "quantum" refers to the fact that in the world of the atom, where particle waves must be considered, the particle energy is quantized, so only certain energies are allowed. To understand the structure of the atom and the phenomena related to it, quantum mechanics is essential, and the Schrödinger equation for calculating the wave function is now widely used. In the next chapter, we will explore the structure of the atom based on the ideas of quantum mechanics.

Check Your Understanding

(*The answers are given at the end of the book.*)

11. A stone is dropped from the top of a building. As the stone falls, does its de Broglie wavelength increase, decrease, or remain the same?

12. An electron and a neutron have different masses. Is it possible, according to Equation 29.8, that they can have the same de Broglie wavelength? **(a)** Yes, provided the magnitudes of their momenta are different. **(b)** Yes, provided their speeds are different. **(c)** No; two particles with different masses always have different de Broglie wavelengths.

13. In **Interactive Figure 29.1b**, replace the electrons with protons that have the same speed. With the aid of Equation 27.1 for the bright fringes in Young's double-slit experiment and Equation 29.8, decide whether the angular separation between the fringes would increase, decrease, or remain the same, relative to the angular separation produced by the electrons.

14. A beam of electrons passes through a single slit, and a beam of protons passes through a second, but identical, slit. The electrons and the protons have the same speed. Which one of the following correctly describes the beam that experiences the greatest amount of diffraction? **(a)** The electrons, because they have the smaller momentum and, hence, the smaller de Broglie wavelength **(b)** The electrons, because they have the smaller momentum and, hence, the larger de Broglie wavelength **(c)** The protons, because they have the smaller momentum and, hence, the smaller de Broglie wavelength **(d)** The protons, because they have the larger momentum and, hence, the smaller de Broglie wavelength **(e)** Both beams experience the same amount of diffraction, because the electrons and protons have the same de Broglie wavelength.

29.6 The Heisenberg Uncertainty Principle

As the previous section discusses, the bright fringes in **Figure 29.14** indicate the places where there is a high probability of an electron striking the screen. Since there are a number of bright fringes, there is more than one place where each electron has some probability of hitting. As a result, it is not possible to specify in advance exactly where on the screen an individual electron will hit. All we can do is speak of the probability that the electron may end up in a number of different places. No longer is it possible to say, as Newton's laws would suggest, that a single electron, fired through the double slit, will travel directly forward in a straight line and strike the screen. This simple model just does not apply when a particle as small as an electron passes through a pair of closely spaced narrow slits. Because the wave nature of particles is important in such circumstances, we lose the ability to predict with 100% certainty the path that a single particle will follow. Instead, only the average behavior of large numbers of particles is predictable, and the behavior of any individual particle is uncertain.

To see more clearly into the nature of the uncertainty, consider electrons passing through a single slit, as in **Figure 29.15**. After a sufficient number of electrons strike the screen, a diffraction pattern emerges. The electron diffraction pattern consists of alternating bright and dark fringes and is analogous to the pattern for light waves shown in **Figure 27.23. Figure 29.15** shows the slit and locates the first dark fringe on either side of the central bright fringe. The central fringe is bright because electrons strike the

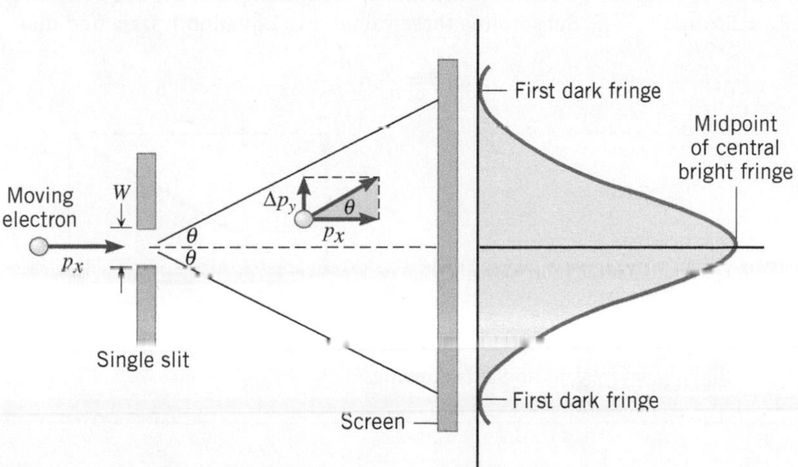

FIGURE 29.15 When a sufficient number of electrons pass through a single slit and strike the screen, a diffraction pattern of bright and dark fringes emerges. (Only the central bright fringe is shown.) This pattern is due to the wave nature of the electrons and is analogous to that produced by light waves.

screen over the entire region between the dark fringes. If the electrons striking the screen outside the central bright fringe are neglected, the extent to which the electrons are diffracted is given by the angle θ in the drawing. To reach locations within the central bright fringe, some electrons must have acquired momentum in the y direction, despite the fact that they enter the slit traveling along the x direction and have no momentum in the y direction to start with. The figure illustrates that the y component of the momentum may be as large as Δp_y. The notation Δp_y indicates the difference between the maximum value of the y component of the momentum after the electron passes through the slit and its value of zero before the electron passes through the slit. Δp_y represents the *uncertainty* in the y component of the momentum, in the sense that the y component may have any value from zero to Δp_y.

It is possible to relate Δp_y to the width W of the slit. To do this, we assume that Equation 27.4, which applies to light waves, also applies to particle waves whose de Broglie wavelength is λ. This equation, $\sin\theta = \lambda/W$, specifies the angle θ that locates the first dark fringe. If θ is small, then $\sin\theta \approx \tan\theta$. Moreover, **Figure 29.15** indicates that $\tan\theta = \Delta p_y/p_x$, where p_x is the x component of the momentum of the electron. Therefore, $\Delta p_y/p_x < \lambda/W$. However, $p_x = h/\lambda$ according to de Broglie's equation, so that

$$\frac{\Delta p_y}{p_x} = \frac{\Delta p_y}{h/\lambda} \approx \frac{\lambda}{W}$$

As a result,

$$\Delta p_y \approx \frac{h}{W} \tag{29.9}$$

which indicates that a smaller slit width leads to a larger uncertainty in the y component of the electron's momentum.

It was Heisenberg who first suggested that the uncertainty Δp_y in the y component of the momentum is related to the uncertainty in the y position of the electron as the electron passes through the slit. To get a feel for this relationship, let's assume that the center of the slit is at $y = 0$ m. Because the width of the slit is W, the electron is somewhere within $\pm\frac{1}{2}W$ from the center of the slit. Thus, we take the uncertainty

Math Skills To see why $\sin\theta \approx \tan\theta$ when θ is small, refer to **Figure 29.16a** and recall that the sine and tangent functions are defined as follows in Section 1.4:

$$\sin\theta = \frac{h_o}{h} \quad (1.1) \quad \text{and} \quad \tan\theta = \frac{h_o}{h_a} \quad (1.3)$$

where h_o is the length of the side of a right triangle opposite the angle θ, h_a is the length of the side adjacent to the angle θ, and h is the hypotenuse. When θ is small, h and h_a become approximately equal, or $h \approx h_a$, and the right-hand sides of Equations 1.1 and 1.3 become approximately the same.

To see why $\tan\theta = \Delta p_y/p_x$, refer to **Figure 29.16b**, which shows in enlarged form the portion of **Figure 29.15** that establishes the angle θ. A comparison of the shaded right triangle in **Figure 29.16b** with the shaded right triangle in **Figure 29.16a** reveals that $h_o = \Delta p_y$ and $h_a = p_x$. Substituting these values into Equation 1.3, we find that

$$\tan\theta = \frac{h_o}{h_a} = \frac{\Delta p_y}{p_x}$$

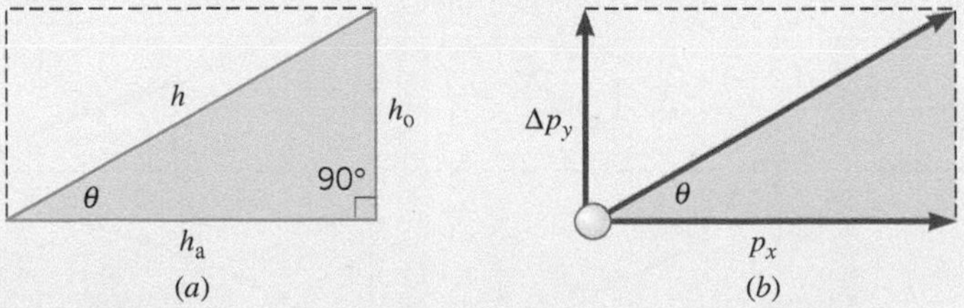

FIGURE 29.16 Math Skills drawing.

in the y position of the electron to be $\Delta y = \frac{1}{2}W$, so that $W = 2\,\Delta y$. Substituting this relation into Equation 29.9 shows that $\Delta p_y \approx h/(2\,\Delta y)$ or $(\Delta p_y)(\Delta y) \approx \frac{1}{2}h$. The result of Heisenberg's more complete analysis is given below in Equation 29.10 and is known as the **Heisenberg uncertainty principle.** Note that the Heisenberg uncertainty principle is a general principle with wide applicability. It does not just apply to the case of single-slit diffraction, which we have used here for the sake of convenience.

THE HEISENBERG UNCERTAINTY PRINCIPLE

Momentum and position

$$(\Delta p_y)(\Delta y) \geq \frac{h}{4\pi} \quad (29.10)$$

Δy = uncertainty in a particle's position along the y direction

Δp_y = uncertainty in the y component of the linear momentum of the particle

Energy and time

$$(\Delta E)(\Delta t) \geq \frac{h}{4\pi} \quad (29.11)$$

ΔE = uncertainty in the energy of a particle when the particle is in a certain state

Δt = time interval during which the particle is in the state

The Heisenberg uncertainty principle places limits on the accuracy with which the momentum and position of a particle can be specified simultaneously. These limits are not just limits due to faulty measuring techniques. They are fundamental limits imposed by nature, and there is no way to circumvent them. Equation 29.10 indicates that Δp_y and Δy cannot both be arbitrarily small at the same time. If one is small, then the other must be large, so that their product equals or exceeds ($\geq$) Planck's constant divided by 4π. For example, if the position of a particle is known exactly, so that Δy is zero, then Δp_y is an infinitely large number, and the momentum of the particle is completely uncertain. Conversely, if we assume that Δp_y is zero, then Δy is an infinitely large number, and the position of the particle is completely uncertain. In other words, the Heisenberg uncertainty principle states that it is impossible to specify precisely both the momentum and position of a particle at the same time.

There is also an uncertainty principle that deals with energy and time, as expressed by Equation 29.11. The product of the uncertainty ΔE in the energy of a particle and the time interval Δt during which the particle remains in a given energy state is greater than or equal to Planck's constant divided by 4π. Therefore, the shorter the lifetime of a particle in a given state, the greater is the uncertainty in the energy of that state.

Example 7 shows that the uncertainty principle has significant consequences for the motion of tiny particles such as electrons but has little effect on the motion of macroscopic objects, even those with as little mass as a Ping-Pong ball.

EXAMPLE 7 | The Heisenberg Uncertainty Principle

Assume that the position of an object is known so precisely that the uncertainty in the position is only $\Delta y = 1.5 \times 10^{-11}$ m. **(a)** Determine the minimum uncertainty in the momentum of the object. Find the corresponding minimum uncertainty in the speed of the object in the case when the object is **(b)** an electron (mass = 9.1×10^{-31} kg) and **(c)** a Ping-Pong ball (mass = 2.2×10^{-3} kg).

Reasoning The minimum uncertainty Δp_y in the y component of the momentum is given by the Heisenberg uncertainty principle as $\Delta p_y = h/(4\pi\Delta y)$, where Δy is the uncertainty in the position of the object along the y direction. Both the electron and the Ping-Pong ball have the same uncertainty in their momenta because they have the same uncertainty in their positions. However, these objects have very different masses. As a result, we will find that the uncertainty in the speeds of these objects is very different.

Problem-Solving Insight **The Heisenberg uncertainty principle states that the product of Δp_y and Δy is greater than or equal to $h/4\pi$. For a given value of Δp_y or Δy, the minimum uncertainty in the other term occurs when the product is equal to $h/4\pi$.**

Solution (a) From Equation 29.10, the minimum uncertainty in the y component of the momentum is given by

$$\Delta p_y = \frac{h}{4\pi\,\Delta y} = \frac{6.63 \times 10^{-34}\ \text{J}\cdot\text{s}}{4\pi(1.5 \times 10^{-11}\ \text{m})} = \boxed{3.5 \times 10^{-24}\ \text{kg}\cdot\text{m/s}}$$

(b) The magnitude p_y of the momentum is $p_y = mv_y$ (Equation 7.2), where m is the mass of the object and v_y is its speed. Therefore, the uncertainty Δp_y is $\Delta p_y = m\Delta v_y$, and the minimum uncertainty in the speed of the electron is

$$\Delta v_y = \frac{\Delta p_y}{m} = \frac{3.5 \times 10^{-24}\ \text{kg} \cdot \text{m/s}}{9.1 \times 10^{-31}\ \text{kg}} = \boxed{3.8 \times 10^{6}\ \text{m/s}}$$

Thus, the small uncertainty in the y position of the electron gives rise to a large uncertainty in the speed of the electron.

(c) The minimum uncertainty in the speed of the Ping-Pong ball is

$$\Delta v_y = \frac{\Delta p_y}{m} = \frac{3.5 \times 10^{-24}\ \text{kg} \cdot \text{m/s}}{2.2 \times 10^{-3}\ \text{kg}} = \boxed{1.6 \times 10^{-21}\ \text{m/s}}$$

Because the mass of the Ping-Pong ball is relatively large, the small uncertainty in its y position gives rise to an uncertainty in its speed that is much smaller than that for the electron. Thus, in contrast to the case for the electron, we can know simultaneously where the ball is and how fast it is moving, to a very high degree of certainty.

Related Homework: Problem 36.

Example 7 emphasizes how the uncertainty principle imposes different uncertainties on the speeds of an electron (small mass) and a Ping-Pong ball (large mass). For objects like the ball, which have relatively large masses, the uncertainties in position and speed are so small that they have no effect on our ability to determine simultaneously where such objects are and how fast they are moving. The uncertainties calculated in Example 7 depend on more than just the mass, however. They also depend on Planck's constant, which is a very small number. It is interesting to speculate about what life would be like if Planck's constant were much larger than 6.63×10^{-34} J · s. Conceptual Example 8 deals with just such speculation.

CONCEPTUAL EXAMPLE 8 | What If Planck's Constant Were Large?

Suppose that you are target shooting at a stationary target. A bullet leaving the barrel of a gun is analogous to an electron passing through the single slit in **Figure 29.15**. With this analogy in mind and assuming that the magnitude of the bullet's momentum is not abnormally large, what would target shooting be like if Planck's constant had a relatively large value instead of its extremely small value of 6.63×10^{-34} J · s? **(a)** It would be more accurate because there would be less uncertainty in where the bullet hits the target. **(b)** It would be less accurate because there would be greater uncertainty in where the bullet hits the target. **(c)** There would be no difference.

Reasoning Let's assume that the bullet is moving down the barrel of the gun in the $+x$ direction and that the target lies on the x axis. When it exits the barrel, the bullet—like the electron passing through a single slit—acquires a momentum component that is perpendicular (in the y direction) to the barrel. This happens even though inside the barrel the bullet travels only along the x direction and has no momentum component in the y direction. Analogous to the discussion relating to **Figure 29.15**, the y component of the momentum may be as large as Δp_y, where Δp_y indicates the difference between the maximum value of the y component of the momentum after the bullet leaves the barrel and its value of zero while the bullet is in the barrel. Δp_y is related to Planck's constant h and the diameter W of the barrel opening via the relation $\Delta p_y \approx h/W$ (Equation 29.9). Since we are now postulating that Planck's constant is large, Δp_y is also large.

Answers (a) and (c) are incorrect. Target shooting becomes more accurate if Planck's constant becomes smaller, not larger. Here's the reason: Inside the barrel the bullet is moving in the $+x$ direction. However, upon exiting the barrel, the bullet acquires a momentum component Δp_y in the y direction and begins to deviate from its original path by moving in the y direction. According to $\Delta p_y \approx h/W$ (Equation 29.9), the smaller the value of h, the smaller is Δp_y. If, in the extreme limit, Planck's constant were zero, Δp_y would also be zero, and the bullet would move only in the $+x$ direction and, thus, would hit the target.

Answer (b) is correct. If the bullet, after leaving the barrel, had only a momentum component that was parallel to the barrel, the bullet would strike the target. However, upon leaving the barrel, the bullet also acquires a momentum component Δp_y that is perpendicular to the barrel. The relation $\Delta p_y \approx h/W$ (Equation 29.9) shows that the larger the value of Planck's constant h, the larger is the value of Δp_y. Since this momentum component is perpendicular to the barrel itself, the bullet can strike at locations other than the target. Thus, target shooting would be less accurate if Planck's constant had a relatively large value.

EXAMPLE 9 | BIO Breathing in Electrons

Our lungs contain approximately 500 million alveoli, which are the small, hollow chambers lined with a membrane that allows the exchange of oxygen and carbon dioxide in our blood. Consider the situation of an electron located in one of these sacks and **(a)** calculate the uncertainty in its momentum using the Heisenberg uncertainty principle. For the uncertainty in the electron's position, use the diameter of the alveoli, which is approximately 0.05 mm. Next, assume that at some instant in time, the momentum of the electron is equal to the uncertainty in its momentum and **(b)** calculate the kinetic energy of the electron.

Reasoning We simply apply the Heisenberg uncertainty principle for position and momentum (Equation 29.10) to calculate the uncertainty in the electron's momentum. We then assume this value to be equal to the electron's momentum and calculate its kinetic energy using Equation 6.2.

Solution (a) From Equation 29.10, we have:

$$\Delta p_y = \frac{h}{(4\pi)(\Delta y)} = \frac{6.63 \times 10^{-34}\ \text{J} \cdot \text{s}}{4\pi(0.05 \times 10^{-3}\ \text{m})} = \boxed{1.1 \times 10^{-30}\ \text{kg} \cdot \text{m/s}}$$

(b) Now assume that $p = \Delta p_y = 1.1 \times 10^{-30}$ kg · m/s and calculate the electron's kinetic energy:

$$\text{KE} = \tfrac{1}{2}mv^2 = \frac{p^2}{2m} = \frac{(1.1 \times 10^{-30}\ \text{kg} \cdot \text{m/s})^2}{2(9.11 \times 10^{-31}\ \text{kg})} = \boxed{6.1 \times 10^{-31}\ \text{J}}$$

While this is an incredibly small kinetic energy, the speed of the electron is still ≈ 1 m/s.

Concept Summary

29.1 The Wave–Particle Duality, 29.2 Blackbody Radiation and Planck's Constant The wave–particle duality refers to the fact that a wave can exhibit particle-like characteristics and a particle can exhibit wave-like characteristics.

At a constant temperature, a perfect blackbody absorbs and reemits all the electromagnetic radiation that falls on it. Max Planck calculated the emitted radiation intensity per unit wavelength as a function of wavelength. In his theory, Planck assumed that a blackbody consists of atomic oscillators that can have only discrete, or quantized, energies. Planck's quantized energies are given by Equation 29.1, where h is Planck's constant (6.63×10^{-34} J · s) and f is the vibration frequency of an oscillator.

$$E = nhf \quad n = 0, 1, 2, 3, \ldots \tag{29.1}$$

29.3 Photons and the Photoelectric Effect All electromagnetic radiation consists of photons, which are packets of energy. The energy of a photon is given by Equation 29.2, where h is Planck's constant and f is the frequency of the photon. A photon has no mass and always travels at the speed of light c in a vacuum.

$$E = hf \tag{29.2}$$

The photoelectric effect is the phenomenon in which light shining on a metal surface causes electrons to be ejected from the surface. The work function W_0 of a metal is the minimum work that must be done to eject an electron from the metal. In accordance with the conservation of energy, the electrons ejected from a metal have a maximum kinetic energy KE_{max} that is related to the energy hf of the incident photon and the work function of the metal by Equation 29.3.

$$hf = \text{KE}_{\text{max}} + W_0 \tag{29.3}$$

29.4 The Momentum of a Photon and the Compton Effect The magnitude p of a photon's momentum is given by Equation 29.6, where h is Planck's constant and λ is the wavelength of the photon.

$$p = \frac{h}{\lambda} \tag{29.6}$$

The Compton effect is the scattering of a photon by an electron in a material, the scattered photon having a smaller frequency and, hence, a smaller energy than the incident photon. Part of the photon's energy and momentum are transferred to the recoiling electron. The difference between the wavelength λ' of the scattered photon and the wavelength λ of the incident photon is related by Equation 29.7 to the scattering angle θ (see **Animated Figure 29.11**), where m is the mass of the electron. The quantity $h/(mc)$ is known as the Compton wavelength of the electron.

$$\lambda' - \lambda = \frac{h}{mc}(1 - \cos\theta) \tag{29.7}$$

29.5 The De Broglie Wavelength and the Wave Nature of Matter The de Broglie wavelength λ of a particle is given by Equation 29.8, where h is Planck's constant and p is the magnitude of the relativistic momentum of the particle. Because of its wavelength, a particle can exhibit wave-like characteristics. The wave associated with a particle is a wave of probability.

$$\lambda = \frac{h}{p} \tag{29.8}$$

29.6 The Heisenberg Uncertainty Principle The Heisenberg uncertainty principle places limits on our knowledge about the behavior of a particle. The uncertainty principle is stated according to Equation 29.10, where Δy is the uncertainty in the particle's position along the y direction, and Δp_y is the uncertainty in the y component of the linear momentum of the particle.

$$(\Delta p_y)(\Delta y) \geq \frac{h}{4\pi} \tag{29.10}$$

The uncertainty principle can also be stated according to Equation 29.11, where ΔE is the uncertainty in the energy of a particle when it is in a certain state, and Δt is the time interval during which the particle is in the state.

$$(\Delta E)(\Delta t) \geq \frac{h}{4\pi} \tag{29.11}$$

Focus on Concepts

Online

Additional questions are available for assignment in WileyPLUS.

Section 29.2 Blackbody Radiation and Planck's Constant

1. An astronomer is measuring the electromagnetic radiation emitted by two stars, which are both assumed to be perfect blackbody emitters. For each star she makes a plot of the radiation intensity per unit wavelength as a function of wavelength. She notices that the curve for star A has a maximum that occurs at a shorter wavelength than does the curve for star B. What can she conclude about the surface temperatures of the two stars? **(a)** Star A has the greater surface temperature. **(b)** Star B has the greater surface temperature. **(c)** Both stars, being perfect blackbody emitters, have the same surface temperature.

(d) There is not enough information to draw a conclusion about the temperatures.

Section 29.3 Photons and the Photoelectric Effect

2. Photons are generated by a microwave oven in a kitchen and by an X-ray machine at a dentist's office. Which type of photon has the greater frequency, the greater energy, and the greater wavelength?

	Greater Frequency	Greater Energy	Greater Wavelength
(a)	X-ray	X-ray	X-ray
(b)	X-ray	X-ray	Microwave
(c)	Microwave	X-ray	Microwave
(d)	Microwave	Microwave	X-ray
(e)	X-ray	Microwave	Microwave

3. A laser emits a beam of light whose photons all have the same frequency. When the beam strikes the surface of a metal, photoelectrons are ejected from the surface. What happens if the laser emits twice the number of photons per second? (a) The photoelectrons are ejected from the surface with twice the maximum kinetic energy. (b) The photoelectrons are ejected from the surface with the same maximum kinetic energy. (c) The number of photoelectrons ejected per second from the surface doubles. (d) Both b and c happen. (e) Both a and c happen.

4. The surface of a metal plate is illuminated with light of a certain frequency. Which of the following conditions determine whether or not photoelectrons are ejected from the metal?

1. The number of photons per second emitted by the light source
2. The length of time that the light is turned on
3. The thermal conductivity of the metal
4. The surface area of the metal illuminated by the light
5. The type of metal from which the plate is made

(a) 1 and 2 (b) 5 (c) 3 and 5 (d) 4 (e) 2 and 3

Section 29.4 The Momentum of a Photon and the Compton Effect

5. Does a photon, like a moving particle such as an electron, have a momentum? (a) No, because a photon is a wave, and a wave does not have a momentum. (b) No, because a photon has no mass, and mass is necessary in order to have a momentum. (c) No, because a photon, always traveling at the speed of light in a vacuum, would have an infinite momentum. (d) Yes, and the magnitude p of the photon's momentum is related to its wavelength λ by $p = h/\lambda$, where h is Planck's constant. (e) Yes, and the magnitude p of the photon's momentum is related to its wavelength λ by $p = h\lambda$, where h is Planck's constant.

Section 29.5 The De Broglie Wavelength and the Wave Nature of Matter

6. Two particles, A and B, have the same mass, but particle A has a charge of $+q$ and B has a charge of $+2q$. The particles are accelerated from rest through the same potential difference. Which one has the longer de Broglie wavelength at the end of the acceleration? (a) Particle A, because it has the greater momentum, and, hence, the longer de Broglie wavelength (b) Particle B, because it has the greater momentum, and, hence, the longer de Broglie wavelength (c) Particle A, because it has the smaller momentum, and, hence, the longer de Broglie wavelength (d) Particle B, because it has the smaller momentum, and, hence, the longer de Broglie wavelength (e) Both particles have the same de Broglie wavelength.

Section 29.6 The Heisenberg Uncertainty Principle

7. Suppose that the momentum of an electron is measured with complete accuracy (i.e., the uncertainty in its momentum is zero). The uncertainty in a simultaneous measurement of the electron's position ________. (a) is also zero (b) is infinitely large (c) is some finite value between zero and infinity (d) cannot be measured, because one cannot measure the position and momentum of a particle, such as an electron, simultaneously

Problems

Online

Additional questions are available for assignment in WileyPLUS.

***Note:** In these problems, ignore relativistic effects unless instructed otherwise, and assume that wavelengths are in a vacuum unless otherwise specified.*

SSM Student Solutions Manual	BIO Biomedical application
MMH Problem-solving help	E Easy
GO Guided Online Tutorial	M Medium
V-HINT Video Hints	H Hard
CHALK Chalkboard Videos	WS Worksheet
	T Team Problem

Section 29.3 Photons and the Photoelectric Effect

1. E The dissociation energy of a molecule is the energy required to break the molecule apart into its separate atoms. The dissociation energy for the cyanogen molecule is 1.22×10^{-18} J. Suppose that this energy is provided by a single photon. Determine the (a) wavelength and (b) frequency of the photon. (c) In what region of the electromagnetic spectrum (see Figure 24.9) does this photon lie?

2. E An AM radio station broadcasts an electromagnetic wave with a frequency of 665 kHz, whereas an FM station broadcasts an electromagnetic wave with a frequency of 91.9 MHz. How many AM photons are needed to have a total energy equal to that of one FM photon?

3. E GO SSM Ultraviolet light with a frequency of 3.00×10^{15} Hz strikes a metal surface and ejects electrons that have a maximum kinetic energy of 6.1 eV. What is the work function (in eV) of the metal?

4. E GO Light is shining perpendicularly on the surface of the earth with an intensity of 680 W/m^2. Assuming that all the photons in the light have the same wavelength (in vacuum) of 730 nm, determine the number of photons per second per square meter that reach the earth.

5. **E** **SSM** Ultraviolet light is responsible for sun tanning. Find the wavelength (in nm) of an ultraviolet photon whose energy is 6.4×10^{-19} J.

6. **E** The maximum wavelength that an electromagnetic wave can have and still eject electrons from a metal surface is 485 nm. What is the work function W_0 of this metal? Express your answer in electron volts.

7. **E** Radiation of a certain wavelength causes electrons with a maximum kinetic energy of 0.68 eV to be ejected from a metal whose work function is 2.75 eV. What will be the maximum kinetic energy (in eV) with which this same radiation ejects electrons from another metal whose work function is 2.17 eV?

8. **E** **GO** Multiple-Concept Example 4 reviews the concepts necessary to solve this problem. Radiation with a wavelength of 238 nm shines on a metal surface and ejects electrons that have a maximum speed of 3.75×10^5 m/s. Which one of the following metals is it, the values in parentheses being the work functions: potassium (2.24 eV), calcium (2.71 eV), uranium (3.63 eV), aluminum (4.08 eV), or gold (4.82 eV)?

9. **M** **SSM** **MMH** An owl has good night vision because its eyes can detect a light intensity as small as 5.0×10^{-13} W/m^2. What is the minimum number of photons per second that an owl eye can detect if its pupil has a diameter of 8.5 mm and the light has a wavelength of 510 nm?

10. **M** **V-HINT** A proton is located at a distance of 0.420 m from a point charge of $+8.30\ \mu$C. The repulsive electric force moves the proton until it is at a distance of 1.58 m from the charge. Suppose that the electric potential energy lost by the system were carried off by a photon. What would be its wavelength?

11. **M** **CHALK** When light with a wavelength of 221 nm is incident on a certain metal surface, electrons are ejected with a maximum kinetic energy of 3.28×10^{-19} J. Determine the wavelength (in nm) of light that should be used to double the maximum kinetic energy of the electrons ejected from this surface.

12. **M** **GO** A glass plate has a mass of 0.50 kg and a specific heat capacity of 840 J/(kg · C°). The wavelength of infrared light is 6.0×10^{-5} m, while the wavelength of blue light is 4.7×10^{-7} m. Find the number of infrared photons and the number of blue photons needed to raise the temperature of the glass plate by 2.0 C°, assuming that all the photons are absorbed by the glass.

13. **H** **SSM** A laser emits 1.30×10^{18} photons per second in a beam of light that has a diameter of 2.00 mm and a wavelength of 514.5 nm. Determine **(a)** the average electric field strength and **(b)** the average magnetic field strength for the electromagnetic wave that constitutes the beam.

Section 29.4 The Momentum of a Photon and the Compton Effect

14. **E** A light source emits a beam of photons, each of which has a momentum of 2.3×10^{-29} kg · m/s. **(a)** What is the frequency of the photons? **(b)** To what region of the electromagnetic spectrum do the photons belong? Consult Figure 24.9 if necessary.

15. **E** A photon of red light (wavelength = 720 nm) and a Ping-Pong ball (mass = 2.2×10^{-3} kg) have the same momentum. At what speed is the ball moving?

16. **E** **CHALK** **SSM** In a Compton scattering experiment, the incident X-rays have a wavelength of 0.2685 nm, and the scattered X-rays have a wavelength of 0.2703 nm. Through what angle θ in **Animated Figure 29.11** are the X-rays scattered?

17. **E** **V-HINT** A sample is bombarded by incident X-rays, and free electrons in the sample scatter some of the X-rays at an angle of θ = 122.0° with respect to the incident X-rays (see **Animated Figure 29.11**). The scattered X-rays have a momentum whose magnitude is 1.856×10^{-24} kg · m/s. Determine the wavelength (in nm) of the incident X-rays. (For accuracy, use $h = 6.626 \times 10^{-34}$ J · s, $c = 2.998 \times 10^8$ m/s, and $m = 9.109 \times 10^{-31}$ kg for the mass of an electron.)

18. **E** **MMH** An incident X-ray photon of wavelength 0.2750 nm is scattered from an electron that is initially at rest. The photon is scattered at an angle of θ = 180.0° in **Animated Figure 29.11** and has a wavelength of 0.2825 nm. Use the conservation of linear momentum to find the momentum gained by the electron.

19. **E** **GO** In the Compton effect, momentum conservation applies, so the total momentum of the photon and the electron is the same before and after the scattering occurs. Suppose that in **Animated Figure 29.11** the incident photon moves in the $+x$ direction and the scattered photon emerges at an angle of θ = 90.0°, which is in the $-y$ direction. The incident photon has a wavelength of 9.00×10^{-12} m. Find the x and y components of the momentum of the scattered electron.

20. **M** **SSM** What is the maximum amount by which the wavelength of an incident photon could change when it undergoes Compton scattering from a nitrogen molecule (N_2)?

21. **M** **GO** **Animated Figure 29.11** shows the setup for measuring the Compton effect. With a fixed incident wavelength, a wavelength of λ'_1 is measured for a scattering angle of θ_1 = 30.0°, whereas a wavelength of λ'_2 is measured for a scattering angle of θ_2 = 70.0°. Find the difference in wavelengths, $\lambda'_2 - \lambda'_1$.

22. **M** **GO** A photon of wavelength 0.45000 nm strikes a free electron that is initially at rest. The photon is scattered straight backward. What is the speed of the recoil electron after the collision?

Section 29.5 The De Broglie Wavelength and the Wave Nature of Matter

23. **E** A bacterium (mass = 2×10^{-15} kg) in the blood is moving at 0.33 m/s. What is the de Broglie wavelength of this bacterium?

24. **E** What are **(a)** the wavelength of a 5.0-eV photon and **(b)** the de Broglie wavelength of a 5.0-eV electron?

25. **E** An electron and a proton have the same speed. Ignore relativistic effects and determine the ratio $\lambda_{electron}/\lambda_{proton}$ of their de Broglie wavelengths.

26. **E** Recall from Section 14.3 that the average kinetic energy of an atom in a monatomic ideal gas is given by $\overline{KE} = \frac{3}{2}kT$, where $k = 1.38 \times 10^{-23}$ J/K and T is the Kelvin temperature of the gas. Determine the de Broglie wavelength of a helium atom (mass = 6.65×10^{-27} kg) that has the average kinetic energy at room temperature (293 K).

27. **E** **GO** In a Young's double-slit experiment that uses electrons, the angle that locates the first-order bright fringes is $\theta_A = 1.6 \times 10^{-4}$ degrees when the magnitude of the electron momentum is $p_A = 1.2 \times 10^{-22}$ kg · m/s. With the same double slit, what momentum magnitude p_B is necessary so that an angle of $\theta_B = 4.0 \times 10^{-4}$ degrees locates the first-order bright fringes?

28. **M** **SSM** A particle has a de Broglie wavelength of 2.7×10^{-10} m. Then its kinetic energy doubles. What is the particle's new de Broglie wavelength, assuming that relativistic effects can be ignored?

29. **M** **GO** From a cliff that is 9.5 m above a lake, a young woman (mass = 41 kg) jumps from rest, straight down into the water. At the instant she strikes the water, what is her de Broglie wavelength?

30. M V-HINT The width of the central bright fringe in a diffraction pattern on a screen is identical when either electrons or red light (vacuum wavelength = 661 nm) pass through a single slit. The distance between the screen and the slit is the same in each case and is large compared to the slit width. How fast are the electrons moving?

31. M GO Particle A is at rest, and particle B collides head-on with it. The collision is completely inelastic, so the two particles stick together after the collision and move off with a common velocity. The masses of the particles are different, and no external forces act on them. The de Broglie wavelength of particle B before the collision is 2.0×10^{-34} m. What is the de Broglie wavelength of the object that moves off after the collision?

32. M SSM An electron, starting from rest, accelerates through a potential difference of 418 V. What is the final de Broglie wavelength of the electron, assuming that its final speed is much less than the speed of light?

33. H The kinetic energy of a particle is equal to the energy of a photon. The particle moves at 5.0% of the speed of light. Find the ratio of the photon wavelength to the de Broglie wavelength of the particle.

Section 29.6 The Heisenberg Uncertainty Principle

34. E SSM An object is moving along a straight line, and the uncertainty in its position is 2.5 m. **(a)** Find the minimum uncertainty in the momentum of the object. Find the minimum uncertainty in the object's velocity, assuming that the object is **(b)** a golf ball (mass = 0.045 kg) and **(c)** an electron.

35. E CHALK A proton is confined to a nucleus that has a diameter of 5.5×10^{-15} m. If this distance is considered to be the uncertainty in the position of the proton, what is the minimum uncertainty in its momentum?

36. E SSM For insight into this problem, review Example 9. In the lungs there are tiny sacs of air, which are called alveoli. An oxygen molecule (mass = 5.3×10^{-26} kg) is trapped within a sac, and the uncertainty in its position is 0.12 mm. What is the minimum uncertainty in the speed of this oxygen molecule?

37. E GO Particles pass through a single slit of width 0.200 mm (see **Figure 29.15**). The de Broglie wavelength of each particle is 633 nm. After the particles pass through the slit, they spread out over a range of angles. Assume that the uncertainty in the position of the particles is one-half the width of the slit, and use the Heisenberg uncertainty principle to determine the minimum range of angles.

38. M GO The minimum uncertainty Δy in the position y of a particle is equal to its de Broglie wavelength. Determine the minimum uncertainty in the speed of the particle, where this minimum uncertainty Δv_y is expressed as a percentage of the particle's speed v_y $\left(\text{Percentage} = \frac{\Delta v_y}{v_y} \times 100\%\right)$. Assume that relativistic effects can be ignored.

39. M V-HINT A subatomic particle created in an experiment exists in a certain state for a time of $\Delta t = 7.4 \times 10^{-20}$ s before decaying into other particles. Apply both the Heisenberg uncertainty principle and the equivalence of energy and mass (see Section 28.6) to determine the minimum uncertainty involved in measuring the mass of this short-lived particle.

Additional Problems

Online

40. E The interatomic spacing in a crystal of table salt is 0.282 nm. This crystal is being studied in a neutron diffraction experiment, similar to the one that produced the photograph in **Figure 29.13*a***. How fast must a neutron (mass = 1.67×10^{-27} kg) be moving to have a de Broglie wavelength of 0.282 nm?

41. E SSM The de Broglie wavelength of a proton in a particle accelerator is 1.30×10^{-14} m. Determine the kinetic energy (in joules) of the proton.

42. E V-HINT Find the de Broglie wavelength of an electron with a speed of 0.88*c*. Take relativistic effects into account.

43. E GO The work function of a metal surface is 4.80×10^{-19} J. The maximum speed of the electrons emitted from the surface is $v_A = 7.30 \times 10^5$ m/s when the wavelength of the light is λ_A. However, a maximum speed of $v_B = 5.00 \times 10^5$ m/s is observed when the wavelength is λ_B. Find the wavelengths λ_A and λ_B.

44. E GO How fast does a proton have to be moving in order to have the same de Broglie wavelength as an electron that is moving with a speed of 4.50×10^6 m/s?

Physics in Biology, Medicine, and Sports

45. E BIO **29.3** Review Example 2 as an aid in understanding this problem. While ultraviolet light from the UV-B band has enough energy to cause sunburns and damage DNA, light from the shorter wavelength UV-C band has an even greater energy. It is so effective at damaging base pairs in DNA that it is utilized in germicidal lamps. Some of these lamps can render inactive 99.99% of bacteria and viruses, so they are used to sterilize workplaces and common objects. The popularity of the lamps surged in 2020 due to the COVID-19 pandemic. The lamps proved effective at deactivating the SARS-CoV-2 coronavirus, assuming that the exposure was long enough. If the energy of the photons produced by the lamp is 4.88 eV, what is the wavelength of the light?

46. E BIO **29.3** Vitamin D is an essential fat-soluble secosteroid that supports many functions in the human body. It facilitates the absorption of calcium in the digestive tract, which leads to strong teeth and bones. It strengthens the immune system and plays an important role in the proper functioning of the body's major organs, such as the muscles, heart, lungs, and brain. The major source of vitamin D is a photochemical reaction between UV-B sunlight and a compound in the skin called

7-dehydrocholesterol. This compound exists in most vertebrate animals, and it reacts with UV-B light at wavelengths between 290 and 315 nm. What is the range of energies of UV-B light that reacts with 7-dehydrocholesterol? Give your answers in eV.

47. M BIO **29.3** If after reviewing Example 2 you still feel like visiting a tanning salon, the use of safety eyewear cannot be overstated. Tanning beds produce mostly UV-A and some UV-B light. Even with your eyes closed, the thin skin of your eyelids does not provide sufficient protection from the ultraviolet light. The photons in the UV-B light deposit their energy in the cornea of the eye, causing damage and an acute condition known as *photokeratitis*. This can also be caused by the flash of a welder's arc or the reflection of sunlight from snow, which can occur after a long day of downhill skiing in bright sunlight without proper eye protection, in which case the photokeratitis is referred to as snow blindness. Symptoms include intense tearing, pain, and the feeling of having sand in the eyes (see the photo). **(a)** Consider a person in a tanning bed whose eyes are exposed to UV-B light with an intensity of 10.0 W/m^2. If we treat their cornea as an opening with a circular cross section and radius of 5.75 mm, how much energy enters their cornea in 20 minutes? **(b)** If the wavelength of the UV light is 300 nm, how many photons are incident on the cornea during this time?

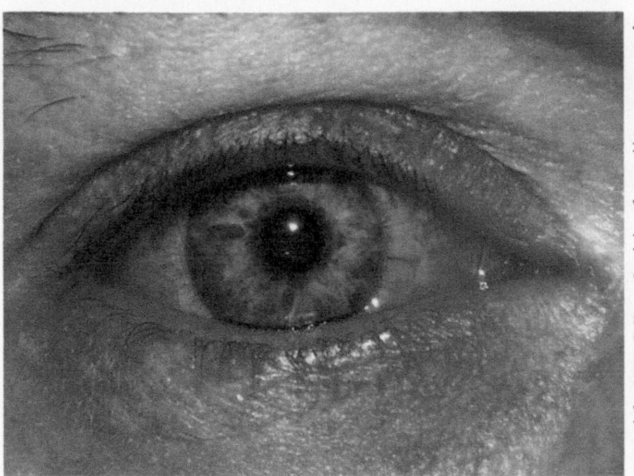

PROBLEM 47

48. E BIO **29.3** One of the primary compounds utilized in many sunscreens is *homosalate*. Its chemical formula is $C_{16}H_{22}O_3$, and it absorbs UV light with wavelengths between 295 and 315 nm. Part of the homosalate compound is salicylic acid, which has a strong absorption at 298 nm. What is the photon energy that is absorbed by the salicylic acid?

49. M BIO **29.2** The curves in **Figure 29.2** follow a relationship known as *Wein's displacement law*. It states that the peak in the radiation intensity per unit wavelength is inversely proportional to the temperature of the blackbody, and it is given by the following:

$$\lambda_{\text{peak}} = \frac{b}{T}$$

where b is the Wein displacement constant, which has a value of 2.90×10^{-3} m · K.

A glowworm (see the photo) creates light through the process of *bioluminescence*, which is the result of a chemical reaction with a light-emitting molecule called a luciferin. Assume the light emitted by the glowworm can be treated as coming from a blackbody. **(a)** Use Wein's displacement law to calculate the temperature of the glowworm, assuming the intensity of the light it emits peaks at a frequency of 5.94×10^{14} Hz. **(b)** Based on your answer in part (a), is the blackbody approximation a good one? Explain.

PROBLEM 49

Concepts and Calculations Problems

Online

In the photoelectric effect, electrons can be emitted from a metal surface when light shines on it. Einstein explained this phenomenon in terms of photons and the conservation of energy. Our discussion of this topic in Section 29.3 deals with a single wavelength. Problem 50, in contrast, considers a mixture of wavelengths and reviews the basic concepts that come into play in this important phenomenon. In this chapter we have seen that moving particles of matter not only possess kinetic energy and momentum but also are characterized by a wavelength that is known as the de Broglie wavelength. Problem 51 reviews kinetic energy and momentum, and focuses on the relation between these fundamental quantities and the de Broglie wavelength.

50. M CHALK Sunlight, whose visible wavelengths range from 380 to 750 nm, is incident on a sodium surface. The work function for sodium is $W_0 = 2.28$ eV. *Concepts:* (i) Will electrons with a greater value of KE_{max} be emitted when the incident photons have a relatively greater or relatively smaller amount of energy? (ii) In the range of visible wavelengths, which wavelength corresponds to incident photons that carry the greatest energy? (iii) What is the smallest value of KE_{max} with which an electron can be ejected from the sodium? *Calculations:* Find the maximum kinetic energy KE_{max} (in joules) of the photoelectrons emitted from the surface and the range of wavelengths that will cause photoelectrons to be emitted.

51. M CHALK SSM An electron and a proton have the same kinetic energy and are moving at speeds much less than the speed of light. *Concepts:* (i) How is the de Broglie wavelength λ related to the magnitude p of the momentum? (ii) How is the magnitude of the momentum related to the kinetic energy of a particle of mass m that is moving at a speed that is much less than the speed of light? (iii) Which has the greater de Broglie wavelength, the electron or the proton? *Calculations:* Determine the ratio of the de Broglie wavelength of the electron to that of the proton.

Team Problems

Online

52. M T WS **A Laser Heater.** You and your team have been tasked with finding a method to heat an object located inside a vacuum chamber. The problem is that there is no way to feed electrical wires into the vessel; thus, an electrical heater is out of the question. There is, however, a window through which infrared (IR) laser light (λ = 715 nm) can pass into the chamber where the object, a 2.20-g aluminum cube, is located. The cube has a small hole that leads to an internal cavity so that all light that enters the hole is absorbed and converted into thermal energy of the cube. **(a)** If the diameter of the laser beam is $D = 1.20$ mm (smaller than the hole in the cube), what must be the intensity of the beam such that the temperature of the aluminum cube increases by 1.00 K in 5.00 s? You can neglect the effects of any heat loss mechanisms. The specific heat capacity of aluminum is 9.00×10^2 J/(kg · C°) **(b)** How many photons are delivered during this time? **(c)** Suppose you instead used ultraviolet light (λ = 248 nm) rather than infrared, but the number of photons absorbed per unit time was the same as that of the infrared light. How long would it take to increase the temperature of the cube by 1.00 K in this case?

53. M T WS **A Light Thruster.** You and your team are given the task of estimating the force provided by a hypothetical "laser thruster." The idea is to use a powerful laser to expel photons out the back of a spacecraft in order to generate a reactive force that propels the ship in the opposite direction. Suppose you intend to use an enormous green laser of wavelength 520.0 nm and a beam with circular cross section of diameter $D = 0.330$ m, which will be mounted on the back of the spaceship. **(a)** What power output of the "laser thruster" would provide a thrust of 1.00 N? **(b)** What is the intensity of the laser beam? **(c)** How much energy is carried away by the laser beam in 5.00 s?

54. M T **Identifying Unknown Metals.** You and your team are tasked with sorting a hundred metallic samples composed of pure silver (Ag), nickel (Ni), indium (In), or cobalt (Co). The problem is that they all look the same (metallic) and are not labeled. The work functions (W_0) of the materials are: Ag (4.50 eV), Ni (5.20 eV), In (4.10 eV), and Co (4.90 eV). You have a monochromatic light source that can be tuned to output wavelengths from 250 nm to 400 nm. You get the idea to use it to measure the threshold frequencies of the samples and, therefore, to identify the metals. **(a)** Calculate the threshold frequencies and wavelengths of all of the metals. **(b)** Can the light source reach the threshold frequencies of all of the metals? If not, can it still be used to sort the samples?

55. M T **Electron Diffraction.** You and your team have set up an experiment where a beam of electrons is accelerated from rest through a potential difference (V), and then passes through a crystalline material that effectively acts as a set of slits separated by 9.62×10^{-11} m. **(a)** If the first order ($m = 1$) bright fringe is located at $\theta = 12.0°$ relative to the initial direction of the electron beam, what is the de Broglie wavelength of the electrons? **(b)** What is the potential difference through which the electrons are accelerated?

Computed axial tomography, or CAT, scanning is an important noninvasive technique that utilizes X-rays to provide image "slices" of the interior of the human body. This colored 3D CAT scan of an adult jaw and part of the skull illustrates the level of detail achievable today. A computer with suitable imaging software assembles the slices into such 3D images. Surgeons can even navigate around the body using rapid animations of CAT data. The production of X-rays is related to the structure of the atom, and that structure is the main topic of this chapter.

The Nature of the Atom

LEARNING OBJECTIVES

After reading this module, you should be able to...

- **30.1** Describe the nuclear atom.
- **30.2** Calculate the wavelengths of hydrogen line spectra.
- **30.3** Use the Bohr model to predict the energy levels in hydrogen.
- **30.4** Explain Bohr's assumption regarding angular momentum.
- **30.5** Apply quantum mechanics to electron energy levels in hydrogen.
- **30.6** Use the Pauli exclusion principle to explain the periodic table.
- **30.7** Apply the Bohr model to X-ray production in atoms.
- **30.8** Explain how a laser operates.
- **30.9** Describe laser applications in medicine.
- **30.10** Explain how holographic images are made and viewed.

30.1 Rutherford Scattering and the Nuclear Atom

An atom contains a small, positively charged nucleus (radius $\approx 10^{-15}$ m), which is surrounded at relatively large distances (radius $\approx 10^{-10}$ m) by a number of electrons, as **Figure 30.1** (not to scale) illustrates. In the natural state, an atom is electrically neutral because the nucleus contains a number of protons (each with a charge of $+e$) that equals the number of electrons (each with a charge of $-e$). This model of the atom is universally accepted now and is referred to as the **nuclear atom**.

The nuclear atom is a relatively recent idea. In the early part of the twentieth century a widely accepted model, developed by the English physicist Joseph J. Thomson (1856–1940), pictured the atom very differently. In Thomson's view there was no nucleus at the center of an atom. Instead, the positive charge was assumed to be spread throughout

Negative electron

Positive nucleus

FIGURE 30.1 In the nuclear atom a small positively charged nucleus is surrounded at relatively large distances by a number of electrons. Drawing is not to scale.

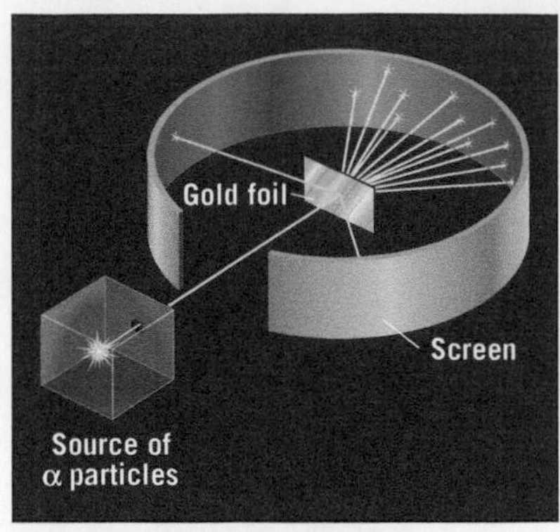

INTERACTIVE FIGURE 30.2 A Rutherford scattering experiment in which α particles are scattered by a thin gold foil. The entire apparatus is located within a vacuum chamber (not shown).

the atom, forming a kind of paste or pudding, in which the negative electrons were suspended like plums.

The "plum-pudding" model was discredited in 1911 when the New Zealand physicist Ernest Rutherford (1871–1937) published experimental results that the model could not explain. As **Interactive Figure 30.2** indicates, Rutherford and his co-workers directed a beam of alpha particles (α particles) at a thin metal foil made of gold. Alpha particles are positively charged particles (the nuclei of helium atoms, although this was not recognized at the time) emitted by some radioactive materials. If the plum-pudding model were correct, the α particles would be expected to pass nearly straight through the foil. After all, there is nothing in this model to deflect the relatively massive α particles, since the electrons have a comparatively small mass and the positive charge is spread out in a "diluted" pudding. Using a zinc sulfide screen, which flashed briefly when struck by an α particle, Rutherford and co-workers were able to determine that not all the α particles passed straight through the foil. Instead, some were deflected at large angles, even backward. Rutherford himself said, "It was almost as incredible as if you had fired a fifteen-inch shell at a piece of tissue and it came back and hit you." Rutherford concluded that the positive charge, instead of being distributed thinly and uniformly throughout the atom, was concentrated in a small region called the nucleus.

But how could the electrons in a nuclear atom remain separated from the positively charged nucleus? If the electrons were stationary, they would be pulled inward by the attractive electric force of the nuclear charge. Therefore, the electrons must be moving around the nucleus in some fashion, like planets in orbit around the sun. In fact, the nuclear model of the atom is sometimes referred to as the "planetary" model. The dimensions of the atom, however, are such that it contains a larger fraction of empty space than our solar system does, as Conceptual Example 1 discusses.

CONCEPTUAL EXAMPLE 1 | Are Atoms Mostly Empty Space?

In the planetary model of the atom, the radius of the nucleus ($\approx 1 \times 10^{-15}$ m) is analogous to the radius of the sun ($\approx 7 \times 10^{8}$ m). The electrons orbit the nucleus at a radial distance ($\approx 1 \times 10^{-10}$ m) that is analogous to the radial distance ($\approx 1.5 \times 10^{11}$ m) at which the earth orbits the sun. Suppose that the dimensions of the sun and the earth's orbit had the same proportions as those of an atomic nucleus and an electron's orbit. What then would be true about the distance between the earth and the sun? **(a)** It would be much greater than it actually is. **(b)** It would be much smaller than it actually is. **(c)** It would be roughly the same as it actually is.

Reasoning The radius of an electron orbit is one hundred thousand times larger than the radius of the nucleus: $(1 \times 10^{-10}\text{ m})/(1 \times 10^{-15}\text{ m}) = 10^{5}$. Using this factor with the radius of the sun will reveal the correct answer.

Answers (b) and (c) are incorrect. Suppose that the earth's orbital radius about the sun were indeed 10^{5} times the sun's radius. The distance between the earth and the sun, then, would be $10^{5} \times (7 \times 10^{8}\text{ m}) = 7 \times 10^{13}$ m, which is neither smaller than nor roughly the same as the actual distance.

Answer (a) is correct. If the earth's orbital radius about the sun were 10^5 times the sun's radius, the distance between the earth and the sun would be $10^5 \times (7 \times 10^8 \text{ m}) = 7 \times 10^{13}$ m, which is more than four hundred times greater than the actual orbital radius of 1.5×10^{11} m. In fact, the earth would be more than ten times farther from the sun than is Pluto, which has an orbital radius of about 6×10^{12} m. An atom, then, contains a much greater fraction of empty space than does our solar system.

Related Homework: Problem 3

Although the planetary model of the atom is easy to visualize, it too is fraught with difficulties. For instance, an electron moving on a curved path has a centripetal acceleration, as Section 5.2 discusses. And when an electron is accelerating, it radiates electromagnetic waves, as Section 24.1 discusses. These waves carry away energy. With their energy constantly being depleted, the electrons would spiral inward and eventually collapse into the nucleus. Since matter is stable, such a collapse does not occur. Thus, the planetary model, although providing a more realistic picture of the atom than the "plum-pudding" model, must be telling only part of the story. The full story of atomic structure is fascinating, and the next section describes another aspect of it.

30.2 Line Spectra

We have seen in Sections 13.3 and 29.2 that all objects emit electromagnetic waves, and we will see in Section 30.3 how this radiation arises. For a solid object, such as the hot filament of a light bulb, these waves have a continuous range of wavelengths, some of which are in the visible region of the spectrum. The continuous range of wavelengths is characteristic of the entire collection of atoms that make up the solid. In contrast, individual atoms, free of the strong interactions that are present in a solid, emit only certain specific wavelengths rather than a continuous range. These wavelengths are characteristic of the atom and provide important clues about its structure. To study the behavior of individual atoms, low-pressure gases are used in which the atoms are relatively far apart.

A low-pressure gas in a sealed tube can be made to emit electromagnetic waves by applying a sufficiently large potential difference between two electrodes located within the tube. With a grating spectroscope like that in Figure 27.36, the individual wavelengths emitted by the gas can be separated and identified as a series of bright fringes. The series of fringes is called a **line spectrum** because each bright fringe appears as a thin rectangle (a "line") resulting from the large number of parallel, closely spaced slits in the grating of the spectroscope.

THE PHYSICS OF . . . neon signs and mercury vapor street lamps. Two familiar examples of low-pressure gases are the neon in neon signs and the mercury in mercury vapor street lamps. **Figure 30.3** shows the visible parts of the line spectra for these two atoms. The specific visible wavelengths that the atoms emit give neon signs and mercury vapor street lamps their characteristic colors.

FIGURE 30.3 The line spectra for neon and mercury.

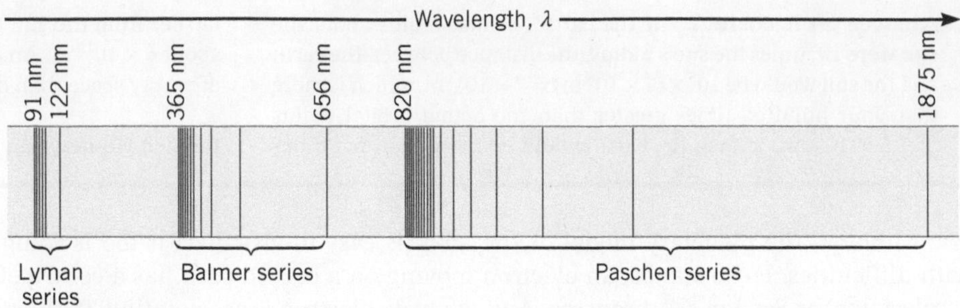

FIGURE 30.4 Line spectrum of atomic hydrogen. Only the Balmer series lies in the visible region of the electromagnetic spectrum.

The simplest line spectrum is that of atomic hydrogen, and much effort has been devoted to understanding the pattern of wavelengths that it contains. **Figure 30.4** illustrates in schematic form some of the groups or series of lines in the spectrum of atomic hydrogen. Only one of the groups is in the visible region of the electromagnetic spectrum; it is known as the **Balmer series**, in recognition of Johann J. Balmer (1825–1898), a Swiss schoolteacher who found an empirical equation that gave the values for the observed wavelengths. This equation is given next, along with similar equations that apply to the **Lyman series** at shorter wavelengths and the **Paschen series** at longer wavelengths, which are also shown in the drawing:

Lyman series
$$\frac{1}{\lambda} = R\left(\frac{1}{1^2} - \frac{1}{n^2}\right) \quad n = 2, 3, 4, \ldots \tag{30.1}$$

Balmer series
$$\frac{1}{\lambda} = R\left(\frac{1}{2^2} - \frac{1}{n^2}\right) \quad n = 3, 4, 5, \ldots \tag{30.2}$$

Paschen series
$$\frac{1}{\lambda} = R\left(\frac{1}{3^2} - \frac{1}{n^2}\right) \quad n = 4, 5, 6, \ldots \tag{30.3}$$

In these equations, the constant term R has the value of $R = 1.097 \times 10^7\ \text{m}^{-1}$ and is called the *Rydberg constant*. An essential feature of each group of lines is that there are long and short wavelength limits, with the lines being increasingly crowded toward the short wavelength limit. **Figure 30.4** also gives these limits for each series, and Example 2 determines them for the Balmer series.

EXAMPLE 2 | The Balmer Series

Find **(a)** the longest and **(b)** the shortest wavelengths of the Balmer series.

Reasoning Each wavelength in the series corresponds to one value for the integer n in Equation 30.2. Longer wavelengths are associated with smaller values of n. The longest wavelength occurs when n has its smallest value of $n = 3$. The shortest wavelength arises when n has a very large value, so that $1/n^2$ is essentially zero.

Solution **(a)** With $n = 3$, Equation 30.2 reveals that for the longest wavelength

$$\frac{1}{\lambda} = R\left(\frac{1}{2^2} - \frac{1}{n^2}\right) = (1.097 \times 10^7\ \text{m}^{-1})\left(\frac{1}{2^2} - \frac{1}{3^2}\right) = 1.524 \times 10^6\ \text{m}^{-1}$$

or $\boxed{\lambda = 656\ \text{nm}}$

(b) With $1/n^2 = 0$, Equation 30.2 reveals that for the shortest wavelength

$$\frac{1}{\lambda} = (1.097 \times 10^7\ \text{m}^{-1})\left(\frac{1}{2^2} - 0\right) = 2.743 \times 10^6\ \text{m}^{-1}$$

or $\boxed{\lambda = 365\ \text{nm}}$

Equations 30.1–30.3 are useful because they reproduce the wavelengths that hydrogen atoms radiate. However, these equations are empirical and provide no insight as to *why* certain wavelengths are radiated and others are not. It was the great Danish physicist, Niels Bohr (1885–1962), who provided the first model of the atom that predicted the discrete wavelengths emitted by atomic hydrogen. Bohr's model started us on the way toward understanding how the structure of the atom restricts the radiated wavelengths to certain values. In 1922 Bohr received the Nobel Prize in Physics for his accomplishment.

30.3 The Bohr Model of the Hydrogen Atom

In 1913 Bohr presented a model that led to equations such as Balmer's for the wavelengths that the hydrogen atom radiates. Bohr's theory begins with Rutherford's picture of an atom as a nucleus surrounded by electrons moving in circular orbits. In his theory, Bohr made a number of assumptions and combined the new quantum ideas of Planck and Einstein with the traditional description of a particle in uniform circular motion.

Adopting Planck's idea of quantized energy levels (see Section 29.2), Bohr hypothesized that in a hydrogen atom there can be only certain values of the total energy (electron kinetic energy plus potential energy). These allowed energy levels correspond to different orbits for the electron as it moves around the nucleus, the larger orbits being associated with larger total energies. **Animated Figure 30.5** illustrates two of the orbits. In addition, Bohr assumed that an electron in one of these orbits *does not* radiate electromagnetic waves. For this reason, the orbits are called **stationary orbits** or **stationary states**. Bohr recognized that radiationless orbits violated the laws of physics, as they were then known. But the assumption of such orbits was necessary, because the traditional laws indicated that an electron radiates electromagnetic waves as it accelerates around a circular path, and the loss of the energy carried by the waves would lead to the collapse of the orbit.

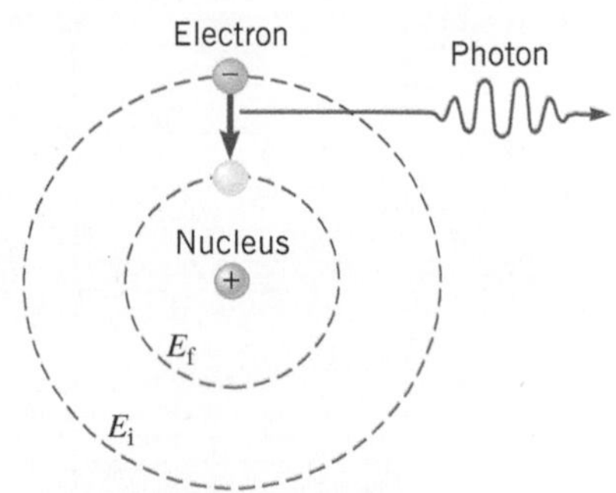

ANIMATED FIGURE 30.5 In the Bohr model, a photon is emitted when the electron drops from a larger, higher-energy orbit (energy = E_i) to a smaller, lower-energy orbit (energy = E_f).

To incorporate Einstein's photon concept (see Section 29.3), Bohr theorized that a photon is emitted only when the electron *changes* orbits from a larger one with a higher energy to a smaller one with a lower energy, as **Animated Figure 30.5** indicates. How do electrons get into the higher-energy orbits in the first place? They get there by picking up energy when atoms collide, which happens more often when a gas is heated, or by acquiring energy when a high voltage is applied to a gas.

When an electron in an initial orbit with a larger energy E_i changes to a final orbit with a smaller energy E_f, the emitted photon has an energy of $E_i - E_f$, consistent with the law of conservation of energy. But according to Einstein, the energy of a photon is hf, where f is its frequency and h is Planck's constant. As a result, we find that

$$E_i - E_f = hf \tag{30.4}$$

Since the frequency of an electromagnetic wave is related to the wavelength by $f = c/\lambda$, Bohr could use Equation 30.4 to determine the wavelengths radiated by a hydrogen atom. First, however, he had to derive expressions for the energies E_i and E_f.

The Energies and Radii of the Bohr Orbits

For an electron of mass m and speed v in an orbit of radius r (see **Figure 30.6**), the total energy is the kinetic energy $\left(\text{KE} = \frac{1}{2}mv^2\right)$ of the electron plus the electric potential energy EPE. The potential energy is the product of the charge $(-e)$ on the electron and the electric potential produced by the positive nuclear charge, in accord with Equation 19.3. We assume that the nucleus contains Z protons,* for a total nuclear charge of $+Ze$. The electric potential at a distance r from a point charge of $+Ze$ is given as $+kZe/r$ by Equation 19.6, where the constant k is $k = 8.988 \times 10^9\ \text{N} \cdot \text{m}^2/\text{C}^2$. The electric potential energy is, then, $\text{EPE} = (-e)(+kZe/r)$. Consequently, the total energy E of the atom is

$$\begin{aligned} E &= \text{KE} + \text{EPE} \\ &= \tfrac{1}{2}mv^2 - \frac{kZe^2}{r} \end{aligned} \tag{30.5}$$

But a centripetal force of magnitude mv^2/r (Equation 5.3) acts on a particle in uniform circular motion. As Figure 30.6 indicates, the centripetal force is provided by the electrostatic force of attraction $\vec{\mathbf{F}}$ that the protons in the nucleus exert on the electron.

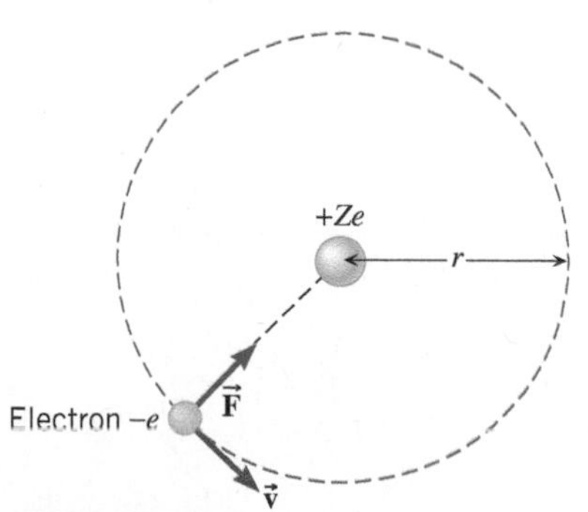

FIGURE 30.6 In the Bohr model, the electron is in uniform circular motion around the nucleus. The centripetal force $\vec{\mathbf{F}}$ is the electrostatic force of attraction that the positive nuclear charge exerts on the electron.

*For hydrogen, $Z = 1$, but we also wish to consider situations in which Z is greater than 1.

According to Coulomb's law (Equation 18.1), the magnitude of the electrostatic force is $F = kZe^2/r^2$. Therefore, $mv^2/r = kZe^2/r^2$, or

$$mv^2 = \frac{kZe^2}{r} \tag{30.6}$$

We can use this relation to eliminate the term mv^2 from Equation 30.5, with the result that

$$E = \frac{1}{2}\left(\frac{kZe^2}{r}\right) - \frac{kZe^2}{r} = -\frac{kZe^2}{2r} \tag{30.7}$$

The total energy of the atom is negative because the negative electric potential energy is larger in magnitude than the positive kinetic energy.

A value for the radius r is needed, if Equation 30.7 is to be useful. To determine r, Bohr made an assumption about the orbital angular momentum of the electron. The magnitude L of the angular momentum is given by Equation 9.10 as $L = I\omega$, where $I = mr^2$ is the moment of inertia of the electron moving on its circular path and $\omega = v/r$ (Equation 8.9) is the angular speed of the electron in radians per second. Thus, the angular momentum is $L = (mr^2)(v/r) = mvr$. Bohr conjectured that the angular momentum can assume only certain discrete values; in other words, L is quantized. He postulated that the allowed values are integer multiples of Planck's constant divided by 2π:

$$L_n = mv_n r_n = n\frac{h}{2\pi} \quad n = 1, 2, 3, \ldots \tag{30.8}$$

Solving this equation for v_n and substituting the result into Equation 30.6 lead to the following expression for the radius r_n of the nth Bohr orbit:

$$r_n = \left(\frac{h^2}{4\pi^2 mke^2}\right)\frac{n^2}{Z} \quad n = 1, 2, 3, \ldots \tag{30.9}$$

Math Skills To obtain Equation 30.9 for the radius r_n, we begin by writing Equation 30.6 with the symbols v_n instead of v and r_n instead of r:

$$m{v_n}^2 = \frac{kZe^2}{r_n} \tag{30.6}$$

Next, we solve $mv_n r_n = n\frac{h}{2\pi}$ (Equation 30.8) for v_n, by dividing both sides by mr_n:

$$\frac{\cancel{m}v_n\cancel{r_n}}{(\cancel{mr_n})} = n\frac{h}{2\pi(mr_n)} \quad \text{or} \quad v_n = \frac{nh}{2\pi mr_n} \tag{1}$$

Substituting Equation 1 into Equation 30.6, we obtain

$$m\underbrace{\left(\frac{nh}{2\pi mr_n}\right)^2}_{{v_n}^2} = \frac{kZe^2}{r_n} \quad \text{or} \quad \frac{n^2h^2}{4\pi^2 m{r_n}^2} = \frac{kZe^2}{r_n} \tag{2}$$

To isolate r_n on one side of the equals sign in Equation 2, we multiply both sides by $\frac{{r_n}^2}{kZe^2}$:

$$\frac{n^2h^2}{4\pi^2 m\cancel{r_n^2}}\left(\frac{\cancel{r_n^2}}{kZe^2}\right) = \frac{\cancel{kZe^2}}{\cancel{r_n}}\left(\frac{r_n\cancel{r_n}}{\cancel{kZe^2}}\right) \quad \text{or} \quad \frac{n^2h^2}{4\pi^2 mkZe^2} = r_n \tag{3}$$

Finally, in Equation 3, we factor out the term $\frac{n^2}{Z}$ to give

$$r_n = \left(\frac{h^2}{4\pi^2 mke^2}\right)\frac{n^2}{Z} \tag{30.9}$$

With $h = 6.626 \times 10^{-34}$ J · s, $m = 9.109 \times 10^{-31}$ kg, $k = 8.988 \times 10^9$ N · m²/C², and $e = 1.602 \times 10^{-19}$ C, this expression reveals that

Radii for Bohr orbits (in meters)

$$r_n = (5.29 \times 10^{-11}\ \text{m})\frac{n^2}{Z} \quad n = 1, 2, 3, \ldots \tag{30.10}$$

Therefore, in the hydrogen atom ($Z = 1$) the smallest Bohr orbit ($n = 1$) has a radius of $r_1 = 5.29 \times 10^{-11}$ m. This particular value is called the **Bohr radius. Figure 30.7** shows the first three Bohr orbits for the hydrogen atom.

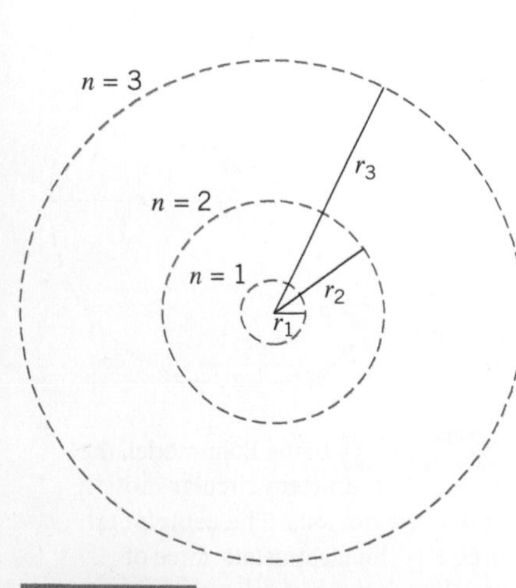

FIGURE 30.7 The first Bohr orbit in the hydrogen atom has a radius $r_1 = 5.29 \times 10^{-11}$ m. The second and third Bohr orbits have radii $r_2 = 4r_1$ and $r_3 = 9r_1$, respectively.

Equation 30.9 for the radius of a Bohr orbit can be substituted into Equation 30.7 to show that the corresponding total energy for the nth orbit is

$$E_n = -\left(\frac{2\pi^2 mk^2e^4}{h^2}\right)\frac{Z^2}{n^2} \quad n = 1, 2, 3, \ldots \tag{30.11}$$

Substituting values for h, m, k, and e into this expression yields

Bohr energy levels in joules

$$E_n = -(2.18 \times 10^{-18}\ \text{J})\frac{Z^2}{n^2} \quad n = 1, 2, 3, \ldots \tag{30.12}$$

Often, atomic energies are expressed in units of electron volts rather than joules. Since $1.60 \times 10^{-19}\ \text{J} = 1\ \text{eV}$, Equation 30.12 can be rewritten as

Bohr energy levels in electron volts

$$E_n = -(13.6\ \text{eV})\frac{Z^2}{n^2} \quad n = 1, 2, 3, \ldots \tag{30.13}$$

Energy Level Diagrams

It is useful to represent the energy values given by Equation 30.13 on an *energy level diagram,* as in **Figure 30.8**. In this diagram, which applies to the hydrogen atom ($Z = 1$), the highest energy level corresponds to $n = \infty$ in Equation 30.13 and has an energy of 0 eV. This is the energy of the atom when the electron is completely removed ($r = \infty$) from the nucleus and is at rest. In contrast, the lowest energy level corresponds to $n = 1$ and has a value of -13.6 eV. The lowest energy level is called the **ground state**, to distinguish it from the higher levels, which are called **excited states**. Observe how the energies of the excited states come closer and closer together as n increases.

The electron in a hydrogen atom at room temperature spends most of its time in the ground state. To raise the electron from the ground state ($n = 1$) to the highest possible excited state ($n = \infty$), 13.6 eV of energy must be supplied. Supplying this amount of energy removes the electron from the atom, producing the positive hydrogen ion H^+. This is the minimum energy needed to remove the electron and is called the **ionization energy**. Thus, the Bohr model predicts that the ionization energy of atomic hydrogen is 13.6 eV, in excellent agreement with the experimental value. In Example 3 the Bohr model is applied to doubly ionized lithium.

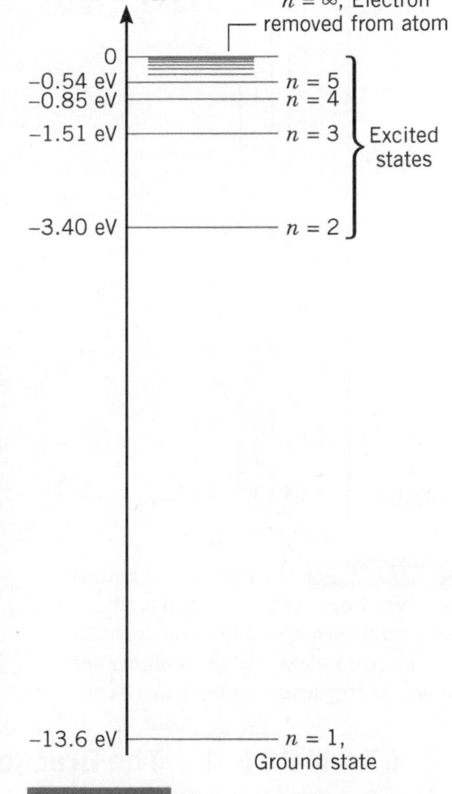

FIGURE 30.8 Energy level diagram for the hydrogen atom.

EXAMPLE 3 | The Ionization Energy of Li^{2+}

The Bohr model does not apply when more than one electron orbits the nucleus because it does not account for the electrostatic force that one electron exerts on another. For instance, an electrically neutral lithium atom (Li) contains three electrons in orbit around a nucleus that includes three protons ($Z = 3$), and Bohr's analysis is not applicable. However, the Bohr model can be used for the doubly charged positive ion of lithium (Li^{2+}) that results when two electrons are removed from the neutral atom, leaving only one electron to orbit the nucleus. Obtain the ionization energy that is needed to remove the remaining electron from Li^{2+}.

Reasoning The lithium ion Li^{2+} contains three times the positive nuclear charge that the hydrogen atom contains. Therefore, the orbiting electron is attracted more strongly to the nucleus in Li^{2+} than to the nucleus in the hydrogen atom. As a result, we expect that more energy is required to ionize Li^{2+} than the 13.6 eV required for atomic hydrogen.

Solution The Bohr energy levels for Li^{2+} are obtained from Equation 30.13 with $Z = 3$; $E_n = -(13.6\ \text{eV})(3^2/n^2)$. Therefore, the ground state ($n = 1$) energy is

$$E_1 = -(13.6\ \text{eV})\frac{3^2}{1^2} = -122\ \text{eV}$$

Removing the electron from Li^{2+} requires 122 eV of energy:

$$\boxed{\text{Ionization energy} = 122\ \text{eV}}$$

This value for the ionization energy agrees well with the experimental value of 122.4 eV and, as expected, is greater than the 13.6 eV required for atomic hydrogen.

The Line Spectra of the Hydrogen Atom

To predict the wavelengths in the line spectrum of the hydrogen atom, Bohr combined his ideas about atoms (electron orbits are stationary orbits and the angular momentum of an electron is quantized) with Einstein's idea of the photon.

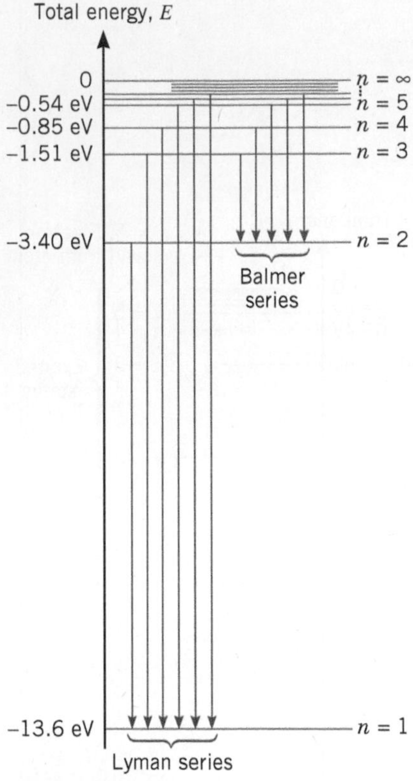

FIGURE 30.9 The Lyman and Balmer series of lines in the hydrogen atom spectrum correspond to transitions that the electron makes between higher and lower energy levels, as indicated here.

As applied by Bohr, the photon concept is inherent in Equation 30.4, $E_i - E_f = hf$, which states that the frequency f of the photon is proportional to the difference between two energy levels of the hydrogen atom. If we substitute Equation 30.11 for the total energies E_i and E_f into Equation 30.4 and recall from Equation 16.1 that $f = c/\lambda$, we obtain the following result:

$$\frac{1}{\lambda} = \frac{2\pi^2 mk^2e^4}{h^3c}(Z^2)\left(\frac{1}{n_f^2} - \frac{1}{n_i^2}\right) \tag{30.14}$$

$$n_i, n_f = 1, 2, 3, \ldots \quad \text{and} \quad n_i > n_f$$

Using known values for h, m, k, e, and c, we find that $2\pi^2mk^2e^4/(h^3c) = 1.097 \times 10^7\ \text{m}^{-1}$, in agreement with the Rydberg constant R that appears in Equations 30.1–30.3. The agreement between the theoretical and experimental values of the Rydberg constant was a major accomplishment of Bohr's theory.

With $Z = 1$ and $n_f = 1$, Equation 30.14 reproduces Equation 30.1 for the Lyman series. Thus, Bohr's model shows that the Lyman series of lines occurs when electrons make transitions from higher energy levels with $n_i = 2, 3, 4, \ldots$ to the first energy level where $n_f = 1$. **Figure 30.9** shows these transitions. Notice that when an electron makes a transition from $n_i = 2$ to $n_f = 1$, the longest wavelength photon in the Lyman series is emitted, since the energy change is the smallest possible. When an electron makes a transition from the highest level where $n_i = \infty$ to the lowest level where $n_f = 1$, the shortest wavelength is emitted, since the energy change is the largest possible. Since the higher energy levels are increasingly close together, the lines in the series become more and more crowded toward the short wavelength limit, as can be seen in **Figure 30.4. Figure 30.9** also shows the energy level transitions for the Balmer series, where $n_i = 3, 4, 5, \ldots$, and $n_f = 2$. In the Paschen series (see **Figure 30.4**) $n_i = 4, 5, 6, \ldots$, and $n_f = 3$. The next example deals further with the line spectrum of the hydrogen atom.

EXAMPLE 4 | The Brackett Series for Atomic Hydrogen

In the line spectrum of atomic hydrogen there is also a group of lines known as the Brackett series. These lines are produced when electrons, excited to high energy levels, make transitions to the $n = 4$ level. Determine **(a)** the longest wavelength in this series and **(b)** the wavelength that corresponds to the transition from $n_i = 6$ to $n_f = 4$. **(c)** Refer to Figure 24.9 and identify the spectral region in which these lines are found.

Reasoning The longest wavelength corresponds to the transition that has the smallest energy change, which is between the $n_i = 5$ and $n_f = 4$ levels in **Figure 30.8**. The wavelength for this transition, as well as that for the transition from $n_i = 6$ to $n_f = 4$, can be obtained from Equation 30.14.

> **Problem-Solving Insight In the line spectrum of atomic hydrogen, all lines in a given series (e.g., the Brackett series) are identified by a single value of the quantum number n_f for the *lower energy level into which an electron falls*. Each line in a given series, however, corresponds to a different value of the quantum number n_i for the higher energy level where an electron originates.**

Solution (a) Using Equation 30.14 with $Z = 1$, $n_i = 5$, and $n_f = 4$, we find that

$$\frac{1}{\lambda} = (1.097 \times 10^7\ \text{m}^{-1})(1^2)\left(\frac{1}{4^2} - \frac{1}{5^2}\right) = 2.468 \times 10^5\ \text{m}^{-1}$$

or $\boxed{\lambda = 4051\ \text{nm}}$

(b) The calculation here is similar to that in part (a):

$$\frac{1}{\lambda} = (1.097 \times 10^7\ \text{m}^{-1})(1^2)\left(\frac{1}{4^2} - \frac{1}{6^2}\right) = 3.809 \times 10^5\ \text{m}^{-1}$$

or $\boxed{\lambda = 2625\ \text{nm}}$

(c) According to Figure 24.9, these lines lie in the $\boxed{\text{infrared region}}$ of the spectrum.

The various lines in the hydrogen atom spectrum are produced when electrons change from higher to lower energy levels and photons are emitted. Consequently, the spectral lines are called **emission lines**. Electrons can also make transitions in the reverse direction, from lower to higher levels, in a process known as *absorption*. In this case, an atom

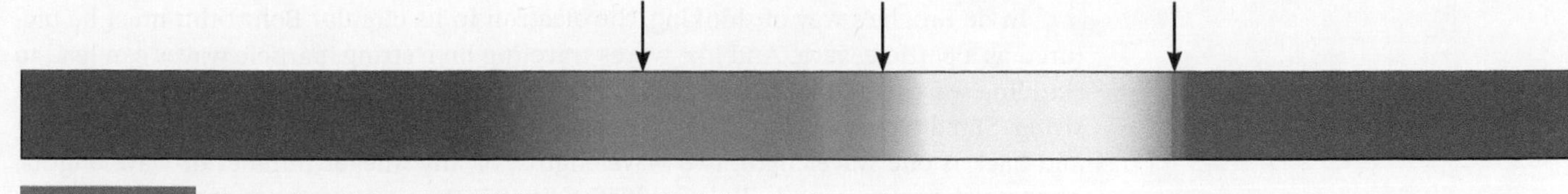

FIGURE 30.10 Illustration of the sun's visible spectrum. Dark lines in the sun's spectrum are absorption lines called Fraunhofer lines, in honor of their discoverer. There are thousands of these absorption lines in the sun's spectrum. The locations of three of the more prominent Fraunhofer lines are marked by arrows.

absorbs a photon that has precisely the energy needed to produce the transition. Thus, if photons with a continuous range of wavelengths pass through a gas and then are analyzed with a grating spectroscope, a series of dark **absorption lines** appears in the continuous spectrum. The dark lines indicate the wavelengths removed by the absorption process.

THE PHYSICS OF . . . absorption lines in the sun's spectrum. Absorption lines can be seen in **Figure 30.10** in the spectrum of the sun, where they are called Fraunhofer lines, after their discoverer. They are due to atoms, located in the outer and cooler layers of the sun, that absorb radiation coming from the interior. The interior portion of the sun is too hot for individual atoms to retain their structures and, therefore, the interior emits a continuous spectrum of wavelengths.

The Bohr model provides a great deal of insight into atomic structure. However, this model is now known to be oversimplified and has been superseded by a more detailed picture provided by quantum mechanics and the Schrödinger equation (see Section 30.5).

Check Your Understanding

(*The answers are given at the end of the book.*)

1. Which one of the following statements is true? **(a)** An atom is less easily ionized when its outermost electron is in an excited state than when it is in the ground state. **(b)** An atom is more easily ionized when its outermost electron is in an excited state than when it is in the ground state. **(c)** The energy state (excited state or ground state) of the outermost electron in an atom has nothing to do with how easily the atom can be ionized.

2. An electron in the hydrogen atom is in the $n = 4$ energy level. When this electron makes a transition to a lower energy level, is the wavelength of the photon emitted in **(a)** the Lyman series only, **(b)** the Balmer series only, **(c)** the Paschen series only, or **(d)** could it be in the Lyman, the Balmer, or the Paschen series?

3. A tube contains atomic hydrogen, and nearly all of the electrons in the atoms are in the ground state or $n = 1$ energy level. Electromagnetic radiation with a continuous spectrum of wavelengths (including those in the Lyman, Balmer, and Paschen series) enters one end of the tube and leaves the other end. The exiting radiation is found to contain strong absorption lines. To which one or more of the series do the wavelengths of these absorption lines correspond? Assume that once an electron absorbs a photon and jumps to a higher energy level, it does not absorb yet another photon and jump to an even higher energy level.

30.4 De Broglie's Explanation of Bohr's Assumption About Angular Momentum

Of all the assumptions Bohr made in his model of the hydrogen atom, perhaps the most puzzling is the assumption about the angular momentum of the electron [$L_n = mv_nr_n = nh/(2\pi)$; $n = 1, 2, 3, \ldots$]. Why should the angular momentum have only those values that are integer multiples of Planck's constant divided by 2π? In 1923, ten years after Bohr's work, de Broglie pointed out that his own theory for the wavelength of a moving particle could provide an answer to this question.

In de Broglie's way of thinking, the electron in its circular Bohr orbit must be pictured as a particle wave. And like waves traveling on a string, particle waves can lead to standing waves under resonant conditions. Section 17.5 discusses these conditions for a string. Standing waves form when the total distance traveled by a wave down the string and back is one wavelength, two wavelengths, or any integer number of wavelengths. The total distance around a Bohr orbit of radius r is the circumference of the orbit or $2\pi r$. By the same reasoning, then, the condition for standing particle waves for the electron in a Bohr orbit would be

$$2\pi r = n\lambda \quad n = 1, 2, 3, \ldots$$

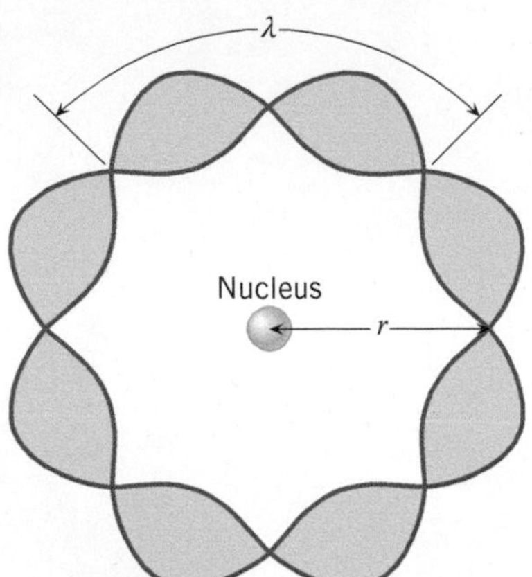

FIGURE 30.11 De Broglie suggested standing particle waves as an explanation for Bohr's angular momentum assumption. Here, a standing particle wave is illustrated on a Bohr orbit where four de Broglie wavelengths fit into the circumference of the orbit.

where n is the number of whole wavelengths that fit into the circumference of the circle. But according to Equation 29.8 the de Broglie wavelength of the electron is $\lambda = h/p$, where p is the magnitude of the electron's momentum. If the speed of the electron is much less than the speed of light, the momentum is $p = mv$, and the condition for standing particle waves becomes $2\pi r = nh/(mv)$. A rearrangement of this result gives

$$mvr = n\frac{h}{2\pi} \quad n = 1, 2, 3, \ldots$$

which is just what Bohr assumed for the angular momentum of the electron. As an example, **Figure 30.11** illustrates the standing particle wave on a Bohr orbit for which $2\pi r = 4\lambda$.

De Broglie's explanation of Bohr's assumption about angular momentum emphasizes an important fact—namely, that particle waves play a central role in the structure of the atom. Moreover, the theoretical framework of quantum mechanics includes the Schrödinger equation for determining the wave function ψ (Greek letter psi) that represents a particle wave. The next section deals with the picture that quantum mechanics gives for atomic structure, a picture that supersedes the Bohr model. In any case, the Bohr expression for the energy levels (Equation 30.11) can be applied when a single electron orbits the nucleus, whereas the theoretical framework of quantum mechanics can be applied, in principle, to atoms that contain an arbitrary number of electrons.

30.5 The Quantum Mechanical Picture of the Hydrogen Atom

The picture of the hydrogen atom that quantum mechanics and the Schrödinger equation provide differs in a number of ways from the Bohr model. The Bohr model uses a single integer number n to identify the various electron orbits and the associated energies. Because this number can have only discrete values, rather than a continuous range of values, n is called a **quantum number**. In contrast, quantum mechanics reveals that four different quantum numbers are needed to describe each state of the hydrogen atom. These four are described as follows:

1. **The principal quantum number n.** As in the Bohr model, this number determines the total energy of the atom and can have only integer values: $n = 1, 2, 3, \ldots$. In fact, the Schrödinger equation predicts* that the energy of the hydrogen atom is identical to the energy obtained from the Bohr model: $E_n = -(13.6 \text{ eV})\, Z^2/n^2$.

2. **The orbital quantum number ℓ.** This number determines the angular momentum of the electron due to its orbital motion. The values that ℓ can have depend on the value of n, and only the following integers are allowed:

$$\ell = 0, 1, 2, \ldots, (n-1)$$

*This prediction requires that small relativistic effects and small interactions within the atom be ignored, and assumes that the hydrogen atom is not located in an external magnetic field.

For instance, if $n = 1$, the orbital quantum number can have only the value $\ell = 0$, but if $n = 4$, the values $\ell = 0, 1, 2,$ and 3 are possible. The magnitude L of the angular momentum of the electron is

$$L = \sqrt{\ell(\ell + 1)}\,\frac{h}{2\pi} \tag{30.15}$$

3. **The magnetic quantum number m_ℓ.** The word "magnetic" is used here because an externally applied magnetic field influences the energy of the atom, and this quantum number is used in describing the effect. Since the effect was discovered by the Dutch physicist Pieter Zeeman (1865–1943), it is known as the *Zeeman effect*. When there is no external magnetic field, m_ℓ plays no role in determining the energy. In either event, the magnetic quantum number determines the component of the angular momentum along a specific direction, which is called the z direction by convention. The values that m_ℓ can have depend on the value of ℓ, with only the following positive and negative integers being permitted:

$$m_\ell = -\ell, \ldots, -2, -1, 0, +1, +2, \ldots, +\ell$$

For example, if the orbital quantum number is $\ell = 2$, then the magnetic quantum number can have the values $m_\ell = -2, -1, 0, +1,$ and $+2$. The component L_z of the angular momentum in the z direction is

$$L_z = m_\ell \frac{h}{2\pi} \tag{30.16}$$

4. **The spin quantum number m_s.** This number is needed because the electron has an intrinsic property called "spin angular momentum." Loosely speaking, we can view the electron as spinning while it orbits the nucleus, analogous to the way the earth spins as it orbits the sun. There are two possible values for the spin quantum number of the electron:

$$m_s = +\tfrac{1}{2} \quad \text{or} \quad m_s = -\tfrac{1}{2}$$

Sometimes the phrases "spin up" and "spin down" are used to refer to the directions of the spin angular momentum associated with the values for m_s.

Table 30.1 summarizes the four quantum numbers that are needed to describe each state of the hydrogen atom. One set of values for n, ℓ, m_ℓ, and m_s corresponds to one state. As the principal quantum number n increases, the number of possible combinations of the four quantum numbers rises rapidly, as Example 5 illustrates.

TABLE 30.1 Quantum Numbers for the Hydrogen Atom

Name	Symbol	Allowed Values
Principal quantum number	n	$1, 2, 3, \ldots$
Orbital quantum number	ℓ	$0, 1, 2, \ldots, (n-1)$
Magnetic quantum number	m_ℓ	$-\ell, \ldots, -2, -1, 0, +1, +2, \ldots, +\ell$
Spin quantum number	m_s	$-\frac{1}{2}$ or $+\frac{1}{2}$

EXAMPLE 5 | Quantum Mechanical States of the Hydrogen Atom

Determine the number of possible states for the hydrogen atom when the principal quantum number is **(a)** $n = 1$ and **(b)** $n = 2$.

Reasoning Each different combination of the four quantum numbers summarized in **Table 30.1** corresponds to a different state. We begin with the value for n and find the allowed values for ℓ. Then, for each ℓ value we find the possibilities for m_ℓ. Finally, m_s may be $+\frac{1}{2}$ or $-\frac{1}{2}$ for each group of values for n, ℓ, and m_ℓ.

Solution **(a)** The diagram below shows the possibilities for ℓ, m_ℓ, and m_s when $n = 1$: Thus, there are two different states for the hydrogen atom. In the absence of an external magnetic field, these two states have the same energy, since they have the same value of n.

(b) When $n = 2$, there are eight possible combinations for the values of n, ℓ, m_ℓ, and m_s, as the diagram below indicates: With the same value of $n = 2$, all eight states have the same energy when there is no external magnetic field.

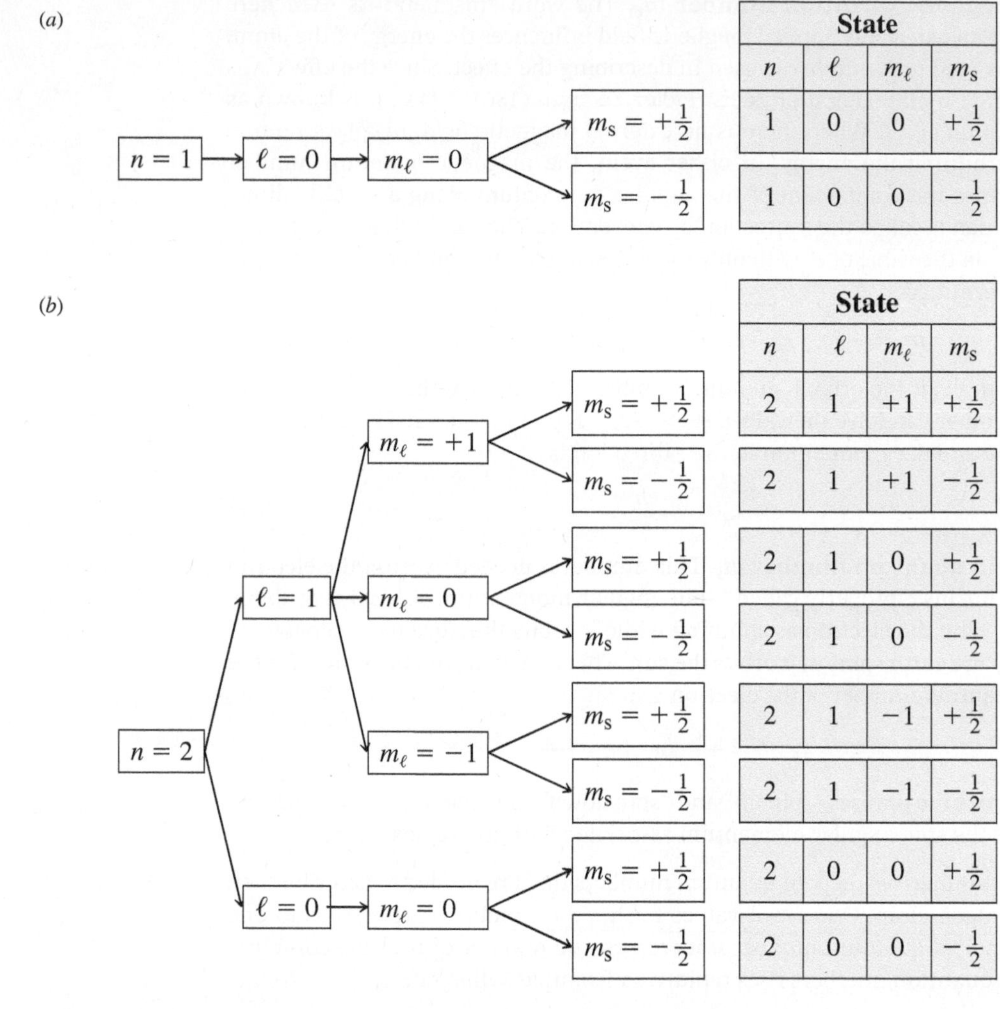

(a)

State			
n	ℓ	m_ℓ	m_s
1	0	0	$+\frac{1}{2}$
1	0	0	$-\frac{1}{2}$

(b)

State			
n	ℓ	m_ℓ	m_s
2	1	+1	$+\frac{1}{2}$
2	1	+1	$-\frac{1}{2}$
2	1	0	$+\frac{1}{2}$
2	1	0	$-\frac{1}{2}$
2	1	−1	$+\frac{1}{2}$
2	1	−1	$-\frac{1}{2}$
2	0	0	$+\frac{1}{2}$
2	0	0	$-\frac{1}{2}$

Quantum mechanics provides a more accurate picture of atomic structure than does the Bohr model. It is important to realize that the two pictures differ substantially, as Conceptual Example 6 illustrates.

CONCEPTUAL EXAMPLE 6 | The Bohr Model Versus Quantum Mechanics

Consider two hydrogen atoms. There are no external magnetic fields present, and the electron in each atom has the same energy. According to the Bohr model and to quantum mechanics, is it possible for the electrons in these atoms **(a)** to have zero orbital angular momentum and **(b)** to have different orbital angular momenta?

Reasoning and Solution **(a)** In both the Bohr model and quantum mechanics, the energy is proportional to $1/n^2$, according to Equation 30.13, where n is the principal quantum number. Moreover, the value of n may be $n = 1, 2, 3, \ldots$, and may not be zero. ***In the Bohr model, the fact that n may not be zero means that it is not possible for the orbital angular momentum to be zero*** because the angular momentum is proportional to n, according to Equation 30.8. In the quantum mechanical picture the magnitude of the orbital angular momentum is proportional to $\sqrt{\ell(\ell+1)}$, as given by Equation 30.15. Here, ℓ is the orbital quantum number and may take on the values $\ell = 0, 1, 2, \ldots, (n-1)$. We note that ℓ [and therefore $\sqrt{\ell(\ell+1)}$] may be zero, no matter what the value for n is. Consequently, ***the orbital angular momentum may be zero according to quantum mechanics***, in contrast to the case for the Bohr model.

(b) If the electrons have the same energy, they have the same value for the principal quantum number n. ***In the Bohr model, this means that they cannot have different values for the orbital angular momentum L_n***, since $L_n = nh/(2\pi)$, according to Equation 30.8. In quantum mechanics, the energy is also determined by n when external magnetic fields are absent, but the orbital angular momentum is determined by ℓ. Since $\ell = 0, 1, 2, \ldots, (n - 1)$, different values of ℓ are compatible with the same value of n. For instance, if $n = 2$ for both electrons, one of them could have $\ell = 0$, while the other could have $\ell = 1$. ***According to quantum mechanics, then, the electrons could have different orbital angular momenta, even though they have the same energy.***

The following table summarizes the discussion from parts (a) and (b):

	Bohr Model	Quantum Mechanics
(a) For a given n, can the angular momentum ever be zero?	No	Yes
(b) For a given n, can the angular momentum have different values?	No	Yes

According to the Bohr model, the nth orbit is a circle of radius r_n, and every time the position of the electron in this orbit is measured, the electron is found exactly at a distance r_n away from the nucleus. This simplistic picture is now known to be incorrect, and the quantum mechanical picture of the atom has replaced it. Suppose that the electron is in a quantum mechanical state for which $n = 1$, and we imagine making a number of measurements of the electron's position with respect to the nucleus. We would find that its position is uncertain, in the sense that there is a probability of finding the electron sometimes very near the nucleus, sometimes very far from the nucleus, and sometimes at intermediate locations. The probability is determined by the wave function ψ, as Section 29.5 discusses. We can make a three-dimensional picture of our findings by marking a dot at each location where the electron is found. More dots occur at places where the probability of finding the electron is higher, and after a sufficient number of measurements, a picture of the quantum mechanical state emerges. **Figure 30.12** shows the spatial distribution for the position of an electron in a state for which $n = 1$, $\ell = 0$, and $m_\ell = 0$. This picture is constructed from so many measurements that the individual dots are no longer visible but have merged to form a kind of probability "cloud" whose density changes gradually from place to place. The dense regions indicate places where the probability of finding the electron is higher, and the less dense regions indicate places where the probability is lower. Also indicated in **Figure 30.12** is the radius where quantum mechanics predicts the greatest probability per unit radial distance of finding the electron in the $n = 1$ state. This radius matches exactly the radius of 5.29×10^{-11} m found for the first Bohr orbit.

For a principal quantum number of $n = 2$, the probability clouds are different than for $n = 1$. In fact, more than one cloud shape is possible because with $n = 2$ the orbital quantum number can be either $\ell = 0$ or $\ell = 1$. Although the value of ℓ does not affect the energy of the hydrogen atom, the value does have a significant effect on the shape of the probability clouds. **Figure 30.13*a*** shows the cloud for $n = 2$, $\ell = 0$, and $m_\ell = 0$. Part *b* of the drawing shows that when $n = 2$, $\ell = 1$, and $m_\ell = 0$, the cloud has a two-lobe shape with the nucleus at the center between the lobes. For larger values of n, the probability clouds become increasingly complex and are spread out over larger volumes of space.

The probability cloud picture of the electron in a hydrogen atom is very different from the well-defined orbit of the Bohr model. The fundamental reason for this difference is to be found in the Heisenberg uncertainty principle, as Conceptual Example 7 discusses.

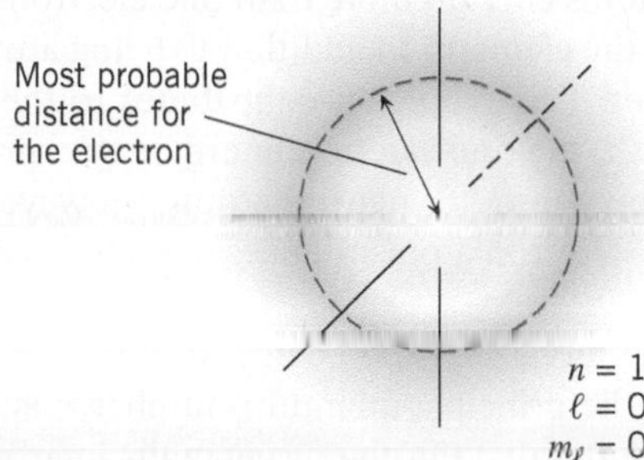

FIGURE 30.12 The electron probability cloud for the ground state ($n = 1$, $\ell = 0$, $m_\ell = 0$) of the hydrogen atom.

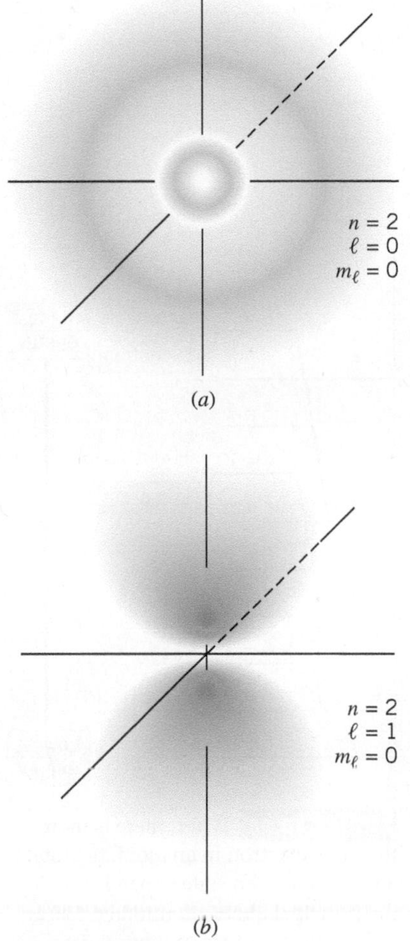

FIGURE 30.13 The electron probability clouds for the hydrogen atom when (*a*) $n = 2$, $\ell = 0$, $m_\ell = 0$ and (*b*) $n = 2$, $\ell = 1$, $m_\ell = 0$.

CONCEPTUAL EXAMPLE 7 | The Uncertainty Principle and the Hydrogen Atom

In the Bohr model of the hydrogen atom, the electron in the ground state ($n = 1$) is in an orbit that has a radius of exactly 5.29×10^{-11} m, so that the uncertainty Δy in its radial position is $\Delta y = 0$ m. According to the Heisenberg uncertainty principle, what does the fact that there is no uncertainty in the electron's radial position imply about the electron's radial speed? The uncertainty principle implies **(a)** nothing about the radial speed, **(b)** that the radial speed has only a small uncertainty, **(c)** that the radial speed has an infinitely large uncertainty.

Reasoning We need to obtain an expression for the uncertainty in the radial speed of the electron to use as a guide. As stated in Equation 29.10, the Heisenberg principle is $(\Delta p_y)(\Delta y) \geq h/(4\pi)$. In the present context, Δy is the uncertainty in the electron's radial position, so that Δp_y is the uncertainty in the electron's radial momentum. According to Equation 7.2, however, the magnitude of the momentum is $p_y = mv_y$, where m is the electron's mass and v_y is the electron's radial speed. As a result, the uncertainty in the momentum is $\Delta p_y = \Delta(mv_y) = m\Delta v_y$. With this substitution for Δp_y, the Heisenberg principle becomes $(m\Delta v_y)(\Delta y) \geq h/(4\pi)$, which can be rearranged to show that

$$\Delta v_y \geq \frac{h}{m(\Delta y)(4\pi)}$$

Answers (a) and (b) are incorrect. Our result for Δv_y shows that the Heisenberg principle does indeed imply something about the uncertainty in the electron's radial speed. Since $\Delta y = 0$ m in the Bohr model, our result for Δv_y shows that the uncertainty in the speed is infinitely large (Δy is in the denominator on the right). Therefore, answers (a) and (b) cannot be correct.

Answer (c) is correct. Our expression for Δv_y shows that, in fact, the uncertainty in the radial speed is infinitely large ($\Delta y = 0$ m in the Bohr model and Δy is in the denominator on the right in the expression above). Such a large uncertainty in the radial speed means that the electron may be moving very rapidly in the radial direction and, therefore, would not remain in its Bohr orbit. Quantum mechanics, with its probability-cloud picture of atomic structure, correctly represents the positional and motional uncertainty that the Heisenberg principle reveals. The Bohr model does not correctly represent this aspect of reality at the atomic level.

Check Your Understanding

(The answers are given at the end of the book.)

4. In the Bohr model for the hydrogen atom, the closer the electron is to the nucleus, the smaller is the total energy of the electron. Is this also true in the quantum mechanical picture of the hydrogen atom?

5. In the quantum mechanical picture of the hydrogen atom, the orbital angular momentum of the electron may be zero in any of the possible energy states. For which energy state *must* the orbital angular momentum be zero?

6. Consider two different hydrogen atoms. The electron in each atom is in a different excited state, so that each electron has a different total energy. Is it possible for the electrons to have the same orbital angular momentum L, according to **(a)** the Bohr model and **(b)** quantum mechanics?

7. The magnitude of the orbital angular momentum of the electron in a hydrogen atom is observed to increase. According to **(a)** the Bohr model and **(b)** quantum mechanics, does this necessarily mean that the total energy of the electron also increases?

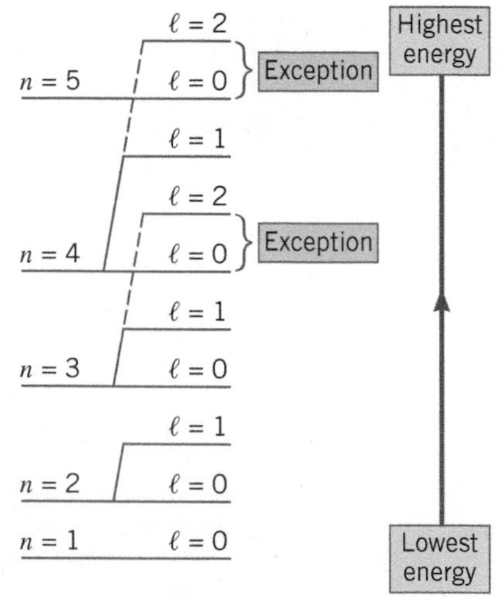

FIGURE 30.14 When there is more than one electron in an atom, the total energy of a given state depends on the principal quantum number n and the orbital quantum number ℓ. The energy increases with increasing n (with some exceptions) and, for a fixed n, with increasing ℓ. For clarity, levels for $n = 6$ and higher are not shown.

30.6 The Pauli Exclusion Principle and the Periodic Table of the Elements

Except for hydrogen, all electrically neutral atoms contain more than one electron, with the number given by the atomic number Z of the element. In addition to being attracted by the nucleus, the electrons repel each other. This repulsion contributes to the total energy of a multiple-electron atom. As a result, the one-electron energy expression for hydrogen, $E_n = -(13.6 \text{ eV})Z^2/n^2$, does not apply to other neutral atoms. However, the simplest approach for dealing with a multiple-electron atom still uses the four quantum numbers n, ℓ, m_ℓ, and m_s.

Detailed quantum mechanical calculations reveal that the energy level of each state of a multiple-electron atom depends on both the principal quantum number n and the orbital quantum number ℓ. **Figure 30.14** illustrates that the energy generally increases as

n increases, but there are exceptions, as the drawing indicates. Furthermore, for a given n, the energy also increases as ℓ increases.

In a multiple-electron atom, all electrons with the same value of n are said to be in the same **shell**. Electrons with $n = 1$ are in a single shell (sometimes called the K shell), electrons with $n = 2$ are in another shell (the L shell), those with $n = 3$ are in a third shell (the M shell), and so on. Those electrons with the same values for both n and ℓ are often referred to as being in the same **subshell**. The $n = 1$ shell consists of a single $\ell = 0$ subshell. The $n = 2$ shell has two subshells, one with $\ell = 0$ and one with $\ell = 1$. Similarly, the $n = 3$ shell has three subshells, one with $\ell = 0$, one with $\ell = 1$, and one with $\ell = 2$.

In the hydrogen atom near room temperature, the electron spends most of its time in the lowest energy level, or ground state—namely, in the $n = 1$ shell. Similarly, when an atom contains more than one electron and is near room temperature, the electrons spend most of their time in the lowest energy levels possible. The lowest energy state for an atom is called the **ground state**. However, when a multiple-electron atom is in its ground state, not every electron is in the $n = 1$ shell in general because the electrons obey a principle discovered by the Austrian physicist Wolfgang Pauli (1900–1958).

THE PAULI EXCLUSION PRINCIPLE

No two electrons in an atom can have the same set of values for the four quantum numbers n, ℓ, m_ℓ, and m_s.

Suppose two electrons in an atom have three quantum numbers that are identical: $n = 3$, $m_\ell = 1$, and $m_s = -\frac{1}{2}$. According to the exclusion principle, it is not possible for each to have $\ell = 2$, for example, since each would then have the same four quantum numbers. Each electron must have a different value for ℓ (for instance, $\ell = 1$ and $\ell = 2$) and, consequently, would be in a different subshell. With the aid of the Pauli exclusion principle, we can determine which energy levels are occupied by the electrons in an atom in its ground state, as the next example demonstrates.

EXAMPLE 8 | Ground States of Atoms

Determine which of the energy levels in **Figure 30.14** are occupied by the electrons in the ground state of hydrogen (1 electron), helium (2 electrons), lithium (3 electrons), beryllium (4 electrons), and boron (5 electrons).

Reasoning In the ground state of an atom the electrons are in the lowest available energy levels. Consistent with the Pauli exclusion principle, they fill those levels "from the bottom up"—that is, from the lowest to the highest energy.

Solution As the colored dot (●) in **Figure 30.15** indicates, the electron in the hydrogen atom (H) is in the $n = 1$, $\ell = 0$ subshell, which has the lowest possible energy. A second electron is present in the helium atom (He), and both electrons can have the quantum numbers $n = 1$, $\ell = 0$, and $m_\ell = 0$. However, in accord with the Pauli exclusion principle, each electron must have a different spin quantum number, $m_s = +\frac{1}{2}$ for one electron and $m_s = -\frac{1}{2}$ for the other. Thus, the drawing shows both electrons in the lowest energy level.

The third electron that is present in the lithium atom (Li) would violate the exclusion principle if it were also in the $n = 1$, $\ell = 0$ subshell, no matter what the value for m_s is. Thus, the $n = 1$, $\ell = 0$ subshell is filled when occupied by two electrons. With this level filled, the $n = 2$, $\ell = 0$ subshell becomes the next lowest energy level available and is where the third electron of lithium is found (see **Figure 30.15**). In the beryllium atom (Be), the fourth electron is in the $n = 2$, $\ell = 0$ subshell, along with the third electron. This is possible, since the third and fourth electrons can have different values for m_s.

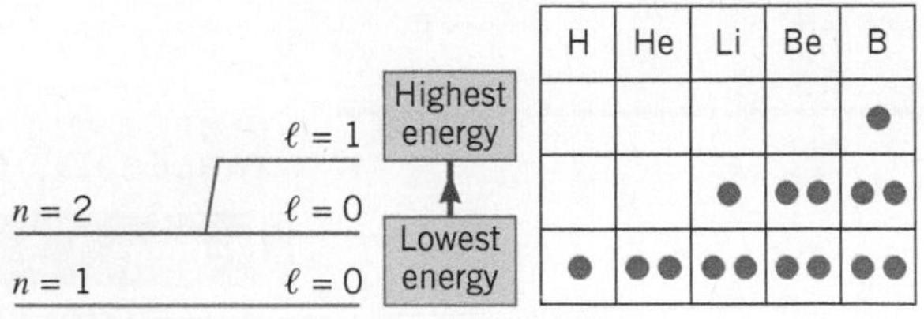

FIGURE 30.15 The electrons (●) in the ground state of an atom fill the available energy levels "from the bottom up"—that is, from the lowest to the highest energy, consistent with the Pauli exclusion principle. The ranking of the energy levels in this figure is meant to apply for a given atom only.

With the first four electrons in place as just discussed, the fifth electron in the boron atom (B) cannot fit into the $n = 1$, $\ell = 0$ or the $n = 2$, $\ell = 0$ subshell without violating the exclusion principle. Therefore, the fifth electron is found in the $n = 2$, $\ell = 1$ subshell, which is the next available energy level with the lowest energy, as **Figure 30.15** indicates. For this electron, m_ℓ can be -1, 0, or $+1$, and m_s can be $+\frac{1}{2}$ or $-\frac{1}{2}$ in each case. However, in the absence of an external magnetic field, all six possibilities correspond to the same energy.

	Subshell	Maximum number of electrons in subshell	
	$\ell = 2$	10	Highest energy
$n = 5$	$\ell = 0$	2	
	$\ell = 1$	6	
	$\ell = 2$	10	
$n = 4$	$\ell = 0$	2	
	$\ell = 1$	6	
$n = 3$	$\ell = 0$	2	
	$\ell = 1$	6	
$n = 2$	$\ell = 0$	2	
$n = 1$	$\ell = 0$	2	Lowest energy

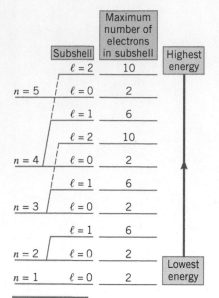

FIGURE 30.16 The maximum number of electrons that the ℓth subshell can hold is $2(2\ell + 1)$.

Because of the Pauli exclusion principle, there is a maximum number of electrons that can fit into an energy level or subshell. Example 8 shows that the $n = 1$, $\ell = 0$ subshell can hold at most two electrons. The $n = 2$, $\ell = 1$ subshell, however, can hold six electrons because with $\ell = 1$, there are three possibilities for m_ℓ (-1, 0, and $+1$), and for each of these choices, the value of m_s can be $+\frac{1}{2}$ or $-\frac{1}{2}$. In general, m_ℓ can have the values $0, \pm1, \pm2, \ldots, \pm\ell$, for $2\ell + 1$ possibilities. Since each of these can be combined with two possibilities for m_s, the total number of different combinations for m_ℓ and m_s is $2(2\ell + 1)$. This, then, is the maximum number of electrons the ℓth subshell can hold, as **Figure 30.16** summarizes.

For historical reasons, there is a widely used convention in which each subshell of an atom is referred to by a letter rather than by the value of its orbital quantum number ℓ. For instance, an $\ell = 0$ subshell is called an s subshell. An $\ell = 1$ subshell and an $\ell = 2$ subshell are known as p and d subshells, respectively. The higher values of $\ell = 3$, 4, and so on, are referred to as f, g, and so on, in alphabetical sequence, as **Table 30.2** indicates.

TABLE 30.2 The Convention of Letters Used to Refer to the Orbital Quantum Number

Orbital Quantum Number ℓ	Letter
0	s
1	p
2	d
3	f
4	g
5	h

This convention of letters is used in a shorthand notation that is convenient for indicating simultaneously the principal quantum number n, the orbital quantum number ℓ, and the number of electrons in the n, ℓ subshell. An example of this notation follows:

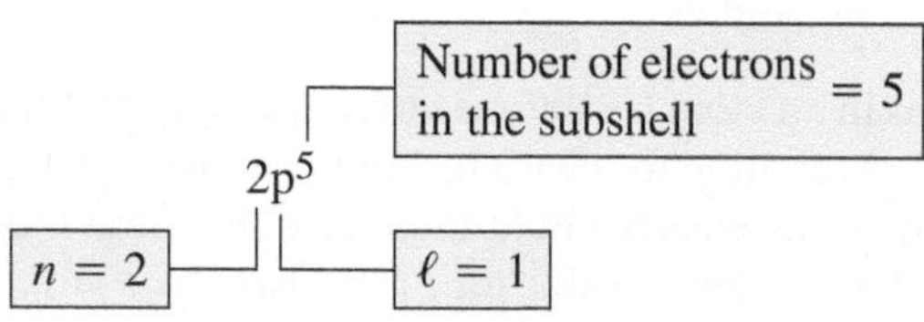

With this notation, the arrangement, or configuration, of the electrons in an atom can be specified efficiently. For instance, in Example 8, we found that the electron configuration for boron has two electrons in the $n = 1$, $\ell = 0$ subshell, two in the $n = 2$, $\ell = 0$ subshell, and one in the $n = 2$, $\ell = 1$ subshell. In shorthand notation this arrangement is expressed as $1s^2\ 2s^2\ 2p^1$. **Table 30.3** gives the ground-state electron configurations written in this fashion for elements containing up to thirteen electrons. The first five entries are those worked out in Example 8.

TABLE 30.3 Ground-State Electronic Configurations of Atoms

Element	Number of Element	Configuration of the Electrons
Hydrogen (H)	1	$1s^1$
Helium (He)	2	$1s^2$
Lithium (Li)	3	$1s^2\ 2s^1$
Beryllium (Be)	4	$1s^2\ 2s^2$
Boron (B)	5	$1s^2\ 2s^2\ 2p^1$
Carbon (C)	6	$1s^2\ 2s^2\ 2p^2$
Nitrogen (N)	7	$1s^2\ 2s^2\ 2p^3$
Oxygen (O)	8	$1s^2\ 2s^2\ 2p^4$
Fluorine (F)	9	$1s^2\ 2s^2\ 2p^5$
Neon (Ne)	10	$1s^2\ 2s^2\ 2p^6$
Sodium (Na)	11	$1s^2\ 2s^2\ 2p^6\ 3s^1$
Magnesium (Mg)	12	$1s^2\ 2s^2\ 2p^6\ 3s^2$
Aluminum (Al)	13	$1s^2\ 2s^2\ 2p^6\ 3s^2\ 3p^1$

Each entry in the periodic table of the elements often includes the ground-state electronic configuration, as **Interactive Figure 30.17** illustrates for argon. To save space, only the configuration of the outermost electrons and unfilled subshells is specified, using the shorthand notation just discussed. Originally the periodic table was developed by the Russian chemist Dmitri Mendeleev (1834–1907) on the basis that certain groups of elements exhibit similar chemical properties. There are eight of these groups, plus the transition elements in the middle of the table, which include the lanthanide series and the actinide series. The similar chemical properties within a group can be explained on the basis of the configurations of the outer electrons of the elements in the group. Thus, quantum mechanics and the Pauli exclusion principle offer an explanation for the chemical behavior of the atoms. The full periodic table can be found immediately before the back cover of this text.

An application that utilizes atomic emission in forensic science is discussed in Example 9.

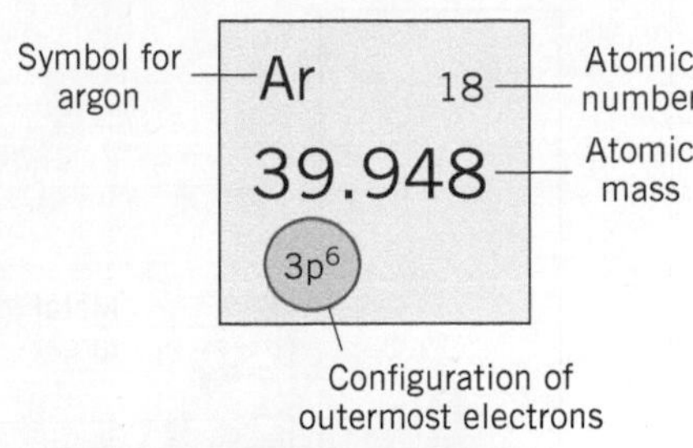

INTERACTIVE FIGURE 30.17 The entries in the periodic table of the elements often include the ground-state configuration of the outermost electrons.

EXAMPLE 9 | BIO Atomic Emission in Forensic Medicine

Since the atomic energy levels of the elements are unique, so are their respective emission spectra. For instance, the line spectrum emitted by excited gold atoms is distinctly different than that emitted by excited lead atoms. Imagine a substance that is a mixture (or alloy) of gold and lead. Such a material, when excited in some way, would simultaneously emit the line spectra of both gold and lead. Atomic emission can therefore be used to identify the elemental compositions of materials, even in small quantities. This has great utility in medicine and forensic science. For example, arsenic has an affinity for keratin and therefore gets trapped in human hair at the time of ingestion. As the hair grows, with a known average rate of about 1.2 cm per month, the trapped arsenic remains. Therefore, not only can atomic emission detect the presence of arsenic, but it can estimate the time at which the person was exposed.

There are various methods that can be used to excite atoms in a material, including corona (high-voltage) discharge, incineration, and X-ray excitation. One commonly used method in medicine is inductively coupled plasma–atomic emission spectroscopy (ICP-AES), which consists of a device which uses electromagnetic induction to create a plasma (a highly energetic ionized gas that emits light) that includes sample atoms and their ions. The emitted light is then analyzed using an optical spectrometer.

Suppose a hair sample is suspected of containing arsenic. **(a)** What is the ground-state atomic configuration of arsenic (Z = 33)? **(b)** In arsenic, suppose an electron in an excited state with energy E_i = 6.770 eV undergoes a transition to a final state with energy E_f = 1.353 eV. Determine the wavelength of the light that is emitted during this transition. **(c)** If arsenic were detected only in a short hair segment, 3.5 cm above the root of the hair, estimate how much time had passed between the ingestion of the arsenic and the removal of the hair from the head.

Reasoning **(a)** Arsenic (As) has 33 electrons. We will follow the style used in Table 30.3 of Section 30.6 to fill the subshells in accordance with the Pauli exclusion principle, the order of which is depicted in Figure 30.16. **(b)** The energy of the emitted photons will be the energy difference between the levels. Knowing the energy of the emitted photons, we can use Equation 30.4 to find the frequency, and then convert to wavelength using $f = c/\lambda$. **(c)** Since the arsenic was first detected 3.5 cm above the root, and we know the approximate rate of growth of the hair, we can estimate how much time had passed between the ingestion of the arsenic and the removal of the hair from the head.

Solution **(a)** Following the style of Table 30.3, the $\ell = 0$ subshell is denoted by s, the $\ell = 1$ subshell is denoted by p, and the $\ell = 2$ subshell is denoted by d. In Figure 30.16 we see that the 4s subshell fills before the 3d subshell, and that the 5s subshell fills before the 4d subshell. We also see in Figure 30.16 that s subshells can contain up to 2 electrons, p subshells can contain up to 6 electrons, and d subshells can contain a maximum of 10 electrons. Following these rules, in accordance with the Pauli exclusion principle, we have the following for the ground state configuration of arsenic:

$$\boxed{1s^2\,2s^2\,2p^6\,3s^2\,3p^6\,4s^2\,3d^{10}\,4p^3}$$

(b) With the difference between energy levels given by $E_i - E_f$, we can apply Equation 30.4, and substitute $f = c/\lambda$ to give the following result:

$$E_i - E_f = hf = \frac{hc}{\lambda}$$

The energies were given in eV, and must first be converted to joules. Using the conversion factor 1 eV = 1.602×10^{-19} J, we have $E_i = 10.846 \times 10^{-19}$ J and $E_f = 2.168 \times 10^{-19}$ J. Solving for the wavelength, we have

$$\lambda = \frac{hc}{E_i - E_f} = \frac{(6.626 \times 10^{-34}\ \text{J}\cdot\text{s})(2.998 \times 10^{8}\ \text{m/s})}{(10.846 \times 10^{-34}\ \text{J} - 2.168 \times 10^{-34}\ \text{J})} = \boxed{228.9\ \text{nm}}$$

where more accurate values of h and c were used to accommodate the number of significant digits in the final answer. Note that this wavelength is in the ultraviolet part of the spectrum. It can therefore be detected by spectrometers, but not by the human eye.

(c) Since the arsenic was detected 3.5 cm above the root, and hair grows at an average rate of 1.2 cm/month, the hair grew for a time t after ingestion given by

$$t = \frac{3.5\ \text{cm}}{1.2\ \text{cm/month}} = \boxed{2.9\ \text{months}}$$

Thus, the person ingested the arsenic 2.9 months before the hair was removed from their head.

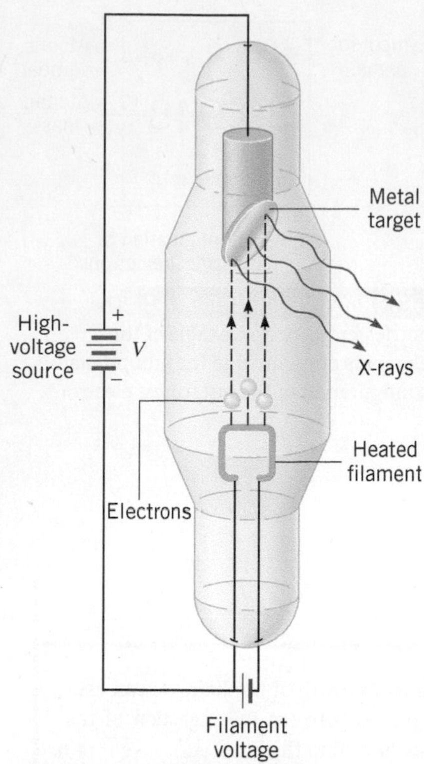

INTERACTIVE FIGURE 30.18 In an X-ray tube, electrons are emitted by a heated filament, accelerate through a large potential difference V, and strike a metal target. The X-rays originate when the electrons interact with the target.

Check Your Understanding

(The answers are given at the end of the book.)

8. Using the convention of letters to refer to the orbital quantum number, write down the ground-state configuration of the electrons in krypton ($Z = 36$).

9. Can a 5g subshell contain **(a)** 22 electrons? **(b)** 17 electrons?

10. An electronic configuration for manganese ($Z = 25$) is written as $1s^2\,2s^2\,2p^6\,3s^2\,3p^6\,4s^2\,3d^4\,4p^1$. Does this configuration represent **(a)** the ground state or **(b)** an excited state?

30.7 X-Rays

THE PHYSICS OF . . . X-rays. X-rays were discovered by the Dutch physicist Wilhelm K. Roentgen (1845–1923), who performed much of his work in Germany. X-rays can be produced when electrons, accelerated through a large potential difference, collide with a metal target made, for example, from molybdenum or platinum. The target is contained within an evacuated glass tube, as **Interactive Figure 30.18** shows. Example 10 discusses the relationship between the wavelength of the emitted X-rays and the speed of the impinging electrons.

Analyzing Multiple-Concept Problems

EXAMPLE 10 | X-Rays and Electrons

The highest-energy X-rays produced by an X-ray tube have a wavelength of 1.20×10^{-10} m. What is the speed of the electrons in **Interactive Figure 30.18** just before they strike the metal target? Ignore the effects of relativity.

Reasoning We can find the electron's speed from a knowledge of its kinetic energy, since the two are related. Moreover, it is the electron's kinetic energy that determines the energy of any photon. The energy of any photon, on the other hand, is directly proportional to its frequency, as discussed in Section 29.3. But we know from our study of waves (see Section 16.2) that the frequency of a wave is inversely proportional to its wavelength. Thus, we will be able to find the speed of an impinging electron from the given X-ray wavelength.

Knowns and Unknowns The following table summarizes what we know and what we seek:

Description	Symbol	Value
Wavelength of highest-energy X-ray photon	λ	1.20×10^{-10} m
Unknown Variable		
Speed of electron just before it strikes target	v	?

Modeling the Problem

STEP 1 Kinetic Energy The kinetic energy of an electron is the energy it has because of its motion. Since we are ignoring the effects of relativity, the kinetic energy KE is given by Equation 6.2 as $\text{KE} = \frac{1}{2}mv^2$, where m and v are the electron's mass and speed. Solving this expression for the speed gives Equation 1 at the right. The mass of the electron is known but its kinetic energy is not, so this will be evaluated in Steps 2 and 3.

$$v = \sqrt{\frac{2(\text{KE})}{m}} \quad (1)$$

?

STEP 2 Energy of a Photon An X-ray photon is a discrete packet of electromagnetic-wave energy. The photon's energy E is given by $E = hf$, where h is Planck's constant and f is the photon's frequency (see Equation 29.2). The energy needed to produce an X-ray photon comes from the kinetic energy of an electron striking the target. We know that the photons have the highest possible energy. This means that all of an electron's kinetic energy KE goes into producing a photon, so KE = E. Substituting KE = E into $E = hf$ gives

$$\boxed{\text{KE} = hf} \quad \textbf{(2)}$$

This result for the energy of the photon can be substituted into Equation 1, as indicated at the right. The frequency is not known, but Step 3 discusses how it can be obtained from the given wavelength.

$$v = \sqrt{\frac{2(\text{KE})}{m}} \quad \textbf{(1)}$$

$$\boxed{\text{KE} = hf} \quad \textbf{(2)}$$

$$\boxed{?}$$

STEP 3 Relation Between Frequency and Wavelength Electromagnetic waves such as X-rays travel at the speed c of light in a vacuum. According to Equation 16.1, this speed is related to the frequency f and the wavelength λ by $c = f\lambda$. Solving for the frequency gives

$$\boxed{f = \frac{c}{\lambda}}$$

All the variables on the right side of this equation are known, so we substitute it into Equation 2 for the kinetic energy, as shown in the right column.

$$v = \sqrt{\frac{2(\text{KE})}{m}} \quad \textbf{(1)}$$

$$\boxed{\text{KE} = hf} \quad \textbf{(2)}$$

$$\boxed{f = \frac{c}{\lambda}}$$

Solution Algebraically combining the results of the three steps, we have

STEP 1 STEP 2 STEP 3

$$v = \sqrt{\frac{2(\text{KE})}{m}} = \sqrt{\frac{2(hf)}{m}} = \sqrt{\frac{2h\left(\frac{c}{\lambda}\right)}{m}} = \sqrt{\frac{2hc}{m\lambda}}$$

The speed of an electron just before it strikes the metal target is

$$v = \sqrt{\frac{2hc}{m\lambda}} = \sqrt{\frac{2(6.63 \times 10^{-34}\ \text{J} \cdot \text{s})(3.00 \times 10^{8}\ \text{m/s})}{(9.11 \times 10^{-31}\ \text{kg})(1.20 \times 10^{-10}\ \text{m})}} = \boxed{6.03 \times 10^{7}\ \text{m/s}}$$

Related Homework: Problem 41

A plot of X-ray intensity per unit wavelength versus the wavelength looks similar to **Figure 30.19** and consists of sharp peaks or lines superimposed on a broad continuous spectrum. The sharp peaks are called characteristic lines or **characteristic X-rays** because they are characteristic of the target material. The broad continuous spectrum is referred to as **Bremsstrahlung** (German for "braking radiation") and is emitted when the electrons decelerate or "brake" upon hitting the target.

In **Figure 30.19** the characteristic lines are marked K_α and K_β because they involve the $n = 1$ or K shell of a metal atom. If an electron with enough energy strikes the target, one of the K-shell electrons can be knocked out. An electron in one of the outer shells can then fall into the K shell, and an X-ray photon is emitted in the process. The K_α line arises when an electron in the $n = 2$ level falls into the vacancy that the impinging electron has created in the $n = 1$ level. Similarly, the K_β line arises when an electron in the $n = 3$ level falls to the $n = 1$ level. Example 11 shows that a large potential difference is needed to operate an X-ray tube so that the electrons impinging on the metal target will have sufficient energy to generate the characteristic X-rays. Example 12 determines an estimate for the K_α wavelength of platinum.

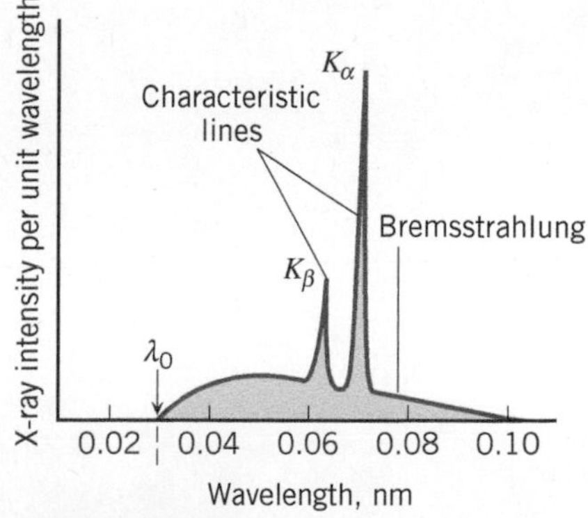

FIGURE 30.19 When a molybdenum target is bombarded with electrons that have been accelerated from rest through a potential difference of 45 000 V, this X-ray spectrum is produced. The vertical axis is not to scale.

EXAMPLE 11 | Operating an X-Ray Tube

Strictly speaking, the Bohr model does not apply to multiple-electron atoms, but it can be used to make estimates. Use the Bohr model to estimate the minimum energy that an incoming electron must have to knock a K-shell electron entirely out of an atom in a platinum ($Z = 78$) target in an X-ray tube.

Reasoning According to the Bohr model, the energy of a K-shell electron is given by Equation 30.13, $E_n = -(13.6 \text{ eV})Z^2/n^2$, with $n = 1$. When striking a platinum target, an incoming electron must have at least enough energy to raise the K-shell electron from this low energy level up to the 0-eV level that corresponds to a very large distance from the nucleus. Only then will the incoming electron knock the K-shell electron entirely out of a target atom.

Problem-Solving Insight Equation 30.13 for the Bohr energy levels [$E_n = -(13.6 \text{ eV})Z^2/n^2$, $n = 1$] can be used in rough calculations of the energy levels involved in the production of K_α X-rays. In this equation, however, the atomic number Z must be reduced by one, to account approximately for the shielding of one K-shell electron by the other K-shell electron.

Solution We will use Equation 30.13 to estimate the minimum energy that an incoming electron must have. However, strictly speaking, this equation applies only to one-electron atoms, because it neglects the repulsive force between electrons in a multiple-electron atom. In the platinum K-shell, each electron exerts on the other electron a repulsive force that balances (approximately) the attractive force of one nuclear proton. In effect, one of the K-shell electrons shields the other from the force of that proton. Therefore, in our calculation we replace Z in Equation 30.13 by $Z - 1$ and find that

$$E_1 = -(13.6 \text{ eV})\frac{(Z-1)^2}{n^2} = -(13.6 \text{ eV})\frac{(78-1)^2}{1^2} = -8.1 \times 10^4 \text{ eV}$$

Hence, to raise the K-shell electron up to the 0-eV level, the minimum energy for an incoming electron is $\boxed{8.1 \times 10^4 \text{ eV}}$. One electron volt is the kinetic energy acquired when an electron accelerates from rest through a potential difference of one volt. Thus, a potential difference of at least 81 000 V must be applied to the X-ray tube.

EXAMPLE 12 | The K_α Characteristic X-Ray for Platinum

Use the Bohr model to estimate the wavelength of the K_α line in the X-ray spectrum of platinum ($Z = 78$).

Reasoning This example is very similar to Example 4, which deals with the emission line spectrum of the hydrogen atom. As in that example, we use Equation 30.14, this time with the initial value of n being $n_i = 2$ and the final value being $n_f = 1$. As in Example 11, a value of 77 rather than 78 is used for Z to account approximately for the shielding effect of the single K-shell electron in canceling out the attraction of one nuclear proton.

Solution Using Equation 30.14, we find that

$$\frac{1}{\lambda} = (1.097 \times 10^7 \text{ m}^{-1})(78-1)^2\left(\frac{1}{1^2} - \frac{1}{2^2}\right) = 4.9 \times 10^{10} \text{ m}^{-1}$$

$$\text{or} \quad \boxed{\lambda = 20 \times 10^{-11} \text{ m}}$$

This answer is close to an experimental value of 1.9×10^{-11} m.

Another interesting feature of the X-ray spectrum in **Figure 30.19** is the sharp cutoff that occurs at a wavelength of λ_0 on the short-wavelength side of the Bremsstrahlung. This cutoff wavelength is independent of the target material but depends on the energy of the impinging electrons. An impinging electron cannot give up any more than all of its kinetic energy when decelerated by the metal target in an X-ray tube. Thus, at most, an emitted X-ray photon can have an energy equal to the kinetic energy KE of the electron and a frequency given by Equation 29.2 as $f = (\text{KE})/h$, where h is Planck's constant. But the kinetic energy acquired by an electron (charge magnitude = e) in accelerating from rest through a potential difference V is e times V, according to earlier discussions in Section 19.2; V is the potential difference applied across the X-ray tube (see **Interactive Figure 30.18**). Thus, the maximum photon frequency is $f_0 = (eV)/h$. Since $f_0 = c/\lambda_0$, a maximum frequency corresponds to a minimum wavelength, which is the cutoff wavelength λ_0:

$$\lambda_0 = \frac{hc}{eV} \quad \textbf{(30.17)}$$

Figure 30.19, for instance, assumes a potential difference of 45 000 V, which corresponds to a cutoff wavelength of

$$\lambda_0 = \frac{(6.63 \times 10^{-34} \text{ J} \cdot \text{s})(3.00 \times 10^8 \text{ m/s})}{(1.60 \times 10^{-9} \text{ C})(45\,000 \text{ V})} = 2.8 \times 10^{-11} \text{ m}$$

The medical profession began using X-rays for diagnostic purposes almost immediately after their discovery. When a conventional X-ray is obtained, the patient is typically positioned in front of a piece of photographic film, and a single burst of radiation is directed through the patient and onto the film. Since the dense structure of bone absorbs X-rays much more than soft tissue does, a shadow-like picture is recorded on the film. As useful as such pictures are, they have an inherent limitation. The image on the film is a superposition of all the "shadows" that result as the radiation passes through one layer of

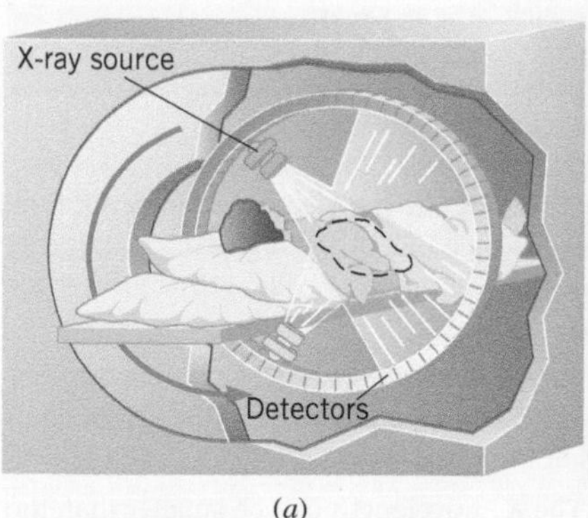

(a)

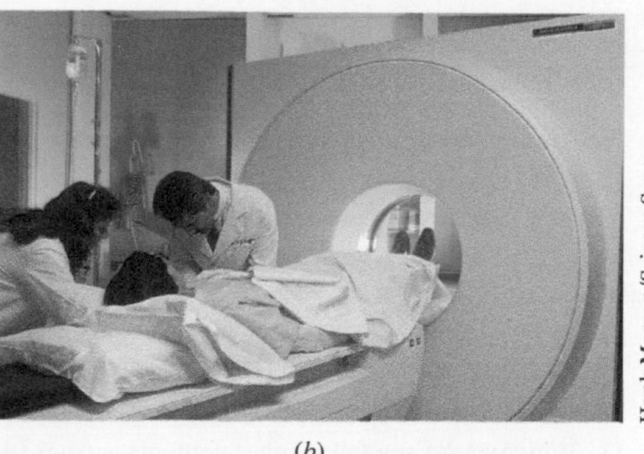

(b)

Hank Morgan/Science Source

FIGURE 30.20 (*a*) In CAT scanning, a "fanned-out" array of X-ray beams is sent through the patient from different orientations. (*b*) A doctor and a nurse prepare a patient for a CAT scan.

body material after another. Interpreting which part of a conventional X-ray corresponds to which layer of body material is very difficult.

BIO THE PHYSICS OF . . . CAT scanning. The technique known as CAT scanning, or CT scanning, has greatly extended the ability of X-rays to provide images of specific locations within the body. The acronym CAT stands for **c**omputerized **a**xial **t**omography or **c**omputer-**a**ssisted **t**omography, and the shorter version CT stands for **c**omputerized **t**omography. In this technique a series of X-ray images are obtained as indicated in **Figure 30.20**. A number of X-ray beams form a "fanned out" array of radiation and pass simultaneously through the patient. Each of the beams is detected on the other side by a detector, which records the beam intensity. The various intensities are different, depending on the nature of the body material through which the beams have passed. The feature of CAT scanning that leads to dramatic improvements over the conventional technique is that the X-ray source can be rotated to different orientations, so that the fanned-out array of beams can be sent through the patient from various directions. **Figure 30.20*a*** singles out two directions for illustration. In reality many different orientations are used, and the intensity of each beam in the array is recorded as a function of orientation. The way in which the intensity of a beam changes from one orientation to another is used as input to a computer. The computer then constructs a highly resolved image of the cross-sectional slice of body material through which the fan of radiation has passed. In effect, the CAT scanning technique makes it possible to take an X-ray picture of a cross-sectional "slice" that is perpendicular to the body's long axis. In fact, the word "axial" in the phrase "computerized axial tomography" refers to the body's long axis. The chapter-opening photograph and **Figure 30.21** show three-dimensional CAT scans of parts of the human anatomy.

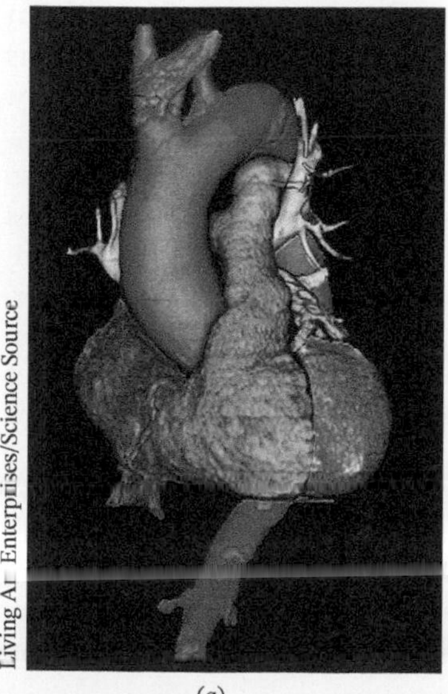

Living Art Enterprises/Science Source

(a)

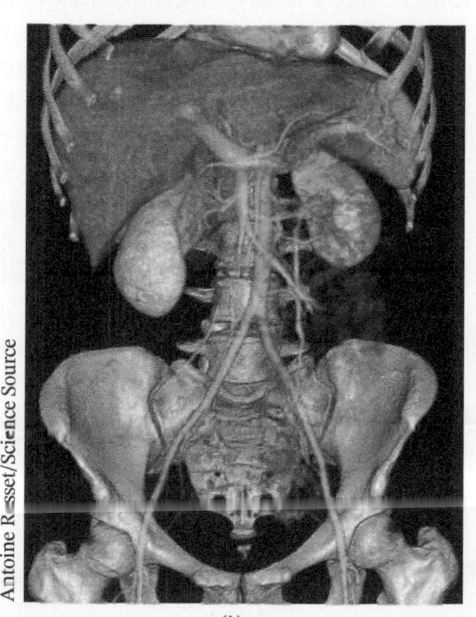

Antoine Rosset/Science Source

(b)

FIGURE 30.21 (*a*) A 3D CAT scan of a human heart. The inferior vena cava and right atrium are purple. The right ventricle and pulmonary outflow tract are blue. The pulmonary veins are yellow. The left ventricle is orange, and the aorta is red. (*b*) A 3D CAT scan of an abdomen in frontal view. Near the top of the image the large mass is the liver (green), and beneath it the two smaller teardrop-shaped objects (green/yellow) are the kidneys.

Check Your Understanding

(The answers are given at the end of the book.)

11. X-ray tube A and X-ray tube B use the same voltage to accelerate the electrons. However, tube A uses a copper target, whereas tube B uses a silver target. Which one of the following statements is true? **(a)** The cutoff wavelength is greater for tube A. **(b)** The cutoff wavelength is greater for tube B. **(c)** Both tubes have the same cutoff wavelengths.

12. Is it possible to adjust the electric potential V used to operate an X-ray tube so that Bremsstrahlung X-rays are created, but characteristic X-rays are not created? **(a)** Yes, if V is small enough. **(b)** Yes, if V is large enough. **(c)** No, regardless of the value of V.

13. Which one of the following statements is true? **(a)** The K_α wavelength can be smaller than the cutoff wavelength λ_0, assuming that both are produced by the same X-ray tube. **(b)** The K_α wavelength is produced when an electron undergoes a transition from the $n = 1$ energy level to the $n = 2$ energy level. **(c)** The K_β wavelength is always smaller than the K_α wavelength for a given metal target.

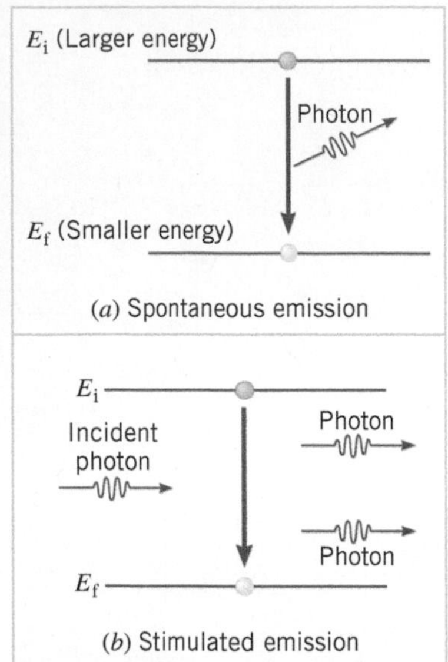

FIGURE 30.22 (*a*) Spontaneous emission of a photon occurs when the electron (●) makes an unprovoked transition from a higher to a lower energy level, the photon departing in a random direction. (*b*) Stimulated emission of a photon occurs when an incoming photon with the correct energy induces an electron to change energy levels, the emitted photon traveling in the same direction as the incoming photon.

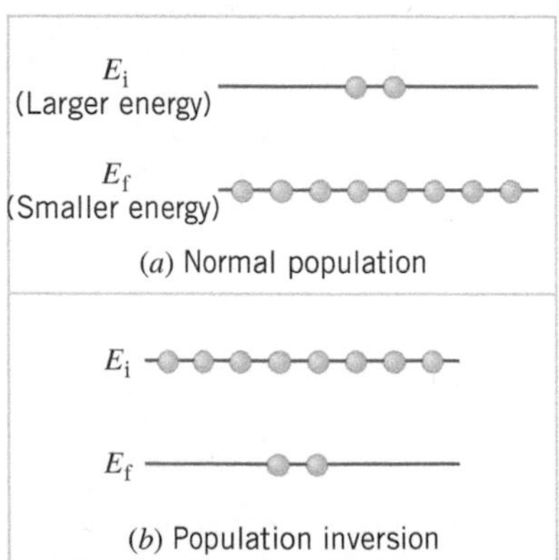

FIGURE 30.23 (*a*) In a normal situation at room temperature, most of the electrons in atoms are found in a lower or ground-state energy level. (*b*) If an external energy source is provided to excite electrons into a higher energy level, a population inversion can be created in which more electrons are in the higher level than in the lower level.

30.8 The Laser

THE PHYSICS OF . . . the laser. The laser is one of the most useful inventions of the twentieth century. Today, there are many types of lasers, and most of them work in a way that depends directly on the quantum mechanical structure of the atom.

When an electron makes a transition from a higher energy state to a lower energy state, a photon is emitted. The emission process can be one of two types, spontaneous or stimulated. In **spontaneous emission** (see Figure 30.22*a*), the photon is emitted spontaneously, in a random direction, without external provocation. In **stimulated emission** (see Figure 30.22*b*), an incoming photon induces, or stimulates, the electron to change energy levels. To produce stimulated emission, however, the incoming photon must have an energy that exactly matches the difference between the energies of the two levels—namely, $E_i - E_f$. Stimulated emission is similar to a resonant process, in which the incoming photon "jiggles" the electron at just the frequency to which it is particularly sensitive and causes the change between energy levels. This frequency is given by Equation 30.4 as $f = (E_i - E_f)/h$. The operation of lasers depends on stimulated emission.

Stimulated emission has three important features. First, one photon goes in and two photons come out (see Figure 30.22*b*). In this sense, the process amplifies the number of photons. In fact, this is the origin of the word "laser," which is an acronym for **l**ight **a**mplification by the **s**timulated **e**mission of **r**adiation. Second, the emitted photon travels in the same direction as the incoming photon. Third, the emitted photon is exactly in step with or has the same phase as the incoming photon. In other words, the two electromagnetic waves that these two photons represent are coherent (see Section 17.2) and are locked in step with one another. In contrast, two photons emitted by the filament of an incandescent light bulb are emitted independently. They are not coherent, since one does not stimulate the emission of the other.

Although stimulated emission plays a pivotal role in a laser, other factors are also important. For instance, an external source of energy must be available to excite electrons into higher energy levels. The energy can be provided in a number of ways, including intense flashes of ordinary light and high-voltage discharges. If sufficient energy is delivered to the atoms, more electrons will be excited to a higher energy level than remain in a lower level, a condition known as a **population inversion**. Figure 30.23 compares a normal energy level population with a population inversion. The population inversions used in lasers involve a higher energy state that is **metastable**, in the sense that electrons remain in the metastable state for a much longer period of time than they do in an ordinary excited state (10^{-3} s versus 10^{-8} s, for example). The requirement of a metastable higher energy state is essential, so that there is more time to enhance the population inversion.

Figure 30.24 shows the widely used helium/neon laser. To sustain the necessary population inversion, a high voltage is discharged across a low-pressure mixture of 15% helium

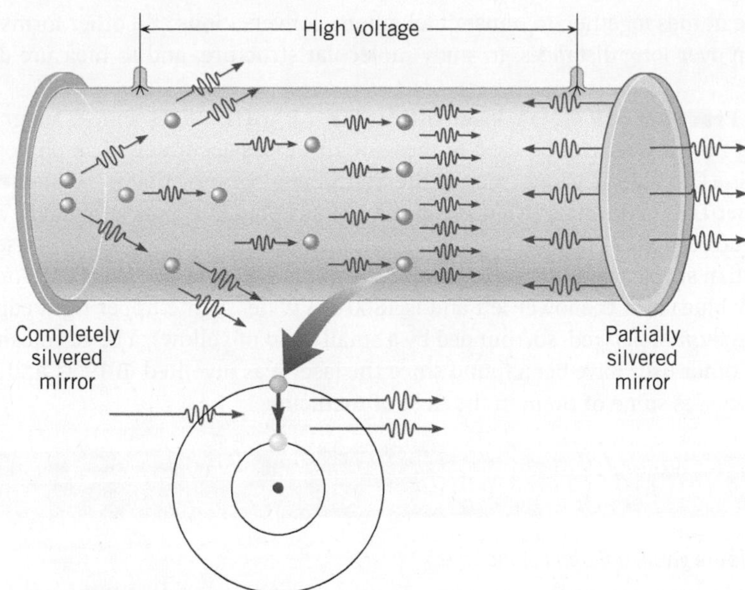

FIGURE 30.24 A schematic drawing of a helium/neon laser. The blow-up shows the stimulated emission that occurs when an electron in a neon atom is induced to change from a higher to a lower energy level.

and 85% neon contained in a glass tube. The laser process begins when an atom, via spontaneous emission, emits a photon parallel to the axis of the tube. This photon, via stimulated emission, causes another atom to emit two photons parallel to the tube axis. These two photons, in turn, stimulate two more atoms, yielding four photons. Four yield eight, and so on, in a kind of avalanche. To ensure that more and more photons are created by stimulated emission, both ends of the tube are silvered to form mirrors that reflect the photons back and forth through the helium/neon mixture. One end is only partially silvered, however, so that some of the photons can escape from the tube to form the laser beam. When the stimulated emission involves only a single pair of energy levels, the output beam has a single frequency or wavelength and is said to be monochromatic.

A laser beam is also exceptionally narrow. The width is determined by the size of the opening through which the beam exits, and very little spreading-out occurs, except that due to diffraction around the edges of the opening. A laser beam does not spread much because any photons emitted at an angle with respect to the tube axis are quickly reflected out the sides of the tube by the silvered ends (see **Figure 30.24**). These ends are carefully arranged to be perpendicular to the tube axis. Since all the power in a laser beam can be confined to a narrow region, the intensity, or power per unit area, can be quite large.

Figure 30.25 shows the pertinent energy levels for a helium/neon laser. By coincidence, helium and neon have nearly identical metastable higher energy states, respectively located 20.61 and 20.66 eV above the ground state. The high-voltage discharge across the gaseous mixture excites electrons in helium atoms to the 20.61-eV state. Then, when an excited helium atom collides inelastically with a neon atom, the 20.61 eV of energy is given to an electron in the neon atom, along with 0.05 eV of kinetic energy from the moving atoms. As a result, the electron in the neon atom is raised to the 20.66-eV state. In this fashion, a population inversion is sustained in the neon, relative to an energy level that is 18.70 eV above the ground state. In producing the laser beam, stimulated emission causes electrons in neon to drop from the 20.66-eV level to the 18.70-eV level. The energy change of 1.96 eV corresponds to a wavelength of 633 nm, which is in the red region of the visible spectrum.

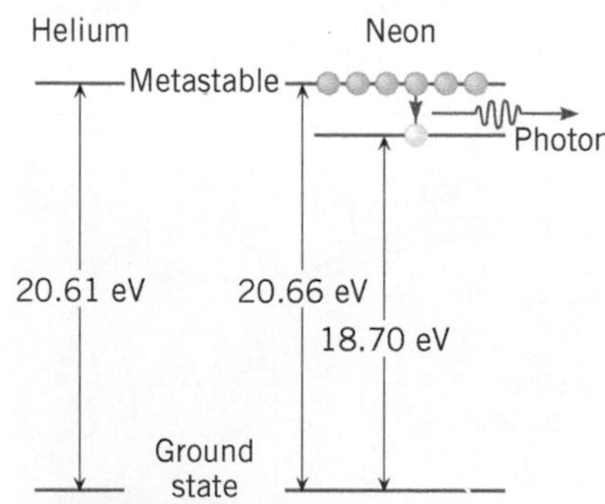

FIGURE 30.25 These energy levels are involved in the operation of a helium/neon laser.

The helium/neon laser is not the only kind of laser. There are many different types, including the ruby laser, the argon-ion laser, the carbon dioxide laser, the gallium arsenide solid state laser, and chemical dye lasers. Depending on the type and whether the laser operates continuously or in pulses, the available beam power ranges from milliwatts to megawatts. Since lasers provide coherent monochromatic electromagnetic radiation that can be confined to an intense narrow beam, they are useful in a wide variety of situations. Today they are used to reproduce music in compact disc players, to weld parts of

FIGURE 30.26 A 3D map of the topography of Mars. The elevation is color-coded from white (highest) through red, yellow, green, blue, and purple (lowest).

automobile frames together, to transmit telephone conversations and other forms of communication over long distances, to study molecular structure, and to measure distances accurately.

THE PHYSICS OF . . . a laser altimeter. Figure 30.26 shows an impressive example of how a laser can be used to measure distances accurately. The photograph in the figure is a three-dimensional map of the Martian topography that was obtained by the Mars Orbiter Laser Altimeter (MOLA) on the Mars Global Surveyor spacecraft. The map was constructed from 27 million height measurements, each made by sending laser pulses to the Martian surface and measuring their return times. The large *Hellas Planitia* impact basin (dark blue) is at the lower left and is 1800 km wide. At the upper right edge of the image is *Elysium Mons* (red, surrounded by a small band of yellow), a large volcano.

Many other uses have been found since the laser was invented in 1960, and the next section discusses some of them in the field of medicine.

Check Your Understanding

(The answers are given at the end of the book.)

14. A certain laser is designed to operate continuously. Which one of the following statements is false? **(a)** The population inversion used in this laser involves a higher energy state and a lower energy state. **(b)** The population inversion used in this laser involves a metastable higher energy state. **(c)** The laser needs an external source of energy to operate. **(d)** The external energy source that the laser uses can be disconnected once the population inversion is established.

15. Laser A produces green light. Laser B produces red light. Which laser utilizes energy levels that have a larger energy difference between them? **(a)** Laser A **(b)** Laser B **(c)** The energy difference between the levels is the same for both lasers.

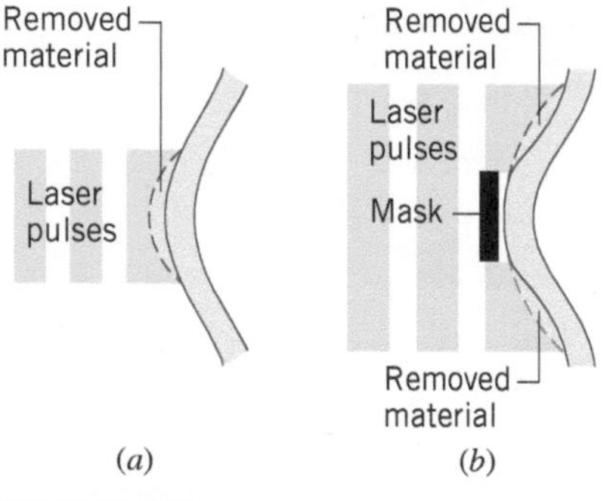

FIGURE 30.27 *(a)* To correct for myopia (nearsightedness) using the PRK procedure, a laser vaporizes tissue (dashed line) on the center of the cornea, thereby flattening it. *(b)* To correct for hyperopia (farsightedness), a laser vaporizes tissue on the peripheral region of the cornea, thereby steepening its contour.

30.9 *Medical Applications of the Laser

One of the medical areas in which the laser has had a substantial impact is in ophthalmology, which deals with the structure, function, and diseases of the eye. Section 26.10 discusses the human eye and the use of contact lenses and eyeglasses to correct nearsightedness and farsightedness. In these conditions, the eye cannot refract light properly and produces blurred images on the retina.

BIO **THE PHYSICS OF . . . PRK eye surgery.** A laser-based procedure known as PRK (**p**hoto**r**efractive **k**eratectomy) offers an alternative treatment for nearsightedness and farsightedness that does not rely on lenses. It involves the use of a laser to remove small amounts of tissue from the cornea of the eye (see Figure 26.33) and thereby changes its curvature. As Section 26.10 points out, light enters the eye through the cornea, and it is at the air/cornea boundary that most of the refraction of the light occurs. Therefore, changing the curvature of that boundary can correct deficiencies in the way the eye refracts light, thus causing the image to be focused onto the retina, where it belongs. Ideally, the cornea is dome-shaped. If the dome is too steep, however, the rays of light are focused in front of the retina and nearsightedness results. As Figure 30.27*a* shows, the laser light removes tissue from the center of the cornea, thereby flattening it and increasing the eye's effective focal length. On the other hand, if the shape of the cornea is too flat, light rays would come to a focus behind the retina if they could, and farsightedness occurs. As part *b* of the drawing illustrates, the center of the cornea is now masked and the laser is used to remove peripheral tissue. This steepens the shape of the cornea, thereby shortening the eye's effective focal length and allowing rays to be focused on the retina.

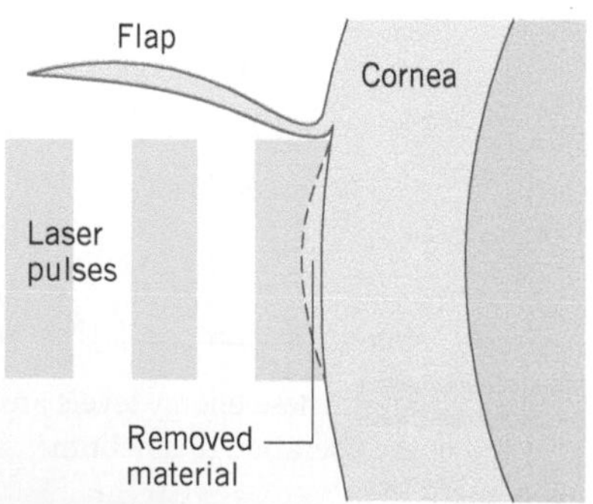

FIGURE 30.28 To correct for myopia (nearsightedness) using the LASIK technique, a laser vaporizes tissue (dashed line) on the cornea, thereby flattening it.

BIO **THE PHYSICS OF . . . LASIK eye surgery.** The LASIK (**l**aser-**a**ssisted **i**n **si**tu **k**eratomileusis) procedure uses a motor-powered blade known as a microkeratome to partially detach a thin flap (about 0.2 mm thick) in the front of the cornea (see Figure 30.28). The flap is pulled back and the laser beam then remodels the corneal tissue underneath by vaporizing cells. Afterward, the flap is folded back into place, with no stitches being required. The laser light in the PRK and LASIK techniques is pulsed and comes from an ultraviolet excimer laser that produces a wavelength of 193 nm. The cornea absorbs this wavelength extremely well, so that weak pulses can be used, leading to highly precise and controllable removal of corneal tissue. Typically, 0.1 to 0.5 μm of tissue is removed by each pulse without damaging adjacent layers.

BIO THE PHYSICS OF . . . removing port-wine stains. Another medical application of the laser is in the treatment of congenital capillary malformations known as port-wine stains, which affect 0.3% of children at birth. These birthmarks are usually found on the head and neck. Preferred treatment for port-wine stains now utilizes a pulsed dye laser, as shown in **Figure 30.29**. The light is absorbed by oxyhemoglobin in the malformed capillaries, which are destroyed in the process without damage to adjacent normal tissue. Eventually the destroyed capillaries are replaced by normal blood vessels, which causes the port-wine stain to fade.

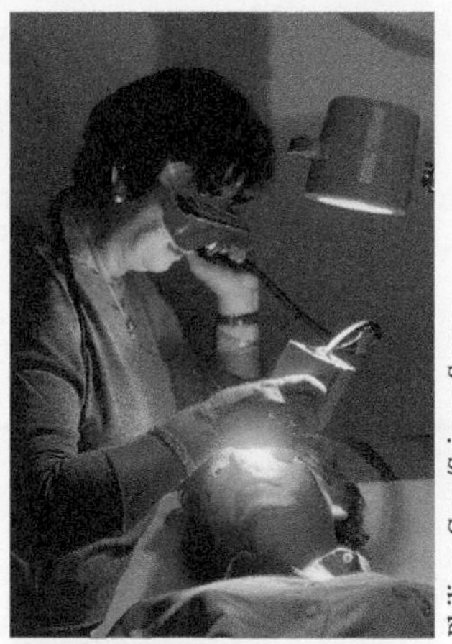
Philippe Garo/Science Source

FIGURE 30.29 A patient receives treatment from a pulsed dye laser. This medical application is effective at removing skin abnormalities, such as port-wine stains and rosacea.

BIO THE PHYSICS OF . . . photodynamic therapy for cancer. In the treatment of cancer, the laser is being used along with light-activated drugs in photodynamic therapy. The procedure involves administering the drug intravenously, so that the tumor can absorb it from the bloodstream, the advantage being that the drug is then located right near the cancer cells. When the drug is activated by laser light, a chemical reaction ensues that disintegrates the cancer cells and the small blood vessels that feed them. In **Figure 30.30** a patient is being treated for cancer of the esophagus. An endoscope that uses optical fibers is inserted down the patient's throat to guide the red laser light to the tumor site and activate the drug. Photodynamic therapy works best with small tumors in their early stages.

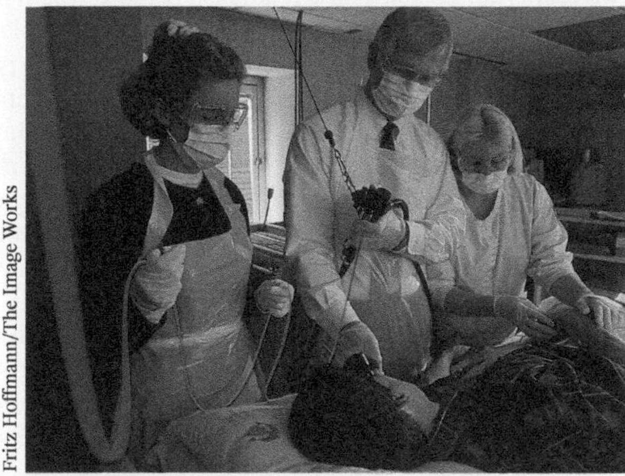
Fritz Hoffmann/The Image Works

FIGURE 30.30 Photodynamic therapy to treat cancer of the esophagus is being administered to this patient. Red laser light is routed to the tumor site with an endoscope that incorporates optical fibers.

EXAMPLE 13 | BIO The Physics of Laser Lithotripsy

Solid masses are sometimes produced in the kidneys, known as kidney stones, and if they are too large, they can become stuck in the ureter, which is the tube leading from the kidney to the bladder. This condition can be extremely painful, and it is often corrected with a medical treatment known as lithotripsy, which is designed to break up the stone, so that it can pass naturally through the urine. One kind of lithotripsy involves the use of a laser, where a fiber optic is inserted into the body near the stone and delivers intense laser pulses to pulverize it. The type of laser often used in this technique is a holmium YAG laser, which is a solid state laser that uses a crystal of yttrium aluminum garnet (YAG), where a small fraction of the yttrium atoms in the crystal have been replaced by holmium atoms. The laser produces infrared light with a wavelength of 2.10 μm and operates with a frequency of 5 Hz, which means it delivers 5 pulses of light to the stone each second. If the energy delivered in each pulse is 0.5 J, how many photons of laser light strike the kidney stone in one second? Round your answer to the nearest whole number.

Reasoning Since we are given the wavelength of the laser light, we can use Equation 29.2 to calculate the energy of an individual photon in the beam. The number of photons striking the stone will be given by the total energy delivered divided by the energy per photon.

Solution (a) From Equation 29.2, we have:

$$E_{\text{photon}} = hf = \frac{hc}{\lambda}$$

The total energy delivered in one second is equal to the number of pulses in one second multiplied by the energy per pulse:

$$E_{\text{tot}} = (\text{number of pulses})\left(\frac{\text{energy}}{\text{pulse}}\right)$$

The number of photons N is then given by:

$$N = \frac{E_{\text{tot}}}{E_{\text{photon}}} = \frac{(\text{number of pulses})\left(\frac{\text{energy}}{\text{pulse}}\right)}{hc/\lambda}$$

$$= \frac{(5)(0.5\text{ J})(2.10\times10^{-6}\text{ m})}{(6.63\times10^{-34}\text{ J}\cdot\text{s})(3.00\times10^{8}\text{ m/s})}$$

$$= \boxed{3\times10^{19}\text{ photons}}$$

30.10 *Holography

THE PHYSICS OF . . . holography. One of the most familiar applications of lasers is in holography, which is a process for producing three-dimensional images. The information used to produce a holographic image is captured on photographic film, which is referred to as a hologram. **Figure 30.31** illustrates how a hologram is made. Laser light strikes a half-silvered mirror, or beamsplitter, which reflects part and transmits part of the light. In the drawing, the reflected part is called the *object beam* because it illuminates the object (a chess piece). The transmitted part is called the *reference beam*. The object beam reflects from the chess piece at points such as A and B and, together with the reference beam, falls on the film. One of the main characteristics of laser light is that it is coherent. Thus, the light from the two beams has a stable phase relationship, like the light from the two slits in Young's double-slit experiment (see Section 27.2). Because of the stable phase relationship and because the two beams travel different distances, an interference pattern is formed on the film. This pattern is the hologram and, although much more complex, is analogous to the pattern of bright and dark fringes formed in the double-slit experiment.

Figure 30.32 shows in greater detail how a holographic interference pattern arises. This drawing considers only the reference beam and the light (wavelength $= \lambda$) coming from point A on the chess piece. As we know from Section 27.1, constructive interference between the two light waves leads to a bright fringe; it occurs when the waves, in reaching the film, travel distances that differ by an integer number m of wavelengths. In the drawing, ℓ_m is the distance between point A and the place on the film where the mth-order bright fringe occurs, and ℓ_0 is the perpendicular distance that the reference beam would travel from point A to the $m = 0$ bright fringe. In addition, r_m is the distance along the film that locates the bright fringe. In terms of these distances, we know that

$$\ell_m - \ell_0 = m\lambda \qquad \text{(condition for constructive interference)}$$

$$\ell_0^{\,2} + r_m^2 = \ell_m^{\,2} \qquad \text{(Pythagorean theorem)}$$

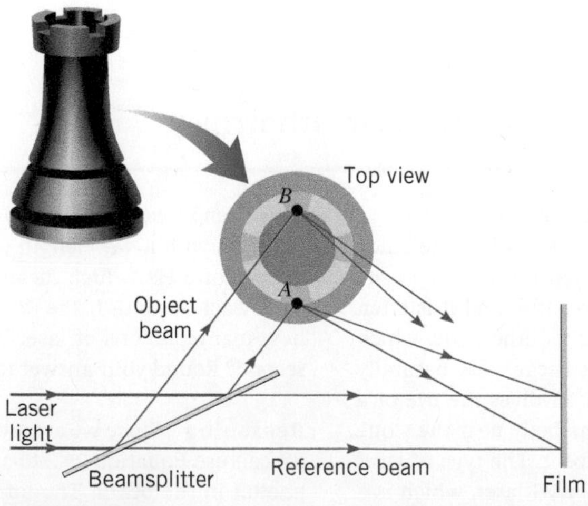

FIGURE 30.31 An arrangement used to produce a hologram.

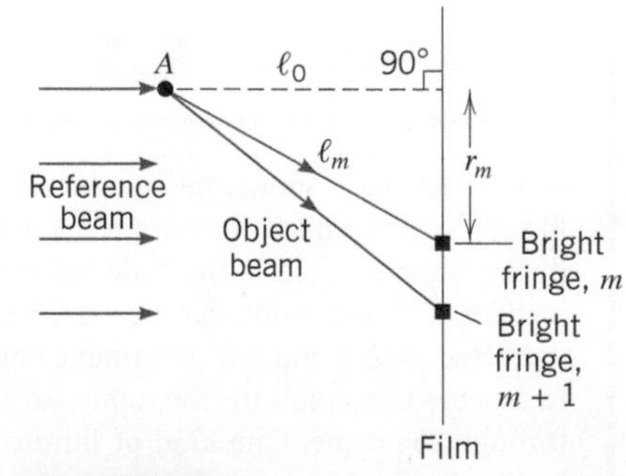

FIGURE 30.32 This drawing helps to explain how the interference pattern arises on the film when light from point A (see **Figure 30.31**) and light from the reference beam combine there.

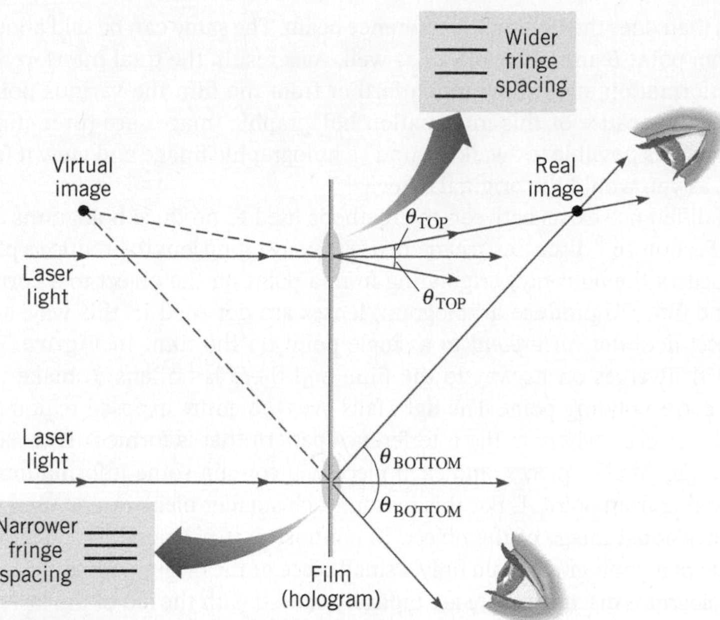

FIGURE 30.33 When the laser light used to produce a hologram is shone through it, both a real and a virtual image of the object are produced.

The first equation indicates that $\ell_m = m\lambda + \ell_0$, which can be substituted into the second equation. The result can be rearranged to show that

$$r_m{}^2 = m\lambda(m\lambda + 2\ell_0)$$

Since ℓ_0 is typically much larger than λ (for instance, $\ell_0 \approx 10^{-1}$ m and $\lambda \approx 10^{-6}$ m), it follows that $r_m \approx \sqrt{m\lambda 2\ell_0}$. In other words, r_m is roughly proportional to $\sqrt{m}$. Therefore, the fringes are farther apart near the top of the film than they are near the bottom. For example, for the $m = 1$ and $m = 2$ fringes, we have $r_2 - r_1 \propto \sqrt{2} - \sqrt{1} = 0.41$ whereas for the $m = 2$ and $m = 3$ fringes, we have $r_3 - r_2 \propto \sqrt{3} - \sqrt{2} = 0.32$.

In addition to the fringe pattern just discussed, the total interference pattern on the hologram includes interference effects that are related to light coming from point *B* and other locations on the object in **Figure 30.31**. The total pattern is very complicated. Nevertheless, the fringe pattern for point *A* alone is sufficient to illustrate the fact that a hologram can be used to produce both a virtual image and a real image of the object, as we will now see.

To produce the holographic images, the laser light is directed through the interference pattern on the film, as in **Figure 30.33**. The pattern can be thought of as a kind of diffraction grating, with the bright fringes analogous to the spaces between the slits of the grating. Section 27.7 discusses how light passing through a grating is split into higher-order bright fringes that are symmetrically located on either side of a central bright fringe. **Figure 30.33** shows the three rays that correspond to the central and first-order bright fringes, as they originate from a spot near the top and a spot near the bottom of the film. The angle θ, which locates the first-order fringes relative to the central fringe, is given by Equation 27.7 (with $m = 1$) as $\sin\theta = \lambda/d$, where *d* is the separation between the slits of the grating. When the slit separation is greater, as it is near the top of the film, the angle is smaller. When the slit separation is smaller, as it is near the bottom, the angle is larger. Thus, **Figure 30.33** has been drawn with θ_{TOP} smaller than θ_{BOTTOM}. Of the three rays emerging from the film at the top and the three at the bottom, we use the uppermost one in each case to locate the real image of point *A* on the chess piece. The real image is located where these two rays intersect, when extended to the right. To locate the virtual image, we use the lower ray in each of the three-ray bundles at the top and bottom of the film. When projected to the left, these rays appear to be originating from the spot where the projections intersect—that is, from the virtual image.

A holographic image differs greatly from a photographic image. The most obvious difference is that a hologram provides a three-dimensional image, whereas photographs are two-dimensional. The three-dimensional nature of the holographic image is inherent in the interference pattern formed on the film. In **Figure 30.31** part of this pattern arises because the light emitted by point *A* travels different distances in reaching different spots on the film than does the light in the reference beam. The same can be said about the light emitted from point B and other places as well. As a result, the total interference pattern contains information about how much farther from the film the various points on the object are, and because of this information holographic images are three-dimensional. Furthermore, it is possible to "walk around" a holographic image and view it from different angles, as you would the original object.

GIPhotoStock/Science Source

FIGURE 30.34 A hologram of a dove on a Visa credit card. When illuminated in white light, the hologram appears to change colors when viewed from different angles.

A vast difference exists between the methods used to produce holograms and photographs. As Section 26.7 discusses, a camera uses a converging lens to produce a photograph. The lens focuses the light rays originating from a point on the object to a corresponding point on the film. To produce a hologram, lenses are not used in this way, and a point on the object *does not correspond* to a single point on the film. In **Figure 30.32**, light from point A diverges on its way to the film, and there is no lens to make it converge to a single corresponding point. The light falls over the entire exposed region of the film and contributes everywhere to the interference pattern that is formed. Thus, a hologram may be cut into smaller pieces, and each piece will contain some information about the light originating from point A. For this reason, each smaller piece can be used to produce a three-dimensional image of the object. In contrast, it is not possible to reconstruct the entire image in a photograph from only a small piece of the original film.

The holograms discussed here are typically viewed with the aid of the laser light used to produce them. There are also other kinds of holograms. Credit cards, for example, use holograms for identification purposes, as shown in **Figure 30.34**. This kind of hologram is called a rainbow hologram and is designed to be viewed in white light that is reflected from it. Other applications of holography include head-up displays for instrument panels in high-performance fighter planes, laser scanners at checkout counters, computerized data storage and retrieval systems, and methods for high-precision biomedical measurements.

Concept Summary

30.1 Rutherford Scattering and the Nuclear Atom The idea of a nuclear atom originated in 1911 as a result of experiments by Ernest Rutherford in which α particles were scattered by a thin metal foil. The phrase "nuclear atom" refers to the fact that an atom consists of a small, positively charged nucleus surrounded at relatively large distances by a number of electrons, whose total negative charge equals the positive nuclear charge when the atom is electrically neutral.

30.2 Line Spectra A line spectrum is a series of discrete electromagnetic wavelengths emitted by the atoms of a low-pressure gas that is subjected to a sufficiently high potential difference. Certain groups of discrete wavelengths are referred to as "series." The following equations can be used to determine the wavelengths in three of the series that are found in the line spectrum of atomic hydrogen:

Lyman series $$\frac{1}{\lambda}=R\left(\frac{1}{1^2}-\frac{1}{n^2}\right)\quad n=2,3,4,\ldots \tag{30.1}$$

Balmer series $$\frac{1}{\lambda}=R\left(\frac{1}{2^2}-\frac{1}{n^2}\right)\quad n=3,4,5,\ldots \tag{30.2}$$

Paschen series $$\frac{1}{\lambda}=R\left(\frac{1}{3^2}-\frac{1}{n^2}\right)\quad n=4,5,6,\ldots \tag{30.3}$$

The constant term R is called the Rydberg constant and has the value $R = 1.097 \times 10^7\ \text{m}^{-1}$.

30.3 The Bohr Model of the Hydrogen Atom The Bohr model applies to atoms or ions that have only a single electron orbiting a nucleus containing Z protons. This model assumes that the electron exists in circular orbits that are called stationary orbits because the electron does not radiate electromagnetic waves while in them. According to this model, a photon is emitted only when an electron changes from an orbit with a higher energy E_i to an orbit with a lower energy E_f. The orbital energies and the photon frequency f are related according to Equation 30.4, where h is Planck's constant. The model also assumes that the magnitude L_n of the orbital angular momentum of the electron can only have the discrete values indicated in Equation 30.8. With these assumptions, it can be shown that the nth Bohr orbit has a radius r_n and is associated with a total energy E_n, as given in Equations 30.10 and 30.13.

$$E_i - E_f = hf \tag{30.4}$$

$$L_n = n\frac{h}{2\pi}\quad n=1,2,3,\ldots \tag{30.8}$$

$$r_n = (5.29\times10^{-11}\ \text{m})\frac{n^2}{Z}\quad n=1,2,3,\ldots \tag{30.10}$$

$$E_n = -(13.6\ \text{eV})\frac{Z^2}{n^2}\quad n=1,2,3,\ldots \tag{30.13}$$

The ionization energy is the minimum energy needed to remove an electron completely from an atom. The Bohr model predicts that the wavelengths comprising the line spectrum emitted by a hydrogen atom can be calculated according to Equation 30.14.

$$\frac{1}{\lambda}=RZ^2\left(\frac{1}{n_f^2}-\frac{1}{n_i^2}\right) \tag{30.14}$$

$$n_i, n_f = 1, 2, 3, \ldots \quad \text{and} \quad n_i > n_f$$

30.4 De Broglie's Explanation of Bohr's Assumption About Angular Momentum Louis de Broglie proposed that the electron in a circular Bohr orbit should be considered as a particle wave and that standing particle waves around the orbit offer an explanation of the angular momentum assumption in the Bohr model.

30.5 The Quantum Mechanical Picture of the Hydrogen Atom Quantum mechanics describes the hydrogen atom in terms of the following four quantum numbers:

(1) The principal quantum number n, which can have the integer values $n = 1, 2, 3, \ldots$

(2) The orbital quantum number ℓ, which can have the integer values $\ell = 0, 1, 2, \ldots, (n-1)$

(3) The magnetic quantum number m_ℓ, which can have the positive and negative values $m_\ell = -\ell, \ldots, -2, -1, 0, +1, +2, \ldots, +\ell$

(4) The spin quantum number m_s, which, for an electron, can be either $m_s = +\frac{1}{2}$ or $m_s = -\frac{1}{2}$

According to quantum mechanics, an electron does not reside in a circular orbit but, rather, has some probability of being found at various distances from the nucleus.

30.6 The Pauli Exclusion Principle and the Periodic Table of the Elements The Pauli exclusion principle states that no two electrons in an atom can have the same set of values for the four quantum numbers n, ℓ, m_ℓ, and m_s. This principle determines the way in which the electrons in multiple-electron atoms are distributed into shells (defined by the value of n) and subshells (defined by the value of ℓ).

The following notation is used to refer to the orbital quantum numbers: s denotes $\ell = 0$, p denotes $\ell = 1$, d denotes $\ell = 2$, f denotes $\ell = 3$, g denotes $\ell = 4$, h denotes $\ell = 5$, and so on.

The arrangement of the periodic table of the elements is related to the Pauli exclusion principle.

30.7 X-Rays X-rays are electromagnetic waves emitted when high-energy electrons strike a metal target contained within an evacuated glass tube. The emitted X-ray spectrum of wavelengths consists of sharp "peaks" or "lines," called characteristic X-rays, superimposed on a broad continuous range of wavelengths called Bremsstrahlung, or "braking radiation." The K_α characteristic X-ray is emitted when an electron in the $n = 2$ level (L shell) drops into a vacancy in the $n = 1$ level (K shell). Similarly, the K_β characteristic X-ray is emitted when an electron in the $n = 3$ level (M shell) drops into a vacancy in the $n = 1$ level (K shell). The minimum wavelength, or cutoff wavelength λ_0, of the Bremsstrahlung is determined by the kinetic energy of the electrons striking the target in the X-ray tube, according to Equation 30.17, where h is Planck's constant, c is the speed of light in a vacuum, e is the magnitude of the charge on an electron, and V is the potential difference applied across the X-ray tube.

$$\lambda_0 = \frac{hc}{eV} \tag{30.8}$$

30.8 The Laser/30.9 Medical Applications of the Laser A laser is a device that generates electromagnetic waves via a process known as stimulated emission. In this process, one photon stimulates the production of another photon by causing an electron in an atom to fall from a higher energy level to a lower energy level. The emitted photon travels in the same direction as the photon causing the stimulation. Because of this mechanism of photon production, the electromagnetic waves generated by a laser are coherent and may be confined to a very narrow beam. Stimulated emission contrasts with the process known as spontaneous emission, in which an electron in an atom also falls from a higher to a lower energy level, but does so spontaneously, in a random direction, without any external provocation.

Focus on Concepts

Online

Additional questions are available for assignment in WileyPLUS.

Section 30.3 The Bohr Model of the Hydrogen Atom

1. Consider applying the Bohr model to a neutral helium atom ($Z = 2$). The model takes into account a number of factors. Which one of the following does it not take into account? (a) The quantization of the orbital angular momentum of an electron (b) The centripetal acceleration of an electron (c) The electric potential energy of an electron (d) The electrostatic repulsion between electrons (e) The electrostatic attraction between the nucleus and an electron

2. According to the Bohr model, what determines the shortest wavelength in a given series of wavelengths emitted by the atom? (a) The quantum number n_i that identifies the higher energy level from which the electron falls into a lower energy level (b) The quantum number n_f that identifies the lower energy level into which the electron falls from a higher energy level (c) The ratio n_f/n_i, where n_f is the quantum number that identifies the lower energy level into which the electron falls and n_i is the quantum number that identifies the higher level from which the electron falls (d) The sum $n_f + n_i$ of two quantum numbers, where n_f identifies the lower energy level into which the electron falls and n_i identifies the higher level from which the electron falls (e) The difference $n_f - n_i$ of two quantum numbers, where n_f identifies the lower energy level into which the electron falls and n_i identifies the higher level from which the electron falls

Section 30.5 The Quantum Mechanical Picture of the Hydrogen Atom

3. According to quantum mechanics, only one of the following combinations of the principal quantum number n and the orbital quantum number ℓ is possible for the electron in a hydrogen atom. Which combination is it? (a) $n = 3, \ell = 3$ (b) $n = 2, \ell = 3$ (c) $n = 1, \ell = 2$ (d) $n = 0, \ell = 0$ (e) $n = 3, \ell = 1$

4. Which one of the following statements is false? (a) The orbits in the Bohr model have precise sizes, whereas in the quantum mechanical picture of the hydrogen atom they do not. (b) In the absence of external magnetic fields, both the Bohr model and quantum mechanics predict the same total energy for the electron in the hydrogen atom. (c) The spin angular momentum of the electron plays a role in both the Bohr model and the quantum mechanical picture of the hydrogen atom. (d) The magnitude of the orbital angular momentum cannot be zero in the Bohr model, but it can be zero in the quantum mechanical picture of the hydrogen atom.

Section 30.6 The Pauli Exclusion Principle and the Periodic Table of the Elements

5. Each of the following answers indicates the quantum mechanical states of two electrons, A and B. Which pair of states could *not* describe two of the electrons in a multiple-electron atom?

(a)

	n	ℓ	m_ℓ	m_s
A	4	1	+1	$-\frac{1}{2}$
B	3	1	+1	$-\frac{1}{2}$

(b)

	n	ℓ	m_ℓ	m_s
A	3	2	−1	$-\frac{1}{2}$
B	3	1	−1	$+\frac{1}{2}$

(c)

	n	ℓ	m_ℓ	m_s
A	2	0	0	$-\frac{1}{2}$
B	2	1	+1	$+\frac{1}{2}$

(d)

	n	ℓ	m_ℓ	m_s
A	5	3	+1	$-\frac{1}{2}$
B	4	1	0	$+\frac{1}{2}$

(e)

	n	ℓ	m_ℓ	m_s
A	3	2	−2	$+\frac{1}{2}$
B	3	2	−2	$+\frac{1}{2}$

6. Consider the 5f and 6h subshells in a multiple-electron atom. Which of these subshells can contain 19 electrons? **(a)** Only the 6h subshell **(b)** Only the 5f subshell **(c)** Both subshells **(d)** Neither subshell

Section 30.7 X-Rays

7. Silver ($Z = 47$), copper ($Z = 29$), and platinum ($Z = 78$) can be used as the target in an X-ray tube. Rank in descending order (largest first) the energies needed for impinging electrons to knock a K-shell electron completely out of an atom in each of these targets. **(a)** Silver, copper, platinum **(b)** Silver, platinum, copper **(c)** Platinum, silver, copper **(d)** Platinum, copper, silver **(e)** Copper, silver, platinum

8. The voltage applied across an X-ray tube is doubled. What happens to the cutoff wavelength in the spectrum of wavelengths emitted by the tube's metal target? **(a)** It also doubles. **(b)** It decreases by a factor of two. **(c)** It increases by a factor of four. **(d)** It decreases by a factor of four. **(e)** Nothing happens to it.

Section 30.8 The Laser

9. Consider two energy levels that characterize the atoms of a material used in a laser. A population inversion between these two levels ________. **(a)** has the lower energy level more populated than it normally is and the higher energy level less populated than it normally is **(b)** is the same thing as a metastable state **(c)** requires no external source of energy to be sustained **(d)** has the higher energy level more populated than it normally is and the lower energy level less populated than it normally is

Problems

Online

Additional questions are available for assignment in WileyPLUS.
Note: *In working these problems, ignore relativistic effects.*

SSM Student Solutions Manual
MMH Problem-solving help
GO Guided Online Tutorial
V-HINT Video Hints
CHALK Chalkboard Videos
BIO Biomedical application
E Easy
M Medium
H Hard
WS Worksheet
T Team Problem

Section 30.1 Rutherford Scattering and the Nuclear Atom

1. **E SSM** The nucleus of the hydrogen atom has a radius of about 1×10^{-15} m. The electron is normally at a distance of about 5.3×10^{-11} m from the nucleus. Assuming that the hydrogen atom is a sphere with a radius of 5.3×10^{-11} m, find **(a)** the volume of the atom, **(b)** the volume of the nucleus, and **(c)** the percentage of the volume of the atom that is occupied by the nucleus.

2. **E** The nucleus of a hydrogen atom is a single proton, which has a radius of about 1.0×10^{-15} m. The single electron in a hydrogen atom normally orbits the nucleus at a distance of 5.3×10^{-11} m. What is the ratio of the density of the hydrogen nucleus to the density of the complete hydrogen atom?

3. **E** Review Conceptual Example 1 and use the information therein as an aid in working this problem. Suppose that you're building a scale model of the hydrogen atom, and the nucleus is represented by a ball that has a radius of 3.2 cm (somewhat smaller than a baseball). How many miles away ($1 \text{ mi} = 1.61 \times 10^5$ cm) should the electron be placed?

4. **E V-HINT** In a Rutherford scattering experiment a target nucleus has a diameter of 1.4×10^{-14} m. The incoming α particle has a mass of 6.64×10^{-27} kg. What is the kinetic energy of an α particle that has a de Broglie wavelength equal to the diameter of the target nucleus? Ignore relativistic effects.

5. **M CHALK SSM** There are Z protons in the nucleus of an atom, where Z is the atomic number of the element. An α particle carries a charge of $+2e$. In a scattering experiment, an α particle, heading directly toward a nucleus in a metal foil, will come to a halt when all the particle's kinetic energy is converted to electric potential energy. In such a situation, how close will an α particle with a kinetic energy of 5.0×10^{-13} J come to a gold nucleus ($Z = 79$)?

6. **M GO** The nucleus of a copper atom contains 29 protons and has a radius of 4.8×10^{-15} m. How much work (in electron volts) is done by the electric force as a proton is brought from infinity, where it is at rest, to the "surface" of a copper nucleus?

Section 30.2 Line Spectra,

Section 30.3 The Bohr Model of the Hydrogen Atom

7. **E SSM** For a doubly ionized lithium atom Li^{2+} ($Z = 3$), what is the principal quantum number of the state in which the electron has the same total energy as a ground-state electron has in the hydrogen atom?

8. **E** A singly ionized helium atom (He^+) has only one electron in orbit about the nucleus. What is the radius of the ion when it is in the second excited state?

9. E Using the Bohr model, determine the ratio of the energy of the nth orbit of a triply ionized beryllium atom (Be^{3+}, $Z = 4$) to the energy of the nth orbit of a hydrogen atom (H).

10. E MMH The electron in a hydrogen atom is in the first excited state, when the electron acquires an additional 2.86 eV of energy. What is the quantum number n of the state into which the electron moves?

11. E SSM Find the energy (in joules) of the photon that is emitted when the electron in a hydrogen atom undergoes a transition from the $n = 7$ energy level to produce a line in the Paschen series.

12. E GO (a) What is the ionization energy of a hydrogen atom that is in the $n = 4$ excited state? (b) For a hydrogen atom, determine the ratio of the ionization energy for the $n = 4$ excited state to the ionization energy for the ground state.

13. E CHALK A hydrogen atom is in the ground state. It absorbs energy and makes a transition to the $n = 3$ excited state. The atom returns to the ground state by emitting two photons. What are their wavelengths?

14. E In the hydrogen atom, what is the total energy (in electron volts) of an electron that is in an orbit that has a radius of 4.761×10^{-10} m?

15. M GO A sodium atom ($Z = 11$) contains 11 protons in its nucleus. Strictly speaking, the Bohr model does not apply, because the neutral atom contains 11 electrons instead of a single electron. However, we can apply the model to the outermost electron as an approximation, provided that we use an effective value $Z_{\text{effective}}$ rather than 11 for the number of protons in the nucleus. (a) The ionization energy for the outermost electron in a sodium atom is 5.1 eV. Use the Bohr model with $Z = Z_{\text{effective}}$ to calculate a value for $Z_{\text{effective}}$. (b) Using $Z = 11$ and $Z = Z_{\text{effective}}$, determine the corresponding two values for the radius of the outermost Bohr orbit.

16. M GO SSM A wavelength of 410.2 nm is emitted by the hydrogen atoms in a high-voltage discharge tube. What are the initial and final values of the quantum number n for the energy level transition that produces this wavelength?

17. M V-HINT MMH A hydrogen atom emits a photon that has momentum with a magnitude of 5.452×10^{-27} kg · m/s. This photon is emitted because the electron in the atom falls from a higher energy level into the $n = 1$ level. What is the quantum number of the level from which the electron falls? Use a value of 6.626×10^{-34} J · s for Planck's constant.

18. E SSM For atomic hydrogen, the Paschen series of lines occurs when $n_f = 3$, whereas the Brackett series occurs when $n_f = 4$ in Equation 30.14. Using this equation, show that the ranges of wavelengths in these two series overlap.

19. M GO CHALK Doubly ionized lithium Li^{2+} ($Z = 3$) and triply ionized beryllium Be^{3+} ($Z = 4$) each emit a line spectrum. For a certain series of lines in the lithium spectrum, the shortest wavelength is 40.5 nm. For the same series of lines in the beryllium spectrum, what is the shortest wavelength?

20. H In the Bohr model of hydrogen, the electron moves in a circular orbit around the nucleus. Determine the angular speed of the electron, in revolutions per second, when it is in (a) the ground state and (b) the first excited state.

Section 30.5 The Quantum Mechanical Picture of the Hydrogen Atom

21. E A hydrogen atom is in its second excited state. Determine, according to quantum mechanics, (a) the total energy (in eV) of the atom, (b) the magnitude of the maximum angular momentum the electron can have in this state, and (c) the maximum value that the z component L_z of the angular momentum can have.

22. E GO The table lists quantum numbers for five states of the hydrogen atom. Which (if any) of them are not possible? For those that are not possible, explain why.

Process	n	ℓ	m_ℓ
(a)	3	3	0
(b)	2	1	−1
(c)	4	2	3
(d)	5	−3	2
(e)	4	0	0

23. E The orbital quantum number for the electron in a hydrogen atom is $\ell = 5$. What is the smallest possible value (the most negative) for the total energy of this electron? Give your answer in electron volts.

24. E GO It is known that the possible values for the magnetic quantum number m_ℓ are −4, −3, −2, −1, 0, +1, +2, +3, and +4. Determine the orbital quantum number ℓ and the smallest possible value of the principal quantum number n.

25. E SSM The maximum value for the magnetic quantum number in state A is $m_\ell = 2$, while in state B it is $m_\ell = 1$. What is the ratio L_A/L_B of the magnitudes of the orbital angular momenta of an electron in these two states?

26. M V-HINT CHALK The electron in a certain hydrogen atom has an angular momentum of 8.948×10^{-34} J · s. What is the largest possible magnitude for the z component of the angular momentum of this electron? For accuracy, use $h = 6.626 \times 10^{-34}$ J · s.

27. M SSM MMH For an electron in a hydrogen atom, the z component of the angular momentum has a *maximum* value of $L_z = 4.22 \times 10^{-34}$ J · s. Find the three smallest possible values (the most negative) for the total energy (in electron volts) that this atom could have.

28. H GO An electron is in the $n = 5$ state. What is the smallest possible value for the angle between the z component of the orbital angular momentum and the orbital angular momentum?

Section 30.6 The Pauli Exclusion Principle and the Periodic Table of the Elements

29. E CHALK Two of the three electrons in a lithium atom have quantum numbers of $n = 1, \ell = 0, m_\ell = 0, m_s = +\frac{1}{2}$ and $n = 1, \ell = 0, m_\ell = 0, m_s = -\frac{1}{2}$. What quantum numbers can the third electron have if the atom is in (a) its ground state and (b) its first excited state?

30. E Following the style used in **Table 30.3**, determine the electronic configuration of the ground state for yttrium Y ($Z = 39$). Refer to **Figure 30.16** to see the order in which the subshells fill.

31. E **Figure 30.16** was constructed using the Pauli exclusion principle and indicates that the $n = 1$ shell holds 2 electrons, the $n = 2$ shell holds 8 electrons, and the $n = 3$ shell holds 18 electrons. These numbers can be obtained by adding the numbers given in the figure for the subshells contained within a given shell. How many electrons can be put into the $n = 5$ shell, which is only partly shown in the figure?

32. E GO Which of the following subshell configurations are not allowed? For those that are not allowed, give the reason why. (a) $3s^1$ (b) $2d^2$ (c) $3s^4$ (d) $4p^8$ (e) $5f^{12}$

33. E SSM When an electron makes a transition between energy levels of an atom, there are no restrictions on the initial and final values of the principal quantum number n. According to quantum mechanics, however, there is a rule that restricts the initial and final values of the orbital quantum number ℓ. This rule is called a *selection rule* and states that $\Delta\ell = \pm 1$. In other words, when an electron makes a transition between energy levels, the value of ℓ can only increase or decrease by one. The value of ℓ may not remain the same nor may it increase or decrease by more than one. According to this rule, which of the following energy level transitions are allowed? (a) 2s → 1s (b) 2p → 1s (c) 4p → 2p (d) 4s → 2p (e) 3d → 3s

34. M In the ground state, the outermost shell ($n = 1$) of helium (He) is filled with electrons, as is the outermost shell ($n = 2$) of neon (Ne). The full outermost shells of these two elements distinguish them as the first two so-called "noble gases." Suppose that the spin quantum number m_s had *three* possible values, rather than two. If that were the case, which elements would be (a) the first and (b) the second noble gases? Assume that the possible values for the other three quantum numbers are unchanged, and that the Pauli exclusion principle still applies.

Section 30.7 X-Rays

35. E MMH By using the Bohr model, decide which element is likely to emit a K_α X-ray with a wavelength of 4.5×10^{-9} m.

36. E GO What is the minimum potential difference that must be applied to an X-ray tube to knock a K-shell electron completely out of an atom in a copper ($Z = 29$) target? Use the Bohr model as needed.

37. E V-HINT In the X-ray spectrum of niobium ($Z = 41$), a K_α peak is observed at a wavelength of 7.462×10^{-11} m. (a) Determine the magnitude of the difference between the observed wavelength of the K_α X-ray for niobium and that predicted by the Bohr model. (b) Express the magnitude of this difference as a percentage of the observed wavelength.

38. E CHALK SSM When a certain element is bombarded with high-energy electrons, K_α X-rays that have an energy of 9890 eV are emitted. Determine the atomic number Z of the element, and identify the element. Use the Bohr model as necessary.

39. M GO The Bohr model, although not strictly applicable, can be used to estimate the minimum energy E_{min} that an incoming electron must have in an X-ray tube in order to knock a K-shell electron entirely out of an atom in the metal target. The K_α X-ray wavelength of metal A is 2.0 times the K_α X-ray wavelength of metal B. What is the ratio of $E_{min, A}$ for metal A to $E_{min, B}$ for metal B?

40. M SSM An X-ray tube contains a silver ($Z = 47$) target. The high voltage in this tube is increased from zero. Using the Bohr model, find the value of the voltage at which the K_α X-ray just appears in the X-ray spectrum.

41. M GO Multiple-Concept Example 10 reviews the concepts that are important in this problem. An electron, traveling at a speed of 6.00×10^7 m/s, strikes the target of an X-ray tube. Upon impact, the electron decelerates to one-quarter of its original speed, an X-ray photon being emitted in the process. What is the wavelength of the photon?

Section 30.8 The Laser

42. E GO A pulsed laser emits light in a series of short pulses, each having a duration of 25.0 ms. The average power of each pulse is 5.00 mW, and the wavelength of the light is 633 nm. Find the number of photons in each pulse.

43. E GO CHALK The drawing shows three energy levels of a laser that are involved in the lasing action. These levels are analogous to the levels in the Ne atoms of a He-Ne laser. The E_2 level is a metastable level, and the E_0 level is the ground state. The difference between the energy levels of the laser is shown in the drawing. (a) What energy (in eV per electron) must an external source provide to start the lasing action? (b) What is the wavelength of the laser light? (c) In what region of the electromagnetic spectrum (see Figure 24.9) does the laser light lie?

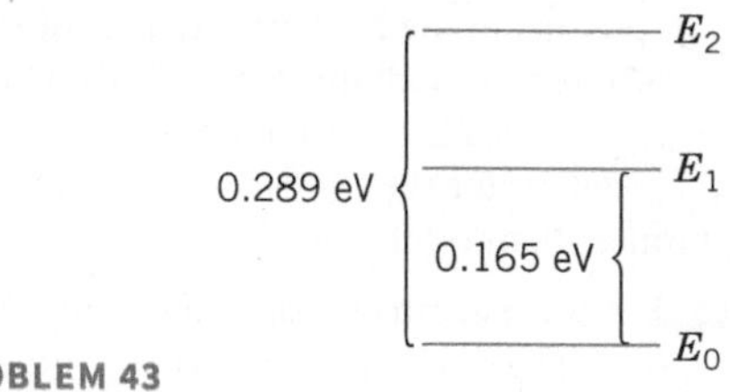

PROBLEM 43

44. M V-HINT Fusion is the process by which the sun produces energy. One experimental technique for creating controlled fusion utilizes a solid-state laser that emits a wavelength of 1060 nm and can produce a power of 1.0×10^{14} W for a pulse duration of 1.1×10^{-11} s. In contrast, the helium/neon laser used in a bar-code scanner at the checkout counter emits a wavelength of 633 nm and produces a power of about 1.0×10^{-3} W. How long (in days) would the helium/neon laser have to operate to produce the same number of photons that the solid-state laser produces in 1.1×10^{-11} s?

Additional Problems

Online

45. E GO (a) What is the minimum energy (in electron volts) that is required to remove the electron from the ground state of a singly ionized helium atom (He^+, $Z = 2$)? (b) What is the ionization energy for He^+?

46. E Molybdenum has an atomic number of $Z = 42$. Using the Bohr model, estimate the wavelength of the K_α X-ray.

47. E SSM In the line spectrum of atomic hydrogen there is also a group of lines known as the Pfund series. These lines are produced when electrons, excited to high energy levels, make transitions to the $n = 5$ level. Determine (a) the longest wavelength and (b) the shortest wavelength in this series. (c) Refer to Figure 24.9 and identify the region of the electromagnetic spectrum in which these lines are found.

48. E GO The voltage across an X-ray tube is 35.0 kV. Suppose that the molybdenum ($Z = 42$) target in the X-ray tube is replaced by a silver ($Z = 47$) target. Determine (a) the tube's cutoff wavelength and (b) the wavelengths of the K_α X-ray photons emitted by the molybdenum and silver targets.

49. **M** **V-HINT** The energy of the $n = 2$ Bohr orbit is -30.6 eV for an unidentified ionized atom in which only one electron moves about the nucleus. What is the radius of the $n = 5$ orbit for this species?

50. **M** Consider a particle of mass m that can exist only between $x = 0$ m and $x = +L$ on the x axis. We could say that this particle is confined to a "box" of length L. In this situation, imagine the standing de Broglie waves that can fit into the box. For example, the drawing shows the first three possibilities. Note in this picture that there are either one, two, or three half-wavelengths that fit into the distance L. Use Equation 29.8 for the de Broglie wavelength of a particle and derive an expression for the allowed energies (only kinetic energy) that the particle can have. This expression involves m, L, Planck's constant, and a quantum number n that can have only the values 1, 2, 3, ...

PROBLEM 50

Physics in Biology, Medicine, and Sports

51. **E** **BIO** **SSM** **30.8** A laser is used in eye surgery to weld a detached retina back into place. The wavelength of the laser beam is 514 nm, and the power is 1.5 W. During surgery, the laser beam is turned on for 0.050 s. During this time, how many photons are emitted by the laser?

52. **E** **BIO** **GO** **30.8** The ultraviolet excimer laser used in the PRK technique (see Section 30.9) has a wavelength of 193 nm. A carbon dioxide laser produces a wavelength of 1.06×10^{-5} m. What is the minimum number of photons that the carbon dioxide laser must produce to deliver at least as much or more energy to a target as does a single photon from the excimer laser?

53. **E** **BIO** **30.8** A laser peripheral iridotomy is a procedure for treating an eye condition known as narrow-angle glaucoma, in which pressure buildup in the eye can lead to loss of vision. A neodymium YAG laser (wavelength = 1064 nm) is used in the procedure to punch a tiny hole in the peripheral iris, thereby relieving the pressure buildup. In one application the laser delivers 4.1×10^{-3} J of energy to the iris in creating the hole. How many photons does the laser deliver?

54. **M** **BIO** **30.7** The tube in a hospital X-ray machine has an operating voltage of 90 kV. **(a)** What is the maximum energy (in eV) of a photon produced by this tube? **(b)** What is the shortest wavelength of an X-ray photon produced by the same tube?

55. **M** **BIO** **30.7** The target element in a veterinary X-ray machine is tungsten. Use the Bohr model to estimate the wavelength of **(a)** the K_α line and **(b)** the K_β line for this X-ray source.

56. **E** **BIO** **30.1** The number of atoms in a single human cell is on the order of the number of cells in the entire human body. Research suggests that a single cell consists of 100 trillion (1.0×10^{14}) atoms. Assuming the radius of these atoms is one angstrom, what would be the volume of a single cell?

57. **M** **BIO** **30.3** A pulsed dye laser that is used for the treatment (removal) of port-wine stains from the skin produces laser light at a wavelength of 486 nm. Assume this light is produced by electrons in hydrogen that undergo a transition in the Balmer series. Use the Bohr model to determine the initial energy state (n_i) of these electrons.

58. **M** **BIO** **30.1** There are several proteins that exist in nature that are *fluorescent*, which means they absorb visible light or other electromagnetic radiation and reemit that light, usually at a longer wavelength. One of the more prevalent fluorescent proteins is GFP, or green fluorescent protein, which emits bright green fluorescent light when exposed to light in the blue to ultraviolet region of the electromagnetic spectrum (see the photo). The main absorption, or excitation, wavelength for one type of GFP is 395 nm. **(a)** What is the energy (in eV) of a photon in this absorbed light? The emission spectrum of the light from the GFP peaks near 509 nm, which is in the shorter wavelength (higher energy) portion of the green part of the visible spectrum. **(b)** What is the energy (in eV) of a photon in the emitted light? **(c)** If the emitted light corresponded to the longest wavelength in the Balmer series from a hydrogen-like atom, what would be the effective value of Z for this atom? (*Note*: Z will not be a whole number in this case.)

The mice on the left and right have been genetically modified to express GFP when illuminated with UV light. Their exposed skin cells (eyes, nose, ears, and tail) glow bright green due to fluorescence and indicate healthy cells. The mice also contain subcutaneous breast cancer tumor cells that glow red (not shown), so they can be easily distinguished from healthy cells. This aids researchers in tracking the metastization process of the cancer. The mouse in the middle has not been modified.

Ingrid Moen, Charlotte Jevne, Jian Wang, Karl-Henning Kalland, Martha Chekenya, Lars A Akslen, Linda Sleire, Per ØEnger, Rolf K Reed, Anne M Øyan and Linda EB Stuhr: *Gene expression in tumor cells and stroma in dsRed 4T1 tumors in eGFP-expressing mice with and without enhanced oxygenation*, https://commons.wikimedia.org/wiki/File:4T1_Mice_01.jpg

PROBLEM 58

Concepts and Calculations Problems

Online

The Bohr model of the hydrogen atom introduces a number of important features that characterize the quantum picture of atomic structure. Among them are the concepts of quantized energy levels and the photon emission that occurs when an electron makes a transition from a higher to a lower energy state. Problem 59 deals with these ideas. Problem 60 reviews the physics of how a K_α X-ray is produced, how its energy is related to the ionization energies of the target atom, and how to determine the minimum voltage needed to produce it in an X-ray tube.

59. M CHALK A hydrogen atom ($Z = 1$) is in the third excited state, and a photon is either emitted or absorbed. *Concepts:* (i) What is the quantum number of the third excited state? (ii) When an atom emits a photon, is the final quantum number n_f of the atom greater than or less than the initial quantum number n_i? (iii) When an atom absorbs a photon, is the final quantum number n_f of the atom greater than or less than the initial quantum number n_i? (iv) How is the wavelength of a photon related to its energy? *Calculations:* Determine the quantum number n_f of the final state and the energy of the photon when the photon is **(a)** emitted with the shortest possible wavelength, **(b)** emitted with the longest possible wavelength, and **(c)** absorbed with the longest possible wavelength.

60. M CHALK SSM The K-shell and L-shell ionization energies of a metal are 8979 eV and 951 eV, respectively. *Concepts:* (i) How is a K_α photon produced, and how much energy does it have? (ii) What must be the minimum voltage across the X-ray tube to produce a K_α photon? (iii) What is meant by the phrases "K-shell ionization energy" and "L-shell ionization energy"? (iv) What does the difference between the K-shell and L-shell ionization energies represent? *Calculations:* **(a)** Assuming that there is a vacancy in the L shell, what must be the minimum voltage across an X-ray tube with a target made from this metal to produce K_α X-ray photons? **(b)** Determine the wavelength of a K_α photon.

Team Problems

Online

61. E T WS **Detecting a Poison Gas.** You and your team are exploring an old laboratory and happen upon a sealed corridor with a sign on the door that reads "Danger: fluorine leak." You use your phone to look up fluorine (F_2) on the Internet and find that it is a halogen gas that is extremely poisonous, and one should be careful not to allow it to make contact with their skin and to not breathe it in, even in small amounts. The problem is that you have to get through the area that is sealed, and therefore must determine whether there is still any gas present in the space. You are fortunate that, in a room that you have already explored, there is a spectrometer that is capable of energizing a vial of gas by subjecting it to a high voltage, and then analyzing the atomic light emission. The spectrometer has a detectable wavelength range of 385 to 750 nm. The only atomic transitions in fluorine that you are able to find on the Internet are the following: $E_{01} = -12.697$ eV to $E_{f1} = -14.683$ eV; $E_{02} = -12.731$ eV to $E_{f2} = -14.683$ eV; $E_{03} = -12.697$ eV to $E_{f3} = -14.505$ eV; and $E_{04} = -0.05010$ eV to $E_{f4} = -13.025$ eV, where the subscripts "0" and "f" correspond to the initial and final states, respectively. **(a)** Determine the energies of the photons emitted during each transition in joules. **(b)** What are the wavelengths of the light emitted during these transitions? **(c)** What wavelengths (emission lines) will be detected by the spectrometer if fluorine gas is in the room?

62. M T WS **An Atomic Emission Device.** You and your team are designing a device that utilizes a large hydrogen-gas discharge lamp as a light source. A ruled grating is used to separate the emission lines emanating from the lamp, and the specific line that you will use is one of the Lyman lines ($n = 4 \rightarrow n = 1$). The light is collimated into a beam of rectangular cross section with length $L = 0.75$ cm and width $W = 0.55$ cm, which illuminates a flat, grounded metallic surface with an intensity of 1250 W/m^2. **(a)** What is the wavelength of the emitted light? **(b)** What is the energy of each emitted photon (in J)? **(c)** How many photons hit the metallic surface per second? **(d)** If 0.75% of the impinging photons result in ejected photoelectrons, what is the magnitude of the resulting electric current in the metal?

63. T M **A Simple Spectrometer.** In the Balmer series for the atomic emission of hydrogen, emission lines with n going from $3 \rightarrow 2$, $4 \rightarrow 2$, $5 \rightarrow 2$, and $6 \rightarrow 2$ are in the visible spectrum. **(a)** What are the wavelengths of the light emitted in these transitions? **(b)** A spectrometer is a device that disperses light into its constituent wavelengths, and provides a means of measuring these wavelengths. You and your team are designing a simple spectrometer using a transmission grating with 1.50×10^3 lines/cm. A collimated beam of light from a hydrogen discharge tube (which emits light from all possible hydrogen atomic transitions) passes through the grating. At what second-order angles ($m = 2$) do you expect each of the emission lines calculated in (a) to appear? **(c)** You place a screen with a coating that fluoresces when exposed to ultraviolet light in the light emerging from the grating, and you observe another line at $\theta = 6.84°$. Calculate the wavelength of this light. Does it come from the Balmer series?

64. T M **An X-ray Source.** You and your team are building an X-ray source for a diffraction instrument used to study the crystalline structure of materials. You decide to use the K_α radiation from a copper ($Z = 29$) target. **(a)** Estimate the minimum energy that an incoming electron must have in order to knock a K-shell electron out of the Cu atom. **(b)** Through what voltage must you accelerate the electrons? **(c)** What is the cutoff wavelength of the Bremsstrahlung radiation for the electron energy found in (a)? **(d)** What is the wavelength of the K_α line? Convert this to the energy of the K_α X-ray photons (in eV) **(e)** If a first-order peak ($m = 1$) appears at an angle of 17.0° when the K_α radiation found in (d) passes through a hypothetical diffraction grating, what would be the "line spacing" of the grating? Note: distances between atoms in solids are on the order of 10^{-10} m.

CHAPTER 31

Nuclear reactors are used for research and power generation. Their operation relies on a fuel that is composed of radioactive elements. The nuclei in the atoms of these elements are unstable, and they break apart over time, releasing different types of radiation, known as decay products. The image above shows spent nuclear fuel being removed from the High Flux Isotope Reactor at Oak Ridge National Lab in Oak Ridge, Tennessee. The intense blue light, known as Cherenkov radiation, is the result of the decay products from the fuel interacting with the surrounding water that is required to keep the assembly cool.

Genevieve Martin/ORNL, U.S. Dept. of Energy

Nuclear Physics and Radioactivity

LEARNING OBJECTIVES

After reading this module, you should be able to...

31.1 Identify and explain the properties of the nucleus.

31.2 Describe the strong nuclear force.

31.3 Calculate nuclear binding energy and mass defect.

31.4 Analyze alpha, beta, and gamma decays.

31.5 Explain the role of the neutrino in weak nuclear decay.

31.6 Solve problems involving radioactive decay.

31.7 Calculate age using radioactive dating.

31.8 Analyze radioactive decay series.

31.9 Explain how radiation detectors operate.

31.1 Nuclear Structure

Atoms consist of electrons in orbit about a central nucleus. As we have seen in Chapter 30, the electron orbits are quantum mechanical in nature and have interesting characteristics. Little has been said about the nucleus, however. Since the nucleus is interesting in its own right, we now consider it in greater detail.

The nucleus of an atom consists of neutrons and protons, collectively referred to as **nucleons**. The **neutron**, discovered in 1932 by the English physicist James Chadwick (1891–1974), carries no electric charge and has a mass slightly larger than that of a proton (see **Table 31.1**).

The number of protons in the nucleus is different in different elements and is given by the **atomic number** Z. In an electrically neutral atom, the number of nuclear protons equals the number of electrons in orbit around the nucleus. The number of neutrons in the nucleus is N.

TABLE 31.1 Properties of Select Particles

		Mass	
Particle	Electric Charge (C)	Kilograms (kg)	Atomic Mass Units (u)
Electron	-1.60×10^{-19}	$9.109\,382 \times 10^{-31}$	$5.485\,799 \times 10^{-4}$
Proton	$+1.60 \times 10^{-19}$	$1.672\,622 \times 10^{-27}$	1.007 276
Neutron	0	$1.674\,927 \times 10^{-27}$	1.008 665
Hydrogen atom	0	$1.673\,534 \times 10^{-27}$	1.007 825

The total number of protons and neutrons is referred to as the **atomic mass number** A because the total nuclear mass is *approximately* equal to A times the mass of a single nucleon:

$$\underbrace{A}_{\substack{\text{Number of protons}\\ \text{and neutrons (atomic}\\ \text{mass number or}\\ \text{nucleon number)}}} = \underbrace{Z}_{\substack{\text{Number of protons}\\ \text{(atomic number)}}} + \underbrace{N}_{\substack{\text{Number of}\\ \text{neutrons}}} \tag{31.1}$$

Sometimes A is also called the **nucleon number.** A shorthand notation is often used to specify Z and A along with the chemical symbol for the element. For instance, the nuclei of all naturally occurring aluminum atoms have $A = 27$, and the atomic number for aluminum is $Z = 13$. In shorthand notation, then, the aluminum nucleus is specified as $^{27}_{13}\text{Al}$. The number of neutrons in an aluminum nucleus is not given explicitly by this shorthand notation. However, it can be determined easily with the aid of Equation 31.1, which indicates that $N = A - Z = 14$. In general, for an element whose chemical symbol is X, the symbol for the nucleus is

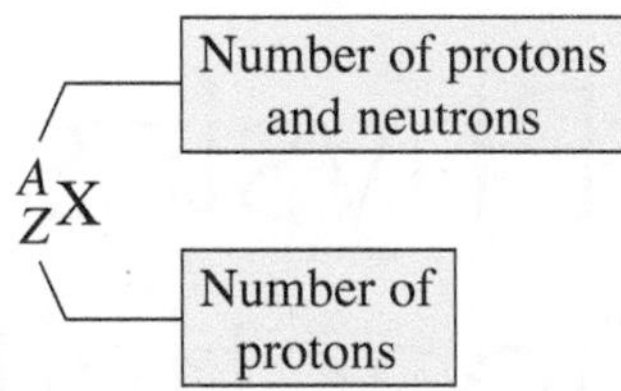

For a proton the symbol is $^{1}_{1}\text{H}$, since the proton is the nucleus of a hydrogen atom. A neutron is denoted by $^{1}_{0}\text{n}$. In the case of an electron we use $^{0}_{-1}\text{e}$, where $A = 0$ because an electron is not composed of protons or neutrons and $Z = -1$ because the electron has a negative charge.

Nuclei that contain the same number of protons, but a different number of neutrons, are known as **isotopes.** Carbon, for example, occurs in nature in two stable forms. In most carbon atoms (98.90%), the nucleus is the $^{12}_{6}\text{C}$ isotope and consists of six protons and six neutrons. A small fraction (1.10%), however, contain nuclei that have six protons and seven neutrons—namely, the $^{13}_{6}\text{C}$ isotope. The percentages given here are the natural abundances of the isotopes. The atomic masses in the periodic table are average atomic masses, taking into account the abundances of the various isotopes.

The protons and neutrons in the nucleus are clustered together to form an approximately spherical region, as **Interactive Figure 31.1** illustrates. Experiment shows that the radius r of the nucleus depends on the atomic mass number A and is given approximately in meters by

$$r \approx (1.2 \times 10^{-15}\ \text{m})A^{1/3} \tag{31.2}$$

This equation indicates that the radius of the aluminum nucleus ($A = 27$), for example, is $r \approx (1.2 \times 10^{-15}\ \text{m})27^{1/3} = 3.6 \times 10^{-15}$ m. Equation 31.2 leads to an important conclusion concerning the nuclear density of different atoms, as Conceptual Example 1 discusses.

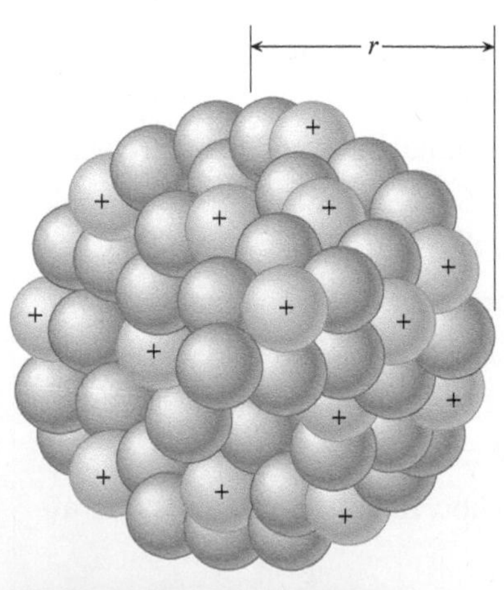

INTERACTIVE FIGURE 31.1 The nucleus in an atom is approximately spherical (radius = r) and contains protons (+) clustered closely together with neutrons (●).

CONCEPTUAL EXAMPLE 1 | Nuclear Density

It is well known that lead and oxygen contain different atoms and that the density of solid lead is much greater than the density of gaseous oxygen. Using the definition of density along with Equation 31.2, decide whether the density of the *nucleus* in a lead atom is **(a)** greater than, **(b)** approximately equal to, or **(c)** less than the density of the *nucleus* in an oxygen atom.

Reasoning The density ρ of an object, such as the nucleus, is defined as its mass M divided by its volume V: $\rho = M/V$ (Equation 11.1). The mass of a nucleus is approximately equal to the number A of nucleons in the nucleus times the mass m of a single nucleon, since the masses of a proton and a neutron are nearly the same. Thus, we have that $M \approx Am$, where A is greater for lead than for oxygen, but m is the same for both. The nucleus is approximately spherical with a radius r, so its volume V is given by $V = \frac{4}{3}\pi r^3$. The radius, however, depends on the number A of nucleons through the relation $r \approx (1.2 \times 10^{-15}\ \text{m})A^{1/3}$ (Equation 31.2). Therefore, we can write the density of a nucleus as follows:

$$\rho = \frac{M}{V} \approx \frac{Am}{\frac{4}{3}\pi r^3} = \frac{Am}{\frac{4}{3}\pi[(1.2 \times 10^{-15}\ \text{m})A^{1/3}]^3} \approx \frac{m}{\frac{4}{3}\pi(1.2 \times 10^{-15}\ \text{m})^3}$$

Note that the nucleon number A has been eliminated algebraically from this result, as a direct consequence of Equation 31.2.

Answers (a) and (c) are incorrect. The result obtained in the Reasoning section for the nuclear density ρ depends only on numerical factors and the value of m, which is the mass of a single nucleon no matter where the nucleon is located. The nuclear density does not depend on the nuclear number A. Thus, the nuclear density of lead, which is the ratio of its mass to its volume, is neither greater than nor less than the nuclear density of oxygen.

Answer (b) is correct. The result obtained in the Reasoning section for the nuclear density ρ indicates that the density of the nucleus in a lead atom is approximately the same as it is in an oxygen atom. In general, because of Equation 31.2, the *nuclear* density has nearly the same value in all atoms. The difference in densities between solid lead and gaseous oxygen, however, arises mainly because of the difference in how closely the atoms are packed together in the solid and gaseous phases.

Check Your Understanding

(The answers are given at the end of the book.)

1. Two nuclei differ in their numbers of protons and their numbers of neutrons. Which one or more of the following statements is/are true? **(a)** They are different isotopes of the same element. **(b)** They have the same electric charge. **(c)** They could have the same radii. **(d)** They have approximately the same nuclear density.

2. A material is known to be an isotope of lead, although the particular isotope is not known. From such limited information, which of the following quantities can you specify? **(a)** Its atomic number **(b)** Its neutron number **(c)** Its atomic mass number

3. Two nuclei have different nucleon numbers A_1 and A_2. Are the two nuclei necessarily isotopes of the same element?

4. Can two nuclei have the same radius, even though they contain different numbers of protons and different numbers of neutrons?

31.2 The Strong Nuclear Force and the Stability of the Nucleus

Two positive charges that are as close together as they are in a nucleus repel one another with a very strong electrostatic force. What, then, keeps the nucleus from flying apart? Clearly, some kind of attractive force must hold the nucleus together, since many kinds of naturally occurring atoms contain stable nuclei. The gravitational force of attraction between nucleons is too weak to counteract the repulsive electric force, so a different type of force must hold the nucleus together. This force is the **strong nuclear force** and is one of only three fundamental forces that have been discovered, fundamental in the sense that all forces in nature can be explained in terms of these three. The gravitational force is also one of these forces, as is the electroweak force (see Section 31.5).

Many features of the strong nuclear force are well known. For example, it is almost independent of electric charge. At a given separation distance, nearly the same nuclear force of attraction exists between two protons, between two neutrons, or between a proton and a neutron. The range of action of the strong nuclear force is extremely short,

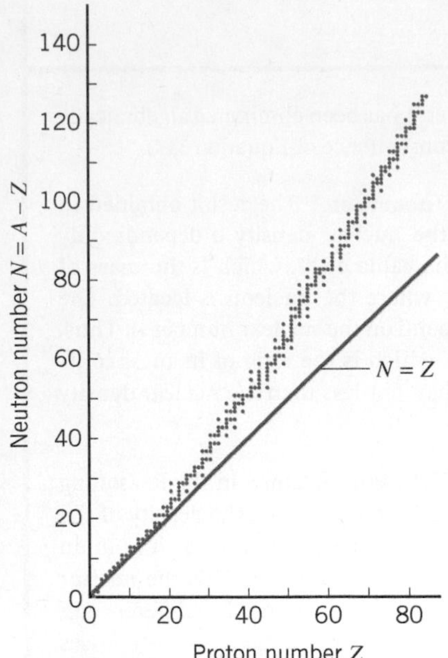

FIGURE 31.2 With few exceptions, the naturally occurring stable nuclei have a number N of neutrons that equals or exceeds the number Z of protons. Each dot in this plot represents a stable nucleus.

with the force of attraction being very strong when two nucleons are as close as 10^{-15} m and essentially zero at larger distances. In contrast, the electric force between two protons decreases to zero only gradually as the separation distance increases to large values and, therefore, has a relatively long range of action.

The limited range of action of the strong nuclear force plays an important role in the stability of the nucleus. For a nucleus to be stable, the electrostatic repulsion between the protons must be balanced by the attraction between the nucleons due to the strong nuclear force. But one proton repels all other protons within the nucleus, since the electrostatic force has such a long range of action. In contrast, a proton or a neutron attracts only its nearest neighbors via the strong nuclear force. As the number Z of protons in the nucleus increases under these conditions, the number N of neutrons has to increase even more, if stability is to be maintained. **Figure 31.2** shows a plot of N versus Z for naturally occurring elements that have stable nuclei. For reference, the plot also includes the straight line that represents the condition $N = Z$. With few exceptions, the points representing stable nuclei fall above this reference line, reflecting the fact that the number of neutrons becomes greater than the number of protons as the atomic number Z increases.

As more and more protons occur in a nucleus, there comes a point when a balance of repulsive and attractive forces cannot be achieved by an increased number of neutrons. Eventually, the limited range of action of the strong nuclear force prevents extra neutrons from balancing the long-range electric repulsion of extra protons. The stable nucleus with the largest number of protons ($Z = 83$) is that of bismuth, $^{209}_{83}\text{Bi}$, which contains 126 neutrons. All nuclei with more than 83 protons (e.g., uranium, $Z = 92$) are unstable and spontaneously break apart or rearrange their internal structures as time passes. This spontaneous disintegration or rearrangement of internal structure is called **radioactivity**, first discovered in 1896 by the French physicist Antoine Becquerel (1852–1908). Section 31.4 discusses radioactivity in greater detail.

31.3 The Mass Defect of the Nucleus and Nuclear Binding Energy

Because of the strong nuclear force, the nucleons in a stable nucleus are held tightly together. Therefore, energy is required to separate a stable nucleus into its constituent protons and neutrons, as **Interactive Figure 31.3** illustrates. The more stable the nucleus is, the greater is the amount of energy needed to break it apart. The required energy is called the **binding energy** of the nucleus.

Two ideas that we have studied previously come into play as we discuss the binding energy of a nucleus. These are the rest energy of an object (Section 28.6) and mass (Section 4.2). In Einstein's theory of special relativity, energy and mass are equivalent; in fact, the rest energy E_0 and the mass m are related via $E_0 = mc^2$ (Equation 28.5), where c is the speed of light in a vacuum. Therefore, a change ΔE_0 in the rest energy of the system is equivalent to a change Δm in the mass of the system, according to $\Delta E_0 = (\Delta m)c^2$. We see, then, that the binding energy used in **Interactive Figure 31.3** to disassemble the nucleus appears as extra mass of the separated and stationary nucleons. In other words, the sum of the individual masses of the separated protons and neutrons is greater by an amount Δm than the mass of the stable nucleus. The difference in mass Δm is known as the **mass defect** of the nucleus.

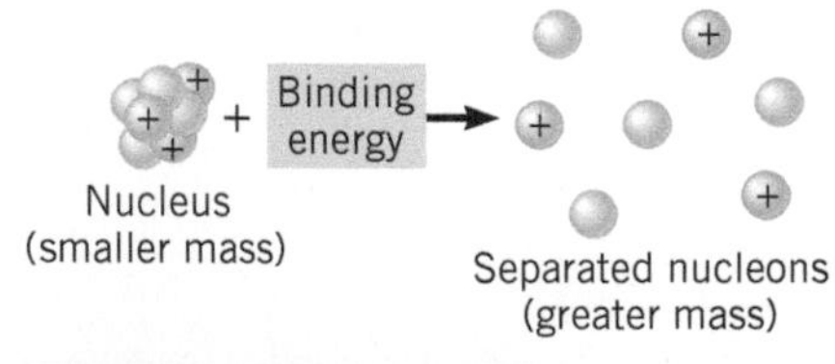

INTERACTIVE FIGURE 31.3 Energy, called the binding energy, must be supplied to break the nucleus apart into its constituent protons and neutrons. Each of the separated nucleons is at rest and out of the range of the forces of the other nucleons.

As Example 2 shows, the binding energy of a nucleus can be determined from the mass defect according to Equation 31.3:

$$\text{Binding energy} = (\text{Mass defect})c^2 = (\Delta m)c^2 \tag{31.3}$$

EXAMPLE 2 | The Binding Energy of the Helium Nucleus

The most abundant isotope of helium has a ^4_2He nucleus whose mass is 6.6447×10^{-27} kg. For this nucleus, find **(a)** the mass defect and **(b)** the binding energy.

Reasoning The symbol ^4_2He indicates that the helium nucleus contains $Z = 2$ protons and $N = 4 - 2 = 2$ neutrons. To obtain the mass defect Δm, we first determine the sum of the individual masses of the separated protons and neutrons. Then we subtract from this sum the mass of the ^4_2He nucleus. Finally, we use Equation 31.3 to calculate the binding energy from the value for Δm.

Solution **(a)** Using data from **Table 31.1**, we find that the sum of the individual masses of the nucleons is

$$\underbrace{2(1.6726 \times 10^{-27}\text{ kg})}_{\text{Two protons}} + \underbrace{2(1.6749 \times 10^{-27}\text{ kg})}_{\text{Two neutrons}} = 6.6950 \times 10^{-27}\text{ kg}$$

This value is greater than the mass of the intact ^4_2He nucleus, and the mass defect is

$$\Delta m = 6.6950 \times 10^{-27}\text{ kg} - 6.6447 \times 10^{-27}\text{ kg} = \boxed{0.0503 \times 10^{-27}\text{ kg}}$$

(b) According to Equation 31.3, the binding energy is

$$\text{Binding energy} = (\Delta m)c^2 = (0.0503 \times 10^{-27}\text{ kg})(3.00 \times 10^8\text{ m/s})^2 = 4.53 \times 10^{-12}\text{ J}$$

Usually, binding energies are expressed in energy units of electron volts instead of joules ($1\text{ eV} = 1.60 \times 10^{-19}$ J):

$$\text{Binding energy} = (4.53 \times 10^{-12}\text{ J})\left(\frac{1\text{ eV}}{1.60 \times 10^{-19}\text{ J}}\right) = 2.83 \times 10^7\text{ eV} = \boxed{28.3\text{ MeV}}$$

In this result, one million electron volts is denoted by the unit MeV. The value of 28.3 MeV is more than two million times greater than the energy required to remove an orbital electron from an atom.

In calculations such as that in Example 2, it is customary to use the **atomic mass unit** (u) instead of the kilogram. As introduced in Section 14.1, the atomic mass unit is one-twelfth of the mass of a $^{12}_6C$ atom of carbon. In terms of this unit, the mass of a $^{12}_6C$ atom is exactly 12 u. **Table 31.1** also gives the masses of the electron, the proton, and the neutron in atomic mass units. For future calculations, the energy equivalent of one atomic mass unit can be determined by observing that the mass of a proton is 1.6726×10^{-27} kg or 1.0073 u, so that

$$1\text{ u} = (1\text{ u})\left(\frac{1.6726 \times 10^{-27}\text{ kg}}{1.0073\text{ u}}\right) = 1.6605 \times 10^{-27}\text{ kg}$$

and

$$\Delta E_0 = (\Delta m)c^2 = (1.6605 \times 10^{-27}\text{ kg})(2.9979 \times 10^8\text{ m/s})^2 = 1.4924 \times 10^{-10}\text{ J}$$

In electron volts, therefore, one atomic mass unit is equivalent to

$$1\text{ u} = (1.4924 \times 10^{-10}\text{ J})\left(\frac{1\text{ eV}}{1.6022 \times 10^{-19}\text{ J}}\right) = 9.315 \times 10^8\text{ eV} = 931.5\text{ MeV}$$

Data tables for isotopes, such as the table in Appendix F, give masses in atomic mass units. Typically, however, the given masses are not nuclear masses. They are *atomic masses*—that is, the masses of neutral atoms, including the mass of the orbital electrons. Example 3 deals again with the ^4_2He nucleus and shows how to take into account the effect of the orbital electrons when using such data to determine binding energies.

EXAMPLE 3 | The Binding Energy of the Helium Nucleus, Revisited

The atomic mass of helium ^4_2He is 4.0026 u, and the atomic mass of hydrogen ^1_1H is 1.0078 u. Using atomic mass units instead of kilograms, obtain the binding energy of the ^4_2He nucleus.

Reasoning To determine the binding energy, we calculate the mass defect in atomic mass units and then use the fact that one atomic mass unit is equivalent to 931.5 MeV of energy. The mass of 4.0026 u for ^4_2He *includes the mass of the two electrons in the neutral helium atom.* To calculate the mass defect, we must subtract 4.0026 u from the sum of the individual masses of the nucleons, including the mass of the electrons. As **Figure 31.4** illustrates, the electron mass will be included if the masses of two hydrogen atoms are used in the calculation instead of the masses of two protons. The mass of a ^1_1H hydrogen atom is given in **Table 31.1** as 1.0078 u, and the mass of a neutron as 1.0087 u.

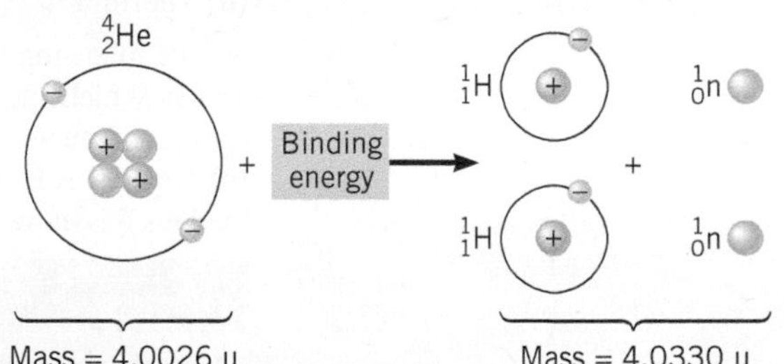

FIGURE 31.4 Data tables usually give the mass of the neutral atom (including the orbital electrons) rather than the mass of the nucleus. When data from such tables are used to determine the mass defect of a nucleus, the mass of the orbital electrons must be taken into account, as this drawing illustrates for the ^4_2He isotope of helium. See Example 3.

Solution The sum of the individual masses is

$$\underbrace{2(1.0078\text{ u})}_{\text{Two hydrogen atoms}} + \underbrace{2(1.0087\text{ u})}_{\text{Two neutrons}} = 4.0330\text{ u}$$

The mass defect is $\Delta m = 4.0330\text{ u} - 4.0026\text{ u} = 0.0304$ u. Since 1 u is equivalent to 931.5 MeV, the binding energy is

$$\text{Binding energy} = (0.0304\text{ u})\left(\frac{931.5\text{ MeV}}{1\text{ u}}\right) = \boxed{28.3\text{ MeV}}$$

which matches the result obtained in Example 2.

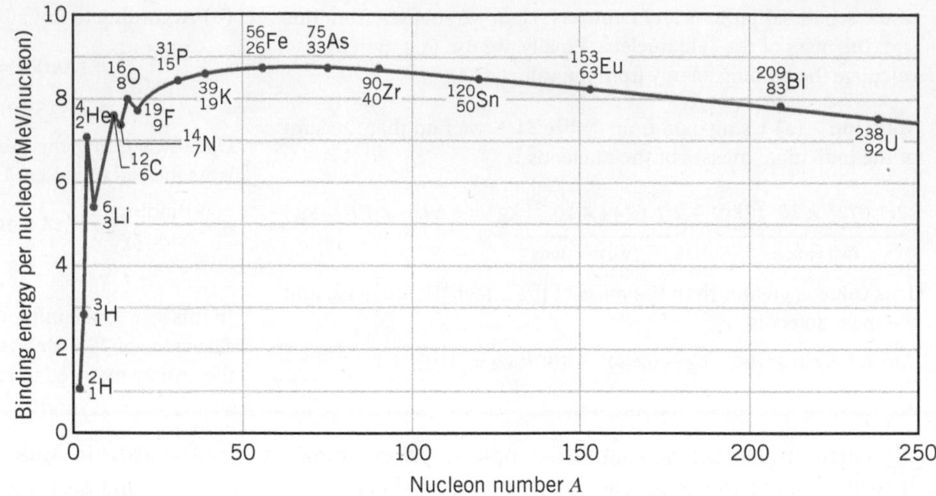

FIGURE 31.5 A plot of binding energy per nucleon versus the nucleon number A.

To see how the nuclear binding energy varies from nucleus to nucleus, it is necessary to compare the binding energy for each nucleus on a per-nucleon basis. The graph in **Figure 31.5** shows a plot in which the binding energy divided by the nucleon number A is plotted against the nucleon number itself. In the graph, the peak for the $^{4}_{2}$He isotope of helium indicates that the $^{4}_{2}$He nucleus is particularly stable. The binding energy per nucleon increases rapidly for nuclei with small masses and reaches a maximum of approximately 8.7 MeV/nucleon for a nucleon number of about $A = 60$. For greater nucleon numbers, the binding energy per nucleon decreases gradually. Eventually, the binding energy per nucleon decreases enough so there is insufficient binding energy to hold the nucleus together. Nuclei more massive than the $^{209}_{83}$Bi nucleus of bismuth are unstable and hence radioactive.

Check Your Understanding

(The answers are given at the end of the book.)

5. Using **Figure 31.5**, rank the following nuclei in ascending order according to the binding energy per nucleon (smallest first): **(a)** Phosphorus $^{31}_{15}$P **(b)** Cobalt $^{59}_{27}$Co **(c)** Tungsten $^{84}_{74}$W **(d)** Thorium $^{232}_{90}$Th

6. The following table gives values for the mass defect Δm for four hypothetical nuclei: A, B, C, and D. Which statement is true regarding the stability of these nuclei? **(a)** Nucleus D is the most stable, and nucleus A is the least stable. **(b)** Nucleus C is stable, whereas nuclei A, B, and D are not. **(c)** Nucleus A is the most stable, and nucleus D is not stable. **(d)** Nuclei A and B are stable, but nucleus B is more stable than nucleus A.

	A	B	C	D
Mass defect, Δm	$+6.0 \times 10^{-29}$ kg	$+2.0 \times 10^{-29}$ kg	0 kg	-6.0×10^{-29} kg

31.4 Radioactivity

When an unstable or radioactive nucleus disintegrates spontaneously, certain kinds of particles and/or high-energy photons are released. These particles and photons are collectively called "rays." Three kinds of rays are produced by naturally occurring radioactivity: **α rays, β rays,** and **γ rays.** They are named according to the first three

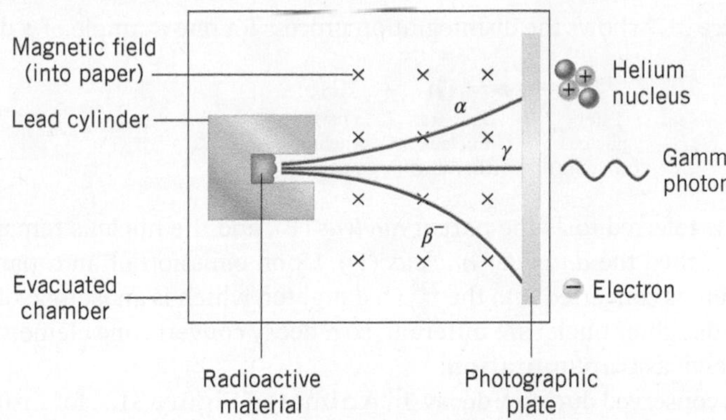

INTERACTIVE FIGURE 31.6 α and β rays are deflected by a magnetic field and, therefore, consist of moving charged particles. γ rays are not deflected by a magnetic field and, consequently, must be uncharged.

letters of the Greek alphabet, alpha (α), beta (β), and gamma (γ), to indicate the extent of their ability to penetrate matter. α rays are the least penetrating, being blocked by a thin ($\approx$0.01 mm) sheet of lead, whereas β rays penetrate lead to a much greater distance ($\approx$0.1 mm). γ rays are the most penetrating and can pass through an appreciable thickness ($\approx$100 mm) of lead.

The nuclear disintegration process that produces α, β, and γ rays must obey the conservation laws of physics. These laws are called "conservation laws," because each of them deals with a property that is conserved, in the sense that it does not change during a process. The following list shows the property with which each law deals:

1. Conservation of energy/mass (Sections 6.8 and 28.6)
2. Conservation of linear momentum (Section 7.2)
3. Conservation of angular momentum (Section 9.6)
4. Conservation of electric charge (Section 18.2)
5. Conservation of nucleon number (Section 31.4)

We have studied the first four of these laws in previous chapters, and to them we now add a fifth, **the conservation of nucleon number.**

In all radioactive decay processes it has been observed that the number of nucleons (protons plus neutrons) present before the decay is equal to the number of nucleons after the decay. Therefore, the number of nucleons is conserved during a nuclear disintegration. As applied to the disintegration of a nucleus, the conservation laws require that the energy, electric charge, linear momentum, angular momentum, and nucleon number that a nucleus possesses must remain unchanged when it disintegrates into nuclear fragments and accompanying α, β, or γ rays.

The three types of radioactivity that occur naturally can be observed in a relatively simple experiment. A piece of radioactive material is placed at the bottom of a narrow hole in a lead cylinder. The cylinder is located within an evacuated chamber, as **Interactive Figure 31.6** illustrates. A magnetic field is directed perpendicular to the plane of the paper, and a photographic plate is positioned to the right of the hole. Three spots appear on the developed plate, which are associated with the radioactivity of the nuclei in the material. Since moving particles are deflected by a magnetic field only when they are electrically charged, this experiment reveals that two types of radioactivity (α and β rays, as it turns out) consist of charged particles, whereas the third type (γ rays) does not.

α Decay

When a nucleus disintegrates and produces α rays, it is said to undergo **α decay**. Experimental evidence shows that α rays consist of positively charged particles, each one being the ${}^{4}_{2}\text{He}$ nucleus of helium. Thus, an α particle has a charge of $+2e$ and a nucleon number of $A = 4$. Since the grouping of 2 protons and 2 neutrons in a ${}^{4}_{2}\text{He}$ nucleus is particularly stable, as we have seen in connection with **Figure 31.5**, it is not surprising that an α particle can be ejected as a unit from a more massive unstable nucleus.

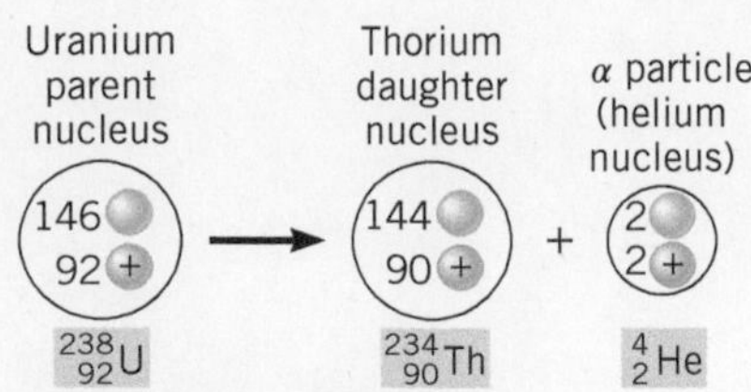

ANIMATED FIGURE 31.7 α decay occurs when an unstable parent nucleus emits an α particle and in the process is converted into a different, or daughter, nucleus.

Animated Figure 31.7 shows the disintegration process for one example of α decay:

$$\underset{\substack{\text{Parent}\\\text{nucleus}\\\text{(uranium)}}}{^{238}_{92}\text{U}} \longrightarrow \underset{\substack{\text{Daughter}\\\text{nucleus}\\\text{(thorium)}}}{^{234}_{90}\text{Th}} + \underset{\substack{\alpha\text{ particle}\\\text{(helium}\\\text{nucleus)}}}{^{4}_{2}\text{He}}$$

The original nucleus is referred to as the *parent nucleus* (P), and the nucleus remaining after disintegration is called the *daughter nucleus* (D). Upon emission of an α particle, the uranium $^{238}_{92}U$ parent is converted into the $^{234}_{90}Th$ daughter, which is an isotope of thorium. The parent and daughter nuclei are different, so α decay converts one element into another, a process known as **transmutation**.

Electric charge is conserved during α decay. In **Animated Figure 31.7**, for instance, 90 of the 92 protons in the uranium nucleus end up in the thorium nucleus, and the remaining 2 protons are carried off by the α particle. The total number of 92, however, is the same before and after disintegration. α decay also conserves the number of nucleons, because the number is the same before (238) and after (234 + 4) disintegration. Consistent with the conservation of electric charge and nucleon number, the general form for α decay is

α Decay

$$\underset{\substack{\text{Parent}\\\text{nucleus}}}{^{A}_{Z}\text{P}} \longrightarrow \underset{\substack{\text{Daughter}\\\text{nucleus}}}{^{A-4}_{Z-2}\text{D}} + \underset{\substack{\alpha\text{ particle (helium}\\\text{nucleus)}}}{^{4}_{2}\text{He}}$$

When a nucleus releases an α particle, the nucleus also releases energy. In fact, the energy released by radioactive decay is responsible, in part, for keeping the interior of the earth hot and, in some places, even molten. The following example shows how the conservation of mass/energy can be used to determine the amount of energy released in α decay.

EXAMPLE 4 | α Decay and the Release of Energy

The atomic mass of uranium $^{238}_{92}U$ is 238.0508 u, that of thorium $^{234}_{90}Th$ is 234.0436 u, and that of an α particle $^{4}_{2}He$ is 4.0026 u. Determine the energy released when α decay converts $^{238}_{92}U$ into $^{234}_{90}Th$.

Reasoning Since energy is released during the decay, the combined mass of the $^{234}_{90}Th$ daughter nucleus and the α particle is less than the mass of the $^{238}_{92}U$ parent nucleus. The difference in mass is equivalent to the energy released. We will determine the difference in mass in atomic mass units and then use the fact that 1 u is equivalent to 931.5 MeV.

Solution The decay and the masses are shown below:

$$\underset{238.0508\text{ u}}{^{238}_{92}\text{U}} \longrightarrow \underbrace{\underset{234.0436\text{ u}}{^{234}_{90}\text{Th}} + \underset{4.0026\text{ u}}{^{4}_{2}\text{He}}}_{238.0462\text{ u}}$$

The decrease in mass, or mass defect for the decay process, is 238.0508 u − 238.0462 u = 0.0046 u. As usual, the masses are atomic masses and include the mass of the orbital electrons. But this causes no error here because the same total number of electrons is included for $^{238}_{92}U$, on the one hand, and for $^{234}_{90}Th$ plus $^{4}_{2}He$, on the other. Since 1 u is equivalent to 931.5 MeV, the released energy is

$$\text{Released energy} = (0.0046\ \cancel{\text{u}})\left(\frac{931.5\text{ MeV}}{1\ \cancel{\text{u}}}\right) = \boxed{4.3\text{ MeV}}$$

When α decay occurs as in Example 4, the energy released appears as kinetic energy of the recoiling $^{234}_{90}Th$ nucleus and the α particle, except for a small portion carried away as a γ ray. Conceptual Example 5 discusses how the $^{234}_{90}Th$ nucleus and the α particle share in the released energy.

CONCEPTUAL EXAMPLE 5 | How Energy Is Shared During the α Decay of $^{238}_{92}U$

In Example 4, the energy released by the α decay of $^{238}_{92}U$ is found to be 4.3 MeV. Since this energy is carried away as kinetic energy of the recoiling $^{234}_{90}Th$ nucleus and the α particle, it follows that $KE_{Th} + KE_{\alpha} = 4.3$ MeV. However, KE_{Th} and KE_{α} are not equal. Which particle carries away more kinetic energy, the $^{234}_{90}Th$ nucleus or the α particle?

Reasoning and Solution Kinetic energy depends on the mass m and speed v of a particle, since $KE = \frac{1}{2}mv^2$ (Equation 6.2). The $^{234}_{90}Th$ nucleus has a much greater mass than the α particle, and since the kinetic energy is proportional to the mass, it is tempting to conclude that the $^{234}_{90}Th$ nucleus has the greater kinetic energy. This conclusion is not correct, however, since it does not take into account the fact that the $^{234}_{90}Th$ nucleus and the α particle have different speeds after the decay. In fact, we expect the thorium nucleus to recoil with the smaller speed precisely *because* it has the greater mass. The decaying $^{238}_{92}U$ is like a parent and their young child on ice skates, pushing off against one another. The more massive parent recoils with much less speed than the child. We can use the principle of conservation of linear momentum to verify our expectation.

As Section 7.2 discusses, the conservation principle states that the total linear momentum of an isolated system remains constant. An isolated system is one for which the vector sum of the external forces acting on the system is zero, and the decaying $^{238}_{92}U$ nucleus fits this description. It is stationary initially, and since momentum is mass times velocity, its initial momentum is zero. In its final form, the system consists of the $^{234}_{90}Th$ nucleus and the α particle and has a final total momentum of $m_{Th}v_{Th} + m_{\alpha}v_{\alpha}$. According to momentum conservation, the initial and final values of the total momentum of the system must be the same, so that $m_{Th}v_{Th} + m_{\alpha}v_{\alpha} = 0$. Solving this equation for the velocity of the thorium nucleus, we find that $v_{Th} = -m_{\alpha}v_{\alpha}/m_{Th}$. Since m_{Th} is much greater than m_{α}, we can see that the speed of the thorium nucleus is less than the speed of the α particle. Moreover, the kinetic energy depends on the square of the speed and only the first power of the mass. As a result of its much greater speed, ***the α particle has the greater kinetic energy***.

Related Homework: Problem 27

THE PHYSICS OF . . . radioactivity and smoke detectors. One widely used application of α decay is in smoke detectors. **Figure 31.8** illustrates how a smoke detector operates. Two small and parallel metal plates are separated by a distance of about one centimeter. A tiny amount of radioactive material at the center of one of the plates emits α particles, which collide with air molecules. During the collisions, the air molecules are ionized to form positive and negative ions. The voltage from a battery causes one plate to be positive and the other negative, so that each plate attracts ions of opposite charge. As a result there is a current in the circuit attached to the plates. The presence of smoke particles between the plates reduces the current, since the ions that collide with a smoke particle are usually neutralized. The drop in current that smoke particles cause is used to trigger an alarm.

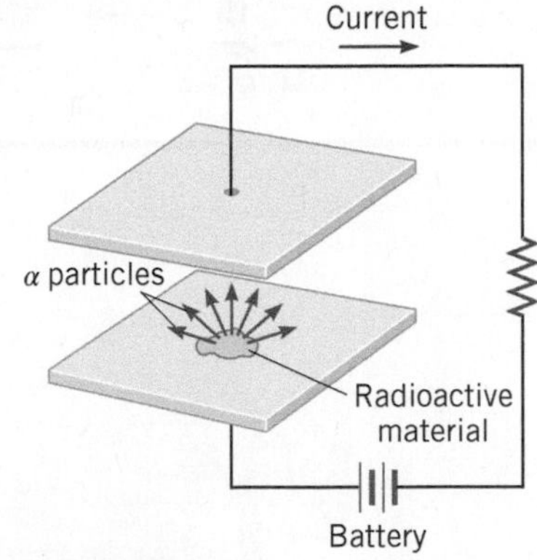

FIGURE 31.8 A smoke detector.

β Decay

The β rays in **Interactive Figure 31.6** are deflected by the magnetic field in a direction opposite to that of the positively charged α rays. Consequently, these β rays, which are the most common kind, consist of negatively charged particles or β^- particles. Experiment shows that β^- particles are electrons. As an illustration of β^- decay, consider the thorium $^{234}_{90}Th$ nucleus, which decays by emitting a β^- particle, as in **Figure 31.9**:

$$\underset{\substack{\text{Parent}\\ \text{nucleus}\\ \text{(thorium)}}}{^{234}_{90}\text{Th}} \longrightarrow \underset{\substack{\text{Daughter}\\ \text{nucleus}\\ \text{(protactinium)}}}{^{234}_{91}\text{Pa}} + \underset{\substack{\beta^- \text{ particle}\\ \text{(electron)}}}{^{0}_{-1}\text{e}}$$

β^- decay, like α decay, causes a transmutation of one element into another. In this case, thorium $^{234}_{90}Th$ is converted into protactinium $^{234}_{91}Pa$. The law of conservation of charge is obeyed, since the net number of positive charges is the same before (90) and after (91 − 1) the β^- emission. The law of conservation of nucleon number is obeyed, since the nucleon number remains at $A = 234$. The general form for β^- decay is

β⁻ decay

$$\underset{\substack{\text{Parent}\\ \text{nucleus}}}{^{A}_{Z}\text{P}} \longrightarrow \underset{\substack{\text{Daughter}\\ \text{nucleus}}}{_{Z+1}^{A}\text{D}} + \underset{\substack{\beta^- \text{ particle}\\ \text{(electron)}}}{^{0}_{-1}\text{e}}$$

The electron emitted in β^- decay does *not* actually exist within the parent nucleus and is *not* one of the orbital electrons. Instead, the electron is created when a neutron decays into a proton and an electron; when this occurs, the proton number of the parent

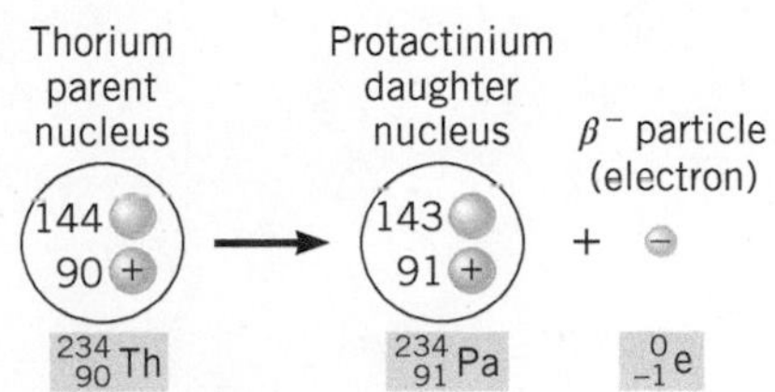

FIGURE 31.9 β^- decay occurs when a neutron in an unstable parent nucleus decays into a proton and an electron, the electron being emitted as the β^- particle. In the process, the parent nucleus is transformed into the daughter nucleus.

nucleus increases from Z to $Z + 1$ and the nucleon number remains unchanged. The newly created electron is usually fast-moving and escapes from the atom, leaving behind a positively charged atom.

Example 6 illustrates that energy is released during β^- decay, just as it is during α decay, and that the conservation of mass/energy applies.

EXAMPLE 6 | β^- Decay and the Release of Energy

The atomic mass of thorium ${}^{234}_{90}\text{Th}$ is 234.043 59 u, and the atomic mass of protactinium ${}^{234}_{91}\text{Pa}$ is 234.043 30 u. Find the energy released when β^- decay changes ${}^{234}_{90}\text{Th}$ into ${}^{234}_{91}\text{Pa}$.

Reasoning To find the energy released, we follow the usual procedure of determining how much the mass has decreased because of the decay and then calculating the equivalent energy.

> **Problem-Solving Insight In β^- decay, be careful not to include the mass of the electron (${}^{0}_{-1}\text{e}$) twice. As discussed here, the atomic mass of the daughter atom (${}^{234}_{91}\text{Pa}$) already includes the mass of the emitted electron.**

Solution The decay and the masses are shown below:

$$\underbrace{{}^{234}_{90}\text{Th}}_{234.043\ 59\ \text{u}} \longrightarrow \underbrace{{}^{234}_{91}\text{Pa} + {}^{0}_{-1}\text{e}}_{234.043\ 30\ \text{u}}$$

When the ${}^{234}_{90}\text{Th}$ nucleus of a thorium atom is converted into a ${}^{234}_{91}\text{Pa}$ nucleus, the number of orbital electrons remains the same, so the resulting protactinium atom is missing one orbital electron. However, the given mass includes all 91 electrons of a neutral protactinium atom. In effect, then, the value of 234.043 30 u for ${}^{234}_{91}\text{Pa}$ already includes the mass of the β^- particle. The mass decrease that accompanies the β^- decay is

$$234.043\ 59\ \text{u} - 234.043\ 30\ \text{u} = 0.000\ 29\ \text{u}$$

Since 1 u is equivalent to 931.5 MeV, the released energy is

$$\text{Released energy} = (0.00029\ \cancel{\text{u}})\left(\frac{931.5\ \text{MeV}}{1\ \cancel{\text{u}}}\right) = \boxed{0.27\ \text{MeV}}$$

This is the maximum kinetic energy that the emitted electron can have.

A second kind of β decay sometimes occurs.* In this process the particle emitted by the nucleus is a **positron** rather than an electron. A positron, also called a β^+ particle, has the same mass as an electron but carries a charge of $+e$ instead of $-e$. The disintegration process for β^+ decay is

β^+ decay

$$\underbrace{{}^{A}_{Z}\text{P}}_{\text{Parent nucleus}} \longrightarrow \underbrace{{}_{Z-1}^{\ \ \ A}\text{D}}_{\text{Daughter nucleus}} + \underbrace{{}^{0}_{+1}\text{e}}_{\beta^+\text{ particle (positron)}}$$

The emitted positron does *not* exist within the nucleus but, rather, is created when a nuclear proton is transformed into a neutron. In the process, the proton number of the parent nucleus decreases from Z to $Z - 1$, and the nucleon number remains the same. As with β^-decay, the laws of conservation of charge and nucleon number are obeyed, and there is a transmutation of one element into another.

γ Decay

The nucleus, like the orbital electrons, exists only in discrete energy states or levels. When a nucleus changes from an excited energy state (denoted by an asterisk*) to a lower energy state, a photon is emitted. The process is similar to the one discussed in Section 30.3 for the photon emission that leads to the hydrogen atom line spectrum. With nuclear energy levels, however, the photon has a much greater energy and is called a γ ray. The γ decay process is written as follows:

γ decay

$$\underbrace{{}^{A}_{Z}\text{P}^*}_{\text{Excited energy state}} \longrightarrow \underbrace{{}^{A}_{Z}\text{P}}_{\text{Lower energy state}} + \underbrace{\gamma}_{\gamma\text{ ray}}$$

γ decay does *not* cause a transmutation of one element into another.

*A third kind of β decay also occurs in which a nucleus pulls in, or captures, one of the orbital electrons from outside the nucleus. The process is called **electron capture,** or **K capture,** since the electron normally comes from the innermost, or K, shell.

Medical Applications of Radioactivity

BIO **THE PHYSICS OF . . . Gamma Knife radiosurgery.** Gamma Knife radiosurgery is becoming a very promising medical procedure for treating certain problems of the brain, including benign and cancerous tumors as well as blood vessel malformations. The procedure, which involves no knife at all, uses powerful, highly focused beams of γ rays aimed at the tumor or malformation. The γ rays are emitted by a radioactive cobalt-60 source. As **Figure 31.10*a*** illustrates, the patient wears a protective metal helmet that is perforated with many small holes. Part *b* of the figure shows that the holes focus the γ rays to a single tiny target within the brain. The target tissue thus receives a very intense dose of radiation and is destroyed, while the surrounding healthy tissue is undamaged. Gamma Knife surgery is a noninvasive, painless, and bloodless procedure that is often performed under local anesthesia. Hospital stays are 70 to 90% shorter than with conventional surgery, and patients often return to work within a few days.

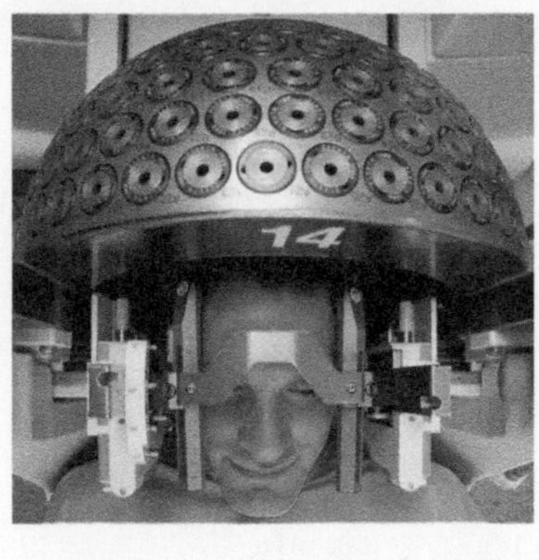

(*a*)

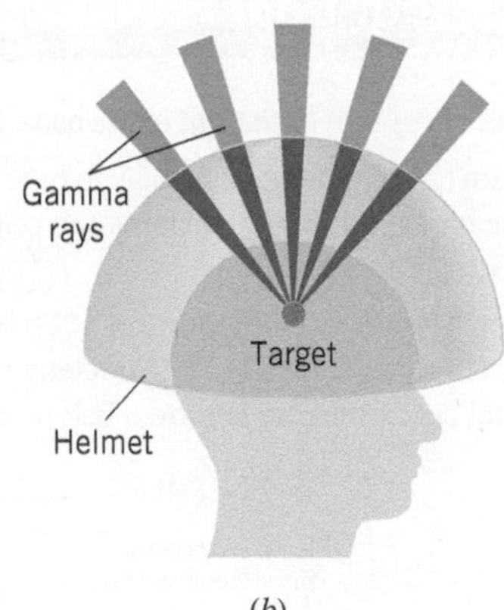

(*b*)

FIGURE 31.10 (*a*) In Gamma Knife radiosurgery, a protective metal helmet containing many small holes is placed on the patient's head. (*b*) The holes focus the beams of γ rays to a tiny target within the brain.

BIO **THE PHYSICS OF . . . an exercise thallium heart scan.** An exercise thallium heart scan is a test that uses radioactive thallium to produce images of the heart muscle. When combined with an exercise test, such as walking on a treadmill, the thallium scan helps identify regions of the heart that do not receive enough blood. The scan is especially useful in diagnosing the presence of blockages in the coronary arteries, which supply oxygen-rich blood to the heart muscle. During the test, a small amount of thallium is injected into a vein while the patient walks on a treadmill. The thallium attaches to the red blood cells and is carried throughout the body. The thallium enters the heart muscle by way of the coronary arteries and collects in heart -muscle cells that come into contact with the blood. The thallium isotope used, ${}^{201}_{81}\text{Tl}$, emits γ rays, which a special camera records. Since the thallium reaches those regions of the heart that have an adequate blood supply, lesser amounts show up in areas where the blood flow has been reduced due to arterial blockages (see **Figure 31.11**). A second set of images is taken several hours later, while the patient is resting. These images help differentiate between regions of the heart that temporarily do not receive enough blood (the blood flow returns to normal after the exercise) and regions that are permanently damaged due to, for example, a previous heart attack (the blood flow does not return to normal).

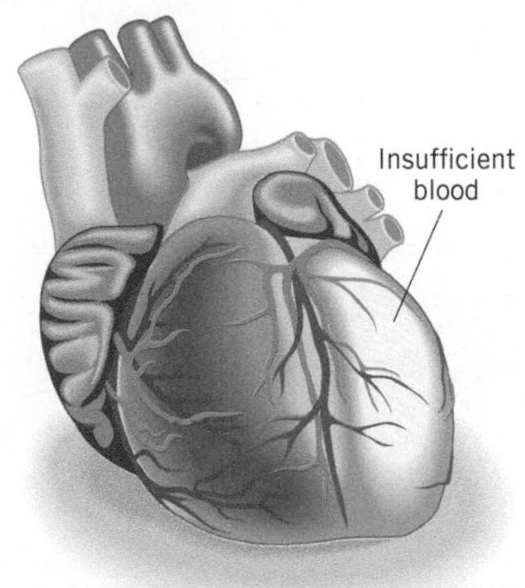

FIGURE 31.11 An exercise thallium heart scan indicates regions of the heart that receive insufficient blood during exercise.

BIO **THE PHYSICS OF . . . brachytherapy implants** The use of radioactive isotopes to deliver radiation to specific targets in the body is an important medical technique. In treating cancer, for example, the method of delivery should ideally apply a high dose of radiation to a malignant tumor in order to kill it, while applying only a small (non-damaging) dose to healthy surrounding tissue. Brachytherapy implants offer such a delivery method. In this type of treatment, radioactive isotopes are formed into small seeds and implanted directly in the tumor according to a predesigned pattern. The energy and type of radiation emitted by the isotopes can be exploited to optimize a treatment design and minimize damage to healthy tissue. Seeds containing iridium $^{192}_{77}\text{Ir}$ are used to treat many cancers, and seeds containing iodine $^{125}_{53}\text{I}$ and palladium $^{103}_{46}\text{Pd}$ are used for prostate cancer. Research has also indicated that brachytherapy implants may have an important role to play in the treatment of atherosclerosis, in which blood vessels become blocked with plaque. Such blockages are often treated using the technique of balloon angioplasty. With the aid of a catheter inserted into an occluded coronary artery, a balloon is inflated to open the artery and place a stent (a metallic mesh that provides support for the arterial wall) at the site of the blockage. Sometimes the arterial wall is damaged in this process, and as it heals, the artery often becomes blocked again. Brachytherapy implants (using iridium $^{192}_{77}\text{Ir}$ or phosphorus $^{32}_{15}\text{P}$, for instance) have been found to inhibit repeat blockages following angioplasty.

Check Your Understanding

(The answers are given at the end of the book.)

7. Polonium $^{216}_{84}\text{Po}$ undergoes α decay to produce a daughter nucleus that itself undergoes β^- decay. Which one of the following nuclei is the one that ultimately results? **(a)** $^{211}_{82}\text{Pb}$ **(b)** $^{211}_{81}\text{Tl}$ **(c)** $^{212}_{81}\text{Tl}$ **(d)** $^{212}_{83}\text{Bi}$ **(e)** $^{213}_{82}\text{Pb}$

8. Uranium $^{238}_{92}\text{U}$ decays into thorium $^{234}_{90}\text{Th}$ by means of α decay, as Example 4 in the text discusses. Another possibility is that the $^{238}_{92}\text{U}$ nucleus just emits a single proton instead of an α particle. This hypothetical decay scheme is shown below, along with the pertinent atomic masses:

$$\underset{\substack{\text{Uranium} \\ 238.050\ 78\ \text{u}}}{^{238}_{92}\text{U}} \longrightarrow \underset{\substack{\text{Protactinium} \\ 237.051\ 14\ \text{u}}}{^{237}_{91}\text{Pa}} + \underset{\substack{\text{Proton} \\ 1.007\ 83\ \text{u}}}{^{1}_{1}\text{H}}$$

For a decay to be possible, it must bring the parent nucleus toward a more stable state by allowing the release of energy. Compare the total mass of the products of this hypothetical decay with the mass of $^{238}_{92}\text{U}$ and decide whether the emission of a single proton is possible for $^{238}_{92}\text{U}$.

31.5 The Neutrino

When a β particle is emitted by a radioactive nucleus, energy is simultaneously released, as Example 6 illustrates. Experimentally, however, it is found that most β particles do not have enough kinetic energy to account for all the energy released. If a β particle carries away only part of the energy, where does the remainder go? The question puzzled physicists until 1930, when Wolfgang Pauli proposed that part of the energy is carried away by another particle that is emitted along with the β particle. This additional particle is called the **neutrino**, and its existence was verified experimentally in 1956. The Greek letter nu (ν) is used to symbolize the neutrino. For instance, the β^- decay of thorium $^{234}_{90}\text{Th}$ (see Section 31.4) is more correctly written as

$$^{234}_{90}\text{Th} \longrightarrow {}^{234}_{91}\text{Pa} + {}^{\ 0}_{-1}\text{e} + \bar{\nu}$$

The bar above the ν is included because the neutrino emitted in this particular decay process is an antimatter neutrino, or antineutrino. A normal neutrino (ν without the bar) is emitted when β^+ decay occurs.

The emission of neutrinos and β particles involves a force called the *weak nuclear force* because it is much weaker than the strong nuclear force. It is now known that the weak nuclear force and the electromagnetic force are two different manifestations of a single, more fundamental force, the **electroweak force**. The theory for the electroweak

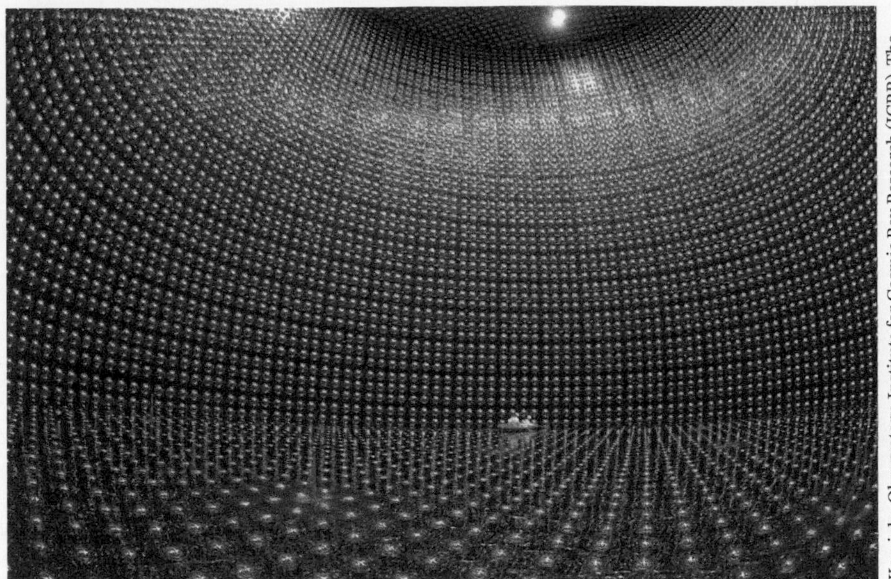

Kamioka Observatory, Institute for Cosmic Ray Research (ICRR), The University of Tokyo

FIGURE 31.12 The Super-Kamiokande neutrino detector in Japan is an underground steel cylindrical tank with its inner wall lined with 11 000 photomultiplier tubes. It is filled with 12.5 million gallons of ultrapure water when operational. In this photograph it is partially full, and the technicians in the boat are inspecting the photomultiplier tubes.

force was developed by Sheldon Glashow (1932–), Abdus Salam (1926–1996), and Steven Weinberg (1933–), who shared a Nobel Prize for their achievement in 1979. The electroweak force, the gravitational force, and the strong nuclear force are the three fundamental forces in nature.

The neutrino has zero electric charge and is extremely difficult to detect because it interacts very weakly with matter. For example, the average neutrino can penetrate one light-year of lead (about 9.5×10^{15} m) without interacting with it. Thus, even though trillions of neutrinos pass through our bodies every second, they have no effect. One of the major scientific questions of our time is whether neutrinos have mass. The question is important because neutrinos are so plentiful in the universe. Even a very small mass could account for a significant portion of the mass in the universe and, possibly, could have an effect on the formation of galaxies.

Although difficult, it is possible to detect neutrinos. **Figure 31.12** shows the Super-Kamiokande neutrino detector in Japan. It is located 915 m underground and consists of a steel cylindrical tank, ten stories tall, whose inner wall is lined with 11 000 photomultiplier tubes (see Section 31.9). The tank is filled with 12.5 million gallons of ultrapure water. Neutrinos colliding with the water molecules produce light patterns that the photomultiplier tubes detect. In 1998 the Super-Kamiokande detector yielded the first strong, but indirect, evidence that neutrinos do indeed have a small mass. (The mass of the electron neutrino is less than 0.0004% of the mass of an electron.) This finding implies that neutrinos travel at less than the speed of light. If the neutrino's mass were zero, like that of a photon, it would travel at the speed of light.

31.6 Radioactive Decay and Activity

The question of which radioactive nucleus in a group of nuclei disintegrates at a given instant is decided like the winning numbers in a state lottery: individual disintegrations occur randomly. As time passes, the number N of parent nuclei decreases, as **Figure 31.13** shows. This graph of N versus time indicates that the decrease occurs in a smooth fashion, with N approaching zero after enough time has passed. To help describe the graph, it is useful to define the **half-life** $T_{1/2}$ of a radioactive isotope

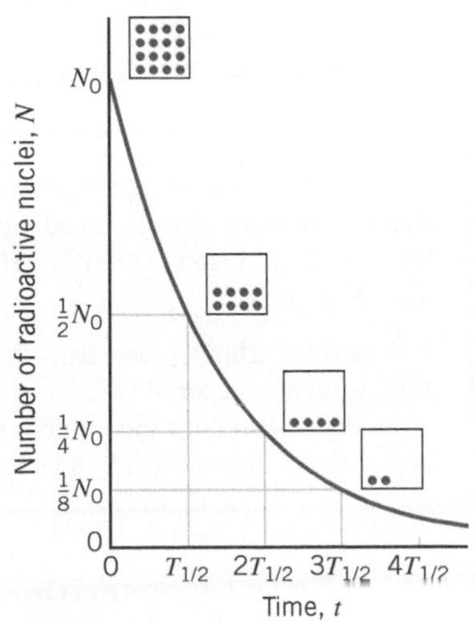

FIGURE 31.13 The half-life $T_{1/2}$ of a radioactive decay is the time in which one-half of the radioactive nuclei disintegrate.

TABLE 31.2 Some Half-Lives for Radioactive Decay

Isotope		Half-Life
Polonium	$^{214}_{84}Po$	1.64×10^{-4} s
Krypton	$^{89}_{36}Kr$	3.16 min
Radon	$^{222}_{86}Rn$	3.83 d
Strontium	$^{90}_{38}Sr$	29.1 yr
Radium	$^{226}_{88}Ra$	1.6×10^{3} yr
Carbon	$^{14}_{6}C$	5.73×10^{3} yr
Uranium	$^{238}_{92}U$	4.47×10^{9} yr
Indium	$^{115}_{49}In$	4.41×10^{14} yr

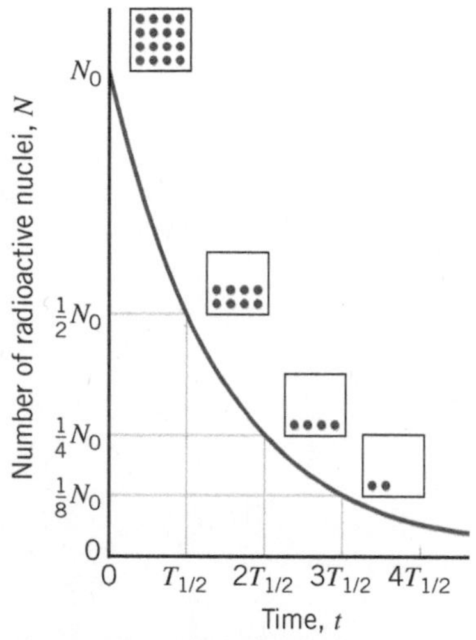

FIGURE 31.13 (REPEATED) The half-life $T_{1/2}$ of a radioactive decay is the time in which one-half of the radioactive nuclei disintegrate.

as the time required for one-half of the nuclei present to disintegrate. For example, radium $^{226}_{88}Ra$ has a half-life of 1600 years, because it takes this amount of time for one-half of a given quantity of this isotope to disintegrate. In another 1600 years, one-half of the remaining radium atoms will have disintegrated, leaving only one-fourth of the original number intact. In **Figure 31.13 (Repeated)**, the number of nuclei present at time $t = 0$ s is $N = N_0$, and the number present at $t = T_{1/2}$ is $N = \frac{1}{2}N_0$. The number present at $t = 2T_{1/2}$ is $N = \frac{1}{4}N_0$, and so on. The value of the half-life depends on the nature of the radioactive nucleus. Values ranging from a fraction of a second to billions of years have been found (see **Table 31.2**).

THE PHYSICS OF . . . radioactive radon gas in houses. Radon $^{222}_{86}Rn$ is a naturally occurring radioactive gas produced when radium $^{226}_{88}Ra$ undergoes α decay. There is a nationwide concern about radon as a health hazard because radon in the soil is gaseous and can enter the basement of homes through cracks in the foundation. (It should be noted, however, that the mechanism of indoor radon entry is not well understood and that entry via foundation cracks is likely only part of the story.) Once inside, the concentration of radon can rise markedly, depending on the type of housing construction and the concentration of radon in the surrounding soil. Radon gas decays into daughter nuclei that are also radioactive. The radioactive nuclei can attach to dust and smoke particles that can be inhaled, and they remain in the lungs to release tissue-damaging radiation. Prolonged exposure to high levels of radon can lead to lung cancer. Since radon gas concentrations can be measured with inexpensive monitoring devices, it is recommended that all homes be tested for radon. Example 7 deals with the half-life of radon $^{222}_{86}Rn$.

EXAMPLE 7 | The Radioactive Decay of Radon Gas

Suppose that 3.0×10^{7} radon atoms are trapped in a basement at the time the basement is sealed against further entry of the gas. The half-life of radon is 3.83 days. How many radon atoms remain after 31 days?

Reasoning During each half-life, the number of radon atoms is reduced by a factor of two. Thus, for each half-life in the period of 31 days, we reduce the number of radon atoms present at the beginning of that half-life by a factor of two.

Solution In a period of 31 days there are (31 days)/(3.83 days) = 8.1 half-lives. In 8 half-lives the number of radon atoms is reduced by a factor of $2^8 = 256$. Ignoring the difference between 8 and 8.1 half-lives, we find that the number of atoms remaining after 31 days is $(3.0 \times 10^{7})/256 = \boxed{1.2 \times 10^{5}}$.

The **activity** of a radioactive sample is the number of disintegrations per second that occur. Each time a disintegration occurs, the number N of radioactive nuclei decreases. As a result, the activity can be obtained by dividing ΔN, the change in the number of nuclei, by Δt, the time interval during which the change takes place; the average activity over the time interval Δt is the magnitude of $\Delta N/\Delta t$, or $|\Delta N/\Delta t|$. Since the decay of any individual nucleus is completely random, the number of disintegrations per second that occurs in a sample is proportional to the number of radioactive nuclei present, so that

$$\frac{\Delta N}{\Delta t} = -\lambda N \tag{31.4}$$

where λ is a proportionality constant referred to as the **decay constant.** The minus sign is present in this equation because each disintegration decreases the number N of nuclei originally present.

The SI unit for activity is the *becquerel* (Bq), named after Antoine Becquerel (1852–1908). One becquerel is one disintegration per second. Activity is also measured using a unit called the *curie* (Ci), honoring Marie (1867–1934) and Pierre (1859–1906) Curie, the discoverers of radium and polonium. Historically, the curie was chosen as a unit because it is roughly the activity of one gram of pure radium. In terms of becquerels,

$$1\ \text{Ci} = 3.70 \times 10^{10}\ \text{Bq}$$

The activity of the radium put into the dial of a watch to make it glow in the dark is about 4×10^4 Bq, and the activity used in radiation therapy for cancer is approximately a billion times greater, or 4×10^{13} Bq.

The mathematical expression for the graph of N versus t shown in **Figure 31.13** can be obtained from Equation 31.4 with the aid of calculus. The result for the number N of radioactive nuclei present at time t is

$$N = N_0 e^{-\lambda t} \tag{31.5}$$

assuming that the number present at $t = 0$ s is N_0. The exponential e has the value $e = 2.718\ldots$, and many calculators provide the value of e^x. We can relate the half-life $T_{1/2}$ of a radioactive nucleus to its decay constant λ in the following manner. By substituting $N = \frac{1}{2}N_0$ and $t = T_{1/2}$ into Equation 31.5, we find that $\frac{1}{2} = e^{-\lambda T_{1/2}}$. Solving this equation for $T_{1/2}$ reveals that

$$T_{1/2} = \frac{\ln 2}{\lambda} = \frac{0.693}{\lambda} \tag{31.6}$$

Examples 8 through 10 illustrate the use of Equations 31.5 and 31.6.

Math Skills To obtain Equation 31.6, we take the natural logarithm of both sides of the equation $\frac{1}{2} = e^{-\lambda T_{1/2}}$, which gives

$$\ln\left(\tfrac{1}{2}\right) = \ln(e^{-\lambda T_{1/2}})$$

According to Equation D-12 in Appendix D, the left side of this result is $\ln\left(\frac{1}{2}\right) = \ln 1 - \ln 2$. According to Equation D-9 in Appendix D, the right side is $\ln(e^{-\lambda T_{1/2}}) = -\lambda T_{1/2}$. Thus, we have

$$\ln 1 - \ln 2 = -\lambda T_{1/2}$$

However, since $\ln 1 = 0$, this result becomes

$$\ln 2 = \lambda T_{1/2} \quad \text{or} \quad T_{1/2} = \frac{\ln 2}{\lambda} \tag{31.6}$$

EXAMPLE 8 | The Activity of Radon $^{222}_{86}\text{Rn}$

As in Example 7, suppose that there are 3.0×10^7 radon atoms ($T_{1/2} = 3.83$ days or 3.31×10^5 s) trapped in a basement. **(a)** How many radon atoms remain after 31 days? Find the activity **(b)** just after the basement is sealed against further entry of radon and **(c)** 31 days later.

Reasoning The number N of radon atoms remaining after a time t is given by $N = N_0 e^{-\lambda t}$ (Equation 31.5), where $N_0 = 3.0 \times 10^7$ is the original number of atoms when $t = 0$ s and λ is the decay constant. The decay constant is related to the half-life $T_{1/2}$ of the radon atoms by $\lambda = 0.693/T_{1/2}$. The activity can be obtained from $\Delta N/\Delta t = -\lambda N$ (Equation 31.4).

Solution (a) The decay constant is

$$\lambda = \frac{0.693}{T_{1/2}} = \frac{0.693}{3.83 \text{ days}} = 0.181 \text{ days}^{-1} \quad \textbf{(31.6)}$$

and the number N of radon atoms remaining after 31 days is

$$N = N_0 e^{-\lambda t} = (3.0 \times 10^7)e^{-(0.181 \text{ days}^{-1})(31 \text{ days})} = \boxed{1.1 \times 10^5} \quad \textbf{(31.5)}$$

This value is slightly less than that found in Example 7 because there we ignored the difference between 8.0 and 8.1 half-lives.

(b) The activity can be obtained from Equation 31.4, provided the decay constant is expressed in reciprocal seconds:

$$\lambda = \frac{0.693}{T_{1/2}} = \frac{0.693}{3.31 \times 10^5 \text{ s}} = 2.09 \times 10^{-6} \text{ s}^{-1} \quad \textbf{(31.6)}$$

Thus, the number of disintegrations per second is

$$\frac{\Delta N}{\Delta t} = -\lambda N = -(2.09 \times 10^{-6} \text{ s}^{-1})(3.0 \times 10^7)$$
$$= -63 \text{ disintegations/s} \quad \textbf{(31.4)}$$

The activity is the magnitude of $\Delta N/\Delta t$, so initially $\boxed{\text{Activity} = 63 \text{ Bq}}$.

(c) From part (a), the number of radioactive nuclei remaining at the end of 31 days is $N = 1.1 \times 10^5$, and reasoning similar to that in part (b) reveals that $\boxed{\text{Activity} = 0.23 \text{ Bq}}$.

EXAMPLE 9 | BIO The Physics of Brachytherapy Implants Revisited

As part of a treatment plan for prostate cancer, a patient undergoes brachytherapy, in which seeds containing iodine $^{125}_{53}\text{I}$ are implanted close to the prostate (see **Figure 31.14**). The iodine atoms have a half-life of 59.5 days, and they decay by electron capture into tellurium $^{125}_{52}\text{Te}^*$ that is in an excited state. The tellurium atoms immediately undergo gamma decay, which releases a gamma ray with a maximum energy of 35 keV. This provides the majority of the dose of radiation that is used to kill the cancer cells. **(a)** During a low-dose treatment, a seed of $^{125}_{53}\text{I}$ is left near the prostate for 48 hours and then removed. What fraction of the $^{125}_{53}\text{I}$ atoms remains in the seed after the treatment? **(b)** What is the correct decay process for $^{125}_{52}\text{Te}^*$?

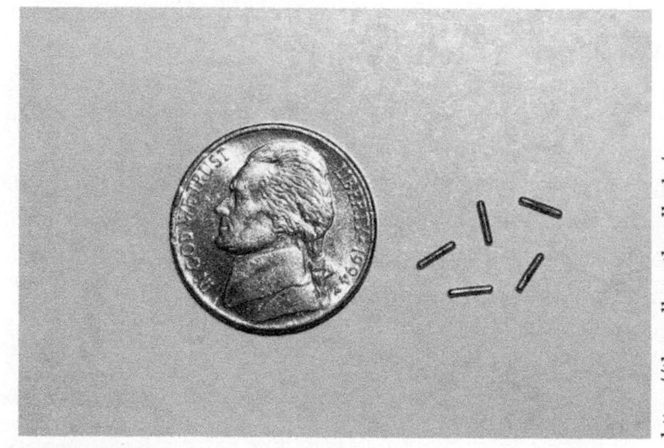

https://thewell.northwell.edu/brachytherapy-right-me

FIGURE 31.14 Photo of the seeds used in brachytherapy next to a nickel to show their size. The titanium capsules contain a radioactive element, such as $^{125}_{53}\text{I}$. The seeds are implanted directly into a tumor to kill cancer cells.

Reasoning (a) As time passes, the number of radioactive atoms of $^{125}_{53}\text{I}$ will decay. The number remaining after a time t is given by Equation 31.5. Thus, N_0 represents the original number of $^{125}_{53}\text{I}$ atoms, and N represents the final number. **(b)** The gamma decay of $^{125}_{52}\text{Te}^*$ will be described by the gamma decay process: $^{A}_{Z}\text{P}^* \rightarrow {}^{A}_{Z}\text{P} + \gamma$.

Solution (a) According to Equation 31.5, the fraction of the sample that remains will be given by

$$\frac{N}{N_0} = e^{-\lambda t}$$

In order to find this ratio, we need the value of the decay constant λ, which is related to the half-life by Equation 31.6:

$$\lambda = \frac{0.693}{T_{1/2}} = \frac{0.693}{59.5 \text{ days}} = 1.16 \times 10^{-2} \text{ days}^{-1}$$

We can now calculate the fraction of $^{125}_{53}\text{I}$ atoms that remains after $t = 48.0$ hours $= 2.00$ days:

$$\frac{N}{N_0} = e^{-\lambda t} = e^{-(1.16 \times 10^{-2} \text{ days}^{-1})(2.00 \text{ days})} = \boxed{0.977}$$

Thus, 97.7% of the radioactive sample remains. **(b)** When $^{125}_{52}\text{Te}^*$ undergoes gamma decay, it will be described by the following decay process:

$$\boxed{^{125}_{52}\text{Te}^* \rightarrow {}^{125}_{52}\text{Te} + \gamma}$$

During gamma decay, no transmutation takes place. A gamma ray is produced, and the parent nucleus returns to a stable state.

EXAMPLE 10 | BIO The Physics of ^{90}Sr in the Bones

Calcium is an essential element in the human body for maintaining proper bone density. Strontium is chemically very similar to calcium, and 99% of strontium in the body is located in the bones. This typically does not cause health problems, if the strontium in the body is a non-radioactive stable isotope. However, if a person ingests a radioactive isotope of strontium, like ^{90}Sr, for example, then the isotope will settle in the bones and deliver harmful radiation to the nearby bone and marrow cells. This can lead to bone cancer and leukemia (**Figure 31.15**). If ^{90}Sr undergoes β^- decay with a half-life of 29.1 years, what percentage of the initial amount is still in the body after 50.0 years?

Reasoning We can use Equation 31.5 to calculate the ratio of remaining ^{90}Sr atoms N to the original number N_0. To find this ratio, we will need the decay constant λ for ^{90}Sr, which is related to its half-life $T_{1/2}$ by Equation 31.6.

Solution Beginning with Equation 31.5, we have:

$$N = N_0 e^{-\lambda t} \quad \text{or} \quad \frac{N}{N_0} = e^{-\lambda t}$$

To find the decay constant, we use Equation 31.6:

$$\lambda = \frac{0.693}{T_{1/2}} = \frac{0.693}{29.1 \text{ yr}} = 0.0238 \text{ yr}^{-1}$$

Substituting this result into the previous expression, we calculate the percentage of remaining radioactive Sr nuclei as follows:

$$\frac{N}{N_0} = e^{-(0.0238 \text{ yr}^{-1})(50.0 \text{ yr})} = 0.304 = \boxed{30.4\%}$$

The half-life of ^{90}Sr is a significant fraction of a typical lifespan. Thus, even after 50 years, there is a considerable percentage of atoms remaining.

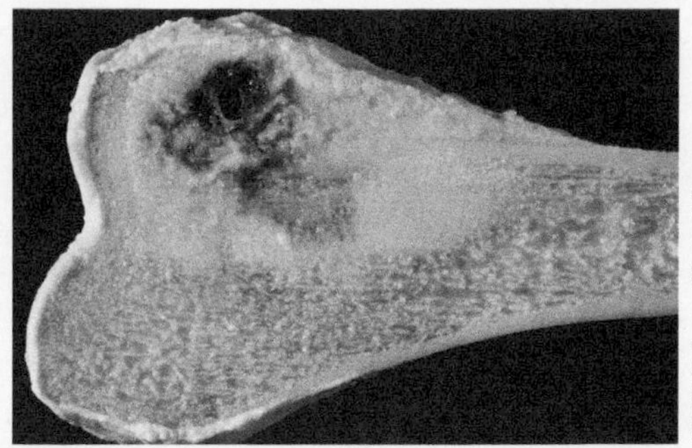

FIGURE 31.15 A cross-section of a human femur bone showing a malignant tumor known as osteosarcoma. This is an aggressive form of cancer that can result from ^{90}Sr replacing Ca in the bone.

Check Your Understanding

(*The answers are given at the end of the book.*)

9. The thallium $^{208}_{81}Tl$ nucleus is radioactive, with a half-life of 3.053 min. At a given instant, the activity of a certain sample of thallium is 2400 Bq. Using the concept of a half-life, and without doing any written calculations, determine whether the activity 9 minutes later is **(a)** a little less than $\frac{1}{8}$(2400 Bq) = 300 Bq, **(b)** a little more than $\frac{1}{8}$(2400 Bq) = 300 Bq, **(c)** a little less than $\frac{1}{3}$(2400 Bq) = 800 Bq, or **(d)** a little more than $\frac{1}{3}$(2400 Bq) = 800 Bq.

10. The half-life of indium $^{115}_{49}In$ is 4.41×10^{14} yr. Thus, one-half of the nuclei in a sample of this isotope will decay in this time, which is very long. Is it possible for any single nucleus in the sample to decay after only one second?

11. Could two different samples of the same radioactive element have different activities?

31.7 Radioactive Dating

THE PHYSICS OF . . . radioactive dating. One important application of radioactivity is the determination of the age of archaeological or geological samples as in the case of the mummified remains of Queen Hatshepsut (see **Figure 31.16**). If an object contains radioactive nuclei when it is formed, then the decay of these nuclei marks the passage of time like a clock, half of the nuclei disintegrating during each half-life. If the half-life is known, a measurement of the number of nuclei present today relative to the number present initially can give the age of the sample. According to Equation 31.4, the activity of a sample is proportional to the number of radioactive nuclei, so one way to obtain the age is to compare present activity with initial activity. A more accurate way is to determine the present number of radioactive nuclei with the aid of a mass spectrometer.

The present activity of a sample can be measured, but how is it possible to know what the original activity was, perhaps thousands of years ago? Radioactive dating methods entail certain assumptions that make it possible to estimate the original activity. For instance, the radiocarbon technique utilizes the $^{14}_{6}C$ isotope of carbon, which undergoes β^- decay with a half-life of 5730 yr. This isotope is currently present in the earth's atmosphere at an equilibrium concentration of about one atom for every 8.3×10^{11} atoms of normal carbon $^{12}_{6}C$. It is often assumed* that this value has remained constant over the years because $^{14}_{6}C$ is created when cosmic rays interact with the earth's upper atmosphere, a production method that offsets the loss via β^- decay. Moreover, nearly all living organisms ingest the equilibrium concentration of $^{14}_{6}C$. However, once an organism dies, metabolism no longer sustains the input of $^{14}_{6}C$, and β^- decay causes half of the $^{14}_{6}C$ nuclei to disintegrate every 5730 years. Example 11 illustrates how to determine the $^{14}_{6}C$ activity of one gram of carbon in a living organism.

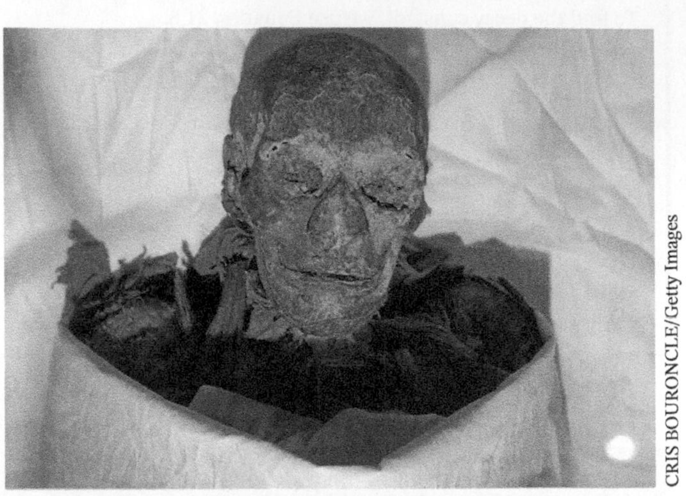

FIGURE 31.16 The mummified remains of Queen Hatshepsut, who ruled ancient Egypt from 1479 to 1458 BC. Radioactive dating is one of the techniques used to determine the age of such remains.

CRIS BOURONCLE/Getty Images

EXAMPLE 11 | $^{14}_{6}C$ Activity per Gram of Carbon in a Living Organism

(a) Determine the number of carbon $^{14}_{6}C$ atoms present for every gram of carbon $^{12}_{6}C$ in a living organism. Find **(b)** the decay constant and **(c)** the activity of this sample.

Reasoning The total number of carbon $^{12}_{6}C$ atoms in one gram of carbon $^{12}_{6}C$ is equal to the corresponding number of moles times Avogadro's number (see Section 14.1). Since there is only one $^{14}_{6}C$ atom for every 8.3×10^{11} atoms of $^{12}_{6}C$, the number of $^{12}_{6}C$ atoms is equal to the total number of $^{12}_{6}C$ atoms divided by 8.3×10^{11}. The decay constant λ for $^{14}_{6}C$ is $\lambda = 0.693/T_{1/2}$, where $T_{1/2}$ is the half-life. The activity is equal to the magnitude of $\Delta N/\Delta t$, which is equal to the decay constant times the number of $^{14}_{6}C$ atoms present, according to Equation 31.4.

Solution **(a)** One gram of carbon $^{12}_{6}C$ (atomic mass = 12 u) is equivalent to 1.0/12 mol. Since Avogadro's number is 6.02×10^{23} atoms/mol and since there is one $^{14}_{6}C$ atom for every 8.3×10^{11} atoms of $^{12}_{6}C$, the number of $^{14}_{6}C$ atoms is

$$\begin{array}{l}\text{Number of } ^{14}_{6}\text{C atoms for every}\\ \text{1.0 grams of carbon } ^{12}_{6}\text{C}\end{array} = \left(\frac{1.0}{12}\text{ mol}\right)\left(6.02 \times 10^{23}\ \frac{\text{atoms}}{\text{mol}}\right)\left(\frac{1}{8.3 \times 10^{11}}\right) = \boxed{6.0 \times 10^{10}\text{ atoms}}$$

(b) Since the half-life of $^{14}_{6}C$ is 5730 yr (1.81×10^{11} s), the decay constant is

$$\lambda = \frac{0.693}{T_{1/2}} = \frac{0.693}{1.81 \times 10^{11}\text{ s}} = \boxed{3.83 \times 10^{-12}\text{ s}^{-1}} \quad \textbf{(31.6)}$$

(c) Equation 31.4 indicates that $\Delta N/\Delta t = -\lambda N$, so the magnitude of $\Delta N/\Delta t$ is λN.

$$\begin{array}{l}\text{Activity of } ^{14}_{6}\text{C for}\\ \text{every 1.0 gram of}\\ \text{carbon } ^{12}_{6}\text{C in a}\\ \text{living organism}\end{array} = \lambda N = (3.83 \times 10^{-12}\text{ s}^{-1})(6.0 \times 10^{10}\text{ atoms}) = \boxed{0.23\text{ Bq}}$$

An organism that lived thousands of years ago presumably had an activity of about 0.23 Bq per gram of carbon. When the organism died, the activity began decreasing. From a sample of the remains, the current activity per gram of carbon can be measured and compared to the value of 0.23 Bq to determine the time that has transpired since death. This procedure is illustrated in Example 12.

*The assumption that the $^{14}_{6}C$ concentration has always been at its present equilibrium value has been evaluated by comparing $^{14}_{6}C$ ages with ages determined by counting tree rings. More recently, ages determined using the radioactive decay of uranium $^{238}_{92}U$ have been used for comparison. These comparisons indicate that the equilibrium value of the $^{14}_{6}C$ concentration has indeed remained constant for the past 1000 years. However, from there back about 30 000 years, it appears that the $^{14}_{6}C$ concentration in the atmosphere was larger than its present value by up to 40%. As a first approximation we ignore such discrepancies.

Analyzing Multiple-Concept Problems

EXAMPLE 12 | The Ice Man

On September 19, 1991, German tourists in the Italian Alps found a Stone-Age traveler, later dubbed the Ice Man, whose body had become trapped in a glacier. **Figure 31.17** shows the well-preserved remains, which were dated using the radiocarbon method. Material found with the body had a $^{14}_{6}\text{C}$ activity of about 0.121 Bq per gram of carbon. Find the age of the Ice Man's remains.

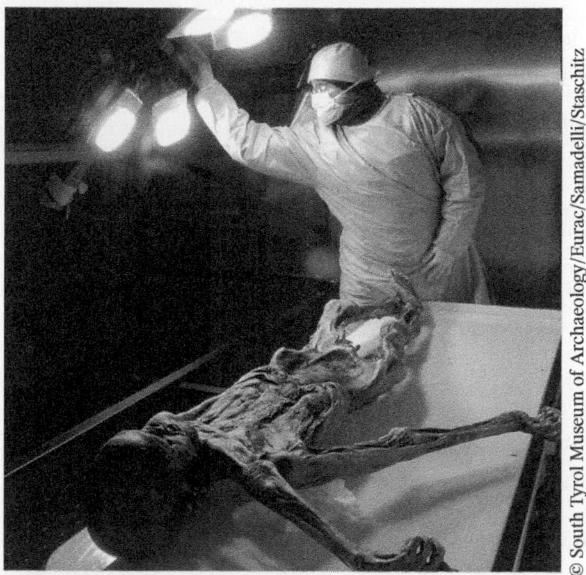

© South Tyrol Museum of Archaeology/Eurac/Samadelli/Staschitz

FIGURE 31.17 The frozen remains of the Ice Man or "Oetzi," as he also is called, were discovered in the ice of a glacier in the Italian Alps in 1991. Radiocarbon dating has revealed his age.

Reasoning In the radiocarbon method, the number of radioactive nuclei remaining at a given instant is related to the number present initially, the time that has passed since the Ice Man died, and the decay constant for $^{14}_{6}\text{C}$. Thus, to determine the age of the remains, we will need information about the number of nuclei present when the body was discovered and the number present initially, which is related to the activity of the material found with the body and the initial activity. To determine the age, we will also need the decay constant, obtainable from the half-life of $^{14}_{6}\text{C}$.

Knowns and Unknowns We have the following data:

Description	Symbol	Value	Comment
Explicit Data			
Activity of material found with body	A	0.121 Bq	This is the activity per gram of carbon.
Implicit Data			
Half-life of $^{14}_{6}\text{C}$	$T_{1/2}$	5730 yr	The radiocarbon dating method is specified.
Initial activity of material found with body	A_0	0.23 Bq	This activity is assumed for one gram of carbon in a living organism.
Unknown Variable			
Age of Ice Man's remains	t	?	

Modeling the Problem

STEP 1 Radioactive Decay The number N of radioactive nuclei present at a time t is

$$N = N_0 e^{-\lambda t} \tag{31.5}$$

where N_0 is the number present initially at $t = 0$ s and λ is the decay constant for $^{14}_{6}\text{C}$. Rearranging terms gives

$$\frac{N}{N_0} = e^{-\lambda t}$$

Taking the natural logarithm of both sides of this result (see Appendix D), we find that

$$\ln\left(\frac{N}{N_0}\right) = -\lambda t$$

Solving for t shows that the age of the Ice Man's remains is given by Equation 1 at the right. To use this result, we need information about the ratio N/N_0 and λ. We deal with N/N_0 in Step 2 and with λ in Step 3.

$$t = -\left(\frac{1}{\lambda}\right) \ln\left(\frac{N}{N_0}\right) \tag{1}$$

? ?

STEP 2 Activity The activity A is the number of disintegrations per second, or $\left|\frac{\Delta N}{\Delta t}\right|$, where ΔN is the number of disintegrations that occur in the time interval Δt.

Noting that $\frac{\Delta N}{\Delta t} = -\lambda N$ according to Equation 31.4, we find for the activity A that

$$A = \left|\frac{\Delta N}{\Delta t}\right| = |-\lambda N| = \lambda N$$

Using this expression, we have that

$$\boxed{\frac{N}{N_0} = \frac{\lambda N}{\lambda N_0} = \frac{A}{A_0}}$$

The substitution of this result into Equation 1 is shown at the right. We turn now to Step 3, in order to evaluate the decay constant λ.

$$t = -\left(\frac{1}{\lambda}\right)\ln\left(\frac{N}{N_0}\right) \quad (1)$$

$$\boxed{?} \qquad \boxed{\frac{N}{N_0} = \frac{A}{A_0}}$$

STEP 3 Decay Constant The decay constant is related to the half-life $T_{1/2}$ according to

$$\boxed{\lambda = \frac{0.693}{T_{1/2}}} \quad \textbf{(31.6)}$$

which we substitute into Equation 1, as shown at the right.

$$t = -\left(\frac{1}{\lambda}\right)\ln\left(\frac{N}{N_0}\right) \quad (1)$$

$$\boxed{\lambda = \frac{0.693}{T_{1/2}}} \qquad \boxed{\frac{N}{N_0} = \frac{A}{A_0}}$$

Solution Combining the results of each step algebraically, we find that

$$\underset{\text{STEP 1}}{t = -\left(\frac{1}{\lambda}\right)\ln\left(\frac{N}{N_0}\right)} = \underset{\text{STEP 2}}{-\left(\frac{1}{\lambda}\right)\ln\left(\frac{A}{A_0}\right)} = \underset{\text{STEP 3}}{-\left(\frac{1}{0.693/T_{1/2}}\right)\ln\left(\frac{A}{A_0}\right)}$$

$$t = -\left(\frac{1}{\lambda}\right)\ln\left(\frac{N}{N_0}\right) \quad (1)$$

$$\boxed{\lambda = \frac{0.693}{T_{1/2}}} \qquad \boxed{\frac{N}{N_0} = \frac{A}{A_0}}$$

This result reveals that the age of the Ice Man's remains is

$$t = -\left(\frac{T_{1/2}}{0.693}\right)\ln\left(\frac{A}{A_0}\right) = -\left(\frac{5730\text{ yr}}{0.693}\right)\ln\left(\frac{0.121\text{ Bq}}{0.23\text{ Bq}}\right) = \boxed{5300\text{ yr}}$$

Note that this solution implies for the activity that

$$A = A_0 e^{-\lambda t}$$

This can be seen by combining the result from Step 2 ($N/N_0 = A/A_0$) with Equation 31.5 ($N = N_0 e^{-\lambda t}$).

Related Homework: Problems 38, 40

Radiocarbon dating is not the only radioactive dating method. For example, other methods utilize uranium $^{238}_{92}\text{U}$, potassium $^{40}_{19}\text{K}$, and lead $^{210}_{82}\text{Pb}$. For such methods to be useful, the half-life of the radioactive species must be neither too short nor too long relative to the age of the sample to be dated, as Conceptual Example 13 discusses.

CONCEPTUAL EXAMPLE 13 | Dating a Bottle of Wine

A bottle of red wine is thought to have been sealed about 5 years ago. The wine contains a number of different atoms, including carbon, oxygen, and hydrogen. Each of these has a radioactive isotope. The radioactive isotope of carbon is the familiar $^{14}_{6}\text{C}$, with a half-life of 5730 yr. The radioactive isotope of oxygen is $^{15}_{8}\text{O}$ and has a half-life of 122.2 s. The radioactive isotope of hydrogen, called tritium, is $^{3}_{1}\text{H}$; its half-life is 12.33 yr. The activity of each of these isotopes is known at the time the bottle was sealed. However, only one of the isotopes is useful for determining the age of the wine accurately from a measurement of its current activity. Which is it? **(a)** $^{14}_{6}\text{C}$ **(b)** $^{15}_{8}\text{O}$ **(c)** $^{3}_{1}\text{H}$

Reasoning To find the age of the wine, it is necessary to determine the ratio of the current activity A to the initial activity A_0 (see Example 11). If the age of the sample is very small relative to the half-life of the nuclei, relatively few of the nuclei would have decayed during the wine's life, and the measured activity would have changed little from its initial value ($A \approx A_0$). To obtain an accurate age from such a small change would require prohibitively precise measurements. On the other hand, if the age of the sample is many times greater than the half-life of the nuclei, virtually all of the nuclei would have decayed, and the current activity would be so small($A \approx 0$) that it would be virtually impossible to detect.

Answer (a) is incorrect. The expected age of the wine is about 5 years. This period is only a tiny fraction of the 5730-yr half-life of $^{14}_{6}\text{C}$. As a result, relatively few of the $^{14}_{6}\text{C}$ nuclei would have decayed during the wine's life, and the current activity would be nearly the same as the initial activity ($A \approx A_0$), thus requiring prohibitively precise measurements.

Answer (b) is incorrect. The $^{15}_{8}\text{O}$ isotope is not very useful either, because of its relatively short half-life of 122.2 s. During a 5-year period, so many half-lives of 122.2 s would have occurred that the current activity would be vanishingly small ($A \approx 0$) and undetectable.

Answer (c) is correct. The only remaining option is the $^{3}_{1}\text{H}$ isotope. The expected age of 5 yr is long enough relative to the half-life of 12.33 yr that a measurable change in activity will have occurred, but not so long that the current activity will have completely vanished for all practical purposes.

Related Homework: Check Your Understanding Question 14

Check Your Understanding

(*The answers are given at the end of the book.*)

12. To which one or more of the following objects, each about 1000 yr old, can the radiocarbon dating technique *not* be applied? **(a)** A wooden box **(b)** A gold statue **(c)** Some well-preserved animal fur

13. Suppose there were a greater number of carbon $^{14}_{6}C$ atoms in a plant living 5000 yr ago than is currently believed. When the seeds of this plant are tested using radiocarbon dating, is the age obtained too small or too large compared to the true age?

14. Review Conceptual Example 13 as an aid in answering this question. Tritium is an isotope of hydrogen and undergoes β^- decay with a half-life of 12.33 yr. Like carbon $^{14}_{6}C$, tritium is produced in the atmosphere because of cosmic rays and can be used in a radioactive dating technique. Can tritium dating be used to determine a reliable date for a sample that is about 700 yr old?

31.8 Radioactive Decay Series

When an unstable parent nucleus decays, the resulting daughter nucleus is sometimes also unstable. If so, the daughter then decays and produces its own daughter, and so on, until a completely stable nucleus is produced. This sequential decay of one nucleus after another is called a **radioactive decay series.** Examples 4–6 discuss the first two steps of a series that begins with uranium $^{238}_{92}U$:

$$\underset{\text{Uranium}}{^{238}_{92}\text{U}} \longrightarrow \underset{\text{Thorium}}{^{234}_{90}\text{Th}} + ^{4}_{2}\text{He}$$

$$^{234}_{90}\text{Th} \longrightarrow \underset{\text{Protactinium}}{^{234}_{91}\text{Pa}} + ^{0}_{-1}\text{e}$$

Furthermore, Examples 7 and 8 deal with radon $^{222}_{86}Rn$, which is formed down the line in the $^{238}_{92}U$ radioactive decay series. **Figure 31.18** shows the entire series. At several points,

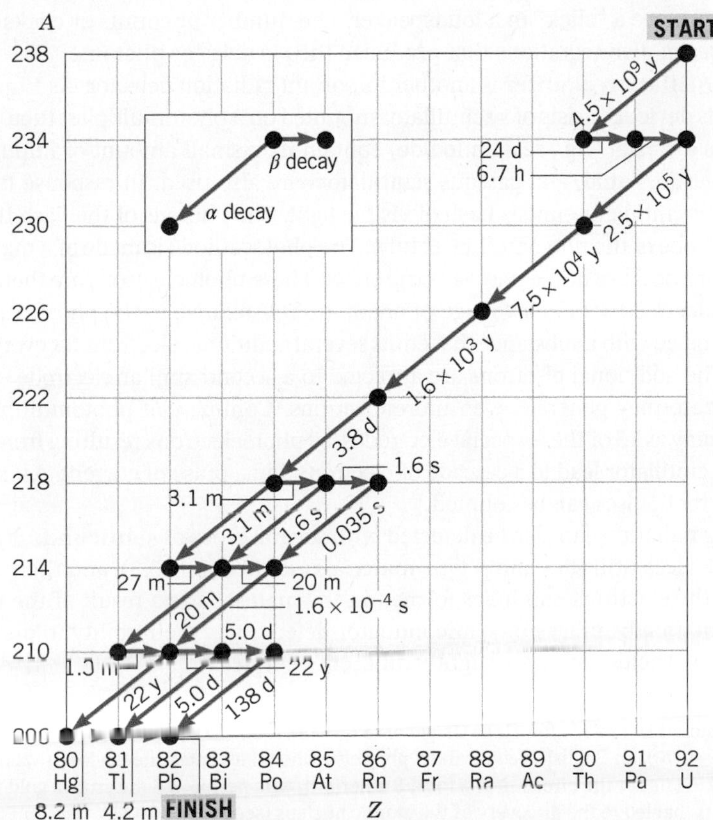

FIGURE 31.18 The radioactive decay series that begins with uranium $^{238}_{92}U$ and ends with lead $^{206}_{82}Pb$. Half-lives are given in seconds (s), minutes (m), hours (h), days (d), or years (y). The inset in the upper left identifies the type of decay that each nucleus undergoes.

branches occur because more than one kind of decay is possible for an intermediate species. Ultimately, however, the series ends with lead ${}^{206}_{82}\text{Pb}$, which is stable.

The ${}^{238}_{92}\text{U}$ series and other such series are the only sources of some of the radioactive elements found in nature. Radium ${}^{226}_{88}\text{Ra}$, for instance, has a half-life of 1600 yr, which is short enough that all the ${}^{226}_{88}\text{Ra}$ created when the earth was formed billions of years ago has now disappeared. The ${}^{238}_{92}\text{U}$ series provides a continuing supply of ${}^{226}_{88}\text{Ra}$, however.

Check Your Understanding

(The answer is given at the end of the book.)

15. Because of radioactive decay, one element can be transmuted into another. Thus, a container of uranium ${}^{238}_{92}\text{U}$ ultimately becomes a container of lead ${}^{206}_{82}\text{Pb}$, as **Figure 31.18** indicates. Roughly, how long does it take for ${}^{238}_{92}\text{U}$ to transmute entirely into ${}^{206}_{82}\text{Pb}$? **(a)** Several decades **(b)** Several centuries **(c)** Thousands of years **(d)** Millions of years **(e)** Billions of years

31.9 Radiation Detectors

THE PHYSICS OF . . . radiation detectors. There are a number of devices that can be used to detect the particles and photons (γ rays) emitted when a radioactive nucleus decays. Such devices detect the ionization that these particles and photons cause as they pass through matter.

The most familiar detector is the **Geiger counter***, which **Figure 31.19** illustrates. The Geiger counter consists of a gas-filled metal cylinder. The α, β, or γ rays enter the cylinder through a thin window at one end. γ rays can also penetrate directly through the metal. A wire electrode runs along the center of the tube and is kept at a high positive voltage (1000–3000 V) relative to the outer cylinder. When a high-energy particle or photon enters the cylinder, it collides with and ionizes a gas molecule. The electron produced from the gas molecule accelerates toward the positive wire, ionizing other molecules in its path. Additional electrons are formed, and an avalanche of electrons rushes toward the wire, leading to a pulse of current through the resistor R. This pulse can be counted or made to produce a "click" in a loudspeaker. The number of counts or clicks is related to the number of disintegrations that produced the particles or photons.

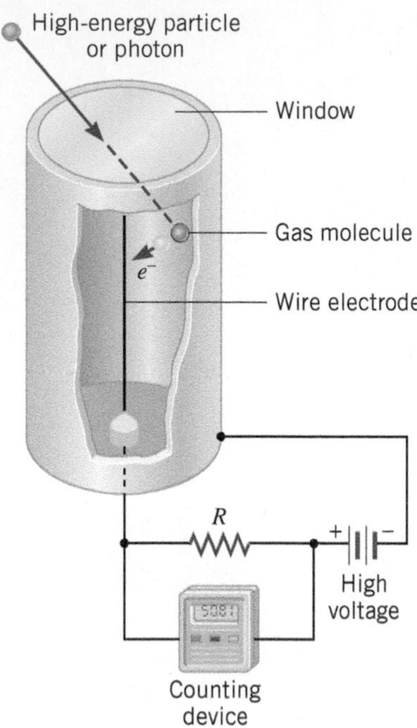

FIGURE 31.19 A Geiger counter.

The **scintillation counter** is another important radiation detector. As **Figure 31.20** indicates, this device consists of a scintillator mounted on a photomultiplier tube. Often the scintillator is a crystal (e.g., cesium iodide) containing a small amount of impurity (thallium), but plastic, liquid, and gaseous scintillators are also used. In response to ionizing radiation, the scintillator emits a flash of visible light. The photons of the flash then strike the photocathode of the photomultiplier tube. The photocathode is made of a material that emits electrons because of the photoelectric effect. These photoelectrons are then attracted to a special electrode kept at a voltage of about +100 V relative to the photocathode. The electrode is coated with a substance that emits several additional electrons for every electron striking it. The additional electrons are attracted to a second similar electrode (voltage = +200 V), where they generate even more electrons. Commercial photomultiplier tubes contain as many as 15 of these special electrodes, so photoelectrons resulting from the light flash of the scintillator lead to a cascade of electrons and a pulse of current. As in a Geiger tube, the current pulses can be counted.

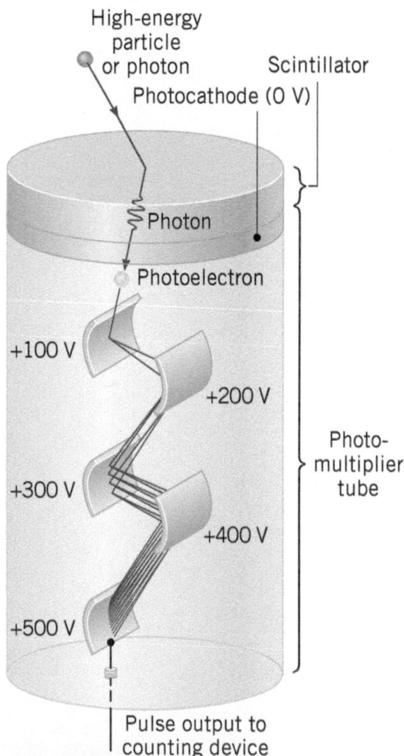

FIGURE 31.20 A scintillation counter.

Ionizing radiation can also be detected with several types of **semiconductor detectors.** Such devices utilize n- and p-type materials (see Section 23.5), and their operation depends on the electrons and holes formed in the materials as a result of the radiation. One of the main advantages of semiconductor detectors is their ability to discriminate between two particles with only slightly different energies.

*The Geiger counter is named after Hans Geiger (1882–1945), a German physicist, along with Sir Ernest Marsden (1889–1970), an English-New Zealand physicist, who co-invented the device. It was Geiger and Marsden, working under the direction of Ernest Rutherford, who performed the crucial gold foil scattering experiments that led to the discovery of the atomic nucleus (see Section 30.1).

A number of instruments provide a pictorial representation of the path that high-energy particles follow after they are emitted from unstable nuclei. In a **cloud chamber,** a gas is cooled just to the point at which it will condense into droplets, provided nucleating agents are available on which the droplets can form. When a high-energy particle, such as an α particle or a β particle, passes through the gas, the ions it leaves behind serve as nucleating agents, and droplets form along the path of the particle. A **bubble chamber** works in a similar fashion, except that it contains a liquid that is just at the point of boiling. Tiny bubbles form along the trail of a high-energy particle passing through the liquid. Paths revealed in a cloud or bubble chamber can be photographed to provide a permanent record of the event. **Figure 31.21** shows a photograph of tracks in a cloud chamber. A **photographic emulsion** also can be used directly to produce a record of the path taken by a particle of ionizing radiation. Ions formed as the particle passes through the emulsion cause silver to be deposited along the track when the emulsion is developed.

Photo Researchers / Science History Images / Alamy Stock Photo

FIGURE 31.21 A photograph showing particle tracks in a cloud chamber.

Concept Summary

31.1 Nuclear Structure The nucleus of an atom consists of protons and neutrons, which are collectively referred to as nucleons. A neutron is an electrically neutral particle whose mass is slightly larger than that of the proton. The atomic number Z is the number of protons in the nucleus. The atomic mass number A (or nucleon number) is the total number of protons and neutrons in the nucleus and is given by Equation 31.1, where N is the number of neutrons. For an element whose chemical symbol is X, the symbol for the nucleus is ${}^{A}_{Z}\text{X}$. Nuclei that contain the same number of protons, but a different number of neutrons, are called isotopes. The approximate radius (in meters) of a nucleus is given by Equation 31.2.

$$A = Z + N \tag{31.1}$$

$$r \approx (1.2 \times 10^{-15}\ \text{m})A^{1/3} \tag{31.2}$$

31.2 The Strong Nuclear Force and the Stability of the Nucleus The strong nuclear force is the force of attraction between nucleons (protons and neutrons) and is one of the three fundamental forces of nature. This force balances the electrostatic force of repulsion between protons and holds the nucleus together. The strong nuclear force has a very short range of action and is almost independent of electric charge.

31.3 The Mass Defect of the Nucleus and Nuclear Binding Energy The binding energy of a nucleus is the energy required to separate the nucleus into its constituent protons and neutrons. The binding energy is given by Equation 31.3, where Δm is the mass defect of the nucleus and c is the speed of light in a vacuum. The mass defect is the amount by which the sum of the individual masses of the protons and neutrons exceeds the mass of the intact nucleus.

When specifying nuclear masses, it is customary to use the atomic mass unit (u). One atomic mass unit has a mass of 1.6605×10^{-27} kg and is equivalent to an energy of 931.5 MeV.

$$\text{Binding energy} = (\Delta m)c^2 \tag{31.3}$$

31.4 Radioactivity Unstable nuclei spontaneously decay by breaking apart or rearranging their internal structure in a process called radioactivity. Naturally occurring radioactivity produces α, β, and γ rays. α rays consist of positively charged particles, each particle being the ${}^{4}_{2}\text{He}$ nucleus of helium. The general form for α decay is

$$\underbrace{{}^{A}_{Z}\text{P}}_{\substack{\text{Parent}\\\text{nucleus}}} \longrightarrow \underbrace{{}^{A-4}_{Z-2}\text{D}}_{\substack{\text{Daughter}\\\text{nucleus}}} + \underbrace{{}^{4}_{2}\text{He}}_{\substack{\alpha\text{ particle}\\\text{(helium nucleus)}}}$$

The most common kind of β ray consists of negatively charged particles, or β^- particles, which are electrons. The general form for β^- decay is

$$\underbrace{{}^{A}_{Z}\text{P}}_{\substack{\text{Parent}\\\text{nucleus}}} \longrightarrow \underbrace{{}_{Z+1}^{A}\text{D}}_{\substack{\text{Daughter}\\\text{nucleus}}} + \underbrace{{}^{0}_{-1}\text{e}}_{\substack{\beta^-\text{particle}\\\text{(electron)}}}$$

β^+ decay produces another kind of β ray, which consists of positively charged particles, or β^+ particles. A β^+ particle, also called a positron, has the same mass as an electron, but carries a charge of $+e$ instead of $-e$.

If a radioactive parent nucleus disintegrates into a daughter nucleus that has a different atomic number, as occurs in α and β decay, one element has been converted into another element, the conversion being referred to as a transmutation.

γ rays are high-energy photons emitted by a radioactive nucleus. The general form for γ decay is

$$\underbrace{{}^{A}_{Z}\text{P}^*}_{\substack{\text{Excited}\\\text{energy state}}} \longrightarrow \underbrace{{}^{A}_{Z}\text{P}}_{\substack{\text{Lower}\\\text{energy state}}} + \underbrace{\gamma}_{\gamma\text{ ray}}$$

γ decay does not cause a transmutation of one element into another.

31.5 The Neutrino The neutrino is an electrically neutral particle that is emitted along with β particles and has a mass that is much, much smaller than the mass of an electron.

31.6 Radioactive Decay and Activity The half-life of a radioactive isotope is the time required for one-half of the nuclei present to disintegrate or decay. The activity is the number of disintegrations per second that occur. Activity is the magnitude of $\Delta N/\Delta t$, where ΔN is the change in the number N of radioactive nuclei and Δt is the time interval during which the change occurs. In other words, activity is $|\Delta N/\Delta t|$. The SI unit for activity is the becquerel (Bq), one becquerel being one disintegration per second. Activity is sometimes also measured in a unit called the curie (Ci); 1 Ci = 3.70×10^{10} Bq.

$$\frac{\Delta N}{\Delta t} = -\lambda N \tag{31.4}$$

Radioactive decay obeys Equation 31.4, where λ is the decay constant. This equation can be solved by the methods of integral calculus to show that N is given by Equation 31.5, where N_0 is the original number of nuclei. The decay constant λ is related to the half-life $T_{1/2}$ according to Equation 31.6.

$$N = N_0 e^{-\lambda t} \tag{31.5}$$

$$\lambda = \frac{0.693}{T_{1/2}} \tag{31.6}$$

31.7 Radioactive Dating If an object contained radioactive nuclei when it was formed, then the decay of these nuclei can be used to determine the age of the object. One way to obtain the age is to relate the present activity A of an object to its initial activity A_0, according to Equation 1, where λ is the decay constant and t is the age of the object. For radiocarbon dating that uses the ${}^{14}_{6}\text{C}$ isotope of carbon, the initial activity is often assumed to be $A_0 = 0.23$ Bq.

31.8 Radioactive Decay Series The sequential decay of one nucleus after another is called a radioactive decay series. A decay series starts with a radioactive nucleus and ends with a completely stable nucleus. **Figure 31.18** illustrates one such series that begins with uranium ${}^{238}_{92}\text{U}$ and ends with lead ${}^{206}_{82}\text{Pb}$.

31.9 Radiation Detectors A number of devices are used to detect α and β particles as well as γ rays. These include the Geiger counter, the scintillation counter, semiconductor detectors, cloud and bubble chambers, and photographic emulsions.

Focus on Concepts

Online

Additional questions are available for assignment in WileyPLUS.

Section 31.1 Nuclear Structure

1. An indium (In) nucleus contains 49 protons and 66 neutrons. Which one of the following symbols describes this nucleus? (a) ${}^{49}_{115}\text{In}$ (b) ${}^{49}_{66}\text{In}$ (c) ${}^{115}_{66}\text{In}$ (d) ${}^{66}_{49}\text{In}$ (e) ${}^{115}_{49}\text{In}$

2. The notation for a particular nucleus is ${}^{85}_{37}\text{Rb}$. In an electrically neutral atom, how many electrons are in orbit about this nucleus? (a) 37 + 85 = 122 (b) 85 (c) 37 (d) 85 − 37 = 48 (e) The number of electrons cannot be determined from the notation.

Section 31.3 The Mass Defect of the Nucleus and Nuclear Binding Energy

3. Suppose that we lived in a hypothetical world in which the mass of each proton and each neutron were exactly 1 u. In this world, the atomic mass of copper ${}^{63}_{29}\text{Cu}$ is 62.5 u. What would be the mass defect for this nucleus? (a) 63 u (b) 29 u (c) 63 u − 29 u = 34 u (d) 0.5 u (e) 63 u + 29 u = 92 u

Section 31.4 Radioactivity

4. Which one or more of the three decay processes (α, β^-, or γ) results in a new element? (a) α and β^- (b) Only α (c) Only β^- (d) β^- and γ (e) Only γ

5. A nucleus can undergo α, β^-, or γ decay. For each type of decay, is the radius of the daughter nucleus greater than, less than, or about the same as the radius of the parent nucleus?

	α Decay	β^- Decay	γ Decay
(a)	Greater than	Greater than	About the same as
(b)	Greater than	Less than	About the same as
(c)	Less than	About the same as	Greater than
(d)	Less than	About the same as	About the same as
(e)	About the same as	Less than	Less than

Section 31.6 Radioactive Decay and Activity

6. Two samples contain different radioactive isotopes. Is it possible for these samples to have the same activity? (a) Yes, if they have the same number of nuclei but different half-lives. (b) Yes, if they have different numbers of nuclei and different half-lives. (c) Yes, if they have different numbers of nuclei but the same half-lives. (d) No, because they can have different half-lives. (e) No, because they can have different numbers of nuclei.

7. The drawing shows the activities of three radioactive samples. Rank the samples according to half-life, largest first. (a) 2, 3, 1 (b) 1, 2, 3 (c) 3, 2, 1 (d) 1, 3, 2 (e) 3, 1, 2

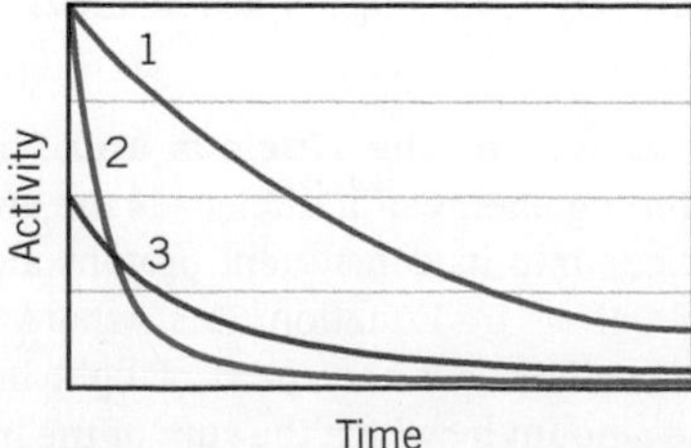

QUESTION 7

Section 31.7 Radioactive Dating

8. The bones from an animal found at an archaeological dig have a $^{14}_{6}C$ activity of 0.10 Bq per gram of carbon. The half-life of the radioactive isotope $^{14}_{6}C$ is 5730 yr. Which one of the following best describes the age of the bones? **(a)** It is less than 2000 years. **(b)** It is between 2000 and 3000 years. **(c)** It is between 3000 and 4000 years. **(d)** It is between 4000 and 5000 years. **(e)** It is more than 5000 years.

Problems

Online

Additional questions are available for assignment in WileyPLUS.

Note: *The data given for atomic masses in these problems include the mass of the electrons orbiting the nucleus of the electrically neutral atom.*

SSM Student Solutions Manual
MMH Problem-solving help
GO Guided Online Tutorial
V-HINT Video Hints
CHALK Chalkboard Videos
BIO Biomedical application
E Easy
M Medium
H Hard
WS Worksheet
T Team Problem

Section 31.1 Nuclear Structure, Section 31.2 The Strong Nuclear Force and the Stability of the Nucleus

1. E SSM For $^{208}_{82}Pb$ find **(a)** the net electrical charge of the nucleus, **(b)** the number of neutrons, **(c)** the number of nucleons, **(d)** the approximate radius of the nucleus, and **(e)** the nuclear density.

2. E A nucleus contains 18 protons and 22 neutrons. What is the radius of this nucleus?

3. E In each of the following cases, what element does the symbol X represent and how many neutrons are in the nucleus? Use the periodic table on the inside of the back cover of the textbook as needed. **(a)** $^{195}_{78}X$ **(b)** $^{32}_{16}X$ **(c)** $^{63}_{29}X$ **(d)** $^{11}_{5}X$ **(e)** $^{239}_{94}X$

4. E By what factor does the nucleon number of a nucleus have to increase in order for the nuclear radius to double?

5. E SSM In electrically neutral atoms, how many **(a)** protons are in the uranium $^{238}_{92}U$ nucleus, **(b)** neutrons are in the mercury $^{202}_{80}Hg$ nucleus, and **(c)** electrons are in orbit about the niobium $^{93}_{41}Nb$ nucleus?

6. E GO The largest stable nucleus has a nucleon number of 209, and the smallest has a nucleon number of 1. If each nucleus is assumed to be a sphere, what is the ratio (largest/smallest) of the surface areas of these spheres?

7. M GO CHALK The ratio r_X/r_T of the radius of an unknown nucleus $^{A}_{Z}X$ to the radius of a tritium nucleus $^{3}_{1}T$ is $\frac{r_X}{r_T} = 1.10$. Both nuclei contain the same number of neutrons. Identify the unknown nucleus in the form $^{A}_{Z}X$. Use the periodic table immediately before the back cover of this text.

8. M V-HINT CHALK An unknown nucleus contains 70 neutrons and has twice the volume of the nickel $^{60}_{28}Ni$ nucleus. Identify the unknown nucleus in the form $^{A}_{Z}X$. Use the periodic table immediately before the back cover of this text.

Section 31.3 The Mass Defect of the Nucleus and Nuclear Binding Energy

(*Note:* The atomic mass for hydrogen $^{1}_{1}H$ is 1.007 825 u; *this includes the mass of one electron.*)

9. E SSM Find the binding energy (in MeV) for lithium $^{7}_{3}Li$ (atomic mass = 7.016 003 u).

10. E The binding energy of a nucleus is 225.0 MeV. What is the mass defect of the nucleus in atomic mass units?

11. E Determine the mass defect (in atomic mass units) for **(a)** helium $^{3}_{2}He$, which has an atomic mass of 3.016 030 u, and **(b)** the isotope of hydrogen known as tritium $^{3}_{1}T$, which has an atomic mass of 3.016 050 u. **(c)** On the basis of your answers to parts (a) and (b), state which nucleus requires more energy to disassemble it into its separate and stationary constituent nucleons. Give your reasoning.

12. E GO A 245-kg boulder is dropped into a mine shaft that is 3.0×10^3 m deep. During the boulder's fall, the system consisting of the earth and the boulder loses a certain amount of gravitational potential energy. It would take an equal amount of energy to "free" the boulder from the shaft by raising it back to the top, so this can be considered the system's binding energy. **(a)** Determine the binding energy (in joules) of the earth–boulder system. **(b)** How much mass does the earth–boulder system lose when the boulder falls to the bottom of the shaft?

13. E CHALK For lead $^{206}_{82}Pb$ (atomic mass = 205.974 440 u) obtain **(a)** the mass defect in atomic mass units, **(b)** the binding energy (in MeV), and **(c)** the binding energy per nucleon (in MeV/nucleon).

14. M **(a)** Energy is required to separate a nucleus into its constituent nucleons, as **Interactive Figure 31.3** indicates; this energy is the *total* binding energy of the nucleus. In a similar way one can speak of the energy that binds a single nucleon to the remainder of the nucleus. For example, separating nitrogen $^{14}_{7}N$ into nitrogen $^{13}_{7}N$ and a neutron takes energy equal to the binding energy of the neutron, as shown below:

$$^{14}_{7}N + \text{Energy} \longrightarrow {}^{13}_{7}N + {}^{1}_{0}n$$

Find the energy (in MeV) that binds the neutron to the $^{14}_{7}N$ nucleus by considering the mass of $^{13}_{7}N$ (atomic mass = 13.005 738 u) and the mass of $^{1}_{0}n$ (atomic mass = 1.008 665 u), as compared to the mass of $^{14}_{7}N$ (atomic mass = 14.003 074 u). **(b)** Similarly, one can speak of the energy that binds a single proton to the $^{14}_{7}N$ nucleus:

$$^{14}_{7}N + \text{Energy} \longrightarrow {}^{13}_{6}C + {}^{1}_{1}H$$

Following the procedure outlined in part (a), determine the energy (in MeV) that binds the proton (atomic mass = 1.007 825 u) to the $^{14}_{7}N$ nucleus. The atomic mass of carbon $^{13}_{6}C$ is 13.003 355 u. **(c)** Which nucleon is more tightly bound, the neutron or the proton?

15. M SSM Two isotopes of a certain element have binding energies that differ by 5.03 MeV. The isotope with the larger binding energy contains one more neutron than the other isotope. Find the difference in atomic mass between the two isotopes.

16. M GO A copper penny has a mass of 3.0 g. Determine the energy (in MeV) that would be required to break all the copper nuclei into their constituent protons and neutrons. Ignore the energy that binds the electrons to the nucleus and the energy that binds one atom to another in the structure of the metal. For simplicity, assume that all the copper nuclei are $^{63}_{29}Cu$ (atomic mass = 62.939 598 u).

Section 31.4 Radioactivity

17. **E** **SSM** Write the β^+ decay process for each of the following nuclei, being careful to include Z and A and the proper chemical symbol for each daughter nucleus: (a) $^{18}_{9}F$ (b) $^{15}_{8}O$

18. **E** Write the β^- decay process for carbon $^{14}_{6}C$, including the chemical symbols as well as the values of Z and A for the parent and daughter nuclei and the β^- particle.

19. **E** **SSM** Osmium $^{191}_{76}Os$ (atomic mass = 190.960 920 u) is converted into iridium $^{191}_{77}Ir$ (atomic mass = 190.960 584 u) via β^- decay. What is the energy (in MeV) released in this process?

20. **E** Find the energy that is released when a nucleus of lead $^{211}_{82}Pb$ (atomic mass = 210.988 735 u) undergoes β^- decay to become bismuth $^{211}_{83}Bi$ (atomic mass = 210.987 255 u).

21. **E** Find the energy (in MeV) released when α decay converts radium $^{226}_{88}Ra$ (atomic mass = 226.025 40 u) into radon $^{222}_{86}Rn$ (atomic mass = 222.017 57 u). The atomic mass of an α particle is 4.002 603 u.

22. **E** **GO** Lead $^{207}_{82}Pb$ is a stable daughter nucleus that can result from either an α decay or a β^- decay. Write the decay processes, including the chemical symbols and values for Z and A of the parent nuclei, for (a) the α decay and (b) the β^- decay.

23. **E** In the form $^{A}_{Z}X$ identify the daughter nucleus that results when (a) plutonium $^{242}_{94}Pu$ undergoes α decay, (b) sodium $^{24}_{11}Na$ undergoes β^- decay, and (c) nitrogen $^{13}_{7}N$ undergoes β^+ decay.

24. **E** **V-HINT** When uranium $^{235}_{92}U$ decays, it emits (among other things) a γ ray that has a wavelength of 1.14×10^{-11} m. Determine the energy (in MeV) of this γ ray.

25. **M** **CHALK** **MMH** Polonium $^{210}_{84}Po$ (atomic mass = 209.982 848 u) undergoes α decay. Assuming that all the released energy is in the form of kinetic energy of the α particle (atomic mass = 4.002 603 u) and ignoring the recoil of the daughter nucleus (lead $^{206}_{82}Pb$, 205.974 440 u), find the speed of the α particle. Ignore relativistic effects.

26. **M** **GO** Radon $^{220}_{86}Rn$ produces a daughter nucleus that is radioactive. The daughter, in turn, produces its own radioactive daughter, and so on. This process continues until lead $^{208}_{82}Pb$ is reached. What are the total number N_α of α particles and the total number N_β of β^- particles that are generated in this series of radioactive decays?

27. **M** Review Conceptual Example 5 as background for this problem. The α decay of uranium $^{238}_{92}U$ produces thorium $^{234}_{90}Th$ (atomic mass = 234.0436 u). In Example 4, the energy released in this decay is determined to be 4.3 MeV. Determine how much of this energy is carried away by the recoiling $^{234}_{90}Th$ daughter nucleus and how much by the α particle (atomic mass = 4.002 603 u). Assume that the energy of each particle is kinetic energy, and ignore the small amount of energy carried away by the γ ray that is also emitted. In addition, ignore relativistic effects.

28. **H** **MMH** An isotope of beryllium (atomic mass = 7.017 u) emits a γ ray and recoils with a speed of 2.19×10^4 m/s. Assuming that the beryllium nucleus is stationary to begin with, find the wavelength of the γ ray.

29. **H** **SSM** Find the energy (in MeV) released when β^+ decay converts sodium $^{22}_{11}Na$ (atomic mass = 21.994 434 u) into neon $^{22}_{10}Ne$ (atomic mass = 21.991 383 u). Notice that the atomic mass for $^{22}_{11}Na$ includes the mass of 11 electrons, whereas the atomic mass for $^{22}_{10}Ne$ includes the mass of only 10 electrons.

Section 31.6 Radioactive Decay and Activity

30. **E** In 9.0 days the number of radioactive nuclei decreases to one-eighth the number present initially. What is the half-life (in days) of the material?

31. **E** The $^{32}_{15}P$ isotope of phosphorus has a half-life of 14.28 days. What is its decay constant in units of s^{-1}?

32. **E** **GO** Two radioactive waste products from nuclear reactors are strontium $^{90}_{38}Sr$ ($T_{1/2} = 29.1$ yr) and cesium $^{134}_{55}Cs$ ($T_{1/2} = 2.06$ yr). These two species are present initially in a ratio of $N_{0,Sr}/N_{0,Cs} = 7.80 \times 10^{-3}$. What is the ratio N_{Sr}/N_{Cs} fifteen years later?

33. **E** **SSM** The number of radioactive nuclei present at the start of an experiment is 4.60×10^{15}. The number present twenty days later is 8.14×10^{14}. What is the half-life (in days) of the nuclei?

34. **M** **GO** A one-gram sample of radium $^{224}_{88}Ra$ (atomic mass = 224.020 186 u, $T_{1/2} = 3.66$ days) contains 2.69×10^{21} nuclei and undergoes α decay to produce radon $^{220}_{86}Rn$ (atomic mass = 220.011 368 u). The atomic mass of an α particle is 4.002 603 u. The latent heat of fusion for water is 33.5×10^4 J/kg. With the energy released in 3.66 days, how many kilograms of ice could be melted at 0 °C?

35. **M** **GO** Outside the nucleus, the neutron itself is radioactive and decays into a proton, an electron, and an antineutrino. The half-life of a neutron (mass = 1.675×10^{-27} kg) outside the nucleus is 10.4 min. On average, over what distance (in meters) would a beam of 5.00-eV neutrons travel before the number of neutrons decreased to 75.0% of its initial value?

36. **M** **SSM** Two radioactive nuclei A and B are present in equal numbers to begin with. Three days later, there are three times as many A nuclei as there are B nuclei. The half-life of species B is 1.50 days. Find the half-life of species A.

Section 31.7 Radioactive Dating

37. **E** A sample has a $^{14}_{6}C$ activity of 0.0061 Bq per gram of carbon. (a) Find the age of the sample, assuming that the activity per gram of carbon in a living organism has been constant at a value of 0.23 Bq. (b) Evidence suggests that the value of 0.23 Bq might have been as much as 40% larger. Repeat part (a), taking into account this 40% increase.

38. **E** **SSM** Review Multiple-Concept Example 11 for help in approaching this problem. An archaeological specimen containing 9.2 g of carbon has an activity of 1.6 Bq. How old (in years) is the specimen?

39. **E** **CHALK** **GO** The half-life for the α decay of uranium $^{238}_{92}U$ is 4.47×10^9 yr. Determine the age (in years) of a rock specimen that contains 60.0% of its original number of $^{238}_{92}U$ atoms.

40. **E** **GO** Multiple-Concept Example 12 reviews most of the concepts that are needed to solve this problem. Material found with a mummy in the arid highlands of southern Peru has a $^{14}_{6}C$ activity per gram of carbon that is 78.5% of the activity present initially. How long ago (in years) did this individual die?

41. **M** **SSM** When any radioactive dating method is used, experimental error in the measurement of the sample's activity leads to error in the estimated age. In an application of the radiocarbon dating technique to certain fossils, an activity of 0.100 Bq per gram of carbon is measured to within an accuracy of ±10.0%. Find the age of the fossils and the maximum error (in years) in the value obtained. Assume that there is no error in the 5730-year half-life of $^{14}_{6}C$ nor in the value of 0.23 Bq per gram of carbon in a living organism.

Additional Problems

Online

42. E GO In a nucleus, each proton experiences a repulsive electrostatic force from each of the other protons. In a nucleus of gold ${}^{197}_{79}\text{Au}$, what is the magnitude of the least possible electrostatic force of repulsion that one proton can exert on another?

43. E V-HINT When a sample from a meteorite is analyzed, it is determined that 93.8% of the original mass of a certain radioactive isotope is still present. Based on this finding, the age of the meteorite is calculated to be 4.51×10^9 yr. What is the half-life (in yr) of the isotope used to date the meteorite?

44. E GO The β^- decay of phosphorus ${}^{32}_{15}\text{P}$ (atomic mass = 31.973 907 u) produces a daughter nucleus that is sulfur ${}^{32}_{16}\text{S}$ (atomic mass = 31.972 070 u), a β^- particle, and an antineutrino. The kinetic energy of the β^- particle is 0.90 MeV. Find the maximum possible energy (in MeV) that the antineutrino could carry away.

45. E Complete the following decay processes by stating what the symbol X represents (X = α, β^-, β^+, or γ):

(a) ${}^{211}_{82}\text{Pb} \rightarrow {}^{211}_{83}\text{Bi} + \text{X}$
(c) ${}^{231}_{90}\text{Th}^* \rightarrow {}^{231}_{90}\text{Th} + \text{X}$
(b) ${}^{11}_{6}\text{C} \rightarrow {}^{11}_{5}\text{B} + \text{X}$
(d) ${}^{210}_{84}\text{Po} \rightarrow {}^{206}_{82}\text{Pb} + \text{X}$

46. M V-HINT A sample of ore containing radioactive strontium ${}^{90}_{38}\text{Sr}$ has an activity of 6.0×10^5 Bq. The atomic mass of strontium is 89.908 u, and its half-life is 29.1 yr. How many grams of strontium are in the sample?

47. M V-HINT In a radioactive decay series similar to that shown in **Figure 31.18**, thorium ${}^{228}_{90}\text{Th}$ (atomic mass = 228.028 715 u) undergoes four successive α decays, producing a daughter nucleus. **(a)** Determine the symbol ${}^{A}_{Z}\text{X}$ for the nucleus produced by four successive α decays of ${}^{228}_{90}\text{Th}$. **(b)** What is the total amount of energy (in MeV) released in this series of α decays? The mass of the daughter nucleus can be obtained by using the result of part (a) and consulting Appendix F at the back of the book. The mass of a single α particle is 4.002 603 u.

48. M GO SSM What is the wavelength (in vacuum) of the 0.186-MeV γ-ray photon emitted by ${}^{226}_{88}\text{Ra}$?

Physics in Biology, Medicine, and Sports

49. E BIO 31.6 Review Example 10 as background for this problem. Strontium ${}^{90}_{38}\text{Sr}$ has a half-life of 29.1 yr. It is chemically similar to calcium, enters the body through the food chain, and collects in the bones. Consequently, ${}^{90}_{38}\text{Sr}$ is a particularly serious health hazard. How long (in years) will it take for 99.9900% of the ${}^{90}_{38}\text{Sr}$ released in a nuclear reactor accident to disappear?

50. E BIO 31.6 Iodine ${}^{131}_{53}\text{I}$ is used in diagnostic and therapeutic techniques in the treatment of thyroid disorders. This isotope has a half-life of 8.04 days. What percentage of an initial sample of ${}^{131}_{53}\text{I}$ remains after 30.0 days?

51. M BIO 31.6 A device used in radiation therapy for cancer contains 0.50 g of cobalt ${}^{60}_{27}\text{Co}$ (59.933 819 u). The half-life of ${}^{60}_{27}\text{Co}$ is 5.27 yr. Determine the activity of the radioactive material.

52. M BIO V-HINT 31.6 The isotope ${}^{198}_{79}\text{Au}$ (atomic mass = 197.968 u) of gold has a half-life of 2.69 days and is used in cancer therapy. What mass (in grams) of this isotope is required to produce an activity of 315 Ci?

53. M BIO 31.6 Positron emission tomography, or PET, is a medical imaging technique that uses radionuclides to visualize and measure biological processes inside the body. A radioactive isotope, called a tracer, is injected into the body and decays by positron emission. Which isotope is chosen will depend on what is being imaged. Once the tracer is in its proper location, the positrons it emits will combine with electrons in the surrounding tissue to create gamma rays that are detected by a scanning ring around the body (see the photo, left). The energy and momentum of the gamma rays are measured, from which a 3-dimensional image of the tracer concentration in the body can be formed. If the tracer is designed to detect cancer, and thus collects in a tumor, the precise location and size of the tumor is revealed (see the photo, right). One of the most common tracers for cancer detection is a glucose compound called fludeoxyglucose. One of the hydroxyl groups in the compound is replaced by radioactive fluorine-18 (${}^{18}\text{F}$). The half-life of ${}^{18}\text{F}$ is only 109.8 minutes, so it remains in the body for a short period of time. **(a)** If ${}^{18}\text{F}$ undergoes β^+ decay, what is the daughter nucleus? **(b)** If the activity of the tracer is 10 500 μCi at the beginning of the procedure, what is the activity 12.0 hours later?

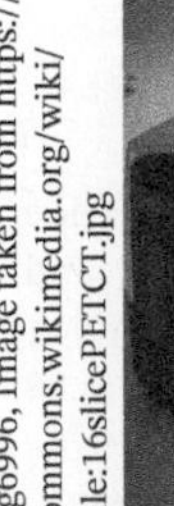

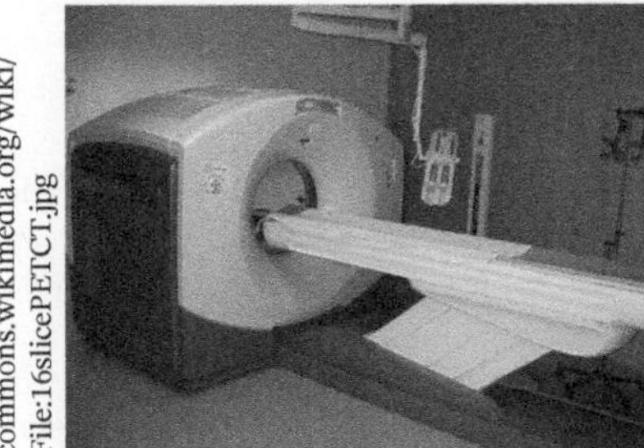

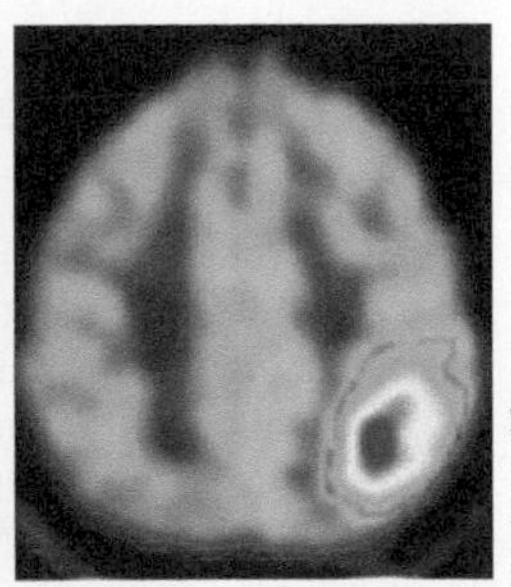

PROBLEM 53

54. M BIO 31.6 Cancer cells in tumors tend to accumulate more phosphate than normal cells. For this reason, a radioactive isotope of phosphorous (${}^{32}\text{P}$) is often used in diagnostic imaging to pinpoint the location of malignant tumors. The half-life of ${}^{32}\text{P}$ is 14.3 days. After a small amount of ${}^{32}\text{P}$ is ingested, how long will it take for 99.99% of the ${}^{32}\text{P}$ to disappear?

55. M BIO 31.3 The human body is composed of 18.5% carbon. This corresponds to 5.0×10^2 moles of carbon atoms for the average human. What would be the total nuclear binding energy of all the carbon atoms in the body? For simplicity, assume all of the carbon atoms are carbon-12 (${}^{12}\text{C}$). In comparison, the energy released by the atomic bomb dropped on Hiroshima, Japan, at the end of World War II was 60 trillion joules (6.0×10^{13} J).

56. M BIO 31.6 A radiation safety officer is using a Geiger counter to measure the activity of a radioactive material. Her initial measurement records 515 counts/min. She returns 2.00 hours later and records an activity of 125 counts/min. **(a)** What is the half-life of the material? **(b)** If the officer takes a third measurement, 2.00 hours after the second measurement, what activity will she measure then?

57. M BIO 31.6 Bananas are an excellent source of potassium. In naturally occurring potassium, a very small fraction, 0.0120%, is composed of potassium-40 (${}^{40}\text{K}$), which is a radioactive isotope with a half-life of 1.251×10^9 years. In fact, ${}^{40}\text{K}$ is responsible for the majority of the naturally occurring radioactivity in all animals, including humans. The activity of ${}^{40}\text{K}$ in the human body is approximately 4.300×10^3 Bq.

What mass of ^{40}K is contained in the body, if its isotopic mass is 39.96 g/mol? *Note:* Your body regulates the amount of potassium you need to maintain homeostasis by excreting excess potassium through sweat and urine. Eating many bananas, or other potassium-rich foods, will not increase the amount of potassium in your body. Thus, you will not increase your annual dose of radiation from ^{40}K.

Concepts and Calculations Problems

Online

Radioactive decay obeys the conservation laws of physics, and we have studied five of these. Two of them are particularly important in understanding the types of radioactivity that occur and the nuclear changes that accompany them. These are the conservation of electric charge and the conservation of nucleon number. Problem 58 emphasizes their importance and reviews how they are applied. Problem 59 explores how radioactive decay can be a source of energy in the form of heat.

58. M CHALK Thorium $^{228}_{90}$Th produces a daughter nucleus that is radioactive. The daughter, in turn, produces its own radioactive daughter, and so on. This process continues until bismuth $^{212}_{83}$Bi is reached. *Concepts:* **(i)** How many of the 90 protons in the thorium nucleus are carried off by the α particles? **(ii)** How many protons are left behind when the β^- particles are emitted? **(iii)** How many of the 228 nucleons in the thorium nucleus are carried off by the α particles? **(iv)** Does the departure of a β^- particle alter the number of nucleons? *Calculations*: What are the total number N_α of α particles and the total number N_β of β^- particles that are generated in this series of radioactive decays?

59. M CHALK SSM A one-gram sample of thorium $^{228}_{90}$Th contains 2.64×10^{21} atoms and undergoes α decay with a half-life of 1.913 yr (1.677×10^4 h). Each disintegration releases an energy of 5.52 MeV (8.83×10^{-13} J). Assume that all of the energy is used to heat a 3.80-kg sample of water. *Concepts*: **(i)** How much heat Q is needed to raise the temperature of a mass m of water by ΔT degrees? **(ii)** The energy released by each disintegration is E. What is the total energy E_{total} released by a number n of disintegrations? **(iii)** What is the number n of disintegrations that occur during a time t? *Calculations:* Find the change in temperature of the 3.80-kg sample of water that occurs in one hour.

Team Problems

Online

60. E T WS **Californium.** Neutrons have applications in a wide variety of disciplines, including medicine, materials science, and nuclear energy. One small-scale source of neutrons is an isotope of the artificially synthesized element californium (Cf). Cf has $Z = 98$ protons in its nucleus, but has a large number of isotopes, so the number of neutrons varies. One particularly useful isotope is ^{252}Cf, which has a half-life of 2.65 years. ^{252}Cf undergoes α decay 96.9% of the time, and spontaneously fissions 3.10% of the time, a process by which it releases a variety of fragments that, on average, include about 3.77 free neutrons. This relatively large rate of spontaneous neutron emission makes ^{252}Cf a viable small-scale neutron source. You and your team have been tasked with characterizing this radioactive isotope for use in an experimental neutron source. **(a)** How many neutrons are in the ^{252}Cf nucleus? **(b)** What is the approximate radius of the ^{252}Cf nucleus? **(c)** What is the daughter nucleus in the α decay of ^{252}Cf? **(d)** How many α particles are emitted per second by 35.0 g of ^{252}Cf? **(e)** Approximately how many neutrons are emitted per second by 35.0 g of ^{252}Cf? Assume the mass per mole of ^{252}Cf is 252.08 g/mol.

61. M T WS **A Nuclear Clock.** You and your team are designing a device that will be used to turn on the communications systems on a space probe 350.0 years after it is launched. Americium-241 (^{241}Am) has a half-life of 432.2 years and decays through α decay, as well as through other processes such as spontaneous fission. The decay products have enough energy to ionize the atoms of a gas, creating charges that can move between the charged plates of a parallel plate capacitor. The gas pressure between the plates is set such that each disintegration of an ^{241}Am nucleus results in an average net charge flow of $1.55 \times 10^4 e$ between the plates, that is, a current flows between the plates. The idea is that when the activity of the radioactive material decays to a level such that the current drops below some predetermined value, the circuit monitoring the current will energize the communications systems of the spacecraft. **(a)** What current will flow between the plates if 1.50 kg of ^{241}Am is placed in the volume between them? **(b)** At what magnitude of the current should the communications system energize if it is to turn on 350.0 years after the launch? The molar mass of Americium-241 is 241.06 g/mol.

62. M T **A Radioisotope Thermoelectric Generator.** You and your team are designing a thermoelectric generator that converts the thermal energy released during the decay of a radioactive material into electrical energy. Such devices are used in deep-space probes such as Voyager 1, which was launched in 1977. Its radioisotope thermoelectric generators are expected to function until the year 2025. The isotope that your team will use is $^{238}_{94}$Pu (atomic mass = 238.049553 u), which has a half-life of 87.7 years. **(a)** If $^{238}_{94}$Pu undergoes alpha decay, what is the daughter nucleus? **(b)** Assuming the daughter nucleus found in (a) has an atomic mass of 234.0409468 u, calculate the energy released in the decay in joules (the atomic weight of an alpha particle is 4.002603 u). **(c)** How many atoms are in one gram of $^{238}_{94}$Pu? **(d)** How many atoms in one gram of $^{238}_{94}$Pu decay after one year? **(e)** How much energy is released from one gram of $^{238}_{94}$Pu during one year? **(f)** What is the average power output of one gram during one year (in watts)?

63. M T **Radioactive Dating of a Mystery Object.** In a science fiction movie, a strange object is discovered buried deep in the ice of western Antarctica. It appears to be a radioisotope thermoelectric device from a spacecraft, the function of which is to convert thermal energy released in the decay of its radioactive contents into electrical energy. The radioactive material is identified as americium-241 ($^{241}_{95}$Am), which has a half-life of 432 years. **(a)** If $^{241}_{95}$Am undergoes alpha decay, what is its daughter nucleus? **(b)** What is the *activity* you would expect for 1.00 g of $^{241}_{95}$Am (in Bq)? **(c)** In the movie, 1.00 g of material is extracted from the device, and the activity is measured to be 4.00×10^{10} Bq. Assuming the device was initially loaded with 100% $^{241}_{95}$Am, how old is the device? Express your answer in years.

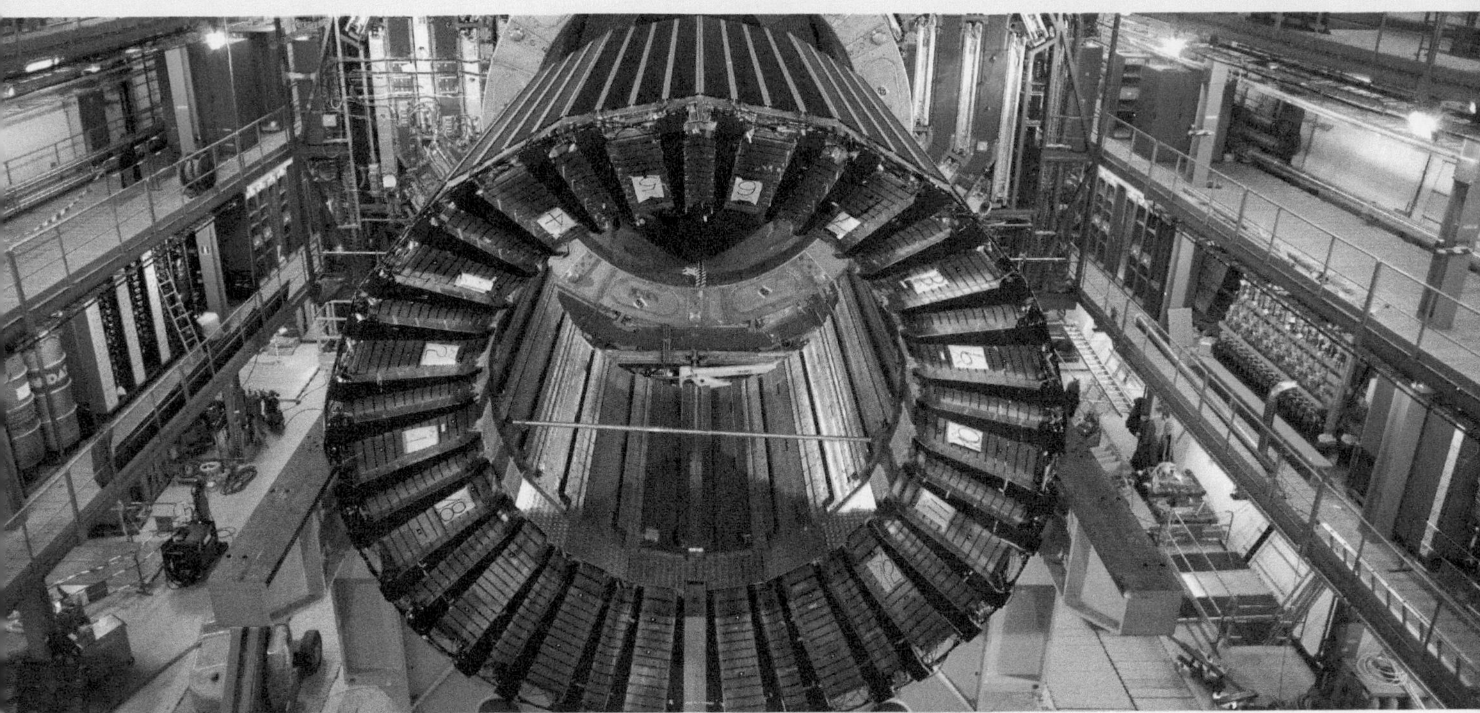

Elementary particles are the basic building blocks for all matter. They are studied by accelerating particles such as protons to high speeds and crashing them together. The Large Hadron Collider is designed to do just such a job. It consists of a large underground ring (diameter 8.6 km or 5.3 mi) and associated facilities located beneath parts of France and Switzerland. High-speed protons travel in opposite directions in the ring, and the elementary particles that result from the collisions between them are observed using specialized detectors. This photograph shows the Compact Muon Solenoid detector for the Large Hadron Collider.

VALERIO MEZZANOTTI/The New York/Redux Pictures

Ionizing Radiation, Nuclear Energy, and Elementary Particles

LEARNING OBJECTIVES

After reading this module, you should be able to...

32.1 Calculate biological effects of ionizing radiation.

32.2 Apply nuclear conservation laws to complete induced nuclear reactions.

32.3 Calculate the energy released in nuclear fission reactions.

32.4 Describe the basic components of nuclear reactors.

32.5 Calculate the energy released in nuclear fusion reactions.

32.6 Describe elementary particle theories leading to the standard model.

32.7 Use Hubble's law to explain the age and size of the universe.

32.8 Discuss recent discoveries in modern physics.

32.1 Biological Effects of Ionizing Radiation

BIO THE PHYSICS OF . . . the biological effects of ionizing radiation. **Ionizing radiation** consists of photons and/or moving particles that have sufficient energy to knock an electron out of an atom or molecule, thus forming an ion. The photons usually lie in the ultraviolet, X-ray, or γ-ray regions of the electromagnetic spectrum (see Figure 24.9), whereas the moving particles can be the α and β particles emitted during radioactive decay. An energy of roughly 1 to 35 eV is needed to ionize an atom or molecule, and the particles and γ rays emitted during nuclear disintegration often have energies of several million eV. Therefore, a single α particle, β particle, or γ ray can ionize thousands of molecules.

Nuclear radiation is potentially harmful to humans because the ionization it produces can significantly alter the structure of molecules within a living cell. The alterations can lead to the death of the cell and even of the organism itself. Despite the potential hazards, however, ionizing radiation is used in medicine for diagnostic and therapeutic purposes, such as locating bone fractures and treating cancer. The hazards can be minimized only if the fundamentals of radiation exposure, including dose units and the biological effects of radiation, are understood.

Exposure is a measure of the ionization produced in air by X-rays or γ rays, and it is defined in the following manner: A beam of X-rays or γ rays is sent through a mass m of dry air at standard temperature and pressure (STP: 0 °C, 1 atm pressure). In passing through the air, the beam produces positive ions whose total charge is q. Exposure is defined as the total charge per unit mass of air: exposure $= q/m$. The SI unit for exposure is coulombs per kilogram (C/kg). However, the first radiation unit to be defined was the *roentgen* (R), and it is still used today. With q expressed in coulombs (C) and m in kilograms (kg), the exposure in roentgens is given by

$$\text{Exposure (in roentgens)} = \left(\frac{1}{2.58 \times 10^{-4}}\right)\frac{q}{m} \tag{32.1}$$

Thus, when X-rays or γ rays produce an exposure of one roentgen, $q = 2.58 \times 10^{-4}$ C of positive charge are produced in $m = 1$ kg of dry air:

$$1\,\text{R} = 2.58 \times 10^{-4}\ \text{C/kg} \qquad \text{(dry air, at STP)}$$

Since the concept of exposure is defined in terms of the ionizing abilities of X-rays and γ rays in air, it does not specify the effect of radiation on living tissue. For biological purposes, the **absorbed dose** is a more suitable quantity because it is the energy absorbed from the radiation per unit mass of absorbing material:

$$\text{Absorbed dose} = \frac{\text{Energy absorbed}}{\text{Mass of absorbing material}} \tag{32.2}$$

The SI unit of absorbed dose is the *gray* (Gy), which is the unit of energy divided by the unit of mass: 1 Gy = 1 J/kg. Equation 32.2 is applicable to all types of radiation and absorbing media. Another unit is often used for absorbed dose—namely, the *rad* (sometimes abbreviated rd). The word "rad" is an acronym for **r**adiation **a**bsorbed **d**ose. The rad and the gray are related by

$$1\ \text{rad} = 0.01\ \text{gray}$$

Example 1 deals with the gray and the rad as units for the absorbed dose.

Analyzing Multiple-Concept Problems

EXAMPLE 1 | Absorbed Dose of γ Rays

Figure 32.1 shows γ rays being absorbed by water. What is the absorbed dose (in rads) of γ rays that will heat the water from 20.0 to 50.0 °C?

Reasoning When γ rays are absorbed by the water, they cause it to heat up. The absorbed dose of γ rays is the energy (heat) absorbed by the water divided by its mass. According to the discussion in Section 12.7, the heat that must be absorbed by the water in order for its temperature to increase by a given amount depends on the mass and specific heat capacity of the water. We will use the concept of specific heat capacity to evaluate the absorbed dose of γ rays.

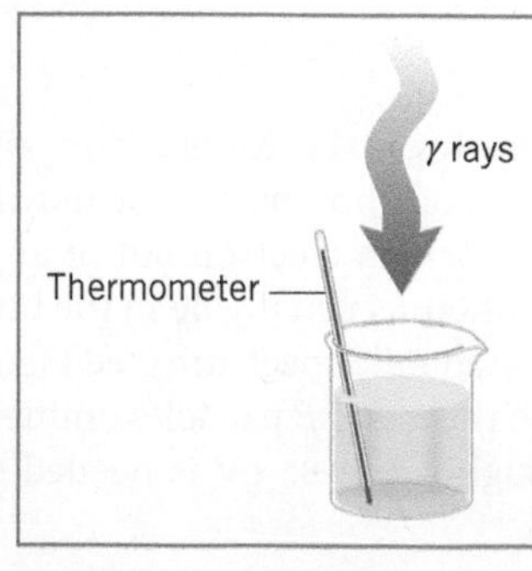

FIGURE 32.1 When the water absorbs the γ rays, its temperature rises.

Knowns and Unknowns The following table summarizes the given information:

Description	Symbol	Value
Initial temperature of water	T_0	20.0 °C
Final temperature of water	T	50.0 °C
Unknown Variable		
Absorbed dose of γ rays (in rads)	Absorbed dose	?

Modeling the Problem

STEP 1 Absorbed Dose The absorbed dose of γ rays is the energy (heat) Q absorbed by the water divided by its mass m (see Equation 32.2), as indicated in the right column. Neither Q nor m is known. Both variables will be dealt with in Step 2.

$$\text{Absorbed dose} = \frac{Q}{m} \quad (32.2)$$

?

STEP 2 Heat Needed to Increase the Temperature of the Water The heat Q that is needed to increase the temperature of a mass m of water by an amount ΔT is $Q = cm\Delta T$ (Equation 12.4), where c is the specific heat capacity of water. The change in temperature ΔT is equal to the higher temperature T minus the lower temperature T_0, or $\Delta T = T - T_0$. Thus, the heat can be expressed as

$$\boxed{Q = cm\Delta T = cm(T - T_0)} \quad (12.4)$$

This expression for the heat absorbed by the water can be substituted into Equation 32.2 for the absorbed dose, as indicated in the right column. Note that the mass m of the water appears in both the numerator and denominator, so it can be eliminated algebraically.

$$\text{Absorbed dose} = \frac{Q}{m} \quad (32.2)$$

$$\boxed{Q = cm(T - T_0)} \quad (12.4)$$

Solution Algebraically combining the results of the two modeling steps gives

$$\text{Absorbed dose} \overset{\text{STEP 1}}{=} \frac{Q}{m} \overset{\text{STEP 2}}{=} \frac{c\cancel{m}(T - T_0)}{\cancel{m}} = c(T - T_0)$$

Taking the specific heat capacity c of water from Table 12.2, we find that the absorbed dose of γ rays [expressed in grays (Gy)] is

$$\text{Absorbed dose} = c(T - T_0) = [4186\ \text{J/(kg} \cdot \text{C°)}](50.0\ \text{°C} - 20.0\ \text{°C}) = 1.26 \times 10^5\ \text{Gy}$$

The problem asks that the absorbed dose be expressed in rads, rather than in grays. To this end, we note that 1 rad = 0.01 Gy, so

$$\text{Absorbed dose} = (1.26 \times 10^5\ \cancel{\text{Gy}})\left(\frac{1\ \text{rad}}{0.01\ \cancel{\text{Gy}}}\right) = \boxed{1.26 \times 10^7\ \text{rad}}$$

Related Homework: Problem 42

The amount of biological damage produced by ionizing radiation is different for different kinds of radiation. For instance, a 1-rad dose of neutrons is far more likely to produce eye cataracts than a 1-rad dose of X-rays. To compare the damage caused by different types of radiation, the **relative biological effectiveness** (RBE) is used.* The relative biological effectiveness of a particular type of radiation is the ratio of the dose of 200-keV X-rays needed to produce a certain biological effect to the dose of the radiation needed to produce the same biological effect:

$$\text{Relative biological effectiveness (RBE)} = \frac{\text{The dose of 200-keV X-rays that produces a certain biological effect}}{\text{The dose of radiation that produces the same biological effect}} \quad (32.3)$$

*The RBE is sometimes called the *quality factor* (QF).

TABLE 32.1 Relative Biological Effectiveness (RBE) for Various Types of Radiation

Type of Radiation	RBE
200-keV X-rays	1
γ rays	1
β^- particles (electrons)	1
Protons	10
α particles	10–20
Neutrons	
Slow	2
Fast	10

The RBE depends on the nature of the ionizing radiation and its energy, as well as on the type of tissue being irradiated. Table 32.1 lists some typical RBE values for different kinds of radiation, assuming that an "average" biological tissue is being irradiated. A value of RBE = 1 for γ rays and β^- particles indicates that they produce the same biological damage as do 200-keV X-rays. The larger RBE values for protons, α particles, and neutrons indicate that they cause substantially more damage. The RBE is often used in conjunction with the absorbed dose to reflect the damage-producing character of the radiation. The product of the absorbed dose in rads (not in grays) and the RBE is the **biologically equivalent dose**:

$$\underset{\text{(in rems)}}{\text{Biologically equivalent dose}} = \underset{\text{(in rads)}}{\text{Absorbed dose}} \times \text{RBE} \tag{32.4}$$

The unit for the biologically equivalent dose is the *rem,* short for **r**oentgen **e**quivalent, **m**an. Example 2 illustrates the use of the biologically equivalent dose.

EXAMPLE 2 | BIO Comparing Absorbed Doses of γ Rays and Neutrons

A biological tissue is irradiated with γ rays that have an RBE of 0.70. The absorbed dose of γ rays is 850 rad. The tissue is then exposed to neutrons whose RBE is 3.5. The biologically equivalent dose of the neutrons is the same as that of the γ rays. What is the absorbed dose of neutrons?

Reasoning The biologically equivalent doses of the neutrons and the γ rays are the same. Therefore, the tissue damage produced in each case is the same. However, the RBE of the neutrons is larger than the RBE of the γ rays by a factor of 3.5/0.70 = 5.0. Consequently, we will find that the absorbed dose of the neutrons is only one-fifth as great as that of the γ rays.

Solution According to Equation 32.4, the biologically equivalent dose is the product of the absorbed dose (in rads) and the RBE; it is the same for the γ rays and the neutrons. Therefore, we have

$$\begin{aligned}\text{Biologically equivalent dose} &= (\text{Absorbed dose})_{\gamma\text{ rays}}\,\text{RBE}_{\gamma\text{ rays}}\\ &= (\text{Absorbed dose})_{\text{neutrons}}\,\text{RBE}_{\text{neutrons}}\end{aligned}$$

Solving for the absorbed dose of the neutrons gives

$$\begin{aligned}(\text{Absorbed dose})_{\text{neutrons}} &= (\text{Absorbed dose})_{\gamma\text{ rays}}\left(\frac{\text{RBE}_{\gamma\text{ rays}}}{\text{RBE}_{\text{neutrons}}}\right)\\ &= (850\text{ rad})\left(\frac{0.70}{3.5}\right) = \boxed{170\text{ rad}}\end{aligned}$$

Everyone is continually exposed to background radiation from natural sources, such as cosmic rays (high-energy particles that come from outside the solar system), radioactive materials in the environment, radioactive nuclei (primarily carbon ${}^{14}_{6}\text{C}$ and potassium ${}^{40}_{19}\text{K}$) within our own bodies, and radon. Table 32.2 lists the average biologically equivalent doses received from these sources by a citizen in the United States. According to this table, radon is a major contributor to the natural background radiation. Radon is an odorless radioactive gas and poses a health hazard because, when inhaled, it can damage the lungs and cause cancer. Radon is found in soil and rocks and enters houses via a mechanism that is not well understood. One possibility for indoor radon entry is through cracks and crevices in the foundation. The amount of radon in the soil varies greatly throughout the country, with some localities having significant amounts and others having virtually none. Accordingly, the dose that any individual receives can vary widely from the average value of 207 mrem/yr given in Table 32.2 (1 mrem = 10^{-3} rem). In many houses, the entry of radon can be reduced significantly by sealing the foundation against entry of the gas and providing good ventilation so it does not accumulate.

To the natural background of radiation, a significant amount of human-made radiation has been added, mostly from medically related sources. Among these sources, CAT scanning (see Section 30.7) is the major contributor, as Table 32.2 indicates.

The effects of radiation on humans can be grouped into two categories, according to the time span between initial exposure and the appearance of physiological symptoms: (1) short-term or acute effects that appear within a matter of minutes, days, or weeks, and (2) long-term or latent effects that appear years, decades, or even generations later.

Radiation sickness is the general term applied to the acute effects of radiation. Depending on the severity of the dose, a person with radiation sickness can exhibit nausea, vomiting, fever, diarrhea, and loss of hair. Ultimately, death can occur. The severity of radiation sickness is related to the dose received, and in the following discussion the biologically equivalent doses quoted are whole-body, single doses. A dose less than 50 rem causes no short-term ill effects. A dose between 50 and 300 rem brings on radiation sickness, the severity increasing with increasing dosage. A whole-body dose in the range of 400–500 rem is classified as an LD_{50} dose, meaning that it is a lethal dose (LD) for about 50% of the people so exposed; death occurs within a few months. Whole-body doses greater than 600 rem result in death for almost all individuals.

Long-term or latent effects of radiation may appear as a result of high-level brief exposure or low-level exposure over a long period of time. Some long-term effects are hair loss, eye cataracts, and various kinds of cancer. In addition, genetic defects caused by mutated genes may be passed on from one generation to the next.

Because of the hazards of radiation, the federal government has established dose limits. The permissible dose for an individual is defined as the dose, accumulated over a long period of time or resulting from a single exposure, that carries negligible probability of a severe health hazard. Federal standards (1991) state that an individual in the general population should not receive more than 500 mrem of human-made radiation each year, *exclusive* of medical sources. A person exposed to radiation in the workplace (e.g., a radiation therapist) should not receive more than 5 rem per year from work-related sources.

TABLE 32.2 Average Biologically Equivalent Doses of Radiation Received by a U. S. Citizen[a]

Source of Radiation	Biologically Equivalent Dose (mrem/yr)[b]
Natural background radiation	
Cosmic rays	33
Radioactive earth and air	21
Internal radioactive nuclei	29
Inhaled radon	207
Human-made radiation	
Consumer products	13
CAT scanning	147
Routine medical/dental diagnostics	33
Nuclear medicine	74

[a]National Council on Radiation Protection and Measurements, Report No. 160, "Ionizing Radiation Exposure of the Population of the United States," 2009.
[b]1 mrem = 10^{-3} rem.

BIO THE PHYSICS OF . . . cosmic rays and space travel. One of the central problems in future long-term space travel is the dangerous radiation to which astronauts are exposed, even while inside their ship or space station. Cosmic rays are highly energetic particles, such as atomic nuclei and protons, which travel near the speed of light. They can originate from the sun, or from distant sources, like other galaxies. Cosmic rays carry enough energy to ionize atoms, and are therefore dangerous to humans. The surface of the earth is mostly protected from cosmic rays due to both the earth's atmosphere and its magnetic field. Astronauts in earth orbit or deep space, however, must be shielded from this dangerous radiation by materials in their spacecraft, or by other means.

One might think that a layer of a dense metal, such as lead or tungsten, built into the walls of the spaceship would suffice, but it is considerably more complicated than that. First, any added weight presents both technical issues and significant added cost (many thousands of dollars per kilogram) for launching the ship into space. More importantly, it turns out that some radiation-shielding materials, such as lead, can make the radiation inside the ship worse due to the secondary radiation that is produced when the highly energetic cosmic rays are absorbed. These secondary particles and photons can penetrate into the ship with lesser energies that can collectively cause more damage than the original cosmic ray. Thus, materials are sought that can absorb cosmic rays in a way that produces less secondary radiation than heavy-element shielding. For instance, the outer walls of the International Space Station (ISS) are composed of aluminum, a metal that is much lighter than lead or tungsten, which results in less secondary radiation and a comparatively reduced net exposure. Hydrogen-rich plastics are also a possible shielding solution for the same reason. Liquid hydrogen is an effective radiation shield that produces a smaller amount of secondary radiation, as is water, but these are both consumables on spacecraft and will diminish over time. Other hypothetical solutions include electrostatic or magnetic deflection, which are currently undeveloped, and flying close to asteroids and using them as shields! Protecting space travelers from cosmic rays and other radiation sources is a critical problem that must be solved before humans will be able to safely venture into deep space.

Example 3 discusses the biologically equivalent dose a patient receives during a CT scan.

EXAMPLE 3 | BIO The Energy Absorbed by a Body During a CT Scan

Consider a patient who receives a biologically equivalent dose, or BED, of 1.7 rem during a full-body CT scan. If the RBE of the X-rays in this procedure is 0.90, how much energy is absorbed by a patient with a mass of 82 kg?

Reasoning We can use Equation 32.4 to calculate the absorbed dose and then Equation 32.2 to calculate the energy absorbed by the patient.

Solution From Equation 32.4, we have:

$$\underset{\text{(in rems)}}{\text{Biologically equivalent dose}} = \underset{\text{(in rads)}}{\text{Absorbed dose}} \times \text{RBE}$$

Rearranging, we find:

$$\underset{\text{dose (in rads)}}{\text{Absorbed}} = \frac{\underset{\text{(in rems)}}{\text{Biologically equivalent dose}}}{\text{RBE}} = \frac{1.7\text{ rem}}{0.90} = 1.9\text{ rad}$$

In order to calculate the energy absorbed in joules, we need to convert the absorbed dose to Gy: 1.9 rad = 0.019 Gy. Using Equation 32.2, we calculate the energy absorbed:

$$\text{Energy absorbed} = (\text{Absorbed dose}) \times (\text{Mass of absorbing material})$$
$$= (0.019\text{ Gy}) \times (82\text{ kg}) = \boxed{1.6\text{ J}}$$

Check Your Understanding

(The answers are given at the end of the book.)

1. Two different types of radiation have the same RBE. Is it possible for these two types of radiation to deliver different biologically equivalent doses of radiation to a given tissue sample?

2. The damage-producing character of a given type of ionizing radiation depends on **(a)** only the RBE of the radiation, **(b)** only the absorbed dose of the radiation, **(c)** both the RBE and the absorbed dose of the radiation.

3. A person faces the possibility of receiving the following absorbed doses of ionizing radiation: 20 rad of γ rays (RBE = 1), 5 rad of neutrons (RBE = 10), and 2 rad of α particles (RBE = 20). Rank the amount of biological damage that these possibilities will cause in decreasing order (greatest damage first).

32.2 Induced Nuclear Reactions

Section 31.4 discusses how a radioactive parent nucleus disintegrates spontaneously into a daughter nucleus. It is also possible to bring about, or induce, the disintegration of a stable nucleus by striking it with another nucleus, an atomic or subatomic particle, or a γ-ray photon. A **nuclear reaction** is said to occur whenever an incident nucleus, particle, or photon causes a change to occur in a target nucleus.

In 1919, Ernest Rutherford observed that when an α particle strikes a nitrogen nucleus, an oxygen nucleus and a proton are produced. This nuclear reaction is written as

$$\underbrace{{}^{4}_{2}\text{He}}_{\substack{\text{Incident}\\ \alpha\text{ particle}}} + \underbrace{{}^{14}_{7}\text{N}}_{\substack{\text{Nitrogen}\\ \text{(target)}}} \longrightarrow \underbrace{{}^{17}_{8}\text{O}}_{\text{Oxygen}} + \underbrace{{}^{1}_{1}\text{H}}_{\text{Proton, }p}$$

Because the incident α particle induces the transmutation of nitrogen into oxygen, this reaction is an example of an **induced nuclear transmutation**.

Nuclear reactions are often written in a shorthand form. For example, the reaction above is designated by ${}^{14}_{7}\text{N}\,(\alpha, p)\,{}^{17}_{8}\text{O}$. The first and last symbols represent the initial and final nuclei, respectively. The symbols within the parentheses denote the incident α particle (on the left) and the small emitted particle or proton p (on the right). Some other induced nuclear transmutations are listed below, together with the equivalent shorthand notations:

Nuclear Reaction	Notation
${}^{1}_{0}\text{n} + {}^{10}_{5}\text{B} \rightarrow {}^{7}_{3}\text{Li} + {}^{4}_{2}\text{He}$	${}^{10}_{5}\text{B}\,(n, \alpha)\,{}^{7}_{3}\text{Li}$
$\gamma + {}^{25}_{12}\text{Mg} \rightarrow {}^{24}_{11}\text{Na} + {}^{1}_{1}\text{H}$	${}^{25}_{12}\text{Mg}\,(\gamma, p)\,{}^{24}_{11}\text{Na}$
${}^{1}_{1}\text{H} + {}^{13}_{6}\text{C} \rightarrow {}^{14}_{7}\text{N} + \gamma$	${}^{13}_{6}\text{C}\,(p, \gamma)\,{}^{14}_{7}\text{N}$

Induced nuclear reactions, like the radioactive decay process discussed in Section 31.4, obey the conservation laws of physics. Each of these laws deals with a property that does not change during a process. The following list shows the property with which each law deals:

1. Conservation of energy/mass (Sections 6.8 and 28.6)
2. Conservation of linear momentum (Section 7.2)

3. Conservation of angular momentum (Section 9.6)
4. Conservation of electric charge (Section 18.2)
5. Conservation of nucleon number (Section 31.4)

In particular, items 4 and 5 on this list indicate that both the total electric charge of the nucleons and the number of nucleons in a nuclear reaction are conserved. The next example illustrates how the conservation of total electric charge and number of nucleons can be used to identify the nucleus produced in a reaction.

EXAMPLE 4 | An Induced Nuclear Transmutation

An α particle strikes an aluminum ${}^{27}_{13}\text{Al}$ nucleus. As a result, an unknown nucleus ${}^{A}_{Z}\text{X}$ and a neutron ${}^{0}_{1}\text{n}$ are produced:

$$ {}^{4}_{2}\text{He} + {}^{27}_{13}\text{Al} \longrightarrow {}^{A}_{Z}\text{X} + {}^{1}_{0}\text{n} $$

Identify the nucleus produced, including its atomic number Z (the number of protons) and its atomic mass number A (the number of nucleons).

Reasoning The total electric charge of the nucleons is conserved, so that we can set the total number of protons before the reaction equal to the total number after the reaction. The total number of nucleons is also conserved, so that we can set the total number before the reaction equal to the total number after the reaction. These two conserved quantities will allow us to identify the nucleus ${}^{A}_{Z}\text{X}$.

Solution The conservation of total electric charge and total number of nucleons leads to the equations listed below:

Conserved Quantity	Before Reaction		After Reaction
Total electric charge (number of protons)	$2 + 13$	$=$	$Z + 0$
Total number of nucleons	$4 + 27$	$=$	$A + 1$

Solving these equations for Z and A gives $Z = 15$ and $A = 30$. Since $Z = 15$ identifies the element as phosphorus (P), the nucleus produced is $\boxed{{}^{30}_{15}\text{P}}$.

Induced nuclear transmutations can be used to produce isotopes that are not found naturally. In 1934, Enrico Fermi (1901–1954) suggested a method for producing elements with a higher atomic number than uranium ($Z = 92$). These elements—neptunium ($Z = 93$), plutonium ($Z = 94$), americium ($Z = 95$), and so on—are known as *transuranium elements,* and none occurs naturally. They are created in a nuclear reaction between a suitably chosen lighter element and a small incident particle, usually a neutron or an α particle. For example, **Figure 32.2** shows a reaction that produces plutonium from uranium. A neutron is captured by a uranium ${}^{239}_{92}\text{U}$ nucleus, producing ${}^{239}_{92}\text{U}$ and a γ ray. The ${}^{239}_{92}\text{U}$ nucleus is radioactive and decays with a half-life of 23.5 min into neptunium ${}^{239}_{93}\text{Np}$. Neptunium is also radioactive and disintegrates with a half-life of 2.4 days into plutonium ${}^{239}_{94}\text{Pu}$. Plutonium is the final product and has a half-life of 24 100 yr.

The neutrons that participate in nuclear reactions can have kinetic energies that cover a wide range. In particular, those that have a kinetic energy of about 0.04 eV or less are called **thermal neutrons.** The name derives from the fact that such a relatively small kinetic energy is comparable to the average translational kinetic energy of a molecule in an ideal gas at room temperature.

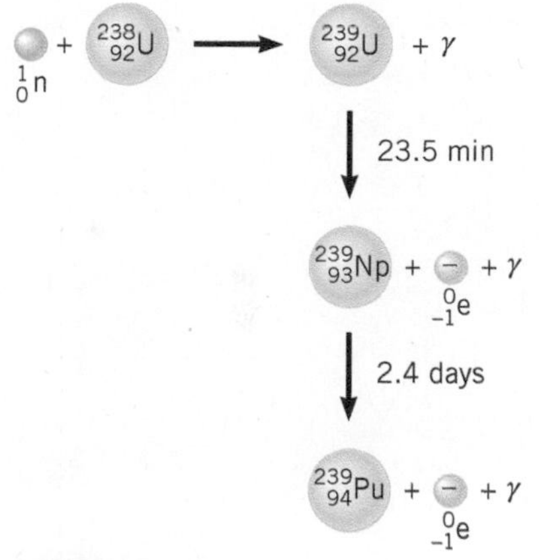

$$ {}^{1}_{0}\text{n} + {}^{238}_{92}\text{U} \longrightarrow {}^{239}_{92}\text{U} + \gamma $$

23.5 min

$$ {}^{239}_{93}\text{Np} + {}^{0}_{-1}\text{e} + \gamma $$

2.4 days

$$ {}^{239}_{94}\text{Pu} + {}^{0}_{-1}\text{e} + \gamma $$

FIGURE 32.2 An induced nuclear reaction is shown in which ${}^{238}_{92}\text{U}$ is transmuted into the transuranium element plutonium ${}^{239}_{94}\text{Pu}$.

Check Your Understanding

(The answers are given at the end of the book.)

4. Which one or more of the following nuclear reactions could possibly occur? **(a)** ${}^{15}_{7}\text{N}\ (\alpha, \gamma)\ {}^{18}_{9}\text{F}$ **(b)** ${}^{14}_{6}\text{C}\ (p, n)\ {}^{15}_{8}\text{O}$ **(c)** ${}^{14}_{7}\text{N}\ (p, \gamma)\ {}^{15}_{8}\text{O}$ **(d)** ${}^{15}_{8}\text{O}\ (n, p)\ {}^{13}_{6}\text{C}$

5. Why is each of the following reactions *not* allowed? **(a)** ${}^{60}_{28}\text{Ni}\ (\alpha, p)\ {}^{62}_{29}\text{Cu}$ **(b)** ${}^{27}_{13}\text{Al}\ (n, n)\ {}^{28}_{13}\text{Al}$ **(c)** ${}^{39}_{19}\text{K}\ (p, \alpha)\ {}^{36}_{17}\text{Cl}$

32.3 Nuclear Fission

In 1939 four German scientists, Otto Hahn, Lise Meitner, Fritz Strassmann, and Otto Frisch, made an important discovery that ushered in the atomic age. They found that a

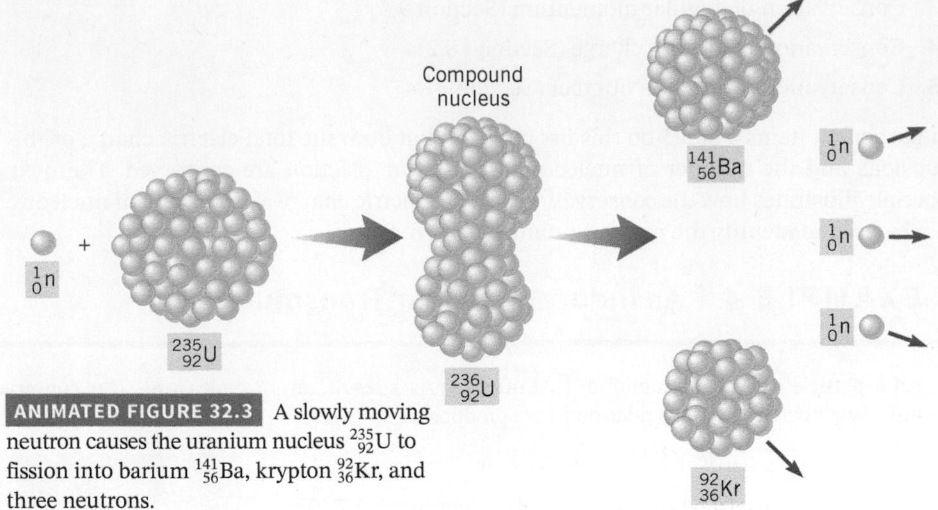

ANIMATED FIGURE 32.3 A slowly moving neutron causes the uranium nucleus $^{235}_{92}\text{U}$ to fission into barium $^{141}_{56}\text{Ba}$, krypton $^{92}_{36}\text{Kr}$, and three neutrons.

uranium nucleus, after absorbing a neutron, splits into two fragments, each with a smaller mass than the original nucleus. The splitting of a massive nucleus into two less massive fragments is known as **nuclear fission**.

Animated Figure 32.3 shows a fission reaction in which a uranium $^{235}_{92}\text{U}$ nucleus is split into barium $^{141}_{56}\text{Ba}$ and krypton $^{92}_{36}\text{Kr}$ nuclei. The reaction begins when $^{235}_{92}\text{U}$ absorbs a slowly moving neutron, creating a "compound nucleus," $^{236}_{92}\text{U}$. The compound nucleus disintegrates quickly into $^{141}_{56}\text{Ba}$, $^{92}_{36}\text{Kr}$, and three neutrons according to the following reaction:

$$^{1}_{0}\text{n} + {}^{235}_{92}\text{U} \longrightarrow \underbrace{^{236}_{92}\text{U}}_{\substack{\text{Compound}\\\text{nucleus}\\\text{(unstable)}}} \longrightarrow \underbrace{^{141}_{56}\text{Ba}}_{\text{Barium}} + \underbrace{^{92}_{36}\text{Kr}}_{\text{Krypton}} + \underbrace{3\,^{1}_{0}\text{n}}_{\text{3 neutrons}}$$

This reaction is only one of the many possible reactions that can occur when uranium fissions. For example, another reaction is

$$^{1}_{0}\text{n} + {}^{235}_{92}\text{U} \longrightarrow \underbrace{^{236}_{92}\text{U}}_{\substack{\text{Compound}\\\text{nucleus}\\\text{(unstable)}}} \longrightarrow \underbrace{^{140}_{54}\text{Xe}}_{\text{Xenon}} + \underbrace{^{94}_{38}\text{Sr}}_{\text{Strontium}} + \underbrace{2\,^{1}_{0}\text{n}}_{\text{2 neutrons}}$$

Some reactions produce as many as 5 neutrons; however, the average number produced per fission is 2.5.

When a neutron collides with and is absorbed by a uranium nucleus, the uranium nucleus begins to vibrate and becomes distorted. The vibration continues until the distortion becomes so severe that the attractive strong nuclear force can no longer balance the electrostatic repulsion between the nuclear protons. At this point, the nucleus bursts apart into fragments, which carry off energy, primarily in the form of kinetic energy. The energy carried off by the fragments is enormous and was stored in the original nucleus mainly in the form of electric potential energy. An average of roughly 200 MeV of energy is released per fission. This energy is approximately 10^8 times greater than the energy released per molecule in an ordinary chemical reaction, such as the combustion of gasoline or coal. Example 5 demonstrates how to estimate the energy released during the fission of a nucleus.

EXAMPLE 5 | The Energy Released During Nuclear Fission

Estimate the amount of energy released when a massive nucleus ($A = 240$) fissions.

Reasoning Figure 31.5 shows that the binding energy of a nucleus with $A = 240$ is about 7.6 MeV per nucleon. We assume that this nucleus fissions into two fragments, each with $A \approx 120$. According to Figure 31.5, the binding energy of the fragments increases to about 8.5 MeV per nucleon. Consequently, when a massive nucleus fissions, there is a release of about 8.5 MeV − 7.6 MeV = 0.9 MeV of energy per nucleon.

Solution Since there are 240 nucleons involved in the fission process, the total energy released per fission is approximately (0.9 MeV/nucleon)(240 nucleons) ≈ $\boxed{200\text{ MeV}}$.

Virtually all naturally occurring uranium is composed of two isotopes. These isotopes and their natural abundances are $^{238}_{92}U$ (99.275%) and $^{235}_{92}U$ (0.720%). Although $^{238}_{92}U$ is by far the most abundant isotope, the probability that it will capture a neutron and fission is very small. For this reason, $^{238}_{92}U$ is not the isotope of choice for generating nuclear energy. In contrast, the isotope $^{235}_{92}U$ readily captures a neutron and fissions, *provided the neutron is a thermal neutron* (kinetic energy) $\approx$ 0.04 eV or less). The probability of a thermal neutron causing $^{235}_{92}U$ to fission is about 500 times greater than the probability for a neutron whose energy is relatively high—say, 1 MeV. Thermal neutrons can also be used to fission other nuclei, such as plutonium $^{239}_{94}Pu$. Conceptual Example 6 deals with one of the reasons why thermal neutrons are useful for inducing nuclear fission.

CONCEPTUAL EXAMPLE 6 | Neutrons Versus Protons or Alpha Particles

A thermal neutron has a relatively small amount of kinetic energy but, nevertheless, can penetrate a nucleus. To penetrate the same nucleus, would a proton or an α particle need **(a)** the same small amount of kinetic energy as the neutron needs, **(b)** a much larger amount of kinetic energy than the neutron needs, or **(c)** much less kinetic energy than the neutron needs?

Reasoning To penetrate a nucleus, a particle such as a neutron, a proton, or an α particle must have enough kinetic energy to do the work of overcoming any repulsive force that it encounters. A repulsive force can arise because protons in the nucleus are electrically charged. Since a neutron is electrically neutral, however, it encounters no electrostatic force of repulsion as it approaches the nuclear protons, and, hence, needs relatively little energy to reach the nucleus.

Answers (a) and (c) are incorrect. A proton and an α particle each carry a positive charge, so that each would encounter an electrostatic force of repulsion as it approached the nuclear protons, a force that a thermal neutron does not encounter. These answers ignore the additional kinetic energy that a proton or an α particle would need to overcome the repulsion.

Answer (b) is correct. A proton and an α particle, each being positively charged, would each require much more kinetic energy than a neutron does, in order to overcome the electrostatic force of repulsion from the nuclear protons. It is true that each would also experience the attractive strong nuclear force from the nuclear protons and neutrons. However, this force has an extremely short range of action and, therefore, would come into play only after an impinging particle reached the target nucleus. In comparison, the electrostatic force has a long range of action and is encountered throughout the entire journey to the target.

Related Homework: Problem 38

The fact that the uranium fission reaction releases 2.5 neutrons, on the average, makes it possible for a self-sustaining series of fissions to occur. As **Figure 32.4** illustrates, each neutron released can initiate another fission event, resulting in the emission of still more neutrons, followed by more fissions, and so on. A **chain reaction** is a series of nuclear fissions whereby some of the neutrons produced by each fission cause additional fissions. During an uncontrolled chain reaction, it would not be unusual for the number of fissions to increase a thousandfold within a few millionths of a second. With an average energy of about 200 MeV being released per fission, an uncontrolled chain reaction can generate an incredible amount of energy in a very short time, as happens in an atomic bomb (which is actually a *nuclear* bomb).

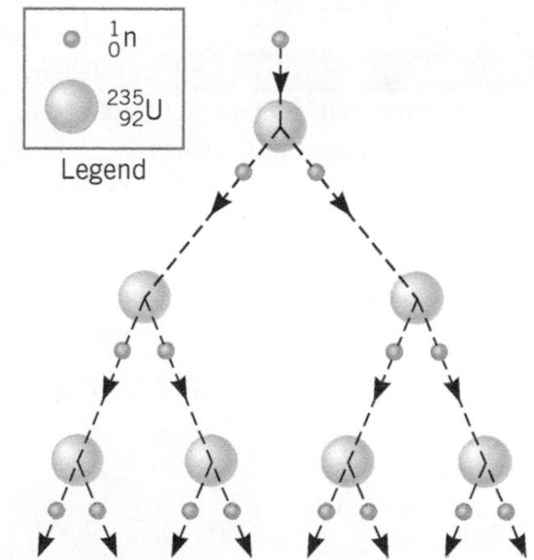

FIGURE 32.4 A chain reaction. For clarity, it is assumed that each fission generates two neutrons (2.5 neutrons are actually liberated on the average). The fission fragments are not shown.

By using a material that can absorb neutrons without fissioning, it is possible to limit the number of neutrons in the environment of the fissile nuclei. In this way, a condition can be established whereby each fission event contributes, on average, only *one neutron* that fissions another nucleus (see **Figure 32.5**). Thus, the chain reaction and the rate of energy production are controlled. The controlled-fission chain reaction is the principle behind nuclear reactors used in the commercial generation of electric power.

Check Your Understanding

(*The answers are given at the end of the book.*)

6. When the nucleus of a certain element absorbs a thermal neutron, fission usually occurs, with the production of nuclear fragments and a number *N* of neutrons. However, in a collection of atoms of this element, a small fraction of the thermal neutrons absorbed by the nuclei does not lead to

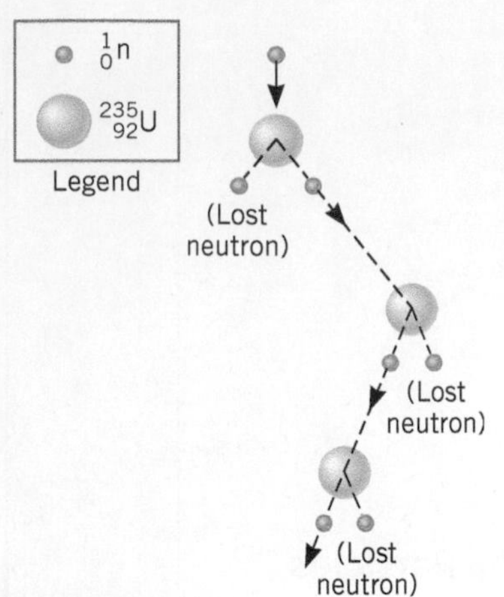

FIGURE 32.5 In a controlled chain reaction, only one neutron, on average, from each fission event causes another nucleus to fission. The "lost neutron" is absorbed by a material (not shown) that does not fission. As a result, energy is released at a steady or controlled rate.

fission. For which one of the following values of N can a self-sustaining chain reaction *not* be produced using this element? **(a)** $N = 4$ **(b)** $N = 3$ **(c)** $N = 2$ **(d)** $N = 1$

7. Thermal neutrons, thermal protons, and thermal electrons all have the same kinetic energy of about 0.04 eV. Rank the speeds of these particles in descending order (greatest speed first).

8. Would a release of energy accompany the fission of a nucleus of nucleon number 25 into two fragments of about equal mass? Consult **Figure 31.5** as needed.

32.4 Nuclear Reactors

THE PHYSICS OF . . . nuclear reactors. A nuclear reactor is a type of furnace in which energy is generated by a controlled-fission chain reaction. The first nuclear reactor was built by Enrico Fermi in 1942, on the floor of a squash court under the west stands of Stagg Field at the University of Chicago. Today, there are a number of kinds and sizes of reactors, and many have the same three basic components: fuel elements, a neutron moderator, and control rods. **Interactive Figure 32.6** illustrates these components.

The **fuel elements** contain the fissile fuel and, for example, may be thin rods about 1 cm in diameter. In a large power reactor there may be thousands of fuel elements placed close together, and the entire region of fuel elements is known as the **reactor core**. Uranium $^{235}_{92}$U is a common reactor fuel. Since the natural abundance of this isotope is only about 0.7%, there are special uranium-enrichment plants to increase the percentage. Most commercial reactors use uranium in which the amount of $^{235}_{92}$U has been enriched to about 3%.

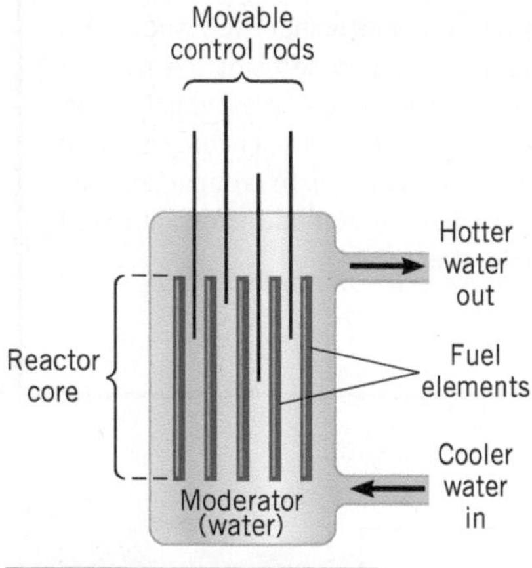

INTERACTIVE FIGURE 32.6 A nuclear reactor consists of fuel elements, control rods, and a moderator (in this case, water).

Whereas neutrons with energies of about 0.04 eV (or less) readily fission $^{235}_{92}$U, the neutrons released during the fission process have significantly greater energies of several MeV or so. Consequently, a nuclear reactor must contain some type of material that will decrease or moderate the speed of such energetic neutrons so they can readily fission additional $^{235}_{92}$U nuclei. The material that slows down the neutrons is called a **moderator**. One commonly used moderator is water. When an energetic neutron leaves a fuel element, the neutron enters the surrounding water and collides with water molecules. With each collision, the neutron loses an appreciable fraction of its energy and slows down. Once slowed down to thermal energy by the moderator, a process that takes less than 10^{-3} s, the neutron is capable of initiating a fission event upon reentering a fuel element.

If the output power from a reactor is to remain constant, only one neutron from each fission event must trigger a new fission, as **Figure 32.5** illustrates. When each fission leads to one additional fission—no more or no less—the reactor is said to be *critical*. A reactor normally operates in a critical condition, because then it produces a steady output of energy. The reactor is *subcritical* when, on average, the neutrons from each fission trigger *less than one* subsequent fission. In a subcritical reactor, the chain reaction is not self-sustaining and eventually dies out. When the neutrons from each fission trigger *more than one* additional fission, the reactor is *supercritical*. During a supercritical condition, the energy released by a reactor increases. If left unchecked, the increasing energy can lead to a partial or total meltdown of the reactor core, with the possible release of radioactive material into the environment.

Clearly, a control mechanism is needed to keep the reactor in its normal, or critical, state. This control is accomplished by a number of **control rods** that can be moved into and out of the reactor core (see **Interactive Figure 32.6**). The control rods contain an element, such as boron or cadmium, that readily absorbs neutrons without fissioning. If the reactor becomes supercritical, the control rods are automatically moved farther into the core to absorb the excess neutrons causing the condition. In response, the reactor returns to its critical state. Conversely, if the reactor becomes subcritical, the control rods are partially withdrawn from the core. Fewer neutrons are absorbed, more neutrons are available for fission, and the reactor again returns to its critical state.

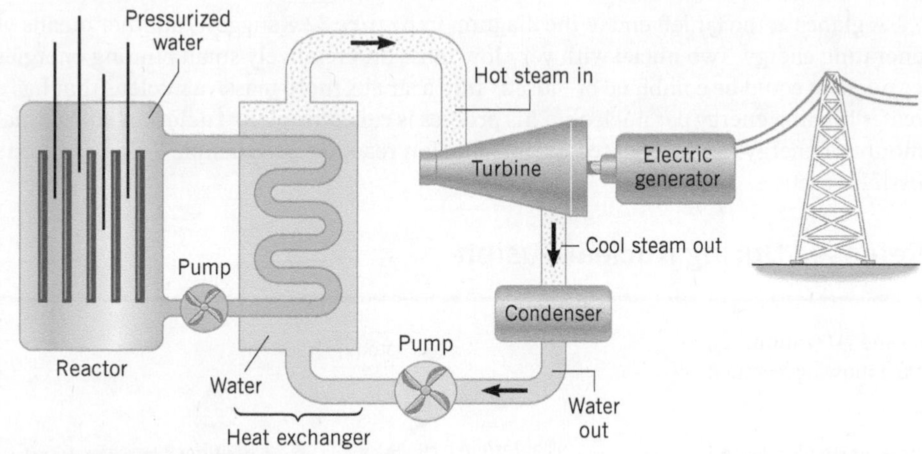

INTERACTIVE FIGURE 32.7 Diagram of a nuclear power plant that uses a pressurized water reactor.

Interactive Figure 32.7 illustrates a pressurized water reactor. In such a reactor, the heat generated within the fuel rods is carried away by water that surrounds the rods. To remove as much heat as possible, the water temperature is allowed to rise to a high value (about 300 °C). To prevent boiling, which occurs at 100 °C at 1 atmosphere of pressure, the water is pressurized in excess of 150 atmospheres. The hot water is pumped through a heat exchanger, where heat is transferred to water flowing in a second, closed system. The heat transferred to the second system produces steam that drives a turbine. The turbine is coupled to an electric generator, whose output electric power is delivered to consumers via high-voltage transmission lines. After exiting the turbine, the steam is condensed back into water that is returned to the heat exchanger.

32.5 Nuclear Fusion

In Example 5 in Section 32.3, the plot of binding energy per nucleon in Figure 31.5 is used to estimate the amount of energy released in the fission process. As summarized in **Figure 32.8**, the massive nuclei at the right end of the curve have a binding energy of about 7.6 MeV per nucleon. The less massive fission fragments are near the center of the curve and have a binding energy of approximately 8.5 MeV per nucleon. The energy released per nucleon by fission is the difference between these two values, or about 0.9 MeV per nucleon.

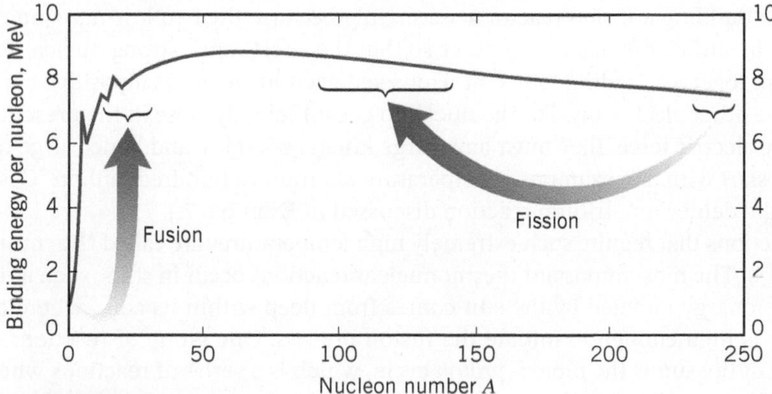

FIGURE 32.8 When fission occurs, a massive nucleus divides into two fragments whose binding energy per nucleon is greater than that of the original nucleus. When fusion occurs, two low-mass nuclei combine to form a more massive nucleus whose binding energy per nucleon is greater than that of the original nuclei.

A glance at the far left end of the diagram in **Figure 32.8** suggests another means of generating energy. Two nuclei with very low mass and relatively small binding energies per nucleon could be combined or "fused" into a single, more massive nucleus that has a greater binding energy per nucleon. This process is called **nuclear fusion**. A substantial amount of energy can be released during a fusion reaction, as Example 7 shows for one possible reaction.

EXAMPLE 7 | The Energy Released During Nuclear Fusion

Two isotopes of hydrogen, ${}^{2}_{1}\text{H}$ (deuterium, D) and ${}^{3}_{1}\text{H}$ (tritium, T), fuse to form ${}^{4}_{2}\text{He}$ and a neutron according to the following reaction:

$$ {}^{2}_{1}\text{H} + {}^{3}_{1}\text{H} \longrightarrow {}^{4}_{2}\text{He} + {}^{1}_{0}\text{n} $$

Determine the energy that is released by this particular fusion reaction, which is illustrated in **Figure 32.9**.

Reasoning Energy is released, so the total mass of the final nuclei is less than the total mass of the initial nuclei. To determine the energy released, we find the amount (in atomic mass units u) by which the total mass has decreased. Then, we use the fact that 1 u is equivalent to 931.5 MeV of energy, as determined in Section 31.3. This approach is the same as that used in Section 31.4 for radioactive decay.

Solution The masses of the initial and final nuclei in this reaction, as well as the mass of the neutron, are

Initial Masses		Final Masses	
${}^{2}_{1}\text{H}$	2.0141 u	${}^{4}_{2}\text{He}$	4.0026 u
${}^{3}_{1}\text{H}$	3.0161 u	${}^{1}_{0}\text{n}$	1.0087 u
Total:	5.0302 u	Total:	5.0113 u

The decrease in mass, or the mass defect, is $\Delta m = 5.0302\text{ u} - 5.0113\text{ u} = 0.0189\text{ u}$. Since 1 u is equivalent to 931.5 MeV, the energy released is

$$ \text{Released energy} = (0.0189\text{ u})\left(\frac{931.5\text{ MeV}}{1\text{ u}}\right) = \boxed{17.6\text{ MeV}} $$

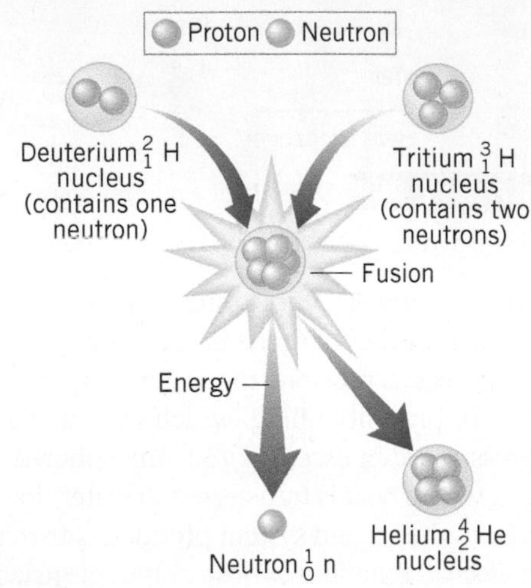

FIGURE 32.9 Deuterium and tritium are fused together to form a helium nucleus (${}^{4}_{2}\text{He}$). The result is the release of an enormous amount of energy, mainly carried by a single high-energy neutron (${}^{1}_{0}\text{n}$).

The deuterium nucleus contains 2 nucleons, and the tritium nucleus contains 3. Thus, there are 5 nucleons that participate in the fusion, so the energy released per nucleon is about 3.5 MeV. This energy per nucleon is greater than the energy released in a fission process ($\approx$0.9 MeV per nucleon).

Because fusion reactions release so much energy, there is considerable interest in fusion reactors, although to date no commercial units have been constructed. The difficulties in building a fusion reactor arise mainly because the two low-mass nuclei must be brought sufficiently near each other so that the short-range strong nuclear force can pull them together, leading to fusion. However, each nucleus has a positive charge and repels the other electrically. For the nuclei to get sufficiently close in the presence of the repulsive electric force, they must have large kinetic energies, and hence large temperatures, to start with. For example, a temperature of around a hundred million °C is needed to start the deuterium–tritium reaction discussed in Example 7.

Reactions that require such extremely high temperatures are called **thermonuclear reactions**. The most important thermonuclear reactions occur in stars, such as our own sun. The energy radiated by the sun comes from deep within its core, where the temperature is high enough to initiate the fusion process. One group of reactions thought to occur in the sun is the *proton–proton* cycle, which is a series of reactions whereby six protons form a helium nucleus, two positrons, two γ rays, two protons, and two neutrinos. The energy released by the proton–proton cycle is about 25 MeV.

Human-made fusion reactions have been carried out in a fusion-type nuclear bomb—commonly called a hydrogen bomb. In a hydrogen bomb, the fusion reaction is ignited by a fission bomb using uranium or plutonium. The temperature produced by the fission bomb is sufficiently high to initiate a thermonuclear reaction where, for example, hydrogen isotopes are fused into helium, releasing even more energy. For fusion to be useful as a commercial energy source, the energy must be released in a steady, controlled manner—unlike the energy in a bomb. To date, scientists have not succeeded in constructing a fusion device that produces more energy on a continual basis than is expended in operating the device. A fusion device uses a high temperature to start a reaction, and under such a condition, all the atoms are completely ionized to form a *plasma* (a gas composed of charged particles, like ${}_{1}^{2}H^{+}$ and e^{-}). The problem is to confine the hot plasma for a long enough time so that collisions among the ions can lead to fusion.

THE PHYSICS OF . . . magnetic confinement and fusion. One ingenious method of confining the plasma is called *magnetic confinement* because it uses a magnetic field to contain and compress the charges in the plasma. Charges moving in the magnetic field are subject to magnetic forces. As the forces increase, the associated pressure builds, and the temperature rises. The gas becomes a superheated plasma, ultimately fusing when the pressure and temperature are high enough.

THE PHYSICS OF . . . inertial confinement and fusion. Another type of confinement scheme, known as *inertial confinement,* is also being developed. Tiny, solid pellets of fuel are dropped into a container. As each pellet reaches the center of the container, a number of high-intensity laser beams strike the pellet simultaneously. The heating causes the exterior of the pellet to vaporize almost instantaneously. However, the inertia of the vaporized atoms keeps them from expanding outward as fast as the vapor is being formed. As a result, high pressures, high densities, and high temperatures are achieved at the center of the pellet, thus causing fusion. A variant of inertial confinement fusion, called "Z pinch," is under development at Sandia National Laboratories in New Mexico. This device would also implode tiny fuel pellets, but without using lasers. Instead, scientists are using a cylindrical array of fine tungsten wires that are connected to a gigantic capacitor. When the capacitor is discharged, a huge current is sent through the wires. The heated wires vaporize almost instantly, generating a hot gas of ions, or plasma. The plasma is driven inward upon itself by the huge magnetic field produced by the current. The compressed plasma becomes superhot and generates a gigantic X-ray pulse. This pulse, it is hoped, would implode the solid fuel pellets to temperatures and pressures at which fusion would occur.

When compared to fission, fusion has some attractive features as an energy source. As we have seen in Example 7, fusion yields more energy per nucleon of fuel than fission does. Moreover, one type of fuel, ${}_{1}^{2}H$ (deuterium), is found in the waters of the oceans and is plentiful, cheap, and relatively easy to separate from the common ${}_{1}^{1}H$ isotope of hydrogen. Fissile materials like naturally occurring uranium ${}_{92}^{235}U$ are much less available, and supplies could be depleted within a century or two. However, the commercial use of fusion to provide cheap energy remains in the future.

Check Your Understanding

(*The answers are given at the end of the book.*)

9. Which one or more of the following statements correctly describe differences between fission and fusion? **(a)** Fission involves the combining of low-mass nuclei to form a more massive nucleus, whereas fusion involves the splitting of a massive nucleus into less massive fragments. **(b)** Fission involves the splitting of a massive nucleus into less massive fragments, whereas fusion involves the combining of low-mass nuclei to form a more massive nucleus. **(c)** More energy per nucleon is released when a fission event occurs than when a fusion event occurs. **(d)** Less energy per nucleon is released when a fission event occurs than when a fusion event occurs.

10. Which one or more of the following statements concerning fission and fusion are true? **(a)** Both fission and fusion reactions are characterized by a mass defect. **(b)** Both fission and fusion reactions always obey the conservation laws of physics. **(c)** Both fission and fusion take advantage of the fact that the binding energy per nucleon varies with the nucleon number of the nucleus.

11. Would the fusion of two nuclei, each with a nucleon number of 60, release energy? Consult **Figure 32.8** as needed.

32.6 Elementary Particles

Setting the Stage

By 1932 the electron, the proton, and the neutron had been discovered and were thought to be nature's three **elementary particles**, in the sense that they were the basic building blocks from which all matter is constructed. Experimental evidence obtained since then, however, shows that several hundred additional particles exist, and scientists no longer believe that the proton and the neutron are elementary particles.

Most of these new particles have masses greater than the electron's mass, and many are more massive than protons or neutrons. Most of the new particles are unstable and decay in times between about 10^{-6} and 10^{-23} s.

Often, new particles are produced by accelerating protons or electrons to high energies and letting them collide with a target nucleus. For example, **Figure 32.10** shows a collision between an energetic proton and a stationary proton. If the incoming proton has sufficient energy, the collision produces an entirely new particle, the *neutral pion* (π^0). The π^0 particle lives for only 8.4×10^{-17} s before it decays into two γ-ray photons. Since the pion did not exist before the collision, it was created from part of the incident proton's energy. Because a new particle such as the neutral pion is often created from energy, it is customary to report the mass of the particle in terms of its equivalent *rest energy* (see Equation 28.5). Often, energy units of MeV are used. For instance, detailed analyses of experiments reveal that the mass of the π^0 particle is equivalent to a rest energy of 135.0 MeV. For comparison, the more massive proton has a rest energy of 938.3 MeV. Analyses of experiments also provide the electric charge and other properties of particles created in high-energy collisions. In the limited space available here, it is not possible to describe all the new particles that have been found. However, we will highlight some of the more significant discoveries.

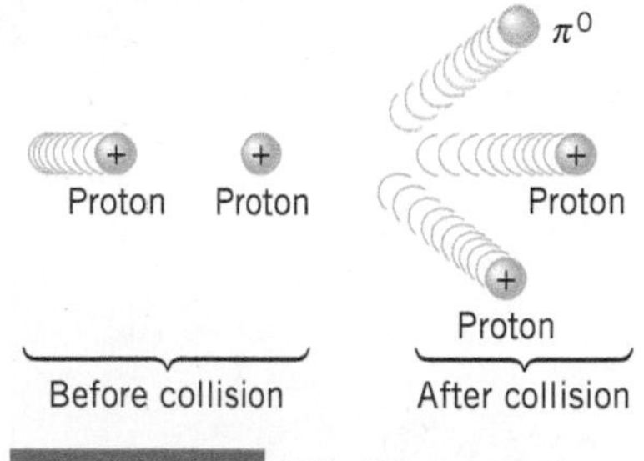

FIGURE 32.10 When an energetic proton collides with a stationary proton, a neutral pion (π^0) is produced. Part of the energy of the incident proton goes into creating the pion.

Neutrinos

In 1930, Wolfgang Pauli suggested that a particle called the **neutrino** (now known as the "electron neutrino") should accompany the β decay of a radioactive nucleus. As Section 31.5 discusses, the neutrino has no electric charge, has a very small mass (a tiny fraction of the mass of an electron), and travels at speeds approaching (but less than) the speed of light. Neutrinos were finally discovered in 1956. Today, neutrinos are created in abundance in nuclear reactors and particle accelerators and are thought to be plentiful in the universe.

Positrons and Antiparticles

The year 1932 saw the discovery of the **positron** (a contraction for "positive electron"). The positron has the same mass as the electron but carries an opposite charge of $+e$. A collision between a positron and an electron is likely to annihilate both particles, converting them into electromagnetic energy in the form of γ rays. For this reason, positrons never coexist with ordinary matter for any appreciable length of time. The mutual annihilation of a positron and an electron lies at the heart of an important medical diagnostic technique, as Conceptual Example 8 discusses.

CONCEPTUAL EXAMPLE 8 | BIO The Physics of PET Scanning

Certain radioactive isotopes decay by positron emission—for example, oxygen $^{15}_{8}O$. In the medical diagnostic technique known as PET scanning (positron emission tomography), such isotopes are injected into the body, where they collect at specific sites. A positron ($^{0}_{1}e$) emitted by the decaying isotope immediately encounters an electron ($^{0}_{-1}e$) in the body tissue, and the resulting mutual annihilation produces two γ-ray photons ($^{0}_{1}e + {}^{0}_{-1}e \rightarrow \gamma + \gamma$), which are detected by devices mounted on a ring around the patient. As **Figure 32.11a** shows, the two photons strike oppositely positioned detectors and, in so doing, reveal the line on which the annihilation occurred. Such information leads to a computer-generated image that can be useful in diagnosing

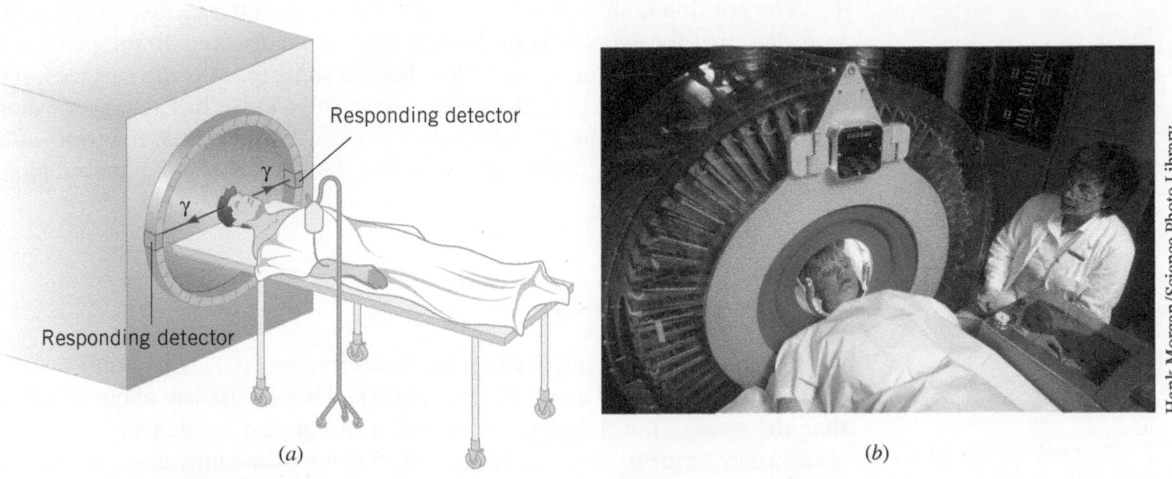

FIGURE 32.11 (*a*) In positron emission tomography, or PET scanning, a radioactive isotope is injected into the body. The isotope decays by emitting a positron, which annihilates an electron in the body tissue, producing two γ-ray photons. These photons strike detectors mounted on opposite sides of a ring that surrounds the patient. (*b*) A patient undergoing a PET scan of the brain.

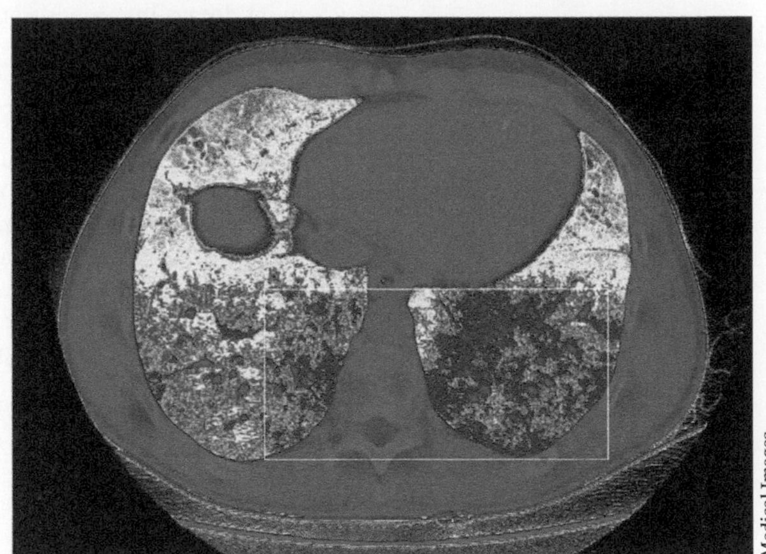

FIGURE 32.12 This PET scan of the lungs reveals pulmonary edema (see the white rectangle), which is an abnormal accumulation of fluid.

abnormalities at the site where the radioactive isotope collects (see Figure 32.12). Which conservation principle accounts for the fact that the photons strike oppositely positioned detectors, the principle of conservation of **(a)** linear momentum or **(b)** energy?

Reasoning Momentum is a vector concept and, therefore, has a direction. The momentum-conservation principle states that the total linear momentum of an isolated system remains constant (see Section 7.2). An isolated system is one on which no net external force acts. Energy is not a vector concept and has no direction associated with it. The energy-conservation principle states that energy can neither be created nor destroyed, but can only be converted from one form to another (see Section 6.8).

Answer (b) is incorrect. Energy is not a vector and has no direction associated with it. Therefore, it cannot, by itself, reveal the directional line along which the γ-ray photons were emitted.

Answer (a) is correct. The positron and the electron constitute an isolated system, so that momentum conservation applies. They do exert electrostatic forces on one another, since they carry electric charges. However, these forces are internal, not external, forces and cannot change the total linear momentum of the two-particle system. The total photon momentum, then, must equal the total momentum of the positron and the electron, which is nearly zero, to the extent that these particles have much less momentum than the photons do. With a total linear momentum of zero, the momentum vectors of the photons must point in opposite directions. Thus, the two photons depart from the annihilation site heading toward oppositely located detectors.

Related Homework: Problem 37

The positron is an example of an antiparticle, and after its discovery scientists came to realize that for every type of particle there is a corresponding type of antiparticle. The antiparticle is a form of matter that has the same mass as the particle but carries an opposite electric charge (e.g., the electron–positron pair) or has a magnetic moment that is oriented in an opposite direction relative to the spin (e.g., the neutrino–antineutrino pair). A few electrically neutral particles, like the photon and the neutral pion (π^0), are their own antiparticles.

Muons and Pions

In 1937, the American physicists S. H. Neddermeyer (1907–1988) and C. D. Anderson (1905–1991) discovered a new charged particle whose mass was about 207 times greater than the mass of the electron. The particle is designated by the Greek letter μ (mu) and is known as a **muon**. There are two muons that have the same mass but opposite charge: the particle μ^- and its antiparticle μ^+. The μ^- muon has the same charge as the electron, whereas the μ^+ muon has the same charge as the positron. Both muons are unstable, and have a lifetime of 2.2×10^{-6} s. The μ^- muon decays into an electron (β^-), a muon neutrino (ν_μ), and an electron antineutrino ($\bar{\nu}_e$) according to the following reaction:

$$\mu^- \longrightarrow \beta^- + \nu_\mu + \bar{\nu}_e$$

The μ^+ muon decays into a positron (β^+), a muon antineutrino ($\bar{\nu}_\mu$), and an electron neutrino (ν_e):

$$\mu^+ \longrightarrow \beta^+ + \bar{\nu}_\mu + \nu_e$$

Muons interact with protons and neutrons via the weak nuclear force (see Section 31.5).

The Japanese physicist Hideki Yukawa (1907–1981) predicted in 1935 that **pions** exist, but they were not discovered until 1947. Pions come in three varieties: one that is positively charged, the negatively charged antiparticle with the same mass, and the neutral pion, mentioned earlier, which is its own antiparticle. The symbols for these pions are, respectively, π^+, π^-, and π^0. The charged pions are unstable and have a lifetime of 2.6×10^{-8} s. The decay of a charged pion almost always produces a muon:

$$\pi^- \longrightarrow \mu^- + \bar{\nu}_\mu$$
$$\pi^+ \longrightarrow \mu^+ + \nu_\mu$$

As mentioned earlier, the neutral pion π^0 is also unstable and decays into two γ-ray photons, the lifetime being 8.4×10^{-17} s. The pions are of great interest because, unlike the muons, the pions interact with protons and neutrons via the strong nuclear force.

Classification of Particles

It is useful to group the known particles into three families—the bosons, the leptons, and the hadrons—as **Table 32.3** summarizes. The **boson family** is composed of particles (all are types of bosons) that play central roles in nature's three fundamental forces (see Section 4.6). The photon is associated with the electromagnetic force, which is one manifestation of the electroweak force. The W^-, W^+, and Z^0 particles are associated with the weak nuclear force, which is the other manifestation of the electroweak force. The gluons are associated with the strong nuclear force, whereas the graviton is thought to be associated with the gravitational force.

The **lepton family** consists of particles that interact by means of the *weak nuclear force.* Leptons can also exert gravitational and (if the leptons are charged) electromagnetic forces on other particles. The four better-known leptons are the electron, the muon, the electron neutrino ν_e, and the muon neutrino ν_μ. **Table 32.3** lists these particles together with their antiparticles. Two other leptons have also been discovered, the tau particle τ and its neutrino ν_τ, bringing the number of particles in the lepton family to six.

The **hadron family** contains the particles that interact by means of *both the strong nuclear force and the weak nuclear force.* Hadrons can also interact by gravitational and electromagnetic forces, but at short distances ($\lesssim 10^{-15}$ m) the strong nuclear force dominates. Among the hadrons are the proton, the neutron, and the pions. As **Table 32.3** indicates, most hadrons are short-lived. The hadrons are subdivided into two groups, the **mesons** and the **baryons**, for a reason that will be discussed in connection with the idea of quarks.

TABLE 32.3 Some Particles and Their Properties

Family	Particle	Particle Symbol	Antiparticle Symbol	Rest Energy (MeV)	Lifetime (s)
Boson[a]	Photon	γ	Self[b]	0	Stable
	$W^{\pm}$	W^-	W^+	8.04×10^4	3×10^{-25}
	Z^0	Z^0	Self[b]	9.12×10^4	3×10^{-25}
	Gluons[c]	g	—	0	—
	Graviton[c]	—	—	0	—
Lepton	Electron	e^- or β^-	e^+ or β^+	0.511	Stable
	Muon	μ^-	μ^+	105.7	2.2×10^{-6}
	Tau	τ^-	τ^+	1777	2.9×10^{-13}
	Electron neutrino	ν_e	$\bar{\nu}_e$	$<2 \times 10^{-6}$	Stable
	Muon neutrino	ν_μ	$\bar{\nu}_\mu$	<0.19	Stable
	Tau neutrino	ν_τ	$\bar{\nu}_\tau$	<18.2	Stable
Hadron					
Mesons					
	Pion	π^+	π^-	139.6	2.6×10^{-8}
		π^0	Self[b]	135.0	8.4×10^{-17}
	Kaon	K^+	K^-	493.7	1.2×10^{-8}
		K^0	$\bar{K}^0$	497.6	—[d]
	Eta	η^0	Self[b]	547.3	$<10^{-18}$
Baryons					
	Proton	p	$\bar{p}$	938.3	Stable
	Neutron	n	$\bar{n}$	939.6	886
	Lambda	Λ^0	$\bar{\Lambda}^0$	1116	2.6×10^{-10}
	Sigma	Σ^+	$\bar{\Sigma}^-$	1189	8.0×10^{-11}
		Σ^0	$\bar{\Sigma}^0$	1193	7.4×10^{-20}
		Σ^-	$\bar{\Sigma}^+$	1197	1.5×10^{-10}
	Omega	Ω^-	Ω^+	1672	8.2×10^{-11}

[a]The particles in this family are types of bosons that are associated with (or mediate) nature's fundamental forces.
[b]The particle is its own antiparticle.
[c]Free gluons and the graviton have not been observed experimentally.
[d]This particle and its antiparticle do not have definite lifetimes.

TABLE 32.4 Quarks and Antiquarks

	Quarks		Antiquarks	
Name	Symbol	Charge	Symbol	Charge
Up	u	$+\frac{2}{3}e$	$\overline{u}$	$-\frac{2}{3}e$
Down	d	$-\frac{1}{3}e$	$\overline{d}$	$+\frac{1}{3}e$
Strange	s	$-\frac{1}{3}e$	$\overline{s}$	$+\frac{1}{3}e$
Charmed	c	$+\frac{2}{3}e$	$\overline{c}$	$-\frac{2}{3}e$
Top	t	$+\frac{2}{3}e$	$\overline{t}$	$-\frac{2}{3}e$
Bottom	b	$-\frac{1}{3}e$	$\overline{b}$	$+\frac{1}{3}e$

Quarks

As more and more hadrons were discovered, it became clear that they were not all elementary particles. The suggestion was made that the hadrons are made up of smaller, more elementary particles called **quarks**. In 1963, a quark theory was advanced independently by M. Gell-Mann (1929–2019) and G. Zweig (1937–). The theory proposed that there are three quarks and three corresponding antiquarks, and that hadrons are constructed from combinations of these. Thus, the quarks are elevated to the status of elementary particles for the hadron family. The particles in the lepton family are considered to be elementary, and as such they are not composed of quarks.

The three quarks were named *up* (u), *down* (d), and *strange* (s), and were assumed to have, respectively, fractional charges of $+\frac{2}{3}e$, $-\frac{1}{3}e$, and $-\frac{1}{3}e$. In other words, a quark possesses a charge magnitude smaller than that of an electron, which has a charge of $-e$. **Table 32.4** lists the symbols and electric charges of these quarks and the corresponding antiquarks. Experimentally, quarks should be recognizable by their fractional charges, but in spite of an extensive search for them, free quarks have never been found.

According to the original quark theory, the mesons are different from the baryons, because each meson consists of only two quarks—a quark and an antiquark—whereas a baryon contains three quarks. For instance, the π^- pion (a meson) is composed of a d quark and a $\overline{u}$ antiquark, $\pi^- = d + \overline{u}$, as **Figure 32.13** shows. These two quarks combine to give the π^- pion a net charge of $-e$. Similarly, the π^+ pion is a combination of the $\overline{d}$ and u quarks, $\pi^+ = \overline{d} + u$. In contrast, protons and neutrons, being baryons, consist of three quarks. A proton contains the combination $d + u + u$, and a neutron contains the combination $d + d + u$ (see **Figure 32.13**). These groups of three quarks give the correct charges for the proton and neutron.

The original quark model was extremely successful in predicting not only the correct charges for the hadrons, but other properties as well. However, in 1974 a new particle, the J/ψ meson, was discovered. This meson has a rest energy of 3097 MeV, much larger than the rest energies of other known mesons. The existence of the J/ψ meson could be explained only if a new quark–antiquark pair existed; this new quark was called *charmed* (c). With the discovery of more and more particles, it has been necessary to postulate a fifth and a sixth quark; their names are *top* (t) and *bottom* (b), although some scientists prefer to call these quarks *truth* and *beauty*. Today, there is firm evidence for all six quarks, each with its corresponding antiquark. All of the hundreds of known hadrons can be accounted for in terms of these six quarks and their antiquarks.

In addition to electric charge, quarks also have other properties. For example, each quark possesses a characteristic called **color**, for which there are three possibilities: blue, green, or red. The corresponding possibilities for the antiquarks are antiblue, antigreen, and antired. The use of the term "color" and the specific choices of blue, green, and red are arbitrary, for the visible colors of the electromagnetic spectrum have nothing to do with quark properties. The quark property of color, however, is important, because it

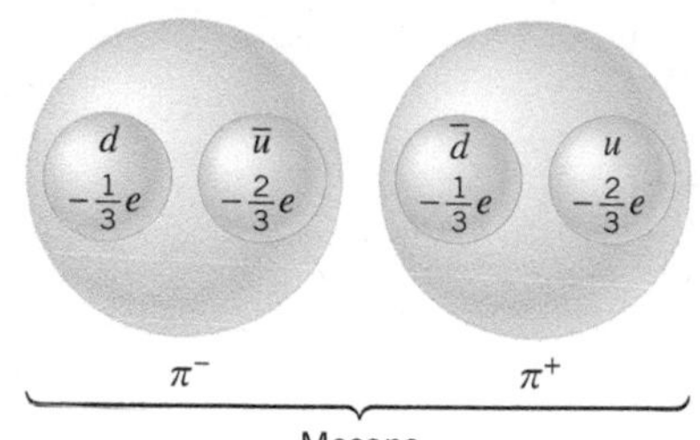

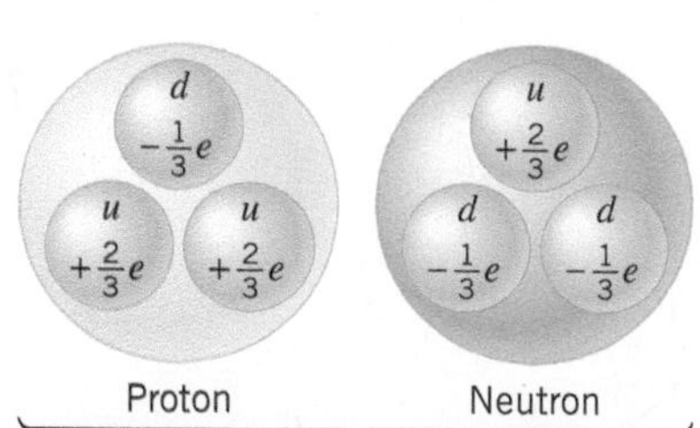

FIGURE 32.13 According to the original quark model of hadrons, all mesons consist of a quark and an antiquark, whereas baryons contain three quarks.

brings the quark model into agreement with the Pauli exclusion principle and enables the model to account for experimental observations that are otherwise difficult to explain.

The Standard Model

The various elementary particles that have been discovered can interact via one or more of the following four forces: the gravitational force, the strong nuclear force, the weak nuclear force, and the electromagnetic force. In particle physics, the phrase "**the standard model**" refers to the currently accepted explanation for the strong nuclear force, the weak nuclear force, and the electromagnetic force. In this model, the strong nuclear force between quarks is described in terms of the concept of color, and the theory is referred to as quantum chromodynamics. According to the standard model, the weak nuclear force and the electromagnetic force are separate manifestations of a single, even more fundamental, force, referred to as the electroweak force, as we have seen in Section 31.5.

In the standard model, our understanding of the building blocks of matter follows the hierarchical pattern illustrated in **Interactive Figure 32.14**. Molecules, such as water (H_2O) and glucose ($C_6H_{12}O_6$), are composed of atoms. Each atom consists of a nucleus that is surrounded by a cloud of electrons. The nucleus, in turn, is made up of protons and neutrons, which are composed of quarks.

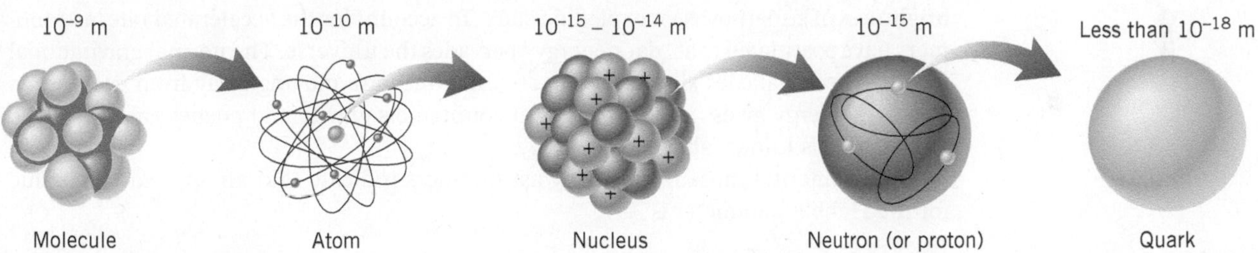

INTERACTIVE FIGURE 32.14 The current view of how matter is composed of basic units, starting with a molecule and ending with a quark. The approximate sizes of each unit are also listed.

Check Your Understanding

(The answers are given at the end of the book.)

12. Of the following particles, which ones are *not* composed of quarks and antiquarks? **(a)** A proton **(b)** An electron **(c)** A neutron **(d)** A neutrino

13. The sigma-minus particle Σ^- has a charge of $-e$. Which one of the following quark combinations corresponds to this particle? **(a)** *dds* **(b)** *ssu* **(c)** *uus*

32.7 Cosmology

Cosmology is the study of the structure and evolution of the universe. In this study, both the very large and the very small aspects of the universe are important. Astronomers, for example, study stars located at enormous distances from the earth, up to billions of light-years away. In contrast, particle physicists focus their efforts on the very small elementary particles (10^{-18} m or smaller) that comprise matter. The synergy between the work of astronomers and that of particle physicists has led to significant advances in our understanding of the universe. Central to that understanding is the belief that the universe is expanding, and we begin by discussing the evidence that justifies this belief.

The Expanding Universe and the Big Bang

THE PHYSICS OF . . . an expanding universe. The idea that the universe is expanding originated with the astronomer Edwin P. Hubble (1889–1953). He found that light reaching the earth from distant galaxies is Doppler-shifted toward greater wavelengths—that is, toward the red end of the visible spectrum. As Section 24.5 discusses, this type of Doppler shift results when the observer and the source of the light are moving away from each other. The speed at which a galaxy is receding from the earth can be determined from the measured Doppler shift in wavelength. Hubble found that a galaxy located at a distance d from the earth recedes from the earth at a speed v given by

Hubble's law

$$v = Hd \tag{32.5}$$

where H is a constant known as the Hubble parameter. In other words, the recession speed is proportional to the distance d, so that more distant galaxies are moving away from the earth at greater speeds. Equation 32.5 is referred to as Hubble's law.

Hubble's picture of an expanding universe does not mean that the earth is at the center of the expansion. In fact, there is no literal center. Imagine a loaf of raisin bread expanding as it bakes. Each raisin moves away from every other raisin, without any single one acting as a center for the expansion. Galaxies in the universe behave in a similar fashion. Observers in other galaxies would see distant galaxies moving away, just as we do.

THE PHYSICS OF . . . "dark energy." Not only is the universe expanding, it is doing so at an accelerated rate, according to recent astronomical measurements of the brightness of supernovas, or exploding stars. To account for the accelerated rate, astronomers have postulated that "dark energy" pervades the universe. The normal gravitational force between galaxies slows the rate at which they are moving away from each other. The dark energy gives rise to a force that counteracts gravity and pushes galaxies apart. As yet, little is known about dark energy.

Experimental measurements by astronomers indicate that an approximate value for the Hubble parameter is

$$H = 0.022 \frac{\text{m}}{\text{s} \cdot \text{light-year}}$$

The value for the Hubble parameter is believed to be accurate within 10%. Scientists are very interested in obtaining an accurate value for H because it can be related to an age for the universe, as the next example illustrates.

EXAMPLE 9 | An Age for the Universe

Determine an estimate of the age of the universe using Hubble's law.

Reasoning Consider a galaxy currently located at a distance d from the earth. According to Hubble's law, this galaxy is moving away from us at a speed of $v = Hd$. At an earlier time, therefore, this galaxy must have been closer. We can imagine, in fact, that in the remote past the separation distance was relatively small and that the universe originated at such a time. To estimate the age of the universe, we calculate the time it has taken the galaxy to recede to its present position. For this purpose, time is simply distance divided by speed, or $t = d/v$.

Solution Using Hubble's law and the fact that a distance of 1 light-year is 9.46×10^{15} m, we estimate the age of the universe to be

$$t = \frac{d}{v} = \frac{d}{Hd} = \frac{1}{H}$$

$$t = \frac{1}{0.022 \dfrac{\text{m}}{\text{s} \cdot \text{light-year}}} = \frac{1}{\left(0.022 \dfrac{\cancel{\text{m}}}{\text{s} \cdot \cancel{\text{light-year}}}\right)\left(\dfrac{1\ \cancel{\text{light-year}}}{9.46 \times 10^{15}\ \cancel{\text{m}}}\right)}$$

$$= 4.3 \times 10^{17}\ \text{s} \quad \text{or} \quad \boxed{1.4 \times 10^{10}\ \text{yr}}$$

The idea presented in Example 9, that our galaxy and other galaxies in the universe were very close together at some earlier instant in time, lies at the heart of the **Big Bang theory**. This theory postulates that the universe had a definite beginning in a cataclysmic event, sometimes called the primeval fireball. Dramatic evidence supporting the theory was discovered in 1965 by Arno A. Penzias (1933–) and Robert W. Wilson (1936–). Using a radio telescope, they discovered that the earth is being bathed in weak electro-magnetic waves in the microwave region of the spectrum (wavelength = 7.35 cm, see **Figure 24.9**). They observed that the intensity of these waves is the same, no matter where in the sky they pointed their telescope, and concluded that the waves originated outside of our galaxy. This microwave background radiation, as it is called, represents radiation left over from the Big Bang and is a kind of cosmic afterglow. Subsequent measurements have confirmed the research of Penzias and Wilson and have shown that the microwave radiation is consistent with a perfect blackbody (see Sections 13.3 and 29.2) radiating at a temperature of 2.7 K, in agreement with theoretical analysis of the Big Bang. In 1978, Penzias and Wilson received the Nobel Prize in Physics for their discovery.

The Standard Model for the Evolution of the Universe

Based on the recent experimental and theoretical research in particle physics, scientists have proposed an evolutionary sequence of events following the Big Bang. This sequence is known as the **standard cosmological model** and is illustrated in **Figure 32.15**.

Immediately following the Big Bang, the temperature of the universe was incredibly high, about 10^{32} K. During this initial period, the three fundamental forces (the gravitational force, the strong nuclear force, and the electroweak force) all behaved as a single unified force. Very quickly, in about 10^{-43} s, the gravitational force took on a separate identity all its own, as **Figure 32.15** indicates. Meanwhile, the strong nuclear force and the electroweak force continued to act as a single force, which is sometimes referred to as the GUT force. GUT stands for the **g**rand **u**nified **t**heory that presumably would explain such a force. Slightly later, at about 10^{-35} s after the Big Bang, the GUT force separated into the strong nuclear force and the electroweak force, the universe expanding and cooling somewhat to a temperature of roughly 10^{28} K (see **Figure 32.15**). From this point on, the strong nuclear force behaved as we know it today, while the electroweak force maintained its identity. In this scenario, note that the weak nuclear force and the electromagnetic force have not yet manifested themselves as separate entities. The disappearance of the electroweak force and the appearance of the weak nuclear force and the electromagnetic force eventually occurred at approximately 10^{-10} s after the Big Bang, when the temperature of the expanding universe had cooled to about 10^{15} K.

From the Big Bang up until the strong nuclear force separated from the GUT force at a time of 10^{-35} s, all particles of matter were similar, and there was no distinction between quarks and leptons. After this time, quarks and leptons became distinguishable. Eventually the quarks and antiquarks formed hadrons, such as protons and neutrons and their

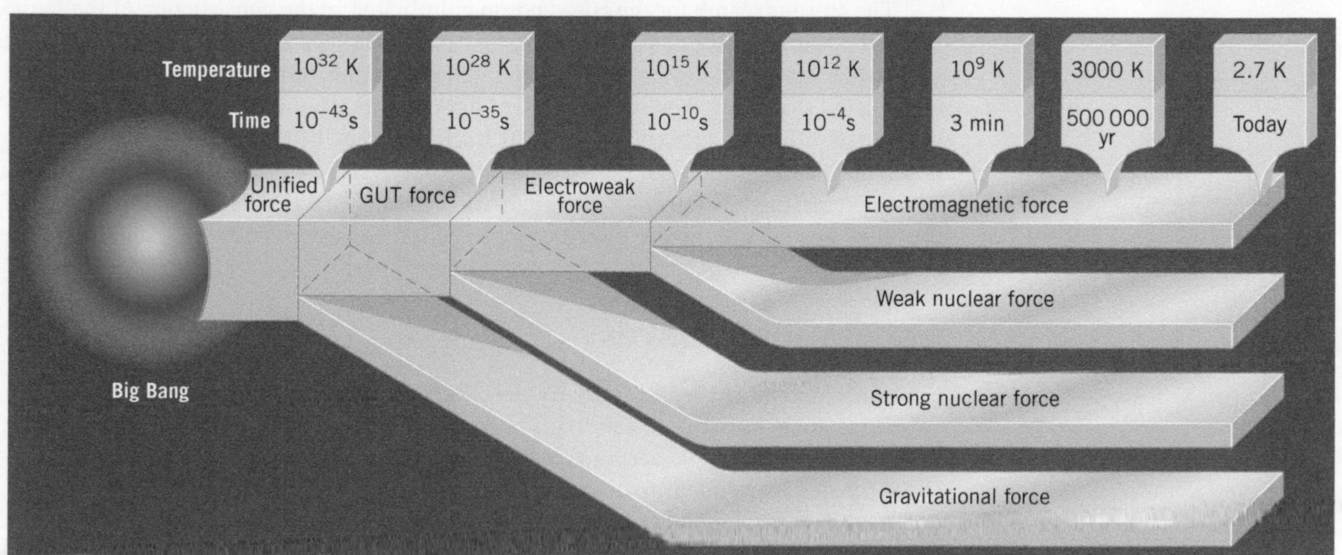

FIGURE 32.15 According to the standard cosmological model, the universe has evolved as illustrated here. In this model, the universe is presumed to have originated with a cataclysmic event known as the Big Bang. The times shown are those following this event.

antiparticles. By a time of 10^{-4} s after the Big Bang, however, the temperature had cooled to approximately 10^{12} K, and the hadrons had mostly disappeared. Protons and neutrons survived only as a very small fraction of the total number of particles, the majority of which were leptons such as electrons, positrons, and neutrinos. Like most of the hadrons before them, most of the electrons and positrons eventually disappeared. However, they did leave behind a relatively small number of electrons to join the small number of protons and neutrons at a time of about 3 min following the Big Bang. At this time the temperature of the expanding universe had decreased to about 10^9 K, and small nuclei such as that of helium began forming. Later, when the universe was about 500 000 years old and the temperature had dropped to near 3000 K, hydrogen and helium atoms began forming. As the temperature decreased further, stars and galaxies formed, and today we find a temperature of 2.7 K characterizing the cosmic background radiation of the universe.

32.8 Recent Discoveries in Modern Physics

The advancement of science is a continuing process that trudges along slowly at times, but sometimes leaps forward with entirely new and unexpected discoveries, or observations that verify or disprove the predictions of longstanding theories. In the past decade, two Nobel Prizes in Physics have been awarded as the result of two important experiments: the discovery of the **Higgs boson** and the observation of **gravitational waves**.

The Higgs Boson

In the **standard model** (see Section 32.6), a type of field is required in order to explain certain symmetry properties of particles, and to give the predicted particles their correct masses. This field, which is assumed to exist in all space, is called the Higgs field, after Peter Ware Higgs (1929–), the British physicist who contributed to its theoretical development and predictions. The Higgs boson, sometimes referred to as the God particle, is a manifestation of the Higgs field, and therefore its detection would constitute strong evidence of the existence of the theoretically predicted field, and confirm the predictions of the standard model.

The 40-year search for the Higgs boson culminated in the construction of the Large Hadron Collider (LHC) at CERN, a particle accelerator located on the border of France and Switzerland. The device produces high-energy beams of particles (e.g., protons and lead nuclei) that travel close to the speed of light and then collide, breaking up into a cornucopia of different particles. Early in 2013, scientists at CERN confirmed that the Higgs boson had been detected. The 2013 Nobel Prize in Physics was awarded to Peter W. Higgs and François Englert (1932–) for their theoretical work that led to the discovery of the Higgs boson, and therefore verified a prediction consistent with the standard model.

Gravitational Waves

Einstein's general theory of relativity describes, among other things, the curvature of spacetime near massive objects. This warping of spacetime amounts to a geometric description of gravity that accounts for the bending of light around stars, which was precisely the one experiment that was used to verify this prediction of the theory. In 1919, the results of the Eddington experiment, organized by the British astronomers Frank Watson Dyson and Arthur Stanley Eddington, were reported, confirming the predicted bending of starlight around the sun.

Another prediction of the general theory of relativity is the creation of disturbances in spacetime by accelerating masses in a way that is analogous to the production of electromagnetic waves by accelerating charges. These ripples in spacetime are called

Courtesy Caltech/MIT/LIGO Laboratory

FIGURE 32.16 The Laser Interferometer Gravitational-Wave Observatory (LIGO) at Livingston, Louisiana.

gravitational waves, and they travel at the speed of light. They can only be detected using very sensitive instrumentation, such as the enormous laser interferometer at the Laser Interferometer Gravitational-Wave Observatory (LIGO) at Livingston, Louisiana (see **Figure 32.16**), and an identical one 3000 miles away at Hanover, Washington. With two perpendicular arms, each 4 kilometers in length, the LIGO interferometers can detect very slight distortions in their lengths, on the order of 1/10 000 the width of a proton, a miniscule effect that would occur as the result of the spacetime distortions produced by a gravitational wave as it passes through. Even though the LIGO detectors are incredibly sensitive, the intensity of gravitational waves is so weak that it requires collisions in the universe that are, well, on an astronomical scale. LIGO can detect gravity waves from events such as the merger of two neutron stars or the collision of two black holes. In order for a gravitational-wave detection to be considered legitimate, it would have to be detected by both facilities.

Albert Einstein predicted the existence of gravitational waves in 1916. Nearly a century later, on September 14, 2015, the LIGO collaboration detected the first gravitational waves due to the merging of two back holes 1.3 billion light-years from earth. Not only did this observation confirm the prediction of Einstein's general theory of relativity, but it provided a new observational tool for astronomers, and it will continue to offer valuable information to astrophysicists and cosmologists in the future. Rainer Weiss (1932–), Kip Thorne (1940–), and Barry Barish (1936–) were awarded the 2017 Nobel Prize in Physics for their role in the direct detection of gravitational waves.

Concept Summary

32.1 Biological Effects of Ionizing Radiation Ionizing radiation consists of photons and/or moving particles that have enough energy to ionize an atom or molecule. Exposure is a measure of the ionization produced in air by X-rays or γ rays. When a beam of X-rays or γ rays is sent through a mass m of dry air (0 °C, 1 atm pressure) and produces positive ions whose total charge is q, the exposure in coulombs per kilogram (C/kg) is

$$\text{Exposure (in coulombs per kilogram)} = \frac{q}{m}$$

With q in coulombs (C) and m in kilograms (kg), the exposure in roentgens is given by Equation 32.1.

$$\begin{matrix}\text{Exposure}\\ \text{(in roentgens)}\end{matrix} = \left(\frac{1}{2.58 \times 10^{-4}}\right)\frac{q}{m} \qquad \textbf{(32.1)}$$

The absorbed dose is the amount of energy absorbed from the radiation per unit mass of absorbing material, as specified by Equation 32.2.

The SI unit of absorbed dose is the gray (Gy); 1 Gy = 1 J/kg. However, the rad is another unit that is often used: 1 rad = 0.01 Gy.

$$\text{Absorbed dose} = \frac{\text{Energy absorbed}}{\text{Mass of absorbing material}} \qquad \textbf{(32.2)}$$

The amount of biological damage produced by ionizing radiation is different for different types of radiation. The relative biological effectiveness (RBE) is the absorbed dose of 200-keV X-rays required to produce a certain biological effect divided by the dose of a given type of radiation that produces the same biological effect, as given by Equation 32.3.

$$\text{RBE} = \frac{\begin{matrix}\text{The dose of 200-keV X-rays that}\\ \text{produces a certain biological effect}\end{matrix}}{\begin{matrix}\text{The dose of radiation that}\\ \text{produces the same biological effect}\end{matrix}} \qquad \textbf{(32.3)}$$

The biologically equivalent dose (in rems) is the product of the absorbed dose (in rads) and the RBE, as shown in Equation 32.4.

$$\text{Biologically equivalent dose (in rems)} = \text{Absorbed dose (in rads)} \times \text{RBE} \quad \textbf{(32.4)}$$

32.2 Induced Nuclear Reactions An induced nuclear reaction occurs whenever a target nucleus is struck by an incident nucleus, an atomic or subatomic particle, or a γ-ray photon and undergoes a change as a result. An induced nuclear transmutation is a reaction in which the target nucleus is changed into a nucleus of a new element.

All nuclear reactions (induced or spontaneous) obey the conservation laws of physics as they relate to mass/energy, electric charge, linear momentum, angular momentum, and nucleon number.

Nuclear reactions are often written in a shorthand form, such as ${}^{14}_{7}\text{N}\ (\alpha, p)\ {}^{17}_{8}\text{O}$. The first and last symbols ${}^{14}_{7}\text{N}$ and ${}^{17}_{8}\text{O}$ denote, respectively, the initial and final nuclei. The symbols within the parentheses denote the incident α particle (on the left) and the small emitted particle or proton p (on the right).

A thermal neutron is one that has a kinetic energy of about 0.04 eV.

32.3 Nuclear Fission Nuclear fission occurs when a massive nucleus splits into two less massive fragments. Fission can be induced by the absorption of a thermal neutron. When a massive nucleus fissions, energy is released because the binding energy per nucleon is greater for the fragments than for the original nucleus. Neutrons are also released during nuclear fission. These neutrons can, in turn, induce other nuclei to fission and lead to a process known as a chain reaction. A chain reaction is said to be controlled if each fission event contributes, on average, only one neutron that fissions another nucleus.

32.4 Nuclear Reactors A nuclear reactor is a device that generates energy by a controlled chain reaction. Many reactors in use today have the same three basic components: fuel elements, a neutron moderator, and control rods. The fuel elements contain the fissile fuel, and the entire region of fuel elements is known as the reactor core. The neutron moderator is a material (water, for example) that slows down the neutrons released in a fission event to thermal energies so they can initiate additional fission events. Control rods contain material that readily absorbs neutrons without fissioning. They are used to keep the reactor in its normal, or critical, state, in which each fission event leads to one additional fission, no more, no less. The reactor is subcritical when, on average, the neutrons from each fission trigger less than one subsequent fission. The reactor is supercritical when, on average, the neutrons from each fission trigger more than one additional fission.

32.5 Nuclear Fusion In a fusion process, two nuclei with smaller masses combine to form a single nucleus with a larger mass. Energy is released by fusion when the binding energy per nucleon is greater for the larger nucleus than for the smaller nuclei. Fusion reactions are said to be thermonuclear because they require extremely high temperatures to proceed. Current studies of nuclear fusion utilize either magnetic confinement or inertial confinement to contain the fusing nuclei at the high temperatures that are necessary.

32.6 Elementary Particles Subatomic particles are divided into three families: the boson family (which includes the photon), the lepton family (which includes the electron), and the hadron family (which includes the proton and the neutron). Elementary particles are the basic building blocks of matter. All members of the boson and lepton families are elementary particles.

The quark theory proposes that the hadrons are not elementary particles but are composed of elementary particles called quarks. Currently, the hundreds of hadrons can be accounted for in terms of six quarks (up, down, strange, charmed, top, and bottom) and their antiquarks.

The standard model consists of two parts: (1) the currently accepted explanation for the strong nuclear force in terms of the quark concept of "color" and (2) the theory of the electroweak interaction.

32.7 Cosmology Cosmology is the study of the structure and evolution of the universe. Our universe is expanding. The speed v at which a distant galaxy recedes from the earth is given by Hubble's law (Equation 32.5), where $H = 0.022$ m/(s · light-year) is called the Hubble parameter and d is the distance of the galaxy from the earth.

$$v = Hd \quad \textbf{(32.5)}$$

The Big Bang theory postulates that the universe had a definite beginning in a cataclysmic event, sometimes called the primeval fireball. The radiation left over from this event is in the microwave region of the electromagnetic spectrum, and it is consistent with a perfect blackbody radiating at a temperature of 2.7 K, in agreement with theoretical analysis of the Big Bang.

The standard cosmological model for the evolution of the universe is summarized in **Figure 32.15**.

32.8 Recent Discoveries in Modern Physics In the last decade, we have witnessed profound discoveries in elementary particle physics. The 2013 Nobel Prize in physics was awarded to Peter W. Higgs and François Englert for their theoretical work that led to the discovery of the Higgs boson, or "God particle" that verified a prediction consistent with the standard model of particle physics that explains the building blocks of matter.

Rainer Weiss, Kip Thorne, and Barry Barish were awarded the 2017 Nobel Prize in physics for their role in the direct detection of gravitational waves, ripples in spacetime that move at the speed of light and were first predicted by Albert Einstein in 1916. These disturbances are created by the motion of matter, such as merging neutron stars, and their detection not only confirmed Einstein's general theory of relativity but demonstrated a new tool to be used by astronomers and astrophysicists in the future.

Focus on Concepts

Online

Additional questions are available for assignment in WileyPLUS.

Section 32.1 Biological Effects of Ionizing Radiation

1. Biologically equivalent doses are specified in units called_______.
(a) rads (b) grays (c) rems (d) J/kg (e) roentgens

Section 32.2 Induced Nuclear Reactions

2. Determine the unknown nuclear species ${}^{A}_{Z}\text{X}$ in the following nuclear reaction:

$${}^{A}_{Z}\text{X} + {}^{14}_{7}\text{N} \longrightarrow {}^{14}_{6}\text{C} + {}^{1}_{1}\text{H}$$

(a) ${}^{2}_{1}\text{H}$ (b) ${}^{1}_{0}\text{n}$ (c) γ ray (d) ${}^{4}_{2}\text{He}$

Section 32.3 Nuclear Fission

3. The fission of $^{235}_{92}U$ can occur via many different reactions. In general, they can be written as follows:

$$^1_0n + ^{235}_{92}U \longrightarrow ^{A_X}_{Z_X}X + ^{A_Y}_{Z_Y}Y + \eta\,^1_0n$$

where X and Y refer to the identities of the fission fragments and η is the number of neutrons produced. Which one or more of the following statements are true?

A. The compound nucleus that is formed to initiate the fission process is the same, no matter what X and Y refer to.

B. The greater the number η of neutrons produced by the reaction, the smaller is the sum of the nucleon numbers A_X and A_Y.

C. The sum of the atomic numbers Z_X and Z_Y is the same, no matter what X and Y refer to.

(a) A (b) A and B (c) A and C (d) B and C (e) A, B, and C

Section 32.5 Nuclear Fusion

4. In each of the following three nuclear fusion reactions, the masses of the nuclei are given beneath each nucleus. Rank the energy produced by each reaction in descending order (greatest first).

Reaction I $\quad \underbrace{^2_1H}_{2.0141\text{ u}} + \underbrace{^2_1H}_{2.0141\text{ u}} \longrightarrow \underbrace{^3_2He}_{3.0160\text{ u}} + \underbrace{^1_0n}_{1.0087\text{ u}}$

Reaction II $\quad \underbrace{^3_2He}_{3.0160\text{ u}} + \underbrace{^3_2He}_{3.0160\text{ u}} \longrightarrow \underbrace{^4_2He}_{4.0026\text{ u}} + \underbrace{^1_1H}_{1.0078\text{ u}} + \underbrace{^1_1H}_{1.0078\text{ u}}$

Reaction III $\quad \underbrace{^{15}_7n}_{15.0001\text{ u}} + \underbrace{^1_1H}_{1.0078\text{ u}} \longrightarrow \underbrace{^{12}_6C}_{12.0000\text{ u}} + \underbrace{^4_2He}_{4.0026\text{ u}}$

(a) I, II, III (b) I, III, II (c) II, III, I (d) II, I, III (e) III, II, I

Section 32.6 Elementary Particles

5. The drawings show four possibilities for hadrons in the quark theory. In each of the possibilities, the symbols for the quarks are shown together with the corresponding electric charges. Note that e stands for the magnitude of the charge on an electron. Which one shows the quark structure for an antiproton? (a) A (b) B (c) C (d) D

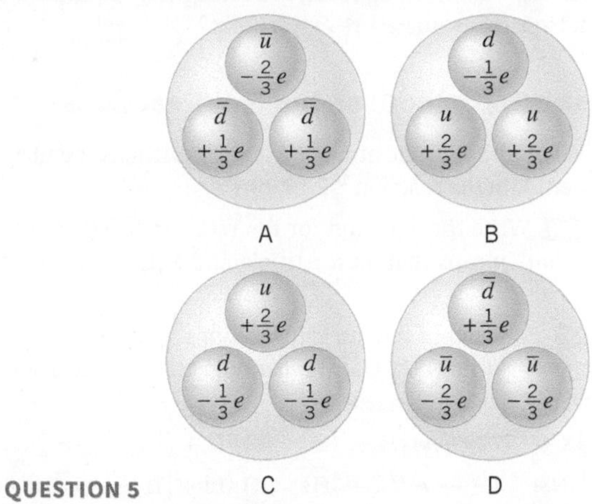

QUESTION 5

Section 32.7 Cosmology

6. Which one of the following statements is *not* accepted as a part of the current picture of the universe? (a) The universe is expanding. (b) Early in the history of the universe, there was only one fundamental force. (c) The universe began with a cataclysmic event known as the Big Bang. (d) The weak electromagnetic radiation in the microwave region of the spectrum that bathes the earth provides evidence of the Big Bang theory. (e) Hubble's law indicates that a galaxy located at a distance d from the earth is moving away from the earth at a speed that is inversely proportional to d.

Problems

Online

Additional questions are available for assignment in WileyPLUS.

SSM Student Solutions Manual

MMH Problem-solving help

GO Guided Online Tutorial

V-HINT Video Hints

CHALK Chalkboard Videos

BIO Biomedical application

E Easy

M Medium

H Hard

WS Worksheet

T Team Problem

Section 32.1 Biological Effects of Ionizing Radiation

1. E SSM Neutrons (RBE = 2.0) and α particles have the same biologically equivalent dose. However, the absorbed dose of the neutrons is six times the absorbed dose of the α particles. What is the RBE for the α particles?

2. E What absorbed dose (in rads) of α particles (RBE = 15) causes as much biological damage as a 60-rad dose of protons (RBE = 10)?

3. E BIO Over a full course of treatment, two different tumors are to receive the same absorbed dose of therapeutic radiation. The smaller of the tumors (mass = 0.12 kg) absorbs a total of 1.7 J of energy. (a) Determine the absorbed dose, in Gy. (b) What is the total energy absorbed by the larger of the tumors (mass = 0.15 kg)?

4. E Over a year's time, a person receives a biologically equivalent dose of 24 mrem (millirems) from cosmic rays, which consist primarily of high-energy protons bombarding earth's atmosphere from space. The relative biological effectiveness of protons is 10. (a) What is the person's absorbed dose in rads? (b) The person absorbs 1.9×10^{-3} J of energy from cosmic rays in a year. What is the person's mass?

5. E BIO GO SSM A beam of particles is directed at a 0.015-kg tumor. There are 1.6×10^{10} particles per second reaching the tumor, and the energy of each particle is 4.0 MeV. The RBE for the radiation is 14. Find the biologically equivalent dose given to the tumor in 25 s.

6. E GO A 75-kg person is exposed to 45 mrem of α particles (RBE = 12). How much energy (in joules) has this person absorbed?

7. E GO The biologically equivalent dose for a typical chest X-ray is 2.5×10^{-2} rem. The mass of the exposed tissue is 21 kg, and it absorbs 6.2×10^{-3} J of energy. What is the relative biological effectiveness (RBE) for the radiation on this particular type of tissue?

8. M BIO A 2.0-kg tumor is being irradiated by a radioactive source. The tumor receives an absorbed dose of 12 Gy in a time of 850 s. Each

disintegration of the radioactive source produces a particle that enters the tumor and delivers an energy of 0.40 MeV. What is the activity $\Delta N/\Delta t$ (see Section 31.6) of the radioactive source?

9. **M** **CHALK** **SSM** A beam of nuclei is used for cancer therapy. Each nucleus has an energy of 130 MeV, and the relative biological effectiveness (RBE) of this type of radiation is 16. The beam is directed onto a 0.17-kg tumor, which receives a biologically equivalent dose of 180 rem. How many nuclei are in the beam?

Section 32.2 Induced Nuclear Reactions

10. **E** What is the atomic number Z, the atomic mass number A, and the element X in the reaction ${}^{10}_{5}\text{B}\ (\alpha, p)\ {}^{A}_{Z}\text{X}$?

11. **E** **SSM** Write the equation for the reaction ${}^{17}_{8}\text{O}\ (\gamma, \alpha n)\ {}^{12}_{6}\text{C}$. The notation "$\alpha n$" means that an α particle and a neutron are produced by the reaction.

12. **E** **GO** For each of the nuclear reactions listed below, determine the unknown particle ${}^{A}_{Z}\text{X}$. Use the periodic table on the inside of the back cover of the text as needed.

(a) ${}^{A}_{Z}\text{X} + {}^{14}_{7}\text{N} \longrightarrow {}^{1}_{1}\text{H} + {}^{17}_{8}\text{O}$ (c) ${}^{1}_{1}\text{H} + {}^{27}_{13}\text{Al} \longrightarrow {}^{A}_{Z}\text{X} + {}^{1}_{0}\text{n}$

(b) ${}^{15}_{7}\text{N} + {}^{A}_{Z}\text{X} \longrightarrow {}^{12}_{6}\text{C} + {}^{4}_{2}\text{He}$ (d) ${}^{7}_{3}\text{Li} + {}^{1}_{1}\text{H} \longrightarrow {}^{4}_{2}\text{He} + {}^{A}_{Z}\text{X}$

13. **E** A neutron causes ${}^{232}_{90}\text{Th}$ to change according to the reaction

$${}^{1}_{0}\text{n} + {}^{232}_{90}\text{Th} \longrightarrow {}^{A}_{Z}\text{X} + \gamma$$

(a) Identify the unknown nucleus ${}^{A}_{Z}\text{X}$, giving its atomic mass number A, its atomic number Z, and the symbol X for the element. (b) The ${}^{A}_{Z}\text{X}$ nucleus subsequently undergoes β^- decay, and its daughter does too. Identify the final nucleus, giving its atomic mass number, atomic number, and name.

14. **E** Write the reactions below in the shorthand form discussed in the text.

(a) ${}^{4}_{2}\text{He} + {}^{27}_{13}\text{Al} \longrightarrow {}^{30}_{15}\text{P} + {}^{1}_{0}\text{n}$

(b) ${}^{1}_{1}\text{H} + {}^{9}_{4}\text{Be} \longrightarrow {}^{6}_{3}\text{Li} + {}^{4}_{2}\text{He}$

(c) ${}^{1}_{0}\text{n} + {}^{55}_{25}\text{Mn} \longrightarrow {}^{56}_{25}\text{Mn} + \gamma$

15. **E** **SSM** Complete the following nuclear reactions, assuming that the unknown quantity signified by the question mark is a single entity: (a) ${}^{34}_{18}\text{Ar}\ (n, \alpha)?$ (b) ${}^{82}_{34}\text{Se}\ (?, n)\ {}^{82}_{35}\text{Br}$ (c) ${}^{58}_{28}\text{Ni}\ ({}^{40}_{18}\text{Ar}, ?)\ {}^{57}_{27}\text{Co}$ (d) $?(\gamma, \alpha)\ {}^{16}_{8}\text{O}$

16. **M** **CHALK** During a nuclear reaction, an unknown particle is absorbed by a copper ${}^{63}_{29}\text{Cu}$ nucleus, and the reaction products are ${}^{62}_{29}\text{Cu}$, a neutron, and a proton. What are the name, atomic number, and nucleon number of the nucleus formed temporarily when the copper ${}^{63}_{29}\text{Cu}$ nucleus absorbs the unknown particle?

Section 32.3 Nuclear Fission, Section 32.4 Nuclear Reactors

17. **E** Determine the atomic number Z, the atomic mass number A, and the element X for the unknown species ${}^{A}_{Z}\text{X}$ in the following reaction for the fission of uranium ${}^{235}_{92}\text{U}$:

$${}^{1}_{0}\text{n} + {}^{235}_{92}\text{U} \longrightarrow {}^{133}_{51}\text{Sb} + {}^{A}_{Z}\text{X} + 4\,{}^{1}_{0}\text{n}$$

Consult the periodic table on the inside of the back cover of the text as needed.

18. **E** **SSM** When a ${}^{235}_{92}\text{U}$ (235.043 924 u) nucleus fissions, about 200 MeV of energy is released. What is the ratio of this energy to the rest energy of the uranium nucleus?

19. **E** **CHALK** How many neutrons are produced when ${}^{235}_{92}\text{U}$ fissions in the following way? ${}^{1}_{0}\text{n} + {}^{235}_{92}\text{U} \rightarrow {}^{132}_{50}\text{Sn} + {}^{101}_{42}\text{Mo} + \text{neutrons}$

20. **E** Neutrons released by a fission reaction must be slowed by collisions with the moderator nuclei before the neutrons can cause further fissions. Suppose a 1.5-MeV neutron leaves each collision with 65% of its incident energy. How many collisions are required to reduce the neutron's energy to at least 0.040 eV, which is the energy of a thermal neutron?

21. **E** **V-HINT** The energy released by each fission within the core of a nuclear reactor is 2.0×10^2 MeV. The number of fissions occurring each second is 2.0×10^{19}. Determine the power (in watts) that the reactor generates.

22. **E** Uranium ${}^{235}_{92}\text{U}$ fissions into two fragments plus three neutrons: ${}^{1}_{0}\text{n} + {}^{235}_{92}\text{U} \rightarrow (2\ \text{fragments}) + 3\,{}^{1}_{0}\text{n}$. The mass of a neutron is 1.008 665 u and the mass of ${}^{235}_{92}\text{U}$ is 235.043 924 u. If 225.0 MeV of energy is released, what is the total mass of the two fragments?

23. **M** **GO** **CHALK** When 1.0 kg of coal is burned, approximately 3.0×10^7 J of energy is released. If the energy released during each ${}^{235}_{92}\text{U}$ fission is 2.0×10^2 MeV, how many kilograms of coal must be burned to produce the same energy as 1.0 kg of ${}^{235}_{92}\text{U}$?

24. **M** **GO** The water that cools a reactor core enters the reactor at 216 °C and leaves at 287 °C. (The water is pressurized, so it does not turn to steam.) The core is generating 5.6×10^9 W of power. Assume that the specific heat capacity of water is 4420 J/(kg · C°) over the temperature range stated above, and find the mass of water that passes through the core each second.

25. **H** **SSM** A nuclear power plant is 25% efficient, meaning that only 25% of the power it generates goes into producing electricity. The remaining 75% is wasted as heat. The plant generates 8.0×10^8 watts of electric power. If each fission releases 2.0×10^2 MeV of energy, how many kilograms of ${}^{235}_{92}\text{U}$ are fissioned per year?

Section 32.5 Nuclear Fusion

26. **E** Two deuterium atoms (${}^{2}_{1}\text{H}$) react to produce tritium (${}^{3}_{1}\text{H}$) and hydrogen (${}^{1}_{1}\text{H}$) according to the following fusion reaction:

$$\underbrace{{}^{2}_{1}\text{H}}_{2.014\,102\ \text{u}} + \underbrace{{}^{2}_{1}\text{H}}_{2.014\,102\ \text{u}} \longrightarrow \underbrace{{}^{3}_{1}\text{H}}_{3.016\,050\ \text{u}} + \underbrace{{}^{1}_{1}\text{H}}_{1.007\,825\ \text{u}}$$

What is the energy (in MeV) released by this deuterium–deuterium reaction?

27. **E** **GO** In one type of fusion reaction a proton fuses with a neutron to form a deuterium nucleus:

$${}^{1}_{1}\text{H} + {}^{1}_{0}\text{n} \longrightarrow {}^{2}_{1}\text{H} + \gamma$$

The masses are ${}^{1}_{1}\text{H}$ (1.0078 u), ${}^{1}_{0}\text{n}$ (1.0087 u), and ${}^{2}_{1}\text{H}$ (2.0141 u). The γ-ray photon is massless. How much energy (in MeV) is released by this reaction?

28. **E** **MMH** The fusion of two deuterium nuclei (${}^{2}_{1}\text{H}$, mass = 2.0141 u) can yield a helium nucleus (${}^{3}_{2}\text{He}$, mass = 3.0160 u) and a neutron (${}^{1}_{0}\text{n}$, mass = 1.0087 u). What is the energy (in MeV) released in this reaction?

29. **E** **V-HINT** Tritium (${}^{3}_{1}\text{H}$) is a rare isotope of hydrogen that can be produced by the following fusion reaction:

$$\underbrace{{}^{1}_{Z}\text{X}}_{1.0087\ \text{u}} + \underbrace{{}^{A}_{1}\text{Y}}_{2.0141\ \text{u}} \longrightarrow \underbrace{{}^{3}_{1}\text{H}}_{3.0161\ \text{u}} + \gamma$$

(a) Determine the atomic mass number A, the atomic number Z, and the names X and Y of the unknown particles. (b) Using the masses given in the reaction, determine how much energy (in MeV) is released by this reaction.

30. **M** **CHALK** **SSM** One proposed fusion reaction combines lithium ${}^{6}_{3}\text{Li}$ (6.015 u) with deuterium ${}^{2}_{1}\text{H}$ (2.014 u) to give helium ${}^{4}_{2}\text{He}$ (4.003 u):

${}_1^2\mathrm{H} + {}_3^6\mathrm{Li} \rightarrow 2\,{}_2^4\mathrm{He}$. How many kilograms of lithium would be needed to supply the energy needs of one household for a year, estimated to be 3.8×10^{10} J?

31. M GO In Example 7 it was determined that 17.6 MeV of energy is released when the following fusion reaction occurs:

$$\underbrace{{}_1^2\mathrm{H}}_{2.0141\ \mathrm{u}} + \underbrace{{}_1^3\mathrm{H}}_{3.0161\ \mathrm{u}} \longrightarrow \underbrace{{}_2^4\mathrm{He}}_{4.0026\ \mathrm{u}} + \underbrace{{}_0^1\mathrm{n}}_{1.0087\ \mathrm{u}}$$

Ignore relativistic effects and determine the kinetic energies of the neutron and the α particle.

32. M Deuterium (${}_1^2\mathrm{H}$) is an attractive fuel for fusion reactions because it is abundant in the oceans, where about 0.015% of the hydrogen atoms in the water (H_2O) are deuterium atoms. **(a)** How many deuterium atoms are there in one kilogram of water? **(b)** If each deuterium nucleus produces about 7.2 MeV in a fusion reaction, how many kilograms of water would be needed to supply the energy needs of the United States for one year, estimated to be 1.1×10^{20} J?

Section 32.6 Elementary Particles

33. E The main decay mode for the negative pion is $\pi^- \rightarrow \mu^- + \bar{\nu}_\mu$. Find the energy (in MeV) released in this decay. Consult **Table 32.3** for rest energies and assume that the rest energy for $\bar{\nu}_\mu$ is ≈ 0 MeV.

34. E V-HINT A neutral pion π^0 (rest energy = 135.0 MeV) produced in a high-energy particle experiment moves at a speed of 0.780 c. After a very short time, it decays into two γ-ray photons. One of the γ-ray photons has an energy of 192 MeV. What is the energy (in MeV) of the second γ-ray photon? Take relativistic effects into account.

35. E GO In addition to its rest energy, a moving proton (p') has kinetic energy. This proton collides with a stationary proton (p), and the reaction forms a stationary neutron (n), a stationary proton (p), and a stationary pion (π^+), according to the following reaction: $p' + p \rightarrow n + p + \pi^+$. The rest energy of each proton is 938.3 MeV, and the rest energy of the neutron is 939.6 MeV. The rest energy of the pion is 139.6 MeV. What is the kinetic energy (in MeV) of the moving proton?

36. E SSM Suppose a neutrino is created and has an energy of 35 MeV. **(a)** Assuming the neutrino, like the photon, has no mass and travels at the speed of light, find the momentum of the neutrino. **(b)** Determine the de Broglie wavelength of the neutrino.

37. M GO Review Conceptual Example 8 as background for this problem. An electron and its antiparticle annihilate each other, producing two γ-ray photons. The kinetic energies of the particles are negligible. Determine the magnitude of the momentum of each photon.

38. M Review Conceptual Example 6 as background for this problem. An energetic proton is fired at a stationary proton. For the reaction to produce new particles, the two protons must approach each other to within a distance of about 8.0×10^{-15} m. The moving proton must have a sufficient speed to overcome the repulsive Coulomb force. What must be the minimum initial kinetic energy (in MeV) of the proton?

Additional Problems

Online

39. E GO Someone stands near a radioactive source and receives doses of the following types of radiation: γ rays (20 mrad, RBE = 1), electrons (30 mrad, RBE = 1), protons (5 mrad, RBE = 10), and slow neutrons (5 mrad, RBE = 2). Rank the types of radiation, highest first, as to which produces the largest biologically equivalent dose.

40. E Identify the unknown species ${}_Z^A\mathrm{X}$ in the following nuclear reaction: ${}_{11}^{22}\mathrm{Na}\,(d, \alpha)\,{}_Z^A\mathrm{X}$. Here, d stands for the deuterium isotope ${}_1^2\mathrm{H}$ of hydrogen.

41. E GO MMH What energy (in MeV) is released by the following fission reaction?

$$\underbrace{{}_0^1\mathrm{n}}_{1.009\ \mathrm{u}} + \underbrace{{}_{92}^{235}\mathrm{U}}_{235.044\ \mathrm{u}} \longrightarrow \underbrace{{}_{54}^{140}\mathrm{Xe}}_{139.922\ \mathrm{u}} + \underbrace{{}_{38}^{94}\mathrm{Sr}}_{93.915\ \mathrm{u}} + \underbrace{2\,{}_0^1\mathrm{n}}_{2(1.009\ \mathrm{u})}$$

42. M V-HINT Multiple-Concept Example 1 discusses some of the physics principles that are used to solve this problem. What absorbed dose (in rads) of γ rays is required to change a block of ice at 0.0 °C into steam at 100.0 °C?

43. M SSM Imagine that your car is powered by a fusion engine in which the following reaction occurs: $3\,{}_1^2\mathrm{H} \rightarrow {}_2^4\mathrm{He} + {}_1^1\mathrm{H} + {}_0^1\mathrm{n}$. The masses are ${}_1^2\mathrm{H}$ (2.0141 u), ${}_2^4\mathrm{He}$ (4.0026 u), ${}_1^1\mathrm{H}$ (1.0078 u), and ${}_0^1\mathrm{n}$ (1.0087 u). The engine uses 6.1×10^{-6} kg of deuterium ${}_1^2\mathrm{H}$ fuel. If one gallon of gasoline produces 2.1×10^9 J of energy, how many gallons of gasoline would have to be burned to equal the energy released by all the deuterium fuel?

44. M GO The energy consumed in one year in the United States is about 1.1×10^{20} J. With each ${}_{92}^{235}\mathrm{U}$ fission, about 2.0×10^2 MeV of energy is released. How many kilograms of ${}_{92}^{235}\mathrm{U}$ would be needed to generate this energy if all the nuclei fissioned?

45. M **(a)** If each fission reaction of a ${}_{92}^{235}\mathrm{U}$ nucleus releases about 2.0×10^2 MeV of energy, determine the energy (in joules) released by the complete fissioning of 1.0 gram of ${}_{92}^{235}\mathrm{U}$. **(b)** How many grams of ${}_{92}^{235}\mathrm{U}$ would be consumed in one year to supply the energy needs of a household that uses 30.0 kWh of energy per day, on the average?

46. H One kilogram of dry air at STP conditions is exposed to 1.0 R of X-rays. One roentgen is defined by Equation 32.1. An equivalent definition can be based on the fact that an exposure of one roentgen deposits 8.3×10^{-3} J of energy per kilogram of dry air. Using the two definitions and assuming that all ions produced are singly charged, determine the average energy (in eV) needed to produce one ion in air.

Physics in Biology, Medicine, and Sports

47. M BIO 32.1 In March 2011, the Fukushima Daiichi nuclear power station, which is located about 130 miles northeast of Tokyo, Japan, experienced a powerful 9.0 earthquake and subsequent tsunami that caused three of its reactor cores to melt down. The damaged reactors leaked radiation into the atmosphere and Pacific Ocean, forcing over 150 000 residents within 20 km of the plant to be

evacuated. Elevated levels of radiation will exist at the plant site for decades.

The torch relay for the 2020 Summer Olympics in Tokyo was to begin at the J-Village Sports Facility in Fukushima in December 2019. However, background radiation measurements at the facility in October of that year measured a value of 7.1 mrem/hr at one location. Businesses in and around Fukushima often display current radiation levels (see the photo). **(a)** If people receive, on average, 630 mrem of radiation per year from natural sources, how many days at the sports facility would it take to receive an annual dose of radiation? **(b)** If a 95-kg person stays at the facility for the number of days in part (a), how much energy does their body absorb from the radiation during that time? Assume the RBE of the radiation is 1. *Note:* The Olympic Torch Relay in 2019, and the entire 2020 Summer Games, were rescheduled due to the COVID-19 pandemic.

Haruhiko Okumura

A hotel in Naraha, near Fukushima, Japan, displays the radiation dose rate in microsieverts (μSv) per hour, five years after the Fukushima disaster. The sievert (Sv) is the SI unit of dose. 1 μSv is equal to 0.1 mrem.
PROBLEM 47

48. **M** **BIO** **32.1** A now-retired Russian particle physicist named Anatoli Bugorski survived accidental exposure to a very high dose of radiation in 1978 while working at the U-70 synchrotron in Protvino, Russia. At the time, the U-70 synchrotron was one of the world's largest and most powerful particle accelerators. It is similar to the present-day Large Hadron Collider (LHC) near Geneva, in that both are proton synchrotrons. In order to repair a piece of malfunctioning equipment, it was necessary for Bugorski to enter the proton beam tunnel. A breakdown in several safety protocols allowed him to enter the tunnel while the beam was still on. The beam of protons is invisible, so once inside, there was no indication that he was in any danger. He leaned over to perform a visual inspection of the malfunctioning equipment and placed his head in the proton beam. Relativistic protons moving near the speed of light with an energy of 76 GeV cut a 2- to 3-mm-wide path through his face, head, and brain in a fraction of a second (see the photo). He recalls seeing a flash of light, "brighter than a thousand suns," but said he felt no pain. He knew immediately what had happened. Surprisingly, he finished the repairs and noted the accident in his lab notebook.

The damage to Bugorski's cells was extensive. Over the next few weeks, his face would swell up, his skin would die and peel off, he would suffer nerve damage to the left side of his face that led to facial paralysis, he would lose hearing in his left ear, and he began to suffer seizures. Although the prognosis from his medical doctors was grim, he remarkably survived and began to recover. The overall damage to his brain was small, but so extensive along the path of the beam that cancer cells were unable to form. He would eventually recover nearly all of his intellectual capacity, finish his PhD, get married, and have children. Sources disagree on the exact absorbed dose of radiation he received, but the value is near 3.0×10^5 rad. **(a)** Assuming this value for the absorbed dose, how much energy was deposited in his head? Take the mass of his head to be 3.7 kg. **(b)** If the RBE of the protons was 10.0, what biologically equivalent dose of radiation did he receive?

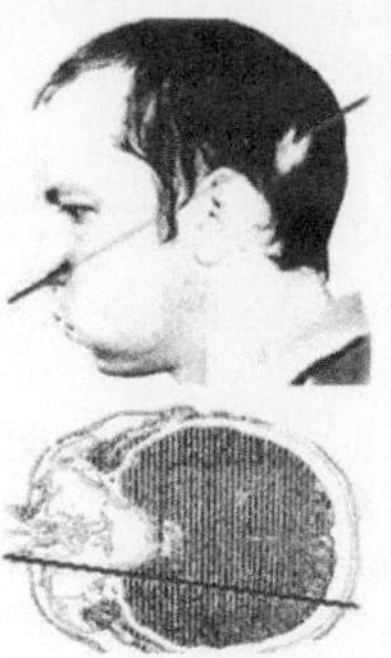

(Left, top image) A picture of Anatoli Bugorski several days after the accident. Notice the large amount of swelling on the left side of his face. The red line shows the path of the proton beam through Bugorski's head. The beam entered through the back of his head above his left ear. Notice the small patch of hair that is missing at the entrance point. (Left, bottom image) A sketch of the cross section of his head, showing how the beam bored through nonessential parts of his brain. The beam exited his face near the left side of his nose. (Right) A more recent picture of Bugorski. Note the slight droop in the left side of his face and a lack of wrinkles on that side due to the paralysis.
PROBLEM 48

49. **E** **BIO** **32.7** Over the past few decades, scientists have discovered over 4000 exoplanets, which are planets that exist outside of our solar system. This is quite remarkable, when one considers that only 30 years ago, we only knew of 9 planets. Current estimates suggest there could be billions of exoplanets in just the Milky Way Galaxy alone. One of these planets is KOI-456.04, which orbits the star Kepler-160 within the constellation Lyra. This exoplanet is being heavily studied, since it is similar to earth and orbits a star like our sun. Measurements suggest that KOI-456.04 is less than twice the size of earth and that it orbits its star at roughly the same distance as the earth orbits the sun. This places the planet in the so-called habitable zone or Goldilocks zone, which means it's neither too hot nor too cold, but just right to potentially support liquid water and possibly life. The Kepler-160 system is located 3.14×10^3 ly from earth. Apply Hubble's law to estimate the speed at which this system is moving away from earth. *Note:* Hubble's law should, technically, only be applied to very distant galaxies. The motion of galaxies that are relatively close to earth is dominated by local forces, such as gravitational attraction, and is not representative of the overall expansion of the universe.

50. **M** **BIO** **32.1** A patient with a brain tumor is undergoing a gamma knife treatment (see Section 31.4). Multiple sources of ^{60}Co emit two gamma rays with a total energy of 2.50 MeV. If the activity of all the sources combined is 5.00 mCi, what is the biologically equivalent dose delivered to the brain tumor in one hour? The mass of the tumor is 0.150 kg. Since the half-life of ^{60}Co is 5.27 yr, we can assume the activity remains constant during the procedure.

51. **M** **BIO** **32.1** A scientist working at a neutron reactor facility accidentally exposes his hand to slow neutrons and receives an absorbed dose of radiation of 2.25 rad. **(a)** How much energy is absorbed by his hand, which has a mass of 0.570 kg? **(b)** If we assume his hand is completely composed of water, what would be the rise in temperature of his hand due to the exposure?

Concepts and Calculations Problems

Online

When considering the biological effects of ionizing radiation, the concept of biologically equivalent dose is especially important. Its importance lies in the fact that the biologically equivalent dose incorporates both the amount of energy per unit mass that is absorbed and the effectiveness of a particular type of radiation in producing a certain biological effect. Problem 52 examines this concept and also reviews the notions of power (Section 6.7) and intensity (Section 16.7) of a wave. Problem 53 illustrates the decay of a particle into two photons, and provides a review of the principles of conservation of energy and conservation of momentum.

52. M CHALK SSM A patient is being given a chest X-ray. The X-ray beam is turned on for 0.20 s, and its intensity is 0.40 W/m^2. The area of the chest being exposed is 0.072 m^2, and the radiation is absorbed by 3.6 kg of tissue. The relative biological effectiveness (RBE) of the X-ray for this tissue is 1.1. *Concepts*: (i) How is the power of the beam related to the beam intensity? (ii) How is the energy absorbed by the tissue related to the power of the beam? (iii) What is the absorbed dose? (iv) How is the biologically equivalent dose related to the absorbed dose? *Calculation*: Calculate the biologically equivalent dose received by the patient.

53. M CHALK The π^0 meson is a particle that has a rest energy of 135.0 MeV (see **Table 32.3**). It lives for a very short time and then decays into two γ-ray photons: $\pi^0 \rightarrow \gamma + \gamma$. Suppose that one of the γ-ray photons travels along the $+x$ axis. *Concepts*: (i) How is the energy E of each γ-ray photon related to the rest energy E_0 of the π_0 particle? (ii) How can the frequency and wavelength of a photon be determined from its energy? (iii) How is the total linear momentum of the photons related to the momentum of the π^0 particle, and what is the momentum of each particle? *Calculations*: If the π^0 is at rest when it decays, find **(a)** the energy (in MeV), **(b)** the frequency and wavelength, and **(c)** the momentum of each γ-ray photon.

Team Problems

Online

54. E T WS **The Safety of a Neutron Source.** Californium-252 is a strong neutron source. You and your team have been assigned to assess the risk associated with storing 1.50 grams of californium-252 that is to be used in an experiment that requires neutrons. Through spontaneous fission processes, the 1.50-g ^{252}Cf source emits 3.47×10^{12} neutrons per second. The material is contained in a shielded container, and your task is to estimate the biologically equivalent dose (in rems) delivered to an unsuspecting person who has opened the shielded vessel without taking proper precautions. The neutrons have an average energy of 2.30 MeV and have an RBE = 10.0. Since the source emits neutrons uniformly in all directions, the geometry of the container is such that a person opening its hatch will only receive a fraction (7.50%) of the total emitted neutrons. The affected area on the body would be the head and chest, and the estimated mass in which the neutrons would be absorbed is 21.0 kg. If a person opens the hatch and is exposed for 5.00 s, **(a)** what is the absorbed dose in rad? **(b)** What is the biologically equivalent dose in rem? **(c)** How long would it take for a person standing in front of the open container to get a dose of neutrons that would result in radiation sickness (i.e., greater than 50 rem)? Note that this result pertains only to the emitted neutrons. Californium-252 also emits other ionizing radiation including α particles, fission fragments, and gamma rays.

55. E T WS **A New Heavy Element.** Scientists can induce nuclear transmutations through collisions, as in the case of an α particle striking a nitrogen nucleus to produce an oxygen nucleus (see Section 32.2). In the case of ${}^{48}_{20}\text{Ca}$ nuclei striking ${}^{243}_{95}\text{Am}$ nuclei, some of the collisions result in the formation of one large nucleus plus two free neutrons. You and your team have been assigned to identify the unknown nucleus and its decay products. **(a)** Assuming the number of protons and total number of nucleons are conserved in the collision, what is the unknown nucleus? **(b)** The unknown nucleus has been observed to undergo α decay. What is the daughter nucleus that results from this process? *Note:* both the parent and daughter nuclei have only recently been discovered. They were artificially synthesized and are not found in nature since they are extremely unstable and decay quickly.

56. M T **A Rough Measure of Exposure.** You and your team are designing a crude device to estimate radiation exposure. The device consists of a set of parallel plates with a large voltage across them. The circular plates have a radius of 3.50 cm and are separated by distance of 1.00 cm. The region between the plates is filled with dry air at standard temperature and pressure (STP: $T = 0°$, $P = 1$). A cubic meter of dry air at STP has a mass of 1.29 kg. When radiation ionizes an air molecule, the stripped electron is accelerated by the electric field between the plates and is registered as a current. When the device is located near a particularly strong source, it registers a current of 1.30×10^{-9} A. What is the exposure between the plates after 5.00 s (in roentgens)?

57. M T **A Hypothetical Fusion Reactor.** You and your team are tasked with evaluating the following deuterium-deuterium fusion reaction for use in a future fusion power reactor:

$$\underbrace{{}^{2}_{1}\text{H}}_{2.014\,102\text{ u}} + \underbrace{{}^{2}_{1}\text{H}}_{2.014\,102\text{ u}} \longrightarrow \underbrace{{}^{3}_{1}\text{H}}_{3.016\,050\text{ u}} + \underbrace{{}^{1}_{1}\text{H}}_{1.007\,825\text{ u}}$$

(a) What is the energy released in this reaction (in joules)? **(b)** How many reactions per second would be required to run a 5000 MW reactor? **(c)** What mass of deuterium (in kg) would be needed to run the reactor for a year with a power output of 5000 MW?

Powers of Ten and Scientific Notation

In science, very large and very small decimal numbers are conveniently expressed in terms of powers of ten, some of which are listed below:

$$10^3 = 10 \times 10 \times 10 = 1000 \qquad 10^{-3} = \frac{1}{10 \times 10 \times 10} = 0.001$$

$$10^2 = 10 \times 10 = 100 \qquad 10^{-2} = \frac{1}{10 \times 10} = 0.01$$

$$10^1 = 10 \qquad 10^{-1} = \frac{1}{10} = 0.1$$

$$10^0 = 1$$

Using powers of ten, we can write the radius of the earth in the following way, for example:

$$\text{Earth radius} = 6\ 380\ 000\ \text{m} = 6.38 \times 10^6\ \text{m}$$

The factor of ten raised to the sixth power is ten multiplied by itself six times, or one million, so the earth's radius is 6.38 million meters. Alternatively, the factor of ten raised to the sixth power indicates that the decimal point in the term 6.38 is to be moved six places *to the right* to obtain the radius as a number without powers of ten.

For numbers less than one, negative powers of ten are used. For instance, the Bohr radius of the hydrogen atom is

$$\text{Bohr radius} = 0.000\ 000\ 000\ 0529\ \text{m} = 5.29 \times 10^{-11}\ \text{m}$$

The factor of ten raised to the minus eleventh power indicates that the decimal point in the term 5.29 is to be moved eleven places *to the left* to obtain the radius as a number without powers of ten. Numbers expressed with the aid of powers of ten are said to be in ***scientific notation.***

Calculations that involve the multiplication and division of powers of ten are carried out as in the following examples:

$$(2.0 \times 10^6)(3.5 \times 10^3) = (2.0 \times 3.5) \times 10^{6+3} = 7.0 \times 10^9$$

$$\frac{9.0 \times 10^7}{2.0 \times 10^4} = \left(\frac{9.0}{2.0}\right) \times 10^7 \times 10^{-4} = \left(\frac{9.0}{2.0}\right) \times 10^{7-4} = 4.5 \times 10^3$$

The general rules for such calculations are

$$\frac{1}{10^n} = 10^{-n} \qquad \textbf{(A-1)}$$

$$10^n \times 10^m = 10^{n+m} \qquad \text{(Exponents added)} \qquad \textbf{(A-2)}$$

$$\frac{10^n}{10^m} = 10^{n-m} \qquad \text{(Exponents subtracted)} \qquad \textbf{(A-3)}$$

where n and m are any positive or negative number.

Scientific notation is convenient because of the ease with which it can be used in calculations. Moreover, scientific notation provides a convenient way to express the significant figures in a number, as Appendix B discusses.

Significant Figures

The number of ***significant figures*** in a number is the number of digits whose values are known with certainty. For instance, a person's height is measured to be 1.78 m, with the measurement error being in the third decimal place. All three digits are known with certainty, so that the number contains three significant figures. If a zero is given as the last digit to the right of the decimal point, the zero is presumed to be significant. Thus, the number 1.780 m contains four significant figures. As another example, consider a distance of 1500 m. This number contains only two significant figures, the one and the five. The zeros immediately to the left of the unexpressed decimal point are not counted as significant figures. However, zeros located between significant figures are significant, so a distance of 1502 m contains four significant figures.

Scientific notation is particularly convenient from the point of view of significant figures. Suppose it is known that a certain distance is fifteen hundred meters, to four significant figures. Writing the number as 1500 m presents a problem because it implies that only two significant figures are known. In contrast, the scientific notation of 1.500×10^3 m has the advantage of indicating that the distance is known to four significant figures.

When two or more numbers are used in a calculation, the number of significant figures in the answer is limited by the number of significant figures in the original data. For instance, a rectangular garden with sides of 9.8 m and 17.1 m has an area of (9.8 m)(17.1 m). A calculator gives 167.58 m^2 for this product. However, one of the original lengths is known only to two significant figures, so the final answer is limited to only two significant figures and should be rounded off to 170 m^2. In general, ***when numbers are multiplied or divided, the number of significant figures in the final answer equals the smallest number of significant figures in any of the original factors.***

The number of significant figures in the answer to an addition or a subtraction is also limited by the original data. Consider the total distance along a biker's trail that consists of three segments with the distances shown as follows:

	2.5 km
	11 km
	5.26 km
Total	18.76 km

The distance of 11 km contains no significant figures to the right of the decimal point. Therefore, neither does the sum of the three distances, and the total distance should not be reported as 18.76 km. Instead, the answer is rounded off to 19 km. In general, ***when numbers are added or subtracted, the last significant figure in the answer occurs in the last column (counting from left to right) containing a number that results from a combination of digits that are all significant.*** In the answer of 18.76 km, the eight is the sum of 2 + 1 + 5, each digit being significant. However, the seven is the sum of 5 + 0 + 2, and the zero is not significant, since it comes from the 11-km distance, which contains no significant figures to the right of the decimal point.

Algebra

C.1 Proportions and Equations

Physics deals with physical variables and the relations between them. Typically, variables are represented by the letters of the English and Greek alphabets. Sometimes, the relation between variables is expressed as a proportion or inverse proportion. Other times, however, it is more convenient or necessary to express the relation by means of an equation, which is governed by the rules of algebra.

If two variables are ***directly proportional*** and one of them doubles, then the other variable also doubles. Similarly, if one variable is reduced to one-half its original value, then the other is also reduced to one-half its original value. In general, if x is directly proportional to y, then increasing or decreasing one variable by a given factor causes the other variable to change in the same way by the same factor. This kind of relation is expressed as $x \propto y$, where the symbol $\propto$ means "is proportional to."

Since the proportional variables x and y always increase and decrease by the same factor, the ratio of x to y must have a constant value, or $x/y = k$, where k is a constant, independent of the values for x and y. Consequently, a proportionality such as $x \propto y$ can also be expressed in the form of an equation: $x = ky$. The constant k is referred to as a ***proportionality constant.***

If two variables are ***inversely proportional*** and one of them increases by a given factor, then the other decreases by the same factor. An inverse proportion is written as $x \propto 1/y$. This kind of proportionality is equivalent to the following equation: $xy = k$, where k is a proportionality constant, independent of x and y.

C.2 Solving Equations

Some of the variables in an equation typically have known values, and some do not. It is often necessary to solve the equation so that a variable whose value is unknown is expressed in terms of the known quantities. ***In the process of solving an equation, it is permissible to manipulate the equation in any way, as long as a change made on one side of the equals sign is also made on the other side.*** For example, consider the equation $v = v_0 + at$. Suppose values for v, v_0, and a are available, and the value of t is required. To solve the equation for t, we begin by subtracting v_0 from *both* sides:

$$\begin{aligned} v &= v_0 + at \\ -v_0 &= -v_0 \\ \hline v - v_0 &= at \end{aligned}$$

Next, we divide both sides of $v - v_0 = at$ by the quantity a:

$$\frac{v - v_0}{a} = \frac{at}{a} = (1)t$$

On the right side, the a in the numerator divided by the a in the denominator equals one, so that

$$t = \frac{v - v_0}{a}$$

It is always possible to check the correctness of the algebraic manipulations performed in solving an equation by substituting the answer back into the original equation. In the previous example, we substitute the answer for t into $v = v_0 + at$:

$$v = v_0 + a\left(\frac{v - v_0}{a}\right) = v_0 + (v - v_0) = v$$

The result $v = v$ implies that our algebraic manipulations were done correctly.

Algebraic manipulations other than addition, subtraction, multiplication, and division may play a role in solving an equation. The same basic rule applies, however: Whatever is done to the left side of an equation must also be done to the right side. As another example, suppose it is necessary to express v_0 in terms of v, a, and x, where $v^2 = v_0^2 + 2ax$. By subtracting $2ax$ from both sides, we isolate v_0^2 on the right:

$$\begin{aligned} v^2 &= v_0^2 + 2ax \\ -2ax &= -2ax \\ \hline v^2 - 2ax &= v_0^2 \end{aligned}$$

To solve for v_0, we take the positive and negative square root of *both* sides of $v^2 - 2ax = v_0^2$:

$$v_0 = \pm\sqrt{v^2 - 2ax}$$

C.3 Simultaneous Equations

When more than one variable in a single equation is unknown, additional equations are needed if solutions are to be found for all of the unknown quantities. Thus, the equation $3x + 2y = 7$ cannot be solved by itself to give unique values for both x and y. However, if x and y also (i.e., simultaneously) obey the equation $x - 3y = 6$, then both unknowns can be found.

There are a number of methods by which such simultaneous equations can be solved. One method is to solve one equation for x in terms of y and substitute the result into the other equation to obtain an expression containing only the single unknown variable y. The equation $x - 3y = 6$, for instance, can be solved for x by adding $3y$ to each side, with the result that $x = 6 + 3y$. The substitution of this expression for x into the equation $3x + 2y = 7$ is shown below:

$$3x + 2y = 7$$
$$3(6 + 3y) + 2y = 7$$
$$18 + 9y + 2y = 7$$

We find, then, that $18 + 11y = 7$, a result that can be solved for y:

$$\begin{array}{rr} 18 + 11y = & 7 \\ -18 \quad & -18 \\ \hline 11y = & -11 \end{array}$$

Dividing both sides of this result by 11 shows that $y = -1$. The value of $y = -1$ can be substituted in either of the original equations to obtain a value for x:

$$\begin{array}{rcl} x - 3y & = & 6 \\ x - 3(-1) & = & 6 \\ x + 3 & = & 6 \\ -3 & & -3 \\ \hline x & = & 3 \end{array}$$

C.4 The Quadratic Formula

Equations occur in physics that include the square of a variable. Such equations are said to be *quadratic* in that variable, and often can be put into the following form:

$$ax^2 + bx + c = 0 \tag{C-1}$$

where a, b, and c are constants independent of x. This equation can be solved to give the ***quadratic formula,*** which is

$$x = \frac{-b \pm \sqrt{b^2 - 4ac}}{2a} \tag{C-2}$$

The $\pm$ in the quadratic formula indicates that there are two solutions. For instance, if $2x^2 - 5x + 3 = 0$, then $a = 2$, $b = -5$, and $c = 3$. The quadratic formula gives the two solutions as follows:

Solution 1: Plus sign

$$x = \frac{-b + \sqrt{b^2 - 4ac}}{2a} = \frac{-(-5) + \sqrt{(-5)^2 - 4(2)(3)}}{2(2)} = \frac{+5 + \sqrt{1}}{4} = \frac{3}{2}$$

Solution 2: Minus sign

$$x = \frac{-b - \sqrt{b^2 - 4ac}}{2a} = \frac{-(-5) - \sqrt{(-5)^2 - 4(2)(3)}}{2(2)} = \frac{+5 - \sqrt{1}}{4} = 1$$

Exponents and Logarithms

Appendix A discusses powers of ten, such as 10^3, which means ten multiplied by itself three times, or $10 \times 10 \times 10$. The three is referred to as an ***exponent.*** The use of exponents extends beyond powers of ten. In general, the term y^n means the factor y is multiplied by itself n times. For example, y^2, or y squared, is familiar and means $y \times y$. Similarly, y^5 means $y \times y \times y \times y \times y$.

The rules that govern algebraic manipulations of exponents are the same as those given in Appendix A (see Equations A-1, A-2, and A-3) for powers of ten:

$$\frac{1}{y^n} = y^{-n} \tag{D-1}$$

$$y^n y^m = y^{n+m} \quad \text{(Exponents added)} \tag{D-2}$$

$$\frac{y^n}{y^m} = y^{n-m} \quad \text{(Exponents subtracted)} \tag{D-3}$$

To the three rules above we add two more that are useful. One of these is

$$y^n z^n = (yz)^n \tag{D-4}$$

The following example helps to clarify the reasoning behind this rule:

$$3^2 5^2 = (3 \times 3)(5 \times 5) = (3 \times 5)(3 \times 5) = (3 \times 5)^2$$

The other additional rule is

$$(y^n)^m = y^{nm} \quad \text{(Exponents multiplied)} \tag{D-5}$$

To see why this rule applies, consider the following example:

$$(5^2)^3 = (5^2)(5^2)(5^2) = 5^{2+2+2} = 5^{2 \times 3}$$

Roots, such as a square root or a cube root, can be represented with fractional exponents. For instance,

$$\sqrt{y} = y^{1/2} \quad \text{and} \quad \sqrt[3]{y} = y^{1/3}$$

In general, the nth root of y is given by

$$\sqrt[n]{y} = y^{1/n} \tag{D-6}$$

The rationale for Equation D-6 can be explained using the fact that $(y^n)^m = y^{nm}$. For instance, the fifth root of y is the number that, when multiplied by itself five times, gives back y. As shown below, the term $y^{1/5}$ satisfies this definition:

$$(y^{1/5})(y^{1/5})(y^{1/5})(y^{1/5})(y^{1/5}) = (y^{1/5})^5 = (y^{1/5})^{\times 5} = y$$

Logarithms are closely related to exponents. To see the connection between the two, note that it is possible to express any number y as another number B raised to the exponent x. In other words,

$$y = B^x \tag{D-7}$$

The exponent x is called the ***logarithm*** of the number y. The number B is called the ***base number.*** One of two choices for the base number is usually used. If $B = 10$, the logarithm is known as the *common logarithm,* for which the notation "log" applies:

Common logarithm $\quad y = 10^x \quad$ or $\quad x = \log y \quad$ **(D-8)**

If $B = e = 2.718\ldots$, the logarithm is referred to as the *natural logarithm,* and the notation "ln" is used:

Natural logarithm $\quad y = e^z \quad$ or $\quad z = \ln y \quad$ **(D-9)**

The two kinds of logarithms are related by

$$\ln y = 2.3026 \log y \tag{D-10}$$

Both kinds of logarithms are often given on calculators.

The logarithm of the product or quotient of two numbers A and C can be obtained from the logarithms of the individual numbers according to the rules below. These rules are illustrated here for natural logarithms, but they are the same for any kind of logarithm.

$$\ln(AC) = \ln A + \ln C \tag{D-11}$$

$$\ln\left(\frac{A}{C}\right) = \ln A - \ln C \tag{D-12}$$

Thus, the logarithm of the product of two numbers is the sum of the individual logarithms, and the logarithm of the quotient of two numbers is the difference between the individual logarithms. Another useful rule concerns the logarithm of a number A raised to an exponent n:

$$\ln A^n = n \ln A \tag{D-13}$$

Rules D-11, D-12, and D-13 can be derived from the definition of the logarithm and the rules governing exponents.

APPENDIX E

Geometry and Trigonometry

E.1 Geometry

Angles

Two angles are equal if

1. They are vertical angles (see **Figure E1**).
2. Their sides are parallel (see **Figure E2**).

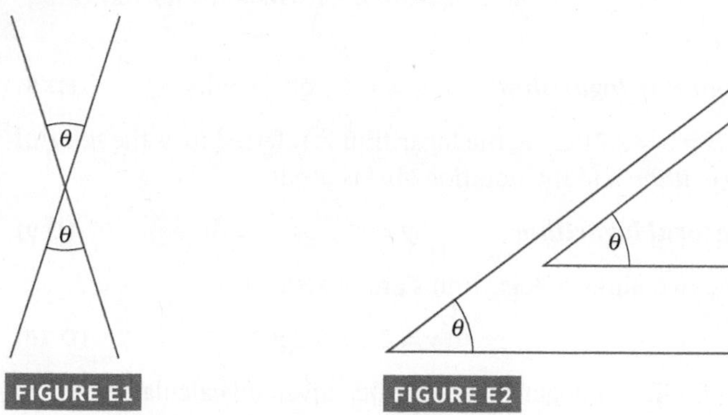

FIGURE E1

FIGURE E2

3. Their sides are mutually perpendicular (see **Figure E3**).

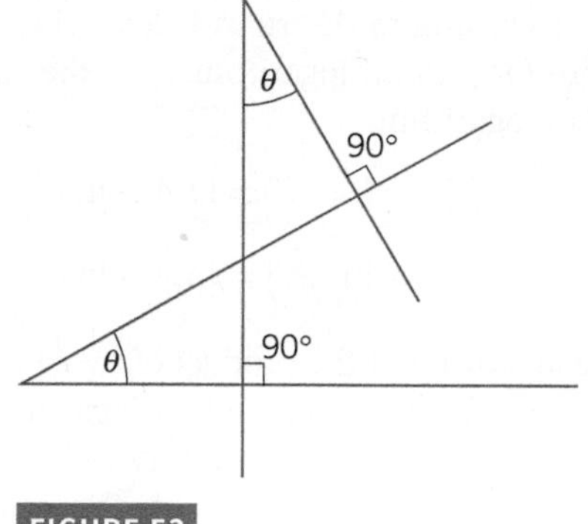

FIGURE E3

Triangles

1. The ***sum of the angles*** of any triangle is 180° (see **Figure E4**).

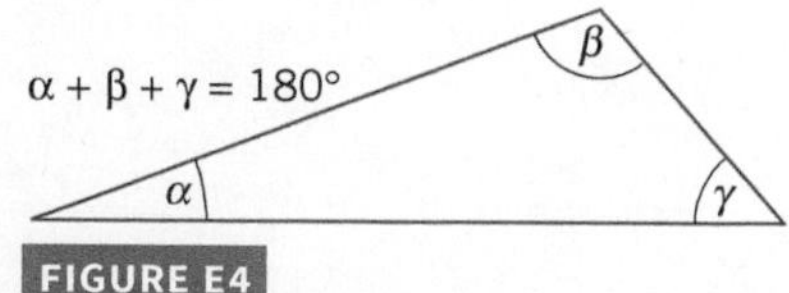

FIGURE E4

2. A ***right triangle*** has one angle that is 90°.
3. An ***isosceles triangle*** has two sides that are equal.
4. An ***equilateral triangle*** has three sides that are equal. Each angle of an equilateral triangle is 60°.
5. Two triangles are ***similar*** if two of their angles are equal (see **Figure E5**). The corresponding sides of similar triangles are proportional to each other:

$$\frac{a_1}{a_2} = \frac{b_1}{b_2} = \frac{c_1}{c_2}$$

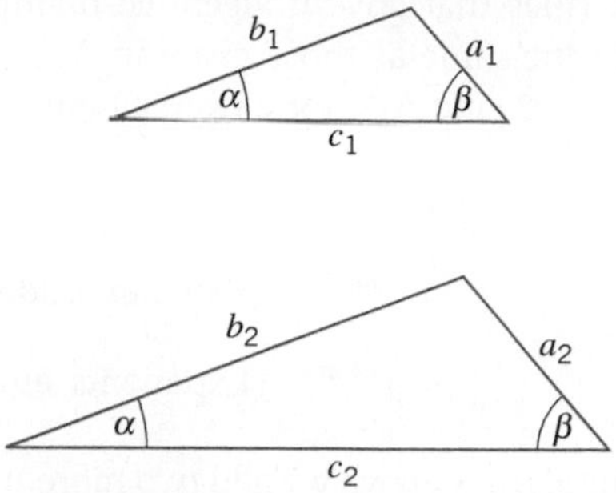

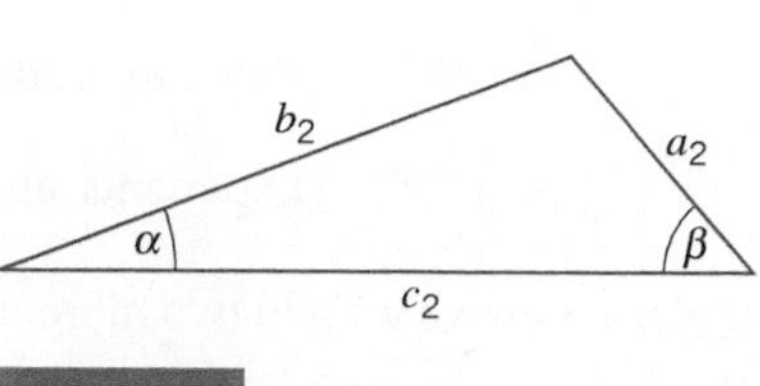

FIGURE E5

6. Two similar triangles are ***congruent*** if they can be placed on top of one another to make an exact fit.

Circumferences, Areas, and Volumes of Some Common Shapes

1. Triangle of base b and altitude h (see **Figure E6**):

$$\text{Area} = \frac{1}{2} bh$$

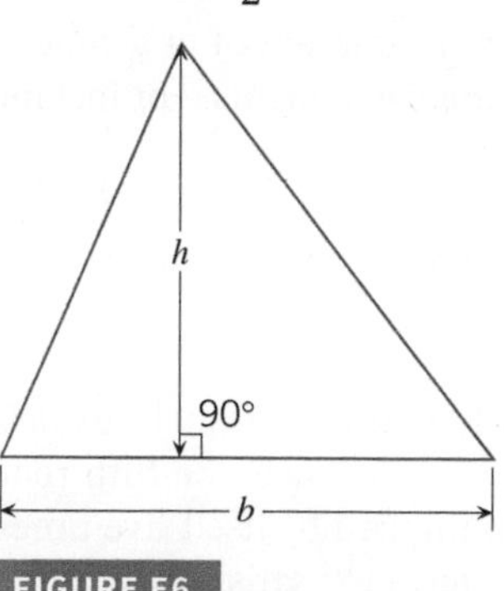

FIGURE E6

2. Circle of radius r:

$$\text{Circumference} = 2\pi r$$

$$\text{Area} = \pi r^2$$

3. Sphere of radius r:

$$\text{Surface area} = 4\pi r^2$$

$$\text{Volume} = \frac{4}{3}\pi r^3$$

4. Right circular cylinder of radius r and height h (see **Figure E7**):

$$\text{Surface area} = 2\pi r^2 + 2\pi rh$$
$$\text{Volume} = \pi r^2 h$$

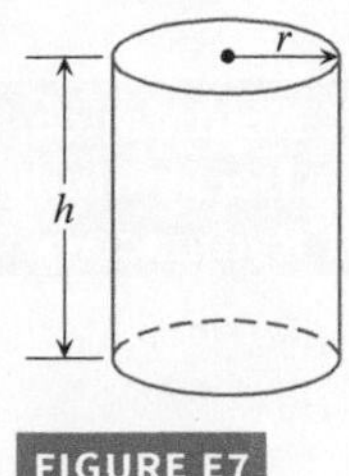

FIGURE E7

E.2 Trigonometry

Basic Trigonometric Functions

1. For a right triangle, the sine, cosine, and tangent of an angle θ are defined as follows (see **Figure E8**):

$$\sin\theta = \frac{\text{Side opposite } \theta}{\text{Hypotenuse}} = \frac{h_o}{h}$$

$$\cos\theta = \frac{\text{Side adjacent } \theta}{\text{Hypotenuse}} = \frac{h_a}{h}$$

$$\tan\theta = \frac{\text{Side opposite } \theta}{\text{Side adjacent to } \theta} = \frac{h_o}{h_a}$$

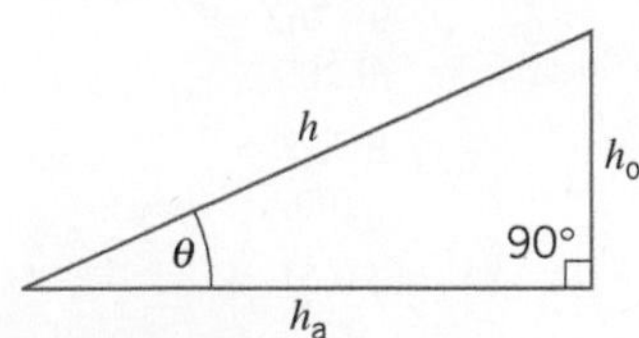

FIGURE E8

2. The secant ($\sec\theta$), cosecant ($\csc\theta$), and cotangent ($\cot\theta$) of an angle θ are defined as follows:

$$\sec\theta = \frac{1}{\cos\theta} \qquad \csc\theta = \frac{1}{\sin\theta} \qquad \cot\theta = \frac{1}{\tan\theta}$$

Triangles and Trigonometry

1. The ***Pythagorean theorem*** states that the square of the hypotenuse of a right triangle is equal to the sum of the squares of the other two sides (see **Figure E8**):

$$h^2 = h_o^{\,2} + h_a^{\,2}$$

2. The ***law of cosines*** and the ***law of sines*** apply to any triangle, not just a right triangle, and they relate the angles and the lengths of the sides (see **Figure E9**):

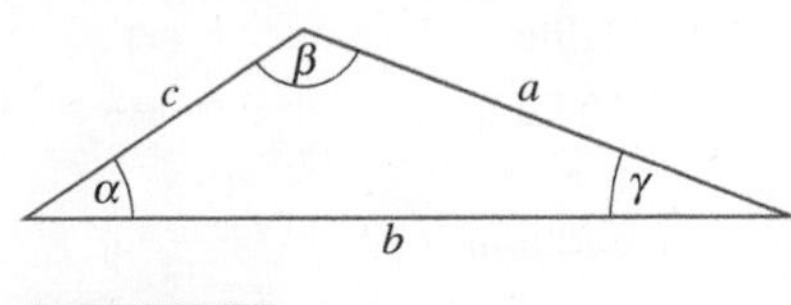

FIGURE E9

Law of cosines $\quad c^2 = a^2 + b^2 - 2ab\cos\gamma$

Law of sines $\quad \dfrac{a}{\sin\alpha} = \dfrac{b}{\sin\beta} = \dfrac{c}{\sin\gamma}$

Other Trigonometric Identities

1. $\sin(-\theta) = -\sin\theta$
2. $\cos(-\theta) = \cos\theta$
3. $\tan(-\theta) = -\tan\theta$
4. $(\sin\theta)/(\cos\theta) = \tan\theta$
5. $\sin^2\theta + \cos^2\theta = 1$
6. $\sin(\alpha \pm \beta) = \sin\alpha\cos\beta \pm \cos\alpha\sin\beta$

   If $\alpha = 90°$, $\sin(90° \pm \beta) = \cos\beta$

   If $\alpha = \beta$, $\sin 2\beta = 2\sin\beta\cos\beta$
7. $\cos(\alpha \pm \beta) = \cos\alpha\cos\beta \mp \sin\alpha\sin\beta$

   If $\alpha = 90°$, $\cos(90° \pm \beta) = \mp\sin\beta$

   If $\alpha = \beta$, $\cos 2\beta = \cos^2\beta - \sin^2\beta = 1 - 2\sin^2\beta$

APPENDIX F

Selected Isotopes[a]

Atomic No. Z	Element	Symbol	Atomic Mass No. A	Atomic Mass (u)	% Abundance, or Decay Mode if Radioactive	Half-Life (if Radioactive)
0	(Neutron)	n	1	1.008 665	β^-	10.37 min
1	Hydrogen	H	1	1.007 825	99.985	
	Deuterium	D	2	2.014 102	0.015	
	Tritium	T	3	3.016 050	β^-	12.33 yr
2	Helium	He	3	3.016 030	0.000 138	
			4	4.002 603	≈100	
3	Lithium	Li	6	6.015 121	7.5	
			7	7.016 003	92.5	
4	Beryllium	Be	7	7.016 928	EC, γ	53.29 days
			9	9.012 182	100	
5	Boron	B	10	10.012 937	19.9	
			11	11.009 305	80.1	
6	Carbon	C	11	11.011 432	β^+, EC	20.39 min
			12	12.000 000	98.90	
			13	13.003 355	1.10	
			14	14.003 241	β^-	5730 yr
7	Nitrogen	N	13	13.005 738	β^+, EC	9.965 min
			14	14.003 074	99.634	
			15	15.000 108	0.366	
8	Oxygen	O	15	15.003 065	β^+, EC	122.2 s
			16	15.994 915	99.762	
			18	17.999 160	0.200	
9	Fluorine	F	18	18.000 937	EC, β^+	1.8295 h
			19	18.998 403	100	
10	Neon	Ne	20	19.992 435	90.51	
			22	21.991 383	9.22	
11	Sodium	Na	22	21.994 434	β^+, EC, γ	2.602 yr
			23	22.989 767	100	
			24	23.990 961	β^-, γ	14.659 h
12	Magnesium	Mg	24	23.985 042	78.99	
13	Aluminum	Al	27	26.981 539	100	
14	Silicon	Si	28	27.976 927	92.23	
			31	30.975 362	β^-, γ	2.622 h
15	Phosphorus	P	31	30.973 762	100	
			32	31.973 907	β^-	14.282 days
16	Sulfur	S	32	31.972 070	95.02	
			35	34.969 031	β^-	87.51 days
17	Chlorine	Cl	35	34.968 852	75.77	
			37	36.965 903	24.23	
18	Argon	Ar	40	39.962 384	99.600	
19	Potassium	K	39	38.963 707	93.2581	
			40	39.963 999	β^-, EC, γ	1.277×10^9 yr

[a]Data for atomic masses are taken from *Handbook of Chemistry and Physics,* 66th ed., CRC Press, Boca Raton, FL. The masses are those for the neutral atom, including the Z electrons. Data for percent abundance, decay mode, and half-life are taken from E. Browne and R. Firestone, *Table of Radioactive Isotopes,* V. Shirley, Ed., Wiley, New York, 1986. α = alpha particle emission, β^- = negative beta emission, β^+ = positron emission, γ = γ-ray emission, EC = electron capture.

Appendix F Selected Isotopes (*continued*)

Atomic No. Z	Element	Symbol	Atomic Mass No. A	Atomic Mass (u)	% Abundance, or Decay Mode if Radioactive	Half-Life (if Radioactive)
20	Calcium	Ca	40	39.962 591	96.941	
21	Scandium	Sc	45	44.955 910	100	
22	Titanium	Ti	48	47.947 947	73.8	
23	Vanadium	V	51	50.943 962	99.750	
24	Chromium	Cr	52	51.940 509	83.789	
25	Manganese	Mn	55	54.938 047	100	
26	Iron	Fe	56	55.934 939	91.72	
27	Cobalt	Co	59	58.933 198	100	
			60	59.933 819	β^-, γ	5.271 yr
28	Nickel	Ni	58	57.935 346	68.27	
			60	59.930 788	26.10	
29	Copper	Cu	63	62.939 598	69.17	
			65	64.927 793	30.83	
30	Zinc	Zn	64	63.929 145	48.6	
			66	65.926 034	27.9	
31	Gallium	Ga	69	68.925 580	60.1	
32	Germanium	Ge	72	71.922 079	27.4	
			74	73.921 177	36.5	
33	Arsenic	As	75	74.921 594	100	
34	Selenium	Se	80	79.916 520	49.7	
35	Bromine	Br	79	78.918 336	50.69	
36	Krypton	Kr	84	83.911 507	57.0	
			89	88.917 640	β^-, γ	3.16 min
			92	91.926 270	β^-, γ	1.840 s
37	Rubidium	Rb	85	84.911 794	72.165	
38	Strontium	Sr	86	85.909 267	9.86	
			88	87.905 619	82.58	
			90	89.907 738	β^-	29.1 yr
			94	93.915 367	β^-,γ	1.235 s
39	Yttrium	Y	89	88.905 849	100	
40	Zirconium	Zr	90	89.904 703	51.45	
41	Niobium	Nb	93	92.906 377	100	
42	Molybdenum	Mo	98	97.905 406	24.13	
43	Technetium	Tc	98	97.907 215	β^-, γ	4.2×10^6 yr
44	Ruthenium	Ru	102	101.904 348	31.6	
45	Rhodium	Rh	103	102.905 500	100	
46	Palladium	Pd	106	105.903 478	27.33	
47	Silver	Ag	107	106.905 092	51.839	
			109	108.904 757	48.161	
48	Cadmium	Cd	114	113.903 357	28.73	
49	Indium	In	115	114.903 880	95.7; β^-	4.41×10^{14} yr
50	Tin	Sn	120	119.902 200	32.59	
51	Antimony	Sb	121	120.903 821	57.3	
52	Tellurium	Te	130	129.906 229	38.8; β^-	2.5×10^{21} yr
53	Iodine	I	127	126.904 473	100	
			131	130.906 114	β^-, γ	8.040 days

Appendix F Selected Isotopes (*continued*)

Atomic No. Z	Element	Symbol	Atomic Mass No. *A*	Atomic Mass (u)	% Abundance, or Decay Mode if Radioactive	Half-Life (if Radioactive)
54	Xenon	Xe	132	131.904 144	26.9	
			136	135.907 214	8.9	
			140	139.921 620	β^-, γ	13.6 s
55	Cesium	Cs	133	132.905 429	100	
			134	133.906 696	β^-, EC, γ	2.062 yr
56	Barium	Ba	137	136.905 812	11.23	
			138	137.905 232	71.70	
			141	140.914 363	β^-, γ	18.27 min
57	Lanthanum	La	139	138.906 346	99.91	
58	Cerium	Ce	140	139.905 433	88.48	
59	Praseodymium	Pr	141	140.907 647	100	
60	Neodymium	Nd	142	141.907 719	27.13	
61	Promethium	Pm	145	144.912 743	EC, α, γ	17.7 yr
62	Samarium	Sm	152	151.919 729	26.7	
63	Europium	Eu	153	152.921 225	52.2	
64	Gadolinium	Gd	158	157.924 099	24.84	
65	Terbium	Tb	159	158.925 342	100	
66	Dysprosium	Dy	164	163.929 171	28.2	
67	Holmium	Ho	165	164.930 319	100	
68	Erbium	Er	166	165.930 290	33.6	
69	Thulium	Tm	169	168.934 212	100	
70	Ytterbium	Yb	174	173.938 859	31.8	
71	Lutetium	Lu	175	174.940 770	97.41	
72	Hafnium	Hf	180	179.946 545	35.100	
73	Tantalum	Ta	181	180.947 992	99.988	
74	Tungsten (wolfram)	W	184	183.950 928	30.67	
75	Rhenium	Re	187	186.955 744	62.60; β^-	4.6×10^{10} yr
76	Osmium	Os	191	190.960 920	β^-, γ	15.4 days
			192	191.961 467	41.0	
77	Iridium	Ir	191	190.960 584	37.3	
			193	192.962 917	62.7	
78	Platinum	Pt	195	194.964 766	33.8	
79	Gold	Au	197	196.966 543	100	
			198	197.968 217	β^-, γ	2.6935 days
80	Mercury	Hg	202	201.970 617	29.80	
81	Thallium	Tl	205	204.974 401	70.476	
			208	207.981 988	β^-, γ	3.053 min
82	Lead	Pb	206	205.974 440	24.1	
			207	206.975 872	22.1	
			208	207.976 627	52.4	
			210	209.984 163	α, β^-, γ	22.3 yr
			211	210.988 735	β^-, γ	36.1 min
			212	211.991 871	β^-, γ	10.64 h
			214	213.999 798	β^-, γ	26.8 min
83	Bismuth	Bi	209	208.980 374	100	
			211	210.987 255	α, β^-, γ	2.14 min
			212	211.991 255	β^-, α, γ	1.0092 h

Appendix F Selected Isotopes (*continued*)

Atomic No. Z	Element	Symbol	Atomic Mass No. *A*	Atomic Mass (u)	% Abundance, or Decay Mode if Radioactive	Half-Life (if Radioactive)
84	Polonium	Po	210	209.982 848	α, γ	138.376 days
			212	211.988 842	α, γ	45.1 s
			214	213.995 176	α, γ	163.69 μs
			216	216.001 889	α, γ	150 ms
85	Astatine	At	218	218.008 684	α, β^-	1.6 s
86	Radon	Rn	220	220.011 368	α, γ	55.6 s
			222	222.017 570	α, γ	3.825 days
87	Francium	Fr	223	223.019 733	α, β^-, γ	21.8 min
88	Radium	Ra	224	224.020 186	α, γ	3.66 days
			226	226.025 402	α, γ	1.6×10^3 yr
			228	228.031 064	β^-, γ	5.75 yr
89	Actinium	Ac	227	227.027 750	α, β^-, γ	21.77 yr
			228	228.031 015	β^-, γ	6.13 h
90	Thorium	Th	228	228.028 715	α, γ	1.913 yr
			231	231.036 298	β^-, γ	1.0633 days
			232	232.038 054	100; α, γ	1.405×10^{10} yr
			234	234.043 593	β^-, γ	24.10 days
91	Protactinium	Pa	231	231.035 880	α, γ	3.276×10^4 yr
			234	234.043 303	β^-, γ	6.70 h
			237	237.051 140	β^-, γ	8.7 min
92	Uranium	U	232	232.037 130	α, γ	68.9 yr
			233	233.039 628	α, γ	1.592×10^5 yr
			235	235.043 924	0.7200; α, γ	7.037×10^8 yr
			236	236.045 562	α, γ	2.342×10^7 yr
			238	238.050 784	99.2745; α, γ	4.468×10^9 yr
			239	239.054 289	β^-, γ	23.47 min
93	Neptunium	Np	239	239.052 933	β^-, γ	2.355 days
94	Plutonium	Pu	239	239.052 157	α, γ	2.411×10^4 yr
			242	242.058 737	α, γ	3.763×10^5 yr
95	Americium	Am	243	243.061 375	α, γ	7.380×10^3 yr
96	Curium	Cm	245	245.065 483	α, γ	8.5×10^3 yr
97	Berkelium	Bk	247	247.070 300	α, γ	1.38×10^3 yr
98	Californium	Cf	249	249.074 844	α, γ	350.6 yr
99	Einsteinium	Es	254	254.088 019	α, γ, β^-	275.7 days
100	Fermium	Fm	253	253.085 173	EC, α, γ	3.00 days
101	Mendelevium	Md	255	255.091 081	EC, α	27 min
102	Nobelium	No	255	255.093 260	EC, α	3.1 min
103	Lawrencium	Lr	257	257.099 480	α, EC	646 ms
104	Rutherfordium	Rf	261	261.108 690	α	1.08 min
105	Dubnium	Db	262	262.113 760	α	34 s

Answers to Check Your Understanding

Chapter 18

CYU 1: c

CYU 2: $+3.2 \times 10^{-13}$ C on object A and -3.2×10^{-13} C on object B

CYU 3: $+1.6 \times 10^{-13}$ C on object A and -3.2×10^{-13} C on object B

CYU 4: b and e

CYU 5: Yes, because the charge on the balloon will induce a slight charge of opposite polarity in the surface of the ceiling, analogous to that in Figure 18.8.

CYU 6: a

CYU 7: b

CYU 8: C, A, B

CYU 9: the electron, because, being less massive, it has the greater acceleration

CYU 10: No, because the force of the spring changes direction when the spring is stretched compared to when it is compressed, while the electrostatic force does not have this characteristic.

CYU 11: d

CYU 12: 0 N/C

CYU 13: (a) corner C
(b) negative
(c) greater

CYU 14: a

CYU 15: (a) No.
(b) No.

CYU 16: For rod A, the field points perpendicularly away from the rod. For rod B, it points parallel to the rod and is directed from the positive toward the negative half.

CYU 17: (a) false
(b) false
(c) true
(d) false
(e) false

CYU 18: d

CYU 19: The flux does not change, as long as the charge remains within the Gaussian surface.

CYU 20: The same flux passes through each, since each encloses the same net charge.

CYU 21: (a) q_1 and q_2
(b) q_1, q_2, and q_3

Chapter 19

CYU 1: (a) Yes.
(b) No.
(c) Yes.
(d) Yes.

CYU 2: The work is the same in all three cases (see Equation 19.4).

CYU 3: The electron arrives at a plate first.

CYU 4: a

CYU 5: b

CYU 6: c

CYU 7: a

CYU 8: d

CYU 9: (a) remains the same
(b) decreases

CYU 10: the electron

CYU 11: (a) +2.0 V
(b) 0 V
(c) +2.0 V

CYU 12: The electric field is zero.

CYU 13: b

CYU 14: a

CYU 15: c

CYU 16: (a) bottom of a valley
(b) top of a mountain

CYU 17: (a) decreases
(b) increases
(c) increases
(d) increases

CYU 18: (a) decreases
(b) increases
(c) remains the same
(d) increases

Chapter 20

CYU 1: d

CYU 2: 0.50 A

CYU 3: b

CYU 4: a

CYU 5: b, d, and e

CYU 6: c (A value for the current is also needed.)

CYU 7: a

CYU 8: c

CYU 9: b and d

CYU 10: The 75-W bulb. See Equation 20.15c.

CYU 11: d

CYU 12: e

CYU 13: in parallel

CYU 14: b

CYU 15: c

CYU 16: a, b, d, and e

CYU 17: There are two ways. One is to form two groups of two parallel resistors and then connect the groups in series. The other is to form two groups of two series resistors and then connect the groups in parallel.

CYU 18: Junction rule: $I_1 + I_3 = I_2$
Loop rule, loop *ABCD:*
$3.0\text{ V} + 7.0\text{ V} + I_3R_3 = I_1R_1$
Loop rule, loop *BEFC:*
$5.0\text{ V} = I_3R_3 + 7.0\text{ V} + I_2R_2$

CYU 19: c

CYU 20: b

CYU 21: ohm × farad
= (volt/ampere)(coulomb/volt)
= coulomb/ampere
= coulomb/(coulomb/second)
= second

CYU 22: e

Chapter 21

CYU 1: c

CYU 2: d

CYU 3: (a) Yes.
(b) No, because the particle could move either parallel or anti-parallel to the magnetic field.

CYU 4: b

CYU 5: d

CYU 6: particle 3

CYU 7: b

CYU 8: c

CYU 9: b

CYU 10: c

CYU 11: (a) The direction of the magnetic force reverses.
(b) The direction of the magnetic force does not change.

CYU 12: B and D (a tie), A, C

CYU 13: a

CYU 14: (a) repelled
(b) repelled

CYU 15: (a) attracted
(b) repelled

CYU 16: c

CYU 17: a

CYU 18: Part *a*: There is a point to the right of both wires where the total magnetic field is zero. Part *b*: There is a point between the wires where the total magnetic field is zero. This point is closer to the wire carrying the current I_2.

CYU 19: A, D, C, B

CYU 20: d

CYU 21: No, because aluminum is a non-ferromagnetic material.

CYU 22: b

Chapter 22

CYU 1: No. With both the magnet and coil moving at the same velocity with respect to the earth, there is no relative motion between the magnet and the coil, which is needed for there to be an induced current in the coil.

CYU 2: d

CYU 3: a

CYU 4: b

CYU 5: c

CYU 6: b

CYU 7: A lightning bolt is a large electric current that changes in time and, thus, produces a magnetic field that also changes in time. When this changing field passes through a coil or loop of wire in an appliance, it can, via Faraday's law, create an induced emf, which can lead to an induced current.

CYU 8: c

CYU 9: a and d

CYU 10: a

CYU 11: b

CYU 12: Answer 1: downward and decreasing
Answer 2: upward and increasing

CYU 13: c

CYU 14: d

CYU 15: b

CYU 16: With the headlights off, the engine does not need to do the work of keeping the battery charged.

CYU 17: a

CYU 18: c

CYU 19: b

CYU 20: b and d

Chapter 23

CYU 1: The ratio decreases by a factor of 3.

CYU 2: (a) the circuit containing the inductor
(b) the circuit containing the resistor

CYU 3: less than

CYU 4: decreases

CYU 5: decreases

CYU 6: d

CYU 7: (a) increases
(b) increases

CYU 8: a

CYU 9: (a) remains the same
(b) decreases

CYU 10: in phase (see Equation 23.8, in which $X_L = X_C$)

CYU 11: (a) Yes.
(b) Yes.

CYU 12: a

CYU 13: (a) left to right
(b) left to right

Chapter 24

CYU 1: d

CYU 2: a

CYU 3: because, according to Faraday's law of electromagnetic induction, the emf depends on how rapidly the magnetic field of the wave is changing and this is determined by the frequency of the wave

CYU 4: e

CYU 5: d

CYU 6: b

CYU 7: No. The same Doppler change results when the star moves away from the earth and when the earth moves away from the star. Only the relative motion between the star and the earth can be detected.

CYU 8: B, A, C

CYU 9: Yes.

CYU 10: The light intensity that is not transmitted is absorbed by the polarizer and the analyzer. The polarizer absorbs one-half of the incident intensity, and the analyzer absorbs four-tenths of the incident intensity.

CYU 11: unpolarized: c
horizontally polarized: b
vertically polarized: c

CYU 12: because the transmission axis of the Polaroid material is nearly horizontal, in the same direction as the polarized light reflected from the lake

Chapter 25

CYU 1: 55°

CYU 2: Yes.

CYU 3: Yes.

CYU 4: $f_{\text{inside}} = +0.30$ m, $f_{\text{outside}} = -0.30$ m

CYU 5: (a) concave
(b) The sodium unit and engine are located at the focal point of the mirror.

CYU 6: Open the surface up to produce a more gently curving shape.

CYU 7: No.

CYU 8: (a) upright
(b) upside down

CYU 9: (a) Yes, provided the object distance is greater than the focal length of the mirror.
(b) It is not possible for a convex mirror to project an image directly onto a screen.

CYU 10: (a) No.
(b) No.

CYU 11: (a) The magnitude of the image distance becomes larger.
(b) The magnitude of the image height becomes larger.

CYU 12: (a) No. You can see yourself anywhere on the principal axis.
(b) You cannot see yourself when you are between the center of curvature and the focal point of the mirror because your image is behind you.

CYU 13: A, D, and E

CYU 14: The image will never be located beyond the focal point (behind the mirror).

Chapter 26

CYU 1: slab B

CYU 2: liquid B

CYU 3: Yes. To see why, apply Snell's law at the air–water interface and at the water–glass interface.

CYU 4: the one filled with water

CYU 5: liquid A

CYU 6: c

CYU 7: b

CYU 8: a

CYU 9: c

CYU 10: The critical angle for a water–air interface is 48.8° (see Equation 26.4). Any light emitted at an angle greater than 48.8° with respect to the vertical is incident on the surface at an angle exceeding the critical angle. It is totally internally reflected and doesn't exit the water.

CYU 11: No. To see why, apply Snell's law at both surfaces of the glass slab and use Equation 26.4.

CYU 12: c (They are most effective when the angle of incidence is the Brewster angle and the reflected light is 100% polarized.)

CYU 13: liquid A

CYU 14: a (Since $n = 1.520$ for red light and $n = 1.538$ for violet-colored light, the critical angle for total internal reflection is greater for red than for violet-colored light.)

CYU 15: b

CYU 16: Yes.

CYU 17: a

CYU 18: the lens

CYU 19: converging lens, $d_o = \frac{1}{2}f$

CYU 20: d

CYU 21: the glasses of the farsighted person, since they use converging lenses

CYU 22: 13 cm

CYU 23: Light normally passes from air ($n = 1.00$) into the cornea ($n = 1.38$), at which time most of the eye's refraction of the light occurs. If water ($n = 1.33$) replaces air, the similarity of the index of refraction of water to that of the cornea reduces the eye's normal refraction and causes blurred vision. Goggles preserve the air–cornea boundary.

CYU 24: b

CYU 25: hawk, kestrel, eagle

CYU 26: a

CYU 27: 0.042 rad

CYU 28: b

CYU 29: the longer telescope

CYU 30: microscope

CYU 31: c, d, e, f

CYU 32: because chromatic aberration is related to the refraction of light and not to the reflection of light

Chapter 27

CYU 1: (a) constructive
(b) destructive
(c) destructive

CYU 2: c

CYU 3: Yes.

CYU 4: a

CYU 5: (a) d_1 and λ_2
(b) d_2 and λ_1

CYU 6: (a) The pattern would be the same.
(b) The positions of the light and dark fringes would be interchanged.

CYU 7: No, because θ in Equations 27.1 and 27.2 approaches 90° as λ becomes larger and larger.

CYU 8: b

CYU 9: c

CYU 10: enhances

CYU 11: (a) A and C
(b) B

CYU 12: c

CYU 13: b

CYU 14: (a) broadens
(b) contracts

CYU 15: a

CYU 16: c

CYU 17: a, c, b

CYU 18: Yes.

CYU 19: small *f*-number setting

CYU 20: (a) the maximum that is closer to the central maximum
(b) away from the central maximum

CYU 21: The distance between the bright fringes would decrease.

Chapter 28

CYU 1: d

CYU 2: a, b

CYU 3: C, B, A

CYU 4: No, because the term v^2/c^2 in Equations 28.1 and 28.2 would then be zero.

CYU 5: c

CYU 6: c, d

CYU 7: No, because the two diagonals are perpendicular, so that diagonal AC is contracted, whereas diagonal BD is not contracted.

CYU 8: greatest mass: c, smallest mass: b

CYU 9: a, because then they have more electric potential energy (see Example 8 in Chapter 19)

CYU 10: b, because the fully charged capacitor stores electric potential energy (see Section 19.5)

CYU 11: a. The work is the change in kinetic energy, which is proportional to the mass (see the work–energy theorem in Section 6.2). The electron has the smaller mass.

CYU 12: c

Chapter 29

CYU 1: No.

CYU 2: (a) red
(b) violet

CYU 3: No.

CYU 4: b

CYU 5: a

CYU 6: (a) increases
(b) increases
(c) remains the same
(d) remains the same

CYU 7: less than

CYU 8: c

CYU 9: a

CYU 10: No. Photon collisions would cause spinning in a direction from the shiny side of a panel toward the black side.

CYU 11: decreases

CYU 12: b

CYU 13: decreases

CYU 14: b

Chapter 30

CYU 1: b

CYU 2: d

CYU 3: The absorption lines belong only to the Lyman series, since very few electrons are present with $n = 2$ or $n = 3$.

CYU 4: No, because the location of the electron in a given quantum mechanical energy state is uncertain.

CYU 5: when the electron is in the $n = 1$ state, because then the only possible value for the orbital quantum number is $\ell = 0$

CYU 6: (a) No, because the Bohr model uses the same quantum number n for the total energy and the orbital angular momentum (see Equations 30.13 and 30.8).
(b) Yes, because quantum mechanics uses the quantum number n for the total energy but the quantum number ℓ for the orbital angular momentum.

CYU 7: (a) Yes, because the Bohr model uses the same quantum number n for the orbital angular momentum and the total energy (see Equations 30.8 and 30.13).
(b) No, because quantum mechanics uses the quantum number ℓ for the orbital angular momentum but the quantum number n for the total energy.

CYU 8: $1s^2\ 2s^2\ 2p^6\ 3s^2\ 3p^6\ 4s^2\ 3d^{10}\ 4p^6$

CYU 9: (a) No.
(b) Yes.

CYU 10: b

CYU 11: c

CYU 12: a

CYU 13: c

CYU 14: d

CYU 15: a

Chapter 31

CYU 1: c, d

CYU 2: a

CYU 3: No, because they could have different numbers of protons (different atomic numbers).

CYU 4: Yes, because the total number A of nucleons could be the same, and it is the value of A that determines the radius.

CYU 5: d, c, a, b

CYU 6: c

CYU 7: d

CYU 8: It is not possible, because the total mass of the decay products is greater than the mass of the parent nucleus, ${}^{238}_{92}U$, indicating that energy would not be released.

CYU 9: b

CYU 10: Yes, because the decay of any single nucleus occurs randomly and can happen at any moment.

CYU 11: Yes.

CYU 12: b, because the gold statue does not contain carbon atoms

CYU 13: too small

CYU 14: No, because in 700 years the activity of a sample would have decreased to an immeasurably small fraction of its initial value.

CYU 15: e

Chapter 32

CYU 1: Yes, if the absorbed dose of the radiation is different for each type of radiation (see Equation 32.4).

CYU 2: c

CYU 3: neutrons, α particles, γ rays

CYU 4: c

CYU 5: (a) because it violates the conservation of nucleon number
(b) because it violates the conservation of nucleon number
(c) because it violates the conservation of electric charge

CYU 6: d

CYU 7: electrons, protons, neutrons

CYU 8: No, because the binding energy per nucleon is greater for the original nucleus than for the two fragments, as indicated in Figure 31.5.

CYU 9: b and d

CYU 10: a, b, and c

CYU 11: No, because the binding energy per nucleon is greater for the original nuclei than for the nucleus resulting from the fusion, as indicated in Figure 32.8.

CYU 12: b and d

CYU 13: a

Answers to Odd-Numbered Problems

Chapter 18

1. $+3.04 \times 10^{-18}$ C
3. (a) $-1.6\ \mu$C (b) 1.0×10^{13} electrons
5. (a) $+1.5q$ (b) $+4q$ (c) $+4q$
7. (a) 3.35×10^{26} electrons (b) -5.36×10^{7} C
9. 8 electrons
11. (a) 0.83 N (a) attractive
13. 1.8×10^{-5} C
15. (a) Like signs (i.e., either both positive or both negative)
    (b) 1.7×10^{-16} C
17. (a) 4.56×10^{-8} C (b) 3.25×10^{-6} kg
19. 7.19×10^{23} m/s^2
21. (a) 0.166 N directed along the $+y$ axis
    (b) 111 m/s^2 directed along the $+y$ axis
23. 1.96×10^{-17} J
25. -3.3×10^{-6} C
27. 1.37
29. 1.8 N due east
31. 54 N/C
33. (a) 7700 N/C (b) 1300 N/C (c) 5500 N/C
35. 6.5×10^{3} N/C directed downward
37. 1.81×10^{2} N/C in Fig. 18.20a and 3.11×10^{2} N/C in Fig. 18.20b
39. 2.2×10^{5} N/C directed along the $-x$ axis
41. 2900 N/C directed 47° above the $+x$ axis
43. (a) $+2.0\ \mu$C (b) $-6.0\ \mu$C
45. 6.5 m
47. 3.25×10^{-8} C
49. (a) 350 N · m^2/C (b) 460 N · m^2/C
51. 58°
53. (a) 7.9×10^{5} N/C, radially outward
    (b) 1.4×10^{6} N/C, radially inward
    (c) 0 N/C
55. (a) Positive (b) 2.53×10^{7} protons
57. 0.80 m
59. $F_A = 430$ N; $F_B = 250$ N; $F_C = 250$ N
61. (a) 8.2×10^{-8} C (b) 8.2×10^{-3} N
63. 8.0×10^{5} N/C (b) 1.3×10^{-13} N
65. 2.6×10^{-9} N
67. 2.7×10^{-14} C
69. 1340 m
71. (a) 7.56×10^{4} N/C (along the $+y$ axis)
    (b) 5.04×10^{2} m/s^2 (along the $+y$ axis)
73. (a) 2.9×10^{-9} C (b) 2.1×10^{-4} kg
75. (a) 1.23×10^{5} N/C (b) -1.94×10^{5} N/C

Chapter 19

1. -2.1×10^{-11} J
3. (a) 4.5×10^{-3} N directed from A toward B
   (b) 3.0×10^{3} N/C directed from A toward B
5. 19 m/s
7. $+4.80 \times 10^{-18}$ C
9. -4.05×10^{4} V
11. 2.4
13. -9.4×10^{3} V
15. $+7.8 \times 10^{6}$ V
17. -3.1×10^{-6} C
19. -0.746 J
21. 1.53×10^{-14} m
23. 0.0342 m
25. 1.41×10^{-2} m
27. 1.6×10^{-8} C
29. 3.5×10^{4} V
31. (a) 179 V (b) 143 V (c) 155 V
33. (a) 0 V/m (b) 1.0×10^{1} V/m (c) 5.0 V/m
35. (a) 0 V (b) +290 V (c) −290 V
37. 36 V
39. 6.1×10^{-5} C
41. 5.66 V
43. 52 V
45. 18 V
47. (a) -5.6×10^{-5} J (b) 0 J
49. 3.82
51. 0.213 J
53. 1.1×10^{-20} J
55. 1.1×10^{3} V
57. 2×10^{-8} F
59. (a) decreases (b) 36
61. (a) 3.75×10^{9} J (b) 1.45×10^{3} kg
63. 0.068 J
65. (a) 2.06×10^{3} V (b) 1.37×10^{-4} W
67. (a) 3657 capacitors (b) 398 capacitors

Chapter 20

1. 174 C
3. 8.6 A
5. 16 Ω
7. (a) 0.600 Ω (for *a*); 0.0375 Ω (for *b*); 0.150 Ω (for *c*)
   (b) 5.00 A (for *a*); 80.0 A (for *b*); 20.0 A (for *c*)
9. Aluminum
11. 58 m
13. 9.3%
15. 1.8 V
17. 6.0×10^{2} Ω
19. \$$5.9 \times 10^{6}$
21. 8.9 hours
23. 3.1×10^{-4} m
25. 250 °C
27. (a) 786 W (b) 1572 W
29. 150 W
31. 21 V
33. (a) 50.0 Hz (b) 2.40×10^{2} Ω (c) 60.0 W
35. 4.0×10^{1} Ω
37. $P(2.0\ \Omega) = 8.0$ W; $P(4.0\ \Omega) = 16$ W; $P(6.0\ \Omega) = 24$ W
39. 85.9 Ω and 242 Ω
41. 140 W
43. 1800 W
45. (a) $I_1 = I_2 = I_3 = 0.282$ A; $V_1 = 14.1$ V; $V_2 = 7.05$ V; $V_3 = 2.82$ V
    (b) $I_1 = 0.480$ A; $I_2 = 0.960$ A; $I_3 = 2.40$ A; $V_1 = V_2 = V_3 = 24.0$ V
47. 2.3 Ω
49. (a) 130 Ω (b) 7.2 W
51. 2.00 Ω and 4.00 Ω
53. 4
55. 9.2 A
57. 4.6 Ω
59. 42 Ω
61. 6.00 Ω, 0.545 Ω, 3.67 Ω, 2.75 Ω, 2.20 Ω, 1.50 Ω, 1.33 Ω, and 0.833 Ω
63. 25 Ω
65. 12.0 V
67. 24.0 V
69. 4.84 A
71. 2.0 A
73. 0.75 V, left end at higher potential
75. (a) 15.0 A (b) 6.5 Ω (c) 3.5 Ω
77. 2.49×10^{4} Ω
79. 820 Ω and 8.00×10^{-3} A
81. 9.0 V
83. 18 μF
85. 2.0 μF

87. Demonstration of the equality is proof
89. 11 V
91. 3.9 s
93. 1.61
95. (a) 24.0 W (for *a*); 12.0 W (for *b*); 96.0 W (for *c*); 48.0 W (for *d*)
(b) 2.00 A (for *a*); 1.00 A (for *b*); 4.00 A (for *c*); 2.00 A (for *d*)
97. −34.6 °C
99. 21 V
101. (a) 1.2 Ω (b) 110 V
103. 0.049 m
105. (a) 1.5×10^{-11} A (b) 4.7×10^{7} ions
107. 4.1×10^{-7} F
109. (a) 860 W (b) 5700 cells (c) 5700
111. 1.92×10^{-15} moles
113. (a) 9.6 W (b) 2.4 W (c) 9.6 W
115. (a) 6.65 A (b) 1.37×10^{4} W
117. (a) 12 Ω
(b) Four 47-Ω resistors in parallel have a resistance of 11.8 Ω, which is within 10% of the 12-Ω requirement.

Chapter 21

1. 5.9×10^{12} m/s²
3. 4.1×10^{-3} m/s
5. 58°
7. 1.1×10^{-2} N
9. 1.3×10^{-10} m, directed out of the page
11. (a) due south (b) 2.55×10^{14} m/s²
13. 1.5×10^{-8} s
15. 2.7×10^{-4} T
17. 0.0904 m
19. 19 cm
21. 1.41
23. 5.25×10^{3} m/s
25. 0.71 m
27. 6.8×10^{5} C/kg
29. 5.1×10^{-5} T
31. 3.4×10^{-3} T
33. 2.7 m
35. (a) left-to-right (b) 1.1×10^{-2} m
37. (a) 37° (b) 0.49 N
39. 3.3 cm
41. 0.053 m
43. (a) 13.4 A · m² (b) 24.1 N · m
45. (a) 170 N · m (b) 35° angle increases
47. 3.1×10^{-4} T
49. 1.8×10^{-4} N·m
51. 190 A
53. 8.71×10^{-3} m
55. 0.800 m
57. 1.04×10^{-2} T
59. 320 A
61. (a) 1.1×10^{-5} T (b) 4.4×10^{-6} T
63. 19.7°
65. 0.062 m
67. 0.19 N
69. 1.7×10^{-3} N
71. (a) $F_E = +1.37 \times 10^{-3}$ N, along $+x$ axis, $F_{Bx} = 0$ N, $F_{By} = 0$ N
(b) $F_E = +1.37 \times 10^{-3}$ N, along the $+x$ axis, $F_{Bx} = 0$ N, $F_{By} = 2.94 \times 10^{-3}$ N, along $+z$ axis
(c) $F_E = +1.37 \times 10^{-3}$ N, along $+x$ axis, $F_{Bx} = 3.78 \times 10^{-3}$ N, along $+y$ axis, $F_{By} = 2.94 \times 10^{-3}$ N, along $-x$ axis
73. 0.14 T
75. (a) 9.1×10^{-7} T (b) 49
77. 4.0×10^{3} A
79. 1.18 T
81. 0.325 kg
83. (a) 4590 turns (b) 519 m (c) −1.52 A
85. (a) ^{1}H: 6.9×10^{5} m/s, ^{2}H: 4.9×10^{5} m/s, ^{3}H: 4.0×10^{5} m/s (b) 0.125 T
(c) ^{1}H: 11.6 cm, ^{2}H: 16.4 cm

Chapter 22

1. (a) 3.7×10^{-5} T (b) East end is positive
3. 7800 V
5. (a) 1 (b) 2 (c) 8
7. 0.15 W
9. (a) 0.23 kg (b) −1.8 J (c) 1.8 J
11. (a) 0.086 m² (b) 0.018 m²
13. −0.094 Wb
15. 8.9×10^{-3} Wb
17. (a) −1.0 V (0 – 3.0 s); 0 V (3.0 – 6.0 s); +0.50 V (6.0 – 9.0 s)
(b) −2.0 A (0 – 3.0 s); +1.0 A (6.0 – 9.0 s)
19. 8.6×10^{-5} T
21. (a) 0.38 V (b) 0.43 m²/s
23. 0.14 V
25. (a) 3.6×10^{-3} V (b) The area of the loop must shrink at a rate of 2.0×10^{-3} m²/s
27. −0.84 A
29. The right end is positive
31. Left-to-right
33. (a) counterclockwise
(b) No induced current
(c) clockwise (d) no induced current
35. (a) Lenz's law explains the behavior
(b) No induced current can flow in the cut ring
37. 0.150 m
39. 3.0×10^{5}
41. 2.0 V
43. 1.5×10^{9} J
45. 2.5×10^{-2} H
47. 1.4 V
49. 220 turns
51. 3.8 V
53. (a) 1:13 (b) 1.7×10^{-2} A
(c) 2.0 W delivered by wall socket
(d) 2.0 W delivered to the battery
55. 41 V
57. 140
59. $R_1 = (N_p/N_s)R_2$
61. 1.5 m²/s
63. 756
65. 0.050 V
67. 6.6×10^{-2} J
69. 1.0×10^{-3} V
71. 38 V
73. 5.2×10^{-5} T
75. (a) 1.1 m (b) 3.2×10^{-5} Wb
77. (a) 0.040 s (b) 160 rad/s (c) 65 V
(d) −49 V
79. (a) 5.01×10^{4} V (b) 0.402 V
(c) 4.54×10^{3} V
(d) Yes, the probe should take a more distant path.
81. (a) 534 V (b) 378 V (c) 0.066 T
(d) $N_p : N_s = 3.44:1$

Chapter 23

1. 9470 Hz
3. 36 Ω
5. 2.7×10^{-6} F
7. (a) 6.4×10^{-6} F (b) 9.0×10^{-4} C
9. 0.44 A
11. 24.6 Ω
13. 176 mH
15. (a) 1.11×10^{4} Hz (b) 6.83×10^{-9} F
(c) 6.30×10^{3} Ω (d) 7.00×10^{2} Ω
17. 0.819
19. $R = 49.7$ Ω, $X_C = 185$ Ω
21. 3.0 W
23. 270 Hz
25. (a) 1.7×10^{-5} H (b) 8.8×10^{-5} H
27. 123 V
29. 1.8 μF
31. 2.8 kHz
33. 521 Hz
35. 2.4%

37. $V_R = 10.5$ V, $V_C = 19.0$ V, $V_L = 29.6$ V
39. 8.0×10^1 Hz
41. 0.075 A
43. (a) 2.94×10^{-3} H (b) 4.84 Ω (c) 0.163
45. $R = 443$ Ω, $X_C = 76.6$ Ω
47. (a) 2.2×10^5 Ω (b) 1.3×10^5 Ω (c) 0.30 Hz
49. (a) 4.20×10^{-8} H (b) 2.89×10^{-12} F
51. 1.8 A
53. (a) 24.5 W (b) 138 W
55. (a) 1.22×10^{-5} H
(b) Two possible configurations: (i) L_1 and L_2 in series. (ii) L_1 in series with $L_3 || L_4$.
(c) 588 kHz $\leq f_0 \leq$ 1019 kHz
(d) Lowest resistance is obtained with the resistors wired in parallel.

Chapter 24

1. 2.73×10^{12} m
3. 4.1×10^{16} m
5. 11.118 m
7. 4.4×10^8 Hz
9. (a) 1.4×10^4 (b) 1.4×10^4
11. 177 N/m
13. 540 rev/s
15. 7.0 bits
17. 1.6×10^6 m
19. 3.8×10^2 W/m²
21. 990 N/C
23. 1.8×10^{-5} J
25. 3.93×10^{26} W
27. 1.7×10^{11} W
29. (a) Receding
(b) 3.1×10^6 m/s
31. (a) 6.175×10^{14} Hz (b) 6.159×10^{14} Hz
33. (a) 0.55 W/m² (b) 3.7×10^{-2} W/m²
35. 21.5°
37. For $\alpha = 0°$ and 35°, respectively:
(a) 3.5 W/m², 3.5 W/m²
(b) 7.0 W/m², 4.7 W/m²
(c) 0 W/m², 2.3 W/m²
39. 720 W/m²
41. Sheet A removed: 3.8 W/m²; Sheet B removed: 0 W/m²; Sheet C removed: 0 W/m²; Sheet D removed: 5.1 W/m²
43. For $E = 315$ N/C and 945 N/C, respectively:
(a) 263 W/m², 2370 W/m²
(b) 1.05×10^{-6} T, 3.15×10^{-6} T
(c) 263 W/m², 2370 W/m²
45. 640 Hz
47. 68 N/C
49. 4.44×10^{-10}
51. 2.1×10^3 Hz
53. (a) 2.1×10^4 W/m² (b) 1.1×10^2 W (c) 18 s
55. (a) 3.0×10^8 m/s (b) 1.30×10^{19} Hz
57. 102.2 mph
59. (a) 2.5×10^8 m (b) 0.011 V/m (c) 3.7×10^{-11} T (d) 5.4×10^{-16} J/m³
61. (a) 4.0 W/m² (b) 6.0 W/m²
63. 4.00×10^{11} m
65. (a) 500 W/m² (b) 8.52 W/m²
(c) −1.5° (clockwise); 1.5° (counterclockwise)

Chapter 25

1. $x = +3.0$ m
3. −1.80 m/s
5. 1.67 m
7. (a) 0.0670° (b) 7 m
9. (a) 70.6° (b) 62.1°
11. (a) 24 cm
(b) Ray diagram similar to Figure 25.16.
13. (a) 7.5 cm in front of the mirror
(b) Image height is 1.0 cm
15. Image located at 10.9 cm. Ray diagram below.
17. (a) $d_o = +62$ cm (b) $m = +0.35$
(c) upright (d) smaller
19. $f = 9.1$ cm
21. (a) 180 cm (b) 6.0×10^1 cm
23. Smaller object distance: $m = 0.750$, Greater object distance: $m = 0.600$
25. −2.0
27. (a) $d_o = 82$ m (b) $h_o = 11$ m
29. 56 cm
31. (a) As object moves closer to the mirror, the magnitude of the image distance becomes smaller.
(b) As object moves closer to the mirror, the magnitude of the image height becomes larger.
33. 74 cm
35. −0.16 cm
37. −17 cm
39. (a) concave (b) +3.1 cm (c) +2.8 (d) upright
41. (a) 70° (b) 20° (c) 225 m (d) 420 m
43. 1.4 m
45. −13 cm
47. (a) 16 (b) above the horizontal
(c) +10.0° (d) $\beta = 5.0°$
49. (a) $d_o = 308.5$ m (b) $h = 147.2$ m
(c) $w_o = 10.3$ m

Chapter 26

1. 2.0×10^{-11} s
3. 1.66×10^8 m/s
5. Ethyl alcohol
7. 1.82
9. (a) 46°
(b) 50° (Drawing wrongly shows an angle of refraction greater than 55°)
(c) 69°
(d) 0° (Drawing wrongly shows an angle of refraction greater than 0°)
11. 0.9°
13. 1.92×10^8 m/s
15. 2.46×10^8 m/s
17. 1.65
19. 2.7 m
21. 1.19 mm
23. The answer is a derivation
25. 37.79°
27. 2.5 m
29. Layer a top surface: 24.7°; Layer b top surface: 14.9°; Layer b bottom surface: 6.0°; Layer c bottom surface: 29.4°
31. $n_b = 1.16$; $n_c = 1.38$
33. 3.36×10^{-8} s
35. 32°
37. 55.0°
39. 0.725 m
41. 1.73
43. 44.6° (red), 45.9° (violet)
45. (a) 1.075 (b) 1.083
47. (a) 1.2 m (b) 1.7 m

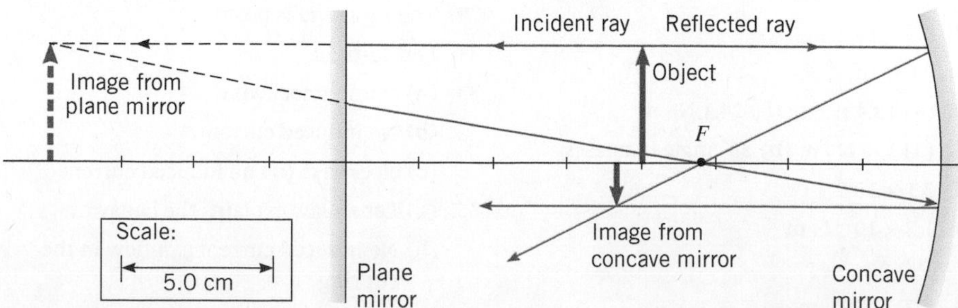

DIAGRAM FOR CHAPTER 25, PROBLEM 15.

49. (a) virtual (b) −8.1

51. (a) −10.0 cm (b) +0.500

53. (a) 60.0 cm (b) 20.0 cm

55. (a) −85.1 cm (b) 134.1 cm

57. 0.333 m

59. 37.3 cm

61. −5.6 cm

63. −12 cm

65. (a) 4.00 cm to the left of the diverging lens (b) −0.167 (c) virtual (d) inverted (e) smaller

67. (a) 18.1 cm (b) real (c) inverted

69. (a) +0.600 (b) +2.00

71. −9.2 m

73. 28.0 cm

75. −2.0 diopters (right eye); −0.15 diopters (left eye)

77. (a) −4.5 m (b) 0.50 m

79. 18

81. (a) 6.88 cm (b) 3.63

83. 13.7 cm

85. 9×10^{-3} rad

87. 3.0 cm

89. (a) −30.0 (b) 4.27 cm (c) −4.57

91. 1.1 m

93. (a) 1.3-dopter lens (b) 0.86 m (c) −8.5

95. −31

97. 1.51

99. (a) 6.74×10^{-7} m^2 (a) 7.86×10^5 W/m^2

101. (a) −24 cm (b) 6.0 mm

103. (a) converging (b) $2f$ (c) $2f$

105. −181

107. (a) 22.4 cm (b) 28.4 cm

109. +2.92 diopters

111. −1100

113. 47.5°

115. (a) 382 J/m^3 (b) 1.15×10^{11} W/m^2 (c) −0.405 (d) 1.04 m (e) 0.422 m

117. (a) 23.8° (b) 26.4° (c) 2.60°

Chapter 27

1. (a) 11° (b) 22° (c) 34° (d) 48°

3. (a) destructive interference is occurring (b) 3.25 m and 0.75 m from one of the sources

5. 4.9×10^{-7} m

7. Version A: 6 bright fringes, Version B: 4 bright fringes

9. 0.0248 m

11. 487 nm

13. 102 nm

15. 6.12×10^{-7} m

17. 1.18

19. 115 nm

21. 1.5

23. 3.3×10^{-6} m

25. 4.9×10^{-7} m

27. 0.447

29. 0.013

31. 1.1×10^{-4} m

33. 644 nm

35. 630 nm

37. 4.0×10^{-6} m

39. 640 nm and 480 nm

41. (a) 7.9° for violet light, 13° for red light
(b) For $m = 2$: 16° for violet light and 26° for red light
(c) For $m = 3$: 24° for violet light and 41° for red light
(d) The second and third orders overlap.

43. 571 nm

45. 1.2×10^{-5} m

47. 3.2×10^3 m

49. 1560 nm

51. (a) 220 nm (b) No

53. (a) 9.7×10^{-3} m = 9.7 mm
(b) The hunter's claim is not reasonable.

55. 5.1 s

57. (a) 0.82 (b) 1.2

59. (a) 740 nm (b) 31°

61. 522 nm

63. 499 nm

65. (a) 22.2×10^3 m (b) 24.8×10^3 m

67. (a) 12.8 μm (b) 40.7 μm
(c) $N = 47$ for sample in (a). $N = 149$ for sample in (b).

Chapter 28

1. (a) 4.9×10^{-9} s (b) 1.5 m

3. 0.999 95c

5. 2.28 s

7. 16

9. 2.6×10^8 m/s

11. 8.1 km

13. 1.3

15. 40.2°

17. 7.5×10^{-19} kg · m/s

19. 1.0 m

21. −2.0 m/s

23. 5.0×10^{-13} J

25. (a) 7.44×10^{-12} kg (b) 5.02×10^{-11} kg

27. 1.1 kg

29. (a) 3.84×10^{-12} J (b) 0.999 781c

31. 0.31c

33. +0.80c

35. Assuming the direction of the cruiser, ions, and laser light is in the positive direction; (a) +c (b) +0.994c (c) +0.200c (d) +0.194c

37. 1.7×10^{-13} J

39. 72 h

41. 0.866c (A); 0.943c (B); 0.745c (C)

43. (a) 0.020c (6.0×10^6 m/s)
(b) $2.83 \times 10^{-5}c$ (8490 m/s)

45. 2.60×10^8 m/s

47. 2.96×10^8 m/s

49. (a) $m_a = 6.64 \times 10^{-27}$ kg; $m_b = 6.64 \times 10^{-27}$ kg; $m_c = 33.2 \times 10^{-27}$ kg
(b) $KE_a = 5.98 \times 10^{-10}$ J; $KE_b = 17.9 \times 10^{-10}$ J; $KE_c = 5.98 \times 10^{-10}$ J

51. (a) 0.501c (b) 0.733c (c) too early

53. (a) 56.6 yr (b) 0.12 min = 7.1 s (c) 0.014:1

Chapter 29

1. (a) 1.63×10^{-7} m (b) 1.84×10^{15} Hz (c) UV region of the electromagnetic spectrum

3. 6.3 eV

5. 310 nm

7. 1.26 eV

9. 73 photons/s

11. 162 nm

13. (a) 7760 N/C (b) 2.59×10^{-5} T

15. 4.2×10^{-25} m/s

17. 0.3533 nm

19. $p_x = 7.37 \times 10^{-23}$ kg·m/s, $p_y = 5.82 \times 10^{-23}$ kg · m/s

21. 1.27×10^{-12} m

23. 1×10^{-18} m

25. 1830

27. 4.8×10^{-23} kg · m/s

29. 1.2×10^{-36} m

31. $\lambda_f = 2.0 \times 10^{-34}$ m

33. 4.0×10^1

35. 9.6×10^{-21} kg · m/s

37. −0.0289° to +0.0289°

39. 7.9×10^{-33} kg

41. 7.77×10^{-13} J

43. $\lambda_A = 2.75 \times 10^{-7}$ m, $\lambda_B = 3.35 \times 10^{-7}$ m

45. 254 nm

47. (a) 1.25 J (b) 1.89×10^{18} photons

49. (a) 5740 K
(b) No. The answer in part (a) is unreasonably hot.

51. 42.8

53. (a) 3.00×10^8 W (b) 3.51×10^9 W/m^2 (c) 1.50×10^9 J

55. (a) 2.00×10^{-11} m (b) 3.80 kV

Chapter 30

1. (a) 6.2×10^{-31} m^3 (b) 4×10^{-45} m^3 (c) 7×10^{-13} %

3. 2 mi

5. 7.3×10^{-14} m

7. $n = 3$

9. 16

11. 1.98×10^{-19} J

13. 6.56×10^{-7} m and 1.22×10^{-7} m

15. (a) 1.8 (b) 4.3×10^{-11} m ($Z = 11$) and 2.6×10^{-10} m ($Z = 1.8$)

17. 2

19. 22.8 nm

21. (a) −1.51 eV (b) 2.58×10^{-34} J · s (c) 2.11×10^{-34} J · s

23. −0.378 eV

25. 1.732

27. −0.278 eV, −0.378 eV, −0.544 eV

29. (a)

n	ℓ	m_ℓ	m_s
2	0	0	1/2
2	0	0	−1/2

(b)

n	ℓ	m_ℓ	m_s
2	1	1	1/2
2	1	1	−1/2
2	1	0	1/2
2	1	0	−1/2
2	1	−1	1/2
2	1	−1	−1/2

31. 50

33. (a) Not allowed (b) Allowed (c) Not allowed (d) Allowed (e) Not allowed

35. Carbon ($Z = 6$)

37. (a) 1.34×10^{-12} m (b) 1.80 %

39. 0.50

41. 1.29×10^{-10} m

43. (a) 0.289 eV (b) 1.00×10^{-5} m (c) infrared

45. (a) 40.8 eV (b) 54.4 eV

47. (a) 7458 nm (b) 2279 nm (c) Infrared region

49. 4.41×10^{-10} m

51. 1.9×10^{17} photons

53. 2.2×10^{16} photons

55. (a) 2.28×10^{-11} m (b) 1.92×10^{-11} m

57. 4

59. (a) $n_f = 1$, 12.8 eV (b) $n_f = 3$, 0.661 eV (c) $n_f = 5$, 0.306 eV

61. (a) $E_1 = 3.182 \times 10^{-19}$ J, $E_2 = 3.127 \times 10^{-19}$ J, $E_3 = 2.897 \times 10^{-19}$ J, $E_4 = 2.079 \times 10^{-18}$ J (b) $\lambda_1 = 624.3$ nm, $\lambda_2 = 635.3$ nm, $\lambda_3 = 685.7$ nm, $\lambda_4 = 95.55$ nm (c) 624.3 nm, 635.3 nm, and 685.7 nm

63. (a) $\lambda_1(3 \to 2) = 656.3$ nm (red); $\lambda_2(4 \to 2) = 486.2$ nm (aqua); $\lambda_3(5 \to 2) = 434.1$ nm (blue); $\lambda_4(6 \to 2) = 410.2$ nm (violet) (b) $\theta_1 = 11.3°$; $\theta_2 = 8.38°$; $\theta_3 = 7.48°$; $\theta_4 = 7.07°$ (c) 3.97 nm, this is the 7→2 transition of the Balmer series

Chapter 31

1. (a) $+1.31 \times 10^{-17}$ C (b) 126 (c) 208 (d) 7.1×10^{-15} m (e) 2.3×10^{17} kg/m^3

3. (a) 117 neutrons (platinum, Pt) (b) 16 neutrons (sulfur, S) (c) 34 neutrons (copper, Cu) (d) 6 neutrons (boron, B) (e) 145 neutrons (plutonium, Pu)

5. (a) 92 (b) 122 (c) 41

7. $^{4}_{2}$He

9. 39.25 MeV

11. (a) 0.008 285 u (b) 0.009 105 u (c) More energy must be supplied to tritium ($^{3}_{1}$T) than to helium ($^{3}_{2}$He).

13. (a) 1.741 670 u (b) 1622 MeV (c) 7.87 MeV/nucleon

15. 1.003 27 u

17. (a) $^{18}_{9}\text{F} \to {}^{18}_{8}\text{O} + {}^{0}_{+1}\text{e}$ (b) $^{15}_{8}\text{O} \to {}^{15}_{7}\text{N} + {}^{0}_{+1}\text{e}$

19. 0.313 MeV

21. 4.87 MeV

23. (a) $^{238}_{92}$U (b) $^{24}_{12}$Mg (c) $^{13}_{6}$C

25. 1.61×10^7 m/s

27. $KE_{Th} = 0.072$ MeV, $KE_\alpha = 4.2$ MeV

29. 1.82 MeV

31. 5.62×10^{-7} s^{-1}

33. 8.00 days

35. 8.01×10^6 m

37. (a) 3.0×10^4 yr (b) 3.3×10^4 yr

39. 3.29×10^9 yr

41. Age of the fossils is 6100 yr. Maximum error is 900 yr.

43. 4.88×10^{10} yr

45. (a) X represents a β^- particle (electron). (b) X represents a β^+ particle (positron). (c) X represents a γ ray. (d) X represents an α particle (helium nucleus).

47. (a) $^{212}_{82}$Pb (b) 24.62 MeV

49. 387 yr

51. 2.1×10^{13} Bq

53. (a) oxygen (b) 112 μCi

55. 2.78×10^{28} MeV

57. 16.24 mg

59. 6.1 C°

61. (a) 4.74×10^{-13} A (b) 2.73×10^{-13} A

63. (a) $^{237}_{93}$Np (b) 1.27×10^{11} Bq (c) 720.6 yr

Chapter 32

1. 12

3. (a) 14 Gy (b) 2.1 J

5. 2.4×10^4 rem

7. 0.85

9. 9.2×10^8 nuclei

11. $\gamma + {}^{17}_{8}\text{O} \to {}^{12}_{6}\text{C} + {}^{4}_{2}\text{He} + {}^{1}_{0}\text{n}$

13. (a) $A = 233$, $Z = 90$, X = Th (Thorium) (b) $A = 233$, $Z = 92$, X = U (Uranium)

15. (a) Sulfur $^{31}_{16}$S (b) proton $^{1}_{1}$H (c) potassium $^{41}_{19}$K (d) neon $^{20}_{10}$Ne

17. $^{A}_{Z}\text{X} = {}^{99}_{41}\text{Nb}$

19. 3

21. 6.4×10^8 W

23. 2.7×10^6 kg

25. 1200 kg

27. 2.2 MeV

29. (a) $A = 2$, $Z = 0$, X = neutron, Y = deuteron nucleus (b) 6.2 MeV

31. $KE_n = 14.1$ MeV, $KE_\alpha = 3.5$ MeV

33. 33.9 MeV

35. 140.9 MeV

37. 2.73×10^{-22} kg · m/s

39. Protons, electrons, gamma rays, slow neutrons

41. 184 MeV

43 1.0 gal

45. (a) 8.2×10^{10} J (b) 0.48 g

47. (a) 89 h = 3.7 days (b) 0.60 J

49. 69 m/s

51. (a) 0.0128 J (b) 5.36×10^{-6} C°

53. (a) 67.5 MeV (b) 1.63×10^{22} Hz; 1.84×10^{-14} m (c) $+3.60 \times 10^{-20}$ kg · m/s, -3.60×10^{-20} kg · m/s

55. (a) $^{289}_{115}$Mc (moscovium) (b) $^{285}_{113}$Nh (nihonium)

57. (a) 6.46×10^{-13} J (b) 7.73×10^{21} reactions/s (c) 1.63×10^3 kg

Index

Fundamental Constants

Quantity	Symbol	Value*
Avogadro's number	N_A	$6.022\,140\,76 \times 10^{23}$ mol^{-1}
Boltzmann's constant	k	$1.380\,649 \times 10^{-23}$ J/K
Electron charge magnitude	e	$1.602\,176\,634 \times 10^{-19}$ C
Permeability of free space	μ_0	$4\pi \times 10^{-7}$ T · m/A
Permittivity of free space	ε_0	$8.854\,187\,8128 \times 10^{-12}$ C^2/(N · m^2)
Planck's constant	h	$6.626\,070\,15 \times 10^{-34}$ J · s
Mass of electron	m_e	$9.109\,383\,7015 \times 10^{-31}$ kg
Mass of neutron	m_n	$1.674\,927\,471 \times 10^{-27}$ kg
Mass of proton	m_p	$1.672\,621\,923\,69 \times 10^{-27}$ kg
Speed of light in vacuum	c	$2.997\,924\,58 \times 10^8$ m/s
Universal gravitational constant	G	$6.674\,30 \times 10^{-11}$ N · m^2/kg^2
Universal gas constant	R	8.314 462 618 J/(mol · K)

*2018 CODATA recommended values.

Useful Physical Data

Acceleration due to earth's gravity	9.80 m /s^2 = 32.2 ft /s^2
Atmospheric pressure at sea level	1.013×10^5 Pa = 14.70 lb/in.^{2}
Density of air (0 °C, 1 atm pressure)	1.29 kg /m^3
Speed of sound in air (20 °C)	343 m/s
Water	
Density (4 °C)	1.000×10^3 kg /m^3
Latent heat of fusion	3.35×10^5 J/ kg
Latent heat of vaporization	2.26×10^6 J/kg
Specific heat capacity	4186 J/(kg · C°)
Earth	
Mass	5.98×10^{24} kg
Radius (equatorial)	6.38×10^6 m
Mean distance from sun	1.50×10^{11} m
Moon	
Mass	7.35×10^{22} kg
Radius (mean)	1.74×10^6 m
Mean distance from earth	3.85×10^8 m
Sun	
Mass	1.99×10^{30} kg
Radius (mean)	6.96×10^8 m

Frequently Used Mathematical Symbols

Symbol	Meaning
$=$	is equal to
$\neq$	is not equal to
$\propto$	is proportional to
$>$	is greater than
$<$	is less than
$\approx$	is approximately equal to
$\lvert x \rvert$	absolute value of x (always treated as a positive quantity)
Δ	the difference between two variables (e.g., ΔT is the final temperature minus the initial temperature)
Σ	the sum of two or more variables (e.g., $\sum_{i=1}^{3} x_i = x_1 + x_2 + x_3$)

Conversion Factors

Length

1 in. = 2.54 cm

1 ft = 0.3048 m

1 mi = 5280 ft = 1.609 km

1 m = 3.281 ft

1 km = 0.6214 mi

1 angstrom (Å) = 10^{-10} m

Mass

1 slug = 14.59 kg

1 kg = 1000 grams = 6.852×10^{-2} slug

1 atomic mass unit (u) = 1.6605×10^{-27} kg

(1 kg has a weight of 2.205 lb where the acceleration due to gravity is 32.174 ft/s^2)

Time

1 d = 24 h = 1.44×10^3 min = 8.64×10^4 s

1 yr = 365.24 days = 3.156×10^7 s

Speed

1 mi/h = 1.609 km/h = 1.467 ft/s = 0.4470 m/s

1 km/h = 0.6214 mi/h = 0.2778 m/s = 0.9113 ft/s

Force

1 lb = 4.448 N

1 N = 10^5 dynes = 0.2248 lb

Work and Energy

1 J = 0.7376 ft · lb = 10^7 ergs

1 kcal = 4186 J

1 Btu = 1055 J

1 kWh = 3.600×10^6 J

1 eV = 1.602×10^{-19} J

Power

1 hp = 550 ft · lb/s = 745.7 W

1 W = 0.7376 ft · lb/s

Pressure

1 Pa = 1 N/m^2 = 1.450×10^{-4} lb/in.^{2}

1 lb/in.^{2} = 6.895×10^3 Pa

1 atm = 1.013×10^5 Pa = 1.013 bar = 14.70 lb/in.^{2} = 760 torr

Volume

1 liter = 10^{-3} m^3 = 1000 cm^3 = 0.03531 ft^3

1 ft^3 = 0.02832 m^3 = 7.481 U.S. gallons

1 U.S. gallon = 3.785×10^{-3} m^3 = 0.1337 ft^3

Angle

1 radian = 57.30°

1° = 0.01745 radian

Standard Prefixes Used to Denote Multiples of Ten

Prefix	Symbol	Factor
Tera	T	10^{12}
Giga	G	10^{9}
Mega	M	10^{6}
Kilo	k	10^{3}
Hecto	h	10^{2}
Deka	da	10^{1}
Deci	d	10^{-1}
Centi	c	10^{-2}
Milli	m	10^{-3}
Micro	μ	10^{-6}
Nano	n	10^{-9}
Pico	p	10^{-12}
Femto	f	10^{-15}

Basic Mathematical Formulas

Area of a circle = πr^2

Circumference of a circle = $2\pi r$

Surface area of a sphere = $4\pi r^2$

Volume of a sphere = $\frac{4}{3}\pi r^3$

Pythagorean theorem: $h^2 = h_o^2 + h_a^2$

Sine of an angle: $\sin\theta = h_o/h$

Cosine of an angle: $\cos\theta = h_a/h$

Tangent of an angle: $\tan\theta = h_o/h_a$

Law of cosines: $c^2 = a^2 + b^2 - 2ab\cos\gamma$

Law of sines: $a/\sin\alpha = b/\sin\beta = c/\sin\gamma$

Quadratic formula:

If $ax^2 + bx + c = 0$, then, $x = (-b \pm \sqrt{b^2 - 4ac})/(2a)$

Units

Quantity	Name of Unit	Symbol	Expression in Terms of Other SI Units
Length	meter	m	Base unit
Mass	kilogram	kg	Base unit
Time	second	s	Base unit
Electric current	ampere	A	Base unit
Temperature	kelvin	K	Base unit
Amount of substance	mole	mol	Base unit
Velocity	—	—	m/s
Acceleration	—	—	m/s^2
Force	newton	N	$kg \cdot m/s^2$
Work, energy	joule	J	$N \cdot m$
Power	watt	W	J/s
Impulse, momentum	—	—	$kg \cdot m/s$
Plane angle	radian	rad	m/m
Angular velocity	—	—	rad/s
Angular acceleration	—	—	rad/s^2
Torque	—	—	$N \cdot m$
Frequency	hertz	Hz	s^{-1}
Density	—	—	kg/m^3

Quantity	Name of Unit	Symbol	Expression in Terms of Other SI Units
Pressure, stress	pascal	Pa	N/m^2
Viscosity	—	—	$Pa \cdot s$
Electric charge	coulomb	C	$A \cdot s$
Electric field	—	—	N/C
Electric potential	volt	V	J/C
Resistance	ohm	Ω	V/A
Capacitance	farad	F	C/V
Inductance	henry	H	$V \cdot s/A$
Magnetic field	tesla	T	$N \cdot s/(C \cdot m)$
Magnetic flux	weber	Wb	$T \cdot m^2$
Specific heat capacity	—	—	$J/(kg \cdot K)$ or $J/(kg \cdot C°)$
Thermal conductivity	—	—	$J/(s \cdot m \cdot K)$ or $J/(s \cdot m \cdot C°)$
Entropy	—	—	J/K
Radioactive activity	becquerel	Bq	s^{-1}
Absorbed dose	gray	Gy	J/kg
Exposure	—	—	C/kg

The Greek Alphabet

Alpha	A	α	Iota	I	ι	Rho	P	ρ
Beta	B	β	Kappa	K	κ	Sigma	Σ	σ
Gamma	Γ	γ	Lambda	Λ	λ	Tau	T	τ
Delta	Δ	δ	Mu	M	μ	Upsilon	ϒ	υ
Epsilon	E	ε	Nu	N	ν	Phi	Φ	ϕ
Zeta	Z	ζ	Xi	Ξ	ξ	Chi	X	χ
Eta	H	η	Omicron	O	o	Psi	ψ	ψ
Theta	Θ	θ	Pi	Π	π	Omega	Ω	ω

Periodic Table of the Elements

Symbol — **Cl** 17 — Atomic number
Atomic mass* — 35.453
$3p^5$ — Electron configuration

Group I	Group II	Transition elements										Group III	Group IV	Group V	Group VI	Group VII	Group 0
H 1 1.00794 $1s^1$																	**He** 2 4.00260 $1s^2$
Li 3 6.941 $2s^1$	**Be** 4 9.01218 $2s^2$											**B** 5 10.81 $2p^1$	**C** 6 12.011 $2p^2$	**N** 7 14.0067 $2p^3$	**O** 8 15.9994 $2p^4$	**F** 9 18.9984 $2p^5$	**Ne** 10 20.180 $2p^6$
Na 11 22.9898 $3s^1$	**Mg** 12 24.305 $3s^2$											**Al** 13 26.9815 $3p^1$	**Si** 14 28.0855 $3p^2$	**P** 15 30.9738 $3p^3$	**S** 16 32.07 $3p^4$	**Cl** 17 35.453 $3p^5$	**Ar** 18 39.948 $3p^6$
K 19 39.0983 $4s^1$	**Ca** 20 40.08 $4s^2$	**Sc** 21 44.9559 $3d^14s^2$	**Ti** 22 47.87 $3d^24s^2$	**V** 23 50.9415 $3d^34s^2$	**Cr** 24 51.996 $3d^54s^1$	**Mn** 25 54.9380 $3d^54s^2$	**Fe** 26 55.845 $3d^64s^2$	**Co** 27 58.9332 $3d^74s^2$	**Ni** 28 58.69 $3d^84s^2$	**Cu** 29 63.546 $3d^{10}4s^1$	**Zn** 30 65.41 $3d^{10}4s^2$	**Ga** 31 69.72 $4p^1$	**Ge** 32 72.64 $4p^2$	**As** 33 74.9216 $4p^3$	**Se** 34 78.96 $4p^4$	**Br** 35 79.904 $4p^5$	**Kr** 36 83.80 $4p^6$
Rb 37 85.4678 $5s^1$	**Sr** 38 87.62 $5s^2$	**Y** 39 88.9059 $4d^15s^2$	**Zr** 40 91.224 $4d^25s^2$	**Nb** 41 92.9064 $4d^45s^1$	**Mo** 42 95.94 $4d^55s^1$	**Tc** 43 (98) $4d^55s^2$	**Ru** 44 101.07 $4d^75s^1$	**Rh** 45 102.906 $4d^85s^1$	**Pd** 46 106.42 $4d^{10}5s^0$	**Ag** 47 107.868 $4d^{10}5s^1$	**Cd** 48 112.41 $4d^{10}5s^2$	**In** 49 114.82 $5p^1$	**Sn** 50 118.71 $5p^2$	**Sb** 51 121.76 $5p^3$	**Te** 52 127.60 $5p^4$	**I** 53 126.904 $5p^5$	**Xe** 54 131.29 $5p^6$
Cs 55 132.905 $6s^1$	**Ba** 56 137.33 $6s^2$	57–71	**Hf** 72 178.49 $5d^26s^2$	**Ta** 73 180.948 $5d^36s^2$	**W** 74 183.84 $5d^46s^2$	**Re** 75 186.207 $5d^56s^2$	**Os** 76 190.2 $5d^66s^2$	**Ir** 77 192.22 $5d^76s^2$	**Pt** 78 195.08 $5d^86s^2$	**Au** 79 196.967 $5d^{10}6s^1$	**Hg** 80 200.59 $5d^{10}6s^2$	**Tl** 81 204.383 $6p^1$	**Pb** 82 207.2 $6p^2$	**Bi** 83 208.980 $6p^3$	**Po** 84 (209) $6p^4$	**At** 85 (210) $6p^5$	**Rn** 86 (222) $6p^6$
Fr 87 (223) $7s^1$	**Ra** 88 (226) $7s^2$	89–103	**Rf** 104 (261) $6d^27s^2$	**Db** 105 (262) $6d^37s^2$	**Sg** 106 (266) $6d^47s^2$	**Bh** 107 (264) $6d^57s^2$	**Hs** 108 (277) $6d^67s^2$	**Mt** 109 (268) $6d^77s^2$	**Ds** 110 (281)	**Rg** 111 (282)	**Cn** 112 (285)	**Nh** 113 (286)	**Fl** 114 (289)	**Mc** 115 (289)	**Lv** 116 (293)	**Ts** 117 (294)	**Og** 118 (294)

Lanthanide series

La 57 138.906 $5d^16s^2$	**Ce** 58 140.12 $4f^26s^2$	**Pr** 59 140.908 $4f^36s^2$	**Nd** 60 144.24 $4f^46s^2$	**Pm** 61 (145) $4f^56s^2$	**Sm** 62 150.36 $4f^66s^2$	**Eu** 63 151.96 $4f^76s^2$	**Gd** 64 157.25 $5d^14f^76s^2$	**Tb** 65 158.925 $4f^96s^2$	**Dy** 66 162.50 $4f^{10}6s^2$	**Ho** 67 164.930 $4f^{11}6s^2$	**Er** 68 167.26 $4f^{12}6s^2$	**Tm** 69 168.934 $4f^{13}6s^2$	**Yb** 70 173.04 $4f^{14}6s^2$	**Lu** 71 174.967 $5d^14f^{14}6s^2$

Actinide series

Ac 89 (227) $6d^17s^2$	**Th** 90 232.038 $6d^27s^2$	**Pa** 91 231.036 $5f^26d^17s^2$	**U** 92 238.029 $5f^36d^17s^2$	**Np** 93 (237) $5f^46d^17s^2$	**Pu** 94 (244) $5f^66d^07s^2$	**Am** 95 (243) $5f^76d^07s^2$	**Cm** 96 (247) $5f^76d^17s^2$	**Bk** 97 (247) $5f^96d^07s^2$	**Cf** 98 (251) $5f^{10}6d^07s^2$	**Es** 99 (252) $5f^{11}6d^07s^2$	**Fm** 100 (257) $5f^{12}6d^07s^2$	**Md** 101 (258) $5f^{13}6d^07s^2$	**No** 102 (259) $6d^07s^2$	**Lr** 103 (262) $6d^17s^2$

* Atomic mass values are averaged over isotopes according to the percentages that occur on the earth's surface. For unstable elements, the mass number of the most stable known isotope is given in parentheses. *Source: IUPAC Commission on Atomic Weights and Isotopic Abundances, 2001.*